FEATURES AND BENEFITS

INTRODUCTORY ANALYSIS

Comprehensive coverage of topics provides solid preparation for college-level courses in calculus, linear algebra, finite mathematical structures, and probability and statistics. pp. iii–xii

The **variety of course options** made possible by the organization and scope of topics makes it possible to plan courses to meet different needs. pp. iii–xii

A **balance of theory and its application,** seen in numerous theorems, proofs, examples, and solutions, enables students to read and write mathematics with understanding and precision. pp. 116–117

The **text design** with its open format and its highlighting of important content enables students to read mathematics with focus and concentration. Numerous diagrams and illustrations also further this aim. pp. 174–175

Abundant exercises reinforce theory, skills, and applications at three levels of difficulty. pp. 294–295

Chapter summaries, tests, and **cumulative reviews** provide continuing reinforcement and opportunities for evaluation of progress. pp. 432–433; pp. 438–439

Applications show students how mathematics is used in related fields. **Extras** enrich the course by extending concepts to other mathematical areas. pp. 38–39; pp. 34–35

Computer exercises and **Computer applications** provide opportunities for students to apply new technologies in a meaningful way. p. 261; pp. 520–521

Compatibility with calculators allows students to use the book successfully with or without calculators. A discussion of interpolation and extensive tables are provided. pp. 292–293; pp. 646–668

A comprehensive **Teacher's Manual with Solutions** and **Tests** (on duplicating masters) offer maximum teaching support.

Introductory Analysis

Mary P. Dolciani
Robert H. Sorgenfrey
John A. Graham
David L. Myers

Editorial Advisers

Andrew M. Gleason
Joshua Barlaz
William Ted Martin

Teacher Consultants

Yvonne Tomlinson Lehr
Barbara E. Nunn
Karl Randall
William E. Raschkow
Carl H. Sweitzer

HOUGHTON MIFFLIN COMPANY / BOSTON

Atlanta Dallas Geneva, Ill. Lawrenceville, N.J. Palo Alto Toronto

AUTHORS

Mary P. Dolciani, formerly Professor of Mathematical Sciences, Hunter College of the City University of New York.

Robert H. Sorgenfrey, Professor of Mathematics, University of California, Los Angeles.

John A. Graham, Mathematics Teacher, Buckingham, Browne and Nichols School, Cambridge, Massachusetts.

David L. Myers, Computer Coordinator and Mathematics Teacher, Winsor School, Boston, Massachusetts.

Editorial Advisers

Andrew M. Gleason, Hollis Professor of Mathematics and Natural Philosophy, Harvard University, Cambridge, Massachusetts.

Joshua Barlaz, Professor of Mathematics, Rutgers, The State University of New Jersey, New Brunswick, New Jersey.

William Ted Martin, Professor of Mathematics and Education, Emeritus, Massachusetts Institute of Technology, Cambridge, Massachusetts.

Teacher Consultants

Yvonne Tomlinson Lehr, Mathematics Teacher, Riverdale High School, New Orleans, Louisiana.

Barbara E. Nunn, Mathematics Teacher and Chairman of the Mathematics Department, Coral Springs High School, Coral Springs, Florida.

Karl Randall, Chairman of the Mathematics Department, Hinsdale High School Central, Hinsdale, Illinois.

William E. Raschkow, Basic Skills Director, Kent Public Schools, Kent, Washington.

Carl H. Sweitzer, Mathematics Teacher, Bernardsville High School, Bernardsville, New Jersey.

Printed in U.S.A.

ISBN: 0-395-40655-2

DEFGHIJ-RM-9543210/898

CONTENTS

Chapter 1 Foundations of Real Analysis

Sets and Sentences

□ 1-1 SETS 1 □ 1-2 OPEN SENTENCES 5 □ 1-3 CONDITIONAL SENTENCES 10

Real Numbers

□ 1-4 ADDITION AND MULTIPLICATION PROPERTIES OF REAL NUMBERS 13
□ 1-5 ORDER AND INEQUALITIES 19 □ 1-6 ABSOLUTE VALUE 23
□ 1-7 SUBSETS OF $\mathcal{R}$ 28

Review and Testing

SUMMARY 32 TEST 33

Features

COMPUTER EXERCISES 5, 9, 19, 27, 31
EXTRA 34 COMPUTER APPLICATION 36 APPLICATION 38

Chapter 2 Analytic Geometry

Coordinate Geometry and Lines

□ 2-1 THE CARTESIAN COORDINATE SYSTEM: DISTANCE AND MIDPOINTS 41
□ 2-2 LINES 46

Conic Sections

□ 2-3 CIRCLES 53 □ 2-4 ELLIPSES 56 □ 2-5 HYPERBOLAS 63
□ 2-6 PARABOLAS 69 □ 2-7 CONIC SECTIONS 73
□ 2-8 INTERSECTIONS OF CONIC SECTIONS 76

Review and Testing

SUMMARY 80 TEST 81

Features

COMPUTER EXERCISES 52, 63, 79
EXTRA 82 COMPUTER APPLICATION 84 APPLICATION 86

Chapter 3 Sequences, Series, and Limits

Mathematical Induction

□ 3-1 MATHEMATICAL INDUCTION 89 □ 3-2 THE BINOMIAL THEOREM 94

Sequences and Series

□ 3-3 RECURSIVE DEFINITIONS: SEQUENCES AND SERIES 99 □ 3-4 ARITHMETIC SEQUENCES 103 □ 3-5 GEOMETRIC SEQUENCES AND SERIES 108

Limits and Sequences

□ 3-6 LIMITS OF SEQUENCES 114 □ 3-7 INFINITE SERIES 119 □ 3-8 THE AXIOM OF COMPLETENESS 125 □ 3-9 INTEGRAL AND RATIONAL EXPONENTS 129

Review and Testing

SUMMARY 133 TEST 135 CUMULATIVE REVIEW 138

Features

COMPUTER EXERCISES 99, 113, 124
APPLICATION 136

Chapter 4 Functions and Limits

Functions

□ 4-1 FUNCTIONS AND FUNCTIONAL NOTATION 141 □ 4-2 THE ALGEBRA OF FUNCTIONS 146 □ 4-3 COMPOSITION AND INVERSION OF FUNCTIONS 151

Limits of Functions

□ 4-4 LIMITS AND THEIR PROPERTIES 157 □ 4-5 CONTINUOUS FUNCTIONS 163 □ 4-6 THE INTERMEDIATE-VALUE THEOREM 167
□ 4-7 LIMITS INVOLVING INFINITY 172

Review and Testing

SUMMARY 177 TEST 178

Features

COMPUTER EXERCISES 146, 156, 171
EXTRA 179 COMPUTER APPLICATION 181 APPLICATION 182

Chapter 5 Theory of Polynomial Equations

Complex Numbers

☐ 5-1 THE COMPLEX FIELD 185 ☐ 5-2 THE COMPLEX PLANE 189
☐ 5-3 SQUARE ROOTS OF COMPLEX NUMBERS 193

General Theory of Polynomials

☐ 5-4 POLYNOMIALS OVER FIELDS 196 ☐ 5-5 THE FUNDAMENTAL THEOREM OF ALGEBRA 201

Polynomials with Real Coefficients

☐ 5-6 CONJUGATE IMAGINARY ROOTS 205 ☐ 5-7 BOUNDS FOR REAL ROOTS; DESCARTES' RULE 208 ☐ 5-8 FINDING RATIONAL ROOTS 211

Review and Testing

SUMMARY 215 TEST 216

Features

COMPUTER EXERCISES 188, 193, 200, 214
COMPUTER APPLICATION 217 APPLICATION 218

Chapter 6 Introduction to Differential Calculus

Differentiation

☐ 6-1 DERIVATIVES OF FUNCTIONS 221 ☐ 6-2 PROPERTIES OF DERIVATIVES 226 ☐ 6-3 NOTATIONS FOR DERIVATIVES 230
☐ 6-4 THE CHAIN RULE 234

Applications

☐ 6-5 USING DERIVATIVES IN GRAPHING 238 ☐ 6-6 CONCAVITY AND THE SECOND DERIVATIVE 242 ☐ 6-7 APPLIED MAXIMA AND MINIMA 245
☐ 6-8 DERIVATIVES AS RATES OF CHANGE 253 ☐ 6-9 NEWTON'S METHOD 258

Review and Testing

SUMMARY 262 TEST 263 CUMULATIVE REVIEW 268

Features

COMPUTER EXERCISES 226, 241, 261
EXTRA 264 COMPUTER APPLICATION 266 APPLICATION 267

Chapter 7 Trigonometric Functions and Triangle Solving

Circular Functions of Numbers

□ 7-1 THE SINE AND COSINE FUNCTIONS 271 □ 7-2 THE OTHER CIRCULAR FUNCTIONS 277

Trigonometric Functions of Angles

□ 7-3 ANGLES AND ANGLE MEASURE 281 □ 7-4 THE TRIGONOMETRIC FUNCTIONS 286 □ 7-5 VALUES OF THE TRIGONOMETRIC FUNCTIONS 290

Solving Triangles

□ 7-6 RIGHT TRIANGLES 292 □ 7-7 THE LAWS OF COSINES AND SINES 296
□ 7-8 SOLVING GENERAL TRIANGLES 300 □ 7-9 AREAS OF TRIANGLES 305

Review and Testing

SUMMARY 308 TEST 309

Features

COMPUTER EXERCISES 281, 295, 305
COMPUTER APPLICATION 310 APPLICATION 311

Chapter 8 Trigonometric Identities and Graphs

Trigonometric Identities

□ 8-1 IDENTITIES IN ONE VARIABLE 313 □ 8-2 TRIGONOMETRIC ADDITION FORMULAS 318 □ 8-3 DOUBLE-ANGLE AND HALF-ANGLE FORMULAS 322
□ 8-4 MORE TRIGONOMETRIC IDENTITIES 326 □ 8-5 TRIGONOMETRIC EQUATIONS 330

Trigonometric Graphs

□ 8-6 PERIODICITY AND OTHER PROPERTIES 334 □ 8-7 SINE AND COSINE GRAPHS 338 □ 8-8 MORE GENERAL SINE CURVES 343 □ 8-9 OTHER TRIGONOMETRIC GRAPHS 348 □ 8-10 INVERSE TRIGONOMETRIC FUNCTIONS 350

Review and Testing

SUMMARY 355 TEST 357

Features

COMPUTER EXERCISES 317, 322, 333, 343, 350
APPLICATION 358

Chapter 9 Applications of Trigonometry

Applications Involving Polar Coordinates

□ 9-1 THE POLAR COORDINATE SYSTEM 361 □ 9-2 GRAPHS OF POLAR EQUATIONS 365 □ 9-3 POLAR FORM OF COMPLEX NUMBERS 369 □ 9-4 POWERS AND ROOTS OF COMPLEX NUMBERS 372 □ 9-5 PARAMETRIC EQUATIONS 375

Calculus of Trigonometric Functions

□ 9-6 SOME TRIGONOMETRIC LIMITS 380 □ 9-7 DERIVATIVES OF THE TRIGONOMETRIC FUNCTIONS 384 □ 9-8 RATES AND EXTREMA 388

Review and Testing

SUMMARY 393 TEST 394

Features

COMPUTER EXERCISES 368, 375, 383
EXTRA 395 APPLICATION 396

Chapter 10 Exponential and Logarithmic Functions

Exponential Functions

□ 10-1 REAL EXPONENTS 399 □ 10-2 EXPONENTIAL FUNCTIONS 403
□ 10-3 THE NATURAL EXPONENTIAL FUNCTION 408

Logarithmic Functions

□ 10-4 LOGARITHMIC FUNCTIONS 412 □ 10-5 LAWS OF LOGARITHMS 416

Derivatives

□ 10-6 DERIVATIVES OF EXPONENTIAL AND LOGARITHMIC FUNCTIONS 422
□ 10-7 EXPONENTIAL GROWTH AND DECAY 427

Review and Testing

SUMMARY 432 TEST 433 CUMULATIVE REVIEW 438

Features

COMPUTER EXERCISES 412, 421, 432
EXTRA 434 COMPUTER APPLICATION 436 APPLICATION 437

Chapter 11 Vectors

Vectors

☐ 11-1 INTRODUCTION TO VECTORS 441 ☐ 11-2 THREE-DIMENSIONAL SPACE 446 ☐ 11-3 THE NORM OF A VECTOR 451 ☐ 11-4 THE DOT PRODUCT 454 ☐ 11-5 BASIS VECTORS 457

Applications

☐ 11-6 APPLICATIONS OF VECTORS 462 ☐ 11-7 VECTORS AND PROOFS 466
☐ 11-8 VECTOR SPACES 469

Review and Testing

SUMMARY 471 TEST 472

Features

COMPUTER EXERCISES 446, 453, 461, 466
EXTRA 474 APPLICATION 476

Chapter 12 Matrices, Determinants, and Systems of Linear Equations

Transformations and Matrices

☐ 12-1 LINEAR TRANSFORMATIONS 479 ☐ 12-2 MATRICES 485
☐ 12-3 MATRICES OF LINEAR TRANSFORMATIONS 490 ☐ 12-4 ROTATION OF AXES 493

Systems of Linear Equations

☐ 12-5 LINEAR SYSTEMS I 497 ☐ 12-6 LINEAR SYSTEMS II 501
☐ 12-7 DETERMINANTS 505 ☐ 12-8 THE INVERSE OF A MATRIX 510
☐ 12-9 CRAMER'S RULE 515

Review and Testing
SUMMARY 518 TEST 519

Features
COMPUTER EXERCISES 484, 489, 514
COMPUTER APPLICATION 520 APPLICATION 522

Chapter 13 Further Vector Topics

Planes and Lines
□ 13-1 THE CROSS PRODUCT 525 □ 13-2 PLANES 530
□ 13-3 LINES 534

Surfaces
□ 13-4 QUADRIC SURFACES 539 □ 13-5 CYLINDRICAL AND SPHERICAL COORDINATES 545

Review and Testing
SUMMARY 549 TEST 551 CUMULATIVE REVIEW 554

Features
COMPUTER EXERCISES 529, 539
COMPUTER APPLICATION 552 APPLICATION 553

Chapter 14 Introduction to Integral Calculus

Antiderivatives and Definite Integrals
□ 14-1 ANTIDERIVATIVES 557 □ 14-2 INTEGRATION BY SUBSTITUTION 561
□ 14-3 DEFINITE INTEGRALS 565

Applications of Integration
□ 14-4 AREAS OF PLANE REGIONS 568 □ 14-5 VOLUMES OF SOLIDS 574
□ 14-6 WORK AND ENERGY 580

Review and Testing
SUMMARY 584 TEST 585

Features

COMPUTER EXERCISES 579, 583
COMPUTER APPLICATION 586 APPLICATION 587

Chapter 15 Probability and Statistics

Probability

☐ 15-1 PERMUTATIONS 589 ☐ 15-2 COMBINATIONS 594
☐ 15-3 PROBABILITY 599 ☐ 15-4 COMPOUND EVENTS 603
☐ 15-5 RANDOM VARIABLES AND PROBABILITY DISTRIBUTIONS 610

Statistics

☐ 15-6 STATISTICAL MEASURES 616 ☐ 15-7 VARIANCE AND STANDARD DEVIATION 621 ☐ 15-8 NORMAL DISTRIBUTIONS 626 ☐ 15-9 CONFIDENCE INTERVALS 632 ☐ 15-10 CORRELATION 635

Review and Testing

SUMMARY 639 TEST 641

Features

COMPUTER EXERCISES 598, 626
COMPUTER APPLICATION 643 APPLICATION 644

Using the Tables 646
Tables 647
Reference Materials 669
Glossary 671
Index 680
Credits 689
Answers to Selected Exercises

SYMBOLS

		PAGE
$\in$	is a member of	1
$/$	is not (used to negate another symbol)	2
$\ldots$	suggested pattern continues	2
N	set of natural numbers	2
Z	set of integers	2
Q	set of rational numbers	2
$\mathcal{R}$	set of real numbers	2
$\{\ \}$	the set with members . . .	2
$\subseteq$	is a subset of	2
$\subset$	is a proper subset of	2
$\emptyset$	null set, or empty set	2
$\cup$	union (of sets)	3
$\cap$	intersection (of sets)	3
A'	complement of A	3
$\{x \in \mathcal{R}: x > 3\}$	the set of all real numbers x such that $x > 3$	6
$p \wedge q$	p and q	6
$p \vee q$	p or q	7
p'	negation of p	7
$p \rightarrow q$	if p, then q	10
$p \leftrightarrow q$	p if and only if q	10
$=$	equals, is equal to	13
$<$	is less than	19
$>$	is greater than	20
$\geq$	is greater than or equal to	20
$\leq$	is less than or equal to	20
$\lvert x \rvert$	absolute value of x	23
$a \mid b$	a divides b	28
(a, b)	ordered pair of real numbers	41
$P(x, y)$	point with coordinates x and y	42
r^m	the mth power of r	90
$n!$	n factorial	94
$\binom{n}{k}$	binomial coefficient	94
a_n	nth term of a sequence	100
S_n	sum of the first n terms of a sequence	101

		PAGE
Σ	sigma (summation symbol)	101
$\lim_{n \to \infty}$	the limit as n increases without bound	114
$\sqrt[n]{a}$	nth root of a	128
$g(x)$	the value of g at x	142
$g: x \rightarrow g(x)$	the function that assigns the value of $g(x)$ to x	142
$f \circ g$	composition of functions f and g	151
f^{-1}	inverse of f	153
$\lim_{x \to c} f(x)$	the limit of $f(x)$ as x approaches c	158
$[a, b]$	closed interval	167
(a, b)	open interval	167
C	the complex field	185, 186
$\text{Re } z$	real part of a complex number z	186
$\text{Im } z$	imaginary part of a complex number z	186
$\bar{z}$	conjugate of a complex number z	189
$\lvert z \rvert$	absolute value, or modulus, of a complex number z	190
i	$\sqrt{-1}$; the pair $(0, 1)$	193
$f'(c)$	the derivative of f at c	223
f'	the derivative of f	227
$\frac{d}{dx}$	the derivative of (derivative operator)	230
$f'(x), y', \frac{dy}{dx}$	notations for derivatives	231
$f'', \frac{d^2y}{dx^2}$	second derivative of f	231
f'''	third derivative of f	231
$f^{(n)}$	nth derivative of f	231
(a, ∞)	the set $\{x: x > a\}$; an open interval	238
$(-\infty, b)$	the set $\{x: x < b\}$; an open interval	238
$\approx$	is approximately equal to	256
$\cos x$	the cosine of x	272

		PAGE
$\sin x$	the sine of x	272
$\tan x$	the tangent of x	277
$\cot x$	the cotangent of x	277
$\sec x$	the secant of x	277
$\csc x$	the cosecant of x	277
1°	one degree	282
$1'$	one minute	290
$1''$	one second	290
$\text{Cos } x$	$\cos x,\ 0 \le x \le \pi$	350
Cos^{-1}	inverse of Cos	351
$\text{Sin } x$	$\sin x,\ -\frac{\pi}{2} \le x \le \frac{\pi}{2}$	351
Sin^{-1}	inverse sine	351
$\text{Tan } x$	$\tan x,\ -\frac{\pi}{2} < x < \frac{\pi}{2}$	352
$\text{Cot } x$	$\cot x,\ 0 < x < \pi$	352
Tan^{-1}	inverse tangent	352
Cot^{-1}	inverse cotangent	352
$\text{Sec } x$	$\sec x,\ 0 \le x \le \pi,\ x \ne \frac{\pi}{2}$	353
$\text{Csc } x$	$\csc x,\ -\frac{\pi}{2} \le x \le \frac{\pi}{2},\ x \ne 0$	353
Sec^{-1}	inverse secant	353
Csc^{-1}	inverse cosecant	353
(r, θ)	polar coordinates of a point	361
$\arg z$	argument of z	369
$x \to b^x$	exponential function with base b	403
e	base of natural exponential function	408
$\log x$	logarithm of x with base 10	410
$\ln x$	natural logarithm of x	410
$\log_b x$	logarithm of x with base b	412
sinh	hyperbolic sine	434
cosh	hyperbolic cosine	434
tanh	hyperbolic tangent	435
$\overrightarrow{PQ}$	directed line segment, vector	441
$\mathbf{u}, \mathbf{v}$	vectors	442
$\mathbf{0}$	zero vector	442
(a, b)	ordered pair denoting a vector	444
(x, y, z)	ordered triple of real numbers	446
$P(x, y, z)$	point in three-space	446

		PAGE
(x, y, z)	vector in three-space	449
$\lVert \mathbf{v} \rVert$	norm, or magnitude, of a vector $\mathbf{v}$	451
$\mathbf{u} \cdot \mathbf{v}$	dot product of two vectors	454
$\{\mathbf{i}, \mathbf{j}\}$	standard basis for vectors in the plane	457
$\{\mathbf{i}, \mathbf{j}, \mathbf{k}\}$	standard basis for vectors in three-space	457
T_α	rotation of the plane through an angle α	481
$\begin{bmatrix} a & b & c \\ d & e & f \end{bmatrix}$	a matrix	485
$\det T$	determinant of T	505
$\begin{vmatrix} a & b \\ c & d \end{vmatrix}$	a determinant	505
A^{-1}	inverse of a matrix A	510
$\text{adj } A$	adjoint of a matrix A	511
A^T	transpose of a matrix A	511
$\mathbf{u} \times \mathbf{v}$	cross product of $\mathbf{u}$ and $\mathbf{v}$	525
$P(r, \theta, z)$	cylindrical coordinates of a point in space	545
(ρ, θ, φ)	spherical coordinates of a point in space	546
$\int$	integral sign	558
$\int f(x)\,dx$	indefinite integral of f	558
$\int_a^b f(x)\,dx$	definite integral of f from a to b	565
$[F(x)]_a^b$	$F(b) - F(a)$	566
${}_nP_r$	the number of permutations of n objects taken r at a time	591
${}_nC_r$	the number of combinations of n objects taken r at a time	594
$P(A)$	the probability of A	600
$E(X)$	expected value of a random variable X	613
$\bar{x}$	mean of a set of data	617
μ	mean of the probability distribution of a random variable	619
σ	standard deviation	622
σ^2	variance	622

USING YOUR TEXTBOOK

Reading a mathematics textbook is different from reading a novel, a history textbook, and even a chemistry or physics textbook. You need to read each discussion slowly and work through each example and its solution carefully. You might use pencil and paper to outline a solution strategy and to fill in missing details in solutions. This technique is particularly useful in studying theorems and their proofs. You might also make sketches as aids to understanding what you are reading.

In order to make full use of your textbook, you need to know its features. The chapter summary is a concise road map for the chapter. Pay careful attention to key vocabulary terms. These appear in bold type. Rely upon the glossary at the back of the book as a brief dictionary. The index with its cross references helps you see how terms are used in various contexts. The symbol list at the front of the book gives brief descriptions of the meanings of important symbols and the pages on which they are introduced. Throughout this book, boxes with displays are used to highlight important facts, methods, and results. These serve as a focal point for review.

Numerous diagrams are provided in examples, discussions, and exercises. These will aid understanding and serve as an example for you when you need to construct your own diagrams.

The exercise sets with their A, B, C delineation provide the opportunity to deepen your understanding of concepts and to master methods. Rely upon the answers to selected exercises at the back of the book to check your progress.

The textbook also provides optional computer exercises. You might try these if you have familiarity with BASIC. The book also provides extra topics, computer applications, and applications. The *Teacher's Manual with Solutions* provides programs in both BASIC and Pascal for the computer applications.

Chapter **1**

Foundations of Real Analysis

The set of real numbers with addition and multiplication is a special mathematical system. In this chapter some basic ideas underlying mathematical systems are discussed. Moreover, properties of the set of real numbers and open sentences involving real numbers are reviewed.

Sets and Sentences

1-1 SETS

Frequently in mathematics we use the notion of a *set*. A **set** is a collection of objects for which there is a definite criterion that enables us to tell what is in the set and what is not in the set. The following are examples of sets.

the set of all even integers

the set of all triangles in the plane

the set of all mathematics teachers at your school

We call the objects in a given set the **members** of that set. For example, 2 is a member of the set of even integers, but 3 is not a member of that set.

Capital letters such as A, B, and C are used to denote sets. Lower-case letters such as a, b, and c are used to denote members of sets. The symbol $\in$ is used to indicate membership in a set. If A is a set and a is a member of A,

---

Computers have been of great value in many kinds of mathematical exploration. The computer-generated display opposite shows some details of the Mandelbrot set. This set is a region of the complex plane, which is discussed in a later chapter.

we write $a \in A$. If a is not a member of A, we write $a \notin A$. For example, if E is the set of even integers,

$$2 \in E$$
$$3 \notin E.$$

In this text we will use the following special symbols to denote important sets of numbers.

N	the set of natural numbers
Z	the set of integers
Q	the set of rational numbers
$\mathcal{R}$	the set of real numbers

We can also write a set by **roster**, or **listing**. For example, we can write the set N of natural numbers as

$$\{1, 2, 3, \ldots\}$$

where . . . indicates that the suggested pattern continues.

All the members of one set might be members of a second set. For example, each mathematics teacher at your school is a teacher of some subject at your school. We say that the set of all mathematics teachers at your school is a *subset* of the set of all teachers at your school. In general, if every member of a set G is also a member of a set H, then G is a **subset** of H, written $G \subseteq H$. If H has a member that G does not have, then G is a **proper subset** of H, written $G \subset H$.

Two sets G and H are **equal** (written $G = H$) if they have exactly the same members. To show that two sets are equal, we can use the fact that $H = G$ if $G \subseteq H$ and $H \subseteq G$.

EXAMPLE 1 Describe any subset relationships for each pair of sets.

a. Z and $\mathcal{R}$

b. the set of odd integers and the set of even integers

c. $\{1, 2, 3, 4\}$ and $\{2, 1, 4, 3\}$

Solution

a. Since every integer is a real number, $Z \subseteq \mathcal{R}$, and because $\frac{1}{2}$ is a real number but not an integer, $Z \subset \mathcal{R}$.

b. No odd integer is an even integer. No even integer is an odd integer. Thus neither set is a subset of the other set.

c. The sets $\{1, 2, 3, 4\}$ and $\{2, 1, 4, 3\}$ have exactly the same members. $\{1, 2, 3, 4\} \subseteq \{2, 1, 4, 3\}$, $\{2, 1, 4, 3\} \subseteq \{1, 2, 3, 4\}$. Furthermore $\{1, 2, 3, 4\} = \{2, 1, 4, 3\}$.

The **null set**, or **empty set**, denoted $\emptyset$, is the set having no members. The null set is a subset of every set.

We can perform operations on sets.

DEFINITION

Let A and B be sets.

The **union** of A and B, written $A \cup B$, is the set consisting of all members of either A or B.

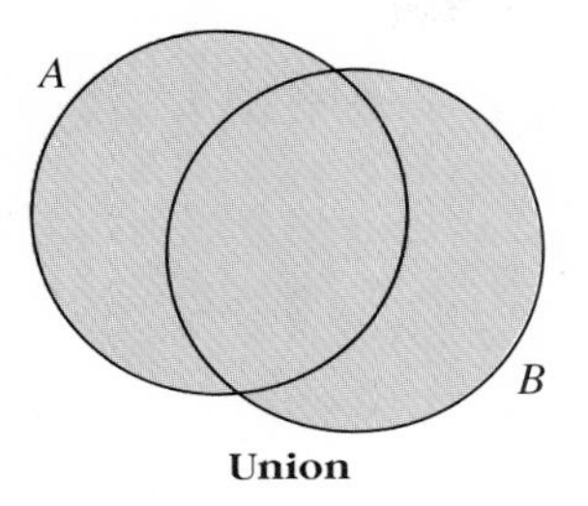

Union

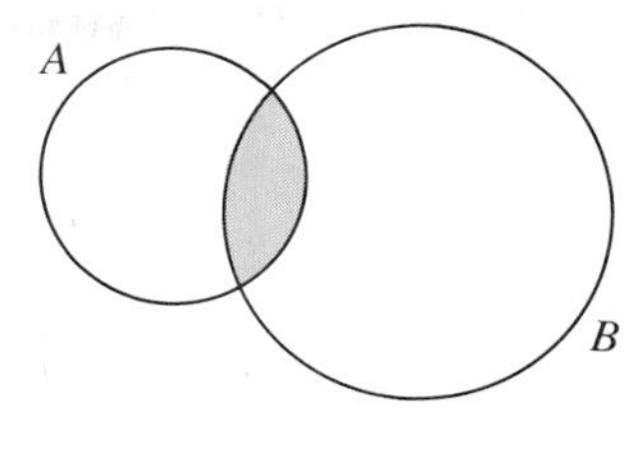

Intersection

The **intersection** of A and B, written $A \cap B$, is the set consisting of all members of both A and B.

EXAMPLE 2 Let $A = \{1, 2, 3, 4, 5\}$, $B = \{3, 4, 5, 6, 7\}$, and $C = \{1, 2, 3, 6, 7\}$. Find each intersection or union.

a. $A \cup B$ **b.** $A \cap B$ **c.** $A \cup (B \cap C)$

Solution

a. The union of A and B is the set of all members of A or B. $A \cup B = \{1, 2, 3, 4, 5, 6, 7\}$.

b. The intersection of A and B is the set of all members of both A and B. $A \cap B = \{3, 4, 5\}$.

c. First find $B \cap C$. $B \cap C = \{3, 6, 7\}$. Then find the union of A and $\{3, 6, 7\}$. $A \cup (B \cap C) = \{1, 2, 3, 4, 5, 6, 7\}$.

In order to define a third operation, *complementation*, we must introduce the concept of the *universal set*. For example, in geometry the universal set might be the set of all points in the plane. In algebra, the universal set might be the set of all real numbers.

DEFINITION

Let A be a set and U be the universal set. The **complement** of A, denoted by A', is the set of all members of U not in A.

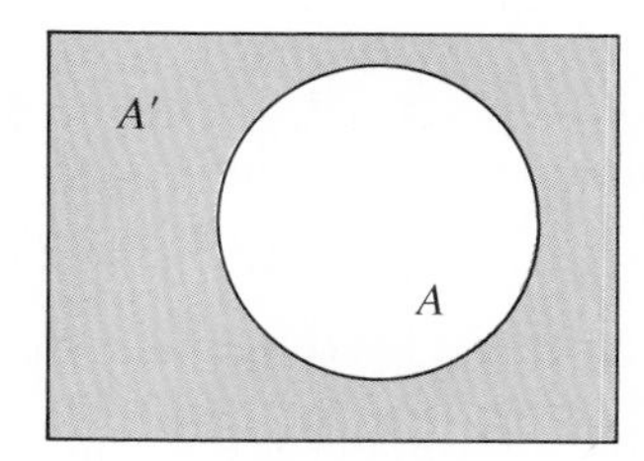

For example, if U is the set of all real numbers and A is the set of all integers, then A' is the set of all real numbers that are not integers. Note that the complement of $\emptyset$ is U and the complement of U is $\emptyset$.

EXAMPLE 3 Let $U = \{1, 2, 3, 4, 5, 6, 7, 8\}$, $A = \{1, 2, 3\}$, $B = \{4, 5, 6\}$, and $C = \{1, 2, 4, 8\}$. Find each complement.

a. A' **b.** $(A \cup C)'$ **c.** $(A \cap B)'$

Solution **a.** $A' = \{4, 5, 6, 7, 8\}$

b. $A \cup C = \{1, 2, 3, 4, 8\}$. Thus $(A \cup C)' = \{5, 6, 7\}$.

c. $A \cap B = \emptyset$. Thus $(A \cap B)' = U = \{1, 2, 3, 4, 5, 6, 7, 8\}$.

EXERCISES

Let $A = \{a, b, c, d\}$, $B = \{a, b, c, d, e\}$, $C = \{a, d\}$, and $D = \{b, c\}$. Describe any subset relationships for each pair of sets.

A **1.** $A; D$ **2.** $C; D$ **3.** $C \cup D; A$ **4.** $A \cap B; D$

Let E = {even integers}, O = {odd integers}, and Z be the universal set. Find each union, intersection, or complement.

5. E' **6.** O' **7.** $E \cup O$ **8.** $E \cap O$

State whether each statement is true or false.

9. A square $\notin$ {all parallelograms}

10. $\{21\} \subset$ {all integral multiples of 7}

11. $\{1, 3\} \subseteq$ {all odd integers}

12. $\emptyset \in \{\{1\}, \{2\}, \emptyset\}$

Let $A = \{1, 2, 3, 4\}$, $B = \{1, 4, 6, 8\}$, $C = \{2, 4, 5, 8, 10\}$, and $U = \{1, 2, 3, 4, 5, 6, 7, 8, 9, 10, 11, 12, 13\}$. Find each union, intersection, or complement.

13. $A \cup B$ **14.** $B \cap C$ **15.** $(A \cup B) \cup C$

16. $A \cap (B \cup C)$ **17.** A' **18.** $(A \cup C)'$

19. $(A \cap B)'$ **20.** $A' \cup B$ **21.** $(A \cup B) \cap C'$

22. $C \cup (B \cup C)'$ **23.** $A' \cap B' \cap C'$ **24.** $(B \cap B') \cap A$

25. $\emptyset \cup A$ **26.** $\emptyset \cap A$ **27.** $U \cap (A \cup B \cup C)'$

List all subsets of each set.

B **28.** $\emptyset$ **29.** $\{4\}$ **30.** $\{1, 3\}$ **31.** $\{1, 2, 3\}$

The *power set* of a set A, denoted by $\mathcal{P}(A)$, is the set of all subsets of A. Tell how many members the power set of each set has.

32. $\emptyset$ **33.** Exercise 29 **34.** Exercise 30 **35.** Exercise 31

36. If a set has five members, how many members does its power set have?

C **37.** Give a formula for the number of members of $\mathcal{P}(A)$ if A has n members.

38. Give a formula for the number of members of the power set of the power set of a set having n members.

39. If the power set of B has eight members more than the power set of A, how many members do A and B have?

In Exercises 40–43, follow the method shown to prove each statement.

EXAMPLE Show that $A \cup B = B \cup A$.

Solution If $x \in A \cup B$, then $x \in A$ or $x \in B$. Thus $x \in B$ or $x \in A$.
Hence $A \cup B \subseteq B \cup A$.
If $x \in B \cup A$, then $x \in B$ or $x \in A$. Thus $x \in A$ or $x \in B$.
Hence $B \cup A \subseteq A \cup B$.
Therefore $A \cup B = B \cup A$.

40. $A \cap B = B \cap A$

41. $\emptyset \cup A = A$

42. $A \cup (B \cap C) = (A \cup B) \cap (A \cup C)$

43. $A \cap (B \cup C) = (A \cap B) \cup (A \cap C)$

COMPUTER EXERCISES

1. Write a program to find any subset relationships between two sets A and B of numbers that the user enters. Assume that the maximum number of members in each set is less than 100. *Note:* For a particular pair of sets A and B it is possible to have: A is a subset of B and B is not a subset of A.

2. Run the program in Exercise 1 for each pair of sets.
a. $A = \{1, 2, 3, 8\}$; $B = \{1, 2, 3, 4, 5, 8\}$
b. $A = \{4, 3, 5, 1\}$; $B = \{5, 4, 1, 3\}$
c. $A = \{7, 5, 9, 3\}$; $B = \{3, 5, 7\}$
d. $A = \{2, 4, 6, 8\}$; $B = \{1, 3, 5, 7\}$

1-2 OPEN SENTENCES

In mathematics we frequently encounter *sentences*. For example, $2 = \frac{6}{3}$ is a sentence that is true, and $1 = 0$ is a sentence that is false. We also encounter sentences of a different type. For example, $x + 3 = 5$ is a sentence that is neither true nor false as it stands. If $x = 2$, then the sentence $x + 3 = 5$ is true. If $x = 4$, then $x + 3 = 5$ is false.

Let D be a set and let x represent any member of D. Any sentence involving x is called an **open sentence** in x; x is called a **variable**; and D is called the **domain**, or **replacement set**, of x. The **solution set**, or **truth set**, of an open sentence in x is the set of all x in D such that the open sentence is true. For example, $x^2 = 4$ is an open sentence in x. If D is the set of all integers, then the solution set of $x^2 = 4$ is $\{-2, 2\}$. If D is the set of all positive integers, then the solution set of $x^2 = 4$ is $\{2\}$.

An **identity** is an open sentence whose solution set is the domain of its variables. For example, $x + 3 = 3 + x$ is an identity over the set of real numbers. A **contradiction** is a sentence whose solution set is empty. For example, $x + 3 = x + 5$ is a contradiction because no real number satisfies $x + 3 = x + 5$.

Often the solution set of an open sentence is written in *set-builder notation*. For example, the solution set of $2x > 6$ over $\mathcal{R}$ can be written as

$\{x \in \mathcal{R}: x > 3\}$. This is read "the set of all real numbers x such that $x > 3$." If the domain of a variable is $\mathcal{R}$, we usually omit $\mathcal{R}$ and write $\{x: x > 3\}$.

The sentence $2 = \frac{6}{3}$ and $1 = 0$ is a compound sentence formed from two sentences and the word *and*. In general, if p represents a sentence and q represents a sentence, then the **conjunction** of p and q (also written $p \wedge q$) is the sentence p *and* q. The conjunction $p \wedge q$ is defined to be true if both p and q are true and is defined to be false otherwise. The definition can be given in a *truth table*.

The truth or falsity of a sentence is called its *truth value*.

p	q	$p \wedge q$
true	true	true
true	false	false
false	true	false
false	false	false

EXAMPLE 1 Give the truth value of each conjunction.

a. $2 < 5$ and $-5 < 2$ **b.** $2 = \frac{6}{3}$ and $1 = 0$

c. $3x + 1 < 5$ and $x < 5$ where x is an integer

Solution

a. Since both $2 < 5$ and $-5 < 2$ are true, the conjunction is true.

b. Since $1 = 0$ is false, the conjunction is false even though $2 = \frac{6}{3}$ is true.

c. The truth value of the conjunction cannot be determined unless specific values of x are given. If $x = 1$, the conjunction is true. If $x = 2$, the conjunction is false.

The solution set of the conjunction of two open sentences is the intersection of the solution sets of the given open sentences.

EXAMPLE 2 If p is the open sentence $2x - 8 < 0$ and q is the open sentence $4 - x \leq 3$, find and graph each solution set over $\mathcal{R}$ on a number line.

a. p **b.** q **c.** $p \wedge q$

Solution

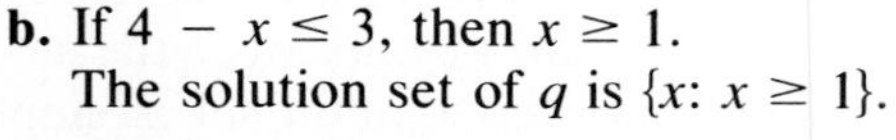

a. If $2x - 8 < 0$, then $x < 4$.
The solution set of p is $\{x: x < 4\}$.

b. If $4 - x \leq 3$, then $x \geq 1$.
The solution set of q is $\{x: x \geq 1\}$.

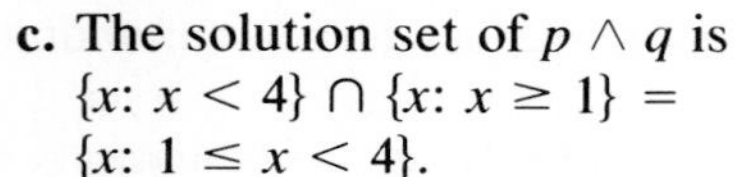

c. The solution set of $p \wedge q$ is
$\{x: x < 4\} \cap \{x: x \geq 1\} =$
$\{x: 1 \leq x < 4\}$.

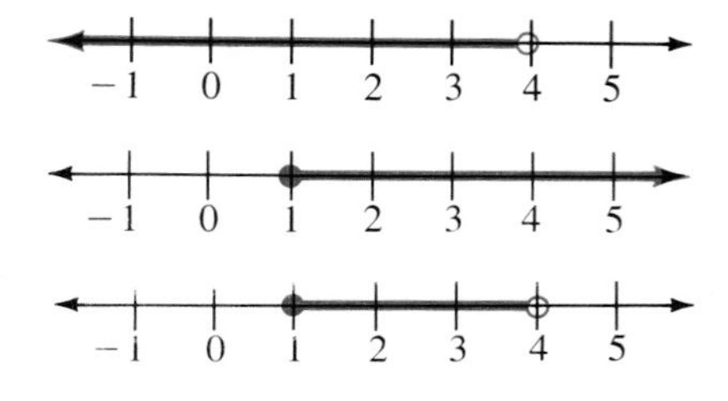

Two sentences may be joined by the word *or*. Let p and q represent sentences. The *disjunction* of p and q is the sentence p *or* q (also written $p \vee q$). The **disjunction** $p \vee q$ is defined to be true if either p or q is true. This definition can be given in a truth table.

p	q	$p \vee q$
true	true	true
true	false	true
false	true	true
false	false	false

EXAMPLE 3 Give the truth value of each disjunction.

a. Jupiter's orbit lies outside the orbit of Mars or it never snows in Minneapolis.

b. $3x + 1 \geq 22$ or $x < 5$.

Solution **a.** Since Jupiter's orbit does lie outside the orbit of Mars, the disjunction is true.

b. The truth value cannot be determined unless specific values of x are given. If $x = 4$, the disjunction is true. If $x = 6$, the disjunction is false.

The solution set of the disjunction of two open sentences is the union of the solution sets of the given open sentences.

EXAMPLE 4 If p is the open sentence $2x - 1 > 7$ and q is the open sentence $x + 4 \leq 2$, find and graph each solution set over $\mathcal{R}$ on a number line.

a. p **b.** q **c.** $p \vee q$

Solution **a.** If $2x - 1 > 7$, then $x > 4$.
The solution set of p is $\{x: x > 4\}$.

b. If $x + 4 \leq 2$, then $x \leq -2$.
The solution set of q is $\{x: x \leq -2\}$.

c. The solution set of $p \vee q$ is
$\{x: x > 4\} \cup \{x: x \leq -2\} =$
$\{x: x \leq -2 \text{ or } x > 4\}$.

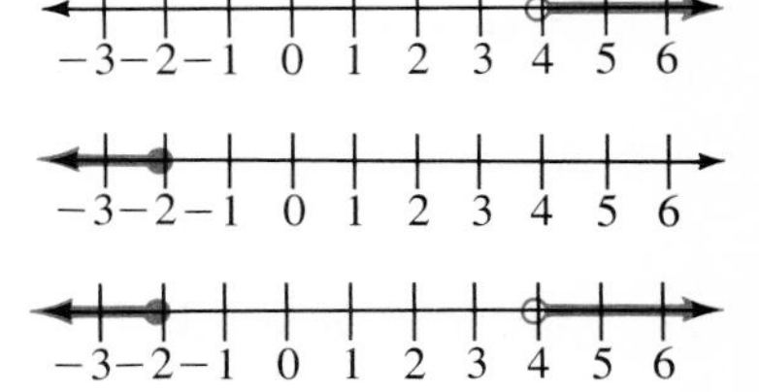

The sentence "$1 = 0$" is false. The sentence "1 is not equal to 0" is true. We call the second sentence the *negation* of the first sentence. In general, if p is a sentence, then the sentence *not* p, also written p', is called the **negation** of p. *Not* p is true when p is false and false when p is true.

EXAMPLE 5 State the negation of each sentence.

a. $a = 0$ or $b = 0$ **b.** $a = 0$ and $b = 0$

Solution **a.** To negate "$a = 0$ or $b = 0$," negate the parts "$a = 0$" and "$b = 0$," and the connective *or*. Thus the negation of the given sentence is "$a \neq 0$ and $b \neq 0$."

b. To negate "$a = 0$ and $b = 0$," negate "$a = 0$," "$b = 0$," and the connective *and*. Thus the negation of the given sentence is "$a \neq 0$ or $b \neq 0$."

In general, if p and q represent sentences, then the negation of "p or q" is "not p and not q"; the negation of "p and q" is "not p or not q".

EXAMPLE 6 State the negation of each sentence.

a. For every real number x, $x^2 \geq 0$.

b. There is a real number x such that $x^2 = -1$.

Solution **a.** The negation of the given sentence is "There is *at least one* real number whose square is negative."

b. The negation of the given sentence is "There are *no* real numbers x such that $x^2 = -1$."

EXERCISES

State whether each statement is true or false.

A

1. 4 is an even number and 5 is an odd number.

2. 3 is negative or 3 is positive.

3. 10 is a multiple of 3 or 10 is a multiple of 4.

4. $-3 < 0$ and $-3 \geq 0$.

Find and graph each solution set over $\mathcal{R}$. **a.** p **b.** q **c.** $p \wedge q$

5. p: $x > 0$
q: $2x < 6$

6. p: $x < 0$
q: $3x > -15$

7. p: $2x \leq 6$
q: $3x \geq 9$

8. p: $4y + 1 > 5$
q: $3y - 5 \leq 7$

9. p: $4t - 5 \geq 3$
q: $3t + 5 \leq 26$

10. p: $2x - 1 \geq 1$
q: $3x + 1 \leq 4$

Find and graph each solution set over $\mathcal{R}$. **a.** p **b.** q **c.** $p \vee q$

11. p: $x > 0$
q: $2x < -4$

12. p: $x < 0$
q: $3x < 12$

13. p: $3w - 1 > 5$
q: $4w + 3 \leq -1$

14. p: $6x \geq 6$
q: $3x - 5 < -14$

15. p: $2x - 5 \geq -1$
q: $3x - 5 \leq 1$

16. p: $4t + 6 \leq 14$
q: $t - 3 \geq -1$

Write the negation of each sentence.

17. There is a positive square root of 2.

18. For every real number x, $x > 0$ or $x < 0$.

19. For every real number x, $x < 5$, $x = 5$, or $x > 5$.

20. There is a square whose sides have length 2.

21. $2 = \frac{6}{3}$ and $\frac{6}{3} = \frac{12}{6}$.

22. $0 < 1$ or $4 - 1 = 3$.

B

23. $2x < -4$ or $3x > 6$.

24. $2x > -4$ and $3x > 6$.

25. Find and graph on a number line the solution set over $\mathcal{R}$ of the negation of the disjunction in Exercise 23.

26. Find and graph on a number line the solution set over $\mathcal{R}$ of the negation of the conjunction in Exercise 24.

27. If p is the open sentence $x > 2$ and q is the open sentence $x > 5$, find and graph the solution set over $\mathcal{R}$ of $p \wedge q'$.

28. If p is the open sentence $x \leq 4$ and q is the open sentence $x > 7$, find and graph the solution set over $\mathcal{R}$ of $p \vee q'$.

State whether each sentence over $\mathcal{R}$ is an identity, a contradiction, or a sentence that is sometimes true and sometimes false.

29. $x^2 + 2x + 1 = (x + 1)^2$

30. $(x - 3)^2 = -4$

31. $x^2 - 5x + 6 = 0$

32. $2x - 1 < 5$ and $3x + 2 > 11$

33. $\sqrt{x^2} = x$

34. $x^2 - y^2 = (x + y)(x - y)$

In Exercises 35–42 let p and q represent sentences and let p be true and let q be false. Give the truth value of each sentence.

C

35. $p \wedge q'$

36. $(p \vee q)'$

37. $(p \wedge q)'$

38. $p \vee (p \wedge q)$

39. $p' \wedge q'$

40. $p \wedge q \wedge p'$

41. $p \vee p'$

42. $(p' \vee q')'$

Let p, q, r, and s represent the following open sentences: p: $2x - 1 > 5$, q: $3x + 1 < 4$, r: $x + 4 < 9$, and s: $3x - 5 > 13$. Find and graph each solution set over $\mathcal{R}$.

43. $(r \wedge p) \vee s$

44. $q' \wedge s' \wedge r'$

45. $(r \wedge s)' \vee q$

46. $(r \vee s) \wedge p'$

COMPUTER EXERCISES

1. Write a computer program to list the members of the intersection of two sets of numbers entered by the user. Assume that the number of members of each set is less than 100. *Note*: The intersection might be empty.

2. Run the program in Exercise 1 for each pair of sets.
   a. $A = \{4, 6, 8, 3, 5\}$; $B = \{6, 5, 1, 2\}$
   b. $A = \{5, 6, 8, 9\}$; $B = \{5, 6, 8, 9\}$
   c. $A = \{3, 8, 5, 2\}$; $B = \{2, 4, 7, 9\}$
   d. $A = \{1, 2, 3, 4, 5\}$; $B = \{1, 2, 3, 4\}$

1-3 CONDITIONAL SENTENCES

Many theorems in mathematics are expressed in the form "if p, then q" where p and q are sentences. For example,

> If two sides of a triangle are congruent,
> then the angles opposite those sides are congruent.

If p and q represent sentences, then the sentence "if p, then q" (also written $p \rightarrow q$) is called a **conditional sentence**. It is defined to be true in every case except when p is true and q is false. This definition can be given in a truth table.

p	q	$p \rightarrow q$
true	true	true
true	false	false
false	true	true
false	false	true

In the conditional sentence $p \rightarrow q$, we call p the *hypothesis* and we call q the *conclusion*. We can summarize rows 3 and 4 of the truth table by saying that any conditional sentence with a false hypothesis is true.

The **converse** of "if p, then q" is the conditional sentence "if q, then p." The **biconditional** sentence "p if and only if q" (also written $p \leftrightarrow q$) is the conjunction of "if p, then q" and its converse. The following truth table summarizes the definition of $p \leftrightarrow q$.

p	q	$p \rightarrow q$	$q \rightarrow p$	$p \leftrightarrow q$
true	true	true	true	true
true	false	false	true	false
false	true	true	false	false
false	false	true	true	true

EXAMPLE 1 If p is the sentence "$a = b$" and q is the sentence "$a^2 = b^2$", state the truth value of each of the following conditional sentences.

a. $p \rightarrow q$ **b.** $q \rightarrow p$ **c.** $p \leftrightarrow q$

Solution **a.** $p \rightarrow q$ is "If $a = b$, then $a^2 = b^2$."
This conditional sentence is true.

b. $q \to p$ is "If $a^2 = b^2$, then $a = b$."
This conditional sentence is false.
For example, $(-2)^2 = (2)^2$ but $2 \neq -2$.

c. $p \leftrightarrow q$ is "$a = b$ if and only if $a^2 = b^2$."
The biconditional sentence is false because $q \to p$ is false.

The conditional sentence $p \to q$ can be stated in a variety of ways.

If p, then q.

p is sufficient for q.

q is necessary for p.

p only if q.

For example, "If two triangles are congruent, then they are similar" can also be stated in the following three ways:

Congruence of two triangles is sufficient for their similarity.

Similarity of two triangles is necessary for their congruence.

Two triangles are congruent only if they are similar.

The sentence p *only if* q means that if q is not true, then p is not true. We call $q' \to p'$ the **contrapositive** of $p \to q$. Thus in the example above the contrapositive of "If two triangles are congruent, then they are similar" is "If two triangles are not similar, then they are not congruent."

EXAMPLE 2 State the contrapositive of "If a quadrilateral is a parallelogram, then opposite sides are congruent."

Solution A quadrilateral has two pairs of opposite sides. The negation of the conclusion of the given conditional sentence is "At least one pair of opposite sides of a quadrilateral are not congruent." The contrapositive of the given statement is "If at least one pair of opposite sides of a quadrilateral are not congruent, then the quadrilateral is not a parallelogram."

Many theorems in mathematics are proved by the strategy of *contraposition*. That is, by proving that $q' \to p'$ is true we prove that $p \to q$ is true.

Other theorems are proved by contradiction. We may prove that $p \to q$ is true by assuming that p is true and q is false. If we logically obtain a false conclusion from these assumptions, we may assert that $p \to q$ is true. A proof using this strategy is called an *indirect proof* or *proof by contradiction*.

EXAMPLE 3 Given the conditional sentence "If $x \neq y$, then $x + 1 \neq y + 1$," what assumptions would be made in order to give an indirect proof?

Solution To prove the given conditional sentence indirectly, assume that $x \neq y$ and $x + 1 = y + 1$.

EXERCISES

State whether each conditional sentence is true or false.

A

1. If $1 = 0$, then $1 = -1$.
2. If 12 is a multiple of 6, then 24 is a multiple of 6.
3. If every square is a rectangle, then every rectangle is a square.
4. A triangle is isosceles only if it is equilateral.

Give the converse of each conditional sentence. State whether the converse is sometimes, always, or never true.

5. If two triangles are congruent, then their corresponding angles are congruent.
6. If 2 is a factor of an integer, then 2 is a factor of the square of that integer.
7. If x is an odd integer, then x^2 is an odd integer.
8. If $x + y = x + z$, then $y = z$.
9. If $ab < ac$, then $b < c$.
10. If $x^2 < 0$, then $x^4 \geq 0$.

State the contrapositive of each conditional sentence.

11. If roses are black, then violets are green.
12. If a triangle is equilateral, then each angle has measure 60°.
13. If an angle of a parallelogram is a right angle, then the parallelogram is a rectangle.
14. If $ab = ac$ and $a \neq 0$, then $b = c$.
15. If $ab = b$ and $b \neq 0$, then $a = 1$.
16. If $x^2 - 9 = 0$, then $x = 3$ or $x = -3$.

State the assumptions that would be made if the given statement is to be proved by contradiction.

17. If $x^2 \neq y^2$, then $x \neq y$.
18. If $x \neq y$, then $-x \neq -y$.
19. $1 \neq 2$.
20. If $x < y$ and $x = u$, then $u < y$.

State each conditional sentence in the if-then form.

B

21. For all sets R and S, $R \cap S = R$ is necessary for $R \subset S$.
22. An integer x is odd only if x^2 is an odd integer.
23. For $a, b \in \mathcal{R}$, $a \leq b$ and $b \leq a$ only if $a = b$.
24. For all sets S and T, $S \subset T$ is sufficient for $S \cup T = T$.
25. For $a > 0$, $b > 0$, $m \neq 0$, $a = b$ is necessary for $a^m = b^m$.
26. For a, b and $c \in \mathcal{R}$, and $a > 0$, $ab > ac$ is sufficient for $b > c$.

C **27.** For all x and y and $a > 0$, $a^x a^y = a^{x+y}$.

28. There is at least one triangle for which the sum of the lengths of two sides is less than the length of the third side.

Use the symbols $\wedge$, $\vee$, and $'$ to form a sentence involving p and q whose truth table is given.

29.

p	q	?
T	T	T
T	F	F
F	T	T
F	F	T

30.

p	q	?
T	T	F
T	F	F
F	T	T
F	F	F

31.

p	q	?
T	T	F
T	F	F
F	T	F
F	F	T

32.

p	q	?
T	T	F
T	F	F
F	T	F
F	F	F

Real Numbers

■ 1–4 ADDITION AND MULTIPLICATION PROPERTIES OF REAL NUMBERS

A **formal mathematical system** consists of *undefined objects*, *postulates* or *axioms*, *definitions*, and *theorems*. Geometry is such a system. Real analysis, the main subject of this text, is based on the system of real numbers. We will accept the real numbers as undefined. We will also accept equality (=), addition (+), and multiplication (×) as undefined. Some of the fundamental properties of this system are stated below.

Axioms of Equality

For all real numbers a, b, and c:

Reflexive Property $a = a$

Symmetric Property If $a = b$, then $b = a$.

Transitive Property If $a = b$ and $b = c$, then $a = c$.

Substitution Axiom

If $a = b$, then in any true sentence involving a we may substitute b for a and obtain another true sentence.

The following five axioms express the basic properties of addition.

Axioms of Addition

Closure For all real numbers a and b, $a + b$ is a unique real number.

Associative For all real numbers a, b, and c,

$$(a + b) + c = a + (b + c).$$

Additive Identity There exists a unique real number 0 (zero) such that $a + 0 = 0 + a = a$ for every real number a.

Additive Inverses For each real number a, there exists a real number $-a$ (the opposite of a) such that

$$a + (-a) = (-a) + a = 0.$$

Commutative For all real numbers a and b, $a + b = b + a$.

Because of the associative axiom we may write $a + b + c$ for either $(a + b) + c$ or $a + (b + c)$. It follows from this axiom that all such groupings lead to the same result. For example, $a + (b + (c + d)) = (a + b) + (c + d)$.

The operation of multiplication has similar properties.

Axioms of Multiplication

Closure For all real numbers a and b, ab is a unique real number.

Associative For all real numbers a, b, and c, $(ab)c = a(bc)$.

Multiplicative Identity There exists a unique nonzero real number 1 (one) such that $1 \cdot a = a \cdot 1 = a$.

Multiplicative Inverses For each nonzero real number a there exists a real number $\frac{1}{a}$ (the reciprocal of a) such that

$$a\left(\frac{1}{a}\right) = \left(\frac{1}{a}\right)a = 1.$$

Commutative For all real numbers a and b, $ab = ba$.

It can be shown that the additive inverse of each real number is unique (Exercise 40) and the multiplicative inverse of each nonzero real number is unique (Exercise 41).

Addition and multiplication are related by the following axiom.

The Distributive Axiom of Multiplication over Addition

For all real numbers a, b, and c, $a(b + c) = ab + ac$.

Subtraction and *division* are defined in terms of addition and multiplication respectively.

DEFINITION

For all real numbers a and b:

Subtraction $\quad a - b = a + (-b)$

Division $\quad \frac{a}{b} = a\left(\frac{1}{b}\right)$ provided $b \neq 0$

Two important types of mathematical structures are *groups* and *fields*. A **group** is a set of objects having one operation satisfying the axioms of closure, associativity, identity, and inverses. If the commutative axiom is also satisfied, the set is a **commutative group**. Both Z and $\mathcal{R}$ are commutative groups. A **field** F is a set of objects with two operations such that

(1) F is a commutative group under the first operation;

(2) its nonzero members form a commutative group under the second operation; and

(3) the second operation is distributive over the first operation.

The set $\mathcal{R}$ with $+$ and $\cdot$ is a field, but Z with $+$ and $\cdot$ is not a field.

Many theorems can be proved from the field axioms. Some useful facts about real numbers are stated in Theorem 1.

THEOREM 1 For all real numbers a, b, and c:

1. $a = b$ if and only if $a + c = b + c$. (Cancellation Law for Addition)

2. $a = b$ if and only if $ac = bc$ $(c \neq 0)$. (Cancellation Law for Multiplication)

3. If $a = b$, $-a = -b$.

4. $-(-a) = a$

5. $a \cdot 0 = 0$

6. $-0 = 0$

7. $-a = -1(a)$

8. $-(ab) = a(-b) = (-a)b$

9. $-(a + b) = -a + (-b)$

10. If $a \neq 0$, $\frac{1}{-a} = \frac{-1}{a} = -\frac{1}{a}$.

We will prove part (5) of the theorem.

Proof:

$a \cdot 0 = a \cdot 0$	reflexive property of equality
$a \cdot (0 + 0) = a \cdot 0$	additive identity; substitution
$a \cdot 0 + a \cdot 0 = a \cdot 0$	distributive axiom
$-(a \cdot 0) + a \cdot 0 + a \cdot 0 = -(a \cdot 0) + a \cdot 0$	part (1) of Theorem 1
$0 + a \cdot 0 = 0$	additive inverse
$a \cdot 0 = 0$	additive identity ■

The proofs of the other parts of the theorem are left as Exercises 21–29 on page 18.

When we solve equations such as $2x + 7 = 6$, we use Theorem 1 and the field axioms.

$2x + 7 = 6$	given equation
$2x + 7 + (-7) = 6 + (-7)$	part (1) of Theorem 1
$2x + 0 = -1$	additive inverse
$2x = -1$	additive identity
$\left(\frac{1}{2}\right)2x = \left(\frac{1}{2}\right)(-1)$	part (2) of Theorem 1
$x = -\frac{1}{2}$	multiplicative inverse; multiplicative identity part (7) of Theorem 1

This shows that if $2x + 7 = 6$ has a solution, then the solution must be $-\frac{1}{2}$. To show that $-\frac{1}{2}$ is indeed a solution, we can substitute $-\frac{1}{2}$ in the original equation or observe that the steps in the solution are reversible.

To solve an equation such as $x^2 + 3x - 10 = 0$, we use the **zero-product property**.

THEOREM 2 For all real numbers a and b:

$ab = 0$ if and only if $a = 0$ or $b = 0$.

Proof: *If part:*

Suppose that $a = 0$ or $b = 0$.

If $a = 0$, then $ab = 0 \cdot b = 0$ by Theorem 1, parts (2) and (5).

If $b = 0$, then $ab = a \cdot 0 = 0$ by Theorem 1, parts (2) and (5).

Only if part:

Suppose that $ab = 0$.

If $a \neq 0$, then $\frac{1}{a}(ab) = \frac{1}{a}(0)$ by Theorem 1, part (2)

$\left(\frac{1}{a}a\right)b = \frac{1}{a}(0)$	associativity
$1 \cdot b = 0$	multiplicative inverse
$b = 0$	multiplicative identity

Therefore $a = 0$ or $b = 0$. ■

Applying this theorem to $x^2 + 3x - 10 = 0$, we can find x.

$$x^2 + 3x - 10 = 0$$
$$(x + 5)(x - 2) = 0$$

Thus by Theorem 2 $\quad x + 5 = 0 \text{ or } x - 2 = 0$
and $\quad x = -5 \text{ or } x = 2.$

EXERCISES

Name the axiom, theorem, or definition that justifies each step in the proof. Variables represent real numbers.

A

1. $(x + y) - y = x$

Proof:

$$\begin{aligned}(x + y) - y &= (x + y) + (-y) \\ &= x + (y + (-y)) \\ &= x + 0 \\ &= x \\ (x + y) - y &= x\end{aligned}$$

2. If $a = b$, then $a^2 = b^2$.

Proof:

$$\begin{aligned}a &= b \\ aa &= ab \\ ab &= bb \\ aa &= bb \\ a^2 &= b^2\end{aligned}$$

3. $(-a)(-b) = ab$

Proof:

$$\begin{aligned}(-a)(-b) &= -((-a)b) \\ &= (-(-a))b \\ &= ab \\ (-a)(-b) &= ab\end{aligned}$$

4. $x(y - z) = xy - xz$

Proof:

$$\begin{aligned}x(y - z) &= x(y + (-z)) \\ &= xy + x(-z) \\ &= xy + (-xz) \\ &= xy - xz \\ x(y - z) &= xy - xz\end{aligned}$$

Solve each equation over $\mathcal{R}$.

5. $7(x - 3) + 2(2 - 3x) = 8$

6. $(3x + 1)(x - 4) = 0$

7. $\frac{1}{2}x + 2 = -\frac{3}{4}x + 1$

8. $\frac{3}{4}(x + 5) - 4x = \frac{3}{4}(2x - 3)$

9. $6x^2 + 5x + 1 = 0$

10. $7x^2(3x - 5) = 0$

State whether each set is closed under **(a)** addition and **(b)** multiplication. If a set is not closed under an operation, give an example that shows this.

11. $\{0\}$ **12.** $\{1, 2\}$ **13.** N **14.** $\{1\}$ **15.** $\{0, 3\}$ **16.** $\{0$ and natural numbers$\}$

In Exercises 17–20 an operation $*$ has been defined over the set of natural numbers. **a.** Find $2 * 3$. **b.** Determine whether or not N is closed under $*$. **c.** Determine whether $*$ is commutative. **d.** Determine whether $*$ is associative.

B

17. $x * y = (1 + x) + y$

18. $x * y = xy^2$

19. $x * y = x(2y)$

20. $x * y = \frac{x + y}{xy}$

In Exercises 21–29 prove each part of Theorem 1.

21. Part (1) **22.** Part (2) **23.** Part (3) **24.** Part (4)

25. Part (6) (*Hint*: $0 + (-0) = 0$)

26. Part (7) (*Hint*: $(-1)a + a = (-1)a + 1 \cdot a$)

27. Part (8) (*Hint*: $-(ab) = (-1)(ab)$)

28. Part (9) (*Hint*: $(a + b) + (-(a + b)) = 0$)

29. Part (10) (*Hint*: Prove that $\frac{1}{-a} = \frac{-1}{a}$ and that $\frac{-1}{a} = -\frac{1}{a}$ and use the transitive property.)

Prove each statement. Variables represent real numbers and are nonzero where necessary.

30. $\dfrac{1}{\frac{1}{a}} = a$

31. $\dfrac{0}{a} = 0$

32. $\dfrac{1}{ab} = \dfrac{1}{a} \cdot \dfrac{1}{b}$

33. $\dfrac{1}{\frac{a}{b}} = \dfrac{b}{a}$

34. $\dfrac{\frac{a}{b}}{\frac{c}{d}} = \dfrac{ad}{bc}$

35. $\dfrac{1}{a} + \dfrac{1}{b} = \dfrac{a + b}{ab}$

36. $\dfrac{a + b}{c} = \dfrac{a}{c} + \dfrac{b}{c}$

37. $\dfrac{ab}{a} = b$

38. $\dfrac{a}{b} = \dfrac{c}{d}$ if and only if $ad = bc$.

39. $\dfrac{ac}{bc} = \dfrac{a}{b}$

C

40. Prove that the additive inverse of a real number is unique. (*Hint*: Assume that a real number a has two distinct additive inverses.)

41. Prove that the multiplicative inverse of a nonzero real number is unique. (*Hint*: Assume that a nonzero real number a has two distinct multiplicative inverses.)

42. Let $K = \{a, b, c, d\}$. Let the operation $\circ$ be defined on K by the table below.

a. Is K closed under $\circ$?
b. Is $\circ$ an associative operation?
c. Is there an identity member for $\circ$ in K?
d. Does each member in K have an inverse under $\circ$?
e. Is K a group under $\circ$?
f. Is K a commutative group under $\circ$?

$\circ$	a	b	c	d
a	a	b	c	d
b	b	a	d	c
c	c	d	a	b
d	d	c	b	a

43. Let $Z_5 = \{0, 1, 2, 3, 4\}$. Define $\oplus$ and $\odot$ for Z_5 as follows:
$a \oplus b =$ the remainder on dividing $a + b$ by 5 and
$a \odot b =$ the remainder on dividing ab by 5.

a. Show that Z_5 is a commutative group under $\oplus$.
b. Show that Z_5 without 0 is a commutative group under $\odot$.
c. Show that Z_5 is a field.

44. Let $Z_6 = \{0, 1, 2, 3, 4, 5\}$. Define $\oplus$ and $\odot$ for Z_6 as follows:
$a \oplus b =$ the remainder on dividing $a + b$ by 6 and
$a \odot b =$ the remainder on dividing ab by 6.
Is Z_6 a field under $\oplus$ and $\odot$?

45. For what n is $Z_n = \{0, 1, 2, \ldots, n - 1\}$ a field?

COMPUTER EXERCISES

1. Let $Z_5 = \{0, 1, 2, 3, 4\}$. Define multiplication in Z_5 as follows: For all a and b in Z_5, ab is the remainder when ab is divided by 5. Write a computer program to display the multiplication table for Z_5.
2. From the table what is the inverse of 4?
3. What algebraic property is reflected by the fact that the entries in the table are symmetric about the diagonal from the upper left to the lower right of the table?

1-5 ORDER AND INEQUALITIES

The real numbers are ordered by a relation less than (<). This is illustrated on the following number line.

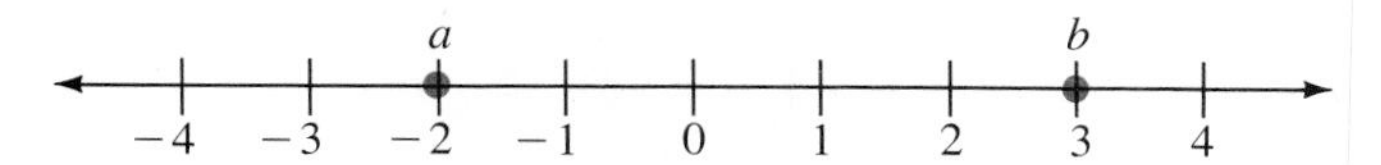

We say that *a is less than b* ($a < b$) if the graph of a is to the left of the graph of b on the number line. This does not constitute a precise definition of order. We take < as an undefined relation and use the number line as a means of picturing order.

The following axioms describe the basic properties of order.

Axiom of Comparison	For all real numbers a and b, $a < b$, $a = b$, or $b < a$.
Transitive Axiom of Order	For all real numbers a, b, and c, if $a < b$ and $b < c$, then $a < c$.
Addition Axiom of Order	For all real numbers a, b, and c, if $a < b$, then $a + c < b + c$.

A real number c is **positive** if $0 < c$ and **negative** if $c < 0$. The number 0 is neither positive nor negative. The following axiom involves multiplication and positive numbers.

Multiplication Axiom of Order For all real numbers a, b and c, if $0 < c$ and $a < b$, then $ac < bc$.

We can define $>$, $\geq$, and $\leq$ in terms of $<$ and $=$.

$a > b$ (a is greater than b) if and only if $b < a$.

$a \geq b$ (a is greater than or equal to b) if and only if $a > b$ or $a = b$.

$a \leq b$ (a is less than or equal to b) if and only if $a < b$ or $a = b$.

All the axioms of order are true when $<$ is replaced by any of the symbols defined above. The symbols $<$, $>$, $\leq$, and $\geq$, together with their negations $\nless$, $\ngtr$, $\nleq$, and $\ngeq$, and the symbol $\neq$ are called *inequality symbols*. An **inequality** is any sentence involving an inequality symbol.

The following theorems are useful in dealing with inequalities.

THEOREM 3 For all real numbers a, b, c, and d:

1. If $a < b$ and $c < d$, then $a + c < b + d$.

2. If $a > 0$, then $-a < 0$ and $\frac{1}{a} > 0$.

3. If $a < 0$, then $-a > 0$ and $\frac{1}{a} < 0$.

We will prove part (1) and leave the proofs of parts (2) and (3) as Exercises 21 and 22.

Proof:

$a < b$ and $c < d$	hypothesis
$a + c < b + c$ and $b + c < b + d$	addition axiom of order
$a + c < b + d$	transitive axiom of order ■

THEOREM 4 For all real numbers a, b, and c, if $c < 0$ and $a < b$, then $ac > bc$.

Proof: Assume that $c < 0$ and $a < b$. Then by Theorem 3, part (3), $-c > 0$.

$a(-c) < b(-c)$	multiplication axiom of order
$-ac < -bc$	Theorem 1, part (8)
$-ac + (ac + bc) < -bc + (ac + bc)$	addition axiom of order
$(-ac + ac) + bc < (-bc + bc) + ac$	commutative and associative axioms of addition
$bc < ac$,	additive inverses and identity
or $ac > bc$ ■	

The axioms and theorems can be used to solve inequalities.

EXAMPLE 1 Solve $-5x + 3 \le 2$ over $\mathcal{R}$ and graph the solution set.

Solution

$$-5x + 3 \le 2$$
$$-5x + 3 + (-3) \le 2 + (-3)$$
$$-5x \le -1$$
$$\left(-\frac{1}{5}\right)(-5x) \ge \left(-\frac{1}{5}\right)(-1)$$
$$x \ge \frac{1}{5}$$

The solution set is $\left\{x: x \ge \frac{1}{5}\right\}$. Its graph is shown below.

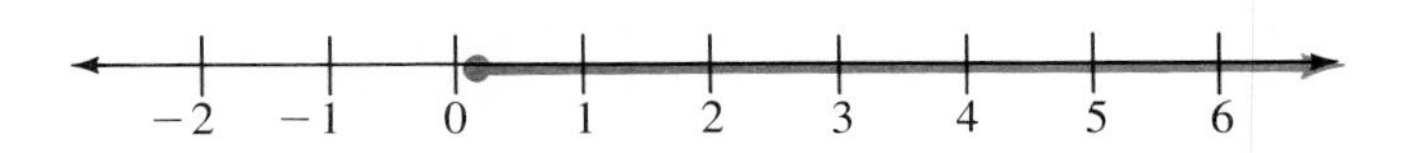

EXAMPLE 2 Solve $3x + 4 \ge 10$ and $2x - 5 < 9$ simultaneously over $\mathcal{R}$ and graph the solution set.

Solution Solve each inequality separately and form the intersection of the solution sets.

$$\begin{array}{rcr} 3x + 4 \ge 10 & \text{and} & 2x - 5 < 9 \\ 3x \ge 6 & | & 2x < 14 \\ x \ge 2 & \text{and} & x < 7 \end{array}$$

The complete solution set is $\{x: 2 \le x < 7\}$. The graph of the solution set is shown below.

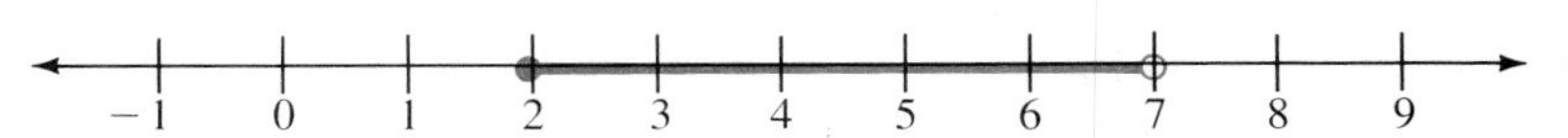

EXAMPLE 3 Solve $-3 \le 2t - 5 \le 7$ and graph the solution set over $\mathcal{R}$.

Solution The given open sentence can be written

$$-3 \le 2t - 5 \text{ and } 2t - 5 \le 7.$$

Solve each inequality and graph the intersection of the solution sets.

$$\begin{array}{rcr} -3 + 5 \le 2t - 5 + 5 & \text{and} & 2t - 5 + 5 \le 7 + 5 \\ 2 \le 2t & | & 2t \le 12 \\ 1 \le t & \text{and} & t \le 6 \end{array}$$

The solution set of both inequalities together is $\{t: 1 \le t \le 6\}$. The graph is shown below.

The following theorem extends the zero-product property to inequalities. You are asked for proof of the theorem in Exercises 27 and 28.

THEOREM 5 For all real numbers a and b:

1. $ab > 0$ if and only if a and b are both positive or both negative.
2. $ab < 0$ if and only if a and b are opposite in sign.

EXAMPLE 4 Solve $x^2 - 3x > 0$ and graph the solution set over $\mathcal{R}$.

Solution

$$x^2 - 3x = x(x - 3)$$

In order for $x(x - 3)$ to be positive, both factors must have the same sign (Theorem 5, part (1)).

$$\begin{array}{ccc} x > 0 \text{ and } x - 3 > 0 & \text{or} & x < 0 \text{ and } x - 3 < 0 \\ x > 0 \text{ and } x > 3 & | & x < 0 \text{ and } x < 3 \\ x > 3 & \text{or} & x < 0 \end{array}$$

The solution set is $\{x\colon x < 0 \text{ or } x > 3\}$. The graph is shown below.

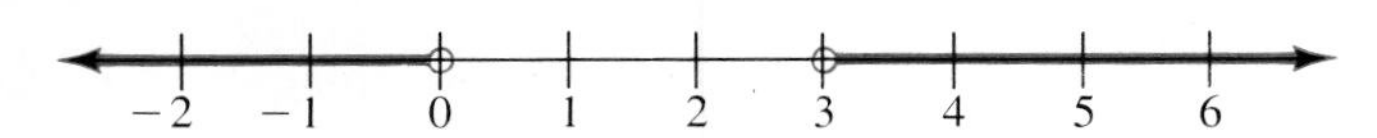

EXERCISES

Solve each inequality over $\mathcal{R}$ and graph each nonempty solution set.

A

1. $2t + 5 \le 6t - 1$
2. $-2 \le 2t - 4 \le 4$
3. $6(4x + 1) \ge 12x + 6$
4. $t^2 - 5t < 0$
5. $r - 6 \le 2r - 5 < r$
6. $20 \ge 6t - 4 \ge 5$
7. $(x - 5)(x - 4) < 0$
8. $(2t - 3)(t + 4) \ge 0$
9. $t^2 < 4t - 4$
10. $n^2 \ge 2n + 3$
11. $2t^2 + 5t - 3 \ge 0$
12. $3x^2 + 10x - 8 < 0$
13. $r^2 + 10r + 25 > 0$
14. $4r^2 + 4r + 1 < 0$

Solve and graph the intersection of the solution sets over $\mathcal{R}$.

15. $2x - 5 < 1$; $-3x + 4 < 10$
16. $4x - 7 < 5$; $4 - 2x > 2$
17. $3x - 6 < 0$; $2x - 5 > 3$
18. $4x - 7 \le 13$; $3x + 2 > 17$

B

19. **a.** If $a < b$, what additional assumptions on a and b will guarantee that $\frac{1}{a} < \frac{1}{b}$?
    b. Prove your assertion.
20. **a.** If $a < b$, what additional assumptions on a and b will guarantee that $a^2 < b^2$?
    b. Prove your assertion.

In Exercises 21 and 22 prove each statement by indirect proof.

21. If $a > 0$, then **(a)** $-a < 0$ and **(b)** $\frac{1}{a} > 0$.

22. If $a < 0$, then **(a)** $-a > 0$ and **(b)** $\frac{1}{a} < 0$.

In Exercises 23–26, prove each statement.

23. If $x > 1$, then $(x - 1)(x^2 + 3x + 1) > 0$.

24. If $x < 2$, then $(x - 2)(x^2 + 1) < 0$.

25. For all real numbers x and y, $x^2 + y^2 \geq 0$.

26. For all real numbers x and y, $x^2 + y^2 \geq 2xy$.

27. Prove Theorem 5, part (1).

28. Prove Theorem 5, part (2).

29. Prove that if $x < y$, then $x < \frac{x + y}{2} < y$.

30. Prove that $\frac{a}{c} > 0$ if and only if $ac > 0$. $\left(\textit{Hint}: \frac{a}{c} = a\left(\frac{1}{c}\right).\right)$

31. Prove that $\frac{a}{c} < 0$ if and only if $ac < 0$.

Solve each inequality and graph each nonempty solution set.

32. $\frac{x}{x - 1} > 0$

33. $\frac{x^2 - 1}{2x + 1} \leq 0$

C **34.** $\frac{x - 3}{2x - 1} > 2(x - 2)$

35. $\frac{(x - 2)(x - 3)}{x + 1} \geq 0$

1-6 ABSOLUTE VALUE

The distance between the points with coordinates 0 and 3 on the number line is 3. The distance between the points with coordinates 0 and -3 is also 3.

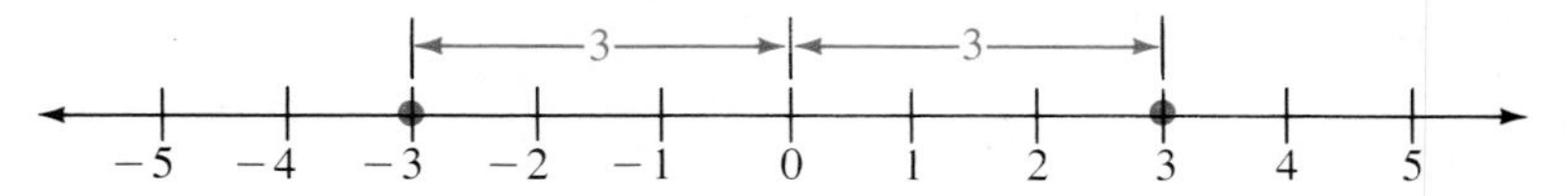

With this in mind we make the following definition of the **absolute value** of a real number x, denoted $|x|$.

DEFINITION

For all real numbers x:

$$|x| = \begin{cases} x \text{ if } x \geq 0 \\ -x \text{ if } x < 0 \end{cases}$$

For example, $|3| = 3$ and $|-3| = -(-3) = 3$.

The following example introduces an important theorem involving absolute value.

EXAMPLE 1 Solve **a.** $|x| = 4$ **b.** $|x| < 4$ **c.** $|x| > 4$.

Solution Think in terms of distance.

a. The only points 4 units from 0 have coordinates 4 and -4. So $x = 4$ or $x = -4$.

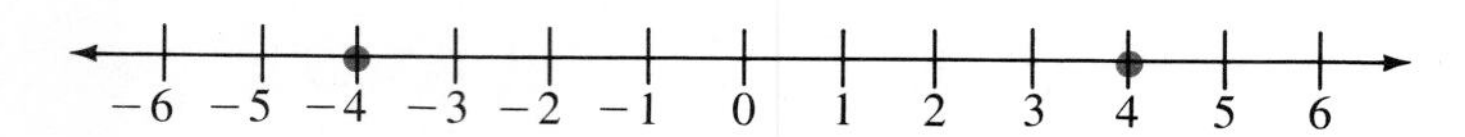

b. The only points less than four units from 0 have coordinates x between -4 and 4. Thus $-4 < x < 4$.

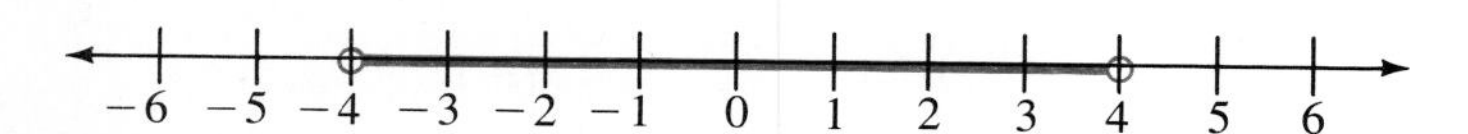

c. Any x satisfying $|x| > 4$ must be located more than four units from 0. Thus $x < -4$ or $x > 4$.

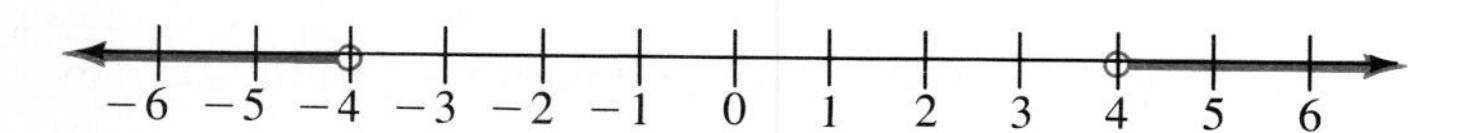

THEOREM 6
1. If $a \geq 0$, $|x| = a$ if and only if $x = -a$ or $x = a$.
2. If $a > 0$, $|x| < a$ if and only if $-a < x < a$.
3. $|x| > a$ if and only if $x < -a$ or $x > a$.

We will prove part (2) of the theorem and leave the proofs of parts (1) and (3) as Exercises 39 and 40.

Proof: Suppose that $a > 0$ and $|x| < a$.
If $x \geq 0$, then $|x| = x$. Thus $x = |x| < a$; $x < a$. Since $a > 0$, $-a < 0 < x$. Hence $-a < x$. Therefore $-a < x$ and $x < a$. That is, $-a < x < a$.
If $x < 0$, then $x < 0 < a$ and $x < a$. Since $x < 0$, $|x| = -x$. Therefore $-x = |x| < a$ and $-x < a$. Hence by Theorem 4, $-a < x$. Therefore $-a < x$ and $x < a$. That is, $-a < x < a$.
In either case $-a < x < a$.

Conversely suppose that $a > 0$ and $-a < x < a$.
If $x \geq 0$, then $|x| = x$. Thus $-a < |x| < a$. Hence $|x| < a$.
If $x < 0$, then $|x| = -x$ and $x = -|x|$. So $-a < -|x| < a$.
Multiply by -1 to obtain $a > |x| > -a$. Hence $|x| < a$.
In either case $|x| < a$. ■

EXAMPLE 2 Solve $|3y + 5| = 2$.

Solution

$$\begin{array}{rcl} 3y + 5 = 2 & \text{or} & 3y + 5 = -2 \\ 3y = -3 & & 3y = -7 \\ y = -1 & \text{or} & y = -\frac{7}{3} \end{array}$$

Therefore $y = -1$ or $-\frac{7}{3}$.

EXAMPLE 3 Solve $|2x - 1| < 5 - x$.

Solution Apply Theorem 6, part (2).

$$-(5 - x) < 2x - 1 < 5 - x$$
$$x - 5 < 2x - 1 < 5 - x$$
$$\begin{array}{rcl} x - 5 < 2x - 1 & \text{and} & 2x - 1 < 5 - x \\ -4 < x & \text{and} & x < 2 \end{array}$$

Therefore $-4 < x < 2$.

EXAMPLE 4 Solve $|x + 2| \geq x - 7$.

Solution

$$\begin{array}{rcl} x + 2 \leq -(x - 7) & \text{or} & x + 2 \geq x - 7 \\ x \leq \frac{5}{2} & \text{or} & 0 \geq -9 \end{array}$$

Therefore the solution set is the set of all real numbers.

THEOREM 7 For all real numbers a, $|a|^2 = a^2$.

EXAMPLE 5 Solve $|2x| = |x + 1|$.

Solution Apply Theorem 7 and the zero-product property.

$$|2x|^2 = |x + 1|^2$$
$$(2x)^2 = (x + 1)^2$$
$$3x^2 - 2x - 1 = 0$$
$$(3x + 1)(x - 1) = 0$$

Thus $3x + 1 = 0$ or $x - 1 = 0$. Hence $x = -\frac{1}{3}$ or 1.

Using absolute value, we can find the *distance* between two points on the number line. If A and B are two points on the number line and a and b are their coordinates respectively, then the **distance** between A and B, denoted AB, is given by

$$|a - b|.$$

In the following theorem many important properties of absolute value are stated.

THEOREM 8 For all real numbers a and b,

1. $|ab| = |a||b|$
2. $\left|\frac{a}{b}\right| = \frac{|a|}{|b|} \quad (b \neq 0)$
3. $|a + b| \leq |a| + |b|$
4. $|a - b| \leq |a| + |b|$

We will prove part (3) of the theorem. The proofs of the other parts are requested in Exercises 45–47.

Proof: Since $|a + b|$ and $|a| + |b|$ are nonnegative, it is sufficient to show that

$$|a + b|^2 \leq (|a| + |b|)^2.$$

or in light of Theorem 7

$$(a + b)^2 \leq (|a| + |b|)^2. \qquad (*)$$

Since for any real number x, $x \leq |x|$, we have

$$ab \leq |ab|.$$

Therefore

$$a^2 + 2ab + b^2 \leq a^2 + 2|ab| + b^2.$$

By Theorem 7 and part (1) of Theorem 8

$$a^2 + 2ab + b^2 \leq |a|^2 + 2|a||b| + |b|^2.$$

This is equivalent to (*). ■

The inequality $|a + b| \leq |a| + |b|$ is called the **triangle inequality**. We will make use of it in later chapters.

EXERCISES

Find and graph each nonempty solution set over $\mathcal{R}$.

A

1. $|a| \leq 2.5$
2. $|3 - 2y| = 5$
3. $|m - 1| \leq 0$
4. $|d + 2| > 1$
5. $|s| = -3$
6. $\left|\frac{1}{2}t\right| \leq 2$
7. $|2x + 7| \geq 3$
8. $|5 - b| = \frac{5}{2}$
9. $|4k - 1| < 15$
10. $|9 - 7c| \geq 0$
11. $|18 + 3y| > 0$
12. $|5j - 4| < -2$
13. $|2 - (4 + z)| = 6$
14. $|7 - 2(2 - x)| \leq 8$
15. $|3y - 1| > y + 9$
16. $|2g + 3| = g - 4$
17. $\left|1 + \frac{1}{2}z\right| = \frac{5}{6}z$
18. $|8 + 3r| \geq 2 - 3r$

19. $|n - 1| \leq n + 2$

20. $\left|\frac{3}{4}t - \frac{1}{3}\right| \leq \frac{1}{2}t - 1$

21. $\left|\frac{1}{x} - 3\right| > 2$

22. $\left|\frac{1}{2} - \frac{2}{y}\right| > 1$

23. $|2(1 + k) - 3| = |0.4k - 3|$

24. $|s - 3(s - 2)| = |2(s - 3)|$

Solve.

25. $|y^2 + 6y + 9| > 0$

26. $|x^2 - 1| \leq 3$

B

27. $\left|\frac{a + 3}{2a - 1}\right| < 1$

28. $\left|\frac{a - 5}{a - 2}\right| \geq 2$

29. $|2d - 7| \leq |3d + 1|$

30. $\left|\frac{1}{2}n - 1\right| > |n + 1|$

31. $|x + 3| = |x| + |5x|$

32. $|2x + 5| = |x| - |4x|$

33. $|2t - 3| = |2t| - |3t|$

34. $|x^2 - 9| = |2x||x + 3|$

35. $|(x - 3)(x - 4)| > 0$

36. $|x^2 - 5x + 6| \geq 0$

37. $|x^2 - 2x + 1| \geq |x - 1|$

38. $|2n^2 + 5n - 3| \geq |n + 3|$

Prove each statement.

39. Part (1) of Theorem 6

40. Part (3) of Theorem 6

41. $a^2 = b^2$ if and only if $|a| = |b|$.

42. Theorem 7

43. For all real numbers a and b, $|a - b| = |b - a|$.

44. If a and b are nonnegative real numbers, then $a \leq b$ if and only if $a^2 \leq b^2$.

45. Part (1) of Theorem 8

46. Part (2) of Theorem 8

C

47. Part (4) of Theorem 8

48. For all real numbers a and b, $|a - b| \geq |a| - |b|$.

COMPUTER EXERCISES

1. Write a program to find the smallest integer n such that $\left|\frac{n}{n + 1} - 1\right| < h$ where h is a positive number supplied by the user.
2. Run the program in Exercise 1 for the following values of h.
   a. 0.1 **b.** 0.05 **c.** 0.01
3. Modify the program in Exercise 1 to find the smallest integer n such that $\left|\frac{2n - 1}{n + 1} - 2\right| < h$ for a positive number h supplied by the user.
4. Run the program in Exercise 3 for each value of h in Exercise 2.

■ 1–7 SUBSETS OF $\mathscr{R}$

We can describe the set N of *natural numbers* as the smallest set of all real numbers that satisfy the following conditions.

$$1 \in N \qquad \text{and} \qquad \text{If } k \in N, \text{ then } k + 1 \in N.$$

Since $1 \in N$, we have that $1 + 1 = 2 \in N$, $2 + 1 = 3 \in N$, and so on. We can describe the set Z of *integers* as N together with 0 and all the additive inverses of N.

For any integers a and b, we say that a is a **factor**, or **divisor**, of b (written $a|b$ and read "a divides b") if there is an integer c such that $ac = b$. A **prime number**, or **prime**, is any integer greater than 1 that has only 1 and itself as factors. Any integer greater than 1 that is not prime is called **composite**.

The following theorem, presented without proof, is used frequently in calculations and theoretical work involving integers.

THEOREM 9 **The Fundamental Theorem of Arithmetic**
Every integer greater than 1 can be expressed as a product $p_1p_2p_3 \ldots p_n$ in which $p_1, p_2, p_3, \ldots, p_n$ are primes. Furthermore the factorization is unique except for the order in which the primes are written.

EXAMPLE 1 Find the prime factorization of 54.

Solution Use the fundamental theorem of arithmetic to factor 54 into primes.

$$54 = 2(27) = 2 \cdot 3(9) = 2 \cdot 3 \cdot 3 \cdot 3, \text{ or } 2 \cdot 3^3$$

If a, b, and c are integers, $c|a$, and $c|b$, we call c a **common factor** of a and b. The **greatest common factor** (GCF) or **greatest common divisor** of a and b is the largest positive integer that is a common factor of both a and b. For example, 3 is a common factor of 12 and 18, but 6 is their greatest common factor. Two integers are **relatively prime** if their GCF is 1. The fundamental theorem of arithmetic provides a way to find the GCF of two integers.

EXAMPLE 2 Find the GCF of 54 and 180.

Solution Write the prime factorization of each number.

$$54 = 2 \cdot 3 \cdot 3 \cdot 3 \qquad 180 = 2 \cdot 2 \cdot 3 \cdot 3 \cdot 5$$

The GCF of 54 and 180 must contain each prime factor common to the factorizations of 54 and 180. Each such prime must be repeated the minimum number of times it occurs in either of the two factorizations. Thus the GCF of 54 and 180 is $2 \cdot 3 \cdot 3 = 18$.

The following theorem, called the *division algorithm*, provides another way to find the GCF of two numbers.

THEOREM 10 The Division Algorithm
Given integers s and t, $t > 0$, there exist unique integers q and r such that $s = tq + r$ and $0 \leq r < t$.

To find the GCF of 54 and 198 by the division algorithm, divide 198 by 54. Repeat the process using the divisor as the new dividend and the remainder as the new divisor.

$$198 = 3 \cdot 54 + 36$$
$$54 = 1 \cdot 36 + 18$$
$$36 = 2 \cdot 18 + 0$$

When we obtain 0 as a remainder, the last divisor, here 18, is the GCF of the given integers. The procedure or algorithm used above is called the *Euclidean algorithm*. It always gives the GCF of two positive integers.

The set Z of integers satisfies all of the addition and multiplication axioms of the real numbers except for the existence of multiplicative inverses. If we include the reciprocals of all nonzero integers and all products of integers and reciprocals, we obtain the set Q of all *rational numbers*.

$$Q = \left\{x\colon x = \frac{p}{q},\ p \text{ and } q \in Z \text{ and } q \neq 0\right\}$$

If we let $+$ and $\cdot$ be defined for Q by

$$\frac{p}{q} + \frac{r}{s} = \frac{ps + qr}{qs} \quad \text{and} \quad \frac{p}{q} \cdot \frac{r}{s} = \frac{pr}{qs},$$

we can show that Q with $+$ and $\cdot$ is a field. For this reason Q is also called the field of quotients of Z.

We will show that Q under $+$ is closed. That is, we will show that the sum of two rational numbers is another rational number. Let $\frac{p}{q}$ and $\frac{r}{s}$ be rational numbers. Then $q \neq 0$ and $s \neq 0$. Since p, q, r, and s are integers, ps, qr, and qs are also integers. Because Z is a group under addition, $ps + qr$ is an integer. Lastly, by the zero-product property, since $q \neq 0$ and $s \neq 0$, $qs \neq 0$. Thus $\frac{ps + qr}{qs}$ is the quotient of two integers with nonzero denominator. Therefore the sum of $\frac{p}{q}$ and $\frac{r}{s}$ is a rational number.

You are asked to show in Exercises 34–36 that Q with $+$ and $\cdot$ is a field. In fact Q with $+$ and $\cdot$ is an ordered field.

There are *real numbers* that are not rational. Such numbers are called *irrational numbers*. Square roots of integers that are not perfect squares are irrational, as will be shown in Chapter 3. The number π is irrational. It is approximately 3.1415926535 to ten decimal places. The number e, to be introduced in Chapter 10, is also an irrational number.

THEOREM 11 There is no rational number whose square is 2.

Proof: Assume to the contrary that there is a rational number $\frac{p}{q}$ such that $\left(\frac{p}{q}\right)^2 = 2$. We may also assume that p and q are relatively prime. Then we have that $\frac{p^2}{q^2} = 2$ and that $p^2 = 2q^2$. Therefore 2 is a factor of p^2. In Exercise 33 you are asked to show that if a is a prime and $a|k^2$, then $a|k$. Using this fact we have that 2 must be a factor of p. Therefore $p = 2c$ for some integer c. Hence $2q^2 = (2c)^2 = 4c^2$. That is, $q^2 = 2c^2$. Thus 2 is a factor of q^2 and 2 must be a factor of q. The conclusion that 2 is a factor of both p and q contradicts the assumption that p and q are relatively prime. Thus there is no rational number whose square is 2. ■

EXERCISES

Find the prime factorization of each integer.

A
1. 729 **2.** 1960 **3.** 153 **4.** 1287
5. 151 **6.** 508 **7.** 968 **8.** 353

Find the GCF of each pair of integers.

9. 675; 570 **10.** 216; 539 **11.** 97; 485 **12.** 644; 345

The **least common multiple** (LCM) of two integers is the smallest positive integer that is a multiple of both integers. Find the LCM of each pair of integers.

13. 24; 32 **14.** 1000; 120 **15.** 5; 13 **16.** $2^3 \cdot 3^2; 5 \cdot 2^2 \cdot 3$

Use the Euclidean algorithm to find the GCF of each pair of integers.

17. 24; 16 **18.** 33; 9 **19.** 18; 25 **20.** 117; 126

A natural number n is a **perfect number** if it equals the sum of all of its factors other than itself. State whether each given integer is perfect.

21. 28 **22.** 126 **23.** 25 **24.** 3^3

Prove each statement. Assume that variables represent integers and are nonzero where necessary.

B
25. $a|b$ if and only if $-a|b$.
26. If $a|(b + c)$ and $a|b$, then $a|c$.
27. If $a|b$ and $a|c$, then $a|(b + c)$.
28. If $a|(b - c)$ and $a|b$, then $a|c$.
29. If $a|(b + c + d)$, $a|b$, and $a|c$, then $a|d$.

30. Every natural number greater than 1 is of the form $2k$ or $2k + 1$ for some natural number k. (*Hint*: Use the division algorithm.)
31. Every quotient of positive integers can be written as a quotient of relatively prime integers.
32. The LCM of two integers a and b equals ab divided by the GCF of a and b.
33. If a is a prime and $a|k^2$, then $a|k$.
34. Prove that the set Q of rational numbers is a commutative group under addition.
35. Prove that the set of nonzero rational numbers is a commutative group under multiplication.
36. Show that the distributive axiom holds in Q.
37. Show that between any two distinct rational numbers there is a third rational number. Explain why between any two distinct rational numbers there are infinitely many rational numbers.

Let a and b be integers and let n be a positive integer. Then a is **congruent** to b modulo n, written $a \equiv b \pmod{n}$, if $n|(a - b)$. For example, $16 \equiv 4 \pmod{3}$ since $3|(16 - 4)$.

C

38. Show that $a \equiv a \pmod{n}$.
39. Show that if $a \equiv b \pmod{n}$, then $b \equiv a \pmod{n}$.
40. Show that if a, b, and c are integers, $a \equiv b \pmod{n}$, and $b \equiv c \pmod{n}$, then $a \equiv c \pmod{n}$.
41. Show that every natural number k is congruent modulo n to some natural number r where $0 \leq r < n$.
42. Show that if $a \equiv x \pmod{n}$ and $b \equiv y \pmod{n}$, then $a + b \equiv x + y \pmod{n}$.
43. Show that if $a \equiv x \pmod{n}$ and $b \equiv y \pmod{n}$, then $ab \equiv xy \pmod{n}$.

COMPUTER EXERCISES

1. Write a program to determine whether a natural number supplied by the user is prime or composite.
2. Run the program in Exercise 1 for the following numbers.
   a. 51 **b.** 73 **c.** 323 **d.** 257
3. Write a program to find three rational numbers between two rational numbers supplied by the user.
4. Run the program in Exercise 3 for each pair of numbers.
   a. 3; 7 **b.** 1.1; 2.1 **c.** 0.4; 0.5 **d.** -1.1; 0.1
5. Write a program to find the greatest common factor of two positive integers supplied by the user.
6. Run the program in Exercise 5 for each pair of integers.
   a. 18; 24 **b.** 7; 24 **c.** 10; 60 **d.** 14; 49

Chapter Summary

1. A *set* is a collection of objects for which there is a definite criterion that enables us to tell what is in the set and what is not. Objects in a given set are called *members* of that set. The symbols $\in$, $\subseteq$, $\subset$ are used to denote *membership*, *subset*, and *proper subset* relationships respectively. The *null* set, or *empty* set, denoted $\emptyset$, is the set having no members.

   Throughout the text the following special symbols are used.

N	the set of natural numbers
Z	the set of integers
Q	the set of rational numbers
$\mathcal{R}$	the set of real numbers

2. The *union* of two sets A and B, written $A \cup B$, is the set consisting of all members of either A or B. The *intersection* of A and B, written $A \cap B$, is the set consisting of all members of both A and B. The *complement* of A, written A', is the set of all members of some universal set that are not in A.
3. An *open sentence* is any sentence involving a *variable*. The *solution set* of an open sentence is the set of all members of the *domain,* or *replacement set*, for which the sentence is true.
4. Two sentences p and q can be joined by the word *and* to form the *conjunction* "p and q," also written $p \wedge q$. The *disjunction* of p and q is the sentence "p or q," also written $p \vee q$. The truth value of a conjunction or disjunction depends on the truth values of p and q. The *negation* of the sentence p is the sentence "not p," also written p'.
5. A *conditional sentence* is any sentence of the form "if p, then q," also written $p \rightarrow q$. Its *converse* is the conditional sentence "if q, then p." The *biconditional sentence* "p if and only if q" is the conjunction of "if p, then q" and its converse. The *contrapositive* of "if p, then q" is "if not q, then not p." To prove that $p \rightarrow q$ is true by *indirect proof*, assume that p and q' are true and then obtain a contradiction.
6. The set $\mathcal{R}$ of real numbers with addition and multiplication satisfies the axioms of equality, addition, and multiplication. It also satisfies the distributive axiom. The axioms and theorems in Section 1-4 are useful in proving theorems and in solving equations. A *field* is a set with two operations such that (1) it is a commutative group under the first operation, (2) its nonzero members form a commutative group under the second operation, and (3) the second operation is distributive over the first operation. The set $\mathcal{R}$ with addition and multiplication is a field.
7. Real numbers are ordered by a relation *less than* ($<$) which can be illustrated by a number line. The symbols $>$, $\geq$, and $\leq$ are defined in terms of $<$ and $=$. The axioms and theorems in Section 1-5 are useful in solving *inequalities* and in proving theorems.

8. The *absolute value* of a real number x, denoted $|x|$, is defined to be x if $x \geq 0$ and $-x$ if $x < 0$. The *distance* between two points A and B on a number line is $|a - b|$ where a and b are the coordinates of A and B respectively.

9. The *fundamental theorem of arithmetic* states that every integer greater than 1 has a unique prime factorization. The *division algorithm* can be used to find the *greatest common factor* of two integers. The set Q of *rational* numbers is defined to be

$$Q = \left\{x: x = \frac{p}{q}, p \text{ and } q \in Z \text{ and } q \neq 0\right\}.$$

The set Q with addition and multiplication is a field. *Irrational* numbers are real numbers that are not rational.

Chapter Test

Let $B = \{1, 3, 5, 7\}$, $C = \{1, 2, 3, 4, 5\}$, and $U = \{1, 2, 3, 4, 5, 6, 7, 8\}$. Find each union, intersection, or complement.

1-1 **1.** $(B \cap C)'$ **2.** $B \cup C$ **3.** C' **4.** $B' \cup C$

1-2 **5.** Let p be the sentence $x + 2 < -5$ and q be the sentence $3x - 1 > 8$. Let the domain of each sentence be $\mathcal{R}$. Find and graph the solution set of $p \vee q$ on a number line.

6. Write the negation of "$7x \neq 14$ and $-2x = 2$."

1-3 **7.** Write the converse of "If $x = 3$, then $x^2 = 9$" and determine whether the converse is true or false.

8. Give the contrapositive of "If $ab < 0$ and $a > 0$, then $b < 0$."

In Exercises 9–11 solve each equation over $\mathcal{R}$.

1-4 **9.** $2(8x + 1) + 3(3x - 2) = 46$ **10.** $2x^2 + 5x - 3 = 0$ **11.** $3x^2(4x - 5) = 0$

12. Prove that for all real numbers a and b, $(ab)^2 = a^2b^2$.

In Exercises 13–15 solve each inequality over $\mathcal{R}$.

1-5 **13.** $x^2 + 3x < 0$ **14.** $(2x - 3)(3x - 6) \geq 0$ **15.** $2(x - 3) + 4(x + 5) \leq 32$

16. Solve $3x - 5 \geq 7$ and $2x + 10 < 12$ simultaneously over $\mathcal{R}$ and graph the solution set on a number line.

In Exercises 17 and 18 solve each open sentence over $\mathcal{R}$ and graph the solution set.

1-6 **17.** $|2x - 1| < 3$ **18.** $|3x + 5| \geq 20$

1-7 **19.** Find the GCF of 576 and 336.

Extra

Groups and Subgroups

Recall that the set Z under addition is a group. The set, E, of even integers contained in Z is also a group under addition. To see this let $E = \{p \in Z: p = 2k \text{ for some } k \in Z\}$.

1. E is closed under addition. Let $p = 2k$ and $q = 2s$ where k and s are integers. Then $p + q = 2k + 2s = 2(k + s)$. Since Z is closed under addition, $k + s \in Z$. Thus $p + q \in E$.
2. Addition in E is associative since addition in Z is associative.
3. The integer 0 serves as the additive identity in E since $0 \in E$ and $0 + p = p$ for all p in E.
4. Every member of E has an additive inverse in E. Let $p \in E$. Let $p' = 2(-k) \in E$. Then $p + p' = 2k + 2(-k) = 2(k + (-k)) = 2(0) = 0$. Thus $p' \in E$ is the additive inverse of p.

We call E a *subgroup* of Z under addition.

In general, a nonempty set H is a **subgroup** of a group G if $H \subseteq G$ and H is a group. In the exercises we will examine other subgroups of Z.

There are many examples of groups and subgroups whose members are not numbers.

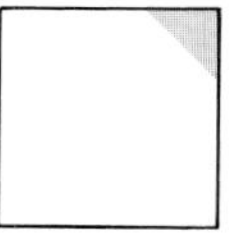

Consider a square with a shaded triangle in one corner, say the upper-right corner. Reflect the square in the horizontal line through the "center" of the square. The shaded triangle then appears in the lower-right corner as shown in the second figure below. In all there are four possible reflections of the square in a line through the "center" of the square, which we name I, A, B, and C.

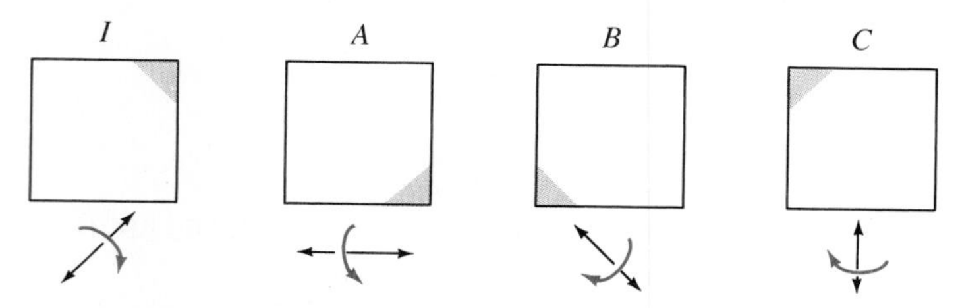

Let $G = \{I, A, B, C\}$. Define $\bullet$ on G as follows: Perform one of the reflections in the set G. On the result, perform another reflection from the set G. For example, $A \bullet C$ means "Perform C. On the result, perform A." We call $A \bullet C$ the *composition* of A with C. Notice that $A \bullet C = B$. The composition of B with C is the same as A. All compositions of members of G are displayed in the following table.

•	I	A	B	C
I	I	A	B	C
A	A	I	C	B
B	B	C	I	A
C	C	B	A	I

We can show that G under • is a group.

1. G is closed under • since the composition of two reflections is another reflection.
2. The operation • is associative.
3. The reflection I serves as the identity. See row 1 and column 1.
4. Every member of G has an inverse under • . For example, $A \bullet A = I$. Thus A is its own inverse. In fact each member of G is its own inverse.

What are the subgroups of $G = \{I, A, B, C\}$? To answer this question, first note that $G \subseteq G$ and G is a group. Thus G is a subgroup of itself. Next note that $\{I\}$ is a group since $I \bullet I = I$. Thus $\{I\}$ is a subgroup of G. The group $\{I\}$ is called the *trivial* subgroup. Let $H = \{I, A\}$. Since A is its own inverse under • , H contains the inverses of each of its members. This observation is sufficient to allow us to conclude that H is a subgroup of G. In all there are five subgroups of G: $\{I\}$ $\{I, A\}$ $\{I, B\}$ $\{I, C\}$ $\{I, A, B, C\}$.

EXERCISES

1. Show that $S = \{p \in Z: p = 7k \text{ for some integer } k\}$ is a subgroup of Z under addition.
2. Let $n \in Z$. Show that $T = \{p \in Z: p = nk \text{ for some integer } k\}$ is a subgroup of Z under addition.
3. Show that every nontrivial subgroup W of Z under addition must be infinite. (*Hint*: Assume that W has a member other than 0. What does closure imply?)
4. Let $U = \left\{\ldots, -\frac{1}{3}, -\frac{1}{2}, -1, 0, 1, \frac{1}{2}, \frac{1}{3}, \ldots\right\}$.

   a. Show that U has an additive identity.

   b. Show that each member of U has an additive inverse.

   c. Show that U is not a subgroup of Q under addition.
5. Let H and K be subgroups of a group G with an operation $*$. Show that $H \cap K$ is a subgroup of G under $*$.
6. Let H and K be subgroups of a group G with an operation $*$. Show that in general $H \cup K$ is not a subgroup of G under $*$. (*Hint*: Let G be the group of integers under addition.)

COMPUTER APPLICATION

TRUTH TABLES

Two sentences are **logically equivalent** (denoted by $\equiv$) if they have the same truth tables. A computer can be used to construct truth tables and show equivalence. Consider

$$p \text{ and } (q \text{ or } r) \qquad (p \text{ and } q) \text{ or } (p \text{ and } r)$$

where p, q, and r have truth values true (1) and false (0). Let addition denote the operation *or*, and let multiplication denote the operation *and*. Then the sentences above become

$$p(q + r) \qquad pq + pr.$$

The following program can be used to make the truth table of $p(q + r)$. A similar program gives the truth table of $pq + pr$.

```
10    FOR P = 1 TO 0 STEP -1
20    FOR Q = 1 TO 0 STEP -1
30    FOR R = 1 TO 0 STEP -1
40    LET A = Q + R
50    IF A > 1 THEN LET A = 1
60    PRINT P;" ";Q;" ";R;" ";P * A
70    NEXT R
80    NEXT Q
90    NEXT P
100     END
```

p	q	r	$p(q + r)$
1	1	1	1
1	1	0	1
1	0	1	1
1	0	0	0
0	1	1	0
0	1	0	0
0	0	1	0
0	0	0	0

To obtain a program for the truth table of $pq + pr$, replace line 40 by LET A = P*Q, and replace line 50 by LET B = P*R. Then insert line 55 IF A + B > 1, LET A + B = 1. Finally replace line 60 by PRINT P; Q; R; A + B.

When the programs are run, it will be clear that the truth tables are identical. That is, $p(q + r)$ is equivalent to $pq + pr$.

$$p \text{ and } (q \text{ or } r) \equiv (p \text{ and } q) \text{ or } (p \text{ and } r)$$

It is also possible to write an expression involving *and*, *or*, and *not* to represent a given truth table. We shall use $'$ to denote *not*. Consider the truth table shown at the right. To represent row 1 of the table, substitute the values $p = 1$ and $q = 1$ into $p'q$. Since $1' = 0$, we have $0 \cdot 1 = 0$. Notice that $p'q$ also represents row 3 of the table since there $p = 0$, $q = 1$, and $0' \cdot 1 = 1 \cdot 1 = 1$. To represent row 2 of the table, consider pq'. In row 2, $p = 1$ and $q = 0$. Hence $1 \cdot 0' = 1 \cdot 1 = 1$. Notice that since $0 \cdot 0' = 0 \cdot 1 = 0$, pq' also represents row 4. The complete table is represented by

p	q	?
1	1	0
1	0	1
0	1	1
0	0	0

$$p'q + pq'.$$

In general, to represent a truth table, find a sum of products representing each row.

EXERCISES *(You may use either BASIC or Pascal.)*

1. Show that the truth value of the negation of p is $1 - p$.
2. Use a computer program to make the truth table of "p or (q and r)."
3. Use a computer program to show that the sentence in Exercise 2 is equivalent to "(p or q) and (p or r)."
4. Use a computer program to show that $(p \text{ or } q)' \equiv p'$ and q'.

Write an expression for each truth table.

5.

p	q	r	?
1	1	1	0
1	1	0	1
1	0	1	1
1	0	0	0
0	1	1	1
0	1	0	0
0	0	1	0
0	0	0	0

6.

p	q	r	?
1	1	1	1
1	1	0	0
1	0	1	0
1	0	0	0
0	1	1	1
0	1	0	1
0	0	1	1
0	0	0	0

APPLICATION

Logic Networks

Figure a shows conjunction, disjunction, and negation as logic networks. Figure b shows a complicated network built from these simpler ones.

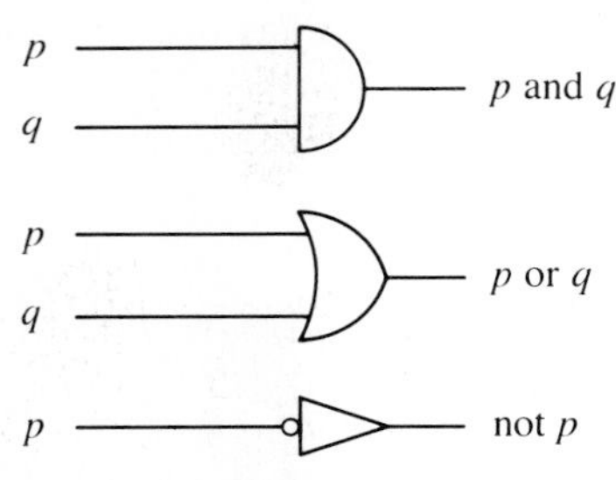

Figure a

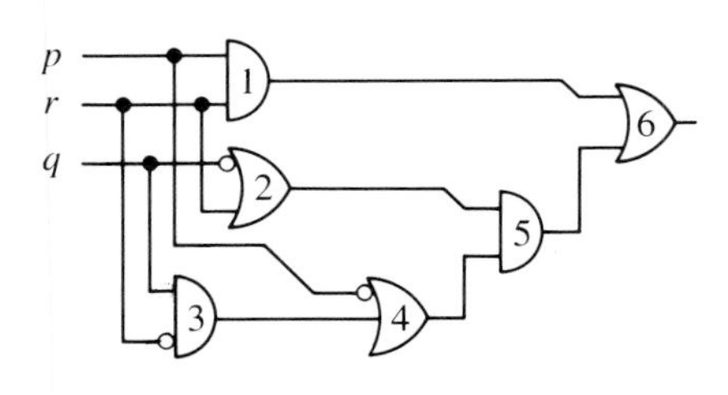

Figure b

In Figure b the third small network is the conjunction: q and not r. The fourth small network is the disjunction: network 3 or not p. Using multiplication for conjunction, addition for disjunction, and prime for negation, the complete network can be written as $pr + (q' + r)(p' + qr')$.

Expressions such as the one above can be simplified by using the following list of facts.

1. $0' = 1$
$1' = 0$

2. $p + 0 = p$
$p \cdot 1 = p$

3. $p + 1 = 1$
$p \cdot 0 = 0$

4. $p + p = p$
$p \cdot p = p$

5. $p + p' = 1$
$p \cdot p' = 0$

6. $(p')' = p$

7. Addition and multiplication are commutative.

8. Addition and multiplication are associative.

9. $p + pq = p$
$p(p + q) = p$

10. $p + p'q = p + q$
$p(p' + q) = pq$

11. $(p + q)' = p'q'$
$(pq)' = p' + q'$

12. $p(q + r) = pq + pr$
$p + qr = (p + q)(p + r)$

Each statement in the foregoing list can be established by the use of truth tables. See pages 36 and 37.

To simplify $pr + (q' + r)(p' + qr')$, begin by using part 12 twice.

$$pr + q'(p' + qr') + r(p' + qr') \quad \text{part 12}$$
$$pr + q'p' + q'qr' + rp' + rqr' \quad \text{part 12}$$
$$pr + q'p' + rp' \quad \text{since } q'qr = 0 \text{ and } rqr' = 0$$

Next use the fact that addition and multiplication are commutative. Then apply the distributive property again (part 12).

$$rp + rp' + q'p'$$
$$r(p + p') + q'p'$$

Since $p + p' = 1$ and $r \cdot 1 = r$, the original expression can simply be written as

$$r + q'p'.$$

The simplified network can be shown as the figure at the right.

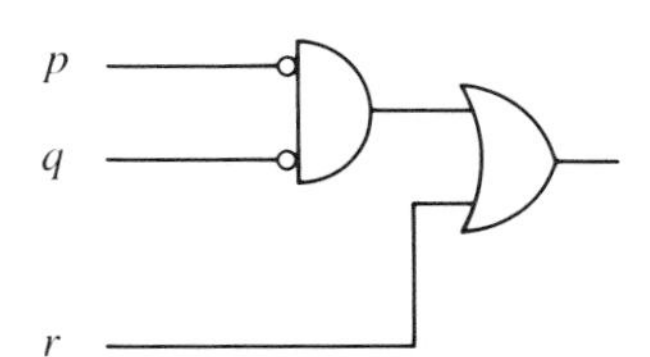

EXERCISES

1. Use truth tables to show that "p or (p and q)" is logically equivalent to p and therefore prove the first statement in part 9.

Simplify each expression.

2. $r + s + r'$
3. $(p + q)(p' + q')$
4. $s + s'(rt)$
5. $(p + q)(p + r)$
6. Write and simplify an expression for the logic network shown below. Use the list of facts on the preceding page.

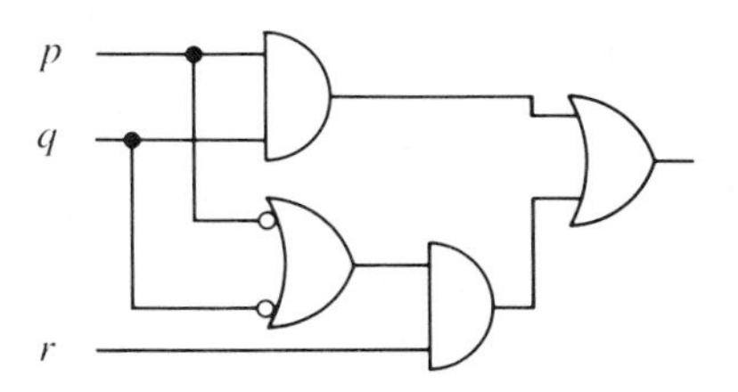

Simplification of electronic circuitry reduces cost of manufacturing.

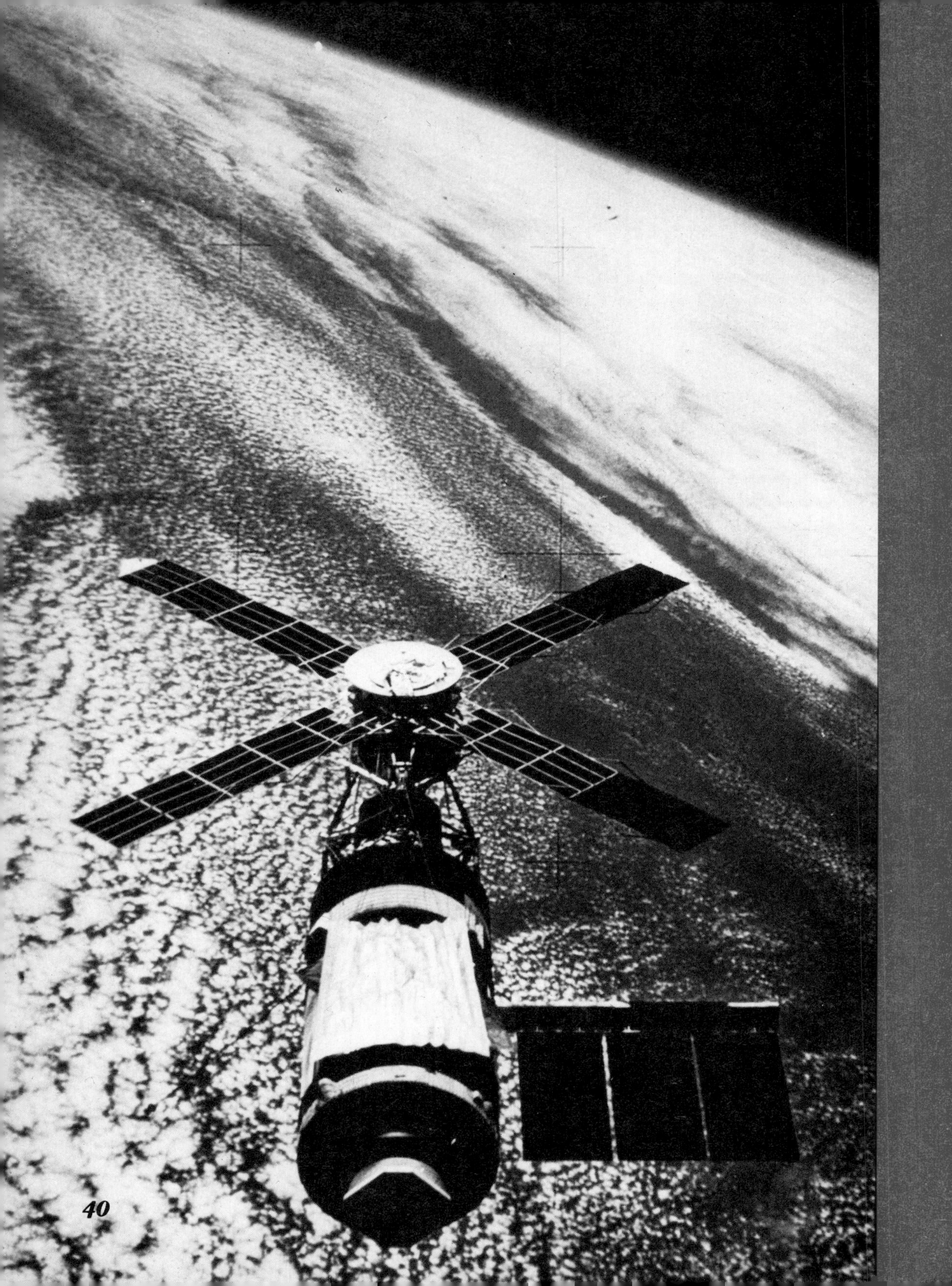

Chapter 2

Analytic Geometry

In this chapter the methods of analytic geometry are introduced. Using these methods we shall discuss lines, circles, ellipses, hyperbolas, and parabolas. Properties and equations of these conic sections *are investigated in detail.*

Coordinate Geometry and Lines

2–1 THE CARTESIAN COORDINATE SYSTEM: DISTANCE AND MIDPOINTS

In Section 1–6 we discussed the real number line and distance. In this section we extend our discussion to the plane.

The **Cartesian product** of $\mathcal{R}$ with $\mathcal{R}$ is defined by

$$\mathcal{R} \times \mathcal{R} = \{(a, b): a, b \in \mathcal{R}\};$$

that is, the set of all *ordered pairs* of real numbers. Two ordered pairs (a, b) and (c, d) are **equal** if their corresponding members are equal. In other words,

$$(a, b) = (c, d) \text{ if and only if } a = c \text{ and } b = d.$$

EXAMPLE 1 Find the values of x and y.

a. $(x^2 - 2x, x + 3) = (0, 5)$ **b.** $(x + 3, 2x - 1) = (y, 1 - y)$

Solution **a.** $x^2 - 2x = 0$ and $x + 3 = 5$

$x(x - 2) = 0$ $x = 2$

$x = 0$ or $x = 2$ and $x = 2.$ Therefore $x = 2$.

Solution continued on following page

The rectangular solar panels of Skylab 4, which always face the sun, are used to generate power for the batteries that operate the satellite's electronic equipment.

b. $x + 3 = y$ and $2x - 1 = 1 - y$.
Substitute $x + 3$ for y in the second equation to obtain

$$2x - 1 = 1 - (x + 3).$$

Therefore $x = -\frac{1}{3}$. Thus $y = -\frac{1}{3} + 3 = \frac{8}{3}$.

To give *coordinates* to points in the plane, draw two perpendicular number lines, or **axes**, that intersect at the origin 0 on each line. It is customary to draw the axes vertically and horizontally, to call the horizontal axis the ***x*-axis**, and to call the vertical axis the ***y*-axis**. The point O where the axes meet is the **origin**. Its coordinates are (0, 0). Any point P in the plane can be assigned coordinates by drawing lines through P parallel to the axes. The coordinates where these lines intersect the axes are the coordinates of P. In the figure at the right, 4 is the ***x*-coordinate** (also called the **abscissa**) of P and 3 is the ***y*-coordinate** (also called the **ordinate**) of P. This system of giving coordinates to points in the plane is called the **rectangular**, or **Cartesian**, **coordinate system** (named after René Descartes).

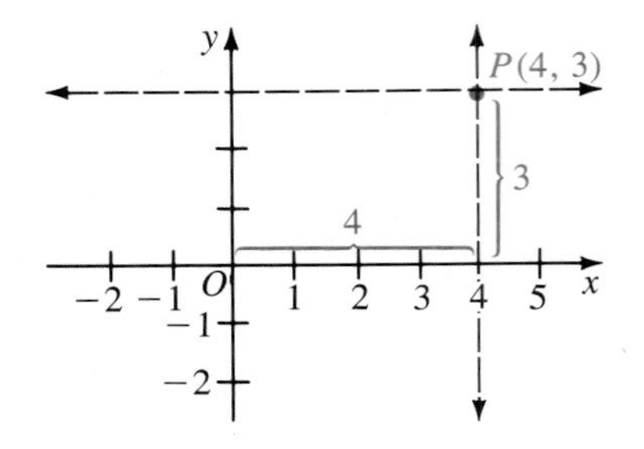

EXAMPLE 2 Give the coordinates of B, C, D, E, and F.

Solution $B = (2, 1)$ $C = (3, 0)$
$D = (0, -1)$ $E = (-3, -2)$
$F = (0, 0)$

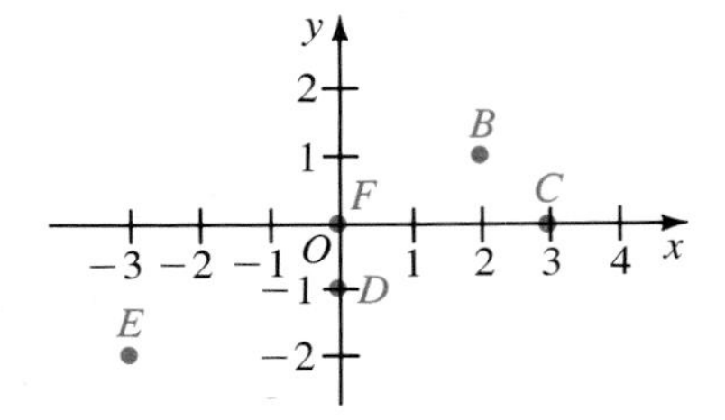

The axes divide the plane into four regions called **quadrants**, numbered as shown in the figure at the right. Notice that a point in the first quadrant has positive coordinates, a point in the second quadrant has a negative x-coordinate and a positive y-coordinate, and so on. Points on the axes are not in any quadrant.

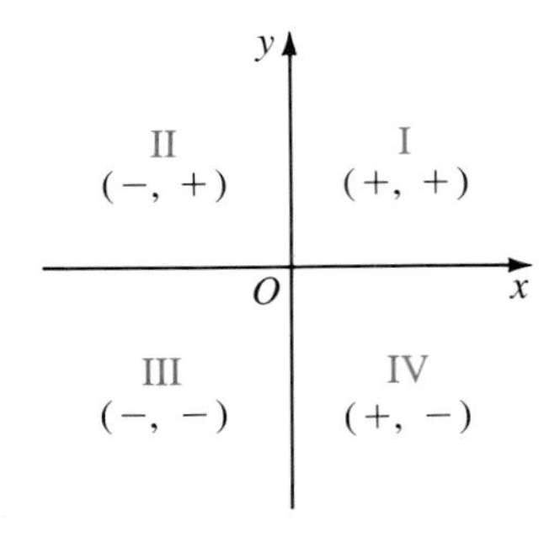

Recall that the distance between two points with coordinates x_1 and x_2 on a number line is given by $|x_1 - x_2|$, or $|x_2 - x_1|$. Using the Pythagorean theorem we can find the distance between any two points in the plane.

EXAMPLE 3 Find the distance between $A(-3, 1)$ and $B(3, 4)$.

Solution Draw lines through A and B parallel to the axes as shown. Then the intersection C of these lines has coordinates (3, 1) and $\triangle ABC$ is a right triangle. Since $\overline{AC}$ and $\overline{BC}$ lie on lines parallel to the axes,

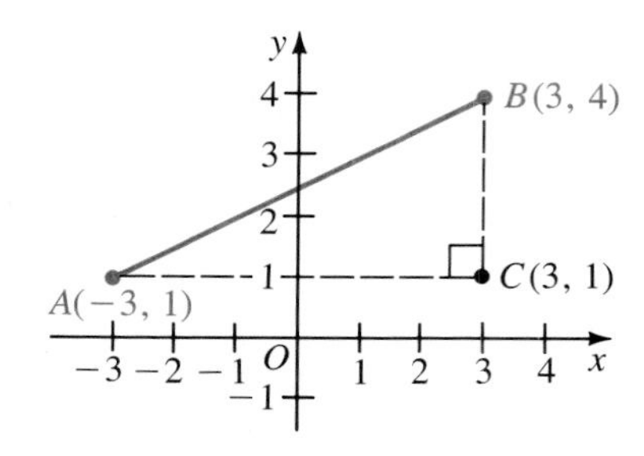

$$AC = |3 - (-3)| = 6 \quad \text{and} \quad BC = |4 - 1| = 3.$$

By the Pythagorean theorem, $(AB)^2 = (AC)^2 + (BC)^2 = 6^2 + 3^2 = 45.$
Therefore $AB = \sqrt{45} = 3\sqrt{5}$. **Answer**

The method used in Example 3 can be generalized to find the distance between any two points in the plane.

THEOREM 1 **The Distance Formula**
The distance d between $A(x_1, y_1)$ and $B(x_2, y_2)$ is given by

$$d = \sqrt{(x_2 - x_1)^2 + (y_2 - y_1)^2}.$$

Every line segment has a *midpoint* whose coordinates can readily be found from the coordinates of the endpoints of the segment. For example, consider the line segment from $A(2, 3)$ to $B(6, -1)$ and its midpoint M in the figures below. The x-coordinate of M is halfway between the x-coordinates of A and B (Figure a), and the y-coordinate of M is halfway between the y-coordinates of A and B (Figure b).

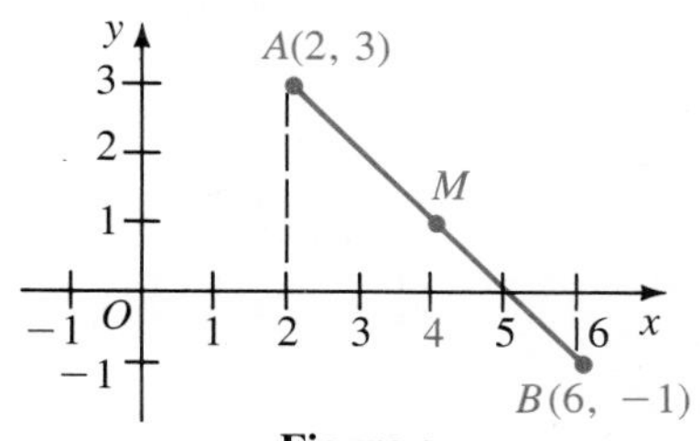

Figure a

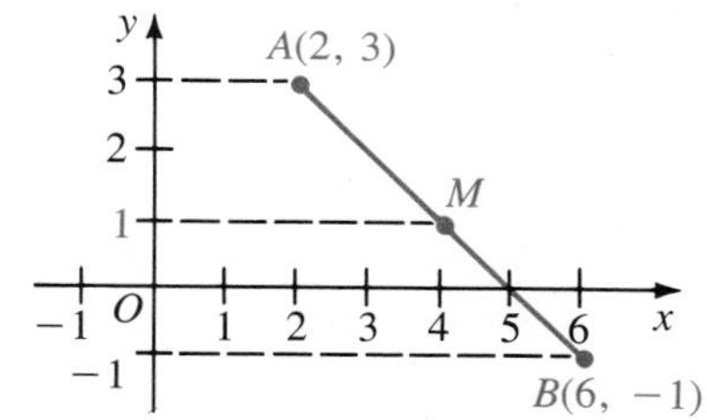

Figure b

The x-coordinate of M is $\frac{2+6}{2} = 4$ and the y-coordinate of M is $\frac{3+(-1)}{2} = 1$.

This method of finding the midpoint of a line segment works in general as stated in the following midpoint formula.

THEOREM 2 **The Midpoint Formula**
The midpoint M of the line segment from $A(x_1, y_1)$ to $B(x_2, y_2)$ is

$$M\left(\frac{x_1 + x_2}{2}, \frac{y_1 + y_2}{2}\right).$$

Proof: If A, B, and M have the coordinates given, calculate AM and BM:

$$AM = \sqrt{\left(x_1 - \left(\frac{x_1 + x_2}{2}\right)\right)^2 + \left(y_1 - \left(\frac{y_1 + y_2}{2}\right)\right)^2}$$

$$= \sqrt{\left(\frac{2x_1 - x_1 - x_2}{2}\right)^2 + \left(\frac{2y_1 - y_1 - y_2}{2}\right)^2}$$

Proof continued on following page

$$AM = \sqrt{\left(\frac{x_1 - x_2}{2}\right)^2 + \left(\frac{y_1 - y_2}{2}\right)^2}$$

$$= \frac{\sqrt{(x_1 - x_2)^2 + (y_1 - y_2)^2}}{2} = \frac{1}{2}AB$$

By similar reasoning,

$$BM = \sqrt{\left(x_2 - \left(\frac{x_1 + x_2}{2}\right)\right)^2 + \left(y_2 - \left(\frac{y_1 + y_2}{2}\right)\right)^2}$$

$$= \frac{\sqrt{(x_1 - x_2)^2 + (y_1 - y_2)^2}}{2} = \frac{1}{2}AB.$$

Since $AM + BM = AB$, M lies on $\overline{AB}$. (See Section 2–2, Exercise 58.) Since $AM = BM$, M must be the midpoint of $\overline{AB}$. ■

The distance and midpoint formulas can be used to prove theorems from geometry.

EXAMPLE 4 Prove that the line segment joining the midpoints of two sides of a triangle is parallel to the third side and is half as long as the third side.

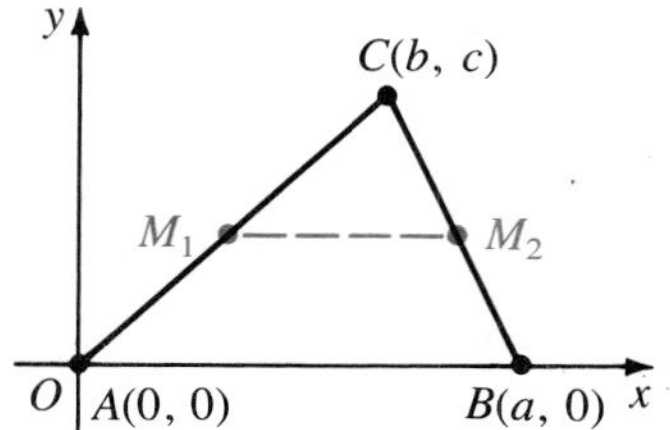

Solution An arbitrary triangle ABC is given. Coordinatize the plane so that A is at the origin and B is along the positive horizontal axis. Let C have coordinates (b, c). Let M_1 and M_2 be the midpoints of $\overline{AC}$ and $\overline{BC}$ respectively. Then

$$M_1 = \left(\frac{b + 0}{2}, \frac{c + 0}{2}\right) = \left(\frac{b}{2}, \frac{c}{2}\right)$$

and

$$M_2 = \left(\frac{a + b}{2}, \frac{0 + c}{2}\right) = \left(\frac{a + b}{2}, \frac{c}{2}\right).$$

Since M_1 and M_2 have the same y-coordinate, $\overline{M_1M_2}$ is parallel to $\overline{AB}$. Furthermore, $AB = a$ and

$$M_1M_2 = \sqrt{\left(\frac{a + b}{2} - \frac{b}{2}\right)^2 + \left(\frac{c}{2} - \frac{c}{2}\right)^2}$$

$$= \sqrt{\left(\frac{a}{2}\right)^2} = \frac{1}{2}a = \frac{1}{2}AB.$$

Geometry that uses a coordinate system is called *analytic geometry*, or *coordinate geometry*.

EXERCISES

A **1.** Find the coordinates of each labeled point.

Find the length of each line segment $\overline{CD}$.

2. $C(2, 3);\ D(5, 1)$

3. $C(-1, 5);\ D(-1, 2)$

4. $C(6, 1);\ D(-5, -2)$

5. $C(-3, -8);\ D(-5, 4)$

6. $C(-2, 2);\ D(-2, 8)$

7. $C(2\sqrt{3}, \sqrt{5});\ D(\sqrt{3}, 2\sqrt{5})$

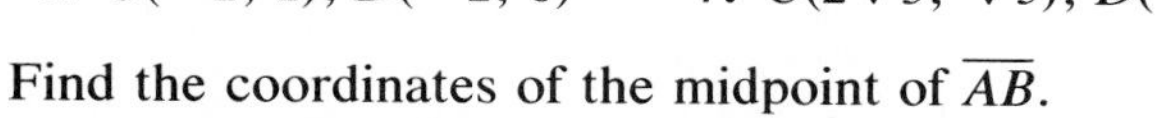

Find the coordinates of the midpoint of $\overline{AB}$.

8. $A(2, 6);\ B(8, 12)$

9. $A(-3, 8);\ B(5, 2)$

10. $A(5, -12);\ B(-9, 3)$

11. $A\left(\frac{7}{2}, \frac{4}{3}\right);\ B\left(3, \frac{2}{5}\right)$

12. $A(\sqrt{5}, 2\sqrt{2});\ B(3\sqrt{5}, 0)$

13. $A\left(\frac{1+\sqrt{2}}{3}, \frac{2-2\sqrt{2}}{2}\right);\ B\left(\frac{3-\sqrt{2}}{2}, \frac{4+\sqrt{2}}{3}\right)$

Find the values of x and y.

14. $(x, 2x) = (3, 6)$

15. $(5x - 1, x^2) = (-11, 4)$

16. $(3x + 1, 2y - 5) = (3, -2)$

17. $(2x^2, 3y + 4) = (x + 1, y^2)$

18. $(2x - 3y, x + 5y) = (13, -13)$

19. $(x + y, x + 5) = (6, y)$

Find the distance from R to S.

20. $R(a, 2);\ S(a, -7)$

21. $R(a, b);\ S(2a, 3b)$

22. $R(b, c);\ S(-3b, c)$

23. $R(a - b, c + d);\ S(a + b, c - d)$

24. If (a, b) is in the third quadrant and $|a| = |b| = 6$, find (a, b).

25. If (x, y) is in the second quadrant and $x^2 = 4$ and $y^2 = 8$, find (x, y).

26. If Sam travels on his snowmobile 5 mi north, then 3 mi west, then 4 mi north, how far is he from his starting point?

B **27.** Find four points in the second quadrant that are 5 units from the origin.

28. Find three points in the fourth quadrant that are 3 units from (2, 0).

29. Find the point on the x-axis that is equidistant from (2, 2) and (6, 6).

30. The converse of the Pythagorean theorem states that if a, b, and c are the lengths of the sides of a triangle and $a^2 + b^2 = c^2$, then the triangle is a right triangle with hypotenuse of length c. Use this theorem to determine if the triangle with the given vertices is a right triangle.

a. $A(2, -3);\ B(5, 1);\ C(1, 4)$

b. $A(3, 1);\ B(6, 0);\ C(5, -3)$

c. $A(\sqrt{3}, \sqrt{5});\ B(2\sqrt{3}, -\sqrt{5});\ C(-\sqrt{3}, -2\sqrt{5})$

d. $A(1.5, 0);\ B(4.5, 3);\ C(13, 9)$

31. Find the length of the median from D of the triangle with vertices $D(2, -1)$, $E(0, 0)$, and $F(2, 5)$.

32. Find the perimeter of the triangle with A, B, and C as vertices.
 a. $A(3, 2)$; $B(0, 6)$; $C(6, 14)$ **b.** $A(2, 5)$; $B(5, 9)$; $C(1, 6)$

33. Use Exercise 30 to show that the triangle with vertices $A(3, 4)$, $B(6, 8)$, and $C(10, 5)$ is a right triangle and find the area.

In Exercises 34–40 write a proof using coordinate geometry.

34. Every point on the perpendicular bisector of a line segment is equidistant from the endpoints of the segment. (*Hint*: Put the segment on the x-axis.)

35. The diagonals of a rectangle have the same length.

36. The length of the line segment with endpoints (a, b) and (x, y) equals the length of the line segment with endpoints $(a + c, b + d)$ and $(x + c, y + d)$.

37. The length of the line segment with endpoints (ka, kb) and (kx, ky) is $|k|$ times the length of the line segment with endpoints (a, b) and (x, y).

C

38. The diagonals of an isosceles trapezoid have the same length.

39. The diagonals of a parallelogram bisect each other. (*Hint*: Use the fact that opposite sides of a parallelogram have the same length to obtain the coordinates of the vertices.)

40. For any points A, B, and C, $AB + BC \geq AC$. (This is the triangle inequality.) (*Hint*: Introduce a coordinate system so that $A = (0, 0)$, $B = (a, b)$, and $C = (c, d)$. Note that $(ad - bc)^2 \geq 0$.)

Exercises 41–43 refer to $\mathcal{R} \times \mathcal{R}$.

41. Let addition $(+)$ be defined by $(a, b) + (c, d) = (a + c, b + d)$. Show that $\mathcal{R} \times \mathcal{R}$ is a commutative group under $+$.

42. Let $*$ be defined by $(a, b) * (c, d) = (ac, bd)$.
 a. Find an identity for $\mathcal{R} \times \mathcal{R}$ under $*$.
 b. Show that not every member of $\mathcal{R} \times \mathcal{R}$ has an inverse under $*$.

43. Let $\cdot$ be defined by $(a, b) \cdot (c, d) = ac + bd$. Show that for every ordered pair (a, b) there exists an ordered pair (c, d) such that $(a, b) \cdot (c, d) = 0$. Is (c, d) unique?

■ 2-2 LINES

The **graph** of an ordered pair (a, b) is the point having (a, b) as its coordinates. The **graph** of an open sentence in two variables is the set of all points in the plane whose coordinates satisfy the open sentence.

EXAMPLE 1 Graph $y = 2x + 1$.

Solution Make a table of values as shown at the top of the following page and graph the ordered pairs.

x	y	(x, y)
-2	$2(-2) + 1 = -3$	$(-2, -3)$
-1	$2(-1) + 1 = -1$	$(-1, -1)$
0	$2(0) + 1 = 1$	$(0, 1)$
1	$2(1) + 1 = 3$	$(1, 3)$
2	$2(2) + 1 = 5$	$(2, 5)$

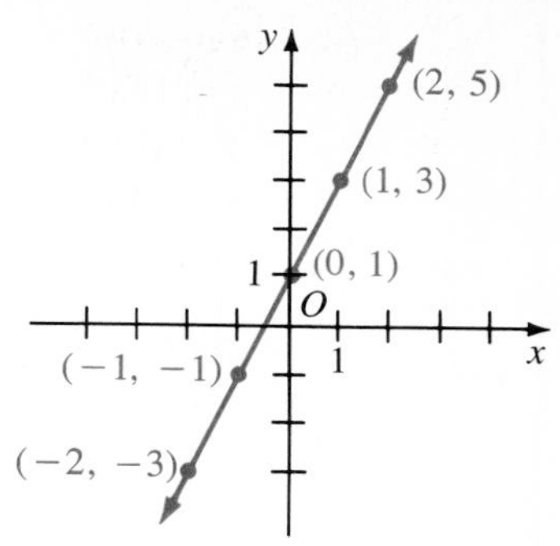

The following theorem, which we accept without proof, relates equations such as $y = 2x + 1$ and lines.

THEOREM 3 Every line in the plane has an equation of the form

$$Ax + By = C \quad (A \neq 0 \text{ or } B \neq 0).$$

Conversely, the graph of an equation of the form $Ax + By = C$ is a line.

Notice from the table of values above that for each increase of 1 unit in x, y increases by 2 units. The following definition of *slope* uses the fact that for two points on a nonvertical line the change in y is proportional to the change in x.

DEFINITION

If $P_1(x_1, y_1)$ and $P_2(x_2, y_2)$ are on a nonvertical line L, then the **slope** of L is given by

$$\frac{y_2 - y_1}{x_2 - x_1}.$$

A vertical line has no slope.

The figure at the right illustrates the definition.

To see that we obtain the same slope for a nonvertical line L regardless of the two points on L used to compute the slope, let $P_1(x_1, y_1)$ and $P_2(x_2, y_2)$ lie on L and let $Ax + By = C$ be an equation of L. Then we have

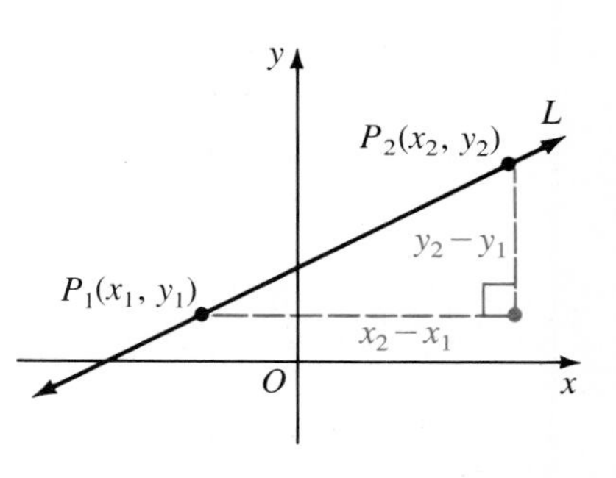

$$Ax_1 + By_1 = C \quad \text{and} \quad Ax_2 + By_2 = C.$$

Therefore

$$\begin{aligned} Ax_1 + By_1 &= Ax_2 + By_2. \\ Ax_1 - Ax_2 &= By_2 - By_1 \\ A(x_1 - x_2) &= B(y_2 - y_1) \end{aligned}$$

Hence

$$-\frac{A}{B} = \frac{y_2 - y_1}{x_2 - x_1} \quad \text{(provided that } x_1 \neq x_2\text{).}$$

Since $-\frac{A}{B}$ is constant, the slope is independent of P_1 and P_2 on L.

We will use the concept of slope to prove that every line in the plane has an equation of the form $Ax + By = C$. Let $P(x_1, y_1)$ be a specific point on L and (x, y) be any other point on L. If L is vertical, then $x = x_1$. This equation is of the form $Ax + By = C$ with $A = 1$, $B = 0$, and $C = x_1$. If L is nonvertical, then $x \neq x_1$ and the slope, which we denote by m, is

$$m = \frac{y - y_1}{x - x_1}.$$

Therefore $$m(x - x_1) = y - y_1.$$

Hence $$(-m)x + y = y_1 - mx_1.$$

Therefore L has an equation of the form $Ax + By = C$ with $A = -m, B = 1$, and $C = y_1 - mx_1$.

The equation $y - y_1 = m(x - x_1)$ is called the **point-slope form** of an equation of the line containing (x_1, y_1) and having slope m. The equation $Ax + By = C$ is called the **standard form** of an equation of a line.

EXAMPLE 2 Find an equation in standard form for the line containing $(-2, 5)$ and $(7, 1)$.

Solution Since $m = \frac{5 - 1}{-2 - 7} = -\frac{4}{9}$, $y - 1 = -\frac{4}{9}(x - 7)$.

Therefore $4x + 9y = 37$. **Answer**

Every nonvertical line crosses the y-axis in some point. The y-coordinate of this point is called the **y-intercept** of the line. If $Ax + By = C$ is an equation of such a line, then $B \neq 0$ and we can write $y = -\frac{A}{B}x + \frac{C}{B}$. When $x = 0$, $y = \frac{C}{B}$. Thus the y-intercept of the line is $\frac{C}{B}$. Letting the slope be denoted by m and the y-intercept be denoted by b, the equation of the line can be written

$$y = mx + b.$$

We call this equation the **slope-intercept form** of an equation of the line. For example, the slope of the line in Example 1 is 2 and its y-intercept is 1.

The ***x*-intercept** of a line is the x-coordinate of the point where the line intersects the x-axis. A line with x-intercept $a \neq 0$ and y-intercept $b \neq 0$ has the following *intercept form*:

$$\frac{x}{a} + \frac{y}{b} = 1$$

If $(a, 0)$ and $(0, b)$ are on the line with equation $Ax + By = C$, then $A = \frac{C}{a}$ and $B = \frac{C}{b}$. Therefore $\frac{C}{a}x + \frac{C}{b}y = C$. Hence $\frac{x}{a} + \frac{y}{b} = 1$.

A vertical line has an equation of the form $x = h$. A horizontal line has an equation of the form $y = k$.

EXAMPLE 3 Graph each equation.

a. $y = 3$ **b.** $x = -3$ **c.** $-3x + 2y = 6$

Solution **a.**

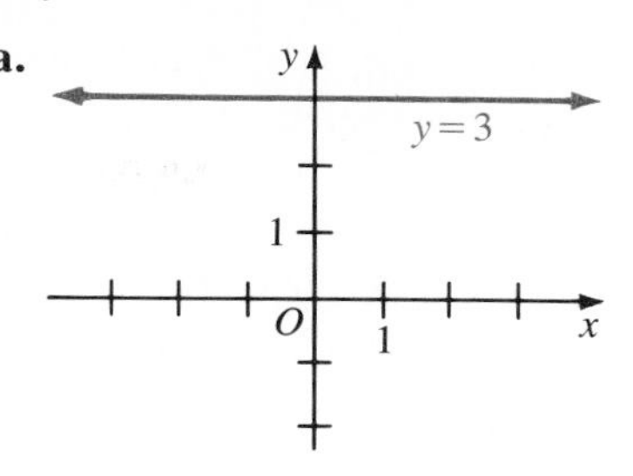

b.

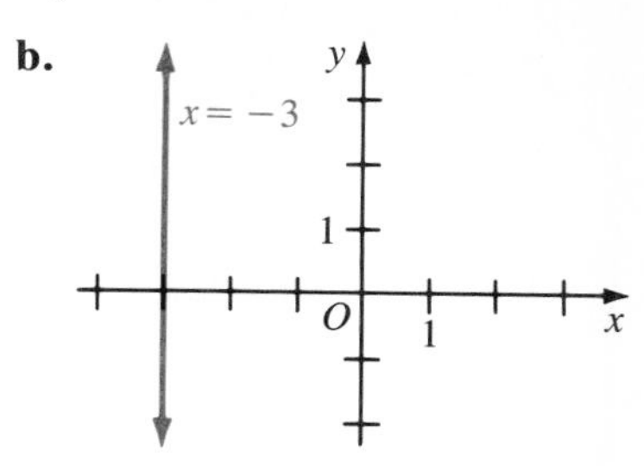

c. The graph is a line. Plot two points and draw a line through them. If $x = 0$, then $y = 3$. If $y = 0$, then $x = -2$. To check, plot a third point. If $x = -1$, $y = \frac{3}{2}$.

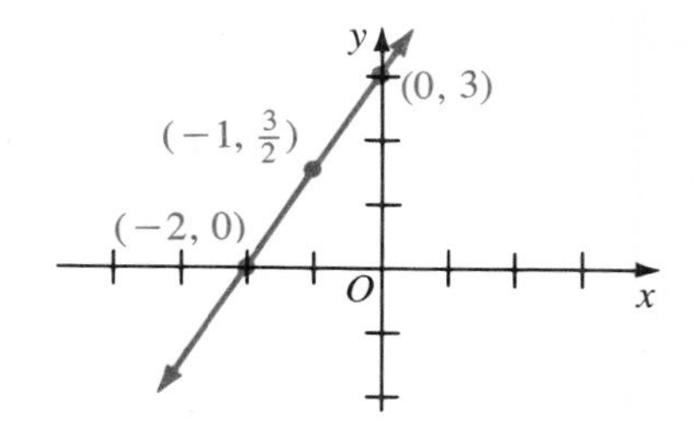

The concept of slope is useful in determining relationships between pairs of lines in the plane.

THEOREM 4 **1.** Two nonvertical lines in the plane are parallel if and only if their slopes are equal.

2. Two nonvertical lines are perpendicular if and only if the product of their slopes is -1.

We will prove that if two nonvertical lines are perpendicular, then the product of their slopes is -1.

Proof: Let L_1 and L_2 be perpendicular lines having slopes m_1 and m_2 respectively. We must show that $m_1 m_2 = -1$. For simplicity we will assume that the lines intersect at the origin. Then $y = m_1 x$ and $y = m_2 x$ are equations for L_1 and L_2 and $P_1\ (1, m_1)$ and $P_2(1, m_2)$ are on the lines. Since L_1 and L_2 are perpendicular, triangle OP_1P_2 is a right triangle. By the Pythagorean theorem,

$$(P_1P_2)^2 = (OP_2)^2 + (OP_1)^2.$$

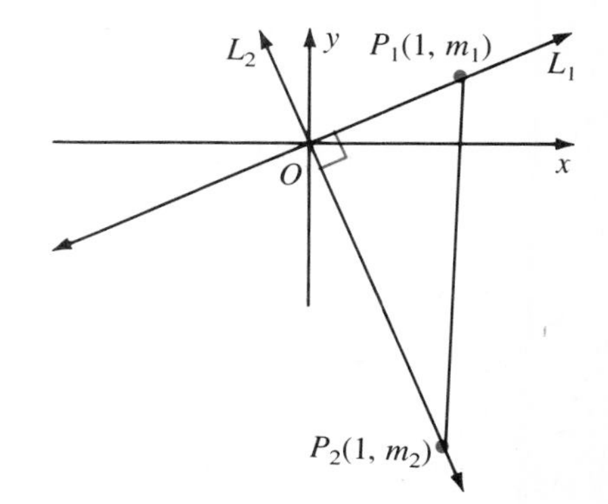

Proof continued on following page.

By the distance formula,

$$(1 - 1)^2 + (m_2 - m_1)^2 = 1^2 + m_1^2 + 1^2 + m_2^2.$$

Simplifying we have

$$-2m_1m_2 = 2.$$

Therefore $m_1m_2 = -1$. ■

You are asked to prove the *if* part of Theorem 4, part (2), in Exercise 59 and to prove part (1) in Exercise 30 in Section 2-8.

EXAMPLE 4 Find an equation in standard form for each line.

a. y-intercept 2; parallel to the line $y = 7x - 4$
b. containing $(2, -7)$; perpendicular to the line $2x - 5y = 3$

Solution **a.** The slope of the required line is 7 by Theorem 4, part (1), and the y-intercept is 2. Therefore $y = 7x + 2$, or $-7x + y = 2$. **Answer**

b. The slope of the given line is $\frac{2}{5}$. Therefore the slope of the required line is $-\frac{5}{2}$. Use the point-slope form to obtain the equation:

$$y - (-7) = -\frac{5}{2}(x - 2)$$

In standard form the equation is $5x + 2y = -4$. **Answer**

EXERCISES

State whether the line has positive, negative, 0, or no slope.

A

1.

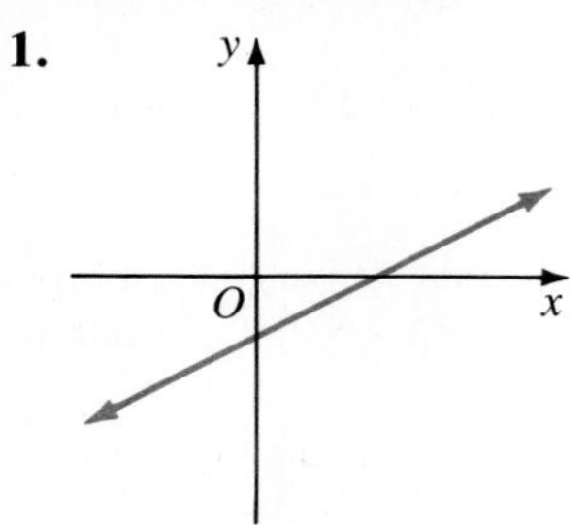

2.

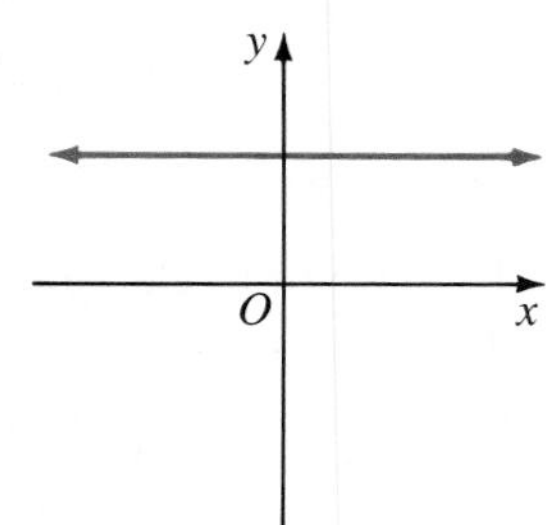

3.

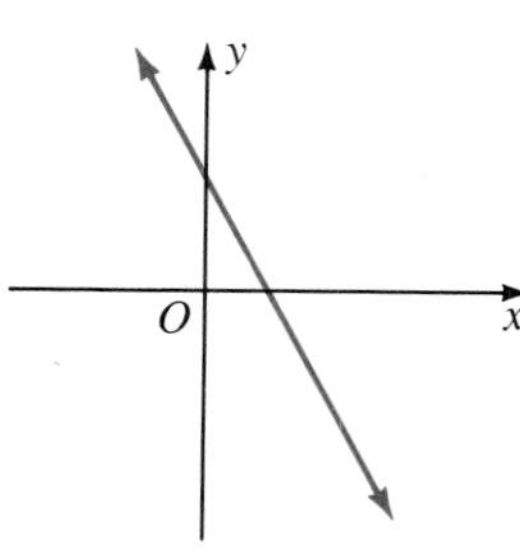

4.

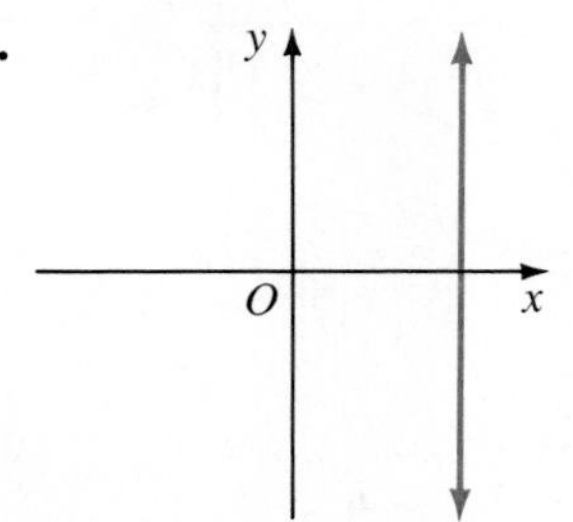

5.

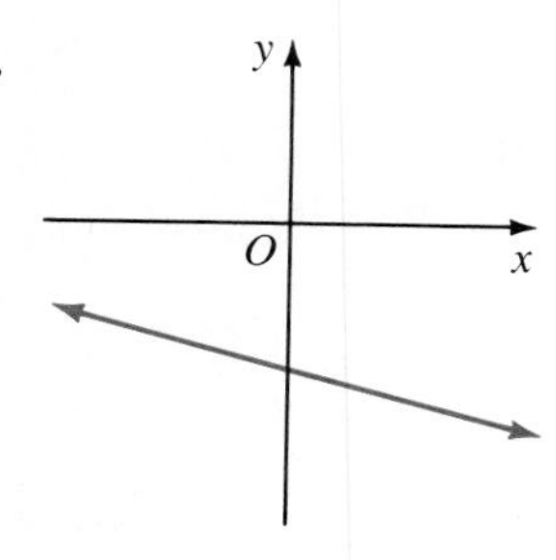

6.

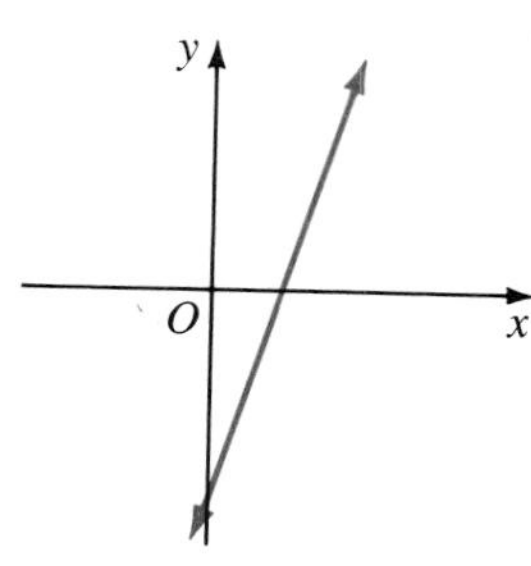

Graph each equation.

7. $y = x - 3$
8. $2y = 3x - 1$
9. $x = y + 2$
10. $2x + 3y = 5$
11. $y = 3x - 2$
12. $x - 2y = 9$
13. $x = 3$
14. $y = -1$
15. $y + 3 = -(x + 4)$

Find the slope (if it exists) of the line containing each pair of points.

16. $(5, 3)$ and $(2, 8)$
17. $(3, -4)$ and $(6, -4)$
18. $(2, 5)$ and $(-2.2, 3.5)$
19. $(-2, 1)$ and $(0, 4)$
20. $\left(\frac{2}{3}, 0\right)$ and $\left(\frac{2}{3}, -18\right)$
21. $(2\sqrt{2}, 3\sqrt{3})$ and $(-\sqrt{2}, 2\sqrt{3})$

Find the slope (if it exists) of each line and the coordinates of any points where the line crosses an axis.

22. $3x - 2y = 7$
23. $2x + 8y = 3$
24. $-5x = 5$
25. $7y = 0$
26. $y = -\frac{11}{5}x + 7$
27. $y + 2 = 5(x + 3)$

Find an equation in standard form for each line.

28. With slope $\frac{1}{2}$ and y-intercept 1
29. With slope $\frac{2}{3}$, containing the point $(7, -3)$
30. Containing the points $(-3, 2)$ and $(7, 4)$
31. Containing the points $(4, 1)$ and $(0, 6)$
32. Containing the points $(-2, 4)$ and $(2, -3)$
33. Parallel to $3x - 2y = 5$, through $(1, 1)$
34. Parallel to $5x = 2y - 6$, through $(-2, 3)$
35. Perpendicular to $y = 4x + 1$, through $(1, 2)$
36. Perpendicular to $5x + 3y = 8$, through $(-5, 23)$
37. Parallel to $y = 5$, through $(-7, 17)$
38. Perpendicular to $x = -2$, through $(2, 4)$
39. Perpendicular to $y = -2$, through $(-5, 4)$

Determine whether the graphs of each pair of equations are parallel, perpendicular, or neither.

40. $y = 2x - 4$; $y = \frac{1}{2}x + 3$
41. $y = 3x + 5$; $y = 3x + 1$
42. $y = \frac{2}{3}x + 1$; $y = -\frac{3}{2}x - 2$
43. $3x + 2y = 7$; $2x + 3y = 3$
44. $5x + 5y = 3$; $2x - 2y = 8$
45. $x + 4 = 0$; $y + 4 = 0$

B 46. By calculating slopes, determine whether A, B, and C are collinear.
   a. $A(2, 1)$; $B(-3, 2)$; $C(-8, 3)$
   b. $A(-5, 1)$; $B(12, -6)$; $C(0, 0)$

47. Given the following slope of a line and point on the line, find two other points on the line.

a. $m = 3, (4, -2)$ **b.** $m = -2, (7, 1)$

c. $m = \frac{1}{2}, (3, 0)$ **d.** $m = \frac{4}{3}, (-2, 3)$

48. Suppose that the cost of owning and operating an automobile is given by an equation of the form $y = mx + b$, where y represents cost and x represents miles driven. If it costs \$210 to drive 300 miles in a month and \$230 to drive 500 miles in a month, what will it cost to drive 1000 miles in a month?

49. Find an equation in standard form of the line through (u, v) perpendicular to the line $Ax + By = C$.

50. Graph the open sentence $|x| + |y| = 1$.

51. Graph the open sentence $|x + 1| - |y| = 3$.

52. Describe the graph of $Ax + By = C$ if both A and B are 0.

53. Find an equation of the line containing $(-2, 5)$ and parallel to the line containing $(1, 6)$ and $(3, 10)$.

54. Find an equation of the perpendicular bisector of the line segment with endpoints $(-2, 4)$ and $(6, 14)$.

55. Find an equation of the perpendicular bisector of the line segment with endpoints $(1, 5)$ and $(7, 12)$.

C **56.** Find an equation of the perpendicular bisector of the line segment with endpoints (x_1, y_1) and (x_2, y_2) where $x_1 \neq x_2$.

57. Prove that if A, B, and C are collinear with B between A and C, then $AB + BC = AC$ by proving (a) and (b).

a. Coordinatize the plane so that $A = (0, 0)$ and $B = (a, b)$. Show for some $k > 1$ that $C = (ka, kb)$.

b. Find $AB + BC$ and AC.

58. Show that if A, B, and C are distinct points such that $AB + BC = AC$, then A, B, and C are collinear with B between A and C.

59. Prove that if the product of the slopes of two lines is -1, then the lines are perpendicular.

COMPUTER EXERCISES

1. Write a program to approximate the shortest distance from a point to a nonvertical line. The user supplies the coordinates of the point and the coefficients of an equation of the line.

2. Run the program in Exercise 1 for each point and line.

a. $(3, 4)$; $y = 6$ **b.** $(4, 8)$; $3x + 4y = -2$

c. $(4, -2)$; $3x + 5y = 2$ **d.** $(-3, -2)$; $x + 5y = 5$

Conic Sections

2-3 CIRCLES

A **circle** is the set of all points in the plane a fixed distance r (the **radius**) from a fixed point (the **center**). An equation of a circle can be found using the distance formula. Suppose a circle has center (h, k) and radius $r > 0$. The point (x, y) is on the circle if and only if

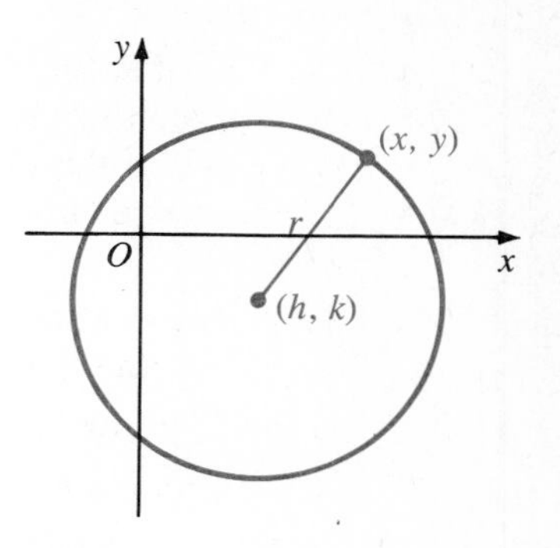

$$\sqrt{(x - h)^2 + (y - k)^2} = r,$$

or

$$(x - h)^2 + (y - k)^2 = r^2.$$

THEOREM 5 An equation of a circle with radius r and center (h, k) is

$$(x - h)^2 + (y - k)^2 = r^2.$$

EXAMPLE 1 Find the center and radius of each circle and draw its graph.

a. $(x - 1)^2 + (y + 2)^2 = 16$ **b.** $x^2 + y^2 - 8x + 12y - 8 = 0$

Solution **a.** The center is $(1, -2)$ and the radius is $\sqrt{16} = 4$. The graph is at the left below.

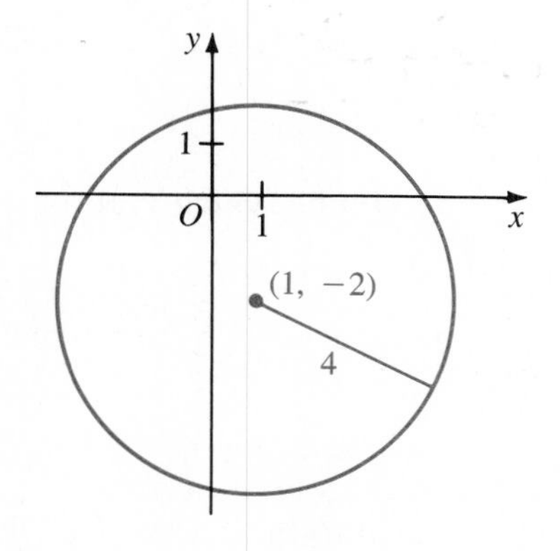

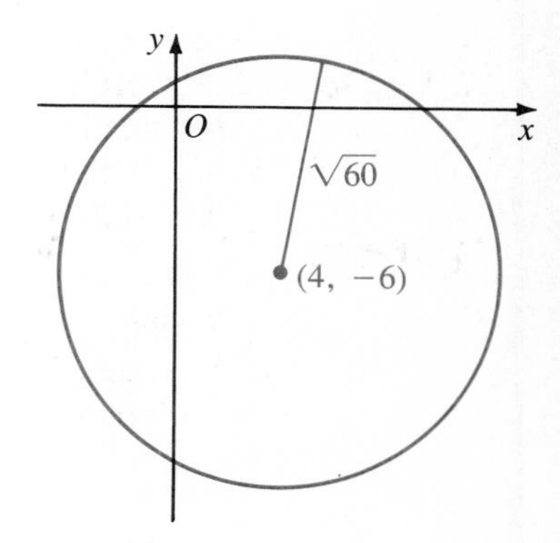

b. Complete the squares in x and y.

$$x^2 + y^2 - 8x + 12y - 8 = 0$$
$$x^2 - 8x + y^2 + 12y = 8$$
$$x^2 - 8x + 16 + y^2 + 12y + 36 = 8 + 16 + 36$$
$$(x - 4)^2 + (y + 6)^2 = 60$$

The center is $(4, -6)$ and the radius is $\sqrt{60} \approx 7.7$. The graph is at the right above.

EXAMPLE 2 Identify the graph of each equation.

a. $x^2 + y^2 + 5x = 0$ **b.** $x^2 + y^2 - 4x + 2y = -5$

c. $x^2 + 2x + y^2 = -3$

Solution **a.**

$$x^2 + y^2 + 5x = 0$$
$$x^2 + 5x + \left(\frac{5}{2}\right)^2 + y^2 = 0 + \left(\frac{5}{2}\right)^2$$
$$\left(x + \frac{5}{2}\right)^2 + y^2 = \left(\frac{5}{2}\right)^2$$

The graph is a circle with center $\left(-\frac{5}{2}, 0\right)$ and radius $\frac{5}{2}$.

b.

$$x^2 + y^2 - 4x + 2y = -5$$
$$x^2 - 4x + 4 + y^2 + 2y + 1 = -5 + 4 + 1$$
$$(x - 2)^2 + (y + 1)^2 = 0$$

The graph is the single point $(2, -1)$ since the sum of two squares is 0 only if both squares are 0.

c.

$$x^2 + 2x + y^2 = -3$$
$$x^2 + 2x + 1 + y^2 = -3 + 1$$
$$(x + 1)^2 + y^2 = -2$$

There is no ordered pair satisfying the equation. The graph is empty.

Example 2 illustrates the following theorem.

THEOREM 6 The graph of an equation of the form $x^2 + y^2 + Ax + By + C = 0$ is a circle, a point, or empty.

The values of A, B, and C jointly determine which type of graph the equation has. (See Exercise 32.)

EXAMPLE 3 Find an equation of the line tangent to the graph of $(x - 3)^2 + (y + 1)^2 = 25$ at the point $(0, -5)$.

Solution Recall that a tangent to a circle intersects the circle in exactly one point and is perpendicular to the radius drawn to the point of tangency. Notice that $(0, -5)$ is on the circle. The y-intercept of the tangent is -5. From the figure, the slope of $\overline{CD}$ is

$$\frac{-1 - (-5)}{3 - 0} = \frac{4}{3}.$$

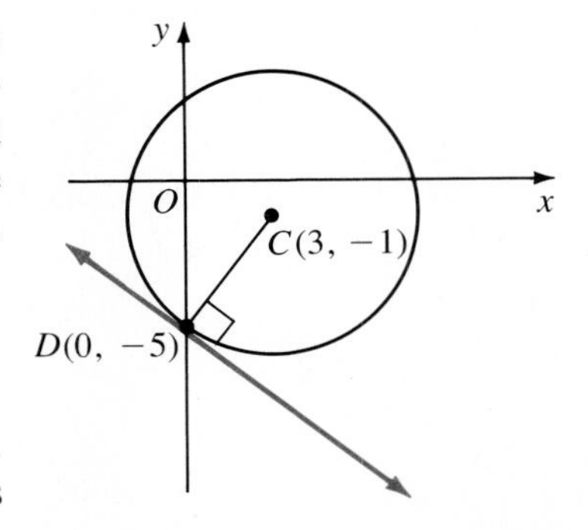

Thus the slope of the tangent line is $-\frac{3}{4}$. The equation of the required tangent line is $y = -\frac{3}{4}x - 5$. **Answer**

EXAMPLE 4 An archaeologist found a piece of broken pottery that appeared to be part of a circular plate. When the piece was placed on a sheet of graph paper marked off in centimeters, the points (4, 6), (7, 7), and (12, 6) lay along the edge of the plate. What was the radius of the plate?

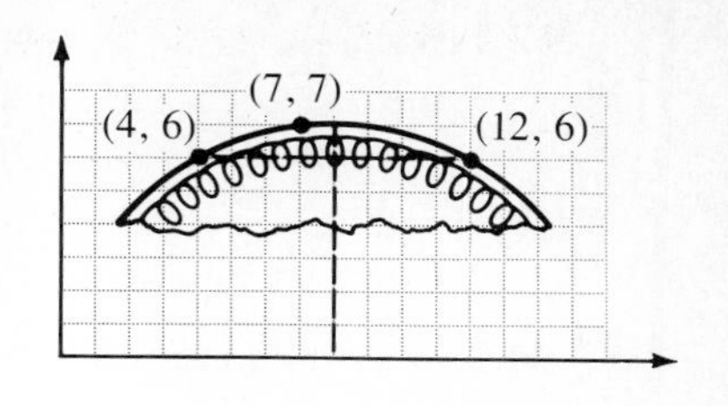

Solution The center must lie on the perpendicular bisector of the segment that joins (4, 6) and (12, 6). The center is $(8, k)$ for some k. So $\sqrt{(4-8)^2+(6-k)^2} = \sqrt{(7-8)^2+(7-k)^2}$ and $16 + 36 - 12k + k^2 = 1 + 49 - 14k + k^2$. Thus $k = -1$ and the radius is $\sqrt{(4-8)^2+(6-(-1))^2} = \sqrt{65} \approx 8.1$ cm.

EXERCISES

Find an equation of each circle.

A

1. Center (2, 3); radius 3
2. Center $(-2, 6)$; radius 5
3. Center (0, 3); radius 12
4. Center (0, 0); radius $2\sqrt{3}$

Graph each equation (if possible). Find the center and radius of each circle.

5. $(x+2)^2 + (y-3)^2 = 1$
6. $x^2 + y^2 = 16$
7. $(x-4)^2 + y^2 = 4$
8. $x^2 + y^2 + 4x = 5$
9. $x^2 + y^2 - 2y = 0$
10. $x^2 + y^2 + 6x - 8y = -9$
11. $x^2 + y^2 + 5x + 7y = 0$
12. $x^2 + y^2 = 0$
13. $x^2 + (y-2)^2 = -7$
14. $x^2 + y^2 + y = -3$
15. $x^2 + y^2 + 8x = -16$
16. $x^2 + y^2 - 3x - 9y = -25$

Find an equation of the line tangent to each circle at the point P.

17. $(x-2)^2 + y^2 = 10$; $P(1, -3)$
18. $(x+3)^2 + (y-4)^2 = 16$; $P(1, 4)$

Determine whether the point A is inside, outside, or on the circle with radius r and center C.

19. $C = (2, -1)$; $r = 3$; $A = (3, 2)$
20. $C = (4, -1)$; $r = 5$; $A = (8, -3)$
21. $C = (0, 0)$; $r = 4$; $A = (2, 2)$
22. $C = (-3, -4)$; $r = 2\sqrt{5}$; $A = (-5, 0)$

Find an equation of each circle.

B

23. Center (3, 5); tangent to the x-axis
24. Center $(5, -3)$; tangent to the y-axis
25. Tangent to the x-axis, the y-axis, and the line $y = 5$ (two answers)
26. Center on the line $y = 2x$, tangent to the x-axis at (3, 0)
27. Center on the line $y = -2x$, tangent to the y-axis at (0, 3)

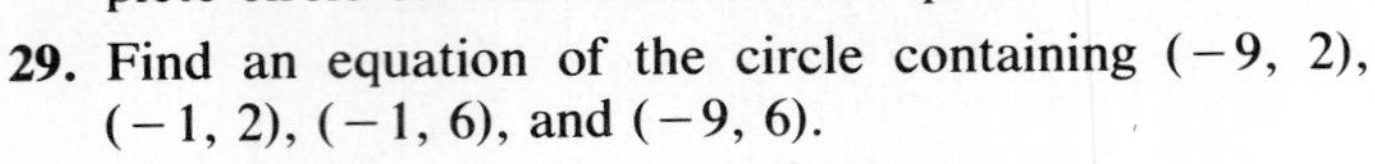

28. The arch shown at the right has height 10 ft and width 30 ft. Use coordinates to find the radius of the complete circle of which the arch is a part.

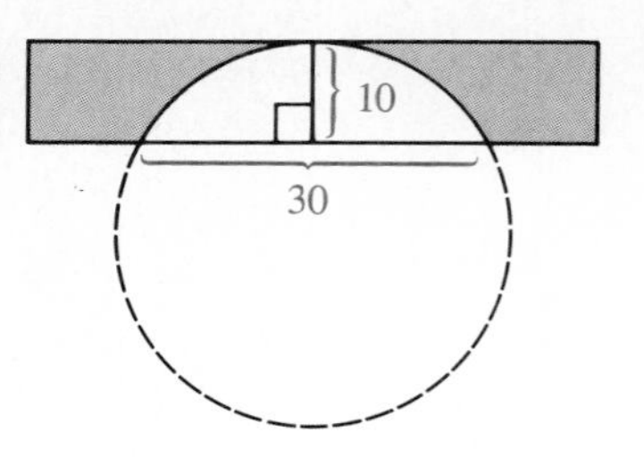

29. Find an equation of the circle containing $(-9, 2)$, $(-1, 2)$, $(-1, 6)$, and $(-9, 6)$.

C **30.** Find an equation of the circle containing $(0, 0)$, $(4, 1)$, and $(3, -3)$.

31. Find an equation of the circle with center $(6, 0)$ and tangent to the line $y = x$.

32. Under what conditions placed on A, B, and C will the graph of $x^2 + y^2 + Ax + By + C = 0$ be **(a)** a circle, **(b)** a point, and **(c)** empty? If the graph is a circle, find a formula for the radius and give the coordinates of the center.

33. Local birdwatchers constructed three observation towers, A, B, and C, in a forest. Tower A is 400 m north and 200 m east of the forest entrance. Tower B is 400 m north and 600 m east of the entrance. Tower C is 300 m directly east of the entrance. If the coordinates of the forest entrance are $(0, 0)$, find the coordinates of the point an equal distance from all three towers.

34. Let $P(x_1, y_1)$ be a point outside the circle of radius r and center (h, k). Draw a line through P tangent to the circle. Show that the distance from point P to the point of tangency is $\sqrt{(x_1 - h)^2 + (y_1 - k)^2 - r^2}$.

35. Show that $x_1x + y_1y = r^2$ is an equation of the line that is tangent to the circle whose equation is $x^2 + y^2 = r^2$ at the point (x_1, y_1).

2-4 ELLIPSES

In Section 2-3 we discussed circles and equations such as $x^2 + y^2 = 4$. If we write this equation as $\frac{x^2}{4} + \frac{y^2}{4} = 1$ and replace the first 4 by 9, we obtain $\frac{x^2}{9} + \frac{y^2}{4} = 1$. We can make a table of values and draw the graph of the equation.

x	y	(x, y)
0	± 2	$(0, \pm 2)$
± 3	0	$(\pm 3, 0)$
± 2	$\pm\frac{2}{3}\sqrt{5}$	$\left(\pm 2, \pm\frac{2}{3}\sqrt{5}\right)$

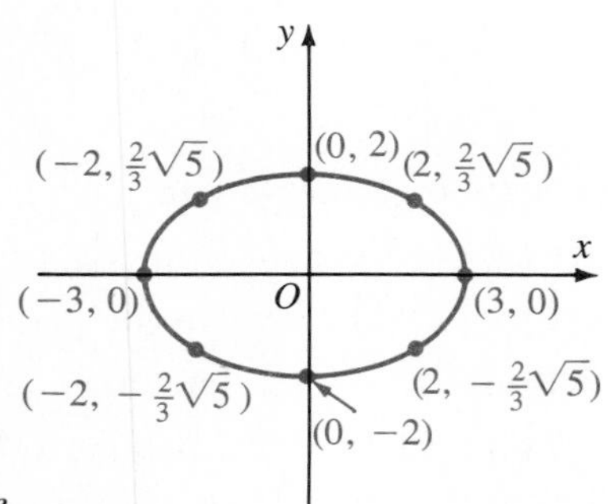

The graph appears to be an oval, or *ellipse*.

#33

DEFINITION

An **ellipse** is a set of all points in the plane such that the sum of the distances from each point to two fixed points called **foci** is a constant.

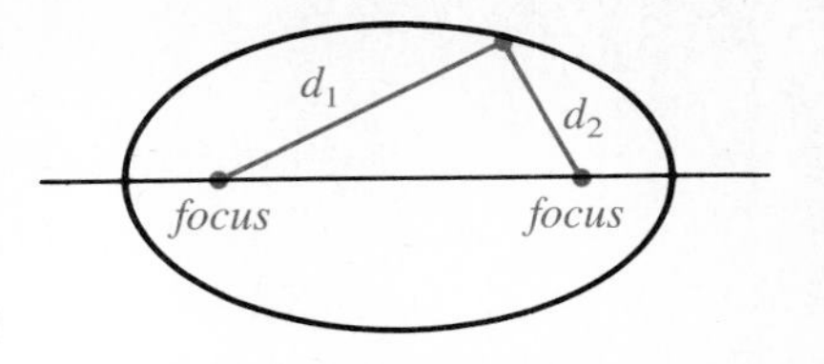

$d_1 + d_2 =$ **a constant**

The distance from each point P on an ellipse to a focus is called a **focal radius** of P.

We can derive an equation for the ellipse with foci $(-c, 0)$ and $(c, 0)$ where $c > 0$ and sum of focal radii is $2a$. Notice that $2a > 2c$. That is, $a > c > 0$. If $P(x, y)$ is any point on the ellipse, then the distance from $P(x, y)$ to $(-c, 0)$ is $\sqrt{(x - (-c))^2 + (y - 0)^2}$ and the distance from $P(x, y)$ to $(c, 0)$ is $\sqrt{(x - c)^2 + (y - 0)^2}$. Therefore

$$\sqrt{(x + c)^2 + y^2} + \sqrt{(x - c)^2 + y^2} = 2a.$$

Then

$$\sqrt{(x + c)^2 + y^2} = 2a - \sqrt{(x - c)^2 + y^2}.$$

Squaring both sides,

$$(x + c)^2 + y^2 = 4a^2 - 4a\sqrt{(x - c)^2 + y^2} + (x - c)^2 + y^2.$$

Simplifying, we have,

$$4xc - 4a^2 = -4a\sqrt{(x - c)^2 + y^2}$$
$$xc - a^2 = -a\sqrt{(x - c)^2 + y^2}.$$

Squaring again,

$$x^2c^2 - 2xca^2 + a^4 = a^2((x - c)^2 + y^2).$$

Writing all variable terms on the left side of the equation and all constant terms on the right side and simplifying, we have

$$x^2(a^2 - c^2) + y^2a^2 = a^2(a^2 - c^2).$$

Since $a > c$ we can let $b^2 = a^2 - c^2$ and write

$$x^2b^2 + y^2a^2 = a^2b^2.$$

Dividing both sides by a^2b^2, we have

$$\frac{x^2}{a^2} + \frac{y^2}{b^2} = 1.$$

Sometimes it is convenient to consider an ellipse with foci $(0, -c)$ and $(0, c)$ and sum of focal radii $2a$. In this case we derive an equation by similar reasoning and obtain

$$\frac{y^2}{a^2} + \frac{x^2}{b^2} = 1 \quad \text{where } b^2 = a^2 - c^2.$$

The graphs of both types of ellipse are shown in the figures at the top of the next page.

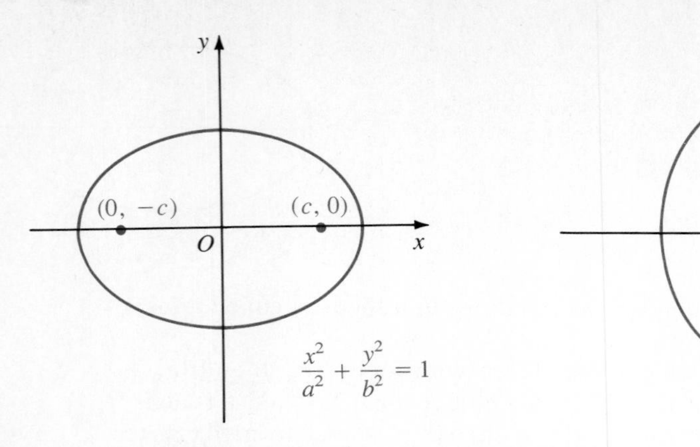

We can graph equations such as $\frac{x^2}{25} + \frac{y^2}{16} = 1$ with the help of the following analysis.

1. If (x, y) lies on the ellipse, then $y = \pm 4\sqrt{1 - \frac{x^2}{25}}$. For y to be a real number we must have $-5 \le x \le 5$. Solving for x in terms of y, we have $x = \pm 5\sqrt{1 - \frac{y^2}{16}}$. For x to be a real number we must have $-4 \le y \le 4$. Therefore the graph is contained within the region bounded by $x = -5$, $x = 5$, $y = -4$, and $y = 4$. Furthermore the graph intersects these lines in the points $(-5, 0)$, $(5, 0)$, $(0, -4)$, and $(0, 4)$. See the figure at the left below.

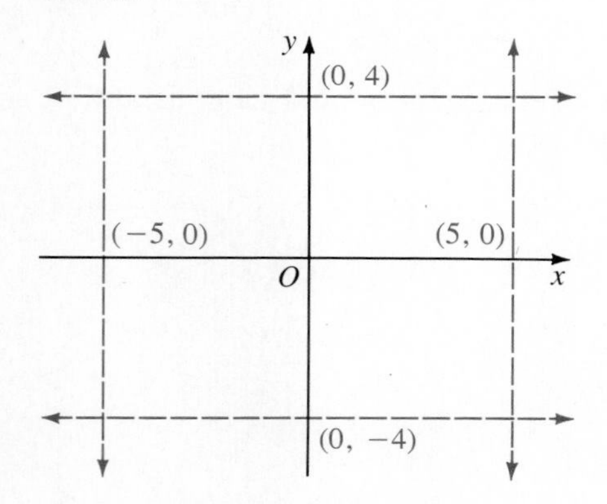

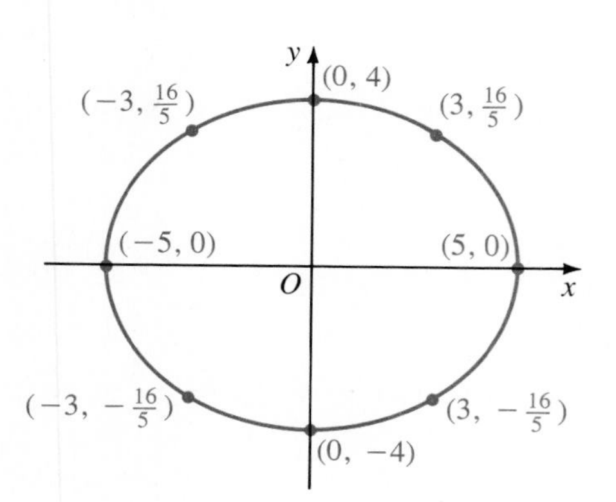

2. The graph is **symmetric** with respect to the x-axis and the y-axis. For example, $\left(3, \frac{16}{5}\right)$ is on the ellipse, so the ellipse also contains the points $\left(-3, \frac{16}{5}\right)$, $\left(-3, -\frac{16}{5}\right)$, and $\left(3, -\frac{16}{5}\right)$.

From this analysis we can draw the graph, shown at the right above.

The chord containing the foci is called the **major axis** of the ellipse. Its endpoints are called the **vertices** of the ellipse. The midpoint of the major axis is the **center** of the ellipse. The chord that lies along the perpendicular bisector of the major axis is the **minor axis**.

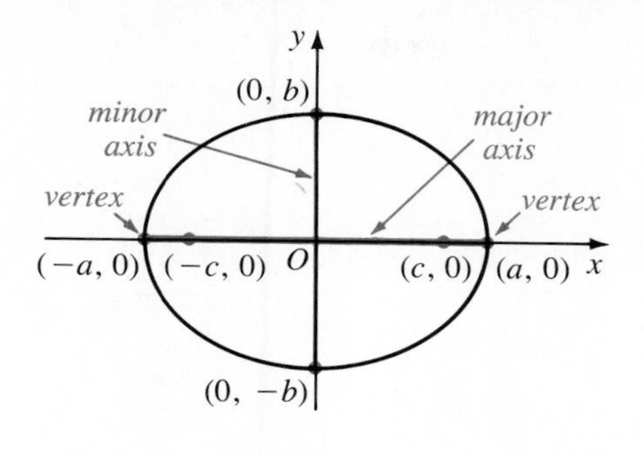

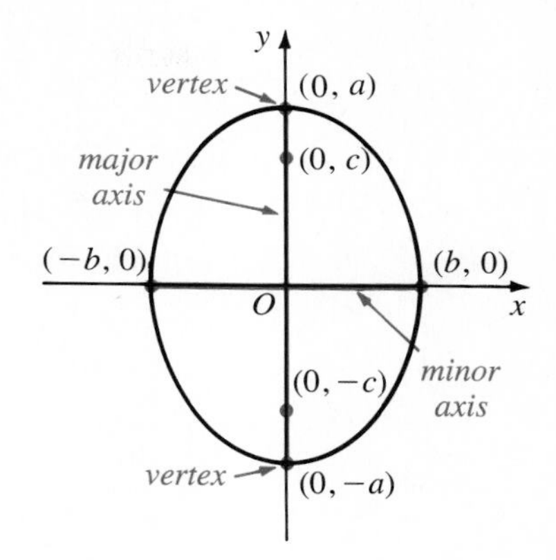

EXAMPLE 1 Find an equation for each ellipse, sketch each graph, and label the foci, the vertices, and the endpoints of the minor axis.

a. foci $(-3, 0)$ and $(3, 0)$
sum of focal radii 12

b. vertices $(0, 7)$ and $(0, -7)$
endpoints of minor axis $(5, 0)$ and $(-5, 0)$

Solution **a.** Since $c = 3$ and $a = 6$, $b^2 = 6^2 - 3^2 = 27$.

So $\frac{x^2}{36} + \frac{y^2}{27} = 1$.

b. Since $a = 7$ and $b = 5$, $\frac{y^2}{49} + \frac{x^2}{25} = 1$. Since $b^2 = a^2 - c^2$, $c^2 = a^2 - b^2 = 49 - 25 = 24$, and $c = 2\sqrt{6}$.

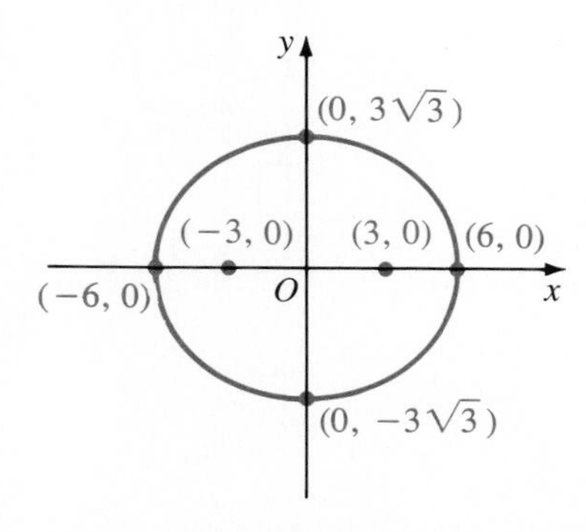

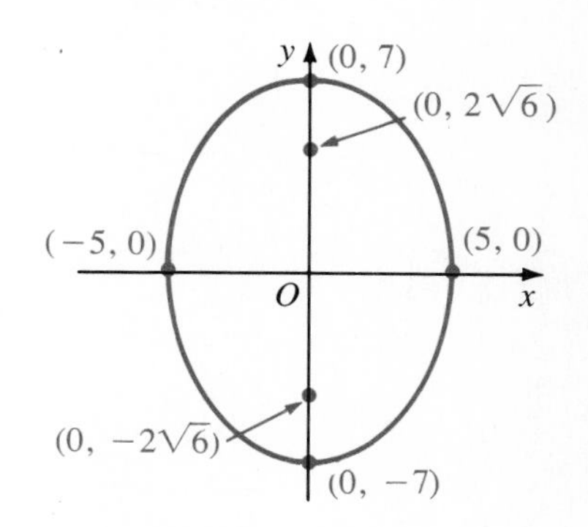

We may also obtain equations for ellipses centered at any point (h, k) with horizontal or vertical major axes. Theorem 7, stated at the top of the following page, generalizes our discussion. Notice that when the center is (h, k), x is replaced by $x - h$ and y is replaced by $y - k$ in the equation for an ellipse.

THEOREM 7 Let $2a$ represent the sum of the focal radii of points of an ellipse.

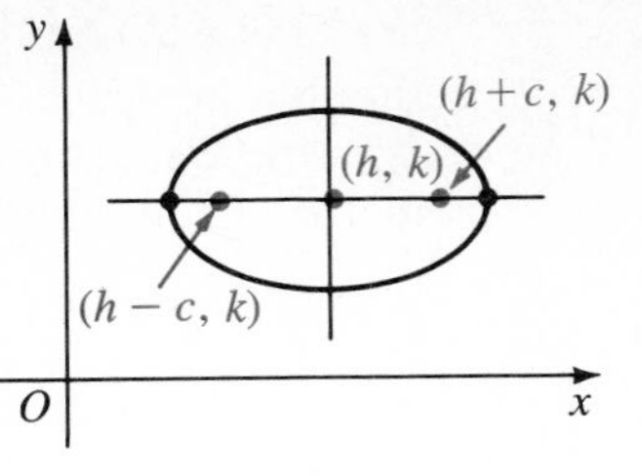

1. If the foci of the ellipse are $(h - c, k)$ and $(h + c, k)$, then

$$\frac{(x - h)^2}{a^2} + \frac{(y - k)^2}{b^2} = 1$$

where $b^2 = a^2 - c^2$.

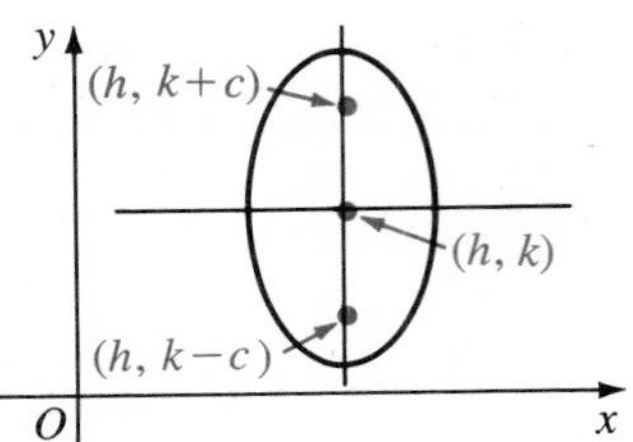

2. If the foci of the ellipse are $(h, k - c)$ and $(h, k + c)$, then

$$\frac{(y - k)^2}{a^2} + \frac{(x - h)^2}{b^2} = 1$$

where $b^2 = a^2 - c^2$.

EXAMPLE 2 Graph $16x^2 + 25y^2 - 32x - 150y = 159$ and label the center, the vertices, and the endpoints of the minor axis.

Solution Complete the square in each variable.

$$\begin{aligned} 16x^2 + 25y^2 - 32x - 150y &= 159 \\ 16(x^2 - 2x) + 25(y^2 - 6y) &= 159 \\ 16(x^2 - 2x + 1) + 25(y^2 - 6y + 9) &= 159 + 16 \cdot 1 + 25 \cdot 9 \\ 16(x - 1)^2 + 25(y - 3)^2 &= 400 \\ \frac{(x - 1)^2}{25} + \frac{(y - 3)^2}{16} &= 1 \end{aligned}$$

The center is $(1, 3)$, $a^2 = 25$, and $b^2 = 16$. So $c = 3$. The major axis is horizontal, so the foci are $(1 - 3, 3) = (-2, 3)$ and $(1 + 3, 3) = (4, 3)$. The vertices are $(1 - 5, 3) = (-4, 3)$ and $(1 + 5, 3) = (6, 3)$.

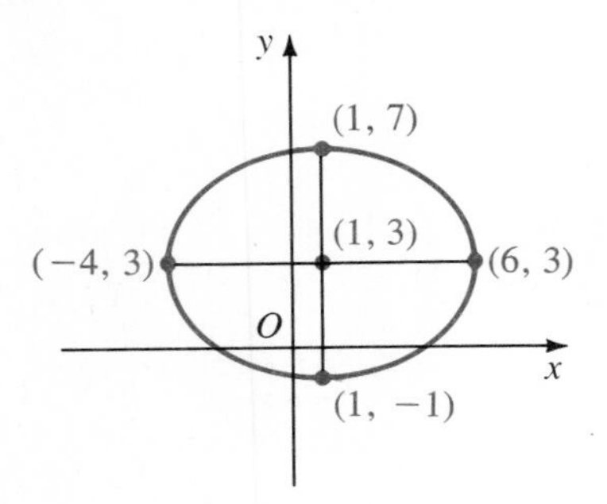

EXAMPLE 3 One focus F of an ellipse is $\left(1, \frac{3}{2}\right)$ and one endpoint E of the minor axis is $\left(\frac{1}{2}, 1\right)$. Find an equation of the ellipse.

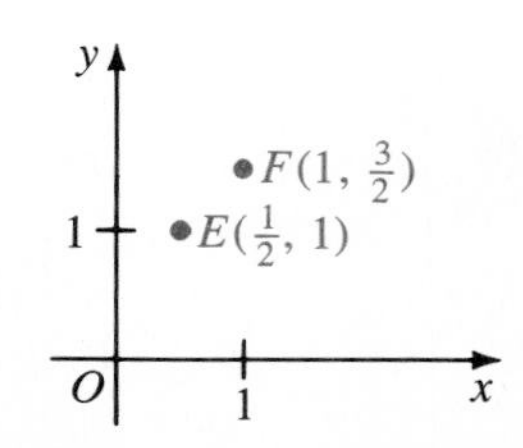

Solution The minor axis might be vertical, horizontal, or tilted. Consider the first two possibilities.

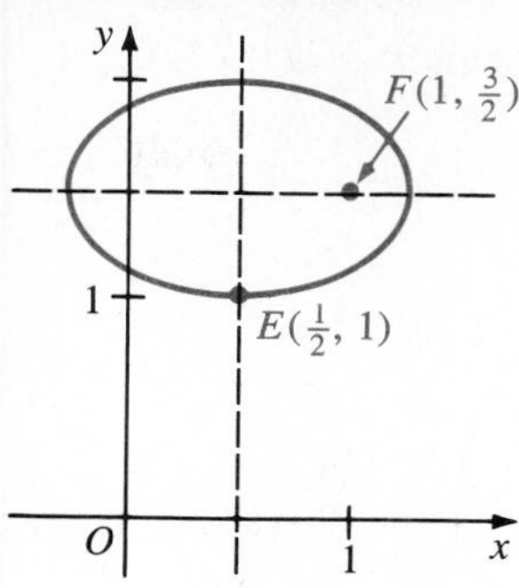

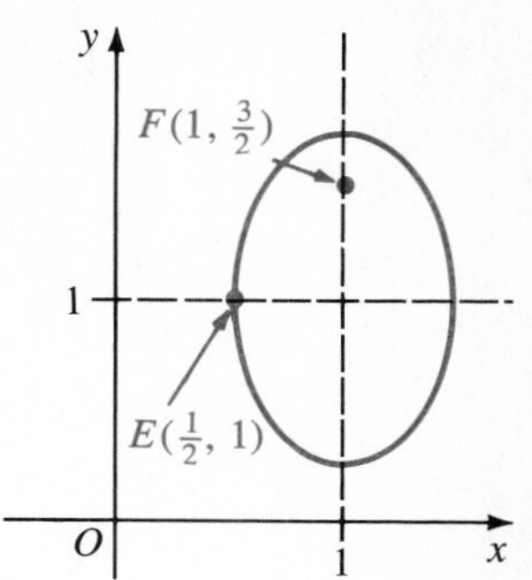

If the minor axis is vertical, the center is $\left(\frac{1}{2}, \frac{3}{2}\right)$. Therefore $c = \frac{1}{2}$, $b = \frac{1}{2}$, and $a^2 = \frac{1}{2}$. The equation of the ellipse is

$$\frac{(x - \frac{1}{2})^2}{\frac{1}{2}} + \frac{(y - \frac{3}{2})^2}{\frac{1}{4}} = 1.$$

If the minor axis is horizontal, the center is (1, 1). Then $c = \frac{1}{2}$ and $b = \frac{1}{2}$. Therefore $\left(\frac{1}{2}\right)^2 = a^2 - \left(\frac{1}{2}\right)^2$ and $a^2 = \frac{1}{2}$. The equation is

$$\frac{(y - 1)^2}{\frac{1}{2}} + \frac{(x - 1)^2}{\frac{1}{4}} = 1.$$

EXERCISES

Sketch the graph of each equation. Label the center and the endpoints of the axes of each ellipse.

A

1. $\frac{y^2}{9} + \frac{x^2}{4} = 1$

2. $x^2 + \frac{y^2}{4} = 1$

3. $\frac{(x + 4)^2}{9} + \frac{(y + 2)^2}{25} = 1$

4. $\frac{(y - 1)^2}{16} + \frac{(x + 3)^2}{4} = 1$

5. $9x^2 + 25y^2 - 225 = 0$

6. $(x - 1)^2 + 20(y + 3)^2 = 100$

7. $x^2 + y^2 - 6x + 10y = 47$

8. $x^2 + y^2 + 10x - 6y = 2$

9. $9x^2 + 16y^2 - 32y = 128$

10. $25x^2 + 16y^2 - 384 = 16$

11. $x^2 + 4y^2 - 4x + 24y = -36$

12. $x^2 + 4y^2 + 4x = 0$

13. $4x^2 + y^2 + 24x - 4y = -36$

14. $x^2 + 81y^2 - 6x + 324y = -252$

Find an equation of each ellipse.

15. Vertices (4, 0) and (−4, 0); endpoints of minor axis (0, 1), (0, −1)

16. Foci (−3, 0) and (3, 0); endpoints of minor axis (0, 4), (0, −4)

Find an equation of each ellipse.

17. Vertices $(-16, 4)$ and $(4, 4)$; endpoints of minor axis $(-6, -4)$, $(-6, 12)$
18. Vertices $(0, 15)$ and $(0, 3)$; endpoints of minor axis $(-2, 9)$, $(2, 9)$
19. Foci $(-7, -7)$ and $(-7, 19)$; vertices $(-7, -9)$, $(-7, 21)$
20. Foci $(-11, 5)$ and $(9, 5)$; vertices $(-19, 5)$, $(17, 5)$
21. One vertex $(6, 0)$; center $(0, 0)$; one focus $(-4, 0)$
22. Major axis of length 8; foci $(2, 0)$ and $(-2, 0)$
23. Minor axis of length 6; foci $(3, 0)$ and $(-3, 0)$
24. Minor axis of length 8; foci $(2\sqrt{6}, 0)$ and $(-2\sqrt{6}, 0)$

Find the coordinates of any points where each ellipse crosses an axis.

25. $x^2 + \frac{y^2}{4} = 1$
26. $\frac{x^2}{16} + y^2 = 1$
27. $\frac{(x-1)^2}{9} + \frac{(y-2)^2}{4} = 1$
28. $\frac{(x+3)^2}{4} + \frac{(y-2)^2}{8} = 1$
29. $x^2 + 4y^2 + 2x + 6y = 1$
30. $x^2 + 2y^2 + 2x + y = -1$

Determine whether P is inside, outside, or on the ellipse.

B

31. $\frac{x^2}{4} + \frac{y^2}{9} = 1$; $P(1, 4)$
32. $\frac{x^2}{9} + \frac{y^2}{4} = 1$; $P(-2, 1)$
33. $\frac{(x+1)^2}{25} + \frac{(y-3)^2}{16} = 1$; $P(4, 3)$
34. $\frac{(x+2)^2}{36} + \frac{(y-2)^2}{32} = 1$; $P(-2, 2)$

In Exercises 35 and 36 find all equations of all ellipses having the given characteristics. Assume that the axes of the ellipses are horizontal or vertical.

35. Focus $(1, 4)$; endpoint of minor axis $(4, 6)$
36. Vertex $(4, 0)$; endpoint of minor axis $(7, 2)$
37. Show that the length of the chord of an ellipse perpendicular to the major axis at a focus is $\frac{2b^2}{a}$. (*Note*: This line segment is called a *latus rectum* of the ellipse.)
38. Prove that the distance from a focus of an ellipse to the nearer vertex is less than the distance from that focus to an endpoint of the nearer latus rectum. (*Note*: The latus rectum is defined in Exercise 37.)

C

39. Derive an equation for the ellipse with foci $(1, 1)$ and $(-1, -1)$ and sum of focal radii equal to 4.
40. Prove that an equation of an ellipse with foci $(0, -c)$ and $(0, c)$ and sum of focal radii $2a$ is $\frac{y^2}{a^2} + \frac{x^2}{b^2} = 1$ where $b^2 = a^2 - c^2$.

41. Prove that if $a > b > 0$ and (x, y) satisfies $\frac{x^2}{a^2} + \frac{y^2}{b^2} = 1$, then the sum of the distances from (x, y) to $(-c, 0)$ and from (x, y) to $(c, 0)$ is $2a$, where $b^2 = a^2 - c^2$.

42. Find an equation for the set of all points (x, y) in the plane such that the distance from (x, y) to the line $y = 6$ is twice the distance from (x, y) to $(2, 0)$.

43. Find an equation for the set of all points (x, y) in the plane such that the distance from (x, y) to $(0, 4)$ is half the distance from (x, y) to the line $y = 10$.

44. Let $Ax^2 + By^2 + Cx + Dy + E = 0$ (A and B positive). Under what conditions on A, B, C, D, and E will the graph of the equation be an ellipse? Find the coordinates of the center.

COMPUTER EXERCISES

1. Write a computer program to approximate the shortest distance from a point to an ellipse. The user supplies the coordinates of the point and an equation of the ellipse. Assume that the origin of the coordinate system is the center of the ellipse.

2. Run the program in Exercise 1 for each point and each equation of the ellipse.

a. $(6, 7)$; $4x^2 + 9y^2 = 36$ **b.** $(0, 4)$; $36x^2 + 20y^2 = 720$
c. $(8, 0)$; $4x^2 + y^2 = 100$ **d.** $(3, 4)$; $x^2 + y^2 = 4$

2-5 HYPERBOLAS

If in the equation $\frac{x^2}{4} + \frac{y^2}{9} = 1$ we replace $+$ by $-$, we obtain $\frac{x^2}{4} - \frac{y^2}{9} = 1$.

The graph of this new equation is not a closed curve but rather a pair of disconnected branches. This can be seen from the following table of values, the graphs of the ordered pairs, and the curves containing them. The graph is called a *hyperbola*.

x	y	(x, y)
± 2	0	$(\pm 2, 0)$
± 3	$\pm\frac{3\sqrt{5}}{2}$	$\left(\pm 3, \pm\frac{3\sqrt{5}}{2}\right)$
± 4	$\pm 3\sqrt{3}$	$(\pm 4, \pm 3\sqrt{3})$

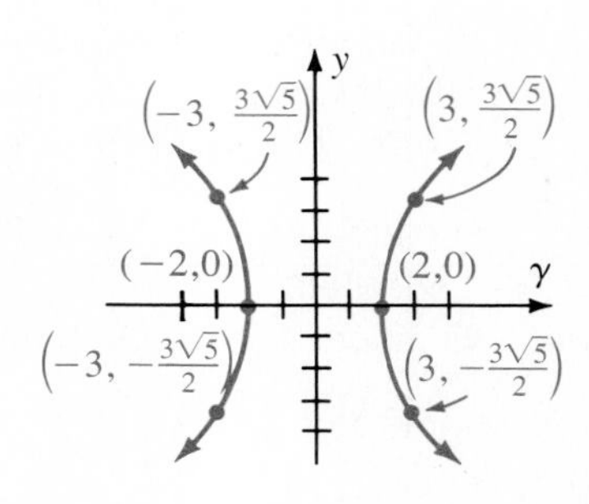

DEFINITION

A **hyperbola** is a set of all points in the plane such that the absolute value of the difference of the distances from each point in the set to two fixed points (the **foci**) is constant.

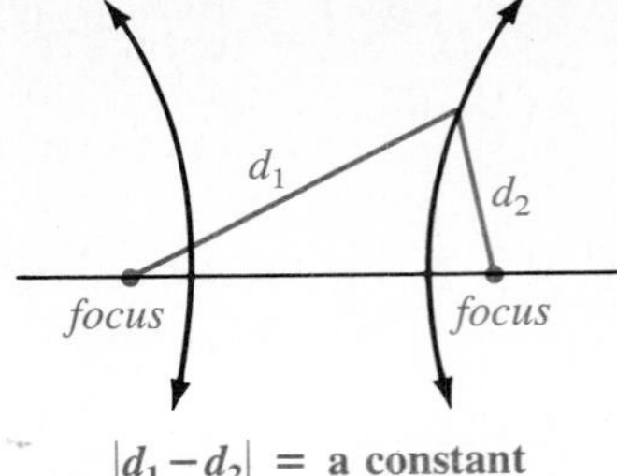

The distance from a point on the hyperbola to a focus is called a **focal radius** of the point.

We can obtain an equation for the hyperbola with foci $(-c, 0)$ and $(c, 0)$ and with the absolute value of the difference of focal radii $2a$. In order to obtain a graph, c must be greater than a.

Let (x, y) be on the hyperbola. Then the distance from (x, y) to $(-c, 0)$ is

$$\sqrt{(x - (-c))^2 + (y - 0)^2} = \sqrt{(x + c)^2 + y^2}.$$

The distance from (x, y) to $(c, 0)$ is $\sqrt{(x - c)^2 + y^2}$.

Therefore $$\left|\sqrt{(x + c)^2 + y^2} - \sqrt{(x - c)^2 + y^2}\right| = 2a.$$

When this is simplified we have

$$(c^2 - a^2)x^2 - a^2y^2 = a^2(c^2 - a^2).$$

Since $c > a$, we let $b^2 = c^2 - a^2$,

$$b^2x^2 - a^2y^2 = a^2b^2.$$

Therefore $$\frac{x^2}{a^2} - \frac{y^2}{b^2} = 1.$$

If instead the foci of a hyperbola are $(0, -c)$ and $(0, c)$ and the absolute value of the difference of the focal radii is $2a$, we obtain the equation

$$\frac{y^2}{a^2} - \frac{x^2}{b^2} = 1 \text{ with } b^2 = c^2 - a^2.$$

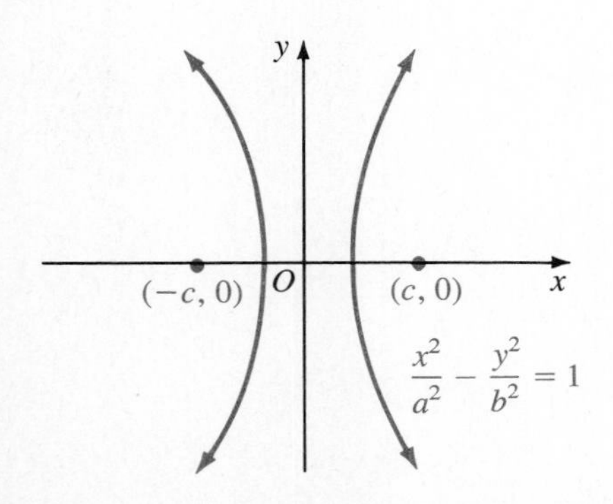

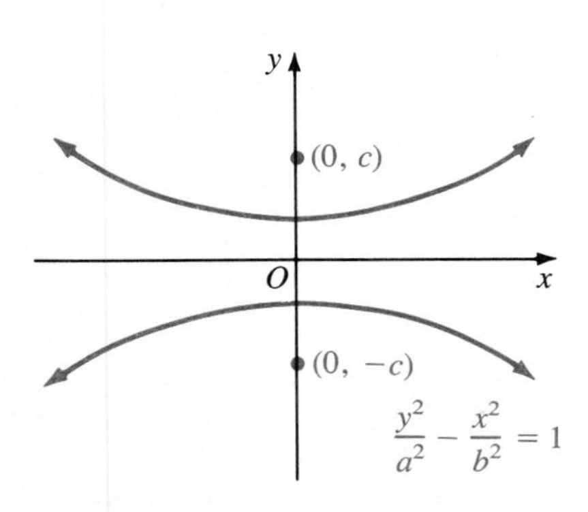

The following analysis provides information about the graph of $\frac{x^2}{a^2} - \frac{y^2}{b^2} = 1$.

1. If $\frac{x^2}{a^2} - \frac{y^2}{b^2} = 1$, then $y = \pm \frac{b}{a}\sqrt{x^2 - a^2}$. If y is to be a real number, then $x \leq -a$ or $x \geq a$. Thus no part of the graph lies inside the region bounded by the lines $x = -a$ and $x = a$. The points $(-a, 0)$ and $(a, 0)$ are on the graph.

2. The graph is symmetric about the x-axis, the y-axis, and the origin. If (r, s) satisfies the equation, then $(r, -s)$, $(-r, s)$, and $(-r, -s)$ also satisfy the equation.

3. By writing $y = \pm\frac{b}{a}\sqrt{x^2 - a^2} = \pm\frac{b}{a}\sqrt{x^2\left(1 - \frac{a^2}{x^2}\right)} = \pm\frac{b}{a}x\sqrt{1 - \frac{a^2}{x^2}}$, we can examine the behavior of y as $|x|$ becomes very large. When $|x|$ is large, $\frac{a^2}{x^2}$ is close to 0, and $1 - \frac{a^2}{x^2}$ is close to 1. Hence y is approximately equal to $\pm\frac{b}{a}x$. The lines $y = \pm\frac{b}{a}x$ are **asymptotes** of the hyperbola.

If the foci of a hyperbola are $(0, -c)$ and $(0, c)$, the foregoing analysis yields the equation $\frac{y^2}{a^2} - \frac{x^2}{b^2} = 1$. In this case the asymptotes are $y = \pm\frac{a}{b}x$.

The **vertices** of a hyperbola are the intersections of the hyperbola with the line containing the foci. The **transverse axis** is the line segment with the vertices as endpoints. The **center** of a hyperbola is the midpoint of the transverse axis.

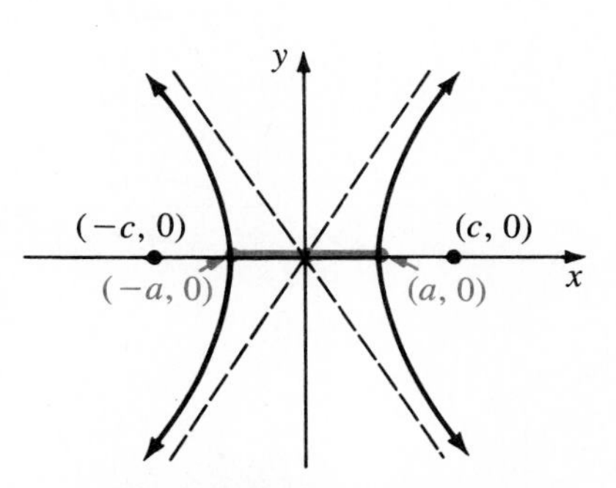

Horizontal transverse axis

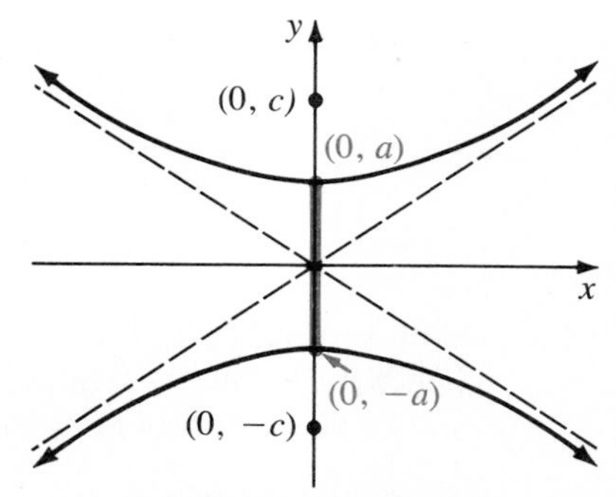

Vertical transverse axis

EXAMPLE 1 Sketch the graph of $\frac{x^2}{16} - \frac{y^2}{9} = 1$.

Show the asymptotes as dashed lines.

Solution on following page

Solution Since $a^2 = 16$ and $b^2 = 9$, $a = 4$ and $b = 3$. The asymptotes are $y = \pm\frac{3}{4}x$.

Plot the points $(-4, 0)$ and $(4, 0)$. Use symmetry to plot additional points.

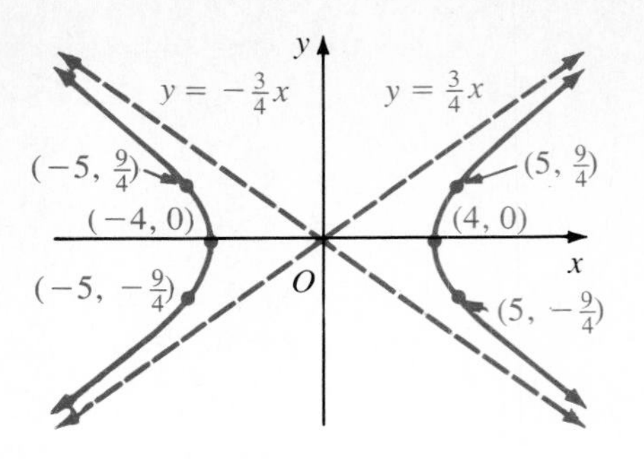

The following theorem describes equations of hyperbolas with center (h, k).

THEOREM 8 Let $2a$ be the absolute value of the difference of the focal radii of a hyperbola, c be the distance from the center to each focus, and $b^2 = c^2 - a^2$.

1. If the transverse axis is horizontal, then

$$\frac{(x-h)^2}{a^2} - \frac{(y-k)^2}{b^2} = 1$$

and the asymptotes are

$$y - k = \pm\frac{b}{a}(x - h).$$

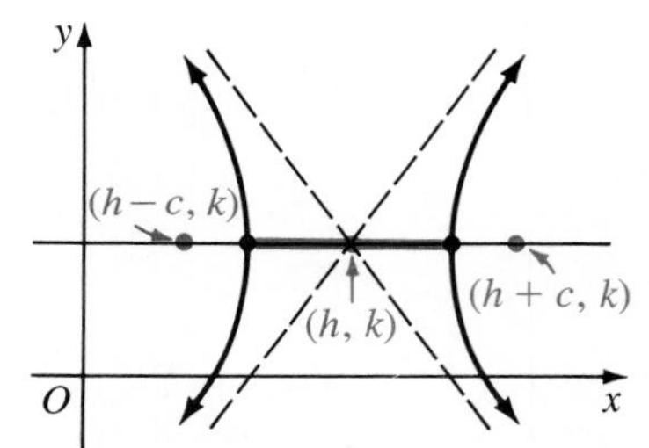

2. If the transverse axis is vertical, then

$$\frac{(y-k)^2}{a^2} - \frac{(x-h)^2}{b^2} = 1$$

and the asymptotes are

$$y - k = \pm\frac{a}{b}(x - h).$$

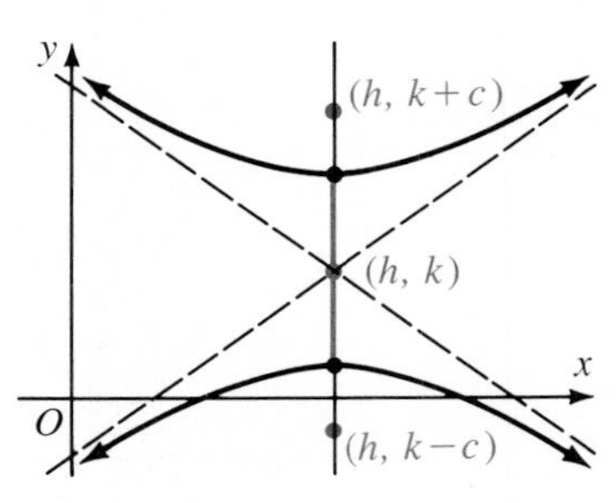

EXAMPLE 2 Sketch the graph of $x^2 - 4y^2 - 6x - 16y + 29 = 0$.
Label the center, vertices, and foci.
Show the asymptotes as dashed lines.

Solution Complete the squares in x and y.

$$x^2 - 4y^2 - 6x - 16y + 29 = 0$$
$$x^2 - 6x + 9 - 4(y^2 + 4y + 4) = -29 + 9 - 16$$
$$(x-3)^2 - 4(y+2)^2 = -36$$
$$\frac{(y+2)^2}{9} - \frac{(x-3)^2}{36} = 1$$

The transverse axis is vertical. The center is $(3, -2)$. Since $a = 3$ and $b = 6$, $c = \sqrt{9 + 36} = 3\sqrt{5}$. The vertices are $(3, -2 + 3) = (3, 1)$ and $(3, -2 - 3) = (3, -5)$. The foci are $(3, -2 + 3\sqrt{5})$ and $(3, -2 - 3\sqrt{5})$. The asymptotes are

$$y + 2 = \pm\frac{1}{2}(x - 3).$$

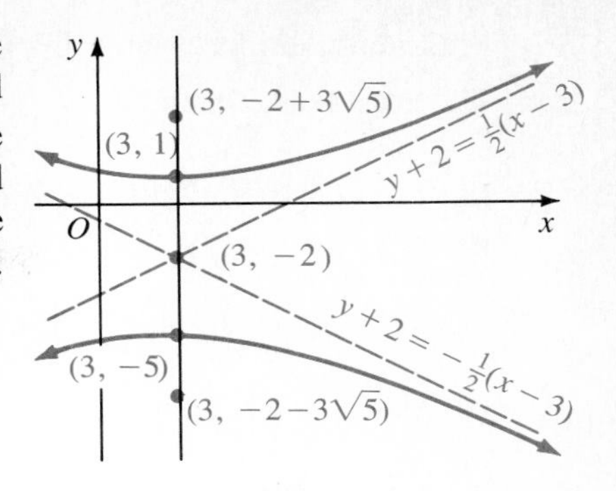

EXAMPLE 3 Find an equation for the hyperbola for which one vertex is $(3, 7)$, one focus is $(3, -1)$, and the center is $(3, 4)$.

Solution Since the center is $(3, 4)$, $h = 3$ and $k = 4$. The transverse axis is vertical, so $c = 4 - (-1) = 5$ and $a = 7 - 4 = 3$. Since $b^2 = c^2 - a^2$, $b^2 = 25 - 9 = 16$. Therefore an equation of the hyperbola is

$$\frac{(y - 4)^2}{9} - \frac{(x - 3)^2}{16} = 1. \quad \textbf{Answer}$$

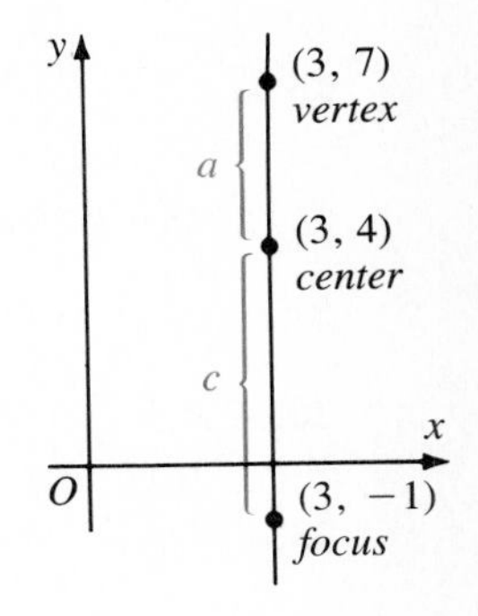

EXERCISES

Sketch the graph of each equation. Draw the asymptotes. Label the center and the endpoints of the transverse axis.

A

1. $\frac{x^2}{25} - \frac{y^2}{16} = 1$

2. $4y^2 - x^2 = 16$

3. $\frac{(x + 2)^2}{9} - \frac{(y - 2)^2}{25} = 1$

4. $\frac{(y + 3)^2}{25} - \frac{(x - 2)^2}{16} = 1$

5. $\frac{(x - 3)^2}{4} - \frac{(y - 4)^2}{16} = 1$

6. $\frac{(y - 1)^2}{4} - \frac{(x + 1)^2}{4} = 1$

7. $x^2 - 4y^2 + 2x + 8y = 7$

8. $4x^2 - 9y^2 + 8x + 36y - 68 = 0$

9. $9y^2 - 4x^2 + 24x - 72 = 0$

10. $3x^2 - 4y^2 - 8y - 16 = 0$

11. $3y^2 - x^2 + 12y + 2x = 1$

12. $9x^2 - y^2 + 18x - 2y = 1$

13. $16x^2 - 36y^2 + 16x + 108y = 221$

14. $16x^2 - 96x - 9y^2 + 36y - 36 = 0$

Find an equation of each hyperbola.

15. Foci $(3, 0)$ and $(-3, 0)$; absolute value of the difference of focal radii $= 2$

16. Foci $(0, 2)$ and $(0, -2)$; absolute value of the difference of focal radii $= 2$

Find an equation of each hyperbola.

17. Foci $(-9, 5)$ and $(-9, 14)$; absolute value of the difference of focal radii $= 3$

18. Foci $(2, 0)$ and $(12, 0)$; vertex at $(4, 0)$

19. Center $(-1, 5)$; vertex $(-1, 6)$; focus $(-1, 9)$

20. Vertices $(8, 0)$ and $(-8, 0)$; slopes of asymptotes $\frac{3}{4}$ and $-\frac{3}{4}$

21. Vertices $(4, -1)$ and $(-12, -1)$; slope of one asymptote is $-\frac{1}{4}$

B **22.** Foci $(4, 0)$ and $(-4, 0)$; slopes of asymptotes 3 and -3

23. Foci $(-3, 5)$ and $(-3, 9)$; slopes of asymptotes 1 and -1

24. Vertices $(3, -2)$ and $(7, -2)$; slopes of asymptotes 2 and -2

25. Focus $(4, 6)$; center $(6, 6)$; slopes of asymptotes 3 and -3

Sketch the graph of each equation.

26. $\frac{x^2}{25} - \frac{y^2}{9} = 0$

27. $y^2 - \frac{x^2}{9} = 0$

28. $\frac{(x - 3)^2}{16} - (y + 1)^2 = 0$

29. $\frac{(y - 1)^2}{16} - (x + 2)^2 = 0$

30. Suppose that a set of all points (x, y) in the plane is such that the absolute value of the difference of the distances from (x, y) to $(0, -c)$ and from (x, y) to $(0, c)$ is $2a$. Prove that the graph has an equation of the form $\frac{y^2}{a^2} - \frac{x^2}{b^2} = 1$. (*Hint*: Use $b^2 = c^2 - a^2$.)

31. Prove that if (x, y) satisfies the equation $\frac{x^2}{a^2} - \frac{y^2}{b^2} = 1$, then the absolute value of the difference of the distances from (x, y) to $(-c, 0)$ and from (x, y) to $(c, 0)$ where $c^2 = a^2 + b^2$ is $2a$ $(a > 0)$.

32. A chord of a hyperbola perpendicular to the transverse axis at a focus is called a *latus rectum* of the hyperbola. Show that the length of a latus rectum of the hyperbola with equation $\frac{x^2}{a^2} - \frac{y^2}{b^2} = 1$ is $\frac{2b^2}{a}$.

33. Prove that when a hyperbola with equation $\frac{x^2}{k^2} - \frac{y^2}{k^2} = 1$ is rotated 45° counterclockwise about the origin, an equation of the resulting figure has the form $xy = m$. Find m in terms of k. (*Hint*: Find the coordinates of the foci of the original hyperbola and the coordinates of the foci after rotation. Then use the definition of a hyperbola.)

34. Find a condition placed on the coefficients of $Ax^2 - By^2 + Cx + Dy + E = 0$, where A and B are positive, that will guarantee that the graph of the equation is a hyperbola. Find the coordinates of the center of the hyperbola.

2-6 PARABOLAS

Equations such as $x^2 - y = 1$ and $y^2 + 3y - x = 5$ have *parabolas* as their graphs. Notice that each of these equations is quadratic in one variable and linear in the other variable.

DEFINITION

A **parabola** is a set of all points in the plane that are equidistant from a fixed line (the **directrix**) and a fixed point (the **focus**) not on the line.

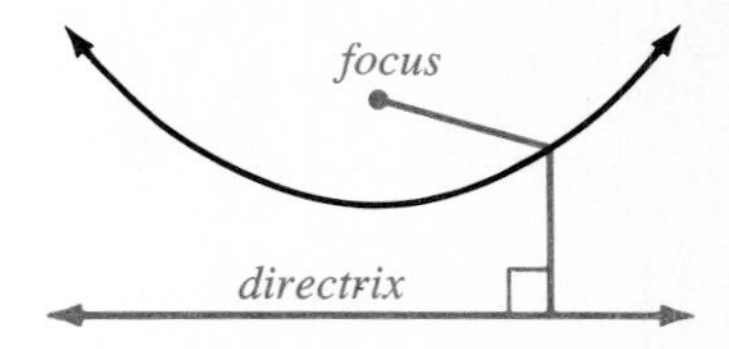

We can find an equation of a parabola with focus $(0, c)$ and directrix $y = -c$. Using the definition of a parabola, we have

$$\sqrt{(x - 0)^2 + (y - c)^2} = |y - (-c)|.$$

Squaring both sides of this equation and simplifying, we obtain

$$x^2 = 4cy, \quad \text{or} \quad y = \frac{1}{4c}x^2.$$

The **axis** (or **axis of symmetry**) of a parabola is the line that contains the focus and is perpendicular to the directrix. The vertex of a parabola is the point where the parabola intersects the axis of symmetry.

If a parabola has focus $(c, 0)$ and directrix $x = -c$, we can obtain the following equations of the parabola:

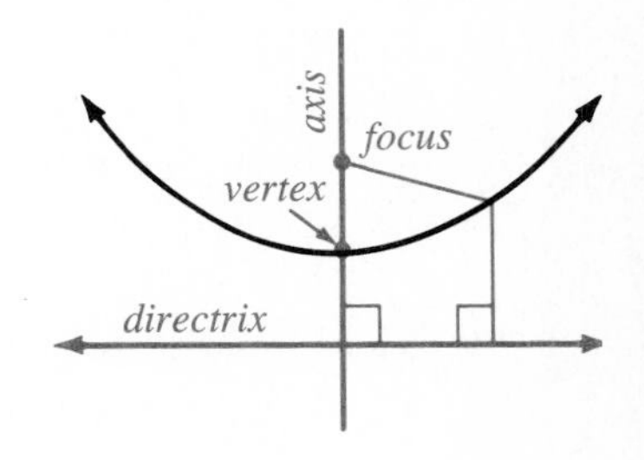

$$y^2 = 4cx, \quad \text{or} \quad x = \frac{1}{4c}y^2$$

The following theorem describes equations of parabolas with vertex (h, k) and horizontal or vertical axis of symmetry.

THEOREM 9 Let (h, k) be the vertex of a parabola.

1. If the directrix is horizontal, the parabola has an equation of the form

$$y - k = \frac{1}{4c}(x - h)^2$$

where $|c|$ is the distance from the focus to the vertex.

2. If the directrix is vertical, the parabola has an equation of the form

$$x - h = \frac{1}{4c}(y - k)^2$$

where $|c|$ is the distance from the focus to the vertex.

The graphs of various types of parabola are shown in the figures at the top of the next page.

$$y - k = \frac{1}{4c}(x - h)^2$$

$c > 0$

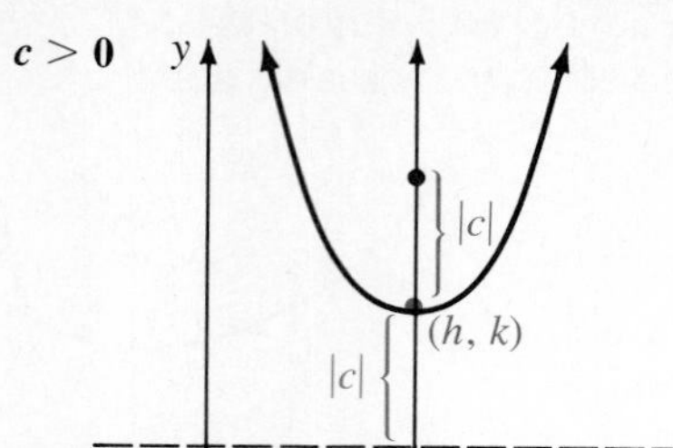

$c < 0$

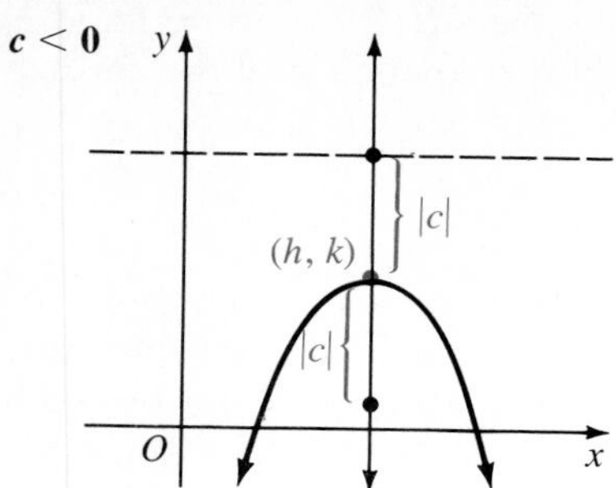

$$x - h = \frac{1}{4c}(y - k)^2$$

$c > 0$

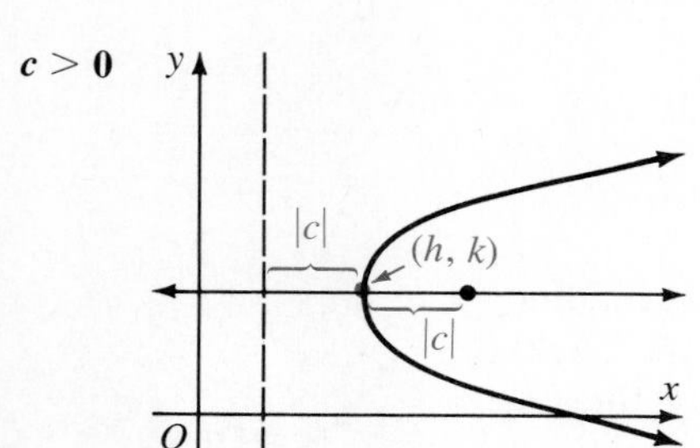

$c < 0$

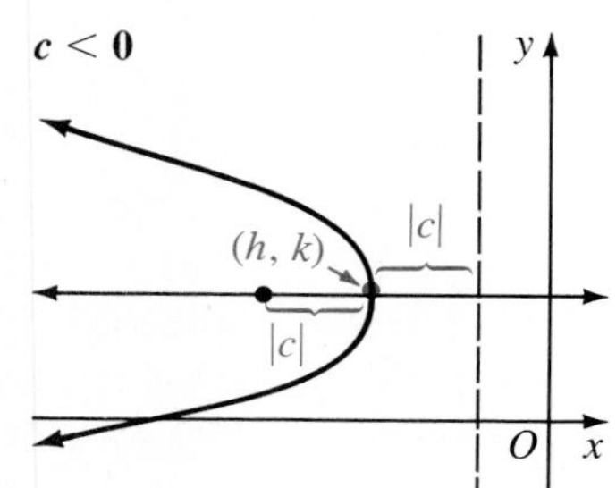

EXAMPLE 1 Graph each equation. Label the vertex, focus, directrix, and axis of symmetry.

a. $x = \frac{1}{4}y^2$ **b.** $x^2 + 2x - 8y = 31$

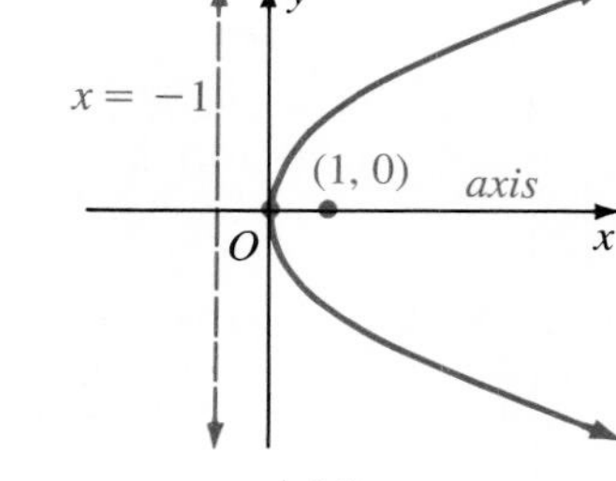

Solution **a.** By Theorem 9, the axis of symmetry is horizontal and the directrix is vertical. The vertex is (0, 0). Since $4c = 4$, $c = 1$. Since $c > 0$, the parabola opens to the right.

b. Complete the square in x.

$$x^2 + 2x - 8y = 31$$
$$x^2 + 2x + 1 = 8y + 32$$
$$(x + 1)^2 = 8(y + 4)$$
$$y + 4 = \frac{1}{8}(x + 1)^2$$

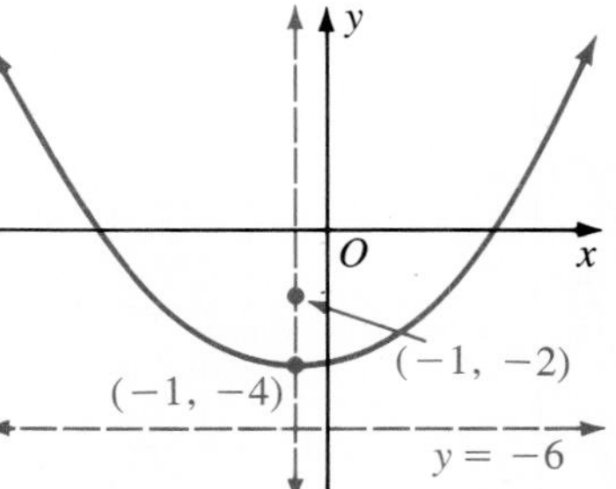

The vertex is $(-1, -4)$. Since $\frac{1}{8} = \frac{1}{4c}$, $c = 2$. Since $c > 0$, the parabola opens upward. The focus is $(-1, -4 + 2) = (-1, -2)$. The directrix is $y = -4 - 2 = -6$. The axis of symmetry is $x = -1$.

EXAMPLE 2 Graph $y = x^2 + x - 6$. Label the vertex, axis of symmetry, and any points where the graph crosses a coordinate axis.

Solution Complete the square in x.

$$y + 6 = x^2 + x$$

$$y + 6 + \frac{1}{4} = x^2 + x + \frac{1}{4}$$

$$y + \frac{25}{4} = \left(x + \frac{1}{2}\right)^2$$

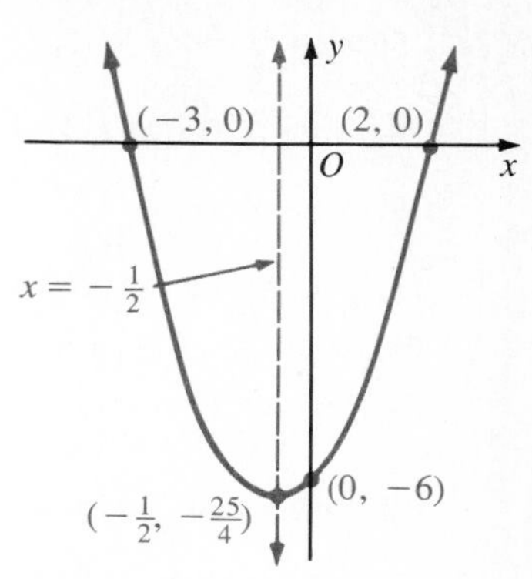

The vertex is $\left(-\frac{1}{2}, -\frac{25}{4}\right)$. Since the directrix is horizontal, the axis of symmetry is $x = -\frac{1}{2}$. To find any y-intercepts let $x = 0$. The y-intercept of the parabola is -6. To find any x-intercepts let $y = 0$. Then $x^2 + x - 6 = 0$. Factoring and solving $x^2 + x - 6 = 0$, $x = 2$ or -3. The x-intercepts are 2 and -3.

EXAMPLE 3 Find an equation of each parabola.

a. Vertex $(-3, 2)$; focus $(-3, 0)$ **b.** Focus $(3, -1)$; directrix $x = -1$

Solution **a.** The distance from the vertex to the focus is $|c| = 2$. Since the vertex is above the focus, $c = -2$. Thus $y - 2 = -\frac{1}{8}(x + 3)^2$. **Answer**

b. Since the directrix is vertical and the focus is to the right of the directrix, the parabola opens to the right. Since the vertex is equidistant from the focus and the directrix, the vertex is $\left(\frac{3 + (-1)}{2}, -1\right) = (1, -1)$. Thus $h = 1$, $k = -1$, and $|c| = |3 - 1| = 2$. Since the parabola opens to the right, $c > 0$. Therefore $x - 1 = \frac{1}{8}(y + 1)^2$. **Answer**

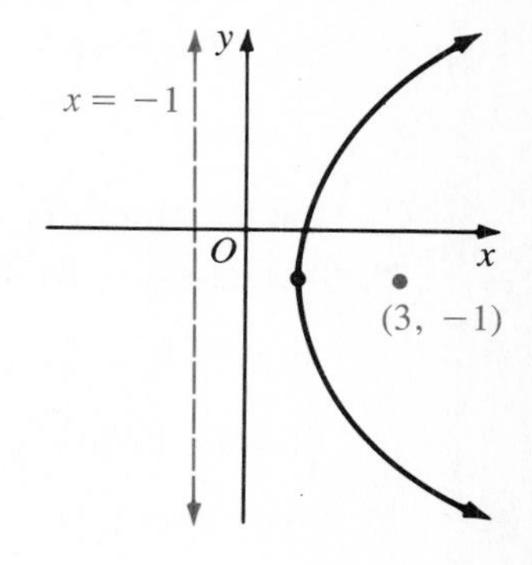

EXERCISES

Graph each equation. Label the vertex and the axis of symmetry. Find the focus, directrix, and any intercepts.

A

1. $y = 4x^2$

2. $y^2 = -12x$

3. $y^2 + x = 0$

4. $y - 2x^2 = 0$

Graph each equation. Label the vertex and the axis of symmetry. Find the focus, directrix, and any intercepts.

5. $y - 3 = (x - 2)^2$

6. $x + 1 = \frac{1}{8}(y - 2)^2$

7. $x - 3 = -(y + 1)^2$

8. $y = (x + 4)^2$

9. $y = -x^2 + 4x - 2$

10. $x = \frac{1}{2}y^2 - 2y - 1$

11. $x = 4y^2 - 2y - 3$

12. $y = 2x^2 - 6x + 1$

13. $y = 3x^2 - 6x + 4$

14. $x = -3y^2 + 12y - 7$

15. $16x - 12y = 4y^2 + 1$

16. $6y + 10x = -x^2 - 13$

Find an equation of each parabola.

17. Vertex (0, 0); focus (0, 12)

18. Focus (7, −3); directrix $y = 1$

19. Vertex (0, 0); directrix $x = -2$

20. Vertex (5, 2); focus (5, −6)

B

21. x-intercepts −2 and 2; containing (0, −5)

22. y-intercepts 4 and 12; containing (6, 8)

23. Containing (2, 3), (2, 9), and (6, 6)

24. Containing (−1, −2), (−1, 8), and (−4, 3)

25. Prove Theorem 9, part (1).

26. Prove Theorem 9, part (2).

27. The chord through the focus of a parabola, perpendicular to the axis of the parabola, and with endpoints on the parabola is called the *latus rectum* of the parabola. If an equation of the parabola is $y = \frac{1}{4c}x^2$, show that the length of the latus rectum is $|4c|$.

28. The distance from a point (x_1, y_1) on a parabola to the focus is called the *focal radius* of the point. If $y = \frac{1}{4c}x^2$ is an equation of the parabola, show that the focal radius of (x_1, y_1) is $|y_1 + c|$.

C

29. Prove the following statement by proving parts (a), (b), and (c).

If $a > 0$, then $y = ax^2 + bx + c$ has its smallest value when $x = -\frac{b}{2a}$.

a. Find y when $x = -\frac{b}{2a}$.

b. Find y when $x = -\frac{b}{2a} + v$ where $v \neq 0$.

c. Show that the result of part (a) is less than the result of part (b).

30. Modify the proof in Exercise 29 to prove that if $a < 0$, then $y = ax^2 + bx + c$ has its maximum value when $x = -\frac{b}{2a}$.

Use the results of Exercises 29 and 30 in Exercises 31–35.

31. Suppose that the gas mileage m (in miles per gallon) of a car depends on speed s (in miles per hour) and is given by $m = -\frac{s^2}{80} + s$. Find the speed at which mileage is a maximum. Find the maximum mileage.

32. Find the dimensions of the rectangle that is inscribed in a right triangle with legs of length 3 and 5 and that has the greatest area. Also find that area.

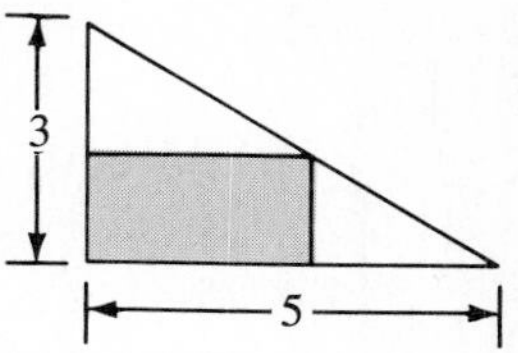

33. Find two real numbers whose difference is 36 and whose product is minimum.

34. A rancher wants to build a rectangular corral with a stream along one side. If he has 500 m of fencing, what are the dimensions that maximize the area of the corral?

35. Find the shortest distance from $(4, -1)$ to the line $y = 2x - 3$. (*Hint*: If d is the distance from $(4, -1)$ to $(x, 2x - 3)$, minimize d^2.)

2–7 CONIC SECTIONS

Conic sections (circles, ellipses, hyperbolas, and parabolas) were known and studied by Greek geometers as early as 350 B.C. The Greeks defined conic sections in terms of an infinite double cone and a cutting plane. As the cutting angle changes, different conics are generated.

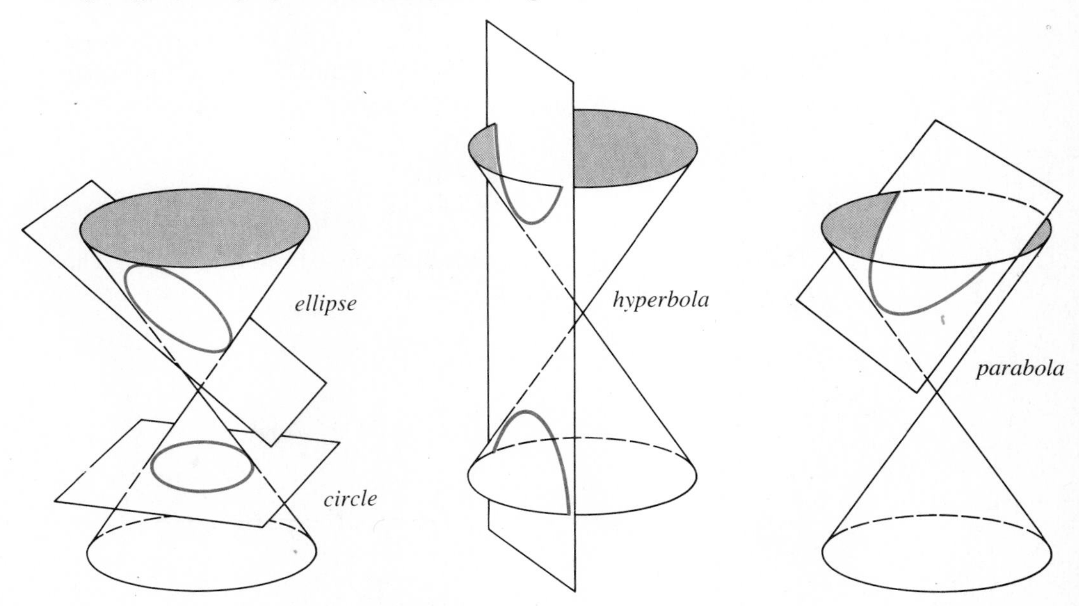

Notice that it is also possible to obtain as the intersection a point, a line, or a pair of intersecting lines. Such intersections are called *degenerate* conics.

Each of the conic sections has an equation of the form

$$Ax^2 + Cy^2 + Dx + Ey + F = 0$$

where the coefficients are real numbers and not all zero. The coefficients determine the conic.

1. If $A = C = 0$, the graph is a line or the empty set.
2. If exactly one of A or C is nonzero, the graph is a parabola, a pair of parallel lines, a line, a point, or the empty set.
3. If A and C have the same sign, the graph is an ellipse, a circle, a point, or the empty set.
4. If A and C are opposite in sign, the graph is a hyperbola or a pair of intersecting lines.

EXAMPLE Identify and draw the graph of each equation.

a. $4y^2 + 8y - x^2 + 2x + 4 = 0$

b. $9x^2 - y^2 + 4y = 4$

c. $9x^2 + 4y^2 + 54x - 16y = -97$

d. $4x^2 - 24x + 3y^2 + 30y = -99$

e. $x^2 + (y - 1)^2 = -3$

Solution **a.** $A = -1$ and $C = 4$. The graph expected is a hyperbola or a pair of intersecting lines. Rearrange the terms and complete the squares.

$$4(y^2 + 2y) - (x^2 - 2x) = -4$$
$$4(y^2 + 2y + 1) - (x^2 - 2x + 1) = -4 + 4 - 1$$
$$\frac{(x - 1)^2}{1} - \frac{(y + 1)^2}{\frac{1}{4}} = 1$$

The graph is a hyperbola with horizontal transverse axis. The center is $(1, -1)$; the vertices are $(0, -1)$ and $(2, -1)$; the asymptotes are

$$y + 1 = \pm\frac{1}{2}(x - 1).$$

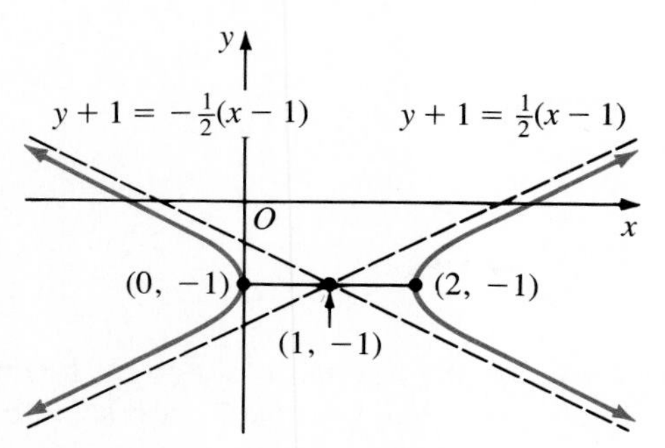

b. The graph is a hyperbola or a pair of intersecting lines. Complete the square in y.

$$9x^2 - (y^2 - 4y + 4) = 4 - 4$$
$$9x^2 - (y - 2)^2 = 0$$

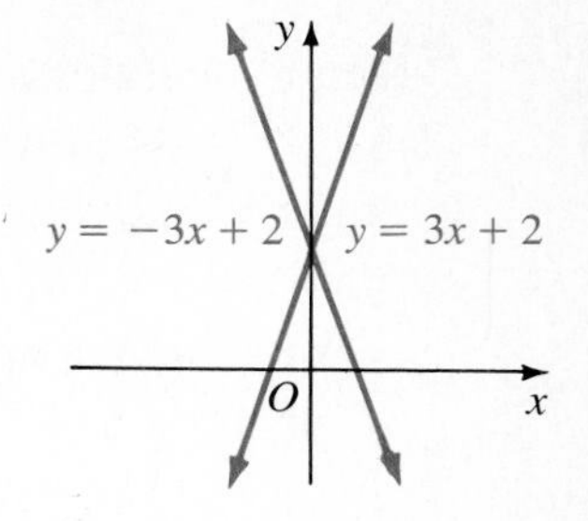

Factor the difference of squares.

$$(3x - (y - 2))(3x + (y - 2)) = 0$$

Thus $3x - (y - 2) = 0$ or $3x + (y - 2) = 0$. So $y = 3x + 2$ or $y = -3x + 2$. The graph is a pair of intersecting lines.

c. A and C have the same sign. Complete the squares.

$$9(x^2 + 6x) + 4(y^2 - 4y) = -97$$
$$9(x^2 + 6x + 9) + 4(y^2 - 4y + 4) = -97 + 81 + 16$$
$$9(x + 3)^2 + 4(y - 2)^2 = 0$$

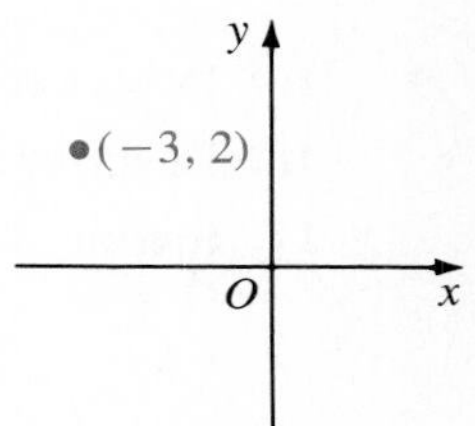

Since the sum of two squares is 0 only when each square is 0, the graph is the point $(-3, 2)$.

d.

$$4x^2 - 24x + 3y^2 + 30y = -99$$
$$4(x^2 - 6x + 9) + 3(y^2 + 10y + 25) = -99 + 36 + 75$$
$$4(x - 3)^2 + 3(y + 5)^2 = 12$$
$$\frac{(x - 3)^2}{3} + \frac{(y + 5)^2}{4} = 1$$

The graph is an ellipse with center $(3, -5)$ and vertical major axis.

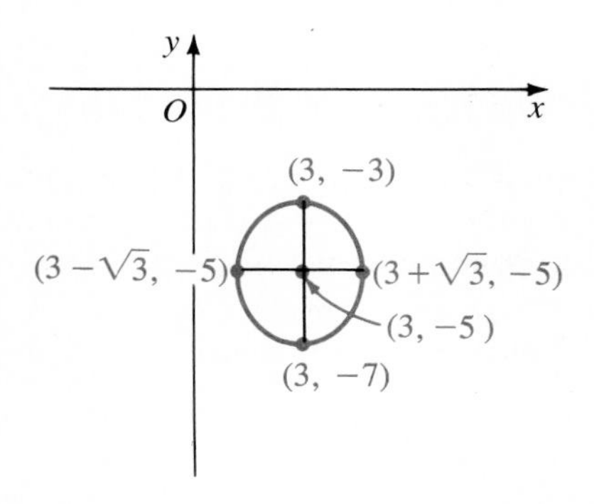

e. Since a sum of squares cannot be negative, the equation is not satisfied by any pair of real numbers. The graph is empty.

A **second-degree equation** in two variables is any equation of the form

$$Ax^2 + Bxy + Cy^2 + Dx + Ey + F = 0$$

where not all coefficients are zero. We have considered only equations for which $B = 0$. In Chapter 12 we shall consider cases where $B \neq 0$.

EXERCISES

Identify and draw the graph of each equation.

A

1. $x^2 + 4y^2 - 6x - 8y + 9 = 0$
2. $x^2 - 4y^2 - 6x + 8y = -1$
3. $x^2 - 16y^2 + 2x = 15$
4. $y^2 + 16x^2 + 2y = 15$
5. $x^2 - y^2 - 4x = -4$
6. $x^2 - y^2 - 2x - 6y = 8$
7. $x^2 + 4y^2 + 8y + 4 = 0$
8. $x^2 + 5y^2 - 6x - 10y + 14 = 0$
9. $x^2 + 3x + 2 = 0$
10. $y^2 - 4y + 4 = 0$
11. $x^2 + 4x = -4$
12. $y^2 + 6y + 12 = 0$
13. $y^2 - x + 8y = 2$
14. $x^2 + 8x + y^2 - 6y = -21$

B

15. Describe the graph of $Ax^2 + Cy^2 = 0$ **(a)** if $AC > 0$; **(b)** if $AC < 0$.
16. Describe the graph of $Dx + Ey + F = 0$ if $DEF = 0$.
17. Identify the graph of $Ax^2 + Cy^2 + Dx + Ey + F = 0$ if $A = 1, C = 0, D = 2, E = 0$, and $F = 1$.
18. Under what conditions on A, D, and F will the graph of $Ax^2 + Dx + F = 0$ be nonempty?

Let $Ax^2 + Cy^2 + Dx + Ey + F = 0$.

C

19. Show that if $AC = 0$, the graph is a parabola or degenerate parabola.
20. Show that if $AC > 0$, the graph is an ellipse, circle, or degenerate ellipse.
21. Show that if $AC < 0$, the graph is a hyperbola or degenerate hyperbola.

2-8 INTERSECTIONS OF CONIC SECTIONS

It is sometimes useful to find points where conic sections intersect. To do so, we can use substitution and linear combination methods to solve the system of corresponding equations.

EXAMPLE 1 Find any points where the graphs of the given equations intersect.

a. $y = 3x + 5$
$2x + 5y = 2$

b. $3x - 4y = 11$
$5x + 3y = 9$

Solution **a.** Substitute $3x + 5$ for y in the second equation to find x:

$$2x + 5(3x + 5) = 2. \text{ So } x = -\frac{23}{17}.$$

Substitute $-\frac{23}{17}$ for x in the first equation to get $y = 3\left(-\frac{23}{17}\right) + 5 = \frac{16}{17}$.

The pair $\left(-\frac{23}{17}, \frac{16}{17}\right)$ satisfies both of the original equations. The point of intersection is $\left(-\frac{23}{17}, \frac{16}{17}\right)$. **Answer**

b. Multiply the first equation by 3 and the second equation by 4.

$$\left.\begin{aligned} 3x - 4y &= 11 \\ 5x + 3y &= 9 \end{aligned}\right\} \longrightarrow \left\{\begin{aligned} 9x - 12y &= 33 \\ 20x + 12y &= 36 \end{aligned}\right.$$

Add the resulting equations to obtain

$$(9x - 12y) + (20x + 12y) = 33 + 36.$$

Therefore $x = \frac{69}{29}$. Substitute $\frac{69}{29}$ for x in either original equation to find y.

$$5\left(\frac{69}{29}\right) + 3y = 9$$

Thus $y = -\frac{28}{29}$. The point of intersection is $\left(\frac{69}{29}, -\frac{28}{29}\right)$. **Answer**

EXAMPLE 2 Find the intersections of the graphs of $x^2 + y^2 = 25$ and $y - 2x = -5$.

Solution Since $y - 2x = -5$, $y = 2x - 5$. By substitution,

$$x^2 + (2x - 5)^2 = 25.$$

Thus

$$\begin{aligned} x^2 + 4x^2 - 20x + 25 &= 25 \\ 5x(x - 4) &= 0 \\ x = 0 \text{ or } x &= 4. \end{aligned}$$

Substitute these values of x in $y - 2x = -5$ to find the corresponding values of y.

$x = 0$: $y - 2(0) = -5$. So $y = -5$.
$x = 4$: $y - 2(4) = -5$. So $y = 3$.
The points of intersection are $(0, -5)$ and $(4, 3)$. **Answer**

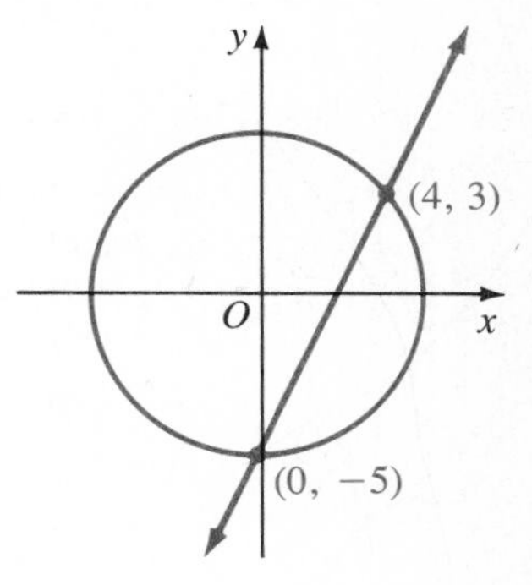

By drawing a sketch of the graphs involved we can usually tell how many solutions to expect and verify the reasonableness of solutions. Note that in the solution of Example 2, $x = 0$ and $x = 4$ might have been substituted into $x^2 + y^2 = 25$.

$$x = 0: (0)^2 + y^2 = 25. \text{ So } y = \pm 5.$$
$$x = 4: (4)^2 + y^2 = 25. \text{ So } y = \pm 3.$$

However this would have given four ordered pairs, two of which are not solutions. The figure shows that $(0, 5)$ and $(4, -3)$ are not solutions.

EXAMPLE 3 Find the points of intersection of the graphs of $x^2 + y^2 = 4$ and $x^2 + y^2 - 4x - 4y = -4$.

Solution Subtract the first equation from the second equation.

$$-4x - 4y = -8.$$

Therefore $y = -x + 2$.

Solution continued on following page

Substitute $-x + 2$ for y in the first equation.

$$\begin{aligned} x^2 + (-x + 2)^2 &= 4 \\ x^2 + x^2 - 4x + 4 &= 4 \\ 2x^2 - 4x &= 0 \\ 2x(x - 2) &= 0 \\ x = 0 \text{ or } x &= 2 \end{aligned}$$

To find y and avoid extra values that are not solutions, substitute $x = 0$ and $x = 2$ in $y = -x + 2$.

$$x = 0\text{: } y = 0 + 2 = 2 \qquad x = 2\text{: } y = -2 + 2 = 0$$

The points of intersection are (0, 2) and (2, 0). **Answer**

Substitution in the original equations serves as a check. A sketch also verifies the solution. Notice that $y = -x + 2$ is an equation of the common chord of the circles.

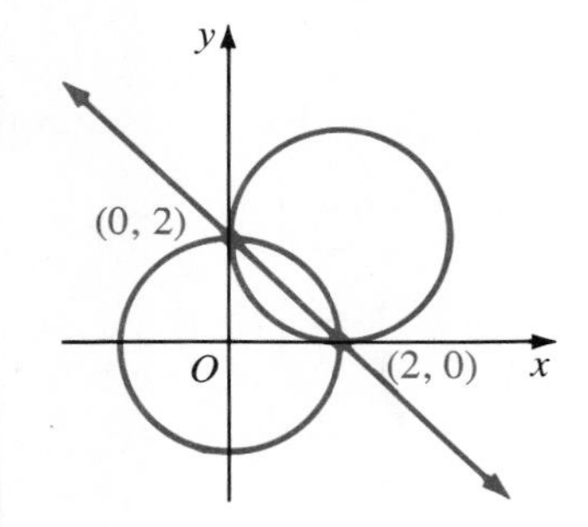

EXERCISES

Find any points of intersection of the graphs of each pair of equations.

A

1. $2x - 5y = 8$
$4x + 6y = -2$

2. $3x + 2y = 4$
$2x - 5y = 2$

3. $y = 2x - 4$
$y = x^2 - 3$

4. $y = 4x - 5$
$y = x^2 - 3x + 1$

5. $y = x + 1$
$x = y^2 - 3y$

6. $x - 2y = 3$
$x^2 - 2y + 3x = 2$

7. $y = 2x - 1$
$x^2 - 3y^2 = 4$

8. $4x^2 + y^2 = 13$
$2x - y = 1$

9. $x^2 + y^2 = 20$
$x^2 - 4x + y^2 = 16$

10. $x^2 + y^2 = 16$
$x^2 + 4x + y^2 = 16$

11. $x^2 + y^2 = 13$
$xy = 6$

12. $x^2 - y^2 = 15$
$xy = 4$

13. $5x^2 - y^2 = 3$
$x^2 + 2y^2 = 5$

14. $x^2 + y^2 = 4$
$x^2 + 4y^2 = 16$

15. $x^2 + y^2 = 25$
$\frac{x^2}{4} + \frac{y^2}{100} = 1$

16. $\frac{x^2}{36} + \frac{y^2}{9} = 1$
$\frac{y^2}{9} - x^2 = 1$

17. $y = (x + 1)^2$

$(x + 1)^2 - \frac{y^2}{4} = 1$

18. $x = (y - 1)^2$

$\frac{x^2}{8} + (y - 1)^2 = -2$

19. $\frac{x^2}{12} + \frac{y^2}{5} = 1$

$\frac{x^2}{12} - y^2 = 1$

20. $y^2 - \frac{x^2}{4} = 1$

$y = x^2 + 1$

21. $y^2 - x^2 = 1$

$x^2 - \frac{y^2}{9} = 1$

22. $(x - 3)^2 + (y - 1)^2 = 1$
$(x - 4)^2 - (y - 1)^2 = 1$

B **23.** Find an equation of the line containing the common chord of the circles whose equations are $x^2 + y^2 = 9$ and $x^2 + y^2 - 6x = 0$.

24. Find an equation of the line containing the common chord of the circles whose equations are $x^2 + y^2 = 16$ and $x^2 + y^2 + 4x + 2y = 4$.

25. Find an equation of the line containing the common chord of the circles whose equations are $(x + 2)^2 + (y - 1)^2 = 9$ and $(x - 1)^2 + (y + 1)^2 = 1$.

26. Find an equation of the perpendicular bisector of the line segment whose endpoints are the common solutions of $x^2 + y^2 = 36$ and $x + y = 6$.

27. Find an equation of the perpendicular bisector of the line segment whose endpoints are the common solutions of $x^2 + y^2 = 6$ and $y = x^2$.

28. Find the distance from $(2, -3)$ to the line $y = x + 1$. (*Hint*: Find an equation for the line containing $(2, -3)$ perpendicular to the line $y = x + 1$. Then find the coordinates of the point where these lines intersect.)

C **29.** Use the method of Exercise 28 to prove that the distance from (u, v) to the line $Ax + By = C$ is given by $\frac{|Au + Bv - C|}{\sqrt{A^2 + B^2}}$.

30. Prove that two nonvertical lines in the plane are parallel if and only if their slopes are equal.

COMPUTER EXERCISES

1. Write a program to find any common solutions of two linear equations in two variables when the user supplies the coefficients of the equations.

2. Run the program in Exercise 1 to solve each system of equations.

a. $2x + 3y = 9$
$5x - 4y = 11$

b. $-3x + y = -2$
$-3x + y = 5$

c. $x = 3$
$y = 4$

d. $12x + 15y = 28$
$24x - 9y = 30$

Chapter Summary

1. In the *Cartesian coordinate system*, each point in the plane is assigned a pair of *coordinates*. The *distance* between two points (x_1, y_1) and (x_2, y_2) is given by $\sqrt{(x_2 - x_1)^2 + (y_2 - y_1)^2}$. The coordinates of the *midpoint* of the line segment with endpoints (x_1, y_1) and (x_2, y_2) are $\left(\frac{x_1 + x_2}{2}, \frac{y_1 + y_2}{2}\right)$.

2. The *graph* of an open sentence in two variables is the set of all points in the plane whose coordinates satisfy the open sentence. Every line in the plane has an equation of the form $Ax + By = C$, called its *standard form*. A nonvertical line containing (x_1, y_1) and (x_2, y_2) has *slope*

$$\frac{y_2 - y_1}{x_2 - x_1}.$$

Every nonvertical line in the plane has an equation of the form:

$y - y_1 = m(x - x_1)$	$y = mx + b$	$\frac{x}{a} + \frac{y}{b} = 1$
point-slope	*slope-intercept*	*intercept* $a, b \neq 0$

Two nonvertical lines are parallel if and only if their slopes are equal. They are perpendicular if and only if the product of their slopes is -1.

3. Below is a summary of facts about circles, ellipses, hyperbolas, and parabolas.

Circle	$(x - h)^2 + (y - k)^2 = r^2$		
center	(h, k)		
radius	r		
Ellipse $(b^2 = a^2 - c^2)$	$\frac{(x - h)^2}{a^2} + \frac{(y - k)^2}{b^2} = 1$	or	$\frac{(y - k)^2}{a^2} + \frac{(x - h)^2}{b^2} = 1$
center	(h, k)		(h, k)
foci	$(h - c, k); (h + c, k)$		$(h, k - c); (h, k + c)$
vertices	$(h - a, k); (h + a, k)$		$(h, k - a); (h, k + a)$
Hyperbola $(b^2 = c^2 - a^2)$	$\frac{(x - h)^2}{a^2} - \frac{(y - k)^2}{b^2} = 1$	or	$\frac{(y - k)^2}{a^2} - \frac{(x - h)^2}{b^2} = 1$
center	(h, k)		(h, k)
vertices	$(h - a, k), (h + a, k)$		$(h, k - a), (h, k + a)$
asymptotes	$y - k = \pm\frac{b}{a}(x - h)$		$y - k = \pm\frac{a}{b}(x - h)$
Parabola	$y - k = \frac{1}{4c}(x - h)^2$	or	$x - h = \frac{1}{4c}(y - k)^2$
vertex	(h, k)		(h, k)
axis of symmetry	$x = h$		$y = k$
focus	$(h, k + c)$		$(h + c, k)$

4. The graph of an equation of the form $Ax^2 + Cy^2 + Dx + Ey + F = 0$, where the coefficients are real numbers and not all zero, is a *conic section* (circle, ellipse, hyperbola, or parabola) or a degenerate conic (point, line, or a pair of lines).
5. Any intersections of a pair of conic sections can be found algebraically from their equations by substitution and linear combination methods.

Chapter Test

2-1

1. Find **(a)** the length and **(b)** the coordinates of the midpoint of the segment whose endpoints are $A(-4, 7)$ and $B(6, 1)$.
2. Use coordinates to prove analytically that the segment joining the midpoints of the nonparallel sides of a trapezoid is parallel to each base and has as its length the average of the lengths of the two bases.

Find an equation in standard form of each line.

2-2

3. Parallel to the line $y = -2x - 6$, through the point $(5, 1)$
4. Containing the points $(-2, -2)$ and $(4, 6)$
5. The perpendicular bisector of the segment from $(-2, 8)$ to $(4, 14)$

2-3

6. Graph $(x + 2)^2 + y^2 = 4$. Label the center and radius.
7. Find an equation of the circle with center $(-2, 3)$ and radius 5.

2-4

8. Graph $x^2 + 4(y - 3)^2 = 4$. Label the center, vertices, and foci.
9. Find an equation of the ellipse with center $(-1, 2)$, one vertex $(-4, 2)$, and foci $(-1 - \sqrt{5}, 2)$ and $(-1 + \sqrt{5}, 2)$.

2-5

10. Graph $\frac{(y + 1)^2}{36} - \frac{(x - 2)^2}{9} = 1$. Draw the asymptotes and label the center, vertices, and foci.
11. Find an equation of the hyperbola with foci $(5, 0)$ and $(-5, 0)$ and absolute value of the difference of the focal radii 8.

2-6

12. Graph $y = -\frac{1}{4}x^2 + 2x - 5$. Label the axis of symmetry and the vertex. Find the focus and the directrix.
13. Find an equation of the parabola with vertex $(-2, -1)$ and focus $(-2, 2)$.

Identify and draw the graph of each equation.

2-7

14. $x^2 - 2y^2 - 2x + 1 = 16$
15. $x^2 + 4y^2 + 6x - 8y = 3$

For each pair of equations find the points of intersection of their graphs.

2-8

16. $2x - 3y = 5, \; -x + 2y = \frac{4}{3}$
17. $x^2 + y^2 - 4x - 6y = 3, \; -2x + 2y = -6$

Extra

Symmetry and Graphs

Recall that the graph of an equation in two variables is the set of all points in the plane whose coordinates satisfy the given equation. We can graph an equation of the form

$$Ax^2 + Cy^2 + Dx + Ey + F = 0$$

by using what we know about conic sections. To graph $|x| + |y| = 1$ we must use other means.

EXAMPLE 1 Graph $|x| + |y| = 1$.

Solution The points $(1, 0)$, $(0, 1)$, $(-1, 0)$, and $(0, -1)$ are on the graph. In order for x and y to satisfy $|x| + |y| = 1$ it must be true that

$$|x| \leq 1 \quad \text{and} \quad |y| \leq 1.$$

That is, $-1 \leq x \leq 1$ and $-1 \leq y \leq 1$.
Thus the graph is completely contained in the square region bounded by the lines $x = 1$, $x = -1$, $y = 1$, and $y = -1$. Obtain the graph of $|x| + |y| = 1$ by considering each quadrant and the signs of x and y in each quadrant.

Quadrant	x	y	Equation
I	+	+	$x + y = 1$
II	−	+	$-x + y = 1$
III	−	−	$-x - y = 1$
IV	+	−	$x - y = 1$

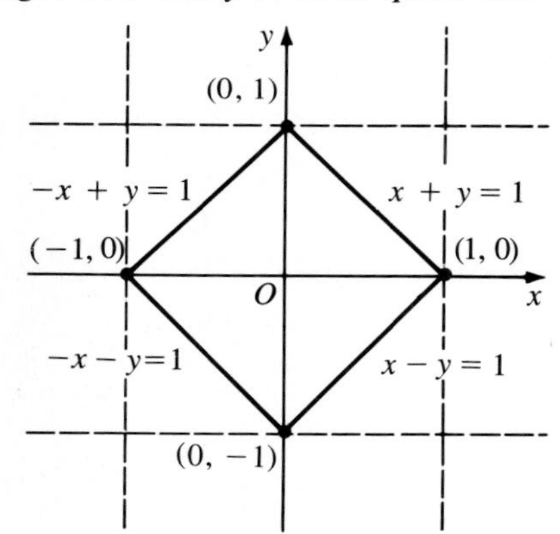

The completed graph is at the right.

EXAMPLE 2 Graph $x^4 + y^4 = 1$.

Solution In order for x and y to satisfy $x^4 + y^4 = 1$, it must be true that

$$x^4 \leq 1 \quad \text{and} \quad y^4 \leq 1.$$

Solve $x^4 - 1 \leq 0$ and $y^4 - 1 \leq 0$:

$$\begin{aligned} (x^2 + 1)(x^2 - 1) &\leq 0 \quad \text{and} \quad & (y^2 + 1)(y^2 - 1) &\leq 0 \\ x^2 - 1 &\leq 0 & y^2 - 1 &\leq 0 \\ -1 \leq x &\leq 1 \quad \text{and} & -1 \leq y &\leq 1 \end{aligned}$$

The graph of $x^4 + y^4 = 1$ is symmetric about the x-axis, the y-axis, and the origin.

x-axis	$(x)^4 + (-y)^4 = x^4 + y^4 = 1$
y-axis	$(-x)^4 + (y)^4 = x^4 + y^4 = 1$
origin	$(-x)^4 + (-y)^4 = x^4 + y^4 = 1$

Obtain the graph of $x^4 + y^4 = 1$ by plotting solutions in the first quadrant and using symmetry.

EXAMPLE 3 Graph $|x| - |y| = 1$.

Solution Write $|x| - |y| = 1$ as $|y| = |x| - 1$, to see that $|x| \geq 1$ and that y can be any real number. Tests for symmetry reveal that the graph is symmetric about the x-axis and the y-axis. The complete graph is shown at the right.

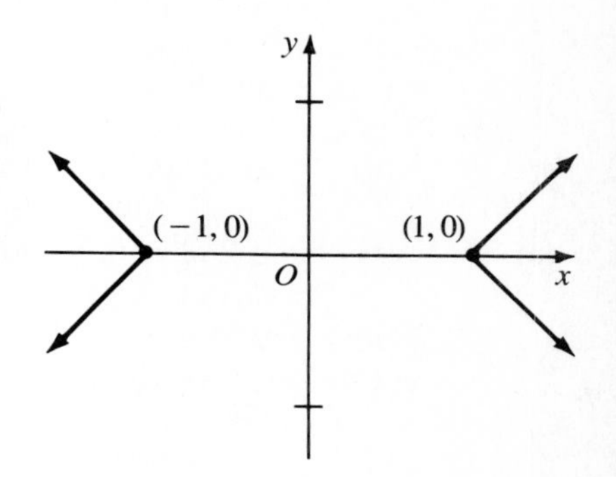

EXERCISES

Graph each equation.

1. $|x| + 2|y| = 1$
2. $3|x| + 2|y| = 1$
3. $|x| - 2|y| = 1$
4. $2|x| - 3|y| = 1$
5. Describe the graph of $x^3 + y^3 = 0$.
6. Describe the graph of $a|x| + b|y| = 1$ if a and b are positive real numbers.
7. Graph $4x^4 + 9y^4 = 36$.
8. Graph $x^3 - y^3 = 1$. (*Hint*: Use a calculator or computer to obtain ordered pairs that satisfy the equation.)

COMPUTER APPLICATION

APPROXIMATING AREAS

It is frequently necessary or useful to use computers and approximation methods to solve problems. Although the formula $A = \pi r^2$ can be used to find the area of a circle, the following method illustrates a strategy that applies widely.

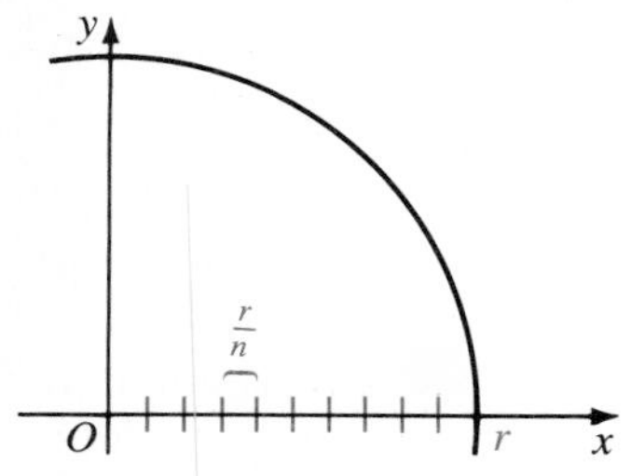

Let C be a circle whose area we wish to find. Let its radius be r. Place a coordinate system on the plane containing the circle in such a way that the origin is at the center of the circle. Then the circle has an equation: $x^2 + y^2 = r^2$. Subdivide the portion of the x-axis from 0 to r into n equal subdivisions.

Construct rectangles as shown in either of the following two figures. Note that in the first figure the rectangles extend beyond the circle and that in the second figure the rectangles lie entirely inside the circular region.

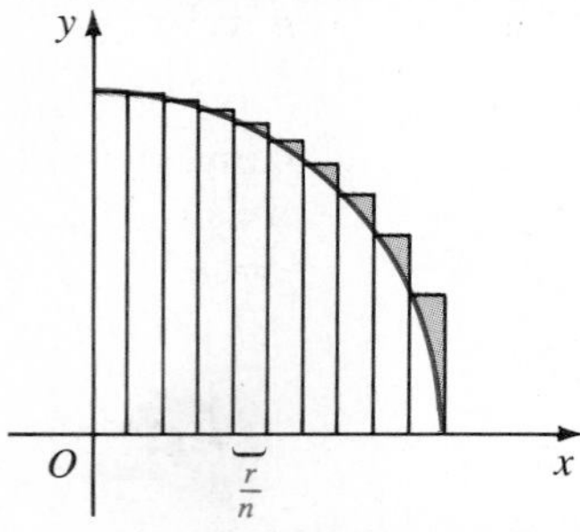

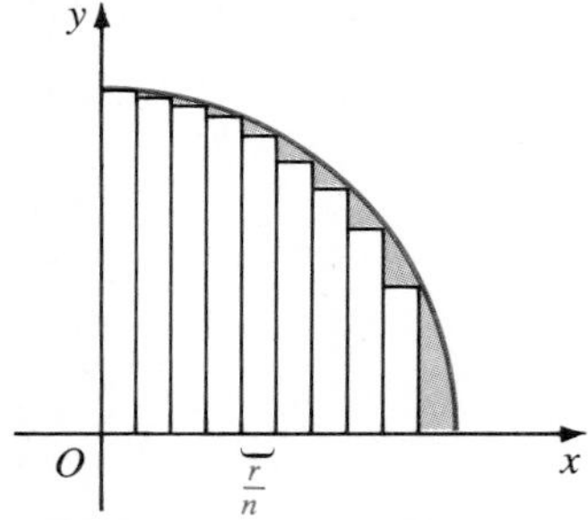

To approximate the area of the circular region in the first quadrant, find the areas of the rectangles and then add the areas. The following solution uses the figure in which the rectangles extend beyond the circle.

Each rectangle has width $\frac{r}{n}$ and height $y = \sqrt{r^2 - x^2}$. The sum of the areas of the 10 rectangles is given by

$$\text{Sum} = \frac{r}{10}\sqrt{r^2 - 0^2} + \frac{r}{10}\sqrt{r^2 - \left(\frac{r}{10}\right)^2} + \cdots + \frac{r}{10}\sqrt{r^2 - \left(\frac{mr}{10}\right)^2}$$

$$+ \cdots + \frac{r}{10}\sqrt{r^2 - \left(\frac{9r}{10}\right)^2}$$

A computer program to approximate the area of a circle in the first quadrant might be designed along the following lines.

Step (1)	Let the sum initially be 0.	$S = 0$
Step (2)	For rectangle $m = 1, 2, 3$, and so on, find the area of the rectangle.	$A = \frac{r}{10}\sqrt{r^2 - \left(\frac{mr}{10}\right)^2}$
Step (3)	Add the area to the sum.	$S = S + A$
Step (4)	Go on to the next rectangle.	Go to $m + 1$

Since a circle with equation $x^2 + y^2 = r^2$ is symmetric about the x-axis, the y-axis, and the origin, an approximation of the area of the circle is $4 \times$ Sum.

EXERCISES

(You may use either BASIC or Pascal.)

1. Write a computer program to find Sum when the user supplies r, the radius of a circle, and n, the number of rectangles in Sum.
2. Run the program in Exercise 1 for each radius and number of rectangles.
   a. $r = 1; n = 10$ **b.** $r = 1, n = 50$
   c. $r = 5; n = 10$ **d.** $r = 5; n = 50$
3. Modify the program in Exercise 1 to approximate the area of a circle when the user supplies a radius r and a number of rectangles n.
4. Run the program in Exercise 3 for each value of r and n in Exercise 2.
5. Approximate π to four decimal places.
6. Find, by using an adapted computer program, an approximation of the surface area of an elliptical swimming pool whose major axis is 75 m and whose minor axis is 30 m. (*Hint*: Consider $b^2x^2 + a^2y^2 = a^2b^2$.)
7. By how much does the approximation in Exercise 6 differ from $\pi\left(\frac{75}{2}\right)\left(\frac{30}{2}\right)$?
8. Use the strategy of constructing rectangles to write a computer program to approximate the area of the region bounded by the graph of $y = x^2$, the x-axis, and the vertical line $x = 3$.

y
O
3
x

9. Write a computer program using rectangles to approximate the area of the two regions formed by a hyperbola with equation $b^2x^2 - a^2y^2 = a^2b^2$ and the vertical lines $x = d$ and $x = -d$. Note that d must be greater than a. The user supplies the length of the transverse axis and the coordinates of a point on the hyperbola.
10. Run the program in Exercise 9 for a hyperbola containing $(10, 4\sqrt{3})$ and having a transverse axis of length 10. Let $d = 9$.

APPLICATION

ORBITS

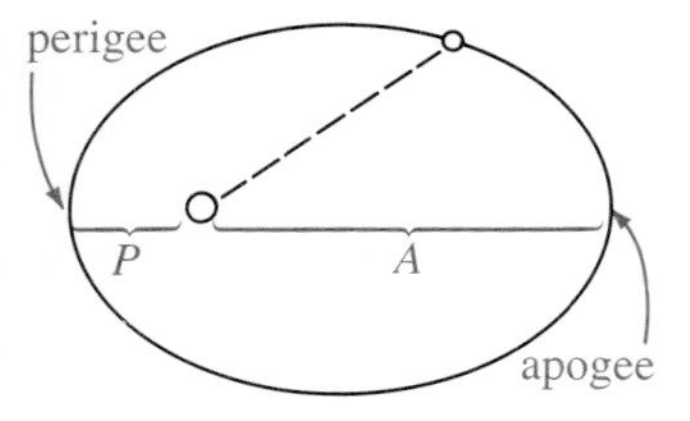

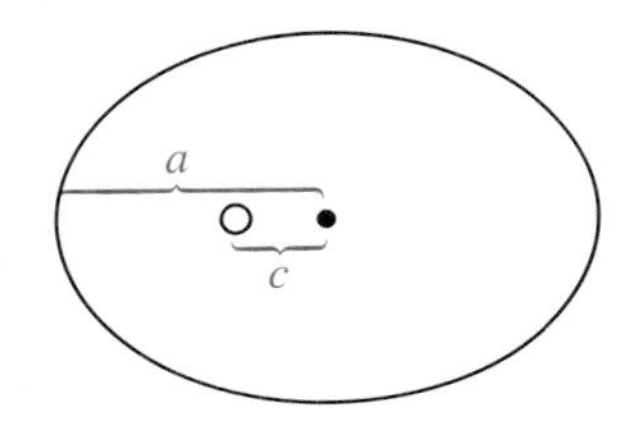

The orbit of a planet around the sun is an ellipse with the sun at one focus. This is Kepler's first law of planetary motion. It applies also to the orbit of a satellite around Earth or the moon. The *apogee* of an orbit is that point on the orbit farthest from the primary body, and the *perigee* is the point closest to the primary body. We denote the distances from the apogee and the perigee to the center of the primary body by A and P respectively.

The eccentricity e of an elliptical orbit is defined by $e = \frac{c}{a}$ where c is the distance from the center of the primary body to the center of the ellipse and a is $\frac{1}{2}$ the length of the major axis of the elliptical orbit. For an ellipse, $0 < e < 1$. The eccentricity can be written in terms of A and P. From the figures above,

$$a = \frac{1}{2}(A + P) \quad \text{and} \quad c = a - P.$$

Hence $c = \frac{1}{2}(A + P) - P = \frac{1}{2}(A - P)$. Therefore $e = \frac{c}{a}$ becomes

$$e = \frac{A - P}{A + P}.$$

A satellite is launched into an elliptical orbit around Earth. The minimum distance above Earth is 110 mi, and the maximum distance above Earth is 140 mi. Using 3960 mi for the radius of Earth, $A = 140 + 3960 = 4100$ and $P = 110 + 3960 = 4070$. Then $A - P = 30$ and $A + P = 8170$. Then the eccentricity of the orbit is found to be 0.00367. Since A and P are nearly equal, the ellipse is nearly circular. This is reflected in the fact that e is nearly 0.

It can be shown that the speed v of a satellite in elliptical orbit and r miles from the center of the primary body is given by

$$v = \sqrt{GM\left(\frac{2}{r} - \frac{1}{a}\right)}$$

where G is the universal gravitational constant and M is the mass of the primary body.

Let v_A be the speed at apogee. Then $r = A$. This speed can be expressed in terms of eccentricity. Since $A = a + c = a + ae = a(1 + e)$,

$$\frac{1}{a} = \frac{1 + e}{A}.$$

Then

$$v_A = \sqrt{GM\left(\frac{2}{A} - \frac{1}{a}\right)} = \sqrt{GM\left(\frac{2}{A} - \frac{1 + e}{A}\right)}$$

Therefore

$$v_A = \sqrt{\frac{GM}{A}(1 - e)}.$$

The speed at perigee, v_P, can be found to be

$$v_P = \sqrt{\frac{GM}{P}(1 + e)}.$$

In calculations involving orbits around Earth, $GM = 1.24 \times 10^{12}$.

Earth rises over the moon's horizon, as seen from the Apollo 11 spacecraft.

EXERCISES

1. The apogee of a satellite orbiting Earth is 5000 mi. The eccentricity of the orbit is 0.55. What is the speed at apogee?

2. The perigee of the satellite in Exercise 1 is 200 mi. What is the speed at perigee?

3. a. Use $P = a - c$ to show that $\frac{1}{a} = \frac{1 - e}{P}$.

b. Use part (a) to show that $v_P = \sqrt{\frac{GM}{P}(1 + e)}$.

4. a. Use $A = a(1 + e)$, $P = a(1 - e)$, and the formulas for v_A and v_P to show that $\frac{v_A}{v_P} = \frac{P}{A}$.

b. Use part (a) to show that in an elliptical orbit the speed at perigee is greater than the speed at apogee; that is $v_P > v_A$.

5. Show that if the orbit of a satellite is a circle with radius r, the speed of the satellite is constant and equal to $\sqrt{\frac{GM}{r}}$.

Chapter 3

Sequences, Series, and Limits

I*n this chapter sequences, series, and limits are discussed. The concept of limit is of fundamental importance in mathematical analysis. In particular, infinite geometric sequences and series are investigated.*

Mathematical Induction

■ 3-1 MATHEMATICAL INDUCTION

In Section 1-7 we described the set N of natural numbers as the smallest set of real numbers such that $1 \in N$ and $k + 1 \in N$ whenever $k \in N$. Since N is the smallest set of real numbers with these properties, any set S contained in N and having these properties must equal N. This fact is formally stated as the *principle of mathematical induction*.

The Principle of Mathematical Induction

If S is a set of natural numbers such that

$$(1)\ 1 \in S \quad \text{and} \quad (2)\ x + 1 \in S \text{ whenever } x \in S,$$

then $S = N$.

One can imagine that the sequence of guard rails shown in the photograph continues infinitely. The concept of infinity is one that has fascinated mathematicians and thinkers in other disciplines for many generations.

This principle can be used to prove many theorems in mathematics. The strategy for proving a statement by mathematical induction is as follows. Let S be the set of all natural numbers for which a specific statement is true. Show that $1 \in S$ and that if $x \in S$ then $x + 1 \in S$. By the principle of mathematical induction we may conclude that $S = N$ and hence that the statement is true for all natural numbers.

EXAMPLE Prove that $2 + 4 + 6 + \cdots + 2n = n(n + 1)$ for all natural numbers n.

Solution Let $S = \{n \in N: 2 + 4 + 6 + \cdots + 2n = n(n + 1)\}$.

(1) Since $2 \cdot 1 = 1(1 + 1)$, the equation is true for $n = 1$. Thus $1 \in S$.

(2) Suppose that $x \in S$. (This assumption is called the **induction hypothesis**.) Use this assumption to prove that $x + 1 \in S$.

$$2 + 4 + \cdots + 2x = x(x + 1) \quad \text{by the induction hypothesis}$$

$$2 + 4 + \cdots + 2x + 2(x + 1) = x(x + 1) + 2(x + 1) \quad \text{by the addition property of equality}$$

$$= (x + 1)(x + 2) \quad \text{by factoring}$$

$$= (x + 1)((x + 1) + 1)$$

This proves that $x + 1 \in S$.

The principle of mathematical induction justifies the conclusion that $S = N$. That is, the formula is true for all natural numbers n.

Mathematical induction can be used to prove many of the laws of elementary algebra. For example, positive integral exponents are defined by $r^1 = r$ and $r^{m+1} = r^m r$ for any real number r and natural number m. With this definition we can prove the *laws of exponents* for positive integral exponents.

THEOREM 1 **Laws of Exponents**

If r and s are real numbers and m and n are natural numbers, then

1. $r^m r^n = r^{m+n}$ **2.** $(r^m)^n = r^{mn}$

3. $\dfrac{r^m}{r^n} = r^{m-n}$ if $m > n$ and $r \neq 0$

4. $\dfrac{r^m}{r^n} = \dfrac{1}{r^{n-m}}$ if $m < n$ and $r \neq 0$

5. $(rs)^m = r^m s^m$

6. $\left(\dfrac{s}{r}\right)^m = \dfrac{s^m}{r^m}$ if $r \neq 0$

7. If $r^m = r^n$ and $r \neq -1, 0,$ or 1, then $m = n$.

We will prove the first law.

Proof: Given r and m fixed, let $S = \{n \in N: r^m r^n = r^{m+n}\}$.

(1) $1 \in S$ because $r^m r^1 = r^{m+1}$ by definition.

(2) Suppose that $x \in S$. That is, $r^m r^x = r^{m+x}$.
We need to prove that $x + 1 \in S$, namely that $r^m r^{x+1} = r^{m+x+1}$. To do this, we can multiply both sides of $r^m r^x = r^{m+x}$ by r.

$$\begin{aligned} (r^m r^x)r &= (r^{m+x})r \\ r^m(r^x r) &= r^{m+x+1} \\ r^m r^{x+1} &= r^{m+x+1} \end{aligned}$$

So $x + 1 \in S$ whenever $x \in S$. Therefore $S = N$ and the first law is true for all natural numbers n. ■

The proofs of some of the other laws are left as Exercises 11–15.

We define the zero exponent by $r^0 = 1$. The laws of exponents stated above remain valid when m or n is 0.

The next two theorems are results that are very useful in analysis.

THEOREM 2 For any real number r and any natural number n,

$$(1 - r)(1 + r + r^2 + \cdots + r^n) = 1 - r^{n+1}.$$

Proof: Given a real number r, let S be the set of all natural numbers for which the statement is true.

(1) $1 \in S$ because $(1 - r)(1 + r) = 1 - r^2 = 1 - r^{1+1}$.

(2) Assuming that $x \in S$, then

$$(1 - r)(1 + r + \cdots + r^x) = 1 - r^{x+1}.$$

To prove that $x + 1 \in S$, we need to show that

$$(1 - r)(1 + r + \cdots + r^x + r^{x+1}) = 1 - r^{x+2}.$$

By the distributive axiom and the induction hypothesis,

$$\begin{aligned} &(1 - r)(1 + r + \cdots + r^x + r^{x+1}) \\ &\quad = (1 - r)(1 + r + \cdots + r^x) + (1 - r)(r^{x+1}) \\ &\quad = 1 - r^{x+1} + (1 - r)(r^{x+1}) \\ &\quad = 1 - r^{x+1} + r^{x+1} - r^{x+2} \\ &\quad = 1 - r^{x+2}. \end{aligned}$$

This shows that $x + 1 \in S$. Therefore $S = N$ and hence, by the principle of mathematical induction, the theorem is proved. ■

Mathematical induction can be used to prove theorems involving inequalities. This is illustrated in Theorem 3.

THEOREM 3 Bernoulli's Inequality
If $n \in N$ and $a \geq -1$, then $(1 + a)^n \geq 1 + na$.

Proof: Let S be the set of all natural numbers for which the inequality is true.

(1) $1 \in S$ since $(1 + a)^1 \geq 1 + 1 \cdot a$.

(2) If $x \in S$, then $(1 + a)^x \geq 1 + xa$.
Since $1 + a \geq 0$,

$$(1 + a)^x(1 + a) \geq (1 + xa)(1 + a)$$
$$(1 + a)^{x+1} \geq (1 + xa)(1 + a) = 1 + xa + a + a^2x$$
$$(1 + a)^{x+1} \geq 1 + (x + 1)a + a^2x.$$

Because $a^2x \geq 0$, $1 + (x + 1)a + a^2x \geq 1 + (1 + x)a$, so

$$(1 + a)^{x+1} \geq 1 + (x + 1)a.$$

Thus if $x \in S$, then $x + 1 \in S$. Thus $S = N$ by the principle of mathematical induction. Hence the theorem is proved. ■

It is important to realize that in proofs by mathematical induction parts (1) and (2) are both necessary. Omitting the proof of either part can lead to erroneous results. (See Exercises 28 and 29.)

EXERCISES #1-14 + 16-24

Use the laws of exponents to simplify each expression.

A

1. $(6xy)^2(-4x^2y^2)^3$ **2.** $(b^6 \cdot 4b^a)^2$

3. $[(3m^3)(2n^2)^2]^3$ **4.** $((5t^2)^2)^3$

5. $(x^n + y^n)(x^m + y^m)$ **6.** $(3cd)(-3d^2) - cd^3$

7. $\dfrac{-2x^{p+q}}{-26x^{p-q}}$ **8.** $\dfrac{35z^{2m-n}}{14z^{m-2n}}$

9. Show by direct calculation that the formula in the example is true for $n = 9$.

10. Show by multiplication that Theorem 2 is true for $n = 4$.

Prove each law of exponents by mathematical induction.

11. Law 2 **12.** Law 3 **13.** Law 4 **14.** Law 5 **15.** Law 6

Prove each statement by mathematical induction.

16. $1 + 3 + 5 + \cdots + (2n - 1) = n^2$

17. $1 + 2 + 3 + \cdots + n = \dfrac{n(n + 1)}{2}$

18. $(a_1 - a_2) + (a_2 - a_3) + \cdots + (a_n - a_{n+1}) = a_1 - a_{n+1}$

19. $\frac{1}{1 \cdot 2} + \frac{1}{2 \cdot 3} + \cdots + \frac{1}{n(n+1)} = \frac{n}{n+1}$

20. $(-1)^1 + (-1)^2 + (-1)^3 + \cdots + (-1)^n = \frac{(-1)^n - 1}{2}$

B 21. $1^2 + 3^2 + \cdots + (2n-1)^2 = \frac{n(2n-1)(2n+1)}{3}$

22. $\frac{1}{2} + \frac{1}{2^2} + \frac{1}{2^3} + \cdots + \frac{1}{2^n} = 1 - \frac{1}{2^n}$

23. $\frac{1}{2} - \frac{1}{4} - \frac{1}{8} - \cdots - \frac{1}{2^n} = \frac{1}{2^n}$

24. $1 - \left[\left(1 - \frac{1}{2}\right) + \left(\frac{1}{2} - \frac{1}{3}\right) + \cdots + \left(\frac{1}{n} - \frac{1}{n+1}\right)\right] = \frac{1}{n+1}$

25. $2n \leq 2^n$

26. $n^3 + 2n$ is an integral multiple of 3.

27. $3^n \geq 2n + 1$

28. Criticize the following fallacious proof that all natural numbers are divisible by 2: Let $S = \{n \in N$: all natural numbers $m \leq n$ are divisible by 2$\}$. Suppose that $x \in S$, so that every natural number $y < x$ is divisible by 2. To prove that $x + 1 \in S$, it is sufficient to show that $x + 1$ is even. If $x \in S$, then $x - 1$ is even. Now $x + 1 = (x - 1) + 2$. Hence $x + 1$ must also be even, since it is the sum of two even integers. Since $x + 1 \in S$ whenever $x \in S$, $S = N$; that is, every natural number is even.

29. Find the error in the following fallacious proof that all flowers are the same color: Let $S = \{n \in N$: in any bouquet of n flowers, all n flowers are the same color$\}$. Clearly $1 \in S$. Assume that $x \in S$ and let an arbitrary bouquet of $x + 1$ flowers be labeled $f_1, f_2, \ldots, f_x, f_{x+1}$. If the last flower is removed, then $\{f_1, f_2, \ldots, f_x\}$ has x flowers and by the induction hypothesis all the flowers are the same color. If the first flower is removed, then $\{f_2, f_3, \ldots, f_{x+1}\}$ has x flowers and all the flowers are the same color. Therefore the entire bouquet of $x + 1$ flowers is the same color. This proves that $x + 1 \in S$. Therefore $S = N$ and all flowers are the same color.

C 30. Let m and n be natural numbers. Prove by induction on n that there exist whole numbers q and r such that $n = mq + r$ and $r < m$.

In Exercises 31 and 32 prove that if $0 < h < 1$ there is a positive integer k for which the inequality is true. (*Hint*: Let M be the positive even integer such that $M - 2 \leq \frac{1}{h} < M$. Choose $k = \frac{M}{2}$ and use Exercise 25 or 27.)

31. $\frac{1}{3^k} < h$

32. $\frac{1}{2^k} < h$

33. If $a_1 = \sqrt{2}$, $a_2 = \sqrt{2 + \sqrt{2}}$, $a_3 = \sqrt{2 + \sqrt{2 + \sqrt{2}}}, \ldots$ prove that $a_n < 2$ for all n.

3-2 THE BINOMIAL THEOREM

In this section we explore efficient methods of calculating powers of binomials: $(a + b)^n$ where n is a natural number. The patterns that emerge are mathematically rich. The proof of their validity uses the principle of mathematical induction.

Expansions of $(a + b)^n$ for $n = 1, 2, 3$, and 4 reveal the patterns for any such expansion.

$$\begin{array}{ll} n = 1 & (a + b)^1 = a + b \\ n = 2 & (a + b)^2 = a^2 + 2ab + b^2 \\ n = 3 & (a + b)^3 = a^3 + 3a^2b + 3ab^2 + b^3 \\ n = 4 & (a + b)^4 = a^4 + 4a^3b + 6a^2b^2 + 4ab^3 + b^4 \end{array}$$

In general, the expansion of $(a + b)^n$ has $n + 1$ terms and the sum of the exponents of a and b in any term in the expansion is n. Furthermore, if we arrange the terms in order of decreasing powers of a and therefore increasing powers of b, then the coefficients can be displayed in a triangular array (called *Pascal's triangle* after the seventeenth-century mathematician and philosopher Blaise Pascal).

$$\begin{array}{lccccccccc} (a + b)^0: & & & & & 1 & & & & \\ (a + b)^1: & & & & 1 & & 1 & & & \\ (a + b)^2: & & & 1 & & 2 & & 1 & & \\ (a + b)^3: & & 1 & & 3 & & 3 & & 1 & \\ (a + b)^4: & 1 & & 4 & & 6 & & 4 & & 1 \end{array}$$

The numbers in each row of the triangle are called **binomial coefficients**. To find a formula for each of these coefficients, we first define $n!$ (read "n factorial"):

$$0! = 1$$
$$n! = n(n - 1)! \text{ for } n \geq 1$$

Thus, for example,
$$\begin{aligned} 1! &= 1 \cdot 0! = 1, \\ 2! &= 2 \cdot 1! = 2 \cdot 1 = 2, \\ 3! &= 3 \cdot 2! = 3 \cdot 2 \cdot 1 = 6, \\ 4! &= 4 \cdot 3! = 4 \cdot 3 \cdot 2 \cdot 1 = 24, \text{ and so forth.} \end{aligned}$$

Next, for nonnegative integers n and k ($k \leq n$), we define $\binom{n}{k}$ by

$$\binom{n}{k} = \frac{n!}{k!(n - k)!}.$$

Notice that the formula gives the following coefficients for the expansion of $(a + b)^4$.

$$\binom{4}{0} = \frac{4!}{0!4!} = 1$$

$$\binom{4}{1} = \frac{4!}{1!3!} = \frac{4 \cdot 3 \cdot 2 \cdot 1}{1 \cdot 3 \cdot 2 \cdot 1} = 4$$

$$\binom{4}{2} = \frac{4!}{2!2!} = \frac{4 \cdot 3 \cdot 2 \cdot 1}{2 \cdot 1 \cdot 2 \cdot 1} = 6$$

$$\binom{4}{3} = \frac{4!}{3!1!} = \frac{4 \cdot 3 \cdot 2 \cdot 1}{3 \cdot 2 \cdot 1 \cdot 1} = 4$$

$$\binom{4}{4} = \frac{4!}{4!0!} = 1$$

It appears that the formula for $\binom{n}{k}$ will enable us to calculate binomial coefficients.

To show that the formula for $\binom{n}{k}$ actually gives binomial coefficients, we first prove the following theorem.

THEOREM 4 Let n and k be nonnegative integers, $1 \leq k \leq n$.

$$\binom{n}{0} = \binom{n}{n} = 1$$

$$\binom{n}{k-1} + \binom{n}{k} = \binom{n+1}{k}$$

Proof

$$\binom{n}{0} = \frac{n!}{0!(n-0)!} = 1 \qquad \binom{n}{n} = \frac{n!}{n!(n-n)!} = 1$$

$$\begin{aligned}\binom{n}{k-1} + \binom{n}{k} &= \frac{n!}{(k-1)!(n-k+1)!} + \frac{n!}{k!(n-k)!} \\ &= \frac{k \cdot n!}{k(k-1)!(n-k+1)!} + \frac{(n-k+1) \cdot n!}{k!(n-k+1)(n-k)!} \\ &= \frac{k \cdot n! + (n-k+1)n!}{k!(n-k+1)!} \\ &= \frac{(n+1) \cdot n!}{k!(n+1-k)!} = \frac{(n+1)!}{k!(n+1-k)!} = \binom{n+1}{k} \quad \blacksquare\end{aligned}$$

We can now prove the binomial theorem.

THEOREM 5 **The Binomial Theorem**

If $n \in N$ and a and $b \in \mathcal{R}$, then

$$(a+b)^n = \binom{n}{0}a^n + \binom{n}{1}a^{n-1}b + \binom{n}{2}a^{n-2}b^2 + \cdots + \binom{n}{n-1}ab^{n-1} + \binom{n}{n}b^n$$

where $\binom{n}{k} = \frac{n!}{k!(n-k)!}$ and $0 \leq k \leq n$.

Proof on following page

Proof: The proof is by induction. Let S be the set of all natural numbers for which the theorem is true.

(1) $1 \in S$ because $(a + b)^1 = 1 \cdot a^1 + 1 \cdot b^1 = \binom{1}{0}a + \binom{1}{1}b.$

(2) If we assume that $x \in S$, then $(a + b)^x =$

$$\binom{x}{0}a^x + \binom{x}{1}a^{x-1}b + \cdots + \binom{x}{j}a^{x-j}b^j + \cdots + \binom{x}{x}b^x.$$

Since $(a + b)^{x+1} = (a + b)^x(a + b)$,

$(a + b)^{x+1} =$

$$a\left[\binom{x}{0}a^x + \binom{x}{1}a^{x-1}b + \cdots + \binom{x}{j}a^{x-j}b^j + \cdots + \binom{x}{x}b^x\right]$$

$$+ b\left[\binom{x}{0}a^x + \binom{x}{1}a^{x-1}b + \cdots + \binom{x}{j}a^{x-j}b^j + \cdots + \binom{x}{x}b^x\right].$$

$(a + b)^{x+1} =$

$$\binom{x}{0}a^{x+1} + \left[\binom{x}{1} + \binom{x}{0}\right]a^xb + \cdots + \left[\binom{x}{j} + \binom{x}{j-1}\right]a^{x+1-j}b^j$$

$$+ \cdots + \left[\binom{x}{x} + \binom{x}{x-1}\right]ab^x + \binom{x}{x}b^{x+1}.$$

Since by Theorem 4,

$$\binom{x}{0} = \binom{x+1}{0} = \binom{x}{x} = \binom{x+1}{x+1} \quad \text{and} \quad \binom{x+1}{j} = \binom{x}{j} + \binom{x}{j-1},$$

$(a + b)^{x+1} =$

$$\binom{x+1}{0}a^{x+1} + \binom{x+1}{1}a^xb + \cdots + \binom{x+1}{j}a^{x+1-j}b^j$$

$$+ \cdots + \binom{x+1}{x+1}b^{x+1}.$$

Therefore $x + 1 \in S$ whenever $x \in S$. Thus $s = N$ and hence by the principle of mathematical induction the theorem is proved. ■

EXAMPLE 1 **a.** Expand $(x - 3)^7$.

b. Find the eighth term in the expansion of $(t^2 + s)^{12}$.

c. In the expansion of $(x^3 - y)^7$, give the term in which the exponent of x is 15.

Solution **a.** $(x - 3)^7 = \binom{7}{0}x^7 + \binom{7}{1}x^6(-3) + \binom{7}{2}x^5(-3)^2 + \binom{7}{3}x^4(-3)^3$

$$+ \binom{7}{4}x^3(-3)^4 + \binom{7}{5}x^2(-3)^5 + \binom{7}{6}x(-3)^6 + \binom{7}{7}(-3)^7$$

$$= x^7 - 21x^6 + 189x^5 - 945x^4 + 2835x^3 - 5103x^2$$
$$+ 5103x^1 - 2187 \quad \textbf{Answer}$$

b. The first term involves s^0, the second involves s^1, the third, s^2, . . . , the eighth, s^7. In that term, the power of t^2 must be $12 - 7 = 5$, so that the term is $\binom{12}{7}(t^2)^5s^7 = \frac{12 \cdot 11 \cdot 10 \cdot 9 \cdot 8}{5 \cdot 4 \cdot 3 \cdot 2 \cdot 1}t^{10}s^7 = 792t^{10}s^7$. **Answer**

c. The required term has x^3 to the fifth power and $(-y)$ to the second power. So the term is $\binom{7}{2}(x^3)^5(-y)^2 = 21x^{15}y^2$. **Answer**

EXAMPLE 2 Estimate $(0.9)^5$ without using a calculator.

Solution $(0.9)^5 = (1 - 0.1)^5$. By the binomial theorem,

$$(1 - 0.1)^5 = \binom{5}{0} \cdot 1^5 (-0.1)^0 + \binom{5}{1} \cdot 1^4(-0.1)^1 + \binom{5}{2} \cdot 1^3 (-0.1)^2$$
$$+ \binom{5}{3} \cdot 1^2 (-0.1)^3 + \binom{5}{4} \cdot 1^1 (-0.1)^4 + \binom{5}{5} \cdot 1^0 (-0.1)^5$$
$$= 1 - 0.5 + 0.1 - 0.01 + 0.0005 - 0.00001.$$

The last two terms are negligible when compared with the first four terms. Therefore $(0.9)^5 \approx 0.59$. **Answer**

EXERCISES

A

1. Verify the formula for $(a + b)^3$ by multiplying out $(a + b)(a + b)^2$.
2. Verify the formula for $(a + b)^4$ by multiplying out $(a + b)(a + b)^3$.

Find the value of each expression.

3. $\frac{7!}{(7-1)!}$ **4.** $\frac{5!}{3!\,2!}$ **5.** $\frac{9!}{5!\,4!}$ **6.** $\frac{4!}{0!}$

7. $\binom{6}{3}$ **8.** $\binom{7}{5}$ **9.** $\binom{9}{3} + \binom{9}{4}$ **10.** $\binom{5}{4} + \binom{5}{5}$

In Exercises 11–16 express the given product or quotient as factorial expressions. Assume that $n > r > 0$.

11. $(n + 1)n!$ **12.** $\frac{(n + 8)!}{n + 8}$

13. $100 \cdot 99 \cdot 98$ **14.** $(n - 1)(n - 2)(n - 3)$

15. $(n - r)(n - r - 1)!$ **16.** $(n - r + 1)(n - r + 1)$

In Exercises 17–22, write the first three terms of the expansion and simplify.

17. $(a + b)^{10}$ **18.** $(x - y)^8$ **19.** $(3c - d^2)^5$

20. $\left(x^3 + \frac{2}{x}\right)^7$ **21.** $\left(6y - \frac{z}{2}\right)^4$ **22.** $\left(4cd^2 - \frac{1}{t}\right)^3$

Write the indicated term in the expansion of the binomial.

23. 7; $(a - b)^{11}$ **24.** 3; $(x + y)^9$

Write the indicated term in the expansion of the binomial.

25. middle; $\left(\frac{4}{y} + \frac{y^2}{4}\right)^6$ **26.** middle; $(x^3 + 2y^2)^{10}$

27. $(n - 3)$; $(a + b)^n$ **28.** $(n - r)$; $(x + y)^n$

29. Write the term in the expansion of $(x^2 - 4y)^8$ in which the exponent of y is 3.

30. Write the term in the expansion of $(2x^4 + yz)^{12}$ in which the exponent of x is 16.

31. If the coefficients of the third and the thirteenth terms in the expansion of $(r + s)^n$ are equal, find the middle term.

32. If the coefficients of the fourth and the fourteenth terms in the expansion of $(c - d)^n$ are equal, find the fifteenth term.

33. Find the term(s) in the expansion of $(a + b)^{15}$ having the coefficient $\binom{15}{3}$.

34. Find the term(s) in the expansion of $(a - b)^{11}$ having the coefficient $\binom{11}{9}$.

In Exercises 35–40, solve each open sentence for n.

B **35.** $n! = 120$ **36.** $\frac{(n + 1)!}{n!} = 7$ **37.** $\binom{n}{2} = 36$

38. $(n - 2)! < 25$ **39.** $10 < \binom{n}{4} + \binom{n}{5} < 100$ **40.** $\frac{(n + 1)!}{(n - 1)!} = 90$

In Exercises 41–43, give the complete expansion in simplified form.

41. $\left(x - \frac{2}{x}\right)^5$ **42.** $(3j + 2k)^4$ **43.** $(2s^2 + t)^3$

In Exercises 44–47, find each value to the nearest thousandth.

44. $(1 + 0.03)^5$ **45.** $(1 - 0.02)^4$ **46.** $(0.97)^6$ **47.** $(1.05)^3$

In Exercises 48 and 49 verify each statement by direct computation.

48. $\binom{3}{0} + \binom{3}{1} + \binom{3}{2} + \binom{3}{3} = 2^3$

49. $\binom{4}{0} + \binom{4}{1} + \binom{4}{2} + \binom{4}{3} + \binom{4}{4} = 2^4$

Prove each statement. Assume that n and k are natural numbers and $k \leq n$.

C **50.** $\binom{n}{k} = \binom{n}{n - k}$ **51.** $\binom{n}{0} + \binom{n}{1} + \cdots + \binom{n}{n} = 2^n$

52. The sum of the coefficients in the expansion of $(a - b)^n$ is 0 $(a, b > 0)$.

53. $\left(1 + \frac{1}{n}\right)^n = 2 + \frac{1}{2!}\left(1 - \frac{1}{n}\right) + \frac{1}{3!}\left(1 - \frac{1}{n}\right)\left(1 - \frac{2}{n}\right) + \cdots + \left(\frac{1}{n}\right)^n$

COMPUTER EXERCISES

1. Write a computer program to find $n!$ for a value of n supplied by the user.
2. What is the largest value of n for which the computer will give $n!$?
3. Write a computer program to find $\binom{n}{k}$ for values of n and k, $n \geq k$, supplied by the user. (*Hint*: Use the result of Exercise 2.)
4. Write a program to find how many ways two 1's and three 2's can be arranged to form 5-digit numbers. Compare the result with $\binom{5}{2}$.

Sequences and Series

3-3 RECURSIVE DEFINITIONS: SEQUENCES AND SERIES

In 1202, Leonardo of Pisa (also called Fibonacci) published his *Liber Abaci* in which he posed and solved problems in arithmetic. One problem had the following *sequence* as its solution: 1, 1, 2, 3, 5, 8, 13, 21, ... This sequence, known as the *Fibonacci sequence*, occurs in nature in the propagation of rabbits and bees and in the arrangement of leaves on plant stems. It occurs also in geometric puzzles and in problems involving the golden ratio $\frac{1 + \sqrt{5}}{2}$.

In general a **sequence** is a set of real numbers in a specified order. Each number in a sequence is called a **term** of the sequence. The order is determined by a pairing of the members of N or a subset $\{1, 2, 3, \ldots, m\}$ of N with the terms of the sequence. The first of these pairings is an **infinite sequence**, and the second pairing is a **finite sequence**. The pairing of the Fibonacci sequence is:

position in list (n):	1	2	3	4	5	6	. . .
	↓	↓	↓	↓	↓	↓	
number (a_n):	1	1	2	3	5	8	. . .

The term corresponding to the natural number n is called the *nth term* of the sequence and is designated by a variable with subscript n, such as a_n.

There is a pattern in the terms of the Fibonacci sequence. The first two terms are 1, and each successive term is the sum of the preceding two terms. That is, $a_1 = 1$, $a_2 = 1$, and $a_n = a_{n-1} + a_{n-2}$ for $n \geq 3$.

A sequence is defined **recursively**, or by a **recursion formula**, when rules are given that

(1) identify the first term or the first several terms of the sequence, and
(2) state how to find subsequent terms from preceding terms.

For example, the recursion formula $a_1 = 3$ and $a_n = 5a_{n-1} + 2$ for $n > 1$ has the following first four terms.

$$a_1 = 3 \qquad a_2 = (5)(3) + 2 = 17$$
$$a_3 = (5)(17) + 2 = 87 \qquad a_4 = (5)(87) + 2 = 437$$

Often, it is useful to define each term a_n of a sequence *explicitly* by a formula in terms of n.

EXAMPLE 1 Find an explicit expression for the nth term of the infinite sequence defined recursively by $a_1 = -1$ and $a_n = a_{n-1} + 2$, $n \in N$, and prove by induction that the formula is correct.

Solution Without simplifying calculations write out the first four terms.

$$\begin{aligned} a_1 &= -1 \\ a_2 &= a_1 + 2 = -1 + 2 = -1 + 1 \cdot 2 \\ a_3 &= a_2 + 2 = -1 + 2 + 2 = -1 + 2 \cdot 2 \\ a_4 &= a_3 + 2 = -1 + 2 + 2 + 2 = -1 + 3 \cdot 2 \end{aligned}$$

It appears that $a_n = -1 + (n - 1)2 = 2n - 3$. To prove this assertion, let $S = \{n \in N: a_n = 2n - 3\}$.

(1) $1 \in S$ since $a_1 = 2 \cdot 1 - 3 = -1$.

(2) Assume $x \in S$; then $a_x = 2x - 3$. By the recursive definition of the sequence,

$$a_{x+1} = a_x + 2 = (2x - 3) + 2 = 2x - 1 = 2(x + 1) - 3.$$

This shows that $x + 1 \in S$ whenever $x \in S$, so that $S = N$ and therefore $a_n = 2n - 3$ for all n.

EXAMPLE 2 Write the terms of the finite sequence whose nth term is given by $2n^2 - 1$ where $n \in \{1, 2, 3, 4, 5\}$.

Solution The formula $a_n = 2n^2 - 1$ yields the table at the right.

$n = 1$	$a_1 = 2(1)^2 - 1 = 1$
$n = 2$	$a_2 = 2(2)^2 - 1 = 7$
$n = 3$	$a_3 = 2(3)^2 - 1 = 17$
$n = 4$	$a_4 = 2(4)^2 - 1 = 31$
$n = 5$	$a_5 = 2(5)^2 - 1 = 49$

The sequence is 1, 7, 17, 31, 49. **Answer**

A **series** is an expression for the sum of the terms of a sequence. For example,

	Sequence	*Series*
finite	3, 7, 11, 15	$3 + 7 + 11 + 15$
	$a_1, a_2, a_3, \ldots, a_n$	$a_1 + a_2 + a_3 + \cdots + a_n$
infinite	$1, 3, 5, \ldots, 2n - 1, \ldots$	$1 + 3 + 5 + \cdots + 2n - 1 + \cdots$
	$a_1, a_2, a_3, \ldots, a_n, \ldots$	$a_1 + a_2 + a_3 + \cdots + a_n + \cdots$

The **sum** of a finite series is just the sum of the terms in the sequence. For example, the sum of the series 3 + 7 + 11 + 15 is 36. In Section 3-7 you will see how it can make sense to talk about the "sum" of an infinite series.

If we let S_n be the sum of the first n terms of a sequence with a_n as the nth term, we can obtain S_n by the following recursive definition.

$$S_0 = 0 \quad \text{and} \quad S_k = S_{k-1} + a_k \text{ where } k = 1, 2, 3, \ldots, n$$

A computer program might use such a definition to compute a sum. For example, the sum $\frac{1}{2} + \frac{1}{4} + \frac{1}{6} + \cdots + \frac{1}{42}$, which we can write as $\frac{1}{2 \cdot 1} + \frac{1}{2 \cdot 2} + \frac{1}{2 \cdot 3} + \frac{1}{2 \cdot 4} + \cdots + \frac{1}{2 \cdot 21}$, can be found by means of the following algorithm. Note that the kth term is $\frac{1}{2k}$.

(1) Let $S = 0$ initially.

(2) As k goes from 1 to 21, increment S by $\frac{1}{2k}$.

The Greek letter **sigma** (Σ), which is called the sum or summation symbol, is used to denote the sum of a series. The sum $\frac{1}{2} + \frac{1}{4} + \frac{1}{6} + \cdots + \frac{1}{42}$ can be compactly written

$$S = \sum_{k=1}^{21} \frac{1}{2k}$$

and read as "the sum of $\frac{1}{2k}$ as k goes from 1 to 21." The letter k is called the **index of summation**. Any other letter could also be used.

EXAMPLE 3 Express the series $\sum_{k=3}^{8} \frac{2^k}{k!}$ in expanded form.

Solution $$\sum_{k=3}^{8} \frac{2^k}{k!} = \frac{2^3}{3!} + \frac{2^4}{4!} + \frac{2^5}{5!} + \frac{2^6}{6!} + \frac{2^7}{7!} + \frac{2^8}{8!}$$

EXAMPLE 4 Find the value of $\sum_{j=1}^{4} [3j + (-1)^j]$.

Solution $$\sum_{j=1}^{4} [3j + (-1)^j] = [3(1) + (-1)^1] + [3(2) + (-1)^2] + [3(3) + (-1)^3] + [3(4) + (-1)^4] = 2 + 7 + 8 + 13 = 30$$

EXAMPLE 5 Write the series 3 − 8 + 15 − 24 + 35 − 48 in sigma notation.

Solution The absolute value of each term is 1 less than a perfect square. There are six terms that alternate in sign. Thus, in sigma notation, the series is

$$\sum_{n=1}^{6} (-1)^{n+1}[(n + 1)^2 - 1], \text{ or } \sum_{n=2}^{7} (-1)^n[n^2 - 1].$$ **Answer**

EXERCISES

A **1.** Write the first five terms of the Fibonacci sequence.

Write the first six terms of each sequence.

2. $a_1 = 7, a_{n+1} = a_n - 3$

3. $a_1 = -20, a_{n+1} = a_n + 6$

4. $a_1 = -3, a_{n+1} = 2a_n$

5. $a_1 = 32, a_{n+1} = -\frac{1}{2}a_n$

6. $a_1 = \frac{3}{4}, a_{n+1} = (-1)^n a_n$

7. $a_1 = \frac{1}{3}, a_{n+1} = (-1)^{n+1} a_n$

8. $a_1 = 1, a_{n+1} = a_n^2 + 1$

9. $a_1 = 1, a_{n+1} = \left(\frac{n}{2n + 1}\right)a_n$

In Exercises 10–15 a formula for a_n is given. Write the first five terms.

10. $a_n = 3\left(\frac{1}{10}\right)^n$

11. $a_n = \frac{(-1)^n}{n}$

12. $a_n = \frac{3 - n}{\sqrt{n + 1}}$

13. $a_n = \frac{\sqrt{n}}{n + 1}$

14. $a_n = \begin{cases} -\frac{1}{n} & \text{if } n \text{ is odd} \\ \frac{1}{n} & \text{if } n \text{ is even} \end{cases}$

15. $a_n = \begin{cases} \sqrt{n} & \text{if } n \text{ is odd} \\ -\sqrt{n} & \text{if } n \text{ is even} \end{cases}$

In Exercises 16–25 find an explicit formula for a_n. In Exercises 16–19 prove by mathematical induction that the formula is correct.

16. $a_1 = 0; a_{n+1} = a_n + 3$

17. $a_1 = -2; a_{n+1} = a_n + 4$

18. $a_1 = 1; a_{n+1} = 3a_n$

19. $a_1 = 1; a_{n+1} = na_n$

20. $\frac{1}{2}, \frac{1}{4}, \frac{1}{8}, \frac{1}{16}, \ldots$

21. $6, 12, 18, 24, \ldots$

22. $\frac{3}{2}, \frac{9}{2}, \frac{27}{2}, \frac{81}{2}, \ldots$

23. $1, \frac{1}{1 \cdot 2}, \frac{1}{1 \cdot 2 \cdot 3}, \frac{1}{1 \cdot 2 \cdot 3 \cdot 4}, \ldots$

24. $4, 16, 36, 64, \ldots$

25. $\frac{2}{2}, \frac{5}{2}, \frac{10}{2}, \frac{17}{2}, \ldots$

Write each series in expanded form.

26. $\sum_{i=3}^{7} 2i$

27. $\sum_{j=0}^{5} j^2$

28. $\sum_{n=1}^{4} |2 - n|$

29. $\sum_{k=1}^{5} (3k - 1)$

30. $\sum_{m=1}^{3} (-1)^m m!$

31. $\sum_{i=1}^{4} (-1)^{i+1}(i^2 + i)$

Write each series in sigma notation.

32. $a_1b_1 + a_2b_2 + a_3b_3 + a_4b_4$

33. $x_1^3 + x_2^3 + x_3^3$

34. $2 + 7 + 12 + 17 + 22$

35. $36 - 12 + 4 - \frac{4}{3}$

B **36.** $\frac{1}{2} - \frac{2}{3} + \cdots - \frac{10}{11}$

37. $\frac{1}{2} - \frac{1}{6} + \frac{1}{12} - \cdots - \frac{1}{110}$

Evaluate each sum.

38. $\sum_{n=1}^{4} n^2$

39. $\sum_{n=1}^{4} n^3$

40. Prove by mathematical induction that $\sum_{k=1}^{n} k^2 = \frac{n(n+1)(2n+1)}{6}$.

41. Prove by mathematical induction that $\sum_{k=1}^{n} k^3 = \left[\frac{n(n+1)}{2}\right]^2$.

Use mathematical induction to prove the statements in Exercises 42–45.

42. $\sum_{i=1}^{n} (a_i + b_i) = \sum_{i=1}^{n} a_i + \sum_{i=1}^{n} b_i$

43. $\sum_{i=1}^{n} ca_i = c\sum_{i=1}^{n} a_i$

44. $\sum_{k=1}^{n} a = na$

45. $\sum_{j=1}^{n} (a_j + b_j)^2 = \sum_{j=1}^{n} a_j^2 + 2\sum_{j=1}^{n} a_j b_j + \sum_{j=1}^{n} b_j^2$

46. Show that $\sum_{k=0}^{x} a_{k+1}$, $\sum_{k=1}^{x+1} a_k$, and $\sum_{k=t}^{x+t} a_{k-t+1}$ represent the same series.

C **47.** Write a formula for the sum of the first n ($n \le 100$) terms of $\sum_{k=1}^{100} (k^2 - 6k)$.

48. If $\sum_{b=2}^{4} (a^2 b - ab) = \sum_{c=3}^{5} (ac + 6)$, determine a.

49. An explicit formula for the terms of the Fibonacci sequence is given by
$$a_n = \frac{1}{\sqrt{5}}\left[\left(\frac{1+\sqrt{5}}{2}\right)^n - \left(\frac{1-\sqrt{5}}{2}\right)^n\right] \text{ for } n \ge 1.$$

a. Verify that the formula is true for $n = 1, 2, 3,$ and 4.

b. Prove that the nth term of the Fibonacci sequence is given by the formula above. (*Hint*: Let $S = \{n \in N: a_m = \frac{1}{\sqrt{5}}\left[\left(\frac{1+\sqrt{5}}{2}\right)^m - \left(\frac{1-\sqrt{5}}{2}\right)^m\right] \text{ for } m \le n\}$.)

50. Let c be a real number such that $c \ne 0$ and $|c| \ne 1$. Let $a_1 = c$ and $a_n = (a_{n-1})^2$ for $n \ge 2$. Show that $a_n = (c)^{2^{n-1}}$ for $n \ge 1$.

3–4 ARITHMETIC SEQUENCES

A grocer wants to make a 10-row display of stacked cans. There is to be one can in the top row, three cans in the next row, five cans in the third row, and so on. How many cans should there be in the tenth row? How many cans will there be in all ten rows?

The display of cans shown at the right begins the sequence 1, 3, 5, ... Notice that the difference between adjacent terms in the sequence is always 2. Such a sequence is called *arithmetic* (a rith MET ik). An **arithmetic sequence** is one that satisfies a recursive definition of the form

$$t_1 = a, \text{ and}$$
$$t_{n+1} = t_n + d, \text{ where } a \text{ and } d \text{ are real numbers.}$$

Since $t_{n+1} - t_n = d$, d is called the **common difference**.

In the can problem, the first term is 1 and the common difference is 2. Since

$$t_1 = 1,$$
$$t_2 = 1 + 2 = 1 + 1 \cdot 2,$$
$$t_3 = 1 + 2 + 2 = 1 + 2 \cdot 2,$$

and

$$t_4 = 1 + 2 + 2 + 2 = 1 + 3 \cdot 2,$$

it seems that the number of cans in the nth row is given by $t_n = 1 + (n - 1)2$. If we assume that this formula is valid for larger values of n, then $t_{10} = 1 + 9 \cdot 2 = 19$ and there will be 19 cans in the tenth row. This formula is indeed correct for all values of n, as the following theorem states.

THEOREM 6 In an arithmetic sequence in which the first term is a and the common difference is d, the nth term is

$$t_n = a + (n - 1)d.$$

You are asked to prove this theorem by induction in Exercise 34.

The formula in Theorem 6 is easy to remember if you realize that there are $n - 1$ intervals of length $|d|$ between $t_1 = a$ and t_n.

$$\underbrace{t_1, \; t_2, \; t_3, \; \ldots, \; t_n}_{|d| \;\; |d| \;\;\;\; |d| \;\;\;\; |d|}$$

$(n - 1)$ intervals

EXAMPLE 1 Find the twelfth term of the arithmetic sequence 2, 9, 16, ...

Solution $a = 2$, $d = 7$, $n = 12$. Since $t_n = a + (n - 1)d$,

$$t_{12} = 2 + (12 - 1) \cdot 7 = 79.$$ **Answer**

EXAMPLE 2 Which term of the arithmetic sequence that begins $-2, 1, 4, \ldots$ equals 40?

Solution $a = -2$, $d = 3$, $t_n = 40$, so that $40 = -2 + (n - 1)(3)$. Thus $n = 15$. Therefore 40 is the 15th term. **Answer**

The terms of an arithmetic sequence are said to be in **arithmetic progression** and the terms between any two given terms are called **arithmetic means** between those terms. For example, in the arithmetic sequence 2, 5, 8, 11, 14 the terms 5, 8, and 11 are the three arithmetic means between 2 and 14, while 5 and 8 are the two arithmetic means between 2 and 11. To insert arithmetic means between numbers in an arithmetic sequence, we need to find the common difference.

EXAMPLE 3 Insert three arithmetic means between 18 and -10.

Solution This problem may be solved by noting that there are to be four intervals, each of length d, between 18 and -10 in an arithmetic sequence:

$$18, \underbrace{\quad}_{|d|} ?, \underbrace{\quad}_{|d|} ?, \underbrace{\quad}_{|d|} ?, \underbrace{\quad}_{|d|} -10$$

Since the difference between 18 and -10 is 28, each interval must have length $\frac{28}{4} = 7$. Since the terms decrease, d is negative. So $d = -7$. Therefore the missing means must be

$$18 - 7 = 11,\ 11 - 7 = 4, \text{ and } 4 - 7 = -3. \qquad \textbf{Answer}$$

(The common difference d can also be found by using the formula in Theorem 6 with $a = 18$, $n = 5$, $t_5 = -10$. Hence, $-10 = 18 + (5 - 1)d$. Thus, $d = -7$.)

A single arithmetic mean between two numbers a and b is called the **average**, or *the* **arithmetic mean**, of the numbers; its value is $\frac{a + b}{2}$.

An **arithmetic series** is one whose terms form an arithmetic sequence. To return to the can problem, the number of cans in all ten rows of the display can be expressed as the arithmetic series:

$$1 + 3 + 5 + \cdots + 17 + 19$$

We could find the sum simply by adding the terms, but the following procedure will lead to a general formula for the sum of any finite arithmetic series.

Let S_{10} be the sum of the ten terms of the can series. Then

$$S_{10} = 1 + 3 + 5 + \cdots + 15 + 17 + 19.$$

Another way of writing this is

$$S_{10} = 19 + 17 + 15 + \cdots + 5 + 3 + 1.$$

Adding the two equations,

$$2S_{10} = 20 + 20 + 20 + \cdots + 20 + 20 + 20 = 10 \cdot 20 = 200.$$

Therefore, $S_{10} = 100$. The display contains 100 cans.

In general, if S_n is the sum of the first n terms of an arithmetic sequence with first term a and common difference d, then we can write S_n two different ways and add the two equations. By doing so, we shall obtain a formula for S_n in terms of a, t_n, and n.

$$\begin{aligned} S_n &= a + (a + d) + (a + 2d) + \cdots + [a + (n - 1)d] \\ S_n &= t_n + (t_n - d) + (t_n - 2d) + \cdots + [t_n - (n - 1)d] \\ 2S_n &= (a + t_n) + (a + t_n) + (a + t_n) + \cdots + (a + t_n), \end{aligned}$$

where $(a + t_n)$ occurs n times. Therefore,

$$S_n = \frac{n(a + t_n)}{2}, \quad \text{or} \quad S_n = n\left(\frac{a + t_n}{2}\right).$$

This formula says that the sum is equal to the product of the number of terms in the series and the average of the first and last terms. This is to be expected because the average value of all the terms is the same as the average of the first and last terms. (See Exercise 38, where you are asked to prove this.)

THEOREM 7 The sum of the first n terms of an arithmetic sequence with first term a and nth term t_n is given by

$$S_n = \frac{n(a + t_n)}{2}.$$

You are asked to prove this in Exercise 35.

EXAMPLE 4 Find the sum of all multiples of 3 between 3 and 468 inclusive.

Solution The problem asks for the sum of the series $3 + 6 + 9 + \cdots + 468$, which is an arithmetic series with $a = 3$ and $d = 3$.
To find the sum, first find n. Since $t_n = 3 + (n - 1)3 = 468$, $n = 156$.

Therefore $S_{156} = 156\,\frac{(3 + 468)}{2} = 36{,}738$. **Answer**

EXERCISES

Find the required term in each arithmetic sequence.

A

1. Twelfth term in 3, 7, 11, 15, . . .
2. Ninth term in $2\sqrt{3}, -\sqrt{3}, -4\sqrt{3}, \ldots$
3. Eighth term in $a - 2b, a, a + 2b, \ldots$
4. Fiftieth term in $\frac{2}{3}, \frac{3}{2}, \frac{7}{3}, \ldots$

Find the first term and the common difference for each arithmetic sequence.

5. $t_n = 5n - 7$ 6. $t_n = 3 - 2n$ 7. $t_n = 4.5n$ 8. $t_n = 1.6n - 2.3$

Complete each arithmetic sequence.

9. 8, ___?___, 14, ___?___, ___?___
10. -5, ___?___, ___?___, ___?___, 3
11. ___?___, 9, ___?___, ___?___, -1
12. ___?___, ___?___, -9, ___?___, -15.2

13. $\underline{\ ?\ }$, x, $\underline{\ ?\ }$, $-2x$, $\underline{\ ?\ }$

14. a, $\underline{\ ?\ }$, $\underline{\ ?\ }$, b, $\underline{\ ?\ }$

15. Which term of 10, 4, −2, . . . is −176?

16. Which term of 4, 16, 28, . . . is 328?

17. Find the sum of the first thirty terms of −2, 3, 8, ...

18. Find the sum of the first twenty positive multiples of 3.

Find the required number in each arithmetic sequence.

19. $d = 7$; $t_{14} = 93$; $a = ?$

20. $t_5 = 12$; $t_{24} = 88$; $t_{12} = ?$

21. $S = -323$; $t_{17} = -43$; $a = ?$

22. $S = 3400$; $n = 100$; $a = 1$; $d = ?$

Find the sum of each series.

B

23. $\sum_{k=1}^{25} (7 - 2k)$

24. $\sum_{i=1}^{40} \left(\frac{3}{2}i + \frac{1}{2}\right)$

25. How many terms of −10, −7, −4, . . . must be added to give a sum of 200?

26. Find the sum of all positive integers less than 500 that are multiples of 11.

27. Find the first three terms of an arithmetic sequence in which $t_7 = 28$ and $t_{18} = 6$.

28. For what value(s) of t will $2t - 7$, $5t - 9$, and $7t + 2$ be consecutive terms in an arithmetic sequence? Find the common difference.

29. For what real values of k will 2, k, and $7k + 2$ be consecutive terms in an arithmetic sequence?

30. If $a_1, a_2, a_3, \ldots$ is an arithmetic sequence and k is a real number, is $ka_1, ka_2, ka_3, \ldots$ an arithmetic sequence? Prove your answer.

31. Can the sum of the terms of an arithmetic sequence be 0? If so what must be the relationship between the first and last terms of the sequence?

32. If $a_1, a_2, a_3, \ldots$ is an arithmetic sequence and t is a real number, is $a_1 + t, a_2 + t, a_3 + t, \ldots$ an arithmetic sequence? Prove your answer.

33. If the terms of an arithmetic sequence are squared, will the resulting sequence be an arithmetic sequence?

C

34. Prove Theorem 6 by mathematical induction.

35. Prove Theorem 7 by mathematical induction.

36. Prove that the sum of the first n terms of an arithmetic sequence is given by $S_n = \frac{n}{2}(2a + (n - 1)d)$.

37. Find a formula for the sum of the terms in an arithmetic sequence from t_k to t_n inclusive ($k < n$).

38. Prove that the average of all the terms in a finite arithmetic sequence is equal to the average of the first and last terms.

39. Derive a formula for $\sum_{i=1}^{n} i$.

40. Prove that the sum of the first n positive odd integers is n^2.

41. Show that if t_n and t_m are distinct terms in an arithmetic sequence, then $\frac{t_m - t_n}{m - n}$ is independent of m and n.

42. Find a formula for the number of diagonals in a convex polygon of n sides. Prove that the formula is correct by mathematical induction.

43. Find a formula for the maximum number of regions into which a circle can be divided by n chords. Prove that the formula is correct by mathematical induction. $\left(\textit{Hint}\text{: Consider the expression } \frac{n^2 + n + 2}{2}.\right)$

3-5 GEOMETRIC SEQUENCES AND SERIES

Some photocopy machines are capable of making reduced copies of original documents. Suppose that a certain machine produces a copy whose dimensions are 90% of those of the original. If a picture originally 10 cm on a side is copied and reduced repeatedly, what will be the length of the side in the tenth copy?

Each side of the original is 10 cm, so each side of the first copy is $(0.9)(10) = 9$ cm. Each side of the second copy is $(0.9)(0.9)(10) = 8.1$ cm, and so forth. The nth copy will be $(0.9)^n(10)$ cm on a side. In particular, the tenth copy will be a picture $(0.9)^{10}(10) \approx 3.49$ cm on a side.

The sequence of the lengths of a side, namely

$$10, (0.9)(10), (0.9)^2(10), (0.9)^3(10), \ldots$$

is an example of a *geometric sequence*. A geometric sequence differs from an arithmetic sequence in that each term after the first term is obtained by multiplying the preceding term by a constant, rather than by adding a constant to the preceding term. In general, a **geometric sequence** is one defined recursively by

$$t_1 = a$$

and

$$t_{n+1} = rt_n \qquad (n \geq 1)$$

The number a is the first term and r is the **common ratio**. For the sequence above, the first term is 10 and the common ratio is 0.9. The formula for the nth term in a geometric sequence is analogous to the formula for the nth term in an arithmetic sequence.

THEOREM 8 In a geometric sequence with first term a and common ratio r, the nth term t_n is given by

$$t_n = ar^{n-1}.$$

You are asked to prove the formula in Exercise 25.

This formula can easily be remembered by visualizing a geometric sequence as in the following figure. The nth term is obtained by multiplying the first term by r, $n - 1$ times.

$$a = t_1 \quad t_2 \quad t_3 \quad t_4 \quad \cdots \quad t_{n-1} \quad t_n$$
$$\times r \quad \times r \quad \times r \quad \cdots \quad \times r$$

EXAMPLE 1 Find the sixth term of the geometric sequence 2, 6, 18, 54, ...

Solution $a = 2$, $r = 3$, $n = 6$. Using $t_n = ar^{n-1}$,

$$t_6 = 2(3)^{6-1} = 2(243) = 486. \quad \textbf{Answer}$$

EXAMPLE 2 Which term of the geometric sequence $\frac{3}{2}$, -3, 6, . . . is equal to 96?

Solution $a = \frac{3}{2}$, $r = -2$, $t_n = 96$. Find n.

$$\begin{aligned} t_n &= ar^{n-1} \\ 96 &= \tfrac{3}{2}(-2)^{n-1} \\ 64 &= (-2)^{n-1} \\ (-2)^6 &= (-2)^{n-1} \\ 6 &= n - 1 \quad \text{(by Theorem 1, part 7)} \\ n &= 7 \quad \textbf{Answer} \end{aligned}$$

In a geometric sequence, terms between any two given terms are called **geometric means**. Thus, in the geometric sequence 3, 6, 12, 24, 48, . . . , the terms 6, 12, and 24 are the three geometric means between 3 and 48. The terms 6 and 12 are the two geometric means between 3 and 24. To insert geometric means between two numbers we must find the common ratio.

EXAMPLE 3 Insert three real geometric means between 1 and 16.

Solution The problem asks for three missing numbers between 1 and 16.

$$1, \ \underline{\ ?\ }, \ \underline{\ ?\ }, \ \underline{\ ?\ }, \ 16$$
$$\times r \quad \times r \quad \times r \quad \times r$$

Evidently, $16 = 1 \cdot r^4$, so that $r^4 - 16 = 0$. Solve for r by factoring to obtain $(r^2 + 4)(r - 2)(r + 2) = 0$ and $r = 2$ or -2. Therefore, there are *two* possible answers:

$$1, 2, 4, 8, 16 \quad \text{or} \quad 1, -2, 4, -8, 16$$

A single *positive* geometric mean between two *positive* real numbers is called *the* **geometric mean**, or **mean proportional**, of the two numbers. If $a > 0$ and $b > 0$, then the geometric mean of a and b is $\sqrt{ab}$. In Exercise 34 you are asked for a proof.

A series whose terms constitute a geometric sequence is a **geometric series**. For example,

$$10 + (0.9)(10) + (0.9)^2(10) + \cdots + (0.9)^{10}(10) + \cdots$$

is such a series.

The following example shows a method for finding the sum of a finite geometric series. Theorem 9 will give a general formula.

EXAMPLE 4 If you deposit 1 cent in a bank the first week and 2 cents the second week, and continue doubling your deposit every week for thirty weeks, how much will you have deposited by the end of that time?

Solution The sequence of deposits is geometric with common ratio 2. In the 30 weeks, your deposits will total $1 + 2 + 4 + 8 + \cdots + 2^{28} + 2^{29}$. Let S equal the sum of this finite geometric series. Write the expression for $2S$ under the expression for S and subtract S from $2S$ to obtain

$$\begin{aligned} S &= 1 + 2 + 4 + 8 + \cdots + 2^{28} + 2^{29} \\ 2S &= \phantom{1 + {}} 2 + 4 + 8 + \cdots + 2^{28} + 2^{29} + 2^{30} \\ \hline S &= 2^{30} - 1 = 1{,}073{,}741{,}823 \end{aligned}$$

The total deposit is 1,073,741,823 cents, or \$10,737,418.23. **Answer**

In general, if S_n is the sum of the first n terms of a geometric sequence with first term a and common ratio r, then we can find a formula for S_n by subtracting r times the series from the series itself, term by term.

$$\begin{aligned} S_n &= a + ar + ar^2 + \cdots + ar^{n-2} + ar^{n-1} \\ rS_n &= \phantom{a + {}} ar + ar^2 + \cdots + ar^{n-2} + ar^{n-1} + ar^n \\ \hline S_n - rS_n &= a + 0 + 0 + \cdots + 0 + 0 - ar^n \end{aligned}$$

Thus, $S_n(1 - r) = a(1 - r^n)$ and

$$S_n = \frac{a(1 - r^n)}{(1 - r)} \quad \text{provided that } 1 - r \neq 0.$$

The following theorem, which summarizes this result, can be proved by induction. See Exercise 26, where a proof is requested.

THEOREM 9 The sum S_n of the first n terms of a geometric series with first term a and common ratio r ($r \neq 1$) is given by

$$S_n = \frac{a(1 - r^n)}{1 - r}.$$

EXAMPLE 5 Find the sum of the first six terms of $5, -10, 20, \ldots$.

Solution $a = 5$, $r = -2$, $n = 6$. By the theorem,

$$S_6 = \frac{5[1 - (-2)^6]}{1 - (-2)} = -105. \quad \textbf{Answer}$$

EXAMPLE 6 Find the sum of the geometric series in which the first term is 3, the last term is $-\frac{3}{128}$, and the common ratio is $-\frac{1}{2}$.

Solution $a = 3$, $r = -\frac{1}{2}$, and $t_n = -\frac{3}{128}$. To apply Theorem 9, first find n.

$$-\frac{3}{128} = 3 \cdot \left(-\frac{1}{2}\right)^{n-1}$$

$$-\frac{1}{128} = \left(-\frac{1}{2}\right)^{n-1}$$

$$\left(-\frac{1}{2}\right)^{7} = \left(-\frac{1}{2}\right)^{n-1}$$

Thus $7 = n - 1$ and $n = 8$.

By Theorem 9, $S_8 = \dfrac{3(1 - (-\frac{1}{2})^8)}{1 - (-\frac{1}{2})} = \dfrac{255}{128}$. **Answer**

EXERCISES

Find the indicated term for each geometric sequence.

A

1. $a = 3; r = -1; t_6 = ?$

2. $a = \frac{1}{5}; r = \frac{2}{5}; t_8 = ?$

3. $\frac{1}{2}, \frac{1}{7}, \frac{2}{49}, \cdots; t_5 = ?$

4. $\frac{3}{5}, 3, 15, \cdots; t_6 = ?$

5. $\frac{3}{64}, -\frac{3}{16}, \frac{3}{4}, \cdots; t_7 = ?$

6. $-1, 1, -1, \cdots; t_{1001} = ?$

Find the value of the specified variable for each geometric sequence.

7. $a = -2; r = 2; t_n = -64; n = ?$

8. $a = \frac{2}{5}; r = \frac{1}{5}; t_n = \frac{2}{625}; n = ?$

9. $t_7 = 243; r = 3; a = ?$

10. $t_8 = \frac{3}{128}; r = \frac{1}{2}; a = ?$

Find the sum of each geometric series.

11. $a = -5; r = \frac{1}{2}; n = 7$

12. $a = \frac{3}{7}; r = -\frac{1}{3}; n = 4$

13. $a = \frac{1}{10}; t_8 = \frac{64}{5}$

14. $a = -\frac{4}{9}; t_5 = -36$

15. $\sum_{k=1}^{7} 3^{k-1}$

16. $\sum_{j=1}^{5} 7(2)^j$

17. $\sum_{i=1}^{4} 6\left(-\frac{1}{2}\right)^{i-1}$

18. $\sum_{r=1}^{6} 135\left(\frac{2}{3}\right)^{r-1}$

Insert the given number of geometric means between the numbers in each pair.

19. 2; between 5 and 625

20. 4; between $\frac{1}{2}$ and $\frac{1}{64}$

21. 3; between 3 and 27

22. 5; between $\frac{1}{9}$ and 81

23. If $t_3 = 15$ and $t_7 = 240$ in a geometric sequence, find t_{10}.

24. If $t_4 = \frac{1}{2}$ and $t_9 = \frac{1}{64}$, find the sum of the first 12 terms of the geometric sequence.

B

25. Prove Theorem 8 by mathematical induction.

26. Prove Theorem 9 **(a)** by mathematical induction, and **(b)** by using Theorem 2 in Section 3-1.

In Exercises 27–32 three of the numbers a, t_n, r, n, and S_n are given. Find the values of the others.

27. $a = -8$; $n = 3$; $S_n = -8$

28. $r = \frac{1}{2}$; $n = 7$; $S_n = \frac{381}{4}$

29. $a = -\frac{3}{2}$; $r = -2$; $S_n = -4.5$

30. $t_n = 768$; $r = 4$; $n = 5$

31. $a = \frac{8}{3}$; $r = \frac{3}{2}$; $t_n = \frac{81}{4}$

32. $a = 75$; $r = 1.4$; $t_n = 288.12$

33. For what value or values of v will $2v + 3$ be the geometric mean of v and $3v + 18$?

34. Prove that if a and b are positive real numbers, then the geometric mean of a and b is $\sqrt{ab}$.

35. Under conditions favorable to the growth of a certain type of bacteria, one organism can divide into two every half-hour. How many times the original number of organisms will there be at the end of a six-hour period?

36. If \$100 is deposited in a savings account paying 5% interest compounded yearly, then after one year the account will have \$100 + 0.05(\$100) = 1.05(\$100). After another year, the account will have 1.05(\$100) + 0.05(1.05)(\$100) = (1.05)(1.05)(\$100). Find a formula for the amount in the account after n years. Find the amount in the account after 17 years.

37. If the population of a certain country is 8 million people, and it is growing at 3% per year, what will be the country's population in 20 years? (*Hint*: See Exercise 36.)

38. A certain kind of light filter reduces the intensity of light going through it by 30%. By what percent is the intensity of light reduced by two such filters, placed one on top of the other? By what percent is intensity reduced by 25 such filters?

39. A certain rubber ball when dropped bounces in such a way that the height of the first bounce is 80% of the initial height, and the height of each bounce after that is 80% of the height of the previous bounce.
   a. Find the height of the fifth bounce if the ball is dropped from an initial height of 20 ft.
   b. Find a formula for the height of the nth bounce if the ball is dropped from an initial height of h ft.

40. A car purchased for $9500 depreciates 15% in value every year. Find the value of the car at the end of a four-year period.

41. One third of the air in a tank is removed with each stroke of a vacuum pump. What part of the original amount of air remains in the tank after five strokes?

42. Show that the sequence formed by the reciprocals of the terms in a geometric sequence is geometric and find its common ratio.

A natural number n can be written in base b, $b > 0$, if there is a positive integer k and there are integers a_i with $0 \le a_i < b$ for $i = 0, 1, \ldots, k$ such that

$$n = a_k b^k + a_{k-1} b^{k-1} + \cdots + a_1 b^1 + a_0 b^0.$$

Such a number is denoted $(a_k a_{k-1} \cdots a_1 a_0)_b$.

C

43. Write 97 as a number in base 3.

44. Write $(13402)_5$ as a decimal number.

45. Prove that the largest number that can be represented in base 2 with k digits is $2^k - 1$.

46. Prove that the largest number that can be represented in base b with k digits is $b^k - 1$. (*Hint*: Since $a_i < b$ for all i, $a_i b^i < b^{i+1}$.)

47. Use a calculator to compute $S_1, S_2, S_3, \ldots, S_{10}$ for $\frac{1}{2} + \frac{1}{4} + \frac{1}{8} + \cdots$.

48. a. Compute S_{100} and S_{1000} for the series in Exercise 47 by formula.
   b. Describe the behavior of S_n as n becomes larger and larger.

COMPUTER EXERCISES

1. Write a computer program to find the nth term of a geometric sequence and the sum of the first n terms of the corresponding geometric series when the user supplies the first term, the common ratio, and the number of terms.

2. Run the program in Exercise 1 for each value of a, r, and n.
   a. $a = 1; r = 2; n = 10$
   b. $a = 1; r = -2; n = 10$
   c. $a = 1; r = 0.1; n = 35$
   d. $a = 1; r = -0.1; n = 35$

Limits and Sequences

3-6 LIMITS OF SEQUENCES

In this section we shall study the behavior of infinite sequences

$$a_1, a_2, a_3, \ldots, a_n, \ldots$$

as n becomes very large.

Consider for example the sequence with nth term $a_n = \frac{n}{n+1}$:

$$\frac{1}{2}, \frac{2}{3}, \frac{3}{4}, \frac{4}{5}, \ldots, \frac{99}{100}, \ldots$$

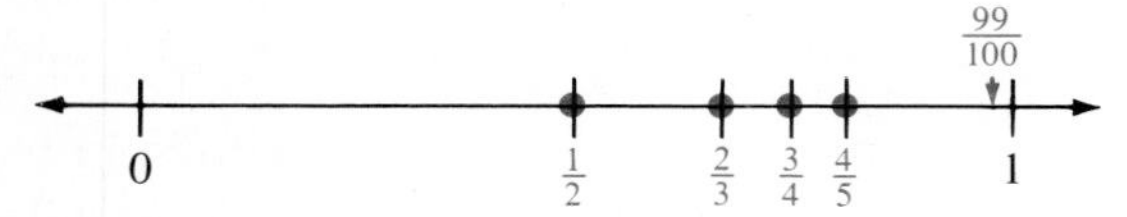

The number line shows the graphs of some of the terms. It appears that the terms "far out" in the sequence cluster near the number 1. That is, it appears that all terms from a_n on are as close to 1 as we please provided that we use large enough values of n.

How large should n be if we wish a_n to be within $\frac{1}{100}$ of 1? That is, for what values of n will $|a_n - 1| < \frac{1}{100}$?

$$|a_n - 1| = \left|\frac{n}{n+1} - 1\right| = \left|\frac{n - 1(n+1)}{n+1}\right| = \left|\frac{-1}{n+1}\right| = \frac{1}{n+1}$$

Now $\frac{1}{n+1} < \frac{1}{100}$ if $n > 99$. Therefore $|a_n - 1| < \frac{1}{100}$ whenever $n > 99$.

We can make $|a_n - 1|$ even smaller by requiring that n be even larger. Indeed, no matter how small a positive number h is given as a bound for $|a_n - 1|$, we can find a number M such that

$$|a_n - 1| < h \text{ whenever } n > M.$$

To see this, note that $|a_n - 1| < h$ is equivalent to $\frac{1}{n+1} < h$. Solving the latter inequality for n, we obtain $n > \frac{1}{h} - 1$. Thus we can let $M = \frac{1}{h} - 1$. We describe this situation by saying that the *limit* of $\frac{n}{n+1}$ as n increases without bound is 1, and we write

$$\lim_{n\to\infty} \frac{n}{n+1} = 1.$$

The limit concept is the basis of that branch of mathematics called *analysis*. The various limits that we will meet in this book all depend on the following basic definition.

DEFINITION

The infinite sequence $a_1, a_2, \ldots, a_n, \ldots$ has A as **limit**, written

$$\lim_{n\to\infty} a_n = A,$$

if for each positive number h, there is a number M such that $|a_n - A| < h$ whenever $n > M$.

A sequence that has a limit is said to be **convergent** and to **converge** to that limit. We speak of *the* limit of such a sequence because it can be shown that a sequence cannot have more than one limit. You are asked to prove this in Exercise 26.

The sequence $1, \frac{1}{2}, \frac{1}{3}, \frac{1}{4}, \ldots, \frac{1}{n}, \ldots$, called the **harmonic sequence**, is of fundamental importance in calculating limits. A **constant sequence** is one in which all the terms are the same. The limits of the harmonic sequence and constant sequences are given in the following theorem.

THEOREM 10 1. $\lim_{n\to\infty} \frac{1}{n} = 0.$

2. If $a_n = c$ for all $n \in N$, then $\lim_{n\to\infty} a_n = c$.

Proof: 1. Given any positive number h, let $M = \frac{1}{h}$. Then whenever $n > M$,

$$\left|\frac{1}{n} - 0\right| = \frac{1}{n} < \frac{1}{M} = h. \quad ■$$

2. Given any positive number h, let M be arbitrary. Then if $n > M$,

$$|a_n - c| = |c - c| = 0 < h. \quad ■$$

Theorems 10 and 11 enable us to find limits of combinations of sequences.

THEOREM 11 If $\lim_{n\to\infty} a_n = A$, $\lim_{n\to\infty} b_n = B$, and c is a real number,

1. $\lim_{n\to\infty} ca_n = c \lim_{n\to\infty} a_n = cA$
2. $\lim_{n\to\infty} (a_n + b_n) = \lim_{n\to\infty} a_n + \lim_{n\to\infty} b_n = A + B$
3. $\lim_{n\to\infty} (a_n - b_n) = \lim_{n\to\infty} a_n - \lim_{n\to\infty} b_n = A - B$
4. $\lim_{n\to\infty} (a_n b_n) = (\lim_{n\to\infty} a_n)(\lim_{n\to\infty} b_n) = AB$
5. $\lim_{n\to\infty} \left(\frac{a_n}{b_n}\right) = \dfrac{\lim_{n\to\infty} a_n}{\lim_{n\to\infty} b_n} = \frac{A}{B}$ provided $B \neq 0$ and $b_n \neq 0$ for all $n \geq 1$.

In particular, according to Theorem 11, "the limit of a sum (difference) is the sum (difference) of the limits."

EXAMPLE Find each limit if it exists.

a. $\lim_{n\to\infty}\left(\frac{3n^2 + 2n + 4}{n^2 + 1}\right)$ **b.** $\lim_{n\to\infty}\left(\frac{n - 5}{n^3 + 2n}\right)$ **c.** $\lim_{n\to\infty}\left(\frac{n^2 - 3n + 1}{n + 8}\right)$

Solution **a.** Divide the numerator and the denominator by the greatest power of n, in this case n^2.

$$\frac{3n^2 + 2n + 4}{n^2 + 1} = \frac{\frac{3n^2 + 2n + 4}{n^2}}{\frac{n^2 + 1}{n^2}} = \frac{3 + \frac{2}{n} + \frac{4}{n^2}}{1 + \frac{1}{n^2}}$$

Therefore $\lim_{n\to\infty}\left(\frac{3n^2 + 2n + 4}{n^2 + 1}\right) = \lim_{n\to\infty}\left(\frac{3 + \frac{2}{n} + \frac{4}{n^2}}{1 + \frac{1}{n^2}}\right)$

Apply part (5) and part (1) of Theorem 11, to obtain

$$\frac{\lim_{n\to\infty} 3 + \lim_{n\to\infty} \frac{2}{n} + \lim_{n\to\infty} \frac{4}{n^2}}{\lim_{n\to\infty} 1 + \lim_{n\to\infty} \frac{1}{n^2}}.$$

Now apply part (1) of Theorem 11:

$$\lim_{n\to\infty} \frac{3n^2 + 2n + 4}{n^2 + 1} = \frac{\lim_{n\to\infty} 3 + 2\lim_{n\to\infty} \frac{1}{n} + 4\lim_{n\to\infty} \frac{1}{n^2}}{\lim_{n\to\infty} 1 + \lim_{n\to\infty} \frac{1}{n^2}} = 3 \quad \textbf{Answer}$$

b. Apply the same technique as in part (a). Divide the numerator and denominator by n^3.

$$\lim_{n\to\infty}\left(\frac{n - 5}{n^3 + 2n}\right) = \lim_{n\to\infty}\left(\frac{\frac{1}{n^2} - \frac{5}{n^3}}{1 + \frac{2}{n^2}}\right) = \frac{0 - 0}{1 + 0} = 0 \quad \textbf{Answer}$$

c. Divide the numerator and the denominator by n^2:

$$\frac{n^2 - 3n + 1}{n + 8} = \frac{1 - \frac{3}{n} + \frac{1}{n^2}}{\frac{1}{n} + \frac{8}{n^2}}$$

Since the limit of the numerator is nonzero and the limit of the denominator is zero, the limit of the sequence does not exist. **Answer**

Long division could also be used to solve the problem.

$$\frac{n^2 - 3n + 1}{n + 8} = n - 11 + \frac{89}{n + 8}$$

So $$\lim_{n\to\infty}\left(\frac{n^2 - 3n + 1}{n + 8}\right) = \lim_{n\to\infty}\left(n - 11 + \frac{89}{n + 8}\right) = \lim_{n\to\infty} n - 11$$

Since $\lim_{n\to\infty} n$ does not exist, the limit of the given quotient does not exist.

A sequence that has no limit is said to **diverge**, or be **divergent**. The sequence in part (c) of the example is divergent.

Geometric sequences play an important role in analysis.

THEOREM 12 If $|r| < 1$, then $\lim_{n\to\infty} r^n = 0$.

Proof: Let h be any positive number. Since $|r| < 1$, there is a positive number a such that $|r| = \frac{1}{1 + a}$. Recall from Theorem 3 that $(1 + a)^n \geq 1 + na$. Therefore

$$|r^n - 0| = |r|^n = \frac{1}{(1 + a)^n} < \frac{1}{1 + na} < \frac{1}{na} < h$$

whenever $n > \frac{1}{ah}$. ■

EXERCISES

Find the limit of each sequence if it exists. If the limit does not exist, so state.

A

1. $a_n = \dfrac{n - 1}{2n}$
2. $a_n = \dfrac{n^2 + 3}{5n - 1}$
3. $a_n = \dfrac{n^3 - 1}{2 + n^2}$
4. $a_n = \dfrac{-4n^2}{5n^2 + 3n}$
5. $a_n = \dfrac{6n^2 - 5n - 4}{8n^2 + n - 7}$
6. $a_n = \left(\dfrac{1}{n} - 4\right)\left(\dfrac{3}{n} - 1\right)$
7. $a_n = \dfrac{3n^2 - 5n + 2}{n + 2}$
8. $a_n = \dfrac{9n^2 - 4}{3n + 2}$
9. $a_n = \left(\dfrac{n + 1}{n + 5}\right)\left(\dfrac{2n - 4}{3n - 1}\right)$
10. $a_n = \dfrac{6 - n}{n + 3} + \dfrac{2n + 5}{3n + 2}$
11. $a_n = \dfrac{2n^2 + 3}{n^3} - \dfrac{3n^3}{n^4 + 1}$
12. $a_n = \left(\dfrac{3n^3 + 4n + 2}{n^2 + 1}\right)\left(\dfrac{5n^2 - 3}{2n^4 + 7}\right)$

In Exercises 13–20 tell whether the given statement is true or false. If it is false, give a counterexample to show that it is false.

13. A term of a convergent sequence can equal the limit of the sequence.
14. If the limit of a convergent sequence is a positive number, all the terms of the sequence are positive numbers.
15. If the terms of a sequence are alternately positive and negative numbers, the sequence cannot be convergent.
16. If a_n is the nth term of a convergent sequence with limit A, then for every two successive terms a_k and a_{k+1}, $|a_k - A| \geq |a_{k+1} - A|$.

Tell whether the given statement is true or false. If it is false, give a counterexample to show that it is false.

17. If $\lim_{n\to\infty} a_n = 0$, then for every n, $a_n = 0$.

18. If $\lim_{n\to\infty} a_n$ exists, then $\lim_{n\to\infty} |a_n|$ exists.

19. If $\lim_{n\to\infty} |a_n|$ exists, then $\lim_{n\to\infty} a_n$ exists.

20. If $a_n \le b_n$ for all n and $\lim_{n\to\infty} b_n$ exists, then $\lim_{n\to\infty} a_n$ exists.

B **21.** If $a_n = \dfrac{n^2}{2n^2 + 1}$, then $\lim_{n\to\infty} a_n = \dfrac{1}{2}$. Find the smallest positive integer n such that $\left|\dfrac{1}{2} - a_n\right| < \dfrac{1}{6}$.

22. If $a_n = \dfrac{3n}{2n^2 + 7}$, then $\lim_{n\to\infty} a_n = 0$. Find the smallest positive integer n such that $|0 - a_n| < \dfrac{1}{5}$.

23. Find two divergent sequences $a_1, a_2, \ldots, a_n, \ldots$ and $b_1, b_2, \ldots, b_n, \ldots$ such that $\lim_{n\to\infty} (a_n + b_n) = 3$.

24. Find two divergent sequences $a_1, a_2, \ldots, a_n, \ldots$ and $b_1, b_2, \ldots, b_n, \ldots$ such that $\lim_{n\to\infty} \left(\dfrac{a_n}{b_n}\right) = 2$.

25. Find one convergent sequence $a_1, a_2, \ldots, a_n, \ldots$ and one divergent sequence $b_1, b_2, \ldots, b_n, \ldots$ such that $a_1b_1, \ldots, a_nb_n, \ldots$ converges.

C **26.** Prove that if a sequence with nth term a_n has a limit, the limit is unique.

27. Prove that $-1, 1, -1, \ldots$ is divergent by completing parts (a) and (b).

a. Let $h = \frac{1}{2}$ and assume that there is a limit A. Show that

$$A - \frac{1}{2} < (-1)^n < A + \frac{1}{2}.$$

b. Show that there is no integer M such that the inequality in part (a) holds for all $n \ge M$.

A sequence $a_1, a_2, \ldots, a_n, \ldots$ increases without bound if for every T (no matter how large) there is an $M \in N$ such that $a_n > T$ for all $n \ge M$. This is written $\lim_{n\to\infty} a_n = +\infty$.

28. Prove that $\lim_{n\to\infty} (n - 11) = +\infty$.

29. State an analogous definition for $\lim_{n\to\infty} a_n = -\infty$.

30. Prove that $\lim_{n\to\infty} (-n^2 + 1) = -\infty$.

31. Prove that if $a_n = \dfrac{s_m n^m + s_{m-1} n^{m-1} + \cdots + s_1 n + s_0}{t_k n^k + t_{k-1} n^{k-1} + \cdots + t_1 n + t_0}$ $(t_k \ne 0)$, then $\lim_{n\to\infty} a_n = 0$ if $m < k$ and $\lim_{n\to\infty} a_n = \dfrac{s_m}{t_k}$ if $m = k$.

3-7 INFINITE SERIES

In Section 3-6 we discussed convergence and divergence of infinite sequences. In this section we shall discuss *infinite series* and see what sense can be made of the sum. We use the notation $\sum_{i=1}^{\infty} a_i$ to denote such a series.

Consider $\sum_{i=1}^{\infty} \frac{1}{i(i+1)} = \frac{1}{1 \cdot 2} + \frac{1}{2 \cdot 3} + \frac{1}{3 \cdot 4} + \cdots + \frac{1}{n(n+1)} + \cdots$.

Let S_n be the sum of the first n terms of the series. Then

$$S_1 = \frac{1}{2},$$

$$S_2 = \frac{1}{2} + \frac{1}{6} = \frac{2}{3},$$

$$S_3 = \frac{2}{3} + \frac{1}{12} = \frac{3}{4},$$

$$S_4 = \frac{3}{4} + \frac{1}{20} = \frac{4}{5}.$$

In Exercise 45 you are asked to show that

$$S_n = \frac{n}{n+1} \text{ and } \lim_{n \to \infty} S_n = 1.$$

It therefore makes sense to say that

$$\sum_{i=1}^{\infty} \frac{1}{i(i+1)} = 1.$$

In general, the **nth partial sum** S_n of an infinite series $\sum_{i=1}^{\infty} a_i$ is $\sum_{i=1}^{n} a_i$. If the sequence of partial sums $S_1, S_2, S_3, \ldots, S_n, \ldots$ converges with limit S, then we define the sum $\sum_{i=1}^{\infty} a_i$ to be the number S.

If the sequence of partial sums of an infinite series converges, then the series **converges**, or is **convergent**. If the sequence of partial sums **diverges**, then the series diverges, or is **divergent**.

THEOREM 13 The series $\sum_{n=1}^{\infty} \frac{1}{n}$, called the **harmonic series**, diverges.

Proof: We will show that the sequence of partial sums diverges.

$$S_1 = 1$$

$$S_4 = 1 + \frac{1}{2} + \left(\frac{1}{3} + \frac{1}{4}\right)$$

$$> 1 + \frac{1}{2} + \left(\frac{1}{4} + \frac{1}{4}\right) = \frac{4}{2}$$

Proof continued on following page

$$S_8 = 1 + \frac{1}{2} + \left(\frac{1}{3} + \frac{1}{4}\right) + \left[\frac{1}{5} + \frac{1}{6} + \frac{1}{7} + \frac{1}{8}\right]$$
$$> 1 + \frac{1}{2} + \left(\frac{1}{4} + \frac{1}{4}\right) + \left[\frac{1}{8} + \frac{1}{8} + \frac{1}{8} + \frac{1}{8}\right] = \frac{5}{2}$$

and $$S_{2^k} > \frac{k+2}{2}.$$

The complete sequence $S_1, S_2, S_3, \ldots, S_n, \ldots$ is unbounded and hence the harmonic series diverges. ■

Unlike the harmonic series, the geometric series $1 + \frac{1}{2} + \frac{1}{4} + \cdots + \left(\frac{1}{2}\right)^{n-1} + \cdots$ converges. Theorem 9 gives a formula for S_n.

$$S_n = \frac{1 - (\frac{1}{2})^n}{1 - \frac{1}{2}} = 2\left(1 - \left(\frac{1}{2}\right)^n\right) = 2 - 2\left(\frac{1}{2}\right)^n$$

Using Theorem 12 with $r = \frac{1}{2}$, we have that $\lim_{n\to\infty}\left(\frac{1}{2}\right)^n = 0$. Thus $\lim_{n\to\infty} S_n = 2$.

Therefore $\sum_{n=1}^{\infty}\left(\frac{1}{2}\right)^{n-1} = 2.$

The general results for convergence or divergence of infinite geometric series are stated below.

THEOREM 14 Given the infinite geometric series

$$\sum_{i=1}^{\infty} ar^{i-1} = a + ar + ar^2 + \cdots + ar^{n-1} + \cdots \quad (a \neq 0).$$

The series converges with sum $\frac{a}{1-r}$ if $|r| < 1$, and the series diverges if $|r| \geq 1$.

Proof: By Theorem 9.

$$S_n = \frac{a(1 - r^n)}{1 - r} = \frac{a}{1 - r} - \left(\frac{a}{1 - r}\right)r^n \quad (r \neq 1).$$

If $|r| < 1$, then by Theorem 12 $\lim_{n\to\infty} r^n = 0$. Therefore $\lim_{n\to\infty} S_n = \frac{a}{1-r}$.

If $|r| > 1$, then $\lim_{n\to\infty} r^n$ does not exist. Therefore the given series diverges.

If $r = 1$, we have $a + a + a + \cdots$. Thus $S_n = na$. Since $\lim_{n\to\infty} na$ does not exist, the series diverges.

If $r = -1$, we have $a - a + a - \cdots$. Thus $S_n = \begin{cases} 1 \text{ if } n \text{ is odd} \\ 0 \text{ if } n \text{ is even.}\end{cases}$

Since $\lim_{n\to\infty} S_n$ does not exist, the series diverges. ■

EXAMPLE 1 Find each sum if it exists.

a. $\sum_{k=1}^{\infty} 5\left(-\frac{1}{6}\right)^{k-1}$ **b.** $\sum_{k=2}^{\infty} \left(\frac{3}{5}\right)^{k}$ **c.** $\sum_{k=1}^{\infty} \left(\frac{1}{1000}\right) \cdot 2^k$

Solution Each series is an infinite geometric series.

a. $\sum_{k=1}^{\infty} 5\left(-\frac{1}{6}\right)^{k-1} = 5 - \frac{5}{6} + \cdots$

Here $a = 5$ and $r = -\frac{1}{6}$. Since $\left|-\frac{1}{6}\right| < 1$, by Theorem 14

$$S = \frac{a}{1 - r} = \frac{5}{1 - \left(-\frac{1}{6}\right)} = \frac{5}{\frac{7}{6}} = \frac{30}{7}. \quad \textbf{Answer}$$

b. $\sum_{k=2}^{\infty} \left(\frac{3}{5}\right)^{k} = \left(\frac{3}{5}\right)^2 + \left(\frac{3}{5}\right)^3 + \left(\frac{3}{5}\right)^4 + \cdots$

Here $a = \left(\frac{3}{5}\right)^2$ and $r = \frac{3}{5}$. Therefore the sum is

$$S = \frac{a}{1 - r} = \frac{\left(\frac{3}{5}\right)^2}{1 - \frac{3}{5}} = \frac{9}{10}. \quad \textbf{Answer}$$

c. Since the common ratio is 2 and $|2| > 1$, the series diverges.

We conclude this section with some remarks on *infinite decimals*. Recall that a **finite decimal** is an expression of the form

$$a_r a_{r-1} \ldots a_1 a_0 . b_1 b_2 \ldots b_n$$

where each a_i and b_i is an integer, $0 \le a_i \le 9$, $0 \le b_i \le 9$. The decimal form represents

$$a_r 10^r + a_{r-1} 10^{r-1} + \cdots + a_1 10 + a_0 + \frac{b_1}{10} + \frac{b_2}{10^2} + \cdots + \frac{b_n}{10^n}.$$

We can extend the notion of finite decimal to the notion of infinite decimal. For example, consider $0.3333\ldots$ (compactly written $0.\overline{3}$, where the bar indicates the digits that repeat). By $0.\overline{3}$ we mean

$$\frac{3}{10} + \frac{3}{10^2} + \frac{3}{10^3} + \cdots.$$

This series is an infinite geometric series with $a = \frac{3}{10}$ and $r = \frac{1}{10}$. By Theorem 14, the series converges and has sum

$$\frac{a}{1 - r} = \frac{\frac{3}{10}}{1 - \frac{1}{10}} = \frac{\frac{3}{10}}{\frac{9}{10}} = \frac{1}{3}.$$

Hence the infinite decimal $0.\overline{3}$ represents the rational number $\frac{1}{3}$.

Using long division, we can represent a rational number as a repeating decimal. For example, $\frac{5}{11} = 0.4545\ldots = 0.\overline{45}$.

THEOREM 15 Every repeating decimal represents a rational number. Conversely, every rational number can be written as a repeating decimal.

EXAMPLE 2 Express $4.1\overline{36}$ as a fraction in the form $\frac{a}{b}$ where a and b are relatively prime.

Solution

$$4.1\overline{36} = 4.13636\ldots$$
$$= 4 + \frac{1}{10} + \frac{3}{10^2} + \frac{6}{10^3} + \frac{3}{10^4} + \frac{6}{10^5} + \cdots$$
$$= 4 + \frac{1}{10} + \left[\frac{36}{10^3} + \frac{36}{10^5} + \frac{36}{10^7} + \cdots\right]$$

The expression in brackets is an infinite geometric series with $a = \frac{36}{10^3}$ and $r = \frac{1}{10^2}$.

$$\text{Therefore } 4.1\overline{36} = \frac{41}{10} + \frac{\frac{36}{10^3}}{1 - \frac{1}{10^2}}$$
$$= \frac{41}{10} + \frac{\frac{36}{1000}}{\frac{99}{100}} = \frac{41}{10} + \frac{2}{55} = \frac{91}{22}.$$ **Answer**

EXERCISES

Find the sum of the infinite geometric series with the specified first term and common ratio.

A

1. $a = 7,\ r = \frac{1}{3}$

2. $a = 6,\ r = -\frac{1}{4}$

3. $a = -11,\ r = \frac{\sqrt{3}}{6}$

4. $a = -\sqrt{2},\ r = -\frac{\sqrt{3}}{2}$

Find the sum of each of the infinite geometric series.

5. $\sum_{k=1}^{\infty} 8\left(-\frac{1}{2}\right)^{k-1}$

6. $\sum_{k=1}^{\infty} 10\left(\frac{2}{3}\right)^{k}$

7. $0.3 - 0.03 + 0.003 - \cdots$

8. $36 + 27 + \frac{81}{4} + \cdots$

9. $35 - \frac{35}{\sqrt{6}} + \frac{35}{6} - \cdots$

10. $\frac{2}{5} + \frac{2}{200} + \frac{2}{8000} + \cdots$

Write the first three terms of the infinite geometric sequence for which

11. $a = \frac{9}{2},\ S = \frac{15}{2}$

12. $a = 0.5,\ S = 0.375$

13. $r = -\frac{3}{4},\ S = 16$

14. $r = 0.1,\ S = \frac{5}{11}$

Write each fraction as a repeating decimal.

15. $\frac{19}{48}$ **16.** $\frac{7}{4}$ **17.** $\frac{33}{25}$ **18.** $\frac{107}{450}$

Write each repeating decimal as a fraction.

19. $0.\overline{14}$ **20.** $0.1\overline{45}$ **21.** $6.\overline{296}$ **22.** $0.1\overline{37}$

In Exercises 23–25 support your answer with an explanation.

23. If two infinite geometric series have the same finite sum, must they be the same series?

24. Can the sum of a convergent infinite geometric series be less than the first term?

25. Can the product of two repeating decimals be a nonrepeating decimal?

In Exercises 26–31 the nth partial sum S_n of a series is given. By investigating $\lim_{n \to \infty} S_n$ determine whether the series is convergent or divergent. When convergent, give the sum.

26. $S_n = \frac{1}{2}(n + 1)(n - 1)$

27. $S_n = 1 - \frac{1}{n}$

28. $S_n = \frac{n}{3n - 1}$

29. $S_n = \frac{2n^2 - n}{3n + 4}$

30. $S_n = \frac{n - 1}{3n^2 - 1}$

31. $S_n = \left(\frac{1}{10}\right)^{2n}$

In Exercises 32–35 find the range of values of x for which each geometric series converges.

B **32.** $8 + 8(x - 3) + 8(x - 3)^2 + \cdots$

33. $1 + 3(x + 1) + 9(x + 1)^2 + \cdots$

34. $1 - \frac{1}{2}x^2 + \frac{1}{4}x^4 + \cdots$

35. $1 + \frac{1}{8}(2 - x)^3 + \frac{1}{64}(2 - x)^6 + \cdots$

For what value(s) of x does each infinite geometric series have the given sum?

36. $\frac{2}{3} = 1 + 3x + 9x^2 + \cdots$

37. $0.8 = 1 - \frac{1}{2}x + \frac{1}{4}x^2 - \cdots$

38. $\frac{x + 2}{2x} = x + x^2 + x^3 + \cdots$

39. $\frac{x + 2}{9x} = x + x^2 + x^3 + \cdots$

40. Find r for the infinite geometric series in which $S = \frac{4 + 3\sqrt{2}}{2}$ and $a = \sqrt{2} + 1$.

41. Prove: If $\sum_{n=1}^{\infty} u_n$ converges with a sum S and k is any constant, then $\sum_{n=1}^{\infty} ku_n$ converges with a sum kS.

42. Prove: If $\sum_{k=1}^{\infty} t_k$ and $\sum_{k=1}^{\infty} p_k$ are convergent series with sums T and P respectively, then $\sum_{k=1}^{\infty} (t_k + p_k)$ and $\sum_{k=1}^{\infty} (t_k - p_k)$ are convergent and $T + P$ and $T - P$ are their respective sums.

43. Given a square of side 20 in. The midpoints of its sides are joined to form an inscribed square. The midpoints of this second square are joined to form a third square. If the process is continued endlessly, find the sum of the perimeters and the sum of the areas of all the squares, including the initial one.

44. Given the infinite geometric series $a + ar + ar^2 + ar^3 + \cdots + ar^{n-1} + \cdots$. $(a > 0)$. If $0 < r < 1$, what fractional part of the sum is the sum of the odd-numbered terms? the even-numbered terms?

C 45. Show that the nth partial sum of $\sum_{i=1}^{\infty} \frac{1}{i(i+1)}$ is given by $S_n = \frac{n}{n+1}$. Use this form of S_n to show that the series is convergent and find its sum. $\left(\textit{Hint: } \frac{1}{i(i+1)} = \frac{1}{i} - \frac{1}{i+1}\right)$

46. Show that the following series converges and find its sum.

$$\left(\frac{1}{2} + \frac{1}{3}\right) + \left(\frac{1}{4} + \frac{1}{9}\right) + \cdots + \left(\frac{1}{2^n} + \frac{1}{3^n}\right) + \cdots$$

(*Hint*: Express S_n as the sum of two separate geometric series.)

47. Prove that if $\sum_{n=1}^{\infty} a_n$ converges, then $\lim_{n \to \infty} a_n = 0$.

COMPUTER EXERCISES

1. Write a computer program to find the value of $\sum_{k=1}^{n} \frac{1}{k^2}$ when the user supplies a value of n.
2. Run the program for each value of n.
   a. $n = 100$ **b.** $n = 120$ **c.** $n = 150$ **d.** $n = 250$
3. Modify the program in Exercise 1 to find the smallest value of n such that two successive sums differ by less than 10^{-5}.
4. Compare the value of either sum obtained from n in Exercise 3 with $\frac{\pi^2}{6}$.
5. Write a computer program to find the sum $1 - \frac{1}{3} + \frac{1}{5} + \cdots + \frac{(-1)^{n-1}}{2n-1}$ when the user supplies a value of n.
6. Run the program for $n = 10$; $n = 20$; $n = 30$; $n = 50$.
7. What is the apparent limit of the series in Exercise 5?

3-8 THE AXIOM OF COMPLETENESS

In this section we discuss the axiom of completeness and thus characterize $\mathcal{R}$ as a complete ordered field. Consider

$$\sum_{i=1}^{\infty} \frac{1}{i!} = 1 + \frac{1}{2 \cdot 1} + \frac{1}{3 \cdot 2 \cdot 1} + \cdots + \frac{1}{i!} + \cdots.$$

$S_1 = 1$	$S_6 \approx 1.718055554$
$S_2 = 1 + \frac{1}{2!} = \frac{3}{2} = 1.5$	$S_7 \approx 1.718253966$
$S_3 = \frac{3}{2} + \frac{1}{3!} = \frac{10}{6} \approx 1.66667$	$S_8 \approx 1.718278767$
$S_4 = \frac{10}{6} + \frac{1}{4!} = \frac{41}{24} \approx 1.70833$	$S_9 \approx 1.718281522$
$S_5 = \frac{41}{24} + \frac{1}{5!} = \frac{206}{120} \approx 1.71667$	$S_{10} \approx 1.718281797$

Although we cannot find a formula for S_n, the table suggests that $\lim_{n \to \infty} S_n$ exists.

Notice that for all n

$$1 + \frac{1}{2 \cdot 1} + \frac{1}{3 \cdot 2 \cdot 1} + \cdots + \frac{1}{n!} \le 1 + \frac{1}{2 \cdot 1} + \frac{1}{3 \cdot 2 \cdot 1} + \cdots + \frac{1}{n!} + \frac{1}{(n+1)!}$$

and therefore that

$$S_n \le S_{n+1}.$$

Furthermore

$$1 + \frac{1}{2 \cdot 1} + \frac{1}{3 \cdot 2 \cdot 1} + \cdots + \frac{1}{n!} < 1 + \frac{1}{2 \cdot 1} + \frac{1}{2 \cdot 2 \cdot 1} + \cdots + \left(\frac{1}{2}\right)^{n-1} + \cdots$$

The infinite geometric series has sum 2. Hence

$$S_n < 2.$$

If we plot points corresponding to the first few nth partial sums on a number line, we can make the following observations.

1. The points move only to the right.
2. The points all lie to the left of 2.

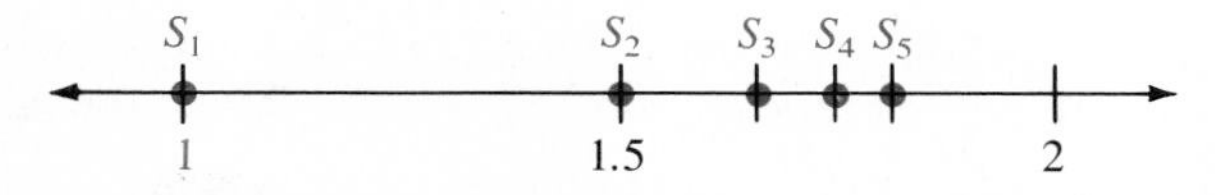

These facts suggest that the sequence of partial sums has a limit. To prove this we need two definitions and the *axiom of completeness*.

DEFINITION

A sequence $a_1, a_2, a_3, \ldots, a_n, \ldots$ is **nondecreasing** if for all n

$$a_{n+1} \geq a_n.$$

A sequence $a_1, a_2, a_3, \ldots, a_n, \ldots$ is **bounded** if there is a real number B such that

$$|a_n| \leq B$$

for all n. We call B a bound for the sequence.

In our example the sequence of partial sums is nondecreasing and 2 is a bound for the sequence.

The Axiom of Completeness

Every nondecreasing and bounded sequence of real numbers converges to a real number.

In Sections 1-4 and 1-5 we stated the axioms for ordered fields. One ordered field is, of course, the system $\mathcal{R}$ of real numbers. The system Q of rational numbers is another such field. There is, however, *only one* complete ordered field, $\mathcal{R}$ itself.

We now know from the axiom of completeness that the series $\sum_{i=1}^{\infty} \frac{1}{i!}$, with which we began this section, has a sum.

Let us return to the question of infinite decimals. Consider the decimal

$$a_r a_{r-1} a_{r-2} \ldots a_1 a_0 . b_1 b_2 \ldots b_n \ldots,$$

which is another way of writing

$$a_r \cdot 10^r + a_{r-1} 10^{r-1} + \cdots + a_1 \cdot 10 + a_0$$
$$+ \frac{b_1}{10} + \frac{b_2}{10^2} + \cdots + \frac{b_n}{10^n} + \cdots.$$

The expression in red is a finite sum and therefore equals a real number, A. The remainder of the expression is an infinite series with nondecreasing partial sums since $b_i \geq 0$ for all i. Furthermore,

$$S_n = \frac{b_1}{10} + \frac{b_2}{10^2} + \cdots + \frac{b_n}{10^n} \leq \frac{9}{10} + \frac{9}{10^2} + \frac{9}{10^3} + \cdots + \frac{9}{10^n}$$
$$\leq \frac{9}{10} + \frac{9}{10^2} + \frac{9}{10^3} + \cdots + \frac{9}{10^n} + \cdots = 1.$$

The infinite series is geometric with sum 1. Therefore S_n is bounded by 1, and thus by the axiom of completeness, $\lim_{n\to\infty} S_n$ exists. Thus the given decimal represents $A + \lim_{n\to\infty} S_n$, a real number.

This argument shows that every decimal represents a real number. The converse is also true.

THEOREM 16 Every decimal represents a real number, and every real number can be expressed as a decimal.

An irrational number is any real number that is not rational. It is possible to construct irrational numbers easily. We need only write infinite decimals that do not repeat. For example,

$$0.1010010001\ldots \qquad 0.2020020002\ldots$$
$$0.1121231234\ldots \qquad 0.1234567891011\ldots$$

are all irrational numbers.

In Section 1-7 we showed that $\sqrt{2}$ could not be written as a quotient of two integers. We will now show that $\sqrt{2}$ is a real number by finding a decimal representation for it. This will also show that $S^2 = 2$ has a unique positive real root.

The largest integer a such that $a^2 \le 2$ is 1. Now we look for the largest integer b_1 such that $\left(1 + \frac{b_1}{10}\right)^2 \le 2$. Since $(1.4)^2 = 1.96 < 2$ and $(1.5)^2 = 2.25 > 2$, b_1 is 4. Next we look for the largest integer b_2 such that $\left(1.4 + \frac{b_2}{10^2}\right)^2 \le 2$. By trial we find that $b_2 = 1$. Continuing in this way we obtain a sequence of finite decimals

$$1.4,\ 1.41,\ 1.414,\ 1.4142,\ \ldots,\ S_n,\ \ldots$$

satisfying $S_n^2 \le 2$. The sequence $S_1, S_2, \ldots, S_n$ is nondecreasing and bounded. By the axiom of completeness, the sequence has a limit S.

The sequence

$$1.5,\ 1.42,\ 1.415,\ 1.4143,\ \ldots,\ T_n,\ \ldots$$

where $T_n = S_n + \frac{1}{10^n}$ is convergent and $T_n^2 \ge 2$.

$$\lim_{n\to\infty} T_n = \lim_{n\to\infty} S_n + \lim_{n\to\infty}\left(\frac{1}{10}\right)^n = S + 0 = S$$

Finally, since $S_n^2 \le 2 \le T_n^2$, we have

$$\lim_{n\to\infty} S_n^2 \le 2 \le \lim_{n\to\infty} T_n^2.$$

Thus

$$S^2 \le 2 \le S^2.$$

Therefore $S^2 = 2$, and we may write $S = \sqrt{2}$.

In general, the existence of positive integral roots of positive real numbers is guaranteed by the following theorem. Its proof can be modeled on the proof given for the existence of $\sqrt{2}$.

THEOREM 17 If a is a positive real number and $n \in N$, then there exists a unique positive real number x such that $x^n = a$.

We call x the **positive nth root** of a and write $x = \sqrt[n]{a}$. The symbol $\sqrt[n]{a}$ is called a **radical,** a is the **radicand,** and n is the **index**. Thus for example, $x^7 = 5$ has a positive real solution and $x = \sqrt[7]{5}$ is a real number. We define $\sqrt[n]{0}$ to be 0 for all natural numbers n.

THEOREM 18 If a is a negative real number and n is an odd positive integer, then there exists a unique negative real number x such that $x^n = a$.

EXERCISES

In Exercises 1–6, write *true* or *false*. If the statement is false, give a counterexample to show that it is false.

A

1. The axiom of completeness permits us to conclude that a sequence which does not have a limit cannot be a nondecreasing sequence.
2. The partial sums of a series must increase without bound as n increases.
3. If each odd-numbered term of a series is greater than zero and each even-numbered term is less than zero, the series cannot be convergent.
4. In the series $\sum_{n=1}^{\infty} a_n$, $S_n - S_{n-1} = a_n$ for $n > 1$.
5. Every nondecreasing sequence of partial sums is bounded.
6. For all real numbers a, $\sqrt{a^2} = a$.

If $a = 3 + \sqrt{5}$ and $b = 3 - \sqrt{5}$, evaluate each expression and state whether it is rational or irrational.

7. $a + b$ 8. ab 9. $a^2 - b^2$
10. $a - b$ 11. $\sqrt{a} \cdot \sqrt{b}$ 12. $\sqrt{a^2 + b^2}$

Find a rational number r and an irrational number s between the two numbers.

13. $\frac{13}{5}$ and $\sqrt{7}$ 14. $\frac{2}{3}$ and $\frac{5}{6}$ 15. $\sqrt{11}$ and π 16. $\frac{3}{5}$ and $\sqrt{0.38}$

Find the positive root(s) of each equation correct to tenths.

17. $3x^2 = 7$ 18. $(5x + 1)^2 = 18$
19. $(x + 3)^2 = 12$ 20. $(x - 4)^3 = 30$

B

21. Prove: If r is a positive irrational number, then $\sqrt[n]{r}$ is irrational.
22. Prove: The reciprocal of an irrational number is an irrational number.
23. Prove $\frac{1}{n!} < \left(\frac{1}{2}\right)^{n-1}$ for all $n \geq 3$ by mathematical induction.
24. Prove that every nonincreasing bounded sequence $a_1, a_2, a_3, \ldots, a_n, \ldots$ converges. (*Hint*: Consider $-a_1, -a_2, -a_3, \ldots, -a_n, \ldots$.)

25. Give an example of a bounded but divergent infinite sequence.

26. Give an example of a nondecreasing but divergent infinite sequence.

27. Prove that if a and b are positive real numbers, then their geometric mean exists.

28. Prove that $\sqrt{\sqrt{a}}$ is real if $a \geq 0$.

29. In the sequence of partial sums S_n that approximates $\sqrt{2}$, why is $S_n^2 \neq 2$ for any n?

30. If a, b, and c are real numbers and $a \neq 0$, under what conditions will $ax^2 + bx + c = 0$ have a real root?

C 31. Let $L = \{a + b\sqrt{2}: a, b \in Q\}$. Let $+$ and $\cdot$ be defined by

$$(a + b\sqrt{2}) + (c + d\sqrt{2}) = (a + c) + (b + d)\sqrt{2};$$
$$(a + b\sqrt{2}) \cdot (c + d\sqrt{2}) = (ac + 2bd) + (bc + ad)\sqrt{2}.$$

a. Show that L is a commutative group under $+$.
b. Find a multiplicative identity for L.
c. Show that L without the additive identity is a commutative group under multiplication.
d. Show that L with $+$ and $\cdot$ is a field.

32. Show that $x^2 + 1 = 0$ has no real roots.

33. Form an equation with integral coefficients that has $2 + \sqrt[3]{3}$ as one of its roots.

34. Show that $1 + x^2 + x^4 + \cdots + x^{2n} + \cdots = 2$ has two real roots and find them.

35. Find the sum of the infinite geometric series $(\sqrt{3} + \sqrt{2}) + 1 + (\sqrt{3} - \sqrt{2}) + \cdots$ and express your result in a form containing no radicals in the denominator.

36. Prove the *comparison test*. Let $a_k \geq b_k > 0$ for all k.

a. If $\sum_{k=1}^{\infty} a_k$ converges, then $\sum_{k=1}^{\infty} b_k$ converges.

b. If $\sum_{k=1}^{\infty} b_k$ diverges, then $\sum_{k=1}^{\infty} a_k$ diverges.

37. Prove that if $\sum_{k=1}^{\infty} a_k$ converges, then $\lim_{n \to \infty} a_n = 0$. Is the converse true?

3-9 INTEGRAL AND RATIONAL EXPONENTS

The laws of exponents for positive integral exponents were presented in Section 3-1. Now we wish to extend the concept of exponents to include negative integers and all rational numbers. Of course we want the previously derived laws for positive integral exponents to remain valid.

If we wish $\frac{a^m}{a^n} = a^{m-n}$ to hold for all natural numbers m and n, then for example, $\frac{3^2}{3^7} = \frac{1}{3^5}$ and $\frac{3^2}{3^7} = 3^{-5}$. This suggests that we define $3^{-5} = \frac{1}{3^5}$.

DEFINITION

For all positive integers n and nonzero real numbers a,

$$a^{-n} = \frac{1}{a^n}.$$

If we wish $a^m \cdot a^n = a^{m+n}$ to hold for all rational exponents, then for example, $a^{\frac{1}{2}} \cdot a^{\frac{1}{2}} = a^{\frac{1}{2}+\frac{1}{2}} = a^1 = a$. This suggests that we define $a^{\frac{1}{2}}$ as $\sqrt{a}$.

DEFINITION

For all positive integers n and all positive real numbers a,

$$a^{\frac{1}{n}} = \sqrt[n]{a}.$$

We can now give meaning to powers with rational exponents.

DEFINITION

For all integers m and n with $n \neq 0$ and any positive real number a,

$$a^{\frac{m}{n}} = (\sqrt[n]{a})^m.$$

For example, $4^{\frac{3}{2}} = (\sqrt{4})^3 = 2^3 = 8$, and $81^{-\frac{1}{4}} = (\sqrt[4]{81})^{-1} = 3^{-1} = \frac{1}{3}$.

If $a < 0$, then the definition makes sense only for some values of m and n. For example, $(-1)^{\frac{1}{3}} = \sqrt[3]{-1}$ is a real number, but $(-1)^{\frac{1}{2}} = \sqrt{-1}$ is not a real number.

To show that the laws of exponents continue to hold when the exponents are rational numbers, we need the following theorem, which you are asked to prove in Exercises 48–51.

THEOREM 19 For all integers m and n with $n \neq 0$ and all positive real numbers a and b,

1. $\sqrt[n]{ab} = \sqrt[n]{a}\sqrt[n]{b}$

2. $\sqrt[n]{a^m} = (\sqrt[n]{a})^m$

3. $\sqrt[n]{\frac{a}{b}} = \frac{\sqrt[n]{a}}{\sqrt[n]{b}}$ $(b \neq 0)$

4. $\sqrt[m]{\sqrt[n]{a}} = \sqrt[mn]{a}$

THEOREM 20 For all rational numbers m and n, and all positive real numbers a and b,

1. $a^m a^n = a^{m+n}$

2. $(a^m)^n = a^{mn}$

3. $\frac{a^m}{a^n} = a^{m-n}$

4. $\left(\frac{a}{b}\right)^m = \frac{a^m}{b^m}$

5. $(ab)^m = a^m b^m$

6. If $a^m = a^n$ and $a \neq 1$, then $m = n$.

We will give a sketch of the proof of Law 1 and leave the proofs of the other laws as Exercises 36–40.

Let $m = \frac{p}{q}$ and let $n = \frac{r}{s}$ where p, q, r, and s are positive integers. Then

$$a^{\frac{p}{q}} = a^{\frac{ps}{qs}} = \sqrt[qs]{a^{ps}} \quad \text{and} \quad a^{\frac{r}{s}} = a^{\frac{rq}{qs}} = \sqrt[qs]{a^{rq}}.$$

Therefore

$$a^{\frac{p}{q}}a^{\frac{r}{s}} = \sqrt[qs]{a^{ps}} \cdot \sqrt[qs]{a^{rq}} = \sqrt[qs]{a^{ps+rq}} = a^{\frac{ps+rq}{qs}} = a^{\frac{p}{q}+\frac{r}{s}}.$$

Therefore, $a^m a^n = a^{m+n}$.

If m or n is negative, say m, then there is a positive integer x such that $m = -x$. Then

$$a^m a^n = a^{-x}a^n = \frac{a^n}{a^x} = a^{n-x} = a^{n+m}.$$

If both m and n are negative, then there are positive integers x and y such that $m = -x$ and $n = -y$. Therefore,

$$a^m a^n = a^{-x}a^{-y} = \frac{1}{a^x a^y} = \frac{1}{a^{x+y}} = a^{-(x+y)} = a^{m+n}.$$

It is clear that $a^m a^n = a^{m+n}$ if either or both of m and n are 0. ■

EXAMPLE 1 Simplify each expression.

a. $\sqrt{8x}\sqrt[3]{4x^2}$ **b.** $\left(\frac{1}{16}\right)^{-\frac{3}{4}}$

Solution **a.**
$$\begin{aligned}\sqrt{8x}\sqrt[3]{4x^2} &= (8x)^{\frac{1}{2}}(4x^2)^{\frac{1}{3}} \\ &= (2^3x)^{\frac{1}{2}}(2^2x^2)^{\frac{1}{3}} \\ &= 2^{\frac{3}{2}}x^{\frac{1}{2}}2^{\frac{2}{3}}x^{\frac{2}{3}} \\ &= 2^{\frac{3}{2}+\frac{2}{3}}x^{\frac{1}{2}+\frac{2}{3}} \\ &= 2^{\frac{13}{6}}x^{\frac{7}{6}} \quad \textbf{Answer}\end{aligned}$$

b.
$$\begin{aligned}\left(\frac{1}{16}\right)^{-\frac{3}{4}} &= \left(\frac{1}{2^4}\right)^{-\frac{3}{4}} \\ &= (2^{-4})^{-\frac{3}{4}} \\ &= 2^{(-4)\left(-\frac{3}{4}\right)} \\ &= 2^3 \\ &= 8 \quad \textbf{Answer}\end{aligned}$$

The laws of exponents, in particular Law 6, are useful in solving exponential equations.

EXAMPLE 2 Solve each exponential equation.

a. $2^{2x-1} = 2^{x+3}$ **b.** $\left(\frac{1}{9}\right)^{x-4} = 27^{1-x}$

Solution **a.** Since $2^{2x-1} = 2^{x+3}$ by Law 6, $2x - 1 = x + 3$. So $x = 4$.

b. To use Law 6, the bases must be equal.

$$\left(\frac{1}{9}\right)^{x-4} = \left(\frac{1}{3^2}\right)^{x-4} = (3^{-2})^{(x-4)} = 3^{-2x+8}$$

$$(27)^{1-x} = (3^3)^{(1-x)} = 3^{3-3x}$$

Therefore $\quad 3^{-2x+8} = 3^{3-3x}$.

By Law 6, $-2x + 8 = 3 - 3x$. So $x = -5$.

Example 2 shows how to solve exponential equations in which the bases can easily be made the same, but not how to solve an equation such as $3^x = 5$. Problems like this and the question of defining irrational exponents will be discussed in Chapter 10.

EXERCISES

Simplify.

A

1. $3^5 \cdot 3^{-2}$

2. $4^0 \cdot 2^3$

3. $\frac{5^2}{5^3}$

4. $\frac{3^6}{3^8}$

5. $(t^2)(2t)^3(2^2t^0)$

6. $(x^3)(2x)(xy^2)^{-1}$

7. $(x^{-2}y^{-3})^{-4}$

8. $(x^2y^{-3})^{-1}$

9. $\frac{(-r^2s)^3}{(2rs^{-1})^2}$

10. $\frac{2^{-1}r^{-2}s}{3r^2s^3}$

11. -2^{-2}

12. $5(-2)^{-3}$

13. $\frac{1}{2^{-1} - 3^{-1}}$

14. $(x^{-1} + y^{-2})^{-1}$

15. $(x + y)^{-1} - x^{-1} - y^{-1}$

16. $1 + r^{-1} - \frac{1}{1 + r}$

17. $8^{-\frac{2}{3}}$

18. $0.25^{\frac{3}{2}}$

19. $9^{\frac{1}{2}} - 9^{-\frac{1}{2}}$

20. $(5^{-\frac{1}{2}})^2 + 3^{\frac{0}{2}}$

21. $\sqrt{x}\sqrt[3]{x}$

22. $x^2\sqrt{2x}$

23. $\frac{2\sqrt{8}}{\sqrt[3]{4}}$

24. $\frac{1}{\sqrt[3]{8x^2}}$

25. $\sqrt{5\sqrt{5x}}$

26. $\sqrt{x^3\sqrt{x\sqrt{x}}}$

Solve each equation.

27. $8^x = 4$

28. $4^{2x-1} = 8^{\frac{1}{2}}$

29. $\left(\frac{1}{4}\right)^{3x+2} = \sqrt[3]{16}$

30. $27^{3-x} = \frac{1}{9}$

31. $x^{\frac{3}{4}} = \frac{1}{8}$

32. $x^{-\frac{2}{3}} = 4$

33. $(2x - 1)^{-\frac{1}{3}} = \frac{1}{2}$

34. $(x + 3)^{\frac{1}{2}} = (\frac{1}{8})^{\frac{1}{3}}$

B

35. Prove that $\frac{1}{a^{-r}} = a^r$ for all nonzero positive real numbers a and rational numbers r.

Prove each of the laws of exponents in Theorem 20.

36. Law 2 **37.** Law 3 **38.** Law 4 **39.** Law 5 **40.** Law 6

Solve each equation.

41. $(2^x)^2 - 3 \cdot 2^x - 4 = 0$

42. $(3^x)^2 - 18 \cdot 3^x + 81 = 0$

43. $(2^x)^2 - 16 = 0$

44. $10 \cdot (3^x)^2 - 2430 = 0$

45. $2^{2x} - 4 \cdot 2^x - 32 = 0$

46. $9^x - 4 \cdot 3^x + 3 = 0$

47. Prove that there is no real square root of -1.

Prove each statement.

48. Part (1) of Theorem 19

49. Part (2) of Theorem 19

50. Part (3) of Theorem 19

51. Part (4) of Theorem 19

C **52.** Prove using the binomial theorem that

$$(1 + x)^n = 1 + \frac{nx}{1} + \frac{n(n-1)x^2}{1 \cdot 2} + \frac{n(n-1)(n-2)x^3}{1 \cdot 2 \cdot 3} + \cdots + \frac{n(n-1)(n-2)\cdots(1)x^n}{1 \cdot 2 \cdot 3 \cdots n}.$$

Although not proved in this text, if r is a rational number and $|x| < 1$, then the infinite geometric series

$$1 + \frac{rx}{1} + \frac{r(r-1)x^2}{1 \cdot 2} + \frac{r(r-1)(r-2)x^3}{1 \cdot 2 \cdot 3} + \cdots + \frac{r(r-1)(r-2)\cdots(r-k+1)x^k}{1 \cdot 2 \cdot 3 \cdots k} + \cdots$$

converges with sum $(1 + x)^r$. This is the *binomial series*.

53. Use the first four terms of the binomial series to approximate $1.01^{\frac{1}{2}}$, $1.05^{\frac{1}{2}}$, and $1.1^{\frac{1}{2}}$.

54. Write the first seven terms of the binomial series for $(1 + x)^{-1}$. Find another series that converges with the same sum.

55. Write the first five terms of the binomial series for $(1 - x^2)^{\frac{2}{3}}$.

Chapter Summary

1. To prove a statement by *mathematical induction*, show (1) that the statement is true for 1, and (2) that whenever the statement is true for x, it is also true for $x + 1$.
   Positive integral exponents are defined by $r^1 = r$ and $r^{m+1} = r^m r$ for any real number r and natural number m. With this definition and mathematical induction, the laws of exponents can be proved for all positive integral exponents.
2. The notation $n!$ (read "n factorial") is defined to be $0! = 1$, and $n! = n(n-1)!$ when n is a positive integer. The *binomial theorem* asserts that

$$(a + b)^n = \binom{n}{0}a^n + \binom{n}{1}a^{n-1}b + \binom{n}{2}a^{n-2}b^2 + \cdots + \binom{n}{n-1}ab^{n-1} + \binom{n}{n}b^n$$

where $\binom{n}{k} = \frac{n!}{k!(n-k)!}$ and $0 \le k \le n$.

3. A *sequence* is a set of real numbers in a specified order. A *series* is an expression for the sum of the terms of a sequence. The Greek letter Σ (sigma) is used to denote the sum of a series.

4. An *arithmetic sequence* is a sequence defined recursively by $t_1 = a$, and $t_{n+1} = t_n + d$, where a is the first term and d is the *common difference*. The nth term of an arithmetic sequence is given by $t_n = a + (n - 1)d$, where a is the first term and d is the common difference. A series formed from the terms of an arithmetic sequence is an *arithmetic series*, and the sum of the first n terms of such a series is $S_n = \frac{n(a + t_n)}{2}$.

5. A *geometric sequence* is a sequence defined recursively by $t_1 = a$, and $t_{n+1} = rt_n$ where a is the first term and r is the *common ratio*. The nth term of a geometric sequence is given by $t_n = ar^{n-1}$. A series formed from the terms of a geometric sequence is a *geometric series*. The sum of the first n terms of a geometric series with first term a and common ratio r, $r \neq 1$, is given by $S_n = \frac{a(1 - r^n)}{1 - r}$.

6. An *infinite sequence* $a_1, a_2, a_3, \ldots, a_n, \ldots$ is said to have a *limit* A (written $\lim_{n \to \infty} a_n = A$), if for each $h > 0$, there exists a number M such that $|a_n - A| < h$ for all $n > M$. The limits of combinations of convergent sequences can be found according to Theorem 11.

7. The *partial sum* S_n of a series with nth term a_n is defined to be the sum of the first n terms of the series and it is denoted $S_n = \sum_{i=1}^{n} a_i$. The *sum* of an infinite series is defined to be the limit, if it exists, of the sequence of its partial sums, $\lim_{n \to \infty} S_n$. The *infinite geometric series* $\sum_{i=1}^{\infty} ar^{i-1} = a + ar + ar^2 + \cdots + ar^{i-1} + \cdots$ $(a \neq 0)$ *converges* to $\frac{a}{1 - r}$ if $|r| < 1$. If $|r| \geq 1$, the series *diverges*.

8. A sequence is *nondecreasing* if every term after the first is greater than or equal to its preceding term. A sequence with nth term a_n is *bounded* if there exists some B such that $|a_n| \leq B$ for all n. The *axiom of completeness* asserts that every nondecreasing, bounded sequence of real numbers converges to a real number.

9. Every decimal represents a real number, and every real number can be expressed as a decimal.

10. If $x^n = a$ where a is a positive number, then x is called the *nth root of a*, written $x = \sqrt[n]{a}$. If n is a positive integer and a is any real number, then a^{-n} is defined as $\frac{1}{a^n}$. If m and n are integers, $n > 0$, and a is a positive number, then $a^{\frac{m}{n}} = (\sqrt[n]{a})^m$.

Chapter Test

3-1 **1.** Prove that $1 + 4 + 7 + \cdots + (3n - 2) = \frac{n(3n - 1)}{2}$ by mathematical induction.

3-2 **2.** Expand $(2x - 3y)^5$.

3-3 **3.** Write the first four terms of the sequence defined recursively by $a_1 = 2$ and $a_n = 2a_{n-1} + 1$ for $n > 1$.

4. Express the series $\sum_{k=1}^{4} (k^2 - k)$ in expanded form.

3-4 **5.** Insert three arithmetic means between 18 and -6.

6. Find the sum of the first twenty positive odd integers.

3-5 **7.** Find the sum of the first eight terms of a geometric sequence whose first term is 3 and whose common ratio is 2.

8. The fifth term of a geometric sequence is -2 and the seventh term is -8. Find the first term.

3-6 **9.** Find $\lim_{n \to \infty} \frac{n(n - 2)}{3n^2 + 2n}$.

10. Find the limit of $\left(\frac{1}{2}\right)\left(\frac{3}{2}\right), \left(\frac{2}{3}\right)\left(\frac{4}{3}\right), \ldots, \left(\frac{n}{n + 1}\right)\left(\frac{n + 2}{n + 1}\right), \ldots$

Find the sum (if it exists) of each geometric series.

3-7 **11.** $4 - 2 + 1 - \cdots + 4\left(-\frac{1}{2}\right)^{n-1} + \cdots$

12. $\sum_{n=1}^{\infty} (5)\left(\frac{1}{3}\right)^n$

13. Express $2.\overline{12}$ as a fraction $\frac{a}{b}$ where a and b are relatively prime.

3-8 **14.** Find a rational number r and an irrational number s between $\sqrt{3}$ and $\sqrt{5}$.

15. Show that the sequence with nth term $a_n = \frac{n - 1}{n + 1}$ is a nondecreasing and bounded sequence for $n \geq 1$. Find the limit of the sequence.

3-9 **16.** Simplify $\sqrt[3]{\frac{16y^2}{2y}}\sqrt[3]{\frac{48y^6}{6y}}$.

17. Solve $9^{-2x} = \frac{1}{3^{x-1}}$.

APPLICATION

ANNUITIES

How do people provide for financial security after their retirement? Since it is not wise to wait until retirement, many people make provision for their post-retirement years as early in their working years as possible. They make such provision by depositing money regularly into some form of retirement account. One form of such an account is an individual retirement account (IRA). Social Security, a mandatory national plan funded by taxes paid through payroll deductions, and company pension programs funded by company and employee contributions also provide retirement income.

How might parents provide for a child's college education? How might state governments pay off a bond issue designed to finance road or bridge construction when the bonds come due? These questions and others can be partially answered by a consideration of annuities. An *annuity* is a sequence of equal amounts of money deposited into an account at equally spaced times.

To understand annuities more clearly, let us consider an example. Suppose that \$1000 is deposited at the end of each year into an account that pays 8% interest compounded annually. (Note that each payment is equal and that each time interval is 1 year.) Let us calculate the amount in the account after 4 years. Each row of the following table shows the growth of each \$1000 deposit.

Year	1	2	3	4
Deposit 1st	1000	(1.08)(1000)	$(1.08)^2(1000)$	$(1.08)^3(1000)$
2nd		1000	(1.08)(1000)	$(1.08)^2(1000)$
3rd			1000	(1.08)(1000)
4th				1000

Notice that each row of the table represents a geometric sequence with common ratio equal to 1.08 since the amount at the end of each year is 1.08 times the amount the year before. Notice also that the amount in the account due to all four deposits is given by the following geometric series.

$$1000 + 1000(1.08) + 1000(1.08)^2 + 1000(1.08)^3$$

This total is a geometric series with first term $a = 1000$, $r = 1.08$, and $n = 4$. By using the formula for the sum of a geometric series, we find that the total amount in the account is about \$4506. This means that a \$4000 investment over four years has grown by about \$506.

The preceding analysis suggests that if P dollars is deposited yearly into an account paying $i\%$, expressed as a decimal, for n years and that if interest is compounded yearly, then the amount, A, in the account after n years is

$$A = \frac{P(1 - (1 + i)^n)}{1 - (i + 1)}, \quad \text{or} \quad \frac{P((1 + i)^n - 1)}{i}.$$

EXERCISES

1. Suppose that \$1500 is deposited at the end of each year into an account that pays 8% interest compounded annually. The depositor makes these payments from age 35 through age 65.
   a. How much money is deposited during that time?
   b. How much money is in the account altogether?
   c. How much interest is earned?
2. Suppose that the money in part (a) of Exercise 1 is invested as a lump sum at age 35. Compare the total interest earned, under the other conditions of Exercise 1, with the interest earned in part (c).
3. Find the annual investment in an annuity if the depositor wants to have \$200,000 in the account after 25 years. Interest is paid at 6% compounded annually.
4. How much more will an annuity be worth in 30 years if \$1000 is deposited per year at 9% rather than at 8% interest compounded annually?

Banks, insurance companies, and other financial organizations rely upon a variety of sources to obtain interest for their customers.

Cumulative Review (*Chapters 1-3*)

1. Show that the circle with center $P(-2, 5)$ and containing point $Q(-3, -2)$ also contains point $R(3, 0)$.
2. Find the range of values of x for which the series $1 - (3x - 1) + (3x - 1)^2 - (3x - 1)^3 + \cdots$ converges. Then find the sum in terms of x.
3. Solve: $81^{-\frac{3}{4}} = 27^{3t-2}$
4. Find an equation in standard form of the perpendicular bisector of the line segment with endpoints $(7, 4)$ and $(3, -6)$.
5. Find **(a)** the GCF and **(b)** the LCM of 245 and 280.
6. State whether the following sentence is an identity, a contradiction, or a sentence that is sometimes true and sometimes false:
$$2x > -8 \text{ or } 1 - x \geq 5$$
7. Solve $2n^2 + 5n - 3 < 0$ over $\mathcal{R}$ and graph the solution set.
8. Verify the following by direct computation: $\binom{7}{4} = \binom{6}{3} + \binom{6}{4}$.
9. Identify and draw the graph of $x^2 + 4y^2 + 2x - 3 = 0$.
10. Show that $S_n = 1 + \frac{1}{2} + \frac{1}{4} + \cdots + \frac{1}{2^{n-1}} = 2 - \frac{1}{2^{n-1}}$.
(*Hint*: Use the formula for the sum of a geometric series.)
11. Prove: If $x \neq 0$, $y \neq 0$, and $\frac{1}{x} = \frac{1}{y}$, then $x = y$.
12. Find any points of intersection of the graphs of $y = x^2 + 4x - 2$ and $2x - y = -1$.
13. Find an equation of the parabola containing $(-3, 11)$, $(1, 11)$, and $(-1, 3)$ and having vertical axis.
14. If $A = \{1, 4, 9\}$, $B = \{2, 4, 6, 8, 10\}$, and the universal set $U = \{1, 2, 3, 4, 5, 6, 7, 8, 9, 10\}$, find **(a)** $A' \cap B$ and **(b)** $A' \cup B$.
15. Use mathematical induction to prove that
$7 + 3 - 1 - 5 - \cdots - (11 - 4n) = n(9 - 2n)$.
16. Find a rational number r and an irrational number s between $\sqrt{2}$ and $\frac{3}{2}$.
17. Find each point on the y-axis that is 5 units from $(3, -1)$.
18. Find and graph the solution set of $\left|\frac{2}{2x - 1}\right| < 1$ over $\mathcal{R}$.
19. Consider the infinite sequence with nth term $a_n = \frac{1}{n}\left(2 - \frac{1}{n}\right)$. Find the limit of the sequence if it exists. If the limit does not exist, so state.
20. State in the if-then form: For an integer x to be divisible by 6, it is necessary that x be divisible by 3.
21. Find an equation of the ellipse with minor axis of length 10 and with foci $(0, 12)$ and $(0, -12)$.

22. Find S_{20} for an arithmetic sequence with $t_9 = 22$ and $t_{15} = 40$.

23. Sketch the graph of $x^2 - y^2 + 4x + 2y - 1 = 0$.

24. Evaluate: $\sum_{n=1}^{3} (n!)^2$

25. Graph the line that is parallel to $3x - 4y = 12$ and contains the point $(-5, -3)$. Find an equation in point-slope form for the line.

26. Insert two geometric means between 18 and $\frac{16}{3}$.

27. If $\sum_{k=1}^{\infty} t_k$ is a convergent series, is $\sum_{k=1}^{\infty} (t_k + 1)$ always, sometimes, or never a convergent series?

28. Solve: $\frac{3x - 2}{x + 1} \leq 0$

29. Show that the line containing the common chord of the circles with equations $(x + 2)^2 + (y - 4)^2 = 25$ and $(x - 2)^2 + y^2 = 1$ is perpendicular to the line containing the centers of the circles.

30. Solve: $4^{2x} - 6 \cdot 4^x + 8 = 0$

31. Which term of the arithmetic sequence 3, 10, 17, . . . is 500?

32. Give the complete expansion of $(3x - y)^5$ in simplified form.

33. Given open sentences p: $\frac{1}{2}x < 2$ and q: $-2x + 3 < 5$, find **(a)** $p \vee q$ and **(b)** $p \wedge q$.

34. Determine whether $P(1, -1)$ is inside, outside, or on the ellipse with equation $\frac{(x + 4)^2}{16} + \frac{y^2}{9} = 1$.

35. Solve: $|2t - 1| = 5$

36. Use the laws of exponents to simplify $[(\frac{1}{2}x^2y)^3(4xy^3)^2] \div x^5y^{10}$.

37. Describe the graph of $\frac{x^2}{4} - y^2 = 0$.

38. State whether {negative integers} is closed under **(a)** addition and **(b)** multiplication.

39. Graph $y = -\frac{1}{4}(x - 1)^2$. Find the focus, directrix, vertex, and points of intersection with the axes.

40. Find the sum of the geometric series $40 - 8 + 1.6 - \cdots$.

41. Write the first six terms of the sequence for which $a_1 = 2$ and $a_{n+1} = 2a_n - 1$.

42. Identify the graph of $2x^2 - 4x - y + 5 = 0$.

43. Show that $\triangle ABC$ with vertices $A(3, -1)$, $B(-5, 3)$, and $C(-5, -7)$ is isosceles.

44. State whether 496 is a perfect number. (*Hint*: It is a perfect number if the sum of all its positive integral factors except 496 is equal to 496.)

45. Give the converse of: If $x = -2$, then $|x| = 2$. State whether the converse is true or not.

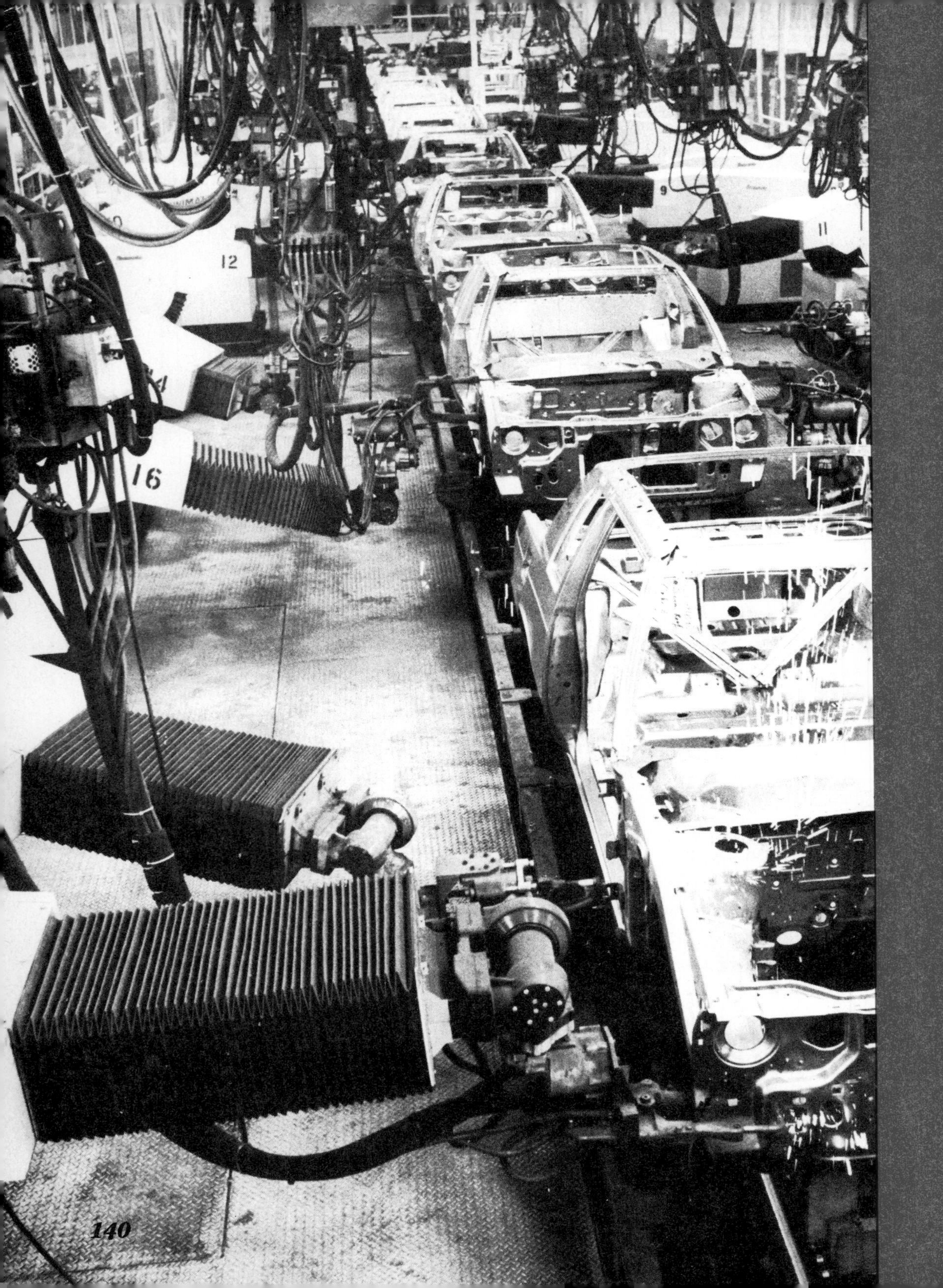
9
11
12
16

Chapter 4

Functions and Limits

In this chapter the concept of function *is reviewed. Then the notion of the* limit of a function *is introduced. In particular,* continuous *functions are discussed.*

Functions

■ 4-1 FUNCTIONS AND FUNCTIONAL NOTATION

In mathematics and its applications we constantly meet situations in which one quantity depends on, or is a *function* of, another quantity. The volume V of a sphere is a function of its radius r; the kinetic energy K of a moving body is a function of its speed v; the cost c of mailing a letter is a function of its weight w. In each such functional relationship there is the idea of pairing: volume with radius, (r, V); energy with speed, (v, K); and cost with weight, (w, c). In general:

DEFINITION

A **function** is a set of ordered pairs in which different pairs have different first members.

The set of all first members is the **domain** of the function and the set of all second members is its **range**. *Any* set of ordered pairs is a **relation**. Thus a function is a special kind of relation.

The cost of manufacturing anything as complicated as an automobile involves many variables. The use of robots, as on the assembly line shown in the photograph, can help to minimize manufacturing costs.

EXAMPLE 1 Which of the following subsets of $\mathcal{R} \times \mathcal{R}$ are functions?

a. $F = \{(x, y): x = |y - 1|\}$ **b.** $G = \{(x, y): y = x - 1\}$

Solution **a.** Since (1, 0) and (1, 2) belong to F and have the same first member, F is not a function.

b. Suppose that two different ordered pairs have the same first member and different second members, and belong to G. Then (x, y_1) and (x, y_2) are in G and $y_1 \neq y_2$. So $y_1 = x - 1$ and $y_2 = x - 1$. Therefore $y_1 = y_2$, and this contradicts the assumption that $y_1 \neq y_2$. Hence G is a function.

Functions in mathematics are usually presented not as sets of ordered pairs, but rather as rules by means of which we can obtain as many ordered pairs as we need. For example, the volume of a sphere is related to the radius by $V = \frac{4}{3}\pi r^3$. From this equation we can find many particular values for V from many particular values for r. Thus a good working definition of a function is the one that follows.

DEFINITION

A **function** is a rule that assigns to each member of a set D, the domain, a unique member of a set R, the range.

In this textbook D and R will be subsets of specified fields; usually both are subsets of $\mathcal{R}$. However in Chapter 5 we shall discuss functions whose domain is C, the field of complex numbers. (A **sequence** can be defined as a function whose domain is N or an initial segment $\{1, 2, 3, \ldots, n\}$ of N.)

Each member of the range is called a **value** of the function. If g is the function that assigns to each real number x the value $4x - x^2$, we write

$$g: x \rightarrow 4x - x^2.$$

To denote the value of g at x, we use **functional notation** and write

$$g(x) = 4x - x^2,$$

read "g of x is $4x - x^2$." For example,

$$g(5) = 4 \cdot 5 - 5^2 = -5.$$

EXAMPLE 2 Given the functions $f: x \rightarrow x^2 - 1$ and $g: x \rightarrow 2x + 1$ with domain $\mathcal{R}$, find each value.

a. $f(2)g(2)$ **b.** $f(g(2))$ **c.** $g(f(2))$

d. $f(x + 1)$ **e.** $\dfrac{f(x + h) - f(x)}{h}$

Solution **a.** $f(2) = 2^2 - 1 = 3$ and $g(2) = 2 \cdot 2 + 1 = 5$.
Therefore $f(2)g(2) = 3 \cdot 5 = 15$. **Answer**

b. From part (a), $g(2) = 5$. So $f(g(2)) = f(5) = 5^2 - 1 = 24$. **Answer**

c. From part (a), $f(2) = 3$. So $g(f(2)) = g(3) = 2 \cdot 3 + 1 = 7$. **Answer**

d. $f(x + 1) = (x + 1)^2 - 1 = (x^2 + 2x + 1) - 1 = x^2 + 2x$ **Answer**

e.
$$\frac{f(x + h) - f(x)}{h} = \frac{[(x + h)^2 - 1] - [x^2 - 1]}{h}$$
$$= \frac{[x^2 + 2xh + h^2 - 1] - [x^2 - 1]}{h}$$
$$= \frac{2xh + h^2}{h} = 2x + h \quad \textbf{Answer}$$

Frequently a function f is described simply by giving a formula for $f(x)$. If no domain is specified, assume that the domain is *the set of all real numbers for which the formula is meaningful*. For example,

Formula	*Domain*
$f(x) = \dfrac{x - 1}{(x + 2)(x - 3)}$	$\{x: x \neq -2, 3\}$
$g(x) = \sqrt{x - 3}$	$\{x: x \geq 3\}$
$\varphi(x) = \dfrac{1}{\sqrt{4 - x^2}}$	$\{x: -2 < x < 2\}$

The **graph** of a function f is the set of all points having coordinates of the form $(x, f(x))$ where x is in the domain of f. Therefore the graph of the function f is the same as the graph of the equation $y = f(x)$.

EXAMPLE 3 Graph each function.

a. $f(x) = \sqrt{4 - x^2}$ **b.** $g(x) = \dfrac{x^2 - 1}{x - 1}$

Solution **a.** The domain of f is $\{x: -2 \leq x \leq 2\}$. Let $y = f(x) = \sqrt{4 - x^2}$. Then $x^2 + y^2 = 4$ and $y \geq 0$. The graph of f is the top half of the circle whose equation is $x^2 + y^2 = 4$, as shown in Figure a.

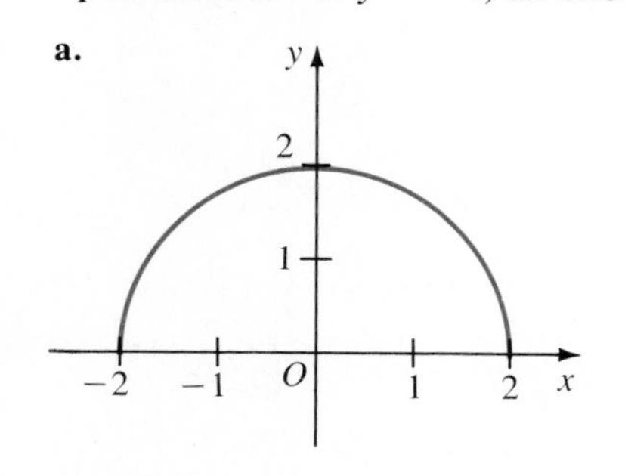

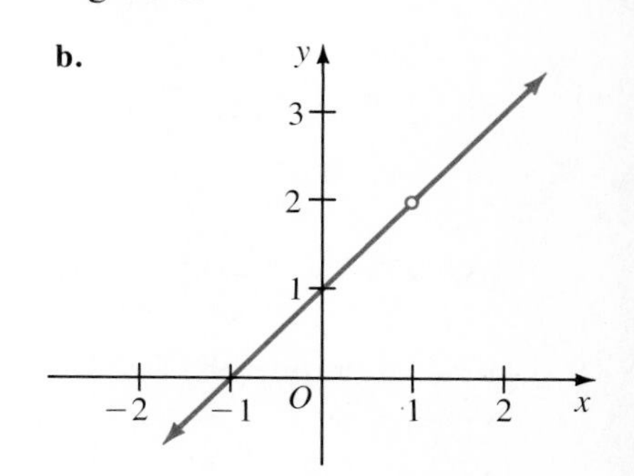

b. The domain of g is $\{x: x \neq 1\}$. If $x \neq 1$, then $y = g(x) = \dfrac{x^2 - 1}{x - 1} = \dfrac{(x + 1)(x - 1)}{x - 1} = x + 1$. The graph is shown in Figure b.

The *vertical line test* is a geometric way to determine whether a curve is the graph of a function. A set in the coordinate plane is the graph of a function if and only if each vertical line intersects it in at most one point.

EXAMPLE 4 Determine whether the graph of each equation is the graph of a function.

a. $\frac{x^2}{4} + \frac{y^2}{9} = 1$ **b.** $y = x^2 + 4x + 5$

Solution **a.** The graph is an ellipse with center at the origin. The vertical line $x = 0$ intersects the graph in two points. Thus the graph is not the graph of a function.

b. The equation $y = x^2 + 4x + 5$ can be written $y - 1 = (x + 2)^2$. Its graph is a parabola with vertex $(-2, 1)$ and axis of symmetry $x = -2$. Any vertical line in the plane intersects the graph in just one point. Thus $y = x^2 + 4x + 5$ is the graph of a function.

EXERCISES

Which of the following subsets of $\mathcal{R} \times \mathcal{R}$ is a function?

A

1. $\{(1, 3), (3, 1), (2, 2)\}$

2. $\{(1, 3), (2, 3), (3, 3)\}$

3. $\{(3, 1), (3, 2), (3, 3)\}$

4. $\{(1, 2), (2, 3), (1, 3)\}$

5. $\{(x, y): y^2 = x\}$

6. $\{(x, y): y^3 = x\}$

7. $\{(x, y): x + y = 4\}$

8. $\{(x, y): x^2 + y^2 = 4\}$

9. $\{(x, y): |x + y| = 4\}$

10. $\{(x, y): |x| + |y| = 4\}$

11. $\{(x, y): x^2 - y^2 = 1\}$

12. $\{(x, y): xy = 1\}$

Find the domain of each function.

13. $f(x) = \frac{x^2 + 1}{x^2 - 1}$

14. $f(x) = \frac{x^2 - 1}{x^2 + 1}$

15. $f(x) = \sqrt{4 - x}$

16. $f(x) = \sqrt{x - 4}$

17. $f(x) = \sqrt{x^2 - 9}$

18. $f(x) = \sqrt{16 - x^2}$

19. $f(x) = \frac{\sqrt{x}}{x^2 - 1}$

20. $f(x) = \sqrt{\frac{x - 1}{x}}$

21. $f(x) = \sqrt{\frac{x + 1}{x - 1}}$

22. $f(x) = \sqrt{\frac{1 - x}{1 + x}}$

Find the range of f.

23. $f(x) = |x - 2|$

24. $f(x) = |x| - 2$

25. $f(x) = x^2 - 1$

26. $f(x) = (x - 1)^2$

27. $f(x) = \frac{1}{\sqrt{x^2 + 1}}$

28. $f(x) = \sqrt{x^2 - 1}$

29. $f(x) = \frac{1}{x^2 + 1}$

30. $f(x) = \sqrt{9 - x^2}$

Find: **(a)** $f(2)g(2)$; **(b)** $f(g(2))$; **(c)** $g(f(2))$; **(d)** $f(x + 2)$; **(e)** $\frac{f(x + h) - f(x)}{h}$

31. $f(x) = 2x + 1,\ g(x) = x - 2$

32. $f(x) = x - 3,\ g(x) = 1 - x$

33. $f(x) = \frac{2}{x},\ g(x) = \frac{x + 2}{x}$

34. $f(x) = x^2,\ g(x) = x - 1$

35. $f(x) = x^2 - 4,\ g(x) = x + 2$

36. $f(x) = \frac{4}{x},\ g(x) = x + 2$

Graph each function or relation.

37. $f\colon x \rightarrow \frac{x^2 - x - 2}{x - 2}$

38. $f\colon x \rightarrow \frac{x^2 - 4}{x - 2}$

39. $f\colon x \rightarrow x^2 - 1$

40. $f\colon x \rightarrow (x - 1)^2$

41. $f\colon x \rightarrow \frac{x^2}{x}$

42. $f\colon x \rightarrow \frac{x^3}{x}$

43. $f\colon x \rightarrow \frac{1}{x}$

44. $f\colon x \rightarrow \sqrt{x^2 + 1}$

B **45.** $\{(x, y)\colon y \geq x + 1\}$

46. $\{(x, y)\colon y - 2x \leq 3\}$

47. $\{(n, y)\colon n \in Z \text{ and } y = (-1)^n 2^n\}$

48. $\{(n, y)\colon n \in Z \text{ and } y = \frac{(-1)^n}{n} + n\}$

Let f be a function having domain $\mathcal{R}$ and range in $\mathcal{R}$ and such that $f(a + b) = f(a) + f(b)$ for all $a, b \in \mathcal{R}$. In Exercises 49–52 prove the following properties of f.

49. $f(0) = 0$

50. $f(a - b) = f(a) - f(b)$

51. $f(3a) = 3f(a)$

52. $f\left(\frac{1}{2}a\right) = \frac{1}{2}f(a)$

Let g be a function having domain $\mathcal{R}$ and range in $\mathcal{R}$ such that $g(1) \neq 0$ and $g(a + b) = g(a)g(b)$. In Exercises 53–56 prove the following properties of g.

53. $g(0) = 1$

54. $g(-b) = \frac{1}{g(b)}$

55. $g(a - b) = \frac{g(a)}{g(b)}$

56. $g(2a) = g(a)g(a)$

Let h be a function with domain the set of all positive real numbers and range in $\mathcal{R}$ such that $h(ab) = h(a) + h(b)$. In Exercises 57–60 prove each property of h.

57. $h(1) = 0$

58. $h\left(\frac{1}{b}\right) = -h(b)$

59. $h\left(\frac{a}{b}\right) = h(a) - h(b)$

60. $h(a^2) = 2h(a)$

61. The equation $RI = E$, known as **Ohm's Law**, relates electrical resistance R to current I and voltage E. If a voltage of 120 v is applied to a circuit and the resistance varies from 10 Ω (Ohms) to 25 Ω, describe the range for current I.

62. The equation $pv = k$, known as **Boyle's Law**, relates pressure p to volume v of a gas at a fixed temperature. If $k = 200$, describe the effect on p when v varies from 11.5 to 22.5.

A definition of "ordered pair": The **pair** a, b is simply the set $\{a, b\}$ (which is the same as $\{b, a\}$). We can define ordered pair in terms of the simpler concept of pair: The **ordered pair** a, b, written (a, b), is the pair having as its two members a and the pair a, b. Thus,

$$(a, b) = \{a, \{a, b\}\} \quad \text{or} \quad (a, b) = \{\{b, a\}, a\}.$$

Thus, in $\{\{3, 5\}, 5\}$, the pair member $\{3, 5\}$ tells you that the pair is either $(3, 5)$ or $(5, 3)$, while the lone member 5 tells you that the first coordinate is 5, so that $\{\{3, 5\}, 5\} = (5, 3)$.

C

63. Express the ordered pair (5, 3) in three other ways.
64. What ordered pair is $\{4, \{1, 4\}\}$?
65. Write (a, b) in two ways different from the ones given above.

COMPUTER EXERCISES

1. Write a program to find $\dfrac{f(3 + h) - f(3)}{h}$ when $h = 0.1, 0.01, 0.001$, and 0.0001 for $f(x) = x^2$.
2. Modify and run the program in Exercise 1 for each function.

   a. $f(x) = x^2 - 1$ **b.** $f(x) = (x - 1)^2$ **c.** $f(x) = \dfrac{4}{x}$ **d.** $f(x) = \sqrt{x - 1}$

4-2 THE ALGEBRA OF FUNCTIONS

If a company can express its income I as a function of goods produced x, and its cost of production C as a function of goods produced, then its profit P can be expressed as a function of goods produced.

$$P(x) = I(x) - C(x)$$

In this section we define *sums*, *differences*, *products*, and *quotients* of functions. We also introduce a variety of special functions.

If two functions f and g have domains with nonempty intersection, we can define the **sum**, $f + g$, and the **product**, fg, in a natural way.

DEFINITION

Let D be the intersection of the domains of f and g and $x \in D$.

$$(f + g)(x) = f(x) + g(x)$$
$$(fg)(x) = f(x) \cdot g(x)$$

EXAMPLE 1 Let the functions f and g be defined by $f(x) = \frac{\sqrt{1-x}}{x}$ and $g(x) = \sqrt{x}$.

a. Find the domain of $f + g$ and fg.

b. Find formulas for $(f + g)(x)$ and $(fg)(x)$.

Solution **a.** The domain of f is $\{x: x \leq 1 \text{ and } x \neq 0\}$. The domain of g is $\{x: x \geq 0\}$. Thus, the domain of $f + g$ and fg is $\{x: x \leq 1 \text{ and } x \neq 0\} \cap \{x: x \geq 0\} = \{x: 0 < x \leq 1\}$.

b. $(f + g)(x) = f(x) + g(x) = \frac{\sqrt{1-x}}{x} + \sqrt{x} = \frac{\sqrt{1-x} + x\sqrt{x}}{x}$

$$(fg)(x) = f(x) \cdot g(x) = \frac{\sqrt{1-x}}{x} \cdot \sqrt{x} = \sqrt{\frac{1-x}{x}}$$

We can define **differences** and **quotients** as well.

DEFINITION

For all x in the domain of g, $(-g)(x) = -g(x)$.

For all x in the intersection of the domains of f and g,

$$(f - g)(x) = f(x) - g(x).$$

For all x in the domain of g such that $g(x) \neq 0$, $\frac{1}{g}(x) = \frac{1}{g(x)}$.

For all x in the intersection of the domains of f and g such that $g(x) \neq 0$,

$$\left(\frac{f}{g}\right)(x) = \frac{f(x)}{g(x)}.$$

The definition of the product of functions suggests the following recursive formulas for a positive integral power of a function. For each positive integer n,

$$f^1 = f,$$
$$f^{n+1} = f \cdot f^n.$$

Thus, if the values of f are defined by the rule $f(x) = x - 1$, then the values of f^2 and f^3 are given by

$$f^2(x) = (x - 1)(x - 1) = (x - 1)^2$$

and

$$f^3(x) = (x - 1)(x - 1)^2 = (x - 1)^3,$$

respectively.

A function may have a **piecewise** definition, that is, it may be defined by different formulas on different parts of its domain. The **absolute value function** is defined by two formulas. The graph consists of the two half-lines shown at the right.

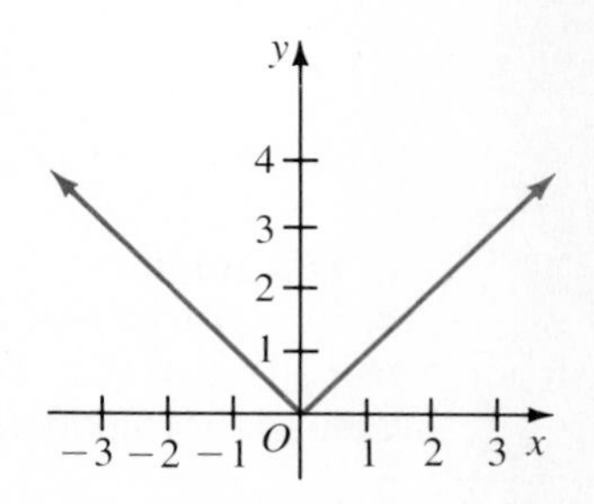

$$x \to |x| = \begin{cases} x \text{ if } x \geq 0 \\ -x \text{ if } x < 0 \end{cases}$$

The **signum** or **sign function** has only three values. The domain $\mathcal{R}$ is divided into three parts. The sign function's graph is shown at the right.

$$x \to \operatorname{sgn} x = \begin{cases} 1 \text{ if } x > 0 \\ 0 \text{ if } x = 0 \\ -1 \text{ if } x < 0 \end{cases}$$

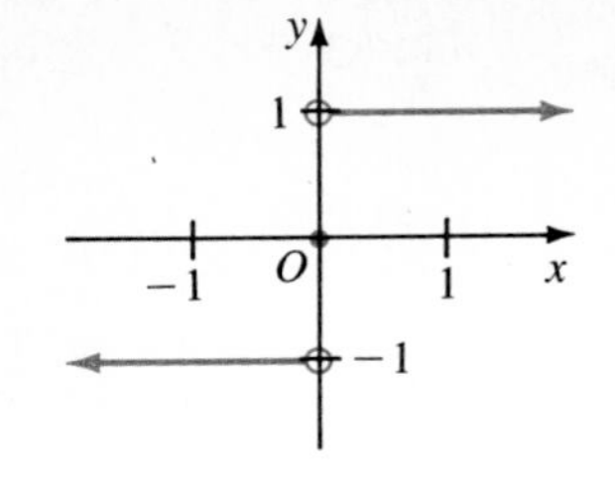

The **greatest integer function**, denoted by $[\![x]\!]$, is defined as follows. For each real number x,

$[\![x]\!]$ = the greatest integer not exceeding x.

It can be thought of as the "rounding down" function. As an example,

$$[\![2.6]\!] = 2 \qquad [\![2]\!] = 2 \qquad [\![-2.6]\!] = -3.$$

The graph of the greatest integer function is shown at the right.

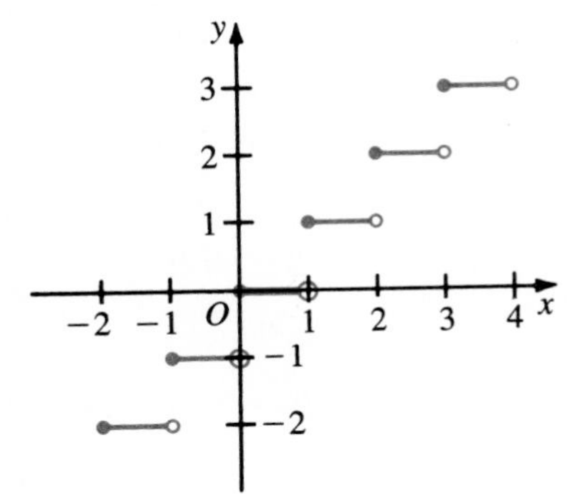

EXAMPLE 2 Graph: **a.** $\varphi: x \to |x| + \operatorname{sgn} x$ **b.** $\chi: x \to [\![x]\!] \operatorname{sgn} x$

Solution The function in (a) is a sum of piecewise functions, and the function in (b) is a product of piecewise functions. Use the definitions of $|x|$, $\operatorname{sgn} x$, and $[\![x]\!]$ to write formulas for φ (phi, pronounced "fī") and χ (chi, pronounced "kī").

a. If $x > 0$, then $\varphi(x) = x + 1$.
If $x = 0$, then $\varphi(x) = 0$.
If $x < 0$, then $\varphi(x) = -x - 1$.

$$\varphi(x) = \begin{cases} x + 1 \text{ if } x > 0 \\ 0 \quad\ \text{ if } x = 0 \\ -x - 1 \text{ if } x < 0 \end{cases}$$

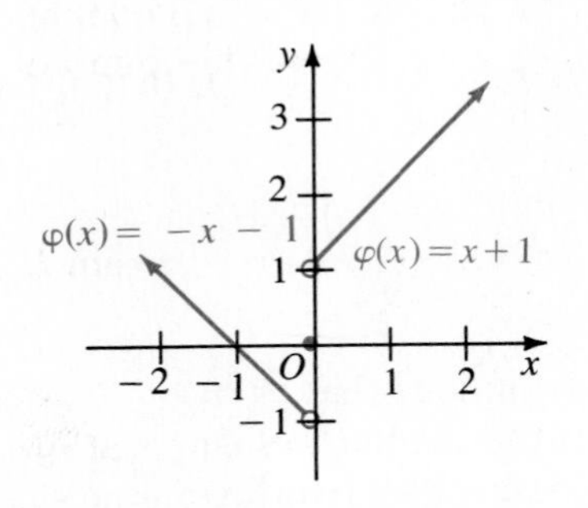

b. If $x \geq 0$, then $\chi(x) = [\![x]\!]$.
If $x < 0$, then $\chi(x) = -[\![x]\!]$.

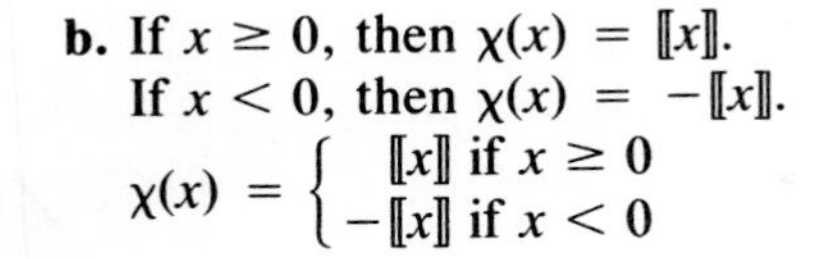

$$\chi(x) = \begin{cases} [\![x]\!] \text{ if } x \geq 0 \\ -[\![x]\!] \text{ if } x < 0 \end{cases}$$

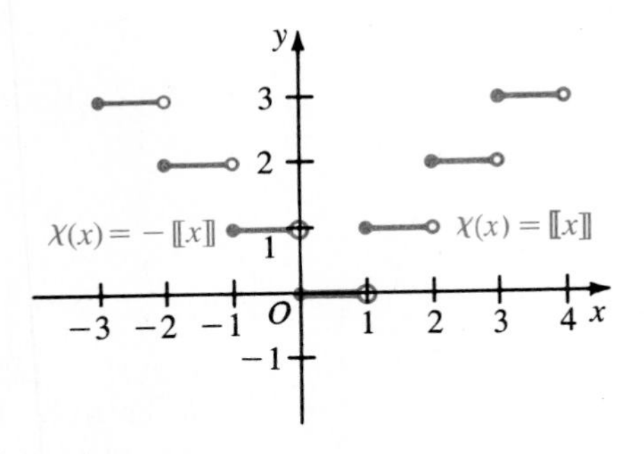

The graph of $y - k = f(x - h)$ can be obtained from the graph of $y = f(x)$ by a **translation** (shift) of the graph of $y = f(x)$ horizontally h units and vertically k units. Therefore many graphs can be obtained from simpler graphs by a translation, or shift.

EXAMPLE 3 Graph $f: x \rightarrow |x + 1| + 2$.

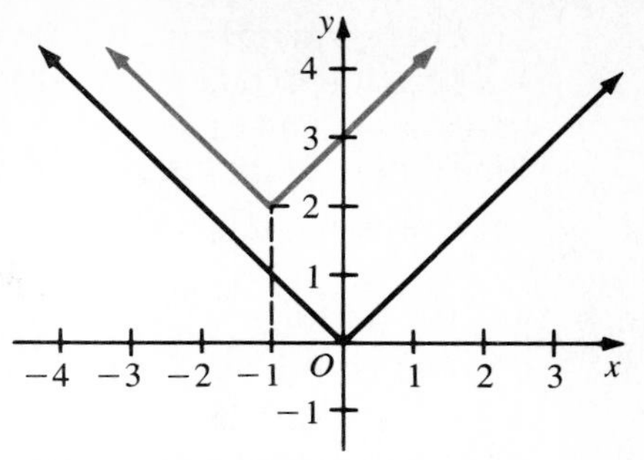

Solution

$$y = |x + 1| + 2$$
$$= |x - (-1)| + 2$$
$$y - 2 = |x - (-1)|$$

The graph shown in red is obtained by translating the graph of $y = |x|$ (shown in black) to the left 1 unit and up 2 units.

The **identity function** on a set X, denoted by I, is defined by $I(x) = x$ for all $x \in X$. The function that assigns to all real numbers x a constant value c is called a **constant function**. For example, the function $f: x \rightarrow 5$ assigns to every real number x the value 5. We usually use the name of the number to serve also as the name of the corresponding constant function. Note the difference between $f(x) = 0$ and $f = 0$. The former means that a certain function f has the value 0 at the particular number x. The latter means that f is the zero function; that is, $f(x) = 0$ for *all* x.

EXAMPLE 4 Find the values of f and g at -1 and 2.

a. $f = \dfrac{I^2 - 4}{I}$ **b.** $g = I^2 \cdot \text{sgn}$

Solution **a.** $f(x) = \dfrac{I^2(x) - 4}{I(x)} = \dfrac{x^2 - 4}{x}$

$\therefore f(-1) = \dfrac{(-1)^2 - 4}{-1} = 3, \quad f(2) = \dfrac{2^2 - 4}{2} = 0$ **Answer**

b. $g(x) = I^2(x) \cdot \text{sgn}\,(x) = x^2 \,\text{sgn}\, x$

$\therefore g(-1) = (-1)^2 \,\text{sgn}\,(-1) = -1, \quad g(2) = 2^2 \,\text{sgn}\, 2 = 4$ **Answer**

Let F be the set of all functions with a common domain D and ranges contained in a field. In Exercises 37–43 it is proved that F is a commutative group under $+$, is closed and associative under $\cdot$, and has the constant function 1 as the multiplicative identity. In Example 5 we show that in F multiplication is distributive over addition.

EXAMPLE 5 Let f, g, and h be functions with a common domain D. Show that $f(g + h) = f \cdot g + f \cdot h$.

Solution For any $x \in D$,

$[f(g + h)](x) = f(x)[(g + h)(x)]$ (Definition of product of functions)
$= f(x)[g(x) + h(x)]$ (Definition of sum of functions)
$= f(x)g(x) + f(x)h(x)$ (Distributive axiom)
$= (fg)(x) + (fh)(x)$ (Definition of product of functions)
$= (fg + fh)(x)$ (Definition of sum of functions)

Thus $f(g + h) = f \cdot g + f \cdot h$.

EXERCISES

Find a rule for **(a)** $f + g$; **(b)** fg; **(c)** $\frac{f}{g}$; and **(d)** $f^2 - g^2$. Specify the domain of each resulting function.

A

1. $f(x) = x + 1;\ g(x) = x$

2. $f(x) = x - 1;\ g(x) = x + 1$

3. $f(x) = \sqrt{1 + x};\ g(x) = \sqrt{1 - x}$

4. $f(x) = \frac{1}{x};\ g(x) = \frac{1}{x - 1}$

5. $f(x) = \frac{x}{x + 1};\ g(x) = \frac{x}{x - 1}$

6. $f(x) = \sqrt{x};\ g(x) = \sqrt{1 - x}$

7. $f(x) = \sqrt{\frac{1 + x}{1 - x}};\ g(x) = \sqrt{1 - x}$

8. $f(x) = \sqrt{\frac{x}{1 - x}};\ g(x) = \sqrt{x}$

Find the value of each function at -1 and at 2.

9. $I^2 - I - 2$

10. $I^2 - 4$

11. $\frac{I^2 - 1}{I}$

12. $\frac{I^2 - 1}{I^3 - 1}$

13. $|2 - I^2|$

14. $[\![2 - I^2]\!]$

15. $\left[\!\!\left[\frac{2 - I^2}{3}\right]\!\!\right]$

16. $\left|\frac{(2 - I^2)}{3}\right|$

In Exercises 17 and 18 state whether or not $f = g$. If not, explain how f and g differ.

17. $f(x) = \frac{x}{|x|};\ g(x) = \text{sgn } x$

18. $f(x) = |x| \text{ sgn } x;\ g(x) = x$

Graph each of the following functions.

19. a. $f\colon x \to |x - 2|$ **b.** $f\colon x \to |x| - 2$

20. a. $f\colon x \to |x + 1|$ **b.** $f\colon x \to |x| + 1$

21. a. $f\colon x \to \text{sgn } (x - 2)$ **b.** $f\colon x \to \text{sgn } x - 2$

22. a. $f\colon x \to [\![x - 1]\!]$ **b.** $f\colon x \to [\![x]\!] - 1$

23. $f\colon x \to |x| + x$

24. $f\colon x \to [\![x]\!] + x$

25. $f\colon x \to x - [\![x]\!]$

26. $f\colon x \to [\![x]\!] - x$

B

27. $f(x) = \begin{cases} 2x - 1 \text{ if } 1 \le x \le 4 \\ -x + 3 \text{ if } 4 < x \le 7 \\ |x| \text{ if } 7 < x \le 10 \end{cases}$

28. $g(x) = \begin{cases} x^2 \text{ if } 0 \le x \le 3 \\ [\![x]\!] \text{ if } 3 < x \le 5 \\ x \text{ if } 5 < x < 8 \end{cases}$

29. $f(x) = [\![x]\!]^2$

30. $g(x) = [\![x - 2]\!]^2$

In Exercises 31–36 graph each equation using translations.

31. $\frac{(x - 2)^2}{4} + \frac{(y - 3)^2}{9} = 1$

32. $(x - 2)(y - 4) = 1$

33. $(x - 3)^2 + (y - 2)^2 = 4\ (y \ge 2)$

34. $(x - 3)^2 - (y - 2)^2 = 1\ (y \ge 0)$

35. $x = (y - 2)^2 + 1\ (y \ge 2)$

36. $x = -2(y + 3)^2 - 2\ (y \le -2)$

In Exercises 37–43 let F be the set of all functions having a common domain D and the operations $+$ and $\cdot$ as defined on page 146. In Example 5 we showed that the distributive law holds in F. Prove that the following hold for all f, g, and h in F.

C

37. $(f + g) + h = f + (g + h)$ (Associative Axiom of Addition)

38. $f + g = g + f$ (Commutative Axiom of Addition)

39. $0 + f = f + 0 = f$ (Existence of Additive Identity)

40. $f + (-f) = (-f) + f = 0$ (Existence of Additive Inverses)

41. $(fg)h = f(gh)$ (Associative Axiom of Multiplication)

42. $fg = gf$ (Commutative Axiom of Multiplication)

43. $1 \cdot f = f \cdot 1 = f$ (Existence of Multiplicative Identity)

44. Show that if D contains more than one number, then there are nonzero members of F that do not have multiplicative inverses.

45. Show that the set of all linear functions, that is, functions of the form $y = mx + b$, is a commutative group under addition.

4-3 COMPOSITION AND INVERSION OF FUNCTIONS

Suppose that a ship starts at a point A 20 km south of a port P and sails due east at 10 km/h. The ship's distance z from P is a function of the distance x it has sailed: $z = \sqrt{400 + x^2}$. The distance x it has sailed is a function of the time t the ship has been sailing: $x = 10t$. Hence $z = \sqrt{400 + x^2} = \sqrt{400 + (10t)^2} = \sqrt{400 + 100t^2} = 10\sqrt{4 + t^2}$. This illustrates how functions can be composed or built up from simpler functions. The two steps in "building" the function $x \to \sqrt{4 - x^2}$ are displayed below.

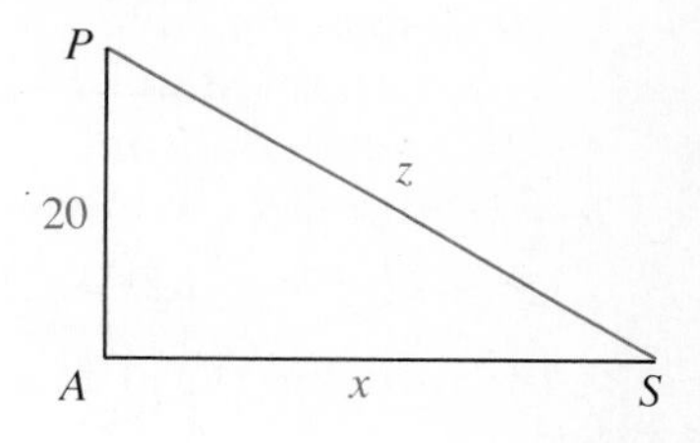

$$x \to 4 - x^2 \to \sqrt{4 - x^2}$$

indicates how we can "build" the function $x \to \sqrt{4 - x^2}$ in two steps: starting with x, first form $4 - x^2$, and then take the square root of $4 - x^2$. We can build up a new function from two functions f and g by first finding the value of g at x and then finding the value of f at $g(x)$.

$$x \to g(x) \to f(g(x))$$

This process is called *composition*. In general:

DEFINITION

The **composition** of the functions f and g is the function, $f \circ g$, defined by

$$(f \circ g)(x) = f(g(x)).$$

Note that the domain of $f \circ g$ is $\{x: g(x) \text{ is in the domain of } f\}$.

Composition is illustrated in the following **mapping diagram**.

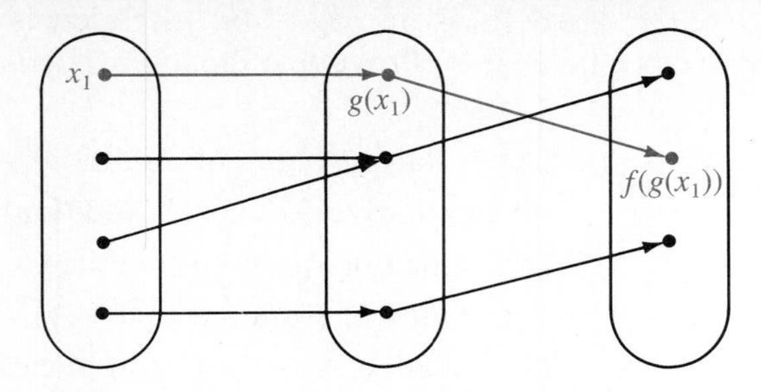

EXAMPLE 1 Given the functions $f\colon x \to \sqrt{4 - x}$ and $g\colon x \to x^2$, find $(f \circ g)(x)$ and $(g \circ f)(x)$. Specify the domains of $f \circ g$ and $g \circ f$.

Solution

$$(f \circ g)(x) = f(g(x)) = f(x^2) = \sqrt{4 - x^2}$$
$$\text{Domain} = \{x\colon -2 \le x \le 2\}$$

$$(g \circ f)(x) = g(f(x)) = g(\sqrt{4 - x}) = (\sqrt{4 - x})^2 = 4 - x$$
$$\text{Domain} = \{x\colon x \le 4\}$$

Note that $\sqrt{4 - x^2} \ne 4 - x$. In general $f \circ g \ne g \circ f$.

We often find it useful to express complicated functions as compositions of simpler functions. We can do this by introducing an "intermediate variable" and writing $(f \circ g)(x)$ or $f(g(x))$ as $f(u)$ where $u = g(x)$.

EXAMPLE 2 Express each function as a composition of two simpler functions.

a. $x \to \sqrt{25 - x^2}$ **b.** $x \to (x + 2)^{10}$

Solution **a.** Let $u = g(x) = 25 - x^2$ and $f(u) = \sqrt{u}$.

b. Let $u = g(x) = x + 2$ and $f(u) = u^{10}$.

The graph of the function in Example 2(a) is the upper half of a circle of radius 5 centered at the origin. Note that when $x = 3$ or $x = -3$, $y = 4$. The mapping diagram and graph that follow illustrate this situation in general.

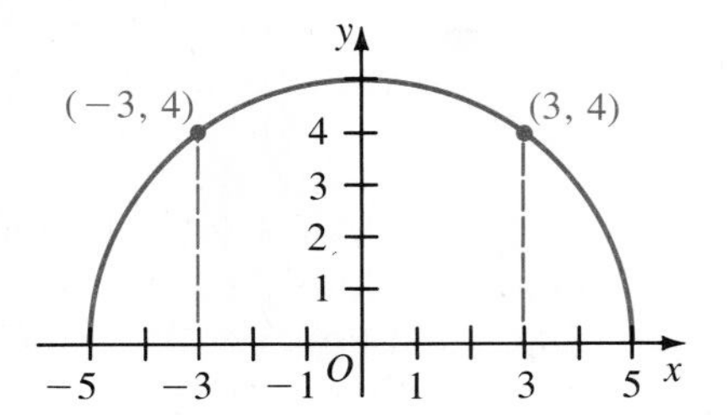

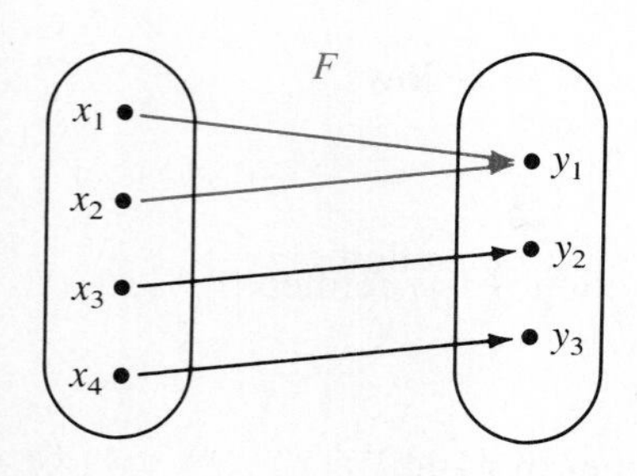

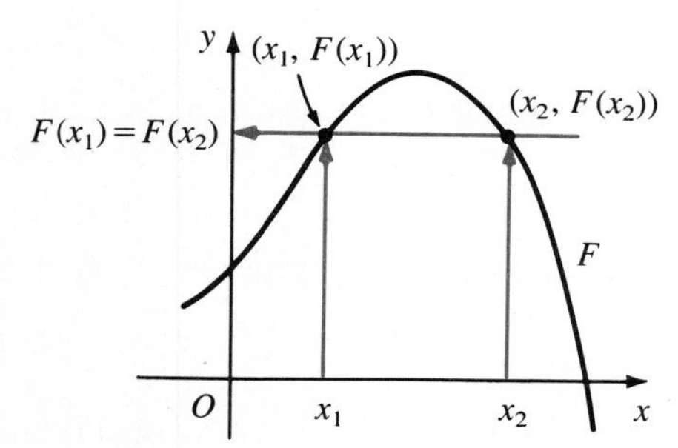

There are many functions for which different members of the domain give different members of the range. This "*one-to-one*" situation is illustrated in the figures below.

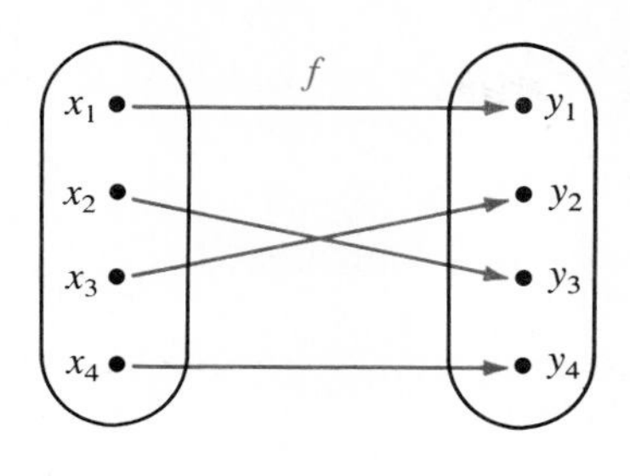

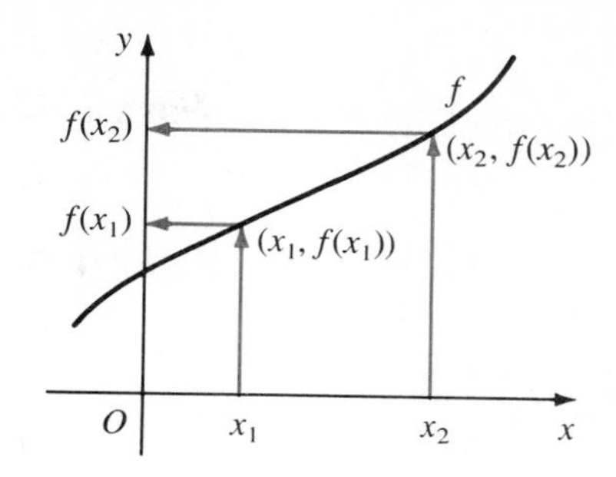

We define a function f to be **one-to-one** if whenever $x_1 \neq x_2$, $f(x_1) \neq f(x_2)$. The horizontal line test is a geometric way to tell whether a function is one-to-one. If every horizontal line intersects the graph of a function f in at most one point, then the function is one-to-one. The graph of the function h: $x \rightarrow \sqrt{25 - x^2}$ is not the graph of a one-to-one function.

If we reverse the arrows in the previous mapping diagram of the function f, we form a new function g which associates each y value with a unique x value. This is shown in the following figures.

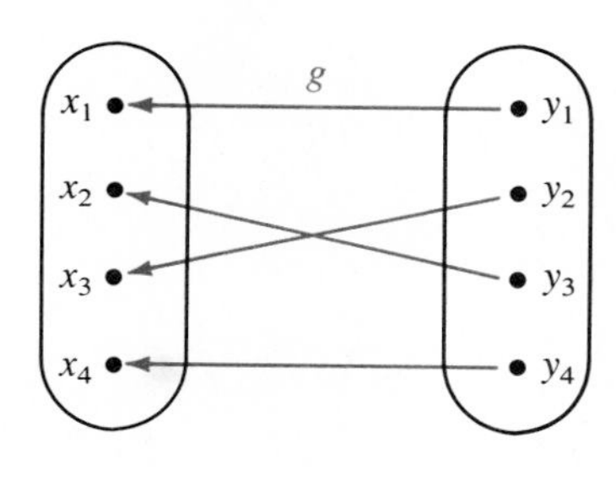

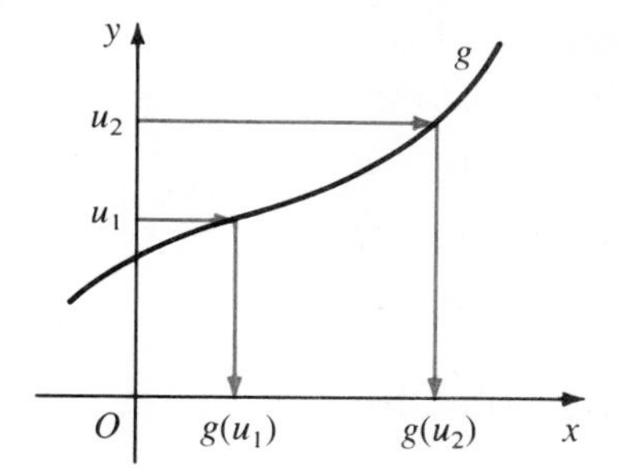

Such a construction of g is always possible if f is one-to-one. The reversal described suggests how to associate with each one-to-one function f a function g that "undoes what f does." We call g the *inverse* of f.

DEFINITION

The **inverse**, g, of the one-to-one function f is defined as follows:

$$g(u) = v \text{ if and only if } f(v) = u.$$

Note that the domain of g is the range of f, and the range of g is the domain of f. We usually denote the inverse of f by f^{-1}. Therefore:

$$f^{-1}(u) = v \text{ if and only if } f(v) = u.$$

When used with functions, f^{-1} means "inverse," not "reciprocal."

The definition of inverse function implies that the inverse of f^{-1} is f, that is, $(f^{-1})^{-1} = f$. Thus the following statements are essentially the same.

$$f^{-1}(f(x)) = x \text{ for each } x \text{ in the domain of } f.$$
$$f(f^{-1}(x)) = x \text{ for each } x \text{ in the domain of } f^{-1}.$$

The equations immediately above can be written as

$$f^{-1} \circ f = I \text{ and } f \circ f^{-1} = I,$$

where I is the identity function on the domain of f and range of f respectively.

EXAMPLE 3 Let f be the function $f\colon x \to \dfrac{1}{x - 2}$. Find f^{-1}.

Solution Let $y = f^{-1}(x)$. Then $x = f(y) = \dfrac{1}{y - 2}$.

Solve for y.

$$x = \frac{1}{y - 2}$$
$$y - 2 = \frac{1}{x}$$
$$y = 2 + \frac{1}{x} = \frac{2x + 1}{x}$$

Therefore $f^{-1}(x) = \dfrac{2x + 1}{x}$. **Answer**

Check $$f^{-1}(f(x)) = f^{-1}\left(\frac{1}{x - 2}\right) = \frac{2\left(\frac{1}{x - 2}\right) + 1}{\frac{1}{x - 2}} = \frac{2 + x - 2}{1} = x$$

$$f(f^{-1}(x)) = f\left(\frac{2x + 1}{x}\right) = \frac{1}{\frac{2x + 1}{x} - 2} = \frac{x}{2x + 1 - 2x} = x$$

The following theorem relates the graph of a one-to-one function f and the graph of its inverse f^{-1}.

THEOREM 1 The graph of a one-to-one function f and the graph of its inverse f^{-1} are reflections of each other in the line $y = x$.

Proof: Recall that the reflection of a point (a, b) in the line $y = x$ is the point (b, a). Thus the theorem is proved by observing that the following statements are equivalent.

(b, a) is on the graph of $y = f^{-1}(x)$.
$a = f^{-1}(b)$.
$b = f(a)$.
(a, b) is on the graph of $y = f(x)$. ■

EXERCISES

Find $(f \circ g)(x)$ and $(g \circ f)(x)$. Specify the domains of $f \circ g$ and $g \circ f$.

A

1. $f(x) = x - 1,\ g(x) = 2x + 1$

2. $f(x) = x + 2,\ g(x) = 2x - 1$

3. $f(x) = x^2 - 1,\ g(x) = x + 1$

4. $f(x) = x^2 - 4,\ g(x) = x - 2$

5. $f(x) = \dfrac{1}{x - 1},\ g(x) = \dfrac{1}{x}$

6. $f(x) = \dfrac{x - 2}{x},\ g(x) = \dfrac{1}{x}$

7. $f(x) = \sqrt{1 - x},\ g(x) = x^2 + 1$

8. $f(x) = \sqrt{1 - x},\ g(x) = x^2 - 1$

In Exercises 9–16, express the given function as a composition of two simpler functions. Use the format of Example 2.

9. $f\colon x \to \sqrt{x + 1}$

10. $f\colon x \to \sqrt{1 - x^2}$

11. $f\colon x \to (x^2 - 4)^8$

12. $f\colon x \to [(x - 1)^2]^3$

13. $f\colon x \to \operatorname{sgn}^2 x$

14. $f\colon x \to \operatorname{sgn} x^2$

15. $f\colon x \to (x + 1)^2$

16. $f\colon x \to \dfrac{1}{x^2 + 4}$

In Exercises 17–24, find $f^{-1}(x)$ and check your result.

17. $f(x) = \dfrac{x - 1}{2}$

18. $f(x) = 3x - 1$

19. $f(x) = \sqrt{x - 1}$

20. $f(x) = \sqrt{1 - x}$

21. $f(x) = \dfrac{1}{x + 1}$

22. $f(x) = \dfrac{x}{x - 2}$

23. $f(x) = (x - 1)^3$

24. $f(x) = \sqrt[3]{x - 1}$

25–28. In Exercises 17–20, graph f and f^{-1} in the same coordinate plane.

B

29. Show that the function $f(x) = x^n$ is one-to-one if n is odd and not one-to-one if n is even.

30. Prove that the line segment joining the points (a, b) and (b, a) has the line $y = x$ as perpendicular bisector.

In Exercises 31 and 32, $f(x) = mx + b$, where $m \neq 0$.

31. Find $f^{-1}(x)$ and verify that $f \circ f^{-1} = I$ and $f^{-1} \circ f = I$.

32. What form must f have if $f^{-1} = f$?

33. A spherical balloon is being inflated in such a way that the radius increases linearly at the rate of 0.5 cm/s. Initially the balloon has a radius of 4 cm. Write an equation to describe the volume of the balloon in terms of time t. What is the volume of the balloon after 1 min?

34. A ship leaves port at 6:00 A.M. at a speed of 22 knots, and a second ship leaves the same port at 7:00 A.M. at a speed of 20 knots. The two ships sail on perpendicular courses. Write an equation to represent the distance between them in terms of hours after 7:00 A.M.

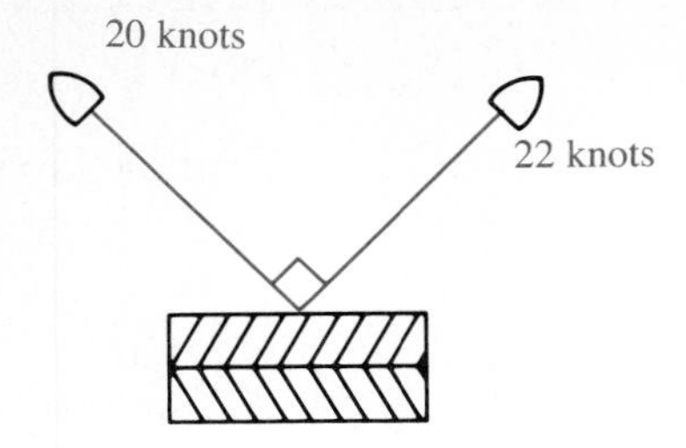

In Exercises 35–38, f is the **linear fractional function** defined by $f(x) = \frac{ax + b}{cx + d}$, where $ad - bc \neq 0$.

35. Find f^{-1}.

36. Verify that $f \circ f^{-1} = I$ and $f^{-1} \circ f = I$.

C 37. Find $f \circ f$. Find a condition on a, b, c, and d to make f its own inverse.

38. Show that f is a constant function if $ad - bc = 0$.

In Exercises 39–42, $G = \{f\colon f(x) = mx + b,\ m \neq 0\}$. Show the following.

39. If $f, g \in G$, then $f \circ g \in G$.

40. If $f, g, h \in G$, then $(f \circ g) \circ h = f \circ (g \circ h)$.

41. For any $f \in G$, $f \circ I = I \circ f = f$.

42. For any $f \in G$, there is an $f^{-1} \in G$ such that $f \circ f^{-1} = f^{-1} \circ f = I$.

43. Exercises 39–42 show that G is an algebraic **group** under the operation $\circ$. Use the group properties in Exercises 39–42 to show that $(f \circ g)^{-1} = g^{-1} \circ f^{-1}$. (*Hint*: $(f \circ g) \circ (f \circ g)^{-1} = I$. Now use associativity and "multiply" both sides on the left first by f^{-1}, then by g^{-1}.)

COMPUTER EXERCISES

1. Write a program to find $f(f(f(x)))$ for $f(x) = \frac{1}{2 + x}$ when x is supplied by the user.

2. Run the program in Exercise 1 for each value of x.
   a. 1 **b.** 7 **c.** 2

3. Modify the program in Exercise 1 to find the composition of $f(x) = \frac{1}{2 + x}$ with itself n times when n and x are supplied by the user.

4. Run the program in Exercise 3 for each value of x and each set of values for n.
   a. $x = 2$; $n = 4, 7, 10, 11$
   b. $x = 7$; $n = 4, 7, 10, 11$

Limits of Functions

4-4 LIMITS AND THEIR PROPERTIES

Analysis can be described as the branch of mathematics that is based on the limit concept. We discussed limits of sequences in Chapter 3 and shall now adapt the notion of limits to functions. In this section we give meaning to the symbol $\lim_{x \to c} f(x)$ (read "the limit of $f(x)$ as x approaches c"). We introduce the subject with an informal discussion that contrasts the behaviors of several functions near $x = 1$. In each example we let x approach 1 but not take on the value 1.

EXAMPLE 1 Discuss the limit of each function at 1.

a. $f(x) = x + 1$

Consider approaching 1 by values of x as given in the table and the corresponding values of $f(x)$.

x	$f(x)$
0.9	$0.9 + 1 = 1.9$
0.99	$0.99 + 1 = 1.99$
0.999	$0.999 + 1 = 1.999$
1.1	$1.1 + 1 = 2.1$
1.01	$1.01 + 1 = 2.01$
1.001	$1.001 + 1 = 2.001$

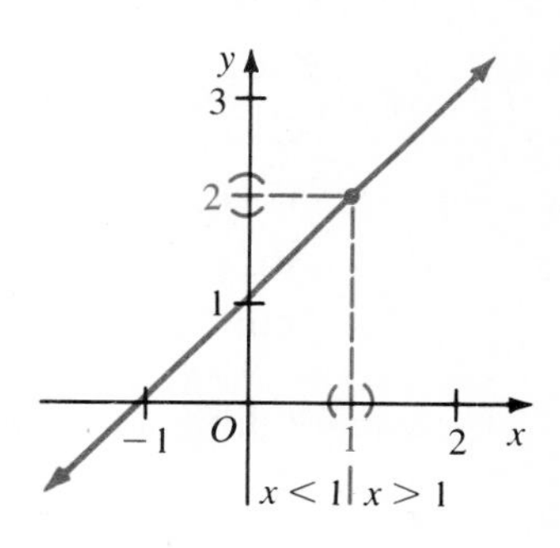

Intuitively, the closer x is to 1, the closer $f(x)$ is to 2. This behavior can be expressed by writing

$$\lim_{x \to 1} f(x) = 2.$$

Note that
$$\lim_{x \to 1} f(x) = f(1).$$

b. $g(x) = \dfrac{x^2 - 1}{x - 1} = \begin{cases} x + 1 & \text{if } x \neq 1 \\ \text{undefined} & \text{if } x = 1 \end{cases}$

The function g behaves the same way as f above since near (but not at) 1, the graph of g is the same as the graph of f. Therefore even though g is not defined at 1,

$$\lim_{x \to 1} g(x) = 2.$$

Example continued on following page

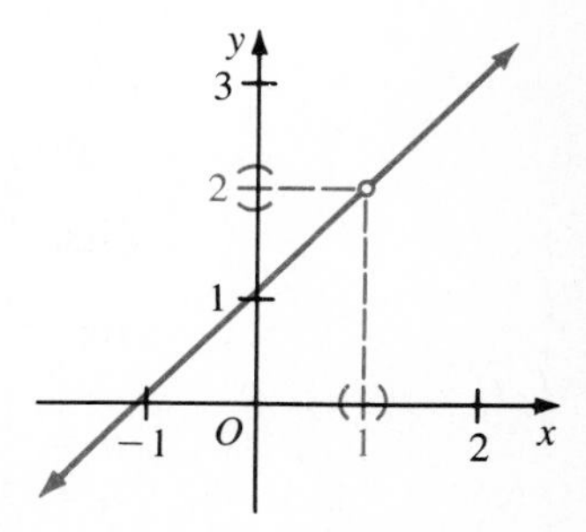

c. $h(x) = x + \operatorname{sgn}^2(x-1) = \begin{cases} x+1 & \text{if } x \neq 1 \\ 1 & \text{if } x = 1 \end{cases}$

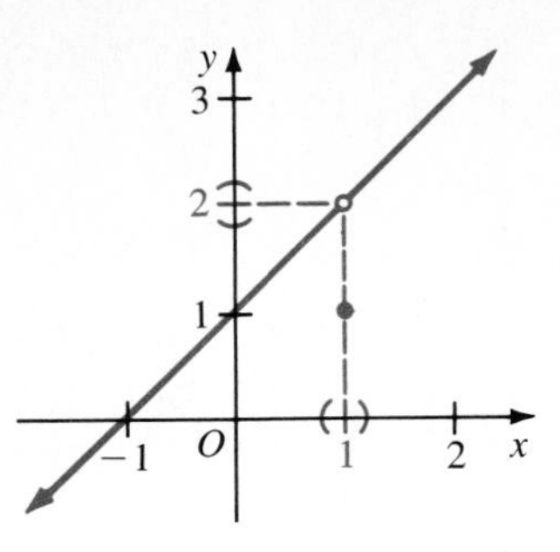

Near $x = 1$, the graph of h is the same as the graph of f or g. As in the first two examples,

$$\lim_{x \to 1} h(x) = 2.$$

Note that h is defined at 1 but that

$$\lim_{x \to 1} h(x) \neq h(1).$$

d. $k(x) = x + 1 + \operatorname{sgn}(x-1) = \begin{cases} x+2 & \text{if } x > 1 \\ 2 & \text{if } x = 1 \\ x & \text{if } x < 1 \end{cases}$

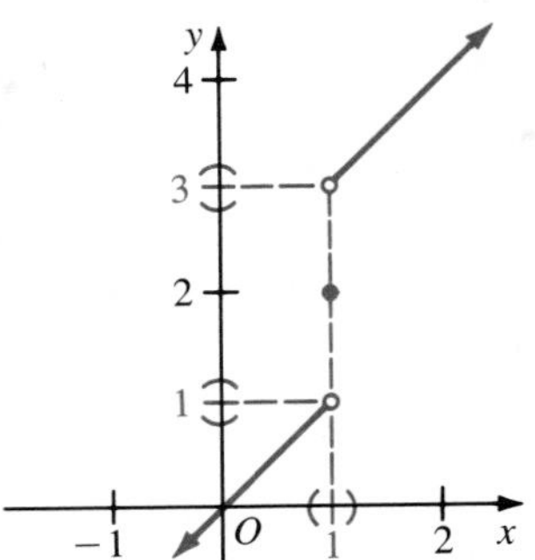

If x is near 1 but $x > 1$, then $k(x) > 3$, while if x is near 1 but $x < 1$, then $k(x) < 1$. Thus, $k(x)$ cannot approach a single number as x approaches 1, and

$$\lim_{x \to 1} k(x) \text{ does not exist.}$$

To give a precise definition of the *limit of a function f* at c, we shall use sequences $x_1, x_2, \ldots, x_n, \ldots$ that approach c and consider the corresponding sequences $f(x_1), f(x_2), \ldots, f(x_n), \ldots$ of functional values.

DEFINITION

The function f has **limit** l at c,

$$\lim_{x \to c} f(x) = l,$$

if and only if $\lim_{n \to \infty} f(x_n) = l$ for every sequence $x_1, x_2, \ldots, x_n, \ldots$ in the domain of f with $\lim_{n \to \infty} x_n = c \quad (x_n \neq c)$.

It is important to realize that for every sequence $x_1, x_2, \ldots$ approaching c the sequence $f(x_1), f(x_2), \ldots$ must approach the same limit l.

EXAMPLE 2 **a.** Let $g(x) = \dfrac{x^2 - 1}{x - 1} = \begin{cases} x+1 & \text{if } x \neq 1 \\ \text{undefined} & \text{if } x = 1 \end{cases}$.

Prove that $\lim_{x \to 1} g(x) = 2$.

b. $k(x) = x + 1 + \operatorname{sgn}(x-1) = \begin{cases} x+2 & \text{if } x > 1 \\ 2 & \text{if } x = 1 \\ x & \text{if } x < 1 \end{cases}$ as in Example 1.

Prove that $\lim_{x \to 1} k(x)$ does not exist.

Solution **a.** Let $x_1, x_2, \ldots, x_n, \ldots$ be an *arbitrary* sequence such that $x_n \neq 1$ and $\lim_{n\to\infty} x_n = 1$. Since $x_n \neq 1$,

$$g(x_n) = \frac{x_n^2 - 1}{x_n - 1} = x_n + 1$$

$$\lim_{n\to\infty} g(x_n) = \lim_{n\to\infty} (x_n + 1).$$

By Theorem 11 in Chapter 3,

$$\begin{aligned}\lim_{n\to\infty} g(x_n) &= \lim_{n\to\infty} x_n + \lim_{n\to\infty} 1 \\ &= 1 + 1 = 2.\end{aligned}$$

Therefore $\lim_{x\to 1} g(x) = 2$.

b. It suffices to find one sequence $x_1, x_2, \ldots, x_n, \ldots$ with $\lim_{n\to\infty} x_n = 1$ and $x_n \neq 1$ such that $\lim_{n\to\infty} k(x_n)$ does not exist. Let $x_n = 1 + \frac{(-1)^n}{n}$. Then

$$\lim_{n\to\infty} x_n = 1.$$

Thus $k(x_n)$ is given by

$$k(x_n) = \begin{cases} x_n + 2 = 3 + \dfrac{(-1)^n}{n} & \text{if } n \text{ is even} \\ x_n = 1 + \dfrac{(-1)^n}{n} & \text{if } n \text{ is odd} \end{cases}$$

for all x_n other than 1. If n is even, $k(x_n) > 3$. If n is odd, $k(x_n) < 1$. Hence $\lim_{n\to\infty} k(x_n)$ cannot approach a limit. Therefore $\lim_{x\to 1} k(x)$ does not exist.

The following theorem provides rules by which many limits can be computed. Since the limit of a function is based on the idea of the limit of a sequence, the truth of the theorem should come as no surprise. In particular, we expect "the limit of a sum is the sum of the limits."

THEOREM 2 Let $\lim_{x\to c} f(x) = l$, $\lim_{x\to c} g(x) = m$, and a be a constant.

1. $\lim_{x\to c} [af(x)] = al$

2. $\lim_{x\to c} [f(x) + g(x)] = l + m$

3. $\lim_{x\to c} [f(x) - g(x)] = l - m$

4. $\lim_{x\to c} [f(x) \cdot g(x)] = lm$

5. $\lim_{x\to c} \left[\dfrac{f(x)}{g(x)}\right] = \dfrac{l}{m}$ provided $g(x), m \neq 0$

The proofs of the parts of Theorem 2 follow from the corresponding properties of sequences and are left as Exercises 47–51.

The following theorem is an immediate consequence of the definition of limit and Theorem 11, page 115.

THEOREM 3 1. $\lim_{x \to c} x = c$ 2. $\lim_{x \to c} k = k$ for every constant k.

EXAMPLE 3 Find the limit of each function as x approaches 2.

a. $f(x) = \dfrac{3x + 4}{x^2}$ **b.** $g(x) = (x - 4)^3$

Solution **a.** (a) $\lim_{x \to 2} x = 2$ Theorem 3 part (1)

(b) $\lim_{x \to 2} 4 = 4$ Theorem 3 part (2)

(c) $\lim_{x \to 2} (3x + 4) = 3 \cdot 2 + 4 = 10$ (a), (b), and Theorem 2 part (1)

(d) $\lim_{x \to 2} x^2 = \lim_{x \to 2} (x \cdot x) = 2 \cdot 2 = 4$ (a) and Theorem 2 part (4)

(e) $\lim_{x \to 2} \dfrac{3x + 4}{x^2} = \dfrac{10}{4} = \dfrac{5}{2}$ (c), (d), and Theorem 2 part (5)

$\lim_{x \to 2} f(x) = \frac{5}{2}$. **Answer**

b. Consider g as a product and apply Theorem 2 part (4) twice.

$$\begin{aligned}\lim_{x \to 2} g(x) &= \lim_{x \to 2} [(x - 4)^2(x - 4)] \\ &= \lim_{x \to 2} (x - 4)^2 \lim_{x \to 2} (x - 4) \\ &= \lim_{x \to 2} (x - 4) \lim_{x \to 2} (x - 4) \lim_{x \to 2} (x - 4)\end{aligned}$$

Since $\lim_{x \to 2} (x - 4) = \lim_{x \to 2} x - \lim_{x \to 2} 4 = 2 - 4 = -2$, $\lim_{x \to 2} g(x) = (-2)(-2)(-2) = -8$. **Answer**

The following theorems state alternative ways to find the limit of a function at c.

THEOREM 4 The following statements are equivalent.

$$\lim_{x \to c} f(x) = l \qquad \lim_{x \to c} (f(x) - l) = 0 \qquad \lim_{x \to c} |f(x) - l| = 0$$

THEOREM 5 If $g(x) \le f(x) \le h(x)$ and $\lim_{x \to c} g(x) = l$ and $\lim_{x \to c} h(x) = l$, then $\lim_{x \to c} f(x) = l$.

For a proof of Theorem 5, sometimes called the *"squeeze" theorem*, see Exercises 52 and 53. We will use these theorems to prove the following theorem.

THEOREM 6 For all $c > 0$, $\lim_{x \to c} \sqrt{x} = \sqrt{c}$.

Proof: For any $x \ge 0$, $\sqrt{x} - \sqrt{c} = \dfrac{\sqrt{x} - \sqrt{c}}{1} \cdot \dfrac{\sqrt{x} + \sqrt{c}}{\sqrt{x} + \sqrt{c}} = \dfrac{x - c}{\sqrt{x} + \sqrt{c}}$.

Note that $|\sqrt{x} - \sqrt{c}| = \left|\dfrac{x - c}{\sqrt{x} + \sqrt{c}}\right| = \dfrac{|x - c|}{\sqrt{x} + \sqrt{c}} \le \dfrac{|x - c|}{\sqrt{c}}$.

Thus $0 \le |\sqrt{x} - \sqrt{c}| \le \dfrac{|x - c|}{\sqrt{c}}$.

By Theorem 2, part (1) and Theorem 3, part (1) $\lim\limits_{x \to c} \dfrac{|x - c|}{\sqrt{c}} = 0$. Therefore $\lim\limits_{x \to c} |\sqrt{x} - \sqrt{c}| = 0$ by Theorem 5 and $\lim\limits_{x \to c} \sqrt{x} = \sqrt{c}$ by Theorem 4. ■

EXAMPLE 4 Find $\lim\limits_{x \to 9} \left(\dfrac{4 - \sqrt{x}}{25 + x^2}\right)$.

Solution Various theorems from this section are used to compute the limit.

$$\lim_{x \to 9} \left(\frac{4 - \sqrt{x}}{25 + x^2}\right) = \frac{\lim\limits_{x \to 9} (4 - \sqrt{x})}{\lim\limits_{x \to 9} (25 + x^2)}$$

$$= \frac{\lim\limits_{x \to 9} 4 - \lim\limits_{x \to 9} \sqrt{x}}{\lim\limits_{x \to 9} 25 + \lim\limits_{x \to 9} x^2} = \frac{4 - \sqrt{9}}{25 + 9^2} = \frac{1}{106}$$

Therefore $\lim\limits_{x \to 9} \left(\dfrac{4 - \sqrt{x}}{25 + x^2}\right) = \dfrac{1}{106}$. **Answer**

EXERCISES

In Exercises 1–6 the graph of a function f is given. State whether or not $\lim\limits_{x \to 2} f(x)$ exists and, if it does, give its value.

A **1.**

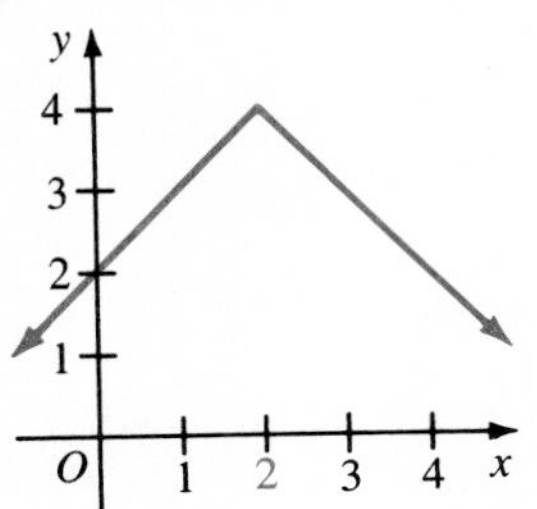

2.

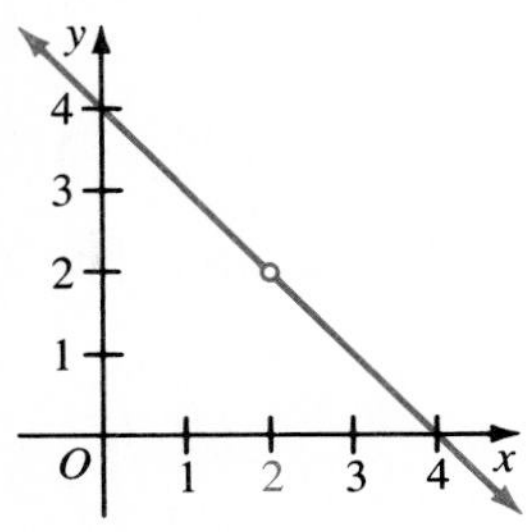

3.

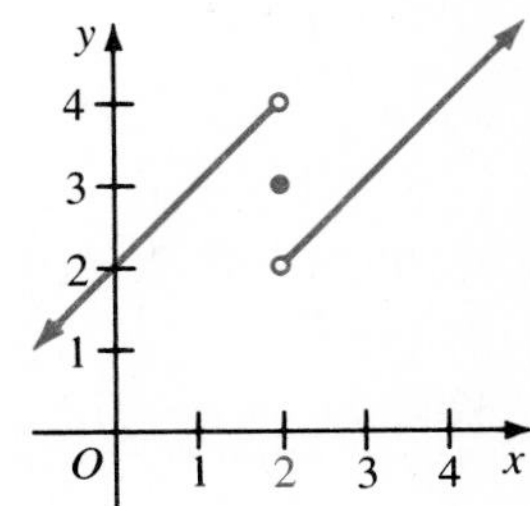

4.

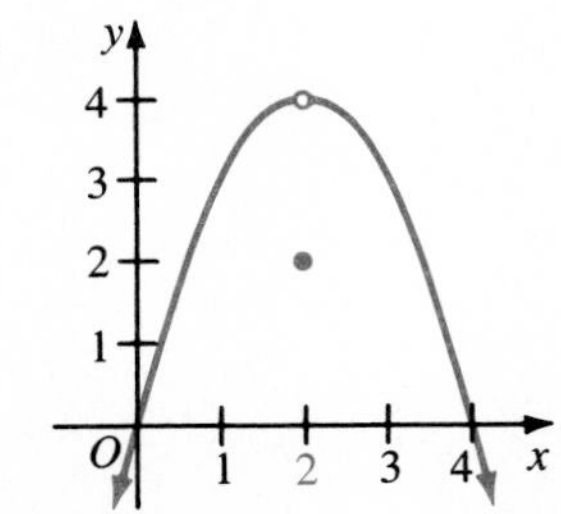

5.

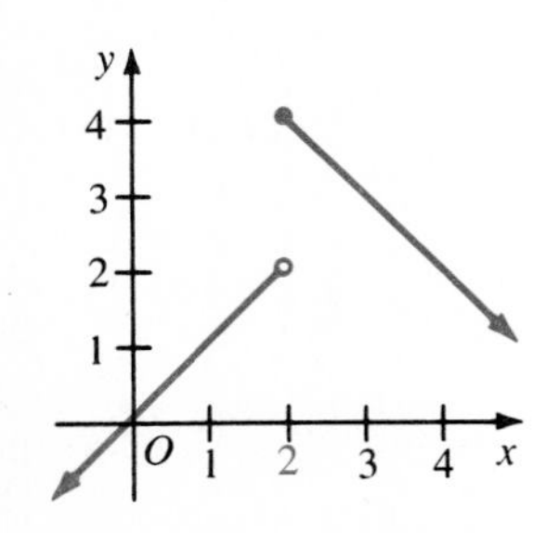

6.

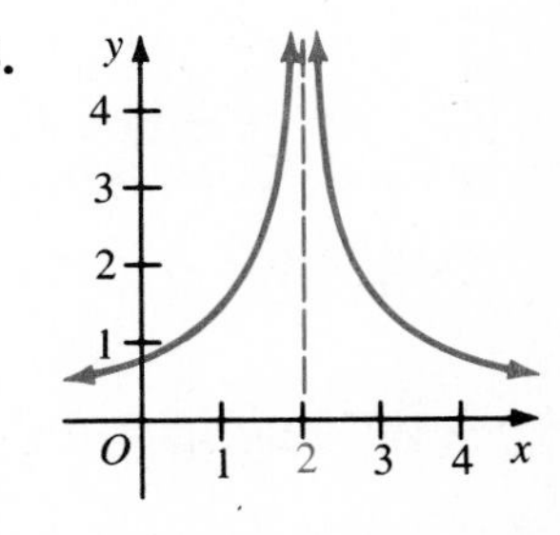

In each exercise state whether or not $\lim_{x \to c} f(x)$ exists and, if it does, give its value.

7. $f(x) = \frac{x}{|x|}$; $c = 0$

8. $f(x) = \frac{x^2}{|x|}$; $c = 0$

9. $f(x) = x[\![x]\!]$; $c = 0$

10. $f(x) = x[\![x]\!]$; $c = 1$

11. $f(x) = \operatorname{sgn}(x - 1)$; $c = 1$

12. $f(x) = \operatorname{sgn}(x - 1)^2$; $c = 1$

13. $f(x) = \frac{x^2 - x - 2}{x - 2}$; $c = 2$

14. $f(x) = \frac{x^2 - x - 2}{x + 1}$; $c = -1$

15. $f(x) = \frac{x^3 + 1}{x + 1}$; $c = -1$

16. $f(x) = \frac{x^3 - 1}{x - 1}$; $c = 1$

Use Theorems 2, 3, and 6 to prove the following limit statements.

17. $\lim_{x \to 2} (3x + 4) = 10$

18. $\lim_{x \to 2} (2x + 5) = 9$

19. $\lim_{x \to 3} (x^2 + 1) = 10$

20. $\lim_{x \to 2} (x^2 - x - 2) = 0$

21. $\lim_{x \to 1} \frac{6}{x + 2} = 2$

22. $\lim_{x \to 2} \frac{x}{x - 1} = 2$

23. $\lim_{x \to 4} (x + \sqrt{x}) = 6$

24. $\lim_{x \to 1} (3x - 2\sqrt{x}) = 1$

25. $\lim_{x \to 2} \frac{x^2 - 1}{x^2 + 2} = \frac{1}{2}$

26. $\lim_{x \to 4} \frac{\sqrt{x}}{x^2 + 9} = \frac{2}{25}$

Find each limit.

27. $\lim_{x \to 2} (4x^3 + 5x + 9)$

28. $\lim_{x \to 4} (7x^2 - 4x + 9)$

29. $\lim_{x \to 0} \sqrt{2}$

30. $\lim_{x \to 5} \pi$

31. $\lim_{x \to \sqrt{2}} (x^2 - 3)(x - 4)$

32. $\lim_{h \to 0} (3h + 5)(7h - 9)$

B

33. $\lim_{x \to \frac{1}{2}} \left(\frac{2x^2 + 5x - 3}{5x^2 - x + 6}\right)$

34. $\lim_{x \to 0} \left(\frac{4x^3 - 2x + 5}{2x^3 + 2x + 5}\right)$

35. $\lim_{x \to 1} \left(\frac{x^2}{x - 1} - \frac{1}{x - 1}\right)$

36. $\lim_{x \to 1} \left(\frac{(x - 1)^5}{x^5 - 1}\right)$

37. $\lim_{x \to 0} \left(\frac{x^3 + 8}{x^3 - 8}\right)$

38. $\lim_{x \to 1} \left(1 + x + \frac{x^2}{2} + \frac{x^3}{3}\right)$

39. $\lim_{h \to 0} \left[\left(\frac{1}{h}\right)\left(\frac{1}{\sqrt{1 + h}} - 1\right)\right]$

40. $\lim_{h \to 0} \left(\frac{5 - \sqrt{25 + h}}{h}\right)$

Use the definition of limit and Theorem 11 on page 115 to prove the following statements.

41. $\lim_{x \to 0} (2x + 1) = 1$

42. $\lim_{x \to -1} \frac{1 - x^2}{1 + x} = 2$

43. $\lim_{x \to 2} \frac{4 - x^2}{2 - x} = 4$

44. $\lim_{x \to 3} (x^2 - x) = 6$

Use Theorem 11, Section 3-6, to prove the following.

C

45. Theorem 3, part (1)
46. Theorem 3, part (2)
47. Theorem 2, part (1)
48. Theorem 2, part (2)
49. Theorem 2, part (3)
50. Theorem 2, part (4)
51. Theorem 2, part (5)

Exercises 52 and 53 refer to this proposition:
(A) If $0 \leq F(x) \leq G(x)$ and $\lim_{x \to c} G(x) = 0$, then $\lim_{x \to c} F(x) = 0$.

52. Explain how to deduce Theorem 5, page 160, from proposition (A).

53. Prove proposition (A) by justifying steps (a) through (e). Let x_1, $x_2, \ldots, x_n, \ldots$ be an arbitrary sequence such that $\lim_{n \to \infty} x_n = c$ and $x_n \neq c$. Let h be any positive number.

a. $\lim_{n \to \infty} G(x_n) = 0$.

b. There is a number N such that

$$G(x_n) = |G(x_n) - 0| < h \text{ if } n \geq N.$$

c. $|F(x_n) - 0| = F(x_n) < h$ if $n \geq N$.

d. $\lim_{n \to \infty} F(x_n) = 0$.

e. $\lim_{x \to c} F(x) = 0$.

54. Let $f(x) = x^3$. Find $\lim_{h \to 0} \left(\frac{f(x + h) - f(x)}{h} \right)$ and $\lim_{x \to c} \left(\frac{f(x) - f(c)}{x - c} \right)$.

4-5 CONTINUOUS FUNCTIONS

Refer to Example 1 in Section 4-4. The graph of f has no break at $x = 1$. However the graphs of functions g, h, and k do have breaks at $x = 1$. We say that f is *continuous* at 1 and that g, h, and k are not continuous at 1. Using limits we can give an analytic definition of continuity at c.

DEFINITION

The function f is **continuous at c** if and only if

$$\lim_{x \to c} f(x) = f(c).$$

This definition requires that

1. f must be defined at c,
2. f must have a limit as x approaches c, and
3. this limit must equal $f(c)$.

To appreciate these requirements consider Example 1 of Section 4-4 again. For f, $f(1)$ exists, $\lim_{x \to 1} f(x)$ exists, and $\lim_{x \to 1} f(x) = f(1)$. For g, since g is not defined for 1, requirements 1 and 3 are not met. The function h is defined at 1 and $\lim_{x \to 1} h(x)$ exists, but $\lim_{x \to 1} h(x) \neq h(1)$. Requirement 2 is not met for the function k since $\lim_{x \to 1} k(x)$ does not exist. In general, if any of these requirements is not met, we say the function is *discontinuous* at c. A function is **continuous** if it is continuous at each member of its domain.

We can prove from the theorems of Section 4-4 that linear combinations, products, and quotients of continuous functions are continuous. (See Exercises 27–29.) (The domain of a quotient function excludes, of course, values at which the denominator is 0.)

A **polynomial function** is any function of the form

$$f(x) = a_0x^n + a_1x^{n-1} + \cdots + a_{n-1}x + a_n.$$

THEOREM 7 Polynomial functions are continuous. In particular, linear functions and quadratic functions are continuous.

The proof of this theorem follows from the continuity of linear combinations and the continuity of products. Chapter 5 is devoted to a study of these important functions.

A *rational function* is continuous at each member of its domain. For example

$$x \to \frac{1}{x}, \quad x \to \frac{x + 1}{x^2 + 1}, \quad \text{and} \quad x \to \frac{x}{x^2 - 4}$$

are all continuous functions. Recall that any value of x that makes the denominator 0 is excluded from the domain.

The **square root function** $x \to \sqrt{x}$ is continuous at each member c of its domain. For $c > 0$, continuity follows from Theorem 6, page 160. The proof of continuity at 0 is left as Exercise 33.

The next theorem shows that the composition of two continuous functions is continuous.

THEOREM 8 If g is continuous at c, and f is continuous at $g(c)$, then $f \circ g$ is continuous at c; that is, $\lim_{x \to c} f(g(x)) = f(g(c))$.

Proof: Let $x_1, x_2, \ldots, x_n, \ldots$ be any sequence such that $\lim_{n \to \infty} x_n = c$. Since g is continuous at c, $\lim_{n \to \infty} g(x_n) = g(c)$. Since f is continuous at $g(c)$ and the sequence $g(x_1), g(x_2), \ldots, g(x_n), \ldots$ has $g(c)$ as limit, we have $\lim_{n \to \infty} f(g(x_n)) = f(g(c))$. Thus, $\lim_{x \to c} f(g(x)) = f(g(c))$. ■

EXAMPLE 1 Prove that $\varphi: x \to \sqrt{\dfrac{x^2 + 1}{x^2 - 1}}$ is continuous.

Solution Let $u = g(x) = \dfrac{x^2 + 1}{x^2 - 1}$ and $f(u) = \sqrt{u}$.

Then $f(g(x)) = \sqrt{g(x)} = \sqrt{\dfrac{x^2 + 1}{x^2 - 1}} = \varphi(x)$ and $\varphi = f \circ g$.

As the quotient of polynomial functions, g is continuous. Since the square root function f is also continuous, the composition of f and g, φ, is continuous.

Since $|x| = \sqrt{x^2}$, the method used in Example 1 can be used to show that the absolute value function $x \rightarrow |x|$ is continuous.

EXAMPLE 2 Show that φ: $x \rightarrow |x^2 - 1| - 1$ is continuous and draw its graph.

Solution The function φ is the composition of simpler functions.

$$x \rightarrow x^2 - 1 \rightarrow |x^2 - 1| \rightarrow |x^2 - 1| - 1$$

Each function in the chain of functions above is continuous. Thus by Theorem 8, φ is continuous.

To graph φ, draw in turn the graphs of $x \rightarrow x^2 - 1$, $x \rightarrow |x^2 - 1|$, and $x \rightarrow |x^2 - 1| - 1 = \varphi(x)$.

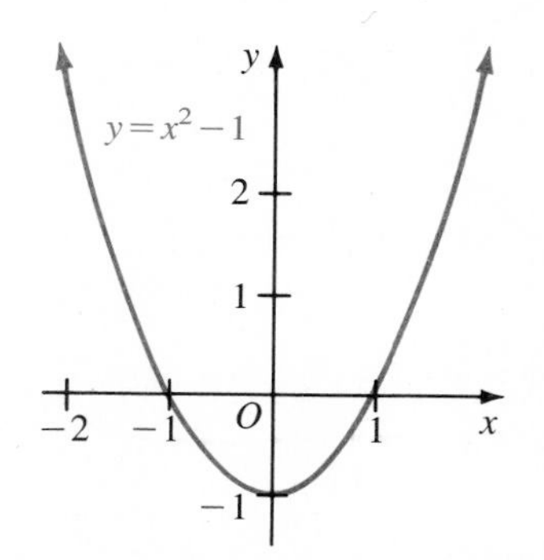

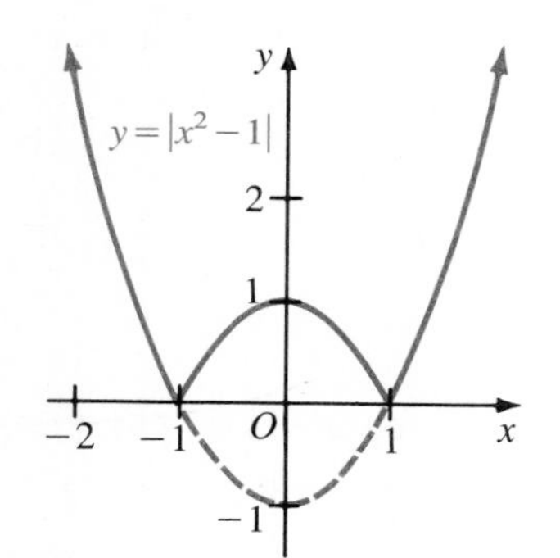

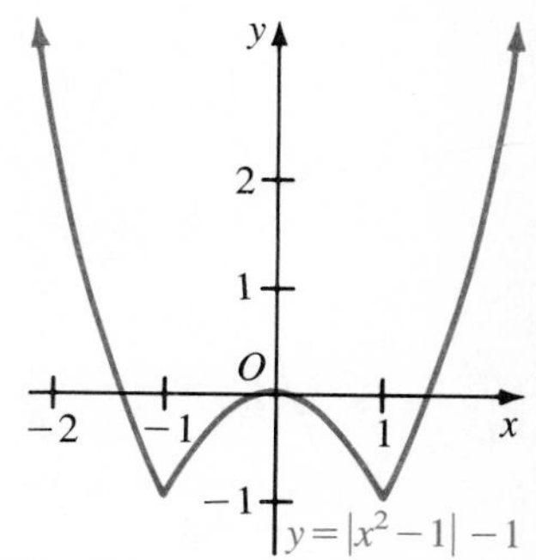

Example 2 illustrates the fact that the graph of a continuous function can have sharp corners.

EXERCISES

In Exercises 1–10 explain how you know that the function φ is continuous.

A

1. $\varphi(x) = |x + 1|$

2. $\varphi(x) = |x - 1| + 1$

3. $\varphi(x) = |1 - x^2|$

4. $\varphi(x) = x|x|$

5. $\varphi(x) = \sqrt{4 - x}$

6. $\varphi(x) = \sqrt{4 - x^2}$

7. $\varphi(x) = x\sqrt{x}$

8. $\varphi(x) = |1 - |x||$

9. $\varphi(x) = \dfrac{x + 1}{\sqrt{x^2 + 1}}$

10. $\varphi(x) = \sqrt{\dfrac{x + 1}{x^2 + 1}}$

11–18. Graph the functions of Exercises 1–8.

Discuss the continuity or discontinuity of each function at the values of x that are given.

19. $f(x) = [\![x]\!]$ at 5; at 4.5

20. $g(x) = \operatorname{sgn} x$ at 0; at -1

21. $f(x) = \begin{cases} 2x - 1 & \text{if } x \geq 3 \\ 8 - x & \text{if } x < 3 \end{cases}$ at 3; at 0

22. $g(x) = \begin{cases} x^2 - 1 & \text{if } x \geq 0 \\ 2x + 4 & \text{if } x < 0 \end{cases}$ at -2; at 0

In Exercises 23 and 24 the symbol max $\{a, b\}$ denotes the larger of the numbers a and b, and min $\{a, b\}$ denotes the smaller of them.

B **23.** Show that $\max\{a, b\} = \frac{1}{2}(a + b + |a - b|)$.

24. Show that $\min\{a, b\} = \frac{1}{2}(a + b - |a - b|)$.

If f and g are functions having the same domain, max $\{f, g\}$ is the function whose value at x is max $\{f(x), g(x)\}$; min $\{f, g\}$ is defined similarly. These ideas are illustrated for $f(x) = |x|$ and $g(x) = 2 - x^2$.

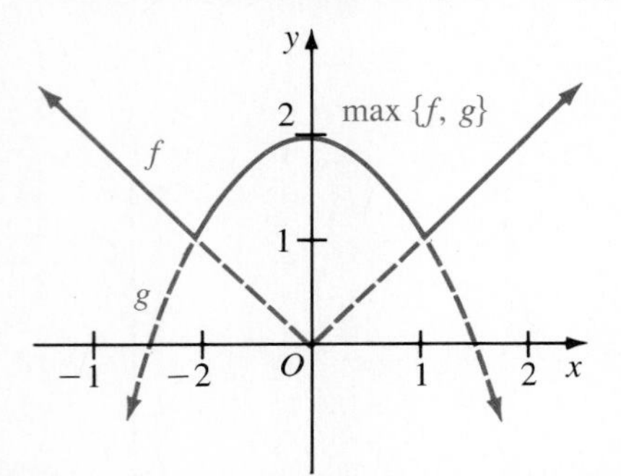

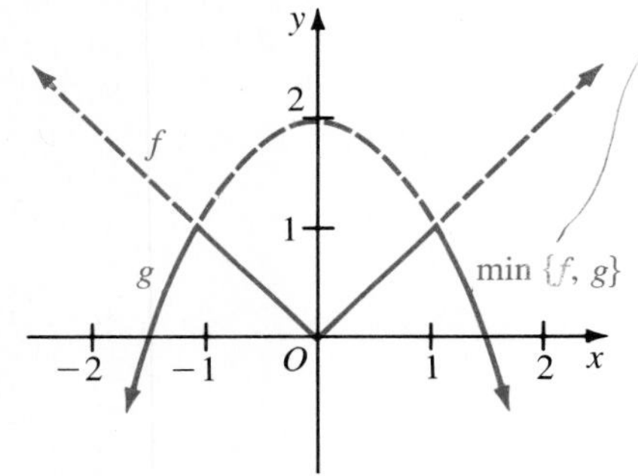

25. Prove: If f and g are continuous, so is max $\{f, g\}$.

26. Prove: If f and g are continuous, so is min $\{f, g\}$.

Let f and g be continuous at c. Use Theorem 11, page 115, to prove the statements in Exercises 27–29.

27. Any linear combination, $af + bg$, is continuous at c.

28. The product function, fg, is continuous at c.

29. The quotient function, $\frac{f}{g}$, is continuous at c, provided that $g(c) \neq 0$.

C **30.** Use mathematical induction to prove that $f(x) = x^n$ is continuous for every positive integer n.

31. Extend Exercise 30 by showing that $f(x) = x^n$ is continuous for every integer n.

32. Show that $f(x) = x^n$ $(x > 0)$ is continuous for every n of the form $k + \frac{1}{2}$, where k is an integer.

33. Prove that $f(x) = \sqrt{x}$ is continuous at 0 by justifying steps (a) through (d):

Let $x_1, x_2, \ldots, x_n, \ldots$ be an arbitrary sequence such that $x_n > 0$ and $\lim_{n \to \infty} x_n = 0$.

Let h be any positive number.

a. There is a number N such that

$$x_n = |x_n - 0| < h^2 \text{ whenever } n > N.$$

b. Therefore $|\sqrt{x_n} - 0| = \sqrt{x_n} < h$ whenever $n > N$.

c. $\lim_{n \to \infty} \sqrt{x_n} = 0.$

d. $\lim_{x \to 0} \sqrt{x} = 0.$

In Exercises 34–36, $Q = \{x \in \mathcal{R}: x \text{ is rational}\}$, and $K = \{x \in \mathcal{R}: x \text{ is irrational}\}$. Both Q and K are **dense** in $\mathcal{R}$, that is, between each two real numbers there are (infinitely many) members both of Q and of K.

34. Let $f(x) = \begin{cases} 1 \text{ if } x \text{ is rational} \\ 2 \text{ if } x \text{ is irrational.} \end{cases}$ Explain why f is continuous *nowhere*.

35. Let $g(x) = \begin{cases} x \text{ if } x \text{ is rational} \\ 2 \text{ if } x \text{ is irrational.} \end{cases}$ Is g continuous anywhere? If so, where?

36. Use Q and K to construct a function that is continuous at exactly two points.

4-6 THE INTERMEDIATE-VALUE THEOREM

Suppose that heat from a source is applied to a horizontal metal bar. If heat is conducted continuously, it seems reasonable that for any temperature between 300°F and 200°F there is a point along the bar having that temperature. For example, there is a point between A and B whose temperature is 240°F. As we shall see in this section, this "*intermediate-value property*" is shared by all continuous functions.

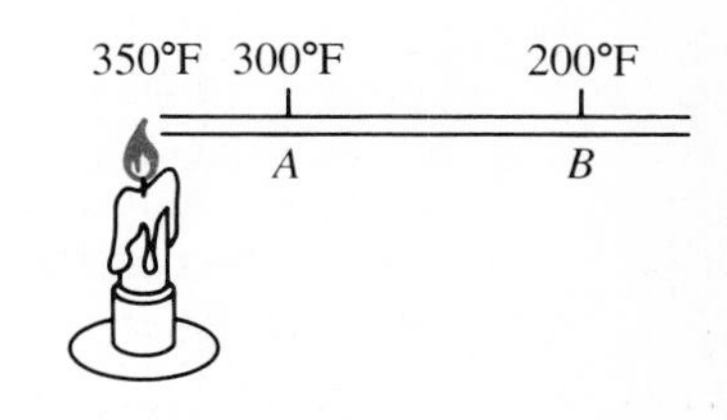

First we define closed and open intervals.

The **closed interval** $[a, b]$ is $[a, b] = \{x: a \leq x \leq b\}$.

The **open interval** (a, b) is $(a, b) = \{x: a < x < b\}$.

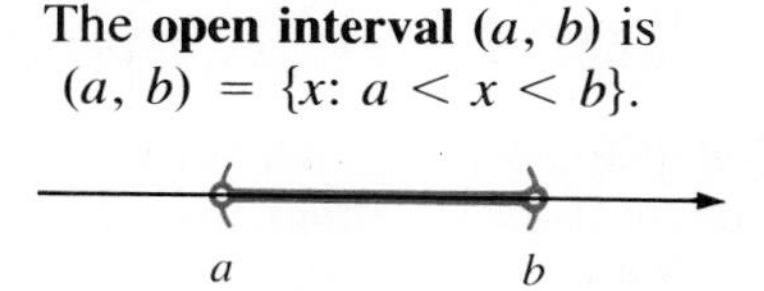

(Whether the symbol (a, b) denotes an interval or the coordinates of a point will be clear from the context.)

The graph of $f(x) = x^2 + 1$ over the interval $[0, 2]$ is shown at the right. Notice that $[0, 2]$ is unbroken, the function f is continuous, and the graph is also unbroken. In general, the graph of any function continuous on an interval has no breaks. This suggests that such a function takes on all values between any two values.

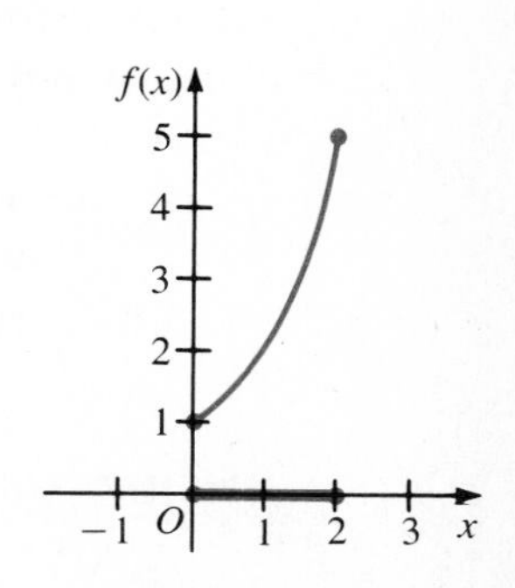

THEOREM 9 **The Intermediate-Value Theorem**

Let f be continuous on $[a, b]$ and let m be any number between $f(a)$ and $f(b)$. Then there is at least one number c in (a, b) such that $f(c) = m$.

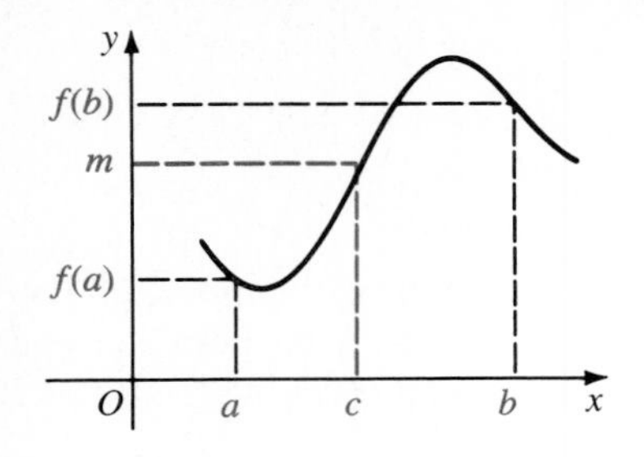

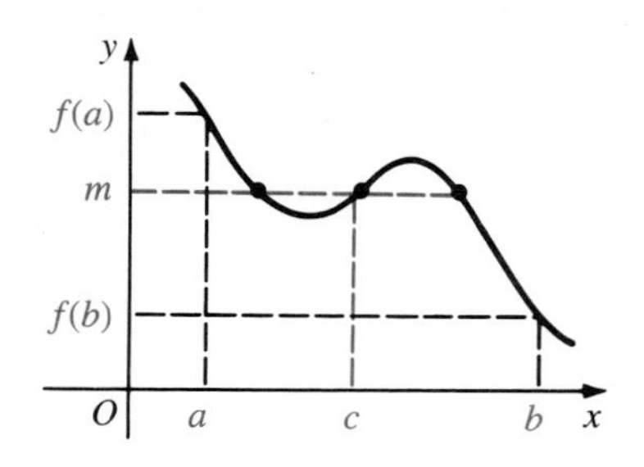

The proof of the intermediate-value theorem depends on the axiom of completeness (page 126) and is given in more advanced courses. The assumption that f is continuous is essential. The figure at the right indicates what can happen if the function is not continuous at even one point in the interval.

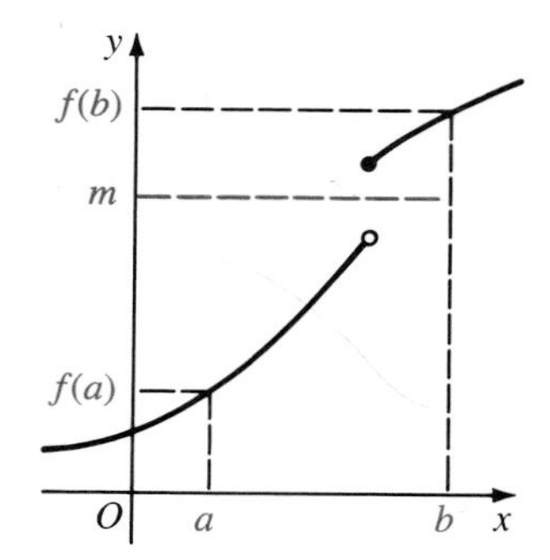

EXAMPLE 1 Find the value of c in Theorem 9 for $f(x) = 4x - x^2$, $a = 0$, $b = 3$, and $m = 2$.

Solution Since f is a polynomial function it is continuous and therefore continuous on $[0, 3]$. So $f(c) = m$ becomes $4c - c^2 = 2$.

By the quadratic formula $c = \dfrac{4 \pm \sqrt{16 - 8}}{2} = 2 \pm \sqrt{2}$.

However, only $2 - \sqrt{2}$ is in $[0, 3]$. Therefore $c = 2 - \sqrt{2}$.

We seldom can find the exact value of c, but in theoretical work we need only its existence to prove other theorems.

Recall that z is a **zero** of a function f if $f(z) = 0$. We can find or approximate zeros of continuous functions by use of the *Location Principle*, a form of the intermediate-value theorem.

THEOREM 10 **The Location Principle**

If f is continuous on $[a, b]$ and $f(a)$ and $f(b)$ have opposite signs, then f has at least one zero between a and b.

The figures on the next page illustrate the relationship between zeros of functions and the location principle. Note that a function f *may* have zeros between a and b even though $f(a)$ and $f(b)$ have the same sign.

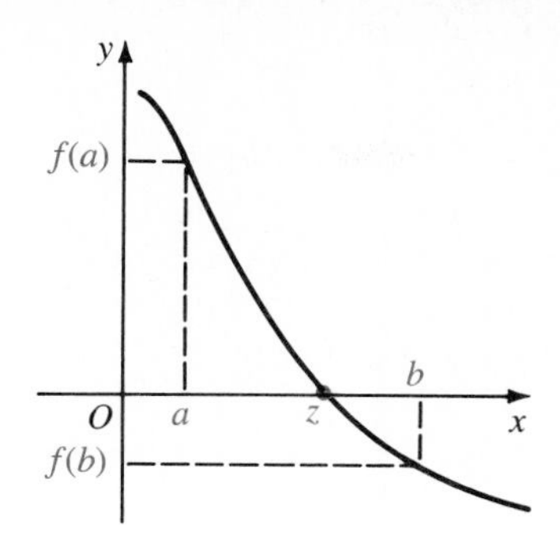

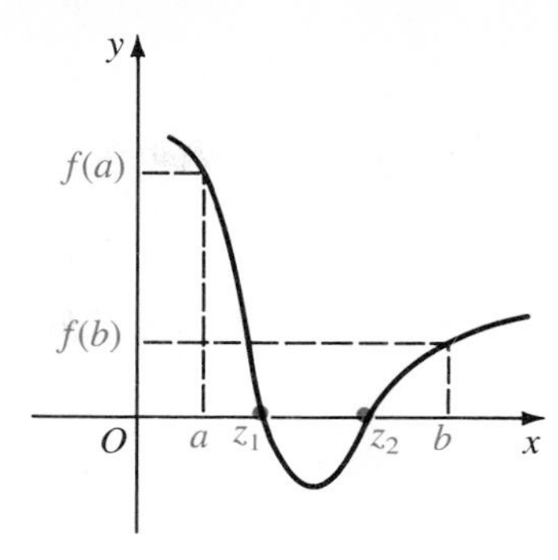

EXAMPLE 2 Verify that the polynomial function $P: x \rightarrow 2x^3 - 7x^2 - 3x + 10$ has a zero between 3 and 4.

Solution Since P is a polynomial function it is continuous and therefore continuous on $[3, 4]$.

$P(3) = 2 \cdot 3^3 - 7 \cdot 3^2 - 3 \cdot 3 + 10 = -8$

$P(4) = 2 \cdot 4^3 - 7 \cdot 4^2 - 3 \cdot 4 + 10 = 14$

Since $P(3)$ and $P(4)$ have opposite signs, P has a zero between 3 and 4.

In Example 2 we calculated $P(3)$ and $P(4)$ by *direct substitution*. Usually, however, it is easier to find values of *polynomial* functions using **synthetic substitution**, a process we will now develop.

You can verify that if $P(x) = a_0x^3 + a_1x^2 + a_2x + a_3$, then

$$\begin{aligned} P(c) &= a_0c^3 + a_1c^2 + a_2c + a_3 \\ &= ((a_0c + a_1)c + a_2)c + a_3. \end{aligned}$$

The last expression can be built up from the coefficients a_0, a_1, a_2, a_3, and c in the following steps.

1. Multiply a_0 by c.	a_0c
2. Add a_1.	$a_0c + a_1$
3. Multiply the result of Step 2 by c.	$(a_0c + a_1)c$
4. Add a_2.	$(a_0c + a_1)c + a_2$
5. Multiply the result of Step 3 by c.	$((a_0c + a_1)c + a_2)c$
6. Add a_3.	$((a_0c + a_1)c + a_2)c + a_3 = P(c)$

Steps 1–6 can be arranged in a three-line format. Circled numbers are used to designate the steps.

c | a_0	a_1	a_2	a_3
	① a_0c	③ $(a_0c + a_1)c$	⑤ $[(a_0c + a_1)c + a_2]c$
a_0	② $a_0c + a_1$	④ $(a_0c + a_1)c + a_2$	⑥ $[(a_0c + a_1)c + a_2]c + a_3 = P(c)$

For example, to find $P(3)$ and $P(4)$ when $P(x) = 2x^3 - 7x^2 - 3x + 10$ we write

$$\begin{array}{r|rrr:r} 3 & 2 & -7 & -3 & 10 \\ & & 6 & -3 & -18 \\ \hline & 2 & -1 & -6 & -8 = P(3) \end{array} \qquad \begin{array}{r|rrr:r} 4 & 2 & -7 & -3 & 10 \\ & & 8 & 4 & 4 \\ \hline & 2 & 1 & 1 & 14 = P(4) \end{array}$$

Synthetic substitution can be applied to polynomials of any degree. The coefficients must be arranged in order of descending powers of the variable and "missing" coefficients 0 and 1 must be supplied (as in Example 3).

EXAMPLE 3 The function $f\colon x \to x^4 - 6x^2 + x - 3$ has one negative zero. Locate it between two consecutive integers.

Solution By direct substitution $f(0) = -3$ and $f(-1) = -9$. The fact that these functional values have the same sign does *not* guarantee that there are no zeros between -1 and 0. However, since $f(-1)$ is *negative*, and, for large negative values of x, $f(x)$ is *positive*, the location principle states that there must be a zero less than -1. Use synthetic substitution to find more values of f.

$$\begin{array}{r|rrrr:r} -2 & 1 & 0 & -6 & 1 & -3 \\ & & -2 & 4 & 4 & -10 \\ \hline & 1 & -2 & -2 & 5 & -13 \end{array} \qquad \begin{array}{r|rrrr:r} -3 & 1 & 0 & -6 & 1 & -3 \\ & & -3 & 9 & -9 & 24 \\ \hline & 1 & -3 & 3 & -8 & 21 \end{array}$$

Since $f(-2)$ and $f(-3)$ have opposite signs, the zero is between -2 and -3. **Answer**

Repeated use of synthetic substitution with the same polynomial can be written in a more compact form. The two synthetic substitutions in Example 3 can be written as:

$$\begin{array}{r|rrrr:r} & 1 & 0 & -6 & 1 & -3 \\ \hline -2 & 1 & -2 & -2 & 5 & -13 \\ -3 & 1 & -3 & 3 & -8 & 21 \end{array}$$

EXERCISES

Find c of the intermediate-value theorem for the given f, a, b, and m.

A

1. $f(x) = 2x - 5,\ a = 1,\ b = 4,\ m = 2$
2. $f(x) = 6 - 2x,\ a = 0,\ b = 3,\ m = 5$
3. $f(x) = \frac{6}{x},\ a = 1,\ b = 3,\ m = 4$
4. $f(x) = x^2,\ a = 1,\ b = 2,\ m = 3$
5. $f(x) = 4x - x^2,\ a = 0,\ b = 2,\ m = 3$

6. $f(x) = x^2 + 3x$, $a = -1$, $b = 2$, $m = 0$
7. $f(x) = (x - 1)^2$, $a = 0$, $b = 3$, $m = 3$
8. $f(x) = x^2 - 2x - 1$, $a = 1$, $b = 4$, $m = 1$
9. $f(x) = \frac{2x - 1}{x - 2}$, $a = 3$, $b = 5$, $m = 4$
10. $f(x) = \frac{2x + 3}{x}$, $a = 1$, $b = 3$, $m = 4$

In Exercises 11–18, f has the indicated number of real zeros. Locate them between consecutive integers.

11. $f(x) = x^3 - 2x^2 - 3$, one
12. $f(x) = x^3 + x^2 + 2$, one
13. $f(x) = x^4 + 4x + 2$, two
14. $f(x) = x^4 - 5x + 3$, two
15. $f(x) = x^3 - 6x + 3$, three
16. $f(x) = x^3 - 4x + 2$, three
17. $f(x) = x^4 - 5x^2 + 3$, four
18. $f(x) = x^4 - 4x^3 + x^2 + 6x - 1$, four

In Exercises 19–22, show that f has two zeros in the specified interval.

B

19. $f(x) = 3x^3 - 9x^2 + 6x + 1$, [1, 2]
20. $f(x) = 4x^3 - 4x + 1$, [0, 1]
21. $f(x) = x^4 - 5x^2 + 6$, [1, 2]
22. $f(x) = x^4 - 11x^2 + 30$, [2, 3]
23. Locate the real root of $f(x) = x^4 - 7x^3 + 10x^2 + 14x - 24$ in [1, 2] in an interval of length 0.1.
24. Locate the real root of $f(x) = x^4 - 16x^2 + 16$ in $[-2, -1]$ in an interval of length 0.1.

C

25. Give an example of a function f defined but *not* continuous on an interval $[a, b]$ and taking on every value between $f(a)$ and $f(b)$.
26. **a.** Prove that if z_0 is the greatest real zero of a function f, continuous on $\mathcal{R}$, then $f(x)$ has the same sign for all $x > z_0$.
    b. State an analogous result involving the least real zero of f.
27. Prove that if z_1 and z_2 are consecutive real zeros of a function f, continuous on $\mathcal{R}$, then $f(x)$ has the same sign for all x between z_1 and z_2.

COMPUTER EXERCISES

1. Write a computer program to approximate to the nearest hundredth a zero of $f(x) = x^2 - 2$ in the interval [1, 2].
2. Modify and run the program in Exercise 1 for each function and interval.
   a. $f(x) = x^3 + x + 1$; $[-1, 1]$
   b. $f(x) = 2x^3 + 3x^2 + 3x + 1$; $[-1, 0]$
3. Use a modification of the program in Exercise 1 to approximate the zeros of $f(x) = 8x^3 - 26x^2 + 23x - 6$ to the nearest hundredth in the interval $[-1, 1]$. (*Hint*: There are two answers.)

4-7 LIMITS INVOLVING INFINITY

Suppose that an object is placed x units from a thin lens with focal length f. Then an image will be formed some y units from the lens. The relationship between the object distance x and the image distance y is given by

$$\frac{1}{x} + \frac{1}{y} = \frac{1}{f}.$$

Solving for y in terms of x we have

$$y = \frac{xf}{x - f}.$$

This is an example of a *rational function* useful in the study of optics.

A quotient of polynomial functions not having common zeros is a **rational function**. A rational function is not defined at points for which the denominator is 0. Such a point is called a *singularity* of the function. For example $x \to \frac{x^2 - 1}{x^2 - 4}$ has singularities at $x = 2$ and at $x = -2$. The behavior of a rational function near its singularities is of great importance in graphing rational functions.

EXAMPLE 1 Graph $g: x \to \frac{x - 3}{x - 2}$.

Solution Investigate the behavior of g to the left and right of its singularity 2. To the left of 2, $g(x) > 0$ and increases without bound when x gets closer to 2. To the right of 2, $g(x) < 0$ and decreases without bound as x gets closer to 2, as shown at the left below.

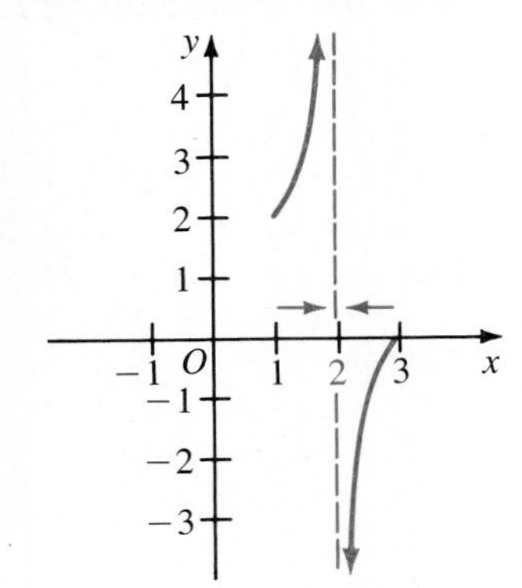

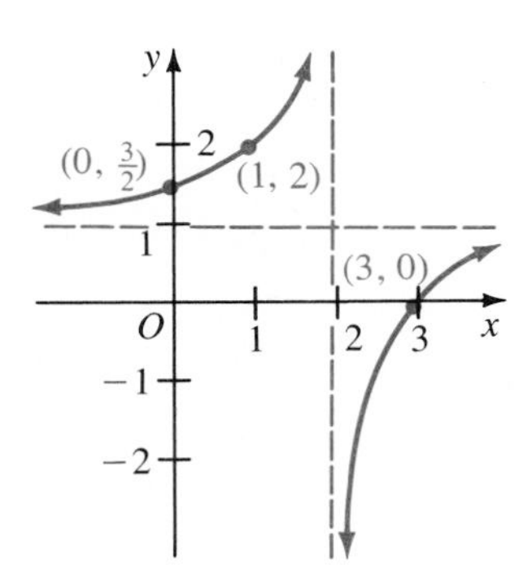

Investigate the behavior of $g(x) = \frac{x - 3}{x - 2}$ for large values of $|x|$. Divide the numerator and denominator by x to obtain

$$g(x) = \frac{1 - \frac{3}{x}}{1 - \frac{2}{x}}.$$

When x is large (x positive or negative), $\frac{3}{x}$ and $\frac{2}{x}$ are near 0. So $g(x)$ is near $\frac{1-0}{1-0}$, or 1. The foregoing analysis and a short table of values enable us to graph g as shown in the right-hand figure on page 172.

x	0	1	3
$g(x)$	$\frac{3}{2}$	2	0

The dashed lines drawn in the graph in Example 1 are called *asymptotes*. In general, a line L is an **asymptote** of a curve C if a point P moving on C becomes arbitrarily close to L as the distance from P to the origin increases without bound. For example, consider the line $y = 1$ in Example 1. If P is placed on the left branch of the curve and is moved farther and farther to the left, then P becomes arbitrarily close to the line $y = 1$.

We can express the behavior of g near the vertical asymptote $x = 2$ by writing $\lim_{x\to 2^-} g(x) = \infty$ (read "$g(x)$ increases without bound as x approaches 2 from the left") and $\lim_{x\to 2^+} g(x) = -\infty$ (read "$g(x)$ increases negatively without bound as x approaches 2 from the right"). Similarly we can write $\lim_{x\to\infty} g(x) = 1$ and $\lim_{x\to-\infty} g(x) = 1$ to describe the behavior of g for large values of $|x|$.

The twelve possible limits involving infinity, illustrated below and on the next page, can be given precise meaning (Exercises 19–29).

(1) $\lim_{x\to c^-} f(x) = \infty$

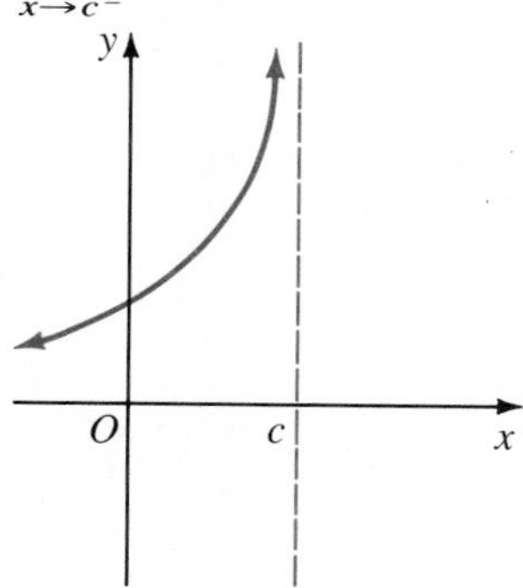

(2) $\lim_{x\to c^-} f(x) = -\infty$

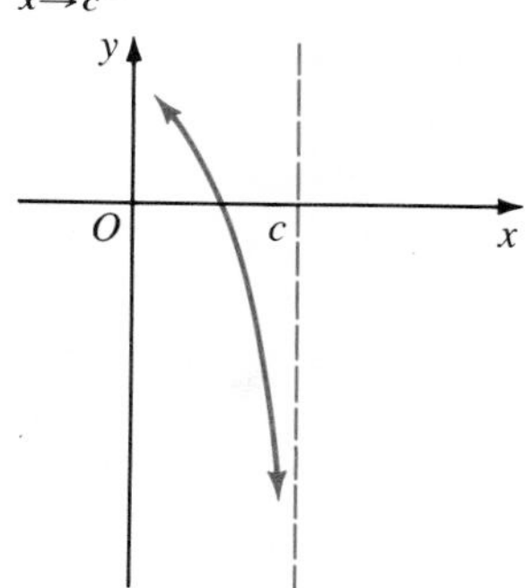

(3) $\lim_{x\to c^+} f(x) = \infty$

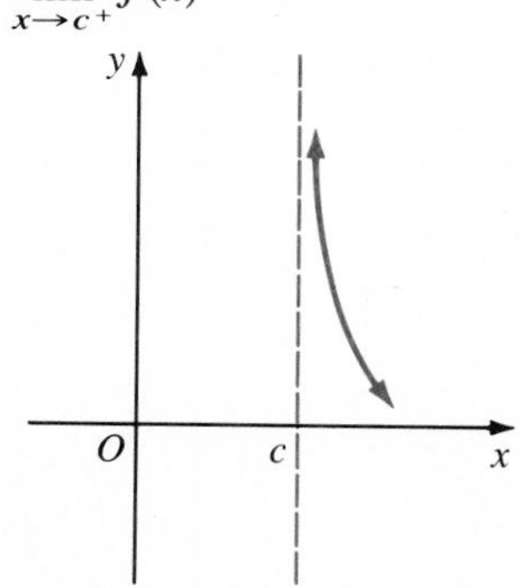

(4) $\lim_{x\to c^+} f(x) = -\infty$

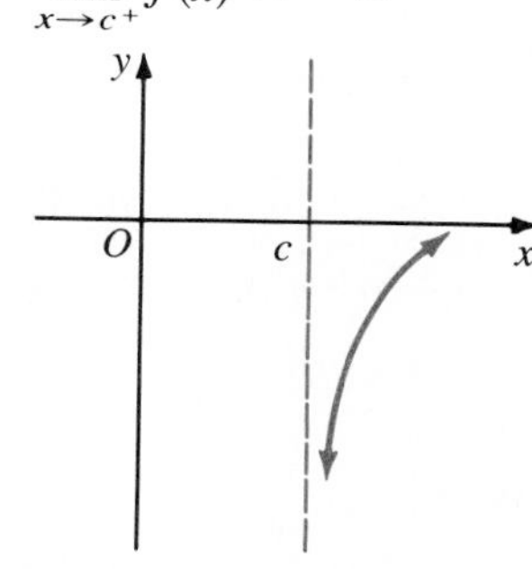

(5) $\lim_{x \to c} f(x) = \infty$

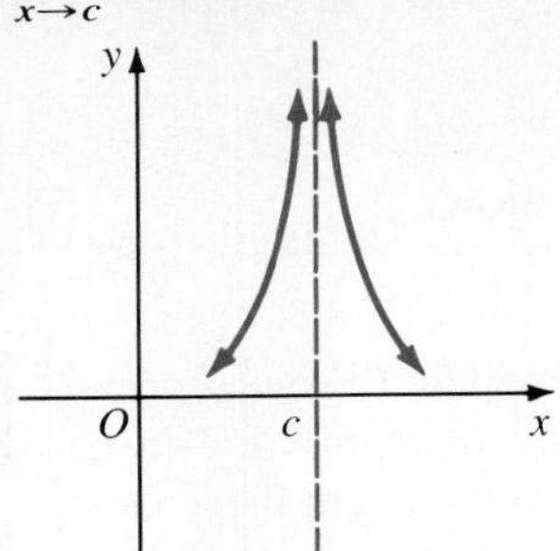

(6) $\lim_{x \to c} f(x) = -\infty$

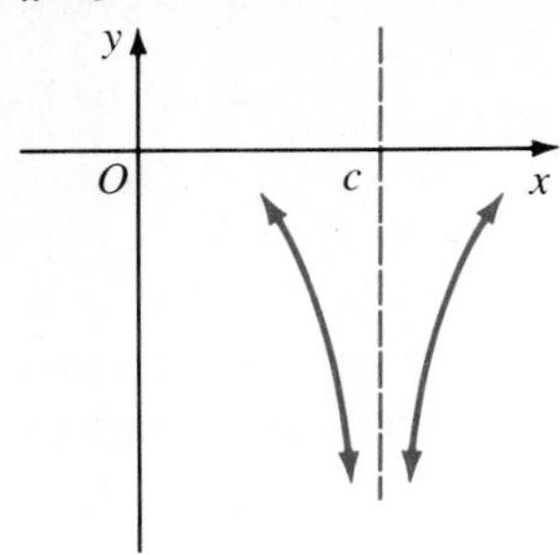

(7) $\lim_{x \to \infty} f(x) = l$

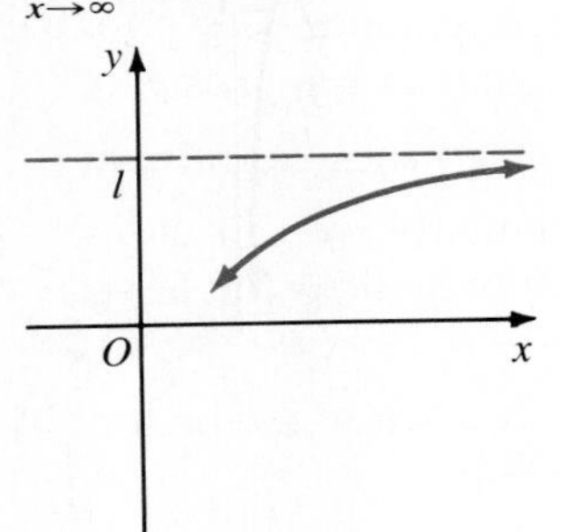

(8) $\lim_{x \to -\infty} f(x) = l$

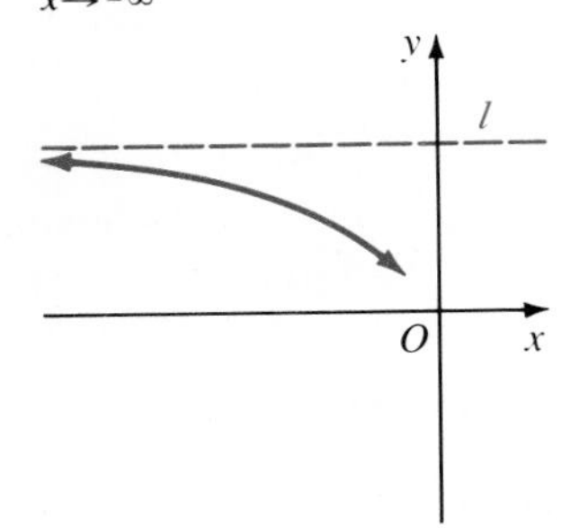

(9) $\lim_{x \to \infty} f(x) = \infty$

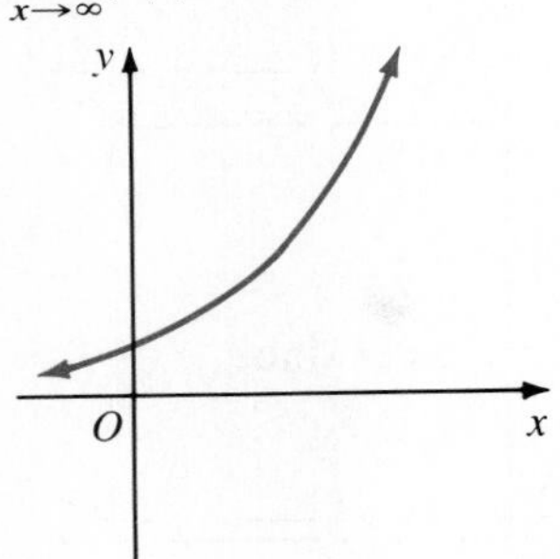

(10) $\lim_{x \to -\infty} f(x) = \infty$

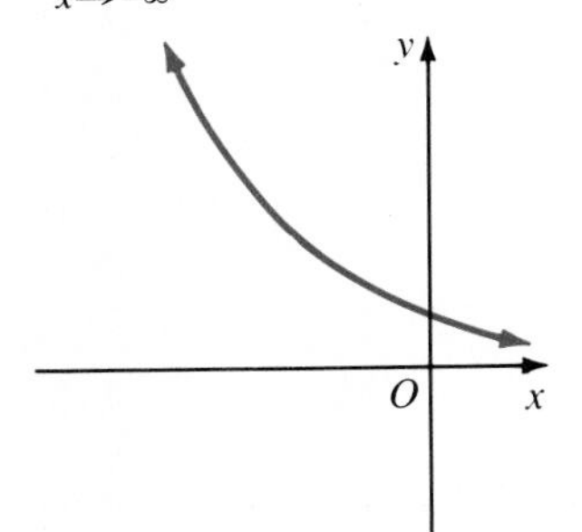

(11) $\lim_{x \to \infty} f(x) = -\infty$

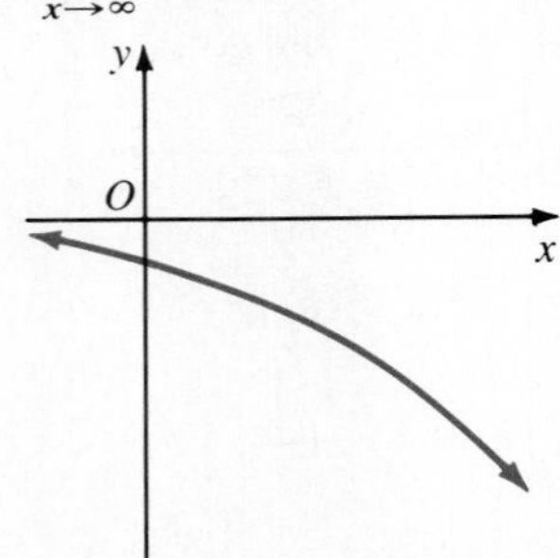

(12) $\lim_{x \to -\infty} f(x) = -\infty$

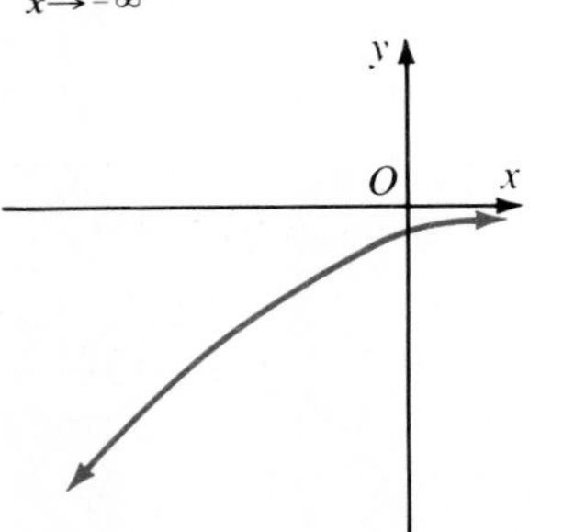

EXAMPLE 2 Graph $\varphi: x \to \dfrac{x^2 - 2}{x^2 - 1}$.

Solution Since $\dfrac{x^2 - 2}{x^2 - 1} = \dfrac{(-x)^2 - 2}{(-x)^2 - 1}$, $\varphi(-x) = \varphi(x)$ and so the graph of φ is symmetric with respect to the y-axis. The behavior of φ near 1 is given by $\lim\limits_{x \to 1^-} \varphi(x) = \infty$ and $\lim\limits_{x \to 1^+} \varphi(x) = -\infty$. Behavior at -1, the other singularity, is obtained by reflection. Use long division to investigate the behavior for large values of x.

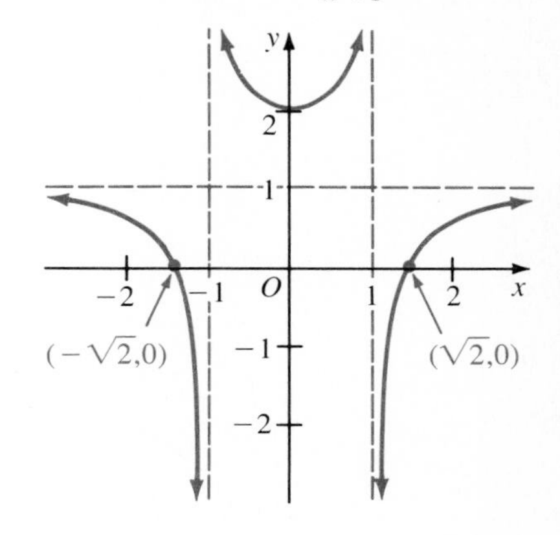

$$\varphi(x) = \frac{x^2 - 2}{x^2 - 1} = 1 - \frac{1}{x^2 - 1}$$

This tells us that $\lim\limits_{x \to \infty} \varphi(x) = 1$ and that if $x > 1$, $\varphi(x) < 1$. Thus the graph of φ approaches the line $y = 1$ from below. The graph crosses the x-axis at $(\sqrt{2}, 0)$ and therefore also at $(-\sqrt{2}, 0)$. It crosses the y-axis at $(0, 2)$.

EXAMPLE 3 Graph $F: x \to \dfrac{x^3 - x^2 + 1}{x^2}$.

Solution The only singularity of F is 0. As $x \to 0$, the numerator $x^3 - x^2 + 1$ is near 1 and the denominator x^2 is near 0 and positive regardless of whether x is to the left or the right of 0. Thus $\lim\limits_{x \to 0} F(x) = \infty$ and the line $x = 0$ is the only vertical asymptote.

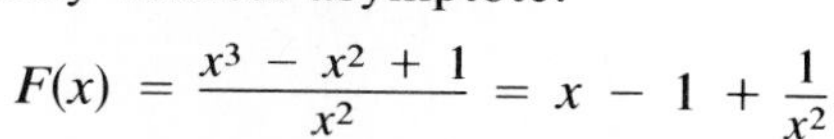

$$F(x) = \frac{x^3 - x^2 + 1}{x^2} = x - 1 + \frac{1}{x^2}$$

So it is clear that when x is large $F(x) \approx x - 1$, since $\dfrac{1}{x^2} \approx 0$. Thus the line $y = x - 1$ is an *oblique asymptote* of the graph.

To graph a rational function F:

1. See if the graph of F is symmetric about the y-axis or the origin.
2. Find the singularities of F and investigate the behavior of the graph to the left and the right of each singularity (vertical asymptotes).
3. Investigate the behavior of the graph as $x \to \infty$ and $x \to -\infty$ (horizontal and oblique asymptotes).
4. Plot the points where the graph crosses the x-axis. Find $F(0)$ to plot any point where the graph crosses the y-axis.
5. Draw smooth curves to fit the facts developed in Steps 1–4. Plot additional points if necessary.

EXERCISES

Write a limit statement that describes the pictured situation.

A

1.

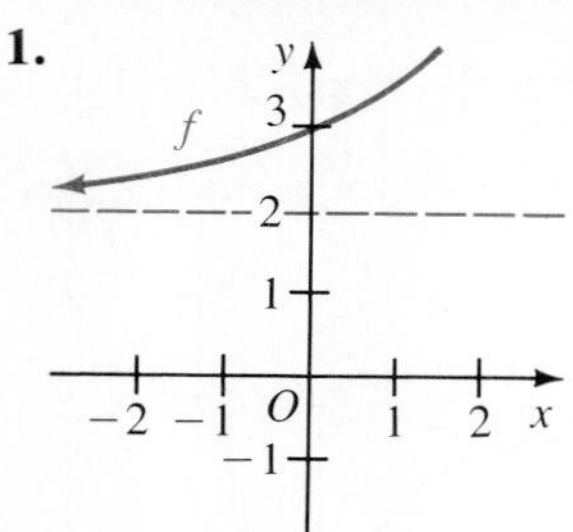

2.

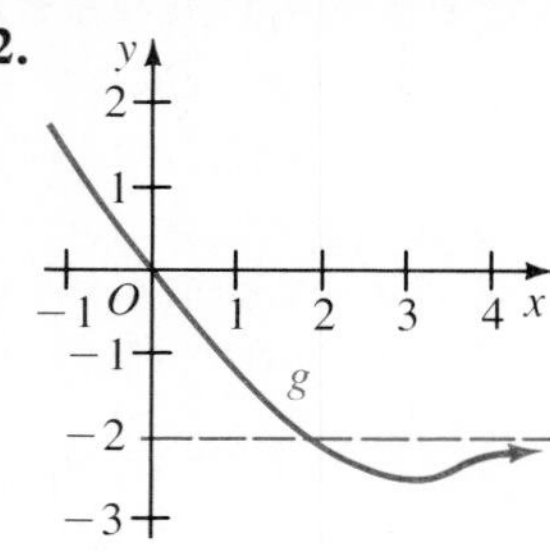

3.

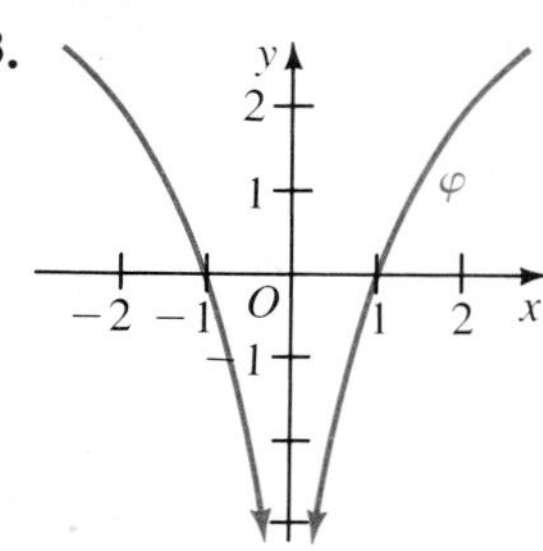

4.

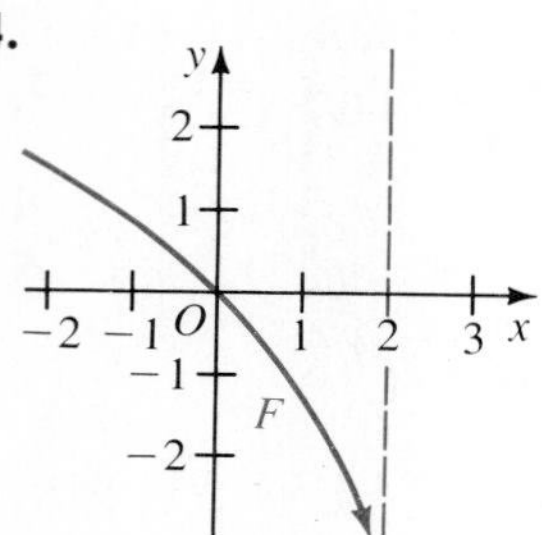

5.

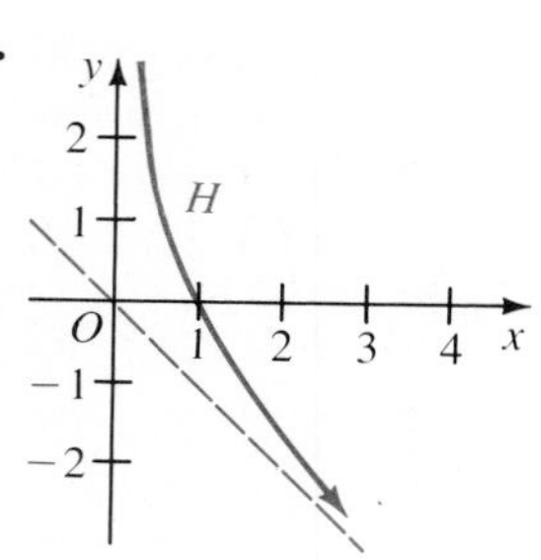

6.

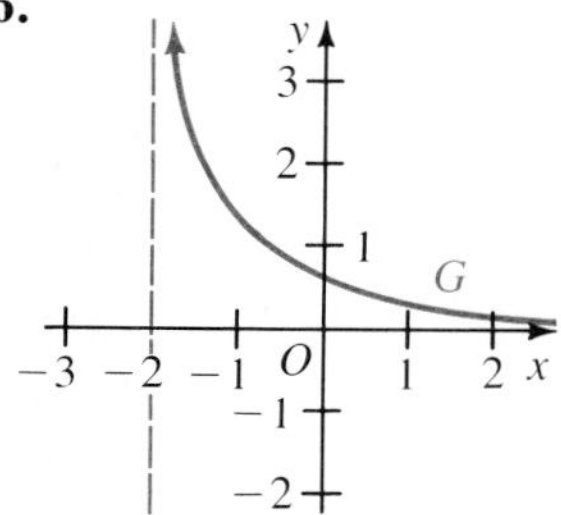

Graph each rational function. Show all asymptotes as dashed lines.

7. $f(x) = \dfrac{x}{x - 2}$ **8.** $f(x) = \dfrac{1 - x}{x}$ **9.** $f(x) = \dfrac{1}{(x - 2)^2}$ **10.** $f(x) = \dfrac{x - 2}{x^2}$

B **11.** $f(x) = \dfrac{x^2 - 1}{x^2 - 4}$ **12.** $f(x) = \dfrac{x}{x^2 + 1}$ **13.** $f(x) = \dfrac{x^2 - 1}{x^2 + 1}$ **14.** $f(x) = \dfrac{x^2}{x^2 - 4}$

15. $f(x) = \dfrac{x^2}{x - 1}$ **16.** $f(x) = \dfrac{x^2 + 1}{x + 1}$ **17.** $f(x) = \dfrac{1}{x} - x^2$ **18.** $f(x) = \dfrac{1}{x} + x^2$

The rest of this exercise set is devoted to definitions of limits involving ∞. Let $x_1, x_2, \ldots, x_n, \ldots$ be a sequence of real numbers. The statement

$$\lim_{n \to \infty} x_n = \infty$$

means that for each number M (no matter how large) there is a number N such that $x_n > M$ provided $n > N$.

C **19.** Give an analogous definition of $\lim_{x \to \infty} x_n = -\infty$.

We can now give precise definitions of the twelve limit statements on pages 173 and 174.

EXAMPLE Define each limit. **a.** $\lim_{x \to c^-} f(x) = \infty$ **b.** $\lim_{x \to \infty} f(x) = l$

Solution **a.** $\lim_{x \to c^-} f(x) = \infty$ if and only if $\lim_{n \to \infty} f(x_n) = \infty$ for every sequence $x_1, x_2, \ldots, x_n, \ldots$ in the domain of f such that $x_n < c$ and $\lim_{n \to \infty} x_n = c$.

b. $\lim_{x \to \infty} f(x) = l$ if and only if $\lim_{n \to \infty} f(x_n) = l$ for every sequence $x_1, x_2, \ldots, x_n, \ldots$ in the domain of f such that $\lim_{n \to \infty} x_n = \infty$.

Define the specified limit statement on pages 173 and 174.

20. (2) **21.** (3) **22.** (4) **23.** (5) **24.** (6)
25. (8) **26.** (9) **27.** (10) **28.** (11) **29.** (12)

Chapter Summary

1. A *function* f is a set of ordered pairs in which different pairs have different first members. The set of all first members of the pairs is the *domain* D of f and the set of all second members is the *range* R of f. Thus, f can be regarded as a rule that assigns to each x in D a unique member of R, denoted by $f(x)$ and called the *value* of f at x. The *graph* of a function f is the set of all points having coordinates $(x, f(x))$.
2. The *sum* $f + g$ of two functions f and g is defined by $(f + g)(x) = f(x) + g(x)$. Its domain is the intersection of the domains of f and g. The *product* fg, *difference* $f - g$, and *quotient* $\frac{f}{g}$ are defined analogously.
3. The *composition* of the functions f and g is the function $f \circ g$ defined by $(f \circ g)(x) = f(g(x))$. The domain of $f \circ g$ is the set of all x such that $g(x)$ is in the domain of f.
4. A function f is *one-to-one* if $f(x_1) \neq f(x_2)$ whenever $x_1 \neq x_2$. A one-to-one function f has an *inverse function* f^{-1}, defined by $f^{-1}(u) = v$ if and only if $f(v) = u$.
5. The *limit* of $f(x)$ as x approaches c, written $\lim_{x \to c} f(x)$, is defined using limits of sequences. We have $\lim_{x \to c} f(x) = l$ provided that $\lim_{n \to \infty} f(x_n) = l$ for every sequence $x_1, x_2, \ldots, x_n, \ldots$ in the domain of f with $\lim_{n \to \infty} x_n = c$ $(x_n \neq c)$.
6. Limits of algebraic combinations of functions behave in a natural way. For example, $\lim_{x \to c} [f(x) \cdot g(x)] = [\lim_{x \to c} f(x)] \cdot [\lim_{x \to c} g(x)]$, with analogous results for sums, differences, and quotients of functions.
7. A function f is *continuous at* c if and only if $\lim_{x \to c} f(x) = f(c)$. If f is continuous at every member of its domain, f is called a *continuous function*. Sums, products, differences, quotients, and compositions of continuous functions are continuous. Polynomial functions are continuous. The graph of a continuous function on an interval is unbroken.
8. The *intermediate-value theorem* asserts that if the function f is continuous on an interval $[a, b]$, then f takes on all values between $f(a)$ and $f(b)$.

Another form of the intermediate-value theorem is the *location principle*: If $f(a)$ and $f(b)$ have opposite signs, then f has at least one zero between a and b.

9. *Synthetic substitution* provides an efficient method of calculating the values of a polynomial function.
10. The quotient of two polynomial functions not having common zeros is called a *rational function*. A rational function has *singularities* at the zeros of the denominator polynomial. Limits are used to investigate the behavior of a rational function f near its singularities and for large values of $|x|$.

Chapter Test

4-1

1. Let $f(x) = x^2 + 1$ and $g(x) = (x + 1)^2$. Find each value.

a. $f(g(1))$ **b.** $g(f(1))$ **c.** $\dfrac{f(x + h) - f(x)}{h}$

2. Graph $\varphi: x \to \dfrac{x^2 - 1}{x - 1}$.

4-2

3. Let $f(x) = \sqrt{x}$ and $g(x) = \sqrt{1 - x}$. Find a rule for **(a)** fg; **(b)** $\dfrac{f}{g}$; and **(c)** $f^2 + g^2$. Specify the domain of each resulting function.

4. Graph each function.

a. $F: x \to x - |x|$ **b.** $G: x \to |x - 1| - 1$

4-3

5. Let $f(x) = \sqrt{x - 4}$ and $g(x) = x^2$. Find $(f \circ g)(x)$ and $(g \circ f)(x)$ and specify the domains of $f \circ g$ and $g \circ f$.

6. Given that $f(x) = \dfrac{x}{x + 1}$, find $f^{-1}(x)$.

4-4

7. Let $f(x) = \dfrac{x^2 - 1}{x^2 - x}$.

a. Does $\lim_{x \to 0} f(x)$ exist? **b.** Does $\lim_{x \to 1} f(x)$ exist?

8. Prove that $\lim_{x \to 2} (x^2 - 4) = 0$. Justify each step using the theorems of Section 4-4.

4-5

9. Prove that $f: x \to |x|$ is continuous.

4-6

10. Let $f(x) = \dfrac{x - 1}{x + 1}$. Find a number c between 1 and 3 such that $f(c) = \frac{1}{2}[f(1) + f(3)]$.

11. Locate the negative zero of $f(x) = x^4 + 2x^3 + 4x - 4$ between two consecutive integers.

4-7

12. Graph $g(x) = \dfrac{x^2 + 4}{x^2 - 4}$, showing all asymptotes as dashed lines.

13. Draw a sketch illustrating the statement: $\lim_{x \to c} f(x) = -\infty$.

Extra

RINGS

In Section 1-4 a group was defined as a set of objects G together with some operation, say $\oplus$, that satisfies the axioms of closure, associativity, identity under $\oplus$, and inverses under $\oplus$.

For many sets of objects we can define two operations, say $\oplus$ and $\odot$. A **ring** is any set of objects G together with two operations such that

1. G is a commutative group under $\oplus$,
2. G is closed under $\odot$ and $\odot$ is associative, and
3. $\odot$ is distributive over $\oplus$.

For example, the set Z of integers is a ring with the familiar operations of $+$ and $\cdot$.

There are special types of rings. A **commutative ring** is a ring that satisfies the commutative axiom of $\odot$. A **ring with identity** is a ring that has an identity under $\odot$. A **commutative ring with identity** is a ring that is both commutative and has an identity under $\odot$. The following diagram shows the relationships among various types of algebraic structures.

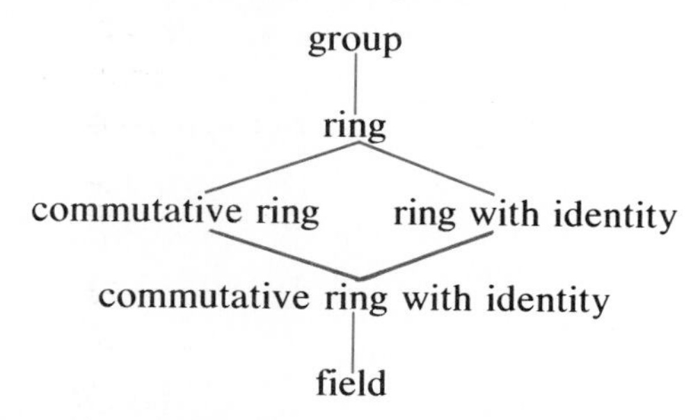

We can show that the set E of even integers is a commutative ring that has no identity under multiplication.

On page 34 it was shown that E with $+$ is a group. It is a commutative group under $+$ since $E \subset \mathcal{R}$. Furthermore $\cdot$ is associative and distributes over $+$ since $E \subset \mathcal{R}$. We need only show that E is closed under $\cdot$ to conclude that E is a commutative ring. Let $a = 2k$ and $b = 2m$, where k and m are integers, be members of E. Then

$$a \cdot b = (2k)(2m) = 2(2km) \in E$$

since $2km$ is an integer and thus $a \cdot b$ is a multiple of 2.

To see that E has no identity under $\cdot$, let us suppose that for all $e \in E$, there is an $I \in E$ such that $e \cdot I = e$. In particular,

$$2 \cdot I = 2.$$

Use of the definition of a group and Theorem 1 of Chapter 1 gives

$$2 \cdot I - 2 = 0;$$
$$2(I - 1) = 0.$$

By Theorem 2 of Chapter 1.

$$2 = 0 \text{ or } I - 1 = 0.$$

Thus $I = 1$. Since 1 is not an even integer, E has no identity under $\cdot$

In Exercises 37–43 and Example 5 of Section 4-2, it was shown that the set F of functions with a common domain D and ranges in a field is a commutative ring with identity.

EXERCISES

In Exercises 1–3, determine whether the set with $+$ and $\cdot$ is a ring. If it is, state what type of ring it is.

1. The set of whole numbers
2. The set of positive real numbers
3. The set of odd integers

In Exercises 4–6, $W = \{(a, b)\text{: } a \text{ and } b \text{ are real numbers}\}$. Define $+$ and $\cdot$ by $(a, b) + (c, d) = (a + c, b + d)$ and $(a, b) \cdot (c, d) = (ac - bd, bc + ad)$.

4. Is W a ring?
5. Show that W is a commutative ring.
6. Does W have an identity under $\cdot$? If it does, what is it?

In Exercises 7–10, $M = \left\{\begin{bmatrix} a & b \\ c & d \end{bmatrix}\text{: } a, b, c, \text{ and } d \text{ are real numbers}\right\}$. Define $+$ and $\cdot$ on M by

$$\begin{bmatrix} a & b \\ c & d \end{bmatrix} + \begin{bmatrix} e & f \\ g & h \end{bmatrix} = \begin{bmatrix} a + e & b + f \\ c + g & d + h \end{bmatrix} \text{ and } \begin{bmatrix} a & b \\ c & d \end{bmatrix} \cdot \begin{bmatrix} e & f \\ g & h \end{bmatrix} = \begin{bmatrix} ae + bg & af + bh \\ ce + dg & cf + dh \end{bmatrix}.$$

It can be shown that M with $+$ and $\cdot$ is a ring.

7. What is the additive identity of M?
8. Is $\cdot$ commutative? If so, prove your answer. If not, give a counterexample.
9. Does M have an identity under $\cdot$? If it does, what is it?
10. Is there a member $\begin{bmatrix} s & t \\ r & v \end{bmatrix}$ of M such that $\begin{bmatrix} 2 & 0 \\ 0 & 2 \end{bmatrix} \cdot \begin{bmatrix} s & t \\ r & v \end{bmatrix} = \begin{bmatrix} 1 & 0 \\ 0 & 1 \end{bmatrix}$?

An *integral domain* is a commutative ring with identity in which $ab = 0$ implies that $a = 0$ or $b = 0$.

11. Show that the set M in Exercises 7–10 is not an integral domain by showing that there are members A and B of M such that $AB = 0$ does not imply that $A = 0$ or $B = 0$. $\left(\textit{Hint}\text{: The 0 member of } M \text{ is } \begin{bmatrix} 0 & 0 \\ 0 & 0 \end{bmatrix}.\right)$

COMPUTER APPLICATION

INTERPOLATION

In many applications numerical data are given as a table of ordered pairs with no given formula or function relating the variables involved. The figure shows the plots of ordered pairs, red line segments joining them in order, and a smooth curve that might represent a function: $f(x) = y$ containing each (x_i, y_i).

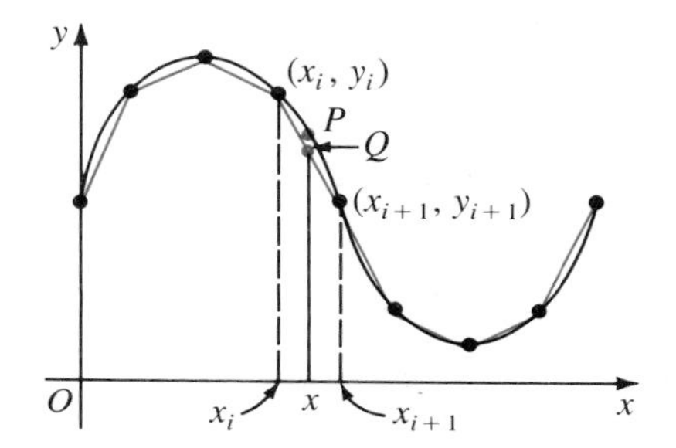

x	x_1	x_2	x_i	$\cdots$	x_n
y	y_1	y_2	y_i	$\cdots$	y_n

If the line segments are close to the curve, we can approximate the y-coordinate of P, a point on the curve, by finding the y-coordinate of Q, a point on a line segment. This process is called *linear interpolation*. We need only have an equation for the line containing (x_i, y_i) and (x_{i+1}, y_{i+1}) and, of course, x. The point-slope form of a line gives

$$y = \left[\frac{y_{i+1} - y_i}{x_{i+1} - x_i}\right](x - x_i) + y_i. \qquad (*)$$

EXERCISES *(You may use either BASIC or Pascal.)*

Write a program using (*) to find y from a table of ordered pairs and a given value of x. Run the program for each table and set of values of x.

1.

x_i	9.5	9.6	9.7	9.8	9.9
y_i	3.082	3.098	3.114	3.130	3.146

$x = 9.56, 9.83$

2.

x_i	1.20	1.21	1.22	1.23	1.24
y_i	2.572	2.650	2.733	2.820	2.912

$x = 1.225, 1.228, 1.235, 1.239$

APPLICATION

Functions in the Sciences

Frequently investigators define and discover functional relationships among quantities.

Celsius temperature is defined as a linear function of Fahrenheit temperature such that 0°C corresponds to 32°F, the freezing point of water, and 100°C corresponds to 212°F, the boiling point of water. From this definition the following formula relating C and F is obtained.

$$C = \tfrac{5}{9}(F - 32)$$

Many functional relationships among physical quantities have been discovered. In the seventeenth century Robert Boyle studied the effect of pressure on the volume of a gas. The table at the left below shows how pressure exerted by a column of mercury affects the height and therefore the volume of a column of air in a glass tube. Boyle discovered that the greater the pressure P, the smaller the volume V. More precisely, volume varies inversely with pressure:

$$V = \frac{k}{P}, \quad \text{or} \quad PV = k,$$

where the constant k depends on the gas. The graph of the data in the table also shows that as pressure increases, volume decreases.

P Total pressure (cm of mercury)	V Height of air column (cm)
88	26
102	22
112	20
142	16
172	13
205	11
285	8

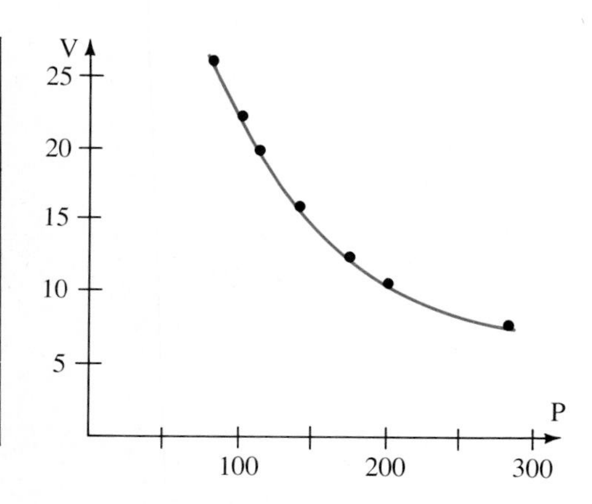

Functional relationships can also be found in the study of economics. The quantity of an item demanded by consumers decreases as the price per item increases. The supply of an item stocked by a store increases as the price consumers are willing to pay increases.

EXERCISES

1. Show that if C is a linear function of F, $C = 0$ when $F = 32$, and $C = 100$ when $F = 212$, then $C = \frac{5}{9}(F - 32)$. Use $C = mF + b$ and two linear equations in two variables.
2. A gas has volume 625 mL when the pressure is 745 mm. Find the volume of the gas when the pressure is increased to 815 mm.
3. A manufacturing firm expects that the demand for one of its products will be given by $D(x) = x^2 - 10x + 43$ and that the supply will be given by $S(x) = x^2 + 3x + 1$ where x is the price of each unit that is sold. Find the number of units that must be sold in order for supply and demand to be equal.

Pressure, volume, temperature and other variables are crucial considerations in the design of ships carrying liquefied natural gas.

Chapter 5

Theory of Polynomial Equations

Solving polynomial equations is the subject of this chapter. In particular, complex numbers *and* the Fundamental Theorem of Algebra *are reviewed. Various techniques for finding roots and information about roots are discussed.*

Complex Numbers

5-1 THE COMPLEX FIELD

We know that the polynomial equation $z^2 + 1 = 0$ has no real solution because, in $\mathcal{R}$, $z^2 + 1 > 0$ for all z. We may ask, "Is it possible to construct a set of numbers containing $\mathcal{R}$ so that in the larger system equations such as $z^2 + 1 = 0$ do have solutions?" Of course we would also like the new system to satisfy the field axioms (pages 14–15). We will show that all of this is possible by constructing a field that contains $\mathcal{R}$ and in which every polynomial equation has a solution.

Denote by C the set of all ordered pairs of real numbers:

$$C = \{(x, y): x, y \in \mathcal{R}\}$$

We will define addition and multiplication in C in the following way.

DEFINITION

Let (u, v) and (x, y) be members of C.

Addition $(u, v) + (x, y) = (u + x, v + y)$

Multiplication $(u, v) \cdot (x, y) = (ux - vy, uy + vx)$

The flow of a stream such as the Nisqually River in Washington produces many interesting effects. Complex variables are used to study many kinds of flow.

With the operations of addition and multiplication as defined, we can show that C is a field. We call C the **complex field**, and each member is called a **complex number**. Two complex numbers (u, v) and (x, y) are **equal** if $u = x$ and $v = y$.

EXAMPLE 1 Show that $(0, 0)$ is an additive identity in C and that $(1, 0)$ is a multiplicative identity in C.

Solution Since $(0, 0) + (x, y) = (0 + x, 0 + y) = (x, y)$ and $(x, y) + (0, 0) = (x + 0, y + 0) = (x, y)$, $(0, 0)$ is an additive identity in C.

Since $(1, 0) \cdot (x, y) = (1 \cdot x - 0 \cdot y, 1 \cdot y + 0 \cdot x) = (x, y)$ and $(x, y) \cdot (1, 0) = (x \cdot 1 - y \cdot 0, y \cdot 1 + x \cdot 0) = (x, y)$, $(1, 0)$ is a multiplicative identity.

Verification that C satisfies the remaining field axioms is left as Exercises 41–44.

It is easy to see that

$$(a, 0) + (b, 0) = (a + b, 0)$$

and that

$$(a, 0) \cdot (b, 0) = (a \cdot b, 0).$$

For example, $(3, 0) + (4, 0) = (3 + 4, 0) = (7, 0)$ and $(3, 0) \cdot (4, 0) = (3 \cdot 4 - 0 \cdot 0, 3 \cdot 0 + 0 \cdot 4) = (12, 0)$. This means that the subset $\{(x, 0): x \in \mathcal{R}\}$ of C behaves algebraically the same as $\mathcal{R}$. In this sense C contains $\mathcal{R}$. We therefore regard $(x, 0)$ as a real number and denote it simply by x. In particular $(1, 0) = 1$.

The special pair $(0, 1)$ is denoted by i. Note that

$$i^2 = i \cdot i = (0, 1) \cdot (0, 1) = (0 \cdot 0 - 1 \cdot 1, 0 \cdot 1 + 1 \cdot 0) = (-1, 0) = -1.$$

THEOREM 1 Every complex number can be written in terms of real numbers and i.

Proof: First note that $(y, 0) \cdot (0, 1) = (y \cdot 0 - 0 \cdot 1, y \cdot 1 + 0 \cdot 0) = (0, y)$. Therefore $(x, y) = (x, 0) + (0, y) = (x, 0) + (y, 0) \cdot (0, 1) = x + yi$. ■

We call x the **real part** and y the **imaginary part** of the complex number $x + yi$. The number $x + yi$ is imaginary if $y \neq 0$ and is **pure imaginary** if $x = 0$ and $y \neq 0$. (The term *imaginary* stems from the 17th-century uneasiness about numbers whose squares are negative.) We often use a single letter, such as z, to denote a complex number and Re z and Im z to denote its real and imaginary parts respectively. For example, if $z = 3 - i$, then Re $z = 3$ and Im $z = -1$. (*Note*: Im z is a real number.)

To operate algebraically with complex numbers in the form $x + yi$, we treat them as ordinary binomials. Then we simplify the results by using the fact that $i^2 = -1$.

EXAMPLE 2 Let $w = 1 - 2i$ and $z = 3 - i$. Express the result of each operation in the form $x + yi$.

a. $w + z$ **b.** $w - z$ **c.** wz **d.** $\frac{w}{z}$

Solution **a.**
$$\begin{aligned} w + z &= (1 - 2i) + (3 - i) \\ &= (1 + 3) + (-2 + (-1))i \\ &= 4 - 3i \quad \textbf{Answer} \end{aligned}$$

b.
$$\begin{aligned} w - z &= (1 - 2i) - (3 - i) \\ &= (1 - 3) + (-2 - (-1))i \\ &= -2 - i \quad \textbf{Answer} \end{aligned}$$

c.
$$\begin{aligned} wz &= (1 - 2i)(3 - i) \\ &= 1 \cdot 3 - (2i)(3) + 1 \cdot (-i) + (-2i)(-i) \\ &= 3 - 7i + 2i^2 = 3 - 7i + 2(-1) = 1 - 7i \quad \textbf{Answer} \end{aligned}$$

d. $\frac{w}{z} = \frac{1 - 2i}{3 - i}$

Because of the presence of i in the denominator, this quotient is not in the form $x + yi$. To obtain the required form multiply the numerator and denominator by $3 + i$.

$$\begin{aligned} \frac{1 - 2i}{3 - i} &= \frac{1 - 2i}{3 - i} \cdot \frac{3 + i}{3 + i} \\ &= \frac{3 - 5i - 2i^2}{9 - i^2} \\ &= \frac{3 - 5i - 2(-1)}{9 - (-1)} = \frac{5 - 5i}{10} = \frac{1 - i}{2} = \frac{1}{2} - \frac{1}{2}i \quad \textbf{Answer} \end{aligned}$$

If we multiply $u + vi$ by $x + yi$ using the method of Example 2, we will obtain $(ux - vy) + (uy + vx)i$, which, in ordered-pair form, is $(ux - vy, uy + vx)$. This motivated the definition of *product* on page 185.

EXERCISES

In Exercises 1–28 express answers in the form $x + yi$.

Find $w + z$ and wz.

A

1. $w = 1 - 2i$, $z = 1 + i$

2. $w = 2 + i$, $z = 1 - i$

3. $w = 2i$, $z = 3 - i$

4. $w = 1 + 2i$, $z = 3$

5. $w = 1 - i$, $z = 4 - 3i$

6. $w = -2 + i$, $z = -1 - i$

7–12. For the w and z of Exercises 1–6, find $w - z$ and $\frac{w}{z}$.

Find the reciprocal (multiplicative inverse) of the given complex number.

13. $2i$ **14.** $1 - i$ **15.** $-1 - i$ **16.** $2 - i$

Express as one of 1, -1, i, or $-i$.

17. i^3 **18.** i^6 **19.** i^{15} **20.** i^{101} **21.** i^{-10} **22.** i^{-21}

Simplify.

23. $(6 + 8i)(6 - 8i) + 100$

24. $(7 - 4i)(7 + 4i) - 65$

25. $\frac{3 - 2i}{6 + 5i} \cdot \frac{4i}{2 + i}$

26. $\frac{5 + 3i}{2i} \cdot \frac{4i}{5 + 2i}$

27. $(4 + 3i)^2 + 5 - 6i$

28. $(5 + 6i) - (7 + 4i)^2$

Find r and s so that the complex numbers are equal. (*Hint*: Equate real and imaginary parts.)

29. $2r + s + (3s - 5r)i = 11 - 11i$

30. $5r - s + (2s + 4r)i = 36 + 26i$

31. $rs + (r + s)i = 6 + 5i$

32. $r + s + (rs)i = 4i$

Solve for z. Assume that the addition and multiplication properties of equality for real numbers apply to complex numbers. Write answers in the form $x + yi$.

B

33. $(1 - i)z = 1 + i$

34. $iz + 2 = 1 + 2i$

35. $2iz - 1 + i = 1 + i$

36. $(1 + i)z + 2i = 1 + i$

37. Verify that $1 + i$ and $1 - i$ are roots of $z^2 - 2z + 2 = 0$.

38. Verify that $2 + 3i$ and $2 - 3i$ are roots of $z^2 - 4z + 13 = 0$.

39. Is the subset $\{(0, b): b \in \mathcal{R}\}$ of C closed under addition? under multiplication?

40. Is the subset $\{(a, b): b \neq 0\}$ of C closed under addition? under multiplication?

In Exercises 41–44 complete the verification that C is a field.

41. Show that $+$ and $\cdot$ are commutative.

C

42. Show that $+$ and $\cdot$ are associative.

43. Show that every nonzero member of C has a multiplicative inverse.

44. Show that multiplication is distributive over addition.

45. Show that $\{i^n: n \in Z\}$ is a group under multiplication by showing that each power of i is 1, -1, i, or $-i$ and by showing that $\{1, -1, i, -i\}$ is a group under multiplication.

COMPUTER EXERCISES

1. Write a computer program to find the quotient of two complex numbers. The quotient should be expressed in the form $x + yi$.

2. Run the program in Exercise 1 to find each quotient.

a. $\frac{1}{2 + 3i}$ **b.** $\frac{2 + i}{2 - i}$ **c.** $\frac{2 - i}{2 + i}$ **d.** $\frac{3 + 4i}{4 + 3i}$

5-2 THE COMPLEX PLANE

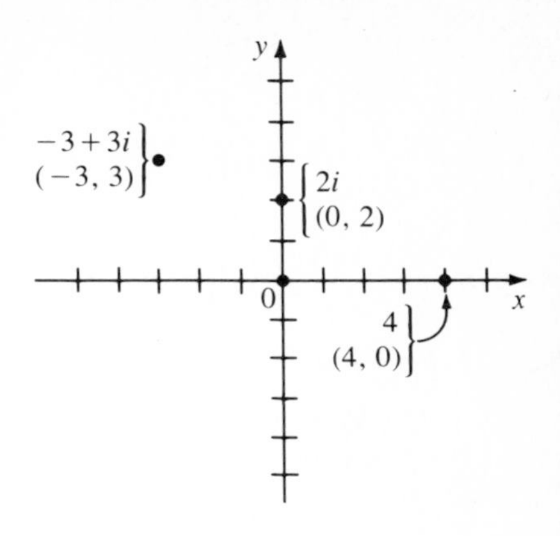

We represent complex numbers in a coordinate plane by letting the **graph** of $x + yi$ be the point whose coordinates are (x, y) as illustrated in the figure. When the plane is used for this purpose, it is called the **complex plane**. Because graphs of real numbers lie on the horizontal axis, it is called the **real axis**. The vertical axis is called the **imaginary axis**.

We can give the operations of addition and subtraction geometric interpretation by considering an example. Let $z = 3 + 4i$ and $w = -5 + i$; then $z + w = -2 + 5i$, $-z = -3 - 4i$, and $z - w = 8 + 3i$. The following figures illustrate $z + w$, $-z$, and $z - w$.

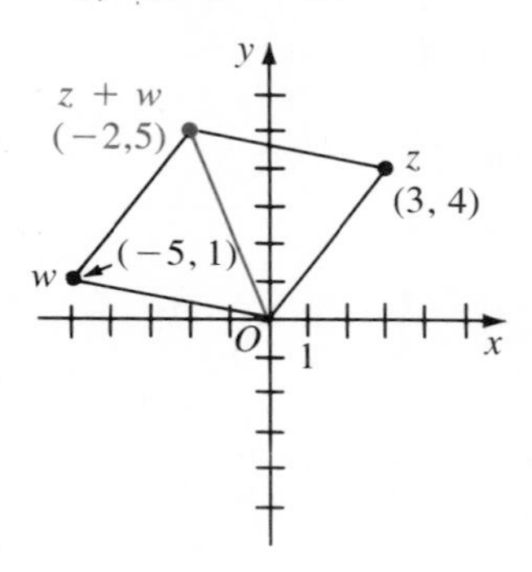

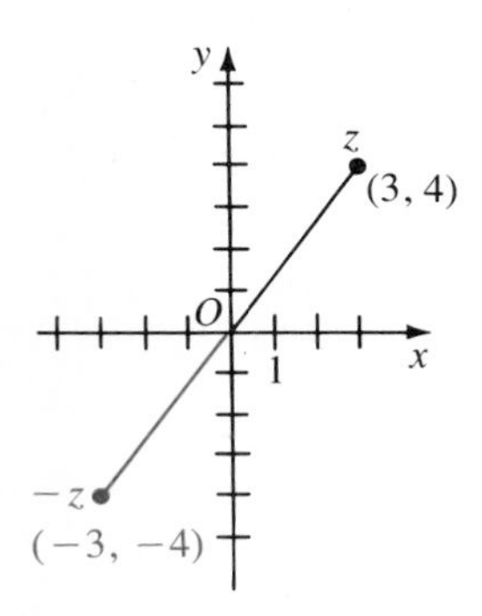

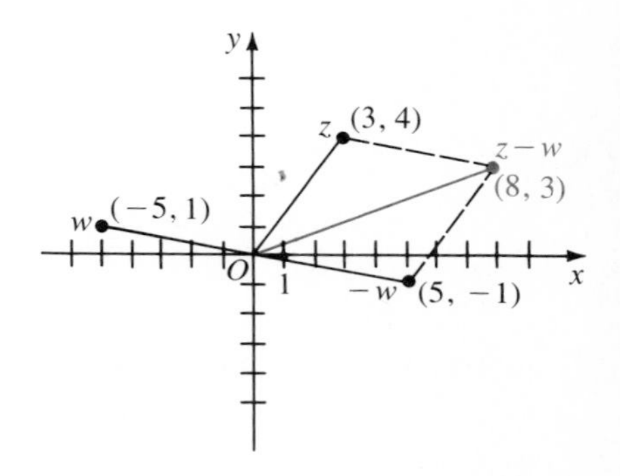

To construct the sum of the complex numbers z and w use the "parallelogram rule." Construct a parallelogram with adjacent sides being line segments from the origin to both z and w. The fourth vertex of the parallelogram is the graph of $z + w$. To obtain $z - w$ first construct $-w$ by reflecting w in the origin. Then construct $z + (-w)$. (The geometry of multiplication and division will be discussed in Section 9-3.)

The **conjugate** of $z = x + yi$, which is denoted $\bar{z}$, is the complex number $\bar{z} = x - yi$. (We used conjugates to find the quotient in Example 2(d) in Section 5-1.) Geometrically, the graph of $\bar{z}$ is the reflection in the real axis of the graph of z.

EXAMPLE 1 Let $z = 3 + 2i$. Find $\bar{z}$, $z + \bar{z}$, and $z - \bar{z}$ and graph all four numbers in the same complex plane.

Solution

$\bar{z} = 3 - 2i$

$z + \bar{z} = (3 + 2i) + (3 - 2i) = 6$

$z - \bar{z} = (3 + 2i) - (3 - 2i) = 4i$

Red is used to show the graphs of $\bar{z}$, $z + \bar{z}$, and $z - \bar{z}$.

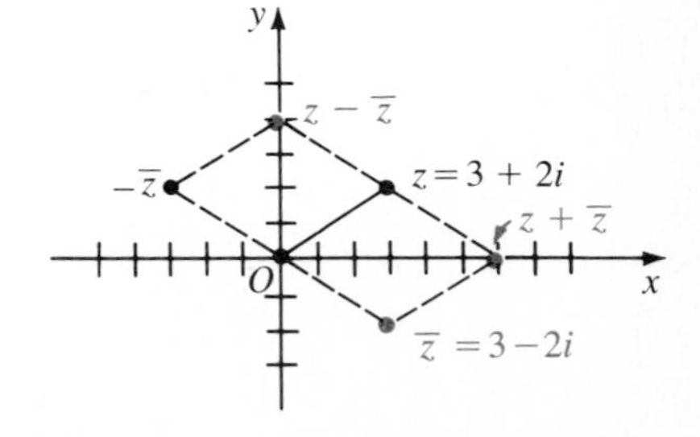

Example 1 illustrates the first two properties of the conjugate as stated in the following theorem.

THEOREM 2 Let w and z be complex numbers. Then

1. $z + \bar{z} = 2 \operatorname{Re} z$ **2.** $z - \bar{z} = 2i \operatorname{Im} z$

3. $\overline{w + z} = \bar{w} + \bar{z}$ **4.** $\overline{w - z} = \bar{w} - \bar{z}$

5. $\overline{wz} = \bar{w}\,\bar{z}$ **6.** $\overline{\left(\dfrac{w}{z}\right)} = \dfrac{\bar{w}}{\bar{z}}$

We will prove the product property (part (5) of the theorem) and leave the proofs of the other parts of the theorem as Exercises 27–30 and 32.

Proof: Let $w = u + vi$ and $z = x + yi$. Then

$$\overline{wz} = \overline{(u + vi)(x + yi)} = \overline{(ux - vy) + (uy + vx)i} = (ux - vy) - (uy + vx)i;$$

and

$$\bar{w}\,\bar{z} = \overline{(u + vi)}\,\overline{(x + yi)} = (u - vi)(x - yi) = (ux - vy) - (uy + vx)i.$$

Therefore $\overline{wz} = \bar{w}\,\bar{z}$. ■

A measure of the "size" of a complex number $z = x + yi$ is the distance from the origin to the graph of z, that is, $\sqrt{x^2 + y^2}$. We call this nonnegative real number the **absolute value**, or **modulus**, of z and denote it by $|z|$. Thus,

$$|z| = |x + yi| = \sqrt{x^2 + y^2}.$$

The following theorem gives some important properties of the absolute value. We shall give a partial proof of part (4) and leave the other proofs as Exercises 26, 46, and 47.

THEOREM 3 Let w and z be complex numbers.

1. $z\bar{z} = |z|^2$ **2.** $|wz| = |w||z|$ **3.** $\left|\dfrac{w}{z}\right| = \dfrac{|w|}{|z|}$

4. $|w + z| \le |w| + |z|$ (the triangle inequality)

Proof: Use the fact that if p and q are nonnegative real numbers and $p^2 \le q^2$, then $p \le q$. Since $|w + z|$ and $|w| + |z|$ are nonnegative, it will be sufficient to show that

$$|w + z|^2 \le (|w| + |z|)^2.$$

$$\begin{aligned} |w + z|^2 &= (w + z)(\overline{w + z}) && \text{by Theorem 3, part (1)} \\ &= (w + z)(\bar{w} + \bar{z}) && \text{by Theorem 2, part (3)} \\ &= w\bar{w} + w\bar{z} + \bar{w}z + z\bar{z} \\ &= |w|^2 + (w\bar{z} + \bar{w}z) + |z|^2 && \text{by Theorem 3, part (1)} \end{aligned}$$

In Exercises 50–53 you are asked to show that $w\bar{z} + \bar{w}z \le 2|w||z|$. Use this inequality to obtain:

$$|w + z|^2 \le |w|^2 + 2|w||z| + |z|^2$$

Hence $\quad |w + z|^2 \le (|w| + |z|)^2.$

Therefore $\quad |w + z| \le |w| + |z|.$ ■

EXAMPLE 2 Let $w = 4 - 3i$ and $z = 2 + i$. Verify that:

a. $\overline{wz} = \bar{w}\,\bar{z}$ **b.** $|w + z| \le |w| + |z|$

Solution **a.** $wz = (4 - 3i)(2 + i) = 11 - 2i$ and $\overline{wz} = 11 + 2i$
$\bar{w}\,\bar{z} = (4 + 3i)(2 - i) = 11 + 2i$
$\therefore \overline{wz} = \bar{w}\,\bar{z}$

b. $w + z = (4 - 3i) + (2 + i) = 6 - 2i$
$|z + w| = \sqrt{36 + 4} = \sqrt{40} = 2\sqrt{10}$
Also $|w| = \sqrt{16 + 9} = 5$ and $|z| = \sqrt{4 + 1} = \sqrt{5}$.
Since $(2\sqrt{10})^2 \le (5 + \sqrt{5})^2$, or $40 \le 30 + 10\sqrt{5}$,
then $2\sqrt{10} \le 5 + \sqrt{5}$. Therefore $|w + z| \le |w| + |z|$.

The following figures give a geometric interpretation of $|w + z|$ and $|w - z|$. The figure at the left indicates why $|w + z| \le |w| + |z|$ is called the triangle inequality. The figure at the right illustrates that $|w - z|$ is the distance between the graphs of w and z.

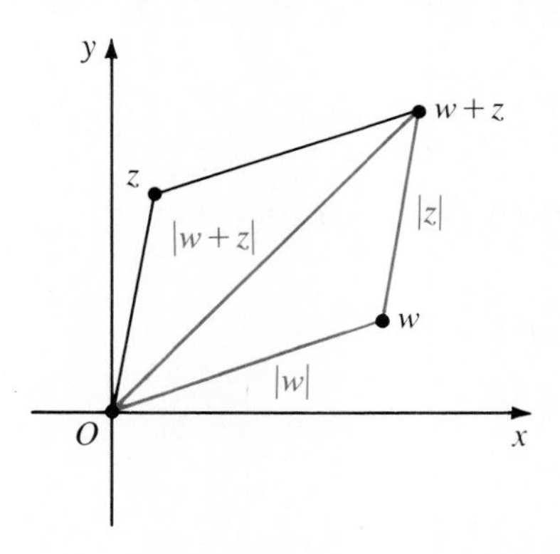

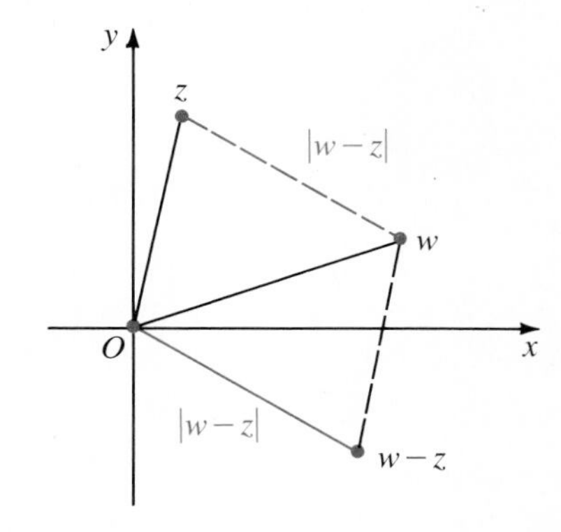

EXAMPLE 3 Graph the equation $|z - i| = 2$.

Solution The complex number z satisfies $|z - i| = 2$ if and only if the distance from the graph of z to the graph of i equals 2. Thus the graph is the circle of radius 2 with center at the graph of i.

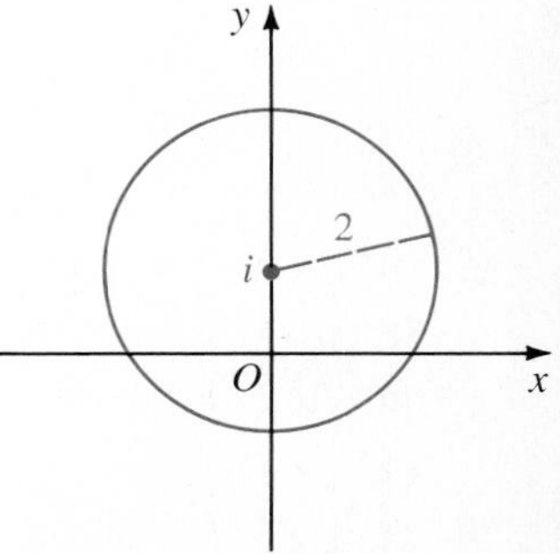

We could also obtain the graph in Example 3 by letting $z = x + yi$. Then $z - i = x + (y - 1)i$ and $|z - i| = \sqrt{x^2 + (y - 1)^2} = 2$. So $z - i = 2$ is equivalent to $x^2 + (y - 1)^2 = 4$.

EXERCISES

In each of Exercises 1–6 graph and label w, z, $w + z$, and $w - z$ in the same complex plane.

A

1. $w = 2 + i,\ z = -1 + i$

2. $w = 4 + 3i,\ z = 3 - 4i$

3. $w = -2 + i,\ z = -2i$

4. $w = 1 - 2i,\ z = i$

5. $w = 12 + 16i,\ z = 9 + 12i$

6. $w = 4 - 2i,\ z = -2 + i$

7–12. Graph and label $\overline{w}$, $\overline{z}$, and $\overline{w + z}$ for Exercises 1–6 in the same complex plane.

13–18. Verify algebraically $\overline{w + z} = \overline{w} + \overline{z}$ for the w and z in Exercises 1–6.

19–24. Verify algebraically $|w + z| \le |w| + |z|$ for the w and z in Exercises 1–6.

Prove the following properties of the conjugate.

25. $\overline{\overline{z}} = z$

26. $z\overline{z} = |z|^2$

27. $z + \overline{z} = 2 \text{ Re } z$

28. $z - \overline{z} = 2i \text{ Im } z$

29. $\overline{w + z} = \overline{w} + \overline{z}$

30. $\overline{w - z} = \overline{w} - \overline{z}$

31. $\overline{\left(\frac{1}{z}\right)} = \frac{1}{\overline{z}}$

32. $\overline{\left(\frac{w}{z}\right)} = \frac{\overline{w}}{\overline{z}}$ (Use Exercise 31.)

B

33. If z is complex and t is real, what can be said about the graphs of z and $t\overline{z}$?

34. If $z = x + yi$ and $z \ne 0$, find $\left|\frac{1}{z}\right|$. Graphically compare $|z|$ and $\left|\frac{1}{z}\right|$.

35. Show that the graphs of -1, $\frac{1}{2} + \frac{\sqrt{3}}{2}i$, and $\frac{1}{2} - \frac{\sqrt{3}}{2}i$ are on the circle of radius 1 centered at the origin.

36. Describe the figure obtained by joining the graphs in Exercise 35. Verify your description by using the distance formula.

37. Use graphical evidence to decide under what conditions $|w + z| = |w| + |z|$.

Graph each equation or inequality in the complex plane.

38. $|z| = 2$

39. $|z - 1| = 1$

40. $\text{Re } z = 3$

41. $\text{Im } z = 2$

C

42. $2 \le |z| \le 3$

43. $2 \le \text{Re } z \le 3$ or $2 \le \text{Im } z \le 3$

44. $\text{Re } z + \text{Im } z = 4$

45. $|z + 1| + |z - 1| = 2\sqrt{10}$ (Recall Sec. 2-4.)

Prove the following properties of the absolute value. (*Hint*: See the beginning of the proof of Theorem 3.)

46. $|wz| = |w||z|$ **47.** $\left|\frac{w}{z}\right| = \frac{|w|}{|z|}$ $\left(\text{First prove } \left|\frac{1}{z}\right| = \frac{1}{|z|}.\right)$

48. Prove: $|w| - |z| \leq |w + z|$
(*Hint*: $|w| = |(w + z) + (-z)| \leq |w + z| + |-z|$.)

49. Prove: $||w| - |z|| \leq |w + z|$

Prove each statement.

50. $\overline{w}\overline{z} = \overline{wz}$ **51.** $\text{Re } z \leq |z|$ **52.** $|\overline{z}| = |z|$

53. Justify each step.

$$\begin{aligned} w\overline{z} + \overline{w}z &= w\overline{z} + \overline{w\overline{z}} \\ &= 2 \text{ Re } (w\overline{z}) \\ &\leq 2\ |w\overline{z}| \\ &\leq 2\ |w||\overline{z}| \\ &\leq 2\ |w||z| \end{aligned}$$

COMPUTER EXERCISES

1. Write a computer program to find $\left|\frac{z}{\overline{z}}\right|$, where $z = a + bi$. Values of a and b are input by the user.
2. Run the program in Exercise 1 for the following values of z.
   a. $2 + 3i$ **b.** -5 **c.** $-3 - 3i$ **d.** $1 - 7i$
3. Write a computer program to find $(bi)^1$, $(bi)^2$, $(bi)^3$, . . . , $(bi)^{n-1}$, and $(bi)^n$ for input values of b and n.
4. Run the program in Exercise 3 to find the first n powers of each complex number.

   a. i; $n = 8$ **b.** $\frac{1}{2}i$; $n = 8$ **c.** $2i$; $n = 8$ **d.** $-2i$; $n = 7$

5-3 SQUARE ROOTS OF COMPLEX NUMBERS

Recall that if r is a nonnegative real number, then $\sqrt{r}$ denotes the nonnegative real number whose square is r. In this section we want to extend the concept of square root to include the case where r is negative. Since $i^2 = -1$, we write $i = \sqrt{-1}$ and call i a square root of -1. Similarly, because $(i\sqrt{2})^2 = -2$, we write $\sqrt{-2} = i\sqrt{2}$. In general, if $r > 0$, then

$$\sqrt{-r} = i\sqrt{r}.$$

Moreover if $r > 0$, the equation $z^2 = -r$ has two roots, or solutions,

$$\sqrt{-r} = i\sqrt{r} \quad \text{and} \quad -\sqrt{-r} = -i\sqrt{r}.$$

For example, if $z^2 = -3$, then $z = \sqrt{-3} = i\sqrt{3}$ or $z = -\sqrt{-3} = -i\sqrt{3}$.

EXAMPLE 1 Solve $(z - 2)^2 = -9$.

Solution Take the square root of each side of $(z - 2)^2 = -9$.

$$z - 2 = \sqrt{-9} \qquad \text{or} \qquad z - 2 = -\sqrt{-9}$$
$$z - 2 = i\sqrt{9} = 3i \qquad z - 2 = -i\sqrt{9} = -3i$$

Therefore $z = 2 + 3i$ or $z = 2 - 3i$. **Answer**

Example 2 illustrates a way to find square roots of complex numbers. (Another method is discussed in Chapter 9.)

EXAMPLE 2 Find the square roots of $3 - 4i$.

Solution The complex number $x + yi$ is a square root of $3 - 4i$ if and only if $(x + yi)^2 = 3 - 4i$. That is,

$$(x^2 - y^2) + 2xyi = 3 - 4i.$$

Equate real and imaginary parts.

$$x^2 - y^2 = 3 \qquad (1)$$
$$2xy = -4 \qquad (2)$$

Also, since $|x + yi|^2 = |3 - 4i| = \sqrt{9 + 16} = 5$,

$$x^2 + y^2 = 5. \qquad (3)$$

Add equations (1) and (3), to solve for x.

$$2x^2 = 8$$
$$x = 2 \text{ or } x = -2$$

From (2), $y = -\frac{4}{2x}$.

If $x = 2$, $y = -1$, and if $x = -2$, $y = 1$.

$\therefore$ the square roots of $3 - 4i$ are $2 - i$ and $-2 + i$. **Answer**

Check: $(2 - i)^2 = (-2 + i)^2 = 4 - 4i + i^2 = 4 - 4i - 1 = 3 - 4i$

We can now extend the quadratic formula learned in earlier courses to solve quadratic equations with complex coefficients. Check the steps in the derivation of the formula, to see that the formula remains valid even if some of the coefficients are imaginary.

THEOREM 4 The roots of $az^2 + bz + c = 0$ $(a \neq 0)$ are given by

$$z = \frac{-b \pm \sqrt{b^2 - 4ac}}{2a}.$$

We take for $\sqrt{b^2 - 4ac}$ either of the square roots of $b^2 - 4ac$.

EXAMPLE 3 Solve $z^2 - iz + (-1 + i) = 0$.

Solution In the quadratic formula let $a = 1$, $b = -i$, and $c = -1 + i$.

$$z = \frac{-(-i) \pm \sqrt{(-i)^2 - 4 \cdot 1 \cdot (-1 + i)}}{2} = \frac{i \pm \sqrt{-1 + 4 - 4i}}{2}$$

$$= \frac{i \pm \sqrt{3 - 4i}}{2}$$

Use Example 2 to substitute $2 - i$ for $\sqrt{3 - 4i}$.

$$z = \frac{i \pm (2 - i)}{2} = \frac{i + (2 - i)}{2} \text{ or } \frac{i - (2 - i)}{2}$$

$\therefore z = 1$ or $-1 + i$. **Answer**

We can also solve quadratic equations with complex coefficients by completing squares rather than by using the quadratic formula.

EXERCISES

Solve each equation.

A

1. $(z - 1)^2 = -1$
2. $(z + 2)^2 = -4$
3. $(z + 2i)^2 = 4$
4. $(z - i)^2 = 9$
5. $(z + 3)^2 = -12$
6. $(z - 3)^2 = -5$
7. $(z - i)^2 = -1$
8. $(z + 2i)^2 = -9$
9. $z^2 - 2z = -10$
10. $z^2 + 4z = -5$

Exercises 11–14 illustrate that such formulas as $\sqrt{a}\sqrt{b} = \sqrt{ab}$, which are true if a and b are nonnegative real numbers, may not be true if they are not.

11. Find **(a)** $\sqrt{-4} \cdot \sqrt{-9}$ and **(b)** $\sqrt{(-4)(-9)}$.
12. Find **(a)** $\sqrt{-1} \cdot \sqrt{-1}$ and **(b)** $\sqrt{(-1)(-1)}$.
13. Find **(a)** $\frac{\sqrt{-16}}{\sqrt{4}}$ and **(b)** $\sqrt{\frac{-16}{4}}$.
14. Find **(a)** $\frac{\sqrt{-8}}{\sqrt{2}}$ and **(b)** $\sqrt{\frac{-8}{2}}$.

Find the two square roots of each complex number.

B

15. $2i$
16. $-2i$
17. $-3 + 4i$
18. $15 - 8i$
19. $2 - 2\sqrt{3}i$
20. $2 + 2\sqrt{3}i$

Solve each equation. Use the indicated exercise(s).

21. $z^2 - 2z + (1 - 2i) = 0$ (Ex. 15)
22. $z^2 - (1 + i)z + i = 0$ (Ex. 16)
23. $z^2 + z + (1 - i) = 0$ (Ex. 17)
24. $z^2 - iz + (-4 + 2i) = 0$ (Ex. 18)
25. $z^4 + 4 = 0$ (Exs. 15, 16)
26. $z^6 + 64 = 0$ (Exs. 19, 20)

C

27. By paralleling the steps used in Example 2, derive general formulas for the square roots of $p + qi$ (p and q real). Consider the two cases $q > 0$ and $q < 0$.

General Theory of Polynomials

5-4 POLYNOMIALS OVER FIELDS

In the rest of this chapter we shall study polynomials over

Q the field of rational numbers,
$\mathcal{R}$ the field of real numbers, and
C the field of complex numbers.

A **polynomial over a field** F is an expression of the form

$$P(x) = a_0x^n + a_1x^{n-1} + \cdots + a_{n-1}x + a_n,$$

where n is a nonnegative integer, and the **coefficients** $a_0, a_1, \ldots, a_n$ are members of F, with the **leading coefficient**, a_0, not 0. The **degree** of $P(x)$ is n. (The zero polynomial has no degree.)

Polynomials of degree less than five have special names.

1. $x + 2i$ is a **linear** polynomial in x.
2. $z^2 + z - \sqrt{3}$ is a **quadratic** polynomial in z.
3. $2t^3 - \frac{1}{2}t + 1$ is a **cubic** polynomial in t.
4. $x^4 + 1$ is a **quartic** polynomial in x.

All four of these are polynomials over C; (2), (3), and (4) are polynomials over $\mathcal{R}$ and over C; and (3) and (4) are polynomials over Q, $\mathcal{R}$, and C.

EXAMPLE 1 Let $P(x) = x^2 - 2x + 5$ and $D(x) = x^3 + 3x^2 - 1$.

a. Find $P(x) + D(x)$. **b.** Find $P(x) \cdot D(x)$.

Solution The work can be arranged in vertical arrays.

a.
$$\begin{array}{r} x^3 + 3x^2 \qquad\quad - 1 \\ x^2 - 2x + 5 \\ \hline x^3 + 4x^2 - 2x + 4 \end{array}$$
Answer

b.
$$\begin{array}{rrrrrr} x^3 & + 3x^2 & - 1 & & & \\ x^2 & - 2x & + 5 & & & \\ \hline x^5 + 3x^4 & & - x^2 & & & \\ - 2x^4 & - 6x^3 & & + 2x & & \\ & 5x^3 & + 15x^2 & & - 5 & \\ \hline x^5 + x^4 & - x^3 & + 14x^2 & + 2x & - 5 & \end{array}$$
Answer

Note that the degree of the product of two nonzero polynomials is the sum of the degrees of the factors. (See Exercise 30.)

Unlike addition and multiplication, division of a polynomial by another polynomial does not yield, in general, a polynomial. We can, however, use the familiar **division algorithm**

$$\text{dividend} = \text{divisor} \times \text{quotient} + \text{remainder}$$

to divide one polynomial by another. The following theorem states the algorithm more precisely.

THEOREM 5 Let $P(x)$ and $D(x)$ be polynomials over a field F, with $D(x)$ not the zero polynomial. Then there are unique polynomials $Q(x)$ and $R(x)$ over F such that

$$P(x) = D(x)Q(x) + R(x),$$

where $R(x)$ is the zero polynomial or else is of degree less than the degree of $D(x)$.

(An inductive proof of this theorem for the case $D(x) = x - c$ is outlined in Exercise 31.)

EXAMPLE 2 Divide $x^3 + 3x^2 - 1$ by $x^2 - 2x + 5$.

Solution The "long division" process is usually arranged as shown below.

$$\begin{array}{r|l}
 & x + 5 \qquad\qquad\qquad = Q(x) \\
x^2 - 2x + 5 & x^3 + 3x^2 \qquad\qquad - 1 \\
 & \underline{x^3 - 2x^2 + 5x} \\
 & \quad 5x^2 - 5x - 1 \\
 & \quad \underline{5x^2 - 10x + 25} \\
 & \qquad\quad 5x - 26 = R(x)
\end{array}$$

Check: $(x^2 - 2x + 5)(x + 5) + (5x - 26)$
$= x^3 + 3x^2 - 5x + 25 + (5x - 26)$
$= x^3 + 3x^2 - 1$

Applying Theorem 5 in the case where $D(x)$ is the linear binomial $x - c$, we find that

$$P(x) = (x - c)Q(x) + R,$$

where R is now a constant. Replacing x by c, we have

$$P(c) = (c - c)Q(c) + R = 0 \cdot Q(c) + R = R.$$

This proves the **remainder theorem**.

THEOREM 6 Let $P(x)$ be a polynomial of positive degree n over a field F. Then for each c in F there is a unique polynomial $Q(x)$ of degree $n - 1$ such that

$$P(x) = (x - c)Q(x) + P(c).$$

We can use synthetic substitution (page 169) to find $P(c)$ in Theorem 6. Further, this process gives the coefficients of $Q(x)$ as well. To see this, let $P(x) = 2x^3 - 9x^2 + 14x - 8$ and $c = 3$. Compare the long division process and the synthetic substitution process, both of which are shown on the following page.

$$
\begin{array}{r}
2x^2 - 3x + 5 \\
x - 3\overline{)2x^3 - 9x^2 + 14x - 8} \\
\underline{2x^3 - 6x^2} \qquad\qquad \\
-3x^2 + 14x \qquad \\
\underline{-3x^2 + 9x} \qquad \\
5x - 8 \\
\underline{5x - 15} \\
7
\end{array}
\qquad
\begin{array}{r|rrrr}
3 & 2 & -9 & 14 & -8 \\
 & & 6 & -9 & 15 \\
\hline
 & 2 & -3 & 5 & 7 = P(3)
\end{array}
$$

Thus we see that the remainder when $P(x)$ is divided by $x - 3$ is indeed $P(3)$. But notice also that the other numbers,

$$2 \quad -3 \quad 5,$$

in the last row of the synthetic substitution process are precisely the coefficients of the quotient,

$$2x^2 - 3x + 5.$$

Since this is always the case (see Exercise 33), we have an easy way of dividing a polynomial by a linear binomial. We call it **synthetic division**.

EXAMPLE 3 Divide $3x^3 + x^2 - 6x + 5$ by $x + 2$.

Solution Use synthetic division with $c = -2$.

$$
\begin{array}{r|rrrr}
-2 & 3 & 1 & -6 & 5 \\
 & & -6 & 10 & -8 \\
\hline
 & 3 & -5 & 4 & -3
\end{array}
$$

Therefore $Q(x) = 3x^2 - 5x + 4$ and $R = -3$. **Answer**

EXAMPLE 4 Divide $2x^4 - 3x^3 - x + 2$ by $2x + 1$.

Solution To use synthetic division the divisor must be of the form $x - c$.

$$\frac{2x^4 - 3x^3 - x + 2}{2x + 1} = \frac{1}{2} \cdot \frac{2x^4 - 3x^3 - x + 2}{x + \frac{1}{2}}$$

Use synthetic division with $c = -\frac{1}{2}$.

$$
\begin{array}{r|rrrrr}
-\frac{1}{2} & 2 & -3 & 0 & -1 & 2 \\
 & & -1 & 2 & -1 & 1 \\
\hline
 & 2 & -4 & 2 & -2 & 3
\end{array}
$$

Therefore $Q(x) = \frac{1}{2}(2x^3 - 4x^2 + 2x - 2) = x^3 - 2x^2 + x - 1$ and $R = 3$. **Answer**

Since $\frac{\text{dividend}}{\text{divisor}} = \text{quotient} + \frac{\text{remainder}}{\text{divisor}}$, we can also write the division as

$$\frac{2x^4 - 3x^3 - x + 2}{2x + 1} = x^3 - 2x^2 + x - 1 + \frac{3}{2x + 1}.$$

EXERCISES

Determine the constants h and k so that the given polynomials will be equal. (Recall that two polynomials are equal if their corresponding coefficients are equal.)

A

1. $x^3 + hx^2 - 2$; $x^3 + 3x^2 + (k + 2)x - 2$
2. $x^3 + (h + 1)x^2 + kx + 4$; $x^3 + (2k + 1)x + 4$
3. $x^4 + (h + k)x^2 + 3$; $x^4 + x^2 + (h - k)$
4. $x^3 + (2h - k)x^2 + 2x + 1$; $x^3 + (2h + k)x + 1$

Find $P(x) + D(x)$, $P(x) - D(x)$, and $P(x) \cdot D(x)$.

5. $P(x) = 6x^3 + 7x^2 - 3x + 2$; $D(x) = 3x^3 + 2x^2 + 2$
6. $P(x) = 2x^3 - 4x^2 + 5x - 6$; $D(x) = 3 + 2x^2$
7. $P(x) = x^4 + 9x - 2x^2 + 4$; $D(x) = x^2 + 2x - 1$
8. $P(x) = 2x^4 - x^3 - 2x^2 + 5$; $D(x) = x^3 - 2x + 3$

9–12. In Exercises 5–8, find the quotient, $Q(x)$, and the remainder, $R(x)$, when $P(x)$ is divided by $D(x)$.

Use synthetic division to divide the first polynomial by the second polynomial.

13. $2x^4 - 9x^2 - x + 2$; $x - 2$
14. $x^3 + 2x^2 - 6x + 5$; $x + 4$
15. $2x^4 + 6x^3 - x^2 + 5$; $x + 3$
16. $x^3 - 9x + 4$; $x - 3$

B

17. $3x^4 + 7x^3 - 7x + 2$; $3x - 2$
18. $4x^4 + 2x^3 + 7x - 1$; $2x - 3$
19. $x^3 + 2x^2 + 9x + 10$; $x - 3i$
20. $x^3 - 2x^2 + 2x - 2$; $x + i$

Determine the constant k so that the first polynomial will be a factor of the second polynomial.

21. $x - 3$; $x^3 - 4x + k$
22. $x + 2$; $x^3 + x + k$
23. $x + 1$; $x^3 + 3x^2 + kx + 3$
24. $x - 2$; $x^3 - 3x^2 + kx + 2$
25. Find a polynomial function whose graph crosses the x-axis only at $(-4, 0)$, $(0, 0)$, and $(4, 0)$.
26. Find a polynomial function whose graph crosses the x-axis only at $(1, 0)$, $(\frac{1}{2}, 0)$ $(\frac{1}{4}, 0)$, and $(\frac{1}{8}, 0)$.
27. Show that if $P(x)$ has exactly k distinct real roots, then the graph of $P(x) = y$ crosses the x-axis in exactly k locations.

In Exercises 28–29, $P(x) = x^3 - (a + b + c)x^2 + (ab + ac + bc)x + abc$.

28. Find $P(a)$.
29. Find $P(b)$.
30. Show that the degree of the product of two nonzero polynomials is the sum of the degrees of the factors. Why does this fact suggest that the zero polynomial has no degree?

Exercise 31 outlines a proof of the existence part of Theorem 5 (the division algorithm), and Exercise 32 indicates a proof of the uniqueness part.

C **31. a.** Show that Theorem 5 is true for polynomials of degree 1.

$$(a_0x + a_1 = a_0x - a_0c + a_0c + a_1)$$

b. Assume that Theorem 5 is true for all polynomials of degree k. Prove that it then is true for $P(x) = a_0x^{k+1} + a_1x^k + \cdots + a_kx + a_{k+1}$ by justifying the following steps:

$$P(x) = x[a_0x^k + a_1x^{k-1} + \cdots + a_k] + a_{k+1}$$

There is a polynomial $Q(x)$ and there is a constant R such that

$$\begin{aligned} P(x) &= x[(x - c)Q(x) + R] + a_{k+1} \\ &= (x - c)[xQ(x)] + Rx - Rc + Rc + a_{k+1} \\ &= (x - c)[xQ(x) + R] + [Rc + a_{k+1}]. \end{aligned}$$

32. Suppose that $P(x) = (x - c)Q_1(x) + R_1 = (x - c)Q_2(x) + R_2$.

a. Show that $R_1 = R_2$.

b. Then show that for all $x \neq c$, $Q_1(x) = Q_2(x)$.

c. Show that $Q_1(c) = Q_2(c)$.

33. To illustrate that the synthetic substitution process (page 169) is also a division process, consider:

$$\begin{aligned} P(x) &= a_0x^3 + a_1x^2 + a_2x + a_3 \\ P(x) - P(c) &= a_0(x^3 - c^3) + a_1(x^2 - c^2) + a_2(x - c) + a_3 - a_3 \\ &= a_0(x - c)(x^2 + cx + c^2) + a_1(x - c)(x + c) + a_2(x - c) \\ &= (x - c)[a_0(x^2 + cx + c^2) + a_1(x + c) + a_2] \\ \therefore P(x) &= (x - c)[a_0x^2 + (a_0c + a_1)x + ((a_0c + a_1)c + a_2)] + P(c) \end{aligned}$$

a. Explain why the polynomial in square brackets is the quotient when $P(x)$ is divided by $x - c$.

b. Compare its coefficients with the numbers in the synthetic substitution process (page 169).

34. Show that if r_1 and r_2 are two consecutive real roots of $P(x) = 0$, then $P(x) > 0$ for all x between r_1 and r_2 or $P(x) < 0$ for all x between r_1 and r_2.

COMPUTER EXERCISES

1. Write a computer program that uses synthetic division to find the quotient and remainder when the polynomial

$$a_0x^n + a_1x^{n-1} + a_2x^{n-2} + \cdots + a_{n-1}x + a_n$$

is divided by $x - k$, where $a_0, a_1, a_2, a_3, \ldots, a_{n-1}, a_n$, and k are input by the user.

2. Run the program for the following polynomials and divisors.

a. $x^2 - 9;\ x + 3$

b. $x^2 - 2x - 30;\ x - 7$

c. $2x^2 + 7x - 15;\ x + 5$

d. $2x^3 - x^2 - 10x + 9;\ x - 2$

e. $x^3 - \frac{1}{2}x^2 - \frac{3}{4}x + 1;\ x - \frac{1}{2}$

f. $3x^3 - 19x^2 + 20x + 10;\ x - 5$

■ 5-5 THE FUNDAMENTAL THEOREM OF ALGEBRA

Our primary objective in this chapter is to find roots of polynomial equations and zeros of polynomial functions. To find roots we rely heavily on factoring.

If $P(x)$, $F(x)$, and $G(x)$ are polynomials over a field F and

$$F(x)G(x) = P(x),$$

then $F(x)$ and $G(x)$ are **factors** of $P(x)$. For example, since

$$(x + 2)(x + 3) = x^2 + 5x + 6,$$

then $x + 2$ and $x + 3$ are factors of $x^2 + 5x + 6$. A polynomial over a field F is **reducible** over F if it is the product of two or more nonconstant polynomials over F; otherwise it is **irreducible** over F. For example,

$x^2 - 4$ is reducible over $\mathcal{R}$ because $x^2 - 4 = (x + 2)(x - 2)$.

On the other hand,

$x^2 + 4$ is irreducible over $\mathcal{R}$, but it is reducible over C because $x^2 + 4 = (x + 2i)(x - 2i)$.

Note that every linear polynomial is irreducible.

A polynomial has been **completely factored** over a field F when it is expressed as a product of polynomials irreducible over F.

EXAMPLE 1 Factor $x^4 - x^2 - 2$ completely over each field.

a. Q **b.** $\mathcal{R}$ **c.** C

Solution **a.** $x^4 - x^2 - 2 = (x^2 + 1)(x^2 - 2)$

b. $x^4 - x^2 - 2 = (x^2 + 1)(x + \sqrt{2})(x - \sqrt{2})$

c. $x^4 - x^2 - 2 = (x + i)(x - i)(x + \sqrt{2})(x - \sqrt{2})$

There is a close relationship between factors and roots, as stated in the following theorem, called *the factor theorem*.

THEOREM 7 **The Factor Theorem**
Let $P(x)$ be a polynomial over the field F. The member r of F is a root of the equation $P(x) = 0$ if and only if $x - r$ is a factor of $P(x)$.

Proof: If r is a root of $P(x) = 0$, then $P(r) = 0$. By Theorem 6, page 197, there is a polynomial $Q(x)$ such that

$$P(x) = (x - r)Q(x) + P(r) = (x - r)Q(x) + 0 = (x - r)Q(x).$$

Hence $x - r$ is a factor of $P(x)$.

Conversely, if $x - r$ is a factor of $P(x)$, then for some polynomial $Q(x)$, $P(x) = (x - r)Q(x)$. Thus, $P(r) = (r - r)Q(r) = 0$, and r is a root of $P(x) = 0$. ■

Once a root r is found, then by Theorem 7

$$P(x) = (x - r)Q(x).$$

It follows that the remaining roots of $P(x) = 0$ must be roots of the **depressed**, or **reduced, equation** $Q(x) = 0$.

EXAMPLE 2 Let $P(x) = 2x^3 + 11x^2 + 12x - 9$.

a. Solve $P(x) = 0$. **b.** Factor $P(x)$ completely over C.

Solution **a.** Use synthetic division to find one root of $P(x) = 0$.

$$\begin{array}{r|rrrr} -3 & 2 & 11 & 12 & -9 \\ & & -6 & -15 & 9 \\ \hline & 2 & 5 & -3 & 0 \end{array}$$

Therefore -3 is a root and the depressed equation is $2x^2 + 5x - 3 = 0$.

$$2x^2 + 5x - 3 = 0$$
$$(x + 3)(2x - 1) = 0$$

The roots of the depressed equation are -3 and $\frac{1}{2}$.
$\therefore$ the roots of the given equation are -3, -3, and $\frac{1}{2}$. **Answer**

b. By the factor theorem,

$$2x^3 + 11x^2 + 12x - 9 = (x + 3)(x + 3)(2x - 1)$$
$$= (x + 3)^2(2x - 1). \quad \textbf{Answer}$$

A root r of $P(x) = 0$ has **multiplicity** k if $(x - r)$ occurs to the power k in the complete factorization of $P(x)$. Thus, in Example 2, -3 is a **double root**, a root of multiplicity 2.

A field F is said to be **algebraically complete**, or **algebraically closed**, if every polynomial equation of positive degree over F has a root in F. We know that neither Q nor $\mathcal{R}$ is algebraically complete because, for example, $x^2 + 1 = 0$ has no root in Q or in $\mathcal{R}$. As we saw in Section 5-3, every quadratic equation over C has a root in C. The fact that C is indeed algebraically complete is the *Fundamental Theorem of Algebra*.

THEOREM 8 **The Fundamental Theorem of Algebra**
Every polynomial equation of positive degree with complex coefficients has a complex root.

This theorem was first proved in 1799 by the great mathematician Karl Friedrich Gauss. Its proof, however, is beyond the scope of this book.

THEOREM 9 Every polynomial of degree n over C is the product of a constant and n factors of the form $x - r$, where $r \in C$.

We will prove Theorem 9 by induction on the degree of the polynomial.

Proof: Let S be the set of all positive integers for which the theorem is true.

(1) $1 \in S$ because $a_0x + a_1 = a_0\left(x - \left(-\frac{a_1}{a_0}\right)\right)$.

(2) Suppose $k \in S$, that is, suppose that every polynomial of degree k can be written in the form $c(x - r_1)(x - r_2) \cdots (x - r_k)$. Now, let $P(x)$ be any polynomial of degree $k + 1$. By Theorem 8, $P(x) = 0$ has a root r, and by Theorem 7,

$$P(x) = (x - r)Q(x),$$

where $Q(x)$ is a polynomial of degree k. By the induction hypothesis, $Q(x) = c(x - r_1)(x - r_2) \cdots (x - r_k)$. Therefore,

$$P(x) = (x - r)Q(x) = c(x - r)(x - r_1)(x - r_2) \cdots (x - r_k).$$

Thus, if $k \in S$, then $k + 1 \in S$.

By (1) and (2) and the principle of mathematical induction, S consists of all positive integers, and the theorem is proved. ■

In Theorem 9 the constant is the leading coefficient of the polynomial

$$a_0x^n + a_1x^{n-1} + \cdots + a_{n-1}x + a_n = a_0(x - r_1)(x - r_2) \cdots (x - r_n).$$

COROLLARY Every polynomial equation of positive degree n over C has exactly n roots if a root of multiplicity k is counted as k roots.

EXAMPLE 3 Solve $x^4 + 2x^3 - 3x^2 - 10x = 0$.

Solution Since $x^4 + 2x^3 - 3x^2 - 10x = x(x^3 + 2x^2 - 3x - 10)$, 0 is a root. Find one root of $x^3 + 2x^2 - 3x - 10 = 0$.

$$\begin{array}{r|rrrr} 1 & 1 & 2 & -3 & -10 \\ & & 1 & 3 & 0 \\ \hline & 1 & 3 & 0 & -10 \end{array} \qquad \begin{array}{r|rrrr} 2 & 1 & 2 & -3 & -10 \\ & & 2 & 8 & 10 \\ \hline & 1 & 4 & 5 & 0 \end{array}$$

Therefore 2 is a root. The remaining roots are solutions of $x^2 + 4x + 5 = 0$. By the quadratic formula,

$$x = \frac{-4 \pm \sqrt{4^2 - 4 \cdot 1 \cdot 5}}{2 \cdot 1} = -2 \pm i.$$

The four roots are 0, 2, $-2 + i$, and $-2 - i$. **Answer**

EXERCISES

Use the factor theorem to determine whether the first polynomial is a factor of the second polynomial.

A

1. $x + 1$; $x^5 + x^4 + x^3 + x^2 + x + 1$

2. $x + 1$; $x^8 - 3x^5 + 2$

3. $x - 1$; $x^{10} + x^9 + x + 1$

4. $x + 2$; $x^8 + 2x^7 + x + 2$

5. $x - 2i$; $x^3 - x^2 + 4x - 4$

6. $x - i$; $x^3 + x^2 + x + 1$

Determine the constant k so that the first polynomial will be a factor of the second polynomial.

7. $x + 2;\ 2x^3 + 3x^2 + kx + k + 1$

8. $x - 3;\ x^3 - 2x^2 + (k - 4)x - (k + 1)$

Solve each equation over C given the indicated root.

9. $x^3 - 3x^2 + 2x - 6 = 0;\ 3$

10. $x^3 + 3x^2 - 4 = 0;\ -2$

11. $x^3 - 2x^2 + 4x - 8 = 0;\ 2$

12. $2x^3 + 5x^2 + 6x + 2 = 0;\ -\frac{1}{2}$

13. $9x^3 - 6x^2 - 20x - 8 = 0;\ -\frac{2}{3}$

14. $4x^3 + 8x^2 - 11x + 3 = 0;\ \frac{1}{2}$

Find a polynomial equation with integral coefficients that has the given numbers as roots. (*Hint*: Use the factor theorem and an appropriately chosen constant to clear of any fractions that result.)

15. $2,\ -2,\ \frac{1}{2}$

16. $3,\ -1,\ -\frac{3}{2}$

17. $2,\ -\frac{3}{2},\ -\frac{1}{2}$

18. $-1,\ \frac{1}{2},\ \frac{1}{3}$

19. $2,\ 2 - i,\ 2 + i$

20. $-3,\ i,\ -i$

Solve each equation over C, given the two indicated roots. (*Hint*: Use the depressed equation.)

B

21. $x^4 + x^3 + 2x - 4 = 0;\ 1,\ -2$

22. $2x^4 - 5x^3 - 11x^2 + 20x + 12 = 0;\ -2,\ 3$

23. $2x^4 - 3x^3 - 3x - 2 = 0;\ 2,\ -\frac{1}{2}$

24. $3x^4 + 5x^3 - 7x^2 - 3x + 2 = 0;\ 1,\ -\frac{2}{3}$

25. $x^4 + 4x^3 - 16x - 16 = 0;\ -2$ is a multiple root.

26. $4x^4 - 4x^3 + 13x^2 + 18x + 5 = 0;\ -\frac{1}{2}$ is a multiple root.

Factor each polynomial completely over **(a)** the rational field, Q; **(b)** the real field, $\mathcal{R}$; and **(c)** the complex field, C. (Two of the three factorizations might be the same.)

27. $x^4 - 2x^2 - 3$

28. $x^4 - 3x^2 - 4$

29. $x^4 - 4$

30. $x^3 + 8$

31. $x^4 - 4x^2 + 4$

32. $x^4 + 4x^2 + 4$

In Exercises 33–40 solve $P(x) = 0$. State the complete factorization of $P(x)$ over C. (*Hint*: In Exercises 39 and 40, find the integral roots first.)

33. $P(x) = x^3 - x^2 - 9x + 9$

34. $P(x) = x^3 - 2x^2 - x + 2$

35. $P(x) = x^3 - 2x^2 - 4x + 8$

36. $P(x) = x^3 - 7x^2 + 15x - 9$

37. $P(x) = 3x^3 + 2x^2 - 3x - 2$

38. $P(x) = 2x^3 - 7x^2 + 4x + 4$

39. $P(x) = x^4 - x^3 + 3x^2 - 9x - 54$

40. $P(x) = x^4 - x^3 + 14x^2 - 16x - 32$

C

41. Show that for each polynomial $P(x)$ there is a circle centered at the origin in the complex plane inside of which the graphs of all the roots of $P(x) = 0$ lie.

42. Use the factor theorem and synthetic division to find all linear factors common to $x^4 + x^3 - 5x^2 - 3x + 6$ and $2x^4 + 2x^3 - 9x^2 - 5x + 10$.

Polynomials with Real Coefficients

5-6 CONJUGATE IMAGINARY ROOTS

When we solve $x^2 - 2x + 5 = 0$ using the quadratic formula, we find that the roots are $1 + 2i$ and $1 - 2i$. These roots are conjugates of each other. This illustrates a property of all polynomial equations with real coefficients:

Imaginary roots occur in conjugate pairs.

This fact follows from Theorem 10.

THEOREM 10 If $a + bi$ (a and b real, $b \neq 0$) is a root of the polynomial equation $P(x) = 0$ with real coefficients, then

$$S(x) = [x - (a + bi)][x - (a - bi)] = x^2 - 2ax + (a^2 + b^2)$$

is a factor of $P(x)$.

Proof: Because $S(x)$ has real coefficients, the division algorithm guarantees that there are polynomials $Q(x)$ and $R(x) = cx + d$ over $\mathcal{R}$ such that

$$P(x) = S(x)Q(x) + cx + d.$$

Substituting $a + bi$ for x, we have

$$P(a + bi) = S(a + bi)Q(a + bi) + c(a + bi) + d.$$

Since $P(a + bi) = 0$ and $S(a + bi) = 0$, this becomes

$$0 = 0 \cdot Q(a + bi) + c(a + bi) + d, \text{ or } (ac + d) + bci = 0.$$

Thus, $ac + d = 0$ and $bc = 0$. Since $b \neq 0$ we have $c = 0$ and therefore $d = 0$. Thus,

$$P(x) = S(x)Q(x). \quad \blacksquare$$

COROLLARY If the imaginary number $a + bi$ (a and b real) is a root of a polynomial equation with real coefficients, then $a - bi$ is also a root.

Proof: Using the notation of Theorem 10 we have $P(x) = S(x)Q(x)$, and since $S(a - bi) = 0$,

$$P(a - bi) = S(a - bi)Q(a - bi) = 0 \cdot Q(a - bi) = 0. \quad \blacksquare$$

EXAMPLE 1 Solve $P(x) = x^4 - 9x^2 + 12x + 10 = 0$, given that $2 - i$ is a root.

Solution By Theorem 10 and its corollary, $2 + i$ is also a root and $S(x) = x^2 - 4x + 5$ is a factor of the polynomial.
Use long division with $S(x)$ as divisor and $P(x)$ as the dividend to find $Q(x)$, as shown on the following page.

$$
\begin{array}{r}
x^2 + 4x + 2 \\
x^2 - 4x + 5\overline{)x^4 \qquad\quad - 9x^2 + 12x + 10} \\
\underline{x^4 - 4x^3 + 5x^2} \qquad\qquad \\
4x^3 - 14x^2 + 12x \qquad \\
\underline{4x^3 - 16x^2 + 20x} \qquad \\
2x^2 - 8x + 10 \\
\underline{2x^2 - 8x + 10} \\
0
\end{array}
$$

The remaining roots are solutions of $x^2 + 4x + 2 = 0$.

$$x = \frac{-4 \pm \sqrt{16 - 8}}{2} = -2 \pm \sqrt{2}$$

$\therefore$ the roots are $2 - i$, $2 + i$, $-2 + \sqrt{2}$, and $-2 - \sqrt{2}$. **Answer**

According to Theorem 9 in Section 5-5 every polynomial $P(x)$ over C can be factored into linear polynomials over C. If instead we restrict attention only to polynomials over $\mathcal{R}$ and look for factors over $\mathcal{R}$ we have the following theorem.

THEOREM 11 Every polynomial of positive degree over $\mathcal{R}$ is the product of linear and irreducible quadratic factors over $\mathcal{R}$.

Proof: Suppose that $P(x)$ is a polynomial of degree n over $\mathcal{R}$. If all the roots $r_1, r_2, \ldots, r_n$ are real, then $x - r_1, x - r_2, \ldots, x - r_n$ are factors of $P(x)$. If $P(x)$ has a complex root c, then by the theorem on conjugate roots, $\overline{c}$ is also a root. Thus $(x - c)(x - \overline{c})$ is a factor. Since

$$(x - c)(x - \overline{c}) = x^2 - (c + \overline{c})x + c\overline{c},$$

$c + \overline{c} = 2 \operatorname{Re} c$, and $c\overline{c} = |c|^2$, then $x^2 - (c + \overline{c})x + c\overline{c}$ is a polynomial over $\mathcal{R}$. It is irreducible since it cannot be factored over $\mathcal{R}$. ■

EXAMPLE 2 Factor $P(x) = x^6 - 3x^2 - 2$ completely over $\mathcal{R}$ given that i is a double root of $P(x) = 0$.

Solution Since i is a double root of $P(x) = 0$, then by Theorem 10 it is easy to see that $-i$ is also a double root. Thus, $(x - i)^2(x + i)^2 = (x^2 + 1)^2 = x^4 + 2x^2 + 1$ is a factor of $P(x)$.

$$
\begin{array}{r}
x^2 \qquad\qquad\quad - 2 \\
x^4 + 2x^2 + 1\overline{)x^6 \qquad\quad - 3x^2 - 2} \\
\underline{x^6 + 2x^4 + x^2} \qquad \\
-2x^4 - 4x^2 - 2 \\
\underline{-2x^4 - 4x^2 - 2} \\
0
\end{array}
$$

$\therefore x^6 - 3x^2 - 2 = (x^2 + 1)^2(x^2 - 2) = (x^2 + 1)^2(x + \sqrt{2})(x - \sqrt{2})$.

EXAMPLE 3 Find a polynomial equation $P(x) = 0$ of minimum degree that has integral coefficients and that has i as a double root and 2 and 3 as roots.

Solution Since i is a double root, so is $-i$. From Example 2, $x^4 + 2x^2 + 1$ is a factor of the desired polynomial. The other factors are $x - 2$ and $x - 3$. Therefore $P(x) = (x^4 + 2x^2 + 1)(x - 2)(x - 3)$

$= x^6 - 5x^5 + 8x^4 - 10x^3 + 13x^2 - 5x + 6.$ **Answer**

EXERCISES

Find a polynomial equation of lowest degree that has the given numbers as roots. Give your answer in the form $a_0x^n + a_1x^{n-1} + \cdots + a_n = 0$ with integral coefficients.

A

1. $2, 1 + i$

2. $-1, 2 - i$

3. $\frac{1}{2}, -\frac{1}{2}, i$

4. $\frac{i}{2}, 1 - i$

5. $1 + i, -1 + i$

6. $\sqrt{2}, -\sqrt{2}, \sqrt{-2}$

7. $1 + i, 1 + i$ (a double root)

8. i, i, i (a triple root)

In Exercises 9–14, solve $P(x) = 0$ over C, given the specified root(s).

9. $P(x) = x^3 - 2x + 4; 1 + i$

10. $P(x) = x^3 - 4x^2 + 3x - 12; i\sqrt{3}$

11. $P(x) = x^4 + 4x^3 + 7x^2 + 8x + 10; -2 - i$

12. $P(x) = x^4 - 3x^3 + 5x^2 - x - 10; 1 - 2i$

13. $P(x) = x^5 + x^3 + x^2 + 1; i$

14. $P(x) = x^6 + 4x^4 + 5x^2 + 2$; i is a double root

15–20. In Exercises 9–14 express $P(x)$ as a product of factors irreducible over $\mathcal{R}$.

In Exercises 21–24, use the following analog of the corollary to Theorem 10: If the irrational number $a + b\sqrt{r}$ (a and b rational) is a root of a polynomial equation with rational coefficients, then $a - b\sqrt{r}$ is also a root.

a. Solve $P(x) = 0$ over C, given the specified root.

b. Express $P(x)$ as a product of factors irreducible over the rational field Q.

B

21. $P(x) = 2x^3 + x^2 - 6x - 3; \sqrt{3}$

22. $P(x) = 2x^3 - 5x^2 + 1; 1 - \sqrt{2}$

23. $P(x) = x^4 - 4x^2 - 8x - 4; 1 + \sqrt{3}$

24. $P(x) = x^4 - 2x^3 - 8x^2 + 12x - 4; -1 - \sqrt{3}$

In Exercises 25 and 26, show that **(a)** the given number is a root of the polynomial and **(b)** its conjugate is not a root.

25. $x^3 - 2x^2 + x + 1 + i; 1 + i$

26. $x^4 + x^2 - ix - 1; i$

C **27.** Prove the proposition about irrational roots stated just before Exercise 21. (*Hint*: Modify the proofs on page 205.)

28. Prove or disprove: Every polynomial equation of positive degree with rational coefficients has an even number of irrational roots.

29. Let $P(x) = 0$ be a polynomial equation over $\mathcal{R}$ with no multiple roots. Prove each statement about the graph of $P(x) = y$.

a. If the degree of P is even, then the graph of P crosses the x-axis an even number of times.

b. If the degree of P is odd, then the graph of P crosses the x-axis an odd number of times.

30. Find all solutions of $x^5 - 1 = 0$. Graph the solution set in the complex plane and describe the figure whose vertices are the graphs of the solutions. (*Hint*: One root is 1. Call the other roots $a + bi$, $a - bi$, $c + di$, and $c - di$. Since each is a root, $a^2 + b^2 = 1$ and $c^2 + d^2 = 1$. This will help simplify calculations used to find a, b, c, and d.)

5-7 BOUNDS FOR REAL ROOTS; DESCARTES' RULE

Various techniques for solving polynomial equations have been presented. In this section we will discuss two additional techniques.

When we search for real roots of a polynomial equation over $\mathcal{R}$, it is helpful to know an interval $[l, m]$ in which all such roots must lie. We call l a **lower bound** and m an **upper bound** for these roots. Theorem 12 provides a way to find such bounds.

THEOREM 12 Let $P(x)$ be a polynomial over $\mathcal{R}$ with positive leading coefficient. Suppose that when $P(x)$ is divided by $x - m$, the coefficients of the quotient and the remainder are all nonnegative. Then no real root of $P(x) = 0$ is greater than m.

You are asked to prove Theorem 12 in Exercise 25.

EXAMPLE 1 Use Theorem 12 to find the least positive integer that is an upper bound for the real roots of $P(x) = 2x^3 - 5x^2 - 8x - 9 = 0$.

Solution Divide $P(x)$ successively by $x - 1$, $x - 2$, ... (Use the compact form of synthetic division.)

	2	−5	−8	−9	
1	2	−3	−11	−20	
2	2	−1	−10	−29	
3	2	1	−5	−24	
4	2	3	4	7	⟵ No negative numbers

$\therefore$ 4 is an upper bound for the real roots of $P(x) = 0$. **Answer**

Note that once a negative number appears in the quotient or remainder, then there is no need to complete the division by that choice of divisor. Thus in Example 1, when -3 appeared in the division by $x - 1$, division by $x - 2$ could begin.

To see why 4 is indeed an upper bound for the real roots of $P(x) = 0$, note that $P(x) = (x - 4)(2x^2 + 3x + 4) + 7$. If $x > 4$, then $x - 4$ and $2x^2 + 3x + 4$ are positive. Therefore if $x > 4$, $P(x)$ cannot be 0.

There are several ways of finding a lower bound for real roots. Suppose that $-r$ is a negative root of $P(x) = 0$. Then r is a positive root of $P(-x) = 0$.

COROLLARY To find a lower bound for the negative roots of $P(x) = 0$, apply Theorem 12 to $P(-x) = 0$ or $-P(-x) = 0$, whichever has positive leading coefficient.

EXAMPLE 2 Use the corollary to find the greatest negative integer that is a lower bound for the real roots of $P(x) = 2x^3 - 5x^2 - 8x - 9 = 0$.

Solution $P(-x) = -2x^3 - 5x^2 + 8x - 9$. Apply Theorem 12 to $-P(-x) = 2x^3 + 5x^2 - 8x + 9$ since $-P(-x)$ has a positive leading coefficient.

	2	5	−8	9	
1	2	7	−1	8	
2	2	9	10	29	⟵No negative numbers

Since 2 is an upper bound for the real roots of $P(-x) = 0$, -2 is a lower bound for the real roots of $P(x) = 0$. **Answer**

Just as Theorem 12 gives information about the *magnitude* of the real roots of a polynomial equation over $\mathcal{R}$, Theorem 13, *Descartes' rule of signs*, gives information about the *number* of real roots. Consider a polynomial $P(x)$ over $\mathcal{R}$ written in order of decreasing powers of x. Whenever the coefficients of two adjacent terms have opposite signs, we say that $P(x)$ has a **variation in sign**. Thus,

$$x^6 - 3x^4 - 5x^3 + 2x - 4$$

1 2 3

has three variations in sign. Descartes' rule, which is stated below without proof, tells us that $x^6 - 3x^4 - 5x^3 + 2x - 4 = 0$ has either three positive roots or one positive root.

THEOREM 13 **Descartes' Rule of Signs**

Let $P(x)$ be a polynomial over $\mathcal{R}$.

The number of positive roots of $P(x) = 0$ either equals the number of variations in sign of $P(x)$ or else is fewer than this by an even number.

The number of negative roots of $P(x) = 0$ either equals the number of variations in sign of $P(-x)$ or else is fewer than this by an even number.

Sometimes Descartes' rule gives complete information about the numbers of roots of various types. For example, $P(x) = x^4 - 3x - 5$ has one variation in sign so that $P(x) = 0$ has exactly one positive root. Since $P(-x) = x^4 + 3x - 5$ also has one variation in sign, $P(x) = 0$ has exactly one negative root. Thus, $P(x) = 0$ has one positive root, one negative root, and two imaginary roots. Usually Descartes' rule leaves us with several possibilities.

EXAMPLE 3 List the possibilities for the nature of the roots (positive, negative, imaginary) of $P(x) = x^5 - 2x^4 + 4x^2 - 3x - 5 = 0$.

Solution $P(x)$ has three variations in sign, so the number of positive roots is 3 or 1. Since $P(-x) = -x^5 - 2x^4 + 4x^2 + 3x - 5$ has two variations in sign, the number of negative roots is 2 or 0. The diagram shows the possibilities.

positive	*negative*	*imaginary*
3	2	$0 = 5 - (3 + 2)$
	0	$2 = 5 - (3 + 0)$
1	2	$2 = 5 - (1 + 2)$
	0	$4 = 5 - (1 + 0)$

EXERCISES

Find the possible numbers of positive, negative, and imaginary roots of each equation. Display multiple possibilities in a diagram as in Example 3.

A

1. $x^3 + 3x - 3 = 0$

2. $x^3 + 2x + 5 = 0$

3. $x^3 + 2x^2 + x + 1 = 0$

4. $3x^3 - 2x^2 - 6 = 0$

5. $2x^4 - 2x - 7 = 0$

6. $x^4 + 3x^2 - 1 = 0$

7. $x^4 + x^3 - 2x - 1 = 0$

8. $x^4 + x^2 - x + 2 = 0$

9. $x^5 + 2x^4 - 2x - 4 = 0$

10. $x^5 - x^4 + x + 1 = 0$

11. $x^5 + 3x^4 - 2x^3 + x^2 - 1 = 0$

12. $x^5 + 2x^4 - x^2 - 3x + 5 = 0$

For each equation use Theorem 12 and its corollary to determine **(a)** the least positive integer that is an upper bound for the real roots and **(b)** the greatest negative integer that is a lower bound.

13. $x^3 - 2x^2 - 3x + 1 = 0$

14. $x^3 + x^2 - 4x - 5 = 0$

15. $2x^3 - 2x^2 - 15x - 20 = 0$

16. $2x^3 - 7x^2 - 21x + 35 = 0$

17. $x^4 + 2x^3 - 6x^2 - 40 = 0$

18. $x^4 - 2x^3 - 30 = 0$

19. $2x^4 - 5x^3 - x^2 + 6x - 12 = 0$

20. $2x^4 - 6x^2 - 3x - 9 = 0$

Show that the specified number is the only real root of the given equation.

B

21. $x^5 - x^4 + 3x^3 - 3x^2 + 2x - 2 = 0$; 1

22. $x^5 + x^4 + x^3 + x^2 + 2x + 2 = 0$; -1

23. $2x^5 + 3x^4 + 4x^3 + 6x^2 + 6x + 9 = 0;\ -\frac{3}{2}$

24. $2x^5 - x^4 + 4x^3 - 2x^2 + 6x - 3 = 0;\ \frac{1}{2}$

C 25. Prove Theorem 12 by paralleling the discussion that follows Example 1 of this section.

5-8 FINDING RATIONAL ROOTS

There is no general method for finding all roots in a field F of a polynomial equation over F. We can, however, find all *rational* roots of any polynomial equation over the *rational* field Q. We make use of the fact that any polynomial equation over Q is equivalent to one having *integral* coefficients. To find rational roots we will use the *rational root theorem.*

THEOREM 14 Rational Root Theorem
Suppose that a polynomial equation with integral coefficients has the root $\frac{h}{k}$, where h and k are relatively prime integers. Then h must be an integral factor of the constant term of the polynomial and k an integral factor of the leading coefficient.

For example, any rational root in lowest terms of

$$6x^3 - 5x^2 + 3x - 3 = 0$$

must have a numerator that is a factor of -3 (namely, ± 1 or ± 3) and a denominator that is a factor of 6 (namely, ± 1, ± 2, ± 3, or ± 6). Thus, the only possible rational roots are

$$1, 3, \frac{1}{2}, \frac{3}{2}, \frac{1}{3}, \frac{1}{6}$$

and their opposites.

EXAMPLE 1 Examine $x^3 - \frac{5}{6}x^2 + \frac{1}{2}x - \frac{1}{2} = 0$ for rational roots.

Solution The given equation is equivalent to

$$P(x) = 6x^3 - 5x^2 + 3x - 3 = 0.$$

By Theorem 14 the only possible rational roots are

$$1, 3, \frac{1}{2}, \frac{3}{2}, \frac{1}{3}, \frac{1}{6}$$

and their opposites. But since $P(-x) = -6x^3 - 5x^2 - 3x - 3$ has no variations in sign, according to Descartes' rule there are no negative roots. Test only the positive candidates.

Solution continued on following page

The test for 1 reveals that 1 is an upper bound for the real roots and not a root itself. Therefore there is no need to test 3 or $\frac{3}{2}$.

	6	−5	3	−3
1	6	1	4	1
$\frac{1}{2}$	6	−2	2	−2
$\frac{1}{3}$	6	−3	2	$-\frac{7}{3}$
$\frac{1}{6}$	6	−4	$\frac{7}{3}$	$-\frac{47}{18}$

Therefore the given equation has no rational root. **Answer**

Note that in Example 1 Descartes' rule and the upper bound theorem were useful in eliminating some of the candidates for roots.

EXAMPLE 2 Find the rational roots of $3x^5 + 11x^4 + 11x^3 + 7x^2 + 8x - 4 = 0$ and solve the equation completely if possible.

Solution By the rational root theorem, the only possible rational roots are

$$\pm 1, \pm 2, \pm 4, \pm\frac{1}{3}, \pm\frac{2}{3}, \pm\frac{4}{3}.$$

Since there is only one variation in sign, by Descartes' rule there is only one positive root.

Look for a positive upper bound. By Theorem 12, no root exceeds 1. Thus 2, 4, and $\frac{4}{3}$ need not be tested.

1	3	11	11	7	8	−4
		3	14	25	32	40
	3	14	25	32	40	36

Test $\frac{1}{3}$ as shown at the right. Since there is only one positive root, $\frac{1}{3}$ must be it. Then only negative candidates for the other roots need to be tested.

$\frac{1}{3}$	3	11	11	7	8	−4
		1	4	5	4	4
	3	12	15	12	12	0

Test −2 in the depressed polynomial after dividing its coefficients by 3. One root is −2. Test the root again in the new depressed polynomial to see if it is a multiple root. Here −2 is a double root.

−2	1	4	5	4	4
		−2	−4	−2	−4
	1	2	1	2	0

−2	1	2	1	2
		−2	0	−2
	1	0	1	0

The remaining roots are roots of $x^2 + 1 = 0$; namely i and $-i$.

The roots of $3x^5 + 11x^4 + 11x^3 + 7x^2 + 8x - 4 = 0$ are $-2, -2, \frac{1}{3}, i$, and $-i$. **Answer**

As a direct result of the rational root theorem we have the following corollary.

COROLLARY If $x^n + a_1x^{n-1} + \cdots + a_{n-1}x + a_n = 0$ has integral coefficients, then its rational roots, if any, must be integers.

Note that the corollary, when applied to the equation $x^2 - 2 = 0$, shows that $\sqrt{2}$ is irrational.

Remarks on solving polynomial equations

We know that linear and quadratic equations can be solved by formula, and indeed there are formulas for solving cubic and quartic equations. (The latter, however, are so complicated that they are little used.) It has been proved that no general formula to solve equations of degree five or higher can exist. It therefore is important to have methods to approximate the roots of polynomial equations. One such method uses the location principle (page 168), and we shall meet a much more powerful one in Section 6-9.

EXERCISES

Find all rational roots of each equation and complete the solution if possible. If the equation has no rational roots, say so. Use the theorems of Section 5-7 to shorten your work.

A

1. $x^3 - 5x^2 + 8x - 6 = 0$

2. $x^3 - 14x + 8 = 0$

3. $x^3 - 2x^2 + 2x - 12 = 0$

4. $x^3 + 3x^2 + 4x + 12 = 0$

5. $2x^3 + 5x^2 + 6x + 2 = 0$

6. $2x^3 + x^2 - 6x - 3 = 0$

7. $x^3 - \frac{2}{3}x^2 + \frac{1}{2}x - \frac{1}{3} = 0$

8. $\frac{1}{3}x^3 - \frac{1}{2}x^2 + \frac{2}{3}x - 1 = 0$

9. $x^4 - 3x^3 - 5x^2 + 13x + 6 = 0$

10. $x^4 - 2x^3 - 7x^2 + 12 = 0$

11. $3x^4 + 16x^3 + 9x^2 - 26x + 8 = 0$

12. $6x^4 + x^3 + 23x^2 + 4x - 4 = 0$

13. $4x^4 - 4x^3 - 11x^2 + 12x - 3 = 0$

14. $4x^4 - 4x^3 + x^2 - 4x - 3 = 0$

15. $6x^5 - 13x^4 + 12x^3 - 19x^2 + 19x - 6 = 0$

16. $12x^5 - 4x^4 + 7x^3 - 2x^2 - 5x + 2 = 0$

Use the rational root theorem to show that the following numbers are irrational.

EXAMPLE $\sqrt[3]{6}$

Solution $\sqrt[3]{6}$ is a root of $x^3 - 6 = 0$, which has 1, 2, 3, and 6 as its only possible rational roots. Since none of these numbers *is* a root, $\sqrt[3]{6}$ is irrational.

17. $\sqrt{15}$

18. $\sqrt[3]{-4}$

19. $\sqrt[5]{-9}$

20. $\sqrt[4]{6}$

21. $\sqrt{5} + \sqrt{3}$

22. $\sqrt{3} - \sqrt{2}$

(*Hint for Exercises 21 and 22*: Use this fact: If p and q are positive numbers, $\sqrt{p} + \sqrt{q}$ and $\sqrt{p} - \sqrt{q}$ are roots of $x^4 - 2(p + q)x^2 + (p - q)^2 = 0$.)

B **23.** Verify the fact stated in the hint for Exercises 21 and 22. What are the other roots of the equation?

24. Prove: If k is an integer and $\sqrt[n]{k}$ is rational, then k is the nth power of an integer.

25. Let $a_0x^n + a_1x^{n-1} + \cdots + a_n$ be a polynomial with integral coefficients. Show that $-|a_n|$ is a lower bound for any rational roots and that $|a_n|$ is an upper bound for any rational roots. Thus all rational roots are in the interval $[-|a_n|, |a_n|]$.

Exercises 26–28 outline a proof of the rational root theorem. We assume that h and k are relatively prime integers and that $\frac{h}{k}$ is a root of

$$a_0x^n + a_1x^{n-1} + \cdots + a_{n-1}x + a_n = 0,$$

where the coefficients $a_0, a_1, \ldots, a_n$ are integers.

26. Show that if h and k are relatively prime and h divides bk, then h divides b.

C **27.** Show that h is a factor of a_n by justifying each step. (*Hint*: Use the field properties and Exercise 26.)

a. $a_0\left(\frac{h}{k}\right)^n + a_1\left(\frac{h}{k}\right)^{n-1} + \cdots + a_{n-1}\left(\frac{h}{k}\right) + a_n = 0$

b. $a_0h^n + a_1h^{n-1}k + \cdots + a_{n-1}hk^{n-1} + a_nk^n = 0$

c. $a_nk^n = -h(a_0h^{n-1} + a_1h^{n-2}k + \cdots + a_{n-1}k^{n-1})$

d. h is a factor of a_nk^n.

e. h is a factor of a_n.

28. Modify the proof in Exercise 27 to show that k is a factor of a_0.

COMPUTER EXERCISES

The computer program of the Computer Exercises following Section 5–4 can be useful here.

1. Write a computer program to find the least positive integer that is an upper bound for the real roots of $P(x) = a_0x^n + a_1x^{n-1} + a_2x^{n-2} + \cdots + a_{n-1}x + a_n = 0$ where the coefficients $a_0, a_1, a_2, \ldots, a_{n-1}$, and a_n are input by the user.

2. Run the computer program of Exercise 1 for the following polynomials.

a. Example 1, Section 5–7 **b.** $P(x) = 5x^2 - 6x - 8$

c. $P(x) = x^2 - 100$ **d.** $P(x) = x^3 + x^2 + x + 1$

e. $P(x) = 4x^3 + 8x^2 - 13x - 3$ **f.** $P(x) = 4x^4 - 3x^3 + 6x^2 - 9x + 2$

3. Write a computer program to find and print all possible rational roots of $P(x) = a_0x^n + a_1x^{n-1} + a_2x^{n-2} + \cdots + a_{n-1}x + a_n = 0$ when the user supplies the coefficients of $P(x)$.

4. Run the computer program of Exercise 3 for the polynomials in (a)–(f) of Exercise 2.

*5. Modify the computer program of Exercise 3 to find and test all possible rational roots of $P(x) = 0$ and report if any are actual roots of the equation. Use the program of Exercise 1 that tests for an upper bound of real roots to eliminate unnecessary tests of possible roots. (*Note*: Check for multiple roots.)

6. Run the program of Exercise 5 for each of the polynomials in parts (a)–(f) of Exercise 2.

Chapter Summary

1. The field C of *complex numbers* is the set of all ordered pairs of real numbers with addition and multiplication defined by $(u, v) + (x, y) = (u + x, v + y)$ and $(u, v) \cdot (x, y) = (ux - vy, uy + vx)$. The pair $(x, 0)$ is identified with the real number x, and the special pair $(0, 1)$ is denoted by i. Then $i^2 = -1$. Since $(x, y) = (x, 0) + (y, 0) \cdot (0, 1)$, the complex number (x, y) can be written as $x + yi$, where $i^2 = -1$.
2. The *conjugate* of the complex number $z = x + yi$ is $\bar{z} = x - yi$. The absolute value, or *modulus*, of z is $|z| = \sqrt{x^2 + y^2}$. These definitions can be illustrated in the complex plane, in which the graph of z, or $x + yi$, is the point (x, y). For example, $|z|$ is the distance from the origin to the graph of z.
3. Some important properties of the conjugate and absolute value are the following:

$$z\bar{z} = |z|^2 \qquad |w + z| \le |w| + |z|$$
$$\overline{w + z} = \bar{w} + \bar{z} \qquad |wz| = |w| \cdot |z|$$
$$\overline{wz} = \bar{w} \cdot \bar{z}$$

4. Every nonzero complex number has two square roots. These can be found using algebraic methods.
5. Polynomials over a field can be added, subtracted, and multiplied. Division of polynomials is possible according to the *division algorithm*:

   dividend = divisor × quotient + remainder,

   where the remainder is zero or of degree less than that of the divisor.
6. As a corollary to the division algorithm we have the *remainder theorem*, which states that when $P(x)$ is divided by $x - c$, the remainder is $P(c)$. *Synthetic division* (the same process as synthetic substitution) reveals the coefficients of the quotient, $Q(x)$, as well as $P(c)$.
7. A polynomial over a field F is *reducible* over F if it is the product of two or more nonconstant polynomials over F; otherwise it is *irreducible* over F. To *factor* a polynomial over F completely, express it as a product of polynomials irreducible over F. The *factor theorem* states that r is a root of $P(x) = 0$ if and only if $x - r$ is a factor of $P(x)$.

8. *The Fundamental Theorem of Algebra* asserts that every polynomial equation of positive degree over the complex field C has a root in C. This implies that every polynomial of degree n over C is the product of a constant and n factors of the form $x - r$, where $r \in C$.
9. Let $P(x)$ be a nonconstant polynomial having *real* coefficients. If the imaginary number $a + bi$ is a root of $P(x) = 0$, then its conjugate, $a - bi$, is also a root. Moreover, $P(x)$ is a product of linear and quadratic factors irreducible over $\mathcal{R}$. Information about the magnitudes and the number of the real roots of $P(x) = 0$ can be obtained by using the theorems on upper and lower bounds and Descartes' rule of signs.
10. Let $P(x)$ be a polynomial with *integral* coefficients. All the possible rational roots of $P(x) = 0$ can be found using the following *rational root theorem*: Let h and k be relatively prime integers and suppose that $\frac{h}{k}$ is a root of $P(x) = 0$. Then h is a factor of the constant term of $P(x)$, and k is a factor of its leading coefficient.

Chapter Test

5-1

1. Simplify $i^{89} + i^{92} + i^{95}$.
2. Let $w = 1 + 2i$ and $z = 2 - i$. Express $w + z$, $w - z$, wz, and $\frac{w}{z}$ in $x + yi$ form.

5-2

3. Graph the equation $|z + 2i| = 2$.
4. Describe a geometric construction to obtain the graph of $\overline{w - z}$ given the graphs of w and z.

5-3

5. Solve $(z + 3i)^2 = -4$.
6. Find the two square roots of $-3 - 4i$.

5-4

7. Use synthetic division to determine the quotient obtained when $x^4 + 2x^3 + 7x - 9$ is divided by $x + 3$.
8. Determine the constant k so that $x + 2$ will be a factor of $x^3 + kx^2 + kx + 2$.

5-5

9. Factor $x^4 - 9$ completely over each field. **a.** Q **b.** $\mathcal{R}$ **c.** C
10. Solve $2x^3 - 3x^2 + 8x - 12 = 0$ over C, given that $\frac{3}{2}$ is a root.

5-6

11. Find a polynomial equation with integral coefficients that has $-1 + i$ as a double root.
12. Solve $x^4 - 2x^3 + 4x - 4 = 0$ over C given that $1 - i$ is a root.

5-7

13. Give the possibilities for the numbers of positive, negative, and imaginary roots of $x^5 + 2x^4 - 9x^2 + x + 5 = 0$.
14. Find the least positive integral upper bound and greatest negative integral lower bound for the real roots of the equation in Exercise 13.

5-8

15. Solve $2x^4 - x^3 + 2x^2 + 19x - 10 = 0$.

COMPUTER APPLICATION

INTEGRAL POWERS OF COMPLEX NUMBERS

The powers of a complex number $a + bi$ are often used in computation, but they also exhibit interesting characteristics of their own.

To see one such characteristic, compute $(a + bi)^2$ and $(a + bi)^3$.

$$(a + bi)^2 = (a + bi)(a + bi) = (a^2 - b^2) + (2ab)i$$
$$\begin{aligned}(a + bi)^3 &= (a + bi)^2(a + bi) \\ &= [a(a^2 - b^2) - b(2ab)] + [b(a^2 - b^2) + a(2ab)]i\end{aligned}$$

Let us use RE(2) and IM(2) to denote the real part and the imaginary part of $(a + bi)^2$ respectively, and let us use RE(3) and IM(3) to denote the real and imaginary parts of $(a + bi)^3$ respectively. Then

$$\text{RE}(2) = a^2 - b^2 \qquad \text{IM}(2) = 2ab$$

and

$$\text{RE}(3) = a\text{RE}(2) - b\text{IM}(2) \qquad \text{IM}(3) = b\text{RE}(2) + a\text{IM}(2).$$

If the pattern continues in this way, then RE(n) and IM(n) are obtained recursively from RE($n - 1$) and IM($n - 1$) for all $n > 1$. That is,

$$\begin{aligned}\text{RE}(n) &= a\text{RE}(n - 1) - b\text{IM}(n - 1) \\ \text{IM}(n) &= b\text{RE}(n - 1) + a\text{IM}(n - 1).\end{aligned}$$

EXERCISES *(You may use either BASIC or Pascal.)*

1. Write a computer program to find the first n positive integral powers of $a + bi$. The user supplies the values of n, a, and b.

Run the program to find the first seven powers of each complex number.

2. $1 + 2i$ **3.** $-3 + 4i$ **4.** $-2 - i$ **5.** $5 - 2i$

6. Modify the program in Exercise 1 to find the absolute value of $(a + bi)^n$ as well as its real and imaginary parts. How does the distance between the origin and the graph of $(a + bi)^n$ change as n increases?

APPLICATION

COMPLEX NUMBERS AND FUNCTIONS

Scientists, engineers, and mathematicians have studied mechanical systems in which an object suspended on a spring bobs up and down once a stretching force has been applied. The object bobs up and down, but, since friction applies, the object's displacement gets smaller and smaller as time goes on. Other studies have involved electrical systems in which applied voltage in a circuit produces a charge. In such a circuit, resistance affects charge and affects the current. Mechanical systems and electrical circuits share many common characteristics.

The mass in a mechanical system is like inductance in an electrical circuit. Friction in a mechanical system is like resistance in an electrical circuit. The displacement of an object is like charge, the impressed force on an object is like voltage, and the velocity of an object is like current.

Problems involving mechanical systems and electrical circuits are often similar. Their solutions often involve similar forms. The solutions involve complex numbers and functions involving complex variables.

Complex numbers and functions of complex variables also arise in problems involving the flow of air over and below the wing of an aircraft. Transformations using complex variables have been used to change circular disks into wing-shaped profiles useful as models for aircraft wing forms.

The use of complex numbers and functions of a complex variable is not restricted to problems involving mechanical systems, electrical circuits, and air flow. Such numbers and functions are useful in the study of heat flow and fluid flow.

The diagrams show the flow of an incompressible fluid such as water around a barrier plate with spaces to the left and right of it, and a flat plate.

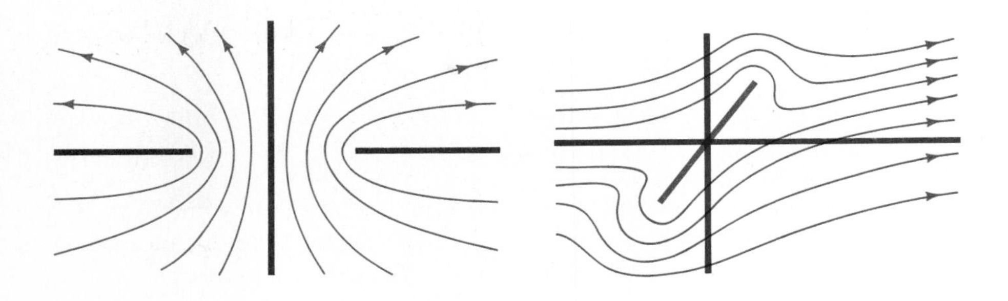

The exercises that follow give some idea of how simple functions of a complex variable work. The functions that are involved in studies of mechanical systems, electrical circuits, air flow, heat flow, and fluid flow are much more involved.

EXERCISES

1. Let $z_1 = 0$, $z_2 = 2$, $z_3 = 2 + i$, and $z_4 = i$.
   a. Graph z_1, z_2, z_3, and z_4.
   b. Let $f(z) = iz$ where $z \in C$. Find $f(z_1)$, $f(z_2)$, $f(z_3)$, and $f(z_4)$.
   c. Graph $f(z_1)$, $f(z_2)$, $f(z_3)$, and $f(z_4)$.
   d. Let P_1, P_2, P_3, and P_4 be the graphs of z_1, z_2, z_3, and z_4 respectively. Let Q_1, Q_2, Q_3, and Q_4 be the graphs of $f(z_1)$, $f(z_2)$, $f(z_3)$, and $f(z_4)$ respectively. Compare P_1P_2 with Q_1Q_2. Are $\overline{P_1P_2}$ and $\overline{P_2P_3}$ perpendicular? Are $\overline{Q_1Q_2}$ and $\overline{Q_2Q_3}$ perpendicular?
2. Let $f(z) = 3z + (5 - 2i)$. Let z_1, z_2, z_3, and z_4 be as in Exercise 1. Graph $f(z_1)$, $f(z_2)$, $f(z_3)$, and $f(z_4)$.

Electron flow through high-voltage power lines as well as through wires in home appliances can be studied with the aid of complex numbers.

Chapter 6

Introduction to Differential Calculus

The concept of the derivative is at the heart of mathematical analysis. In this chapter the derivative is introduced and then applied to various problems such as sketching curves and finding rates of change. Both the English mathematician Newton (1642-1727) and the German mathematician Leibniz (1646-1716) are credited with the discovery of calculus.

Differentiation

6-1 DERIVATIVES OF FUNCTIONS

In preceding chapters we have studied polynomial functions but have done very little graphing of them. In this section we begin to develop techniques for graphing these and other functions by introducing *derivatives* to find *slopes* of graphs.

The figure at the right shows a curve C and arrows at various points on C. The arrows show the direction in which the curve is headed as it is traced out from left to right. The lines containing these arrows are *tangents* to the curve, and their study is an important part of analyzing functions.

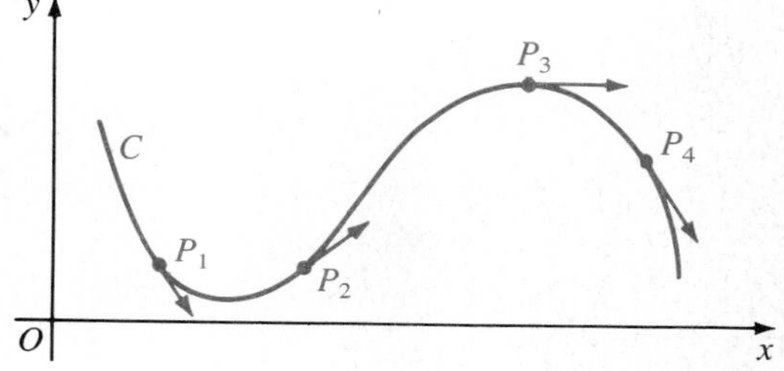

This stroboscopic photograph shows the movement of a tennis racket through a complete stroke—an excellent illustration of continuous change.

The following figures show what is and what is not a tangent. Although L in Figure a intersects C in one point, L is not a tangent to C at P. The line in Figure b is a tangent at P. Even though L in Figure c intersects C in two points, it *is* a tangent to C at P.

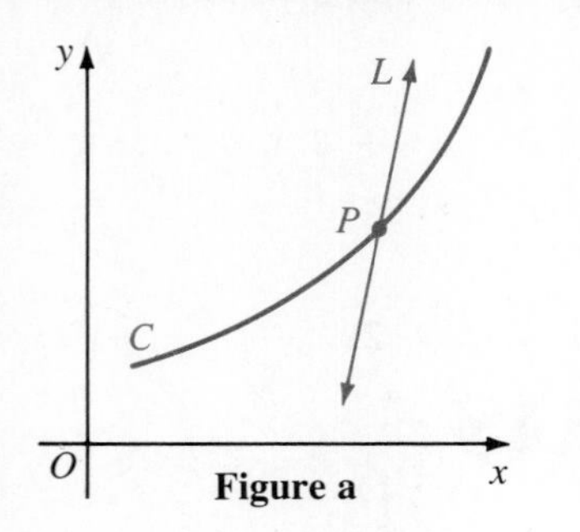

Figure a

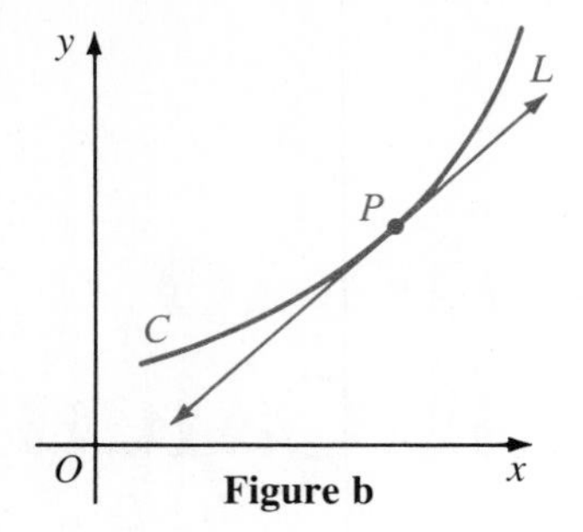

Figure b

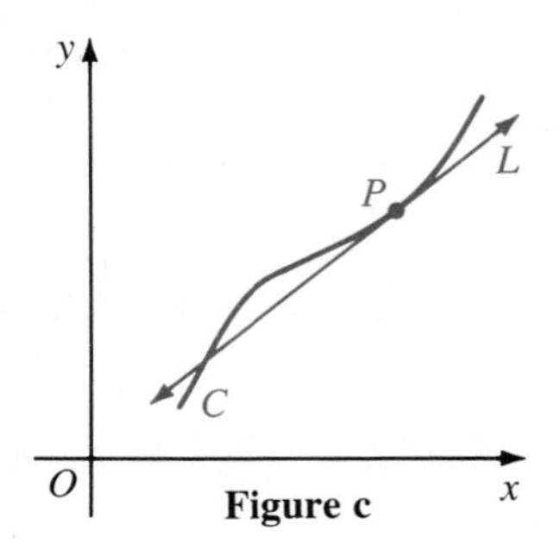

Figure c

Problem: Let C be the graph of a continuous function f and let $P(c, f(c))$ be a point of C. Find the slope m of the line L that is tangent to C at P.

Analysis: Choose a point $Q(x, f(x))$ on C near P and let L_Q be the line through P and Q. See the figures that follow, in which the figure at the right is an enlargement of a portion of the figure at the left.

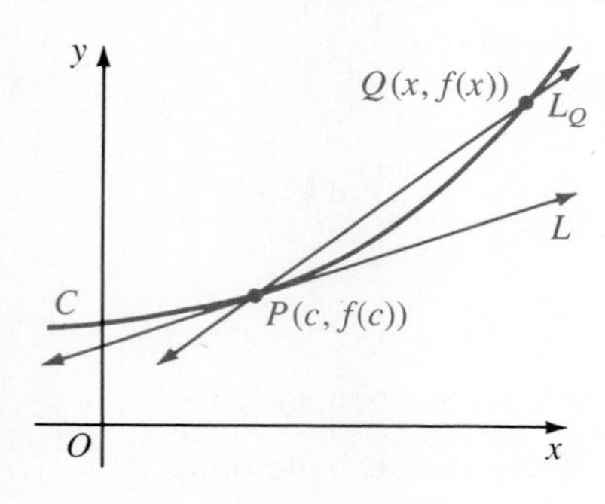

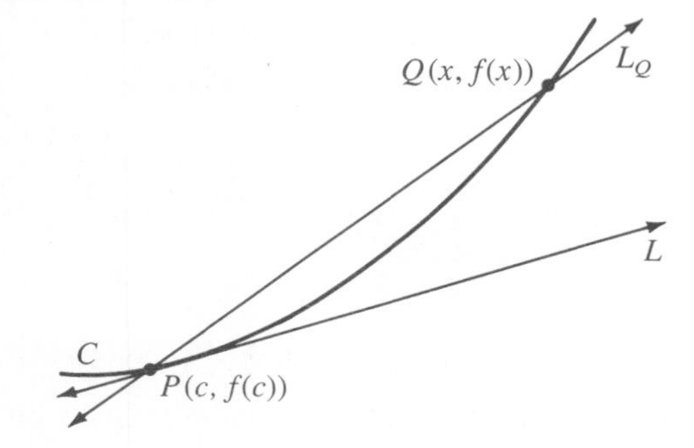

Solution: The closer Q is to P, the closer the slope of L_Q is to the slope of L. (Draw several lines L_Q for points Q closer and closer to P.) That is,

(slope of L_Q) approaches (slope of L) as Q approaches P.

Recall that the slope of the line through $P(c, f(c))$ and $Q(x, f(x))$ is given by

$$m_Q = \frac{f(x) - f(c)}{x - c}. \tag{1}$$

Since Q approaches P if and only if x approaches c, $\frac{f(x) - f(c)}{x - c}$ approaches a number m as x approaches c. In terms of limits,

$$\lim_{x \to c} \frac{f(x) - f(c)}{x - c} = m. \tag{2}$$

The number m thus defined is called the **slope** of the curve C at P. If the limit in (1) does not exist, then C has no slope and no tangent line at P.

EXAMPLE 1 **a.** Find the slope m of the curve C: $y = x^2 - 4x + 2$ at $P(3, -1)$.

b. Find an equation of the line L tangent to C at P.

Solution **a.** Let $f(x) = x^2 - 4x + 2$ and $c = 3$.

$$\begin{aligned} f(x) - f(c) &= (x^2 - 4x + 2) - (3^2 - 4 \cdot 3 + 2) \\ &= x^2 - 4x + 3 \\ &= (x - 3)(x - 1) \end{aligned}$$

So $\dfrac{f(x) - f(c)}{x - c} = \dfrac{(x - 3)(x - 1)}{x - 3} = x - 1.$

Hence $\displaystyle\lim_{x \to 3} \frac{f(x) - f(c)}{x - c} = \lim_{x \to 3} (x - 1)$

$= 2$ **Answer**

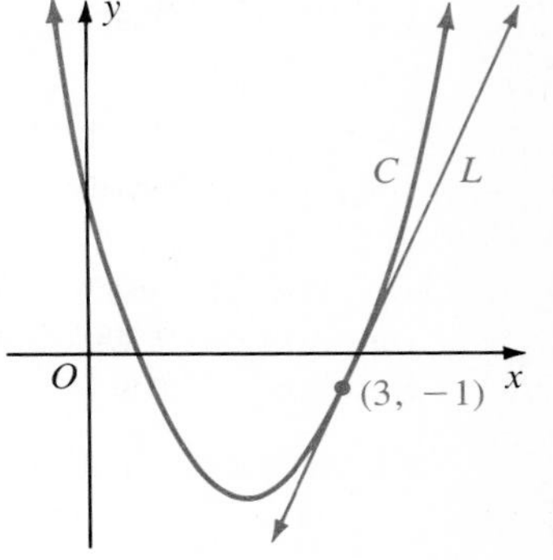

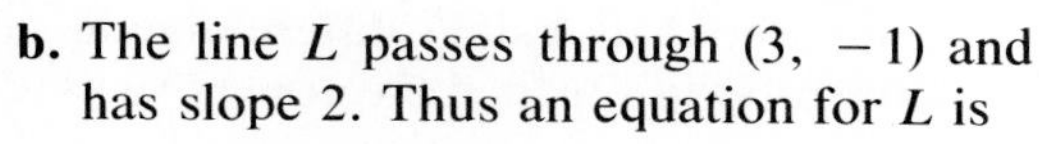

b. The line L passes through $(3, -1)$ and has slope 2. Thus an equation for L is

$y - (-1) = 2(x - 3)$

or $y = 2x - 7.$ **Answer**

(The adjoining figure shows C and L.)

As we shall see, the limit in formula (2) occurs in many different kinds of problems. It forms the basis of *differential calculus*.

DEFINITION

The function f is **differentiable** at c if f is defined in an open interval containing c, and

$$\lim_{x \to c} \frac{f(x) - f(c)}{x - c}$$

exists. If the limit does exist, it is called the **derivative of f at c** and is denoted by $f'(c)$.

Notice in the definition that c is a fixed number and x is a variable.

EXAMPLE 2 If $f(x) = \dfrac{1}{x^2}$ and $c \neq 0$, find $f'(c)$.

Solution $f(x) - f(c) = \dfrac{1}{x^2} - \dfrac{1}{c^2} = \dfrac{c^2 - x^2}{c^2x^2} = \dfrac{(c + x)(c - x)}{c^2x^2}$

Solution continued on following page

$$\text{Therefore } \frac{f(x) - f(c)}{x - c} = \frac{(c + x)(c - x)}{c^2x^2(x - c)} = -\frac{c + x}{c^2x^2}.$$

$$\lim_{x \to c} \frac{f(x) - f(c)}{x - c} = \lim_{x \to c}\left[-\frac{c + x}{c^2x^2}\right] = -\frac{c + c}{c^2c^2} = -\frac{2}{c^3}$$

Therefore $f'(c) = -\frac{2}{c^3}$. **Answer**

If in the definition we let $h = x - c$ so that $x = c + h$, then x approaches c if and only if h approaches 0. Thus we have

$$\lim_{h \to 0} \frac{f(c + h) - f(c)}{h} = f'(c).$$

Using other symbols for the variables, we may write the foregoing limit statements as

$$f'(x) = \lim_{t \to x} \frac{f(t) - f(x)}{t - x} \quad \text{and} \quad f'(x) = \lim_{h \to 0} \frac{f(x + h) - f(x)}{h}.$$

In the equations for $f'(x)$, x is a fixed number in the domain of f and t and h are variables.

The process of finding a derivative is called **differentiation**, and the function $x \to f'(x)$ obtained by the process is called the **derivative** (or **derived**) **function**.

EXAMPLE 3 Differentiate $f(x) = \sqrt{x}$ $(x > 0)$.

Solution The second form for the derivative is useful.

$$\frac{f(x + h) - f(x)}{h} = \frac{\sqrt{x + h} - \sqrt{x}}{h}.$$

Rationalize the numerator.

$$\frac{f(x + h) - f(x)}{h} = \frac{\sqrt{x + h} - \sqrt{x}}{h} \cdot \frac{\sqrt{x + h} + \sqrt{x}}{\sqrt{x + h} + \sqrt{x}}$$

$$= \frac{(x + h) - x}{h(\sqrt{x + h} + \sqrt{x})}$$

$$= \frac{\cancel{h}}{\cancel{h}(\sqrt{x + h} + \sqrt{x})}$$

$$= \frac{1}{\sqrt{x + h} + \sqrt{x}}$$

Since the function $x \to \sqrt{x}$ is continuous (page 164), $\lim_{h \to 0} \sqrt{x + h} = \sqrt{x}$.

$$\text{Thus } f'(x) = \lim_{h \to 0} \frac{f(x + h) - f(x)}{h} = \lim_{h \to 0} \frac{1}{\sqrt{x + h} + \sqrt{x}}$$

$$= \frac{1}{\lim_{h \to 0}(\sqrt{x + h}) + \sqrt{x}}$$

$$= \frac{1}{2\sqrt{x}}. \quad \textbf{Answer}$$

EXAMPLE 4 Show that $f(x) = |x^2 - 4|$ is not differentiable at $x = 2$.

Solution

$$\frac{f(x) - f(2)}{x - 2} = \frac{|x^2 - 4| - 0}{x - 2} = \frac{|x - 2||x + 2|}{x - 2} = \frac{|x - 2|}{x - 2} \cdot |x + 2|$$

$$\frac{f(x) - f(2)}{x - 2} = \begin{cases} |x + 2| \text{ if } x > 2 \\ -|x + 2| \text{ if } x < 2 \end{cases}$$

Now $\lim_{x \to 2} |x + 2| = 4$ for $x > 2$ and $\lim_{x \to 2} (-|x + 2|) = -4$ for $x < 2$.

Thus $\lim_{x \to 2} \frac{f(x) - f(2)}{x - 2}$ does not exist.

Hence f is not differentiable at $x = 2$.

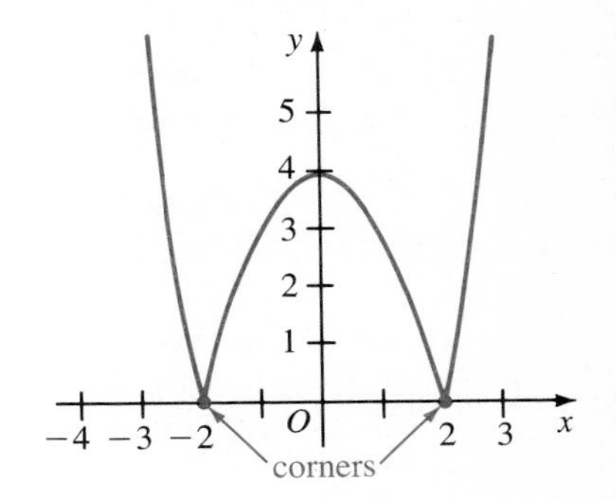

The figure at the right shows the graph of f in Example 4. Roughly speaking, continuity of a function implies that its graph has no breaks, while differentiability implies, in addition, that the graph has no corners.

The following theorem concerning differentiable functions is easy to prove (Exercise 25).

THEOREM 1 If the function f is differentiable at c, then f is continuous at c.

EXERCISES

Use the method of Example 1 **(a)** to find the slope of the curve C at the point P, and **(b)** to find an equation of the line L tangent to C at P.

A

1. C: $y = x^2$; $P(2, 4)$

2. C: $y = \frac{1}{x}$; $P(1, 1)$

3. C: $y = 2x - x^2$; $P(1, 1)$

4. C: $y = x^2 + x$; $P(-1, 0)$

5. C: $y = \frac{1}{1 - x}$; $P(2, -1)$

6. C: $y = x^3 + 1$; $P(2, 9)$

7. C: $y = 1 - x^3$; $P(-1, 2)$

8. C: $y = \frac{6}{x + 1}$; $P(1, 3)$

Use the definition of the derivative to differentiate f.

9. $f(x) = x^4$

10. $f(x) = \frac{1}{x^3}$

11. $f(x) = \frac{1}{\sqrt{x}}$

12. $f(x) = \sqrt{1 - x}$

13. $f(x) = \frac{x}{x + 1}$

14. $f(x) = \frac{1}{x^2 + 1}$

For which value or values of x does g fail to have a derivative?

15. $g(x) = |x| - 2$

16. $g(x) = |x - 2|$

17. $g(x) = |9 - x^2|$

18. $g(x) = |4x - x^2|$

B **19, 20.** Prove your answers to Exercises 15 and 16.

Which of the following functions are differentiable at 0? For those that are, find $f'(0)$.

21. $f(x) = x|x|$ **22.** $f(x) = \sqrt{x^2}$ **23.** $f(x) = \sqrt{|x|}$ **24.** $f(x) = x(1 - |x|)$

C **25.** Prove that if the function f is differentiable at c, then f is continuous at c. $\left(\textit{Hint}: f(x) = f(c) + \frac{f(x) - f(c)}{x - c}(x - c).\right)$

26. Show that the line tangent to the parabola $y = ax^2$ at the point (x_0, y_0) has the equation $y - y_0 = 2ax_0(x - x_0)$.

COMPUTER EXERCISES

1. Write and run a program to approximate the slope of the line tangent to the graph of the function $f(x) = x^3$ for input values of x. Use the formula $S(x) = \frac{f(x + h) - f(x)}{h}$, with $h = 1, 0.1, 0.01, 0.001$, and 0.0001.

2. Modify and run the program in Exercise 1 for each function f and the following values of x: $-2, -1, 0, 1, 2$.
a. $f(x) = -x^2$ **b.** $f(x) = 4x$ **c.** $f(x) = x^3 - 2x$

3. Modify the program in Exercise 1 to calculate the absolute value of the difference between $S(x)$ and $3x^2$ when $h = 0.0001$.

4. Run the program in Exercise 3 for x-values $-2, -1, 0, 1$, and 2.

6-2 PROPERTIES OF DERIVATIVES

Finding derivatives directly from the definition, as we did in Section 6-1, can be very laborious. In this section we shall develop rules to simplify the process. These rules are used so much that they should be learned thoroughly.

Theorem 2 is called the *power rule*. Although it is true for all real exponents n, we shall, for the time being, prove it and use it only for integral values of n and, in view of Example 3, page 224, exponents involving $\frac{1}{2}$.

THEOREM 2 **The Power Rule**
If $f(x) = x^n$, then $f'(x) = nx^{n-1}$.

Here are some examples of the use of the power rule.

1. If $f(x) = x^5$, then $f'(x) = 5x^4$.

2. If $f(x) = \frac{1}{x^2} = x^{-2}$, then $f'(x) = -2x^{-3} = -\frac{2}{x^3}$.

3. If $f(x) = \sqrt{x} = x^{\frac{1}{2}}$, then $f'(x) = \frac{1}{2}x^{-\frac{1}{2}} = \frac{1}{2\sqrt{x}}$.

Proof: (for n a positive integer) Using the binomial theorem, we have

$$\begin{aligned} f(x + h) - f(x) &= (x + h)^n - x^n \\ &= \left(x^n + nx^{n-1}h + \frac{n(n-1)}{2}x^{n-2}h^2 + \cdots + h^n\right) - x^n \\ &= nx^{n-1}h + \frac{n(n-1)}{2}x^{n-2}h^2 + \cdots + h^n. \end{aligned}$$

Now using the second form of the definition of derivative,

$$\frac{f(x + h) - f(x)}{h} = nx^{n-1} + \frac{n(n-1)}{2}x^{n-2}h + \cdots + h^{n-1}.$$

Except for the first term, each term on the right has h as a factor and therefore approaches 0 as h approaches 0.

$$\therefore\ f'(x) = \lim_{h \to 0} \frac{f(x + h) - f(x)}{h} = nx^{n-1}. \quad ■$$

A proof for the case where n is nonpositive is outlined in Exercise 23. Recall from Section 4-2 the following definitions:

1. $(f + g)(x) = f(x) + g(x)$ **2.** $(f - g)(x) = f(x) - g(x)$

3. $(fg)(x) = f(x)g(x)$ **4.** $\left(\frac{f}{g}\right)(x) = \frac{f(x)}{g(x)}$

If a and b are real numbers, we can combine the first two definitions into a single statement:

$$(af + bg)(x) = af(x) + bg(x),$$

called a **linear combination** of f and g.

The next theorem gives rules for differentiating linear combinations, products, and quotients of functions. In rules 3 and 4, g does not have 0 as a value.

THEOREM 3 Let f and g be differentiable functions and a and b be real numbers. Then

1. $(af + bg)' = af' + bg'$ **Linear Combination Rule**

2. $(fg)' = fg' + gf'$ **Product Rule**

3. $\left(\frac{1}{g}\right)' = -\frac{g'}{g^2}$ **Reciprocal Rule**

4. $\left(\frac{f}{g}\right)' = \frac{gf' - fg'}{g^2}$ **Quotient Rule**

Before proving any of the parts of Theorem 3, let us apply the theorem.

EXAMPLE 1 If $P(x) = 2x^3 - 3x^2$, find $P'(x)$.

Solution P is a linear combination of $f(x) = x^3$ and $g(x) = x^2$. Use the power rule and the linear combination rule.

$$f'(x) = 3x^2 \quad \text{and} \quad g'(x) = 2x$$

$$P'(x) = 2 \cdot 3x^2 - 3 \cdot 2x = 6x^2 - 6x \qquad \textbf{Answer}$$

The linear combination rule, whose proof is requested in Exercise 20, can be extended to any finite number of functions. In particular we have:

COROLLARY Every polynomial function is differentiable.

EXAMPLE 2 Differentiate $F(x) = (x^2 + 1)\sqrt{x}$.

Solution Use the product rule with $f(x) = x^2 + 1$ and $g(x) = \sqrt{x}$.

$$f'(x) = 2x \quad \text{and} \quad g'(x) = \frac{1}{2}x^{-\frac{1}{2}} = \frac{1}{2\sqrt{x}}.$$

$$F'(x) = (x^2 + 1)\frac{1}{2\sqrt{x}} + \sqrt{x} \cdot 2x$$

$$= \frac{x^2 + 1 + 2\sqrt{x} \cdot \sqrt{x} \cdot 2x}{2\sqrt{x}} = \frac{5x^2 + 1}{2\sqrt{x}} \qquad \textbf{Answer}$$

EXAMPLE 3 Find the derivative of $G(x) = \dfrac{1 + x^2}{1 - x^2}$.

Solution Use the quotient rule with $f(x) = 1 + x^2$ and $g(x) = 1 - x^2$.

$$f'(x) = 2x \quad \text{and} \quad g'(x) = -2x$$

$$G'(x) = \frac{(1 - x^2)(2x) - (1 + x^2)(-2x)}{(1 - x^2)^2}$$

$$= \frac{2x[(1 - x^2) + (1 + x^2)]}{(1 - x^2)^2}$$

$$= \frac{4x}{(1 - x^2)^2} \qquad \textbf{Answer}$$

EXAMPLE 4 Differentiate $\varphi(x) = \left(\dfrac{1}{\sqrt{x}}\right)^3$.

Solution You could use the product rule and the reciprocal rule. However, since $\left(\dfrac{1}{\sqrt{x}}\right)^3 = (x^{-\frac{1}{2}})^3 = x^{-\frac{3}{2}}$, use the power rule.

$$\varphi(x) = x^{-\frac{3}{2}} \quad \text{and} \quad \varphi'(x) = -\frac{3}{2}x^{-\frac{5}{2}} \qquad \textbf{Answer}$$

We conclude with a proof of part 2 of Theorem 3 (the product rule).

Proof: We shall use the first form of the definition of derivative. At the second step below we use the device of subtracting and adding the same quantity.

$$\begin{aligned}(fg)(t) - (fg)(x) &= f(t)g(t) - f(x)g(x)\\ &= f(t)g(t) - f(t)g(x) + f(t)g(x) - f(x)g(x)\\ &= f(t)[g(t) - g(x)] + g(x)[f(t) - f(x)]\end{aligned}$$

$$\frac{(fg)(t) - (fg)(x)}{t - x} = f(t)\frac{g(t) - g(x)}{t - x} + g(x)\frac{f(t) - f(x)}{t - x}$$

Now let t approach x. Since f is continuous at x, by Theorem 1, $f(t)$ approaches $f(x)$. Therefore

$$\lim_{t\to x}\frac{(fg)(t) - (fg)(x)}{t - x} = f(x)\lim_{t\to x}\frac{g(t) - g(x)}{t - x} + g(x)\lim_{t\to x}\frac{f(t) - f(x)}{t - x}$$

$$\therefore (fg)'(x) = f(x)g'(x) + g(x)f'(x) \quad ■$$

The proofs of the reciprocal and quotient rules are outlined in Exercises 21 and 22.

EXERCISES

In Exercises 1–6, **(a)** differentiate $P(x)$ and **(b)** find an equation of the line tangent to the curve $y = P(x)$ at the point whose x-coordinate is given.

EXAMPLE $P(x) = x^3 - 3x^2 + 5x - 9;\ 2$

Solution **a.** $P'(x) = 3x^2 - 6x + 5$

b. $P(2) = 2^3 - 3 \cdot 2^2 + 5 \cdot 2 - 9 = -3;\ P'(2) = 3 \cdot 2^2 - 6 \cdot 2 + 5 = 5$
$\therefore y - (-3) = 5(x - 2)$, or $5x - y = 13$

A

1. $P(x) = 2x^2 + 5x - 1;\ -3$

2. $P(x) = x^2 - 3x - 1;\ 2$

3. $P(x) = x^3 - 3x^2 + 8;\ 2$

4. $P(x) = x^3 + x^2 + x + 2;\ -1$

5. $P(x) = x^4 - x^2 + 2x + 1;\ 1$

6. $P(x) = x^4 - 4x^3 + 20;\ 3$

Use the rules of Theorem 3 to differentiate each of the following functions.

7. $F(x) = (x - 1)\sqrt{x}$

8. $G(x) = (1 - x^2)\sqrt{x}$

9. $\varphi(x) = \dfrac{1}{x^2 + 1}$

10. $f(x) = \dfrac{1}{2x + 1}$

11. $g(x) = \sqrt{x}\,(x - 1)^2$

12. $\psi(x) = 3x\sqrt{x}$

13. $G(x) = \dfrac{x^2}{1 - x^2}$

14. $\varphi(x) = \dfrac{x^2 + 1}{x^2}$

B

15. $f(x) = \dfrac{1 + \sqrt{x}}{1 - \sqrt{x}}$

16. $g(x) = \dfrac{\sqrt{x}}{1 + \sqrt{x}}$

Use the rules of Theorem 3 to differentiate each of the following functions.

17. $\varphi(x) = \left(x\sqrt{x} - \dfrac{1}{\sqrt{x}}\right)^2$ **18.** $k(x) = (1 + \sqrt{x})^3$

19. Prove that if $\varphi(x) = k$, a constant, then $\varphi'(x) = 0$ for all x.

C **20.** Prove the linear combination rule.

21. Prove the reciprocal rule. $\left(\textit{Hint}\text{: Let } \varphi(x) = \dfrac{1}{g(x)} \text{ so that } \varphi(t) - \varphi(x) = \dfrac{1}{g(t)} - \dfrac{1}{g(x)} = -\dfrac{g(t) - g(x)}{g(t)g(x)}.\right)$

22. Prove the quotient rule. $\left(\textit{Hint}\text{: } \dfrac{f}{g} = \dfrac{1}{g} \cdot f\text{. Use the product and reciprocal rules.}\right)$

23. **a.** Prove Theorem 2 for the case $n = 0$. (*Hint*: Use Exercise 19.)
b. Prove Theorem 2 for the case $n < 0$. (*Hint*: Let $n = -m$ where m is a positive integer. Then use the reciprocal rule.)
c. Prove Theorem 2 for the case $n = k + \frac{1}{2}$, k an integer. (*Hint*: $x^n = x^{k+\frac{1}{2}} = x^k x^{\frac{1}{2}} = x^k\sqrt{x}$. Now use the power rule and simplify.)

6-3 NOTATIONS FOR DERIVATIVES

We often use the **derivative operator notation** $\dfrac{d}{dx}$ (read "the derivative of") to indicate differentiation. Certain aspects of our study are simplified by the use of this notation. For example, the statement,

If $f(x) = \sqrt{x}$, then $f'(x) = \dfrac{1}{2\sqrt{x}}$ can be simplified to $\dfrac{d}{dx}\sqrt{x} = \dfrac{1}{2\sqrt{x}}$.

The notation $\dfrac{d}{dx}$ indicates that the function following it is to be differentiated. Notice that we used $\sqrt{x}$ to abbreviate the function f such that $f(x) = \sqrt{x}$. If the variable involved is t instead of x, the operator would be written $\dfrac{d}{dt}$.

Thus, $\dfrac{d}{dt}\sqrt{t} = \dfrac{1}{2\sqrt{t}}$.

EXAMPLE 1 Differentiate the function $\dfrac{x^2 + 1}{\sqrt{x}}$.

Solution Use the quotient rule and the operator $\dfrac{d}{dx}$.

$$\frac{d}{dx}\left(\frac{x^2 + 1}{\sqrt{x}}\right) = \frac{\sqrt{x}\dfrac{d}{dx}(x^2 + 1) - (x^2 + 1)\dfrac{d}{dx}\sqrt{x}}{(\sqrt{x})^2}$$

$$\frac{d}{dx}\left(\frac{x^2 + 1}{\sqrt{x}}\right) = \frac{\sqrt{x} \cdot 2x - (x^2 + 1) \cdot \dfrac{1}{2\sqrt{x}}}{x}$$

$$= \frac{2\sqrt{x} \cdot \sqrt{x} \cdot 2x - (x^2 + 1)}{2x\sqrt{x}}$$

$$\frac{d}{dx}\left(\frac{x^2 + 1}{\sqrt{x}}\right) = \frac{3x^2 - 1}{2x\sqrt{x}} \quad \textbf{Answer}$$

If a function f is defined by

$$y = f(x),$$

then we can write the derivative in several ways:

$$f'(x), \quad y', \quad \frac{d}{dx} f(x), \quad \frac{d}{dx} y, \quad \text{and} \quad \frac{dy}{dx}$$

EXAMPLE 2 If $y = \dfrac{1}{4 - x}$, find $\dfrac{dy}{dx}$ and write it in terms of y.

Solution Use the reciprocal formula.

$$\frac{dy}{dx} = \frac{d}{dx}\left(\frac{1}{4 - x}\right) = -\frac{-1}{(4 - x)^2} = \left(\frac{1}{4 - x}\right)^2 = y^2$$

If the derivative function f' is itself differentiable, $(f')'$, the derivative of f', exists. It is written f'' and is called the **second derivative** of f. If $y = f(x)$, we can also denote the second derivative by

$$y'' \quad \text{or} \quad \frac{d^2y}{dx^2}.$$

$\left(\dfrac{d^2y}{dx^2}\right.$ is a condensation of $\left.\dfrac{d}{dx}\left(\dfrac{dy}{dx}\right).\right)$

EXAMPLE 3 Find y' and y'' if $y = x^4 - 4x^3 + 6$.

Solution $y' = 4x^3 - 12x^2$ **Answer**

To find y'' take the derivative of y'.

$y'' = 12x^2 - 24x$ **Answer**

Continuation of the process described above leads to the higher derivatives

$$f''', f^{(4)}, \ldots, f^{(n)}, \ldots \quad \text{or} \quad y''', y^{(4)}, \ldots, y^{(n)}, \ldots$$

where n is the **order** of the derivative. For example, the higher derivatives of the function of Example 3 are

$$y''' = 24x - 24, \quad y^{(4)} = 24, \quad y^{(5)} = y^{(6)} = \cdots = 0.$$

Be careful not to confuse the order of a derivative with the power of a function; the order does not denote an exponent.

Suppose that we wish to differentiate $y = (x^2 + 1)^3$. At this stage we do not have a rule for handling powers of functions. We could expand $(x^2 + 1)^3$ to obtain

$$(x^2 + 1)^3 = x^6 + 3x^4 + 3x^2 + 1.$$

Therefore
$$\frac{d}{dx}(x^2 + 1)^3 = 6x^5 + 12x^3 + 6x$$
$$= 6x(x^4 + 2x^2 + 1) = 6x(x^2 + 1)^2.$$

Is there a shorter way to do this? We might be tempted to use the power rule and write $3(x^2 + 1)^2$. However, this is not the correct answer.

THEOREM 4 **The Generalized Power Rule**

If u is a differentiable function of x and n is an integer, then

$$\frac{d}{dx} u^n = nu^{n-1} \frac{du}{dx}.$$

When we apply the generalized power rule to the case where $u = x^2 + 1$ and $n = 3$, we obtain

$$\frac{d}{dx}(x^2 + 1)^3 = 3(x^2 + 1)^2 \frac{d}{dx}(x^2 + 1) = 6x(x^2 + 1)^2.$$

This agrees with the result found previously.

EXAMPLE 4 Find the derivative of each function.

a. $u(x) = (x^3 + 1)^{10}$ **b.** $u(x) = \dfrac{1}{(1 - x^2)^3}$

Solution **a.** $\dfrac{d}{dx}(x^3 + 1)^{10} = 10(x^3 + 1)^9 \cdot 3x^2 = 30x^2(x^3 + 1)^9$ **Answer**

b.
$$\frac{d}{dx}\frac{1}{(1 - x^2)^3} = \frac{d}{dx}(1 - x^2)^{-3}$$
$$= -3(1 - x^2)^{-4} \frac{d}{dx}(1 - x^2)$$
$$= -3(1 - x^2)^{-4}(-2x) = \frac{6x}{(1 - x^2)^4}$$ **Answer**

We can verify the generalized power rule for $n = 2$ and $n = 3$ with the help of the product rule.

If $n = 2$, $\dfrac{d}{dx} u^2 = \dfrac{d}{dx}(u \cdot u) = u \dfrac{du}{dx} + u \dfrac{du}{dx} = 2u \dfrac{du}{dx}.$

If $n = 3$, $\dfrac{d}{dx} u^3 = \dfrac{d}{dx}(u^2 \cdot u) = u^2 \dfrac{du}{dx} + u \dfrac{d}{dx} u^2$

$$= u^2 \frac{du}{dx} + u\left(2u \frac{du}{dx}\right) = 3u^2 \frac{du}{dx}.$$

You are asked to prove the generalized power rule in Exercise 27.

The differentiation rules thus far available to us are listed below.

Let u and v be differentiable functions. Then:

$$\frac{d}{dx}(au + bv) = a\frac{du}{dx} + b\frac{dv}{dx} \qquad \frac{d}{dx}(uv) = u\frac{dv}{dx} + v\frac{du}{dx}$$

$$\frac{d}{dx}\left(\frac{1}{v}\right) = -\frac{\frac{dv}{dx}}{v^2} \qquad \frac{d}{dx}\left(\frac{u}{v}\right) = \frac{v\frac{du}{dx} - u\frac{dv}{dx}}{v^2}$$

$$\frac{d}{dx}u^n = nu^{n-1}\frac{du}{dx}$$

EXERCISES

In Exercises 1–12, perform the indicated differentiation.

A

1. $\frac{d}{dx}(2x + 3)^{10}$ **2.** $\frac{d}{dx}(1 - x)^6$ **3.** $\frac{d}{dt}(1 - t^2)^4$ **4.** $\frac{d}{ds}(1 + s^2)^5$

5. $\frac{d}{dx}\frac{1}{(x^2 + 1)^3}$ **6.** $\frac{d}{dx}\frac{1}{(1 - x^2)^2}$ **7.** $\frac{d}{du}[u(1 - u^2)^3]$ **8.** $\frac{d}{dx}[x^2(1 + x^2)^2]$

9. $\frac{d}{dx}\frac{x}{(1 + x^2)^2}$ **10.** $\frac{d}{dx}\frac{x^2}{(1 - x^2)^2}$ **11.** $\frac{d}{dx}\frac{(x + 1)^4}{x^3}$ **12.** $\frac{d}{dt}t^2(t^2 - 1)^{-3}$

In Exercises 13–18, find y' and y''.

13. $y = x^3 - 3x^2 + 6$ **14.** $y = x^4 - 2x^2 + 2$

15. $y = (2x - 1)^5$ **16.** $y = (4 - x)^4$

B

17. $y = (x^2 - 1)^4$ **18.** $y = \frac{1}{x^2 + 1}$

In Exercises 19–22, find y' and write y' in terms of y.

19. $y = \frac{1}{1 - x}$ **20.** $y = \frac{1}{(1 - x)^2}$ **21.** $y = (1 - x)^2$ **22.** $y = (1 + x)^2$

In Exercises 23 and 24, u, v, and w are differentiable functions of x.

23. If $y = uvw$, show that $y' = uvw' + uv'w + u'vw$.

24. If $y = uv$, find expressions for y', y'', and y'''.

C

25. Let $y = (1 - x)^{-1}$. Show by induction that $y^{(n)} = n!\,(1 - x)^{-n-1}$.

26. Let $y = (1 + x)^{-1}$. Show by induction that $y^{(n)} = (-1)^n n!\,(1 + x)^{-n-1}$.

27. Use mathematical induction to prove the generalized power rule.

28. Let P and Q be the points where a line containing (0, 1) intersects the graph of $4y = x^2$. Show that the tangents at P and Q are perpendicular.

6-4 THE CHAIN RULE

In this section we will discuss and prove the *chain rule*, used to differentiate compositions of functions. The generalized power rule is a special case of it.

$$\text{generalized power rule: } \frac{d}{dx}u^n = nu^{n-1}\frac{du}{dx}$$

$$\text{chain rule: } \frac{d}{dx}f(u) = \frac{d}{du}f(u) \cdot \frac{du}{dx}$$

Before proving the chain rule, let us see how we can use it to differentiate some functions.

EXAMPLE 1 Differentiate $y = \sqrt{1 + x^2}$.

Solution Let $y = \sqrt{u}$ where $u = 1 + x^2$. Apply the chain rule.

$$\frac{dy}{dx} = \frac{d}{dx}\sqrt{u} = \frac{d}{du}\sqrt{u} \cdot \frac{du}{dx}$$

$$= \frac{1}{2\sqrt{u}} \cdot 2x = \frac{x}{\sqrt{u}}$$

$$\frac{dy}{dx} = \frac{x}{\sqrt{1 + x^2}} \quad \textbf{Answer}$$

Notice that in the final answer in Example 1 we eliminated u because it did not appear in the statement of the problem.

By applying the chain rule with formulas presented earlier, we can differentiate a large variety of functions.

EXAMPLE 2 Differentiate $y = \dfrac{x^2}{\sqrt{1 - x}}$.

Solution Use the quotient rule.

$$\frac{dy}{dx} = \frac{\sqrt{1 - x} \cdot \frac{d}{dx}x^2 - x^2\frac{d}{dx}\sqrt{1 - x}}{(\sqrt{1 - x})^2}$$

To find $\frac{d}{dx}\sqrt{1 - x}$, let $u = 1 - x$ and use the chain rule.

$$\frac{d}{dx}\sqrt{1 - x} = \frac{1}{2\sqrt{u}} \cdot \frac{d}{dx}(1 - x)$$

$$= \frac{1}{2\sqrt{u}}(-1) = \frac{-1}{2\sqrt{1 - x}}$$

Therefore

$$\frac{dy}{dx} = \frac{2x\sqrt{1 - x} - x^2\left(\frac{-1}{2\sqrt{1 - x}}\right)}{1 - x}$$

Simplify by multiplying the numerator and the denominator by $2\sqrt{1 - x}$.

$$\frac{dy}{dx} = \frac{4x(1 - x) + x^2}{2(1 - x)^{\frac{3}{2}}} = \frac{4x - 3x^2}{2(1 - x)^{\frac{3}{2}}} \quad \textbf{Answer}$$

EXAMPLE 3 Differentiate $y = \sqrt{\frac{1 + x^2}{1 - x^2}}$.

Solution After introducing the exponent $\frac{1}{2}$, use the generalized power rule, then the quotient rule.

$$\frac{d}{dx}\sqrt{\frac{1 + x^2}{1 - x^2}} = \frac{d}{dx}\left(\frac{1 + x^2}{1 - x^2}\right)^{\frac{1}{2}}$$

$$= \frac{1}{2}\left(\frac{1 + x^2}{1 - x^2}\right)^{-\frac{1}{2}} \frac{d}{dx}\left(\frac{1 + x^2}{1 - x^2}\right)$$

$$= \frac{1}{2}\left(\frac{1 - x^2}{1 + x^2}\right)^{\frac{1}{2}} \frac{(1 - x^2) \cdot 2x - (1 + x^2)(-2x)}{(1 - x^2)^2}$$

$$= \frac{1}{2} \cdot \frac{(1 - x^2)^{\frac{1}{2}}}{(1 + x^2)^{\frac{1}{2}}} \cdot \frac{4x}{(1 - x^2)^2}$$

$$\frac{dy}{dx} = \frac{2x}{(1 + x^2)^{\frac{1}{2}}(1 - x^2)^{\frac{3}{2}}} \quad \textbf{Answer}$$

The chain rule is a formula for differentiating composite functions. The following alternate form is often useful. If we let $f(u) = y$, then

$$\frac{d}{dx}f(u) = \frac{dy}{dx} \quad \text{and} \quad \frac{d}{du}f(u) = \frac{dy}{du}.$$

The chain rule,

$$\frac{d}{dx}f(u) = \frac{d}{du}f(u) \cdot \frac{du}{dx},$$

then becomes

$$\frac{dy}{dx} = \frac{dy}{du} \cdot \frac{du}{dx}.$$

The alternate form is easy to remember because it looks like an algebraic identity. It isn't, of course.

EXAMPLE 4 Find $\frac{dy}{dx}$ if $y = 3u^2 - 2u + 3$ and $u = 4x^2 - 6$.

Solution Use the alternate form for the chain rule.

$$\frac{dy}{du} = 6u - 2 \quad \text{and} \quad \frac{du}{dx} = 8x$$

$$\text{Therefore } \frac{dy}{dx} = (6u - 2)(8x)$$

$$= (6(4x^2 - 6) - 2)(8x)$$

$$\frac{dy}{dx} = 192x^3 - 304x \quad \textbf{Answer}$$

To prove the chain rule we consider the composite function $\varphi(x) = f(g(x))$.

THEOREM 5 The Chain Rule

Let g be differentiable at x and f be differentiable at $g(x)$.

Let $\quad \varphi(x) = f(u), \quad$ where $u = g(x)$.

Then $\quad \varphi'(x) = f'(u) \cdot g'(x), \quad$ where $u = g(x)$.

Proof: (Optional)

We shall prove the theorem in the case where $g(x + h) \neq g(x)$ for all sufficiently small $|h| \neq 0$.

$$\frac{\varphi(x + h) - \varphi(x)}{h} = \frac{f(g(x + h)) - f(g(x))}{h}$$
$$= \frac{f(g(x + h)) - f(g(x))}{g(x + h) - g(x)} \cdot \frac{g(x + h) - g(x)}{h}.$$

With x fixed let $k(h) = g(x + h) - g(x)$ and let $u = g(x)$. Then $k(h) = g(x + h) - u$ and $g(x + h) = u + k(h)$. Thus

$$\frac{\varphi(x + h) - \varphi(x)}{h} = \frac{f(u + k(h)) - f(u)}{k(h)} \cdot \frac{g(x + h) - g(x)}{h}.$$

$$\lim_{h\to 0} \frac{\varphi(x + h) - \varphi(x)}{h} = \lim_{h\to 0} \left[\frac{f(u + k(h)) - f(u)}{k(h)} \cdot \frac{g(x + h) - g(x)}{h}\right]$$
$$= \left(\lim_{h\to 0} \frac{f(u + k(h)) - f(u)}{k(h)}\right)\left(\lim_{h\to 0} \frac{g(x + h) - g(x)}{h}\right)$$

Since g is differentiable at x, g is continuous at x. Thus $\lim_{h\to 0} k(h) = \lim_{h\to 0} [g(x + h) - g(x)] = 0$. That is, when $h \to 0$, $k(h) \to 0$.

$$\lim_{h\to 0} \frac{\varphi(x + h) - \varphi(x)}{h} = \left(\lim_{k(h)\to 0} \frac{f(u + k(h)) - f(u)}{k(h)}\right)\left(\lim_{h\to 0} \frac{g(x + h) - g(x)}{h}\right)$$

Therefore $\varphi'(x) = f'(u) \cdot g'(x)$. ■

EXERCISES

Differentiate each function.

A

1. $(2x - x^2)^5$

2. $(x^3 + 3x)^{10}$

3. $\sqrt{2x + 3}$

4. $\sqrt{1 - x}$

5. $\sqrt{4 - x^2}$

6. $\sqrt{x^2 - 9}$

7. $\dfrac{1}{\sqrt{x^2 + 1}}$

8. $(1 - x^2)^{\frac{3}{2}}$

9. $\sqrt{\dfrac{x}{1 - x}}$

10. $\sqrt{\dfrac{x + 2}{x - 2}}$

11. $x^2\sqrt{4 - x}$

12. $\dfrac{x}{\sqrt{2x - 1}}$

Find $\dfrac{dy}{dx}$.

13. $y = \sqrt{u};\ u = 3x^2 - 9$

14. $y = \sqrt{u} + u^2;\ u = \sqrt{x}$

15. $y = \frac{u + 1}{u - 1}$; $u = \frac{x - 1}{x + 1}$

16. $y = \frac{2u}{u^2 + 1}$; $u = x^3 - 8$

17. $y = (u^2 - 1)^{-2}$; $u = (x^2 - 1)^{-2}$

18. $y = u^3 - 4\sqrt{u}$; $u = 1 - \frac{1}{x} + \frac{1}{x^2}$

In Exercises 19 and 20, prove each formula.

19. $\frac{d}{dx}\sqrt{x^2 \pm a^2} = \frac{x}{\sqrt{x^2 \pm a^2}}$

20. $\frac{d}{dx}\sqrt{a^2 - x^2} = \frac{-x}{\sqrt{a^2 - x^2}}$

In Exercises 21–24, find each derivative. Use Exercises 19 and 20.

B

21. $x\sqrt{4 - x^2}$

22. $\frac{x}{\sqrt{x^2 - 9}}$

23. $\frac{x^2}{\sqrt{x^2 - 1}}$

24. $x^2\sqrt{1 - x^2}$

In Exercises 25 and 26, prove each statement. Use Exercises 19 and 20.

25. If $y = (x + \sqrt{x^2 + 1})^n$, then $y' = \frac{ny}{\sqrt{x^2 + 1}}$.

26. If $y = (x - \sqrt{x^2 - 1})^n$, then $y' = \frac{-ny}{\sqrt{x^2 - 1}}$.

In Exercises 27–30, find y'.

27. $y = \frac{(3x^2 - 9)^5}{(2x^2 + 1)^3}$

28. $y = (4x^2 + 5)^3(x + 9)^2$

29. $y = (5x^2 + 2x + 1)^{10}(3x^2 + 6)^4$

30. $y = \frac{(2x^2 + 3x)^3}{(x^4 - 9)^4}$

31. Prove that $\frac{d}{dx}u^n = nu^{n-1}\frac{du}{dx}$ when $n = k + \frac{1}{2}$, k an integer. $\left(\textit{Hint: } u^n = u^k \cdot u^{\frac{1}{2}}.\right)$

In Exercises 32–36 assume that there are functions E, S, and C with $\mathcal{R}$ as their domains such that

$$E' = E \qquad S' = C \qquad C' = -S$$

for all real numbers.

C

32. Show that if $y = E(kx)$, then $y'' - k^2y = 0$.

33. Show that if $y = aS(kx) + bC(kx)$, then $y'' + k^2y = 0$.

34. Show that $\frac{d}{dx}[S^2(x) + C^2(x)] = 0$.

35. The function E has an inverse function, L, that is differentiable. Show that $L'(x) = \frac{1}{x}$. [*Hint*: Differentiate both sides of $E(L(x)) = x$.]

36. Assuming that the chain rule holds over the complex field C, show that if $y = E(kxi)$, then $y'' + k^2y = 0$.

Applications

6-5 USING DERIVATIVES IN GRAPHING

In the rest of this chapter we discuss some applications of differential calculus. In this section and the next section we shall use derivatives as a tool for graphing functions.

We first extend the discussion, begun in Section 4-6, of properties of functions defined on intervals. We shall consider the sets $\{x: x > a\}$ and $\{x: x < b\}$, denoted (a, ∞) and $(-\infty, b)$ respectively, to be **open intervals**, along with $(a, b) = \{x: a < x < b\}$. We shall use I to denote an interval.

When we graph a function, we need to know whether the graph rises or falls as we move along it from left to right. We therefore make these definitions.

DEFINITION

A function f is **increasing** on an interval I if

$$f(p) < f(q) \text{ for all } p \text{ and } q \text{ in } I \text{ with } p < q.$$

A function f is **decreasing** on an interval I if

$$f(p) > f(q) \text{ for all } p \text{ and } q \text{ in } I \text{ with } p < q.$$

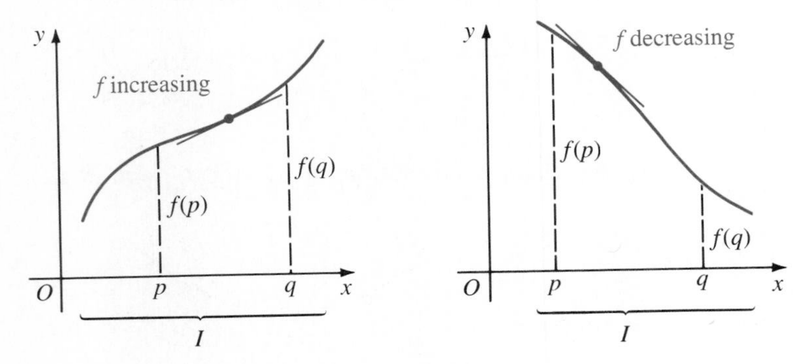

The preceding figures suggest that if tangents to the graph have positive slopes, then f is increasing, while if the tangents have negative slopes, then f is decreasing. Since these slopes are given by $f'(x)$, these remarks suggest Theorem 6, whose proof is discussed in the Extra, pages 264–265.

THEOREM 6 The function f is increasing on an interval I if $f'(x) > 0$ on I; f is decreasing on I if $f'(x) < 0$ on I.

EXAMPLE 1 Let $g(x) = x^3 - 3x^2 - 9x + 22$. **a.** Find the intervals on which g is increasing and on which g is decreasing. **b.** Use (a) to graph g.

Solution **a.** $g'(x) = 3x^2 - 6x - 9 = 3(x + 1)(x - 3)$
$g'(x) > 0$ when $x < -1$ or $x > 3$.
$g'(x) < 0$ when $-1 < x < 3$.
So g is increasing on $(-\infty, -1)$ and on $(3, \infty)$, and g is decreasing on $(-1, 3)$.

b. $g(-1) = 27$ and $g(3) = -5$. Plot the points $(-1, 27)$ and $(3, -5)$ and use (a). Notice that $\lim_{x\to\infty} g'(x) = \infty$ and $\lim_{x\to-\infty} g'(x) = \infty$, so that the graph becomes steeper and steeper the larger $|x|$ is.

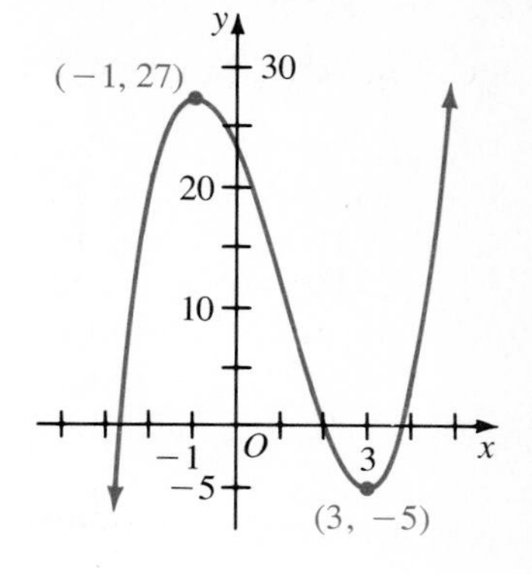

In Example 1 the point $(-1, 27)$ is higher than all other nearby points on the graph. That is, $g(-1) = 27$ is greater than $g(x)$ for all x near -1. We say that g has a **relative**, or **local**, **maximum** value of 27 at -1. Similarly, g has a **relative**, or **local**, **minimum** value of -5 at 3.

In general, f has a relative maximum at c if $f(c) \geq f(x)$, and f has a relative minimum at c if $f(c) \leq f(x)$ for all x in some open interval containing c. **Extremum** (plural: *extrema*) indicates either a maximum or a minimum.

The function g of Example 1 has extrema at -1 and 3. Note that $g'(-1) = 0$ and $g'(3) = 0$. This illustrates the next result.

THEOREM 7 If f is differentiable at c and has a relative extremum at c, then $f'(c) = 0$.

If $f'(c) = 0$, then c is called a **stationary point** of the function f. *At* a stationary point the graph of f has a horizontal tangent line. *Near* a stationary point the graph can behave in various ways.

relative maximum at c

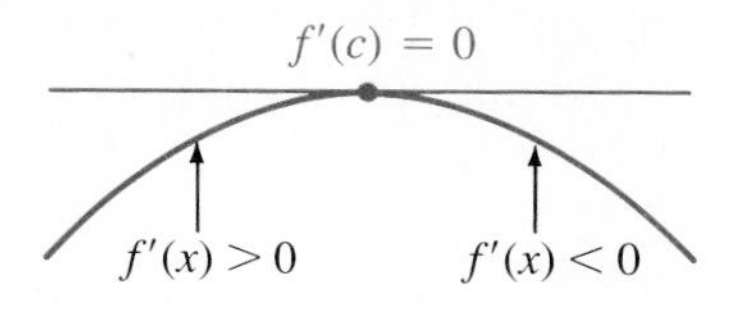

relative minimum at c

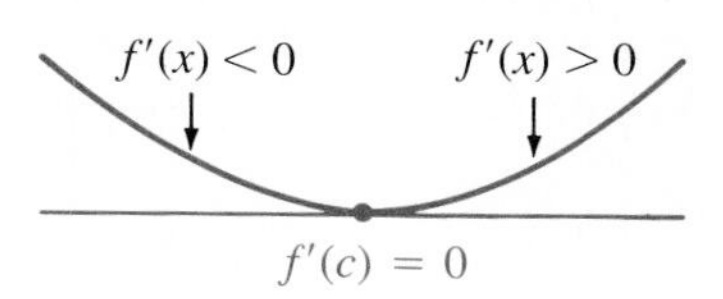

no extremum at c

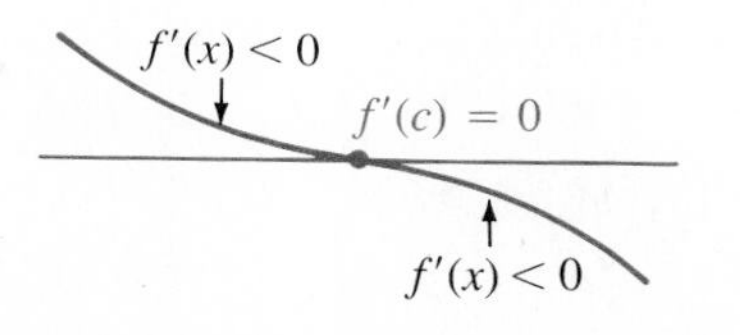

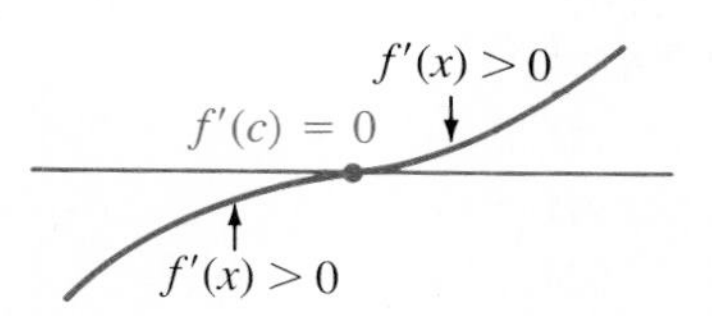

Inspection of the signs of $f'(x)$ on both sides of a stationary point of f enables us to decide whether f has an extremum there and, if so, what kind.

THEOREM 8 **The First-Derivative Test**

Let $f'(c) = 0$ and suppose that for some open interval (a, b) containing c,

1. $f'(x) > 0$ in (a, c) and $f'(x) < 0$ in (c, b). Then f has a relative maximum at c.

2. $f'(x) < 0$ in (a, c) and $f'(x) > 0$ in (c, b). Then f has a relative minimum at c.

3. $f'(x)$ has the same sign in both (a, c) and (c, b). Then f has no relative extremum at c.

EXAMPLE 2 Let $\varphi(x) = x^4 - 4x^3 + 12$. **a.** Find and classify the stationary points of φ. **b.** Graph φ.

Solution **a.** $\varphi'(x) = 4x^3 - 12x^2 = 4x^2(x - 3)$.
When $\varphi'(x) = 0$, $x = 0$ or 3.
The stationary points are 0 and 3.

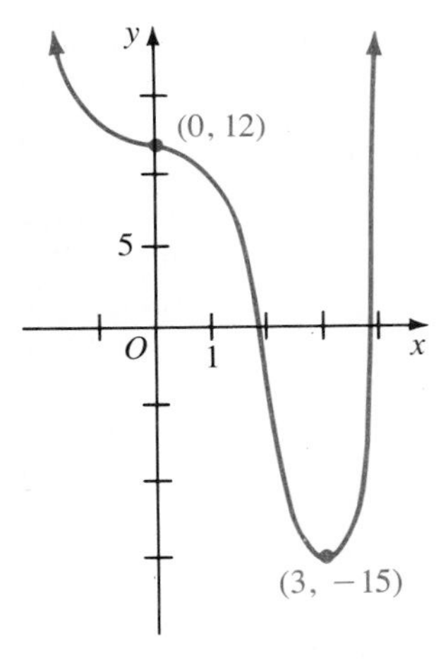

Test $x = 0$: Just to the left of 0, $\varphi'(x) < 0$; just to the right of 0, $\varphi'(x) < 0$. $\therefore$ φ has no extremum at 0.

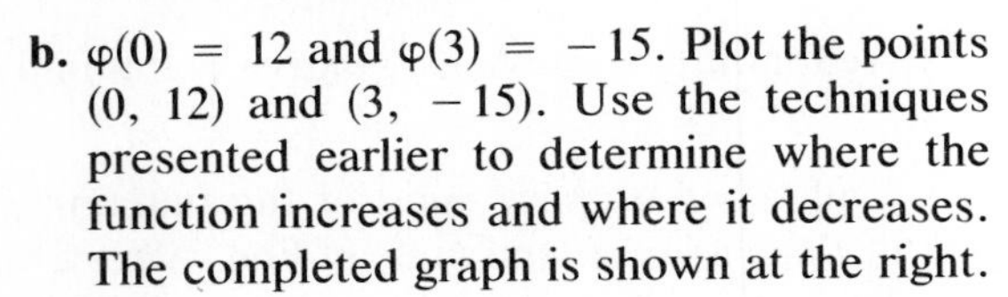

Test $x = 3$: Just to the left of 3, $\varphi'(x) < 0$; just to the right of 3, $\varphi'(x) > 0$. $\therefore$ φ has a relative minimum at 3.

b. $\varphi(0) = 12$ and $\varphi(3) = -15$. Plot the points $(0, 12)$ and $(3, -15)$. Use the techniques presented earlier to determine where the function increases and where it decreases. The completed graph is shown at the right.

Notice in Example 2 that $\varphi(3) = -15$ is less than $\varphi(x)$ for *all* other x, not just those near 3. We say that $\varphi(x)$ has an **absolute**, or **global**, **minimum** value of -15 at 3. An **absolute**, or **global**, **maximum** is defined similarly. Note that neither g in Example 1 nor φ in Example 2 has an absolute maximum.

EXERCISES

In Exercises 1–11: **a.** Find the intervals on which f is increasing and on which f is decreasing. **b.** Find and classify the relative extrema of f. **c.** Graph f.

A

1. $f(x) = x^2 - 4x + 2$ **2.** $f(x) = 2x^2 - 8x + 10$ **3.** $f(x) = 4x - x^2$

4. $f(x) = x^2 + 6x + 5$ **5.** $f(x) = x^3 - 3x^2 + 3$ **6.** $f(x) = 3x - x^3$

7. $f(x) = 6 + 12x - x^3$ **8.** $f(x) = x^3 - 3x^2$ **9.** $f(x) = 8x^2 - x^4$

Hint for 10 and 11: Do not expand; use the product rule for derivatives.

10. $f(x) = (x - 3)^2(x + 1)^2$ **11.** $f(x) = (x + 1)^3(x - 3)$

In Exercises 12–17 find and classify the relative extrema of f if there are any. You need not graph f.

12. $f(x) = \frac{3 - x}{1 + x}$ **13.** $f(x) = \frac{x + 1}{x - 1}$ **14.** $f(x) = \frac{x^2}{x + 1}$ **15.** $f(x) = \frac{x}{x^2 + 1}$

B **16.** $f(x) = x\sqrt{4 - x}$ **17.** $f(x) = \frac{x}{\sqrt{x - 1}}$

Exercises 18 and 19 refer to the cubic function $F(x) = x^3 + ax^2 + bx + c$.

18. Show that the number of stationary points of F is two if $a^2 > 3b$, one if $a^2 = 3b$, and none if $a^2 < 3b$.

19. Show that F has exactly one real root if $a^2 < 3b$.

C **20.** Show that if r is a multiple zero of a polynomial function f, then r is a zero of the function f'.

21. By drawing its graph, show that $g(x) = ||x| - 2|$ has three relative extrema but that there is no well-defined tangent line at any of them.

22. This exercise provides an example of a function that has an extremum at a point where its graph has a *vertical* tangent line.

a. Graph $f(x) = x^3$. Note that the graph of f has a horizontal tangent line at (0, 0).

b. Graph $g(x) = x^{\frac{1}{3}}$, the inverse of f. Note that the graph of g has a vertical tangent line at (0, 0).

c. Graph $\varphi(x) = |g(x)| = |x^{\frac{1}{3}}|$.

COMPUTER EXERCISES

1. Write and run a program to approximate the slope of the line tangent to the graph of $f(x) = x^3 - 4x^2$ at $x = -4, -3.5, -3, \ldots, 4$. Use $S(x) = \frac{f(x + h) - f(x)}{h}$ with $h = 0.0001$.

2. Modify and run the program in Exercise 1 for each function f to print a table of values in three columns. The columns should include x, the calculated slope at x, and an indication of whether the slope is positive, negative, or zero. Use the table to help determine intervals where f increases, where f decreases, and where $f'(x) = 0$.

a. $f(x) = \frac{1}{4}x^3 - x$ **b.** $f(x) = \frac{x^4}{2} - 3x^2$

c. $f(x) = x^5 - 2x^4 + x^3 + x^2 - 2x + 1$ **d.** $f(x) = -x^5 + 10x$

3. Modify and run the program in Exercise 1 for $f(x) = \sqrt{x^2 + 1}$.

6-6 CONCAVITY AND THE SECOND DERIVATIVE

In Example 1 of Section 6-5 we graphed

$$g(x) = x^3 - 3x^2 - 9x + 22.$$

The graph is reproduced in the figure at the right. Notice that as we move to the right along the black part of the curve, the slope of the graph *decreases*; we say that this part of the graph is *concave downward*. On the other hand, as we move to the right along the red part, the slope *increases*; this part of the graph is *concave upward*. Since the slopes in question are given by $g'(x)$, we are led to this general definition.

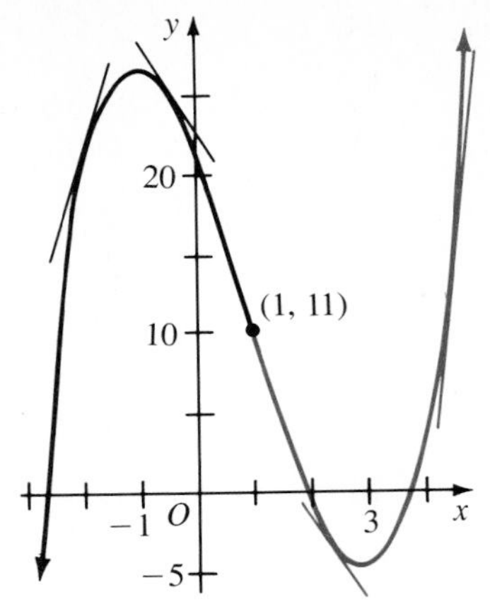

DEFINITION

The graph of a differentiable function f is **concave upward** on an interval I if f' is increasing on I. The graph is **concave downward** if f' is decreasing on I.

By Theorem 6, page 238, the function f' is increasing if $\frac{d}{dx} f'(x) > 0$ and is decreasing if $\frac{d}{dx} f'(x) < 0$. Since $\frac{d}{dx} f'(x) = f''(x)$, we have the following test for concavity.

THEOREM 9 The graph of a function f is concave upward on I if $f''(x) > 0$ on I. The graph of a function f is concave downward on I if $f''(x) < 0$ on I.

For example, if g is the function discussed in the first paragraph,

$$g'(x) = 3x^2 - 6x - 9 \quad \text{and} \quad g''(x) = 6x - 6 = 6(x - 1).$$

Since $g''(x) < 0$ if $x < 1$, and $g''(x) > 0$ if $x > 1$, the theorem tells us that the graph of g is concave downward on $(-\infty, 1)$ and concave upward on $(1, \infty)$. Note that the concavity changes sense at the point $(1, 11)$; such a point is called an **inflection point**.

EXAMPLE 1 Let φ be the function in Example 2, page 240: $\varphi(x) = x^4 - 4x^3 + 12$. Find the intervals on which the graph of φ is **(a)** concave upward and **(b)** concave downward, and **(c)** find the inflection points of the graph.

Solution

$$\varphi'(x) = 4x^3 - 12x^2$$
$$\varphi''(x) = 12x^2 - 24x = 12x(x - 2)$$

$\varphi''(x) > 0$ when $x < 0$ and when $x > 2$.
$\varphi''(x) < 0$ when $0 < x < 2$.

a. The graph is concave upward on $(-\infty, 0)$ and on $(2, \infty)$.
b. The graph is concave downward on $(0, 2)$.
c. The inflection points are $(0, 12)$ and $(2, -4)$.

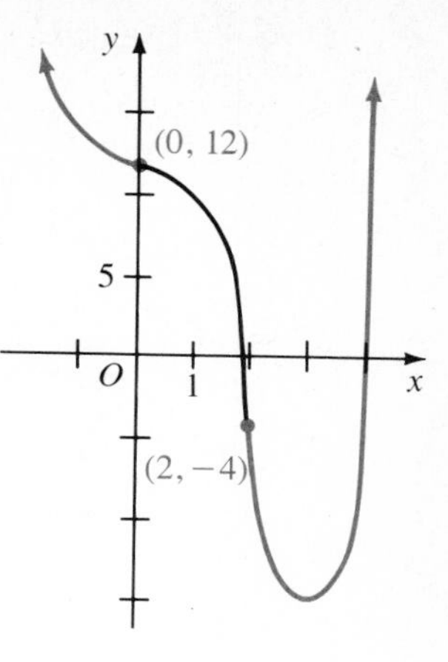

The figure that accompanies Example 1 shows the concave-upward parts of the graph in red and the concave-downward parts of the graph in black.

The second derivative can also be used to test for extrema. The following statements comprise the *second-derivative test*.

THEOREM 10 The Second-Derivative Test

1. If $f'(c) = 0$ and $f''(c) < 0$, then f has a relative maximum at c.
2. If $f'(c) = 0$ and $f''(c) > 0$, then f has a relative minimum at c.

For example, 3 is a stationary point of the function φ of Example 1, and $\varphi''(3) = 36 > 0$. Thus the second-derivative test tells us that φ has a relative minimum at 3, which we already knew from Section 6-5.

Warning: If $f'(c) = 0$ and $f''(c) = 0$, then the second-derivative test gives no information. (See Exercise 21.)

EXAMPLE 2 Use the second-derivative test to find the relative maxima and minima of $F(x) = \dfrac{x^2 + 4}{x}$.

Solution

$$F'(x) = \frac{x \cdot 2x - (x^2 + 4) \cdot 1}{x^2} = \frac{x^2 - 4}{x^2}$$

$$F''(x) = \frac{x^2 \cdot 2x - (x^2 - 4) \cdot 2x}{x^4} = \frac{8}{x^3}$$

$F'(x) = 0$ when $x = -2$ and when $x = 2$.

$F''(-2) = \dfrac{8}{(-2)^3} = -1 < 0$. $\therefore$ F has a relative maximum at -2.

$F''(2) = \dfrac{8}{2^3} = 1 > 0$. $\therefore$ F has a relative minimum at 2.

In Example 2 note that $F(-2) = -4$ and $F(2) = 4$. Thus the relative minimum value of F is greater than the relative maximum value of F. To see how this can be, graph F.

Relationships between a function f and its first two derivative functions, f' and f'', are shown in the figure on the following page. A • indicates a stationary point, a ∘ indicates an inflection point, and × indicates a point that is both a stationary point and an inflection point.

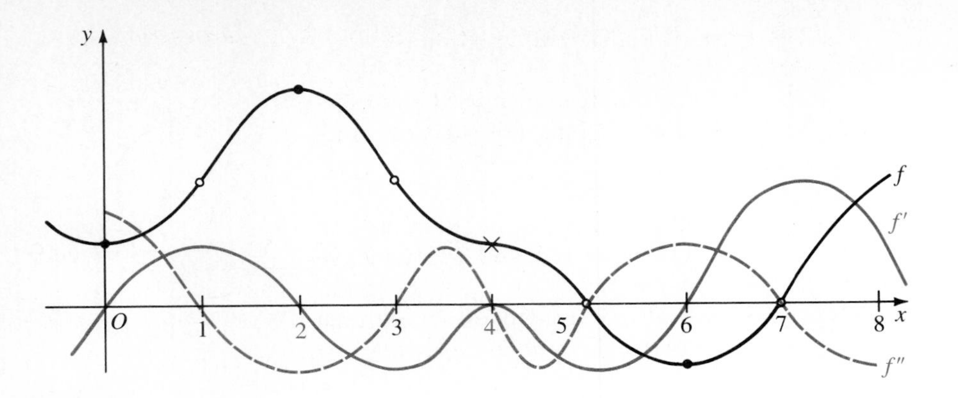

Note that 2 is a stationary point of f and a zero of f'. Moreover, 4 is the x-coordinate of an inflection point of f and a zero of f''.

EXERCISES

In Exercises 1–10, find the intervals on which the graph of f is **(a)** concave upward and **(b)** concave downward, and **(c)** find the inflection points of the graph. Then **(d)** sketch the graph, distinguishing between the concave-upward parts and concave-downward parts.

A

1. $f(x) = x^3 - 6x^2 + 9x$
2. $f(x) = 12x - x^3$
3. $f(x) = x^3 + 3x$
4. $f(x) = x^3 - 3x^2 - 9x + 12$
5. $f(x) = 6x^2 - x^4$
6. $f(x) = x^4 - 4x^3$
7. $f(x) = x^4 - 6x^2 + 8x$
8. $f(x) = 4x^3 - x^4 - 9$
9. $f(x) = 3x^5 - 10x^3$
10. $f(x) = 2x^6 - 5x^4$

Use the second-derivative test to find the relative maxima and relative minima of g.

11. $g(x) = x + \dfrac{2}{\sqrt{x}}$
12. $g(x) = x^2 - \dfrac{2}{x}$
13. $g(x) = \dfrac{x^3 + 4}{x^2}$
14. $g(x) = x^{\frac{1}{2}} + x^{-\frac{1}{2}}$

B

15. Show that at a stationary point c of $f(x) = x^2 + \dfrac{2}{x}$, $f''(c) = 6$, and hence f has a relative minimum but no relative maximum.

16. Show that at a stationary point c of $f(x) = x^2 + \dfrac{2}{x^2}$, $f''(c) = 8$, and hence f has relative minima but no relative maxima.

In Exercises 17–20, draw a coordinate system and reproduce the graph of f on it. Then, in the same coordinate system, sketch the graphs of f' and f'' as was done in the figure on page 244. (A • indicates a stationary point of f, a ∘ indicates an inflection point, and × indicates both.)

17.

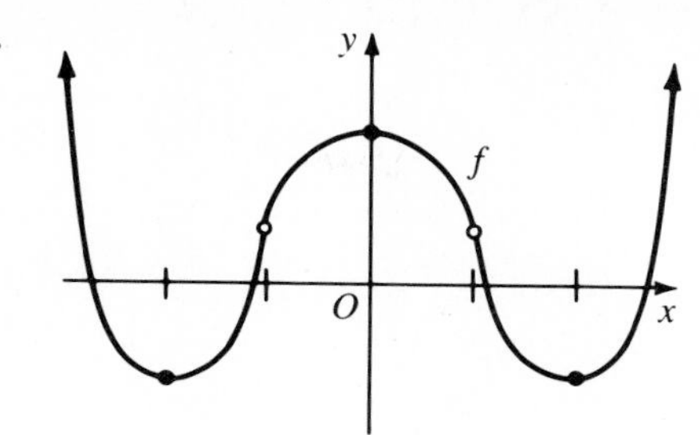

18.

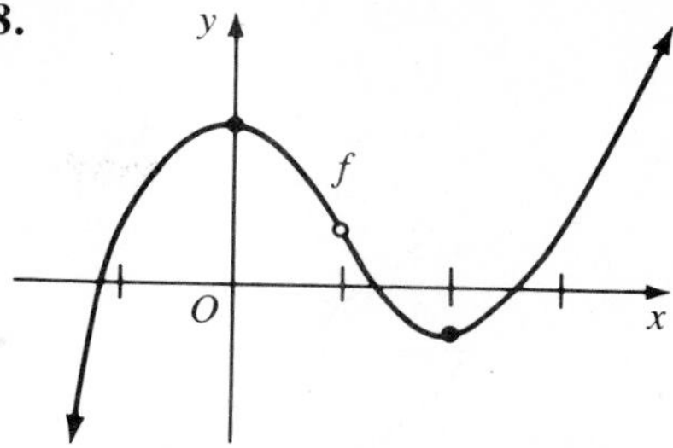

19.

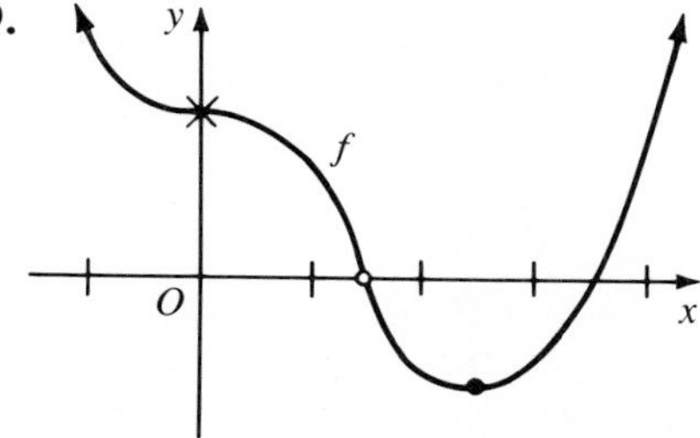

20.

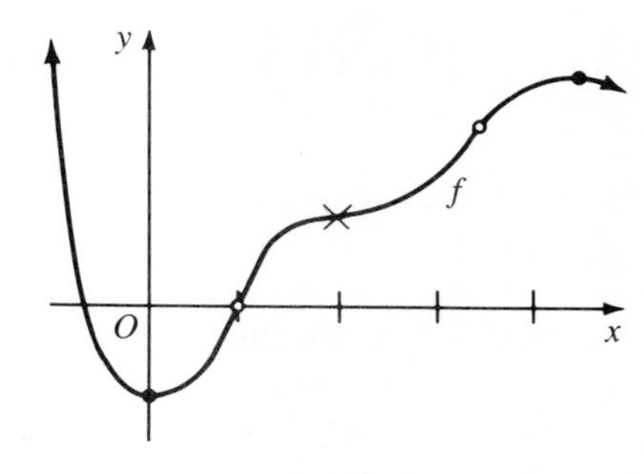

21. Show that each function has the property that $f'(0) = 0$ and that $f''(0) = 0$. Then show that one function has a maximum at 0, one function has a minimum at 0, and one function has neither a maximum nor a minimum at 0.

a. $f(x) = x^4 - 1$ **b.** $f(x) = x^3 - 1$ **c.** $f(x) = 1 - x^4$

Prove the statements made in Exercises 22–24.

22. The graph of a quadratic function is concave upward everywhere or concave downward everywhere.

23. The graph of a cubic function has exactly one inflection point.

24. The graph of a quartic function may have two inflection points or no inflection points.

25. Give examples of the two cases in Exercise 24.

6-7 APPLIED MAXIMA AND MINIMA

Many practical problems require that some quantity (for example, cost of manufacture or fuel consumption) be **minimized**, that is, be made as small as possible. Other problems require that some quantity (for example, profit on sales or attendance at a concert) be **maximized**, that is, be made as large as possible. We can use differential calculus to solve many of these so-called **optimization** problems.

EXAMPLE 1 A box is to be made from a 10 inch by 16 inch sheet of metal by cutting equal squares out of the corners and bending up the flaps to form the sides. Find the dimensions of the box if its volume is to be maximized.

Solution If x inch by x inch squares are cut from the corners, the dimensions of the box will be x, $10 - 2x$, and $16 - 2x$. The volume will be

$$V = x(10 - 2x)(16 - 2x).$$

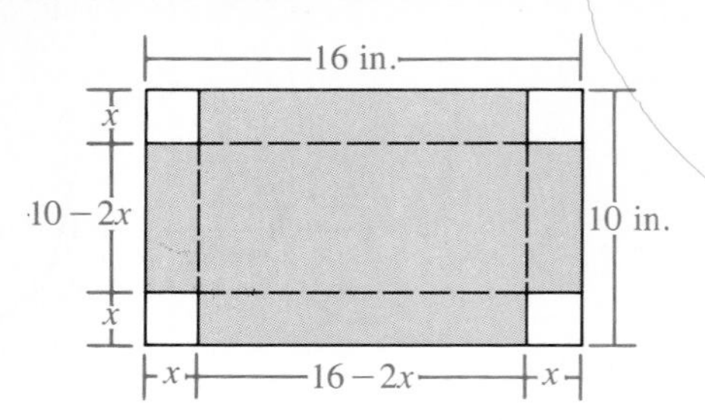

To see where V has its maximum value, find its stationary points.

$$V = 4(x^3 - 13x^2 + 40x)$$

$$\frac{dV}{dx} = 4(3x^2 - 26x + 40)$$

$$= 4(x - 2)(3x - 20).$$

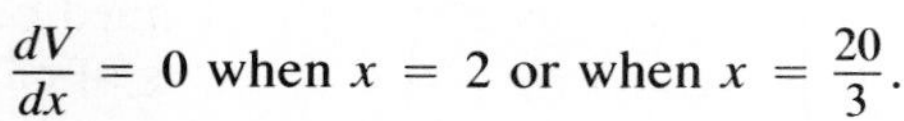

$$\frac{dV}{dx} = 0 \text{ when } x = 2 \text{ or when } x = \frac{20}{3}.$$

Eliminate $\frac{20}{3}$ because it is greater than half the width of the metal sheet. Since physical intuition says that there must *be* a box of maximum volume, it must correspond to $x = 2$. Thus the dimensions of the box are 2 inches by 6 inches by 12 inches. **Answer**

Use of geometrical or physical intuition, as in Example 1, often eliminates the need to use a formal extremum test. Use of the first-derivative test (Theorem 8) shows that $\frac{dV}{dx} < 0$ when $x < \frac{20}{3}$ and $\frac{dV}{dx} > 0$ when $x > \frac{20}{3}$. Thus $x = \frac{20}{3}$ gives a relative minimum not a maximum volume.

EXAMPLE 2 At noon the captain of a fishing trawler sees two supertankers approaching. One of them is 7 miles due west and traveling east at 4 mph. The other is 4 miles due north traveling south at 3 mph. At what time will the tankers be closest together and how far apart will they be then?

Solution The sketch shows the positions of the ships at noon (•) and at t hours after noon (∘). Let s be the distance between the ships t hours after noon. Apply the Pythagorean theorem to the right triangle in the sketch.

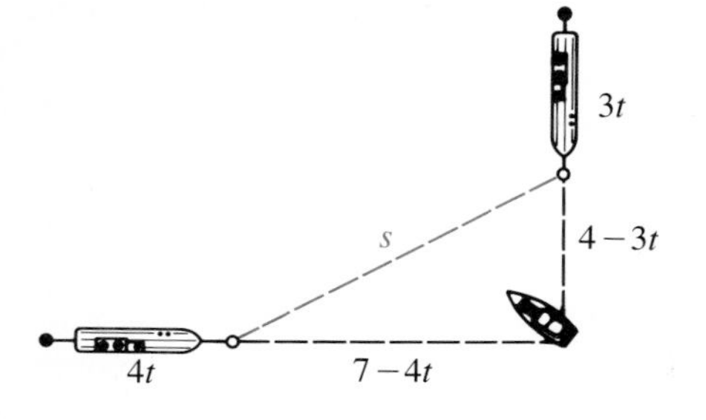

$$s^2 = (7 - 4t)^2 + (4 - 3t)^2$$
$$s^2 = 25t^2 - 80t + 65$$

To minimize the distance s between the ships, find $\frac{ds}{dt}$.

$$\frac{d}{dt}(s^2) = \frac{d}{dt}(25t^2 - 80t + 65)$$

Use the generalized power rule on the left side of the equation above.

$$2s\frac{ds}{dt} = 50t - 80$$

$\frac{ds}{dt} = 0$ when $t = \frac{80}{50} = \frac{8}{5}$. Therefore the tankers will be closest together $\frac{8}{5}$ hours after noon, that is, at 1:36 P.M. At that time

$$s^2 = 25\left(\frac{8}{5}\right)^2 - 80\left(\frac{8}{5}\right) + 65 = 1,$$

so that $s = 1$. The tankers will be 1 mile apart. **Answer**

Notice that in Example 2 the use of the generalized power rule enabled us to avoid radicals in the process of finding the minimum s.

In the next example the function to be maximized is defined only for positive *integers*. However, in order to use the powerful tools of calculus, we assume that it is defined for all positive real numbers.

EXAMPLE 3 A builder of customized cars finds that he can sell n cars a month if he sets the price per car at $p(n) = \frac{2 \times 10^7}{375 + n^2}$ dollars. What price per car maximizes his total revenue?

Solution The total revenue $R(n)$ is the number of cars sold times the price per car.

$$\begin{aligned} R(n) &= np(n) \\ &= 2 \times 10^7\left(\frac{n}{375 + n^2}\right) \end{aligned}$$

Find $R'(n)$ and solve $R'(n) = 0$ for n.

$$\begin{aligned} R'(n) &= 2 \times 10^7 \left[\frac{(375 + n^2) \cdot 1 - n \cdot 2n}{(375 + n^2)^2}\right] \\ &= 2 \times 10^7\left(\frac{375 - n^2}{(375 + n^2)^2}\right) \\ R'(n) &= 0 \text{ if } n = \sqrt{375} \approx 19.4 \end{aligned}$$

Since 19.4 is not an integer, determine whether 19 or 20 gives the greater revenue.

$$R(19) = 2 \times 10^7\left(\frac{19}{375 + 19^2}\right) = \$516{,}304$$

$$R(20) = 2 \times 10^7\left(\frac{20}{375 + 20^2}\right) = \$516{,}129$$

The manufacturer should sell 19 cars a month by pricing each at $p(19) = \frac{2 \times 10^7}{375 + 19^2} = \$27{,}174$. **Answer**

EXAMPLE 4 Find the least area of a triangle in the first quadrant formed by the coordinate axes and a line L through the point $(3, 8)$.

Solution The diagram shows the triangle whose area A is to be minimized.

$$A = \frac{1}{2}ab$$

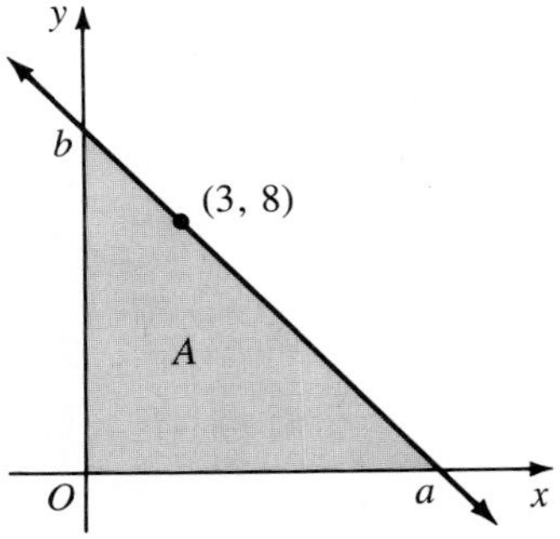

where a and b are the intercepts of L. Next write A in terms of a single variable, say a. Since L has the equation

$$y = -\frac{b}{a}x + b$$

and since $(3, 8)$ is on L, $8 = -\frac{b}{a}(3) + b$.

Thus $$b = \frac{8a}{a - 3}.$$

So the area A in terms of a is given by

$$A = \frac{1}{2}ab = \frac{1}{2}a\left(\frac{8a}{a - 3}\right) = 4\left(\frac{a^2}{a - 3}\right).$$

$$\frac{dA}{da} = 4\left(\frac{(a - 3) \cdot 2a - a^2 \cdot 1}{(a - 3)^2}\right) = 4\left(\frac{a^2 - 6a}{(a - 3)^2}\right) = 4\left(\frac{a(a - 6)}{(a - 3)^2}\right)$$

Solve $\frac{dA}{da} = 0$. If $\frac{dA}{da} = 0$, $a = 0$ or $a = 6$. There is no triangle if $a = 0$. Hence $a = 6$. With $a = 6$ the area is

$$A = 4\left(\frac{a^2}{a - 3}\right) = 4\left(\frac{36}{3}\right) = 48. \qquad \textbf{Answer}$$

Use the first-derivative test to see that $a = 6$ gives a minimum area.

In Example 4 we expressed the quantity A in terms of the x-intercept a. We could have used the y-intercept of L instead. In any case we need to express A as a function of some single variable. The preceding examples suggest the following steps in solving extremum problems.

1. Name the quantity to be maximized or minimized and the other quantities entering into the problem.
2. If appropriate, draw and label a figure to illustrate the problem.
3. Express the quantity to be maximized or minimized as a function of one or more variables in the problem. If more than one such variable is used, eliminate all but one of them by using relationships among them. (We did this in Example 4.)
4. Use Theorem 7 to find the stationary points. Determine which of these gives the required extremum. The first- or second-derivative test is useful.
5. Calculate the maximum or minimum value if required.

EXAMPLE 5 Find the greatest possible volume of a right circular cone inscribed in a sphere of radius a.

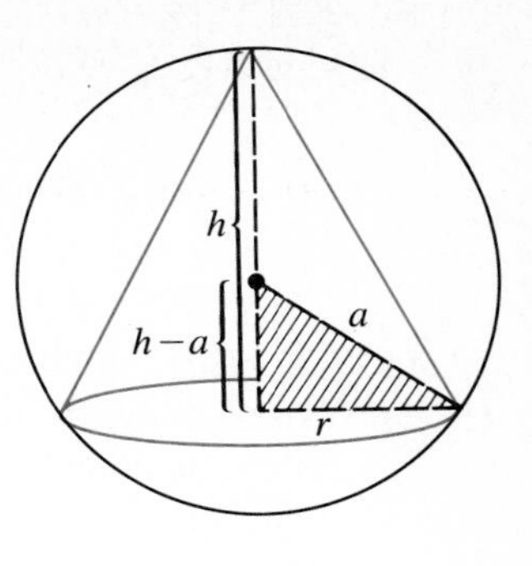

Solution

1. Maximize V, the volume of a cone with base radius r and height h. The radius of the sphere is a given constant a.
2. The cone inscribed in the sphere is shown, labeled, at the right.
3. The volume of the cone is given by

$$V = \frac{1}{3}\pi r^2 h.$$

Write V as a function of one variable, h. Apply the Pythagorean theorem to the shaded right triangle in the figure.

$$r^2 + (h - a)^2 = a^2 \quad \text{or} \quad r^2 + h^2 - 2ah + a^2 = a^2$$

Thus $$r^2 = 2ah - h^2$$

and $$V = \frac{1}{3}\pi(2ah - h^2)h$$

$$= \frac{1}{3}\pi(2ah^2 - h^3) \qquad (*)$$

4. Find $\frac{dV}{dh}$ and solve $\frac{dV}{dh} = 0$.

$$\frac{dV}{dh} = \frac{1}{3}\pi(4ah - 3h^2)$$

$$= \frac{1}{3}\pi h(4a - 3h)$$

$\frac{dV}{dh} = 0$ when $h = 0$ and when $h = \frac{4a}{3}$. Since there is no cone when $h = 0$, use $h = \frac{4a}{3}$.

Use the first-derivative test to see that $\frac{dV}{dh} > 0$ when $h < \frac{4a}{3}$ and $\frac{dV}{dh} < 0$ when $h > \frac{4a}{3}$. Hence $h = \frac{4a}{3}$ gives a relative maximum.

5. The problem asks for the volume of the cone. Substitute for h in (*).

$$V = \frac{1}{3}\pi\left(2a\left(\frac{4a}{3}\right)^2 - \left(\frac{4a}{3}\right)^3\right)$$

$$= \frac{1}{3}\pi \cdot \frac{32}{27}a^3$$

$\therefore$ the greatest V is $\frac{32}{81}\pi a^3$. **Answer**

Since $\frac{32}{81}\pi a^3 = \frac{8}{27} \cdot \frac{4}{3}\pi a^3$, the largest cone takes up $\frac{8}{27}$ of the volume of the sphere.

EXERCISES

A

1. If a ball is thrown vertically upward from the top of an 80-foot tower with initial speed 72 ft/s, its height h above the ground t seconds later is $80 + 72t - 16t^2$ feet. Find the greatest height reached by the ball.
2. A rental agent estimates that the monthly profit p from a building s stories high is given by $p = 4600s - 100s^2$. What height building would she consider most profitable?
3. Find the dimensions of the rectangle having perimeter 24 cm and area as large as possible.
4. Find the dimensions of the rectangle having area 36 cm^2 and perimeter as small as possible.
5. Determine two positive numbers whose sum is 12 and such that the product of one of them and the square of the other is a maximum.
6. Determine two positive numbers whose sum is 12 and such that the product of one and the cube of the other is a maximum.

In Exercises 7 and 8, a contractor is to fence off a rectangular field along a straight river, the side along the river requiring no fence.

7. What is the least amount of fencing needed to fence off 30,000 m^2?
8. What is the greatest area the contractor can fence off using 500 m of fencing?

In Exercises 9 and 10, a rancher wishes to fence in a rectangular plot of land and divide it into two corrals with a fence parallel to two of the sides.

9. If she has 300 yd of fencing available, how large an area can the plot have?
10. What is the least length of fencing needed to enclose a plot of 5400 yd^2?
11. At noon, a tortoise moving south at 8 in./min is 25 in. due north of a snail that is moving west at 6 in./min. When will they be closest together and how far apart will they be then?
12. An open box is to be formed from a square sheet of metal 3 feet on a side by cutting equal squares out of the corners and bending up the flaps. Find the dimensions of the box if its volume is to be as large as possible.
13. The strength of a wooden beam of rectangular cross section is proportional to the product of its width and the square of its height. Find the dimensions of the cross section of the strongest beam that can be cut from a log 12 inches in diameter. (*Hint*: Let S be a measure of strength. Then for some constant k, $S = kwh^2$.)

B **14.** Find two positive numbers whose sum is 4 and such that the sum of the cube of the first and the square of the second is as small as possible.

15. Find two positive numbers whose sum is 1 and such that the product of the cube of the first and the square of the second is as large as possible.

In Exercises 16 and 17, two sides of a rectangle lie along the coordinate axes, and a vertex lies on the first-quadrant part of a curve C as shown at the right. Find the dimensions of the rectangle of greatest area for the given curve C.

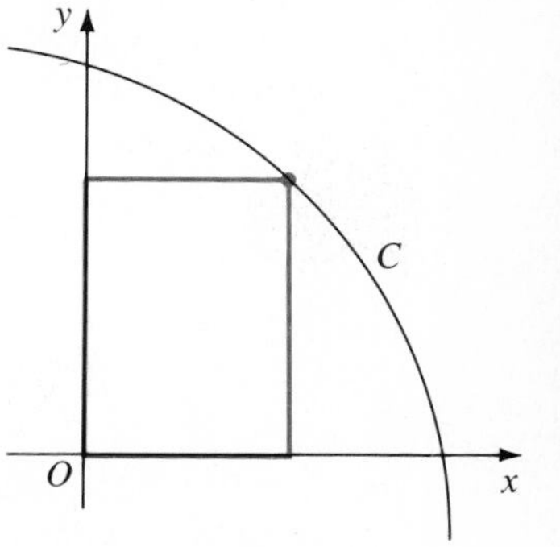

ex 4 **16.** C is the line $3x + 2y = 12$.

17. C is the parabola $y = 12 - x^2$.

18. Rework Example 4 by expressing the area of the triangle as a function of the slope of the line.

19. Find the point on the parabola $y = x^2$ that is closest to the point $P(3, 0)$.

20. A merchant finds that she can sell n units a day of a certain product if she prices it at $\frac{c}{a^2 + n^2}$ dollars per unit $(a, c > 0)$. At what price per unit does she receive the greatest revenue from sales?

21. A chartering company will provide a plane for a fare of \$300 per person for 100 or fewer passengers (but more than 60 passengers). For each passenger over 100, the fare is decreased \$2 per person for everyone. What number of passengers would provide the greatest revenue?

C **22.** The cost of fuel to run an excursion boat is proportional to the cube of its speed and is \$40 per hour when the speed is 10 mph. Expenses are \$270 per hour. Find the most economical speed at which to make a trip of any specified distance.

23. Find the greatest volume of a cylinder inscribed in a cone of base radius 6 and height 9. (*Hint*: Use similar triangles.)

24. Find the greatest volume of a cylinder that can be inscribed in a sphere of radius a.

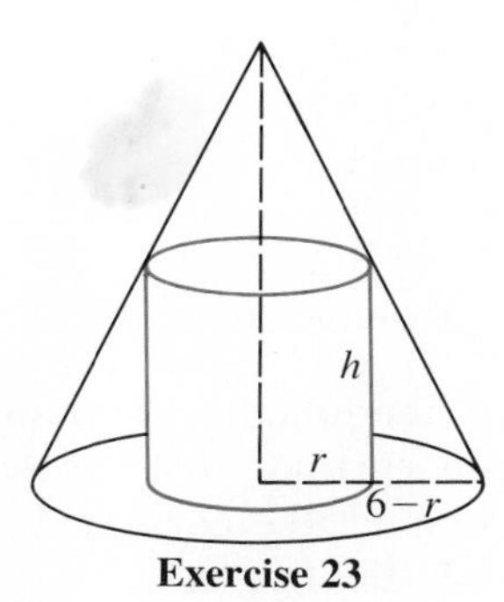

Exercise 23

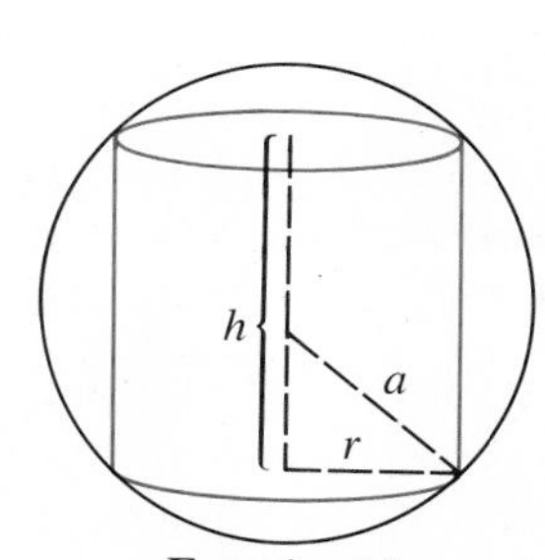

Exercise 24

25. The figure at the right shows a sphere of radius $CT = a$. About it is to be circumscribed a right circular cone with base radius $AB = r$ and height $BV = h$. Find the least volume of such a cone. (*Hint*: Use the fact that triangle CTV is similar to triangle ABV.)

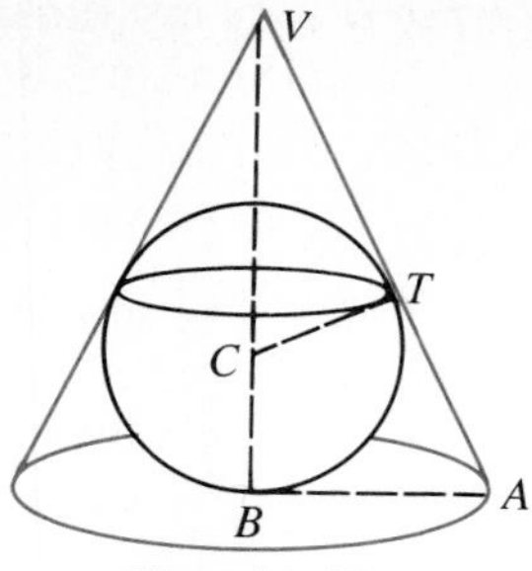

Exercise 25

The remaining exercises involve expressions of the form $\sqrt{a^2 \pm x^2}$ and $\sqrt{x^2 - a^2}$. To differentiate these, recall Section 6-4.

26. A man is in a boat at point B, 2 miles from the nearest point A of a straight shoreline. He wishes to reach a point C on the shore 6 miles from A in the least time possible. How far from A should he land if he can row at 3 mph and jog at 5 mph?

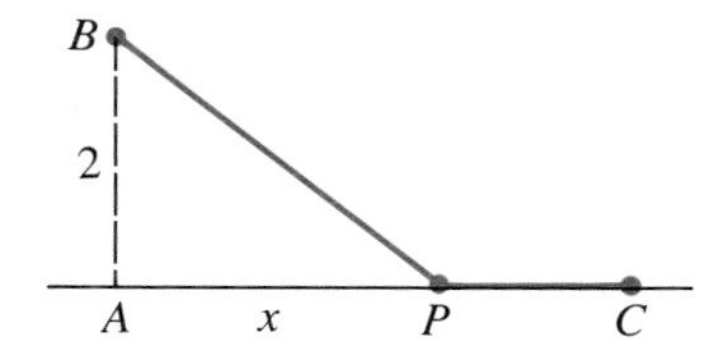

27. The stiffness of a wooden beam of rectangular cross section is proportional to the product of its width and the cube of its height. Find the dimensions of the cross section of the stiffest beam that can be cut from a log 12 inches in diameter. (*Hint*: See Exercise 13.)

28. Follow the instructions preceding Exercise 16 for the case where C is the ellipse $\frac{x^2}{18} + \frac{y^2}{8} = 1$.

29. The figure at the left below shows an isosceles trapezoid inscribed in a semicircle of radius a. Find the maximum area of such a trapezoid.

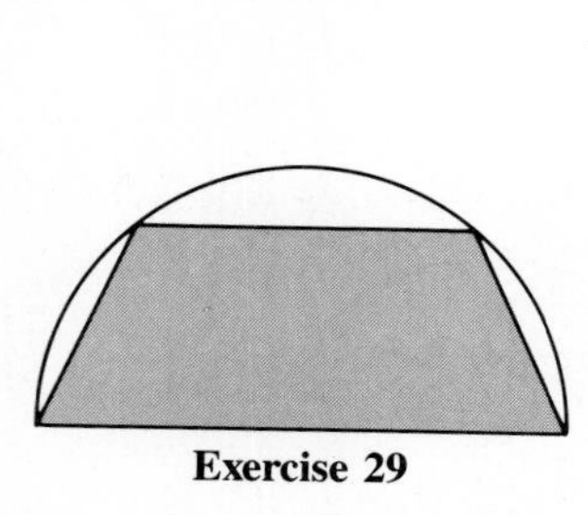
Exercise 29

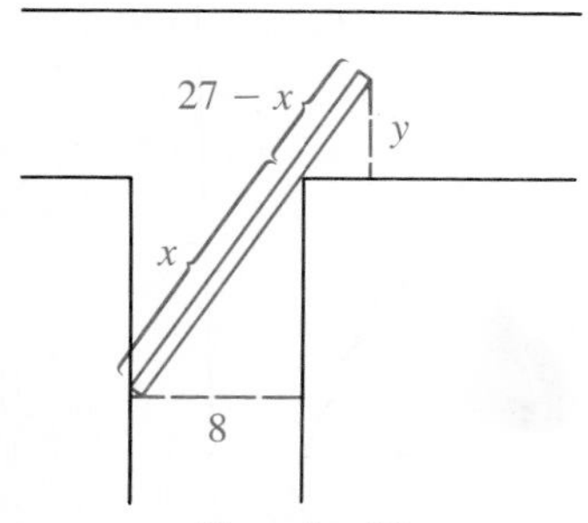

Exercise 30

30. Workers wish to move a 27-foot pole, carried horizontally, from an 8-foot-wide passageway around a corner into a corridor that the passageway intersects at right angles. How wide must the corridor be if they are to do this? (*Hint*: Maximize y shown in the figure.)

6-8 DERIVATIVES AS RATES OF CHANGE

A well-known rate is *speed*, the rate of change of distance with respect to time. For constant speeds we have the familiar formula

$$\text{distance} = \text{rate} \times \text{time}.$$

Related to speed is *velocity*, which tells not only the speed of an object, but also the direction in which it is moving.

> The **velocity** of a moving particle is the rate of change of its position with respect to time.

In this section we will discuss **rectilinear motion**, motion of a particle P along a straight line L. We coordinatize L by placing the origin at the point at which P begins its movement. The velocity of P is positive if P is moving to the right and negative if P is moving to the left.

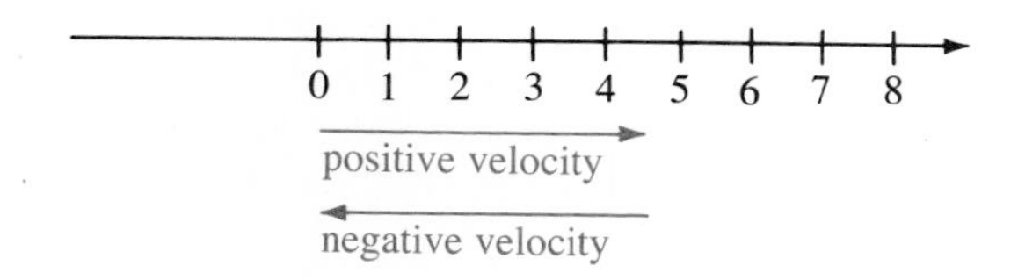

We can often describe the position of P by a law of motion,

$$s = f(t),$$

that gives the coordinate s of P in terms of time t.

Knowing the law of motion of a particle P, we obtain an expression for its velocity as follows: At time c, P's position is $f(c)$, and at a later time t its position is $f(t)$, so the change in position is $f(t) - f(c)$. Since the elapsed time is $t - c$, the average rate of change of the position of P over the time interval $[c, t]$ is

$$\frac{f(t) - f(c)}{t - c}.$$

Now, the shorter the time interval, the better this average rate will approximate the velocity v *at* time c. Thus

$$v = \lim_{t \to c} \frac{f(t) - f(c)}{t - c} = f'(c), \quad \text{or} \quad v = \frac{ds}{dt}.$$

The velocity v of a moving particle P may itself change.

> The **acceleration** of a moving particle is the rate of change of its velocity with respect to time.

Since the velocity of P is given by $v = f'(t) = \frac{ds}{dt}$, the acceleration a is

$$a = \frac{dv}{dt} = f''(t), \quad \text{or} \quad a = \frac{d^2s}{dt^2}.$$

EXAMPLE 1 The law of motion of a particle P is

$$s = t^3 - 6t^2 + 9t \quad (t \geq 0),$$

s being measured in meters and t in seconds. **a.** Find the velocity v (in m/s) **b.** Find the acceleration a (in m/s^2) of P at time t. **c.** When is P moving to the right and when to the left? **d.** Find the positions and accelerations of P when it is instantaneously at rest. **e.** Indicate the motion of P in a diagram.

Solution **a.** $v = \frac{ds}{dt} = 3t^2 - 12t + 9$ m/s **b.** $a = \frac{dv}{dt} = 6t - 12$ m/s^2

c. $v = 3(t^2 - 4t + 3) = 3(t - 1)(t - 3)$

$v > 0$ when $t < 1$ and when $t > 3$. P moves to the right during these times.

$v < 0$ when $1 < t < 3$. P moves to the left during these times.

d. P is at rest when $v = 0$, that is, when $t = 1$ or $t = 3$.
At $t = 1$: $s(1) = 1^3 - 6 \cdot 1^2 + 9 \cdot 1 = 4$ m; $a(1) = 6 \cdot 1 - 12 = -6$ m/s^2
At $t = 3$: $s(3) = 3^3 - 6 \cdot 3^2 + 9 \cdot 3 = 0$ m; $a(3) = 6 \cdot 3 - 12 = 6$ m/s^2

e.

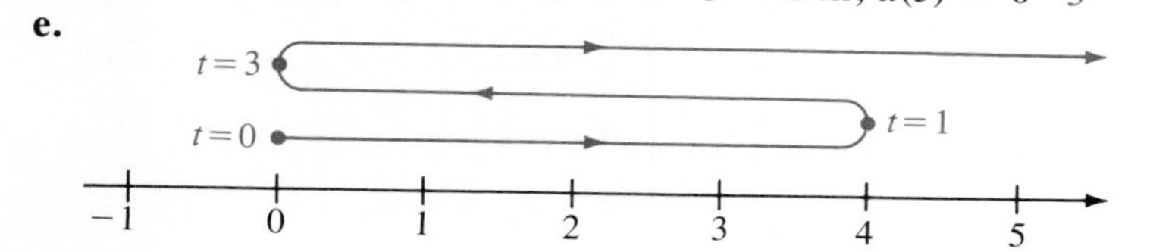

If two or more quantities that vary with time are related by an equation, then a relationship between their rates of change can be found. For example, if a spherical balloon is being inflated, its radius r and volume V are both functions of time, $r = r(t)$ and $V = V(t)$. These functions are related by the equation $V(t) = \frac{4}{3}\pi(r(t))^3$. For simplicity we usually omit the letter t and in this case would write $V = \frac{4}{3}\pi r^3$. Differentiating both sides of this equation yields a relationship between the rates $\frac{dV}{dt}$ and $\frac{dr}{dt}$. Problems involving rates that are related are called **related-rate problems.**

EXAMPLE 2 A car traveling north at 55 mph is approaching an intersection with a street on which a truck, having crossed the intersection, is traveling east at 40 mph. At the moment when the car is 0.3 mile from the intersection, and the truck is 0.4 mile from it, is the distance between the vehicles increasing or decreasing, and how fast?

Solution Let u and w be the distances from the intersection to the car and the truck respectively and let s be the distance between them. It is given that

$\frac{du}{dt} = -55$ and $\frac{dw}{dt} = 40$ since the car is approaching the intersection and the truck is past it and moving away. By the Pythagorean theorem, $s^2 = u^2 + w^2$. Differentiate s^2 and $u^2 + v^2$ with respect to time. Use the generalized power rule.

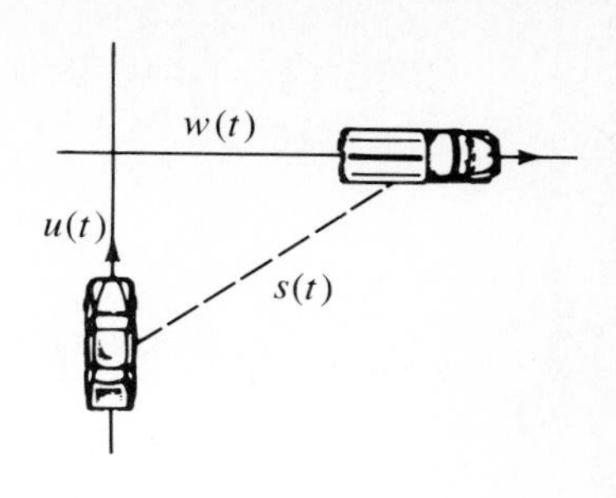

$$2s\frac{ds}{dt} = 2u\frac{du}{dt} + 2w\frac{dw}{dt}$$

$$s\frac{ds}{dt} = u\frac{du}{dt} + w\frac{dw}{dt} \qquad (*)$$

The goal is to find $\frac{ds}{dt}$ when $u = 0.3$ and $w = 0.4$. If $u = 0.3$ and $w = 0.4$, then $s^2 = (0.3)^2 + (0.4)^2 = 0.25$ and $s = 0.5$.
Substitute into (*).

$$0.5\frac{ds}{dt} = 0.3(-55) + 0.4(40)$$
$$= -0.5$$

Hence

$$\frac{ds}{dt} = \frac{-0.5}{0.5}$$
$$= -1.$$

∴ the distance between the vehicles is decreasing at the rate of 1 mph.

Answer

Example 2 suggests the following strategy for solving related-rate problems.

1. *Translate* the problem into mathematical language. This will involve naming quantities in the problem and expressing their rates of change as derivatives.
2. *Relate* the variables (other than time) that were introduced at step (1). Often this will involve geometric relationships. In Example 2 we related s and u and w with the Pythagorean theorem.
3. *Differentiate* the members of the relationship found in (2) with respect to time, using the chain rule.
4. *Substitute* numerical data. Solve for the required rate.

EXAMPLE 3 A tank is in the form of an inverted cone 8 m across the top and 6 m deep. Water is flowing into the tank at the rate of 10 m³/h. At what rate is the depth of the water in the tank changing when it is 5 m deep?

Solution on following page

Solution 1. At any instant the water in the tank forms a cone. Let its volume be V, its base radius be r, and its height be h. It is given that $\frac{dV}{dt} = 10$. The goal is to find $\frac{dh}{dt}$.

2. Relate V and h.

$$V = \frac{1}{3}\pi r^2 h$$

From similar triangles, $\frac{r}{4} = \frac{h}{6}$, or $r = \frac{2h}{3}$.

$$\therefore V = \frac{1}{3}\pi\left(\frac{2h}{3}\right)^2 h = \frac{4\pi}{27}h^3$$

3. Differentiate with respect to time t.

$$\frac{dV}{dt} = \frac{4\pi}{27} \cdot 3h^2 \frac{dh}{dt} = \frac{4\pi}{9}h^2 \frac{dh}{dt}$$

4. Substitute 10 for $\frac{dV}{dt}$ and 5 for h.

$$10 = \frac{4\pi}{9} \cdot 5^2 \frac{dh}{dt} = \frac{100\pi}{9}\frac{dh}{dt}$$

$$\therefore \frac{dh}{dt} = \frac{9}{10\pi} \approx 0.2865 \text{ m/h} \approx 29 \text{ cm/h} \qquad \textbf{Answer}$$

EXERCISES

In each of Exercises 1–6 the position s at time t ($t \geq 0$) of a particle P is given, s being measured in meters and t in seconds.

a. Find the velocity v.
b. Find the acceleration a of P at time t.
c. When is P moving to the right and when to the left?
d. Find the positions and accelerations of P at the times when it is instantaneously at rest.
e. Indicate the motion of P in a diagram like the one in Example 1.

A

1. $s = 2 + 4t - t^2$ **2.** $s = t^2 - 6t + 5$

3. $s = 2t^3 - 9t^2 + 12t$ **4.** $s = 2 - 9t + 6t^2 - t^3$

5. $s = (t - 2)^2(t^2 - 4t + 2)$ **6.** $s = t^4 - 8t^3 + 18t^2 - 16t$

7. A rock dropped into still water sends out concentric ripples. If the radius of the outer ripple increases at the rate of 2 ft/s, how fast is the area of the disturbed surface increasing when it is 6 ft in diameter?

8. Helium is being pumped into a spherical rubber balloon at the rate of 200 cubic inches per second. At what rate is the radius increasing when the balloon is 40 inches in diameter?

9. A boat is being pulled toward a dock by means of a cable attached to a windlass 9 feet above the deck of the boat. The cable is being wound in at the rate of 6 feet per second. How fast is the boat approaching the dock when its horizontal distance to the dock is 12 feet?

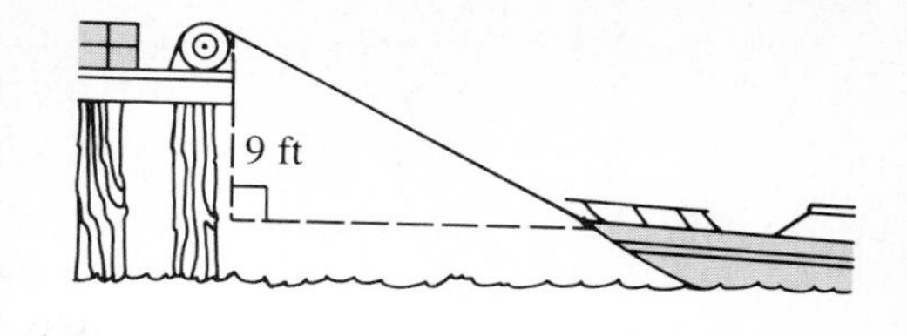

10. The foot of a 13-foot ladder leaning against a high wall is pulled away from the wall at the rate of 4 feet per minute. How fast is the top of the ladder moving when it is **(a)** 12 feet above the ground? **(b)** 5 feet above the ground?

B **11.** A plane takes off from airport A and flies east at 480 mph. At the moment of takeoff, a second plane, approaching A from the north at 320 mph, is still 240 miles from A. Fifteen minutes later is the distance between the planes increasing or decreasing, and how fast?

12. The base of a right triangle is increasing at the rate of 5 cm/min, and its height is decreasing at 4 cm/min. At the moment when the base is 3 cm and the height is 4 cm, **(a)** is the area increasing or decreasing, and how fast? **(b)** is the hypotenuse increasing or decreasing, and how fast?

13. A point (x, y) moves on the top half of the circle $x^2 + y^2 = 25$ in such a way that its x-coordinate increases at the rate of 2 units/s. At what rate is its y-coordinate changing as the point passes through $(4, 3)$?

14. A point moves in the plane of two perpendicular lines so that the product of its distances from the lines always is 12. When the point is 4 units from one of the lines, its distance from that line is decreasing at the rate of 2 units/min. At what rate is its distance from the other line changing at that moment?

15. A man 6 feet tall walks with speed 8 ft/s directly away from a lamppost on which the light is 18 feet high. Find **(a)** the speed of the tip of his shadow and **(b)** the rate at which his shadow is lengthening at the moment when he is 36 feet from the post.

16. Let A be the area of a variable circle having radius r at time t and circumference C. Show that $\frac{dA}{dt} = C\frac{dr}{dt}$.

17. Let V be the volume of a variable sphere having radius r at time t and surface area S. Show that $\frac{dV}{dt} = S\frac{dr}{dt}$.

18. The thin-lens law asserts that $\frac{1}{x} + \frac{1}{y} = \frac{1}{f}$, where x and y are the respective distances from the lens of an object and its image and f is a constant called the focal length of the lens. If an object is moving toward a lens of focal length 10 cm at the rate of 6 cm/s, how fast is its image moving when the object is 30 cm from the lens?

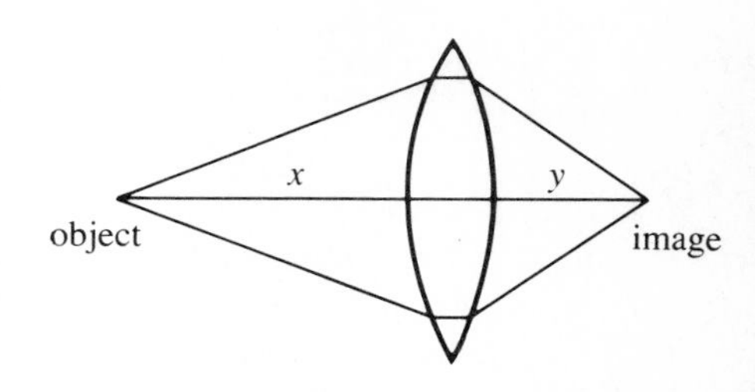

19. Sand is falling from a chute at the rate of one cubic yard per minute to form a conical pile whose height is always one-third its diameter. How fast is the height increasing when the pile is 6 feet high?

20. A V-shaped trough is 8 feet long, 3 feet wide, and 2 feet deep. Water is flowing into the trough at the rate of 5 cubic feet per minute. How fast is the depth of the water increasing just as it is about to overflow?

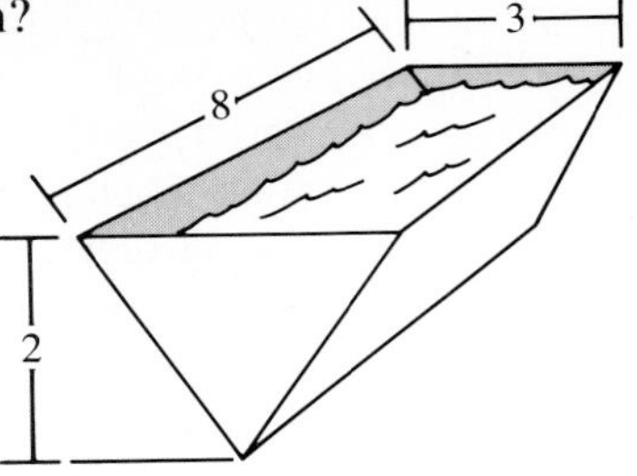

C 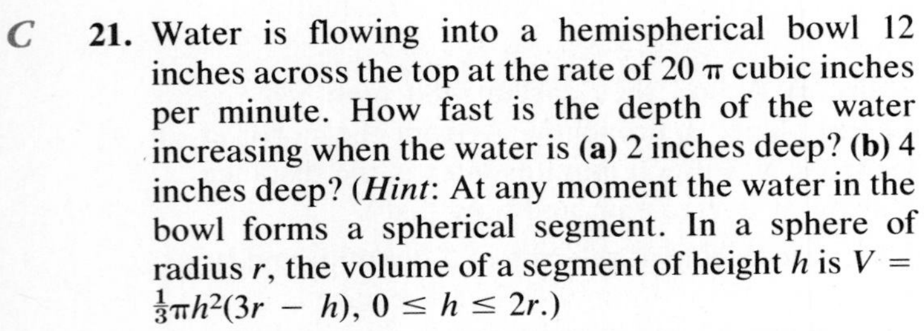

21. Water is flowing into a hemispherical bowl 12 inches across the top at the rate of 20 π cubic inches per minute. How fast is the depth of the water increasing when the water is **(a)** 2 inches deep? **(b)** 4 inches deep? (*Hint*: At any moment the water in the bowl forms a spherical segment. In a sphere of radius r, the volume of a segment of height h is $V = \frac{1}{3}\pi h^2(3r - h)$, $0 \le h \le 2r$.)

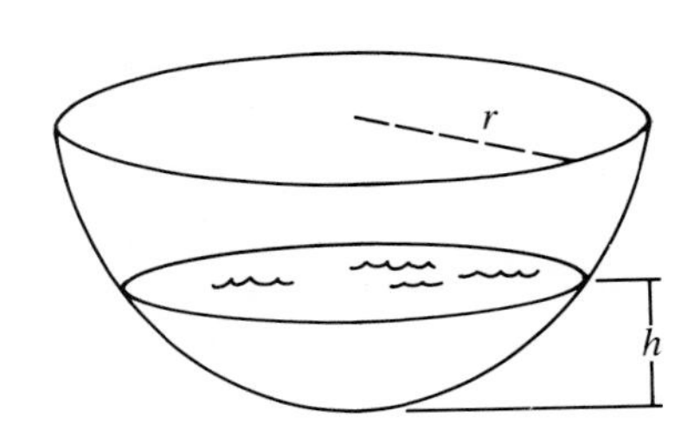

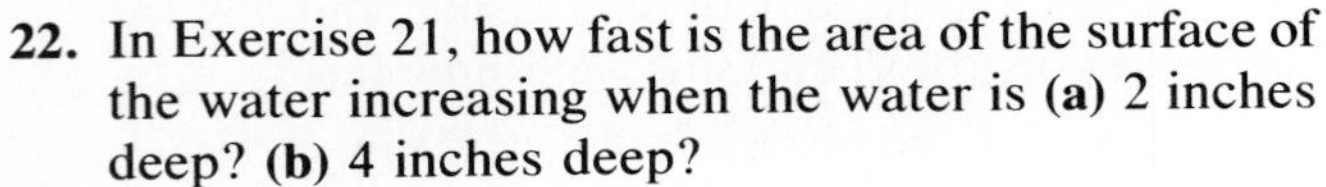

22. In Exercise 21, how fast is the area of the surface of the water increasing when the water is **(a)** 2 inches deep? **(b)** 4 inches deep?

■ 6-9 NEWTON'S METHOD

Many applications require us to find roots of equations or zeros of functions. Although we can sometimes find exact values, we often must be content with approximate answers. In this section we discuss Newton's method, which is used to approximate zeros of differentiable functions.

The figure at the right shows a portion of a graph of a differentiable function f near one of its real zeros r. Let x_1 be a first approximation to r. (One way to obtain such an estimate is to apply the location principle stated on page 168.) To obtain additional approximations to r we reason as follows: The line L tangent to the graph of f at $(x_1, f(x_1))$ will cross the x-axis near the point where the graph itself crosses the axis. That is, the x-intercept, x_2, of L will approximate r.

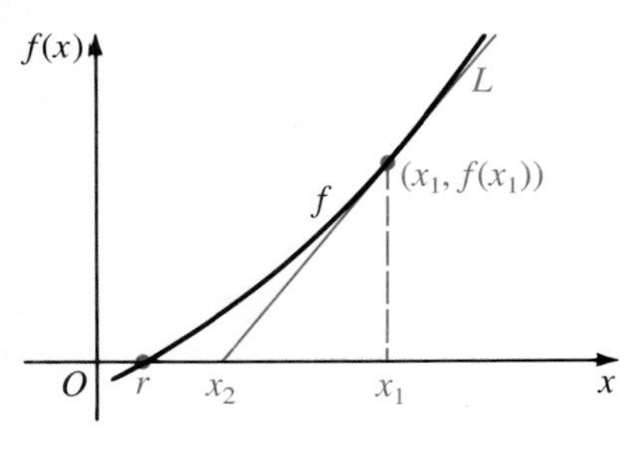

We can find x_2 as follows. Since the slope of L is $f'(x_1)$, L has the equation

$$y - f(x_1) = f'(x_1)(x - x_1).$$

Since L passes through $(x_2, 0)$, we have

$$0 - f(x_1) = f'(x_1)(x_2 - x_1)$$

$$\text{or} \quad -\frac{f(x_1)}{f'(x_1)} = x_2 - x_1.$$

From this we obtain x_2.

$$x_2 = x_1 - \frac{f(x_1)}{f'(x_1)}$$

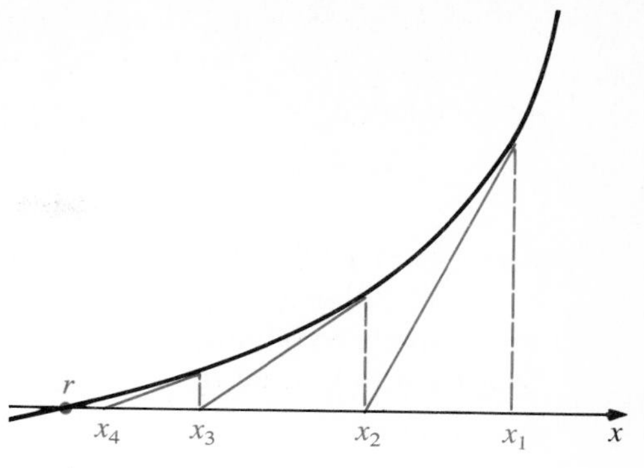

The figure at the right indicates how we can **iterate (repeat)** the method to obtain a third approximation x_3 from x_2 just as we obtained x_2 from x_1.

$$x_3 = x_2 - \frac{f(x_2)}{f'(x_2)}$$

In general we have **Newton's method.**

> If r is a real zero of a differentiable function f and x_n is an approximation to r, then the next approximation x_{n+1} is given by
>
> $$x_{n+1} = x_n - \frac{f(x_n)}{f'(x_n)}, \quad n = 1, 2, 3, \ldots$$

EXAMPLE 1 Approximate the positive zero of $f(x) = x^3 - 3x - 6$ to four decimal places.

Solution

$$f(x) = x^3 - 3x - 6 \quad \text{and} \quad f'(x) = 3x^2 - 3$$

Since $f(2) = -4$ and $f(3) = 12$, the zero lies between 2 and 3, by the location principle. Moreover, $f(2.3) = -0.733$ and $f(2.4) = 0.624$, so the zero lies between 2.3 and 2.4. Take

$$x_1 = 2.4$$

$$f(x_1) = 0.624 \quad \text{and} \quad f'(x_1) = 14.28$$

$$x_2 = x_1 - \frac{f(x_1)}{f'(x_1)} = 2.4 - \frac{0.624}{14.28} = 2.356$$

$$f(x_2) = 0.00953 \quad \text{and} \quad f'(x_2) = 13.652$$

$$x_3 = x_2 - \frac{f(x_2)}{f'(x_2)} = 2.356 - \frac{0.00953}{13.652} = 2.3553$$

Since $\frac{f(x_3)}{f'(x_3)} < 0.0000014$, we assume that to four decimal places the zero is 2.3553. **Answer**

Newton's method can be used to find $\sqrt{a}$ for any positive number a by approximating the positive zero of $f(x) = x^2 - a$. Since $f'(x) = 2x$, we have the following formulas for x_2 and x_{n+1}.

$$x_2 = x_1 - \frac{f(x_1)}{f'(x_1)} = x_1 - \frac{x_1^2 - a}{2x_1} = \frac{2x_1^2 - (x_1^2 - a)}{2x_1} = \frac{x_1^2 + a}{2x_1}.$$

This equation can be written as

$$x_2 = \frac{1}{2}\left(x_1 + \frac{a}{x_1}\right),$$

providing us with the **divide-and-average method** for finding square roots. Make an estimate, x_1, of $\sqrt{a}$, divide it into a, and average the quotient and the estimate to obtain a second approximation. The method can, of course, be iterated by using the recursion formula

$$x_{n+1} = \frac{1}{2}\left(x_n + \frac{a}{x_n}\right).$$

EXAMPLE 2 Use the divide-and-average method to approximate $\sqrt{14}$.

Solution Guess $x_1 = 3.8$ and use $x_2 = \frac{1}{2}\left(x_1 + \frac{14}{x_1}\right)$.

$$x_2 = \frac{1}{2}\left(3.8 + \frac{14}{3.8}\right) = \frac{1}{2}(3.8 + 3.684) = 3.742$$

$$x_3 = \frac{1}{2}\left(3.742 + \frac{14}{3.742}\right) = \frac{1}{2}(3.742 + 3.74131) = 3.74166$$

$\therefore \sqrt{14} \approx 3.74166$ **Answer**

Nowadays the divide-and-average method is mainly of historical interest because of the availability of electronic calculators. Newton's method, however, is still an important one and is suitable for use in digital computers. As we shall see in later chapters, the method is not restricted to the simple polynomial functions we consider in this section.

When you apply Newton's method, try to make the first estimate, x_1, a good one. The following figures indicate some situations that can occur if x_1 is not close enough to the desired zero r. (There are criteria for the sequence of approximations, $x_1, x_2, x_3, \ldots$, to approach r, but they are beyond the scope of this book.)

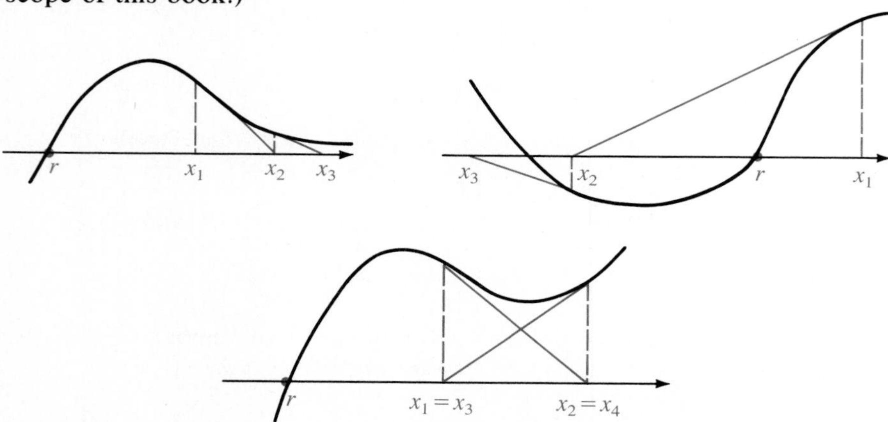

EXERCISES

Use the divide-and-average method or Newton's method to find the indicated roots to four decimal places.

A

1. $\sqrt{19}$
2. $\sqrt{7}$
3. $\sqrt[3]{10}$
4. $\sqrt[3]{-5}$
5. $\sqrt[4]{5}$
6. $\sqrt[4]{15}$
7. $\sqrt[5]{-10}$
8. $\sqrt[5]{8}$

Use Newton's method to approximate (to four decimal places) the indicated zero, r, of the function f.

9. $f(x) = x^3 + 2x - 6;\ 1 < r < 2$
10. $f(x) = x^3 - 5x + 3;\ 0 < r < 1$
11. $f(x) = x^3 - 2x^2 + 2;\ -1 < r < 0$
12. $f(x) = x^3 - x^2 - 2x - 4;\ 2 < r < 3$

B

13. $f(x) = x^4 - 2x - 3;\ 1 < r < 2$
14. $f(x) = x^4 - x^3 - 2;\ 1 < r < 2$
15. If $f(x) = ax + b$ $(a \neq 0)$, and x_1 is any real number, what is the x_2 of Newton's method?
16. Obtain a formula, similar to the divide-and-average formula, for $\sqrt[3]{a}$.
17. If $f(x) = x^2 - 6x + 9$, and x_1 is any real number, to what limit will the sequence $x_1, x_2, x_3, \ldots$ of Newton's method converge?
18. Let $f(x) = \dfrac{x}{x^2 + 1}$.
    - **a.** Find Newton's formula for f.
    - **b.** Show that if $x_n > 1$, then $x_{n+1} > 2x_n$ and hence that the sequence $x_1, x_2, x_3, \ldots$ does not converge.
    - **c.** Use the graph of f to explain this.

COMPUTER EXERCISES

1. Write and run a program using Newton's method to approximate to the nearest ten-thousandth a zero of $f(x) = x^2 - 3$ in $[1, 4]$. The user provides the first guess.
2. Modify and run the program in Exercise 1 for each function.
    - **a.** $f(x) = x^3 - 3$ in $[1, 4]$
    - **b.** $f(x) = x^4 + x - 3$ in $[0.5, 2.5]$
    - **c.** $f(x) = \dfrac{1}{x}$
    - **d.** $f(x) = 6x^4 - 23x^3 - 19x^2 + 4x$ in $[3, 5]$

Chapter Summary

1. The derivative of a function f at c is defined to be

$$f'(c) = \lim_{x \to c} \frac{f(x) - f(c)}{x - c} = \lim_{h \to 0} \frac{f(c + h) - f(c)}{h}$$

if the limit exists. If $f'(c)$ exists, f is said to be differentiable at c. The slope of the graph of f at the point $(c, f(c))$ is given by $f'(c)$. The function defined by $x \to f'(x)$ is called the *derivative function*, or *derived function*, of f. If f is differentiable at c, it is continuous at c.

2. The *power rule* $\frac{d}{dx} x^n = nx^{n-1}$,

$$(af + bg)' = af' + bg',\ (fg)' = fg' + gf', \text{ and } \left(\frac{f}{g}\right)' = \frac{gf' - fg'}{g^2}$$

can be used to differentiate many functions including polynomial functions.

3. If $y = f(x)$, the derivative can be written in several ways:

$$f'(x),\ y',\ \frac{d}{dx} f(x),\ \frac{d}{dx} y,\ \frac{dy}{dx}$$

Derivatives of higher order are written as:
$f''(x), f'''(x), f^{(4)}(x), \ldots, f^{(n)}(x), \ldots$ or $y'', y''', y^{(4)}, \ldots, y^{(n)}, \ldots$

4. The *generalized power rule*, $\frac{d}{dx} u^n = nu^{n-1} \frac{du}{dx}$, is a special case of the *chain rule*:

If $\varphi(x) = f(u)$ and $u = g(x)$, then $\varphi'(x) = f'(u) \cdot g'(x)$.

That is: If $y = f(u)$ and $u = g(x)$, then $\frac{dy}{dx} = \frac{dy}{du} \cdot \frac{du}{dx}$.

5. The derivative is a useful graphing tool. For example, a differentiable function f is *increasing (decreasing)* on an interval I if $f'(x)$ is positive (negative) on I. And $f'(x) = 0$ where f has a relative *extremum* (*maximum* or *minimum*). If $f'(c) = 0$, then c is a *stationary point* of f.

6. The graph of a differentiable function f is *concave upward (downward)* on an interval I if $f''(x)$ is positive (negative) on I. A point where the concavity changes sense is called an *inflection point*. The *second-derivative test* asserts that
If $f'(c) = 0$ and $f''(c) < 0$, then f has a relative maximum at c.
If $f'(c) = 0$ and $f''(c) > 0$, then f has a relative minimum at c.

7. Derivatives can be used to solve many practical *optimization problems*, in which some quantity is to be maximized or minimized. The steps listed on page 248 comprise a plan for solving optimization problems.

8. Many rates of change are given by derivatives. If the position of a particle is given by $s = f(t)$, then the velocity of the particle is given by $\frac{ds}{dt}$. If $\frac{ds}{dt} > 0$, the particle is moving to the right. If $\frac{ds}{dt} < 0$, the particle is moving

to the left. The acceleration of the particle is given by $\frac{d^2s}{dt^2}$. The list of steps on page 255 provides a strategy for solving related-rate problems.

9. To obtain a sequence of approximations to a zero of a function using *Newton's method*, start with a first approximation, x_1, and then use the formula $x_{n+1} = x_n - \frac{f(x_n)}{f'(x_n)}$ for $n = 1, 2, 3$, and so on.

Chapter Test

6-1 **1.** Let $f(x) = \frac{1}{2x + 1}$.

a. Use the definition of derivative to find $f'(x)$.
b. Find an equation of the line tangent to the graph of f at $(0, 1)$.

6-2 **2.** Differentiate each function.

a. $f(x) = 5x^4 - 3x^2 + 1$ **b.** $g(x) = (1 - x^2)\sqrt{x}$ **c.** $f(x) = \frac{x}{1 + x^2}$

6-3 **3.** Find $\frac{d}{dx}\frac{1}{(1 - x^2)^3}$.

4. Let $y = \frac{x}{1 - x}$.

a. Find y'. **b.** Write y' in terms of y.

6-4 **5.** Differentiate each function. In Exercise 5(b) write the answer in a form in which the numerator is rationalized.

a. $f(x) = (2x - x^2)^{10}$ **b.** $f(x) = \frac{\sqrt{x^2 - 1}}{x}$

6-5 **6.** Let $f(x) = x^3 + 3x^2 - 1$.

a. Find the intervals on which f is increasing and the intervals on which f is decreasing.
b. Find and classify the relative extrema of f.
c. Graph f.

6-6 **7.** Find the inflection point(s) of the graph in Exercise 6.

8. Use the second-derivative test to find the relative maxima and minima (if any) of $g(x) = \frac{x^2 + 4}{x}$.

6-7 **9.** An open-topped box having square horizontal cross section is to hold 108 cubic feet of grain when level full. What dimensions should it have if its total area (base plus four vertical sides) is to be as small as possible?

6-8 **10.** Wheat is falling at the rate of 6π ft³/min onto a level floor to form a conical pile whose height is always half its base radius. How fast is the base radius increasing when the pile is 3 feet high?

6-9 **11.** Use Newton's method to find the real zero of $f(x) = x^3 + 2x - 5$ correct to the nearest hundredth.

Extra

A Mean-Value Theorem

The figure at the right shows the chord $\overline{PQ}$ joining points $P(a, f(a))$ and $Q(b, f(b))$ of the graph of a differentiable function f. It seems clear from the figure that there is some point M on the graph of f at which the tangent line is parallel to $\overline{PQ}$. (To convince yourself of this, imagine $\overline{PQ}$ being translated either upward or downward until the chord touches the graph in exactly one point.)

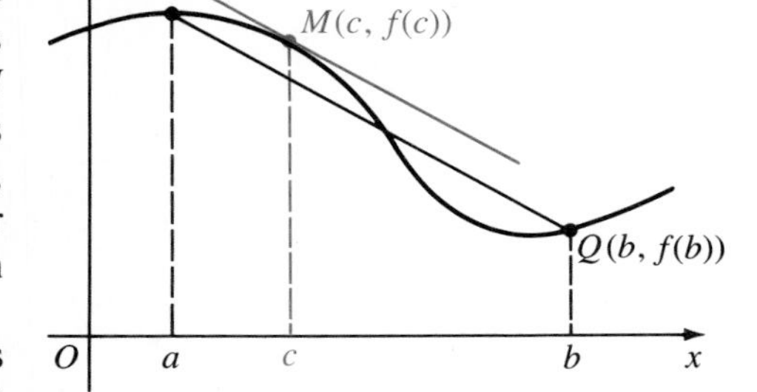

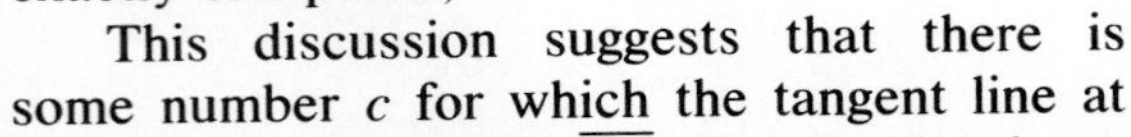

This discussion suggests that there is some number c for which the tangent line at $M(c, f(c))$ is parallel to $\overline{PQ}$. That is, the slope of the tangent at M, namely $f'(c)$, equals the slope of $\overline{PQ}$, namely $\dfrac{f(b) - f(a)}{b - a}$.

$$\frac{f(b) - f(a)}{b - a} = f'(c)$$

This result is called the *mean-value theorem of differential calculus.*

PROPOSITION 1 **The Mean-Value Theorem of Differential Calculus**
Let f be differentiable on (a, b) and continuous on $[a, b]$. Then there is a c in (a, b) such that

$$\frac{f(b) - f(a)}{b - a} = f'(c).$$

Although the truth of the mean-value theorem is obvious geometrically, its proof is too long for this book.

Among the many results that follow from the mean-value theorem is Theorem 6, on page 238, which we state again here.

PROPOSITION 2 The function f is increasing on an interval I if $f'(x) > 0$ on I; f is decreasing on I if $f'(x) < 0$ on I.

Proof: Suppose that $f'(x) > 0$ on I and let p and q be any two numbers in I with $p < q$. Since $f'(x)$ exists on I, the hypotheses of the mean-value theorem are satisfied. Therefore there is a c in (p, q) and hence in I

such that

$$\frac{f(q) - f(p)}{q - p} = f'(c), \quad \text{or} \quad f(q) - f(p) = (q - p)f'(c).$$

Since $q - p > 0$ and $f'(c) > 0$, $f(q) - f(p) > 0$ and therefore $f(p) < f(q)$. Thus f is increasing on I. The other part of the theorem can be proved in a similar way. ■

PROPOSITION 3 If $f'(x) = 0$ on an interval I, then f is constant on I.

Proof: If f is not constant on I, there must be numbers p and q of I such that $f(p) \neq f(q)$. Since the hypotheses of the mean-value theorem are satisfied, there is a number c between p and q, and hence in I, such that $\dfrac{f(q) - f(p)}{q - p} = f'(c)$. Since the numerator of the fraction is not 0, $f'(c) \neq 0$. This contradiction establishes the theorem. ■

PROPOSITION 4 If $F'(x) = G'(x)$ on an interval I, then F and G differ by a constant on I. That is, there is a constant C such that $G(x) = F(x) + C$ for each x in I.

EXERCISES

1. Let f' exist on $\mathcal{R}$. Show that if f has at least two distinct zeros, then there is a zero of f'.
2. Let f'' exist on $\mathcal{R}$. Show that if f has at least three distinct zeros, then f'' has at least one zero.
3. Show that if f is a quadratic function, then (in the notation of the mean-value theorem) $c = \dfrac{a + b}{2}$.
4. Prove Proposition 4. (*Hint*: Apply Proposition 3 to $F - G$.)
5. Give an indirect proof of Proposition 2.
6. Give a proof of Proposition 3 that does not involve drawing a contradiction.

Rolle's theorem states the following: Let f be differentiable on (a, b) and continuous on $[a, b]$, and suppose that $f(a) = f(b)$. Then there is a c in (a, b) such that $f'(c) = 0$.

7. Prove Rolle's theorem by using the mean-value theorem.
8. Prove the mean-value theorem by using Rolle's theorem. (*Hint*: Let $F(x) = f(x) - f(a) - \left[\dfrac{f(b) - f(a)}{b - a}\right](x - a)$. Note that $F(a) = F(b)$.)

COMPUTER APPLICATION

MINIMUM DISTANCE

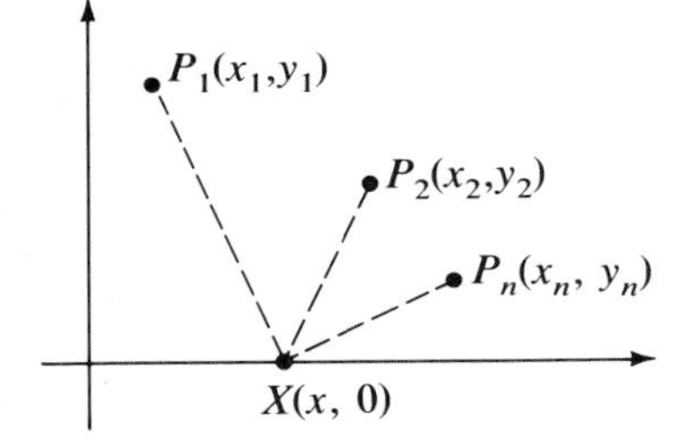

Suppose that a manufacturer wants to build a warehouse along a highway so that the sum of the distances from the warehouse to several stores is minimum. The figure at the right and the following analysis help locate the site of the warehouse.

Let $S(x)$ be the sum of the distances from the warehouse to the stores.

$$S(x) = \sqrt{(x - x_1)^2 + y_1^2} + \cdots + \sqrt{(x - x_n)^2 + y_n^2}$$

To find x such that $S(x)$ is minimum, solve $S'(x) = 0$.

$$S'(x) = \frac{x - x_1}{\sqrt{(x - x_1)^2 + y_1^2}} + \cdots + \frac{x - x_n}{\sqrt{(x - x_n)^2 + y_n^2}} = 0$$

This equation can be solved by applying Newton's method. To do so, $S''(x)$ must be found.

$$S''(x) = \frac{y_1^2}{((x - x_1)^2 + y_1^2)^{\frac{3}{2}}} + \cdots + \frac{y_n^2}{((x - x_n)^2 + y_n^2)^{\frac{3}{2}}}$$

The mth approximation to x, a_m, is given by $a_m = a_{m-1} - \dfrac{S'(a_{m-1})}{S''(a_{m-1})}$.

EXERCISES *(You may use either BASIC or Pascal.)*

1. Write a computer program to approximate x. The user supplies the coordinates of $(x_1, y_1), \ldots, (x_n, y_n)$.
2. Run the program in Exercise 1 for each set of points.
   a. $P_1(0, 4)$; $P_2(4, 4)$
   b. $P_1(2, 3)$; $P_2(8, 16)$
   c. $P_1(0, 5)$; $P_2(3, 8)$; $P_3(7, 5)$
   d. $P_1(0, 2)$; $P_2(2, 4)$; $P_3(3, 3)$; $P_4(4, 1)$

22. Does g fail to have a derivative anywhere? If so, where?

In Exercises 23 and 24, graph the function and state where it fails to be continuous. (Recall that $[\![x]\!]$ = greatest integer $\leq x$ and that $\operatorname{sgn} x = -1, 0,$ or 1 according as $x < 0$, $x = 0$, or $x > 0$.)

23. $x \to \operatorname{sgn} [\![x]\!]$ **24.** $x \to x - \operatorname{sgn} x$

25. State whether or not the indicated limit exists. If it does, give its value.
a. $\lim_{x\to 0} |\operatorname{sgn} x|$ **b.** $\lim_{x\to 0} (x \operatorname{sgn} x)$ **c.** $\lim_{x\to 0} (x + \operatorname{sgn} x)$

26. Prove the statement $\lim_{x\to 2} \frac{8 - x}{2} = 3$. Use the basic properties of limits to justify the steps in your proof.

27. Graph the function $g(x) = \frac{x^2}{1 - x^2}$. Show all asymptotes as dashed lines.

28. Use the location principle to locate the one positive root of $5x^3 - 13x^2 + 5x - 13 = 0$ between two consecutive integers.

29. Use the definition of derivative to find $f'(x)$ if $f(x) = \frac{1}{1 - x}$. Show all steps.

Differentiate, using appropriate formulas. Simplify.

30. $(x^2 + 1)\sqrt{x}$ **31.** $\frac{x}{1 - x^2}$

32. $(x^2 + 1)^{10}$ **33.** $x\sqrt{1 - x^2}$

34. Let $y = \frac{1}{1 - x}$. **a.** Find y', y'', and y'''. **b.** Write a formula for $y^{(n)}$. **c.** Use mathematical induction to prove that your answer to part (b) is correct.

In Exercises 35 and 36 find the intervals on which f is **(a)** increasing, **(b)** decreasing, **(c)** concave up, and **(d)** concave down. **(e)** Find and classify the stationary points of f. **(f)** Graph f.

35. $f(x) = x^3 - 3x^2 - 9x$ **36.** $f(x) = 4x^3 - x^4$

37. Use the second-derivative test to classify the relative extremum of $f(x) = \frac{x^3 - 2}{x}$.

38. Use Newton's method to approximate (to the nearest hundredth) the real zero of $f(x) = x^3 + 2x - 5$.

39. Find the dimensions of a cylinder having volume 54π cubic inches and having the least possible total area.

40. At noon a private plane takes off from airport A and flies due north at 200 mph. At 1:00 P.M. a second plane leaves A and flies due east at 300 mph. At what rate are the planes separating at 2:00 P.M.?

Chapter 7

Trigonometric Functions and Triangle Solving

The basic concepts of trigonometry are introduced in this chapter. A great deal of the chapter is devoted to applying trigonometry to solving triangles.

Circular Functions of Numbers

■ 7-1 THE SINE AND COSINE FUNCTIONS

You have probably used trigonometry to solve problems involving angles, such as finding the height of the tree shown at the right. Later we shall review and extend such geometric applications. It is a fact, however, that many modern applications of trigonometry have little or nothing to do with angles. We therefore start by defining some *circular functions*.

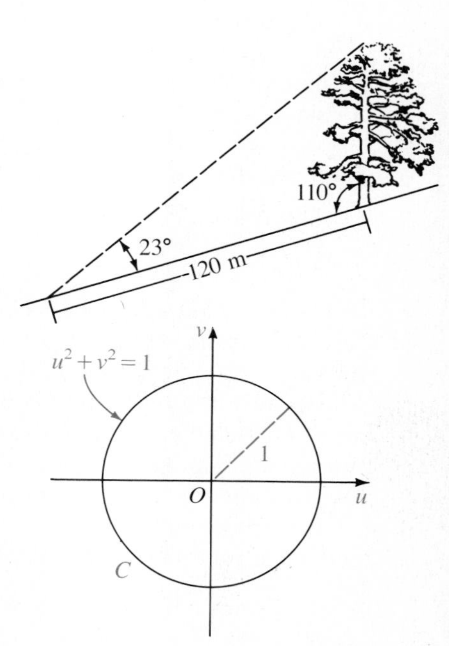

In giving the definitions, we shall use a coordinate plane in which the axes are named by the letters u and v. Thus the **unit circle**, C, the circle of radius 1 centered at (0, 0), has the equation

$$u^2 + v^2 = 1.$$

We shall assume that every arc of a circle has a definite length and that the circumference of the unit circle is 2π.

As suggested by this picture of the Trans-America Tower in San Francisco, the triangle is one of the forms most frequently used in the design of buildings.

We now associate with each real number x a unique point P_x of C as follows:

Start at $U(1, 0)$. Measure $|x|$ units along C in the counterclockwise direction if $x \geq 0$ or in the clockwise direction if $x < 0$. The point reached is P_x.

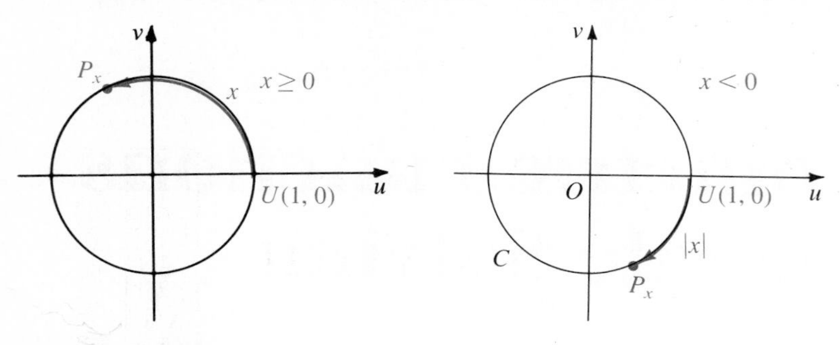

The correspondence

$$x \rightarrow P_x$$

between the set of all real numbers x and the points P_x on the unit circle is shown at the right. Notice that the x-axis has its origin at U. If this axis were flexible and wrapped around C, the graph of x would fall on P_x. Now let x be any real number.

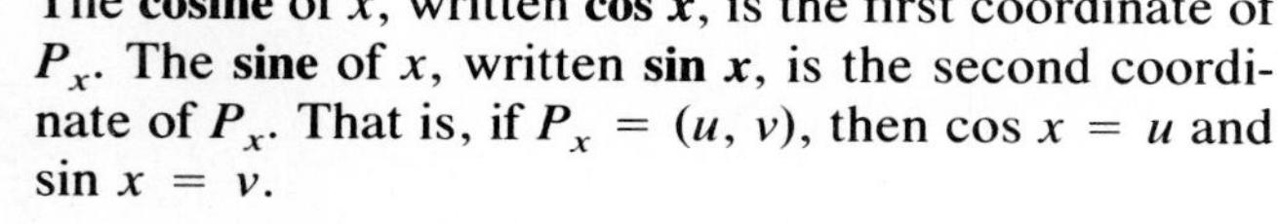

The **cosine** of x, written **cos *x***, is the first coordinate of P_x. The **sine** of x, written **sin *x***, is the second coordinate of P_x. That is, if $P_x = (u, v)$, then $\cos x = u$ and $\sin x = v$.

It follows that the coordinates of P_x are $(\cos x, \sin x)$.

The functions

$$x \rightarrow \sin x \quad \text{and} \quad x \rightarrow \cos x$$

have as their domains the set $\mathcal{R}$ of all real numbers. Since the values of these functions are coordinates of points on the unit circle, the range of each function is the closed interval $[-1, 1]$.

Because the unit circle is used in their definitions, the sine and cosine are often called **circular functions**, and we shall follow this practice for the time being. The relationship of these functions to trigonometric functions you have met in the past will be discussed in Section 7–4.

For some values of x, $\sin x$ and $\cos x$ are easy to find. The figure and tables that follow show P_x, $\sin x$, and $\cos x$ for various values of x. Since the shorter arc of C from $U(1, 0)$ to $(0, 1)$ has length $\frac{1}{4} \cdot 2\pi$, or $\frac{\pi}{2}$, $P_{\frac{\pi}{2}} = (0, 1)$. Thus $\cos \frac{\pi}{2} = 0$ and $\sin \frac{\pi}{2} = 1$. The other entries in Table 1 are obtained similarly.

v (0, 1) C (−1, 0) (1, 0) O u (0, −1)

Table 1

x	0	$\frac{\pi}{2}$	π	$\frac{3\pi}{2}$	2π
$\cos x$	1	0	-1	0	1
$\sin x$	0	1	0	-1	0

Table 2

x	0	$\frac{\pi}{6}$	$\frac{\pi}{4}$	$\frac{\pi}{3}$	$\frac{\pi}{2}$
$\cos x$	1	$\frac{\sqrt{3}}{2}$	$\frac{\sqrt{2}}{2}$	$\frac{1}{2}$	0
$\sin x$	0	$\frac{1}{2}$	$\frac{\sqrt{2}}{2}$	$\frac{\sqrt{3}}{2}$	1

The values given in Table 2 are derived in Exercises 31–34.

EXAMPLE 1 Find the sine and cosine of each number.

a. $\frac{2\pi}{3}$ **b.** $\frac{4\pi}{3}$ **c.** $-\frac{\pi}{3}$ **d.** $\frac{7\pi}{3}$

Solution Draw C for each number and use the fact that $\cos \frac{\pi}{3} = \frac{1}{2}$ and $\sin \frac{\pi}{3} = \frac{\sqrt{3}}{2}$.

a.

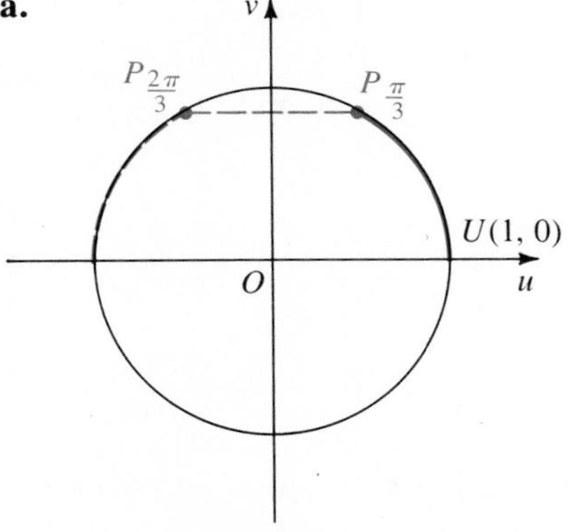

The points $P_{\frac{\pi}{3}}$ and $P_{\frac{2\pi}{3}}$ are symmetric with respect to the v-axis.

$\therefore \cos \frac{2\pi}{3} = -\frac{1}{2}, \sin \frac{2\pi}{3} = \frac{\sqrt{3}}{2}$

Answer

b.

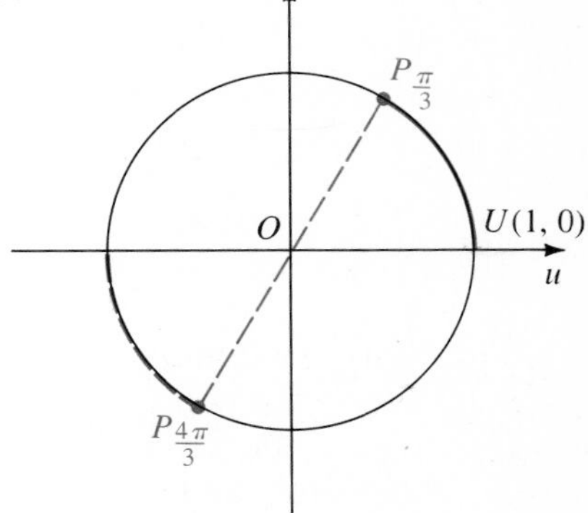

The points $P_{\frac{\pi}{3}}$ and $P_{\frac{4\pi}{3}}$ are symmetric with respect to the origin.

$\therefore \cos \frac{4\pi}{3} = -\frac{1}{2}, \sin \frac{4\pi}{3} = -\frac{\sqrt{3}}{2}$

Answer

Solution continued on following page

c.

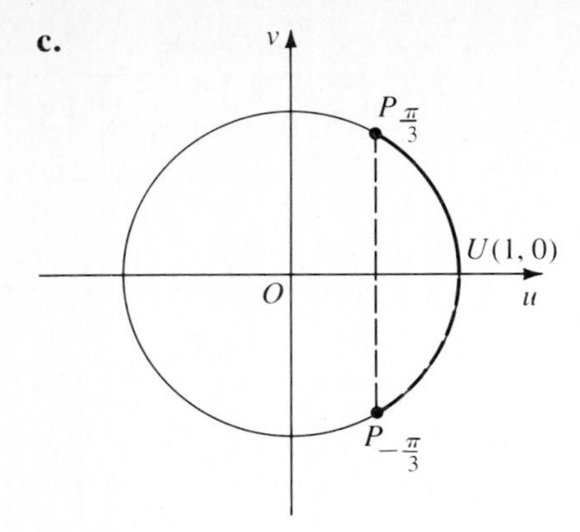

The points are symmetric with respect to the u-axis.

$\therefore \cos\left(-\frac{\pi}{3}\right) = \frac{1}{2}, \sin\left(-\frac{\pi}{3}\right) = -\frac{\sqrt{3}}{2}$. **Answer**

d.

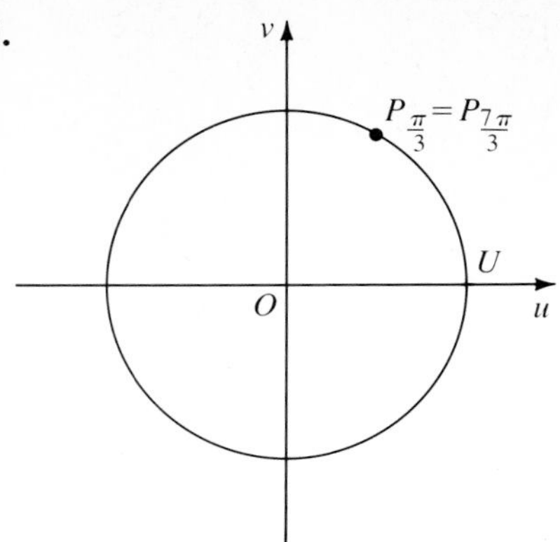

The points $P_{\frac{\pi}{3}}$ and $P_{\frac{7\pi}{3}}$ are the same since $\frac{7\pi}{3} = \frac{\pi}{3} + 2\pi$.

$\therefore \cos\frac{7\pi}{3} = \frac{1}{2}, \sin\frac{7\pi}{3} = \frac{\sqrt{3}}{2}$.

Answer

A number x is said to be a **first-quadrant number** if P_x is in the first quadrant, a **second-quadrant number** if P_x is in the second quadrant, and so on. The quadrant of x thus determines the signs of the coordinates of P_x and therefore of $\cos x$ and $\sin x$. If P_x lies on a coordinate axis, x is called a **quadrantal number**.

Quadrant II	Quadrant I
$\cos x < 0$	$\cos x > 0$
$\sin x > 0$	$\sin x > 0$
Quadrant III	Quadrant IV
$\cos x < 0$	$\cos x > 0$
$\sin x < 0$	$\sin x < 0$

Because for each x, the point $(\cos x, \sin x)$ is on the circle whose equation is $u^2 + v^2 = 1$, we see that

$$\cos^2 x + \sin^2 x = 1.$$

($\cos^2 x$ means $(\cos x)^2$.)

EXAMPLE 2 If $\cos x = \frac{5}{6}$ and $3\pi < x < 4\pi$, find **(a)** the quadrant number of x and **(b)** $\sin x$.

Solution **a.** Since $3\pi = 2\pi + \pi$ and $4\pi = 2\pi + 2\pi$, x is either a third- or a fourth-quadrant number. Since $\cos x > 0$, x is a fourth-quadrant number.

b. Because $\cos x = \frac{5}{6}$ and $\cos^2 x + \sin^2 x = 1$,

$$\sin^2 x = 1 - \cos^2 x = 1 - \frac{25}{36} = \frac{11}{36}$$

$$\sin x = \pm\frac{\sqrt{11}}{6}.$$

Since x is a fourth-quadrant number, $\sin x < 0$. $\therefore \sin x = -\frac{\sqrt{11}}{6}$.

EXERCISES

In Exercises 1–8, find $\cos x$ and $\sin x$.

A **1.**

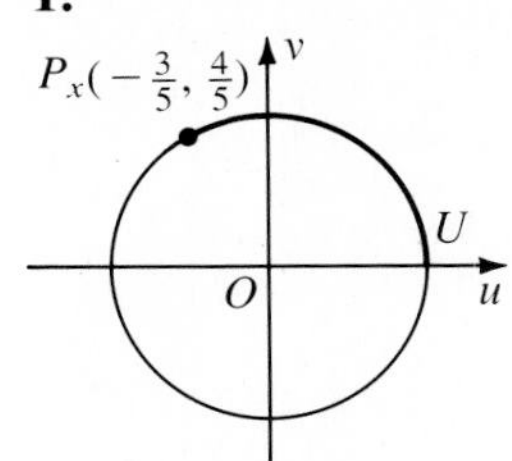

2.

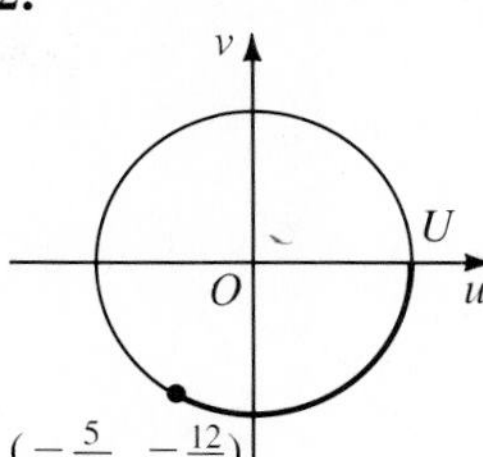

3. **4.**

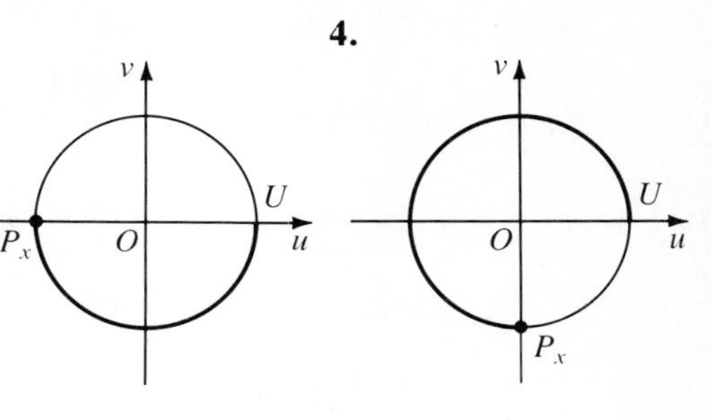

5.

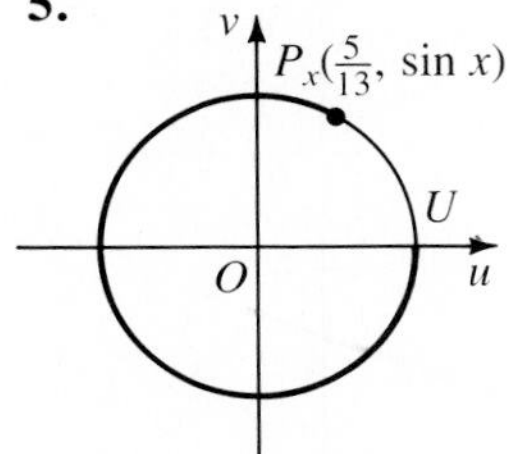

6.

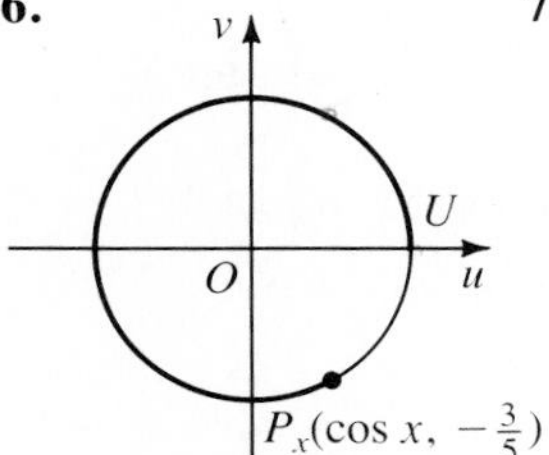

7. **8.**

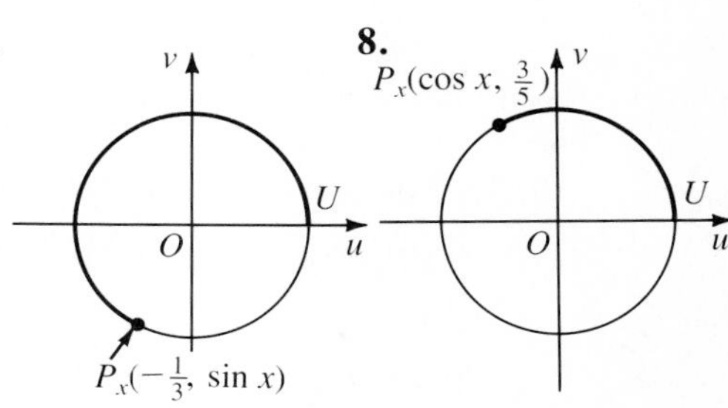

Find the cosine and sine of each of the following numbers.

9. $\frac{3\pi}{4}$ **10.** $\frac{5\pi}{3}$ **11.** $\frac{11\pi}{6}$ **12.** $\frac{7\pi}{6}$

13. $-\frac{\pi}{3}$ **14.** $-\frac{5\pi}{4}$ **15.** $-\frac{11\pi}{6}$ **16.** $\frac{16\pi}{3}$

In Exercises 17–26, either $\cos x$ or $\sin x$ is given. Find the other.

17. $\sin x = \frac{4}{5}$, x a second-quadrant number.

18. $\sin x = -\frac{5}{13}$, x a fourth-quadrant number.

19. $\cos x = -\frac{2}{\sqrt{5}}$, x a third-quadrant number.

20. $\cos x = -\frac{\sqrt{5}}{3}$, x a second-quadrant number.

21. $\sin x = -\frac{2}{3}$, $\cos x > 0$ **22.** $\cos x = \frac{3}{4}$, $\sin x < 0$

B **23.** $\sin x = -\frac{1}{3}$, $|x| < \frac{\pi}{2}$ **24.** $\cos x = -\frac{2}{5}$, $\pi < x < 2\pi$

25. $\cos x = \frac{3\sqrt{5}}{7}$, $\frac{\pi}{2} < x < 2\pi$ **26.** $\sin x = -\frac{2\sqrt{10}}{11}$, $-\frac{\pi}{2} < x < \pi$

Show that the following statements are true.

27. **a.** $\sin \frac{\pi}{3} = 2 \sin \frac{\pi}{6} \cos \frac{\pi}{6}$ **b.** $\sin \frac{2\pi}{3} = 2 \sin \frac{\pi}{3} \cos \frac{\pi}{3}$

28. **a.** $\cos \frac{4\pi}{3} = \cos^2 \frac{2\pi}{3} - \sin^2 \frac{2\pi}{3}$ **b.** $\cos \left(-\frac{2\pi}{3}\right) = \cos^2 \left(-\frac{\pi}{3}\right) - \sin^2 \left(-\frac{\pi}{3}\right)$

29. **a.** $\cos \frac{5\pi}{3} = 2 \cos^2 \frac{5\pi}{6} - 1$ **b.** $\cos \frac{3\pi}{2} = 2 \cos^2 \frac{3\pi}{4} - 1$

30. **a.** $\cos \frac{3\pi}{2} = 1 - 2 \sin^2 \frac{3\pi}{4}$ **b.** $\cos \frac{5\pi}{3} = 1 - 2 \sin^2 \frac{5\pi}{6}$

31. Show that $\sin \frac{\pi}{6} = \frac{1}{2}$ as follows:

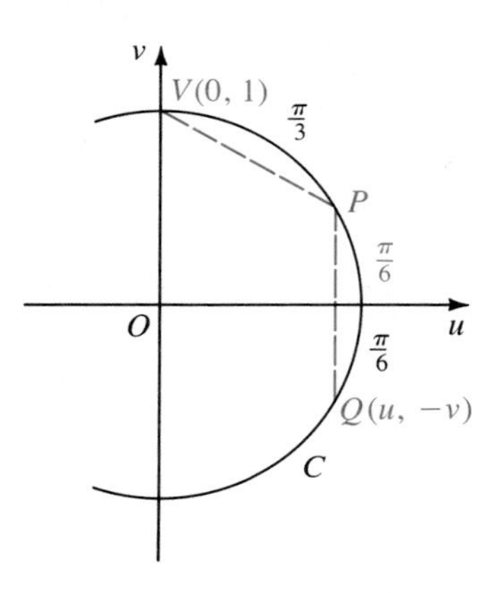

In the adjacent figure let

$$P = P_{\frac{\pi}{6}} = (u, v),$$

so that $v = \sin \frac{\pi}{6}$. Let $Q = P_{-\frac{\pi}{6}} = (u, -v)$ and $V =$ (0, 1). Since arcs QP and PV both have measures $\frac{\pi}{3}$, $QP = VP$. Justify the following steps.

$$(2v)^2 = u^2 + (v - 1)^2$$
$$4v^2 = u^2 + v^2 - 2v + 1 = 1 - 2v + 1$$
$$2v^2 + v - 1 = 0$$
$$(2v - 1)(v + 1) = 0; \quad \therefore v = \tfrac{1}{2} \text{ or } -1$$

Since $v > 0$, $v = \sin \frac{\pi}{6} = \frac{1}{2}$.

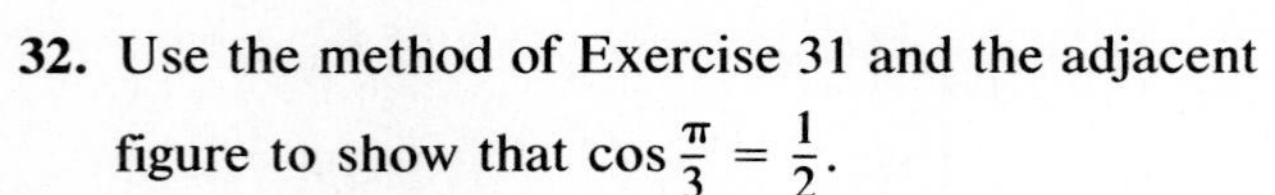

32. Use the method of Exercise 31 and the adjacent figure to show that $\cos \frac{\pi}{3} = \frac{1}{2}$.

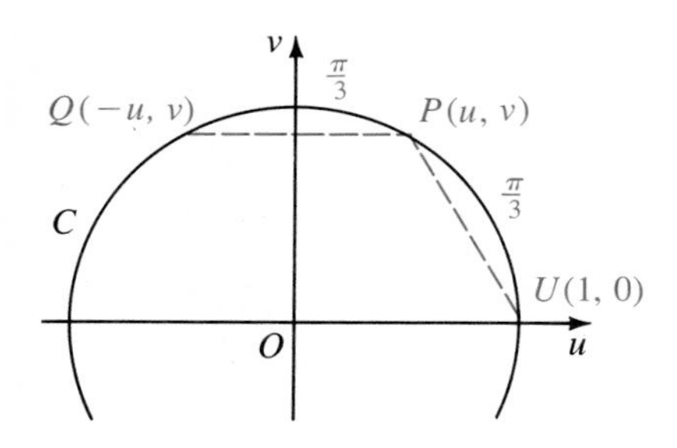

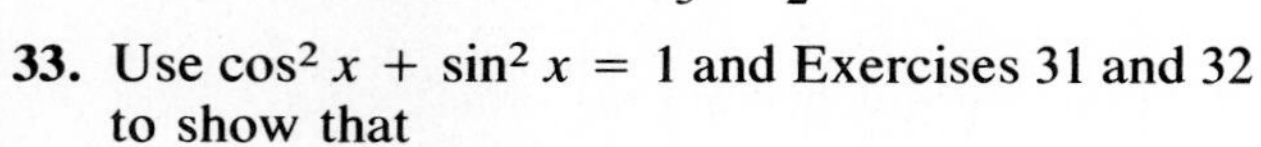

33. Use $\cos^2 x + \sin^2 x = 1$ and Exercises 31 and 32 to show that

(a) $\cos \frac{\pi}{6} = \frac{\sqrt{3}}{2}$ and **(b)** $\sin \frac{\pi}{3} = \frac{\sqrt{3}}{2}$.

34. Show that $\cos \frac{\pi}{4} = \sin \frac{\pi}{4} = \frac{\sqrt{2}}{2}$. (*Hint*: The line $v = u$ intersects the circle C: $u^2 + v^2 = 1$ in the point $P_{\frac{\pi}{4}}$.)

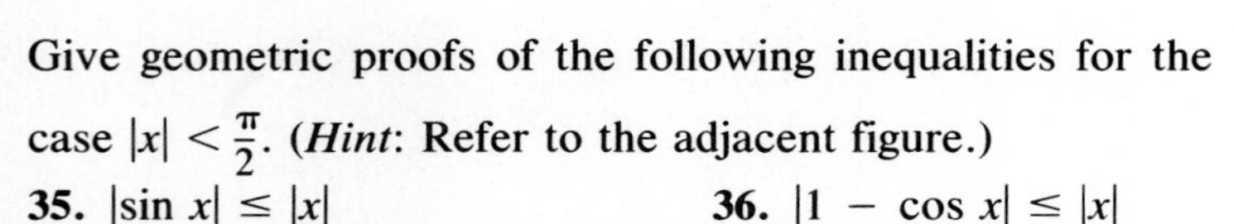

Give geometric proofs of the following inequalities for the case $|x| < \frac{\pi}{2}$. (*Hint*: Refer to the adjacent figure.)

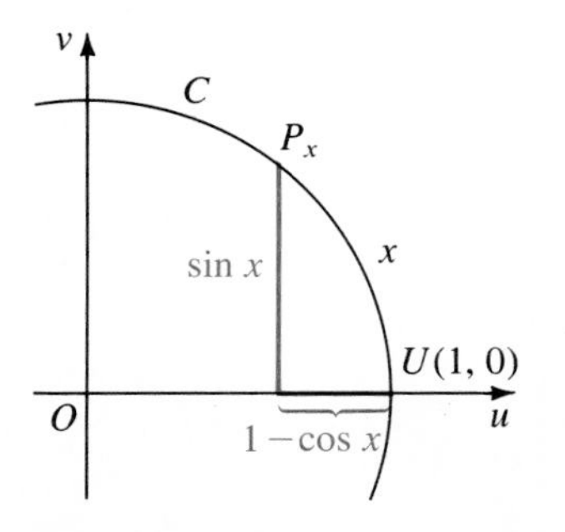

C **35.** $|\sin x| \leq |x|$ **36.** $|1 - \cos x| \leq |x|$

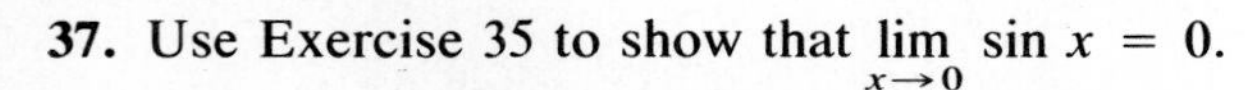

37. Use Exercise 35 to show that $\lim_{x \to 0} \sin x = 0$.

38. Use Exercise 36 to show that $\lim_{x \to 0} \cos x = 1$.

7–2 THE OTHER CIRCULAR FUNCTIONS

The remaining circular functions, the **tangent**, the **cotangent**, the **secant**, and the **cosecant**, can be defined in terms of the sine and cosine.

$$\tan x = \frac{\sin x}{\cos x} \text{ if } \cos x \neq 0 \qquad \cot x = \frac{\cos x}{\sin x} \text{ if } \sin x \neq 0$$

$$\sec x = \frac{1}{\cos x} \text{ if } \cos x \neq 0 \qquad \csc x = \frac{1}{\sin x} \text{ if } \sin x \neq 0$$

EXAMPLE 1 Find $\tan \frac{2\pi}{3}$, $\cot \frac{2\pi}{3}$, $\sec \frac{2\pi}{3}$, and $\csc \frac{2\pi}{3}$.

Solution From Example 1(a), page 273, $\cos \frac{2\pi}{3} = -\frac{1}{2}$ and $\sin \frac{2\pi}{3} = \frac{\sqrt{3}}{2}$.

Therefore

$$\tan \frac{2\pi}{3} = \frac{\sin \frac{2\pi}{3}}{\cos \frac{2\pi}{3}} = \frac{\frac{\sqrt{3}}{2}}{-\frac{1}{2}} = -\sqrt{3}; \quad \cot \frac{2\pi}{3} = \frac{\cos \frac{2\pi}{3}}{\sin \frac{2\pi}{3}} = \frac{-\frac{1}{2}}{\frac{\sqrt{3}}{2}} = -\frac{\sqrt{3}}{3};$$

$$\sec \frac{2\pi}{3} = \frac{1}{\cos \frac{2\pi}{3}} = \frac{1}{-\frac{1}{2}} = -2; \quad \csc \frac{2\pi}{3} = \frac{1}{\sin \frac{2\pi}{3}} = \frac{1}{\frac{\sqrt{3}}{2}} = \frac{2\sqrt{3}}{3}.$$

It is easy to see why the secant and cosine are called reciprocal functions, as are the cosecant and sine and the tangent and cotangent. Two functions that are reciprocals of each other have the same algebraic sign wherever both are defined. The following table gives the signs of the six circular functions.

Quadrant of x →	I	II	III	IV
$\sin x$ and $\csc x$	+	+	−	−
$\cos x$ and $\sec x$	+	−	−	+
$\tan x$ and $\cot x$	+	−	+	−

In Section 7–1 we showed that for all x,

$$\cos^2 x + \sin^2 x = 1.$$

If $\cos x \neq 0$, we may divide both sides of this equation by $\cos^2 x$ to obtain $1 + \frac{\sin^2 x}{\cos^2 x} = \frac{1}{\cos^2 x}$ or

$$1 + \tan^2 x = \sec^2 x \quad (\cos x \neq 0).$$

By dividing each side of $\cos^2 x + \sin^2 x = 1$ by $\sin^2 x$, we can show that

$$1 + \cot^2 x = \csc^2 x \quad (\sin x \neq 0)$$

EXAMPLE 2 Find $\sin x$ and $\cos x$ given that $\tan x = -\frac{12}{5}$ and $\sin x > 0$.

Solution Use $\sec^2 x = 1 + \tan^2 x$.

$$\begin{aligned}\sec^2 x &= 1 + \tan^2 x\\ &= 1 + \left(-\frac{12}{5}\right)^2\\ &= \frac{169}{25}\\ \sec x &= \pm\frac{13}{5}\end{aligned}$$

Since $\tan x < 0$ and $\sin x > 0$, x must be a second-quadrant number. From the table of signs, $\sec x < 0$. Thus $\sec x = -\frac{13}{5}$. Therefore

$$\cos x = \frac{1}{\sec x} = -\frac{5}{13}. \qquad \textbf{Answer}$$

Since $\sin x > 0$,

$$\sin x = \sqrt{1 - \cos^2 x} = \sqrt{1 - \frac{25}{169}} = \frac{12}{13}. \qquad \textbf{Answer}$$

Notice that since the remaining circular functions are defined in terms of the sine and cosine functions, we could go on to find the values of the cotangent and the cosecant functions in Example 2.

Since $(\cos x, \sin x)$ and $(\cos(-x), \sin(-x))$ are points symmetric with respect to the u-axis, we see that

$$\cos(-x) = \cos x \text{ and } \sin(-x) = -\sin x.$$

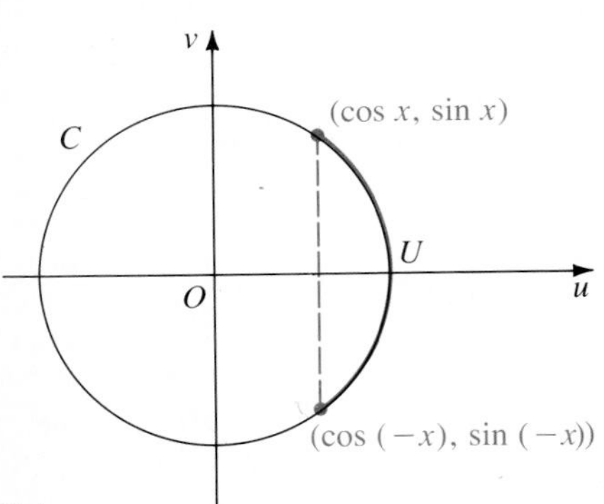

For example,

$$\cos\left(-\frac{\pi}{3}\right) = \cos\frac{\pi}{3} \text{ and } \sin\left(-\frac{\pi}{3}\right) = -\sin\frac{\pi}{3}.$$

We say that the cosine is an *even* function and that the sine is an *odd* function. In general,

f is an **even function** if $f(-x) = f(x)$

and f is an **odd function** if $f(-x) = -f(x)$

for every x in the domain of f.

Since $\tan(-x) = \frac{\sin(-x)}{\cos(-x)} = \frac{-\sin x}{\cos x} = -\tan x$, the tangent is an odd function. It is easy to show that the cosine and secant functions are even functions and that the other four circular functions are odd functions.

Using some geometry we can find the ranges of the other circular functions. Consider the tangent function over the interval $0 < x < \frac{\pi}{2}$. Let

$U = (1, 0)$, $P = (\cos x, \sin x)$, and $M = (\cos x, 0)$. Let Q be the point where the line $\overleftrightarrow{OP}$ meets the line $u = 1$. Then triangles OUQ and OMP are similar.

Thus $\quad \dfrac{UQ}{MP} = \dfrac{OU}{OM}$ or $\dfrac{UQ}{\sin x} = \dfrac{1}{\cos x}$.

Hence $\quad UQ = \dfrac{\sin x}{\cos x} = \tan x.$

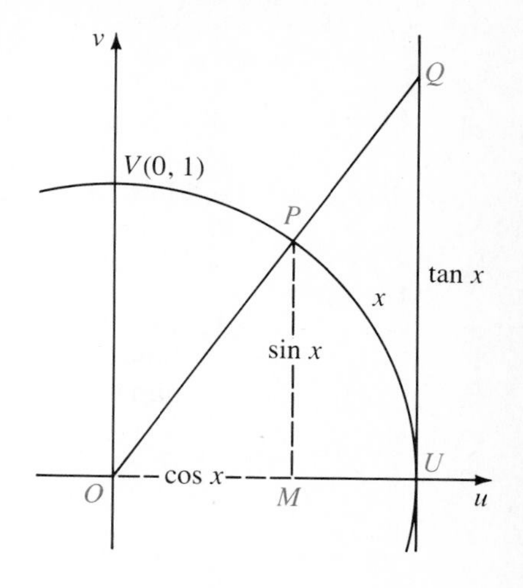

From the figure we see that as x increases from 0 to $\frac{\pi}{2}$, P moves along C from U to $V(0, 1)$, and $UQ = \tan x$ increases from 0 through all positive values. On the other hand, as x decreases from 0 to $-\frac{\pi}{2}$, $-x$ increases from 0 to $\frac{\pi}{2}$. Therefore $\tan x = -\tan(-x)$ decreases from 0 through all negative values. Thus, the range of the tangent function is the set, $\mathcal{R}$, of all real numbers.

The following table gives the domains and the ranges of the six circular

Function	*Domain*	*Range*
Sine	$\mathcal{R}$	$\{y: \lvert y\rvert \le 1\}$
Cosine	$\mathcal{R}$	$\{y: \lvert y\rvert \le 1\}$
Tangent	$\{x: \cos x \neq 0\}$	$\mathcal{R}$
Cotangent	$\{x: \sin x \neq 0\}$	$\mathcal{R}$
Secant	$\{x: \cos x \neq 0\}$	$\{y: \lvert y\rvert \ge 1\}$
Cosecant	$\{x: \sin x \neq 0\}$	$\{y: \lvert y\rvert \ge 1\}$

Notice that the tangent function is not defined for any value of x such that $\cos x = 0$. This means that $x = \dfrac{(2k + 1)\pi}{2}$ for any integer k is excluded from the domain. In Section 8–9 we shall see that the graph of the tangent function has a vertical asymptote at each value of x excluded from the domain of that function.

EXERCISES

Give the quadrant of the number x.

A

1. $\tan x > 0$, $\cos x < 0$
2. $\cos x > 0$, $\csc x < 0$
3. $\sin x > 0$, $\cot x < 0$
4. $\csc x > 0$, $\sec x > 0$
5. $\sec x > 0$, $\sin x < 0$
6. $\cos x < 0$, $\csc x < 0$
7. $\sec x < 0$, $\csc x > 0$
8. $\tan x < 0$, $\cos x > 0$
9. $\cot x > 0$, $\sin x < 0$

Evaluate $\tan x$, $\cot x$, $\sec x$, and $\csc x$ at the given values of x. ($\sin x$ and $\cos x$ were evaluated in Exercises 9–16, page 275.)

10. $\frac{3\pi}{4}$ **11.** $\frac{5\pi}{3}$ **12.** $\frac{11\pi}{6}$ **13.** $\frac{7\pi}{6}$

14. $-\frac{\pi}{3}$ **15.** $-\frac{5\pi}{4}$ **16.** $-\frac{11\pi}{6}$ **17.** $\frac{16\pi}{3}$

Show that the following statements are true.

18. $\tan \frac{\pi}{3} = \dfrac{2 \tan \frac{\pi}{3}}{1 - \tan^2 \frac{\pi}{6}}$

19. $\tan \frac{4\pi}{3} = \dfrac{2 \tan \frac{2\pi}{3}}{1 - \tan^2 \frac{2\pi}{3}}$

20. $\tan \frac{2\pi}{3} = \dfrac{\sin \frac{4\pi}{3}}{1 + \cos \frac{4\pi}{3}}$

21. $\tan \frac{\pi}{3} = \dfrac{1 - \cos \frac{\pi}{6}}{\sin \frac{\pi}{6}}$

Use the given information to find the values of the remaining five circular functions of x.

22. $\tan x = \frac{4}{3}$, $\cos x < 0$

23. $\sec x = \frac{5}{3}$, $\sin x < 0$

24. $\cos x = -\frac{5}{13}$, $\tan x < 0$

25. $\sin x = \frac{7}{25}$, $\sec x < 0$

26. $\csc x = \sqrt{5}$, $\cos x > 0$

27. $\csc x = -\sqrt{10}$, $\cot x > 0$

28. $\sec x = 3$, $\pi < x < 2\pi$

29. $\tan x = -2$, $0 < x < \pi$

Determine whether the function described is even, odd, or neither.

EXAMPLE **a.** $f(x) = x^3 \sin x$ **b.** $g(x) = 2x^2 - 4x$

Solution **a.** $f(-x) = (-x)^3 \sin(-x) = (-x^3)(-\sin x) = x^3 \sin x = f(x)$
$\therefore f$ is even.

b. $g(-x) = 2(-x)^2 - 4(-x) = 2x^2 + 4x \neq g(x)$ or $-g(x)$
$\therefore g$ is neither even nor odd.

B

30. a. $f(x) = x \cos x$ **b.** $g(x) = x^2 + \sin x$

31. a. $f(x) = \sin x + \cos x$ **b.** $g(x) = \sin x \cos x$

32. a. $f(x) = x \sin^3 x$ **b.** $g(x) = x\sqrt{x^2 + 1}$

33. a. $f(x) = \dfrac{\sin x}{1 + x^2}$ **b.** $g(x) = x^2 + \sec^3 x$

Show that the following inequalities are true.

34. $\sin x \le \tan x \le \sec x$ if $0 < x < \frac{\pi}{2}$.

35. $\cos x \le \cot x \le \csc x$ if $0 < x < \frac{\pi}{2}$.

C **36.** Give an argument, based in part on the figure on page 279, to show that the range of the secant function is $\{y: |y| \geq 1\}$ by proving that $\sec x = OQ$.

37. Construct a figure similar to the figure on page 279 to show the relationship of $\cot x$ and $\csc x$ $\left(0 < x < \frac{\pi}{2}\right)$ to the unit circle C. (*Hint*: Let R be the point where the line tangent to C at the point $P(\cos x, \sin x)$ intersects the y-axis. Consider triangle OPR.)

Give arguments, based in part on the figure obtained in Exercise 37, of the following statements.

38. The range of the cotangent function is $\mathcal{R}$.

39. The range of the cosecant function is $\{y: |y| \geq 1\}$.

COMPUTER EXERCISES

1. Write a computer program to evaluate $\cos x - x$ for an input value of x.

2. Write and run a computer program to approximate a solution to $\cos x - x = 0$, where $0 < x < \frac{\pi}{2}$. The program should ask the user to enter a small tolerance T, say $T = 0.01$, and continue to increment x by 0.001 until $|\cos x - x| < T$. A value of x such that $|\cos x - x| < T$ should be printed as an approximate solution to the equation $\cos x - x = 0$.

Trigonometric Functions of Angles

7-3 ANGLES AND ANGLE MEASURE

In order to compare the circular functions and the traditional trigonometric functions we must first discuss angles and angle measure.

In Euclidean geometry an **angle** is simply the union of two rays having a common endpoint. This point is the **vertex** of the angle, and the rays are its **sides**. In trigonometry, however, we often must regard an angle as being generated by a rotation of one side, called the **initial side**, into the other, the **terminal side**. Such rotations can be indicated by curved arrows as shown in the following figures.

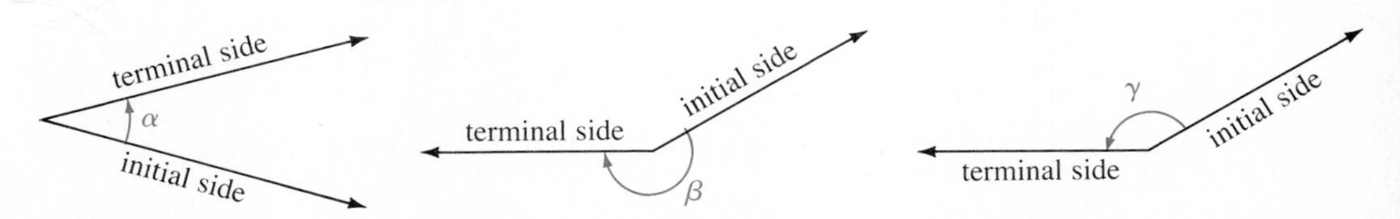

In a given plane, angles having counterclockwise rotations such as α and γ are considered *positive angles*. An angle such as β having a clockwise rotation is considered a *negative angle*.

We now define *radian measure*, which is the basic system of angle measurement in advanced mathematics. (In Chapter 9 we will see that radian measure has definite advantages over other measures.) Given any angle θ let C be a circle of radius r centered at the vertex of θ. Let s be the length of the arc of C intercepted by θ. We take s to be positive if θ is positive, and we take s to be negative if θ is negative. The **radian measure** of θ is $\frac{s}{r}$. We express this by the formula

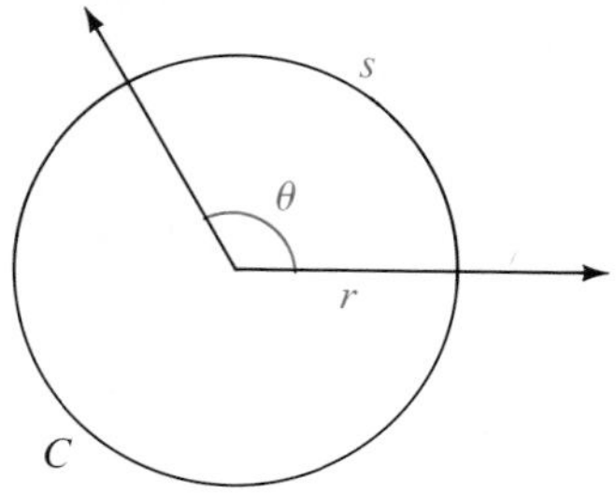

$$\theta = \frac{s}{r}.$$

The symbol θ is used here to denote the measure of an angle as well as the angle itself. It is customary to do this when there is no chance of confusion.

The radian measure of θ does not depend on the size of the circle C used in the definition. Because in similar figures corresponding lengths are proportional, $\frac{s'}{s} = \frac{r'}{r}$. Thus $\frac{s'}{r'} = \frac{s}{r}$ and the ratio is constant for a given θ.

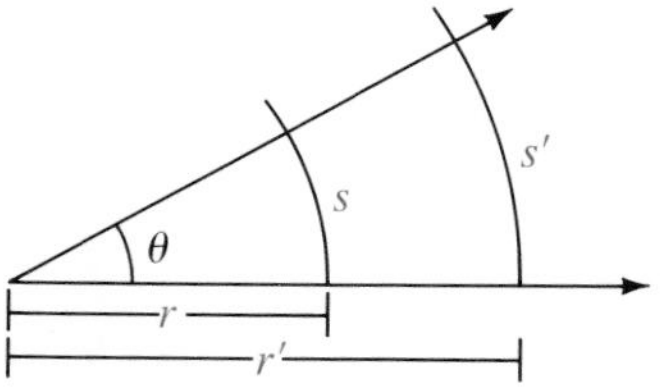

Since the circumference of a circle of radius r is $2\pi r$, the radian measure of a complete revolution is $\frac{2\pi r}{r} = 2\pi$. Measures of other angles are proportional to 2π. In particular the radian measure of a right angle is $\frac{1}{4}(2\pi) = \frac{\pi}{2}$. We usually express radian measure in terms of π. For example, we would write $\theta = \frac{\pi}{2}$ rather than $\theta = 1.571$, which is only an approximation to $\frac{\pi}{2}$.

In many applications it is traditional and convenient to measure angles in **degrees**. To do this, we arbitrarily decide that a right angle shall have measure 90 degrees, written 90°, with the degree measure of any other angle in proportion. To relate degree measure to radian measure, we may take the following point of view: Since a right angle has measure 90° and also has measure $\frac{\pi}{2}$, we may regard 90° as just another way of writing the number $\frac{\pi}{2}$, so that $90° = \frac{\pi}{2}$. Thus, we have the useful conversion formula

$$180° = \pi.$$

When no unit of angle measure is specified, radian measure is being used.

EXAMPLE 1 **a.** Convert each degree measure to radian measure.

$$1°, \quad 60°, \quad -100°$$

b. Convert each radian measure to degree measure.

$$1, \quad \frac{3\pi}{4}, \quad -\frac{\pi}{6}$$

Solution Use the formula $180° = \pi$.

a. $1° = \frac{\pi}{180}$; $60° = 60 \cdot \frac{\pi}{180} = \frac{\pi}{3}$; $-100° = -100 \cdot \frac{\pi}{180} = -\frac{5\pi}{9}$

b. $1 = \frac{180°}{\pi}$; $\frac{3\pi}{4} = \frac{3\pi}{4} \cdot \frac{180°}{\pi} = 135°$; $-\frac{\pi}{6} = -\frac{\pi}{6} \cdot \frac{180°}{\pi} = -30°$

Notice from Example 1(b) that 1 radian is approximately 57.3°.

Radian measure enables us to write simple formulas for **circular-arc length** (s) and **circular-sector area** (A) illustrated in the following figures.

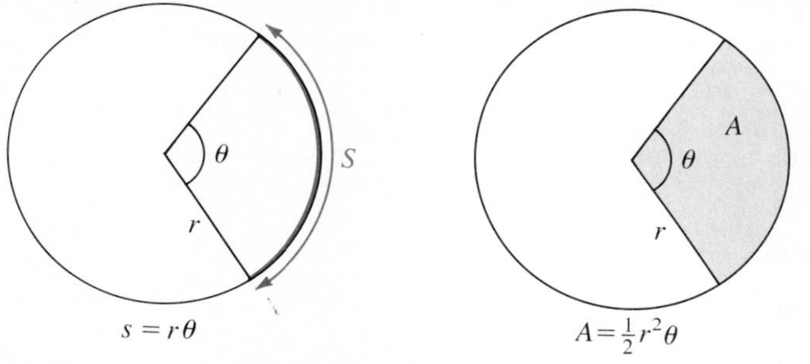

To prove the area formula we use the fact that the areas of sectors of a circle are proportional to the measures of their *central angles*. Since the area of the entire circle is πr^2, we have $\frac{A}{\theta} = \frac{\pi r^2}{2\pi}$. Thus

$$A = \frac{1}{2}r^2\theta.$$

EXAMPLE 2 A central angle of a circle of radius 6.50 cm measures 72°. Find **(a)** the length of the arc it intercepts and **(b)** the area of the sector it defines. Give answers to three significant digits.

Solution $r = 6.50$ and $\theta = 72 \times \frac{\pi}{180} = \frac{2\pi}{5} \approx 1.257$.

a. $s = r\theta = 6.50 \times 1.257 = 8.17$ cm **Answer**

b. $A = \frac{1}{2}r^2\theta = \frac{1}{2}(6.50)^2 \times 1.257 = 26.6$ cm^2 **Answer**

A point P that moves with constant speed v in a circular path of radius r is said to describe **uniform circular motion**. As shown in the figure at the top of the next page, let $\overline{OA}$ be a fixed radius of the path, let θ be the radian

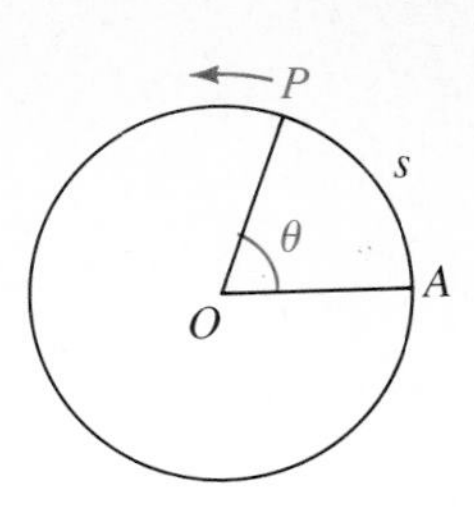

measure of $\angle AOP$, and let s be the length of the arc $\overset{\frown}{AP}$ t seconds after P was at A. Since $s = r\theta$, we have $\frac{s}{t} = r\frac{\theta}{t}$, or

$$v = r\omega$$

where $v = \frac{s}{t}$ is the (linear) speed of P, and $\omega = \frac{\theta}{t}$ is its **angular speed** in radians per second. Of course other units can be used to measure angular speed; revolutions per minute (rpm), for example.

EXAMPLE 3 Find the speed in miles per hour of a point on the rim of a record 12 inches in diameter turning at the rate of $33\frac{1}{3}$ rpm.

Solution $\omega = 33\frac{1}{3}\text{ rpm} = \frac{100}{3} \cdot 2\pi = \frac{200\pi}{3}$ radians/min

Hence $v = r\omega = 6 \cdot \frac{200\pi}{3} = 400\pi$ in./min

$$= \frac{400\pi}{12 \times 5280} \text{ mi/min}$$

$$v = \frac{400\pi}{12 \times 5280} \cdot 60 = 1.19 \text{ mi/h}$$ **Answer**

EXERCISES

A

1. Convert each degree measure to radian measure. Leave each answer in terms of π.
a. $30°$ **b.** $135°$ **c.** $210°$ **d.** $-60°$ **e.** $300°$

2. Convert each radian measure to degree measure.
a. $\frac{3\pi}{2}$ **b.** $\frac{\pi}{3}$ **c.** 3π **d.** $-\frac{2\pi}{3}$ **e.** $\frac{5\pi}{3}$

3. Convert each radian measure to degree measure. Give each answer to the nearest $0.1°$.
a. 2 **b.** -1 **c.** 0.4 **d.** 7.2 **e.** 0.05

4. Convert each degree measure to radian measure. Give each answer to the nearest 0.001 radian.
a. $20°$ **b.** $100°$ **c.** $-60°$ **d.** $-75°$ **e.** $18°$

In Exercises 5–17, r, s, θ, and A are as shown in the adjacent figure. Radian measure of θ is used. Find the unknown quantities.

5. $r = 2,\ \theta = 1.5,\ s = ?,\ A = ?$

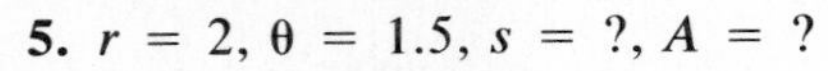

6. $r = 1.2,\ \theta = 0.5,\ s = ?,\ A = ?$

7. $r = 3,\ s = 6,\ \theta = ?,\ A = ?$

8. $r = 2,\ s = 3,\ \theta = ?,\ A = ?$

9. $r = 10$, $A = 20$, $\theta = ?$, $s = ?$

10. $r = 2$, $A = \pi$, $\theta = ?$, $s = ?$

11. $\theta = 1.2$, $s = 1.8$, $r = ?$, $A = ?$

12. $\theta = 1$, $s = 1$, $r = ?$, $A = ?$

13. $s = 6$, $A = 15$, $r = ?$, $\theta = ?$

14. $s = 0.8$, $A = 4.8$, $r = ?$, $\theta = ?$

15. $A = 3\pi$, $\theta = \frac{\pi}{6}$, $r = ?$, $s = ?$

16. $A = 10$, $\theta = 0.8$, $r = ?$, $s = ?$

17. Show that $A = \frac{1}{2}rs$.

Other Angle Measures. Sometimes angles are measured in *revolutions*, especially in connection with angular speed, as in Example 3. In the little-used *centesimal system*, a right angle is assigned the measure 100 *grads*. Thus, 1 revolution = 400 grads.

Write similar formulas relating the following measures.

18. radians and revolutions

19. radians and grads

20. degrees and grads

21. degrees and revolutions

22. A ferris wheel 60 feet in diameter makes one revolution every two minutes. What is the speed of a seat on the rim of the wheel?

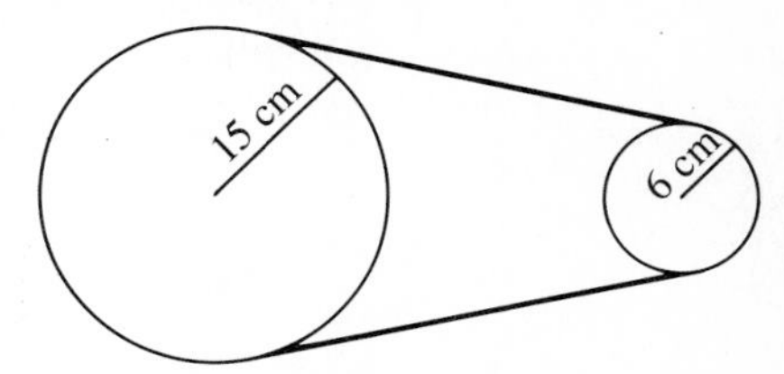

23. The angular speed of the larger wheel shown in the diagram is 100 rpm. Find the angular speed and rim speed of the smaller wheel.

24. A clock pendulum one meter long makes a back-and-forth round trip each two seconds. If the greatest angle the pendulum makes with the vertical is 12°, how many kilometers does its bottom end travel in one day?

The **latitude** of point P shown in the adjacent sketch of Earth is the measure of $\angle POE$. The radius of Earth is about 4000 miles.

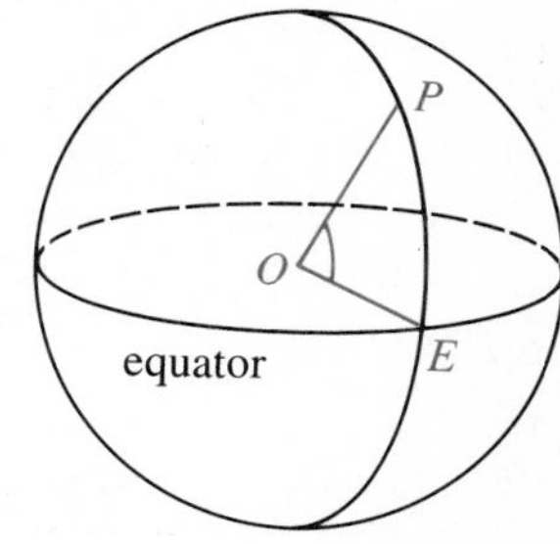

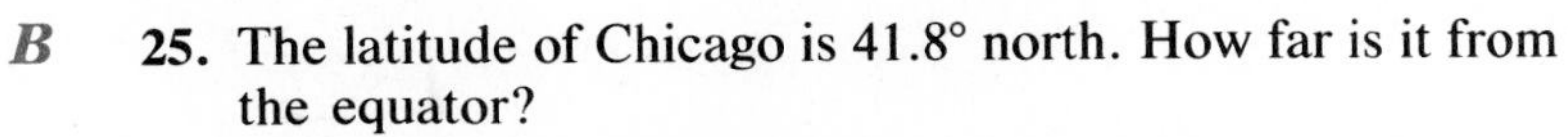

B 25. The latitude of Chicago is 41.8° north. How far is it from the equator?

26. The latitude of Dallas is 32.8° north. How far is it from the North Pole?

27. A cylindrical water tank 4 feet in diameter is lying on its side as shown at the right. What percent of the tank is filled when it contains water to a depth of one foot?

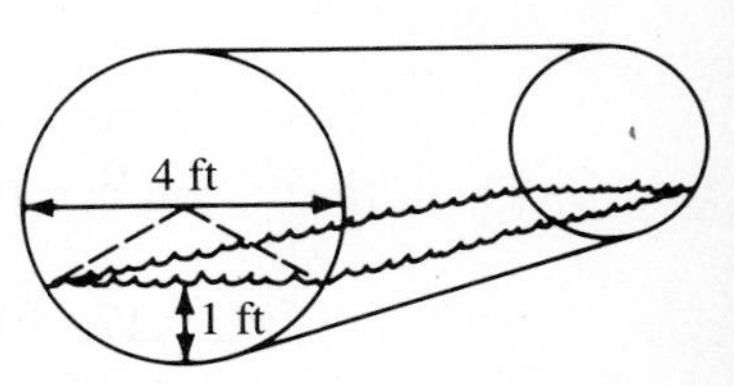

28. The angle subtended by the moon, 240,000 miles away, at the eye of an observer on Earth is about 0.518°. Approximate the diameter of the moon by finding the length of the red arc as shown below.

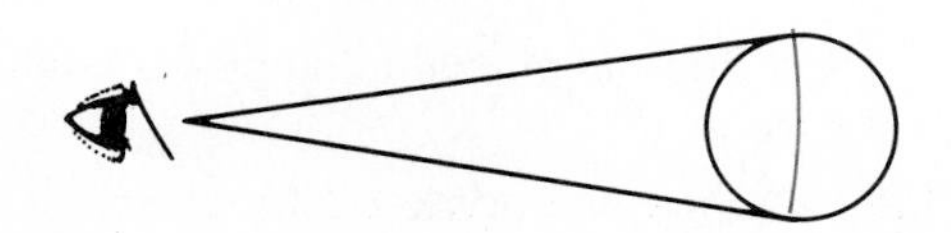

29. A car's wheel (with tire) is 26 inches in diameter and its center is 12 inches above the road, as shown at the left below. If the car is traveling at 60 feet per second, how fast is the highest point of the tire moving relative to the ground?

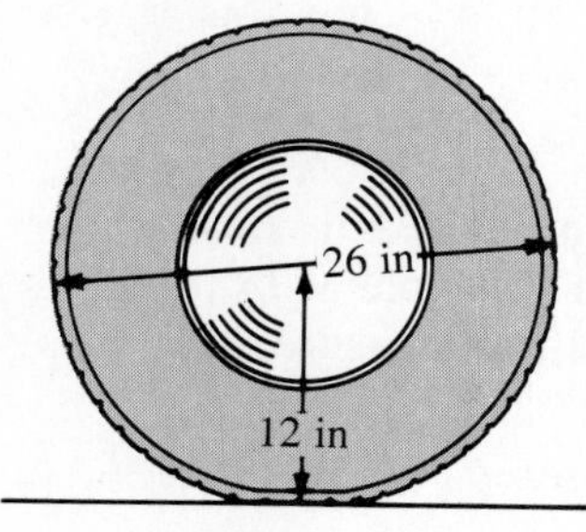

Exercise 29

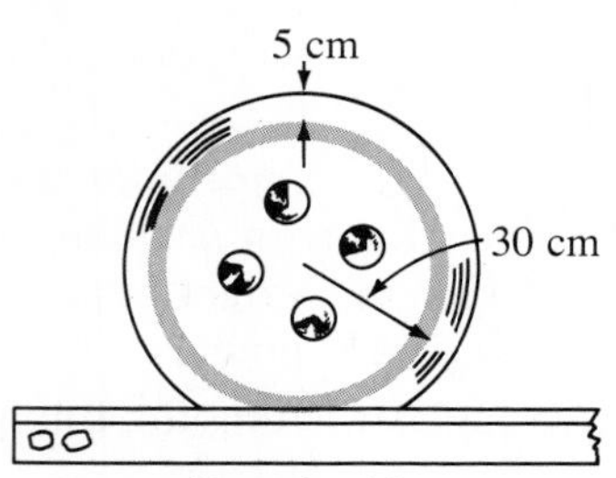

Exercise 30

30. A wheel of a train traveling at 90 km/h is 30 cm in radius and has a 5 cm flange that extends below the top of the rail, as indicated at the right above. How fast is the lowest point of the flange moving and in what direction?

7-4 THE TRIGONOMETRIC FUNCTIONS

Many applications involve *trigonometric* functions. These are closely related to the circular functions, and have the same names, but are defined for *angles* rather than numbers. These functions are defined below.

Let θ be any angle. Introduce a uv-coordinate system so that θ is in **standard position**, that is, so that the initial side of θ coincides with the positive u-axis. Choose any point $P(u, v)$ other than the origin on the terminal side of θ and let

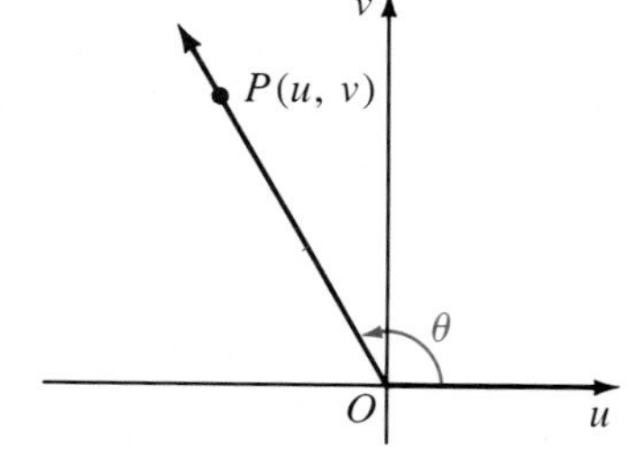

$$r = OP = \sqrt{u^2 + v^2}.$$

Then the **trigonometric cosine** of θ and **trigonometric sine** of θ are defined by

$$\cos\theta = \frac{u}{r} \quad \text{and} \quad \sin\theta = \frac{v}{r}.$$

The other trigonometric functions are defined by

$$\tan\theta = \frac{\sin\theta}{\cos\theta} \qquad \cot\theta = \frac{\cos\theta}{\sin\theta} \qquad \sec\theta = \frac{1}{\cos\theta} \qquad \csc\theta = \frac{1}{\sin\theta}$$

with appropriate restrictions placed on $\sin\theta$ and $\cos\theta$.

EXAMPLE 1 When an angle θ, $0° < \theta < 360°$, is placed in standard position, its terminal side passes through the point $(5, -12)$.

a. Draw θ. **b.** Find the six trigonometric functions of θ.

Solution **a.**

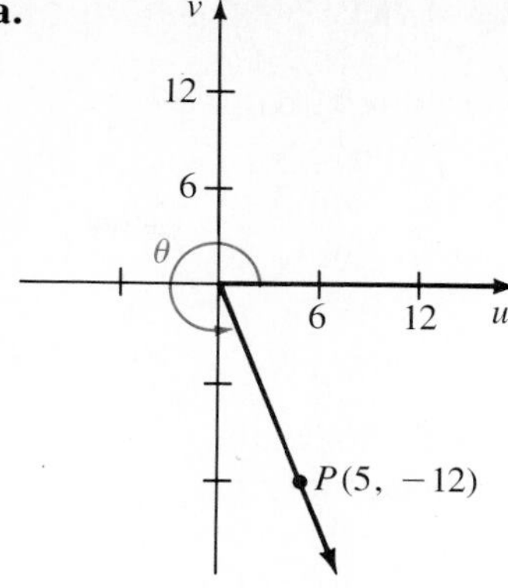

b. $r = OP = \sqrt{5^2 + (-12)^2} = \sqrt{169} = 13$

$$\therefore \cos\theta = \frac{5}{13}, \ \sin\theta = -\frac{12}{13}$$

$$\tan\theta = \frac{\sin\theta}{\cos\theta} = -\frac{12}{5}$$

$$\cot\theta = \frac{\cos\theta}{\sin\theta} = -\frac{5}{12}$$

$$\sec\theta = \frac{1}{\cos\theta} = \frac{13}{5}$$

$$\csc\theta = \frac{1}{\sin\theta} = -\frac{13}{12}$$

We say that θ is an ***n*th-quadrant angle** ($n = 1, 2, 3, 4$) if when θ is in standard position, its terminal side lies in the nth quadrant. We say that θ is a **quadrantal angle** if the terminal side lies along a coordinate axis. Notice that θ is an nth-quadrant angle if and only if its radian measure is an nth-quadrant number. Therefore the table on page 277 shows also how the quadrant of an *angle* θ determines the signs of its trigonometric functions. This follows from the fact that a trigonometric function of an angle equals the corresponding circular function of the radian measure of the angle. For example, if θ measures 60°, then since $60° = \frac{\pi}{3}$, $\sin 60° = \frac{\sqrt{3}}{2}$ and $\cos 60° = \frac{1}{2}$.

The fact just mentioned also guarantees that the following equations, proved in Sections 7–1 and 7–2, continue to hold when θ denotes an angle.

$$\cos^2\theta + \sin^2\theta = 1 \qquad 1 + \tan^2\theta = \sec^2\theta \qquad \cot^2\theta + 1 = \csc^2\theta$$

EXAMPLE 2 Let φ be a second-quadrant angle with $\tan\varphi = -2$. Find the other trigonometric functions of φ.

Solution $\sec^2\varphi = 1 + \tan^2\varphi = 1 + (-2)^2 = 5$. $\therefore \sec\varphi = \pm\sqrt{5}$. Since φ is a second-quadrant angle, $\sec\varphi < 0$.
$\therefore \sec\varphi = -\sqrt{5}$.

$$\cot\varphi = \frac{1}{\tan\varphi} = -\frac{1}{2}$$

$$\cos\varphi = \frac{1}{\sec\varphi} = \frac{1}{-\sqrt{5}} = -\frac{\sqrt{5}}{5}$$

$$\sin\varphi = \frac{\sin\varphi}{\cos\varphi}\cdot\cos\varphi = \tan\varphi\cos\varphi$$

$$= (-2)\left(-\frac{\sqrt{5}}{5}\right) = \frac{2\sqrt{5}}{5}$$

$$\csc\varphi = \frac{1}{\sin\varphi} = \frac{5}{2\sqrt{5}} = \frac{\sqrt{5}}{2}$$

The values of the sine and the cosine for $\theta = 30°, 60°$, and $90°$ occur so often that you should memorize them. We can express the trigonometric functions of many other angles in terms of these angles with the help of reference angles.

Let θ be any nonquadrantal angle in standard position. The **reference angle** of θ is the acute angle α that the terminal side of θ makes with the horizontal axis. Since the unit circle is symmetric about the v-axis, the origin, and the u-axis,

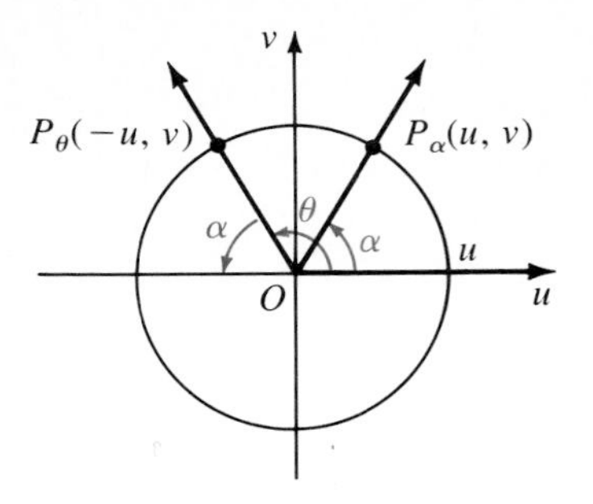

$$\sin\theta = \pm\sin\alpha \quad \cos\theta = \pm\cos\alpha \quad \tan\theta = \pm\tan\alpha$$

and so on. The quadrant of θ determines which sign to use. For example, if θ is in the second quadrant, then $\sin\theta = \sin\alpha$, $\cos\theta = -\cos\alpha$, and $\tan\theta = -\tan\alpha$.

EXAMPLE 3 Find: **(a)** $\sin 150°$ **(b)** $\cos 225°$ **(c)** $\tan 300°$ **(d)** $\sin(-30°)$

Solution Use reference angles and the table of signs.

a. The reference angle of 150° is 180° − 150°, or 30°.
Since 150° is a second-quadrant angle, $\sin 150° > 0$.
$\therefore \sin 150° = +\sin 30° = \frac{1}{2}$. **Answer**

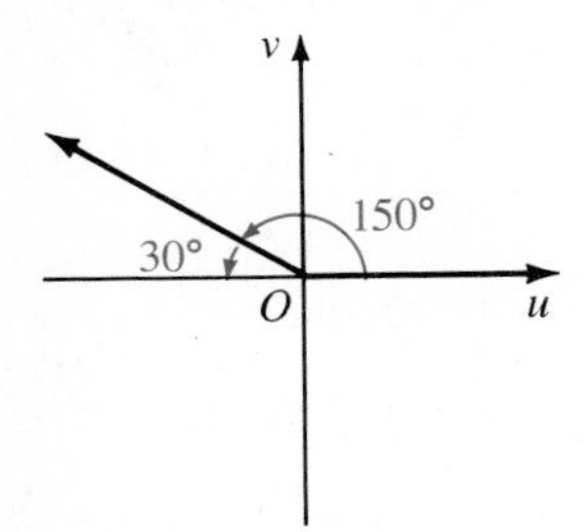

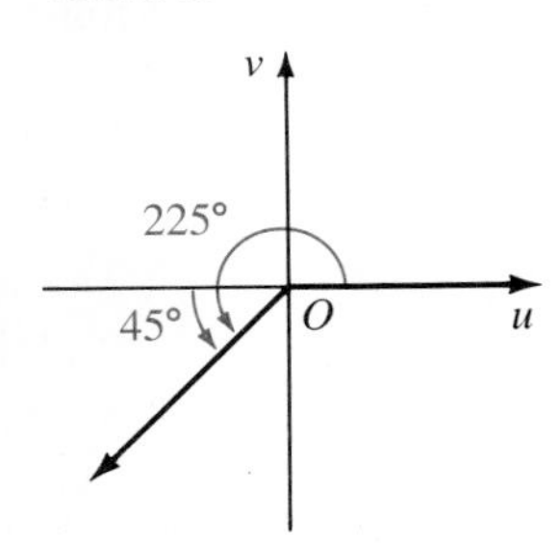

b. The reference angle of 225° is 225° − 180°, or 45°.
Since 225° is a third-quadrant angle, $\cos 225° < 0$.

$\therefore \cos 225° = -\cos 45° = -\frac{\sqrt{2}}{2}$. **Answer**

c. The reference angle of 300° is 360° − 300°, or 60°.
Since 300° is a fourth-quadrant angle, $\tan 300° < 0$.
$\therefore \tan 300° = -\tan 60° = -\sqrt{3}$. **Answer**

d. The reference angle for −30° is 0° − (−30°), or 30°.
Since −30° is a fourth-quadrant angle, $\sin(-30°) < 0$.

$\therefore \sin(-30°) = -\sin 30° = -\frac{1}{2}$. **Answer**

Two angles are **coterminal** if their terminal sides coincide when the angles are placed in standard position. For example, 300° and −60° are measures of coterminal angles. Trigonometric functions of coterminal angles are equal; that is, $\sin 300° = \sin(-60°)$, $\cos 300° = \cos(-60°)$, and so on.

EXERCISES

When an angle θ is placed in standard position, its terminal side passes through the given point. Find all the trigonometric functions of θ that are defined.

A

1. $(4, 3)$ **2.** $(3, -4)$ **3.** $(-5, 12)$

4. $(12, 5)$ **5.** $(-4, 0)$ **6.** $(0, -1)$

7. $(15, -8)$ **8.** $(8, -15)$ **9.** $(-2, 1)$

10. $(1, -1)$ **11.** $(-2, \sqrt{5})$ **12.** $(-\sqrt{3}, -\sqrt{6})$

Find the quadrant of θ given that:

13. $\sin\theta > 0, \cos\theta > 0$ **14.** $\sin\theta < 0, \tan\theta > 0$

15. $\cos\theta < 0, \tan\theta > 0$ **16.** $\sin\theta < 0, \sec\theta > 0$

17. $\csc\theta > 0, \cot\theta < 0$ **18.** $\cos\theta < 0, \csc\theta < 0$

19. $\sin\theta < 0, \sec\theta < 0$ **20.** $\sec\theta < 0, \csc\theta > 0$

Find the values of the six trigonometric functions of each angle. Leave answers in simplest radical form. Certain functions of quadrantal angles are not defined; indicate this fact with a —.

21. $0°$ **22.** $90°$ **23.** $180°$ **24.** $270°$ **25.** $120°$

26. $135°$ **27.** $150°$ **28.** $-60°$ **29.** $-135°$ **30.** $330°$

In Exercises 31–40, the quadrant of an angle φ and one of its trigonometric-function values are given. Find the other trigonometric functions of φ.

31. Quadrant II, $\sin\varphi = \frac{3}{5}$ **32.** Quadrant III, $\tan\varphi = \frac{5}{12}$

33. Quadrant IV, $\cot\varphi = -\frac{8}{15}$ **34.** Quadrant II, $\cos\varphi = -\frac{15}{17}$

35. Quadrant III, $\tan\varphi = 1$ **36.** Quadrant IV, $\sec\varphi = 2$

37. Quadrant IV, $\cos\varphi = \frac{1}{3}$ **38.** Quadrant I, $\sin\varphi = \frac{2}{3}$

39. Quadrant II, $\cos\varphi = -\frac{1}{3}$ **40.** Quadrant II, $\sin\varphi = \frac{2}{3}$

B

41. Use reference angles to show that if $0° < \theta < 90°$, then $\sin(\theta + 90°) = \cos\theta$.

42. Use reference angles to show that if $0° < \theta < 90°$, then $\cos(\theta + 90°) = -\sin\theta$.

43. Prove that if α and θ are coterminal angles, then $\sin\alpha = \sin\theta$ and $\cos\alpha = \cos\theta$.

44. Show that if θ is a third-quadrant angle with reference angle α, then $\sin\theta = -\sin\alpha$.

■ 7-5 VALUES OF THE TRIGONOMETRIC FUNCTIONS

In previous sections we were able to find exact values of trigonometric functions of multiples of $\frac{\pi}{6}$ and $\frac{\pi}{4}$ and hence of angles whose measures are multiples of 30° and 45°. In general, however, we must be content with approximate values obtained from calculators or tables.

To find the value of a trigonometric function of θ, $0° \leq \theta \leq 90°$, from Table 1, page 647, proceed as follows.

1. If $0° \leq \theta \leq 45°$, find the function name at the *top* and read *down* until opposite the measure of θ at the extreme *left*.
2. If $45° \leq \theta \leq 90°$, find the function name at the *bottom* and read *up* until opposite the measure of θ at the extreme *right*.

To find θ given the value of a trigonometric function of θ, locate the given value in the body of the table in the column for that trigonometric function. Then find θ in the extreme left or right column. These processes are illustrated in the following example.

EXAMPLE 1 Find: **(a)** cos 76.3° **(b)** tan 0.92

(c) θ if $\sin \theta = 0.8300$ and $0 < \theta < \frac{\pi}{2}$

Solution **a.** Read Table 1 upward in the column labeled "cos θ" at the bottom.

$$\cos 76.3° = 0.2368$$

b. Read the column labeled "Radians" and the column labeled "tan θ."

$$\tan 0.92 = 1.313$$

c. Use Table 1: $\theta = 56.1°$ or 0.9791

If a calculator is used in Example 1, we obtain (a) 0.2368381, (b) 1.3132637, and (c) 56.098738° or 0.9791077. Even though table entries and calculator results are approximate, it is customary to use = signs, as we did in Example 1, rather than ≈. Notice that although 0.92 was not an entry in the "tan θ" column, we used the entry in that column closest to 0.92.

Traditionally the degree was subdivided into *minutes* and *seconds*.

$$60 \text{ minutes } (60') = 1° \text{ and } 60 \text{ seconds } (60'') = 1'$$

The newer trend, however, is to use decimal degrees as we did in Example 1. We can convert easily from one form to the other. For example, $25°12' = 25° + 12' = 25° + (\frac{12}{60})° = 25° + (0.2)° = 25.2°$. On the other hand, $47.75° = 47° + (0.75)(60') = 47°45'$. In this book we will use decimal degrees in calculations and in giving answers.

To find the values of the trigonometric functions of angles not in the table we use reference angles and then the table.

EXAMPLE 2 Find sin 158.6° and cos 158.6°.

Solution The reference angle of a 158.6° angle measures 180° − 158.6°, or 21.4°. Since 158.6° is a second-quadrant number, its sine is positive and its cosine is negative. Thus,

$$\sin 158.6° = \sin 21.4° = 0.3649$$

and

$$\cos 158.6° = -\cos 21.4° = -0.9311.$$

EXAMPLE 3 Find sin 4.0 to four significant digits.

Solution Since $\pi < 4.0 < \frac{3\pi}{2}$, 4.0 is a third-quadrant number.

Because $\sin 4.0 < 0$, and $\alpha = 4.0 - \pi = 4.0000 - 3.1416 = 0.8584$, $\sin 4.0 = -\sin \alpha = -\sin 0.8584 = -0.7568$.

EXAMPLE 4 Find to the nearest 0.1° the two numbers θ such that $\cos \theta = -0.6215$ and such that $0° < \theta < 360°$.

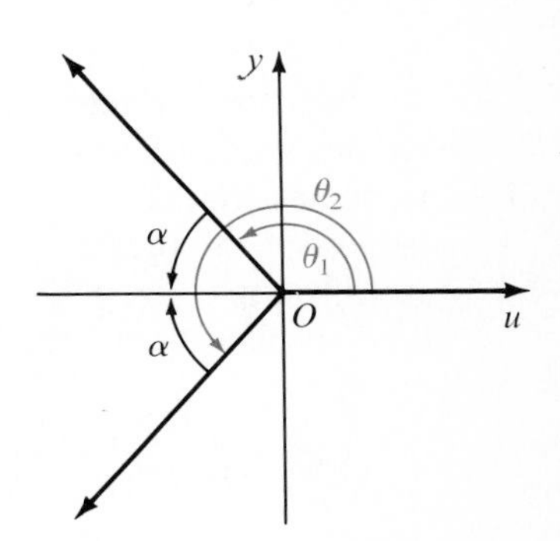

Solution First find the reference angle α of θ. Since $\cos \alpha = 0.6215$, $\alpha = 51.6°$. (Since 0.6215 is close to 0.6211 choose 51.6° for α.) Because $\cos \theta < 0$, θ is a second- or third-quadrant angle. Therefore

$$\theta_1 = 180° - \alpha = 180° - 51.6° = 128.4°$$
$$\theta_2 = 180° + \alpha = 180° + 51.6° = 231.6°$$

EXERCISES

Use the nearest table entry, linear interpolation, or a calculator.

Find each function value to four significant digits.

A

1. cos 36.3° **2.** sin 42.7° **3.** tan 63.5° **4.** cos 72.2°

5. sin 0.26 **6.** tan 1.12 **7.** sin 14.32° **8.** cos 46.48°

9. cos 1.363 **10.** sin 0.548 **11.** cos 163.4° **12.** tan 260°

13. sin (−132°) **14.** cos (−133.5°) **15.** tan 2.45 **16.** cos 5.00

Find the first-quadrant value of θ (to the nearest 0.01°) or of x (to the nearest 0.001).

17. $\sin \theta = 0.5376$ **18.** $\cos \theta = 0.8870$ **19.** $\tan \theta = 1.444$ **20.** $\sin \theta = 0.8415$

21. $\cos x = 0.9001$ **22.** $\tan x = 3.010$ **23.** $\cos \theta = 0.8256$ **24.** $\tan \theta = 0.3317$

25. $\sin \theta = 0.7710$ **26.** $\cos \theta = 0.1855$ **27.** $\tan x = 1.5000$ **28.** $\cos x = 0.6000$

Find the two values of θ (in $0° < \theta < 360°$) or x (in $0 < x < 2\pi$) such that:

29. $\sin \theta = 0.6231$ **30.** $\tan \theta = 1.462$ **31.** $\cos \theta = -0.8484$

32. $\cos \theta = -0.4420$ **33.** $\tan x = -0.1040$ **34.** $\sin x = -0.7120$

Solving Triangles

■ 7–6 RIGHT TRIANGLES

Historically, the first uses of *trigonometry* were in connection with triangles. The word itself is derived from *trigon* (triangle) and *metron* (measurement), and the remainder of this chapter is devoted to "measuring triangles."

Given a triangle ABC, we shall denote the length of the side opposite vertex A by a and the angle at A by α. The symbols b and β are similarly associated with B, and c and γ with C. We say that the sides $\overline{AC}$ and $\overline{AB}$ are **adjacent** to α and that α is **included** by these sides.

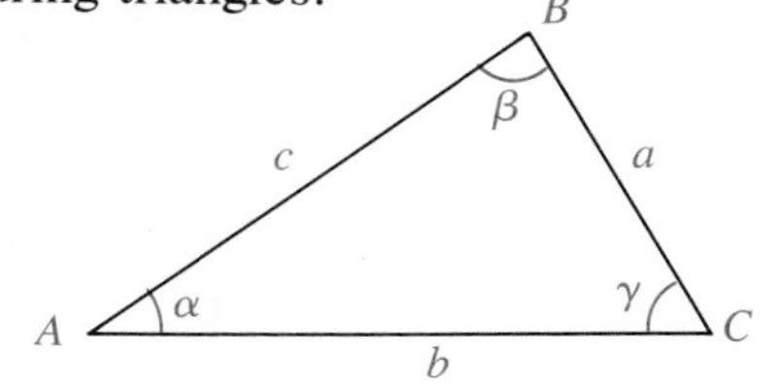

Solving a triangle consists of finding the measures of its **parts**, that is, of its sides and angles. In this section we shall solve right triangles and then isosceles triangles. As we shall see in Section 7–8, a triangle can be solved in general if the measures of any three parts (not all angles) are given. Since a *right* triangle has a 90° angle, the measures of only two other parts (not both angles) need be given. We use the formulas that will now be derived.

Let α be an acute angle of a right triangle ABC with $\gamma = 90°$ and introduce a coordinate system as shown in the figure at the right. Since B has coordinates (b, a) and $c = \sqrt{a^2 + b^2}$, it follows from the definitions of the trigonometric functions on page 286 that:

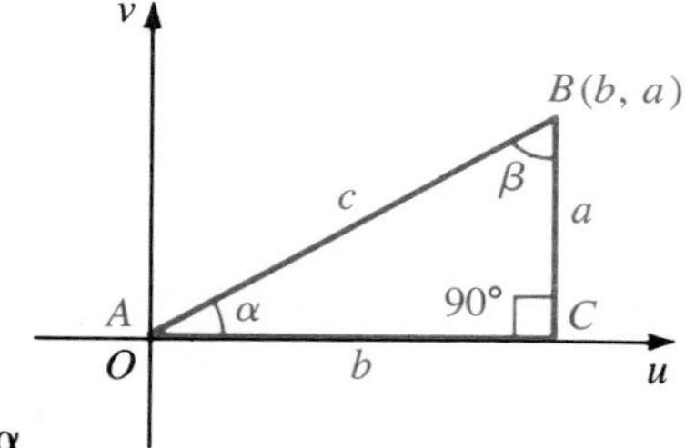

$$\sin \alpha = \frac{a}{c} = \frac{\text{length of the side opposite } \alpha}{\text{length of the hypotenuse}}$$

$$\cos \alpha = \frac{b}{c} = \frac{\text{length of the side adjacent to } \alpha}{\text{length of the hypotenuse}}$$

$$\tan \alpha = \frac{a}{b} = \frac{\text{length of the side opposite } \alpha}{\text{length of the side adjacent to } \alpha}$$

Similarly, $\sin \beta = \frac{b}{c}$, $\cos \beta = \frac{a}{c}$, and $\tan \beta = \frac{b}{a}$.

In our work we will also use the fact that in any triangle ABC,

$$\alpha + \beta + \gamma = 180°.$$

EXAMPLE 1 Solve triangle ABC given $a = 26.0$, $\alpha = 32.5°$, $\gamma = 90°$.

Solution Since the sum of the measures of the angles of any triangle is 180° and $\gamma = 90°$, $\alpha + \beta = 90°$.

$\therefore \beta = 90° - \alpha = 90° - 32.5° = 57.5°$

$$\tan\alpha = \frac{a}{b}$$

$$\tan 32.5^\circ = \frac{26.0}{b}$$

$$b = \frac{26.0}{\tan 32.5^\circ} = 40.8$$

$$\sin\alpha = \frac{a}{c}$$

$$\sin 32.5^\circ = \frac{26.0}{c}$$

$$c = \frac{26.0}{\sin 32.5^\circ} = 48.4$$

Therefore $\beta = 57.5^\circ$, $b = 40.8$, and $c = 48.4$ **Answer**

In Example 1 it was assumed that the given data were correct to three significant digits, so the answers were given to the same accuracy. A scientific calculator was used. If tables were to be used instead, we could use other functions to avoid long division. For example,

$$b = 26.0 \cot 32.5 = 26.0 \times 1.5697 = 40.8.$$

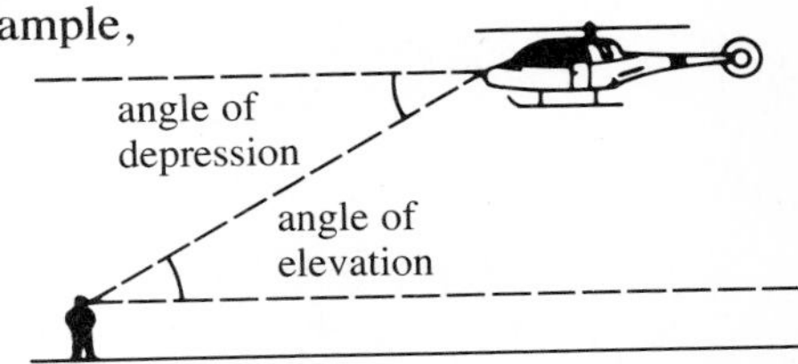

The two angles made by a line of sight with the horizontal are the **angle of elevation** and the **angle of depression**, as shown in the figure at the right.

EXAMPLE 2 A lighthouse 25 m tall stands at the top of a vertical cliff. A boatman directly offshore finds that, to the nearest degree, the angles of elevation of the top and bottom of the lighthouse are 28° and 24°, respectively. How far is he from the bottom of the cliff?

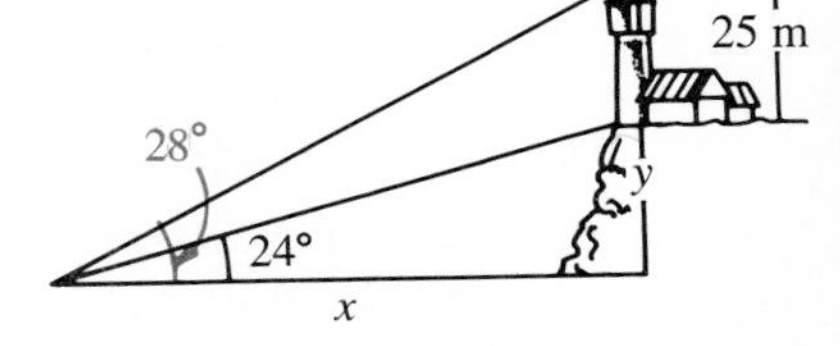

Solution The rough sketch at the right shows the given data and the desired quantity x. From the two right triangles $\tan 24^\circ = \frac{y}{x}$ and $\tan 28^\circ = \frac{y + 25}{x}$.

Therefore, $y = x\tan 24^\circ$ and $y = x\tan 28^\circ - 25$.

Equate these expressions for y.

$$x\tan 28^\circ - 25 = x\tan 24^\circ \quad \text{or} \quad x(\tan 28^\circ - \tan 24^\circ) = 25$$

$$\therefore x = \frac{25}{\tan 28^\circ - \tan 24^\circ} = 290 \text{ m (to two significant digits)}$$

The boatman is about 290 m from the bottom of the cliff.

EXAMPLE 3 The base of an isosceles triangle is 48.8 m long and its vertex angle measures 38.6°. Find the length of each leg.

Solution Bisect the vertex angle and let triangle ABC be one of the two congruent right triangles thus formed. (See the adjoining figure.) The goal is to find c. Note that $\alpha = \frac{1}{2} \cdot 38.6^\circ = 19.3^\circ$ and $a = \frac{1}{2} \cdot 48.8 = 24.4$.

Since $\sin\alpha = \frac{a}{c}$,

$$c = \frac{a}{\sin\alpha} = \frac{24.4}{\sin 19.3^\circ} = 73.8 \text{ m}$$ **Answer**

A, α, c, B, a, C

EXERCISES

In Exercises 1–16, solve triangle ABC, giving lengths to three significant digits and angles to 0.1°.

In Exercises 1–8, triangle ABC is a right triangle with $\gamma = 90°$.

A

1. $a = 24.0$, $\alpha = 42.0°$

2. $a = 14.0$, $\beta = 42.0°$

3. $b = 8.15$, $\alpha = 66.5°$

4. $b = 562$, $\beta = 65.4°$

5. $a = 340$, $b = 530$

6. $b = 6.45$, $c = 13.2$

7. $a = 21.6$, $c = 36.2$

8. $a = 0.560$, $b = 0.445$

In Exercises 9–16, triangle ABC is an isosceles triangle with $a = b$.

9. $b = 16.0$, $\beta = 73.0°$

10. $b = 42.0$, $\alpha = 44.0°$

11. $a = 2.65$, $\gamma = 134.0°$

12. $a = 627$, $\alpha = 72.5°$

13. $a = 0.600$, $c = 1.30$

14. $b = 100$, $c = 80$

15. $c = 264$, $\gamma = 36.6°$

16. $c = 4.25$, $\gamma = 126°$

17. What is the angle of elevation of the sun when a 25 m flagpole casts a shadow 42 m long?

18. A vertically directed searchlight shines a spot of light onto the bottom of the cloud cover over an airport. The angle of elevation of the spot from a point 1500 m from the searchlight is 56.7°. How high is the cloud cover?

19. In two minutes a plane descending at a constant angle of depression of 10.5° travels 3800 m along its flight path. How much altitude has it lost?

20. A television camera on a blimp is focused on a football field with angle of depression 23.2°. A range finder on the blimp determines that the field is 1740 m away. How high is the blimp?

21. What is the greatest latitude from which a signal can be sent (in a straight line) to a communication satellite 23,300 miles above the equator? (Recall the paragraph preceding Exercise 25, page 285.)

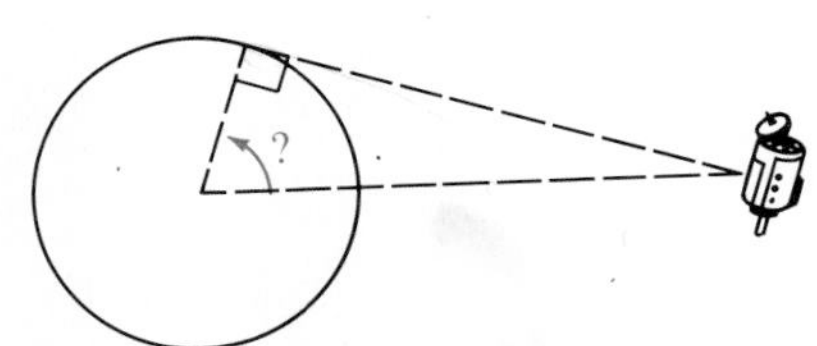

22. A small rectangular park is crossed by two diagonal paths, each 280 m long, that intersect at a 34° angle. Find the dimensions of the park.

23. From a point 100 m from a building the angles of elevation of the top and bottom of a flagpole atop the building are 54.5° and 51.8°. How tall is the flagpole?

24. Two observers 1800 m apart on a straight, level road measure the angles of elevation of a helicopter hovering over the road between them. If these angles are 31.5° and 52.0°, how high is the helicopter?

25. A pilot approaching a 10,000-foot runway finds that the angles of depression of the ends of the runway are 17.0° and 12.5°. How high is his plane and how far is it from the nearer end of the runway?

B **26.** An 8-foot pole and an 18-foot pole are braced by two guy wires, each extending from the top of one pole to the bottom of the other as shown. How far apart are the poles if the guy wires intersect at right angles?

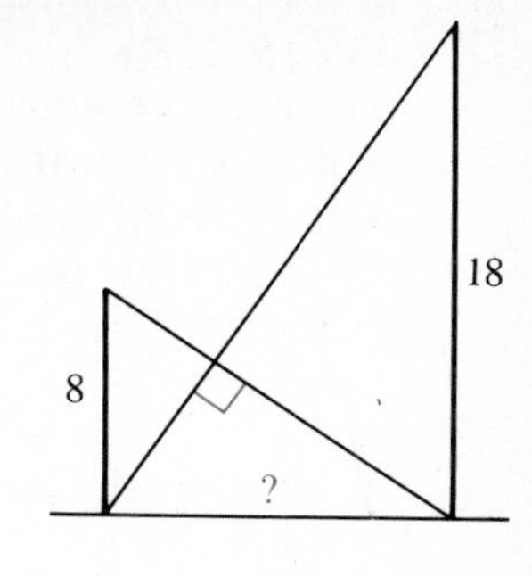

27. Repeat Exercise 26 for two poles that are a feet and b feet tall.

28. Find the perimeters of the regular pentagons **(a)** inscribed in and **(b)** circumscribed about the unit circle. (See the figures below.)

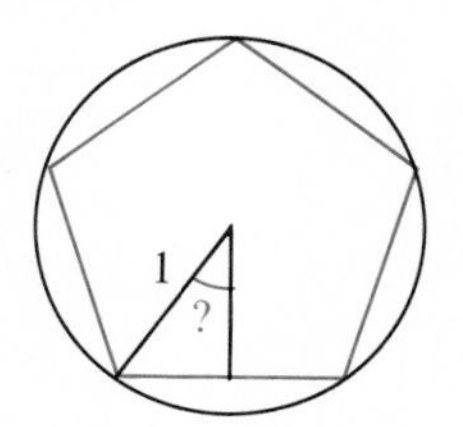

29. Repeat Exercise 28 for a regular polygon having 10 sides.

30. Repeat Exercise 28 for a regular polygon having 20 sides.

C **31.** Explain why the number 2π lies between the two perimeters in parts (a) and (b) of the answers to Exercises 28, 29, and 30.

32. Repeat Exercise 28 for a regular polygon having n sides.

33. What value do the preceding exercises suggest for the following limits?

$$\lim_{n \to \infty} \left(n \sin \frac{180°}{n} \right) \quad \text{and} \quad \lim_{n \to \infty} \left(n \tan \frac{180°}{n} \right)$$

COMPUTER EXERCISES

1. Write a computer program to calculate the area of a triangle ABC whose vertex B is the center of a circle of radius 1 and whose sides $\overline{BA}$ and $\overline{BC}$ are radii of the circle. (See the diagram at the right.) Assume the user enters the measure of $\angle B$. (*Hint*: The segment from the vertex A, perpendicular to the base $\overline{BC}$ of the triangle, can be calculated using a trigonometric function.)

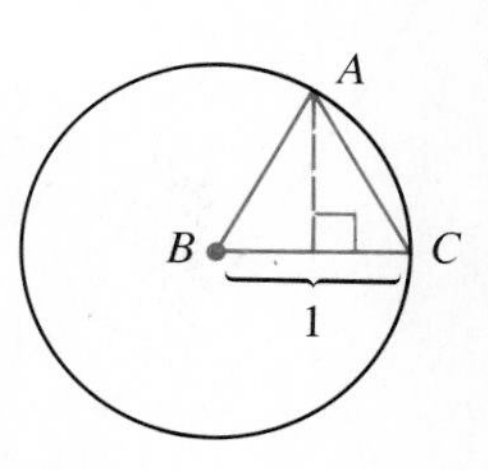

2. Modify the program in Exercise 1 to calculate the area of a regular n-sided polygon inscribed in a circle of radius 1, when the user enters the value of n.

3. Run the program in Exercise 2 for each value of n.
a. 3 **b.** 5 **c.** 8 **d.** 10

4. **a.** Run the program in Exercise 2 for n = 15, 30, 45, and 50.
b. Does the area of a regular n-sided polygon inscribed in a circle of radius 1 appear to have a limit as n increases without bound? If so, use part (a) to make a conjecture as to what that limit is.

7-7 THE LAWS OF COSINES AND SINES

The laws presented in this section enable us to solve triangles in general. We will make use of the following fact, which is a direct consequence of the definitions of the trigonometric functions given on page 286.

If θ is an angle in standard position and P is the point on its terminal side r units from O, then the coordinates of P are

$$(r \cos \theta, r \sin \theta).$$

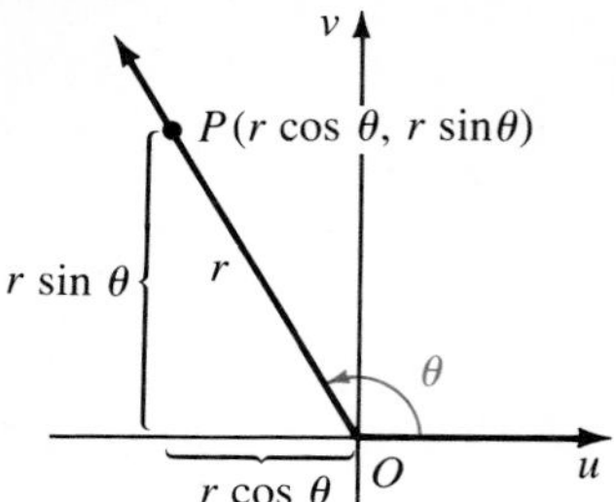

We use the fact just stated in the proofs of the theorems of this section.

THEOREM 1 **The Law of Cosines**

In any triangle ABC,

$$c^2 = a^2 + b^2 - 2ab \cos \gamma$$
$$b^2 = a^2 + c^2 - 2ac \cos \beta$$
$$a^2 = b^2 + c^2 - 2bc \cos \alpha$$

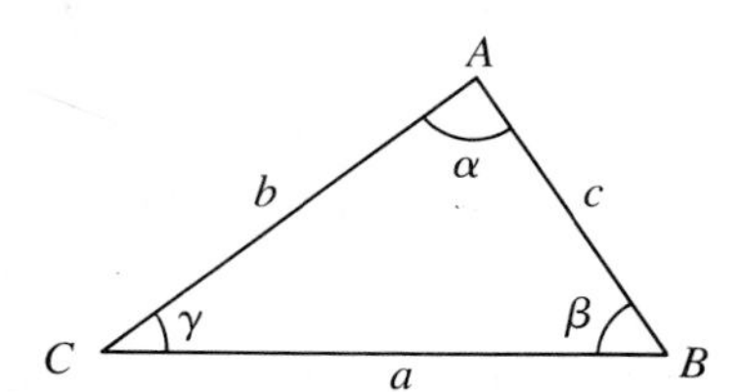

We shall prove the first equation: $c^2 = a^2 + b^2 - 2ab \cos \gamma$.

Proof: Introduce a coordinate system so that γ is in standard position and B is on the u-axis as shown in the following figures. Then A and B have coordinates $(b \cos \gamma, b \sin \gamma)$ and $(a, 0)$, respectively.

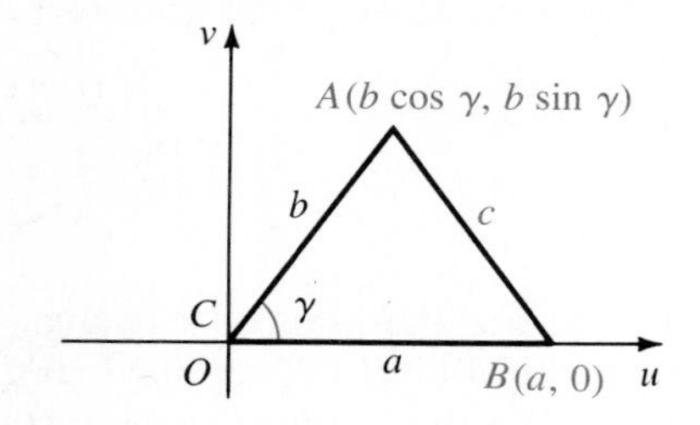

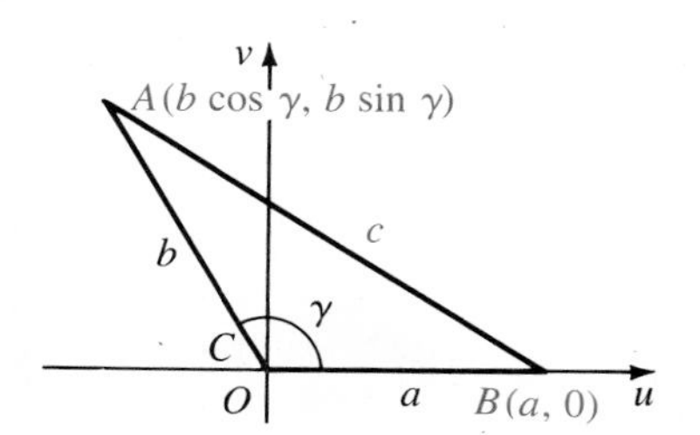

Apply the distance formula to $\overline{AB}$.

$$\begin{aligned} c^2 &= (b \cos \gamma - a)^2 + (b \sin \gamma - 0)^2 \\ &= b^2 \cos^2 \gamma - 2ab \cos \gamma + a^2 + b^2 \sin^2 \gamma \\ &= a^2 + b^2(\cos^2 \gamma + \sin^2 \gamma) - 2ab \cos \gamma \\ &= a^2 + b^2 - 2ab \cos \gamma \quad \blacksquare \end{aligned}$$

EXAMPLE 1 In triangle ABC, $a = 12$, $b = 15$, and $\gamma = 70°$. Find c to three significant digits.

Solution $c^2 = a^2 + b^2 - 2ab \cos \gamma = 12^2 + 15^2 - 2 \cdot 12 \cdot 15 \cdot \cos 70°$
$= 144 + 225 - (360)(0.3420) = 245.9$

$c = \sqrt{245.9} = 15.7$ **Answer**

By putting the equation $c^2 = a^2 + b^2 - 2ab \cos \gamma$ into the form

$$\cos \gamma = \frac{a^2 + b^2 - c^2}{2ab}$$

we can find an angle of a triangle if we know the lengths of its sides.

EXAMPLE 2 A triangular course for a 30 km yacht race has sides 7 km, 9 km, and 14 km long. Find the largest angle of the course to the nearest degree.

Solution The largest angle of a triangle is opposite its longest side. Let $a = 7$, $b = 9$, and $c = 14$ and find γ.

$$\cos \gamma = \frac{7^2 + 9^2 - 14^2}{2 \cdot 7 \cdot 9} = \frac{-66}{126} = -0.5238$$

$\therefore \gamma = 122°$ to the nearest degree. **Answer**

As a companion to the law of cosines we have the following theorem.

THEOREM 2 **The Law of Sines**

In any triangle ABC,

$$\frac{\sin \alpha}{a} = \frac{\sin \beta}{b} = \frac{\sin \gamma}{c}.$$

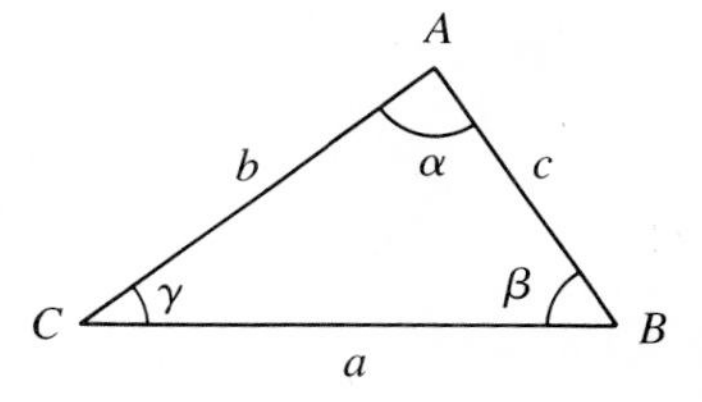

Proof: We first derive a formula for the area, K, of triangle ABC. Let α be in standard position and h be the height measured from B.

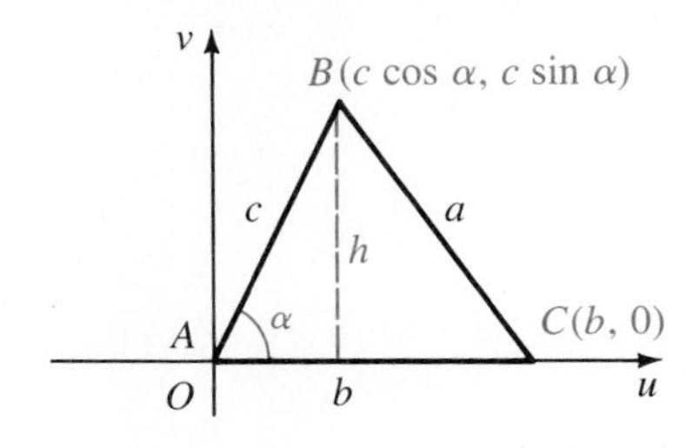

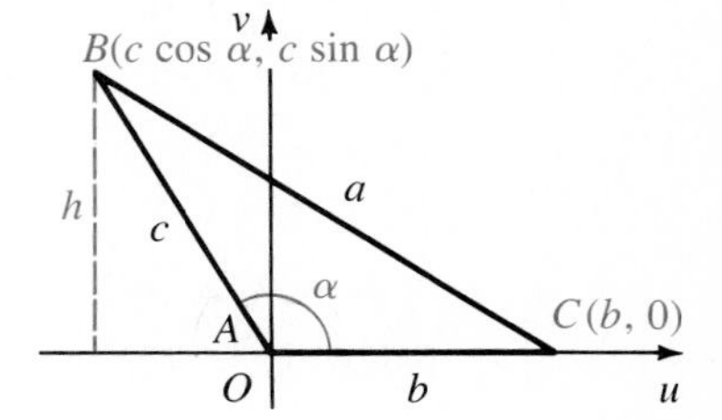

Thus $\qquad K = \frac{1}{2}(\text{base})(\text{height}) = \frac{1}{2}bh.$

Since h is the v-coordinate of B, $h = c \sin \alpha$. Therefore

$$K = \frac{1}{2}bc \sin \alpha.$$

By similar arguments we have

$$K = \frac{1}{2}ac \sin \beta \quad \text{and} \quad K = \frac{1}{2}ab \sin \gamma.$$

Thus,

$$\frac{1}{2}bc \sin \alpha = \frac{1}{2}ac \sin \beta = \frac{1}{2}ab \sin \gamma.$$

Proof continued on following page

When we divide each member of $\frac{1}{2}bc \sin \alpha = \frac{1}{2}ac \sin \beta = \frac{1}{2}ab \sin \gamma$ by $\frac{1}{2}abc$, we obtain

$$\frac{\sin \alpha}{a} = \frac{\sin \beta}{b} = \frac{\sin \gamma}{c}. \quad \blacksquare$$

EXAMPLE 3 Find the height of the tree pictured on page 271.

Solution The given data are shown in the adjacent sketch. Find c.

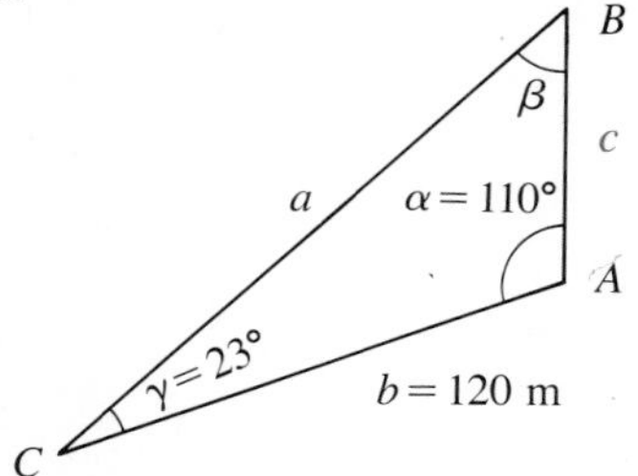

First find β and then apply the law of sines.

$$\beta = 180° - \alpha - \gamma$$
$$= 180° - 110° - 23° = 47°$$

Now use the law of sines.

$$\frac{\sin \beta}{b} = \frac{\sin \gamma}{c}$$

Thus $c = \dfrac{b \sin \gamma}{\sin \beta} = \dfrac{120 \sin 23°}{\sin 47°} = 64.11$

Therefore the tree is 64.1 m tall. **Answer**

EXERCISES

In this exercise set give lengths to three significant digits and angles to the nearest tenth of a degree.

In Exercises 1–12, find the indicated parts of triangle ABC.

A

1. $a = 180$, $\alpha = 35°$, $\beta = 65°$; $b = ?$
2. $a = 11$, $b = 16$, $\gamma = 70°$; $c = ?$
3. $a = 12$, $b = 15$, $c = 20$; $\beta = ?$
4. $a = 5$, $b = 3$, $\alpha = 35°$, $\beta = ?$
5. $a = 0.120$, $\beta = 41.5°$, $\gamma = 38.5°$; $c = ?$
6. $a = 1.6$, $b = 2.5$, $c = 1.9$; smallest angle = ?
7. $a = 12.6$, $b = 15.5$, $c = 25.0$; largest angle = ?
8. $a = 10$, $\alpha = 45°$, $\beta = 60°$, $b = ?$
9. $a = 25$, $b = 24$, $\alpha = 83.5°$, $c = ?$
10. $c = 62.0$, $\alpha = \beta = 58.3°$; $a = ?$
11. $\gamma = 156°$, $a = b = 42.5$; $c = ?$.
12. $a = 2b = 24$, $c = 15$; $\alpha = ?$
13. A tunnel is to be dug from point A to point B. The distances from a third point C to A and B are 2.63 miles and 1.84 miles respectively, and $\angle ACB = 52.2°$. How long will the tunnel be?

14. A pine tree stands on an 18° slope. From a point 100 yards down the slope the angle of elevation of the top of the tree is 32°. How tall is the tree?

15. A pilot approaching a 10,000-foot runway finds that the angles of depression of the ends of the runway are 12° and 15°. How far is she from the nearer end of the runway?

16. A cruise ship and a freighter leave port at the same time and follow straight-line courses at 30 km/h and 10 km/h, respectively. Two hours later they are 50 km apart. What is the angle between their courses?

17. A triangular lot has frontages of 30 m and 50 m on two streets, and its third side is 60 m long. At what angle do the streets intersect?

18. From the top of a building 25 m high the angle of elevation of a weather balloon is 54°, and from the bottom of the building it is 61°. How high is the balloon above the ground?

19. A rhombus has perimeter 24 and an angle of 65°. Find the lengths of its diagonals.

20. Find the lengths of the sides of a parallelogram whose diagonals intersect at a 40° angle and have lengths 8 and 12.

Show that the following formulas hold in any triangle ABC.

21. $\dfrac{\sin \alpha + \sin \beta}{\sin \beta} = \dfrac{a + b}{b}$

22. $\dfrac{\sin \alpha - \sin \beta}{\sin \beta} = \dfrac{a - b}{b}$

23. $\dfrac{\sin \alpha - \sin \beta}{\sin \gamma} = \dfrac{a - b}{c}$

24. $\dfrac{\sin \alpha + \sin \beta}{\sin \gamma} = \dfrac{a + b}{c}$

B

25. A guy wire bracing a transmission tower is 20 m long and makes a 50° angle with the ground. It is to be replaced by a 30-meter wire starting from the same point on the ground. How much farther up the tower will the new wire reach?

26. Two streets intersect at a 60° angle. A triangular lot has three times as much frontage on one of the streets as on the other. How long are these frontages if the third side of the lot is 21 m long?

27. Prove the formula $\cos 2\theta = 1 - 2 \sin^2 \theta$ for $0° < \theta < 90°$ as follows: In the triangle shown express b^2 in two ways: (1) by using the law of cosines, and (2) by first finding $\frac{b}{2}$ from the shaded right triangle.

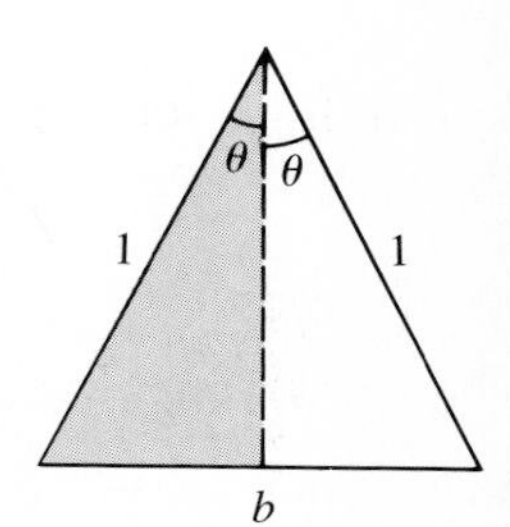

28. Prove the formula $\sin 2\theta = 2 \sin \theta \cos \theta$ for $0° < \theta < 90°$ by following the pattern of Exercise 27, but using the law of sines.

29. Prove that the sum of the squares of the lengths of the diagonals of any parallelogram is equal to the sum of the squares of the lengths of its four sides.

C **30.** Prove that in any triangle the sum of the squares of the lengths of any two sides equals twice the square of the length of the median to the third side plus half the square of the length of the third side.

31. Prove that in any triangle the sum of the squares of the lengths of the medians is equal to three fourths of the sum of the squares of the lengths of the sides. (*Hint*: Use Exercise 30.)

7-8 SOLVING GENERAL TRIANGLES

Triangle-solving problems fall naturally into four cases, depending on which measurements are given:

SSS	Given three sides
SAS	Given two sides and the included angle
SSA	Given two sides and the angle opposite one of them
SAA	Given one side and two angles.

Recall that to solve a triangle means to find all of its unknown parts. We illustrate by solving an SSS case, an SAS case, and an SAA case.

EXAMPLE 1 Solve triangle ABC, given that $a = 4.8$, $b = 6.7$, and $c = 3.9$.

Solution Since b is greater than a and c, β is the largest angle. First use the law of cosines to find β.

$$\cos \beta = \frac{a^2 + c^2 - b^2}{2ac} = \frac{(4.8)^2 + (3.9)^2 - (6.7)^2}{2 \times 4.8 \times 3.9} = -0.1774$$

Thus $\beta = 100.2°$.

Next, use the law of sines to find α.

$$\sin \alpha = \frac{a \sin \beta}{b} = \frac{4.8 \sin 100.2°}{6.7} = 0.7051$$

Since β is the largest angle, α must be acute.
Therefore $\alpha = 44.8°$.

Lastly $\gamma = 180° - \alpha - \beta = 180° - 44.8° - 100.2° = 35.0°$.
Therefore $\beta = 100.2°$, $\alpha = 44.8°$, and $\gamma = 35.0°$. **Answer**

A solution of a triangle can be checked by using a formula that contains all six parts of the triangle. For example, we could use the formula

$$\frac{\sin \alpha + \sin \beta}{\sin \gamma} = \frac{a + b}{c}.$$

Checking Example 1, we find that

$$\frac{\sin \alpha + \sin \beta}{\sin \gamma} = \frac{\sin 44.8° + \sin 100.2°}{\sin 35.0°} = \frac{0.7046 + 0.9842}{0.5736} = 2.944$$

$$\frac{a + b}{c} = \frac{4.8 + 6.7}{3.9} = 2.949$$

This is a satisfactory check.

EXAMPLE 2 Solve triangle ABC, given that $\alpha = 62.5°$, $b = 43$, and $c = 26$.

Solution First find a by using the law of cosines.

$$\begin{aligned} a^2 &= b^2 + c^2 - 2bc \cos \alpha \\ &= 43^2 + 26^2 - 2 \times 43 \times 26 \cos 62.5° = 1493 \end{aligned}$$

Therefore $a = \sqrt{1493} = 38.64$.

Next find γ by using the law of sines.

$$\sin \gamma = \frac{c \sin \alpha}{a} = \frac{26 \sin 62.5°}{38.64} = 0.5969$$

Hence $\gamma = 36.6°$.

Finally, $\beta = 180° - \alpha - \gamma = 180° - 62.5° - 36.6° = 80.9°$.

Therefore $a = 38.64$, $\gamma = 36.6°$, and $\beta = 80.9°$. **Answer**

EXAMPLE 3 Solve triangle ABC, given that $\alpha = 34°$, $\beta = 42°$, and $c = 42.5$.

Solution Find γ. $\gamma = 180° - \alpha - \beta = 180° - 34° - 42° = 104°$

Then find a and b by using the law of sines.

$$a = \frac{c \sin \alpha}{\sin \gamma} = \frac{42.5 \sin 34°}{\sin 104°} = 24.5$$

$$b = \frac{c \sin \beta}{\sin \gamma} = \frac{42.5 \sin 42°}{\sin 104°} = 29.3$$

Therefore $\gamma = 104°$, $a = 24.5$, and $b = 29.3$ **Answer**

SSA is called the ambiguous case because there may be two, one, or no solutions. A geometric explanation of this is given in the following figures for the case where a, b, and α are given, and α is acute.

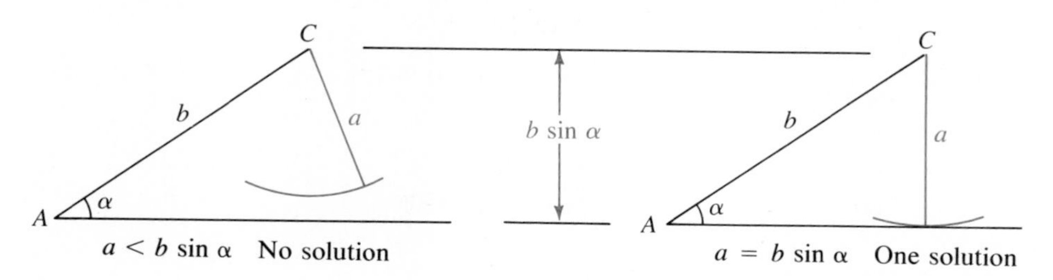

$a < b \sin \alpha$ No solution

$a = b \sin \alpha$ One solution

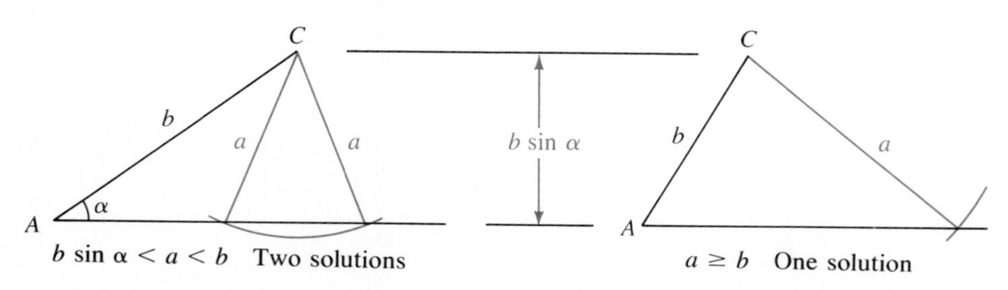

$b \sin \alpha < a < b$ Two solutions

$a \geq b$ One solution

If α is obtuse, there is one solution if $a > b$ and no solution if $a \leq b$.

EXAMPLE 4 In triangle ABC, $\alpha = 42°$ and $b = 35$. Find β given that

a. $a = 20$ **b.** $a = 30$ **c.** $a = 40$.

Solution Use $\frac{\sin \alpha}{a} = \frac{\sin \beta}{b}$ in the form $\sin \beta = \frac{b \sin \alpha}{a}$.

a. $\sin \beta = \frac{35 \sin 42°}{20}$

$= 1.1710$

There is no solution because $\sin \beta$ cannot be greater than 1.

b. $\sin \beta = \frac{35 \sin 42°}{30}$

$= 0.7807$

There are two solutions: $\beta = 51.3°$ and $\beta = 128.7°$.

c. $\sin \beta = \frac{35 \sin 42°}{40} = 0.5855$; $\beta = 35.8°$ or $144.2°$.

But if $\beta = 144.2°$, $\alpha + \beta = 42° + 144.2° = 186.2° > 180°$.
There is one solution: $\beta = 35.8°$.

In most applications involving triangle solving, it is not necessary to find a complete solution. Example 5 is an exception, however, because all parts are found in order to find the one part needed.

EXAMPLE 5 A derrick at the edge of a dock has an arm 25 m long that makes a 122° angle with the floor of the dock. The arm is to be braced with a cable 40 m long from the end of the arm back to the dock. How far from the edge of the dock will the cable be fastened?

Solution The sketch names the parts of the triangle in this SSA problem. The goal is to find c.

First, find β.

$\sin \beta = \frac{b \sin \alpha}{a} = \frac{25 \sin 122°}{40} = 0.5300$

$\beta = 32°$

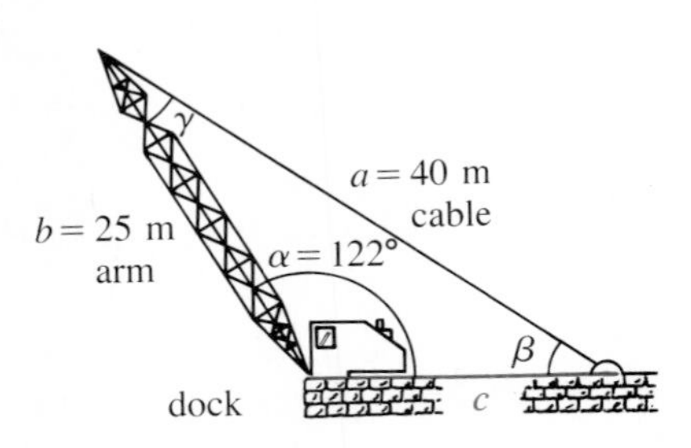

Next, find γ.

$\gamma = 180° - \alpha - \beta$

$= 180° - 122° - 32°$

$\gamma = 26°$

Finally, find c.

$c = \frac{a \sin \gamma}{\sin \alpha} = \frac{40 \sin 26°}{\sin 122°} = 20.677$

$c = 20.7$

Therefore the cable is fastened 20.7 m from the edge of the dock.

EXERCISES

In Exercises 1–12, outline a strategy for solving triangle ABC.

EXAMPLE $b = 18$, $c = 26$, $\alpha = 35°$

Solution **Step 1.** Use the law of cosines to find a.
Step 2. Use the law of sines to find β.
Step 3. Use $\gamma = 180° - \alpha - \beta$ to find γ.

A

1. $a = 16$, $\beta = 32°$, $\gamma = 50°$

2. $a = 21$, $c = 30$, $\beta = 42°$

3. $a = 5$, $b = 8$, $c = 10$

4. $b = 14$, $\beta = 25°$, $\gamma = 110°$

5. $b = 120$, $c = 145$, $\alpha = 100°$

6. $a = 2.3$, $b = 3.7$, $c = 5.0$

7. $a = 32.5$, $b = 21.5$, $\alpha = 130°$

8. $b = 20.3$, $c = 15.8$, $\beta = 116°$

9. $a = 12.0$, $b = 7.8$, $\beta = 35°$

10. $a = 123$, $b = 155$, $\alpha = 55.2°$

11. $b = 110$, $c = 180$, $\beta = 40°$

12. $b = 15.1$, $c = 13.5$, $\gamma = 48.8°$

13–24. Solve the triangles in Exercises 1–12. Give lengths to three significant digits and angles to the nearest tenth of a degree. If there are two solutions, find them both. If there are no solutions, so state.

B

25. In triangle ABC, $a = 4$, $b = 6$, and $\gamma = 120°$. Find the length of the median to the longest side.

26. In triangle ABC, $c = 12$ and $\alpha = \beta = 55°$. Find the length of the median to $\overline{AC}$.

27. Find the lengths of the diagonals of the trapezoid shown below.

28. Find the lengths of the diagonals of the quadrilateral shown below.

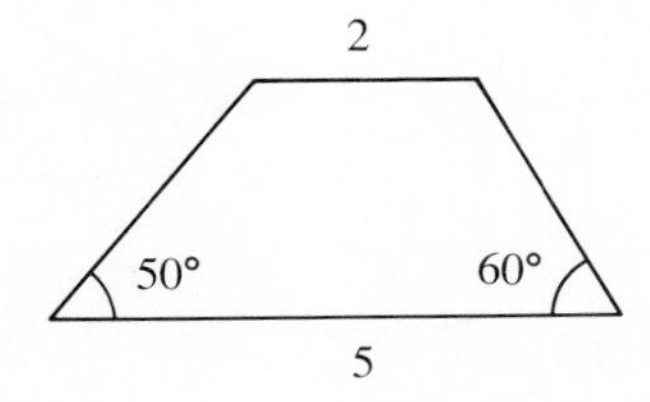

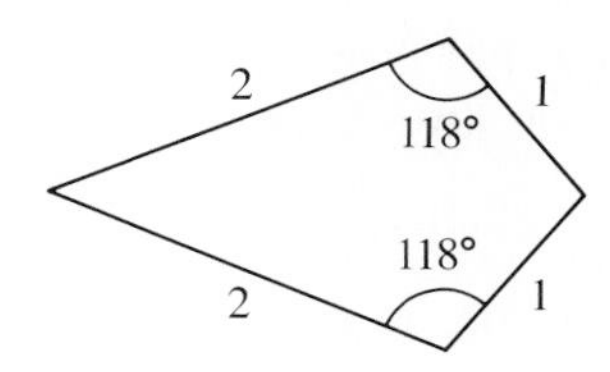

29. A communication satellite is 35,800 km above the equator. Find its angle of elevation from Houston, whose latitude is 29.7°N. Take the radius of Earth to be 6380 km.

30. A monument consists of a 16-meter flagpole standing at the top of a conical mound with vertex angle 140°. How long a shadow does the pole cast on the mound when the angle of elevation of the sun is 63°?

31. Show that in any triangle ABC, $c = a \cos \beta + b \cos \alpha$ (*Hint*: Consider separately the cases where both α and β are acute and where one of them is obtuse.)

C **32.** Show that if $\frac{\cos \alpha}{\cos \beta} = \frac{b}{a}$ (*), then triangle ABC is either isosceles or right.

(*Hint*: Explain why (*) implies that α and β are acute. Then combine (*) and the law of sines and use Exercise 28, page 299.)

33. Suppose that a, b, and α are given, and that there are two solutions, triangle AB_1C and triangle AB_2C, having areas K_1 and K_2 respectively. Show that $\frac{K_1}{K_2} = \frac{\sin \gamma_1}{\sin \gamma_2}$.

Exercises 34 and 35 outline derivatives of Mollweide's equations, (*) and (†). These equations are well suited for checking solutions of triangles because they involve all six parts of triangle ABC.

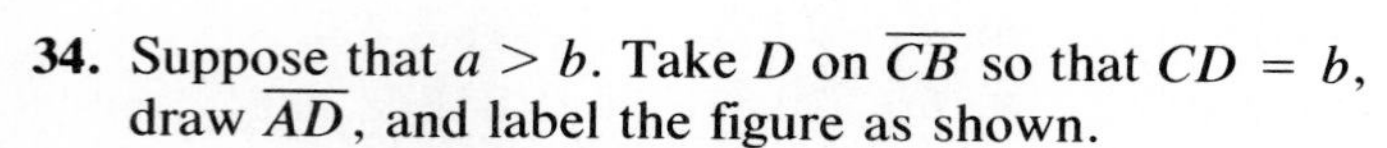

34. Suppose that $a > b$. Take D on $\overline{CB}$ so that $CD = b$, draw $\overline{AD}$, and label the figure as shown.

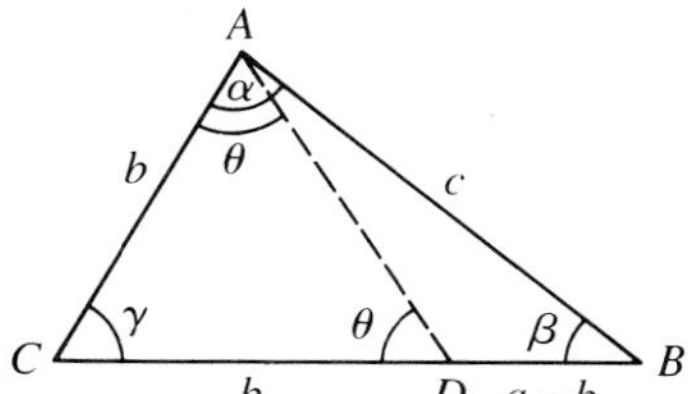

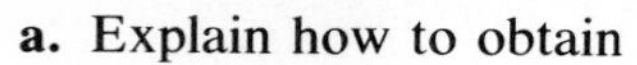

a. Explain how to obtain

$$\frac{a - b}{c} = \frac{\sin (\alpha - \theta)}{\sin (180^\circ - \theta)} = \frac{\sin (\alpha - \theta)}{\sin \theta}. \quad (1)$$

b. Justify the following steps to establish formulas (2) and (3).

$$2\theta + \gamma = 180^\circ$$
$$\theta = 90^\circ - \tfrac{1}{2}\gamma \quad (2)$$

$$\alpha - \theta = \alpha - (90^\circ - \tfrac{1}{2}\gamma) = \alpha - 90^\circ + \tfrac{1}{2}\gamma$$
$$= \alpha - 90^\circ + \tfrac{1}{2}(180^\circ - \alpha - \beta)$$
$$\alpha - \theta = \tfrac{1}{2}(\alpha - \beta) \quad (3)$$

c. Combine (1), (2), and (3) to obtain

$$\frac{a - b}{c} = \frac{\sin \frac{1}{2}(\alpha - \beta)}{\cos \frac{1}{2}\gamma} \quad (*)$$

d. Explain why (*) holds even if $a \leq b$.

35. Take D on $\overrightarrow{BC}$ so that $CD = b$, draw $\overline{AD}$, and label the figure as shown.

a. Explain how to obtain

$$\frac{a + b}{c} = \frac{\sin (\alpha + \theta)}{\sin \theta} \quad (1)$$

b. Justify the following steps to obtain formulas (2) and (3).

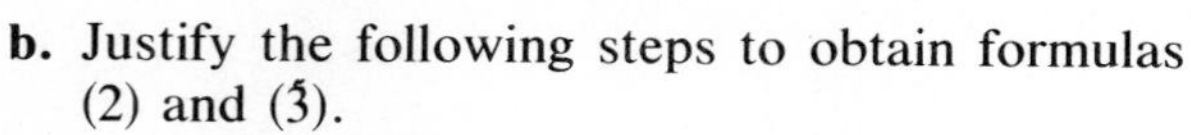

$$\gamma = 2\theta$$
$$\theta = \tfrac{1}{2}\gamma \quad (2)$$

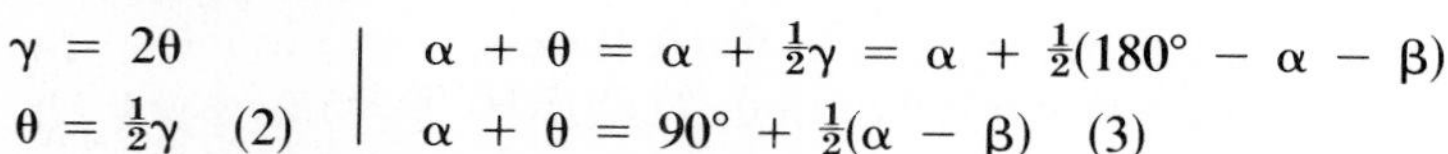

$$\alpha + \theta = \alpha + \tfrac{1}{2}\gamma = \alpha + \tfrac{1}{2}(180^\circ - \alpha - \beta)$$
$$\alpha + \theta = 90^\circ + \tfrac{1}{2}(\alpha - \beta) \quad (3)$$

c. Combine (1), (2), and (3) to obtain

$$\frac{a + b}{c} = \frac{\cos \frac{1}{2}(\alpha - \beta)}{\sin \frac{1}{2}\gamma} \quad (\dagger)$$

36. Use Exercises 34 and 35 to prove the **law of tangents**:

$$\frac{a - b}{a + b} = \frac{\tan \frac{1}{2}(\alpha - \beta)}{\tan \frac{1}{2}(\alpha + \beta)}$$

COMPUTER EXERCISES

1. Write a program to find the number of triangles that the given measurements determine.
2. Run the program for each set of measurements.
   a. $\alpha = 90°$, $a = 23$, $b = 20$
   b. $\alpha = 120°$, $a = 10$, $b = 18$
   c. $\alpha = 45°$, $a = 50$, $b = 60$
   d. $\alpha = 60°$, $a = 12$, $b = 12$
3. Write a program to solve triangle ABC when the measures of two angles and the length of the included side are given.
4. Run the program in Exercise 3 for each set of measurements.
   a. $\alpha = 50°$, $c = 10$, $\beta = 60°$
   b. $\alpha = 50°$, $c = 10$, $\beta = 70°$
   c. $\alpha = 90°$. $c = 10$, $\beta = 60°$

7-9 AREAS OF TRIANGLES

In deriving the law of sines (page 297) we showed that the area K of the triangle ABC is given by $K = \frac{1}{2}bc \sin \alpha$. Similar arguments are used to obtain the other formulas displayed in the first row below.

THEOREM 3 The area K of triangle ABC is given by each of the following formulas.

$$K = \tfrac{1}{2}bc \sin \alpha \qquad K = \tfrac{1}{2}ca \sin \beta \qquad K = \tfrac{1}{2}ab \sin \gamma$$

$$K = \tfrac{1}{2}a^2 \frac{\sin \beta \sin \gamma}{\sin \alpha} \qquad K = \tfrac{1}{2}b^2 \frac{\sin \gamma \sin \alpha}{\sin \beta} \qquad K = \tfrac{1}{2}c^2 \frac{\sin \alpha \sin \beta}{\sin \gamma}$$

$$K = \sqrt{s(s - a)(s - b)(s - c)}, \quad \text{where } s = \tfrac{1}{2}(a + b + c)$$

The law of sines can be put into the form $b = \dfrac{a \sin \beta}{\sin \alpha}$. If this is used to eliminate b from $K = \frac{1}{2}ab \sin \gamma$, we obtain $K = \frac{1}{2}a^2 \dfrac{\sin \beta \sin \gamma}{\sin \alpha}$. The other formulas in the second row are obtained similarly. The last formula, known as **Hero's formula**, is attributed to Hero (or Heron) of Alexandria. Its proof is outlined in the exercises.

EXAMPLE 1 Find the area of triangle ABC given $a = 35$, $c = 42$, and $\beta = 115°$.

Solution $K = \frac{1}{2}ca \sin \beta = \frac{1}{2}(42)(35)\sin 115° = 666$
$K = 666$ square units

EXAMPLE 2 Find the area of the jib sail shown at the right.

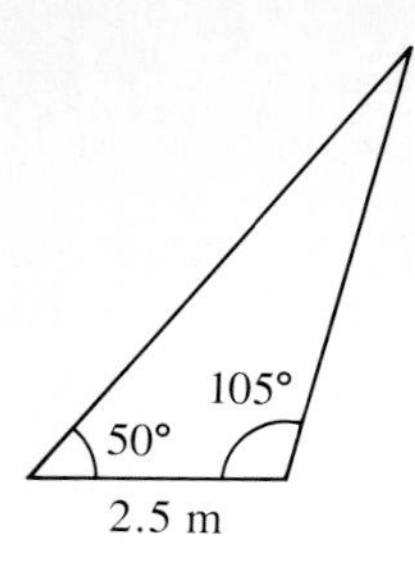

Solution Let $\alpha = 50°$, $\beta = 105°$, and $c = 2.5$.
Then $\gamma = 180° - \alpha - \beta = 25°$.

$$K = \frac{1}{2}c^2 \frac{\sin \alpha \sin \beta}{\sin \gamma}$$

$$= \frac{1}{2}(2.5)^2 \frac{\sin 50° \sin 105°}{\sin 25°} = 5.47$$

The area of the sail is 5.47 m^2. **Answer**

EXAMPLE 3 A surveyor finds that the edges of a triangular lot measure 42.5 m, 37.0 m, and 28.5 m. What is the area of the lot?

Solution Since the lengths of the three sides of the lot are given, use Hero's formula.
Let $a = 42.5$, $b = 37.0$, and $c = 28.5$.

$$s = \frac{1}{2}(a + b + c) = \frac{1}{2}(42.5 + 37.0 + 28.5) = 54.0$$

$$s - a = 11.5 \qquad s - b = 17.0 \qquad s - c = 25.5$$

Therefore $K = \sqrt{s(s - a)(s - b)(s - c)}$

$$= \sqrt{(54.0)(11.5)(17.0)(25.5)} = \sqrt{269204} = 519$$

The area of the lot is 519 m^2. **Answer**

Examples 1, 2, and 3 illustrate how to find areas in cases SAS, SAA, and SSS, respectively. In the ambiguous case, SSA, we first need to find out whether the given information determines a triangle.

EXAMPLE 4 Find the area of triangle ABC, given that $b = 19.2$, $c = 24.6$, and $\beta = 29.4°$.

Solution $\sin \gamma = \frac{c \sin \beta}{b} = \frac{24.6 \sin 29.4°}{19.2} = 0.62897$

There are two values of γ: $\gamma_1 = 39.0°$ or $\gamma_2 = 141.0°$.
From $\alpha = 180° - (\beta + \gamma)$ we find two values of α.

$$\alpha_1 = 111.6° \quad \text{or} \quad \alpha_2 = 9.6°.$$

Now since b, c, and α are known, use the formula

$$K = \frac{1}{2}bc \sin \alpha.$$

$K_1 = \frac{1}{2}(19.2)(24.6)\sin 111.6°$	$K_2 = \frac{1}{2}(19.2)(24.6)\sin 9.6°$
$K_1 = 220$ square units	$K_2 = 39.4$ square units

EXERCISES

A

1–12. Find the areas of the triangles described in Exercises 1–12 on page 303. If no triangle exists, so state.

13. Find the area of a parallelogram that has an angle of 70° and sides of lengths 8 cm and 10 cm.

14. Find the area of a parallelogram whose diagonals are 15 cm and 25 cm long and intersect at a 60° angle.

15. A triangle has area 24 and its shorter sides have lengths 7 and 10. Find its largest angle.

16. A triangle has an area of 10 square units and angles measuring 20°, 50° and 110°. Find the length of its longest side.

B

17. Let P be any point on side $\overline{AB}$ of triangle ABC. Let $\overline{CP}$ have length l and make an angle φ with $\overline{AB}$. Show that the area A of triangle ABC is given by $A = \frac{1}{2} lc \sin \varphi$.

18. The diagonals of a convex quadrilateral have lengths p and q and meet at an angle φ. Derive a formula for the area of the quadrilateral.

19. Explain how to find the area of a convex quadrilateral knowing the lengths of its sides and one of its diagonals.

20. Explain how to find the area of a convex pentagon knowing the lengths of its sides and the two diagonals from some vertex.

21. Explain how to find the area of a convex quadrilateral knowing the lengths of its sides and the measure of one of its angles.

22. Explain how to find the area of a convex pentagon knowing the lengths of its sides and the measures of two of its nonadjacent angles.

Exercise 23 is used in Exercise 24, which outlines a proof of Hero's formula. Both exercises refer to a triangle ABC.

23. Recall that $a + b + c = 2s$. Hence

$$(a + b) - c = a + b + c - 2c = 2s - 2c = 2(s - c).$$

Show that

$$c + (a - b) = 2(s - b) \quad \text{and} \quad c - (a - b) = 2(s - a).$$

24. Justify the following statements.

$$\begin{aligned}
K &= \tfrac{1}{2}ab \sin \gamma \\
4K^2 &= a^2b^2 \sin^2 \gamma \\
&= a^2b^2(1 - \cos^2 \gamma) \\
&= a^2b^2\left[1 - \left(\frac{a^2 + b^2 - c^2}{2ab}\right)^2\right] \\
&= a^2b^2\,\frac{4a^2b^2 - (a^2 + b^2 - c^2)^2}{4a^2b^2} \\
16K^2 &= 4a^2b^2 - (a^2 + b^2 - c^2)^2 \\
&= [2ab + (a^2 + b^2 - c^2)][2ab - (a^2 + b^2 - c^2)] \\
&= [(a + b)^2 - c^2][c^2 - (a - b)^2] \\
&= [(a + b) + c][(a + b) - c][c + (a - b)][c - (a - b)] \\
&= [2s][2(s - c)][2(s - b)][2(s - a)] \\
K^2 &= s(s - a)(s - b)(s - c) \\
K &= \sqrt{s(s - a)(s - b)(s - c)}
\end{aligned}$$

Chapter Summary

1. Given a real number x, let P_x be the point on the *unit circle* C obtained by starting at $(1, 0)$ and measuring $|x|$ units along C in the counterclockwise direction if $x \geq 0$ and in the clockwise direction if $x < 0$. Then:

$$\textit{cosine} \text{ of } x = \cos x = u \quad \text{and} \quad \textit{sine} \text{ of } x = \sin x = v,$$

where (u, v) are the coordinates of P_x. The cosine and sine are often called *circular functions*. The *quadrant* of x is the quadrant in which P_x lies.

2. The *tangent*, *cotangent*, *secant*, and *cosecant* of x are defined by

$$\tan x = \frac{\sin x}{\cos x} \qquad \cot x = \frac{\cos x}{\sin x}$$

$$\sec x = \frac{1}{\cos x} \qquad \csc x = \frac{1}{\sin x}$$

The functions defined above are related in many ways. For example:

$$\cos^2 x + \sin^2 x = 1$$

$$1 + \tan^2 x = \sec^2 x$$

$$\cot^2 x + 1 = \csc^2 x$$

3. Given an angle θ, let C be a circle of radius r centered at the vertex of θ and let s be the length of the arc of C intercepted by θ. Then the *radian measure* of θ is $\frac{s}{r}$. We often use the same symbol for an angle and its measure; thus we may write $\theta = \frac{s}{r}$. *Degree measure* is related to radian measure by the equation $180° = \pi$ (radians).

4. *Trigonometric functions* have the same names as the circular functions, and a trigonometric function of a given angle equals the corresponding circular function of the angle's radian measure. For historical reasons circular functions are usually referred to as trigonometric functions. The *reference angle* α of an angle θ in *standard position* is the acute angle the terminal side of θ makes with the horizontal axis. Any trigonometric function of θ has the same absolute value as the same function of α.

5. Values of the trigonometric functions can be obtained from tables or from scientific calculators.

6. To *solve a triangle* means to find the measures of its sides (a, b, c) and angles (α, β, γ), given any three of these (not all angles). Right triangles can be solved by using the definitions of the trigonometric functions. To solve oblique triangles we may use the *law of cosines*,

$$c^2 = a^2 + b^2 - 2ab \cos \gamma,$$

and the *law of sines*,

$$\frac{\sin \alpha}{a} = \frac{\sin \beta}{b} = \frac{\sin \gamma}{c}.$$

7. The *area* K of a triangle can be found by using such formulas as

$$K = \frac{1}{2}ab \sin \gamma \quad \text{and} \quad K = \frac{1}{2}a^2 \frac{\sin \beta \sin \gamma}{\sin \alpha}$$

or by using *Hero's formula*,

$$K = \sqrt{s(s - a)(s - b)(s - c)}, \text{ where } s = \frac{1}{2}(a + b + c).$$

Chapter Test

7-1 **1.** Find **(a)** $\cos \frac{5\pi}{6}$ and **(b)** $\sin\left(-\frac{5\pi}{4}\right)$.

2. Find $\sin x$ given that $\pi < x < 2\pi$ and $\cos x = \frac{2}{3}$.

7-2 **3.** Find all six circular functions of x given that $\cot x = \frac{3}{4}$ and $\cos x < 0$.

7-3 **4. a.** $\frac{7\pi}{6} = (?)^\circ$ **b.** $165^\circ = (?)$ radians

5. Find the area of a sector of a circle of radius 60 that has central angle 150°.

7-4 **6.** When an angle θ is placed in standard position its terminal side passes through the point $(-1, 2)$. Find the values of the six trigonometric functions of θ.

7. Find $\sin \theta$, $\cos \theta$, and $\tan \theta$ given that θ is a fourth-quadrant angle and the tangent of the reference angle of θ is $\frac{12}{5}$.

7-5 **8.** Use a table or a calculator to find **(a)** $\cos 142.6^\circ$ and **(b)** θ given that $\sin \theta = 0.6312$ and $\frac{\pi}{2} < \theta < \pi$.

7-6 **9.** An observer on level ground finds that the angle of elevation of a balloon tethered 1450 feet above the ground is 52°. How far is the balloon from the observer?

7-7 **10.** A triangle has sides of length 16 m, 26 m, and 38 m. Find its largest angle.

11. A tower stands on a 20° slope. From a point 45 m directly up the slope the angle of elevation of the top of the tower is 10°. How tall is the tower?

7-8 **12.** In triangle ABC, $\alpha = 40^\circ$ and $b = 55$. For what values of a will there be two solutions?

7-9 **13.** Find the area of the triangle of Exercise 10.

AREA

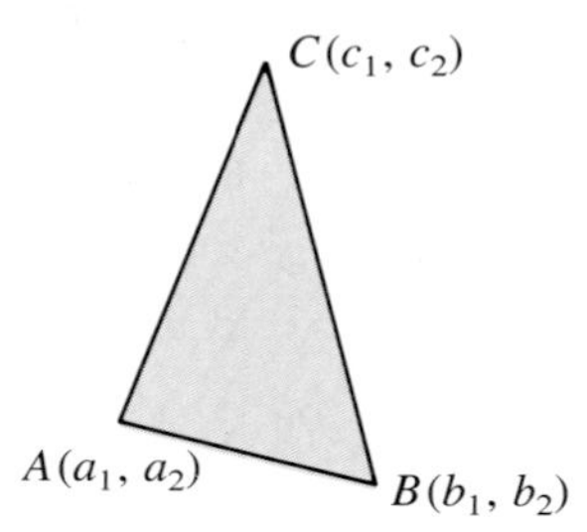

The following computer program can be used to find the area of a triangle when the coordinates of its vertices are given. The program relies upon the distance formula and Hero's formula. If the vertices are A, B, and C, then

$$\text{Area} = \sqrt{S(S - AB)(S - BC)(S - AC)}$$

where $S = \frac{1}{2}(AB + BC + AC)$. The program also serves as a test for collinearity. If, for example, A, B, and C are collinear with B between A and C, then $AB + BC = AC$. Hence $S = AC$, $S - AC = 0$, and Area $= 0$.

```
10  INPUT A1,A2,B1,B2,C1,C2
20  LET AB = SQR ((A1 - B1) * (A1 - B1) + (A2 - B2) * (A2 - B2))
30  LET BC = SQR ((B1 - C1) * (B1 - C1) + (B2 - C2) * (B2 - C2))
40  LET AC = SQR ((A1 - C1) * (A1 - C1) + (A2 - C2) * (A2 - C2))
50  LET S = (AB + BC + AC) / 2
60  LET H = S * (S - AB) * (S - BC) * (S - AC)
70  IF H = 0 THEN PRINT "COLLINEAR POINTS"
80  LET A = SQR (H)
90  LET A = INT (A * (10 ^ 5) + .5) / (10 ^ 5)
100  PRINT "THE AREA IS;"A "."
110  END
```

EXERCISES *(You may use either BASIC or Pascal.)*

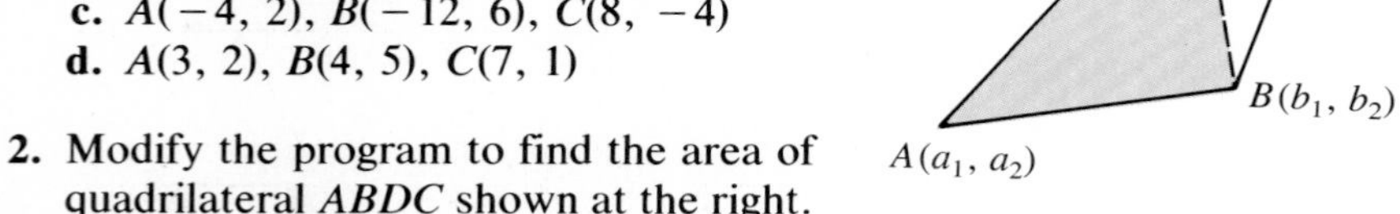

1. Run the program for:
a. $A(0, 0)$, $B(4, 0)$, $C(4, 4)$
b. $A(5, 4)$, $B(9, 4)$, $C(9, 8)$
c. $A(-4, 2)$, $B(-12, 6)$, $C(8, -4)$
d. $A(3, 2)$, $B(4, 5)$, $C(7, 1)$

2. Modify the program to find the area of quadrilateral $ABDC$ shown at the right.

3. Run the program in Exercise 2 for $A(4, 1)$, $B(5, 3)$, $D(4, 6)$, and $C(3, 9)$.

APPLICATION

TRIANGLES IN STRUCTURES

During the restoration of the Statue of Liberty, scaffolding was erected to support workers as they made necessary repairs. Notice that the rectangular elements of the scaffolding are strengthened by diagonal members to form triangles. It is this addition of cross braces that makes the scaffolding stable. Recall from the study of plane geometry that the triangle is a rigid figure.

Theoretically, one diagonal would be enough to stabilize a rectangle. Often a garden gate is braced in this way. However, in the scaffolding shown, both diagonals are used for additional support.

EXERCISES

1. A roof truss is in the shape of an isosceles triangle with base length 16 ft and base angles measuring 45°. Find the height.
2. A truss is in the shape of an isosceles triangle whose base length is 18 ft. It is braced at equally spaced locations. Find the lengths of the vertical and slant braces.
3. A bridge truss is in the shape of an isosceles trapezoid whose base length is 75 m. It is braced at equally spaced locations along the top and the bottom of the truss. Find the lengths of the braces.

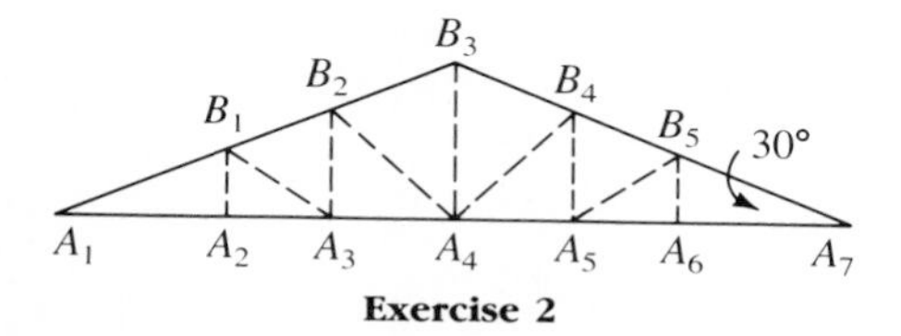

Exercise 2

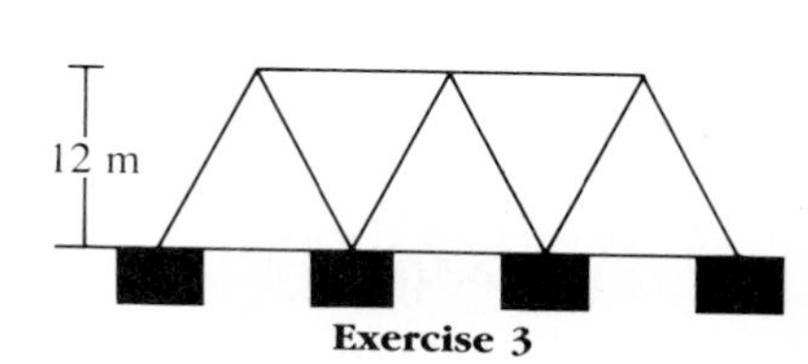

Exercise 3

Chapter 8

Trigonometric Identities and Graphs

In this chapter the study of trigonometric identities is continued. Then the graphs of the trigonometric functions and variations on them are discussed. The chapter concludes with a section on inverse trigonometric functions.

Trigonometric Identities

8-1 IDENTITIES IN ONE VARIABLE

Recall that an identity is an equation that is true for all values of its variable(s) for which it is defined. Thus

$$\frac{1 - x^2}{1 - x} = 1 + x$$

is an identity even though its left side is defined only if $x \neq 1$. Similarly,

$$1 + \tan^2 x = \sec^2 x$$

is an identity because it is true for every x such that $\cos x \neq 0$. From now on restrictions such as $x \neq 1$ and $\cos x \neq 0$ will be understood to apply and thus will not be explicitly stated.

You have already met most of the basic identities listed in the box at the top of the next page. Each of them can be proved by using the definitions of the trigonometric functions.

Wave motion, involved in the transmission of light as well as in the movement of the ocean, is a periodic function of the type discussed in this chapter. The photograph shows Heceta Head lighthouse on the coast of Oregon.

The Reciprocal Identities

$\sin x = \dfrac{1}{\csc x}$	$\sin x \csc x = 1$	$\csc x = \dfrac{1}{\sin x}$
$\cos x = \dfrac{1}{\sec x}$	$\cos x \sec x = 1$	$\sec x = \dfrac{1}{\cos x}$
$\tan x = \dfrac{1}{\cot x}$	$\tan x \cot x = 1$	$\cot x = \dfrac{1}{\tan x}$
$\tan x = \dfrac{\sin x}{\cos x}$		$\cot x = \dfrac{\cos x}{\sin x}$

The Pythagorean Identities

$\sin^2 x + \cos^2 x = 1 \qquad 1 + \tan^2 x = \sec^2 x \qquad 1 + \cot^2 x = \csc^2 x$

We can use these identities to simplify trigonometric expressions and to discover, derive, and prove other identities.

EXAMPLE 1 Simplify $\dfrac{\tan x + \cot x}{\csc^2 x}$.

Solution Begin by writing all the parts of the given expression in terms of $\sin x$ and $\cos x$. Apply basic identities.

$$\frac{\tan x + \cot x}{\csc^2 x} = \frac{\dfrac{\sin x}{\cos x} + \dfrac{\cos x}{\sin x}}{\dfrac{1}{\sin^2 x}}$$

$$= \sin^2 x \left(\frac{\sin x}{\cos x} + \frac{\cos x}{\sin x}\right)$$

$$= \sin^2 x \, \frac{\sin^2 x + \cos^2 x}{\cos x \sin x}$$

$$= \sin^2 x \, \frac{1}{\cos x \sin x} = \frac{\sin x}{\cos x}$$

$$\frac{\tan x + \cot x}{\csc^2 x} = \tan x \qquad \textbf{Answer}$$

EXAMPLE 2 Express $\sin \alpha$ in terms of $\tan \alpha$.

Solution $\sin \alpha = \dfrac{\sin \alpha}{\cos \alpha} \cdot \cos \alpha = \dfrac{\sin \alpha}{\cos \alpha} \cdot \dfrac{1}{\sec \alpha} = \dfrac{\dfrac{\sin \alpha}{\cos \alpha}}{\sec \alpha}$

Since $\dfrac{\sin \alpha}{\cos \alpha} = \tan \alpha$ and $\sec \alpha = \pm\sqrt{1 + \tan^2 \alpha}$,

$$\sin \alpha = \pm \frac{\tan \alpha}{\sqrt{1 + \tan^2 \alpha}}. \qquad \textbf{Answer}$$

We usually do not rationalize the denominator in identities such as the one in Example 2. Furthermore, if α is a first- or fourth-quadrant number, we use $+$. In the other quadrants, we use $-$.

Usually one of two general strategies can be used to prove an identity.

1. Transform the more complicated side of the identity into the simpler side by using reversible steps.
2. Transform both sides of the identity into the same expression by using reversible steps.

Note: It is logically unsound to "work across the = sign." For example, do not cross multiply. To do so *assumes* the truth of what is being proved.

Some special strategies that can be used are listed below.

1. Express all functions in terms of sines and cosines.
2. Look for combinations to which the Pythagorean identities can be applied. For example, substitute $\sec^2 x$ for $1 + \tan^2 x$.
3. Use factoring. For example, $\sin^2 x = 1 - \cos^2 x = (1 + \cos x)(1 - \cos x)$.
4. Combine the terms on each side of the identity into a single fraction.

EXAMPLE 3 Prove each identity.

a. $\dfrac{\tan^2 t}{1 + \sec t} = \sec t - 1$ **b.** $\sin \theta \tan \theta + \cos \theta = \sec \theta$

Solution Use general strategy 1.

a. $$\frac{\tan^2 t}{1 + \sec t} = \frac{\sec^2 t - 1}{\sec t + 1} = \frac{(\sec t + 1)(\sec t - 1)}{\sec t + 1} = \sec t - 1$$

b. $$\sin \theta \tan \theta + \cos \theta = \sin \theta \frac{\sin \theta}{\cos \theta} + \cos \theta$$
$$= \frac{\sin^2 \theta + \cos^2 \theta}{\cos \theta} = \frac{1}{\cos \theta} = \sec \theta$$

EXAMPLE 4 Prove that $\tan^4 x - 1 = \sec^4 x - 2 \sec^2 x$.

Solution Let L.S. denote the left side of the identity and R.S. denote the right side. Use general strategy 2.

L.S. $= (\tan^2 x + 1)(\tan^2 x - 1)$	R.S. $= \sec^2 x(\sec^2 x - 2)$
$= \sec^2 x(\tan^2 x - 1)$	$= \sec^2 x((1 + \tan^2 x) - 2)$
	$= \sec^2 x(\tan^2 x - 1)$

Since all of the steps are reversible, the proof is complete.

EXAMPLE 5 Prove that $\frac{1 + \sin u}{\cos u} = \frac{\cos u}{1 - \sin u}$.

Solution Do not cross multiply. Transform the left side into the right side.

$$\frac{1 + \sin u}{\cos u} = \frac{1 + \sin u}{\cos u} \cdot \frac{1 - \sin u}{1 - \sin u}$$

$$= \frac{1 - \sin^2 u}{\cos u(1 - \sin u)}$$

$$= \frac{\cos^2 u}{\cos u(1 - \sin u)} = \frac{\cos u}{1 - \sin u}$$

EXERCISES

Simplify.

A

1. $\csc^2 x - 1$
2. $\sec^2 x - \tan^2 x$
3. $\sin \alpha \sec \alpha$
4. $\csc t \sin t$
5. $\cos x \csc x \tan x$
6. $\sin \theta \sec \theta \cot \theta$
7. $\frac{\csc t}{\sin t} - \frac{\cot t}{\tan t}$
8. $\frac{\cos x}{\sec x} + \frac{\sin x}{\csc x}$
9. $\frac{\cos u}{\sec u} - \cos u \sec u$
10. $\frac{\sec v}{\cos v} - \sec v \cos v$
11. $\frac{1 + \tan^2 \varphi}{\tan^2 \varphi}$
12. $\frac{\sec^2 \varphi - 1}{\sec^2 \varphi}$
13. $\frac{1 - \cos^2 x}{1 - \cos x} - 1$
14. $\frac{\sin x \cos x}{1 - \cos^2 x}$
15. $\sec x - \sin x \tan x$
16. $\sin x (\csc x - \sin x)$

Express $\sin x$, $\cos x$, $\tan x$, $\cot x$, $\sec x$, and $\csc x$ in terms of:

17. $\sin x$
18. $\tan x$
19. $\sec x$
20. $\cos x$
21. $\cot x$
22. $\csc x$

Prove the following identities.

23. $\tan x(\tan x + \cot x) = \sec^2 x$
24. $\cos^2 x(1 + \cot^2 x) = \cot^2 x$
25. $\cos^2 \theta - \sin^2 \theta = 1 - 2 \sin^2 \theta$
26. $\cos^2 \alpha - \sin^2 \alpha = 2 \cos^2 \alpha - 1$
27. $\sec t - \cos t = \sin t \tan t$
28. $\csc z - \sin z = \cos z \cot z$
29. $\tan^2 x - \sin^2 x = \tan^2 x \sin^2 x$
30. $\sec^2 x + \csc^2 x = \sec^2 x \csc^2 x$
31. $\sin^4 u - \cos^4 u = \sin^2 u - \cos^2 u$
32. $\sec^4 v - \tan^4 v = \sec^2 v + \tan^2 v$

Simplify.

B

33. $\frac{\tan t + \cot t}{\sec^2 t}$
34. $\frac{\csc \alpha - \sin \alpha}{\cot^2 \alpha}$
35. $\frac{\sec x - \cos x}{\tan^2 x}$
36. $\frac{\sec x + \csc x}{1 + \tan x}$

37. $\dfrac{\sin^2 \theta}{1 - \cos \theta} - 1$

38. $\dfrac{\tan^2 t}{\sec t + 1} + 1$

39. $\cot x(\cos x \tan x + \sin x)$

40. $\sin u + \cos u \cot u$

41. $(1 + \cos x)(\csc x - \cot x)$

42. $(1 - \sin t)(\sec t + \tan t)$

43. $(\sin t + \cos t)^2 + (\sin t - \cos t)^2$

44. $(1 + \tan \alpha)^2 + (1 - \tan \alpha)^2$

45. $(3 - \cot x)^2 + (1 + 3 \cot x)^2$

46. $(4 \cos x - 3 \sin x)^2 + (3 \cos x + 4 \sin x)^2$

Prove the following identities.

47. $\dfrac{1}{1 - \sin t} + \dfrac{1}{1 + \sin t} = 2 \sec^2 t$

48. $\dfrac{1}{1 - \sin t} - \dfrac{1}{1 + \sin t} = 2 \tan t \sec t$

49. $\dfrac{\sin x}{1 - \cos x} = \dfrac{1 + \cos x}{\sin x}$

50. $\dfrac{\tan z}{\sec z + 1} = \dfrac{\sec z - 1}{\tan z}$

51. $\dfrac{\cos u}{\sec u + \tan u} = 1 - \sin u$

EC 52. $\dfrac{\sec x + 1}{\sec^2 x} = \dfrac{\sin^2 x}{\sec x - 1}$

53. $(\sin t + \cos t)(\tan t + \cot t) = \sec t + \csc t$

54. $(a \cos x + b \sin x)^2 + (b \cos x - a \sin x)^2 = a^2 + b^2$

C EC 55. $\dfrac{\sin^2 x + 2 \cos x - 1}{\sin^2 x + 3 \cos x - 3} = \dfrac{1}{1 - \sec x}$

EC 56. $\dfrac{\cos^2 x + 3 \sin x - 1}{\cos^2 x + 2 \sin x + 2} = \dfrac{1}{1 + \csc x}$

57. $\sqrt{\dfrac{1 - \cos x}{1 + \cos x}} = |\csc x - \cot x|$

COMPUTER EXERCISES

1. Write a computer program to approximate the value of $\sin x$ by calculating the sum S_n of the first n terms of the series:

   $S = x - \dfrac{x^3}{3!} + \dfrac{x^5}{5!} - \dfrac{x^7}{7!} + \cdots$, where x and n are supplied by the user.

2. Modify the program in Exercise 1 and run it for the following values of x to find n such that S_n is within $\pm$ 0.0001 of the value of $\sin x$ calculated by using the built-in sine function.

   a. $x = \dfrac{\pi}{2}$ **b.** $x = \dfrac{\pi}{4}$

3. Modify the program in Exercise 1 to approximate the value of $\cos x$ by calculating the sum S_n of the first n terms of the series:

   $S_n = 1 - \dfrac{x^2}{2!} + \dfrac{x^4}{4!} - \dfrac{x^6}{6!} + \cdots$, where x and n are input by the user.

4. Modify the program of Exercise 3 and run it for the following values of x to find n such that S_n is within $\pm$ 0.0001 of the value of $\cos x$ calculated by using the built-in cosine function.

   a. $x = \dfrac{\pi}{4}$ **b.** $x = \dfrac{\pi}{6}$

8-2 TRIGONOMETRIC ADDITION FORMULAS

The following identities in two variables are of basic importance.

Addition Formulas for the Sine and Cosine

$$\sin(s+t) = \sin s \cos t + \cos s \sin t$$
$$\sin(s-t) = \sin s \cos t - \cos s \sin t$$
$$\cos(s+t) = \cos s \cos t - \sin s \sin t$$
$$\cos(s-t) = \cos s \cos t + \sin s \sin t$$

We shall prove these identities after illustrating a few ways in which they are used.

EXAMPLE 1 Find the exact value of each function.

a. $\sin 105°$ **b.** $\cos \frac{\pi}{12}$

Solution **a.** Since $105° = 60° + 45°$, use the first formula.

$$\sin 105° = \sin(60° + 45°) = \sin 60° \cos 45° + \cos 60° \sin 45°$$
$$= \frac{\sqrt{3}}{2} \cdot \frac{\sqrt{2}}{2} + \frac{1}{2} \cdot \frac{\sqrt{2}}{2}$$
$$\sin 105° = \frac{\sqrt{6} + \sqrt{2}}{4} \quad \textbf{Answer}$$

b.
$$\cos \frac{\pi}{12} = \cos\left(\frac{\pi}{3} - \frac{\pi}{4}\right) = \cos \frac{\pi}{3} \cos \frac{\pi}{4} + \sin \frac{\pi}{3} \sin \frac{\pi}{4}$$
$$= \frac{1}{2} \cdot \frac{\sqrt{2}}{2} + \frac{\sqrt{3}}{2} \cdot \frac{\sqrt{2}}{2}$$
$$\cos \frac{\pi}{12} = \frac{\sqrt{2} + \sqrt{6}}{4} \quad \textbf{Answer}$$

EXAMPLE 2 Prove the identity $\frac{\cos(\alpha + \beta)}{\sin \alpha \cos \beta} = \cot \alpha - \tan \beta$.

Solution
$$\frac{\cos(\alpha + \beta)}{\sin \alpha \cos \beta} = \frac{\cos \alpha \cos \beta - \sin \alpha \sin \beta}{\sin \alpha \cos \beta}$$
$$= \frac{\cos \alpha \cos \beta}{\sin \alpha \cos \beta} - \frac{\sin \alpha \sin \beta}{\sin \alpha \cos \beta} = \cot \alpha - \tan \beta$$

EXAMPLE 3 Show that $\cos(\pi + \theta) = -\cos \theta$.

Solution Let $s = \pi$ and $t = \theta$ in the third formula.

$$\cos(s + t) = \cos s \cos t - \sin s \sin t$$
$$\cos(\pi + \theta) = \cos \pi \cos \theta - \sin \pi \sin \theta$$
$$= -1 \cdot \cos \theta - 0 \cdot \sin \theta$$
$$= -\cos \theta$$

EXAMPLE 4 Let $\sin p = \frac{4}{5}$ and $\cos q = -\frac{7}{25}$, where p and q are second- and third-quadrant numbers, respectively. Find **(a)** $\sin(p + q)$, **(b)** $\cos(p + q)$, and **(c)** the quadrant of $p + q$.

Solution Use $\cos^2 x + \sin^2 x = 1$ to find $\cos p$ and $\sin q$.

$$\cos p = -\sqrt{1 - \left(\frac{4}{5}\right)^2} = -\frac{3}{5} \qquad \sin q = -\sqrt{1 - \left(-\frac{7}{25}\right)^2} = -\frac{24}{25}$$

a.
$$\begin{aligned}\sin(p + q) &= \sin p \cos q + \cos p \sin q \\ &= \left(\frac{4}{5}\right)\left(-\frac{7}{25}\right) + \left(-\frac{3}{5}\right)\left(-\frac{24}{25}\right) \\ &= \frac{44}{125} \quad \textbf{Answer}\end{aligned}$$

b.
$$\begin{aligned}\cos(p + q) &= \cos p \cos q - \sin p \sin q \\ &= \left(-\frac{3}{5}\right)\left(-\frac{7}{25}\right) - \left(\frac{4}{5}\right)\left(-\frac{24}{25}\right) \\ &= \frac{117}{125} \quad \textbf{Answer}\end{aligned}$$

c. Since $\sin(p + q) > 0$ and $\cos(p + q) > 0$, $p + q$ is a first-quadrant number. **Answer**

We now prove the formula for $\cos(s - t)$, where $0 \le s - t < 2\pi$. The adjoining figure shows the points we need. The circle C is the unit circle.

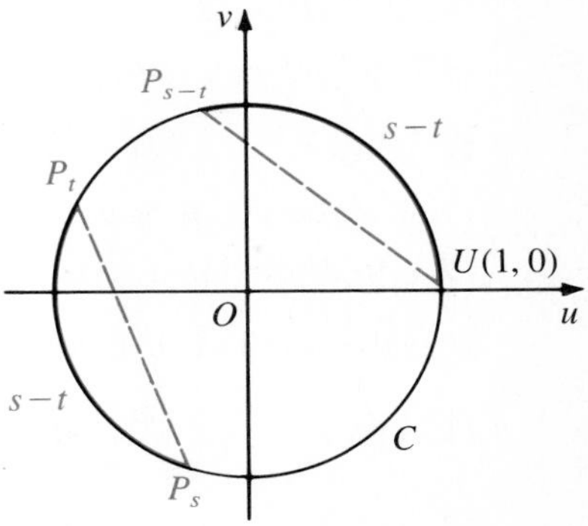

$U(1, 0)$ $\qquad$ $P_{s-t}(\cos(s - t), \sin(s - t))$

$P_t(\cos t, \sin t)$ $\qquad$ $P_s(\cos s, \sin s)$

Because chords $\overline{UP_{s-t}}$ and $\overline{P_tP_s}$ both subtend arcs of length $s - t$, their lengths are equal. Hence

$$UP_{s-t} = P_tP_s.$$

Applying the distance formula and squaring, we have:

$$\begin{aligned}(UP_{s-t})^2 &= [\cos(s - t) - 1]^2 + [\sin(s - t) - 0]^2 \\ &= \cos^2(s - t) - 2\cos(s - t) + 1 + \sin^2(s - t) \\ &= 2 - 2\cos(s - t)\end{aligned}$$

$$\begin{aligned}(P_tP_s)^2 &= [\cos s - \cos t]^2 + [\sin s - \sin t]^2 \\ &= \cos^2 s - 2\cos s \cos t + \cos^2 t + \sin^2 s - 2\sin s \sin t + \sin^2 t \\ &= 2 - 2(\cos s \cos t + \sin s \sin t)\end{aligned}$$

Since $UP_{s-t} = P_tP_s$,

$$2 - 2\cos(s - t) = 2 - 2(\cos s \cos t + \sin s \sin t).$$

Therefore $\qquad \cos(s - t) = \cos s \cos t + \sin s \sin t.$

The restriction $0 \le s - t < 2\pi$ is removed in Exercise 47.

In proving the formula for $\sin(s + t)$ we shall use the identities

$$\cos\left(\frac{\pi}{2} - z\right) = \sin z \quad \text{and} \quad \sin\left(\frac{\pi}{2} - z\right) = \cos z$$

whose proofs are requested in Exercises 38 and 39. Now we have with $z = s + t$

$$\begin{aligned}\sin(s + t) &= \cos\left(\frac{\pi}{2} - (s + t)\right) = \cos\left(\left(\frac{\pi}{2} - s\right) - t\right) \\ &= \cos\left(\frac{\pi}{2} - s\right)\cos t + \sin\left(\frac{\pi}{2} - s\right)\sin t \\ &= \sin s \cos t + \cos s \sin t.\end{aligned}$$

Proofs of the formulas for $\cos(s + t)$ and $\sin(s - t)$ are left as Exercises 40 and 41.

EXERCISES

Evaluate each expression by using the addition formulas.

A

1. $\sin 20° \cos 70° + \cos 20° \sin 70°$
2. $\cos 40° \cos 50° - \sin 40° \sin 50°$
3. $\cos 80° \cos 20° + \sin 80° \sin 20°$
4. $\sin 130° \cos 100° - \cos 130° \sin 100°$
5. $\cos 70° \cos 50° - \sin 70° \sin 50°$
6. $\sin 75° \cos 60° + \cos 75° \sin 60°$

Simplify.

7. $\cos 2\alpha \cos \alpha - \sin 2\alpha \sin \alpha$
8. $\sin \theta \cos 2\theta + \cos \theta \sin 2\theta$
9. $\sin 2t \cos 3t + \cos 2t \sin 3t$
10. $\cos 3z \cos z + \sin 3z \sin z$

Use the method of Example 1 to find the exact value of each function.

11. $\cos 75°$
12. $\sin 75°$
13. $\sin \frac{\pi}{12}$
14. $\cos \frac{7\pi}{12}$
15. $\cos 255°$
16. $\sin 285°$
17. $\sin \frac{11\pi}{12}$
18. $\cos\left(-\frac{5\pi}{12}\right)$

Prove each identity.

19. $\frac{\sin(\alpha + \beta)}{\cos \alpha \cos \beta} = \tan \alpha + \tan \beta$
20. $\frac{\sin(\alpha + \beta)}{\sin \alpha \sin \beta} = \cot \alpha + \cot \beta$
21. $\frac{\cos(\alpha - \beta)}{\sin \alpha \cos \beta} = \cot \alpha + \tan \beta$
22. $\frac{\cos(\alpha + \beta)}{\cos \alpha \cos \beta} = 1 - \tan \alpha \tan \beta$

(*Note*: The identities in Example 3 and Exercises 23–26 are called **reduction formulas**. They can be used to express the sine or cosine of any number in terms of a function of a first-quadrant number.)

23. $\sin(\pi \pm \theta) = \mp\sin \theta$
24. $\sin\left(\frac{\pi}{2} \pm \theta\right) = \cos \theta$

25. $\cos\left(\frac{\pi}{2} \pm \theta\right) = \mp\sin\theta$ 26. $\cos\left(\frac{3\pi}{2} \pm \theta\right) = \pm\sin\theta$

27. Repeat Example 4, page 319, with $p + q$ replaced by $p - q$.

28. Let $\cos p = \frac{4}{5}$ and $\sin q = \frac{5}{13}$, where p and q are fourth- and second-quadrant numbers, respectively. Find **(a)** $\sin(p - q)$, **(b)** $\cos(p - q)$, and **(c)** the quadrant of $p - q$.

29. Repeat Exercise 28 with $p - q$ replaced by $p + q$.

Write each of the following in terms of $\sin\theta$, $-\sin\theta$, $\cos\theta$, or $-\cos\theta$, where $0° < \theta < 45°$.

30. $\cos 100°$ 31. $\sin 200°$ 32. $\sin(-160°)$

33. $\cos 295°$ 34. $\sec 250°$ 35. $\csc(-125°)$

Give specific examples to show that the following are *not* identities.

36. $\sin(\alpha + \beta) = \sin\alpha + \sin\beta$ 37. $\cos(\alpha + \beta) = \cos\alpha + \cos\beta$

In Exercises 38–41 prove each identity.

38. $\cos\left(\frac{\pi}{2} - z\right) = \sin z$

39. $\sin\left(\frac{\pi}{2} - z\right) = \cos z$ (Use Exercise 38.)

40. $\cos(s + t) = \cos s \cos t - \sin s \sin t$ (*Hint*: $s + t = s - (-t)$)

41. $\sin(s - t) = \sin s \cos t - \cos s \sin t$ (*Hint*: $s - t = s + (-t)$)

The formulas of Exercises 38 and 39 can be written as

$$\cos(90° - \theta) = \sin\theta \quad \text{and} \quad \sin(90° - \theta) = \cos\theta.$$

Use these identities to show that

42. $\cot(90° - \theta) = \tan\theta$ and $\tan(90° - \theta) = \cot\theta$

43. $\csc(90° - \theta) = \sec\theta$ and $\sec(90° - \theta) = \csc\theta$

(*Note*: We say that the sine and cosine are **cofunctions**, as are the tangent and cotangent, and the secant and cosecant. Since θ and $90° - \theta$ are measures of complementary angles, the foregoing identities assert that *cofunctions of complementary angles are equal*.)

Prove the following identities. Use the addition formulas.

B 44. $\sin\alpha\sin\beta = \frac{1}{2}[\cos(\alpha - \beta) - \cos(\alpha + \beta)]$

45. $\cos\alpha\cos\beta = \frac{1}{2}[\cos(\alpha - \beta) + \cos(\alpha + \beta)]$

46. $\sin\alpha\cos\beta = \frac{1}{2}[\sin(\alpha - \beta) + \sin(\alpha + \beta)]$

C 47. Prove that $\cos(s - t) = \cos s \cos t + \sin s \sin t$ for any difference, $s - t$. (*Hint*: For any difference, $s - t$, there is an integer k such that $0 \le (s - t) - 2k\pi < 2\pi$.)

COMPUTER EXERCISES

1. Write a computer program to verify numerically that
$$\frac{\sin x}{1 + \cos x} + \frac{1 + \cos x}{\sin x} = \frac{2}{\sin x}$$
is an identity. Run the program for ten values of x produced by the computer's random-number generator.
2. For what values of x is the equation of Exercise 1 undefined?
3. **a.** Write a computer program using random numbers to show that there are real numbers x and y such that $\sin (x - y) \neq \sin x - \sin y$.
   b. Run the program for ten randomly generated values of x and y.
4. Find at least one pair of nonzero, unequal, real numbers x and y for which $\sin (x - y) = \sin x - \sin y$.

8-3 DOUBLE-ANGLE AND HALF-ANGLE FORMULAS

If we replace both s and t by α in the formulas for $\sin (s + t)$ and $\cos (s + t)$, we obtain

$$\sin (\alpha + \alpha) = \sin \alpha \cos \alpha + \cos \alpha \sin \alpha = 2 \sin \alpha \cos \alpha$$

and $$\cos (\alpha + \alpha) = \cos \alpha \cos \alpha - \sin \alpha \sin \alpha = \cos^2 \alpha - \sin^2 \alpha.$$

Thus

$$\sin 2\alpha = 2 \sin \alpha \cos \alpha \qquad \cos 2\alpha = \cos^2 \alpha - \sin^2 \alpha$$

These identities are called **double-angle** formulas. Two other useful formulas for $\cos 2\alpha$ can be obtained with the help of the identity $\sin^2 \alpha + \cos^2 \alpha = 1$.

$$\begin{aligned} \cos 2\alpha &= \cos^2 \alpha - \sin^2 \alpha \\ &= \cos^2 \alpha - (1 - \cos^2 \alpha) \\ &= 2 \cos^2 \alpha - 1 \end{aligned} \qquad \begin{aligned} \cos 2\alpha &= \cos^2 \alpha - \sin^2 \alpha \\ &= (1 - \sin^2 \alpha) - \sin^2 \alpha \\ &= 1 - 2 \sin^2 \alpha \end{aligned}$$

$$\cos 2\alpha = 2 \cos^2 \alpha - 1 \qquad \cos 2\alpha = 1 - 2 \sin^2 \alpha$$

EXAMPLE 1 **a.** Find $\sin 2\theta$ and $\cos 2\theta$ if $\cos \theta = \frac{3}{5}$ and $\sin \theta < 0$.

b. What is the quadrant of 2θ?

Solution **a.** Use $\cos \theta = \frac{3}{5}$ and $\sin^2 \theta + \cos^2 \theta = 1$ to find $\sin \theta$.

$$\sin \theta = -\sqrt{1 - \cos^2 \theta} = -\sqrt{1 - \frac{9}{25}} = -\sqrt{\frac{16}{25}} = -\frac{4}{5}.$$

Next use the double-angle formulas to find sin 2θ and cos 2θ.

$$\sin 2\theta = 2 \sin \theta \cos \theta = 2\left(-\frac{4}{5}\right)\left(\frac{3}{5}\right) = -\frac{24}{25} \qquad \textbf{Answer}$$

$$\cos 2\theta = 2 \cos^2 \theta - 1 = 2 \cdot \left(\frac{9}{25}\right) - 1 = -\frac{7}{25} \qquad \textbf{Answer}$$

b. Since sin 2θ and cos 2θ are both negative, 2θ is a third-quadrant angle. **Answer**

EXAMPLE 2 Prove the identity $\cot \theta - \tan \theta = 2 \cot 2\theta$.

Solution

$$\begin{aligned}\cot \theta - \tan \theta &= \frac{\cos \theta}{\sin \theta} - \frac{\sin \theta}{\cos \theta} \\ &= \frac{\cos^2 \theta - \sin^2 \theta}{\sin \theta \cos \theta} \\ &= 2\,\frac{\cos^2 \theta - \sin^2 \theta}{2 \sin \theta \cos \theta} \\ &= 2\,\frac{\cos 2\theta}{\sin 2\theta} = 2 \cot 2\theta\end{aligned}$$

EXAMPLE 3 Express cos 3*x* in terms of cos *x*.

Solution

$$\begin{aligned}\cos 3x &= \cos (x + 2x) = \cos x \cos 2x - \sin x \sin 2x \\ &= \cos x (\cos^2 x - \sin^2 x) - \sin x (2 \sin x \cos x) \\ &= \cos^3 x - 3 \cos x \sin^2 x \\ &= \cos^3 x - 3 \cos x (1 - \cos^2 x) \\ \cos 3x &= 4 \cos^3 x - 3 \cos x\end{aligned}$$

Formulas for $\sin \frac{\theta}{2}$ and $\cos \frac{\theta}{2}$ can be obtained by replacing α by $\frac{\theta}{2}$ in the last two of the formulas for cos 2α:

$$\begin{aligned}\cos 2\alpha &= 1 - 2 \sin^2 \alpha \\ \cos 2 \cdot \frac{\theta}{2} &= 1 - 2 \sin^2 \frac{\theta}{2} \\ \cos \theta &= 1 - 2 \sin^2 \frac{\theta}{2} \\ \sin^2 \frac{\theta}{2} &= \frac{1 - \cos \theta}{2}\end{aligned} \qquad \begin{aligned}\cos 2\alpha &= 2 \cos^2 \alpha - 1 \\ \cos 2 \cdot \frac{\theta}{2} &= 2 \cos^2 \frac{\theta}{2} - 1 \\ \cos \theta &= 2 \cos^2 \frac{\theta}{2} - 1 \\ \cos^2 \frac{\theta}{2} &= \frac{1 + \cos \theta}{2}\end{aligned}$$

$$\sin \frac{\theta}{2} = \pm \sqrt{\frac{1 - \cos \theta}{2}} \qquad \cos \frac{\theta}{2} = \pm \sqrt{\frac{1 + \cos \theta}{2}}$$

These are called **half-angle** formulas. In each formula, the algebraic sign is determined by the quadrant of $\frac{\theta}{2}$.

EXAMPLE 4 Find $\cos \frac{\theta}{2}$ if $\cos \theta = -\frac{31}{81}$ and $180° < \theta < 360°$.

Solution

$$\cos \frac{\theta}{2} = \pm\sqrt{\frac{1 + \cos \theta}{2}}$$

$$= \pm\sqrt{\frac{1 + \left(-\frac{31}{81}\right)}{2}}$$

$$= \pm\sqrt{\frac{\frac{50}{81}}{2}} = \pm\sqrt{\frac{25}{81}} = \pm\frac{5}{9}$$

Since $180° < \theta < 360°$, $90° < \frac{\theta}{2} < 180°$.

Thus $\cos \frac{\theta}{2}$ is negative and $\cos \frac{\theta}{2} = -\frac{5}{9}$. **Answer**

EXAMPLE 5 Use half-angle formulas to find the exact values of sin 105° and cos 105°.

Solution Since 105° is a second-quadrant number, its sine is positive, and its cosine is negative. Hence

$$\sin 105° = \sin \frac{1}{2} \cdot 210°$$

$$= +\sqrt{\frac{1 - \cos 210°}{2}}$$

$$= \sqrt{\frac{1 - \left(-\frac{\sqrt{3}}{2}\right)}{2}}$$

$$= \sqrt{\frac{2 + \sqrt{3}}{4}}$$

$$\sin 105° = \frac{\sqrt{2 + \sqrt{3}}}{2}$$ **Answer**

$$\cos 105° = \cos \frac{1}{2} \cdot 210°$$

$$= -\sqrt{\frac{1 + \cos 210°}{2}}$$

$$= -\sqrt{\frac{1 + \left(-\frac{\sqrt{3}}{2}\right)}{2}}$$

$$= -\sqrt{\frac{2 - \sqrt{3}}{4}}$$

$$\cos 105° = -\frac{\sqrt{2 - \sqrt{3}}}{2}$$ **Answer**

In Example 1(a), page 318, we found that $\sin 105° = \frac{\sqrt{6} + \sqrt{2}}{4}$ by using an addition formula. You can verify that this result agrees with the result obtained by using the half-angle formula in Example 5.

EXERCISES

Given that $90° < \theta < 180°$ and $\sin \theta = \frac{3}{5}$. Find each value.

A **1.** $\sin 2\theta$ **2.** $\cos 2\theta$ **3.** $\cos \frac{\theta}{2}$ **4.** $\sin \frac{\theta}{2}$

Given that $\frac{3\pi}{2} < t < 2\pi$ and $\sin t = -\frac{12}{13}$. Find each value.

5. $\cos 2t$ **6.** $\sin 2t$ **7.** $\sin \frac{t}{2}$ **8.** $\cos \frac{t}{2}$

In Exercises 9–14, **(a)** find $\sin 2\theta$ and $\cos 2\theta$. **(b)** What is the quadrant of 2θ?

9. $\sin \theta = -\frac{4}{5}, \cos \theta < 0$

10. $\cos \theta = \frac{3}{5}, \sin \theta < 0$

11. $\cos \theta = -\frac{5}{13}, \sin \theta > 0$

12. $\sin \theta = \frac{7}{25}, \cos \theta < 0$

13. $\cos \theta = \frac{2}{3}, \sin \theta > 0$

14. $\sin \theta = -\frac{1}{3}, \cos \theta < 0$

Find exact values of each of the following **(a)** by using a half-angle formula and **(b)** by using an addition formula from Section 8-2. Then **(c)** show that your answers in (a) and (b) are equal.

15. $\cos 15°$

16. $\sin 75°$

17. $\sin \frac{13\pi}{12}$

18. $\cos \frac{5\pi}{12}$

19. $\cos 255°$

20. $\sin 210°$

21. $\sin \frac{7\pi}{12}$

22. $\cos \frac{11\pi}{12}$

Prove the following identities.

23. $\csc 2\alpha = \frac{1}{2} \sec \alpha \csc \alpha$

24. $\sec 2\beta = \frac{\sec^2 \beta}{2 - \sec^2 \beta}$

25. $\cos^4 \theta - \sin^4 \theta = \cos 2\theta$

26. $(\sin t + \cos t)^2 = 1 + \sin 2t$

27. $\cot x + \tan x = 2 \csc 2x$

28. $\frac{1 + \tan^2 u}{1 - \tan^2 u} = \sec 2u$

B

29. $1 - 3 \sin^2 \theta + 2 \sin^4 \theta = \cos^2 \theta \cos 2\theta$

30. $\cos^4 \alpha - 2 \sin^2 \alpha \cos^2 \alpha + \sin^4 \alpha = \cos^2 2\alpha$

31. Express $\sin 3x$ in terms of $\sin x$.

32. Express $\cos 4x$ in terms of $\cos x$.

33. Use a half-angle formula to show that $\cos \frac{\pi}{4} = \frac{\sqrt{2}}{2}$.

C

34. Prove or disprove: If α and β differ by an integral multiple of 2π, then $\sin \frac{\alpha}{2} = \sin \frac{\beta}{2}$ and $\cos \frac{\alpha}{2} = \cos \frac{\beta}{2}$.

35. (Challenge) Prove by mathematical induction: For any positive integer n, $\cos nx$ is a polynomial in $\cos x$, and $\sin nx$ is the product of $\sin x$ and a polynomial in $\cos x$. (*Hint*: Prove the two parts together.)

8-4 MORE TRIGONOMETRIC IDENTITIES

We can obtain addition formulas for the tangent by using the addition formulas for the sine and the cosine. If $\tan(s + t)$, $\tan s$, and $\tan t$ are defined, we have

$$\tan(s + t) = \frac{\sin(s + t)}{\cos(s + t)}$$

$$= \frac{\sin s \cos t + \cos s \sin t}{\cos s \cos t - \sin s \sin t}$$

$$= \frac{\dfrac{\sin s \cos t}{\cos s \cos t} + \dfrac{\cos s \sin t}{\cos s \cos t}}{\dfrac{\cos s \cos t}{\cos s \cos t} - \dfrac{\sin s \sin t}{\cos s \cos t}}$$

Divide numerator and denominator by $\cos s \cos t$.

$$\tan(s + t) = \frac{\tan s + \tan t}{1 - \tan s \tan t}$$

By replacing t by $-t$ and using the fact that $\tan(-t) = -\tan t$, we obtain a formula for $\tan(s - t)$.

$$\tan(s + t) = \frac{\tan s + \tan t}{1 - \tan s \tan t} \qquad \tan(s - t) = \frac{\tan s - \tan t}{1 + \tan s \tan t}$$

EXAMPLE 1 Find the exact value of $\tan 195°$.

Solution

$$\tan 195° = \tan(135° + 60°) = \frac{\tan 135° + \tan 60°}{1 - \tan 135° \tan 60°}$$

$$= \frac{-1 + \sqrt{3}}{1 - (-1)\sqrt{3}}$$

$$\tan 195° = \frac{\sqrt{3} - 1}{\sqrt{3} + 1} \quad \textbf{Answer}$$

Rationalize the denominator to obtain $\tan 195° = 2 - \sqrt{3}$.

When we substitute α for both s and t in the formula for $\tan(s + t)$, we obtain a double-angle formula.

$$\tan 2\alpha = \frac{2 \tan \alpha}{1 - \tan^2 \alpha}$$

Two half-angle identities for the tangent are given below. We shall prove the first identity. The proof of the second identity is left as Exercise 19.

$$\tan \frac{\theta}{2} = \frac{1 - \cos \theta}{\sin \theta} \qquad \tan \frac{\theta}{2} = \frac{\sin \theta}{1 + \cos \theta}$$

Use the sine and cosine double-angle formulas.

$\tan \alpha = \dfrac{\sin \alpha}{\cos \alpha} = \dfrac{2 \sin^2 \alpha}{2 \sin \alpha \cos \alpha}$. Since $2 \sin^2 \alpha = 1 - \cos 2\alpha$,

$$\tan \alpha = \frac{1 - \cos 2\alpha}{\sin 2\alpha}.$$

Let $\frac{\theta}{2} = \alpha$. $\quad \tan \frac{\theta}{2} = \dfrac{1 - \cos\left(2\left(\frac{\theta}{2}\right)\right)}{\sin\left(2\left(\frac{\theta}{2}\right)\right)} = \dfrac{1 - \cos \theta}{\sin \theta}$

EXAMPLE 2 Find the exact value of $\tan \frac{7\pi}{8}$.

Solution Use the formula $\tan \frac{\theta}{2} = \dfrac{1 - \cos \theta}{\sin \theta}$ with $\theta = \frac{7\pi}{4}$.

$$\tan \frac{7\pi}{8} = \tan\left[\frac{1}{2}\left(\frac{7\pi}{4}\right)\right] = \frac{1 - \cos \frac{7\pi}{4}}{\sin \frac{7\pi}{4}} = \frac{1 - \frac{\sqrt{2}}{2}}{-\frac{\sqrt{2}}{2}} = 1 - \sqrt{2}$$ **Answer**

The remainder of this section is devoted to the product-to-sum and sum-to-product formulas.

Product-to-Sum Formulas

$$\sin \alpha \sin \beta = \frac{1}{2}[\cos (\alpha - \beta) - \cos (\alpha + \beta)]$$

$$\cos \alpha \cos \beta = \frac{1}{2}[\cos (\alpha - \beta) + \cos (\alpha + \beta)]$$

$$\sin \alpha \cos \beta = \frac{1}{2}[\sin (\alpha - \beta) + \sin (\alpha + \beta)]$$

We shall prove the identity for $\sin \alpha \sin \beta$.

Use the formulas for $\cos (\alpha - \beta)$ and $\cos (\alpha + \beta)$.

$$\cos (\alpha - \beta) = \cos \alpha \cos \beta + \sin \alpha \sin \beta$$
$$\cos (\alpha + \beta) = \cos \alpha \cos \beta - \sin \alpha \sin \beta$$

Subtract the second equation from the first equation.

$$\cos (\alpha - \beta) - \cos (\alpha + \beta) = 2 \sin \alpha \sin \beta$$

Thus $\sin \alpha \sin \beta = \frac{1}{2}[\cos (\alpha - \beta) - \cos (\alpha + \beta)]$.

Derivations of the other identities are similar. (See Exercises 37 and 38.)

EXAMPLE 3 Write $\sin 2x \cos 3x$ as a sum or difference.

Solution Use the third product-to-sum formula.

$$\sin 2x \cos 3x = \frac{1}{2}[\sin (2x - 3x) + \sin (2x + 3x)]$$
$$= \frac{1}{2}[\sin (-x) + \sin 5x]$$
$$= \frac{1}{2}(\sin 5x - \sin x)$$
$$\sin 2x \cos 3x = \frac{1}{2}\sin 5x - \frac{1}{2}\sin x \qquad \textbf{Answer}$$

The following formulas enable us to write certain sums and differences as products.

Sum-to-Product Formulas

$$\sin s + \sin t = 2 \sin \frac{s+t}{2} \cos \frac{s-t}{2}$$
$$\sin s - \sin t = 2 \cos \frac{s+t}{2} \sin \frac{s-t}{2}$$
$$\cos s + \cos t = 2 \cos \frac{s+t}{2} \cos \frac{s-t}{2}$$
$$\cos s - \cos t = -2 \sin \frac{s+t}{2} \sin \frac{s-t}{2}$$

EXAMPLE 4 Simplify $\dfrac{\sin 3x - \sin x}{\cos 3x + \cos x}$.

Solution Use the second sum-to-product formula with $s = 3x$ and $t = x$ for the numerator and use the third sum-to-product formula with $s = 3x$ and $t = x$ for the denominator.

$$\frac{\sin 3x - \sin x}{\cos 3x + \cos x} = \frac{2 \cos \frac{3x+x}{2} \sin \frac{3x-x}{2}}{2 \cos \frac{3x+x}{2} \cos \frac{3x-x}{2}}$$
$$= \frac{\sin x}{\cos x}$$
$$\frac{\sin 3x - \sin x}{\cos 3x + \cos x} = \tan x \qquad \textbf{Answer}$$

Note: The product-to-sum and sum-to-product formulas are seldom needed today for calculations. These formulas have theoretical value, however. For our purposes it is enough to know that they exist and can be derived from the sum formulas. See Exercises 39–42.

EXERCISES

Exercises involving the tangent identities

Simplify.

A

1. $\dfrac{\tan 20° + \tan 40°}{1 - \tan 20° \tan 40°}$

2. $\dfrac{\tan \frac{5\pi}{8} - \tan \frac{3\pi}{8}}{1 + \tan \frac{5\pi}{8} \tan \frac{3\pi}{8}}$

3. $\dfrac{2 \tan \frac{\pi}{12}}{1 - \tan^2 \frac{\pi}{12}}$

4. $\dfrac{2 \tan 75°}{1 - \tan^2 75°}$

Find the exact value of each tangent.

5. $\tan 15°$
6. $\tan 75°$
7. $\tan \frac{11\pi}{12}$
8. $\tan \frac{13\pi}{12}$
9. $\tan 22.5°$
10. $\tan 67.5°$
11. $\tan \frac{11\pi}{8}$
12. $\tan \frac{5\pi}{8}$

In Exercises 13–16, $\tan \alpha = -\frac{4}{3}$, $\tan \beta = \frac{5}{12}$, and α is a second-quadrant number. Find each tangent.

13. $\tan (\alpha + \beta)$
14. $\tan (\alpha - \beta)$
15. $\tan \frac{\alpha}{2}$
16. $\tan 2\beta$

B

17. Express $\cot (s + t)$ in terms of $\cot s$ and $\cot t$.
18. Express $\cot (s - t)$ in terms of $\cot s$ and $\cot t$.

Prove each of the following identities.

19. $\tan \frac{\theta}{2} = \dfrac{\sin \theta}{1 + \cos \theta}$

20. $\cot \frac{\theta}{2} = \dfrac{\sin \theta}{1 - \cos \theta}$

21. $\tan \frac{\theta}{2} = \pm\sqrt{\dfrac{1 - \cos \theta}{1 + \cos \theta}}$

22. $\cot \frac{\theta}{2} = \pm\sqrt{\dfrac{1 + \cos \theta}{1 - \cos \theta}}$

23. $\tan \left(\theta + \frac{\pi}{4}\right) = \dfrac{\cos \theta + \sin \theta}{\cos \theta - \sin \theta}$

24. $\tan \left(\theta - \frac{\pi}{4}\right) = \dfrac{\sin \theta - \cos \theta}{\sin \theta + \cos \theta}$

Exercises involving the product-to-sum and sum-to-product formulas

Express as a sum or a difference.

A

25. $2 \cos t \cos 2t$
26. $2 \sin 4t \sin 3t$
27. $\sin 2\alpha \cos 4\alpha$
28. $\cos 3\beta \sin 5\beta$
29. $\sin \left(\frac{\pi}{4} + x\right) \sin \left(\frac{\pi}{4} - x\right)$
30. $\cos (p + q) \cos (p - q)$

Express each sum or difference as a product.

31. $\sin \alpha + \sin 3\alpha$
32. $\cos 3t - \cos 5t$
33. $\cos 4x + \cos 6x$
34. $\sin 2\theta - \sin 4\theta$

Express each sum or difference as a product.

35. $\cos\left(\alpha + \frac{\pi}{3}\right) + \cos\left(\alpha - \frac{\pi}{3}\right)$

36. $\sin\left(\alpha + \frac{\pi}{6}\right) - \sin\left(\alpha - \frac{\pi}{6}\right)$

Prove the following identities.

37. $\cos\alpha \cos\beta = \frac{1}{2}[\cos(\alpha - \beta) + \cos(\alpha + \beta)]$

38. $\sin\alpha \cos\beta = \frac{1}{2}[\sin(\alpha - \beta) + \sin(\alpha + \beta)]$

B **39.** $\sin s + \sin t = 2\sin\frac{s + t}{2}\cos\frac{s - t}{2}$
(*Hint*: Let $\alpha - \beta = s$ and $\alpha + \beta = t$. Use the third product-to-sum formula.)

40. $\sin s - \sin t = 2\cos\frac{s + t}{2}\sin\frac{s - t}{2}$

41. $\cos s + \cos t = 2\cos\frac{s + t}{2}\cos\frac{s - t}{2}$

42. $\cos s - \cos t = -2\sin\frac{s + t}{2}\sin\frac{s - t}{2}$

43. $\frac{\cos 3x - \cos x}{\sin 3x + \sin x} = -\tan x$

44. $\frac{\cos 4t + \cos 2t}{\sin 4t - \sin 2t} = \cot t$

45. $\frac{\sin u - \sin v}{\cos u + \cos v} = \tan\frac{u - v}{2}$

46. $\frac{\cos u - \cos v}{\sin u + \sin v} = -\tan\frac{u - v}{2}$

47. $\frac{\sin x + \sin 2x + \sin 3x}{\cos x + \cos 2x + \cos 3x} = \tan 2x$
(*Hint*: First work with $\sin x + \sin 3x$ and $\cos x + \cos 3x$.)

48. $\tan\frac{x \pm y}{2} = \frac{\sin x \pm \sin y}{\cos x + \cos y}$

49. $\frac{\sin x + \sin y}{\sin x - \sin y} = \frac{\tan \frac{1}{2}(x + y)}{\tan \frac{1}{2}(x - y)}$

C **50.** $\cos 0x + \cos 2x + \cos 4x + \cos 6x = 4\cos x \cos 2x \cos 3x$

51. $\sin 0x + \sin 2x + \sin 4x + \sin 6x = 4\cos x \cos 2x \sin 3x$

8-5 TRIGONOMETRIC EQUATIONS

In the preceding sections we have discussed trigonometric identities, that is, equations that are true for *all* allowable values of the variables involved. We now consider trigonometric equations that are true only for certain values of the variables; and our problem is to find those values. In solving such equations we usually use a combination of algebraic methods and trigonometric identities.

EXAMPLE 1 Solve $3\sin^2 x = \cos^2 x$ for x such that $0 \le x < 2\pi$.

Solution Divide both sides by $3\cos^2 x$ and then use the identity $\frac{\sin x}{\cos x} = \tan x$.

$$\frac{\sin^2 x}{\cos^2 x} = \frac{1}{3} \quad \text{or} \quad \tan^2 x = \frac{1}{3}$$

Division by $3\cos^2 x$ is valid if $3\cos^2 x \neq 0$, that is, if $x \neq \frac{\pi}{2}$ or $\frac{3\pi}{2}$. Since neither of these numbers satisfies the given equation, no solutions have been lost.

Thus $\tan x = \pm\sqrt{\frac{1}{3}}$. That is, $\tan x = \frac{\sqrt{3}}{3}$ or $-\frac{\sqrt{3}}{3}$.

If $\tan x = \frac{\sqrt{3}}{3}$, then $x = \frac{\pi}{6}$ or $\frac{7\pi}{6}$.

If $\tan x = -\frac{\sqrt{3}}{3}$, then $x = \frac{5\pi}{6}$ or $\frac{11\pi}{6}$.

$\therefore$ the solutions in $0 \leq x < 2\pi$ are $\frac{\pi}{6}, \frac{5\pi}{6}, \frac{7\pi}{6}$, and $\frac{11\pi}{6}$. **Answer**

Solutions of trigonometric equations are often given in degrees.

EXAMPLE 2 Let $\cos 2\theta = 3\sin\theta - 1$.

a. Find the solutions in the interval $0° \leq \theta < 360°$.

b. Find expressions for all the solutions.

Solution Of the three formulas for $\cos 2\theta$, use $\cos 2\theta = 1 - 2\sin^2\theta$ so that $\sin\theta$ will be the only trigonometric function in the transformed equation.

$$\cos 2\theta = 3\sin\theta - 1$$
$$1 - 2\sin^2\theta = 3\sin\theta - 1$$
$$2\sin^2\theta + 3\sin\theta - 2 = 0$$

a. The preceding equation is quadratic in $\sin\theta$. Solve first for $\sin\theta$ and then for θ.

$$2\sin^2\theta + 3\sin\theta - 2 = 0$$
$$(2\sin\theta - 1)(\sin\theta + 2) = 0$$
$$2\sin\theta - 1 = 0 \quad \text{or} \quad \sin\theta + 2 = 0$$

$\sin\theta = \frac{1}{2}$	$\sin\theta = -2$
$\theta = 30°$ or $150°$	No solution since $-1 \leq \sin\theta \leq 1$.

$\therefore$ the solutions in $0 \leq \theta < 360°$ are $30°$ and $150°$. **Answer**

b. All other solutions of the given equation differ from the ones in (a) by integral multiples of $360°$. Thus, all solutions are given by

$\theta = 30° + k \cdot 360°$ and $\theta = 150° + k \cdot 360°$, k an integer. **Answer**

Solutions in a specified interval, usually $0 \leq x < 2\pi$ or $0° \leq \theta < 360°$, as in Examples 1 and 2(a), are sometimes called **principal solutions.** Formulas such as the ones in Example 2(b) give the **general solution.**

We sometimes must use a calculator or table to complete the solution of a trigonometric equation.

EXAMPLE 3 Solve $\sin 2t = 5\cos^2 t$, $0 \le t < 2\pi$.

Solution Use a double-angle formula and factoring.

$$\begin{aligned} \sin 2t &= 5\cos^2 t \\ 2\sin t\cos t &= 5\cos^2 t \\ 2\sin t\cos t - 5\cos^2 t &= 0 \\ \cos t\,(2\sin t - 5\cos t) &= 0 \\ \cos t = 0 \text{ or } 2\sin t - 5\cos t &= 0 \end{aligned}$$

If $\cos t = 0$, then $t = \frac{\pi}{2}$ or $\frac{3\pi}{2}$.

If $2\sin t - 5\cos t = 0$, then $\frac{\sin t}{\cos t} = \frac{5}{2}$, or $\tan t = 2.5$

A calculator or Table 3 gives $t = 1.19, 1.19 + \pi$.

$\therefore$ the solutions are $1.19, \frac{\pi}{2}, 4.33, \frac{3\pi}{2}$. **Answer**

Notice that in Example 3 we did not divide both sides of $2\sin t\cos t = 5\cos^2 t$ by $\cos t$. Had we done so, we would have lost the solutions for which $\cos t = 0$. Substitution of the values obtained into the original equation shows that the solution is correct.

EXAMPLE 4 Solve $\cos(3\varphi - 75°) = \frac{1}{2}$, $0° \le \varphi < 360°$.

Solution Since $\cos 60° = \frac{1}{2}$ and $\cos 300° = \frac{1}{2}$, then where k is an integer:

$3\varphi - 75° = 60° + k \cdot 360°$	$3\varphi - 75° = 300° + k \cdot 360°$
$3\varphi = 135° + k \cdot 360°$	$3\varphi = 375° + k \cdot 360°$
$\varphi = 45° + k \cdot 120°$	$\varphi = 125° + k \cdot 120°$

Substitute $k = -1, 0, 1$, and 2 to obtain

$$\varphi = -75°, 45°, 165°, 285° \quad | \quad \varphi = 5°, 125°, 245°, 365°$$

$\therefore$ in $0° \le \varphi < 360°$, $\varphi = 5°, 45°, 125°, 165°, 245°$, and $285°$. **Answer**

EXERCISES

If the variable involved is α, θ, or φ, use degree measure. If the variable involved is t, u, or x, use radian measure.

In Exercises 1–10, find the solutions between 0° and 360° or between 0 and 2π.

A

1. $\sin\theta = \cos\theta$

2. $\sin\alpha + \cos\alpha = 0$

3. $4\cos^2 x = 1$

4. $2\sin^2 t = 1$

5. $4\sin^2\varphi = 3$

6. $3\sec^2\theta = 4$

7. $2\sin(\theta - 10°) = 1$

8. $\tan(\alpha + 20°) = 1$

9. $2 \cos\left(t + \frac{\pi}{9}\right) = \sqrt{3}$

10. $\sqrt{2} \sin\left(u + \frac{\pi}{6}\right) = -1$

In Exercises 11–20, find the general solution.

11. $\sin \theta = 1$

12. $\cos \alpha = 0$

13. $2 \cos x = \sec x$

14. $4 \sin t = 3 \csc t$

15. $\sin^2 \alpha = 3 \cos^2 \alpha$

16. $\sec \varphi = 2 \cos \varphi$

17. $\cos 2t = \frac{1}{2}$

18. $\sin 2t = 1$

19. $\tan (\theta + 10°) = 1$

20. $\cos (\varphi - 10°) = \frac{1}{2}$

In Exercises 21–30, find the solutions in $0° \leq \theta < 360°$ or $0 \leq x < 2\pi$.

21. $\tan \theta = 2 \sin \theta$

22. $\sqrt{2} \sin x = \cot x$

23. $\sin 2x + \cos 2x = 0$

24. $\sin 3x = \cos 3x$

B

25. $\sin 2x + \cos x = 0$

26. $\sin 2\theta = \sin \theta$

27. $\cos 2\theta = \cos \theta$

28. $\cos 2x = \sin x$

29. $\cos 2x = 3 \cos x + 1$

30. $3 \sin \theta = 1 + \cos 2\theta$

In Exercises 31–46, find all solutions in $0° \leq \theta < 360°$ or $0 \leq x < 2\pi$. Give answers to the nearest 0.1° or 0.001 radian.

31. $\sin^2 \theta = 2 \cos^2 \theta$

32. $2 \sin \theta + 3 \cos \theta = 0$

33. $2 \sin 2x = \cos x$

34. $2 \cos 2x = 2 + \sin x$

35. $\cos 2\theta = 2 \cos \theta$

36. $\cos 2\theta = 2 \sin \theta$

37. $\tan (2\theta - 20°) = 1$

38. $\cos (2\theta + 40°) = 0$

39. $\sin\left(3x + \frac{\pi}{12}\right) = \frac{1}{2}$

40. $\sin\left(3x - \frac{\pi}{6}\right) = -\frac{1}{2}$

C

41. $4 \sin \theta \cos \theta = \sqrt{3}$

42. $4 \sin x \cos x = 1$

43. $\sin \theta \cos 2\theta + \cos \theta \sin 2\theta = \frac{1}{2}$

44. $\cos x \cos 3x - \sin x \sin 3x = 1$

45. $\sqrt{2} (\sin \theta + \cos \theta) = \sqrt{3}$. (*Hint*: Square both sides.)

46. $\sqrt{2} (\sin x - \cos x) = 1$ (*Hint*: Square both sides.)

COMPUTER EXERCISES

1. Write a program to find, to two decimal places, each x such that $|\tan 3x + 2 \cos x| < 0.005$ in the interval $\left[-\frac{\pi}{6}, \frac{\pi}{6}\right]$.
2. Write a program to find, to two decimal places, each x such that $|\sin x - \cos x - 0.2| < 0.005$ in the interval $\left[\pi, \frac{3\pi}{2}\right]$.

Trigonometric Graphs

■ 8-6 PERIODICITY AND OTHER PROPERTIES

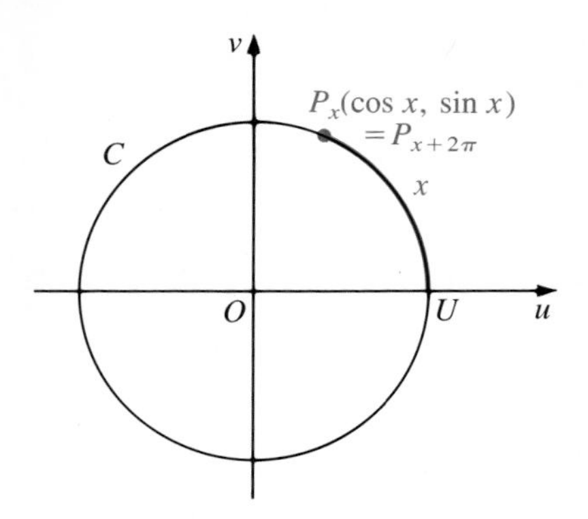

In Section 7-1 we defined $\cos x$ and $\sin x$ to be the first and second coordinates, respectively, of the point P_x on the unit circle C. Because the circumference of C is 2π,

$$P_{x+2\pi} = P_x,$$

$\cos(x + 2\pi) = \cos x$, and $\sin(x + 2\pi) = \sin x$. We say that $\sin x$ and $\cos x$ have *period* 2π.

In general, a function f is **periodic** if for some positive constant p,

$$f(x + p) = f(x)$$

for every x in the domain of f. The smallest such p is the (fundamental) **period** of f. Any portion of the graph of a periodic function f having endpoints of the form $(c, f(c))$ and $(c + p, f(c + p))$ is a **cycle** of the graph.

EXAMPLE 1 One cycle of the graph of a function f having period 2 is shown. Graph f in the interval $-3 \le x \le 5$.

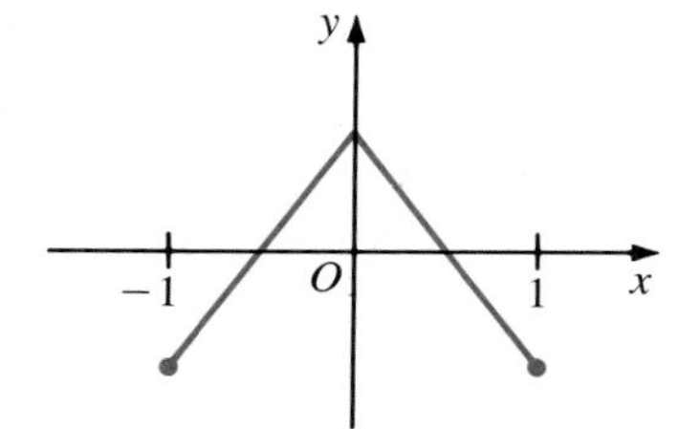

Solution The graph is given on $[-1, 1]$. Copy the graph over the intervals $[-3, -1]$, $[1, 3]$, and $[3, 5]$.

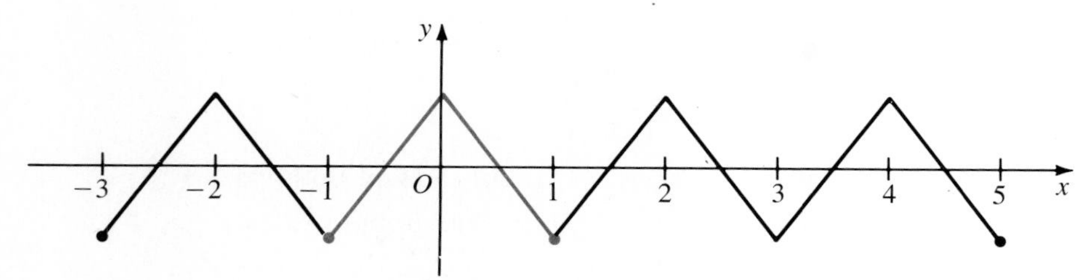

Recall from Section 2-3 that the circle with equation $(x - h)^2 + (y - k)^2 = r^2$ can be obtained by translating the circle whose equation is $x^2 + y^2 = r^2$ horizontally h units and vertically k units.

The graph of

$$y - k = f(x - h), \quad \text{or} \quad y = f(x - h) + k$$

can be obtained from the graph of $y = f(x)$ in the same way. See the figures that follow. If f is periodic, the horizontal translation h is often called the **phase shift**. Example 2(a) illustrates a phase shift of $-\frac{1}{2}$.

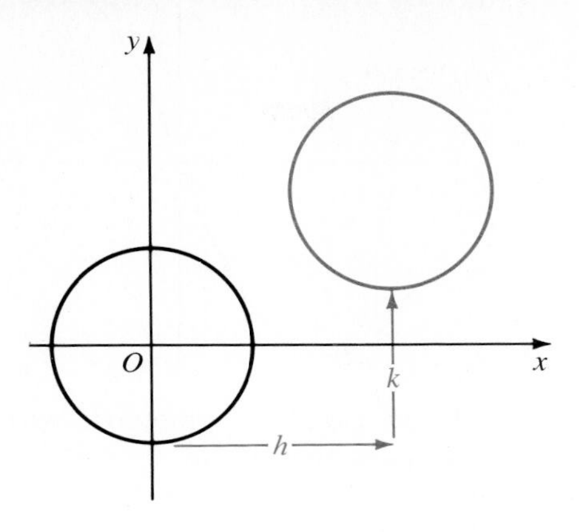

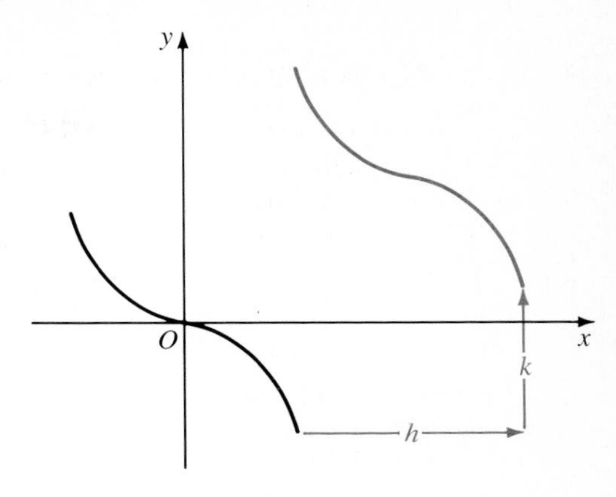

EXAMPLE 2 Let f be the periodic function of Example 1. Graph **(a)** $y = f(x + \frac{1}{2})$ and **(b)** $y = f(x) + \frac{1}{2}$.

Solution **a.** Note that $f(x + \frac{1}{2}) = f(x - (-\frac{1}{2}))$. So $h = -\frac{1}{2}$. Shift the graph of $y = f(x)$ to the left $\frac{1}{2}$ unit.

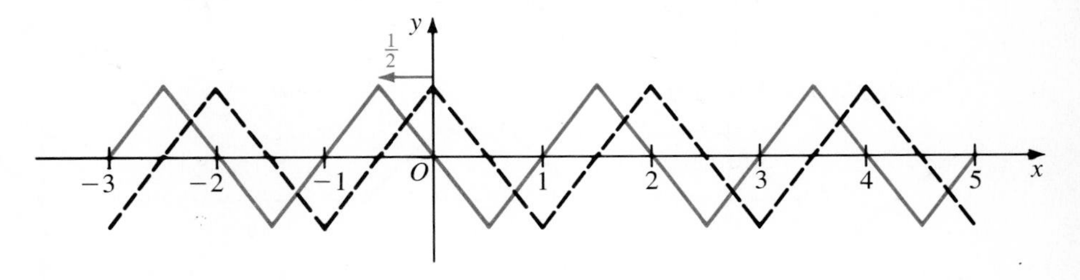

b. Shift the graph of $y = f(x)$ up $\frac{1}{2}$ unit.

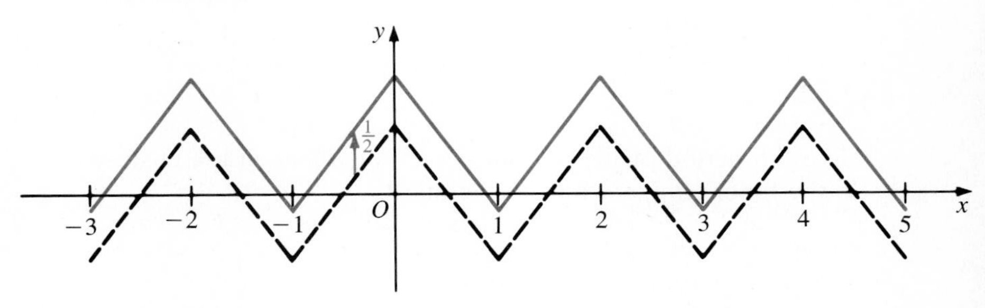

Recall from Section 7-2 that a function f is **even** if $f(-x) = f(x)$; and is **odd** if $f(-x) = -f(x)$, for all x in the domain of f. The following figures illustrate graphs of even and odd functions.

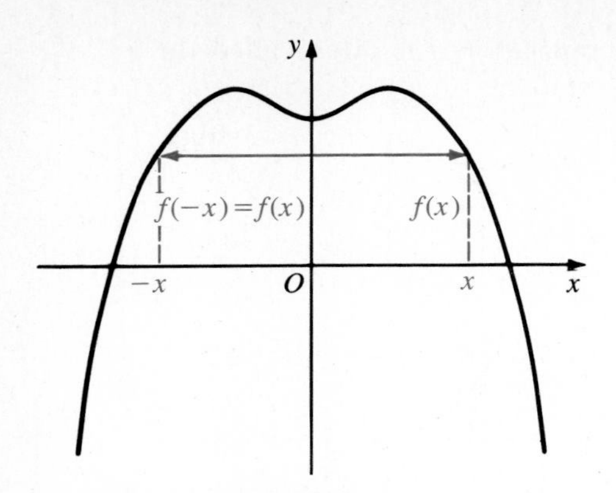

f even: graph is symmetric with respect to the *y*-axis.

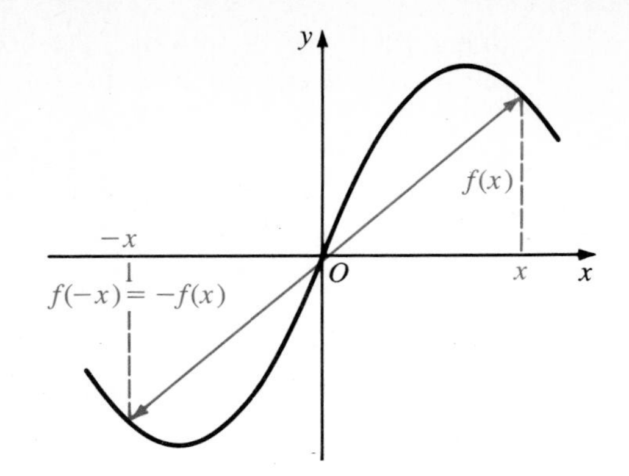

f odd: graph is symmetric with respect to the origin.

EXAMPLE 3 Part of the graph of a periodic function f is shown at the right. Graph f in the interval $-3 \leq x \leq 3$ given that **(a)** f is even; **(b)** f is odd.

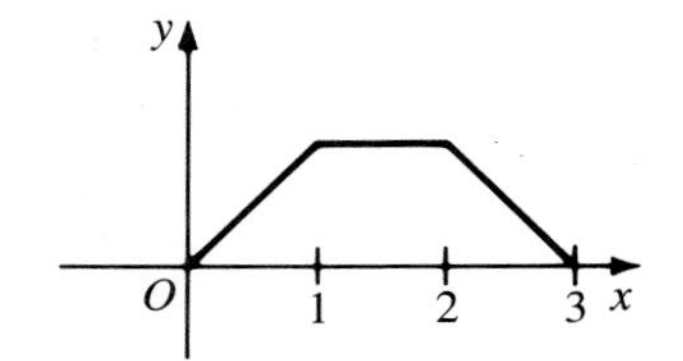

Solution **a.** Reflect the graph of f in the y-axis.

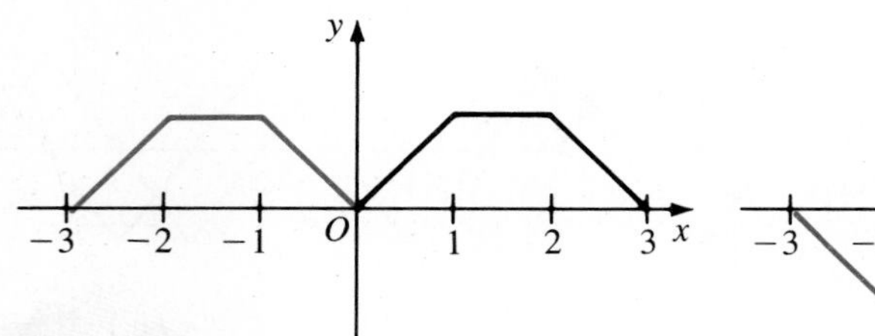

b. Reflect the graph of f in the origin.

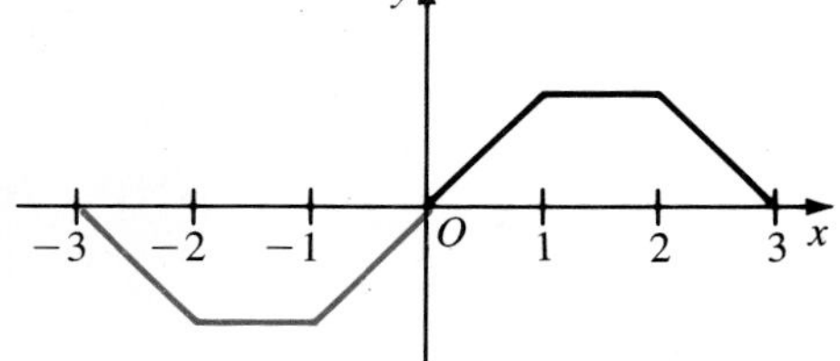

EXERCISES

a. Give the period of the periodic function whose graph is shown.
b. Is the function even, odd, or neither?

A **1.**

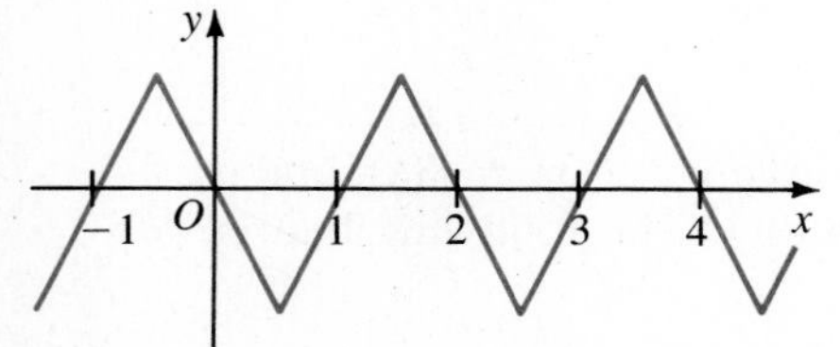

2.

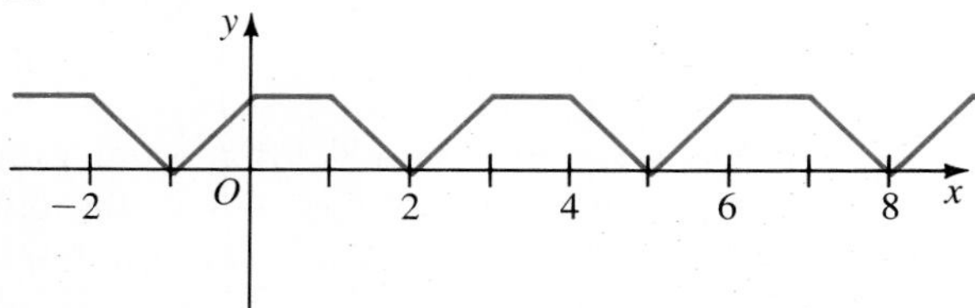

3.

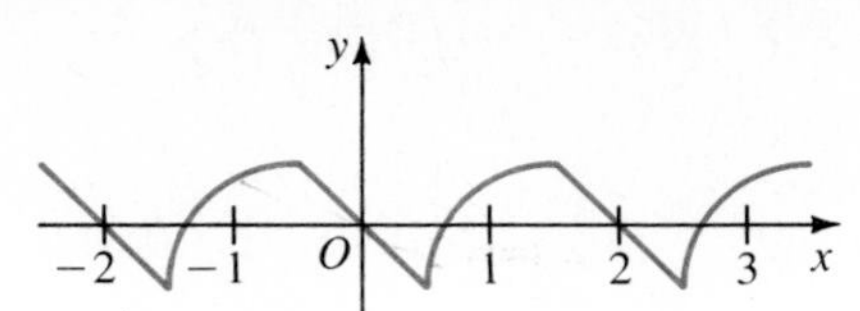

4.

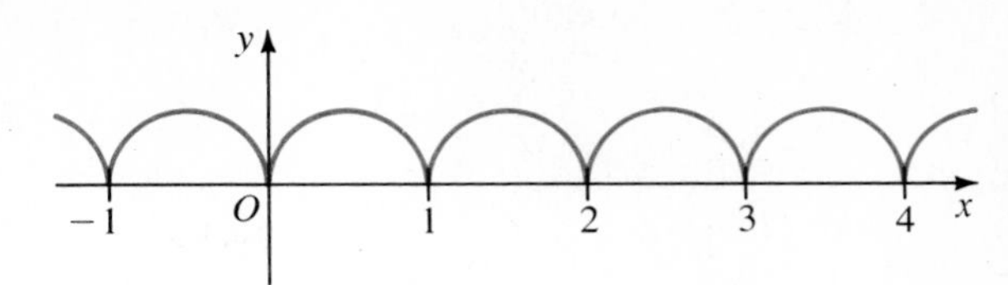

In Exercises 5–10, one cycle of the graph of a periodic function f is given. **a.** Draw three cycles. **b.** Is f even, odd, or neither?

5.

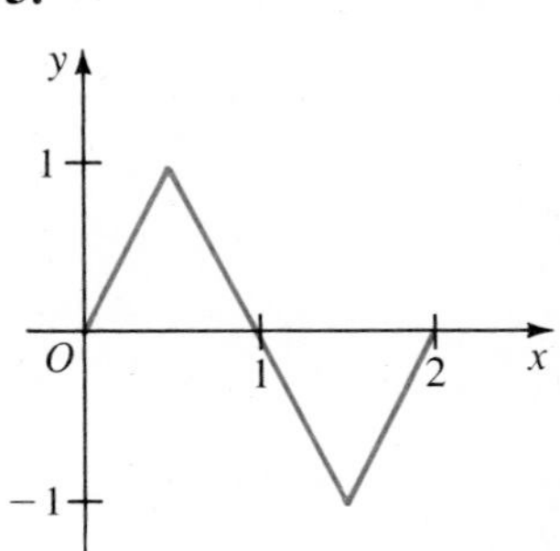

6.

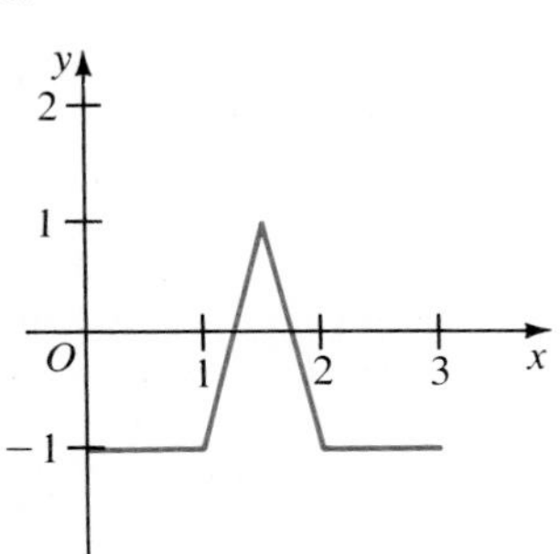

7.

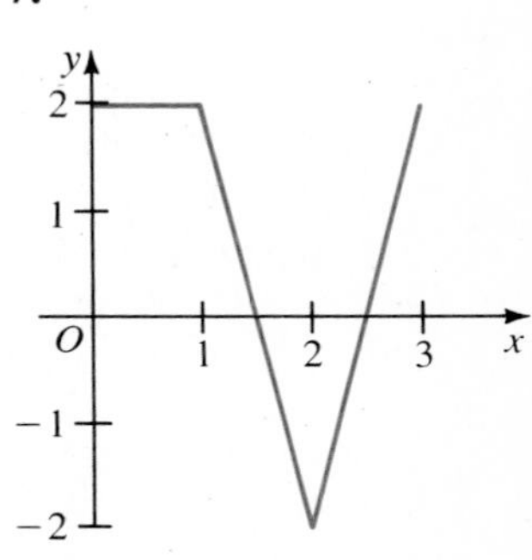

8.

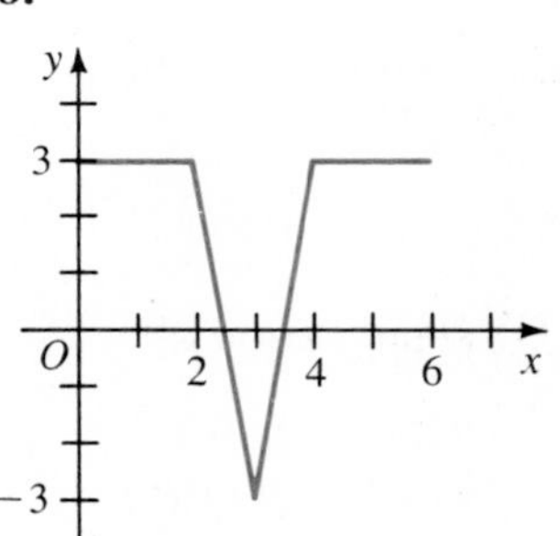

9.

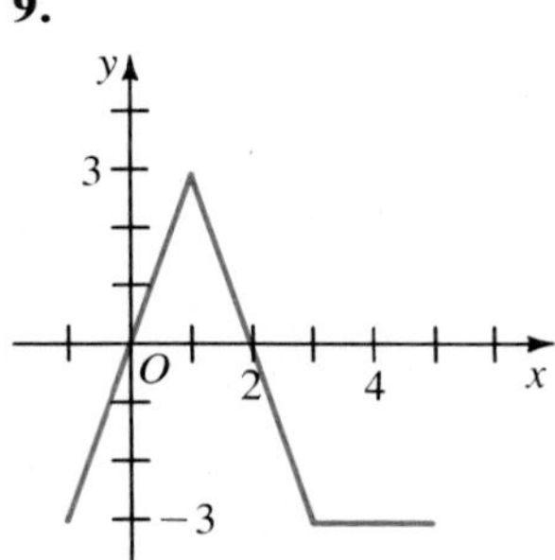

10.

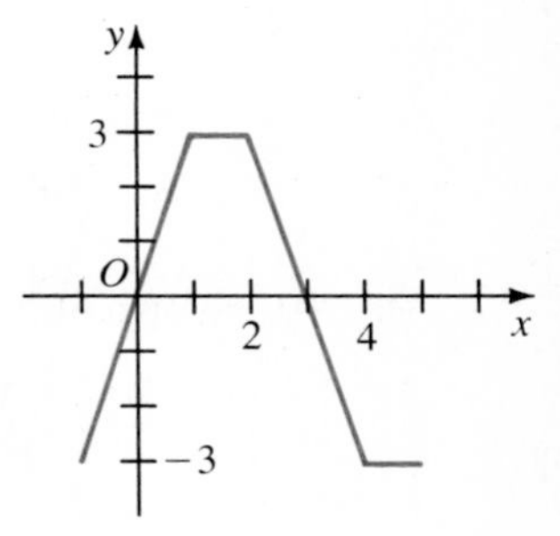

In Exercises 11 and 12, part of the graph of a function g is given. Graph g in the interval $-3 \leq x \leq 3$, given that g is **(a)** even; **(b)** odd.

11.

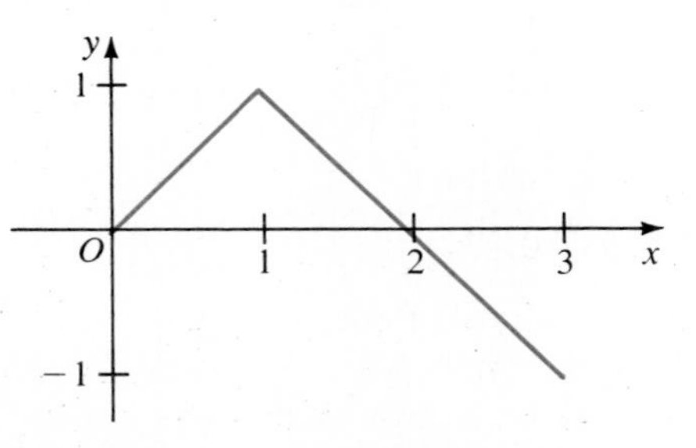

12.

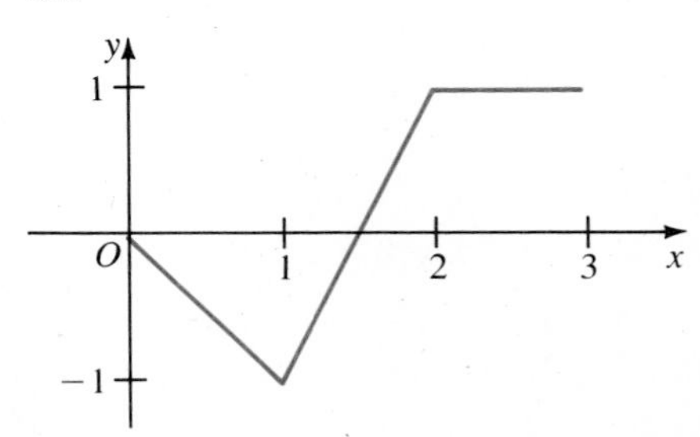

In Exercises 13 and 14, part of the graph of a function g is given. Graph g in the interval $-3 \le x \le 3$, given that g is **(a)** even; **(b)** odd.

13.

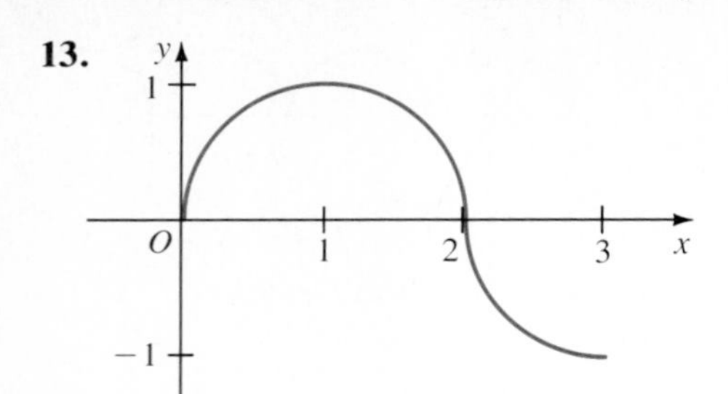

14.

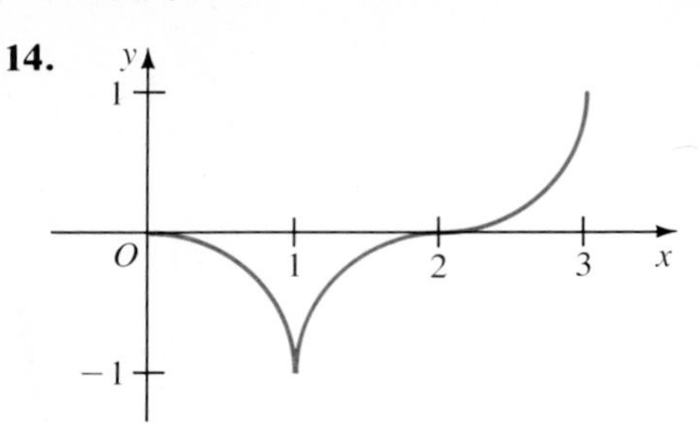

Exercises 15–20 refer to Exercises 5–10, respectively. Draw three cycles of the graph of each equation.

15. a. $y = f(x - \frac{1}{2})$ **b.** $y = f(x) + 1$ **16. a.** $y = f(x + 1)$ **b.** $y = f(x) - 2$

B **17. a.** $y = f(x + 2)$ **b.** $y = -f(x)$ **18. a.** $y = f(x - 2)$ **b.** $y = -f(x)$

19. a. $y = f(x - 1)$ **b.** $y = 3 - f(x)$ **20. a.** $y = f(x + 1)$ **b.** $y = 3 - f(x)$

In each of Exercises 21–26, state whether the result is even, odd, or neither. (Assume that neither function is identically zero.) Prove your answer.

21. The product of two even functions **22.** The product of two odd functions

23. The product of an even function and an odd function

24. The sum of two even functions **25.** The sum of two odd functions

26. The sum of an even function and an odd function

27. Prove that if $f(x) = y$ has period p and $k \neq 0$, then $f(kx)$ has period $\frac{p}{k}$.

28. Prove that if $f(x) = y$ has period p and $k \neq 0$, then $f\left(\frac{x}{k}\right)$ has period kp.

29. Show that if p is a positive constant and $\sin(x + p) = \sin x$ for all x, then p is one of the numbers $2\pi, 4\pi, 6\pi, \ldots$. (*Hint*: Use an addition formula, and by choosing particular values of x, show that $\cos p = 1$ and $\sin p = 0$.)

30. Repeat Exercise 29 with *sin* replaced by *cos*.

8-7 SINE AND COSINE GRAPHS

In Chapter 7 we found, for example, that $\sin \frac{\pi}{6} = \frac{1}{2}$ and $\cos \frac{5\pi}{6} = -\frac{\sqrt{3}}{2}$.

If two-decimal-place approximations are used, these become

$$\sin 0.52 = 0.50 \quad \text{and} \quad \cos 2.62 = -0.87.$$

The following table gives other such approximate values of the sine and cosine functions. The graphs at the right of the table show the results of plotting the entries in the table and joining the points with smooth curves.

x		$\sin x$	$\cos x$
0	0.00	0.00	1.00
$\frac{\pi}{6}$	0.52	0.50	0.87
$\frac{\pi}{3}$	1.05	0.87	0.50
$\frac{\pi}{2}$	1.57	1.00	0.00
$\frac{2\pi}{3}$	2.09	0.87	−0.50
$\frac{5\pi}{6}$	2.62	0.50	−0.87
π	3.14	0.00	−1.00
$\frac{7\pi}{6}$	3.67	−0.50	−0.87
$\frac{4\pi}{3}$	4.19	−0.87	−0.50
$\frac{3\pi}{2}$	4.71	−1.00	0.00
$\frac{5\pi}{3}$	5.24	−0.87	0.50
$\frac{11\pi}{6}$	5.76	−0.50	0.87
2π	6.28	0.00	1.00

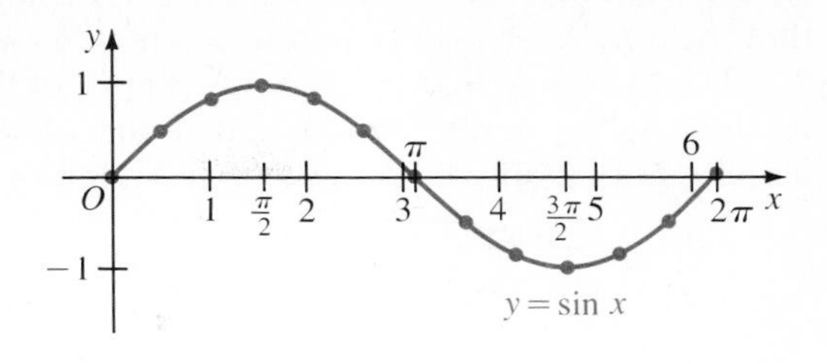

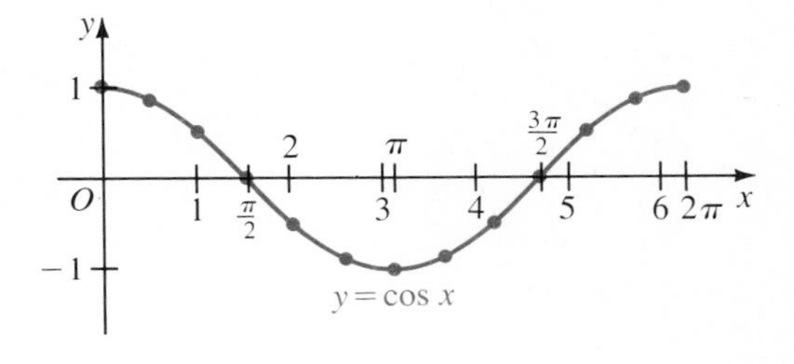

Both sine and cosine have period 2π. Thus, the graphs of $y = \sin x$ and $y = \cos x$ consist of the cycles shown above repeated over and over, as shown in the next two figures.

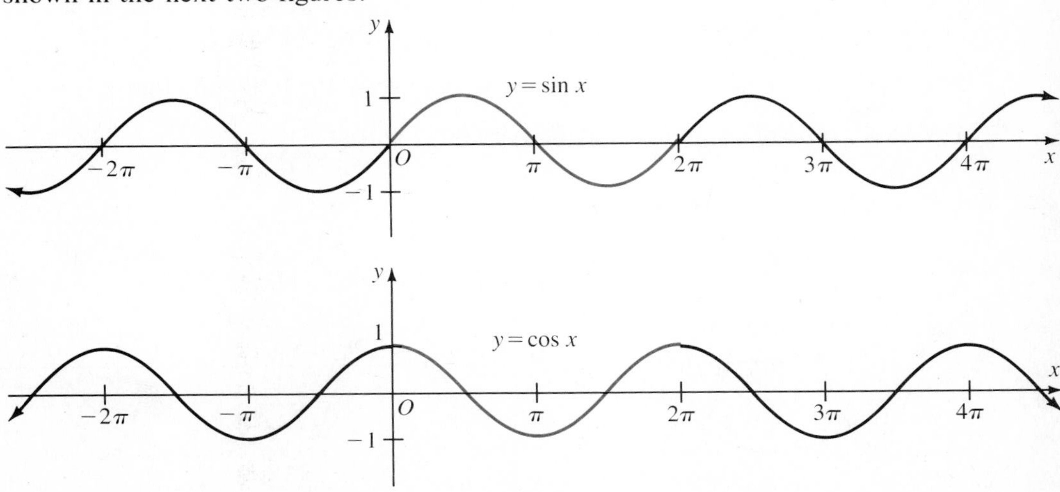

Notice that the graphs of the sine and cosine are symmetric with respect to the origin and the y-axis respectively. This symmetry illustrates the fact that the sine is an odd function and the cosine is an even function.

The following figure shows the graphs of three functions of the form $y = A \sin x$. The number $|A|$ is called the *amplitude* of $A \sin x$. Thus, the amplitude of $2 \sin x$ is 2, of $\sin x$ is 1, and of $\frac{1}{2} \sin x$ is $\frac{1}{2}$.

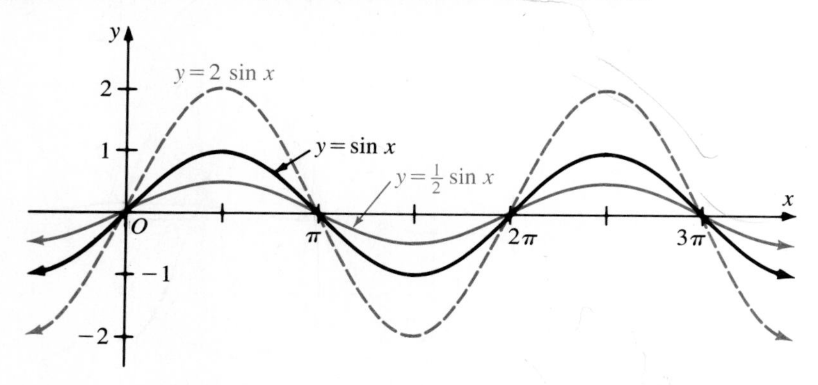

> If a periodic function has maximum value M and minimum value m, then its **amplitude** is $\frac{M - m}{2}$.

Whereas the A in $y = A \sin x$ affects the amplitude, or the *vertical* stretch or compression of the graph of $y = \sin x$, the ω in $y = \sin \omega x$ produces a *horizontal* stretch or compression. That is, the coefficient ω affects the period.

For example, let us find the period, p, of $f(x) = \sin 3x$. We have that $f(x + p) = \sin 3(x + p) = \sin (3x + 3p)$. Thus, $f(x + p) = f(x)$ becomes $\sin (3x + 3p) = \sin 3x$, and this will hold if $3p = 2\pi$. Therefore, $p = \frac{2\pi}{3}$. The graphs of $y = \sin 3x$ and $y = \sin x$ are shown in the following figure. In general both $\sin \omega x$ and $\cos \omega x$ ($\omega > 0$) have period $\frac{2\pi}{\omega}$.

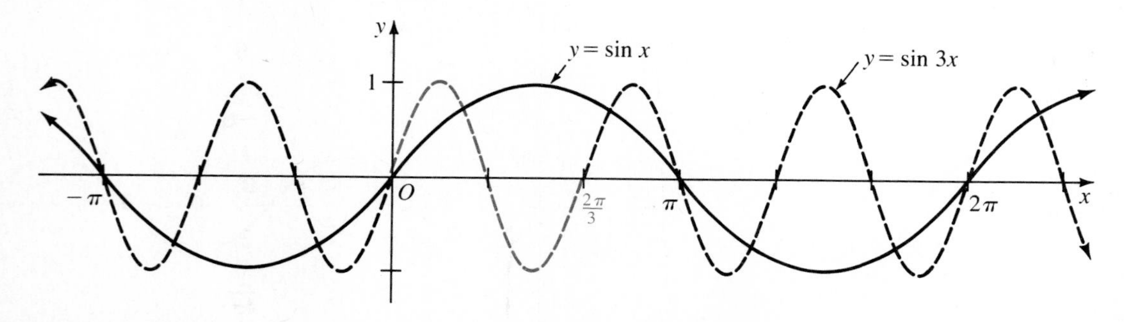

EXAMPLE Graph $y = 2 \sin \frac{\pi}{2}x$.

Solution The amplitude is 2 and the period is $\frac{2\pi}{\frac{\pi}{2}}$, or 4. Since the period is 4, the x-axis is labeled in multiples of 1 rather than of $\frac{\pi}{2}$.

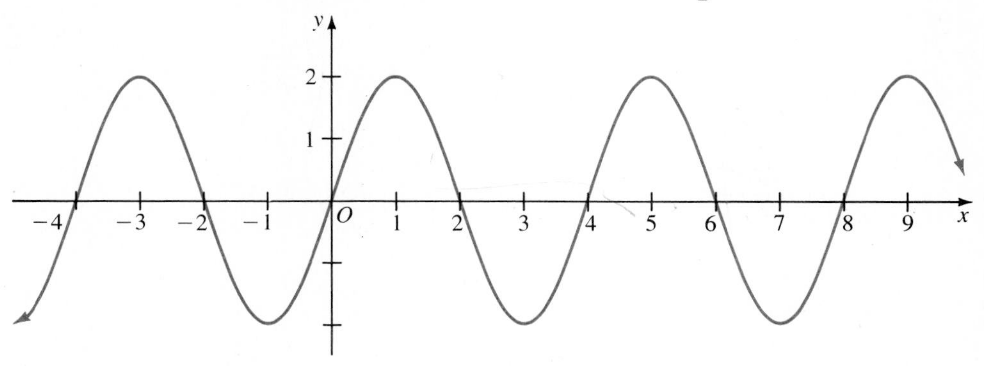

In the function $A \sin \omega t$, where t denotes time measured in seconds, the period $\frac{2\pi}{\omega}$ is the number of *seconds per cycle*. In the example it takes 4 seconds to complete one cycle. The **frequency** of a periodic function is the reciprocal of its period and is measured in **hertz** (cycles per second). Thus the frequency of $A \sin \omega t$ is $\frac{\omega}{2\pi}$ hertz. An equation giving the voltage of ordinary electrical current is $E = 115 \sin (120 \pi t)$ and has frequency $\frac{120\pi}{2\pi}$ or 60 hertz.

EXERCISES

Some characteristics of a sine curve $y = g(x)$ are given using these abbreviations: A = amplitude, p = period, f = frequency, m = minimum value, and M = maximum value. Find an expression for $g(x)$.

A

1. $A = 3, p = 2\pi, g(0) = 0$

2. $A = 2, p = 2\pi, g(0) = 2$

3. $A = 1, p = \pi, g(0) = 1$

4. $A = 1, p = 4\pi, g(0) = 0$

5. $A = 5, f = \frac{2}{\pi}, g(0) = 0$

6. $A = 10, f = \frac{4}{\pi}, g(0) = 0$

7. $A = 4, p = 3, g(0) = -4$

8. $A = 6, p = \frac{1}{2}, g(0) = -6$

9. $A = 20, f = 10, g(0) = 0$

10. $A = 5, f = \frac{1}{10}, g(0) = 0$

11. $M = 6, m = -6, p = \pi$

12. $M = 5, m = -5, p = 3$

13. $M = 4, m = -4, f = 1$

14. $M = 3, m = -3, f = \frac{1}{4\pi}$

Graph each equation, showing at least two cycles of the curve. Label the axes clearly.

15. $y = 2 \sin x$ **16.** $y = 3 \cos x$

17. $y = \sin 2x$ **18.** $y = \cos 3x$

19. $y = 3 \cos \frac{x}{3}$ **20.** $y = 2 \sin \frac{x}{2}$

B **21.** $y = -\sin 2\pi x$ **22.** $y = -2 \sin \pi x$

23. $y = \cos \frac{\pi x}{2}$ **24.** $y = 2 \cos \frac{2\pi x}{3}$

Find an equation of each sine or cosine curve.

25.

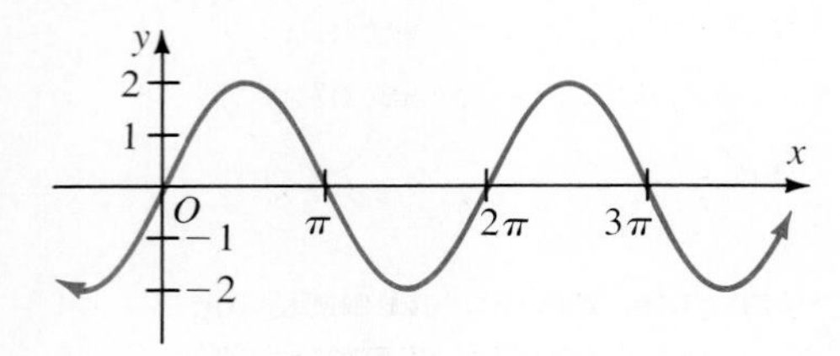

26.

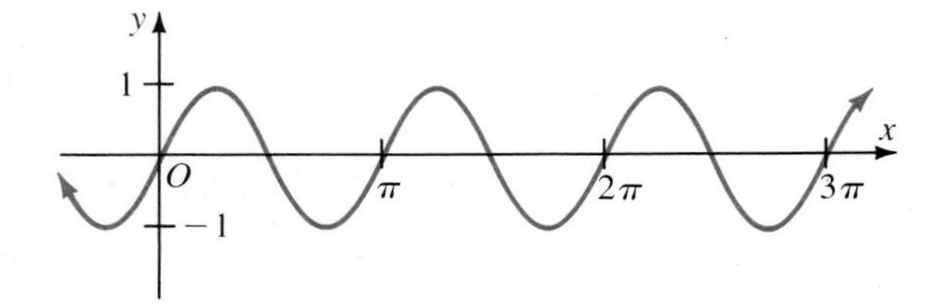

27.

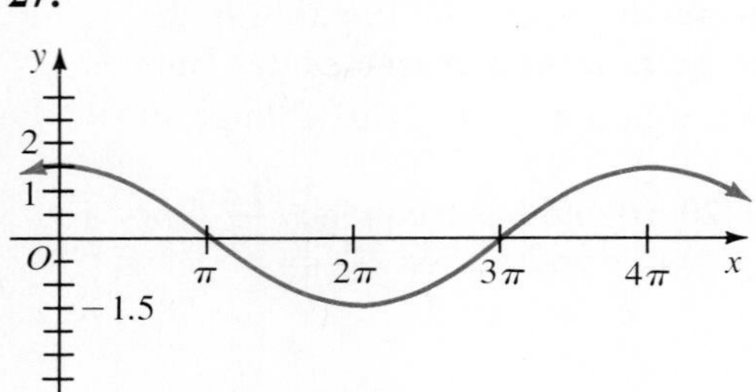

28.

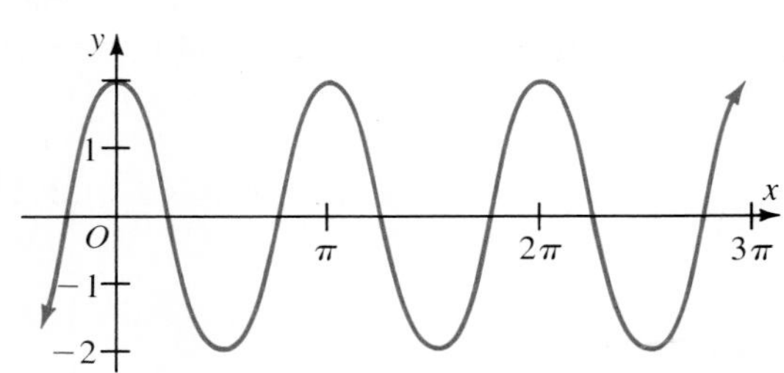

29.

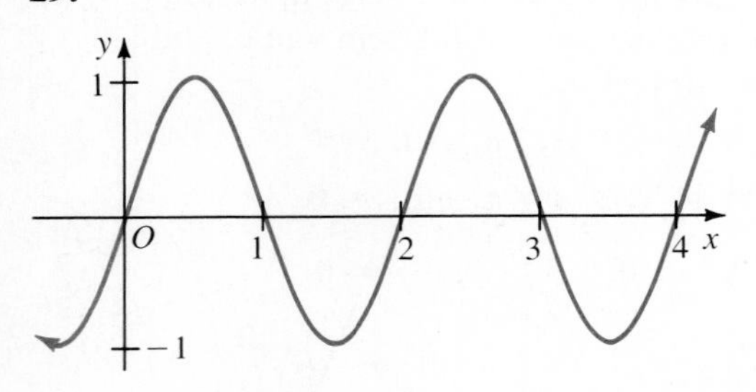

30.

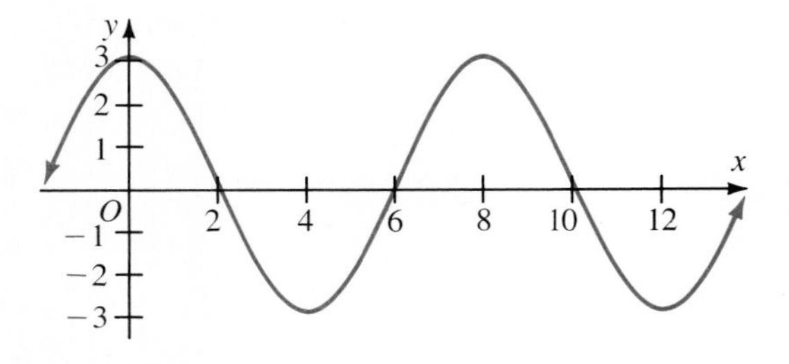

Draw each pair of curves in the same coordinate plane. Then solve a trigonometric equation to find the coordinates of their points of intersection.

31. $y = \sin x$, $y = \sin 2x$ **32.** $y = \cos x$, $y = \sin 2x$

33. $y = \sin x$, $y = \cos 2x$ **34.** $y = \cos x$, $y = \cos 2x$

COMPUTER EXERCISES

1. Let x and y be real numbers and let θ be an angle measured in degrees and input by the user. Write a computer program to make a table of 5 ordered pairs (x, y) that satisfy a given trigonometric equation involving x, y, and θ, such as $x \sin \theta - y \cos \theta = 1$.

In Exercises 2–4, run the program in Exercise 1 for each equation and for $\theta = 45°$, $120°$, and $0°$.

2. $x \sin \theta - y \cos \theta = 1$
3. $x \sin \theta - y^2 \sin \theta = 1$
4. $xy = \cos \theta$
5. Plot the three sets of ordered pairs in Exercise 2 on the same set of axes.
6. Plot the three sets of ordered pairs in Exercise 4 on the same set of axes.

8-8 MORE GENERAL SINE CURVES

Any curve congruent to the graph of an equation of the form $y = A \sin \omega x$ or to the translation of such a graph is called a **sine curve**, **sinusoid**, or **sinusoidal curve**.

EXAMPLE 1 Graph each equation in the same coordinate plane.

a. $y = \frac{3}{2} \sin \frac{\pi}{2}x$ **b.** $y = \frac{3}{2} \sin \left(\frac{\pi}{2}x - \frac{\pi}{3}\right)$

Solution **a.** This sine curve has amplitude $\frac{3}{2}$ and period $\frac{2\pi}{\frac{\pi}{2}} = 4$.

b. Since $\frac{3}{2} \sin \left(\frac{\pi}{2}x - \frac{\pi}{3}\right) = \frac{3}{2} \sin \frac{\pi}{2}\left(x - \frac{2}{3}\right)$, this sine curve has the same amplitude and period as the one in (a). However, there is a phase shift of $\frac{2}{3}$ unit relative to curve (a).

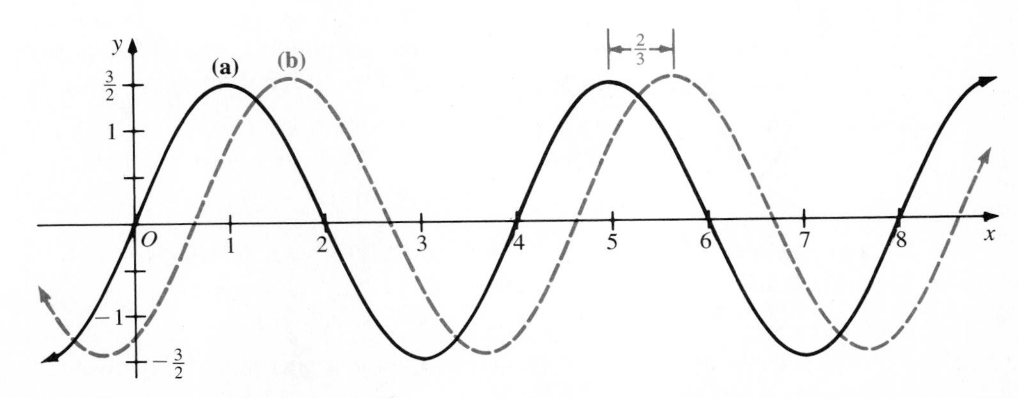

Sine curve (b) in Example 1 is said to **lead** curve (a) by $\frac{2}{3}$ unit. On the other hand, (a) **lags** (b) by the same amount.

In general we have the following results.

> The graph of $\qquad y = A \sin(\omega x - \beta) \quad (\omega > 0)$
>
> is a sine curve of amplitude $|A|$, period $\frac{2\pi}{\omega}$, and phase shift $\frac{\beta}{\omega}$ relative to $y = A \sin \omega x$. These results are also true when sin is replaced by cos.

For example, the graph of $y = \cos x$ is a sine curve because $\cos x = \sin\left(x + \frac{\pi}{2}\right)$.

Suppose that an object attached to the end of a coil spring is pulled down and released. It will then oscillate up and down. Its displacement s from its rest position O is given by a sinusoidal function of time t as indicated in the figure. (We are neglecting the effects of friction.) The resulting **simple harmonic motion** will occur when the net force acting on an object is negatively proportional to its displacement from some central position.

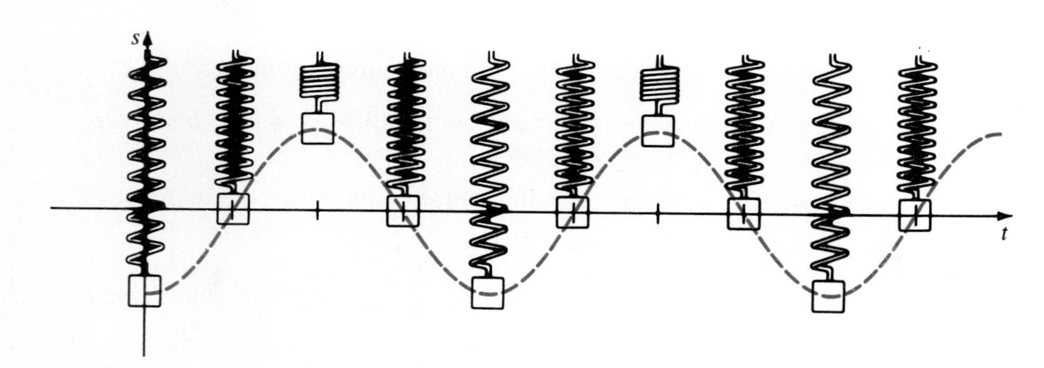

EXAMPLE 2 An object attached to the end of a coil spring is pulled down 20 cm and released. It then describes simple harmonic motion having period 6 seconds. Find an equation that describes the motion.

Solution Introduce an s-axis in the line of the spring with the origin at the rest position of the object. The equation will then have the form

$$s = A \sin(\omega t - \beta).$$

The amplitude $A = 20$, and the period $\frac{2\pi}{\omega} = 6$. Hence $\omega = \frac{\pi}{3}$.

Therefore $$s = 20 \sin\left(\frac{\pi}{3}t - \beta\right).$$

When $t = 0$, $s = -20$ since the object was pulled downward.

Therefore $-20 = 20 \sin(-\beta)$, or $\sin \beta = 1$. Thus we may take $\beta = \frac{\pi}{2}$ and obtain

$$s = 20 \sin\left(\frac{\pi}{3}t - \frac{\pi}{2}\right), \text{ or } s = -20 \cos \frac{\pi t}{3}. \qquad \textbf{Answer}$$

EXAMPLE 3 Graph $y = \sin^2 x$.

Solution Since $\cos 2x = 1 - 2 \sin^2 x$, the given equation is equivalent to

$$y = \frac{1}{2} - \frac{1}{2} \cos 2x = -\frac{1}{2} \cos 2x + \frac{1}{2}.$$

The amplitude is $\frac{1}{2}$ and the period is π. First draw the graph of $y = \frac{1}{2} \cos 2x$. Reflect the graph in the x-axis. The required graph is the graph of the sine curve $y = -\frac{1}{2} \cos 2x$ shifted upward $\frac{1}{2}$ unit.

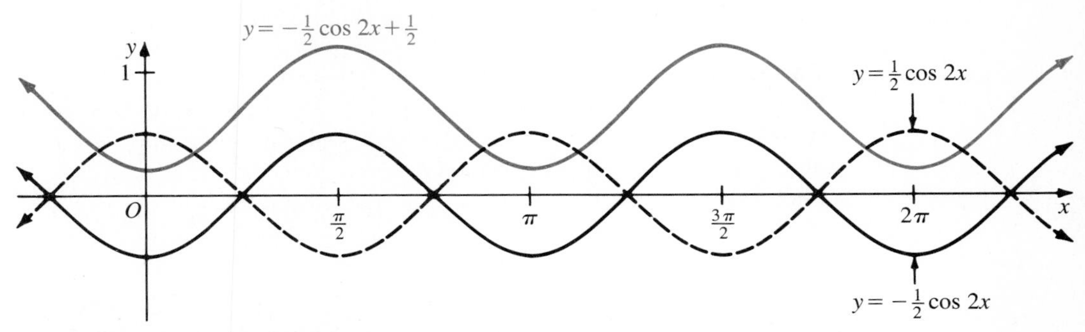

In certain applications it is advantageous to write $a \cos \theta + b \sin \theta$ in terms of a single trigonometric function.

THEOREM $a \cos \theta + b \sin \theta = c \cos(\theta - \beta)$

where $c = \sqrt{a^2 + b^2}$, and $\cos \beta = \frac{a}{c}$, $\sin \beta = \frac{b}{c}$

Proof: Assuming that not both a and b are 0, we have

$$a \cos \theta + b \sin \theta = \sqrt{a^2 + b^2}\left(\frac{a}{\sqrt{a^2 + b^2}} \cos \theta + \frac{b}{\sqrt{a^2 + b^2}} \sin \theta\right)$$

$$= c\left(\frac{a}{c} \cos \theta + \frac{b}{c} \sin \theta\right), \qquad (*)$$

where $c = \sqrt{a^2 + b^2}$. Since $\left(\frac{a}{c}, \frac{b}{c}\right)$ is a point on the unit circle, there is a number β such that $(\cos \beta, \sin \beta) = \left(\frac{a}{c}, \frac{b}{c}\right)$. Thus, (*) becomes

$$a \cos \theta + b \sin \theta = c(\cos \beta \cos \theta + \sin \beta \sin \theta)$$
$$= c \cos(\theta - \beta). \quad \blacksquare$$

We usually choose β so that $-\pi < \beta < \pi$ or $-180° < \beta < 180°$.

EXAMPLE 4 Solve $2 \cos \theta - 3 \sin \theta = 1$, $0° \le \theta < 360°$.

Solution Use the theorem on page 345. Let $a = 2$ and $b = -3$. Then $c = \sqrt{2^2 + (-3)^2} = \sqrt{13}$.

$$\cos \beta = \frac{2}{\sqrt{13}} = 0.5547, \quad \sin \beta = \frac{-3}{\sqrt{13}} = -0.8321$$

From a calculator or Table 1, $\beta = -56.3°$. The given equation becomes

$$\sqrt{13} \cos (\theta - (-56.3°)) = 1.$$

$$\cos (\theta + 56.3°) = \frac{1}{\sqrt{13}} = 0.2774$$

$$\theta + 56.3° = 73.9° \quad \text{or} \quad \theta + 56.3° = 286.1°$$

$$\theta = 17.6° \text{ or } 229.8° \qquad \textbf{Answer}$$

EXAMPLE 5 Graph each equation in the same coordinate plane.

a. $y = 4 \cos x$ **b.** $y = 3 \sin x$ **c.** $y = 4 \cos x + 3 \sin x$

Solution Write equation (c) in the form $y = c \cos (x - \beta)$. Let $a = 4$ and $b = 3$. Then $c = \sqrt{3^2 + 4^2} = 5$, $\cos \beta = \frac{4}{5} = 0.8$, and $\sin \beta = \frac{3}{5} = 0.6$. Hence, $\beta = 0.64$. Equation (c) becomes

$$y = 5 \cos (x - 0.64).$$

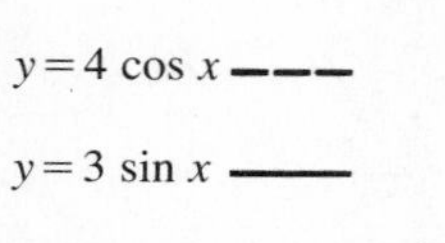

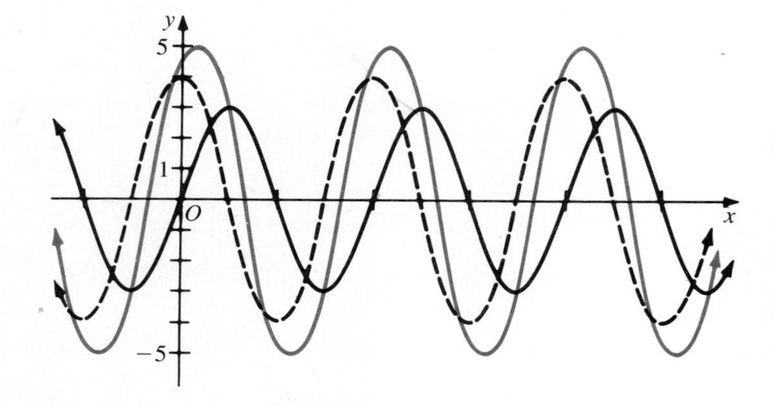

EXERCISES

By how much does curve (a) lead or lag curve (b)?

A

1. a. $y = \sin \left(x - \frac{\pi}{6}\right)$
b. $y = \sin x$

2. a. $y = \cos \left(x + \frac{\pi}{4}\right)$
b. $y = \cos x$

3. a. $y = 2 \cos 2\left(x + \frac{\pi}{3}\right)$
b. $y = 2 \cos 2x$

4. a. $y = 2 \sin \frac{1}{2}\left(x - \frac{\pi}{3}\right)$
b. $y = 2 \sin \frac{1}{2}x$

5. a. $y = 2 \sin (4x - \pi)$
   b. $y = 2 \sin 4x$

6. a. $y = \cos \left(2x + \frac{\pi}{2}\right)$
   b. $y = \cos 2x$

7. a. $y = \cos \frac{\pi}{2}(x - 1)$
   b. $y = \cos \frac{\pi x}{2}$

8. a. $y = \sin \frac{\pi}{3}(x + 1)$
   b. $y = \sin \frac{\pi x}{3}$

9. a. $y = 2 \sin (\pi x + \pi)$
   b. $y = 2 \sin \pi x$

10. a. $y = 3 \cos \left(\frac{\pi}{2}x - \pi\right)$
    b. $y = 3 \cos \frac{\pi x}{2}$

11–20. Graph two cycles of each equation in part (a) of Exercises 1–10.

In Exercises 21–24 graph each equation.

21. $y = 1 - \cos x$

22. $y = 2 + 2 \sin x$

B **23.** $y = 2 + 2 \sin \frac{\pi}{2}(x - 1)$

24. $y = -1 + \cos \left(x - \frac{\pi}{4}\right)$

Solve the given equation in either $0 \leq \theta < 360°$ or $0 \leq x < 2\pi$. Give answers to the nearest 0.1° or 0.01 radian.

25. $\cos \theta + \sqrt{3} \sin \theta = 1$

26. $\sqrt{2}(\cos \theta - \sin \theta) = 1$

27. $2(\sin x - \cos x) = \sqrt{6}$

28. $\sqrt{3} \cos x + \sin x = \sqrt{2}$

29. $3 \cos \theta + \sin \theta = 2$

30. $\cos \theta - 2 \sin \theta = 2$

31. $2 \sin x - 4 \cos x = 3$

32. $3 \cos x + 4 \sin x = 2$

Graph equations (a), (b), and (c) in the same coordinate plane.

33. a. $y = \sin x$ b. $y = \cos x$ c. $y = \sin x + \cos x$

34. a. $y = \cos x$ b. $y = -\sqrt{3} \sin x$ c. $y = \cos x - \sqrt{3} \sin x$

35. a. $y = 4 \cos x$ b. $y = -3 \sin x$ c. $y = 4 \cos x - 3 \sin x$

36. a. $y = 3 \cos x$ b. $y = -4 \sin x$ c. $y = 3 \cos x - 4 \sin x$

In Exercises 37 and 38, find an equation describing the motion of Example 2, page 344, under the given conditions.

37. Amplitude 20 cm; period 2 seconds; at time $t = 0$ the object is at O and moving downward.

38. Amplitude 10 cm; period 4 seconds; at time $t = 0$ the object is 5 cm above O and moving upward.

39. Graph $y = \cos^2 x$.

C **40.** Show that the expression $y = a \cos \theta + b \sin \theta$ can be put into the form $y = c \sin (\theta + \gamma)$. Explain how to find c and γ.

41. Suppose a hole is drilled through the center of the earth. A golf ball dropped into the hole would reach the other end, 8000 miles away, in about 42.5 minutes. The ball would describe simple harmonic motion because within the earth the force of gravity is directly proportional to the distance from the center. Find an equation describing the ball's motion.

8-9 OTHER TRIGONOMETRIC GRAPHS

To graph $y = \tan x$ we first plot the entries of a short table of values. We obtain Figure a by joining these points by a smooth curve. Recall from Section 7-2 that *as x increases toward* $\frac{\pi}{2}$, *tan x increases without bound*. The line $x = \frac{\pi}{2}$ is an asymptote to the graph of $y = \tan x$.

x		$\tan x$
0	0.00	0.00
$\frac{\pi}{12}$	0.26	0.27
$\frac{\pi}{6}$	0.52	0.58
$\frac{\pi}{4}$	0.79	1.00
$\frac{\pi}{3}$	1.05	1.73
$\frac{5\pi}{12}$	1.31	3.73
$\frac{\pi}{2}$	1.57	—

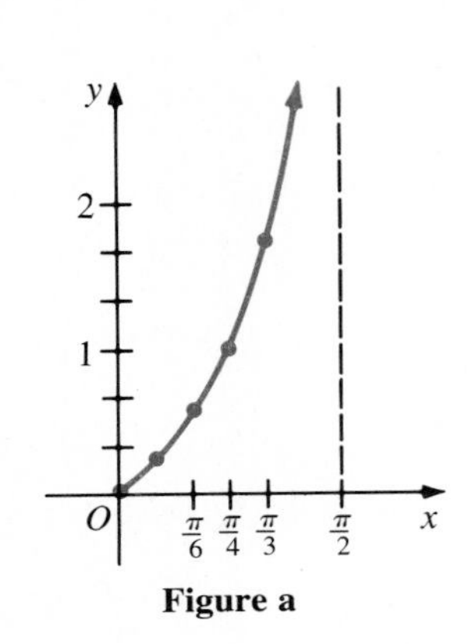

Figure a

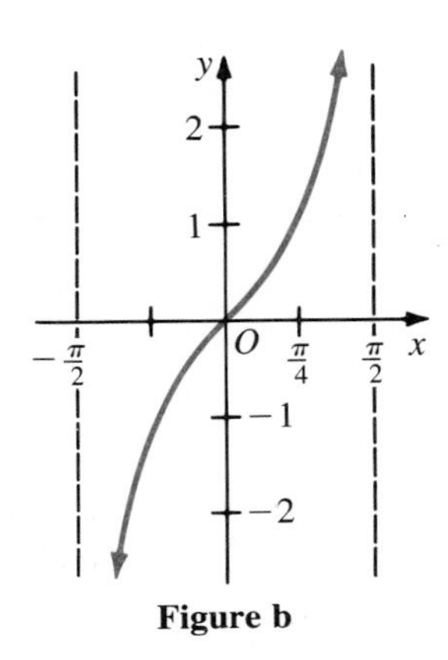

Figure b

We now use the fact that *the tangent is an odd function* (Section 7-2) and reflect Figure a in the origin to obtain Figure b.

Using an addition formula from Section 8-4, we have

$$\tan(x + \pi) = \frac{\tan x + \tan \pi}{1 - \tan x \tan \pi} = \frac{\tan x + 0}{1 - \tan x \cdot 0} = \tan x.$$

Thus, *the tangent function has period* π, and the portion of the graph shown in Figure b is a complete cycle. A larger portion of the graph is shown in Figure c.

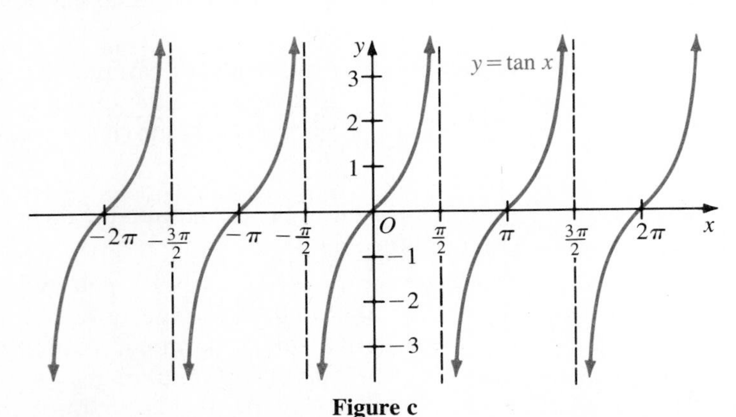

Figure c

Using methods similar to those just discussed, we obtain graphs of the cotangent, secant, and cosecant.

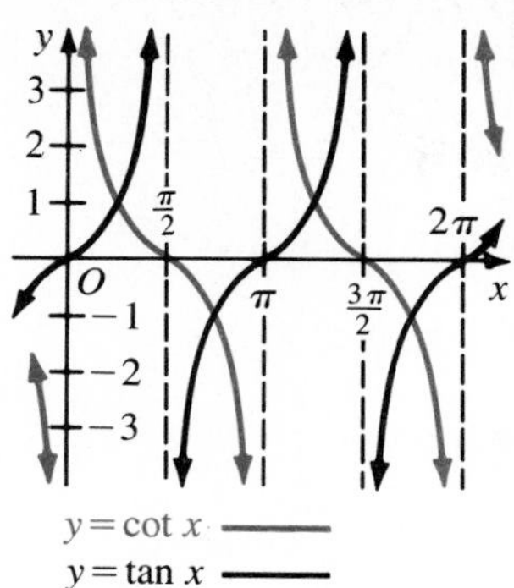

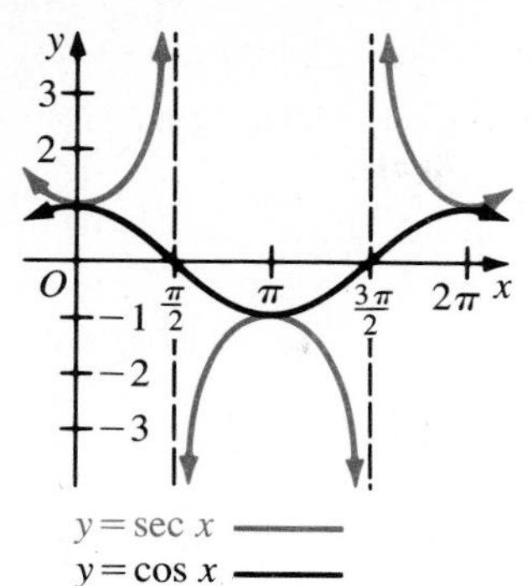

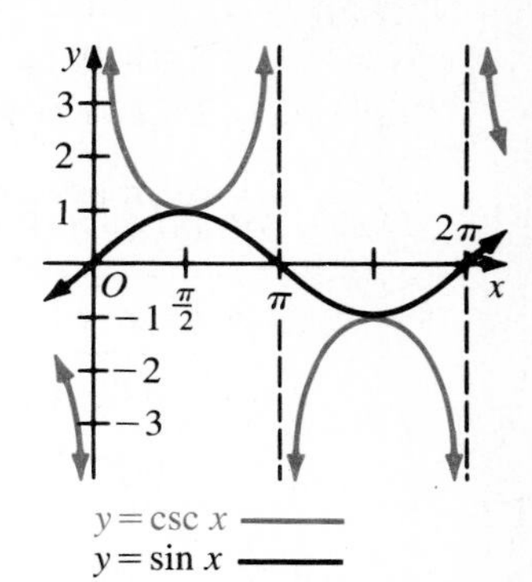

EXAMPLE Graph $y = \tan 2x + 1$.

Solution The asymptotes occur where $\tan 2x$ is undefined. Since the period of $\tan 2x$ is $\frac{\pi}{2}$, the asymptotes are the lines $x = \frac{\pi}{4} + k\left(\frac{\pi}{2}\right)$ for all integers k. Use $y = \tan x$ to graph $y = \tan 2x$. Shift the resulting graph up 1 unit.

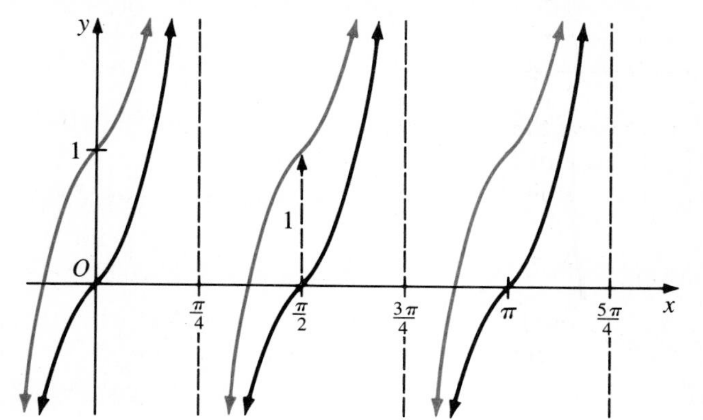

EXERCISES

Give the period of each function f.

A

1. $f(x) = \tan \frac{1}{2}x$

2. $f(x) = \sec 2x$

3. $f(x) = \frac{1}{2} \csc x$

4. $f(x) = \frac{1}{2} \cot \frac{1}{2}x$

5. $f(x) = 2 \cot \pi x$

6. $f(x) = \tan \frac{\pi x}{2}$

Give the period of each function f.

7. $f(x) = 1 + \sec \frac{x}{2}$

8. $f(x) = -1 + \csc 2x$

9. $f(x) = \tan \left(x - \frac{\pi}{2}\right)$

10. $f(x) = \cot \left(x + \frac{\pi}{2}\right)$

11–20. Graph each function f of Exercises 1–10.

B **21.** Graph $x = \tan y$, $-\frac{\pi}{2} < y < \frac{\pi}{2}$.

22. Graph $x = \cot y$, $0 < y < \pi$.

C **23.** Graph **(a)** $y = \frac{1}{2}(\tan x + \cot x)$ and **(b)** $y = \csc 2x$ in the interval $0 < x < \frac{\pi}{2}$. How do the two graphs compare?

24. Graph **(a)** $y = \frac{1}{2}(\cot x - \tan x)$ and **(b)** $y = \cot 2x$ in the interval $0 < x < \frac{\pi}{2}$. How do the two graphs compare?

COMPUTER EXERCISES

1. Write a computer program to make a table of ordered pairs (x, y) that satisfy $y = \dfrac{1}{\tan x + \cot x}$, where $-2\pi < x < 2\pi$.
2. Use the table to draw the graph. Where is the function not defined?

8–10 INVERSE TRIGONOMETRIC FUNCTIONS

The cosine is certainly not a one-to-one function. For example, . . . , $-\frac{\pi}{3}$, $\frac{\pi}{3}$, $\frac{5\pi}{3}$, . . . all have the same cosine, namely, $\frac{1}{2}$. See Figure a. If, however, we restrict the domain of cos x so that $0 \leq x \leq \pi$, then different values of x have different cosines. In doing this, we define a new, *one-to-one* function which we shall denote by Cos with $\{x: 0 \leq x \leq \pi\}$ as domain and $\{y: -1 \leq y \leq 1\}$ as range. See Figure b.

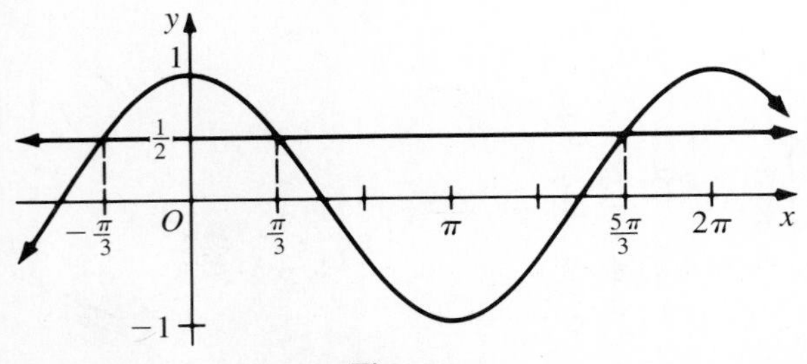

Figure a

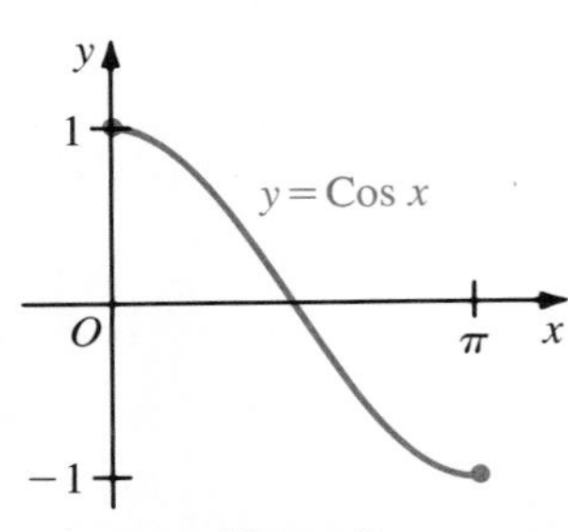

Figure b

Since Cos is one-to-one, it has an inverse function, Cos^{-1}, the **inverse cosine**. Its domain is $\{x: -1 \le x \le 1\}$, and its range is $\{y: 0 \le y \le \pi\}$.

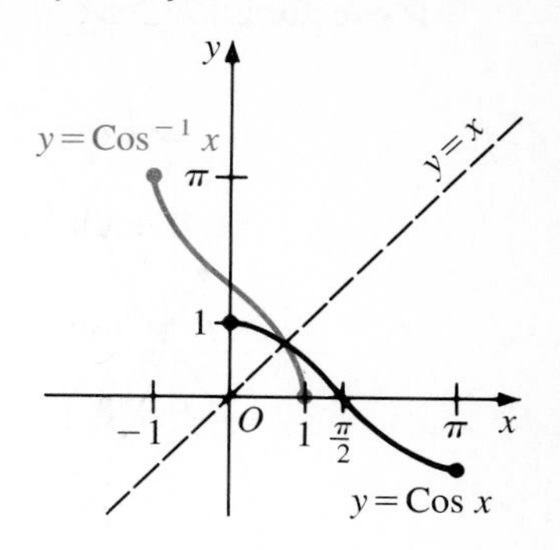

$$y = \text{Cos}^{-1} x, \quad -1 \le x \le 1,$$
$$\text{if and only if}$$
$$\text{Cos } y = x \quad \text{and} \quad 0 \le y \le \pi.$$

The graphs of Cos^{-1} and Cos are reflections of each other in the line $y = x$.

EXAMPLE 1 Evaluate each expression.

a. $\text{Cos}^{-1} 0$ **b.** $\text{Cos}^{-1}\left(-\frac{\sqrt{3}}{2}\right)$ **c.** $\text{Cos}^{-1}\left(\cos\left(\frac{4\pi}{3}\right)\right)$

Solution **a.** Let $y = \text{Cos}^{-1} 0$. Then $\cos y = 0$ and $0 \le y \le \pi$.

$\therefore y = \frac{\pi}{2}$ and $\text{Cos}^{-1} 0 = \frac{\pi}{2}$. **Answer**

b. Similarly, $\text{Cos}^{-1}\left(-\frac{\sqrt{3}}{2}\right) = \frac{5\pi}{6}$. **Answer**

c. Since $\cos \frac{4\pi}{3} = -\frac{1}{2}$, $\text{Cos}^{-1}\left(\cos\frac{4\pi}{3}\right) = \text{Cos}^{-1}\left(-\frac{1}{2}\right) = \frac{2\pi}{3}$. **Answer**

Example 1(c) shows that $\text{Cos}^{-1}(\cos z) = z$ does *not* always hold. It is true, however, that $\cos(\text{Cos}^{-1} z) = z$.

In defining the **inverse sine**, denoted Sin^{-1}, we proceed as we did for Cos^{-1}. However, to obtain a one-to-one function Sin^{-1}, we restrict the domain of the sine to the interval $-\frac{\pi}{2} \le x \le \frac{\pi}{2}$. Therefore

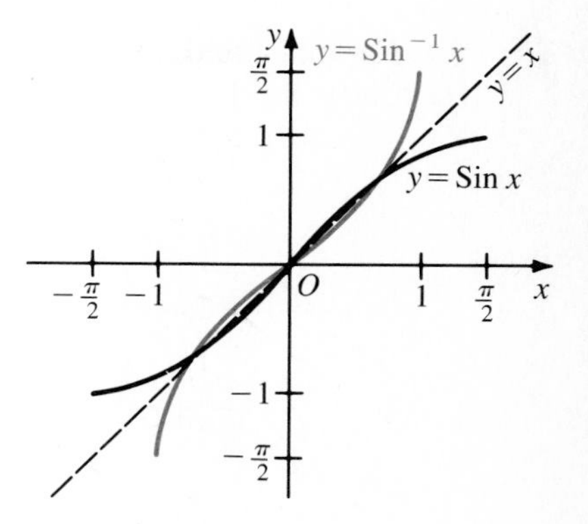

$$y = \text{Sin}^{-1} x, \quad -1 \le x \le 1,$$
$$\text{if and only if}$$
$$\sin y = x \quad \text{and} \quad -\frac{\pi}{2} \le y \le \frac{\pi}{2}.$$

EXAMPLE 2 Show that $\cos(\text{Sin}^{-1} x) = \sqrt{1 - x^2}$.

Solution Let $y = \text{Sin}^{-1} x$. Then $\cos(\text{Sin}^{-1} x) = \cos y$, where

$$\sin y = x \quad \text{and} \quad -\frac{\pi}{2} \le y \le \frac{\pi}{2}.$$

Since y is a first- or fourth-quadrant number, $\cos y \ge 0$.
Use $\cos^2 y + \sin^2 y = 1$ to obtain

$$\cos(\text{Sin}^{-1} x) = \cos y = +\sqrt{1 - \sin^2 y} = \sqrt{1 - x^2}.$$

Figures a and b are the graphs of Tan and Cot respectively. Notice that these functions are defined on $-\frac{\pi}{2} < x < \frac{\pi}{2}$ and $0 < x < \pi$ respectively.

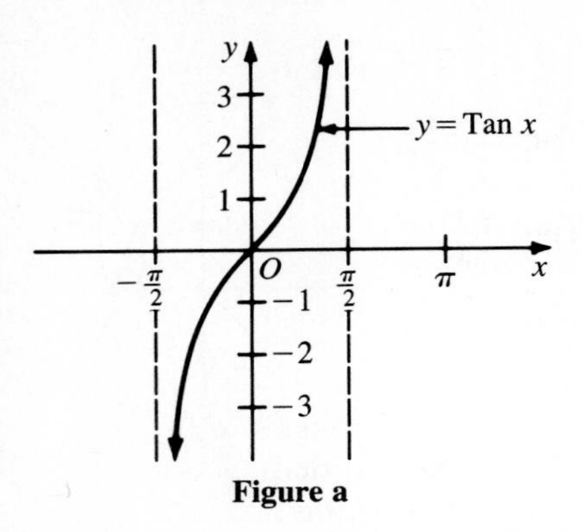

Figure a

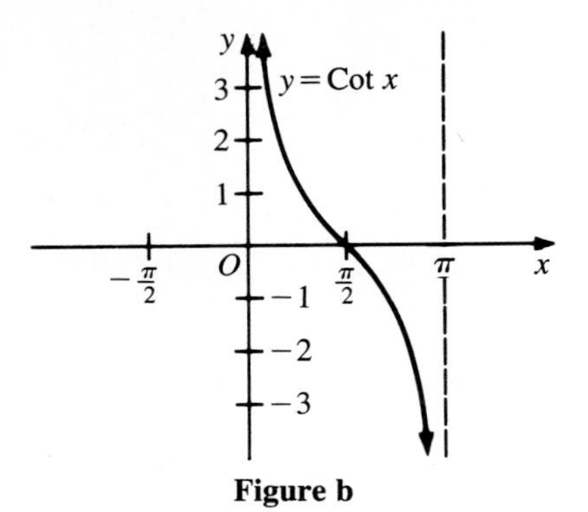

Figure b

These figures suggest the following definitions of the **inverse tangent** (denoted $\mathbf{Tan^{-1}}$) and **inverse cotangent** (denoted $\mathbf{Cot^{-1}}$) functions.

$y = \mathrm{Tan}^{-1}\, x, \quad -\infty < x < \infty,$ if and only if $\tan y = x \quad \text{and} \quad -\frac{\pi}{2} < y < \frac{\pi}{2}.$	$y = \mathrm{Cot}^{-1}\, x, \quad -\infty < x < \infty,$ if and only if $\cot y = x \quad \text{and} \quad 0 < y < \pi.$

Figures c and d are the graphs of Tan^{-1} and Cot^{-1}. They were obtained by reflecting the graphs in Figures a and b in the line $y = x$.

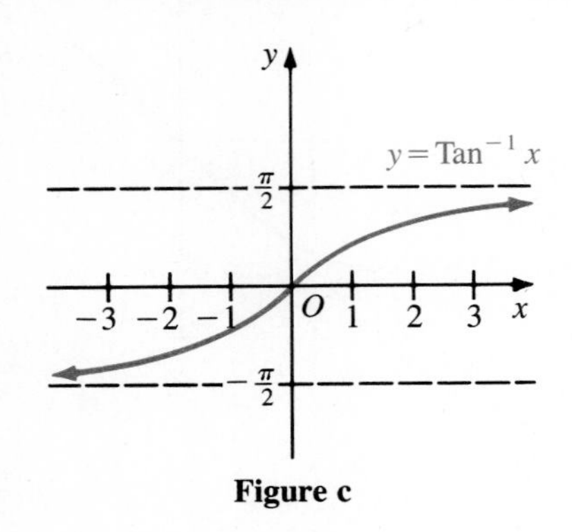

Figure c

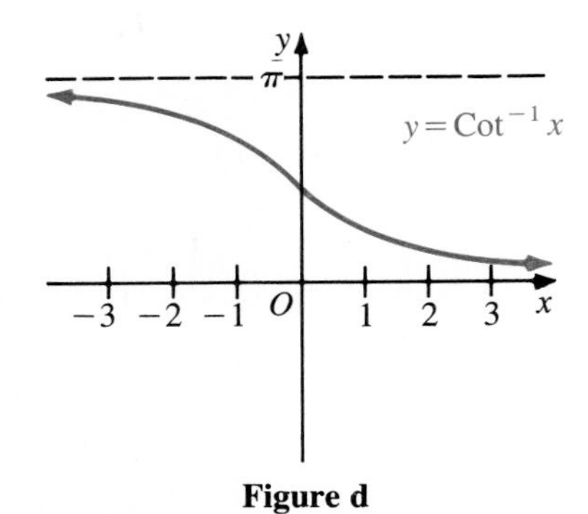

Figure d

The definitions of the inverse trigonometric functions are useful in proving statements about these functions.

EXAMPLE 3 Show that if $x > 0$, then $\mathrm{Cot}^{-1}\, x = \mathrm{Tan}^{-1}\, \frac{1}{x}$.

Solution Let $y = \text{Cot}^{-1} x$. Then $\cot y = x$ and $0 < y < \frac{\pi}{2}$.

$\left(\text{From Figure d, if } x > 0, \text{ then } 0 < y < \frac{\pi}{2}.\right)$

Hence

$$\tan y = \frac{1}{\cot y} = \frac{1}{x},$$

$$\text{Tan}^{-1}(\tan y) = \text{Tan}^{-1}\left(\frac{1}{x}\right),$$

$$y = \text{Tan}^{-1}\frac{1}{x}.$$

Thus

$$\text{Cot}^{-1} x = \text{Tan}^{-1}\frac{1}{x}.$$

The result of Example 3 depended on the fact that the tangent and cotangent are reciprocals of each other. This suggests that we define the **inverse secant (Sec^{-1})** and **inverse cosecant (Csc^{-1})** in terms of the inverse cosine and inverse sine, respectively.

If $|x| \geq 1$,

$$\text{Sec}^{-1} x = \text{Cos}^{-1}\frac{1}{x} \quad \text{and} \quad \text{Csc}^{-1} x = \text{Sin}^{-1}\frac{1}{x}.$$

We sometimes give values of inverse trigonometric functions in degrees. We may think of $\text{Cos}^{-1}(-\frac{1}{2})$ as "the angle between 0° and 180° whose cosine is $-\frac{1}{2}$" and write $\text{Cos}^{-1}(-\frac{1}{2}) = 120°$.

EXAMPLE 4 Evaluate $\cos(2\,\text{Tan}^{-1}(-\frac{3}{4}))$.

Solution Let $\alpha = \text{Tan}^{-1}(-\frac{3}{4})$. Use the formula for $\cos 2\alpha$.

$$\begin{aligned}\cos(2\,\text{Tan}^{-1}(-\tfrac{3}{4})) &= \cos 2\alpha \\ &= \cos^2\alpha - \sin^2\alpha\end{aligned}$$

To find $\cos\alpha$ and $\sin\alpha$, think of α as an angle whose tangent is $\frac{-3}{4}$. Use the sketch to obtain

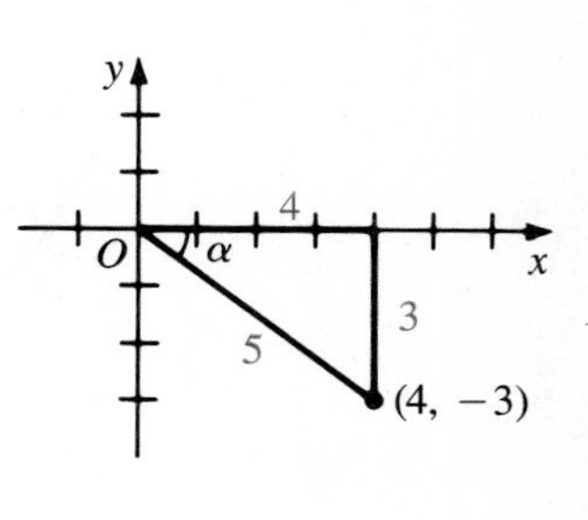

$$\cos\alpha = \frac{4}{5} \quad \text{and} \quad \sin\alpha = -\frac{3}{5}.$$

$$\therefore \cos\left(2\,\text{Tan}^{-1}\left(-\frac{3}{4}\right)\right) = \left(\frac{4}{5}\right)^2 - \left(-\frac{3}{5}\right)^2 =$$

$\frac{7}{25}$. **Answer**

Note: Other notations for the inverse trigonometric functions are in common use. For example, $\text{Sin}^{-1} x$ is also written as $\sin^{-1} x$, Arcsin x, or arcsin x.

EXAMPLE 5 Express $\sin(2\,\text{Cos}^{-1} x)$ in terms of x.

Solution Let $y = \text{Cos}^{-1} x$.
Then $\cos y = x$ and $0 \le y \le \pi$.
Thus, $\sin y = +\sqrt{1 - \cos^2 y} = \sqrt{1 - x^2}$.
$\therefore \sin(2\,\text{Cos}^{-1} x) = \sin 2y = 2 \cos y \sin y = 2x\sqrt{1 - x^2}$

EXERCISES

Evaluate.

A

1. $\text{Sin}^{-1} \frac{\sqrt{3}}{2}$
2. $\text{Cos}^{-1} 1$
3. $\text{Tan}^{-1} 1$
4. $\text{Sin}^{-1} \left(-\frac{1}{2}\right)$
5. $\text{Cos}^{-1} (-1)$
6. $\text{Tan}^{-1} (-\sqrt{3})$
7. $\text{Cot}^{-1} (-1)$
8. $\text{Cos}^{-1} \left(-\frac{\sqrt{2}}{2}\right)$
9. $\text{Sec}^{-1} 2$
10. $\text{Csc}^{-1} \sqrt{2}$
11. $\text{Sec}^{-1} (-\sqrt{2})$
12. $\text{Cot}^{-1} (-\sqrt{3})$
13. $\text{Sin}^{-1} \left(\sin \frac{2\pi}{3}\right)$
14. $\text{Cos}^{-1} \left(\cos \frac{4\pi}{3}\right)$
15. $\cos \left(\text{Sin}^{-1} \frac{3}{5}\right)$
16. $\sin \left(\text{Tan}^{-1} \frac{4}{3}\right)$
17. $\tan \left(\text{Cos}^{-1} \frac{2}{3}\right)$
18. $\sin (\text{Cot}^{-1} (-3))$
19. $\sin \left(2\,\text{Sin}^{-1} \frac{3}{5}\right)$
20. $\cos \left(2\,\text{Cos}^{-1} \frac{4}{5}\right)$
21. $\tan (2\,\text{Tan}^{-1} 2)$
22. $\tan (2\,\text{Cot}^{-1} 2)$
23. $\tan (\text{Tan}^{-1} 2 + \text{Tan}^{-1} 3)$
24. $\cot (\text{Tan}^{-1} 3 - \text{Tan}^{-1} 2)$

B

25. $\cos \left(\text{Cos}^{-1} \frac{5}{13} + \text{Cos}^{-1} \frac{3}{5}\right)$
26. $\sin \left(\text{Sin}^{-1} \frac{12}{13} + \text{Cos}^{-1} \frac{4}{5}\right)$
27. $\tan \left(\text{Sin}^{-1} \frac{2}{3} + \text{Cos}^{-1} \frac{1}{3}\right)$
28. $\cot \left(2\,\text{Cos}^{-1} \frac{2}{5}\right)$
29. $\sin (\text{Tan}^{-1} 3 - \text{Cot}^{-1} 2)$
30. $\cos \left(\text{Sin}^{-1} \frac{1}{3} - \text{Tan}^{-1} \frac{1}{3}\right)$

Express without using trigonometric or inverse trigonometric functions. (See Example 2.)

31. $\sin (\text{Cos}^{-1} x)$
32. $\tan (\text{Cot}^{-1} x)$
33. $\sec \left(\text{Cos}^{-1} \frac{1}{x}\right)$
34. $\csc \left(\text{Sin}^{-1} \frac{1}{x}\right)$

35. $\cos(2\ \mathrm{Cos}^{-1} x)$

36. $\sin(2\ \mathrm{Sin}^{-1} x)$

37. $\tan(2\ \mathrm{Cot}^{-1} x)$

38. $\tan(2\ \mathrm{Tan}^{-1} x)$

39. $\cos(2\ \mathrm{Sec}^{-1} x)$

40. $\sin(2\ \mathrm{Csc}^{-1} x)$

C

41. $\tan(\mathrm{Cot}^{-1} u - \mathrm{Cot}^{-1} v)$

42. $\cot(\mathrm{Tan}^{-1} u + \mathrm{Tan}^{-1} v)$

43. $\sin(\mathrm{Sin}^{-1} u + \mathrm{Sin}^{-1} v)$

44. $\cos(\mathrm{Cos}^{-1} u - \mathrm{Cos}^{-1} v)$

Graph in the same coordinate plane.

45. $y = \mathrm{Sec}^{-1} x$ and $y = \mathrm{Cos}^{-1} x$

46. $y = \mathrm{Csc}^{-1} x$ and $y = \mathrm{Sin}^{-1} x$

Prove the following statements.

47. $\mathrm{Cos}^{-1} \frac{3}{5} - \mathrm{Cos}^{-1} \frac{4}{5} = \mathrm{Cos}^{-1} \frac{24}{25}$

48. $\mathrm{Cos}^{-1} \frac{3}{5} - \mathrm{Sin}^{-1} \frac{3}{5} = \mathrm{Sin}^{-1} \frac{7}{25}$

49. $\mathrm{Sin}^{-1} x + \mathrm{Cos}^{-1} x = \frac{\pi}{2}$, $|x| \leq 1$. (*Hint*: Find the sine of the left side of the equation.)

50. $\mathrm{Tan}^{-1} x + \mathrm{Cot}^{-1} x = \frac{\pi}{2}$, all x. (*Hint*: Find the cotangent of the left side of the equation.)

Chapter Summary

1. An *identity* is an equation that is true for all values of its variables for which it is defined. To prove a trigonometric identity one can (1) transform one side into the other by using reversible steps, or (2) transform both sides into the same expression by using reversible steps. The following basic identities are frequently useful.

   The reciprocal identities:

   $\sin x = \frac{1}{\csc x}$ $\quad\cos x = \frac{1}{\sec x}$ $\quad\tan x = \frac{1}{\cot x}$

   $\csc x = \frac{1}{\sin x}$ $\quad\sec x = \frac{1}{\cos x}$ $\quad\cot x = \frac{1}{\tan x}$

   The Pythagorean identities:

   $\sin^2 x + \cos^2 x = 1$ $\quad 1 + \tan^2 x = \sec^2 x$ $\quad 1 + \cot^2 x = \csc^2 x$

2. Some of the most important trigonometric identities in two variables are the *addition formulas*:

$$\sin(s \pm t) = \sin s \cos t \pm \cos s \sin t$$
$$\cos(s \pm t) = \cos s \cos t \mp \sin s \sin t$$

Special cases of these are the *double-angle formulas*,

$$\sin 2\alpha = 2 \sin \alpha \cos \alpha \quad \text{and} \quad \cos 2\alpha = \cos^2 \alpha - \sin^2 \alpha,$$

and the *half-angle formulas*,

$$\sin \frac{\theta}{2} = \pm \sqrt{\frac{1 - \cos \theta}{2}} \quad \text{and} \quad \cos \frac{\theta}{2} = \pm \sqrt{\frac{1 + \cos \theta}{2}}.$$

3. Formulas involving the tangent include

$$\tan(s \pm t) = \frac{\tan s \pm \tan t}{1 \mp \tan s \tan t},$$

$$\tan 2\alpha = \frac{2 \tan \alpha}{1 - \tan^2 \alpha}, \quad \text{and} \quad \tan \frac{\theta}{2} = \frac{\sin \theta}{1 + \cos \theta}.$$

4. Both algebraic methods and trigonometric identities are used to solve trigonometric equations. The solutions in a specified interval, usually $0 \le x < 2\pi$ or $0° \le \theta < 360°$ are sometimes called *principal solutions*. Formulas are used to give the *general solution*.

5. The following properties are helpful in graphing trigonometric functions.
   (1) A function f is *even* if, for each x in its domain, $f(-x) = f(x)$. A function f is *odd* if, for each x in its domain, $f(-x) = -f(x)$. The graph of an even function is symmetric about the y-axis. The graph of an odd function is symmetric about the origin.
   (2) A function f is *periodic* and has *period* p if for each x in its domain $f(x + p) = f(x)$. The graph of such a function repeats in cycles of length p.

6. The functions $A \sin \omega x$ and $A \cos \omega x$ $(\omega > 0)$, are odd and even, respectively, have amplitude $|A|$, and have period $\frac{2\pi}{\omega}$. Their graphs are called *sine curves*, or *sinusoids*. Both $A \sin(\omega x - \beta)$ and $A \cos(\omega x - \beta)$ have phase shifts of $\frac{\beta}{\omega}$. *Simple harmonic motion*, which is sinusoidal, occurs when an object is subject to a force that is negatively proportional to the object's displacement from some central position.

7. The functions $\tan x$ and $\cot x$ have period π, and their graphs have vertical asymptotes where the functions are undefined.

8. The principal inverse trigonometric functions are defined as follows:

$$y = \text{Cos}^{-1}\, x \text{ if and only if } \cos y = x \text{ and } 0 \le y \le \pi$$

$$y = \text{Sin}^{-1}\, x \text{ if and only if } \sin y = x \text{ and } -\frac{\pi}{2} \le y \le \frac{\pi}{2}$$

$$y = \text{Tan}^{-1}\, x \text{ if and only if } \tan y = x \text{ and } -\frac{\pi}{2} < y < \frac{\pi}{2}$$

Chapter Test

8–1 **1.** Simplify $\frac{1 + \cot^2 x}{\cot^2 x}$.

2. Prove the identity $\frac{1}{1 - \cos\theta} + \frac{1}{1 + \cos\theta} = 2\csc^2\theta$.

8–2 **3.** Simplify $\cos 3x \cos 2x - \sin 3x \sin 2x$.

4. Given: $\cos\alpha = -\frac{12}{13}$, $\sin\beta = \frac{3}{5}$, $\frac{\pi}{2} < \alpha < \pi$, $0 < \beta < \frac{\pi}{2}$. Find each value.

a. $\sin(\alpha + \beta)$ **b.** $\cos(\alpha - \beta)$

8–3 **5.** Find the exact value of $\sin 165°$ **(a)** by using a half-angle formula and **(b)** by using an addition formula. Then **(c)** show that the answers obtained in (a) and (b) are equal.

6. Express $\sec 2\theta$ in terms of $\sec\theta$.

8–4 **7.** For the angles α and β of Exercise 4, find each value.

a. $\tan(\alpha + \beta)$ **b.** $\tan(\alpha - \beta)$ **c.** $\tan\frac{\alpha}{2}$

8–5 **8.** Find the solution of each equation in $0 \le x < 2\pi$.

a. $4\sin^2 x = 3$ **b.** $\cos 2x + \sin x = 0$

8–6 **9.** Is the given function even, odd, or neither?

a. $f(x) = x\cos x$ **b.** $g(x) = x + \cos x$ **c.** $h(x) = x\sin x$

10. If $f(x)$ has period 6, state the period of each function.

a. $2f(\frac{1}{2}x)$ **b.** $f(x - 2)$
c. $2f(x)$ **d.** $\frac{1}{2}f(2x)$

8–7 **11.** Graph the following equations for $0 \le x \le 2\pi$ in the same coordinate plane. Label the graphs.

a. $y = \cos x$ **b.** $y = 2\cos x$ **c.** $y = \cos 2x$

8–8 **12.** Graph the following equations for $-1 \le x \le 3$ in the same coordinate plane. Label the graphs.

a. $y = \sin\pi x$ **b.** $y = 1 + \sin\pi x$ **c.** $y = \sin\pi(x + 1)$

8–9 **13.** Find the period of $\tan\pi x$.

8–10 **14.** Evaluate.

a. $\text{Cos}^{-1}\left(-\frac{\sqrt{2}}{2}\right)$ **b.** $\text{Sin}^{-1}(\sin\pi)$

15. Write each expression without using trigonometric or inverse trigonometric functions.

a. $\cos(\text{Sin}^{-1} x)$ **b.** $\cos(2\,\text{Sin}^{-1} x)$

APPLICATION

ALTERNATING CURRENT

Many portable radios are powered by batteries, a source of direct current. Most electrical energy, however, is produced by alternating-current generators. The source of home lighting, for example, is *alternating current*.

The figure at the left below shows a magnet with its magnetic field, an electrical conductor in the form of a loop, for example a copper wire, and a direction of rotation of the loop.

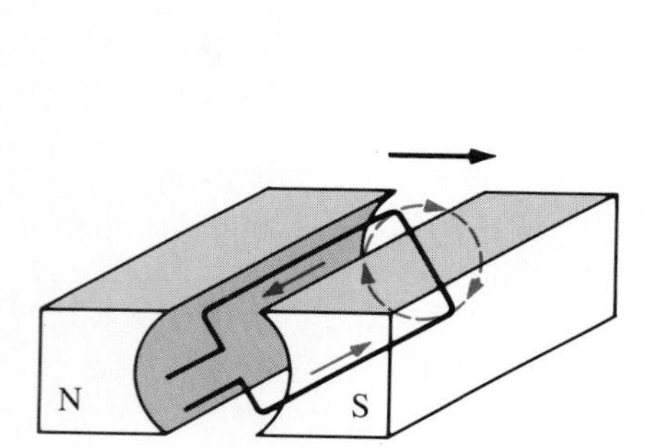

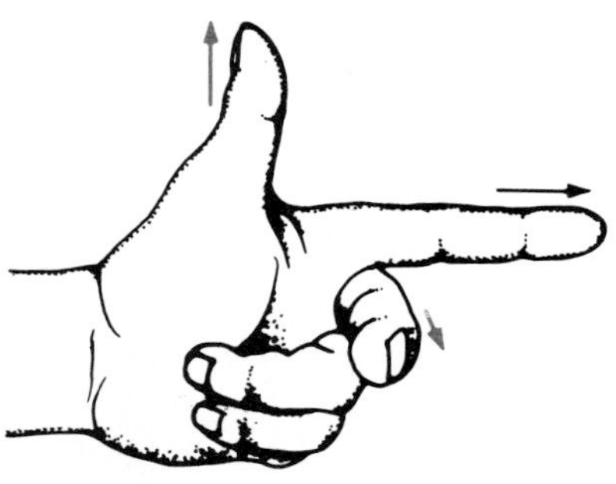

Left-hand rule

As the loop rotates in the magnetic field, the motion causes a flow of electrons in the loop. The direction of flow is found by the so-called generator rule, or left-hand rule. (See the figure at the right above.) The black arrows show the direction of the magnetic field. The dashed arrows show the direction of motion of the loop. Notice that as one side of the loop rotates up, the other side rotates down. The result is electron flow in the loop with directions shown by the red arrows. If wires are attached at the rotating end of the loop and the other ends of the wires are attached to an electrical appliance, then that appliance is powered by **alternating current**.

The following figure shows the plane of the loop at various angles of rotation. At 0° the induced voltage in the loop is 0 V. At 90° the induced voltage is a maximum, E_{max}. At 180° voltage is again 0 V, and so forth.

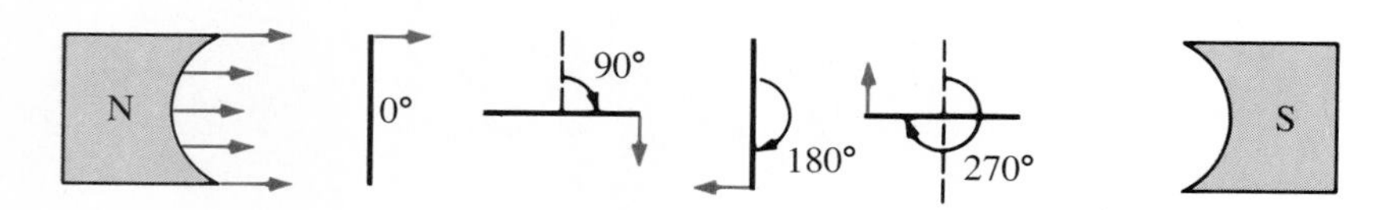

If the loop is rotating at a constant rate in a uniform magnetic field, then the induced voltage, e, is a sinusoidal function of the angle of rotation, θ.

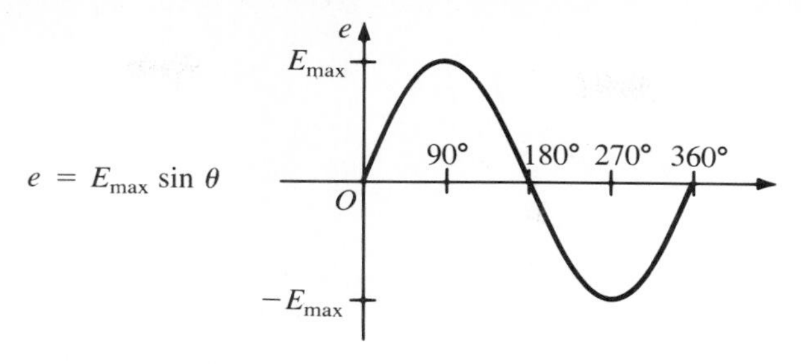

EXERCISES

1. An alternating-current circuit has a voltage of 100 V at the given angle of rotation. Find the maximum voltage in the circuit.
   a. 90° **b.** 30° **c.** 45°
2. An alternating-current circuit has a maximum voltage of 500 V. At what angle θ, $0° \leq \theta \leq 90°$, will the voltage be the given percent of maximum?
   a. 25% **b.** 50% **c.** 90%
3. A loop in a magnetic field rotates with angular speed ω radians per second. Find an equation for e in terms of ω and t. Assume that θ is measured in radians.

Many subway systems are powered by electrical current.

Chapter 9

Applications of Trigonometry

Applications of trigonometry are investigated in this chapter. These applications include the polar coordinate system and the use of limits to find derivatives and solve problems.

Applications Involving Polar Coordinates

9–1 THE POLAR COORDINATE SYSTEM

In Section 2–1 we introduced the rectangular coordinate system in which a point in the plane is represented by (x, y) where x and y are real numbers. The *polar coordinate system* for the plane is another system with many applications. If we know how far away an object is and in what direction it is from us, we can locate the object. This is the principle of the polar coordinate system, the system used to portray an object as a blip on a radar screen such as the one shown at the right.

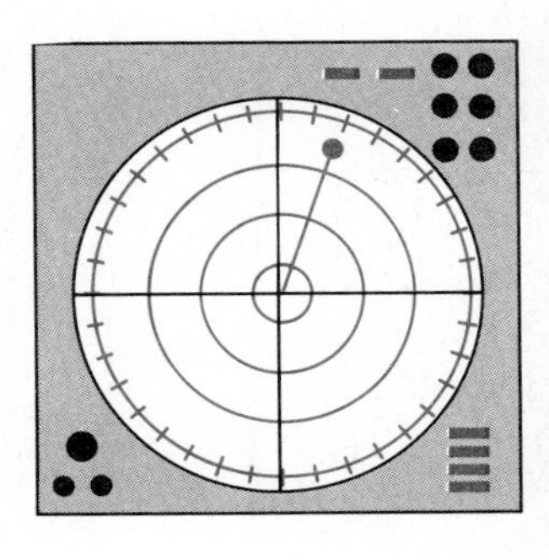

The reference system for polar coordinates consists of a point O, called the **pole**, and a ray, the **polar axis**, having O as its endpoint. We can now describe the position of a point P other than O by giving its **polar coordinates** (r, θ), where $r = OP$ and θ is the measure of an angle from the polar axis to $\overline{OP}$. The pole, O, has polar coordinates $(0, \theta)$, where θ is arbitrary.

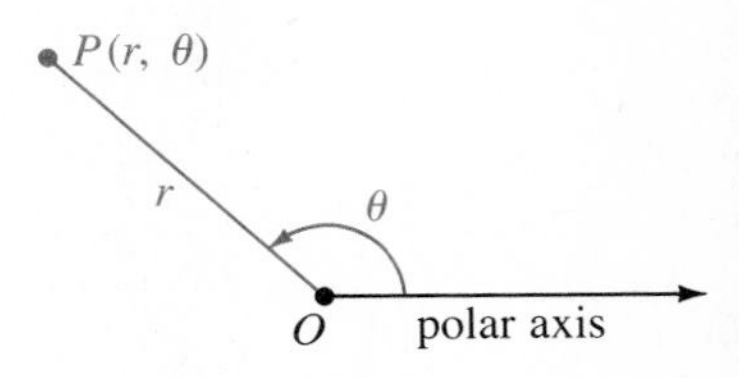

Every kind of musical instrument produces a distinctive pattern of sound waves. The study of such patterns is one of the many applications of trigonometry and the calculus of trigonometric functions.

Each point has many sets of polar coordinates. For example (4, 210°), (4, −150°), and (4, 570°) are all polar coordinates of the point P in the figure at the right. Moreover, negative values of r are used to indicate that P is on the ray opposite the terminal side of θ. Thus, (−4, 30°) and (−4, −330°) are also polar coordinates of P.

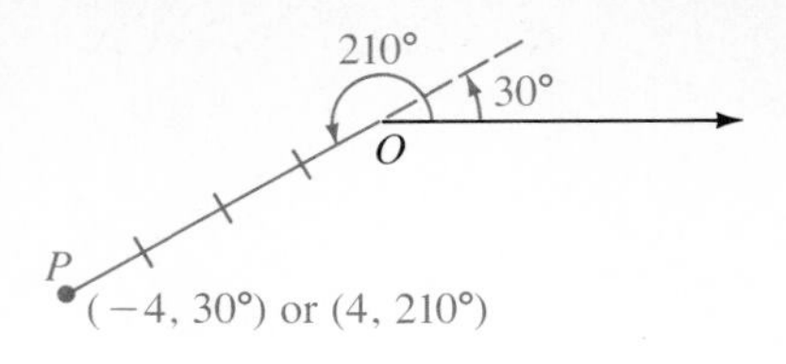

EXAMPLE 1 Plot (4, −30°), (2, π), (−2, 120°), and $\left(3, \frac{\pi}{4}\right)$ in the same polar coordinate system.

Solution The graph is shown below. Notice that θ can be measured in degrees or radians.

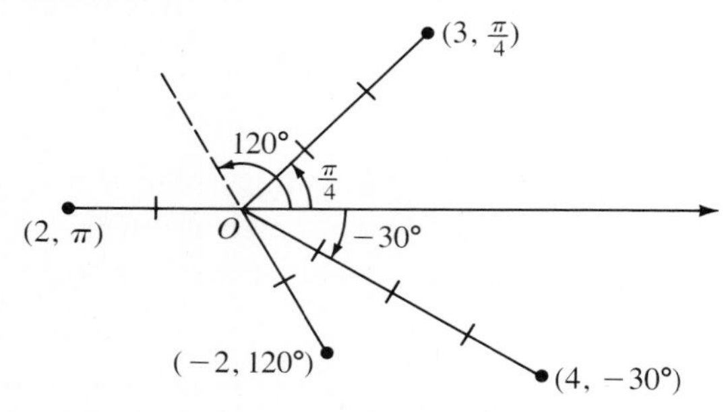

When polar coordinates (r, θ) and rectangular coordinates (x, y) are used together, the polar axis is taken to coincide with the nonnegative x-axis. The equations in the following table enable us to change from one coordinate system to the other. The equations follow directly from the definitions of sin and cos.

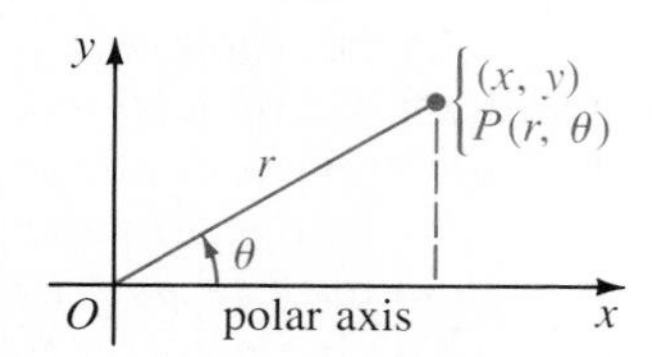

Coordinate-System Changes

From polar to rectangular	*From rectangular to polar*
$x = r\cos\theta$	$r = \pm\sqrt{x^2 + y^2}$
$y = r\sin\theta$	$\cos\theta = \frac{x}{r}, \sin\theta = \frac{y}{r}$

EXAMPLE 2 **a.** Find a set of polar coordinates of $P(-2, 2)$.

b. Find the rectangular coordinates of $Q(4, 300°)$.

Solution **a.** Since $r = \pm\sqrt{(-2)^2 + 2^2} = \pm 2\sqrt{2}$, take $r = 2\sqrt{2}$.

Then $\cos\theta = \frac{-2}{2\sqrt{2}} = -\frac{\sqrt{2}}{2}$ and $\sin\theta = \frac{2}{2\sqrt{2}} = \frac{\sqrt{2}}{2}$. Take $\theta = 135°$.

$\therefore (2\sqrt{2}, 135°)$ is a set of polar coordinates of P. **Answer**

b. $x = 4\cos 300° = 4 \cdot \frac{1}{2} = 2$ and $y = 4\sin 300° = 4\left(-\frac{\sqrt{3}}{2}\right) = -2\sqrt{3}$.

$\therefore (2, -2\sqrt{3})$ are the rectangular coordinates of Q. **Answer**

The **graph** of a polar equation in r and θ is the set of all points whose polar coordinates (r, θ) satisfy the equation.

EXAMPLE 3 **a.** Graph $r = 2$. **b.** Graph $\theta = \frac{\pi}{3}$.

Solution **a.** There is no restriction on θ but $r = 2$. **b.** There is no restriction on r but $\theta = \frac{\pi}{3}$.

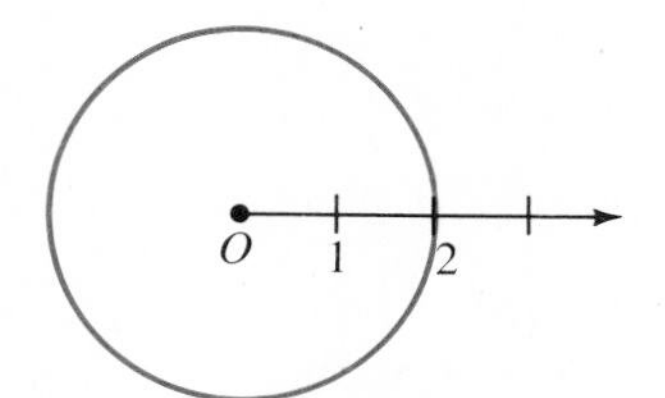

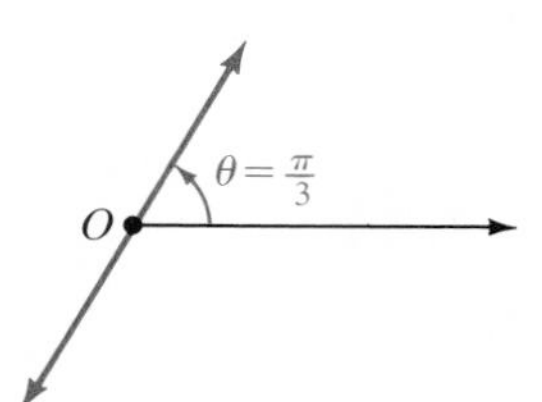

Sometimes we can graph a polar equation by first transforming it into rectangular form by using the equations in the box preceding Example 2.

EXAMPLE 4 **a.** Transform the polar equation $r = 6\cos\theta$ into rectangular form.

b. Graph the equation.

Solution **a.** Multiply both sides of $r = 6\cos\theta$ by r.

$$r^2 = 6r\cos\theta$$

Since $r^2 = x^2 + y^2$ and $r\cos\theta = x$, the equation becomes

$$x^2 + y^2 = 6x.$$

b. By completing a square on the terms involving x, the last equation can be written as

$$(x - 3)^2 + y^2 = 9.$$

Thus, the graph is the circle of radius 3 shown at the right.

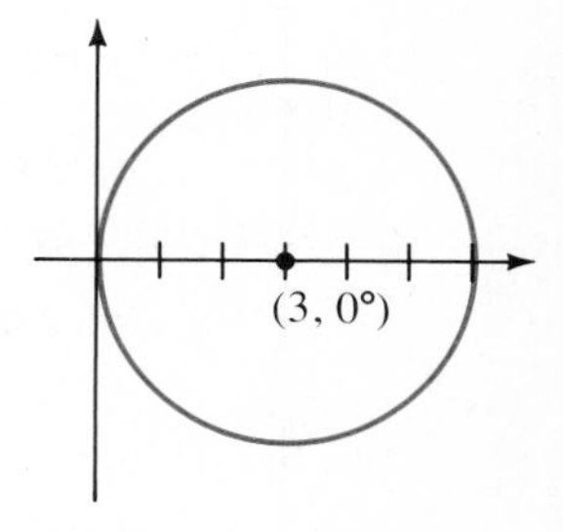

Note: In Example 4, since the origin is a point of the graph of $r = 6\cos\theta$, we add no new points to the graph when we multiply both sides of the original equation by r.

EXERCISES

Plot each point whose polar coordinates are given and find the rectangular coordinates of each point.

A

1. $(2, 60°)$
2. $(5, 180°)$
3. $(4, -45°)$
4. $\left(-6, -\frac{3\pi}{4}\right)$
5. $(-4, 300°)$
6. $(2, -210°)$
7. $\left(-6, -\frac{\pi}{3}\right)$
8. $(-8, 60°)$

Find polar coordinates of the point whose rectangular coordinates are given.

9. $(1, \sqrt{3})$
10. $(0, -3)$
11. $(-5, 0)$
12. $(2, -2)$
13. $(-\sqrt{2}, \sqrt{2})$
14. $(-\frac{1}{2}, -\frac{1}{2})$
15. $(\sqrt{2}, \sqrt{6})$
16. $(\sqrt{6}, -\sqrt{2})$

Write each equation as a polar equation.

17. $y = 3$
18. $x + 2 = 0$
19. $y = x$
20. $y = \sqrt{3}x$
21. $x + y = 0$
22. $x + y = 1$
23. $x^2 + y^2 = 1$
24. $x^2 + y^2 = 4$

Graph the following polar equations.

25. $r = 1$
26. $r = 3$
27. $\theta = \frac{\pi}{4}$
28. $\theta = 120°$
29. $r \cos \theta = 3$
30. $r \sin \theta = -1$
31. $r = 2 \csc \theta$
32. $r = \sec \theta$

Transform each equation into rectangular form and then graph it.

B

33. $r = 2 \sin \theta$
34. $r = -2 \cos \theta$
35. $r = 6 \cos \theta + 8 \sin \theta$
36. $r = \tan \theta \sec \theta$
37. Prove the following **distance formula**.
The distance between points $P_1(r_1, \theta_1)$ and $P_2(r_2, \theta_2)$ is
$$P_1P_2 = \sqrt{r_1^2 + r_2^2 - 2r_1r_2 \cos(\theta_2 - \theta_1)}.$$
(*Hint:* Use the law of cosines.)
38. Use Exercise 37 to show that:
a. if $\theta_1 = \theta_2$, then $P_1P_2 = |r_1 - r_2|$;
b. if $r_1 = r_2 = r_0$, then $P_1P_2 = 2|r_0 \sin \frac{1}{2}(\theta_2 - \theta_1)|$.

In Exercises 39–41, L is a line not passing through the pole O (origin). Let polar coordinates of the foot of the perpendicular segment from O to L be (p, ω) with $p > 0$.

39. Show that a polar equation of L is
$$r \cos(\omega - \theta) = p.$$
40. Show that a rectangular equation of L is
$$(\cos \omega)x + (\sin \omega)y = p. \qquad (*)$$

C

41. Show that if a rectangular equation
$$ax + by = c$$

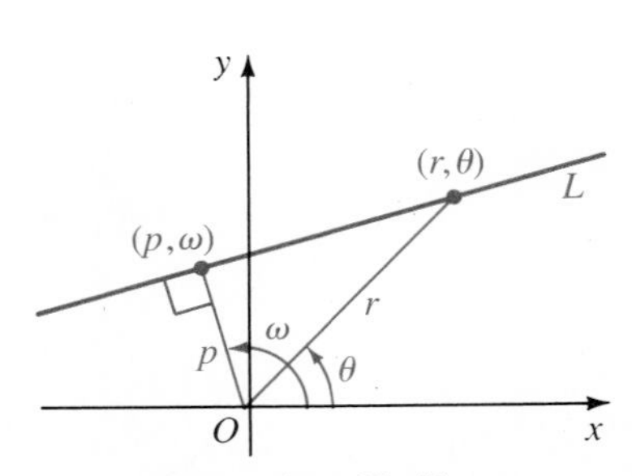

Exercises 39–41

of the line L is put into the form of (*), then the resulting equation is

$$\frac{a}{\pm\sqrt{a^2 + b^2}}x + \frac{b}{\pm\sqrt{a^2 + b^2}}y = \frac{c}{\pm\sqrt{a^2 + b^2}}, \qquad (\dagger)$$

where the sign of the radical is chosen so that the right side of (†) will be positive. [(*) and (†) are called the **normal form** of the equation of L.] Note that $\frac{|c|}{\sqrt{a^2 + b^2}}$ is the distance from O to L.

9–2 GRAPHS OF POLAR EQUATIONS

We can often graph polar-coordinate equations by using our knowledge of how the trigonometric functions behave.

EXAMPLE 1 Graph $r = 2\cos 3\theta$.

Solution Construct a table showing how r varies as θ increases. Choose θ-intervals to restrict 3θ to no more than a quarter of a period of the cosine. By doing so the graph can be drawn quadrant by quadrant.

θ	3θ	$\cos 3\theta$	$r = 2\cos 3\theta$
$0° \to 30°$	$0° \to 90°$	$1 \to 0$	$2 \to 0$
$30° \to 60°$	$90° \to 180°$	$0 \to -1$	$0 \to -2$
$60° \to 90°$	$180° \to 270°$	$-1 \to 0$	$-2 \to 0$
$90° \to 120°$	$270° \to 360°$	$0 \to 1$	$0 \to 2$
$120° \to 150°$	$360° \to 450°$	$1 \to 0$	$2 \to 0$
$150° \to 180°$	$450° \to 540°$	$0 \to -1$	$0 \to -2$

1. As θ increases from 0° to 30°, r decreases from 2 to 0. See row 1 of the table and Figure a.
2. As θ continues to increase from 30° to 60°, r becomes negative and decreases from 0 to -2. See row 2 and Figure b.
3. The complete graph, called a *three-leaved rose*, appears in Figure c. (As θ increases from 180°, the curve is repeated.)

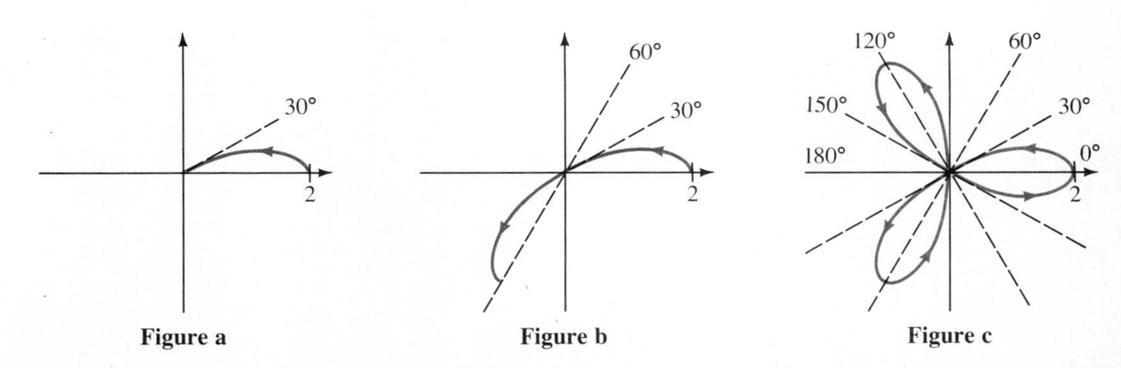

Figure a **Figure b** **Figure c**

Several tests for symmetry in polar graphs can be deduced from the figure at the right. For example, a curve is symmetric with respect to the line $\theta = 0°$ if replacing θ by $-\theta$ leaves its equation unchanged. This was the case in Example 1.

$$2 \cos 3(-\theta) = 2 \cos 3\theta$$

See Figure c for the symmetry in the line $\theta = 0°$.

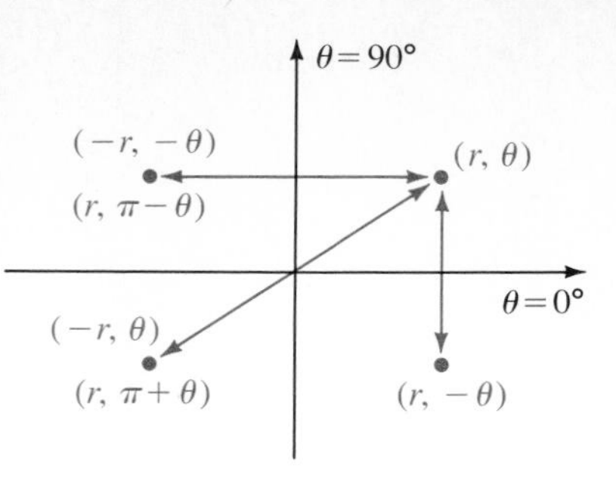

EXAMPLE 2 Graph $r^2 = 4 \sin 2\theta$.

Solution Since $(-r)^2 = r^2 = 4 \sin 2\theta$, the graph is symmetric with respect to the pole. Draw the part of the graph lying above the line of the polar axis, using the equation $r = 2\sqrt{\sin 2\theta}$.

θ	2θ	$\sin 2\theta$	$4 \sin 2\theta$	$r = 2\sqrt{\sin 2\theta}$
$0° \to 45°$ $45° \to 90°$ $90° \to 180°$	$0° \to 90°$ $90° \to 180°$ $180° \to 360°$	$0 \to 1$ $1 \to 0$ negative	$0 \to 4$ $4 \to 0$ negative	$0 \to 2$ $2 \to 0$ r is not real

From the table, draw the part of the graph shown at the left below. Use symmetry about the pole to obtain the complete curve, called a *lemniscate*.

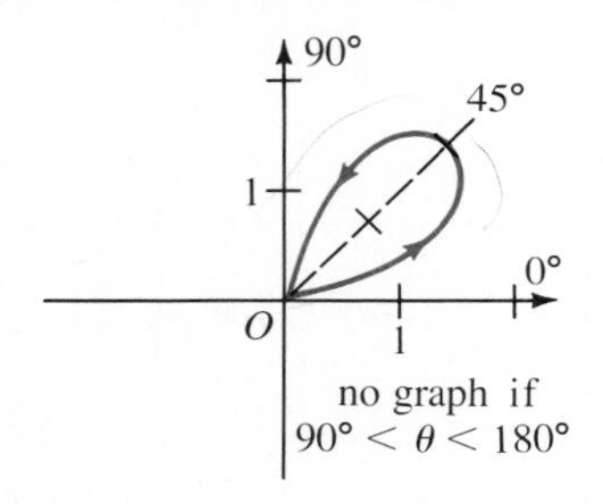

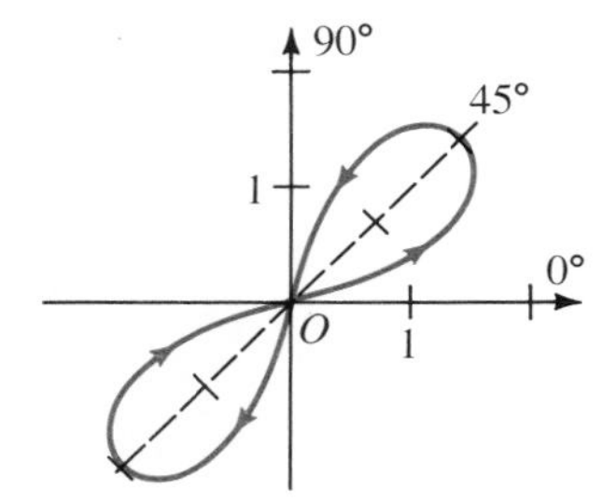

EXAMPLE 3 Graph $r = 1 - 2 \sin \theta$.

Solution Replacing θ by $\pi - \theta$ leaves the equation unchanged, so the graph is symmetric with respect to the vertical line $\theta = 90°$. First draw the part of the graph corresponding to $-90° \le \theta \le 90°$ and then obtain the rest of the graph by reflection in the line $\theta = 90°$. Note that r is sometimes positive and sometimes negative. To find when $r = 0$, solve $1 - 2 \sin \theta = 0$ and obtain $\theta = 30°$. Make θ-intervals in which r has the same sign.

θ	$\sin \theta$	$-2 \sin \theta$	$r = 1 - 2 \sin \theta$
$-90° \to 0°$ $0° \to 30°$ $30° \to 90°$	$-1 \to 0$ $0 \to \frac{1}{2}$ $\frac{1}{2} \to 1$	$2 \to 0$ $0 \to -1$ $-1 \to -2$	$3 \to 1$ $1 \to 0$ $0 \to -1$

From the table we obtain the part of the graph shown at the left below. The complete curve, called a *limaçon*, is shown at the right below.

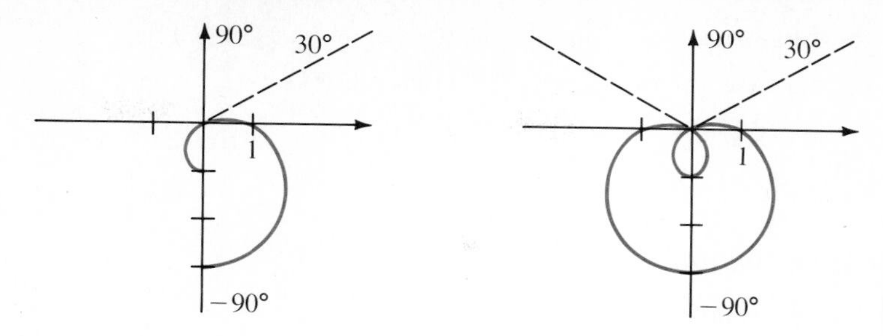

It is frequently useful to test for symmetry before constructing a table of values. For example the graph of $r^2 = \cos 2\theta$ is symmetric with respect to the pole, the line $\theta = 0°$, and the line $\theta = 90°$.

Sometimes, as in the next example, the variable θ appears independently of any trigonometric function. In such cases, θ denotes *radian* measure.

EXAMPLE 4 Graph $r = \frac{1}{\theta}$, $\theta > 0$.

Solution First, notice that as θ increases, the value of r decreases. Construct a table of values with increasing values of θ and approximate values of r.

$\theta =$	$\frac{\pi}{6}$	$\frac{\pi}{3}$	$\frac{\pi}{2}$	$\frac{3\pi}{4}$	π	$\frac{3\pi}{2}$	2π
$r =$	$\frac{6}{\pi}$	$\frac{3}{\pi}$	$\frac{2}{\pi}$	$\frac{4}{3\pi}$	$\frac{1}{\pi}$	$\frac{2}{3\pi}$	$\frac{1}{2\pi}$
$r \approx$	1.91	0.95	0.64	0.42	0.32	0.21	0.16

Lastly plot the points and draw a smooth curve through them. The curve, which is a type of *spiral*, winds around the origin infinitely many times.

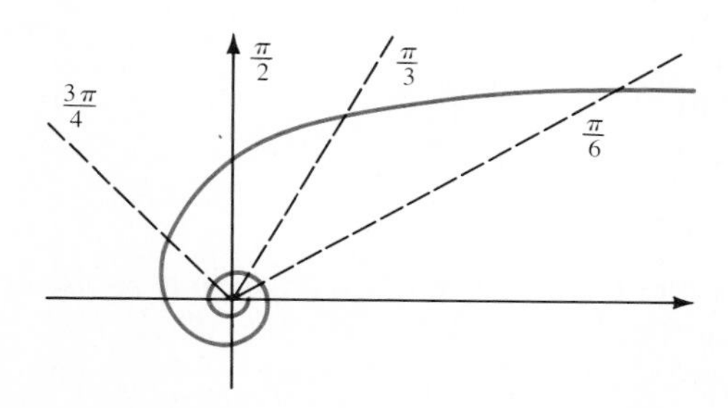

EXERCISES

Graph the following polar equations.

A

1. $r = 1 - \sin\theta$ (*cardioid*, heart-shaped curve)
2. $r = 1 - \cos\theta$ (cardioid)
3. $r = 1 + \cos\theta$ (cardioid)
4. $r = 1 + \sin\theta$ (cardioid)
5. $r = 2\sin 3\theta$ (three-leaved rose)
6. $r = 2\cos 2\theta$ (four-leaved rose)
7. $r = \sin 2\theta$ (four-leaved rose)
8. $r = 2\cos 5\theta$ (five-leaved rose)
9. $r^2 = 4\cos 2\theta$ (lemniscate)
10. $r = 1 + 2\cos\theta$ (limaçon with small loop)
11. $r = 1 + 2\sin\theta$ (limaçon with small loop)
12. $r = 2 + \cos\theta$ (limaçon with no loop)
13. $r = 2 - \sin\theta$ (limaçon with no loop)
14. $r = \theta,\ \theta \geq 0$ (spiral)

B

15. $r = 1 + \frac{1}{\theta},\ \theta > 0$ (spiral)
16. $r = |\cos\theta|$ (two circles)
17. $r = |\sin\theta|$ (two circles)
18. $r^2 - 3r + 2 = 0$ (two circles)
19. $r^2 - r - 2 = 0$ (two circles)
20. $r = \sin\theta + \cos\theta$ (circle)
21. $r = 2\cos\frac{1}{2}\theta$
22. $r = 2\sin\frac{1}{2}\theta$
23. The graph C of $r = a + b\cos\theta$, $a > b$, is a limaçon with no small loop. Show that all chords of C that pass through O have the same length, $2a$. (*Hint:* Let the polar coordinates of one endpoint of the chord be (r_1, α). Write polar coordinates of the other endpoint.)

C

24. Let A and B be the points $(1, 0°)$ and $(1, 180°)$ respectively. Show that $\{P\colon PA \cdot PB = 1\}$ is a lemniscate like the one in Exercise 9.
25. A line segment moves so that one endpoint is on the x-axis and the other is on the y-axis. Let P be the foot of the perpendicular dropped from O to the segment. Show that P moves on a four-leaved rose.

COMPUTER EXERCISE

Write and run a computer program that approximates the coordinates of the point(s) of intersection of the polar graphs of $r_1 = 2\sin\theta$ and $r_2 = 1 + \cos 2\theta$ by constructing a table of values of θ, r_1, and r_2.

9-3 POLAR FORM OF COMPLEX NUMBERS

In Section 5-2 we represented the complex number $z = x + yi$ as the point (x, y) in the plane. Let (r, θ), with $r \geq 0$, be a set of polar coordinates for (x, y). Since $x = r \cos \theta$ and $y = r \sin \theta$, $z = x + yi$ becomes

$$z = r \cos \theta + (r \sin \theta)i,$$

or

$$z = r(\cos \theta + i \sin \theta).$$

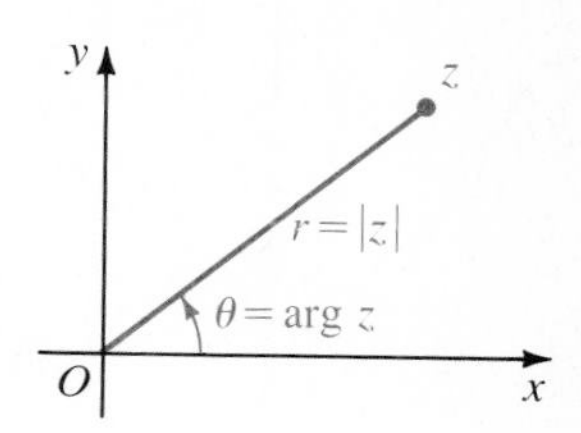

This is the **polar**, or **trigonometric**, **form** of z. It is easy to see that $r = |z|$, the absolute value of z (also called the **modulus** of z). The angle θ is an **argument** of z, denoted by arg z. If $z = 0$, then θ is any angle.

EXAMPLE 1 **a.** Express each complex number in the form $x + yi$:
$3(\cos 270° + i \sin 270°)$, $4(\cos 120° + i \sin 120°)$, and $\sqrt{2}(\cos 315° + i \sin 315°)$

b. Express each number in polar form:
$2i$, $1 + i$, and -3

Solution **a.** $3(\cos 270° + i \sin 270°) = 3(0 + i(-1)) = -3i$

$$4(\cos 120° + i \sin 120°) = 4\left(-\frac{1}{2} + i\frac{\sqrt{3}}{2}\right) = -2 + (2\sqrt{3})i$$

$$\sqrt{2}(\cos 315° + i \sin 315°) = \sqrt{2}\left(\frac{1}{\sqrt{2}} + i\left(-\frac{1}{\sqrt{2}}\right)\right) = 1 - i$$

b. Choose values of θ such that $0° \leq \theta < 360°$.

$$2i = 2(0 + i \cdot 1) = 2(\cos 90° + i \sin 90°)$$

$$1 + i = \sqrt{2}\left(\frac{1}{\sqrt{2}} + i \cdot \frac{1}{\sqrt{2}}\right) = \sqrt{2}(\cos 45° + i \sin 45°)$$

$$-3 = 3(-1 + i \cdot 0) = 3(\cos 180° + i \sin 180°)$$

In Section 5-2 we discussed the geometry of adding and subtracting complex numbers. The polar form enables us to give geometric interpretations of multiplication and division. The following theorem tells us, for example, that to multiply two complex numbers, we multiply their absolute values and add their arguments. (See also Exercises 15-18.)

THEOREM 1 Let $w = a(\cos \alpha + i \sin \alpha)$ and $z = b(\cos \beta + i \sin \beta)$. Then

$$wz = ab[\cos (\alpha + \beta) + i \sin (\alpha + \beta)]$$

and if $b \neq 0$,

$$\frac{w}{z} = \frac{a}{b}[\cos (\alpha - \beta) + i \sin (\alpha - \beta)].$$

Proof on following page

Proof: To prove the product formula we use the trigonometric addition formulas.

$$\begin{aligned} wz &= a(\cos \alpha + i \sin \alpha) \cdot b(\cos \beta + i \sin \beta) \\ &= ab(\cos \alpha \cos \beta + i \cos \alpha \sin \beta + i \sin \alpha \cos \beta + i^2 \sin \alpha \sin \beta) \\ &= ab[(\cos \alpha \cos \beta - \sin \alpha \sin \beta) + i(\sin \alpha \cos \beta + \cos \alpha \sin \beta)] \\ &= ab[\cos (\alpha + \beta) + i \sin (\alpha + \beta)] \end{aligned}$$

For the quotient formula we first obtain a formula for $\frac{1}{z}$ using the fact that $z\bar{z} = |z|^2$ from Section 5–2.

$$\frac{1}{z} = \frac{\bar{z}}{z\bar{z}} = \frac{\bar{z}}{|z|^2} = \frac{b(\cos \beta - i \sin \beta)}{b^2} = \frac{1}{b}(\cos (-\beta) + i \sin (-\beta))$$

Combine this with the product formula.

$$\begin{aligned} \frac{w}{z} &= w \cdot \frac{1}{z} = a(\cos \alpha + i \sin \alpha) \cdot \frac{1}{b}(\cos (-\beta) + i \sin (-\beta)) \\ &= \frac{a}{b}(\cos (\alpha - \beta) + i \sin (\alpha - \beta)) \end{aligned}$$ ■

EXAMPLE 2 Let $w = 6(\cos 55° + i \sin 55°)$ and $z = 4(\cos 95° + i \sin 95°)$.

a. Find wz. **b.** Find $\frac{w}{z}$.

Solution **a.** $wz = 6 \cdot 4(\cos (55° + 95°) + i \sin (55° + 95°))$
$= 24(\cos 150° + i \sin 150°)$ **Answer**

b. $\frac{w}{z} = \frac{6}{4}(\cos (55° - 95°) + i \sin (55° - 95°))$
$= 1.5(\cos (-40°) + i \sin (-40°))$
$= 1.5(\cos 320° + i \sin 320°)$ **Answer**

It is customary to write the argument of a complex number as an angle, θ, such that $0° \le \theta < 360°$.

In $x + yi$ form, the answers in Example 2 are:

$$wz = 24\left(-\frac{\sqrt{3}}{2} + i \cdot \frac{1}{2}\right) = -12\sqrt{3} + 12i \approx -20.8 + 12i$$

and $$\frac{w}{z} \approx 1.5(0.766 - i \cdot 0.643) \approx 1.15 - 0.96i$$

EXERCISES

Convert to $x + yi$ form. In Exercises 7 and 8 give answers to two decimal places.

A

1. $5(\cos 0° + i \sin 0°)$
2. $3\left(\cos \frac{\pi}{2} + i \sin \frac{\pi}{2}\right)$
3. $6(\cos 135° + i \sin 135°)$
4. $4(\cos 300° + i \sin 300°)$

5. $10\left(\cos \frac{4\pi}{3} + i \sin \frac{4\pi}{3}\right)$

6. $2\left(\cos \frac{5\pi}{4} + i \sin \frac{5\pi}{4}\right)$

7. $2(\cos 100° + i \sin 100°)$

8. $12(\cos 200° + i \sin 200°)$

Convert to polar form. Use degree measure.

9. $4 - 4i$

10. $-3i$

11. -4

12. $-\sqrt{3} + i$

13. $\sqrt{3} - i$

14. $-\sqrt{2} + i\sqrt{2}$

In Exercises 15–18:

a. Use Theorem 1 to find wz in polar form.
b. Convert the answer in (a) to $x + yi$ form.
c. Convert w and z to $x + yi$ form.
d. Multiply w and z obtained in (c).
Leave answers in (b) and (d) in simplest radical form.

15. $w = 6(\cos 120° + i \sin 120°)$; $z = 3(\cos 150° + i \sin 150°)$

16. $w = 6(\cos 60° + i \sin 60°)$; $z = 2(\cos 30° + i \sin 30°)$

17. $w = 4(\cos 135° + i \sin 135°)$; $z = 2(\cos 315° + i \sin 315°)$

18. $w = 12(\cos 120° + i \sin 120°)$; $z = 4(\cos 210° + i \sin 210°)$

19–22. Repeat Exercises 15–18 with wz replaced by $\frac{w}{z}$ and "multiply" by "divide."

B

23. Refer to Figure a. Show that the shaded triangles are similar.

24. Describe a geometric (straightedge and compass) construction to locate the point wz given the points w and z. (*Hint*: Use Exercise 23.)

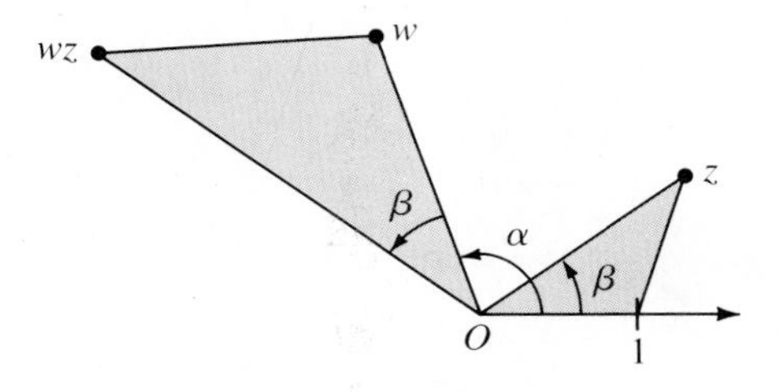

Figure a

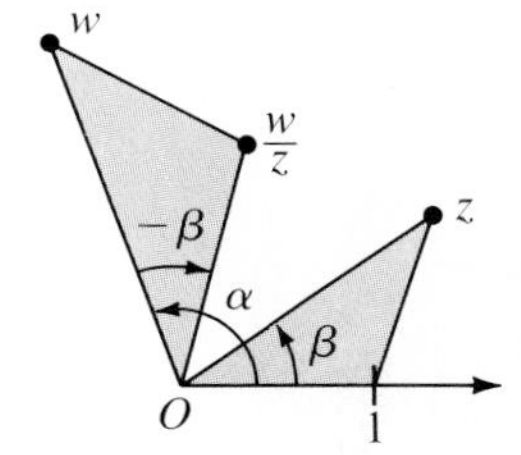

Figure b

25. Repeat Exercise 23 using Figure b.

26. Repeat Exercise 24 with wz replaced by $\frac{w}{z}$.

27. Draw a figure with shaded triangles (similar to Figures a and b) to show the relationship between z and $\frac{1}{z}$.

C

28. Let $\mathcal{U} = \{z: |z| = 1\}$. Show that $\mathcal{U}$ is a commutative group under multiplication.

9-4 POWERS AND ROOTS OF COMPLEX NUMBERS

The theorem on page 369 can be used to find the square of a complex number given in polar form. If $z = r(\cos\theta + i\sin\theta)$, then

$$z^2 = z \cdot z = r \cdot r\,(\cos(\theta + \theta) + i\sin(\theta + \theta))$$
$$\therefore\quad z^2 = r^2(\cos 2\theta + i\sin 2\theta)$$

$$z^3 = z^2 \cdot z = (r^2(\cos 2\theta + i\sin 2\theta))(r(\cos\theta + i\sin\theta))$$
$$= r^3(\cos 2\theta\cos\theta - \sin\theta\sin 2\theta + i(\sin 2\theta\cos\theta + \sin\theta\cos 2\theta))$$
$$\therefore\quad z^3 = r^3(\cos 3\theta + i\sin 3\theta)$$

These equations suggest a general formula for finding powers (and roots) of complex numbers in polar form. The formula is known as *De Moivre's theorem*.

THEOREM 2 **De Moivre's Theorem**

If $z = r(\cos\theta + i\sin\theta)$ and n is an integer, then

$$z^n = r^n(\cos n\theta + i\sin n\theta). \qquad (*)$$

Proof: If $n = 0$, then $z^0 = 1$ and $r^0(\cos 0\cdot\theta + i\sin 0\cdot\theta) = 1$.

The proof for positive integers is by induction. Let S be the set of all positive integers n such that (*) holds.

(1) When $n = 1$, (*) becomes $z^1 = r^1(\cos 1\cdot\theta + i\sin 1\cdot\theta)$, which, since $z = r(\cos\theta + i\sin\theta)$, certainly is true. Thus $1 \in S$.

(2) Now, assume that $k \in S$, that is, that

$$z^k = r^k(\cos k\theta + i\sin k\theta)$$

is true. Then,

$$z^{k+1} = z^k \cdot z = r^k(\cos k\theta + i\sin k\theta)\cdot r(\cos\theta + i\sin\theta).$$

By Theorem 1,

$$z^{k+1} = r^{k+1}(\cos(k + 1)\theta + i\sin(k + 1)\theta).$$

Thus, $k \in S$ implies that $k + 1 \in S$.

By the principle of mathematical induction, (1) and (2) show that (*) holds for all positive integers n. You are asked to prove the theorem for negative integers n in Exercise 24. ■

EXAMPLE 1 Find $(-\sqrt{3} + i)^{10}$.

Solution First convert $-\sqrt{3} + i$ to polar form.

$$-\sqrt{3} + i = 2\left(-\frac{\sqrt{3}}{2} + \frac{1}{2}i\right) = 2(\cos 150^\circ + i\sin 150^\circ)$$

Now use De Moivre's theorem.

$$(-\sqrt{3} + i)^{10} = 2^{10}(\cos(10\cdot 150^\circ) + i\sin(10\cdot 150^\circ))$$

$$
\begin{aligned}
(-\sqrt{3} + i)^{10} &= 1024(\cos 1500^\circ + i \sin 1500^\circ) \\
&= 1024(\cos (60^\circ + 4 \cdot 360^\circ) + i \sin (60^\circ + 4 \cdot 360^\circ)) \\
&= 1024(\cos 60^\circ + i \sin 60^\circ) \\
&= 1024\left(\frac{1}{2} + i\frac{\sqrt{3}}{2}\right) \\
\therefore (-\sqrt{3} + i)^{10} &= 512 + 512i\sqrt{3} \quad \textbf{Answer}
\end{aligned}
$$

De Moivre's theorem is useful in finding roots of complex numbers. As we shall see, every nonzero complex number has n nth roots.

EXAMPLE 2 Find the cube roots of $-8i$.

Solution The number $z = r(\cos \theta + i \sin \theta)$ is a cube root of $-8i$ if and only if

$$z^3 = -8i. \qquad (\dagger)$$

Since $z^3 = r^3(\cos 3\theta + i \sin 3\theta)$ and $-8i = 8(\cos 270^\circ + i \sin 270^\circ)$, equation ($\dagger$) is equivalent to

$$r^3(\cos 3\theta + i \sin 3\theta) = 8(\cos 270^\circ + i \sin 270^\circ).$$

Now solve for r and θ.

From this equation $r^3 = 8$. Since r is a positive real number, $r = 2$. Also

$$\cos 3\theta = \cos 270^\circ \quad \text{and} \quad \sin 3\theta = \sin 270^\circ.$$

Hence

$$
\begin{aligned}
3\theta &= 270^\circ + k \cdot 360^\circ, \ k \text{ an integer.} \\
\therefore \quad \theta &= 90^\circ + k \cdot 120^\circ, \ k \text{ an integer.}
\end{aligned}
$$

Let $k = 0$, 1, and 2. Then

$$\theta = 90^\circ, 210^\circ, \text{ and } 330^\circ.$$

Substitute into $z = r(\cos \theta + i \sin \theta)$. The three cube roots of $-8i$ are

$$
\begin{aligned}
z_1 &= 2(\cos 90^\circ + i \sin 90^\circ) = 2(0 + i \cdot 1) = 2i \\
z_2 &= 2(\cos 210^\circ + i \sin 210^\circ) = 2\left(-\frac{\sqrt{3}}{2} + i\left(-\frac{1}{2}\right)\right) = -\sqrt{3} - i \\
z_3 &= 2(\cos 330^\circ + i \sin 330^\circ) = 2\left(\frac{\sqrt{3}}{2} + i\left(-\frac{1}{2}\right)\right) = \sqrt{3} - i
\end{aligned}
$$

You can check the solution in Example 2 by cubing z_1, z_2, and z_3. Notice that other values of k give the same solution set.

In general we have the following theorem, whose proof is patterned on the solution to Example 2.

THEOREM 3 The n nth roots of $r(\cos \theta + i \sin \theta)$ are given by

$$r^{\frac{1}{n}}\left(\cos \frac{\theta + k \cdot 360^\circ}{n} + i \sin \frac{\theta + k \cdot 360^\circ}{n}\right),$$

where $k = 0, 1, 2, \ldots, n - 1$.

The roots found in Example 2 are evenly spaced around the circle $r = 2$. In general, the n nth roots of any nonzero complex number z are evenly spaced around the circle $r = \sqrt[n]{|z|}$.

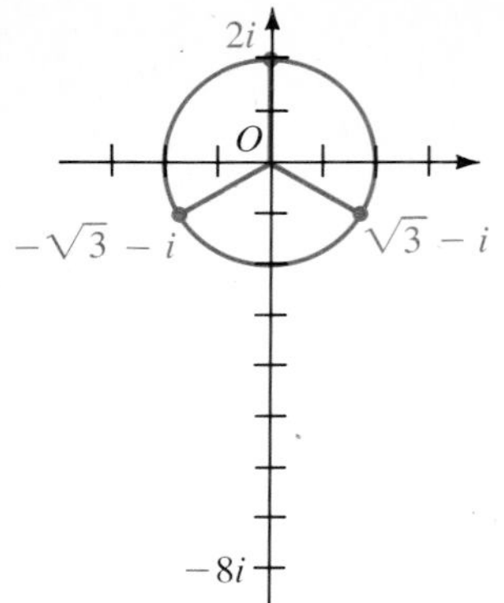

The nth roots of the number 1 are referred to as the **nth roots of unity.** By following the pattern of Example 2, we can show that the nth roots of unity are

$$\cos\left(\frac{k \cdot 360^\circ}{n}\right) + i \sin\left(\frac{k \cdot 360^\circ}{n}\right) \text{ where } k = 0, 1, 2, \ldots, n - 1.$$

EXERCISES

Use De Moivre's theorem to express the following powers in $x + yi$ form.

A

1. $(1 + i)^8$ **2.** $(1 - i)^{10}$ **3.** $(-1 + i)^6$ **4.** $(-1 - i)^{12}$

5. $(\sqrt{3} - i)^7$ **6.** $(1 + i\sqrt{3})^5$ **7.** $(-1 + i\sqrt{3})^9$ **8.** $(\sqrt{3} + i)^7$

Find the specified roots of unity in $x + yi$ form.

9. The fourth roots **10.** The cube roots

11. The sixth roots **12.** The eighth roots

Find the specified roots of unity in polar form.

13. The fifth roots **14.** The tenth roots

In Exercises 15–18, **(a)** find the specified roots in $x + yi$ form and **(b)** draw a figure showing the number, its roots, and the circle containing the roots.

B

15. The cube roots of -1 **16.** The square roots of i

17. The 4th roots of $-2 + 2i\sqrt{3}$ **18.** The cube roots of $-\sqrt{2} + i\sqrt{2}$

In Exercises 19–22 let $\omega = \cos \frac{2\pi}{n} + i \sin \frac{2\pi}{n}$. Express ω^h as an integral power of ω^k.

19. $n = 8, h = 7, k = 3$ **20.** $n = 5, h = 2, k = 3$

21. $n = 6, h = 2, k = 5$ **22.** $n = 9, h = 4, k = 7$

23. Use De Moivre's theorem to express $\cos 3\theta$ and $\sin 3\theta$ in terms of $\cos \theta$ and $\sin \theta$, respectively. (*Hint*: Use De Moivre's theorem and the binomial theorem to express $(\cos \theta + i \sin \theta)^3$ in two different ways.)

24. Show that the formula in De Moivre's theorem is valid if n is a negative integer. (*Hint*: Let $n = -m$, where m is a positive integer. Then $z^n = z^{-m} = \frac{1}{z^m}$. Now use De Moivre's theorem.)

C **25.** Show that for each positive integer n, the nth roots of unity form a commutative group under multiplication by completing parts (a)–(c).

a. Show that the product of two nth roots of unity is another nth root of unity.

b. Show that the multiplicative inverse of each nth root of unity is an nth root of unity.

c. Show that the set of nth roots of unity satisfies the other parts of the definition of commutative group.

In Exercises 26 and 27, $\omega = \cos \frac{2\pi}{n} + i \sin \frac{2\pi}{n}$ (n a positive integer).

26. Show that the nth roots of unity are $\omega, \omega^2, \omega^3, \ldots, \omega^n = 1$.

27. Show that $\omega^p = 1$ if and only if p is an integral multiple of n.

An nth root of unity is *primitive* if every nth root of unity is an integral power of it.

28. Let $\omega = \cos \frac{2\pi}{9} + i \sin \frac{2\pi}{9}$.

a. Show that ω is a primitive 9th root of unity.

b. Show that ω is an integral power of ω^2 and hence that ω^2 is a primitive 9th root of unity.

c. Show that ω is not an integral power of ω^3. (*Hint*: Use Exercise 27.)

d. For what integers k ($1 \leq k \leq 9$) is ω^k a primitive 9th root of unity?

29. Let $\omega = \cos \frac{2\pi}{8} + i \sin \frac{2\pi}{8}$. For what integers k is ω^k a primitive 8th root of unity?

COMPUTER EXERCISES

1. Write a computer program that calculates all n complex nth roots of an input complex number of the form $x + yi$. Use Theorem 3 on page 373.

2. Run the program for the following numbers and given values of n.

a. fourth roots of $3 - i$ **b.** cube roots of $-5 - 2i$

c. eighth roots of $1 + i$ **d.** fifth roots of $2 + 3i$

9-5 PARAMETRIC EQUATIONS

Most of the curves we have met thus far can be described by an equation, such as $x^2 + 4y^2 = 16$. In this equation the variables x and y are related in a direct way. Another way of describing a curve in the plane is to give the x- and y-coordinates of a point on the curve in terms of a third variable, say t, called a **parameter.** A pair of equations

$$x = f(t) \quad \text{and} \quad y = g(t)$$

that give the coordinates of the points of a curve are called **parametric equations** of the curve.

EXAMPLE 1 Sketch the curve defined by $x = 4 \cos t$ and $y = 2 \sin t$.

Solution If $t = 0$, $x = 4 \cos 0 = 4$ and $y = 2 \sin 0 = 0$. If $t = \frac{\pi}{6}$, $x = 4 \cos \frac{\pi}{6} = 4 \cdot \frac{\sqrt{3}}{2} \approx 3.5$, and $y = 2 \sin \frac{\pi}{6} = 2 \cdot \frac{1}{2} = 1$. The following table shows values of t, $0 \le t \le 2\pi$, and the corresponding values of x and y.

t	0	$\frac{\pi}{6}$	$\frac{\pi}{3}$	$\frac{\pi}{2}$	$\frac{2\pi}{3}$	$\frac{5\pi}{6}$	π	$\frac{7\pi}{6}$	$\frac{4\pi}{3}$	$\frac{3\pi}{2}$	$\frac{5\pi}{3}$	$\frac{11\pi}{6}$	2π
x	4	3.5	2	0	−2	−3.5	−4	−3.5	−2	0	2	3.5	4
y	0	1	1.7	2	1.7	1	0	−1	−1.7	−2	−1.7	−1	0

Plot the points (x, y) from the table and join them with a smooth curve to obtain the graph.

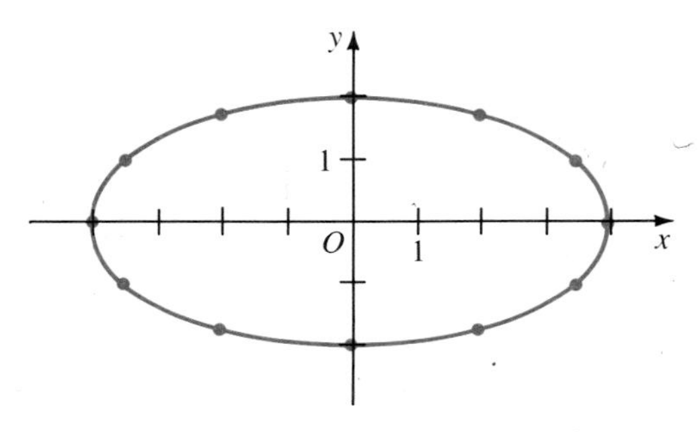

We can show that the curve in Example 1 is an ellipse by eliminating the parameter t. From $x = 4 \cos t$ and $y = 2 \sin t$, we obtain

$$\frac{x}{4} = \cos t \text{ and } \frac{y}{2} = \sin t.$$

Hence $$\left(\frac{x}{4}\right)^2 + \left(\frac{y}{2}\right)^2 = \cos^2 t + \sin^2 t = 1$$

or $$\frac{x^2}{16} + \frac{y^2}{4} = 1.$$

EXAMPLE 2 Identify the curve C defined by $x = \cos t$ and $y = \cos 2t$ by eliminating the parameter t. Then draw C.

Solution Use a double-angle formula for $\cos 2t$.

$$\begin{aligned} y &= \cos 2t \\ &= 2 \cos^2 t - 1 \end{aligned}$$

Substitute x for $\cos t$.

$$y = 2x^2 - 1$$

The graph of $y = 2x^2 - 1$ is the parabola shown at the right. However, the range of the cosine function is $[-1, 1]$. Thus $x \in [-1, 1]$ and the curve C consists only of the arc of the parabola that is shown in red.

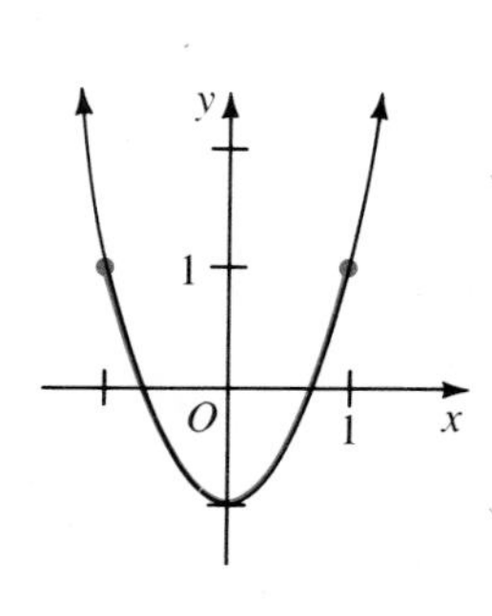

It is frequently useful to find a pair of parametric equations for a given curve.

EXAMPLE 3 Find a pair of parametric equations to describe the position of a point P that starts at $(3, 0)$ and moves in a circular path around a circle of radius 3 and with center at the origin.

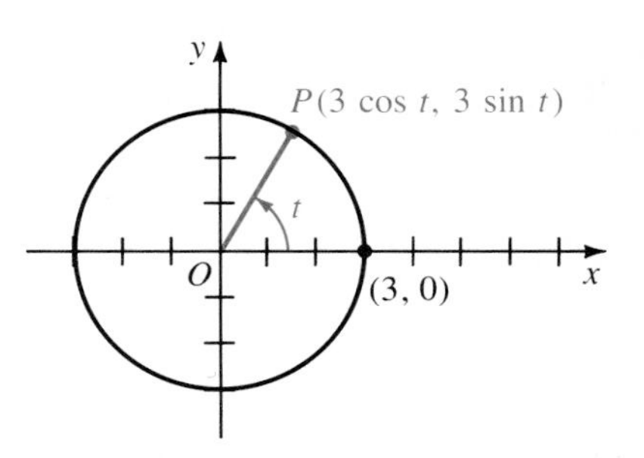

Solution The figure at the right shows the position of P after it has moved through t radians. The coordinates of P are given by

$$x = 3 \cos t \quad \text{and} \quad y = 3 \sin t.$$

In many applications the parameter t might represent time. Thus, the equations in Example 1 might describe the position (x, y) at time t of a moving particle as it starts (at time $t = 0$) at the point $(4, 0)$ and repeatedly traverses the ellipse in a counterclockwise sense. In Example 2 the particle would start at $(1, 1)$ and oscillate back and forth along the parabolic arc C. In Example 3, t represents the measure of the angle through which P moves.

Many curves that do not have simple equations relating x and y directly can be conveniently described parametrically. As an example, we shall derive parametric equations of a *cycloid*, the curve traced out by a point P on the rim of a wheel of radius a as it rolls along the x-axis.

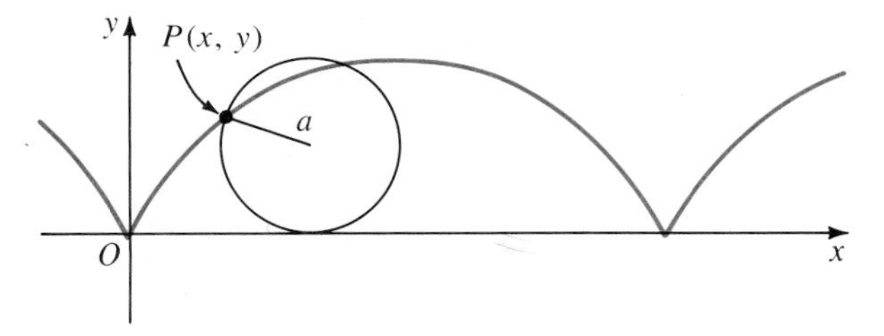

We shall take as the parameter the radian measure of an angle θ through which the wheel has turned since P was at the origin. From the figure at the right, in which θ is acute,

$$x = OQ - PK \quad \text{and} \quad y = QM - KM.$$

Since the wheel rolls without slipping, $OQ =$ length of arc $PQ = a\theta$. Also, $QM = a$, and from triangle PKM, $PK = a \sin \theta$ and $KM = a \cos \theta$.

Thus $x = a\theta - a \sin \theta$ and $y = a - a \cos \theta$.

$$x = a(\theta - \sin \theta) \quad \text{and} \quad y = a(1 - \cos \theta).$$

In Exercise 33 you are asked to show that these equations are valid for all θ.

EXERCISES

In Exercises 1–20, parametric equations of a curve are given. **a.** Identify C (if possible) by eliminating the parameter. **b.** Draw the curve C. (Be sure to show only the curve given by the parametric equations.)

A

1. $x = 1 + t;\ y = 1 - t$

2. $x = \frac{1}{2}t;\ y = t - 2$

3. $x = s^2;\ y = 1 - s^2$

4. $x = \sqrt{t};\ y = 1 - \sqrt{t}$

5. $x = 2\cos^2 t;\ y = 2\sin^2 t$

6. $x = \sec^2 u;\ y = \tan^2 u$

7. $x = 2\cos\theta;\ y = 2\sin\theta$

8. $x = \sqrt{t};\ y = \sqrt{1 - t}$

9. $x = \sqrt{t};\ y = \sqrt{1 + t}$

10. $x = \sec\theta;\ y = \tan\theta$

11. $x = \cos u;\ y = 2\sin u$

12. $x = 3\cos t;\ y = 2\sin t$

13. $x = \tan\theta;\ y = \sec\theta$

14. $x = \csc s;\ y = \cot s$

B

15. $x = 1 + \cos t;\ y = 1 - \sin t$

16. $x = \sin t;\ y = 1 + \cos t$

17. $x = \cos t;\ y = \sec t$

18. $x = \tan u;\ y = \cot u$

19. $x = \sin 2\theta;\ y = 2\sin^2\theta$

20. $x = 2\cos^2\theta;\ y = \sin 2\theta$

21. Find a pair of parametric equations to describe the position P of a particle moving in a counterclockwise direction around a circle of radius r with center at $(1, 1)$.

22. A particle P starts at $(0, 1)$ and moves 2 units to the right and 3 units up every second along a line. Find a pair of parametric equations to describe the position of P after t seconds.

Find a pair of parametric equations of the form $x = f(\theta)$ and $y = g(\theta)$ for each polar curve C.

23. C: $r = \cos 2\theta$

24. C: $r = 1 - \cos\theta$

Find a polar equation for each curve C and name the curve.

25. C: $x = \theta\cos\theta$ and $y = \theta\sin\theta$

26. C: $x = \dfrac{\cos\theta}{\theta}$ and $y = \dfrac{\sin\theta}{\theta}$

In Exercises 27–30 sketch the curve having the given parametric equations by plotting the points (x, y) obtained from the given values of t and then joining the points with a smooth curve.

27. $x = t^2$ and $y = t^3$; $t = 0, \pm\frac{1}{2}, \pm 1, \pm\frac{3}{2}$

28. $x = t^3$ and $y = t^2$; $t = 0, \pm\frac{1}{2}; \pm 1; \pm\frac{3}{2}$

29. $x = t(t^2 - 4),\ y = t^2 - 4$; $t = 0, \pm 1, \pm\frac{3}{2}, \pm 2, \pm\frac{5}{2}$

30. $x = t^2 - 4,\ y = t(t^2 - 4)$; $t = 0, \pm 1, \pm\frac{3}{2}, \pm 2, \pm\frac{5}{2}$

C **31.** Refer to the figure at the left below. Two concentric circles have radii a and b where $a > b$. For each θ, triangle APB is a right triangle.

a. Find parametric equations for the path of P.

b. Eliminate the parameter θ and identify the curve.

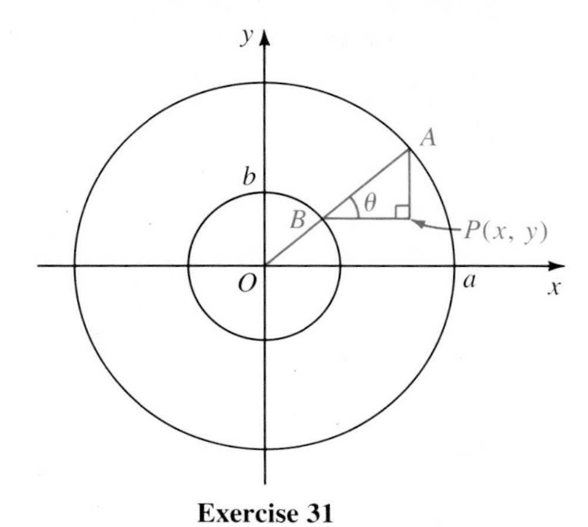

Exercise 31

Exercise 32

32. Refer to the figure at the right above. A projectile P is fired in a vacuum from the origin. Parametric equations of its trajectory are

$$x = (v_0 \cos \alpha)t, \quad y = (v_0 \sin \alpha)t - \frac{1}{2}gt^2,$$

where v_0 is the initial speed of the projectile, α is its initial angle, and g is the acceleration of gravity.

a. Show that the trajectory is an arc of a parabola.

b. Find OQ. For what α is OQ greatest?

33. a. Refer to the figure at the right below. Let (h, k) in an xy-coordinate system be the origin of a uv-coordinate system oriented as shown. Show that the xy-coordinates and the uv-coordinates are related by $x = h - v$ and $y = k - u$.

b. Show that the equations of the cycloid are valid for all θ. (*Hint:* Introduce a uv-coordinate system as in part (a) with $(h, k) = (a\theta, a)$ and with θ in standard position with respect to the uv-coordinate system.)

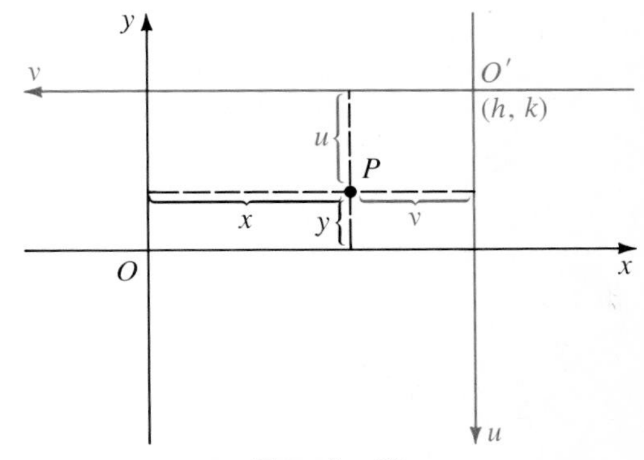

Exercise 33

34. Refer to the figure at the right. A taut string is unrolled from a spool of radius a. Find parametric equations of the spiral curve swept out by the end of the string. The parameter is θ. (The curve is called the *involute* of the circle.)

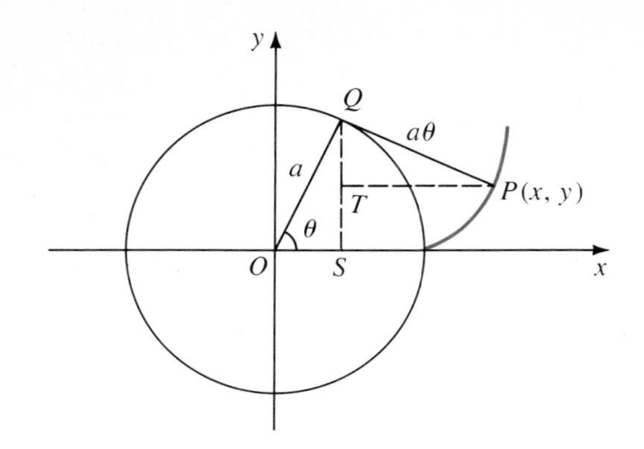

Calculus of Trigonometric Functions

9-6 SOME TRIGONOMETRIC LIMITS

Some of the most important applications of the trigonometric functions involve their derivatives. In order to find these, we need first to establish some trigonometric limits.

The following figures show the unit circle C and the points

$$P(\cos x, \sin x),\ U(1, 0),\ \text{and}\ M(\cos x, 0)$$

for $0 < |x| < \frac{\pi}{2}$.

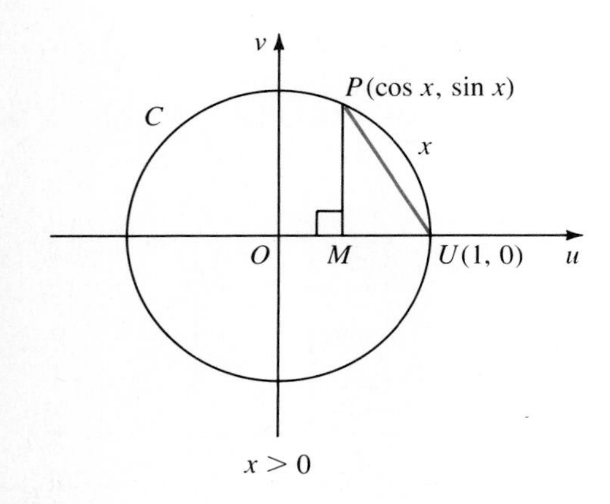

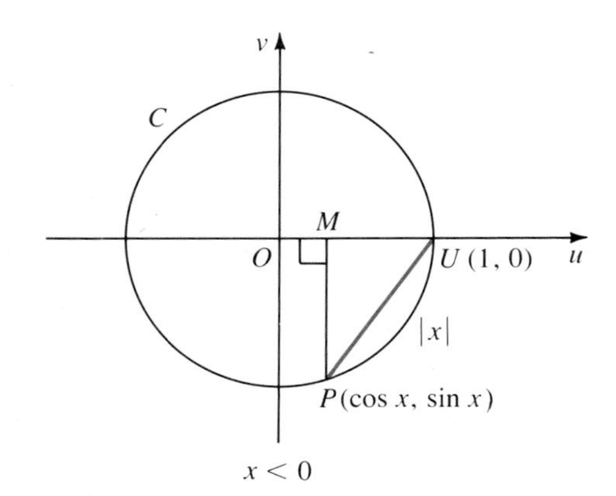

We shall use the fact that the length of the chord $\overline{UP}$ is less than the length, $|x|$, of the arc $\overset{\frown}{UP}$ that it subtends. Therefore

$$|\sin x| = MP < UP < |x| \quad \text{and} \quad |1 - \cos x| = MU < UP < |x|;$$
$$0 < |\sin x| < |x| \quad \text{and} \quad 0 < |1 - \cos x| < |x|.$$

Applying the "squeeze theorem," page 160, we find that

$$\lim_{x \to 0} \sin x = 0 \quad \text{and} \quad \lim_{x \to 0} \cos x = 1.$$

We use these limits to prove the following theorem.

THEOREM 4 The sine and cosine functions are continuous everywhere.

Proof: Let c be any real number. For any h,

$$\sin (c + h) = \sin c \cos h + \cos c \sin h.$$

Using the addition formula for sin, $\lim_{h \to 0} \cos h = 1$ and $\lim_{h \to 0} \sin h = 0$, we have:

$$\begin{aligned} \lim_{h \to 0} \sin (c + h) &= \lim_{h \to 0} (\sin c \cos h + \cos c \sin h) \\ &= \sin c \lim_{h \to 0} \cos h + \cos c \lim_{h \to 0} \sin h \\ &= \sin c \cdot 1 + \cos c \cdot 0 \\ &= \sin c. \end{aligned}$$

This implies that $\lim_{x \to c} \sin x = \sin c$ for all real numbers c. Therefore the sine is continuous everywhere. The proof for the cosine is similar. See Exercise 7. ■

Because the other four trigonometric functions can be expressed as quotients or reciprocals of the sine and cosine, we have the following result.

COROLLARY Each trigonometric function is continuous on its domain.

The following limits are used in Section 9–7.

THEOREM 5 **1.** $\lim_{x \to 0} \frac{\sin x}{x} = 1$ **2.** $\lim_{x \to 0} \frac{1 - \cos x}{x} = 0$

Proof: **1.** Let x be any number between 0 and $\frac{\pi}{2}$. We see that

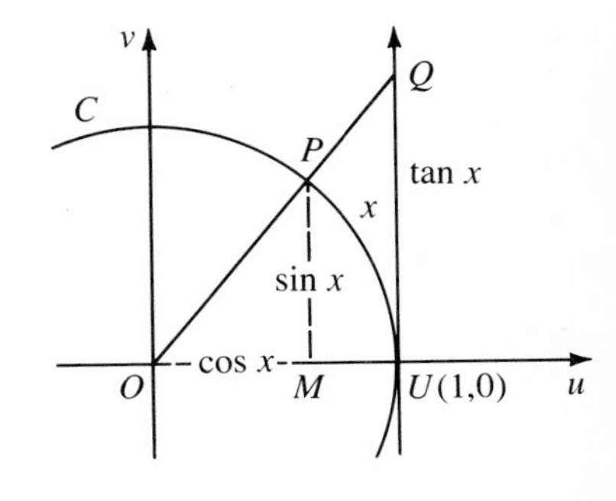

$$\begin{aligned} \text{Area } \triangle OMP &= \frac{1}{2} OM \cdot MP \\ &= \frac{1}{2} \cos x \sin x \end{aligned}$$

$$\begin{aligned} \text{Area } \triangle OUQ &= \frac{1}{2} \cdot 1 \cdot UQ \\ &= \frac{1}{2} \frac{\sin x}{\cos x} \end{aligned}$$

Proof continued on following page

Area sector $OUP = \frac{1}{2} \cdot 1 \cdot x = \frac{1}{2}x$ (recall Exercise 17, page 285).

Because area $\triangle OMP <$ area sector $OUP <$ area $\triangle OUQ$,

$$\frac{1}{2} \cos x \sin x < \frac{1}{2}x < \frac{1}{2} \frac{\sin x}{\cos x}.$$

Multiplying each member by $\frac{2}{\sin x}$, we obtain

$$\cos x < \frac{x}{\sin x} < \frac{1}{\cos x}. \qquad (*)$$

Because $\cos x$, $\frac{x}{\sin x}$, and $\frac{1}{\cos x}$ are all even functions, (*) holds also if $-\frac{\pi}{2} < x < 0$.

Thus $$\lim_{x \to 0} \cos x < \lim_{x \to 0} \frac{x}{\sin x} < \lim_{x \to 0} \frac{1}{\cos x}.$$

We know that $\lim_{x \to 0} \cos x = 1$ and $\lim_{x \to 0} \frac{1}{\cos x} = 1$.

Therefore by the "squeeze theorem," $\lim_{x \to 0} \frac{x}{\sin x} = 1$

and thus $\lim_{x \to 0} \frac{\sin x}{x} = 1$. ■

2. $$\frac{1 - \cos x}{x} = \frac{1 - \cos x}{x} \cdot \frac{1 + \cos x}{1 + \cos x}$$
$$= \frac{1 - \cos^2 x}{x(1 + \cos x)}$$
$$= \frac{\sin^2 x}{x(1 + \cos x)}$$

Hence $\frac{1 - \cos x}{x} = \frac{\sin x}{x} \cdot \frac{\sin x}{1 + \cos x}$.

Take the limit and use the continuity of sine and cosine.

$$\lim_{x \to 0} \frac{1 - \cos x}{x} = \lim_{x \to 0} \frac{\sin x}{x} \cdot \lim_{x \to 0} \frac{\sin x}{1 + \cos x} = 1 \cdot \frac{0}{1 + 1} = 0$$ ■

EXAMPLE Evaluate each limit. **a.** $\lim_{x \to 0} \frac{\sin 3x}{x}$ **b.** $\lim_{x \to \pi} \frac{\sin x}{x - \pi}$

Solution **a.** $\lim_{x \to 0} \frac{\sin 3x}{x} = \lim_{x \to 0} 3 \frac{\sin 3x}{3x} = 3 \lim_{3x \to 0} \frac{\sin 3x}{3x} = 3 \cdot 1 = 3$ **Answer**

b. Let $t = x - \pi$. Then $t \to 0$ as $x \to \pi$.
Also $\sin x = \sin (t + \pi) = \sin t \cos \pi + \cos t \sin \pi = -\sin t$.

Therefore $\lim_{x \to \pi} \frac{\sin x}{x - \pi} = \lim_{t \to 0} \frac{-\sin t}{t} = -\lim_{t \to 0} \frac{\sin t}{t} = -1$. **Answer**

EXERCISES

Evaluate the following limits.

A

1. $\lim_{x \to 0} \frac{\sin \frac{1}{2}x}{x}$
2. $\lim_{x \to 0} x \csc x$
3. $\lim_{x \to 0} \frac{\sin 2x}{\sin x}$
4. $\lim_{x \to 0} \frac{\sin ax}{x}$, $a \neq 0$
5. $\lim_{x \to 0} \frac{\tan x}{x}$
6. $\lim_{x \to 0} \frac{\sin 3x}{\sin 2x}$

B

7. Complete the proof of Theorem 4 by showing that the cosine function is continuous.

Prove the corollary to Theorem 4 by giving specific reasons why each of the following functions is continuous at each number in its domain.

8. The tangent
9. The secant
10. The cosecant
11. The cotangent

Exercises 12 and 13 indicate why, in analysis, radian measure is used rather than degree measure.

12. With a calculator, evaluate $\frac{\sin x}{x}$ for $x = 1.00$, 0.10, and 0.01 with the calculator set in **(a)** the degree mode; **(b)** the radian mode.

C

13. Show that $\lim_{x \to 0} \frac{\sin x^{\circ}}{x} = \frac{\pi}{180}$.

Prove the following statements; x is in radians.

14. $\lim_{x \to 0} \frac{1 - \cos x}{\sin x} = 0$
15. $\lim_{x \to 0} \frac{x - \sin x}{x} = 0$
16. $\lim_{x \to 0} \frac{1 - \cos^2 x}{x^2} = 1$
17. $\lim_{x \to 0} \frac{1 - \cos ax}{x} = 0$
18. $\lim_{x \to \pi} \frac{\pi - x}{\sin x} = 1$
19. $\lim_{x \to \frac{\pi}{2}} \frac{\cos x}{2x - \pi} = -\frac{1}{2}$
20. $\lim_{x \to 0} \frac{1 - \cos x}{x^2} = \frac{1}{2}$
21. $\lim_{x \to 0} \frac{1 - \cos x}{\sin^2 x} = \frac{1}{2}$

COMPUTER EXERCISES

1. Write a computer program to help find $\lim_{x \to 0} f(x)$, if it exists, for a given function f. Let x take on values of a sequence approaching zero through positive values $1, \frac{1}{2}, \frac{1}{4}, \ldots$ and through negative values $-1, -\frac{1}{2}, -\frac{1}{4}, \ldots$.
2. Run the program of Exercise 1 for the following functions.

   a. $f(x) = \frac{\sin x}{x}$ **b.** $f(x) = \frac{1 - \cos x}{x}$ **c.** $f(x) = x \sin \frac{1}{x}$

9-7 DERIVATIVES OF THE TRIGONOMETRIC FUNCTIONS

To find the derivative of the sine function, we use the definition of the derivative in the form

$$f'(x) = \lim_{h \to 0} \frac{f(x + h) - f(x)}{h}$$

(from page 224) and let $f(x) = \sin x$.

$$\begin{aligned} f(x + h) - f(x) &= \sin(x + h) - \sin x \\ &= \sin x \cos h + \cos x \sin h - \sin x \\ &= \sin h \cos x - (1 - \cos h)\sin x \\ \frac{f(x + h) - f(x)}{h} &= \frac{\sin h}{h} \cos x - \frac{1 - \cos h}{h} \sin x \end{aligned}$$

$$\text{Therefore } \lim_{h \to 0} \frac{f(x + h) - f(x)}{h} = \left(\lim_{h \to 0} \frac{\sin h}{h}\right)\cos x - \left(\lim_{h \to 0} \frac{1 - \cos h}{h}\right)\sin x$$

$$= 1 \cdot \cos x - 0 \cdot \sin x = \cos x.$$

Since by Theorem 5, $\lim_{h \to 0} \frac{\sin h}{h} = 1$ and $\lim_{h \to 0} \frac{1 - \cos h}{h} = 0$.

Thus, if $f(x) = \sin x$, then

$$f'(x) = \cos x, \text{ or } \frac{d}{dx} \sin x = \cos x.$$

This proves the first part of Theorem 6. The proof of the second part of the theorem is left as Exercise 1.

THEOREM 6 For all x, $\frac{d}{dx} \sin x = \cos x$ and $\frac{d}{dx} \cos x = -\sin x$.

The formulas of Theorem 6 can, of course, be combined with the differentiation formulas developed in Chapter 6.

EXAMPLE 1 Differentiate each function. **a.** $y = x^2 \cos x$ **b.** $y = \tan x$

Solution **a.** Use the product rule (page 227).

$$\begin{aligned} \frac{d}{dx}(x^2 \cos x) &= x^2 \frac{d}{dx} \cos x + (\cos x) \frac{d}{dx} x^2 \\ &= x^2(-\sin x) + (\cos x)(2x) \end{aligned}$$

$$\therefore \frac{d}{dx}(x^2 \cos x) = 2x \cos x - x^2 \sin x \qquad \textbf{Answer}$$

b. Write $\tan x$ as $\frac{\sin x}{\cos x}$ and use the quotient rule (page 227).

$$\frac{d}{dx} \tan x = \frac{d}{dx} \frac{\sin x}{\cos x}$$

$$\frac{d}{dx}\tan x = \frac{\cos x \frac{d}{dx}\sin x - \sin x \frac{d}{dx}\cos x}{(\cos x)^2}$$

$$= \frac{\cos x(\cos x) - \sin x(-\sin x)}{\cos^2 x}$$

$$= \frac{\cos^2 x + \sin^2 x}{\cos^2 x} = \frac{1}{\cos^2 x} = \sec^2 x$$

$$\therefore \frac{d}{dx}\tan x = \sec^2 x \quad \textbf{Answer}$$

Example 1(b) establishes the first formula stated in Theorem 7. The others can be proved in a similar fashion. See Exercises 2–4.

THEOREM 7

1. $\frac{d}{dx}\tan x = \sec^2 x$
2. $\frac{d}{dx}\cot x = -\csc^2 x$
3. $\frac{d}{dx}\sec x = \sec x \tan x$
4. $\frac{d}{dx}\csc x = -\csc x \cot x$

EXAMPLE 2 Find an equation of the line tangent to the graph of $y = \tan x$ at the point $\left(\frac{\pi}{4}, 1\right)$.

Solution The slope of the curve is given by $y' = \sec^2 x$. When $x = \frac{\pi}{4}$, $y' = \sec^2 \frac{\pi}{4} = (\sqrt{2})^2 = 2$.

$$\therefore y - 1 = 2\left(x - \frac{\pi}{4}\right), \quad \text{or} \quad 2x - y = \frac{\pi}{2} - 1 \quad \textbf{Answer}$$

The chain rule (Section 6–4) enables us to differentiate a great variety of functions.

EXAMPLE 3 Find $\frac{dy}{dx}$. **a.** $y = \sin x^2$ **b.** $y = \sin^5 x$.

Solution **a.** Let $y = \sin u$, where $u = x^2$.

$$\frac{dy}{du} = \cos u \quad \text{and} \quad \frac{du}{dx} = 2x$$

$$\frac{dy}{dx} = \frac{dy}{du} \cdot \frac{du}{dx} = \cos u \cdot 2x$$

$$= 2x \cos u$$

Now eliminate u since the problem involves x.

$$\frac{dy}{dx} = 2x \cos x^2 \quad \textbf{Answer}$$

Solution continued on following page

b. Since $y = \sin^5 x = (\sin x)^5$, let $y = u^5$, where $u = \sin x$.

$$\frac{dy}{du} = 5u^4 \quad \text{and} \quad \frac{du}{dx} = \cos x$$

$$\frac{dy}{dx} = \frac{dy}{du} \cdot \frac{du}{dx}$$

$$= 5u^4 \cos x$$

Therefore $\qquad \dfrac{dy}{dx} = 5 \sin^4 x \cos x.$ **Answer**

With practice you will be able to use the chain rule without introducing the "intermediate variable" u. Thus, the solution of Example 3(b) could be written as

$$\frac{d}{dx} \sin^5 x = 5 \sin^4 x \frac{d}{dx} \sin x = 5 \sin^4 x \cos x.$$

EXAMPLE 4 Find the absolute maximum and minimum values of $f(x) = 2 \sin x + \cos 2x$ where $0 \le x \le 2\pi$.

Solution Find $f'(x)$ and solve $f'(x) = 0$.

$$f(x) = 2 \sin x + \cos 2x$$

$$f'(x) = 2 \cos x - 2 \sin 2x$$

$$= 2(\cos x - \sin 2x)$$

$$= 2(\cos x - 2 \sin x \cos x)$$

$$= 2 \cos x(1 - 2 \sin x)$$

$f'(x) = 0$ when $\cos x = 0$ or $\sin x = \frac{1}{2}$. The solutions of these equations in $0 \le x \le 2\pi$ are

$$x = \frac{\pi}{2}, \frac{3\pi}{2}, \frac{\pi}{6}, \frac{5\pi}{6}.$$

$$f\left(\frac{\pi}{2}\right) = 2 \sin \frac{\pi}{2} + \cos \pi = 2 \cdot 1 + (-1) = 1$$

$$f\left(\frac{3\pi}{2}\right) = 2 \sin \frac{3\pi}{2} + \cos 3\pi = 2(-1) + (-1) = -3$$

$$f\left(\frac{\pi}{6}\right) = 2 \sin \frac{\pi}{6} + \cos \frac{\pi}{3} = 2 \cdot \frac{1}{2} + \frac{1}{2} = \frac{3}{2}$$

$$f\left(\frac{5\pi}{6}\right) = 2 \sin \frac{5\pi}{6} + \cos \frac{5\pi}{3} = 2 \cdot \frac{1}{2} + \frac{1}{2} = \frac{3}{2}$$

To find the absolute maximum and the absolute minimum of f, also find the values of f at 0 and 2π.

$$f(0) = 2 \sin 0 + \cos 0 = 1$$

$$f(2\pi) = 2 \sin 2\pi + \cos 4\pi = 1$$

$\therefore$ the absolute maximum value of f is $\frac{3}{2}$ and the absolute minimum value of f is -3. **Answer**

EXERCISES

A

1. Complete the proof of Theorem 6 by showing that $\frac{d}{dx} \cos x = -\sin x$.

$\left(\textit{Hint}: \cos x = \sin\left(\frac{\pi}{2} - x\right).\right)$

Complete the proof of Theorem 7 by showing that

2. $\frac{d}{dx} \cot x = -\csc^2 x$

3. $\frac{d}{dx} \sec x = \sec x \tan x$

4. $\frac{d}{dx} \csc x = -\csc x \cot x$

In Exercises 5–8 find an equation of the line tangent to the curve C at the point whose x-coordinate is given.

5. $C: y = \cos x; x = \frac{\pi}{6}$

6. $C: y = \sin x; x = \frac{2\pi}{3}$

7. $C: y = 2 \cos x - \cos 2x; x = \frac{\pi}{3}$

8. $C: y = 2 \cos 2x + \sin x; x = 0$

Differentiate each of the following functions. Simplify each answer.

9. $\sec 3x$

10. $\cot 2x$

11. $\sin at$

12. $\tan at$

13. $\cos x^\circ$

14. $\sin x^\circ$

15. $\tan x^2$

16. $\cos x^2$

17. $\tan^2 x$

18. $\cos^2 x$

19. $\sin \theta \cos \theta$

20. $\sin^2 \theta - \cos^2 \theta$

21. $\sin t - t \cos t$

22. $\cos t + t \sin t$

B

23. $2x \cos x + (x^2 - 2)\sin x$

24. $2x \sin x - (x^2 - 2)\cos x$

25. $2x^2 - 2x \sin 2x - \cos 2x$

26. $2x^2 + 2x \cos 2x + \sin 2x$

In Exercises 27 and 28, find $f''(x)$.

27. $f(x) = \tan x$

28. $f(x) = \sec x$

In Exercises 29–32 find the maximum and minimum values of f.

29. $f(x) = 2 \cos x - \cos 2x$

30. $f(x) = 2 \sin x - \cos 2x$

31. $f(x) = 2 \cos x + \sin 2x$

32. $f(x) = 2 \sin x + \sin 2x$

33. Show that if $y = a \cos kt + b \sin kt$, then $\frac{d^2y}{dt^2} + k^2y = 0$.

In Exercises 34 and 35, **(a)** find $f'(x), f''(x), f'''(x)$, and $f^{(4)}(x)$. Then **(b)** give a formula for $f^{(n)}(0)$.

C

34. $f(x) = \sin x$

35. $f(x) = \cos x$

■ 9–8 RATES AND EXTREMA

In this section we shall apply the calculus of the trigonometric functions to problems involving rates and maxima or minima. We shall follow the steps for solving such problems stated in Sections 6–7 and 6–8.

EXAMPLE 1 A 27-foot ladder leans against an 8-foot wall with its top projecting over the wall. The foot of the ladder is being pulled away from the wall at the rate of 5 feet per second. How fast is the ladder rotating when its foot is 6 feet from the wall?

Solution **1.** Let θ be the angle $\left(0 < \theta < \frac{\pi}{2}\right)$ the ladder makes with the ground when its foot is x feet from the wall. It is given that $\frac{dx}{dt} = 5.$ The goal is to find $\frac{d\theta}{dt}$.

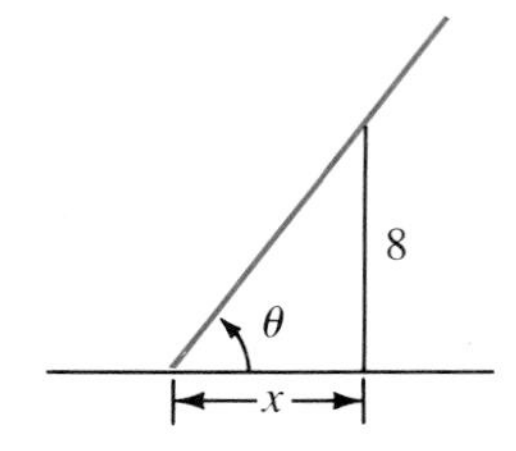

2. Use the sketch at the right.

3. Differentiate with respect to time, t.

$$\frac{dx}{dt} = \frac{dx}{d\theta} \cdot \frac{d\theta}{dt} \qquad (*)$$

Next find $\frac{dx}{d\theta}$ from $x = 8 \cot \theta$.

$$\begin{aligned} \frac{dx}{d\theta} &= -8 \csc^2 \theta \\ &= -8(1 + \cot^2 \theta) \\ &= -8\left(1 + \left(\frac{x}{8}\right)^2\right) \\ &= -\frac{64 + x^2}{8} \end{aligned}$$

Substitute into (*) to obtain

$$\frac{dx}{dt} = -\frac{64 + x^2}{8} \frac{d\theta}{dt}.$$

Hence:

$$\frac{d\theta}{dt} = -\frac{8}{64 + x^2} \frac{dx}{dt}$$

4. Substitute $x = 6$ and $\frac{dx}{dt} = 5$:

$$\begin{aligned} \frac{d\theta}{dt} = \frac{-8}{64 + 36} \cdot 5 &= -\frac{2}{5} \text{ radians per second} \\ &= -\frac{2}{5} \cdot \frac{180°}{\pi} \\ &\approx -22.9° \text{ per second} \end{aligned}$$ **Answer**

EXAMPLE 2 Find the greatest horizontal distance that the ladder of Example 1 projects over the wall.

Solution Follow the steps listed on page 248.

1. Let z be the distance that is to be maximized and θ be the angle $\left(0 < \theta < \frac{\pi}{2}\right)$ the ladder makes with the ground. Let p be the length of the part of the ladder projecting over the wall and q be the length of the rest of the ladder.

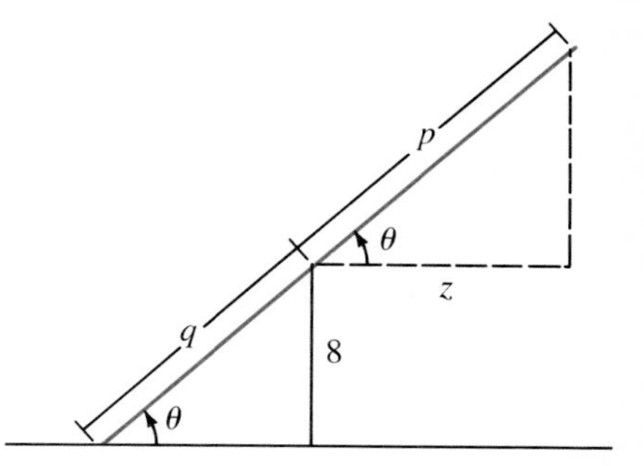

2. A labeled diagram is shown at the right.

3. From the diagram

$$\sec\theta = \frac{p}{z} \text{ and } \csc\theta = \frac{q}{8}.$$

Hence $$p = z\sec\theta \quad \text{and} \quad q = 8\csc\theta.$$

Since $p + q = 27$,

$$z\sec\theta + 8\csc\theta = 27, \quad \text{or} \quad z\sec\theta = 27 - 8\csc\theta.$$

Multiply both sides of $z\sec\theta = 27 - 8\csc\theta$ by $\cos\theta$.

$$z = 27\cos\theta - 8\cot\theta \qquad (*)$$

4. Find $\frac{dz}{d\theta}$. Then set $\frac{dz}{d\theta}$ equal to 0.

$$\frac{dz}{d\theta} = -27\sin\theta + 8\csc^2\theta$$

$\frac{dz}{d\theta} = 0$ when $27\sin\theta = 8\csc^2\theta = \frac{8}{\sin^2\theta}$. Hence

$$\sin^3\theta = \frac{8}{27}, \text{ or } \sin\theta = \frac{2}{3}.$$

There is only one value of θ such that $0 < \theta < 90°$ and $\sin\theta = \frac{2}{3}$.

Use the sketch at the right to find $\cos\theta$ and $\cot\theta$.

$$\cos\theta = \frac{\sqrt{5}}{3} \quad \text{and} \quad \cot\theta = \frac{\sqrt{5}}{2}$$

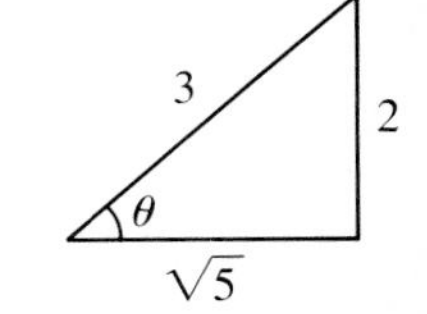

5. Substitute $\frac{\sqrt{5}}{3}$ for $\cos\theta$ and $\frac{\sqrt{5}}{2}$ for $\cot\theta$ into (*).

$$\text{maximum } z = 27 \cdot \frac{\sqrt{5}}{3} - 8\,\frac{\sqrt{5}}{2} = 5\sqrt{5} \text{ feet} \qquad \textbf{Answer}$$

Use identities to simplify your work whenever possible. See step 3 in Example 1. Notice how a double-angle formula is used in step 3 of the next example.

EXAMPLE 3 A cone having vertex angle 2θ is inscribed in a sphere of radius a. Find θ for the cone having greatest volume.

Solution

1. Let V be the volume of the cone and r, h, and s be its base radius, height, and slant height, respectively. The goal is to find θ such that V is maximum.

2. A labeled diagram is shown at the right.

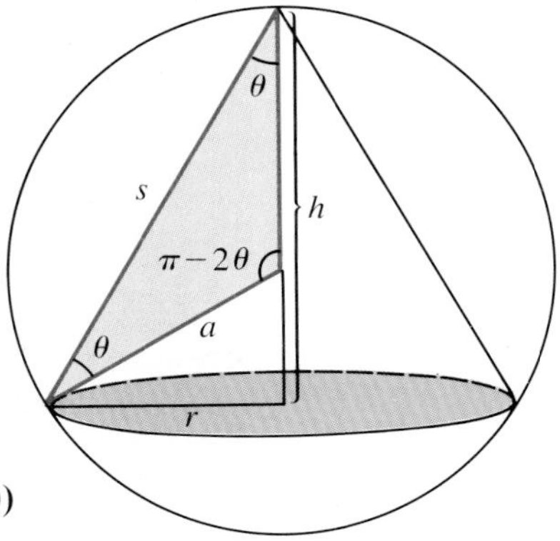

3. Apply the law of cosines to the shaded triangle in the figure to obtain:

$$\begin{aligned} s^2 &= a^2 + a^2 - 2a \cdot a \cdot \cos(\pi - 2\theta) \\ &= 2a^2 + 2a^2 \cos 2\theta \\ &= 2a^2(1 + \cos 2\theta) \\ &= 2a^2(1 + (2\cos^2\theta - 1)) \\ &= 4a^2 \cos^2\theta \end{aligned}$$

Hence $s = 2a \cos \theta$.

Write the volume $V = \frac{1}{3}\pi r^2 h$ in terms of θ:

$$r = s \sin\theta = 2a \cos\theta \sin\theta$$
$$h = s \cos\theta = 2a \cos^2\theta$$
$$V = \frac{1}{3}\pi r^2 h = \frac{1}{3}\pi(2a \cos\theta \sin\theta)^2(2a \cos^2\theta)$$

Therefore $V = \frac{8}{3}\pi a^3 \cos^4\theta \sin^2\theta$.

4. Find $\frac{dV}{d\theta}$ by use of the product rule.

$$\frac{dV}{d\theta} = \frac{8}{3}\pi a^3[\cos^4\theta \cdot 2\sin\theta\cos\theta + \sin^2\theta \cdot 4\cos^3\theta(-\sin\theta)]$$
$$= \frac{16}{3}\pi a^3 \sin\theta \cos^3\theta(\cos^2\theta - 2\sin^2\theta)$$

Solve $\frac{dV}{d\theta} = 0$ for θ.

If $\sin\theta = 0$ or $\cos\theta = 0$, $\theta = 0°$ or $90°$, and these do not give a cone of maximum volume. Therefore $\frac{dV}{d\theta} = 0$ when $\cos^2\theta - 2\sin^2\theta = 0$.

Thus $\cos^2\theta = 2\sin^2\theta$, or $\tan^2\theta = \frac{1}{2}$. Hence $\theta = 35.3°$.

5. The cone with vertex angle 2θ inscribed in a sphere of radius a has maximum volume when $\theta = 35.3°$. **Answer**

EXERCISES

A

1. A plane flying at 600 km/h on a straight horizontal course 5 km above the ground passes directly over an observer who is watching the plane through a telescope. How fast is the telescope rotating **(a)** at that moment? **(b)** one minute later?

2. The foot of a 5-meter ladder leaning against a tall wall is being pulled away from the wall at the rate of 2 m/s. At what rate is the ladder rotating when its foot is **(a)** 3 meters from the wall? **(b)** 4 meters from the wall?

3. The cross section of an aqueduct is to be an isosceles trapezoid, three of whose sides are 2 m long, as shown. Find θ so that the cross-sectional area will be as great as possible.

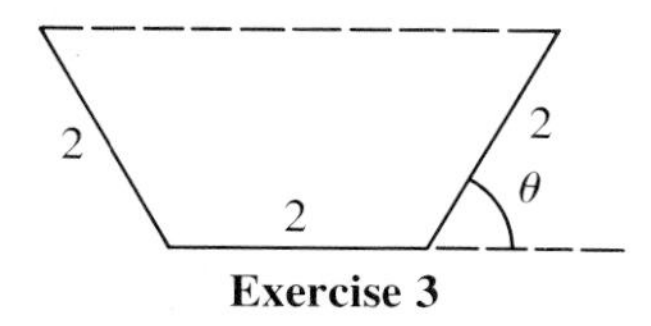

Exercise 3

4. The force magnitude in newtons necessary to drag an object of mass m at constant speed along a rough surface is given by

$$F = \frac{9.8\ mk}{\cos\theta + k\sin\theta},$$

where θ is the angle the force makes with the horizontal and k is the coefficient of friction. For what θ is F least? (*Hint:* Write $\cos\theta + k\sin\theta$ as a single cosine.)

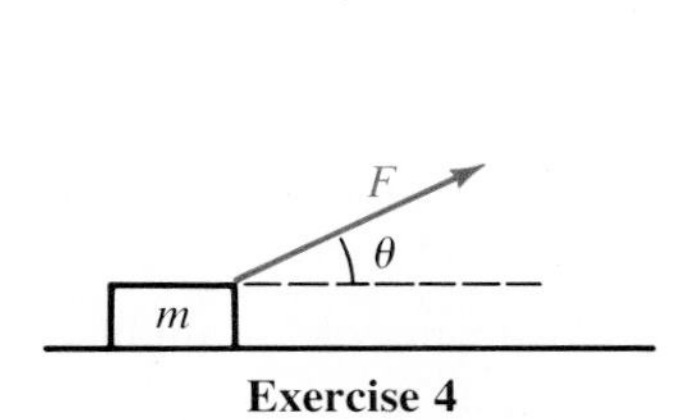

Exercise 4

30
5
50

Exercise 5

5. A ferris wheel 60 feet in diameter has its center 35 feet above the ground. It makes one revolution every 10 seconds. How fast is the height of a point on the rim increasing or decreasing relative to the ground when the point is 50 feet above the ground?

6. Two sides of a triangle are 10 cm long and their included angle is increasing at the rate of 15° per minute. How fast is the third side lengthening at the moment when it is 10 cm long?

7. Refer to Example 3. Find θ for the cone having greatest lateral (curved-surface) area.

Exercises 8–10 refer to a cylinder inscribed in a sphere of radius a, as shown at the right. Find θ for which each of the following is greatest.

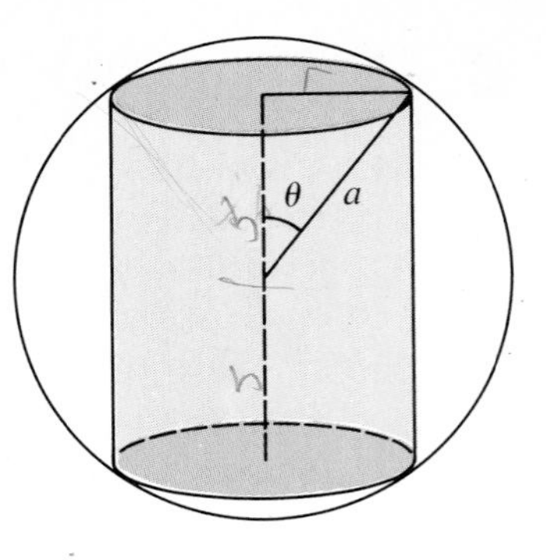

8. The volume of the cylinder.

9. The lateral area of the cylinder.

10. The total area (lateral area plus the area of the two bases).

11. A lighthouse located on an island two miles from the nearest point A of a straight shoreline has a light that rotates once every 20 seconds. How fast is the spot of light it casts on the shore moving when the spot is 4 miles from A?

B **12.** The figure shows a derrick having a 26-foot vertical spar and a 24-foot boom. A cable passing over the top of the spar and attached to the boom is being reeled in at the rate of 3 feet per second. How fast is the angle between the spar and the boom changing when its tangent is $\frac{5}{12}$?

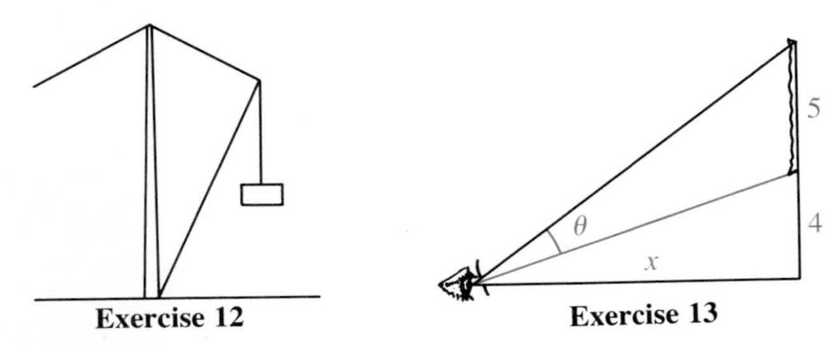

Exercise 12 Exercise 13

13. The bottom of a picture five feet tall is four feet above the level of an observer's eyes. How far from the wall on which the picture hangs should the observer stand in order that the picture subtends the greatest possible angle at the observer's eyes?

14. Solve Exercise 30, page 252, by expressing y in terms of an angle.

15. A wall of a tall building is to be braced by a steel girder that must pass over a four-foot wall 13.5 ft from the building. At least how long must the girder be?

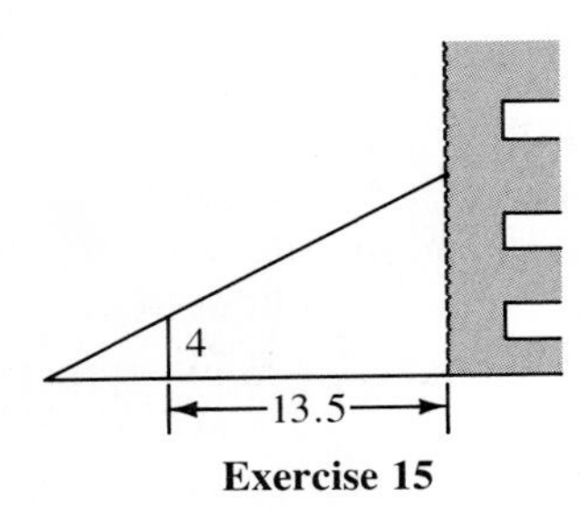

Exercise 15

16. A man walks with a constant speed v along a diameter D of a circular courtyard. A light at one end of the diameter perpendicular to D casts his shadow on the wall of the courtyard. Show that the shadow is moving with speed $\frac{8}{5}v$ at the moment when the man is three fourths of the way across the courtyard.

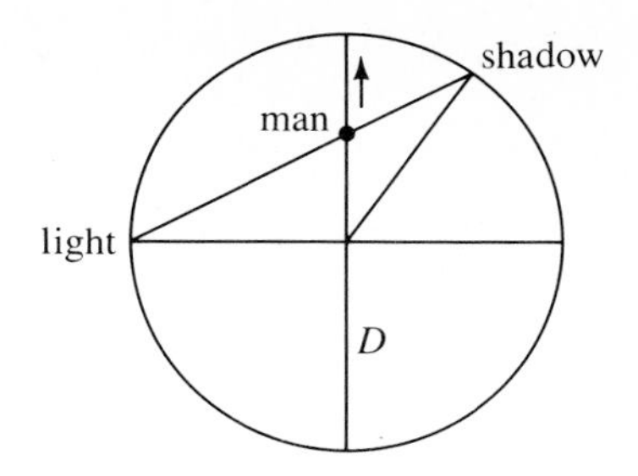

C **17.** Show that the length of the shortest line segment that passes through the point (a, b) and has its ends on the coordinate axes is $(a^{\frac{2}{3}} + b^{\frac{2}{3}})^{\frac{3}{2}}$.

18. The lengths of the sides of a quadrilateral Q are specified. Show that if Q has maximum area, then its opposite angles are supplementary.

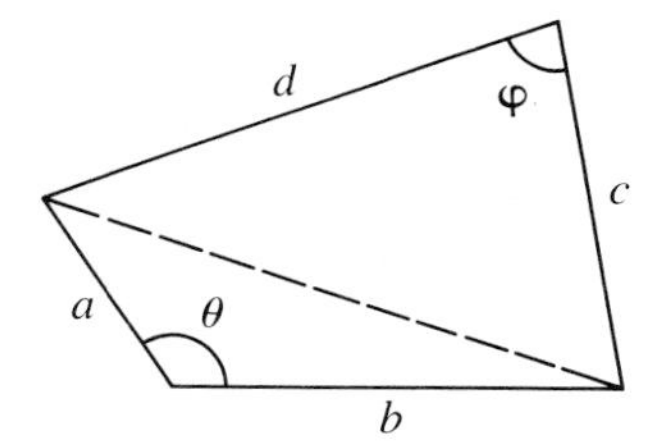

Chapter Summary

1. *Polar coordinates* of a point P are (r, θ), where $r = OP$, the distance from the *pole* O to P, and θ is an angle having the polar axis as initial side and the ray $\overrightarrow{OP}$ as terminal side. Polar and rectangular coordinates of P are related by the equations
$$x = r \cos \theta \qquad y = r \sin \theta.$$
2. In graphing polar equations we make use of our knowledge of how the trigonometric functions vary.
3. The geometry of multiplying and dividing complex numbers makes use of their polar forms: If
$$w = a(\cos \alpha + i \sin \alpha) \quad \text{and} \quad z = b(\cos \beta + i \sin \beta),$$
then:
$$wz = ab(\cos (\alpha + \beta) + i \sin (\alpha + \beta))$$
and if $b \neq 0$, $\quad \frac{w}{z} = \frac{a}{b}(\cos (\alpha - \beta) + i \sin (\alpha - \beta)).$
4. *De Moivre's theorem* is:
$$\text{If } z = r(\cos \theta + i \sin \theta), \text{ then } z^n = r^n(\cos n\theta + i \sin n\theta).$$
This theorem enables us to find powers and roots of complex numbers.

5. *Parametric equations* of the form

$$x = f(t), \quad y = g(t)$$

often describe curves. Many of these equations involve trigonometric functions.

6. Using the basic trigonometric limits,

$$\lim_{x \to 0} \frac{\sin x}{x} = 1 \quad \text{and} \quad \lim_{x \to 0} \frac{1 - \cos x}{x} = 0,$$

we can prove the derivative formulas

$$\frac{d}{dx} \sin x = \cos x \quad \text{and} \quad \frac{d}{dx} \cos x = -\sin x.$$

The derivative formulas for the other trigonometric functions are:

$$\frac{d}{dx} \tan x = \sec^2 x \qquad \frac{d}{dx} \cot x = -\csc^2 x$$

$$\frac{d}{dx} \sec x = \sec x \tan x \qquad \frac{d}{dx} \csc x = -\csc x \cot x$$

7. Many applied and optimization problems can be solved by using trigonometric functions and their derivatives.

Chapter Test

9-1

1. Transform each equation to polar coordinates and simplify the result.
   a. $x = \sqrt{3}y$ **b.** $x^2 + y^2 = 4$ **c.** $x^2 + y^2 = 4x$
2. Graph each equation after transforming it to rectangular coordinates.
   a. $r(\cos \theta + \sin \theta) = 1$ **b.** $r + 2 \cos \theta = 0$

9-2

3. Graph the polar equation $r = 2 \sin 2\theta$.
4. Graph the polar equation $r = 1 - \frac{1}{\theta}$, $\theta > 0$.

9-3

5. **a.** Convert $\sqrt{2}(\cos 45° + i \sin 45°)$ to $x + yi$ form.
   b. Convert -1 to polar form.
6. Express $\dfrac{6(\cos 105° + i \sin 105°)}{3(\cos 60° + i \sin 60°)}$ in **(a)** polar form; **(b)** $x + yi$ form.

9-4

7. Find $(-\sqrt{3} + i)^6$.
8. Find the five fifth roots of unity.

9-5

9. Identify and draw the curve defined parametrically by $x = \cos 2t$ and $y = \sin t$.

9-6

10. Evaluate. **a.** $\lim\limits_{x \to 0} \dfrac{\sin 2x}{x}$ **b.** $\lim\limits_{x \to 0} \dfrac{1 - \sec x}{x}$

9-7

11. Differentiate. **a.** $\sin x^3$ **b.** $\sin^2 x$
12. Find $f''(x)$ given that $f(x) = \cot x$.

9-8

13. A conical funnel is to have slant height 10 cm. What should its vertex angle be to make its volume as great as possible?

Extra

Polar Equations of Conics

We can use eccentricity to define ellipses, parabolas, and hyperbolas and to find polar equations of them. To do so, let D be a line and F be a point not on D. Let e be a positive number. We define the set S of all points P such that

$$\frac{PF}{PD} = e$$

to be an ellipse if $0 < e < 1$, a parabola if $e = 1$, and a hyperbola if $e > 1$. The point F is a focus and the line D is a directrix of S with eccentricity e.

To find a polar equation of S, introduce a polar coordinate system with pole F and polar axis pointing directly away from D. Let $P(r, \theta)$ be any point of S. Let p be the distance between F and D. Then

$$PF = r \quad \text{and} \quad PD = p + r\cos\theta.$$

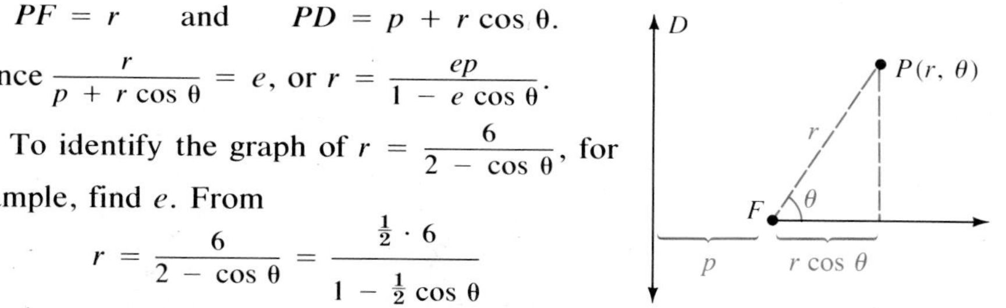

Hence $\dfrac{r}{p + r\cos\theta} = e$, or $r = \dfrac{ep}{1 - e\cos\theta}$.

To identify the graph of $r = \dfrac{6}{2 - \cos\theta}$, for example, find e. From

$$r = \frac{6}{2 - \cos\theta} = \frac{\frac{1}{2}\cdot 6}{1 - \frac{1}{2}\cos\theta}$$

$e = \frac{1}{2}$. Thus the graph is an ellipse.

EXERCISES

Let $r = \dfrac{6}{2 - \cos\theta}$. Let $P(r, \theta)$ be any point on the ellipse.

1. Show that the distance between $F'(4, 0°)$ and $P(r, \theta)$ is $8 - r$. (*Hint:* Use the polar distance formula (page 364) and then the fact that $r\cos\theta = 2r - 6$.)
2. Let D' be the line with equation $r\cos\theta = 10$. Show that $\dfrac{PF'}{PD'} = \dfrac{1}{2}$. This shows that F' is also a focus and that D' is also a directrix.
3. Use the fact that $F(0, 0°)$ is a focus to show that $PF + PF' = 8$.
4. Use $\cos\theta = \dfrac{x}{r}$ to show that $\dfrac{(x - 2)^2}{16} + \dfrac{y^2}{12} = 1$.
5. Use Exercise 4 to show that the ellipse has foci $(0, 0)$ and $(4, 0)$ and that the sum of the focal radii of P is 8.

APPLICATION

Two Laws of Optics

The speed of light in air is approximately 300,000 km/s. When a ray of light enters a second medium, such as water, the ray is bent, or **refracted**, and its speed is reduced, in the case of water to about 225,000 km/s. **Fermat's principle** states that light traveling from one point, A, to another point, B, seeks a path so as to minimize time. Suppose that the incoming ray, or **incident ray**, has speed u and makes an angle α with a perpendicular to the surface and the refracted ray has speed v and makes an angle β with that perpendicular. **Snell's law** provides a relationship among u, v, α, and β:

$$\frac{\sin \alpha}{\sin \beta} = \frac{u}{v}$$

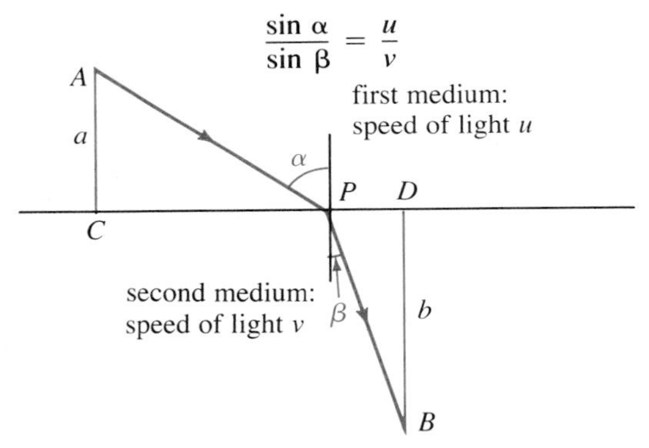

A medium such as water tends to bend light rays and create illusion.

By completing Exercises 1–7 we can make a proof of Snell's law. A short sketch of the proof is as follows.

1. Write $AP + PB$ in terms of u, v, α, and β.
2. Express the time T needed to go from A to P and from P to B in terms of u, v, α, and β.
3. Determine what must be true of u, v, α, and β for $T' = 0$ to be true.

EXERCISES

1. Use the figure to show that $AP + PB = a \sec \alpha + b \sec \beta$.

2. Show that the time T to go from A to P to B is given by $T = \frac{a}{u} \sec \alpha + \frac{b}{v} \sec \beta$. (*Hint:* Recall that distance = rate × time.)

3. Show that α and β are related by $a \tan \alpha + b \tan \beta = k$, $k = CP + PD$.

4. From Exercise 3 both α and β are related variables. Consider α a variable and β a variable that is a function of α. Use Exercise 2 and the chain rule to show that:

$$\frac{dT}{d\alpha} = \frac{a}{u} \sec \alpha \tan \alpha + \frac{b}{v} \sec \beta \tan \beta \frac{d\beta}{d\alpha}$$

$$\frac{dT}{d\alpha} = \frac{a}{u} \frac{\sin \alpha}{\cos^2 \alpha} + \frac{b}{v} \frac{\sin \beta}{\cos^2 \beta} \frac{d\beta}{d\alpha}$$

5. Use Exercise 3 to find $\frac{d\beta}{d\alpha}$ and show that $\frac{d\beta}{d\alpha} = -\frac{a \cos^2 \beta}{b \cos^2 \alpha}$.

6. Substitute the result of Exercise 5 for $\frac{d\beta}{d\alpha}$ in the second equation in Exercise 4 to show that $\frac{dT}{d\alpha} = \frac{a}{\cos^2 \alpha}\left(\frac{\sin \alpha}{u} - \frac{\sin \beta}{v}\right)$.

7. Use Exercise 6 and $\frac{dT}{d\alpha} = 0$ to complete the derivation of Snell's law.

In the case where a ray of light is **reflected** from a mirror, only one medium is involved, but Fermat's principle continues to apply. The angle α is called the **angle of incidence** and the angle β is called the **angle of reflection**. The **reflection law** states that $\alpha = \beta$.

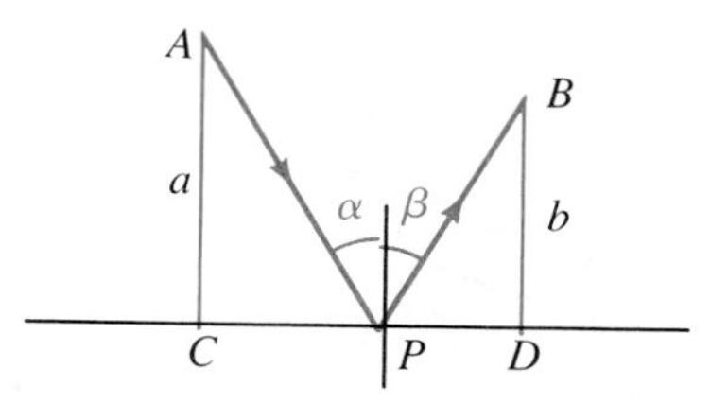

8. Prove the reflection law by proving that $AP + PB$ is least when $\alpha = \beta$. (*Hint:* In this case $u = v$. This fact will simplify the proof.)

Chapter 10

Exponential and Logarithmic Functions

In this chapter exponential and logarithmic functions are discussed. Particular emphasis is placed on base e. Derivatives of these functions are developed and applied to problems involving growth and decay.

Exponential Functions

10-1 REAL EXPONENTS

Recall from Section 3-9 the definition of a rational exponent. If $b > 0$ and $x = \frac{p}{q}$ where p and q are integers ($q \neq 0$), then

$$b^x = b^{\frac{p}{q}} = (\sqrt[q]{b})^p.$$

We now wish to give meaning to b^x where x is any real number. From the definition, we shall see that the laws of exponents remain valid. Furthermore we shall be able to define a function $x \to b^x$ that will be an increasing function if $b > 1$ and that will be a decreasing function if $0 < b < 1$. Each such function will be continuous.

What meaning can be given to the expression $3^{\sqrt{2}}$? Since $\sqrt{2}$ is irrational, we cannot define $3^{\sqrt{2}}$ by use of powers and roots. Since $\sqrt{2} \approx 1.4$, we might consider $3^{\sqrt{2}}$ to be approximately $3^{1.4} \approx 4.65553$. A more accurate approximation to $3^{\sqrt{2}}$ can be obtained by a more accurate approximation to $\sqrt{2}$. The table on the next page shows several such approximations.

The growth of a population such as this colony of king penguins is of interest to biologists. Growth and decay of populations are exponential functions.

r	3^r
1.4	4.65553
1.41	4.70696
1.414	4.72769
1.4142	4.72873
1.41421	4.72878
$\rightarrow \sqrt{2}$	$\rightarrow 3^{\sqrt{2}}$

Notice from the table that

$$1.4 < 1.41 < 1.414 < 1.4142 < 1.41421 < \cdots < 2$$

and that

$$3^{1.4} < 3^{1.41} < 3^{1.414} < 3^{1.4142} < 3^{1.41421} < \cdots < 3^2 = 9$$

The sequence of these powers of 3 is a nondecreasing sequence of real numbers bounded by 9. By the axiom of completeness, the sequence has a real number as a limit.

It can be shown that if $r_1, r_2, r_3, \ldots, r_n, \ldots$ and $s_1, s_2, s_3, \ldots, s_n, \ldots$ are *any* two sequences of rational numbers such that

$$\lim_{n \to \infty} r_n = \sqrt{2} \quad \text{and} \quad \lim_{n \to \infty} s_n = \sqrt{2},$$

then

$$\lim_{n \to \infty} 3^{r_n} = \lim_{n \to \infty} 3^{s_n}.$$

The foregoing discussion means that $3^{\sqrt{2}}$ depends only on 3 and $\sqrt{2}$. The number $3^{\sqrt{2}}$ does not depend on the particular sequence of rational numbers used to approximate $\sqrt{2}$. We can now define b^x.

DEFINITION

Let $b > 0$ and let x be any real number. If $r_1, r_2, r_3, \ldots, r_n, \ldots$ is any sequence of rational numbers such that

$$\lim_{n \to \infty} r_n = x,$$

then

$$b^x = \lim_{n \to \infty} b^{r_n}.$$

We can use the definition to show that the laws of exponents stated in Section 3-9 are valid for all real numbers as exponents.

THEOREM 1 Let a and b be positive real numbers and let m and n be real numbers.

1. $a^m a^n = a^{m+n}$
2. $(a^m)^n = a^{mn}$
3. $\frac{a^m}{a^n} = a^{m-n}$
4. $(ab)^m = a^m b^m$
5. $\left(\frac{a}{b}\right)^m = \frac{a^m}{b^m}$

We shall prove the first law of exponents.

Proof: Let m and n be real numbers. Let $r_1, r_2, r_3, \ldots, r_k, \ldots$ and $s_1, s_2, s_3, \ldots, s_k, \ldots$ be sequences of rational numbers such that

$$\lim_{k\to\infty} r_k = m \quad \text{and} \quad \lim_{k\to\infty} s_k = n.$$

Then by the definition of real exponents,

$$a^m = \lim_{k\to\infty} a^{r_k} \quad \text{and} \quad a^n = \lim_{k\to\infty} a^{s_k}.$$

Since $\lim_{k\to\infty} (r_k + s_k) = \lim_{k\to\infty} r_k + \lim_{k\to\infty} s_k = m + n$,

$a^{m+n} = \lim_{k\to\infty} a^{r_k + s_k}$ by the definition of real exponents.

Thus $a^{m+n} = \lim_{k\to\infty} (a^{r_k} a^{s_k})$ by law 1 for rational exponents.

Hence $a^{m+n} = (\lim_{k\to\infty} a^{r_k})(\lim_{k\to\infty} a^{s_k})$ by Theorem 11, page 115.

Therefore $a^{m+n} = a^m a^n$ by the definition of real exponents. ■

You are asked to prove laws 2 through 5 in similar fashion in Exercises 25–28.

The laws of exponents are useful in simplifying expressions involving real exponents.

EXAMPLE 1 Simplify each expression.

a. $\dfrac{4^{\sqrt{2}} \cdot 8^{\frac{\sqrt{2}}{3}}}{16^{-\frac{1}{2}} \cdot 2^{\sqrt{2}}}$ **b.** $((\sqrt{b})^{\sqrt{6}})^{\sqrt{6}}$

Solution **a.** $\dfrac{4^{\sqrt{2}} \cdot 8^{\frac{\sqrt{2}}{3}}}{16^{-\frac{1}{2}} \cdot 2^{\sqrt{2}}} = \dfrac{(2^2)^{\sqrt{2}} \cdot (2^3)^{\frac{\sqrt{2}}{3}}}{(2^4)^{-\frac{1}{2}} \cdot 2^{\sqrt{2}}} = \dfrac{2^{2\sqrt{2}} \cdot 2^{\sqrt{2}}}{2^{-2} \cdot 2^{\sqrt{2}}} = \dfrac{2^{2\sqrt{2}}}{2^{-2}} = 2^{2\sqrt{2}+2}$ **Answer**

b. $((\sqrt{b})^{\sqrt{6}})^{\sqrt{6}} = (\sqrt{b})^{\sqrt{6}\cdot\sqrt{6}} = (\sqrt{b})^6 = (b^{\frac{1}{2}})^6 = b^3$ **Answer**

The laws are also useful in solving equations involving exponents.

EXAMPLE 2 Solve each equation.

a. $(3x - 10)^{\frac{2}{3}} = 2$ **b.** $(x^2 + 4x + 4)^{\frac{3}{2}} = 27$

Solution on following page

Solution **a.** Raise each side of the given equation to the power $\frac{3}{2}$.

$$((3x - 10)^{\frac{2}{3}})^{\frac{3}{2}} = \pm 2^{\frac{3}{2}}$$
$$3x - 10 = \pm 2\sqrt{2}$$
$$x = \frac{10 \pm 2\sqrt{2}}{3} \quad \textbf{Answer}$$

b. Raise each side of the given equation to the power $\frac{2}{3}$.

$$(x^2 + 4x + 4)^{\frac{3}{2}\cdot\frac{2}{3}} = 27^{\frac{2}{3}}$$
$$x^2 + 4x + 4 = 27^{\frac{2}{3}}$$
$$= 9$$

Solve $x^2 + 4x - 5 = 0$.
Hence $x = 1$ or $x = -5$. **Answer**

You can check the solutions to Example 2(b) by substituting 1 and -5 for x in the original equation.

EXERCISES

In Exercises 1–4 use a calculator to complete each table of values.

A

1.

r_n	1	1.4	1.41	1.414
2^{r_n}				

2.

r_n	1	1.7	1.73	1.732
3^{r_n}				

3.

r_n	1	0.1	0.01	0.001
4^{r_n}				

4.

r_n	0.9	0.99	0.999	0.9999
4^{r_n}				

Use the laws of exponents to simplify each expression.

5. $(3^{\sqrt{2}})^2$
6. $(3^{\sqrt{2}})^{\sqrt{2}}$
7. $2^{\sqrt{2}}2^{\sqrt{2}}$
8. $2^2 2^{\sqrt{2}}$
9. $(x^3)^{\sqrt[3]{3}}$
10. $\sqrt{x^{\sqrt{2}}}$
11. $\frac{x^{\sqrt{2}}}{x^{\sqrt{3}}}$
12. $\frac{x^2}{x^{\sqrt{2}}}$
13. $(x^{\sqrt{2}} - y^{\sqrt{2}})(x^{\sqrt{2}} + y^{\sqrt{2}})$
14. $(x^{\sqrt{2}} - y^{\sqrt{2}})^2$
15. $\left(\frac{x^{\sqrt{5}}}{y^{2\sqrt{5}}}\right)^{\sqrt{5}}$
16. $\frac{x^{2\sqrt{3}}y^{3\sqrt{2}}}{x^{\sqrt{3}}y^{-\sqrt{2}}}$

Solve each equation.

B

17. $x^{\frac{2}{3}} = 4$
18. $y^{-\frac{2}{3}} = 25$
19. $(x - 1)^{\frac{1}{2}} = 4$
20. $(3t + 1)^{\frac{3}{4}} = 8$

21. $(x^2 - 1)^{\frac{3}{5}} = 8$

22. $(x^2 - 6x + 9)^{\frac{3}{2}} = 64$

23. $(x^2 - 10x + 25)^{-\frac{1}{2}} = 8$

24. $(x^2 - 4)^{-\frac{3}{2}} = 125$

C **25.** Prove part (3) of Theorem 1.

26. Prove part (4) of Theorem 1.

27. Prove part (5) of Theorem 1.

28. Prove part (2) of Theorem 1.

10–2 EXPONENTIAL FUNCTIONS

In Section 10–1 we defined b^x for all real numbers x $(b > 0)$. We can now define the *exponential function with base b.*

DEFINITION

Let $b > 0$ and let x be any real number. Then the function

$$x \rightarrow b^x$$

is the **exponential function with base *b*.**

The short tables of values below will enable us to graph exponential functions such as $f(x) = 2^x$ and $g(x) = (\frac{1}{2})^x$.

x	2^x
-2	$\frac{1}{4}$
-1	$\frac{1}{2}$
0	1
1	2
2	4

x	$(\frac{1}{2})^x$
-2	4
-1	2
0	1
1	$\frac{1}{2}$
2	$\frac{1}{4}$

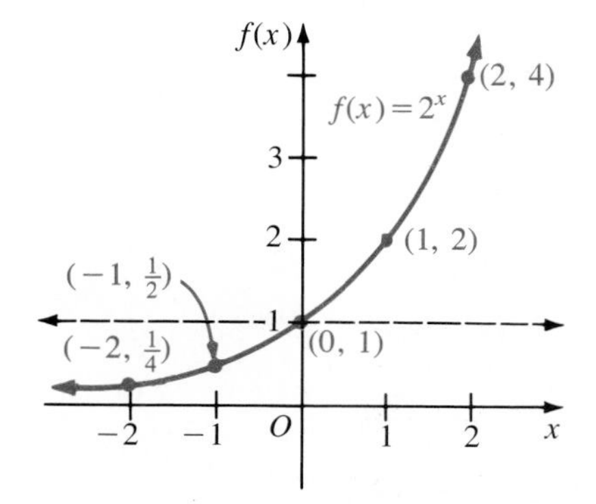

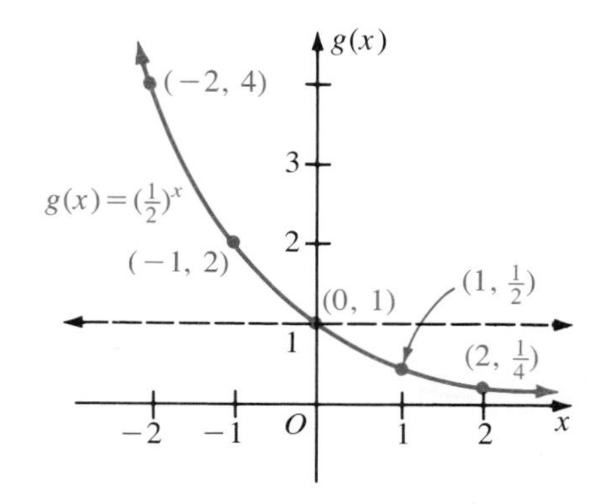

The definition of real exponent from Section 10–1 allows us to draw a smooth curve through the points plotted from the table.

The following theorem lists several important properties of exponential functions.

THEOREM 2 Let b be positive and x and y be real numbers.

1. If $b > 1$, then $b^x > 1$ for all $x > 0$.
2. If $b > 1$, and $x < y$, then $b^x < b^y$.
3. If $0 < b < 1$ and $x < y$, then $b^x > b^y$.
4. For all x, $b^x > 0$.
5. If $b \neq 1$, then $b^x = b^y$ if and only if $x = y$.

Proof:

1. Since $x > 0$, there is an increasing sequence $r_1, r_2, \ldots, r_n, \ldots$ of positive rational numbers such that $\lim_{n \to \infty} r_n = x$. Since $0 < r_1 < r_n$, $1 < b^{r_1} < b^{r_n}$ by Exercise 43. Thus

$$b^x = \lim_{n \to \infty} b^{r_n} \geq b^{r_1} > 1.$$

Therefore $b^x > 1$ for all $x > 0$. ■

2. By part (3) of Theorem 1, $\dfrac{b^y}{b^x} = b^{y-x}$. By part (1) of Theorem 2, since $y - x > 0$, $\dfrac{b^y}{b^x} > 1$. Therefore $b^y > b^x$. That is, $b^x < b^y$. ■

3. Since $0 < b < 1$, $\dfrac{1}{b} > 1$. By part (2) of Theorem 2,

$$\left(\frac{1}{b}\right)^y > \left(\frac{1}{b}\right)^x.$$

Hence by part (5) of Theorem 1,

$$\frac{1}{b^y} > \frac{1}{b^x}.$$

Therefore $b^y < b^x$, or $b^x > b^y$. ■

4. There is an integer n such that $x + n > 0$. By part (3) of Theorem 1,

$$b^x = \frac{b^{x+n}}{b^n}.$$

If $b > 1$, then from part (1) of Theorem 2, $b^{x+n} > 1$. Hence $b^x > \dfrac{1}{b^n}$. Thus $b^x > 0$. If $0 < b < 1$, then $\dfrac{1}{b} > 1$. Hence $b^x = \left(\dfrac{1}{b}\right)^{-x} > 0$. ■

5. Suppose that $b > 1$ and $b^x = b^y$ but $x \neq y$. Then $x < y$ or $x > y$, say $x < y$. Then by part (2), $b^x < b^y$. This provides a contradiction. The proof for $0 < b < 1$ is similar but uses part (3). ■

EXAMPLE 1 Graph each function.

a. $f(x) = 2^x + 2$

b. $f(x) = 2 \cdot 2^x$

Solution **a.** The graph of $f(x) = 2^x + 2$ can be obtained by sketching the graph of the function $f(x) = 2^x$ and then translating that graph upward two units.

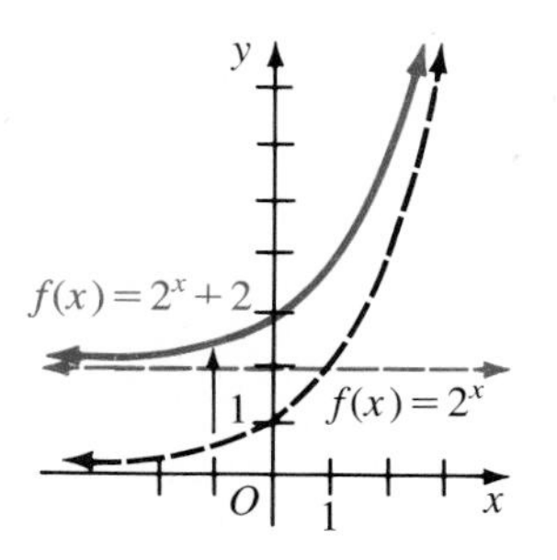

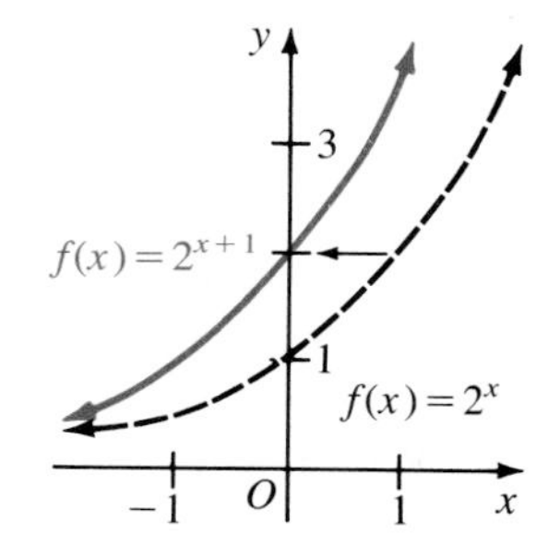

b. Write $f(x) = 2 \cdot 2^x$ as $f(x) = 2^{x+1}$ since $2 \cdot 2^x = 2^1 \cdot 2^x = 2^{x+1}$. The desired graph can be obtained by translating the graph of $f(x) = 2^x$ to the left 1 unit.

Part (5) of Theorem 2 is useful in solving many exponential equations.

EXAMPLE 2 Solve each equation.

a. $3^{2x+5} = 27^{x+1}$

b. $(\frac{1}{2})^{2x-x^2} = 8^{2-x}$

Solution **a.** Write both sides of the equation as powers of 3.

$$3^{2x+5} = 27^{x+1}$$
$$3^{2x+5} = (3^3)^{x+1}$$
$$3^{2x+5} = 3^{3x+3}$$

Hence $\quad 2x + 5 = 3x + 3 \quad$ by part (5) of Theorem 2.

Therefore $x = 2$.

b. Write both sides of the equation as powers of 2.

$$(\tfrac{1}{2})^{2x-x^2} = 8^{2-x}$$
$$2^{x^2-2x} = 2^{6-3x}$$

Use part (5) of Theorem 2.

$$x^2 - 2x = 6 - 3x$$
$$x^2 + x - 6 = 0$$

Therefore $x = 2$ or $x = -3$.

In addition to Theorem 2 we have the following important theorem that states analytic properties of exponential functions.

THEOREM 3 Let $b > 0$ and $b \neq 1$.

1. The exponential function with base b is continuous.
2. The graph of the exponential function with base b is concave up.
3. The graph of the exponential function with base b is asymptotic to the horizontal axis.

The proof of the continuity of the exponential function with base b is too involved to be included in this book. Thus we accept continuity without proof. The proof of part (2) will be delayed until Section 10–6, where derivatives of exponential functions are discussed. The proof of part (3) is quite involved and will be omitted.

EXAMPLE 3 Find each limit.

a. $\lim_{x \to 4} 3^x$ **b.** $\lim_{x \to 1} 2^{3x^2 - 2}$

Solution Use the facts that exponential functions are continuous and that compositions of continuous functions are continuous.

a. $\lim_{x \to 4} 3^x = 3^4$

$= 81$ **Answer**

b. $\lim_{x \to 1} 2^{3x^2 - 2} = 2^{\lim_{x \to 1} (3x^2 - 2)}$

$= 2^1$

$\lim_{x \to 1} 2^{3x^2 - 2} = 2$ **Answer**

Exponential functions are useful in solving a great number of applied problems. For example, if P dollars is deposited in an account that earns interest compounded n times per year at an annual rate r (expressed as a decimal), the amount A in the account after t years is given by

$$A = P\left(1 + \frac{r}{n}\right)^{nt}.$$

The formula for A is called the **compound interest formula.**

EXAMPLE 4 Find the amount of money in the bank after 12.5 years if an original investment of \$1200 is made at 12% interest compounded quarterly.

Solution Find A if $P = 1200$, $t = 12.5$, $r = 0.12$, and $n = 4$.

$$\begin{aligned} A &= P\left(1 + \frac{r}{n}\right)^{nt} \\ &= 1200\left(1 + \frac{0.12}{4}\right)^{4 \cdot 12.5} \\ &= 1200(1.03)^{50} \\ &= 1200(4.3839) \\ A &= 5260.68 \end{aligned}$$

After 12.5 years the account will contain approximately \$5260.68. **Answer**

EXERCISES

Graph each equation.

A

1. $y = 2^x - 1$ **2.** $y = 2^{x-1}$ **3.** $y = 2^{2x}$

4. $y = \frac{1}{2} \cdot 2^x$ **5.** $y = -2^x$ **6.** $y = 3^{x-1} + 2$

7. $y = 2^{-x}$ **8.** $y = -2^{-x}$ **9.** $y = 3 - 3^x$

Solve each equation.

10. $3^{2x-5} = 27$ **11.** $2^{x^2} = 4$ **12.** $5^{2x^2-1} = 25$

13. $\sqrt{125} = 5^x$ **14.** $2^{x^2-2x+1} = 16$ **15.** $(\sqrt{2})^{2x+1} = 2$

16. $x^{\sqrt{2}} = 3$ **17.** $x^{\sqrt{3}} = 2$ **18.** $x^{-\frac{2}{3}} = 4$

Find each limit.

19. $\lim_{x \to 0} (2x + 2^{-x})$ **20.** $\lim_{x \to 2} 2^x x^3$ **21.** $\lim_{x \to 3} 2^{1-x}$

22. $\lim_{x \to 1} 2^{x^2+3x-5}$ **23.** $\lim_{x \to \sqrt{7}} 3^{x^2-4}$ **24.** $\lim_{x \to \sqrt{3}} (0.5)^{x^4-x^2+1}$

B

25. Solve $x^{\sqrt{2}+1} = 2$. (*Hint:* Use $\sqrt{2} - 1$ as an exponent.)

26. Solve $(x + 2)^{1-\sqrt{3}} = 1$.

Solve each equation.

27. $2^{2x} - 12 \cdot 2^x + 32 = 0$ (*Hint:* Let $t = 2^x$.) **28.** $3^{2x} - 82 \cdot 3^x + 81 = 0$

29. $5^{2x} - 20 \cdot 5^x - 125 = 0$ **30.** $3^{2x} - 6 \cdot 3^x - 27 = 0$

In Exercises 31–36 use the compound interest formula. Find the value of the indicated variable from the given values of the other variables.

31. $P = \$1100$; $r = 0.06$; $n = 2$; $t = 4.5$; A

32. $A = \$1338.23$; $r = 0.06$; $n = 1$; $t = 5$; P

33. $A = \$1696.63$; $r = 0.08$; $n = 1$; $t = 4.5$; P

34. $P = \$2500$; $r = 0.075$; $n = 4$; $t = 8$; A

35. $P = \$750$; $n = 1$; $t = 6$; $A = \$1125.55$; r

36. $P = \$820$; $n = 2$; $t = 6$; $A = \$1312.85$; r

An initial population of 12,000 bacteria grows according to the law $P(t) = 12{,}000\ (1.02)^t$ where t is the growing time measured in days and $P(t)$ is the bacteria population after t days. Use this formula and a calculator in Exercises 37 and 38.

37. What is the population after 10.5 days?

38. Approximate the number of days needed for the population to double. (*Hint:* Approximate t such that $P(t) = 24{,}000$.)

An antibacterial agent is introduced into the bacteria population described for Exercises 37 and 38. This agent is introduced at the end of the twentieth day. Use this fact in Exercises 39 and 40.

39. What is the bacteria population when the agent is introduced?

40. The population $Q(t)$ after the time the agent is introduced is given by $Q(t) = P(20)(0.90)^t$. What is the population twenty days after the agent is introduced?

C

41. Sketch the graph of $y(t) = Pb^t$ if $P > 0$ and $b > 1$.

42. Sketch the graph of $y(t) = Pb^t$ if $P > 0$ and $0 < b < 1$.

43. Prove that if $b > 1$ and r and s are rational numbers such that $r < s$, then $b^r < b^s$.

44. Prove that if $0 < b < 1$ and r and s are any rational numbers such that $r < s$, then $b^r > b^s$.

10–3 THE NATURAL EXPONENTIAL FUNCTION

In Section 10–2 we defined the exponential function with base b. There is a particular value of b that plays a central role in mathematical analysis. This base is e (named after the great Swiss mathematician Leonard Euler (1707–1783)). The exponential function with base e,

$$x \to e^x,$$

is called the **natural exponential function.**

We define e as $\lim_{n \to \infty} \left(1 + \frac{1}{n}\right)^n$. At the end of this section we shall show that $\lim_{n \to \infty} \left(1 + \frac{1}{n}\right)^n$ exists.

Use of a calculator or computer provides a decimal approximation of e.

n	100	1000	10,000	100,000
$\left(1 + \frac{1}{n}\right)^n$	2.70481	2.71692	2.71815	2.71825

A decimal approximation of e accurate to 12 decimal places is

$$e \approx 2.718281828459.$$

Since $e > 1$, the function given by $f(x) = e^x$ is an increasing function. The graphs of

$$g(x) = 2^x,$$
$$f(x) = e^x,$$
and
$$h(x) = 3^x$$

are shown at the right. Notice that, like $g(x) = 2^x$ and $h(x) = 3^x$, the function $f(x) = e^x$ is continuous everywhere and is one-to-one.

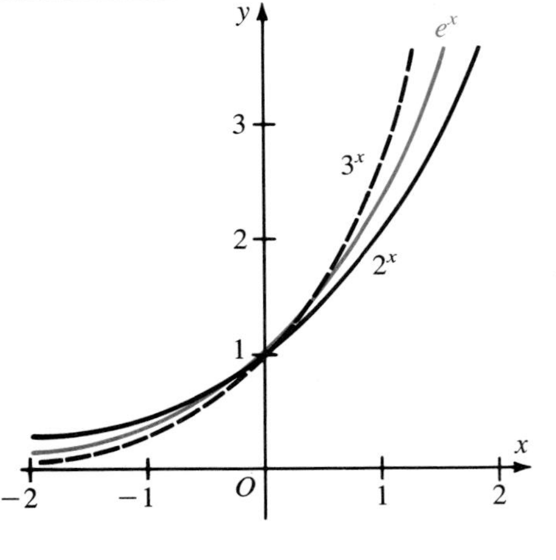

EXAMPLE 1 Graph $y = e^{-x} + 2$.

Solution The graph of $y = e^{-x}$ can be obtained by reflecting the graph of $y = e^x$ in the y-axis. Then the graph of $y = e^{-x} + 2$ can be obtained by translating the graph of $y = e^{-x}$ up 2 units.

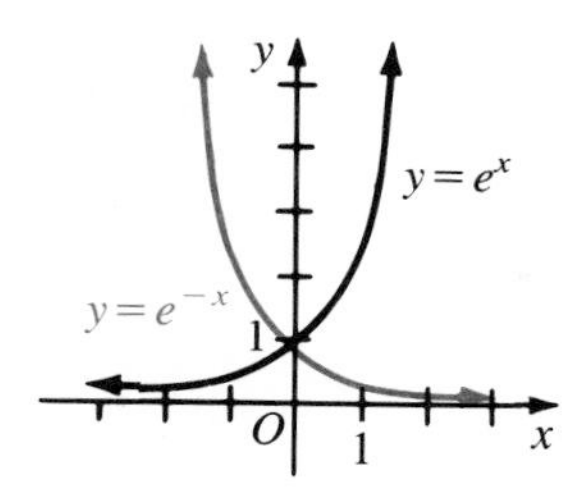

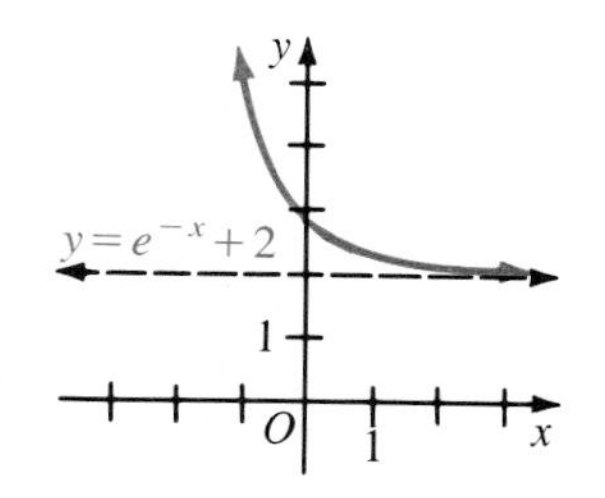

EXAMPLE 2 Suppose that \$1 is deposited in a bank and remains on deposit for t years and that the bank pays 100% interest per year compounded n times per year.

a. Find the amount in the bank after 1 year if interest is paid every minute.

b. Find the amount in the bank after 2 years if interest is paid every minute.

Solution Use the compound interest formula $A = P\left(1 + \frac{r}{n}\right)^{nt}$ with $P = 1$ and $r = 1$.

$$A = \left(1 + \frac{1}{n}\right)^{nt}$$

a. In one year there are $365 \cdot 24 \cdot 60 = 525{,}600$ minutes.

Therefore: $$A = \left(1 + \frac{1}{525{,}600}\right)^{525{,}600}$$
$$= (1.0000019)^{525{,}600} = \$2.7182 \quad \textbf{Answer}$$

b. If $t = 2$, then: $A = (1.0000019)^{525{,}600 \cdot 2} = \$7.3690$ **Answer**

If P dollars are invested at an annual rate r (expressed as a decimal) for t years and interest is compounded not every minute but rather every instant (continuously), then $n \to \infty$ and

$$A = \lim_{n\to\infty} P\left(1 + \frac{r}{n}\right)^{nt}.$$

Letting $m = \frac{n}{r}$, then
$$A = \lim_{m\to\infty} P\left(1 + \frac{1}{m}\right)^{mrt}.$$

Hence
$$A = P\left[\lim_{m\to\infty}\left(1 + \frac{1}{m}\right)^{m}\right]^{rt}.$$

Using the fact that $e = \lim_{m\to\infty}\left(1 + \frac{1}{m}\right)^{m}$, we obtain

$$A = Pe^{rt}.$$

The formula $A = Pe^{rt}$ is called the **continuous compound interest formula.**

EXAMPLE 3 If $1000 is invested at an annual rate of 6% interest compounded continuously, how much money is in the investment after 5 years?

Solution Use the formula $A = Pe^{rt}$ with $P = 1000$, $r = 0.06$, and $t = 5$ to find A.

$$\begin{aligned} A &= 1000e^{(0.06)(5)} \\ &= 1000e^{0.3} \\ &= (1000)(1.3499) \\ A &= 1349.90 \end{aligned}$$

The investment will contain approximately $1350. **Answer**

We shall conclude this section with a proof that $\lim_{n\to\infty}\left(1 + \frac{1}{n}\right)^n$ exists. First we show that the sequence with nth term $e_n = \left(1 + \frac{1}{n}\right)^n$ is nondecreasing. Then we show that the sequence is bounded. Finally we can use the axiom of completeness to establish the existence of $\lim_{n\to\infty}\left(1 + \frac{1}{n}\right)^n$.

Let $e_n = \left(1 + \frac{1}{n}\right)^n$ for all natural numbers n. Then $e_{n+1} = \left(1 + \frac{1}{n+1}\right)^{n+1}$. To show that the sequence with nth term e_n is nondecreasing, we must show that

$$e_n \le e_{n+1}. \text{ That is, } \frac{e_{n+1}}{e_n} \ge 1.$$

$$\text{Now } \frac{e_{n+1}}{e_n} = \frac{\left(1 + \frac{1}{n+1}\right)^{n+1}}{\left(1 + \frac{1}{n}\right)^n} = \left(1 + \frac{1}{n}\right)\left[\frac{1 + \frac{1}{n+1}}{1 + \frac{1}{n}}\right]^{n+1}$$

$$= \left(1 + \frac{1}{n}\right)\left[\frac{n^2 + 2n}{(n+1)^2}\right]^{n+1}$$

$$\frac{e_{n+1}}{e_n} = \left(1 + \frac{1}{n}\right)\left[1 - \frac{1}{(n+1)^2}\right]^{n+1}$$

Using Bernoulli's inequality, $(1 + a)^k \ge 1 + ka$, proved on page 92, with $a = -\frac{1}{(n+1)^2}$ and $k = n + 1$,

$$\left[1 - \frac{1}{(n+1)^2}\right]^{n+1} \ge 1 - \frac{n+1}{(n+1)^2} = 1 - \frac{1}{n+1} = \frac{n}{n+1}.$$

$$\text{Therefore } \frac{e_{n+1}}{e_n} \ge \left(1 + \frac{1}{n}\right)\left(\frac{n}{n+1}\right) = \frac{n+1}{n} \cdot \frac{n}{n+1} = 1.$$

Hence $\frac{e_{n+1}}{e_n} \ge 1$. That is, $e_n \le e_{n+1}$.

To see that the sequence with nth term e_n is bounded, we can use the binomial theorem.

$$e_n = \left(1 + \frac{1}{n}\right)^n$$
$$= 1^n + \frac{n}{1!}\left(\frac{1}{n}\right)^1 + \frac{n(n-1)}{2!}\left(\frac{1}{n}\right)^2 + \cdots + \frac{n(n-1)(n-2)\cdots 1}{n!}\left(\frac{1}{n}\right)^n$$

That is,

$$\left(1 + \frac{1}{n}\right)^n = 1 + \frac{1}{1!} + \frac{\left(1 - \frac{1}{n}\right)}{2!} + \cdots + \frac{\left(1 - \frac{1}{n}\right)\left(1 - \frac{2}{n}\right)\cdots\left(\frac{1}{n}\right)}{n!}.$$

Each numerator of each term of the sum on the right side of the equation above is less than or equal to 1. Therefore we have

$$\left(1 + \frac{1}{n}\right)^n < 1 + \frac{1}{1!} + \frac{1}{2!} + \frac{1}{3!} + \cdots + \frac{1}{n!}.$$

In Section 3–8 it was shown that $\frac{1}{1!} + \frac{1}{2!} + \frac{1}{3!} + \cdots + \frac{1}{n!} < 2$. Therefore $e_n < 3$. Since the sequence with nth term e_n is a nondecreasing and bounded sequence, by the axiom of completeness, $\lim_{n\to\infty}\left(1 + \frac{1}{n}\right)^n$ exists.

EXERCISES

Use the fact that $e = \lim_{n\to\infty}\left(1 + \frac{1}{n}\right)^n$ to find each limit.

A

1. $\lim_{n\to\infty} 2\left(1 + \frac{1}{n}\right)^n$

2. $\lim_{n\to\infty}\left(1 + \frac{1}{n}\right)^{2n}$

3. $\lim_{n\to\infty}\left(1 + \frac{1}{n}\right)^{3n}$

4. $\lim_{n\to\infty}\left(1 + \frac{1}{n}\right)^{2.5n}$

5. $\lim_{n\to\infty}\left(1 + \frac{1}{2n}\right)^{6n}$

6. $\lim_{n\to\infty}\left(1 + \frac{1}{2n}\right)^{n}$

7. $\lim_{n\to 0}(1 + n)^{\frac{1}{n}}$

8. $\lim_{n\to 0}(1 + 3n)^{\frac{1}{3n}}$

Graph each equation.

9. $y = -e^{-x}$

10. $y = e^{-x} - 2$

11. $y = 1 - e^{-x}$

12. $y = e^{x+2}$

In Exercises 13–16 use the formula $A = Pe^{rt}$ to find the value of the indicated variable from the given values of the other variables.

13. $P = \$1000$; $r = 0.08$; $t = 4$; A

14. $P = \$1200$; $r = 0.65$; $t = 3.5$; A

15. $A = \$10{,}000$; $r = 0.75$; $t = 5$; P

16. $A = \$1850$; $r = 0.1$; $t = 8$; P

B **17.** Find the value of $\frac{e^h - 1}{h}$ for each value of h.

a. $h = 1$ **b.** $h = 0.1$ **c.** $h = 0.01$ **d.** $h = 0.001$ **e.** $h = 0.0001$

18. From part (e) of Exercise 17, write an equation of the line tangent to the graph of $y = e^x$ at $x = 0$. (*Hint*: Use $y = mx + b$ and an approximate value of m.)

COMPUTER EXERCISES

1. Write a program to approximate e^x by each of the following methods.

a. $e^x \approx \left(1 + \frac{x}{n}\right)^n$ **b.** $e^x \approx 1 + x + \frac{x^2}{2!} + \cdots + \frac{x^k}{k!}$

2. Run the program of Exercise 1 for $x = 2$, $n = 1000$, and $k = 12$.

3. Run the program of Exercise 1 for $x = 3$, $n = 1000$, and $k = 12$.

Logarithmic Functions

10-4 LOGARITHMIC FUNCTIONS

The exponential function with base b, $b \neq 1$, is a one-to-one function since

$$\text{if } x_1 \neq x_2, \text{ then } b^{x_1} \neq b^{x_2}.$$

Therefore the exponential function with base b has an inverse function, called the *logarithmic function with base b*, denoted by $\log_b$. As the figure at the right shows, the graph of the logarithmic function can be obtained by reflecting the graph of the exponential function in the line $y = x$. Thus if the point (v, u) is on the graph of $y = b^x$, then the point (u, v) is on the graph of $y = \log_b x$.

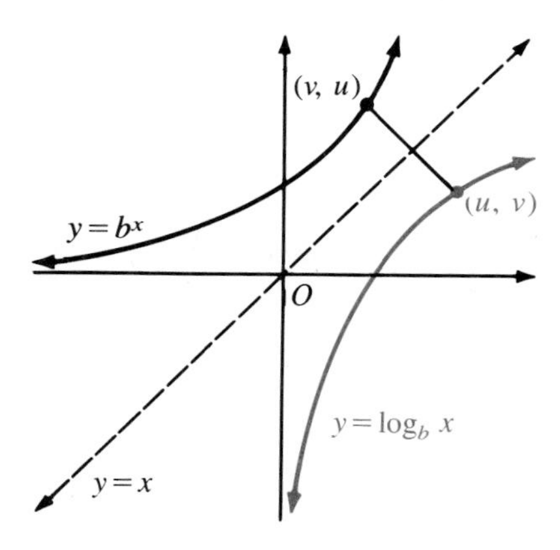

DEFINITION

Let $b > 0$ and $b \neq 1$. The **logarithmic function with base *b*** is the inverse of the exponential function with base b. That is,

$$\log_b x = y \text{ if and only if } b^y = x.$$

Since exponential and logarithmic functions with the same base are inverses of each other, we have the following fundamental identities.

For all real numbers x, $\log_b b^x = x$.
For all positive numbers x, $b^{\log_b x} = x$.

EXAMPLE 1 Find each logarithm. **a.** $\log_3 \frac{1}{27}$ **b.** $\log_{27} \sqrt[3]{9}$ **c.** $\log_3 3$

Solution Use the definition of logarithm and let y be the logarithm to be found.

a. $\log_3 \frac{1}{27} = y$ if and only if $3^y = \frac{1}{27}$. Since $\frac{1}{27} = 3^{-3}$, $3^y = 3^{-3}$. Hence $y = -3$.

b. $\log_{27} \sqrt[3]{9} = y$ if and only if $27^y = \sqrt[3]{9}$. Because $27^y = 3^{3y}$ and $\sqrt[3]{9} = 3^{\frac{2}{3}}$,

$$3^{3y} = 3^{\frac{2}{3}}$$

Hence $3y = \frac{2}{3}$ and $y = \frac{2}{9}$.

c. Since the solution of $3^y = 3$ is 1, $\log_3 3 = 1$.

The logarithmic function with base b has the following properties.

THEOREM 4 Let $b > 0$ and $b \neq 1$.

1. The domain of $f(x) = \log_b x$ is the set of all positive numbers.
2. The range of $f(x) = \log_b x$ is the set of all real numbers.
3. $\log_b 1 = 0$
4. $\log_b b = 1$
5. If $b > 1$, then the logarithmic function is an increasing function.
6. For all positive numbers x_1 and x_2, $x_1 = x_2$ if and only if $\log_b x_1 = \log_b x_2$.

Each part of the theorem can be proved by use of the definition of logarithmic function. The proofs are requested in Exercises 40–45.

EXAMPLE 2 Graph each equation.

a. $y = \log_2 x$ **b.** $y = \log_2 (x + 2)$ **c.** $y = \log_2 x + 2$

Solution **a.** Use the fact that $y = \log_2 x$ if and only if $2^y = x$ and the properties of $\log_2$.

$\log_2 1 = 0$

$\log_2 2 = 1$

$\log_2 8 = y$ if and only if $2^y = 8$. Thus $y = 3$.

$\log_2 16 = y$ if and only if $2^y = 16$. Thus $y = 4$.

Plot the points (1, 0), (2, 1), (8, 3), and (16, 4). Draw a smooth curve through the points plotted.

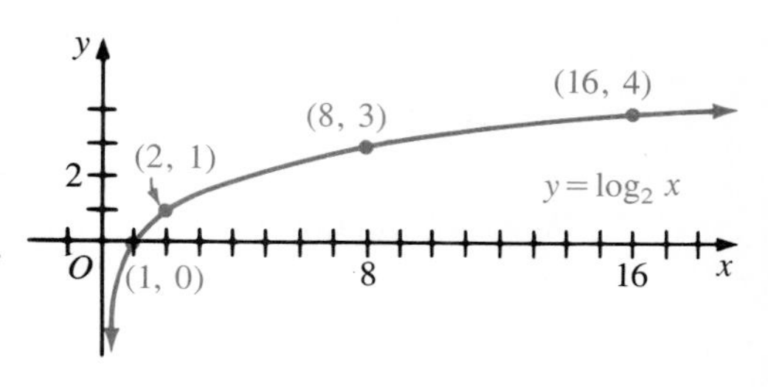

Solution continued on following page

b. To obtain the graph of $y = \log_2 (x + 2)$, translate the graph of $y = \log_2 x$ to the left 2 units. The graph is shown at the left below.

c. The graph of $y = \log_2 x + 2$ can be obtained by translating the graph of $y = \log_2 x$ up 2 units. The graph is shown at the right below.

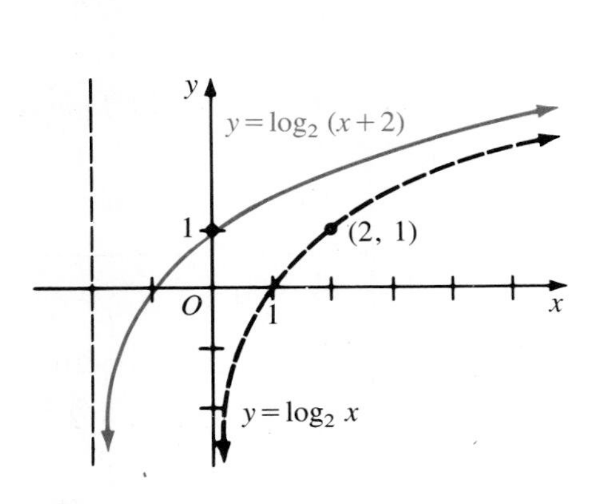

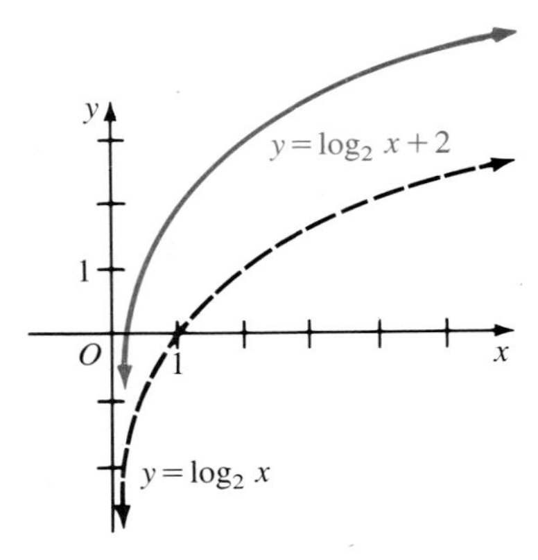

Many logarithmic equations can be solved by using the fact that logarithmic functions are one-to-one.

EXAMPLE 3 Solve each equation.

a. $\log_{10} x = 0.4871$ **b.** $(\log_2 x)^2 + \log_2 x - 6 = 0$

Solution **a.** A scientific calculator can be used to obtain $x = 3.0697$. The value of x can also be obtained by using the fact that $10^{\log_{10} x} = x$.

$$10^{\log_{10} x} = x = 10^{0.4871}$$

Then a calculator gives $x = 3.0697$. **Answer**

b. The equation $(\log_2 x)^2 + \log_2 x - 6 = 0$ is quadratic in $\log_2 x$. Let $t = \log_2 x$. Then solve $t^2 + t - 6 = 0$.

$$t^2 + t - 6 = 0$$
$$(t + 3)(t - 2) = 0$$

Hence $t = 2$ or $t = -3$.

Therefore x can be found by solving $2 = \log_2 x$ and $-3 = \log_2 x$.
$2 = \log_2 x$ if and only if $2^2 = x$. Therefore $x = 4$.
$-3 = \log_2 x$ if and only if $2^{-3} = x$. Therefore $x = \frac{1}{8}$.
Hence $x = 4$ or $x = \frac{1}{8}$. **Answer**

Logarithmic functions play an important role in applications.

EXAMPLE 4 A formula for the intensity level α of a sound is

$$\alpha = 10 \log_{10} \left(\frac{I}{I_0}\right)$$

where I is the intensity of the sound and I_0 is the intensity of a sound barely audible to the human ear. Find α if the intensity level of a sound is 100,000 times the intensity I_0.

Solution Find α if $I = 10^5(I_0)$.

$$\begin{aligned} \alpha &= 10 \log_{10} \left(\frac{10^5(I_0)}{I_0}\right) \\ &= 10 \log_{10} (10^5) \\ &= (10)(5) \qquad \text{since } \log_b b^x = x \\ \alpha &= 50 \end{aligned}$$

The intensity level α is measured in *decibels*. Thus the sound measured in Example 4 has an intensity level of 50 decibels.

EXERCISES

Find the value of each expression.

A

1. $\log_2 16$ **2.** $\log_3 1$ **3.** $\log_3 3$ **4.** $\log_4 \sqrt{2}$

5. $\log_{25} 5$ **6.** $3^{\log_3 \sqrt{2}}$ **7.** $2^{\log_2 \sqrt{3}}$ **8.** $5^{\log_5 7}$

9. $\log_2 (\frac{1}{2})$ **10.** $\log_7 (\frac{1}{7})$ **11.** $\log_3 3^{-5}$ **12.** $10^{\log_{10} 10^2}$

Graph each equation.

13. $y = 2 \log_2 x$ **14.** $y = -\log_2 x$ **15.** $y = 2 - 2 \log_2 x$

16. $y = \log_2 (x - 2) + 1$ **17.** $y = \log_2 |x|$ **18.** $y = |\log_2 x|$

Solve each equation.

19. $(\log_2 x)^2 - \log_2 x - 6 = 0$ **20.** $(\log_3 x)^2 - 6 \log_3 x - 7 = 0$

21. $\log_4 x + 10(\log_4 x)^{-1} - 7 = 0$ **22.** $\log_x 8 = 3$

23. $\log_4 x = \frac{1}{4}$ **24.** $\log_3 \frac{1}{27} = x$

25. $\log_3 x = -4$ **26.** $\log_{\frac{2}{3}} (\frac{9}{4}) = x$

27. $\log_{\frac{1}{2}} x = 32$ **28.** $\log_x \frac{1}{16} = -\frac{4}{3}$

Find the inverse of each function.

B

29. $\log_2 2x$ **30.** $f(x) = 3 \cdot 2^{4-x}$

31. **a.** Graph $y = 2 \log_2 x$ and $y = \log_2 x^2$.
b. Is it always true that $\log_2 x^2 = 2 \log_2 x$?

32. Find the intensity level of a sound one million times as intense as a barely audible sound.

33. How intense must a sound be relative to barely audible sound if the intensity level is 120 decibels?

The magnitude R of an earthquake with an intensity I (as measured on the *Richter scale*) is given by $R = \log_{10} \left(\frac{I}{I_0}\right)$ where I_0 is a certain minimum intensity. Use the formula and a calculator in Exercises 34–36.

34. Find R if I is $(2.2)(10^{12})$ times I_0.

35. How intense is an earthquake relative to the minimum intensity if $R = 8.3$?

36. If one earthquake has intensity level $(10^{5.5})(I_0)$ and a second earthquake has intensity level $(10^{4.4})(I_0)$, what is the difference in the corresponding values of R?

How acidic or alkaline a solution is depends on the concentration of hydrogen ions, $[H^+]$, in it. A measure of acidity or alkalinity is pH. This measure is given by $pH = -\log_{10} [H^+]$. Use this formula in Exercises 37–39.

37. Pure water has a hydrogen ion concentration of 10^{-7} moles per liter. What is the pH of pure water?

38. Vinegar has a hydrogen ion concentration of approximately 6.3×10^{-3} moles per liter. What is the approximate pH of vinegar?

39. A solution is acidic if its pH is less than 7. If the hydrogen ion concentration in an apple is about 10^{-3} moles per liter, is an apple acidic?

In Exercises 40–45, prove each part of Theorem 4.

40. Part (1)	**41.** Part (2)	**42.** Part (3)
43. Part (4)	**44.** Part (5)	**45.** Part (6)

10–5 LAWS OF LOGARITHMS

Recall that since $b^{\log_b a} = a$, $\log_b a$ is the exponent to which b is raised to obtain a. This relationship between logarithms and exponents is used to establish the *laws of logarithms*.

Let $b > 0$ and $b \neq 1$. Let a and c be positive real numbers. Then

$$b^{\log_b ac} = ac$$

$$b^{\log_b a} = a \quad \text{and} \quad b^{\log_b c} = c$$

Hence $b^{\log_b ac} = ac = (b^{\log_b a})(b^{\log_b c}) = b^{\log_b a + \log_b c}$ by the first law of exponents. Therefore $\log_b ac = \log_b a + \log_b c$ by Theorem 2, part (6). This proves the first of the laws of logarithms.

THEOREM 5 **The Laws of Logarithms**

Let $b > 0$ and $b \neq 1$. Let a and c be positive real numbers.

1. $\log_b ac = \log_b a + \log_b c$
2. $\log_b \left(\frac{a}{c}\right) = \log_b a - \log_b c$
3. $\log_b a^c = c \log_b a$ where c is any real number

You are asked to prove laws 2 and 3 in Exercises 33 and 34. We use the laws of logarithms for a variety of purposes.

EXAMPLE 1 Write $\frac{1}{2}\log_{10} 81 - \log_{10} 3$ as a single logarithm.

Solution Use law 2 and law 3.

$$\begin{aligned} \tfrac{1}{2}\log_{10} 81 - \log_{10} 3 &= \log_{10} 81^{\frac{1}{2}} - \log_{10} 3 && \text{by law 3} \\ &= \log_{10} 9 - \log_{10} 3 \\ &= \log_{10} \left(\tfrac{9}{3}\right) && \text{by law 2} \end{aligned}$$

Therefore $\frac{1}{2}\log_{10} 81 - \log_{10} 3 = \log_{10} 3$. **Answer**

EXAMPLE 2 Write $\log_b \left(\frac{M^3N}{P^2}\right)$ as a sum of logarithms.

Solution

$$\begin{aligned} \log_b \left(\frac{M^3N}{P^2}\right) &= \log_b (M^3N) - \log_b (P^2) && \text{by law 2} \\ &= \log_b M^3 + \log_b N - \log_b P^2 && \text{by law 1} \\ &= 3\log_b M + \log_b N - 2\log_b P && \text{by law 3} \end{aligned}$$

EXAMPLE 3 Simplify $5^{\log_{25} 7}$.

Solution Use the fact that $5 = 25^{\frac{1}{2}}$.

$$\begin{aligned} 5^{\log_{25} 7} &= (25^{\frac{1}{2}})^{\log_{25} 7} \\ &= 25^{(\frac{1}{2}\log_{25} 7)} \\ &= 25^{\log_{25} (7^{\frac{1}{2}})} && \text{by law 3} \\ &= 7^{\frac{1}{2}} \text{ or } \sqrt{7} && \text{since } b^{\log_b a} = a \end{aligned}$$

Therefore $5^{\log_{25} 7} = \sqrt{7}$. **Answer**

EXAMPLE 4 Solve each equation.

a. $3^x = 25$ **b.** $3^{x-3} = 5^{2x}$

Solution In each solution take the logarithm with base 10 of each side of the equation.

a.

$$\begin{aligned} \log_{10} 3^x &= \log_{10} 25 \\ x \log_{10} 3 &= \log_{10} 25 \end{aligned}$$

Therefore $x = \dfrac{\log_{10} 25}{\log_{10} 3}$.

By calculator or logarithm table,

$\log_{10} 25 \approx 1.3979$ and $\log_{10} 3 \approx 0.4771$.

Therefore $x \approx 2.9300$. **Answer**

b.

$$\begin{aligned} \log_{10} 3^{x-3} &= \log_{10} 5^{2x} \\ (x - 3)\log_{10} 3 &= 2x \log_{10} 5 && \text{by law 3} \\ x\log_{10} 3 - 3\log_{10} 3 &= x(2\log_{10} 5) \\ x\log_{10} 3 - 2x\log_{10} 5 &= 3\log_{10} 3 \\ x(\log_{10} 3 - 2\log_{10} 5) &= 3\log_{10} 3 \\ x &= \frac{3\log_{10} 3}{\log_{10} 3 - 2\log_{10} 5} \end{aligned}$$

By calculator or table, $x \approx \dfrac{3(0.4771)}{0.4771 - 2(0.6990)} \approx -1.554$ **Answer**

EXAMPLE 5 Solve $\log_{10} x = 1 - \log_{10}(x - 3)$.

Solution Use the fact that $1 = \log_{10} 10$. Then apply the second law of logarithms.

$$\log_{10} x = \log_{10}\left(\frac{10}{x - 3}\right).$$

Therefore $\qquad x = \dfrac{10}{x - 3}$ by part (6) of Theorem 4.

Hence $\qquad x^2 - 3x = 10.$

Solve the quadratic equation for x to get $x = 5$ or $x = -2$.

Since the logarithm function is defined for positive real numbers only, x and $x - 3$ must be positive. Hence $x > 3$. Reject -2. Therefore the only solution is 5. **Answer**

Logarithms with base 10 are called **common logarithms.** We usually omit the subscript 10 and simply write $\log x$. Logarithms with base e are called **natural logarithms,** denoted $\ln x$.

Traditionally the calculation of a common logarithm was done by first writing the number in *scientific notation.* For example $248 = 2.48 \cdot 10^2$. Then use of the first law of logarithms gives

$$\log 248 = \log(2.48 \cdot 10^2) = \log 10^2 + \log 2.48 = 2 + \log 2.48.$$

Consulting a table of common logarithms for numbers between 1 and 10, we obtain $\log 2.48 \approx 0.3945$. Thus $\log 248 \approx 2.3945$. The common logarithm of any positive real number is the sum of an integer part, called the *characteristic,* and a decimal part, called the *mantissa.* Today the use of tables to find logarithms has been largely replaced by the use of scientific calculators. If $\log x = N$, we call x the *antilogarithm* of N. For example, 248 is the antilogarithm of 2.3945.

In Example 4 we solved $3^x = 25$ by using common logarithms to find

$$x = \frac{\log_{10} 25}{\log_{10} 3}.$$

If we solve $3^x = 25$ by using logarithms with base 3, we have

$$\log_3 3^x = \log_3 25;$$
$$x \log_3 3 = \log_3 25;$$
$$x = \frac{\log_3 25}{\log_3 3};$$
$$x = \log_3 25.$$

From the foregoing discussion of the solution of $3^x = 25$, we may write

$$\log_3 25 = \frac{\log_{10} 25}{\log_{10} 3}.$$

That is, the logarithm of 25 with base 3 can be written as a quotient of the common logarithm of 25 and the common logarithm of 3.

THEOREM 6 **The Change-of-Base Formula**

Let a and b be positive numbers other than 1. For all positive real numbers x,

$$\log_b x = \frac{\log_a x}{\log_a b}.$$

Proof: Let $y = \log_b x$. Then $b^y = x$. Taking the logarithm with base a of each side of $b^y = x$, we have

$$\log_a b^y = \log_a x.$$
$$y \log_a b = \log_a x \quad \text{by the third law of logarithms.}$$

Hence $y = \dfrac{\log_a x}{\log_a b}$. Since $y = \log_b x$, $\log_b x = \dfrac{\log_a x}{\log_a b}$. ■

EXAMPLE 6 Find $\log_5 7$.

Solution Use the change-of-base formula with either common or natural logarithms. This solution relies on common logarithms.

$$\log_5 7 = \frac{\log_{10} 7}{\log_{10} 5} \approx \frac{0.8451}{0.6990} \approx 1.2090 \qquad \textbf{Answer}$$

Many scientific calculators have keys for computing common and natural logarithms, so it becomes convenient, even necessary, to convert logarithms in other bases to logarithms in base 10 or base e.

Frequently it is useful to use logarithms to solve applied problems.

EXAMPLE 7 Light with intensity I_0 in the atmosphere enters ocean water, where it is absorbed. The intensity I of the light d centimeters below the surface is given by $I = I_0(4^{-0.0101d})$. At what depth will the intensity of the light be 25% of the intensity of light in the atmosphere?

Solution Find d such that $0.25I_0 = I_0(4^{-0.0101d})$.

$$\begin{aligned} 0.25 &= 4^{-0.0101d} \\ \log 0.25 &= (-0.0101d) \log 4 \\ -0.6021 &= (-0.0101d)(0.6021) \\ 99.01 &= d \end{aligned}$$

Approximately 99 cm below the surface the intensity of the light will be 25% of the intensity of light in the atmosphere. **Answer**

EXERCISES

Write each expression as a single logarithm.

A

1. $3 \log 6 + \log 5$

2. $\log_3 4 - \log_3 6 + \log_3 w$

3. $\frac{1}{2} \log 3 - \frac{1}{3} \log 8$

4. $\log_6 w + \frac{1}{2} \log_6 5 + \frac{1}{2} \log_6 t$

5. $3 \log_5 t - 1.5 \log_5 t$

6. $\log G + \log m + \log M - 2 \log d$

Write each logarithm as a sum or difference of logarithms.

7. $\log\left(\frac{I}{I_0}\right)$ **8.** $\log\left(\frac{GmM}{d^2}\right)$ **9.** $\log_2 (N \cdot 2^t)$ **10.** $\log (P(1 + r)^t)$

Simplify.

11. $e^{3 \ln 4}$ **12.** $e^{\ln 3 - \ln 6}$ **13.** $16^{2 \log_{16} 5}$ **14.** $5^{\log_3 6 - \log_3 2}$

Solve each equation.

15. $\log x = 0.8395$

16. $\log x = -0.4786$

17. $\log x = 1.8395$

18. $\log x = -1.4786$

19. $\log_3 (2x + 1) + \log_3 5 = 9$

20. $\log_4 (3x - 5) - \log_4 x = -\frac{1}{2}$

21. $3^x = 98$

22. $98^x = 3$

23. $3^{x+2} = 7^{2x-1}$

24. $3^{4-x} = 5^x$

25. $(\ln x)^2 - 5 \ln x + 6 = 0$

26. $(\ln x)^2 - 13 \ln x + 42 = 0$

Find each logarithm to four decimal places.

27. $\log_3 7$ **28.** $\log_7 3$ **29.** $\log_4 12$

30. $\log_3 10$ **31.** $\log_2 18.1$ **32.** $\log_2 0.0081$

In Exercises 33 and 34 prove each law of logarithms.

B

33. Law 2 **34.** Law 3

35. a. Find values of a and c to show that in general $\log (a + c) \neq \log a + \log c$.

b. Find values of a and c for which $\log (a + c) = \log a + \log c$.

36. a. Find particular values of a and c to show that in general $\log (ac) \neq (\log a)(\log c)$.

b. Find values of a and c for which $\log (ac) = (\log a)(\log c)$.

37. a. Find a formula to express $\log_{\frac{1}{b}} x$ in terms of $\log_b x$.

b. Prove that the formula from part (a) is correct.

38. a. Find a formula to express $\log_b a$ in terms of $\log_a b$.

b. Prove that the formula from part (a) is correct.

39. Show that if $b > 0$, $b \neq 1$, and $a > 0$, then the horizontal distance between the graph of $y = b^x$ and the graph of $y = ab^x$ is constant.

40. Show that if $x_1, x_2, \ldots, x_n, \ldots$ is a geometric sequence with positive terms, then $\log x_1, \log x_2, \ldots, \log x_n, \ldots$ is an arithmetic sequence.

In Exercises 41 and 42 use the compound interest formula on page 406.

41. How long would it take for \$500 to grow to \$1500 if the annual rate is 8% and interest is compounded quarterly (four times a year)?

42. How long would it take for a deposit to double if interest is compounded monthly at an annual rate of 12%?

43. Use the formula on page 409 to find how long it would take for a deposit to double if interest is compounded continuously at an annual rate of 10%.

The percent $P(t)$, expressed as a decimal, of doctors who accept a new medicine t months after its introduction to the marketplace is given by the formula $P(t) = 1 - e^{-0.02t}$. Use this formula in Exercises 44–46.

44. Can $P(t)$ ever be greater than 1? Explain your answer.

45. How many months will it take for acceptance to reach 99%?

46. Show that $P(t)$ can never equal 1.

A certain electrical circuit has an electromotive force, or voltage, a resistor, and an inductor. The current I at time t measured in seconds is given by $I = \frac{E}{R}(1 - e^{-\frac{Rt}{L}})$ where E measures voltage, R measures resistance, and L measures inductance. Use the formula in Exercises 47 and 48.

C 47. Is the function I increasing or decreasing for $t \geq 0$?

48. Find $\lim_{t \to \infty} I$.

COMPUTER EXERCISES

1. Write a program based on parts (a) and (b) to approximate the area A of the region bounded by $y = \frac{1}{x}$, the x-axis, and the vertical lines $x = 1$ and $x = b$ where $b > 1$ and is input by the user.

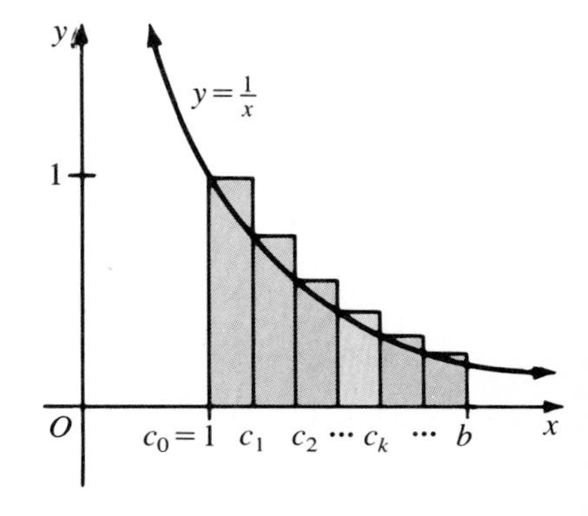

   a. Divide the interval $[1, b]$ into n subintervals of equal length. Let $c_0 = 1$ and let each subdivision point of $[1, b]$ be denoted c_k where $k = 1, 2, \ldots, n - 1$. Write c_k in terms of b and n.

   b. Use c_k to write a formula to find the area of the kth rectangle. Sum the areas of all the rectangles to approximate the area of the region.

Run the program of Exercise 1 for each value of b and n.

2. a. $b = 2, n = 10$
   b. $b = 2, n = 100$

3. a. $b = e, n = 10$
   b. $b = e, n = 100$

4. a. $b = 4, n = 10$
   b. $b = 4, n = 100$

5. Modify the program of Exercise 1 and run it to find $A - \ln b$ in each instance.

   a. Exercise 2(b)   b. Exercise 3(b)   c. Exercise 4(b)

10-6 DERIVATIVES OF EXPONENTIAL AND LOGARITHMIC FUNCTIONS

In this section we shall find formulas for the derivatives of $\ln x$ and e^x. First notice that

$$\lim_{n\to\infty}\left(1+\frac{1}{n}\right)^n = \lim_{n\to 0}(1+n)^{\frac{1}{n}} = e.$$

It can be shown that if t is any real number approaching 0, then

$$\lim_{t\to 0}(1+t)^{\frac{1}{t}} = e. \qquad (*)$$

We shall use (*) to find the derivative of $\ln x$.

Let $y = \ln x$, let $x > 0$ be fixed, and let $h > 0$.

$$\frac{\ln(x+h)-\ln x}{h} = \frac{\ln\left(\frac{x+h}{x}\right)}{h}$$

$$= \frac{\ln\left(1+\frac{h}{x}\right)}{\left(\frac{h}{x}\right)x}$$

$$= \frac{1}{x}\ln\left(1+\frac{h}{x}\right)^{\frac{x}{h}} \quad \text{by law 3 of logarithms}$$

Therefore

$$\frac{dy}{dx} = \lim_{h\to 0}\frac{\ln(x+h)-\ln x}{h}$$

$$= \lim_{h\to 0}\frac{1}{x}\ln\left(1+\frac{h}{x}\right)^{\frac{x}{h}}.$$

$$= \frac{1}{x}\lim_{h\to 0}\ln\left(1+\frac{h}{x}\right)^{\frac{x}{h}}$$

Let $t = \frac{h}{x}$. Then as $h \to 0$, $t \to 0$. Hence $\frac{dy}{dx} = \frac{1}{x}\lim_{t\to 0}\ln(1+t)^{\frac{1}{t}}$

Using equation (*) we have

$$\frac{dy}{dx} = \frac{1}{x}\ln e = \frac{1}{x}\cdot 1 = \frac{1}{x}.$$

The argument just given shows that the natural logarithm function is differentiable for all positive real numbers and provides us with a formula for the derivative.

THEOREM 7 If $y = \ln x$, then $\dfrac{dy}{dx} = \dfrac{1}{x}$.

EXAMPLE 1 Find the derivative of $s(x) = x^2 \ln x$.

Solution Use the product rule.

$$\frac{ds}{dx} = x^2\left[\frac{d}{dx}(\ln x)\right] + \left[\frac{d}{dx}(x^2)\right]\ln x$$

$$= x^2\left(\frac{1}{x}\right) + 2x \cdot \ln x$$

$$\frac{ds}{dx} = x + 2x \ln x \qquad \textbf{Answer}$$

We can obtain a formula for the derivative of the natural exponential function. To do so we shall make use of the fact, which we assume without proof, that the inverse of a differentiable function is also differentiable. Graphical evidence makes this fact reasonable.

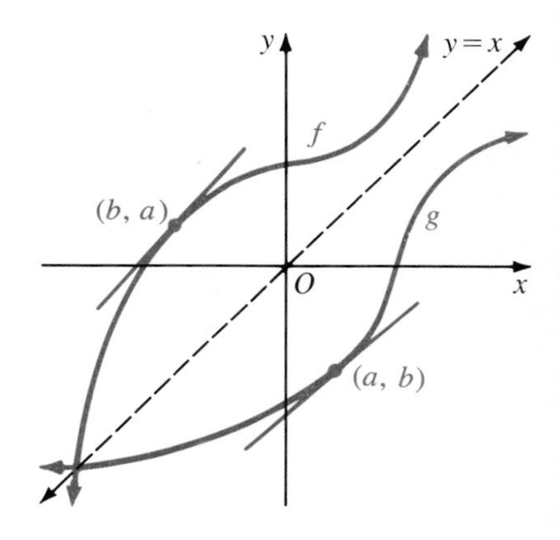

If f and g are inverse functions, then their graphs are reflections of each other in the line $y = x$. The graph of g will have a nonvertical tangent line at the point (a, b) if and only if the graph of f has a nonhorizontal tangent line at the point (b, a). This suggests that g is differentiable at a if $f'(b) = f'(g(a))$ exists and is nonzero.

THEOREM 8 If $y = e^x$, then $\dfrac{dy}{dx} = e^x$.

Proof: Let f be the function given by $f(x) = \ln x$. Then the inverse function g is given by $g(x) = e^x$. Since $f'(x) = \dfrac{1}{x}$ and $\dfrac{1}{x}$ is never 0, we may assume that g' exists. That is, g is differentiable. To find g', apply the chain rule to

$$\ln(e^x) = x.$$

$$\frac{d}{dx}(\ln(e^x)) = \frac{dx}{dx}$$

$$\frac{1}{e^x} \cdot \frac{d}{dx}(e^x) = 1$$

Since e^x is not 0, we can multiply both sides by e^x to obtain

$$\frac{d}{dx}(e^x) = e^x. \quad \blacksquare$$

EXAMPLE 2 Find the derivative of each function.

a. $f(x) = e^x x^3$ **b.** $f(x) = \frac{\ln x}{x}$ **c.** $f(x) = e^{2x^2 + 3x}$

Solution Use the rules of differentiation from Sections 6–3 and 6–4.

a. Use the product rule.

$$\frac{d}{dx}(e^x x^3) = e^x \frac{d}{dx}(x^3) + x^3 \frac{d}{dx}(e^x) = e^x(3x^2) + x^3(e^x)$$
$$= 3e^x x^2 + e^x x^3, \text{ or } x^2 e^x(x + 3)$$

b. Use the quotient rule.

$$\frac{d}{dx}\left(\frac{\ln x}{x}\right) = \frac{x\frac{d}{dx}(\ln x) - \ln x \frac{dx}{dx}}{x^2}$$
$$= \frac{x\left(\frac{1}{x}\right) - (\ln x)(1)}{x^2}$$
$$\frac{d}{dx}\left(\frac{\ln x}{x}\right) = \frac{1 - \ln x}{x^2}$$

c. Since $f(u) = e^u$ and $u = 2x^2 + 3x$ are differentiable functions, the chain rule applies.

$$\frac{d}{dx}(e^{2x^2 + 3x}) = (e^{2x^2 + 3x})\frac{d}{dx}(2x^2 + 3x)$$
$$= (4x + 3)e^{2x^2 + 3x}$$

EXAMPLE 3 Find an equation of the line tangent to the graph of $y = e^{-\frac{x^2}{2}}$ at $x = 0$.

Solution The slope of the tangent line is the value of the derivative.

$$y' = \left(-\frac{2x}{2}\right)e^{-\frac{x^2}{2}} = (-x)e^{-\frac{x^2}{2}}$$

When $x = 0$, $y'(0) = 0$ and $y(0) = 1$. The point-slope form of a line can be used to obtain $y - 1 = 0(x - 0)$. The tangent line has equation $y = 1$.
Answer

EXAMPLE 4 Graph $y = e^{-\frac{x^2}{2}}$.

Solution Since $y > 0$ for all x, the graph is completely above the x-axis. Furthermore, because $e^{-\frac{(-x)^2}{2}} = e^{-\frac{x^2}{2}}$, the graph is symmetric about the y-axis. The results of Example 3 indicate that a relative maximum of 1 occurs at 0. (You can verify this by using the first-derivative test (page 240).) Indeed 1 is an absolute maximum.

To find any points of inflection, set the second derivative equal to 0.

$$y'' = (-x)\frac{d}{dx}\left(e^{-\frac{x^2}{2}}\right) + (-1)e^{-\frac{x^2}{2}}$$
$$= (x^2 - 1)e^{-\frac{x^2}{2}}$$

Then $y'' = 0$ when $x = 1$ or $x = -1$.

When $x = 1$, $y = e^{\frac{-(1)^2}{2}} = e^{-\frac{1}{2}}$, and when $x = -1$, $y = e^{\frac{-(-1)^2}{2}} = e^{-\frac{1}{2}}$. The points of inflection are $(1, e^{-\frac{1}{2}})$ and $(-1, e^{-\frac{1}{2}})$. (You can verify that these points are inflection points by using Theorem 9 of Chapter 6. When $x < -1$, $y'' > 0$. When $-1 < x < 1$, $y'' < 0$, and when $x > 1$, $y'' > 0$.)

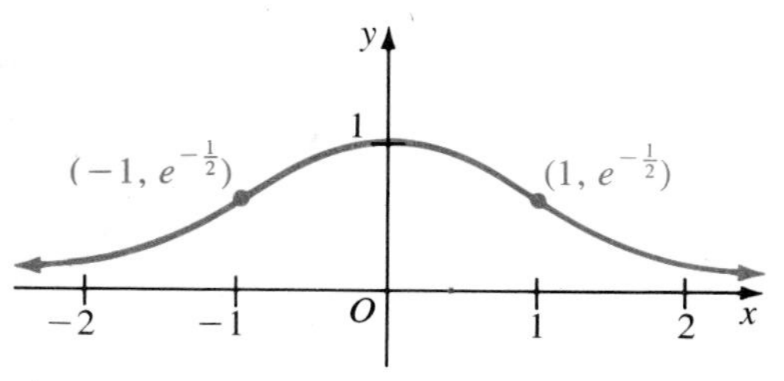

In Exercise 43 you are asked to use the fact that $b^x = (e^{\ln b})^x$ to show that the derivative of b^x is $(\ln b)b^x$. In Exercise 44 you are asked to use the change-of-base formula to show that the derivative of $\log_b x$ is $\frac{1}{x \ln b}$.

Using the derivative for b^x, we can establish the fact that the graph of $f(x) = b^x$, $b \neq 1$, is concave up everywhere.

$$\begin{aligned} f'(x) &= (\ln b)b^x \\ f''(x) &= (\ln b)\frac{d}{dx}(b^x) \\ &= (\ln b)(\ln b)b^x \\ &= (\ln b)^2 b^x \end{aligned}$$

By Theorem 2, $b^x > 0$ for all x. Hence $f''(x)$ is always positive. Therefore the graph of $f(x) = b^x$ is concave up everywhere.

EXERCISES

Find the derivative of each function.

A

1. e^{-5x} **2.** e^{3x} **3.** $\ln(3x)$

4. $x^2 \ln x$ **5.** $x^5 e^x$ **6.** $(x^2 + 1)\ln x$

7. $\frac{e^x}{x^2 - 1}$ **8.** $\frac{\ln x}{x^3}$ **9.** $\frac{x^2}{\ln x}$

10. $\ln(x - 3)^2$ **11.** $\ln(2x^2 + 3x + 1)$ **12.** $(3x^2 - 5)e^{2x}$

13. $\frac{e^{2x} - 1}{e^{2x} + 1}$ **14.** $\frac{e^x - e^{-x}}{e^x + e^{-x}}$ **15.** $\frac{\ln(x + 1)}{\ln(x - 1)}$

16. $e^{4x^3} - 5$ **17.** $\ln(2x^5 + 4x + 1)$ **18.** $\ln(2x)\, e^{2x}$

19. $\sqrt{\ln x}$ **20.** $\sqrt{e^{3x}}$ **21.** $(\ln x)^{-2}$

22. $\sqrt{\ln(x^2 + 1)}$ **23.** $\sqrt{xe^x}$ **24.** $\sqrt{x \ln x}$

In Exercises 25–30 find the value of f' at the given value of x.

25. $(\ln x)^2$; $x = e^2$ **26.** $\ln \sqrt{x}$; $x = 1$ **27.** $e^{\sqrt{x}}$; $x = 1$

28. x^2e^{-x}; $x = 2$ **29.** xe^{x^2}; $x = 1$ **30.** $\ln (\ln x)$; $x = e$

Find an equation of the line tangent to the graph of each function at the given value of x.

31. Exercise 26 **32.** Exercise 27 **33.** Exercise 29

34. $\frac{e^x + e^{-x}}{2}$; $x = 0$ **35.** $\frac{e^x - e^{-x}}{2}$; $x = 0$ **36.** x^2e^{2x}; $x = 1$

B Find the derivative of each function.

37. $\ln (\sin x)$ **38.** $\cos (e^x)$ **39.** $e^{-x} \sin x$

40. $e^{\tan x}$ **41.** $(\cos x)(\ln x)$ **42.** $(\tan x)(\ln x)$

43. Use the fact that $b^x = e^{x \ln b}$ to prove that if $f(x) = b^x$, then $f'(x) = (\ln b)b^x$.

44. Use $\log_b x = \frac{\ln x}{\ln b}$ to prove that if $f(x) = \log_b x$, then $f'(x) = \frac{1}{x \ln b}$.

Use the formulas in Exercises 43 and 44 to find the derivative of each function.

45. $x^2 \log_3 x$ **46.** $x^3 3^x$ **47.** $\frac{10^x}{\log_{10} x}$

48. $\log_5 (x^2 + 3)$ **49.** $10^{x^2 - 3x}$ **50.** $(\log_{10} (2x))\, 10^{3x}$

Graph each equation. Find any relative maxima, relative minima, and points of inflection.

51. $y = \ln x^2$ **52.** $y = \ln\sqrt{x}$ **53.** $y = \frac{e^x + e^{-x}}{2}$ **54.** $y = \frac{e^x - e^{-x}}{2}$

C **55.** Show that if $b > 0$ and $b \neq 1$, then the graph of $f(x) = \log_b x$ is concave down for all $x > 0$.

56. Show that if $x \neq 0$, $\frac{d}{dx} (\ln |x|) = \frac{1}{x}$.

57. Show that $\frac{d}{dx} (x^r) = rx^{r-1}$ for all *real* numbers r and all positive numbers x by justifying parts (a) through (e).

a. $x^r > 0$ for all positive numbers x and all real numbers r.

b. If $y = x^r$, then $\ln y = r \ln x$.

c. Then $\frac{d}{dx} \ln y = r\left(\frac{1}{x}\right)$.

d. $\frac{1}{y}\frac{dy}{dx} = \frac{r}{x}$

e. $\frac{dy}{dx} = rx^{r-1}$

The process of finding a derivative using $\ln y$ and following steps (a) through (e) in Exercise 57 is called *logarithmic differentiation*. Use this process to find y' in Exercises 58–61.

58. $y = x^{2x}$

59. $y = x^{4 \ln x}$

60. $y = (\ln x)^{2x}$

61. $y = (x^3 + 4)^{14}(x^5 - 1)^2 \sqrt{x + 8}$

10–7 EXPONENTIAL GROWTH AND DECAY

Recall from Section 3–5 the discussion of geometric sequences. The following example illustrates a geometric sequence that gives rise to an exponential function.

> A certain city with population 10,000 grows 3% per year. That is, each year the population is 103% of the population the year before.

If $P(n)$ is the population at the end of the nth year, then we can use

$$P(n) = 10{,}000(1.03)^n \quad (n \text{ a natural number})$$

to approximate the population n years after the population is 10,000.

If we assume that the population grows continuously, then we can replace n by t where t is a nonnegative real number.

$$P(t) = 10{,}000(1.03)^t \quad (t \geq 0)$$

In general, if an initial population P_0 grows at $100r\%$ per period of time t, then the population at time t, $P(t)$, is given by the exponential function

$$P(t) = P_0(1 + r)^t.$$

We can also study populations that decrease. In this case, $-1 < r < 0$.

If $r > 0$, we say that the population **grows exponentially** and we call $P(t) = P_0(1 + r)^t$ an **exponential growth model**. If $-1 < r < 0$, we say that the population **decays exponentially** and that $P(t) = P_0(1 + r)^t$ is an **exponential decay model**. The following figures illustrate the two situations.

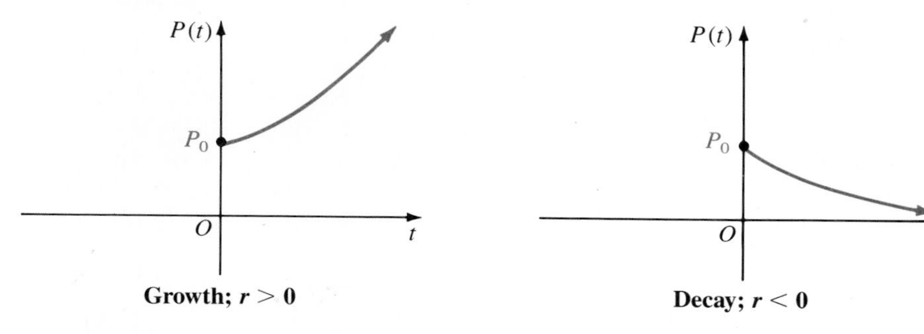

Growth; $r > 0$ **Decay; $r < 0$**

EXAMPLE 1 A city's population grows exponentially. In 3.5 years the population grew from 12,000 people to 14,000 people.

a. Approximate to the nearest thousand the population of the city 6.5 years from the time the population was 12,000.

b. Approximately when will the population double?

Solution First find the exponential growth model. The city's population is given by $P(t) = P_0(1 + r)^t$ where $P_0 = P(0) = 12{,}000$.
Since $P(3.5) = 14{,}000$,

$$14{,}000 = 12{,}000(1 + r)^{3.5}.$$

Thus
$$1.1667 = (1 + r)^{3.5}.$$

Take the common logarithm of each side of the equation.

$$\log(1.1667) = \log(1 + r)^{3.5}$$
$$0.06696 = 3.5 \log(1 + r)$$
$$0.01913 = \log(1 + r)$$

Hence by calculator or table $1 + r = 1.045$.
Therefore $P(t) = 12{,}000(1.045)^t$.

a. $P(6.5) = 12{,}000(1.045)^{6.5} \approx 15{,}975$, or 16,000 **Answer**

b. If d is the time it takes for the population to double, then

$$P(d) = 2P_0.$$

Therefore
$$P_0(1.045)^d = 2P_0.$$

Use logarithms to solve the equation for d.

$$d \log(1.045) = \log 2$$
$$d = \frac{\log 2}{\log(1.045)}$$
$$\approx 15.7$$

Thus it will take approximately 16 years for the population to double. **Answer**

It is often useful to write an exponential growth or decay model in terms of the natural exponential function. Writing $1 + r = e^{\ln(1+r)}$, we have

$$P(t) = P_0(1 + r)^t = P_0 e^{[\ln(1+r)]t}.$$

That is, $P(t) = P_0 e^{kt}$ for some real number k.

EXAMPLE 2 Twenty weeks ago the presence of a noxious chemical pollutant was discovered in a small lake. At that time the level of concentration of the pollutant was 240 parts per million; and at that time a purifying agent was introduced into the lake. The level of concentration of the pollutant decays exponentially and has a half-life of 50 weeks. (The **half-life** of a substance is the time it takes for half of the substance to decay.)

a. What is the level of concentration now?

b. How long will it take for the level to reach 96 parts per million?

Solution First find the exponential decay model. That is, use $P(t) = P_0e^{kt}$ to find k. Since the half-life is 50 weeks,

$$\tfrac{1}{2}(240) = 240e^{k(50)}$$

That is,

$$0.5 = e^{k(50)}$$

Use natural or common logarithms to find k.

$$\ln 0.5 = 50k$$
$$-0.0139 = k$$

Hence $P(t) = 240e^{-0.0139t}$.

a. The level of concentration now is $P(20)$.

$$P(20) = 240e^{(-0.0139)(20)} = 240e^{-0.278} = (240)(0.76) \approx 180$$

Thus the present level of concentration is approximately 180 parts per million. **Answer**

b. To find when the level of concentration will be 96 parts per million, solve $96 = 240e^{-0.0139t}$ for t.

$$96 = 240e^{-0.0139t}$$
$$0.4 = e^{-0.0139t}$$
$$\ln 0.4 = -0.0139t$$
$$-0.9163 = -0.0139t$$
$$65.9 \approx t$$

The level of concentration will be 96 parts per million in approximately 66 weeks from the time the purifying agent is introduced. **Answer**

Note: The equation $P(t) = P_0(1 + r)^t$ could have been used to solve Example 2. Use the half-life information to find $1 + r$.

$$0.5 = (1 + r)^{50}$$
$$\log 0.5 = 50 \log (1 + r)$$
$$-0.006 = \log (1 + r)$$
$$0.9862 = 1 + r$$

The decay model is $P(t) = 240(0.9862)^t$.

The rate of change of a population that grows or decays exponentially can be found by use of Theorem 8 from Section 10–6. If $P(t) = P_0e^{kt}$, then

$$\frac{d}{dt}P(t) = \frac{d}{dt}P_0e^{kt};$$

$$\frac{d}{dt}P(t) = kP_0e^{kt};$$

$$\frac{d}{dt}P(t) = kP(t).$$

The last equation states that the rate of change of population (with respect to time) growing or decaying exponentially is directly proportional to the population at time t. For instance, if one population is twice the size of a second population, then the rate of change in the first population would be twice that of the second population.

Throughout this discussion, we have assumed that both food and space were ample to support growth. We have also assumed that there was no competition for resources. A refinement of the exponential growth or decay model would be necessary to take into account the above-mentioned factors.

EXERCISES

Use the compound interest formula on page 406 for Exercises 1–8. Find the value of the indicated variable from the given values of the other variables.

A

1. $P = \$1500;\ r = 0.09;\ n = 2;\ t = 10;\ A$
2. $P = \$1800;\ r = 0.065;\ n = 4;\ t = 14;\ A$
3. $P = \$2000;\ r = 0.08;\ n = 1;\ A = \$2400;\ t$
4. $P = \$2500;\ r = 0.075;\ n = 1;\ A = \$3000;\ t$
5. $P = \$16{,}000;\ r = 0.12;\ n = 12;\ A = \$20{,}000;\ t$
6. $P = \$1500;\ r = 0.09;\ n = 4;\ A = \$3000;\ t$
7. $P = \$3500;\ n = 1;\ A = \$3850;\ t = 5;\ r$
8. $P = \$8000;\ n = 1;\ A = \$10{,}000;\ t = 6;\ r$

Use $\alpha = 10 \log_{10} \left(\frac{I}{I_0}\right)$ for Exercises 9–13.

9. How intense is a sound relative to I_0 that has a decibel reading of 105?
10. What intensity range relative to I_0 does a sound have if the decibel reading is between 83 and 95?
11. What decibel range corresponds to a sound whose intensity is between I_0 and $10^{12}I_0$?
12. What intensity range does a sound have if the decibel reading is between 25 and 50?
13. If $I = 10^8 I_0$ and $I_1 = 2 \cdot 10^8 I_0$, is the decibel reading for I_1 twice that of I?
14. If a population of bacteria numbers 2000 members at noon and increases 20% per hour, what is the population at 2:00 P.M.? at 11:30 P.M.?
15. A population of trees in a forest is declining at the rate of 4% per year because of a blight. If at present there are 4000 trees, how many trees will there be five years from now? If the same rate of decline applied to years past, how many trees were there five years ago?
16. A colony of bacteria doubles each day. If at first count there were 10^5 members in the colony, how many members were there in the colony five days after the first count? How many members were there after 3.5 days? How many members were there after 17 hours?

17. If a population doubles every 32 years, how long will it take for an initial population of one million to become 3.2 million?

18. If a quantity P increases by 3% per year, what percent increase of P is the population after five years?

19. If a quantity P decreases by 1% per year, what percent decrease of P is the population after 100 years?

The half-life of a radioactive substance is the time it takes for the substance to decay to 50% of its original amount.

20. If 100 g of a radioactive substance decays to 20 g in 16 years, how long will it take to decay to 16 g?

21. If 24 g of a radioactive substance decays to 20 g in 10 years, what is the half-life of the substance?

22. After 12 years a radioactive substance is 18 g and after an additional 4 years the substance is 15 g. What is the half-life of the substance?

If a quantity P grows or decays exponentially, then its rate of change P' is directly proportional to the quantity P. That is $P' = kP$ for some constant k.

B

23. Show that $y = 10{,}000(2^t)$ satisfies $y' = ky$ and find k.

24. Show that $y = 500(\frac{1}{2})^t$ satisfies $y' = ky$ and find k.

25. Show that $y = 0.4e^{-0.2t}$ satisfies $y' = ky$ and find k.

26. Show that $y = 0.12e^t - 10t$ does not satisfy $y' = ky$.

27. Show that $y = e^{x^2}$ does not satisfy $y' = ky$.

28. Show that $y = e^{ax+b}$ satisfies $y' = ky$ for any constants a and b. Find k.

29. **a.** Does $y = 1 - e^t$ satisfy $y' = ky$?
    b. Graph $y = 1 - e^t$ for $t \geq 0$.

30. **a.** Graph $y = P(1 - e^{at})$ for $t \geq 0$. Both P and a are positive constants.
    b. Does $y = P(1 - e^{at})$ satisfy $y' = ky$?

When competition and availability of resources affect population growth, then a limited growth model more properly describes population growth. For instance an initial population $N(0)$ might grow to a population $N(t)$ according to $N(t) = \dfrac{K}{1 + ce^{-2t}}$ where $c = \dfrac{K - N(0)}{N(0)}$ and K is a constant.

C

31. Graph N over the interval $(0, \infty)$ for the case where $K = 1000$ and $N(0) = 500$.

32. Graph N over the interval $(0, \infty)$ for the case where $K = 1000$ and $N(0) = 1000$.

33. Graph N over the interval $(0, \infty)$ for the case where $K = 500$ and $N(0) = 1000$.

34. Show that $N(t) = \dfrac{K}{1 + ce^{-2t}}$ does not satisfy $N' = aN$ for any constant a.

COMPUTER EXERCISES

1. Write a program to find a population $P(t)$ for an input time t, if (1) the population grows exponentially, (2) at $t_1 = 2$, $P(2) = 580$, and (3) at $t_2 = 5$, $P(5) = 700$. (*Hint*: Use 2, 5, $P(2)$, and $P(5)$ in $P(t) = P_0e^{kt}$ to find k and P_0.)
2. Run the program of Exercise 1 to find $P(t)$ for the given values of t.
   a. $t = 3$ **b.** $t = 10$
3. Modify the program of Exercise 1 for t_1, t_2, $P(t_1)$, $P(t_2)$, input by the user, to find $P(t)$ for a t also input by the user.
4. Run the program of Exercise 3 for the following values to find $P(t)$ when $t = 6$.
   a. $t_1 = 3.5$ $\quad t_2 = 4.25$
   $P(t_1) = 12{,}000$ $\quad P(t_2) = 8500$
   b. $t_1 = 4$ $\quad t_2 = 5$
   $P(t_1) = 125$ $\quad P(t_2) = 160$

Chapter Summary

1. Let $b > 0$ and let x be any real number. If $r_1, r_2, r_3, \ldots, r_n, \ldots$ is any sequence of rational numbers such that $\lim_{n\to\infty} r_n = x$, then
$$b^x = \lim_{n\to\infty} b^{r_n}.$$
2. Let $b > 0$ and let x be any real number. The function $x \to b^x$ is called the *exponential function with base b*. If $b > 1$, it is continuous and strictly increasing on $\mathcal{R}$. If $0 < b < 1$, it is continuous and strictly decreasing on $\mathcal{R}$. The exponential function with base b ($b \neq 1$) has a graph that is concave up everywhere and asymptotic to the x-axis.

   The number e is defined as $\lim_{n\to\infty}\left(1 + \frac{1}{n}\right)^n$ and is approximately 2.718. This number is the base of the *natural exponential function*, $x \to e^x$.
3. The *compound interest formula* is
$$A = P\left(1 + \frac{r}{n}\right)^{nt}$$
   where P is an initial deposit, r is the annual interest rate as a decimal, n is the number of interest periods in a year, and t is time in years. If interest is compounded continuously, the formula becomes the *continuous compound interest formula*:
$$A = Pe^{rt}$$
4. If $b > 0$ and $b \neq 1$, then the *logarithmic function with base b* is defined by: $\log_b x = y$ if and only if $b^y = x$. The domain of the logarithmic function is the set of all positive real numbers. The logarithmic function with base b is continuous, and if $b > 1$ it is strictly increasing. The logarithmic function with base e, denoted $\ln x$, is called the *natural logarithmic function*.

5. Let $b > 0$ and $b \neq 1$. If a and c are positive real numbers and d is any real number, then:

$$\log_b ac = \log_b a + \log_b c$$

$$\log_b \left(\frac{a}{c}\right) = \log_b a - \log_b c$$

$$\log_b a^d = d \cdot \log_b a$$

These equations are called the *laws of logarithms.*

The *change-of-base formula* for logarithms is $\log_b c = \dfrac{\log_a c}{\log_a b}$.

6. The natural logarithmic and exponential functions are differentiable and

$$\frac{d}{dx} \ln x = \frac{1}{x} \qquad \frac{d}{dx} e^x = e^x$$

7. Exponential functions are important in studying problems involving growth or decay. An *exponential growth* or *decay model* can be written as $P(t) = P_0 e^{kt}$ where P_0 is the initial population and $P(t)$ is the population after time t. If the population grows exponentially, $k > 0$, and if it decays exponentially, $k < 0$.

Chapter Test

10-1 1. Simplify $((\sqrt[3]{3})^{\sqrt{3}})^{\sqrt{3}}$.

2. Solve $(x^2 + 2)^{\frac{3}{2}} = 27$.

10-2 3. Graph $y - 1 = 3^{-x}$.

Solve each equation.

4. $5^{2x-1} = 125^{\frac{2}{3}}$

5. $3(3^{2x}) + 2(3^x) = 1$

6. A deposit of \$1600 is made. Interest is paid at an annual rate of 8% and interest is compounded quarterly. Find the amount after 12 years.

10-3 7. Find $\lim\limits_{n\to\infty} \left(1 + \frac{1}{n}\right)^{-3n}$.

8. Solve $e^{3x-2} = \sqrt{e}$.

9. A deposit of \$1000 is made. Interest is paid continuously at an annual rate of 7.5%. How much is in the account after 7 years?

10-4 10. Find $\log_9 27$.

11. Graph $y = 1 + \log_2 (x + 1)$.

12. Solve $\log_2 x = -4$.

13. Solve $\log_x 64 = 2$.

10-5 14. Simplify $e^{3 \ln 2 - \ln 3}$.

15. Solve $\log_2 (x + 1) - \log_2 x = 3$.

16. Simplify $(\log_{10} 3)(\log_9 10)$.

17. Solve $3^x = 90$.

10-6 Find each derivative.

18. $e^{-x} \cos x$

19. $\ln (x^2 - 4)$

20. $e^{3x} \ln x$

21. Find an equation of the line tangent to the graph of $y = \sin (\ln x)$ at $x = e^{\pi}$.

10-7 22. If 24 g of a radioactive substance decays to 20 g in 10 years, how long will it take to decay to 16 g?

Extra

Hyperbolic Functions

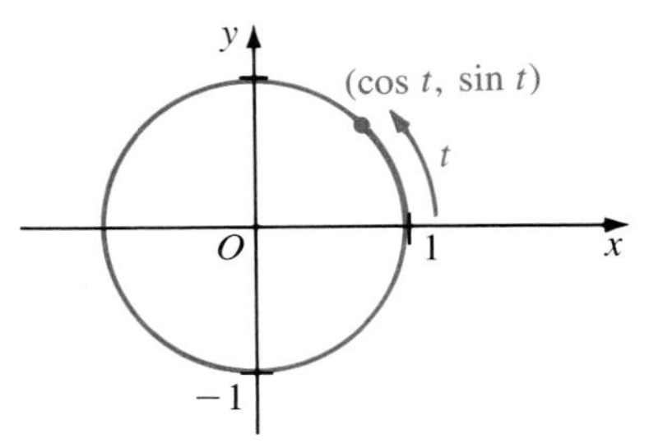

We can interpret the identity $\cos^2 t + \sin^2 t = 1$ as saying that the point with coordinates $(\cos t, \sin t)$ lies on the unit circle. This is the reason we often call cos and sin circular functions.

The **hyperbolic sine**, denoted sinh, and the **hyperbolic cosine**, (denoted cosh), defined by

$$\sinh t = \frac{e^t - e^{-t}}{2} \quad \text{and} \quad \cosh t = \frac{e^t + e^{-t}}{2},$$

are important in many applications of mathematical analysis, physics, and engineering. The following reasoning indicates why these functions might be called hyperbolic.

$$\cosh^2 t - \sinh^2 t = \left(\frac{e^t + e^{-t}}{2}\right)^2 - \left(\frac{e^t - e^{-t}}{2}\right)^2$$

$$= \frac{e^{2t} + 2 + e^{-2t}}{4} - \frac{e^{2t} - 2 + e^{-2t}}{4} = 1$$

We can interpret the identity $\cosh^2 t - \sinh^2 t = 1$ as saying that the point with coordinates $(\cosh t, \sinh t)$ lies on the hyperbola with equation $x^2 - y^2 = 1$.

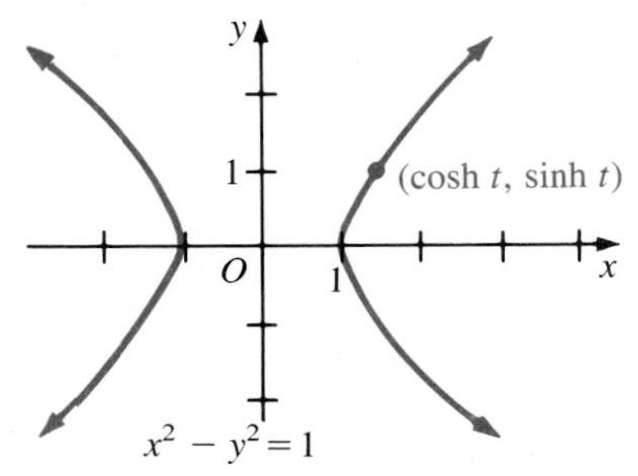

Various properties of hyperbolic functions can be obtained from their definitions. For example, since $\sinh(-t) = \frac{e^{-t} - e^{-(-t)}}{2} = -\frac{e^t - e^{-t}}{2} = -\sinh t$, sinh is an odd function. Recall that sin is also an odd function.

It might come as no surprise that sinh and cosh are differentiable because they are defined in terms of e^t and e^{-t}.

$$\frac{d}{dt}(\sinh t) = \frac{d}{dt}\left(\frac{e^t - e^{-t}}{2}\right) = \frac{e^t - (-1)e^{-t}}{2} = \frac{e^t + e^{-t}}{2} = \cosh t$$

Since $\cosh t > 0$ for all real numbers t and $\frac{d}{dt}(\sinh t) = \cosh t$, we can conclude that sinh is an increasing function on its entire domain. The graphs of sinh and cosh shown below can be obtained from information about the functions and their derivatives. Notice, for example, that the graph of sinh is symmetric about the origin (odd function) and that the graph always rises to the right. (The first derivative is positive for all t.)

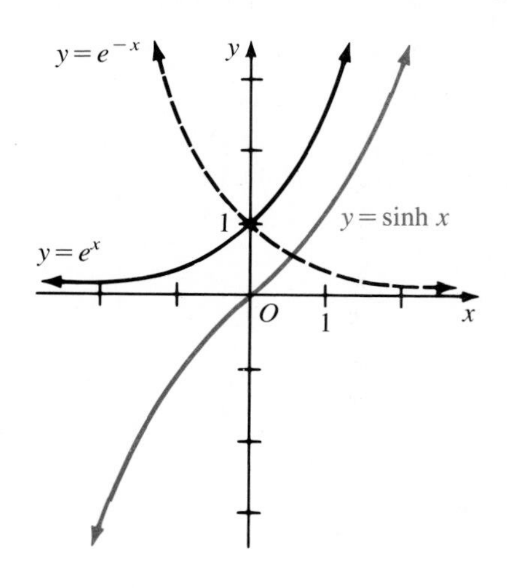

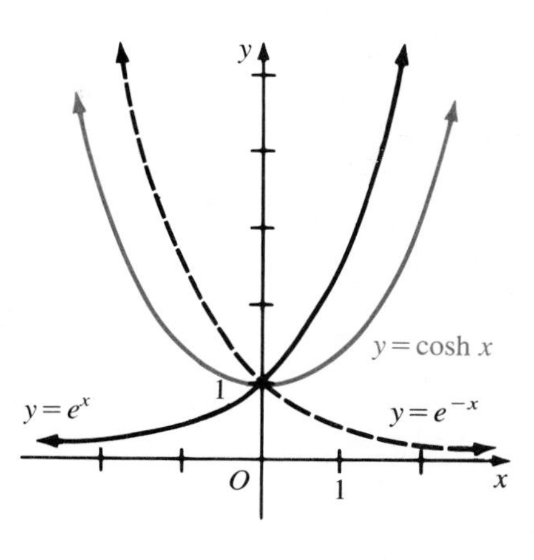

EXERCISES

Use the definitions of sinh and cosh to find each value.

1. $\cosh 0$ **2.** $\sinh 0$ **3.** $\sinh 1$ **4.** $\cosh 1$

Use a calculator to approximate each value.

5. $\sinh 4.8$ **6.** $\cosh(-3.2)$

7. Show that cosh is an even function. **8.** Show that $\frac{d}{dt}(\cosh t) = \sinh t$.

9. a. Find the second, third, and fourth derivatives of sinh.
b. Write a formula for the nth derivative of sinh.

10. Prove that $\sinh(x + y) = (\sinh x)(\cosh y) + (\cosh x)(\sinh y)$ for all real x and y.

11. Prove that $\cosh(x + y) = (\cosh x)(\cosh y) + (\sinh x)(\sinh y)$ for all real x and y.

12. Prove that the graph of cosh is concave up on its domain.

13. Prove that the graph of sinh is concave up on $(0, \infty)$ and concave down on $(-\infty, 0)$.

14. The **hyperbolic tangent**, denoted tanh, is defined by $\tanh t = \frac{\sinh t}{\cosh t}$. Use this definition to find $\frac{d}{dt}(\tanh t)$.

COMPUTER APPLICATION

POWERS OF *e*

In Chapters 9 and 10 it was stated that cos x, sin x, and e^x are all differentiable. Indeed their derivatives of all orders exist for all real values of x. These facts and an important theorem of calculus (*Taylor's theorem*) enable us to write these functions as infinite series involving powers of x.

$$e^x = 1 + x + \frac{x^2}{2!} + \cdots + \frac{x^n}{n!} + \cdots \qquad (1)$$

$$\cos x = 1 - \frac{x^2}{2!} + \frac{x^4}{4!} - \cdots + (-1)^n \frac{x^{2n}}{(2n)!} + \cdots \qquad (2)$$

$$\sin x = x - \frac{x^3}{3!} + \frac{x^5}{5!} - \cdots + (-1)^n \frac{x^{2n+1}}{(2n+1)!} + \cdots \qquad (3)$$

We can use the series for e^x to find e^z when z is a pure imaginary number, that is, when $z = ix$ where x is real.

$$e^z = e^{ix} = 1 + ix + \frac{(ix)^2}{2!} + \cdots + \frac{(ix)^n}{n!} + \cdots$$

$$= \left(1 - \frac{x^2}{2!} + \frac{x^4}{4!} - \cdots\right) + i\left(x - \frac{x^3}{3!} + \frac{x^5}{5!} - \cdots\right)$$

$$e^{ix} = \cos x + i \sin x \qquad (4)$$

Equation (4), known as *Euler's formula*, gives values of e^{ix}. In particular, if $x = \pi$,

$$e^{i\pi} = \cos \pi + i \sin \pi = -1 + i(0) = -1$$

EXERCISES

(You may use either BASIC or Pascal.)

1. Write a program to approximate cos x. Use (2).

2. Write a program to approximate sin x. Use (3).

3. Modify the programs of Exercises 1 and 2 to approximate e^{ix}. Use (4).

4. Run the program in Exercise 3 to find each value of e^{ix}.

a. $x = \pi$ **b.** $x = 0$ **c.** $x = 1$ **d.** $x = 2\pi$

e. $x = \frac{\pi}{2}$ **f.** $x = 3\pi$ **g.** $x = -\pi$ **h.** $x = -\frac{\pi}{2}$

APPLICATION

MAGNITUDES OF STARS

The brightness of a star depends on the amount of energy in the form of light emitted from the star's surface, *luminosity*, and the distance between the star and an observer.

Early Greek astronomers described the brightest stars as stars of the first magnitude and the faintest stars as stars of the sixth magnitude. In the nineteenth century, astronomers noted that first-magnitude stars appeared to be about 100 times as bright as sixth-magnitude stars. This observation gave rise to a scale for measuring and comparing star magnitudes. The scale is a geometric sequence in which the first term is 100 and the sixth term is 1.

If two stars, A and B, have magnitudes m_a and m_b respectively and luminosities l_a and l_b respectively, then, according to the scale, we have $l_a = (\sqrt[5]{100})^{-m_a}$ and $l_b = (\sqrt[5]{100})^{-m_b}$. Thus l_a and l_b can be compared by using $l_a = (\sqrt[5]{100})^{(-1)(m_a - m_b)}l_b$.

EXERCISES

1. Show that a geometric sequence whose first term is 100 and whose sixth term is 1 has common ratio $(\sqrt[5]{100})^{-1}$.
2. Show that $m_a - m_b = -\frac{5}{2} \log \left(\frac{l_a}{l_b}\right)$.
3. If star A is twice as bright as star B and star A has magnitude 3, what is the magnitude of star B?
4. The north star, Polaris, has magnitude 2. If Alpha Centauri is 6.31 times as bright as Polaris, what is its magnitude?
5. The difference in magnitude between Alpha Centauri and the faintest star observable with a 200-inch telescope is about 24. How many times brighter than the faintest star is Alpha Centauri?

Modern astronomy uses computer-controlled telescopes such as the one at the University of Arizona.

Cumulative Review (*Chapters 7–10*)

In Exercises 1 and 2 find all six circular functions of x.

1. $x = \frac{5\pi}{4}$

2. $\cos x = \frac{5}{12}$, $\tan x < 0$

3. a. Convert 420° to radians.
b. Convert 4 radians to degrees.

4. A record is 30 cm in diameter and revolves at $33\frac{1}{3}$ rpm. Find the speed in kilometers per hour of a point on its rim.

5. The terminal side of an angle θ in standard position passes through the point $(1, -2)$. Find the six trigonometric functions of θ.

6. Find two values of θ such that $0° < \theta < 360°$ and $\cos \theta = -0.7880$.

7. The latitude of Chicago is 41.8°N. Find the radius of the circle of latitude through Chicago. (Take the radius of Earth to be 4000 miles.)

8. Find the largest angle of triangle ABC given that $\alpha = 24.5°$, $b = 6.20$, and $c = 4.70$.

9. Find the area of triangle ABC in Exercise 8.

10. Alice runs at 5 m/s and Lyda at 4 m/s. They start from the same point at the same time and run along straight paths that diverge at 120°. How far apart are they after 1 minute?

11. Express $\tan^2 \alpha - \cot^2 \alpha$ in terms of $\cos \alpha$.

12. Find the exact value of $\sin^2 75° - \cos^2 75°$.

13. When angle θ is in standard position, its terminal side passes through the point $(3, -4)$. Find $\tan \frac{1}{2}\theta$.

14. Simplify $\left(\sin \frac{x}{2} + \cos \frac{x}{2}\right)^2 - 1$.

15. Find the solutions of $2 \cos^2 (\theta - 15°) = 1$ in $0° \le \theta < 360°$.

16. The function f has period 8, and

$$f(x) = \begin{cases} -x & \text{if } 0 \le x \le 1 \\ x - 2 & \text{if } 1 < x \le 3 \\ 4 - x & \text{if } 3 < x \le 4. \end{cases}$$

Graph f in $-4 \le x \le 12$ given that f is odd.

In Exercises 17–19 graph each pair of equations in the same coordinate plane.

17. $y = 2 \cos x$; $y = \cos 2x$

18. $y = \sin \pi x$; $y = \sin \left(\pi x - \frac{\pi}{4}\right)$

19. $y = \tan \frac{x}{2}$; $y = \cot \frac{x}{2}$

20. Evaluate $\cos(\text{Sin}^{-1}(-\frac{1}{3}))$.

21. Express $\sin(2\,\text{Tan}^{-1} x)$ without using trigonometric or inverse trigonometric functions.

22. Transform $x^2 + y^2 + 4y = 0$ to polar form.

23. Transform $r(\cos\theta + \sin\theta) = 4$ to rectangular form.

24. Graph the polar curve $r = 4\cos 2\theta$.

In Exercises 25–27, $w = -2 - 2\sqrt{3}i$ and $z = 2(\cos 60° + i \sin 60°)$.

25. Convert w to polar form.

26. Convert z to $x + yi$ form.

27. Express $\frac{w}{z}$ in **(a)** polar form, **(b)** $x + yi$ form.

28. Express the three cube roots of -8 in $x + yi$ form.

29. Identify the curve C: $x = \sin^2 t$, $y = \cos 2t$ by eliminating the parameter. Then graph C.

30. Evaluate $\lim_{x \to 0} \frac{1 - \cos x}{\sin^2 x}$.

In Exercises 31 and 32 differentiate and simplify.

31. $\sin x \cos x$

32. $\tan^2 x$

33. A blimp 1000 m above the ground is traveling toward a football field at 20 km/h. A television camera on the blimp is kept trained on the field. How fast in degrees per minute is the camera rotating when a range finder indicates that the field is 200 m away?

34. Graph $y = (\frac{1}{2})^x$.

35. How much should be invested at 8% per year compounded quarterly to amount to $10,000 after 10 years?

36. Express $\lim_{n \to \infty} \left(\frac{n + 1}{n}\right)^{-2n}$ in terms of e. $\left(\text{Recall that } e = \lim_{n \to \infty} \left(1 + \frac{1}{n}\right)^n.\right)$

37. Evaluate: **a.** $\log_9 \sqrt{3}$ **b.** $e^{\ln 6 - \ln 2}$

38. Solve: **a.** $\log_x 64 = -3$ **b.** $\log_2 5 = \frac{\log_3 5}{\log_3 x}$

39. How many years will it take a deposit to double if it is invested at an annual rate of 8% compounded **(a)** quarterly? **(b)** continuously?

40. A 2.4 mg sample of a radioactive substance decays to 1.8 mg in 5 hours. Find the half-life of the substance.

Chapter 11

Vectors

In this chapter vectors and vector operations are introduced and discussed. Applications of vectors to work, navigation, and mathematical proofs are then investigated.

Vectors

11-1 INTRODUCTION TO VECTORS

Suppose that a power boat whose speed in still water is 18 km/h heads due east across a river flowing due south at 3 km/h. These velocities are called *vector quantities*. Each velocity has a magnitude and a direction. The velocity of the speed boat, for example, has magnitude, 18 km/h, and direction, due east.

A **directed line segment** is a line segment to which a direction has been assigned. If P and Q are the endpoints of the segment, then we call $\overrightarrow{PQ}$ the directed line segment with P as **initial point**, or **tail**, and Q as **terminal point**, or **head**. A directed line segment has both magnitude and direction.

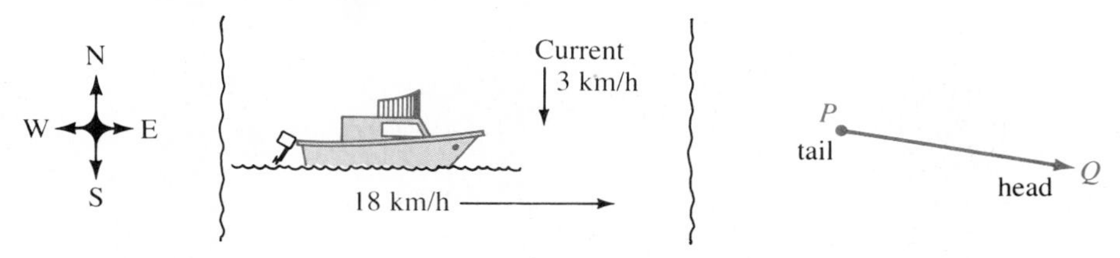

In general, a **vector** is a quantity that has both magnitude and direction. The velocities of the boat and river current as well as the directed line segment $\overrightarrow{PQ}$ are vectors.

The forces of wind, water, and gravity, which can be thought of as vectors, continually act on large structures such as this bridge over the Cape Cod Canal.

Two vectors are **equivalent** if they have the same magnitude and direction. In the following figures $\overrightarrow{RS}$ and $\overrightarrow{TU}$ are equivalent. However, $\overrightarrow{AB}$ and $\overrightarrow{CD}$ are not equivalent since they do not have the same direction. Although $\overrightarrow{EF}$ and $\overrightarrow{GH}$ have the same direction, they are not equivalent since they do not have the same magnitude.

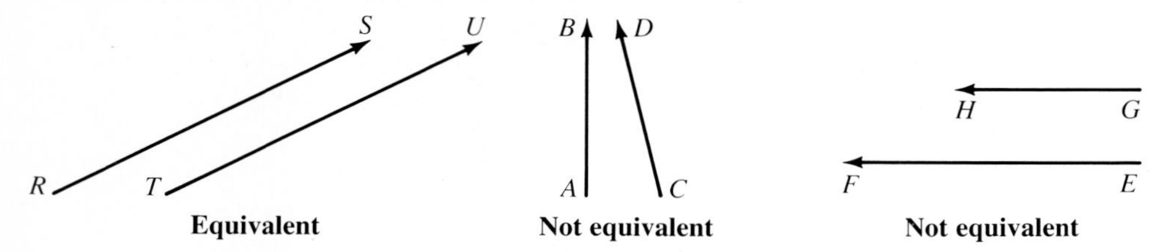

To make our work with vectors easier, we shall use lower-case bold letters such as **v** and **u** to denote vectors. We shall also use italic capital letters with an arrowhead above them, for example $\overrightarrow{PQ}$, to denote vectors. In handwritten work a vector can be written with a small bar or arrow over a letter, such as $\bar{v}$ and $\vec{v}$.

The **sum** of two vectors **u** and **v** is the vector obtained by placing the tail of **v** at the head of **u** and then drawing the arrow whose tail is the tail of **u** and whose head is the head of **v**.

The **zero vector**, denoted **0**, is the vector such that **u** + **0** = **u** for all vectors **u**. The **additive inverse** of **u**, denoted −**u**, is the vector such that **u** + (−**u**) = **0**. The zero vector has magnitude 0 and no direction. The additive inverse of **u** has the same magnitude as **u** but opposite direction.

Subtraction is defined in terms of addition. The **difference** of two vectors **u** and **v**, denoted **u** − **v**, is defined as **u** + (−**v**). This operation is illustrated in the following figures.

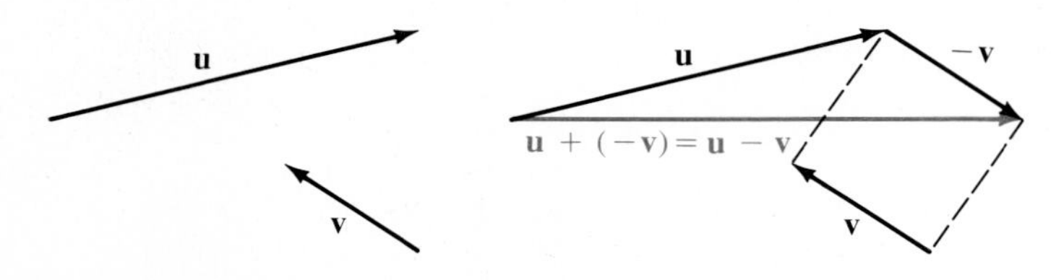

Addition can be extended to any finite number of vectors, as shown in the figure at the right. In other words, addition is associative.

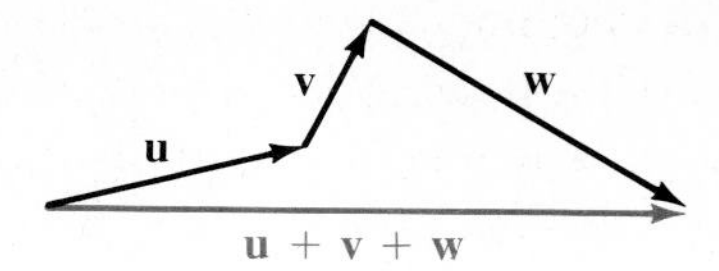

We can define a second vector operation, **scalar multiplication**. If **v** is a vector and k is a real number, then $k\mathbf{v}$, called a **scalar multiple** of **v**, is the vector whose magnitude is $|k|$ times the magnitude of **v** and whose direction is the same as that of **v** if $k > 0$ and whose direction is opposite that of **v** if $k < 0$. If $k = 0$, then $k\mathbf{v} = \mathbf{0}$ and has no direction.

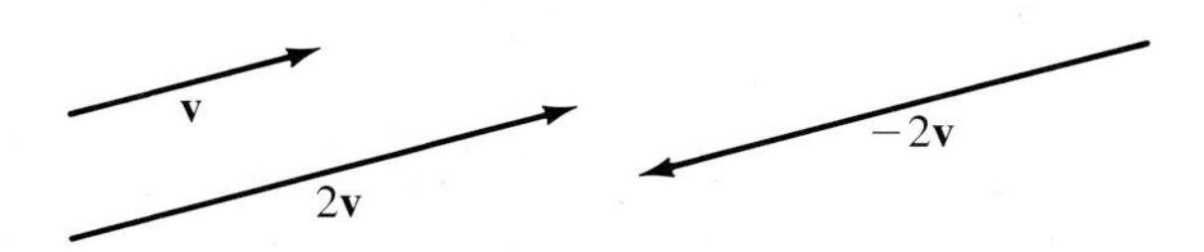

If **u** and **v** are vectors and a and b are both scalars, then $a\mathbf{u} + b\mathbf{v}$ is called a **linear combination** of **u** and **v**.

Two vectors **u** and **v** are **parallel** if $\mathbf{u} = k\mathbf{v}$ for some scalar k. Note that the zero vector is parallel to every vector since $\mathbf{0} = 0\mathbf{v}$ for every vector **v**.

EXAMPLE 1 Use the vectors **u** and **v** shown at the right. **u** **v**
a. Draw $2\mathbf{u} + 3\mathbf{v}$. **b.** Draw $\frac{3}{2}\mathbf{u}$.

Solution **a.** Draw $2\mathbf{u}$ and $3\mathbf{v}$. Then add the resulting pair of vectors.

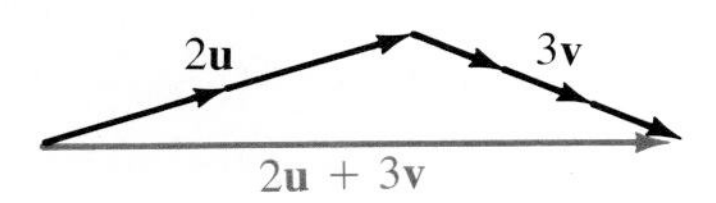

b. First draw $\mathbf{u}'$ whose magnitude is $\frac{1}{2}$ that of **u** and whose direction is the same as **u**. Then draw $3\mathbf{u}'$.

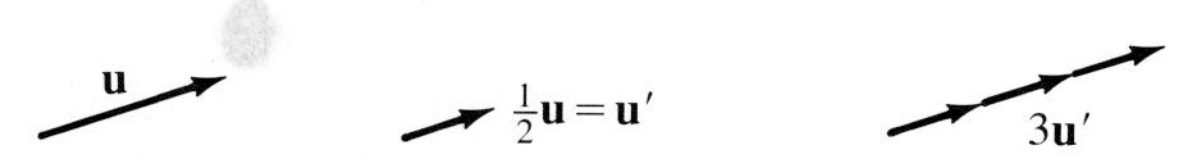

The introduction of a coordinate system enables us to study vectors algebraically.

EXAMPLE 2 If $P(1, 4)$ is the initial point and $T(5, -2)$ is the terminal point of vector $\overrightarrow{PT}$, find the terminal point of a vector equivalent to $\overrightarrow{PT}$ with initial point O, the origin.

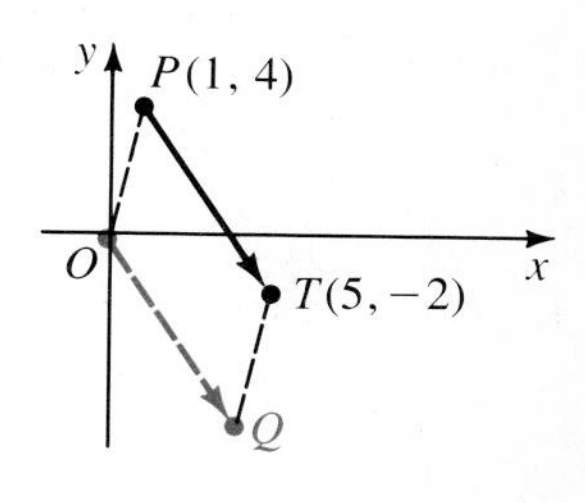

Solution Sketch $\overrightarrow{PT}$. The dashed arrow in the figure at the right shows the vector $\overrightarrow{OQ}$. To find Q, translate T to the left 1 unit and down 4 units. $Q = (5 - 1, -2 - 4) = (4, -6)$.

Example 2 suggests that every vector in the plane has a unique vector equivalent to it with initial point at the origin and terminal point at some point with coordinates (a, b). Such a vector provides a convenient notation for all vectors equivalent to a given vector. In Example 2 we can write $(4, -6)$ for $\overrightarrow{PT}$. When we denote a vector by an ordered pair (a, b), we call a the **first component**, or first coordinate, of the vector and b the **second component**, or second coordinate, of the vector.

EXAMPLE 3 Find the coordinates of $\mathbf{u} + \mathbf{v}$ if $\mathbf{u} = (1, 5)$ and $\mathbf{v} = (3, -2)$.

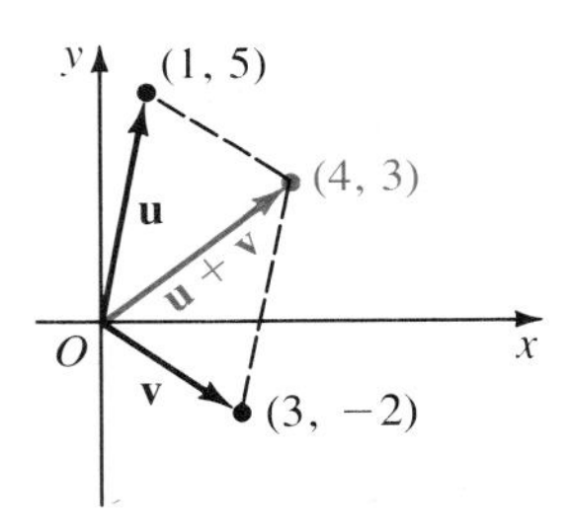

Solution Draw **u** and **v**. Then use the definition of addition: Place the tail of **v** at the head of **u**. Locate the head of **v** to the right 3 units and down 2 units from the tail of **v**. Then the coordinates of $\mathbf{u} + \mathbf{v}$ are $(4, 3)$.

Notice in Example 3 that $\mathbf{u} + \mathbf{v} = (4, 3) = (1 + 3, 5 + (-2))$. In general, if $\mathbf{u} = (a, b)$ and $\mathbf{v} = (c, d)$, then $\mathbf{u} + \mathbf{v} = (a + c, b + d)$. Furthermore, $k\mathbf{u} = (ka, kb)$.

EXAMPLE 4 Find each linear combination if $\mathbf{u} = (-6, 7)$, $\mathbf{v} = (5, 2)$, and $\mathbf{w} = (3, -5)$.

a. $4\mathbf{v} + 2\mathbf{w}$ **b.** $5\mathbf{u} - 6\mathbf{w}$

c. $3(\mathbf{u} + \mathbf{v}) - 3\mathbf{w}$ **d.** $3\mathbf{v} + 8\mathbf{u} - 5\mathbf{w}$

Solution **a.** $4\mathbf{v} = 4(5, 2) = (4 \cdot 5, 4 \cdot 2) = (20, 8)$ and $2\mathbf{w} = 2(3, -5) = (2 \cdot 3, 2 \cdot (-5)) = (6, -10)$

Therefore $4\mathbf{v} + 2\mathbf{w} = (20 + 6, 8 + (-10)) = (26, -2)$. **Answer**

b. $5\mathbf{u} = (5 \cdot (-6), 5 \cdot 7) = (-30, 35)$ and $-6\mathbf{w} = ((-6) \cdot 3, (-6)(-5)) = (-18, 30)$

Therefore $5\mathbf{u} - 6\mathbf{w} = (-30 + (-18), 35 + 30) = (-48, 65)$. **Answer**

c. $3(\mathbf{u} + \mathbf{v}) = 3(-6 + 5, 7 + 2) = 3(-1, 9) = (-3, 27)$ and $-3\mathbf{w} = ((-3)3, (-3)(-5)) = (-9, 15)$

Therefore $3(\mathbf{u} + \mathbf{v}) - 3\mathbf{w} = (-3, 27) + (-9, 15) = (-12, 42)$. **Answer**

d. The first coordinate of $3\mathbf{v} + 8\mathbf{u} - 5\mathbf{w}$ is $3 \cdot 5 + (8)(-6) - 5 \cdot 3 = -48$. The second coordinate is $3 \cdot 2 + 8 \cdot 7 - (5)(-5) = 87$. Hence $3\mathbf{v} + 8\mathbf{u} - 5\mathbf{w} = (-48, 87)$. **Answer**

EXERCISES

In Exercises 1–25 graph paper can be useful.

In Exercises 1–4 draw $\overrightarrow{PT}$. Then find the coordinates of the initial point and the coordinates of the terminal point of the vector that results from translating P and T in the given way.

A

1. $P(-3, 5)$, $T(8, 2)$; to the right 1 unit and down 1 unit

2. $P(-1, 0)$, $T(4, -5)$; to the left 5 units and down 2 units

3. $P(2, 3)$, $T(-6, -1)$; to the left 3 units and up 3 units

4. $P(7, 1)$, $T(3, -8)$; to the right 3 units and up 4 units

In Exercises 5–10 draw $\overrightarrow{PQ}$. Find the coordinates of the vector $\overrightarrow{OQ'}$ having the origin as its initial point and equivalent to $\overrightarrow{PQ}$.

5. $P(4, 2)$, $Q(5, -3)$

6. $P(0, 4)$, $Q(0, -3)$

7. $P(5, 1)$, $Q(8, 1)$

8. $P(2, 2)$, $Q(-2, -2)$

9. $P(-3, -3)$, $Q(0, 0)$

10. $P(-8, 8)$, $Q(8, -8)$

In Exercises 11–18 draw each linear combination of **u** and **v**.

11. $\mathbf{u} - \mathbf{v}$

12. $2\mathbf{u} + \mathbf{v}$

13. $2\mathbf{u} + 2\mathbf{v}$

14. $3\mathbf{u} - 2\mathbf{v}$

15. $\mathbf{u} - \frac{1}{2}\mathbf{v}$

16. $-\mathbf{u} - \mathbf{v}$

17. $2(\mathbf{u} + \mathbf{v})$

18. $2(\mathbf{u} - \mathbf{v})$

u

v

Exercises 11-18

In Exercises 19–22 draw each linear combination of **u** and **v**. Find the coordinates of each linear combination.

19. $\mathbf{u} = (3, 5)$, $\mathbf{v} = (-1, 3)$; $\mathbf{u} + \mathbf{v}$

20. $\mathbf{u} = (4, 0)$, $\mathbf{v} = (0, 6)$; $\mathbf{u} + \mathbf{v}$

21. $\mathbf{u} = (-2, 3)$, $\mathbf{v} = (5, -3)$; $\mathbf{u} - 2\mathbf{v}$

22. $\mathbf{u} = (-4, 5)$, $\mathbf{v} = (4, 0)$; $2\mathbf{u} - \mathbf{v}$

Draw each scalar multiple of $\mathbf{v} = (15, 12)$.

23. $\frac{2}{3}\mathbf{v}$ **24.** $\frac{4}{3}\mathbf{v}$ **25.** $-\frac{1}{3}\mathbf{v}$

Let $\mathbf{u} = (6, 5)$, $\mathbf{v} = (-7, 2)$, $\mathbf{w} = (7, 3)$, and $\mathbf{t} = (-3, -5)$. Find each linear combination.

26. $2\mathbf{u} - 3\mathbf{t}$

27. $-2(5\mathbf{t} - 2\mathbf{w})$

28. $4(\mathbf{u} - \mathbf{v}) + 3(\mathbf{u} + \mathbf{v})$

29. $\mathbf{u} - \mathbf{t} - \mathbf{w}$

30. $5\mathbf{w} - \mathbf{t} + \mathbf{w}$

31. $\mathbf{w} - 2(\mathbf{w} + 3\mathbf{u})$

32. $3\mathbf{u} + 3(\mathbf{t} - \mathbf{v})$

33. $5(\mathbf{w} + \mathbf{t}) - 6(\mathbf{u} - \mathbf{v})$

B

34. Show graphically that if $\mathbf{u} + \mathbf{v} = \mathbf{w}$, then $\mathbf{u} = \mathbf{w} - \mathbf{v}$.

35. Show graphically that $2(\mathbf{u} + \mathbf{v}) = 2\mathbf{u} + 2\mathbf{v}$.

36. Show graphically that $\mathbf{u} + (\mathbf{v} + \mathbf{w}) = (\mathbf{u} + \mathbf{v}) + \mathbf{w}$.

37. Find the sum $\mathbf{u} + \mathbf{v} + \mathbf{w} + \mathbf{t}$ of the vectors $\mathbf{u}$, $\mathbf{v}$, $\mathbf{w}$, and $\mathbf{t}$ shown at the right.

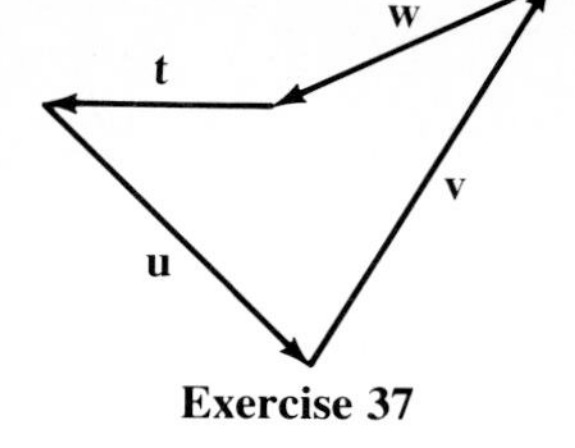

Exercise 37

38. Show that if $\mathbf{v}$ is a nonzero scalar multiple of $\mathbf{u}$, then $\mathbf{u}$ is a nonzero scalar multiple of $\mathbf{v}$. What is the relationship between the two scalars?

In Exercises 39–42, figure $ABCD$ is a parallelogram and E and F are midpoints of $\overline{AB}$ and $\overline{CD}$ respectively. Write each vector in terms of $\overrightarrow{AB}$ and $\overrightarrow{AD}$.

C

39. $\overrightarrow{AC}$

40. $\overrightarrow{DB}$

41. $\overrightarrow{AE} + \overrightarrow{EF}$

42. $\overrightarrow{CF} - \overrightarrow{CB}$

Let $\mathbf{u} = (a, 5b)$ and $\mathbf{v} = (-2b, -3a)$. Solve each vector equation for a and b.

43. $\mathbf{u} + \mathbf{v} = (8, -21)$

44. $3\mathbf{u} - 4\mathbf{v} = (0, 0)$

45. $4\mathbf{u} - 5\mathbf{v} = 2\mathbf{u} + 3\mathbf{v}$

46. $\mathbf{u} + \mathbf{v} = 6\mathbf{u}$

COMPUTER EXERCISES

1. Write a computer program that finds the linear combination, $\mathbf{v}$, of vectors and scalars input by the user.

2. Run the program of Exercise 1 to find each linear combination.

a. $\mathbf{v} = 2(1.5, 4) + 1.3(6, 3.2) - 2(1.5, 4)$

b. $\mathbf{v} = -8(2, 5) + 5.5(0.2, 9.8)$

c. $\mathbf{v} = 3.6(-2, 4.5) + 2.9(-2, 4.5) - (4, 3) + 7.8(0.5, -1.125)$

d. $\mathbf{v} = 0.75(-3, 5) - 6(2, 3) - 4(-1, 0.8) + 7(11, 9) + 0.57(9, 1.4)$

11–2 THREE-DIMENSIONAL SPACE

To locate a point in **three-dimensional space**, also called **three-space**, we draw three mutually perpendicular coordinate axes whose origins coincide. This point of intersection is called the **origin** of the coordinate system. The coordinate axes are usually designated as the x-axis, y-axis, and z-axis. By convention, the positive x-axis points toward the reader, the positive y-axis points to the right, and the positive z-axis points upward (See Figure a.)

The correspondence between ordered triples (x, y, z) of real numbers and points P in three-space is illustrated in Figure b. We accept this correspondence as an axiom. The numbers x, y, and z in (x, y, z) are called the ***x*-**, ***y*-**, and ***z*-coordinates** of P, and P is called the **graph** of (x, y, z).

Each pair of coordinate axes determines a plane. In the xy-plane each point has z-coordinate 0. In the xz-plane each point has y-coordinate 0. In the

yz-plane each point has x-coordinate 0. These **coordinate planes** divide three-space into eight spatial regions called **octants**. The octant in which all three coordinates are positive is called the *first octant*.

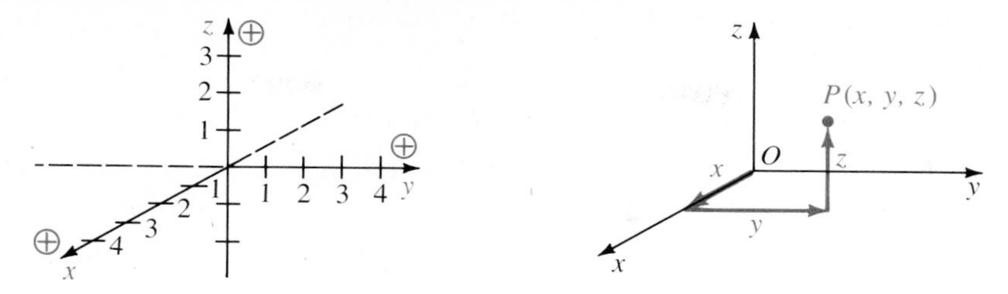

Figure a **Figure b**

EXAMPLE 1 Graph $P(-1, 4, 2)$, $Q(2, -3, 0)$, and $R(0, 0, 3)$.

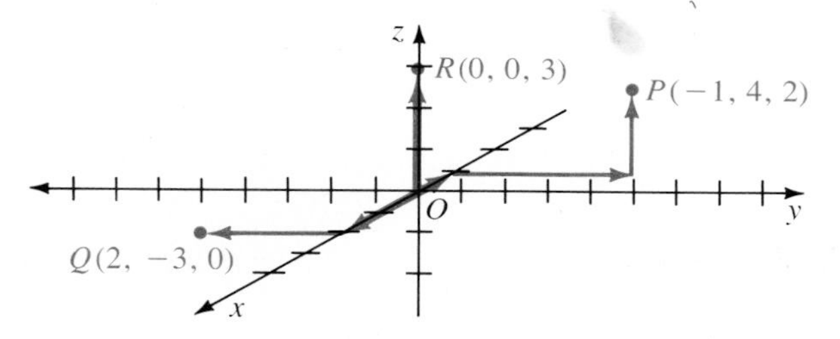

Any two points $P_1(x_1, y_1, z_1)$ and $P_2(x_2, y_2, z_2)$ in three-space determine a coordinate box whose edges are parallel to the coordinate axes. The coordinate box and the distance formula from Chapter 2 will enable us to find the distance between $P_1(x_1, y_1, z_1)$ and $P_2(x_2, y_2, z_2)$. The coordinate box and the vertices we need are shown in Figure a. The two right triangles we need are shown in Figure b.

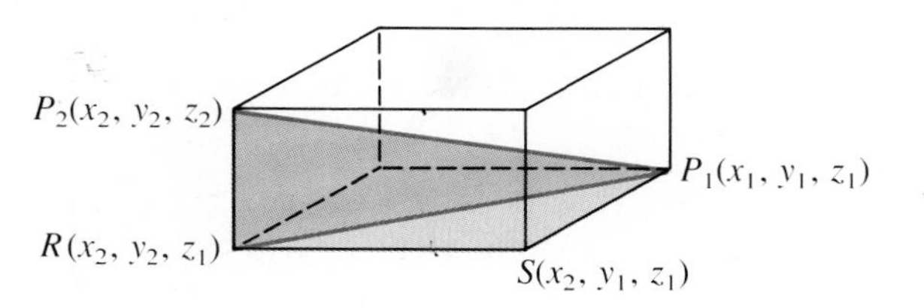

Figure a **Figure b**

Triangle P_1SR is a right triangle in a plane parallel to the xy-plane. Therefore

$$\begin{aligned}(P_1R)^2 &= (P_1S)^2 + RS^2\\ &= |x_2 - x_1|^2 + |y_2 - y_1|^2.\end{aligned}$$

Next we use triangle P_1P_2R and $(P_1R)^2$ to find P_1P_2.

Triangle P_1P_2R is a right triangle. Therefore:

$$(P_1P_2)^2 = (P_1R)^2 + (P_2R)^2$$
$$(P_1P_2)^2 = (P_1R)^2 + |z_2 - z_1|^2$$

Since $(P_1R)^2 = |x_2 - x_1|^2 + |y_2 - y_1|^2$,

$$(P_1P_2)^2 = |x_2 - x_1|^2 + |y_2 - y_1|^2 + |z_2 - z_1|^2.$$

Since $|x|^2 = x^2$ for all real numbers x, we have the following formula for the distance P_1P_2 betweeen P_1 and P_2.

$$P_1P_2 = \sqrt{(x_2 - x_1)^2 + (y_2 - y_1)^2 + (z_2 - z_1)^2}.$$

This argument proves the following theorem, the *distance formula* for three-space.

THEOREM 1 **The Distance Formula for Three-Space**
The distance between $P_1(x_1, y_1, z_1)$ and $P_2(x_2, y_2, z_2)$ is given by

$$P_1P_2 = \sqrt{(x_2 - x_1)^2 + (y_2 - y_1)^2 + (z_2 - z_1)^2}.$$

The midpoint formula for points in the plane can be extended to points in three-space.

THEOREM 2 **The Midpoint Formula for Three-Space**
The midpoint of the line segment with endpoints $P_1(x_1, y_1, z_1)$ and $P_2(x_2, y_2, z_2)$ has coordinates

$$\left(\frac{x_1 + x_2}{2}, \frac{y_1 + y_2}{2}, \frac{z_1 + z_2}{2}\right).$$

EXAMPLE 2 Let points $P(-2, 10, -3)$ and $Q(1, 4, -5)$ be given.

a. Find PQ.

b. Find the midpoint of $\overline{PQ}$.

Solution **a.** Use the distance formula.

$$PQ = \sqrt{(-2 - 1)^2 + (10 - 4)^2 + (-3 - (-5))^2}$$
$$= \sqrt{(-3)^2 + 6^2 + 2^2}$$
$$= \sqrt{49}$$
$$PQ = 7 \quad \textbf{Answer}$$

b. Use the midpoint formula. The coordinates of the midpoint of $\overline{PQ}$ are

$$\left(\frac{-2 + 1}{2}, \frac{10 + 4}{2}, \frac{-3 - 5}{2}\right) = \left(-\frac{1}{2}, 7, -4\right). \quad \textbf{Answer}$$

Just as a vector in the plane, or two-space, can be represented by an ordered pair, each vector in three-space can be represented by an ordered triple. For example, if $\mathbf{v} = \overrightarrow{AB}$ where A is the point $(1, -2, 5)$ and B is $(2, 4, 3)$, then

$$\mathbf{v} = (2 - 1, 4 - (-2), 3 - 5) = (1, 6, -2).$$

In general, given $P_1(x_1, y_1, z_1)$ and $P_2(x_2, y_2, z_2)$, we have

$$\overrightarrow{P_1P_2} = (x_2 - x_1, y_2 - y_1, z_2 - z_1).$$

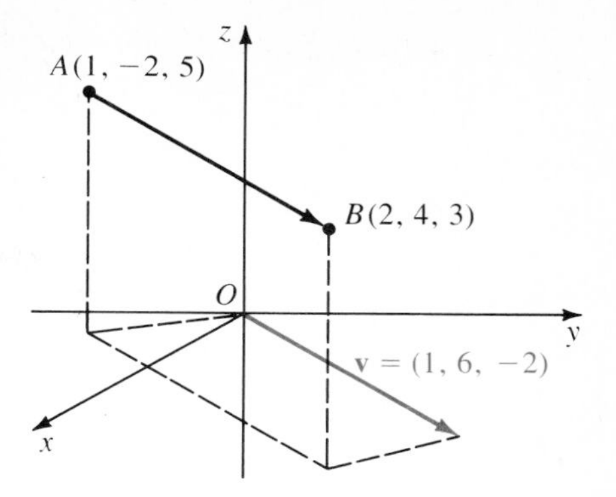

Addition and scalar multiplication of vectors in three-space can be carried out coordinate by coordinate.

EXAMPLE 3 Let $\mathbf{u} = (3, 2, -3)$, $\mathbf{v} = (0, 8, 10)$, and $\mathbf{w} = (-4, 3, 7)$. Find each linear combination.

a. $2\mathbf{u} + 3\mathbf{v}$ **b.** $4\mathbf{u} - 5\mathbf{w}$ **c.** $3\mathbf{w} + 4(\mathbf{u} + \mathbf{v})$

Solution

a. $2\mathbf{u} = 2(3, 2, -3) = (2 \cdot 3, 2 \cdot 2, (2)(-3)) = (6, 4, -6)$
$3\mathbf{v} = 3(0, 8, 10) = (3 \cdot 0, 3 \cdot 8, 3 \cdot 10) = (0, 24, 30)$
Therefore $2\mathbf{u} + 3\mathbf{v} = (6 + 0, 4 + 24, -6 + 30) = (6, 28, 24)$. **Answer**

b. $4\mathbf{u} = 4(3, 2, -3) = (4 \cdot 3, 4 \cdot 2, (4)(-3)) = (12, 8, -12)$
$-5\mathbf{w} = -5(-4, 3, 7) = ((-5)(-4), (-5)(3), (-5)(7)) = (20, -15, -35)$
Therefore $4\mathbf{u} - 5\mathbf{w} = (12 + 20, 8 + (-15), -12 + (-35)) = (32, -7, -47)$. **Answer**

c. $\mathbf{u} + \mathbf{v} = (3 + 0, 2 + 8, -3 + 10) = (3, 10, 7)$. Thus $4(\mathbf{u} + \mathbf{v}) = (12, 40, 28)$. Therefore $3\mathbf{w} + 4(\mathbf{u} + \mathbf{v}) = (-12 + 12, 9 + 40, 21 + 28) = (0, 49, 49)$. **Answer**

EXERCISES

Graph paper can be useful in Exercises 1–16.
In Exercises 1–12 graph each ordered triple.

A

1. $(5, -2, -3)$ **2.** $(4, -1, 5)$ **3.** $(0, 5, -1)$ **4.** $(-3, 6, 0)$

5. $(4, 0, 3)$ **6.** $(1, -4, 7)$ **7.** $(0, 0, -2)$ **8.** $(-3, -4, 5)$

9. $(5, 9, 2)$ **10.** $(3, 7, -6)$ **11.** $(1, 8, 0)$ **12.** $(-1, 8, 3)$

Graph P, Q, and R and draw the triangle PQR.

13. $P(3, 0, 0)$, $Q(0, 5, 0)$, $R(0, 0, -4)$ **14.** $P(6, 0, 0)$, $Q(0, 6, 0)$, $R(0, 0, 6)$

15. $P(2, 3, 0)$, $Q(0, 0, 7)$, $R(0, 6, 0)$ **16.** $P(-3, 5, 0)$, $Q(6, 6, 6)$, $R(0, 0, 0)$

Find PQ and the coordinates of the midpoint of $\overline{PQ}$.

17. $P(6, -4, 3)$, $Q(2, 0, 1)$ **18.** $P(8, 1, 2)$, $Q(9, -3, -7)$

19. $P(4, -3, 12)$, $Q(0, -1, 3)$ **20.** $P(-3, 10, 11)$, $Q(2, 4, 9)$

21. $P(13, -4, 9)$, $Q(1, 3, -4)$ **22.** $P(1, 3, -4)$, $Q(7, 10, 2)$

Let $\mathbf{u} = (1, -3, 2)$, $\mathbf{v} = (-7, 6, 0)$, $\mathbf{t} = (3, 4, -1)$, and $\mathbf{w} = (7, -2, -5)$. Find each linear combination.

23. $4\mathbf{u} + \mathbf{v}$ **24.** $3\mathbf{u} - 2\mathbf{v}$

25. $3(\mathbf{u} + \mathbf{t})$ **26.** $\mathbf{u} + 4\mathbf{v}$

27. $2(\mathbf{u} - \mathbf{v} + \mathbf{t})$ **28.** $3\mathbf{w} + 2(\mathbf{t} + \mathbf{v})$

29. $(\mathbf{t} - \mathbf{w}) + 2\mathbf{t}$ **30.** $2(3\mathbf{u} - \mathbf{t}) + \mathbf{u}$

Classify triangle ABC as isosceles, equilateral, or scalene.

B **31.** $A(2, 0, 1)$, $B(-1, 5, 0)$, $C(-1, 0, -3)$

32. $A(-3, 3, -1)$, $B(1, 3, 2)$, $C(-1, 5, 1)$

33. $A(1, 6, -4)$, $B(7, 0, -4)$, $C(1, 0, 2)$

34. $A(0, 0, 12)$, $B(3, 4, 0)$, $C(-4, 3, 0)$

Use the Pythagorean theorem to determine whether triangle ABC is a right triangle or is not a right triangle.

35. $A(10, 5, 1)$, $B(-2, 2, 5)$, $C(-2, 5, 1)$

36. $A(0, 0, 12)$, $B(0, 9, 0)$, $C(8, 9, 0)$

37. Given $R(a, 0, 0)$, $S(0, a, 0)$, and $T(0, 0, a)$, show that the triangle whose vertices are the midpoints of the sides of triangle RST is equilateral.

38. Find the coordinates of two points on the x-axis that are 7 units from $(4, 6, 3)$.

39. Find the coordinates of two points on the y-axis that are 11 units from $(6, 7, -9)$.

40. Write an equation for the set of all points in three-space that are r units from (c_1, c_2, c_3). What is that set of points?

C **41.** Let $P(x_1, y_1, z_1)$ and $Q(x_2, y_2, z_2)$ be two points and let a and b be positive integers with $a < b$. Show that the coordinates of a point R between P and Q dividing $\overline{PQ}$ such that $\overrightarrow{PR} = \frac{a}{b}\overrightarrow{PQ}$ are

$$\left(\frac{a(x_2 - x_1) + bx_1}{b}, \frac{a(y_2 - y_1) + by_1}{b}, \frac{a(z_2 - z_1) + bz_1}{b}\right).$$

42. Show that if $A(2, 2, 0)$, $B(2, -2, 6)$, $C(4, 0, 0)$, and $D(0, 0, 6)$ are four points in three-space, then $AB = CD$ and $AD = BC$. Is the figure formed by joining A, B, C, and D in that order a parallelogram?

11-3 THE NORM OF A VECTOR

The magnitude of a vector $\mathbf{v}$ is also called its **norm** and is denoted by $\|\mathbf{v}\|$. The distance formulas for the plane and three-space can be used to find the norm of a vector.

DEFINITION

If $\mathbf{v} = (v_1, v_2)$, then $\|\mathbf{v}\| = \sqrt{v_1^2 + v_2^2}$.

If $\mathbf{v} = (v_1, v_2, v_3)$, then $\|\mathbf{v}\| = \sqrt{v_1^2 + v_2^2 + v_3^2}$.

Notice that the norm of a vector is a nonnegative scalar.

EXAMPLE 1 Let $\mathbf{u} = (4, -3)$, $\mathbf{v} = (3, 4)$, $\mathbf{w} = (6, -8)$, $\mathbf{t} = (4, 4, 7)$. Find the norm of each vector.

a. $\mathbf{u}$ **b.** $\mathbf{t}$ **c.** $\mathbf{v} + \mathbf{w}$

Solution **a.** $\|\mathbf{u}\| = \sqrt{4^2 + (-3)^2} = \sqrt{25} = 5$

b. $\|\mathbf{t}\| = \sqrt{4^2 + 4^2 + 7^2} = \sqrt{81} = 9$

c. $\mathbf{v} + \mathbf{w} = (3 + 6, 4 + (-8)) = (9, -4)$

Thus $\|\mathbf{v} + \mathbf{w}\| = \sqrt{9^2 + (-4)^2} = \sqrt{97}$.

A vector with norm 1 is called a **unit vector**.

EXAMPLE 2 Find a unit vector parallel to $\mathbf{v} = (3, 4)$ and having direction opposite to that of $\mathbf{v}$.

Solution The figure at the right shows $\mathbf{v}$ and a unit vector $\mathbf{v}'$ having direction opposite to that of $\mathbf{v}$. If $\mathbf{v}'$ and $\mathbf{v}$ are parallel, then

$$\mathbf{v}' = k\mathbf{v}$$
$$\|\mathbf{v}'\| = \|k\mathbf{v}\|$$
$$1 = \sqrt{(3k)^2 + (4k)^2} = 5|k|.$$

Hence $|k| = \frac{1}{5}$.

Since $\mathbf{v}'$ and $\mathbf{v}$ have opposite directions, choose $k = -\frac{1}{5}$. Then

$$\mathbf{v}' = -\tfrac{1}{5}(3, 4) = \left(-\tfrac{3}{5}, -\tfrac{4}{5}\right). \quad \textbf{Answer}$$

Notice that in part (c) of Example 1 the sum was found before the norm was computed. It is not true in general that $\|\mathbf{u} + \mathbf{v}\| = \|\mathbf{u}\| + \|\mathbf{v}\|$. A relationship that does hold in general is given in part 4 of the next theorem.

THEOREM 3 Let $\mathbf{u}$ and $\mathbf{v}$ be vectors in a vector space V and let r be a scalar.

1. $\|\mathbf{v}\| \geq 0$

2. $\|\mathbf{v}\| = 0$ if and only if $\mathbf{v} = \mathbf{0}$.

3. $\|r\mathbf{v}\| = |r| \cdot \|\mathbf{v}\|$

4. $\|\mathbf{u} + \mathbf{v}\| \leq \|\mathbf{u}\| + \|\mathbf{v}\|$

For vectors in the plane, the following figures illustrate parts 3 and 4 of the theorem. Part 4 is called the **triangle inequality** for vectors. This inequality states in vector terms that the length of one side of a triangle is less than the sum of the lengths of the other two sides.

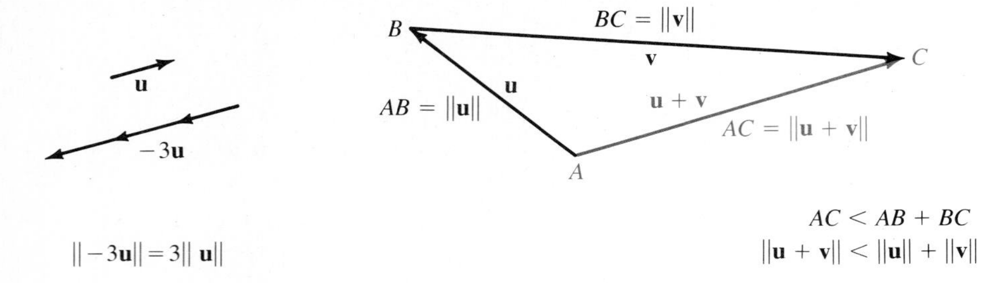

A proof of the triangle inequality is outlined in Exercises 29–32.

EXAMPLE 3 Verify the triangle inequality for $\mathbf{u} = (4, 0)$ and $\mathbf{v} = (5, 12)$.

Solution $\|\mathbf{u}\| = \sqrt{4^2 + 0^2} = 4$ and
$\|\mathbf{v}\| = \sqrt{5^2 + 12^2} = 13$
$\mathbf{u} + \mathbf{v} = (4 + 5, 0 + 12) = (9, 12)$. Thus $\|\mathbf{u} + \mathbf{v}\| = \sqrt{9^2 + 12^2} = 15$.
Since $15 \le 4 + 13$, $\|\mathbf{u} + \mathbf{v}\| \le \|\mathbf{u}\| + \|\mathbf{v}\|$.

EXERCISES

In Exercises 1–12, $\mathbf{u} = (3, 4)$, $\mathbf{v} = (-2, 6)$, $\mathbf{t} = (-5, 12)$, and $\mathbf{w} = (1, 1)$. Find the norm of each vector.

A

1. $\mathbf{t}$ **2.** $\mathbf{w}$ **3.** $\mathbf{u} - \mathbf{v}$

4. $\mathbf{u} + \mathbf{t}$ **5.** $3\mathbf{u}$ **6.** $-2\mathbf{t}$

7. $-5\mathbf{w}$ **8.** $-\frac{1}{2}\mathbf{t}$ **9.** $\frac{1}{2}\mathbf{u} - 3\mathbf{t}$

10. $3\mathbf{w} + \mathbf{u}$ **11.** $\dfrac{\mathbf{t}}{\|\mathbf{t}\|}$ **12.** $\dfrac{\mathbf{v}}{\|\mathbf{v}\|}$

In Exercises 13–16, $\mathbf{u} = (11, -2, 5)$ and $\mathbf{v} = (-1, 12, 5)$. Find the norm of each vector.

13. $3\mathbf{u}$ **14.** $2\mathbf{u} - \mathbf{v}$

15. $\mathbf{u} + 3\mathbf{v}$ **16.** $3\mathbf{v} - 2\mathbf{u}$

17. If $\mathbf{u} = (3, 5, -8)$, verify that $\|2\mathbf{u}\| = 2\|\mathbf{u}\|$ and that $\|-3\mathbf{u}\| = 3\|\mathbf{u}\|$.

18. If $\mathbf{u} = (-3, 7, 1)$ and $\mathbf{v} = (5, -2, 8)$, verify that $\|\mathbf{u} + \mathbf{v}\| \le \|\mathbf{u}\| + \|\mathbf{v}\|$.

19. Find a unit vector parallel to $\mathbf{v} = (5, 12)$ and having the same direction as $\mathbf{v}$.

20. Find a vector with norm 10 parallel to $\mathbf{v} = (8, -3)$ but having direction opposite to that of $\mathbf{v}$.

B

21. Prove that for all vectors in three-space $\mathbf{v} = \mathbf{0}$ if and only if $\|\mathbf{v}\| = 0$.
22. Prove for all vectors $\mathbf{v}$ in three-space and all scalars r that $\|r\mathbf{v}\| = |r| \cdot \|\mathbf{v}\|$.
23. Prove that if $\mathbf{u}$ and $\mathbf{v}$ are vectors in three-space, then $\|\mathbf{u} - \mathbf{v}\| = \|\mathbf{v} - \mathbf{u}\|$.
24. Use the triangle inequality to show that $\|\mathbf{u}\| - \|\mathbf{v}\| \leq \|\mathbf{u} - \mathbf{v}\|$.
25. Use the triangle inequality to show that $\|\mathbf{u} - \mathbf{w}\| \leq \|\mathbf{u} - \mathbf{v}\| + \|\mathbf{v} - \mathbf{w}\|$ for every vector $\mathbf{u}$, $\mathbf{v}$, and $\mathbf{w}$ in three-space. (*Hint:* Write $\mathbf{u} - \mathbf{w}$ in terms of $\mathbf{u}$, $\mathbf{v}$, and $\mathbf{w}$. Apply the triangle inequality.)
26. Under what conditions placed on the vectors $\mathbf{u}$ and $\mathbf{v}$ is it true that $\|\mathbf{u} + \mathbf{v}\| = \|\mathbf{u}\| + \|\mathbf{v}\|$?
27. Prove that if $\mathbf{u}$ and $\mathbf{v}$ are parallel and r and s are scalars, then $r\mathbf{u} + s\mathbf{v}$ is parallel to both $\mathbf{u}$ and $\mathbf{v}$.
28. Let $\mathbf{u}$ and $\mathbf{v}$ be vectors in three-space with the same initial point and let θ be the angle between them. Use the law of cosines to write $\|\mathbf{u} - \mathbf{v}\|^2$ in terms of $\|\mathbf{u}\|$, $\|\mathbf{v}\|$, and θ.

Exercises 29–32 outline a proof of the triangle inequality for vectors in the plane and vectors in three-space.

C

29. Write $\|\mathbf{u} + \mathbf{v}\|^2$ in terms of $\|\mathbf{u}\|$, $\|\mathbf{v}\|$, and θ, the angle between $\mathbf{u}$ and $\mathbf{v}$. (*Hint:* Use the law of cosines.)
30. Use the fact that $\cos \theta \leq 1$ to show that $\|\mathbf{u} + \mathbf{v}\|^2 \leq (\|\mathbf{u}\| + \|\mathbf{v}\|)^2$.
31. Show that if x and y are nonnegative real numbers, then $x \leq y$ if and only if $x^2 \leq y^2$.
32. Show that $\|\mathbf{u} + \mathbf{v}\| \leq \|\mathbf{u}\| + \|\mathbf{v}\|$.

COMPUTER EXERCISES

1. Write a computer program to find the magnitude and direction (in degrees) of the sum of two vectors whose magnitude and direction (in degrees) are input by the user.
2. Run the program for the given vectors.

   a. magnitude = 3, direction 0°
   magnitude = 4, direction 90°

   b. magnitude = 5, direction 30°
   magnitude = 2, direction 125°

   c. magnitude = 8, direction 15°
   magnitude = 8, direction 85°

   d. magnitude = 10, direction 65°
   magnitude = 7, direction 20°

11-4 THE DOT PRODUCT

Operations defined on vectors thus far include:

Addition	$\mathbf{u}, \mathbf{v} \longrightarrow \mathbf{u} + \mathbf{v}$
Scalar multiplication	$k, \mathbf{v} \longrightarrow k\mathbf{v}$
Norm	$\mathbf{v} \longrightarrow \|\mathbf{v}\|$

In this section we shall define the *dot product* of two vectors, also called the *inner product*, or *scalar product*, of two vectors.

Consider vectors **u** and **v** and the angle θ between them, as shown at the right. A triangle whose sides have lengths $\|\mathbf{u}\|$, $\|\mathbf{v}\|$, and $\|\mathbf{u} - \mathbf{v}\|$ is formed from the vectors **u**, **v**, and $\mathbf{u} - \mathbf{v}$. From the law of cosines, we have

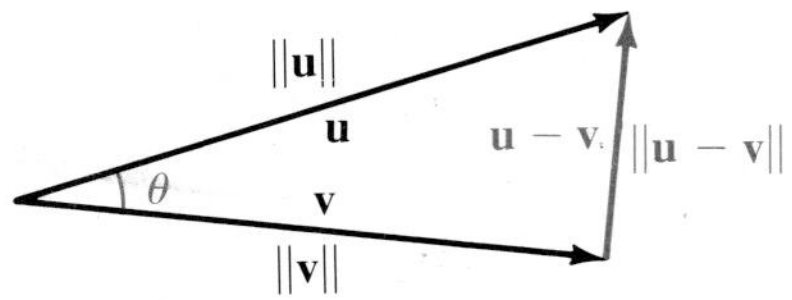

$$\|\mathbf{u} - \mathbf{v}\|^2 = \|\mathbf{u}\|^2 + \|\mathbf{v}\|^2 - 2\,\|\mathbf{u}\|\,\|\mathbf{v}\| \cos \theta.$$

If $\theta = 90°$, then $\cos \theta = 0$ and we have that

$$\|\mathbf{u} - \mathbf{v}\|^2 = \|\mathbf{u}\|^2 + \|\mathbf{v}\|^2.$$

This last equation is the Pythagorean theorem. With this discussion in mind we make the following definition.

DEFINITION

Let **u** and **v** be vectors and let θ be the angle between them. The **dot product** of **u** with **v**, denoted by $\mathbf{u} \cdot \mathbf{v}$, is defined as

$$\mathbf{u} \cdot \mathbf{v} = \|\mathbf{u}\|\,\|\mathbf{v}\| \cos \theta.$$

Thus according to this definition:

$$\|\mathbf{u} - \mathbf{v}\|^2 = \|\mathbf{u}\|^2 - 2\mathbf{u} \cdot \mathbf{v} + \|\mathbf{v}\|^2$$

We now prove formulas that enable us to find the dot product of two vectors from their components.

THEOREM 4 If $\mathbf{u} = (u_1, u_2)$ and $\mathbf{v} = (v_1, v_2)$, then

$$\mathbf{u} \cdot \mathbf{v} = u_1v_1 + u_2v_2.$$

If $\mathbf{u} = (u_1, u_2, u_3)$ and $\mathbf{v} = (v_1, v_2, v_3)$, then

$$\mathbf{u} \cdot \mathbf{v} = u_1v_1 + u_2v_2 + u_3v_3.$$

We shall prove the formula for vectors in the plane. The proof for vectors in three-space is similar.

Proof: We use the diagram above. From the law of cosines,

$$\|\mathbf{u} - \mathbf{v}\|^2 = \|\mathbf{u}\|^2 + \|\mathbf{v}\|^2 - 2\|\mathbf{u}\|\,\|\mathbf{v}\| \cos \theta.$$

Therefore

$$2\|\mathbf{u}\|\,\|\mathbf{v}\| \cos\theta = \|\mathbf{u}\|^2 + \|\mathbf{v}\|^2 - \|\mathbf{u} - \mathbf{v}\|^2.$$

Now $\|\mathbf{u} - \mathbf{v}\|^2 = (u_1 - v_1)^2 + (u_2 - v_2)^2$, $\|\mathbf{u}\|^2 = u_1{}^2 + u_2{}^2$, and $\|\mathbf{v}\|^2 = v_1{}^2 + v_2{}^2$. Hence

$$2\|\mathbf{u}\|\,\|\mathbf{v}\| \cos\theta = u_1{}^2 + u_2{}^2 + v_1{}^2 + v_2{}^2 - [(u_1 - v_1)^2 + (u_2 - v_2)^2].$$

After expanding the squares on the right side of the equation above, we have $2\|\mathbf{u}\|\,\|\mathbf{v}\| \cos\theta = 2u_1v_1 + 2u_2v_2$. Therefore $\mathbf{u} \cdot \mathbf{v} = u_1v_1 + u_2v_2$. ■

EXAMPLE 1 Compute the dot product of each pair of vectors.

a. $\mathbf{u} = (-3, 4)$ and $\mathbf{v} = (8, 6)$ **b.** $\mathbf{t} = (-3, 2, 5)$ and $\mathbf{w} = (5, 3, 7)$

Solution Since ordered pairs and triples are given, use Theorem 4.

a. $\mathbf{u} \cdot \mathbf{v} = (-3)(8) + 4 \cdot 6 = 0$

b. $\mathbf{t} \cdot \mathbf{w} = (-3)(5) + 2 \cdot 3 + 5 \cdot 7 = 26$

We say that two vectors are **orthogonal** if the angle between them is 90° or if either of them is $\mathbf{0}$. Thus $\mathbf{u}$ and $\mathbf{v}$ are orthogonal if and only if $\mathbf{u} \cdot \mathbf{v} = 0$.

EXAMPLE 2 Determine whether $\mathbf{u}$ and $\mathbf{v}$ are orthogonal.

a. $\mathbf{u} = (-4.5, 6)$; $\mathbf{v} = (10.2, 6.5)$

b. $\mathbf{u} = (-4, 6, 3)$; $\mathbf{v} = (3, 1, 2)$

Solution **a.** $\mathbf{u} \cdot \mathbf{v} = (-4.5)(10.2) + 6 \cdot 6.5 = -6.9$ No

b. $\mathbf{u} \cdot \mathbf{v} = (-4)(3) + 6 \cdot 1 + 3 \cdot 2 = 0$ Yes

The following theorem states some of the properties of the dot product.

THEOREM 5 Let $\mathbf{u}$, $\mathbf{v}$, and $\mathbf{w}$ be vectors and let r be a scalar.

1. $\mathbf{u} \cdot \mathbf{v} = \mathbf{v} \cdot \mathbf{u}$ (Commutative Property)
2. $r(\mathbf{u} \cdot \mathbf{v}) = (r\mathbf{u}) \cdot \mathbf{v} = \mathbf{u} \cdot (r\mathbf{v})$ (Associative Property)
3. $\mathbf{u} \cdot (\mathbf{v} + \mathbf{w}) = \mathbf{u} \cdot \mathbf{v} + \mathbf{u} \cdot \mathbf{w}$ (Distributive Property)
4. $|\mathbf{u} \cdot \mathbf{v}| \le \|\mathbf{u}\|\,\|\mathbf{v}\|$ **(Cauchy-Schwarz Inequality)**
5. $\mathbf{v} \cdot \mathbf{v} = \|\mathbf{v}\|^2$ (Norm Property)

We shall prove the Cauchy-Schwarz inequality. The proofs of the other parts of Theorem 5 are requested in Exercises 29–32.

Proof: By the definition of the dot product,

$$|\mathbf{u} \cdot \mathbf{v}| = \big|\|\mathbf{u}\|\,\|\mathbf{v}\| \cos\theta\big|$$

Hence $|\mathbf{u} \cdot \mathbf{v}| = \|\mathbf{u}\|\,\|\mathbf{v}\|\,|\cos\theta|$. Since for all θ, $|\cos\theta| \le 1$,

$$|\mathbf{u} \cdot \mathbf{v}| \le \|\mathbf{u}\|\,\|\mathbf{v}\|. \quad ■$$

Letting $\mathbf{u} = (u_1, u_2)$ and $\mathbf{v} = (v_1, v_2)$, we can obtain another form of the Cauchy-Schwarz inequality:

$$(u_1v_1 + u_2v_2)^2 \le (u_1^2 + u_2^2)(v_1^2 + v_2^2)$$

The dot product has many applications. In particular we can use it to find the measure of an angle between two given vectors.

EXAMPLE 3 Find the measure of the angle between $\mathbf{u} = (3, 4, 5)$ and $\mathbf{v} = (-3, 2, 0)$ to the nearest degree.

Solution Use $\mathbf{u} \cdot \mathbf{v} = \|\mathbf{u}\| \, \|\mathbf{v}\| \cos \theta$.

$\|\mathbf{u}\| = \sqrt{3^2 + 4^2 + 5^2} = \sqrt{50} = 5\sqrt{2}$ and $\|\mathbf{v}\| = \sqrt{(-3)^2 + 2^2 + 0^2} = \sqrt{13}$

$\mathbf{u} \cdot \mathbf{v} = (3)(-3) + 4 \cdot 2 + 5 \cdot 0 = -1$. Thus $-1 = (5\sqrt{2})(\sqrt{13}) \cos \theta$ and

$$\cos \theta = \frac{-1}{(5\sqrt{2})(\sqrt{13})} \approx -0.0392.$$

To the nearest degree the measure of θ is 92°. **Answer**

In general, if $\mathbf{u}$ and $\mathbf{v}$ are vectors in the plane or in three-space and neither vector is $\mathbf{0}$, then the measure of the angle θ between them is given by

$$\cos \theta = \frac{\mathbf{u} \cdot \mathbf{v}}{\|\mathbf{u}\| \, \|\mathbf{v}\|}.$$

EXERCISES

Find the dot product of each pair of vectors. State whether or not the vectors are orthogonal.

A

1. $(-1, 3), (7, 5)$ **2.** $(8, -2), (5, 20)$

3. $(4, -9), (-18, 8)$ **4.** $(6, -3), (-2, 7)$

5. $(12, 9), (3, -4)$ **6.** $(9, 6), (-6, 4)$

7. $(0, -5), (17, 8)$ **8.** $(9, 0), (0, -15)$

9. $(6a, 4b), (-2b, 3a)$ **10.** $(a - b, c), (c, b - a)$

11. $(3, 4, 1), (4, -2, -4)$ **12.** $(0, 8, -4), (9, 7, 15)$

13. $(-8, 0, 6), (5, 3, 7)$ **14.** $(9, 1, 7), (-2, -5, 6)$

Find the measure of the angle between the given pairs of vectors to the nearest degree.

15. $(-3, 5, 6), (1, 1, 3)$ **16.** $(-3, 3, 6), (4, 2, 1)$

17. $(3a, 2a), (2a, -3a)$ **18.** $(3, 6), (9, 18)$

Verify the Cauchy-Schwarz inequality for each pair of vectors.

19. $\mathbf{u} = (7, 24)$; $\mathbf{v} = (12, -5)$ **20.** $\mathbf{u} = (4, -3)$; $\mathbf{v} = (8, -6)$

21. $\mathbf{u} = (1, 2, -2)$; $\mathbf{v} = (3, 6, 2)$ **22.** $\mathbf{u} = (-6, 9, 4)$; $\mathbf{v} = (-2, 10, 11)$

Find k so that the given vectors are **(a)** parallel and **(b)** perpendicular.

23. $(-4, 7), (8, k)$ **24.** $(6, k), (-1, -4)$

25. $(2k, 10), (-3, 6)$ **26.** $(5, 8), (-2, -k)$

27. $(k, 4), (k, -9)$ **28.** $(1, k), (6k, -k)$

In Exercises 29–32 prove each part of Theorem 5 for vectors in three-space.

B **29.** Part 1 **30.** Part 2 **31.** Part 3 **32.** Part 5

33. Prove that if **u** and **v** are vectors in three-space and r is a nonzero scalar, then **u** and **v** are orthogonal if and only if r**u** and **v** are orthogonal.

34. Prove that vectors **u** and **v** in three-space are orthogonal if and only if $\|\mathbf{u} + \mathbf{v}\| = \|\mathbf{u} - \mathbf{v}\|$.

35. Prove that $(\mathbf{v} + \mathbf{t}) \cdot (\mathbf{v} - \mathbf{t}) = \|\mathbf{v}\|^2 - \|\mathbf{t}\|^2$.

C **36.** Use the figure at the right to show that

$$\tfrac{1}{4}\|\mathbf{u} + \mathbf{v}\|^2 - \tfrac{1}{4}\|\mathbf{u} - \mathbf{v}\|^2 = \mathbf{u} \cdot \mathbf{v}.$$

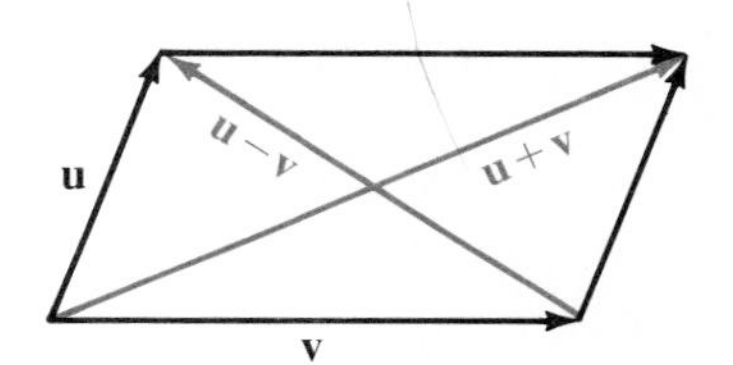

11-5 BASIS VECTORS

Every vector **v** in the plane can be written uniquely as a linear combination of the two unit vectors $\mathbf{i} = (1, 0)$ and $\mathbf{j} = (0, 1)$. For example, if $\mathbf{v} = (3, 2)$, then

$$\mathbf{v} = (3, 2) = 3(1, 0) + 2(0, 1) = 3\mathbf{i} + 2\mathbf{j}.$$

We call $\{\mathbf{i}, \mathbf{j}\}$ the **standard basis** for vectors in the plane.

Every vector **v** in three-space can be uniquely written as a linear combination of the three unit vectors $\mathbf{i} = (1, 0, 0)$, $\mathbf{j} = (0, 1, 0)$, and $\mathbf{k} = (0, 0, 1)$. If $\mathbf{v} = (2, 3, 4)$, then

$$\begin{aligned}\mathbf{v} = (2, 3, 4) &= (2, 0, 0) + (0, 3, 0) + (0, 0, 4)\\ &= 2(1, 0, 0) + 3(0, 1, 0) + 4(0, 0, 1)\\ &= 2\mathbf{i} + 3\mathbf{j} + 4\mathbf{k}.\end{aligned}$$

We call $\{\mathbf{i}, \mathbf{j}, \mathbf{k}\}$ the **standard basis** for vectors in three-space. The context will make it clear when, for example, $\mathbf{i} = (1, 0)$ or $\mathbf{i} = (1, 0, 0)$.

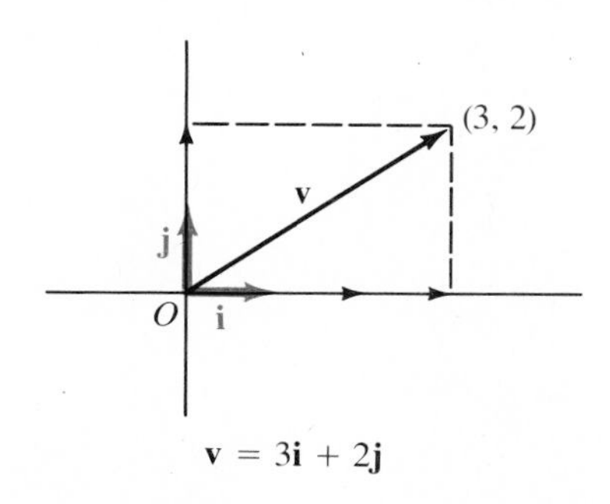

$\mathbf{v} = 3\mathbf{i} + 2\mathbf{j}$

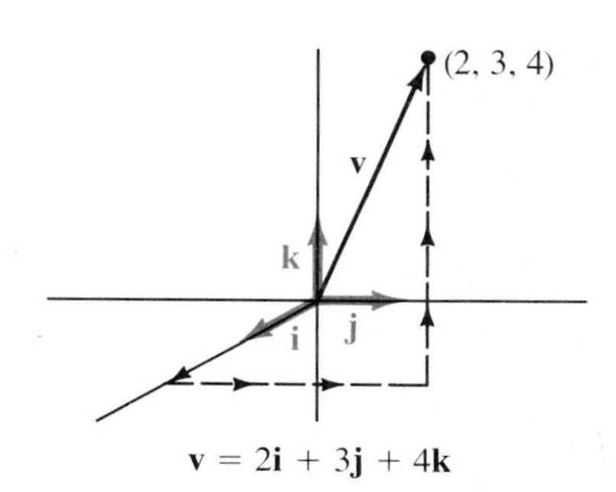

$\mathbf{v} = 2\mathbf{i} + 3\mathbf{j} + 4\mathbf{k}$

A set B of nonzero vectors in two-space or three-space is a **basis** for that space if every vector in that space can be written uniquely as a linear combination of the members of B. Is the set whose members are $\mathbf{t} = (3, 2)$ and $\mathbf{w} = (-4, 6)$ a basis for the set of all vectors in the plane? To answer this question let $\mathbf{v} = (x, y)$. Then the answer is yes if there are unique scalars a and b such that

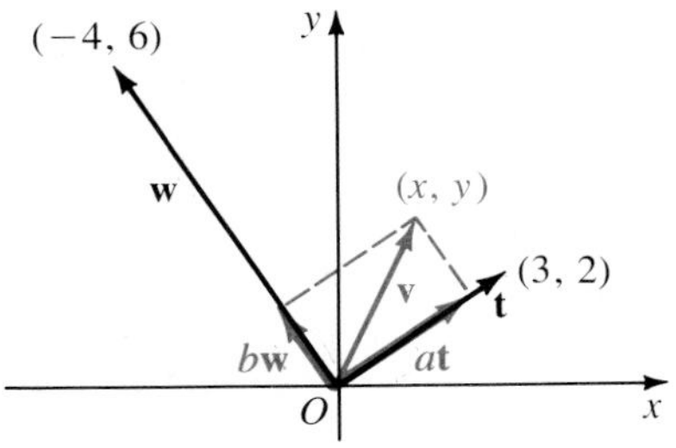

$$\mathbf{v} = a\mathbf{t} + b\mathbf{w}.$$

The figure at the right suggests that a and b can be found. The following argument shows that a and b exist and are unique. It also provides a way to find them.

The following equations are equivalent.

$$\begin{aligned}(x, y) &= a(3, 2) + b(-4, 6)\\ &= (3a, 2a) + (-4b, 6b)\\ &= (3a - 4b, 2a + 6b)\end{aligned}$$

$$\begin{cases} 3a - 4b = x \\ 2a + 6b = y \end{cases}$$

Solving these equations for a and b, we have

$$a = \frac{3x + 2y}{13} \quad \text{and} \quad b = \frac{3y - 2x}{26}.$$

This proves both the existence and the uniqueness of a and b.

Notice that the vectors $\mathbf{t} = (3, 2)$ and $\mathbf{w} = (-4, 6)$ are not parallel since $(3, 2) \neq (-4k, 6k)$ for any k. Theorem 6 is a generalization of the foregoing discussion.

THEOREM 6 If $\mathbf{u}$ and $\mathbf{v}$ are nonparallel vectors in the plane, then every vector in the plane can be written as a unique linear combination of $\mathbf{u}$ and $\mathbf{v}$.

You are asked to prove Theorem 6 in Exercise 40. The proof is based on the argument given above.

EXAMPLE 1 Write $\mathbf{t} = (1, 9)$ as a linear combination of $\mathbf{u} = (1, 2)$ and $\mathbf{v} = (2, -3)$.

Solution Verify that $\mathbf{u}$ and $\mathbf{v}$ are nonparallel. Then $\mathbf{t} = a\mathbf{u} + b\mathbf{v}$ for some scalars a and b.

$$\begin{aligned}\mathbf{t} &= a\mathbf{u} + b\mathbf{v}\\ (1, 9) &= (1a, 2a) + (2b, -3b)\end{aligned}$$

Hence: $$\begin{cases} a + 2b = 1 \\ 2a - 3b = 9 \end{cases}$$

Solve the system of equations to find that $a = 3$ and $b = -1$.
Therefore $\mathbf{t} = 3\mathbf{u} - \mathbf{v}$. **Answer**

Check: $3\mathbf{u} - \mathbf{v} = 3(1, 2) - (2, -3) = (3, 6) - (2, -3) = (1, 9) = \mathbf{t}$

When a vector **t** is written as a linear combination of two nonparallel vectors **u** and **v**, that is, when $\mathbf{t} = a\mathbf{u} + b\mathbf{v}$, we call $a\mathbf{u}$ and $b\mathbf{v}$ the **vector components** of **t** with respect to **u** and **v**. In Example 1 these components are $3\mathbf{u}$ and $-\mathbf{v}$. If **u** and **v** are unit vectors, a and b are called the **scalar components** of **t**. The method used in Example 1 can be used to find a and b. However, if **u** and **v** are orthogonal, the following theorem provides an easier way to find a and b.

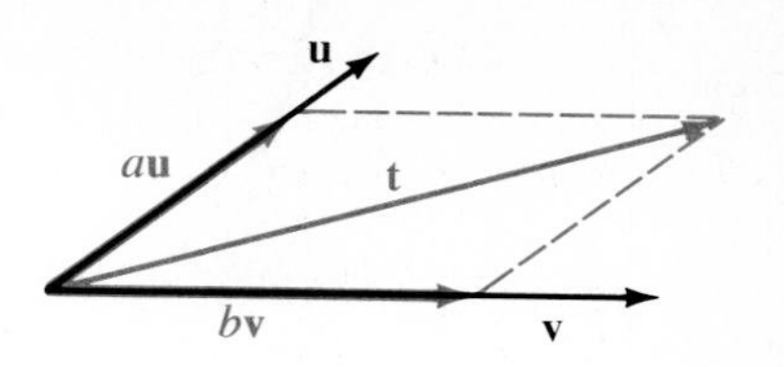

THEOREM 7 If **u** and **v** are nonzero orthogonal vectors in the plane and **t** is any vector in the plane, then

$$\mathbf{t} = \frac{\mathbf{u} \cdot \mathbf{t}}{\|\mathbf{u}\|^2}\mathbf{u} + \frac{\mathbf{v} \cdot \mathbf{t}}{\|\mathbf{v}\|^2}\mathbf{v} = \frac{\mathbf{u} \cdot \mathbf{t}}{\mathbf{u} \cdot \mathbf{u}}\mathbf{u} + \frac{\mathbf{v} \cdot \mathbf{t}}{\mathbf{v} \cdot \mathbf{v}}\mathbf{v}.$$

Proof: Since **u** and **v** are orthogonal, then $\frac{\mathbf{u}}{\|\mathbf{u}\|}$ and $\frac{\mathbf{v}}{\|\mathbf{v}\|}$ are orthogonal unit vectors in the directions of **u** and **v** respectively. Let $\mathbf{u}'$ and $\mathbf{v}'$ be the vectors obtained by projecting **t** onto the lines containing **u** and **v** $\left(\text{and therefore containing } \frac{\mathbf{u}}{\|\mathbf{u}\|} \text{ and } \frac{\mathbf{v}}{\|\mathbf{v}\|}\right)$ as shown in the figure at the right. Then $\mathbf{t} = \mathbf{u}' + \mathbf{v}'$.

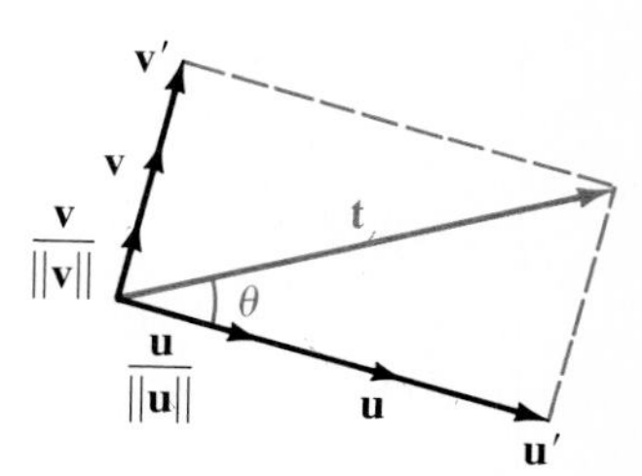

Since $\cos\theta = \frac{\|\mathbf{u}'\|}{\|\mathbf{t}\|}$,

$$\mathbf{u}' = (\|\mathbf{t}\| \cos\theta)\frac{\mathbf{u}}{\|\mathbf{u}\|} = \left[\frac{\|\mathbf{u}\|\,\|\mathbf{t}\|\cos\theta}{\|\mathbf{u}\|^2}\right]\mathbf{u} = \frac{\mathbf{u} \cdot \mathbf{t}}{\|\mathbf{u}\|^2}\mathbf{u}, \text{ or } \frac{\mathbf{u} \cdot \mathbf{t}}{\mathbf{u} \cdot \mathbf{u}}\mathbf{u}$$

Similarly, $\mathbf{v}' = \frac{\mathbf{v} \cdot \mathbf{t}}{\|\mathbf{v}\|^2}\mathbf{v}$, or $\frac{\mathbf{v} \cdot \mathbf{t}}{\mathbf{v} \cdot \mathbf{v}}\mathbf{v}$.

Therefore $\mathbf{t} = \frac{\mathbf{u} \cdot \mathbf{t}}{\mathbf{u} \cdot \mathbf{u}}\mathbf{u} + \frac{\mathbf{v} \cdot \mathbf{t}}{\mathbf{v} \cdot \mathbf{v}}\mathbf{v}$. ■

EXAMPLE 2 Express $\mathbf{t} = (5, 3)$ as a linear combination of $\mathbf{u} = (1, -3)$ and $\mathbf{v} = (3, 1)$.

Solution First note that **u** and **v** are orthogonal: $\mathbf{u} \cdot \mathbf{v} = 1 \cdot 3 + (-3) \cdot 1 = 0$. Apply Theorem 7.

$\mathbf{u} \cdot \mathbf{t} = 1 \cdot 5 + (-3) \cdot 3 = -4$	$\mathbf{v} \cdot \mathbf{t} = 3 \cdot 5 + 1 \cdot 3 = 18$
$\|\mathbf{u}\|^2 = 1^2 + (-3)^2 = 10$	$\|\mathbf{v}\|^2 = 3^2 + 1^2 = 10$

Therefore $\mathbf{t} = \frac{-4}{10}\mathbf{u} + \frac{18}{10}\mathbf{v}$, or $\frac{-2}{5}\mathbf{u} + \frac{9}{5}\mathbf{v}$. **Answer**

Check: $-\frac{2}{5}(1, -3) + \frac{9}{5}(3, 1) = \left(-\frac{2}{5} + \left(\frac{9}{5}\right)(3), \left(-\frac{2}{5}\right)(-3) + \frac{9}{5} \cdot 1\right) = (5, 3)$

EXAMPLE 3 Express $\mathbf{t} = (3, 2)$ as a sum of components in the directions of $\mathbf{u} = (-1, 4)$ and a vector $\mathbf{v}$ orthogonal to $\mathbf{u}$.

Solution Let $\mathbf{v} = (x, y)$. Then since $\mathbf{u} \cdot \mathbf{v} = -x + 4y$, choose any values of x and y that together satisfy $-x + 4y = 0$. Choose $x = 4$ and $y = 1$. Then $\mathbf{v} = (4, 1)$. Apply Theorem 7 with $\mathbf{u} = (-1, 4)$ and $\mathbf{v} = (4, 1)$.

$$\begin{array}{c|c} \mathbf{u} \cdot \mathbf{t} = 5 & \mathbf{v} \cdot \mathbf{t} = 14 \\ \|\mathbf{u}\|^2 = 17 & \|\mathbf{v}\|^2 = 17 \end{array}$$

Therefore $\mathbf{t} = \frac{5}{17}\mathbf{u} + \frac{14}{17}\mathbf{v}$. **Answer**

EXERCISES

Write each vector in the plane as a linear combination of $\mathbf{i}$ and $\mathbf{j}$.

A

1. $(-5, 3)$ **2.** $(6, -3)$ **3.** $(2, 1)$

4. $(-9, -8)$ **5.** $(5, -11)$ **6.** $(-4, -4)$

Write each vector as a linear combination of $\mathbf{i}$, $\mathbf{j}$, and $\mathbf{k}$.

7. $(-1, 0, 5)$ **8.** $(0, 8, 13)$ **9.** $(8, 9, 0)$

10. $(a, 0, 0)$ **11.** $(a, b, 0)$ **12.** (a, b, c)

Write each linear combination as an ordered triple.

13. $3\mathbf{i} + 4\mathbf{j} - \mathbf{k}$ **14.** $7\mathbf{i} - 2\mathbf{j} - 3\mathbf{k}$ **15.** $13\mathbf{i} - 6\mathbf{k}$

16. Show that the set $\{(3, 0), (0, -1)\}$ is a basis for the set of all vectors in the plane.

17. Show that the set $\{(2, 0), (1, 1)\}$ is a basis for the set of all vectors in the plane.

Use the method of Example 1 to write $\mathbf{t}$ as a linear combination of $\mathbf{u} = (1, -2)$ and $\mathbf{v} = (6, 3)$.

18. $\mathbf{t} = (9, -3)$ **19.** $\mathbf{t} = (7, 16)$ **20.** $\mathbf{t} = (-2, -11)$

21. $\mathbf{t} = (8, -16)$ **22.** $\mathbf{t} = (-18, -9)$ **23.** $\mathbf{t} = (8, -1)$

Let $\mathbf{t} = (-3, 1)$. Find the vector components of $\mathbf{t}$ in the direction of $\mathbf{u}$ and the direction of $\mathbf{v}$.

24. $\mathbf{u} = (1, 2)$; $\mathbf{v} = (-4, 2)$ **25.** $\mathbf{u} = (3, -2)$; $\mathbf{v} = (-4, -6)$

26. $\mathbf{u} = (-1, 5)$; $\mathbf{v} = (5, 1)$ **27.** $\mathbf{u} = (a, b)$; $\mathbf{v} = (-b, a)$, $a, b \neq 0$

Let $\mathbf{t} = (5, -2)$. Find the vector components of $\mathbf{t}$ in the direction of $\mathbf{u}$ and the direction of $\mathbf{v}$.

28. $\mathbf{u} = (4, 1)$; $\mathbf{v} = (-2, 8)$ **29.** $\mathbf{u} = (1, 0)$; $\mathbf{v} = (0, -8)$

30. $\mathbf{u} = (7, 0)$; $\mathbf{v} = (0, 5)$ **31.** $\mathbf{u} = (3, -3)$; $\mathbf{v} = (3, 3)$

Let $\mathbf{t} = (4, -3)$. Write $\mathbf{t}$ as a sum of components in the direction of $\mathbf{u}$ and a vector $\mathbf{v}$ orthogonal to $\mathbf{u}$.

B

32. $\mathbf{u} = (1, 5)$

33. $\mathbf{u} = (-2, 6)$

34. $\mathbf{u} = (0, 8)$

35. $\mathbf{u} = (-4, 0)$

36. $\mathbf{u} = (5, 12)$

37. $\mathbf{u} = (5, 5)$

38. Prove that a set of two parallel vectors in the plane cannot be a basis for the set of all vectors in the plane.

39. Prove or disprove: The set $\{(1, 0), (0, 1), (2, 2)\}$ of vectors is a basis for the set of all vectors in the plane. (*Hint*: Recall the uniqueness requirement.)

C

40. Prove Theorem 6. (*Hint*: Let $\mathbf{u} = (m, n)$ and $\mathbf{v} = (r, s)$. Let $\mathbf{t} = (x, y)$ be any vector in the plane. Solve the equation $a\mathbf{u} + b\mathbf{v} = \mathbf{t}$ for the scalars a and b.)

41. Prove that if $\{\mathbf{u}, \mathbf{v}\}$ is a basis for the set of all vectors in the plane and $a\mathbf{u} + b\mathbf{v} = \mathbf{0}$, then a and b are both 0.

42. Prove that the set $\{(1, 1, 1), (1, 0, 0), (0, 1, 0)\}$ of vectors in three-space is a basis for the set of all vectors in three-space.

43. Prove that the set $\{(1, 0, 0), (1, 1, 0), (0, 1, 0)\}$ of vectors in three-space is not a basis for the set of all vectors in three-space.

44. Let $B = \{\mathbf{u}, \mathbf{v}, \mathbf{w}\}$ be a set of vectors in the plane and let $\mathbf{t}$ be any vector in the plane. Show that if $\mathbf{t} = a\mathbf{u} + b\mathbf{v} + c\mathbf{w}$ for some scalars a, b, and c, then at least one of $\mathbf{u}$, $\mathbf{v}$, and $\mathbf{w}$ is a linear combination of the other members of B.

COMPUTER EXERCISES

1. Write a computer program that will print an input vector $\mathbf{t}$ as a sum of vector components in the direction of $\mathbf{u}$ and $\mathbf{v}$, also input by the user. Use Theorem 7 on page 459.

2. Run the program of Exercise 1 for the following vectors.
a. $\mathbf{t} = (14, -22)$; $\mathbf{u} = (-1, 4)$; $\mathbf{v} = (8, 2)$
b. $\mathbf{t} = (-1, 5)$; $\mathbf{u} = (3, -2)$; $\mathbf{v} = (8, 12)$

3. Modify the program of Exercise 1 to print an input vector $\mathbf{t}$ as a sum of vector components in the direction of $\mathbf{u}$, $\mathbf{v}$, and $\mathbf{w}$, also input by the user. Base your modification on a generalization of Theorem 7 for three-space.

4. Run the program of Exercise 3 for the following values of $\mathbf{t}$, $\mathbf{u}$, $\mathbf{v}$, and $\mathbf{w}$.
a. $\mathbf{t} = (-14, 20, -9)$; $\mathbf{u} = (5, -1, 4)$;
$\mathbf{v} = (2, 8, -0.5)$; $\mathbf{w} = (6, -2, -8)$
b. $\mathbf{t} = (-2, 3, -1)$; $\mathbf{u} = (1, -2, 1)$;
$\mathbf{v} = (5, 4, 3)$; $\mathbf{w} = (-3, 0.6, 4.2)$

Applications

11-6 APPLICATIONS OF VECTORS

In many applications the direction of a vector **v** is described by giving the angle θ that **v** makes with due north, $0° \leq \theta < 360°$. This angle is called the **bearing** of **v**.

Because of wind, a plane's **ground speed**, its actual speed relative to the ground, might differ from its **air speed**, its speed in still air, and a plane's **true course**, the direction in which it actually travels, might differ from its **heading**, the direction in which it is pointed. In describing a wind, it is customary to give the direction *from* which it blows. Thus the velocity vector of a west wind points east.

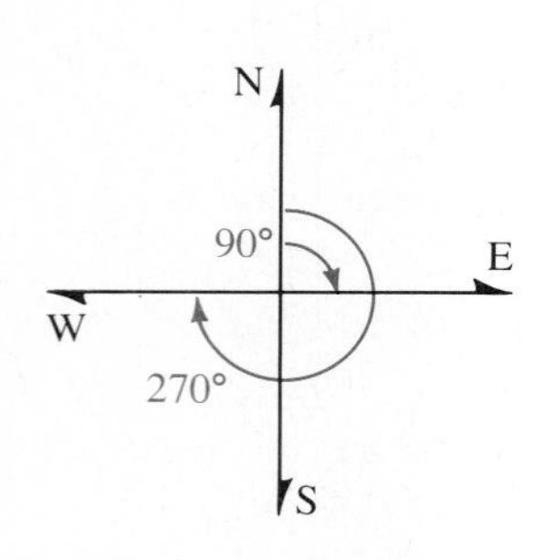

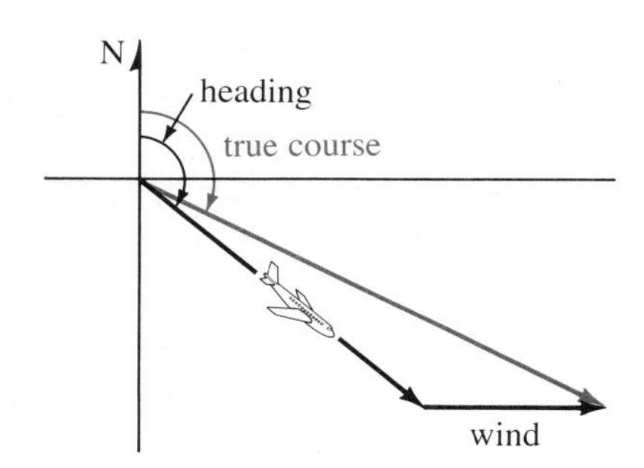

EXAMPLE 1 A plane's heading is 160° and its air speed is 350 mph. If a west wind is blowing at 20 mph, what are the plane's ground speed and true course?

Solution Draw a sketch of the vectors involved. The red arrow indicates the plane's true course. Use the law of cosines and the fact that $\theta = 110°$ to find the ground speed, $\|\mathbf{v}\|$.

$$\|\mathbf{v}\|^2 = 350^2 + 20^2 - 2 \cdot 350 \cdot 20 \cos 110° \approx 127{,}688$$

Hence $\|\mathbf{v}\| \approx 357$.

Use the law of sines to find the measure of the angle α between **u** and **v**.

$$\frac{\sin \alpha}{20} = \frac{\sin 110°}{357}$$

To the nearest degree $\alpha = 3°$. Hence the true course is $160° - 3° = 157°$. Therefore the ground speed is 357 mph and the true course is 157°. **Answer**

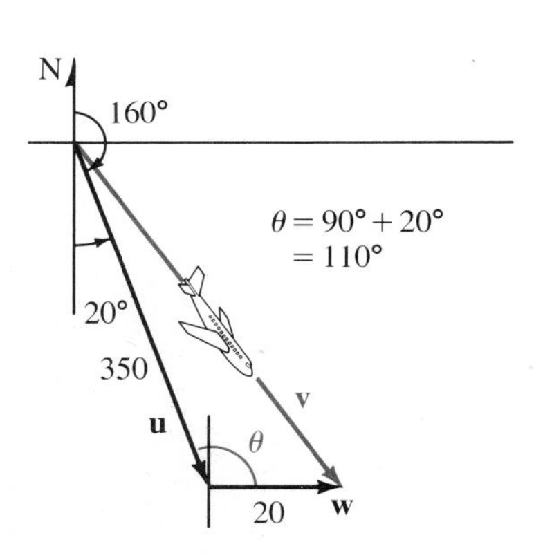

The **resultant** of two vectors **u** and **v** is simply their sum, **u** + **v**. When a vector is written as the sum of two nonparallel vectors, we say that the vector is resolved.

EXAMPLE 2 A force **F** of 24 N (Newtons) is applied at an angle of 30° with the horizontal. Resolve **F** into its horizontal and vertical components.

Solution Write **F** as a linear combination of **i** and **j**.

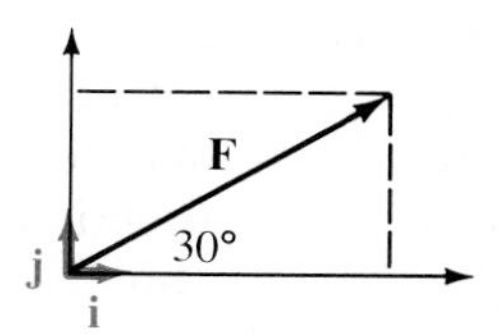

$$\mathbf{F} = (24 \cos 30°)\mathbf{i} + (24 \sin 30°)\mathbf{j}$$
$$= (24)\left(\frac{\sqrt{3}}{2}\right)\mathbf{i} + (24)\left(\frac{1}{2}\right)\mathbf{j}.$$

Therefore $\mathbf{F} = 12\sqrt{3}\mathbf{i} + 12\mathbf{j}$. **Answer**

EXAMPLE 3 Two forces **u** and **v** act on an object at a point P. The force **u** has magnitude 5 N and acts at an angle of 32° with the horizontal. The vector **v** has magnitude 8 N and acts at an angle of 72° with the horizontal. Find the magnitude and direction (with respect to the horizontal) of the resultant vector, **R**.

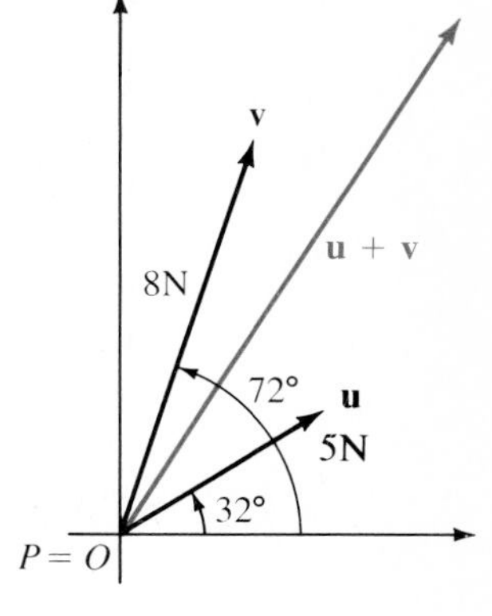

Solution Let P be the origin of a coordinate system with horizontal and vertical axes. Resolve **u** and **v** into horizontal and vertical components.

$$\mathbf{u} = (5 \cos 32°, 5 \sin 32°)$$
$$= (4.2402, 2.6496)$$
$$\mathbf{v} = (8 \cos 72°, 8 \sin 72°)$$
$$= (2.4721, 7.6085)$$

The resultant is

$$(4.2402, 2.6496) + (2.4721, 7.6085) = (6.7123, 10.2581).$$

To the nearest tenth,

$$\|\mathbf{R}\| = \sqrt{(6.7123)^2 + (10.2581)^2} = 12.3.$$

If **R** makes an angle θ with the horizontal, then $\tan \theta = \frac{10.2581}{6.7123}$. To the nearest degree $\theta = 57°$. The force has magnitude 12.3 N and direction 57°. **Answer**

When a force acts to move an object, it does *work* and expends energy. The simplest case occurs when the force **F** moves the object from A to B along a straight path. If $\overrightarrow{AB}$ is denoted by **d**, then the **work** done by **F** is

$$\text{Work} = \|\mathbf{F}\| \, \|\mathbf{d}\| \cos \theta$$

where θ is the angle between **F** and **d**.

EXAMPLE 4 Find the work done by **F** in moving the box shown from A to B.

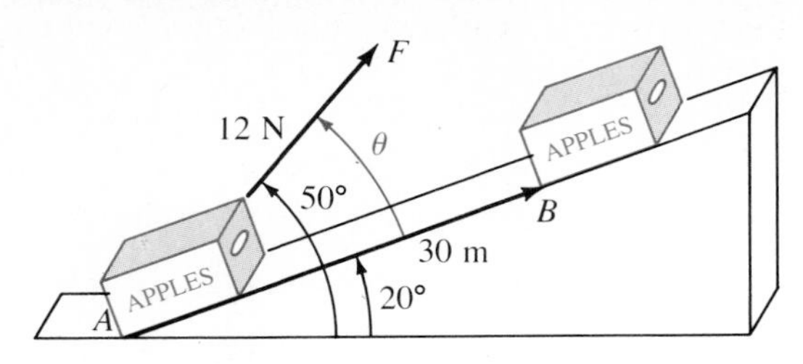

Solution The angle θ measures $50° - 20°$, or $30°$. Therefore Work $= (12)(30)\cos 30°$, or about 312 units of work.

Note: The basic unit of work and energy is the joule (J). One **joule** is the work done by a force of one newton in moving an object one meter. Thus the work done in Example 4 is approximately 312 J.

The **position vector** of a point P is the vector $\mathbf{r} = \overrightarrow{OP}$, where O is the origin. If P_0 is a fixed point having position vector $\mathbf{r}_0$, then a vector equation of the line through P_0 parallel to the fixed vector $\mathbf{m}$ is

$$\mathbf{r} = \mathbf{r}_0 + t\mathbf{m}.$$

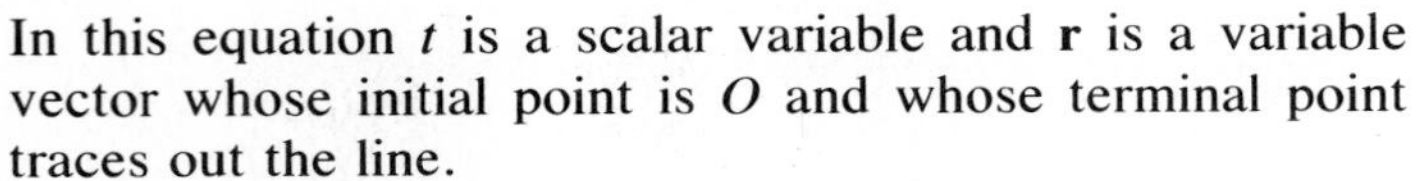

In this equation t is a scalar variable and $\mathbf{r}$ is a variable vector whose initial point is O and whose terminal point traces out the line.

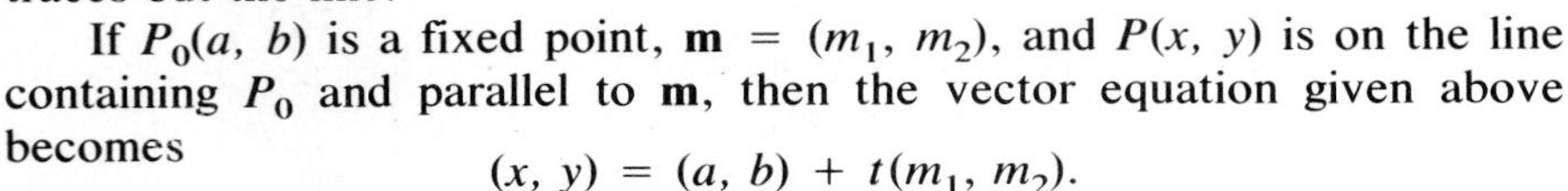

If $P_0(a, b)$ is a fixed point, $\mathbf{m} = (m_1, m_2)$, and $P(x, y)$ is on the line containing P_0 and parallel to $\mathbf{m}$, then the vector equation given above becomes

$$(x, y) = (a, b) + t(m_1, m_2).$$

A set of **parametric equations** of the line is

$$x = a + tm_1 \quad \text{and} \quad y = b + tm_2.$$

EXAMPLE 5 Find a set of parametric equations of the line containing $P_0(6, 4)$ and parallel to $\mathbf{m} = (1, 2)$.

Solution If $P(x, y)$ is any point on the line and $\mathbf{r}$ is the position vector of P, then $(x, y) = (6, 4) + t(1, 2)$. Hence $x = 6 + t$ and $y = 4 + 2t$. **Answer**

EXERCISES

Resolve each vector with given magnitude and making the given angle with the horizontal into its horizontal and vertical components.

A

1. 6, 80° **2.** 13, 45° **3.** 5.5, 120° **4.** 10, 90°

5. 8, 180° **6.** 24, 135° **7.** 7.5, 10° **8.** 100, 140°

Find a set of parametric equations of the line containing the given point P and parallel to the given vector $\mathbf{v}$.

9. $P(4, 7)$; $\mathbf{v} = (1, 3)$

10. $P(-3, 2)$; $\mathbf{v} = (6, -5)$

11. $P(-3, -3)$; $\mathbf{v} = (3, 3)$

12. $P(0, 0)$; $\mathbf{v} = (5, 5)$

13. $P(0, 0)$; $\mathbf{v} = (-6, 8)$

14. $P(-4, -4)$; $\mathbf{v} = (7, -5)$

A force **F** with given magnitude and direction with respect to the horizontal is applied to an object. Resolve **F** into its horizontal and vertical components.

15. 35 N, 45° **16.** 18.5 N, 30° **17.** 65.4 N, −30° **18.** 15 N, 135°

Write the given vector **t** as a sum of components parallel and perpendicular to the unit vector with given direction.

19. **t**: magnitude 20, direction 80°; 45°

20. **t**: magnitude 13.5, direction 120°; 30°

21. **t**: magnitude 1, direction 60.5°; 120°

22. **t**: magnitude 36, direction 110°; 72°

Find the work done by the force **F** with given magnitude and direction in moving an object the given distance at the given angle.

23. force 15 N at 45° along a ramp 60 m long at 30°

24. force 91 N at 10° along a ramp 100 m long at 10°

25. force 15 N at 24° up a vertical cable a distance of 25 m

26. force 42.36 N at 82° up a vertical cable a distance of 1 m

B **27.** Find the work done by pulling a sled 20 m along level ground if a force of 1.5 N is exerted at an angle of 27°.

In Exercises 28–31, find a set of parametric equations of each line.

28. Containing (3, 4) and parallel to $\mathbf{m} = (4, 5)$

29. Containing (−3, 6) and parallel to $\mathbf{m} = (7, 10)$

30. Containing (−3, 5) and (6, 9)

31. Containing (4, 7) and (0, 6)

32. A downstream current and an across-stream wind current at 5 km/h give a sailboat an effective speed of 13 km/h. What is the speed of the downstream current and the angle between the wind current and the path of the sailboat?

33. A plane with a heading of 50° has an air speed of 400 mph. If a 35 mph wind is blowing from the north, what are the plane's ground speed and true course?

34. Repeat Exercise 33 if a plane's heading is 130° and its air speed is 350 mph.
35. An ocean liner has a heading of 250° and a speed of 12 knots. If the ship's true course is 235° and its speed relative to land is 10 knots, what are the speed and direction of the water current?
36. Repeat Exercise 35 if the ship's heading is 100° and its true course is 120°.

In Exercises 37 and 38 use the fact that gravity **G** tends to accelerate an object downward at the rate of 9.8 m/s^2. Also use the fact that the force needed to overcome the effect of gravity is an upward force with magnitude $\|\mathbf{F}\| = 9.8m$ N where m is the mass of the object in kilograms.

C 37. Find the work done in lifting an elevator car weighing 1200 kg up 70 m.
38. Find the work done by the engine of a car weighing 2400 kg if the car travels up a 15% grade over a horizontal distance of 3.5 km.
39. Show that if $\mathbf{F} = (a, b)$ gives an object a displacement $\mathbf{D} = (c, d)$, then the work done is $ac + bd$.

COMPUTER EXERCISES

1. Write a computer program that will resolve an input vector with given magnitude and direction into its horizontal and vertical components.
2. Run the program of Exercise 1 for the following vectors.
   a. 5, 60° **b.** 16, 135°
3. Write and run a computer program to find the work done in pulling a box up a ramp of length 18 m inclined at an angle of 45°, given a force of 20 N in a direction of 60° and a force of 14 N in a direction of 75°. (*Hint*: Find the magnitude and direction of the resultant vector.)

11-7 VECTORS AND PROOFS

We can use vectors to prove theorems from geometry. Often a vector proof is easier to construct than a proof using methods of plane geometry. In order to write a vector proof we shall use the following vector interpretations of geometric terminology.

Geometric term	*Vector interpretation*
$\overline{AB}$ is parallel to $\overline{CD}$.	$\overrightarrow{AB} = k\overrightarrow{CD}$ for some scalar k
$\overline{AB}$ is perpendicular to $\overline{CD}$.	$\overrightarrow{AB} \cdot \overrightarrow{CD} = 0$
$\overline{AB}$ is congruent to $\overline{CD}$.	$\|\overrightarrow{AB}\| = \|\overrightarrow{CD}\|$
P is the midpoint of $\overline{AB}$.	$\overrightarrow{AP} = \frac{1}{2}\overrightarrow{AB}$ and $\overrightarrow{PB} = \frac{1}{2}\overrightarrow{AB}$

EXAMPLE 1 Write a vector proof: The line segment joining the midpoints of two sides of a triangle is parallel to the third side and is half as long as the third side.

Solution The diagram at the right shows the vectors needed to write the proof. Since P and Q are midpoints of $\overrightarrow{AC}$ and $\overrightarrow{CB}$ respectively,

$$\overrightarrow{PC} = \tfrac{1}{2}\overrightarrow{AC} \text{ and } \overrightarrow{CQ} = \tfrac{1}{2}\overrightarrow{CB}.$$

Since $\overrightarrow{PC} + \overrightarrow{CQ} = \overrightarrow{PQ}$,

$$\tfrac{1}{2}\overrightarrow{AC} + \tfrac{1}{2}\overrightarrow{CB} = \overrightarrow{PQ}.$$

Hence $\overrightarrow{PQ} = \frac{1}{2}(\overrightarrow{AC} + \overrightarrow{CB})$. Because $\overrightarrow{AC} + \overrightarrow{CB} = \overrightarrow{AB}$, $\overrightarrow{PQ} = \frac{1}{2}\overrightarrow{AB}$. Thus the segment represented by $\overrightarrow{PQ}$ is parallel to the segment represented by $\overrightarrow{AB}$ and is half as long.

EXAMPLE 2 Write a vector proof: The diagonals of a rhombus are perpendicular.

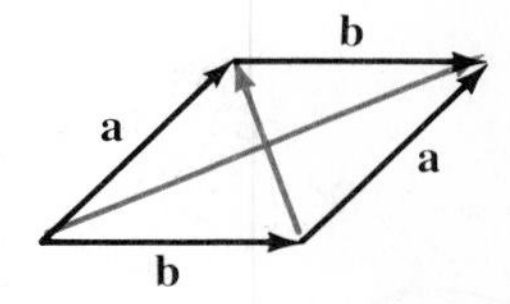

Solution Since the opposite sides of a rhombus are equal in length and parallel, use the same letter for opposite sides. The diagonals are $\mathbf{a} + \mathbf{b}$ and $\mathbf{a} - \mathbf{b}$.

$$\begin{aligned}(\mathbf{a} + \mathbf{b}) \cdot (\mathbf{a} - \mathbf{b}) &= \mathbf{a} \cdot \mathbf{a} - \mathbf{a} \cdot \mathbf{b} + \mathbf{b} \cdot \mathbf{a} - \mathbf{b} \cdot \mathbf{b}\\ &= \mathbf{a} \cdot \mathbf{a} - \mathbf{b} \cdot \mathbf{b} \quad \text{since } \mathbf{a} \cdot \mathbf{b} = \mathbf{b} \cdot \mathbf{a}\\ &= \|\mathbf{a}\|^2 - \|\mathbf{b}\|^2 \quad \text{by the norm property}\end{aligned}$$

Since the given figure is a rhombus, $\|\mathbf{a}\| = \|\mathbf{b}\|$. Therefore $(\mathbf{a} + \mathbf{b}) \cdot (\mathbf{a} - \mathbf{b}) = 0$. Hence the diagonals are perpendicular.

EXAMPLE 3 Write a vector proof: The median to the hypotenuse of a right triangle is half as long as the hypotenuse.

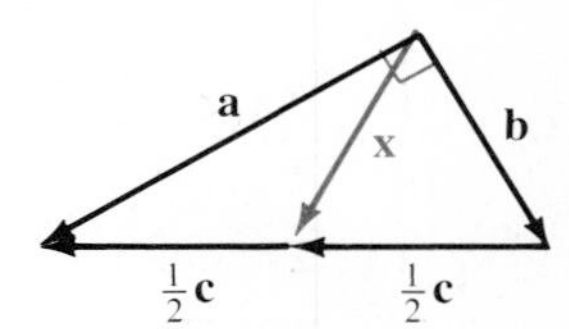

Solution In the diagram $\mathbf{x} + \frac{1}{2}\mathbf{c} = \mathbf{a}$ and thus $\mathbf{x} = \mathbf{a} - \frac{1}{2}\mathbf{c}$. Furthermore from the diagram, $\mathbf{x} = \mathbf{b} + \frac{1}{2}\mathbf{c}$. Thus

$2\mathbf{x} = \mathbf{a} - \frac{1}{2}\mathbf{c} + \mathbf{b} + \frac{1}{2}\mathbf{c} = \mathbf{a} + \mathbf{b}$. Hence $\mathbf{x} = \dfrac{\mathbf{a} + \mathbf{b}}{2}$.

Find $\|\mathbf{x}\|^2$.

$$\begin{aligned}\|\mathbf{x}\|^2 = \mathbf{x} \cdot \mathbf{x} &= \tfrac{1}{4}(\mathbf{a} + \mathbf{b}) \cdot (\mathbf{a} + \mathbf{b})\\ &= \tfrac{1}{4}(\mathbf{a} \cdot \mathbf{a} + 2\,\mathbf{a} \cdot \mathbf{b} + \mathbf{b} \cdot \mathbf{b})\end{aligned}$$

Since $\mathbf{a}$ and $\mathbf{b}$ are perpendicular, $\mathbf{a} \cdot \mathbf{b} = 0$. Therefore

$$\|\mathbf{x}\|^2 = \tfrac{1}{4}(\mathbf{a} \cdot \mathbf{a} + \mathbf{b} \cdot \mathbf{b})$$

and $\|\mathbf{x}\|^2 = \frac{1}{4}(\|\mathbf{a}\|^2 + \|\mathbf{b}\|^2)$ by the norm property.

Since $\|\mathbf{c}\|^2 = \|\mathbf{a}\|^2 + \|\mathbf{b}\|^2$ by the Pythagorean theorem, then

$$\|\mathbf{x}\|^2 = \tfrac{1}{4}\|\mathbf{c}\|^2.$$

Therefore $\|\mathbf{x}\| = \frac{1}{2}\|\mathbf{c}\|$ and the proof is complete.

EXERCISES

Use the diagram at the left below for Exercises 1–6. In the diagram, T is the midpoint of $\overline{PR}$ and $\overline{SQ}$. Write each vector in terms of $\mathbf{w}$ and $\mathbf{x}$.

A **1.** $\overrightarrow{PR}$ **2.** $\overrightarrow{QS}$ **3.** $\overrightarrow{QT}$ **4.** $\overrightarrow{TR}$ **5.** $\overrightarrow{SQ}$ **6.** $\overrightarrow{RP}$

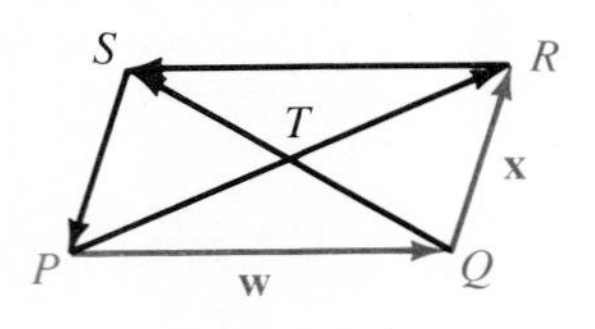

Exercises 1-6

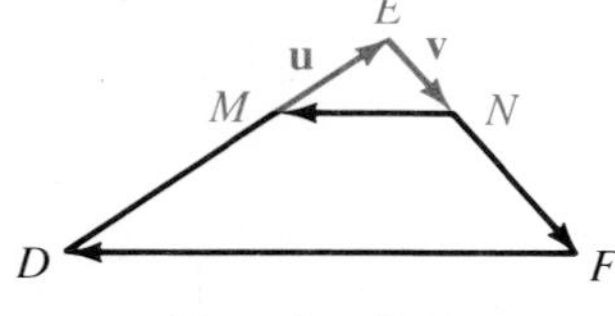

Exercises 7-10

Use the diagram at the right above for Exercises 7–10. In the diagram $\overrightarrow{DM} = 2\,\overrightarrow{ME}$ and $\overrightarrow{NF} = 2\overrightarrow{EN}$. Write each vector in terms of $\mathbf{u}$ and $\mathbf{v}$.

7. $\overrightarrow{DF}$ **8.** $\overrightarrow{DN}$ **9.** $\overrightarrow{MF}$ **10.** $\overrightarrow{DM} + \overrightarrow{MN} + \overrightarrow{NF}$

In Exercises 11 and 12 write a vector proof of each statement by completing each part of the exercise.

B **11.** The sum of the squares of the lengths of the four sides of a parallelogram is equal to the sum of the squares of the lengths of the diagonals.
 a. First write $\overrightarrow{AC}$ and $\overrightarrow{BD}$ in terms of $\mathbf{u}$ and $\mathbf{v}$.
 b. Use the fact that $\|\mathbf{u}\|^2 = \mathbf{u} \cdot \mathbf{u}$ to finish the proof.

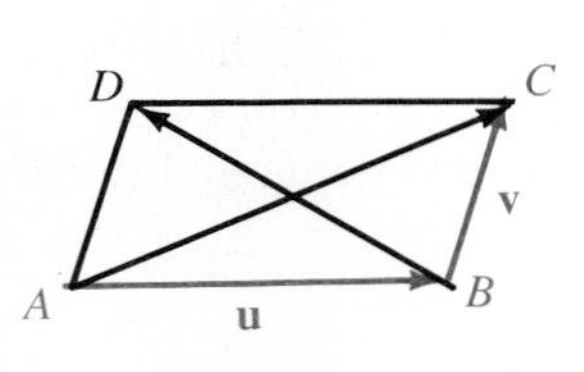

Exercise 11

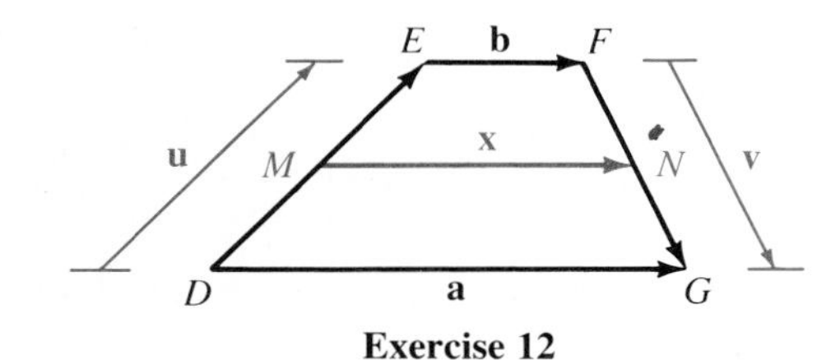

Exercise 12

12. The median of a trapezoid is parallel to the bases and its length is one-half the sum of the lengths of the bases. Let $\overrightarrow{DE} = \mathbf{u}$ and $\overrightarrow{FG} = \mathbf{v}$. (In the diagram M and N are the midpoints of $\overline{DE}$ and $\overline{FG}$ respectively.)
 a. Write $\overrightarrow{DM}$, $\overrightarrow{ME}$, $\overrightarrow{FN}$, and $\overrightarrow{NG}$ in terms of $\mathbf{u}$ and $\mathbf{v}$.
 b. Write $\mathbf{a}$ in terms of $\mathbf{u}$, $\mathbf{b}$, and $\mathbf{v}$ only.
 c. Write $\mathbf{a}$ in terms of $\mathbf{u}$, $\mathbf{x}$, and $\mathbf{v}$ only.
 d. Use parts (b) and (c) to finish the proof.

Write a vector proof for each statement. Begin your proof by drawing a vector diagram.

C **13.** If the midpoints of the sides of a quadrilateral are joined in order, then the quadrilateral so formed is a parallelogram.

14. If the midpoints of three sides of a rhombus are joined in order, then the resulting triangle is a right triangle.

15. Let P be the midpoint of $\overline{AB}$. If C is equidistant from A and B, then $\overline{CP}$ is perpendicular to $\overline{AB}$.

In Exercises 16 and 17 use the definition of the dot product: If **u** and **v** are vectors in the plane, then $\mathbf{u} \cdot \mathbf{v} = \|\mathbf{u}\| \, \|\mathbf{v}\| \cos \theta$ where θ is the measure of the angle between **u** and **v**. Prove each statement.

16. A diagonal of a rhombus bisects two opposite angles.

17. Opposite angles of a parallelogram have equal measure.

11–8 VECTOR SPACES

In Section 11–1 we introduced the operations of addition and scalar multiplication. We introduced the zero vector and the additive inverse of a vector. The fact that addition of vectors is associative was stated and used.

The set of all vectors in two-space (and three-space as well) together with addition and scalar multiplication satisfies the axioms of a mathematical system known as a *vector space*.

DEFINITION

Let V be a set of objects and F be a field, here $\mathcal{R}$. Then V is a **vector space** if for all vectors **u, v,** and **w** in V and scalars r and s in F, $\mathbf{u} + \mathbf{v}$ and $r\mathbf{u}$ are in V and

1. $\mathbf{u} + \mathbf{v} = \mathbf{v} + \mathbf{u}$ **2.** $\mathbf{u} + (\mathbf{v} + \mathbf{w}) = (\mathbf{u} + \mathbf{v}) + \mathbf{w}$

3. There is a unique vector $\mathbf{0} \in V$, called the zero vector, such that, for all $\mathbf{u}$, $\mathbf{0} + \mathbf{u} = \mathbf{u}$.

4. For every $\mathbf{u} \in V$, there is a $-\mathbf{u} \in V$ such that $\mathbf{u} + (-\mathbf{u}) = \mathbf{0}$.

5. $1\mathbf{u} = \mathbf{u}$ **6.** $(rs)\mathbf{u} = r(s\mathbf{u})$

7. $(r + s)\mathbf{u} = r\mathbf{u} + s\mathbf{u}$ and $r(\mathbf{u} + \mathbf{v}) = r\mathbf{u} + r\mathbf{v}$

According to axiom 1 of the definition, addition is commutative, and according to axiom 2, addition is associative.

EXAMPLE 1 Show that addition of vectors in three-space is commutative.

Solution Every vector in three-space has a coordinate representation. Let $\mathbf{u} = (u_1, u_2, u_3)$ and let $\mathbf{v} = (v_1, v_2, v_3)$. Then

$$\begin{aligned} \mathbf{u} + \mathbf{v} &= (u_1, u_2, u_3) + (v_1, v_2, v_3) \\ &= (u_1 + v_1, u_2 + v_2, u_3 + v_3) && \text{definition of addition} \\ &= (v_1 + u_1, v_2 + u_2, v_3 + u_3) && \text{commutative axiom of real numbers} \\ &= (v_1, v_2, v_3) + (u_1, u_2, u_3) && \text{definition of addition} \end{aligned}$$

Therefore $\mathbf{u} + \mathbf{v} = \mathbf{v} + \mathbf{u}$.

In Exercises 1–6 you are asked to show that the set of all vectors in three-space satisfies the other axioms of a vector space.

The vector space as a mathematical concept is important because there are many systems that are vector spaces, some of which have nothing to do with vectors in the plane or in three-space. For example, the set of all polynomials of degree 2 or less, which we shall denote by P_2, with ordinary addition and multiplication by a real number is a vector space.

EXAMPLE 2 Show that P_2 satisfies the first part of axiom 7 of the definition of a vector space.

Solution Every member $\mathbf{u}$ of P_2 is an expression of the form $ax^2 + bx + c$ for some real numbers a, b, and c. Let r and s be real numbers. Then

$$\begin{aligned}(r + s)\mathbf{u} &= (r + s)(ax^2 + bx + c) \\ &= [(r + s)(a)]x^2 + [(r + s)(b)]x + (r + s)(c) \\ &= (ra + sa)x^2 + (rb + sb)x + rc + sc \\ &= rax^2 + rbx + rc + sax^2 + sbx + sc \\ &= r(ax^2 + bx + c) + s(ax^2 + bx + c)\end{aligned}$$

Hence $(r + s)\mathbf{u} = r\mathbf{u} + s\mathbf{u}$.

The following theorem summarizes some properties of any vector space. The theorem can be proved by using the definition of a vector space.

THEOREM 8 Let V be a vector space and let $\mathbf{u}$, $\mathbf{v}$, and $\mathbf{w}$ be any members of V.

1. $0\mathbf{v} = \mathbf{0}$
2. $(-1)\mathbf{v} = -\mathbf{v}$
3. $\mathbf{u} = \mathbf{v}$ if and only if $\mathbf{u} + \mathbf{w} = \mathbf{v} + \mathbf{w}$.
4. $\mathbf{u} + (\mathbf{v} - \mathbf{u}) = \mathbf{v}$
5. $-(-\mathbf{u}) = \mathbf{u}$
6. $-(\mathbf{u} + \mathbf{v}) = -\mathbf{u} + (-\mathbf{v}) = -\mathbf{u} - \mathbf{v}$

EXAMPLE 3 Use the vector space axioms to show **(a)** that $0\mathbf{v} = \mathbf{0}$ and **(b)** that $(-1)\mathbf{v} = -\mathbf{v}$ for every vector $\mathbf{v}$ in a vector space V.

Solution **a.** By the first part of axiom 7,

$$0\mathbf{v} = (0 + 0)\mathbf{v} = 0\mathbf{v} + 0\mathbf{v}$$

Hence $0\mathbf{v} + (-0\mathbf{v}) = 0\mathbf{v} + 0\mathbf{v} + (-0\mathbf{v})$.
Therefore, by axiom 4, $\mathbf{0} = 0\mathbf{v} + \mathbf{0}$.
Thus, by axiom 3, $\mathbf{0} = 0\mathbf{v}$.

b.

	$0\mathbf{v} = \mathbf{0}$	Theorem 8, part 1
	$((-1) + 1)\mathbf{v} = \mathbf{0}$	
	$(-1)\mathbf{v} + 1\mathbf{v} = \mathbf{0}$	Axiom 7 of the definition
	$(-1)\mathbf{v} + \mathbf{v} = \mathbf{0}$	Axiom 5 of the definition
Hence	$(-1)\mathbf{v} = -\mathbf{v}$	Axiom 4 of the definition

EXERCISES

In Exercises 1–6 show that the set of all vectors in three-space satisfies each axiom in the definition of a vector space.

A

1. Axiom 2 **2.** Axiom 3 **3.** Axiom 4

4. Axiom 5 **5.** Axiom 6 **6.** Axiom 7

In Exercises 7–12 show that P_2 with ordinary addition and multiplication by a real number satisfies each axiom in the definition of a vector space.

7. Axiom 1 **8.** Axiom 2 **9.** Axiom 3

10. Axiom 4 **11.** Axiom 5 **12.** Axiom 6

In Exercises 13–16 prove each part of Theorem 8. You may use parts 1 and 2 of the theorem and the axioms of a vector space.

B

13. Part 3 **14.** Part 4

15. Part 5 **16.** Part 6

17. Let D be the set of all differentiable functions on an interval I. Show that under ordinary addition of functions and multiplication of a function by a real number D is a vector space.

Chapter Summary

1. A *vector* is a quantity with magnitude, or length, and direction. A vector can be represented by an arrow or a *directed line segment*. The *sum* of two vectors **u** and **v** is the vector obtained by placing the tail of **v** at the head of **u** and then drawing the arrow whose tail is the tail of **u** and whose head is the head of **v**. A *scalar multiple* of **v** is a vector $k\mathbf{v}$ where k is a scalar. For all vectors **u**, there is a vector, the *zero vector* denoted **0**, such that $\mathbf{u} + \mathbf{0} = \mathbf{u}$. For each vector **u**, there is a vector $-\mathbf{u}$, called the *additive inverse* of **u**, such that $\mathbf{u} + (-\mathbf{u}) = \mathbf{0}$.

   Two vectors **u** and **v** are *parallel* if and only if $\mathbf{u} = k\mathbf{v}$ for some scalar k.

   Every vector in the plane can be represented by an ordered pair (a, b) with a the *first component* and b the *second component*.

2. Let $P_1(x_1, y_1, z_1)$ and $P_2(x_2, y_2, z_2)$ be two points in three-space. The length of $\overline{P_1P_2}$ is given by the *distance formula*:

$$\sqrt{(x_2 - x_1)^2 + (y_2 - y_1)^2 + (z_2 - z_1)^2}$$

   and the coordinates of the midpoint of $\overline{P_1P_2}$ are given by the *midpoint formula*:

$$\left(\frac{x_1 + x_2}{2}, \frac{y_1 + y_2}{2}, \frac{z_1 + z_2}{2}\right)$$

3. The length of a vector $\mathbf{v}$ is its *norm*, denoted $\|\mathbf{v}\|$.
If $\mathbf{v} = (v_1, v_2, v_3)$, then $\|\mathbf{v}\| = \sqrt{v_1^2 + v_2^2 + v_3^2}$. For all vectors $\mathbf{u}$ and $\mathbf{v}$, $\|\mathbf{u} + \mathbf{v}\| \le \|\mathbf{u}\| + \|\mathbf{v}\|$. This is the *triangle inequality*.
4. The *dot product* of $\mathbf{u}$ and $\mathbf{v}$, denoted $\mathbf{u} \cdot \mathbf{v}$, is defined by
$$\mathbf{u} \cdot \mathbf{v} = \|\mathbf{u}\| \, \|\mathbf{v}\| \cos \theta$$
where θ is the measure of the angle between $\mathbf{u}$ and $\mathbf{v}$. If $\mathbf{u} = (u_1, u_2, u_3)$ and $\mathbf{v} = (v_1, v_2, v_3)$, then $\mathbf{u} \cdot \mathbf{v} = u_1v_1 + u_2v_2 + u_3v_3$.
Two vectors $\mathbf{u}$ and $\mathbf{v}$ are *orthogonal* if the angle between them is 90° or if one of them is $\mathbf{0}$, that is, if $\mathbf{u} \cdot \mathbf{v} = 0$.
5. Any vector $\mathbf{t}$ in the plane can be uniquely expressed as a linear combination of two nonzero nonparallel vectors $\mathbf{u}$ and $\mathbf{v}$.
$$\mathbf{t} = \frac{\mathbf{u} \cdot \mathbf{t}}{\|\mathbf{u}\|^2}\mathbf{u} + \frac{\mathbf{v} \cdot \mathbf{t}}{\|\mathbf{v}\|^2}\mathbf{v}$$
6. The *resultant* of two vectors is their sum. The *work* done by a force $\mathbf{F}$ effecting a displacement $\mathbf{d}$ is given by $\|\mathbf{F}\| \, \|\mathbf{d}\| \cos \theta$ where θ is the measure of the angle between $\mathbf{F}$ and $\mathbf{d}$. A vector equation of the line containing P_0 and parallel to $\mathbf{m}$ is $\mathbf{r} = \mathbf{r}_0 + t\mathbf{m}$ where $\mathbf{r}_0$ is the *position vector* of P_0 and $\mathbf{r}$ is the position vector of any point P on the line.
7. Vectors can be used to prove theorems in geometry. To do so, vector interpretations of geometric properties are used.
8. A *vector space* is a set of objects for which addition and scalar multiplication are defined. Addition is commutative and associative. There is an additive identity and each member has an additive inverse. Scalar multiplication is associative and distributes over both scalars and vectors. Furthermore, for all vectors $\mathbf{v}$, $1\mathbf{v} = \mathbf{v}$.

Chapter Test

11-1

1. Find the vector whose initial point is the origin and that is equivalent to $\overrightarrow{PQ}$, where $P = (-1, 7)$ and $Q = (4, 5)$.
2. Let $\mathbf{u} = (-1, 3)$, $\mathbf{v} = (4, -9)$, and $\mathbf{w} = (5, -2)$. Find each linear combination. **a.** $\mathbf{u} + \mathbf{v} - 2\mathbf{w}$ **b.** $2(\mathbf{u} - \mathbf{w}) + 3\mathbf{v}$
3. Illustrate that $2(\mathbf{u} - \mathbf{v}) = 2\mathbf{u} - 2\mathbf{v}$ by means of a diagram.

11-2

4. Let points $P(-3, 7, -5)$ and $Q(2, -1, 9)$ be given.
   a. Find PQ. **b.** Find the midpoint of $\overline{PQ}$.
5. Let $\mathbf{u} = (4, -6, 2)$, $\mathbf{v} = (-8, 3, 5)$, and $\mathbf{w} = (-3, 6, 4)$. Find the vector $\frac{3}{2}\mathbf{u} - \mathbf{w} + \mathbf{v}$.

11-3

6. Find the unit vector having direction opposite that of $\mathbf{v} = (-7, 24)$.
7. Use the triangle inequality to show that for all vectors $\mathbf{u}$, $\mathbf{v}$, and $\mathbf{w}$
$$\|\mathbf{u} + \mathbf{w}\| \le \|\mathbf{u} - \mathbf{v}\| + \|\mathbf{v} + \mathbf{w}\|.$$

11-4 **8.** Find the measure of the angle between $\mathbf{u} = (3, 6, -2)$ and $\mathbf{v} = (-3, 4, 0)$.

9. Find k such that $\mathbf{u} = (-5, k)$ and $\mathbf{v} = (7, -10)$ are orthogonal.

11-5 **10.** Write $\mathbf{t} = (-9, 1)$ as a linear combination of $\mathbf{u} = (-1, 5)$ and $\mathbf{v} = (3, 7)$.

11. Find the components of $\mathbf{t} = (-3, 4)$ in the direction of $\mathbf{u} = (2, -1)$ and $\mathbf{v} = (0, 5)$.

11-6 **12.** A plane's heading is 43° and its air speed is 290 mph. If a west wind is blowing at 15 mph, what are the plane's ground speed and true course?

13. Resolve the vector with magnitude 14 and direction 120° with the horizontal into horizontal and vertical components.

14. Use the figure at the left below. Find the work done by **F** in moving the block from A to B.

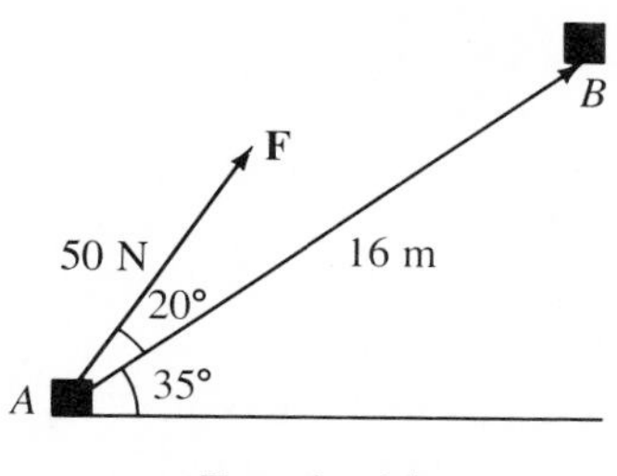

Exercise 14

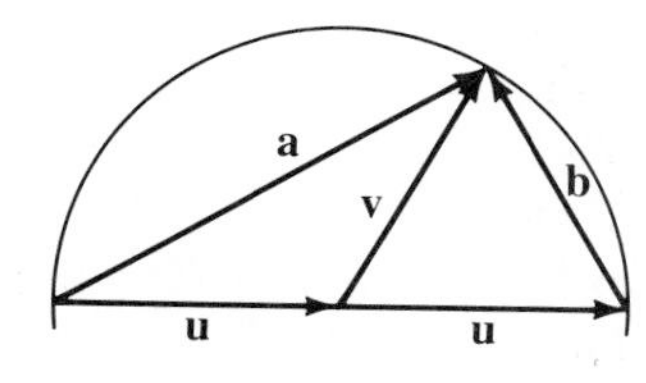

Exercise 15

11-7 **15.** Give a vector proof that the measure of an angle inscribed in a semicircle is 90°. Use the figure at the right above and the fact that **u** and **v** represent radii of the circle.

11-8 **16.** Use the vector space axioms to show for any vector **v** and any scalar r that $(-r)\mathbf{v} = -(r\mathbf{v})$.

Vector Spaces and Subspaces

A nonempty subset U of a vector space V is a **subspace** of V if U is also a vector space.

Let V be the set of all vectors in the plane. Then with addition and scalar multiplication, V is a vector space. Let $\mathbf{v}$ be a nonzero member of V and let $U = \{k\mathbf{v}: k \in \mathcal{R}\}$. Then U is a nonempty subset of V. It is easy to see that U is closed under addition and scalar multiplication. Let $k\mathbf{v}$ and $m\mathbf{v}$ be members of U. Then

$$k\mathbf{v} + m\mathbf{v} = (k + m)\mathbf{v}.$$

Since $k + m \in \mathcal{R}$, $k\mathbf{v} + m\mathbf{v} \in U$. Therefore U is closed under addition. Let $a \in \mathcal{R}$. Then

$$a(k\mathbf{v}) = (ak)\mathbf{v}.$$

Since $ak \in \mathcal{R}$, $a(k\mathbf{v}) \in U$. Hence U is closed under scalar multiplication.

The figure below shows U and V. Notice that $\mathbf{0} \in U$ since $0\mathbf{v} = \mathbf{0}$. The fact that U is a subspace of V can be stated as "Every line containing the origin is a subspace of the plane."

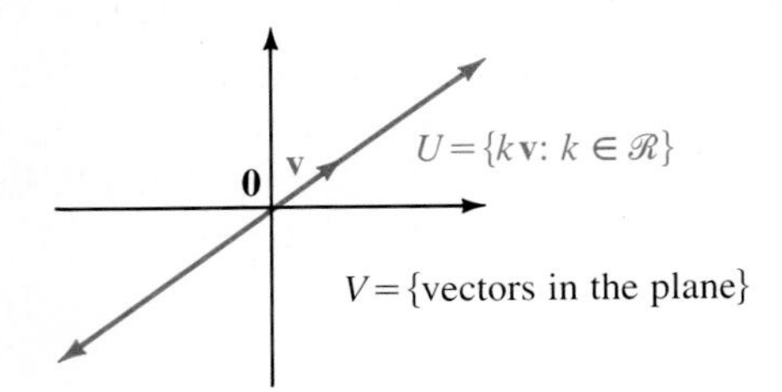

EXERCISES

Let V be the set of all vectors in the plane with addition and scalar multiplication. Let $\mathbf{v}$ be a nonzero member of V and let $U = \{k\mathbf{v}: k \in \mathcal{R}\}$.

1. Show that $-k\mathbf{v}$ is the additive inverse of $k\mathbf{v}$ by giving a reason for each step.

a. $k\mathbf{v} + (-k\mathbf{v}) = k\mathbf{v} + (-k)\mathbf{v}$
b. $= (k + (-k))\mathbf{v}$
c. $= 0\mathbf{v}$
d. $= \mathbf{0}$

2. Suppose that $\mathbf{v} = (1, 3)$ and that $U = \{k(1, 3): k \in \mathcal{R}\}$. Is $(2, 3)$ a member of U?

3. Is $\{\mathbf{0}\}$ a subspace of V?

4. Let $\mathbf{u}$ and $\mathbf{v}$ be nonparallel vectors in the plane and let $U = \{a\mathbf{u} + b\mathbf{v}: a, b \in \mathcal{R}\}$. Show that U equals the set of all vectors in the plane.

In Exercises 5–8, let V be the set of all vectors in three-space with addition and scalar multiplication. Let $\mathbf{i} = (1, 0, 0)$, let $\mathbf{j} = (0, 1, 0)$, and let $\mathbf{k} = (0, 0, 1)$.

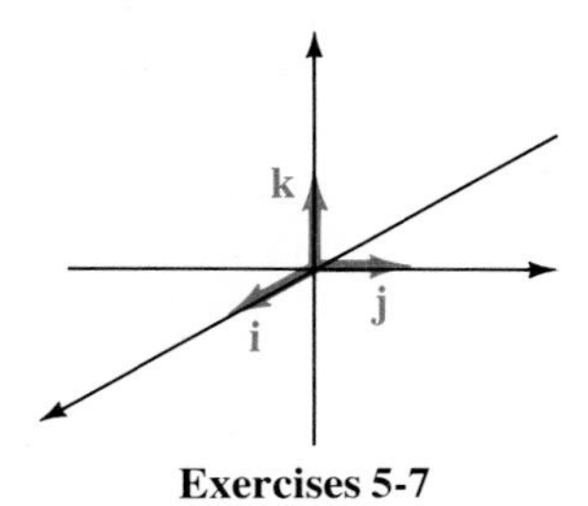

Exercises 5-7

5. Describe geometrically the subspace $U = \{k\mathbf{j}: k \in \mathcal{R}\}$.

6. Describe geometrically $W = \{a\mathbf{i} + b\mathbf{j}: a, b \in \mathcal{R}\}$.

7. Is $\mathbf{k}$ a member of W?

8. Let $\mathbf{u}$ and $\mathbf{v}$ be nonparallel vectors in three-space. Let

$$T = \{a\mathbf{u} + b\mathbf{v}: a, b \in \mathcal{R}\}.$$

Give a geometric description of T and justify your answer.

9. Let $P_2 = \{ax^2 + bx + c: a, b, c \in \mathcal{R}\}$. Show that the subset $P_1 = \{ax + b: a, b \in \mathcal{R}\}$ is a subspace of P_2. Addition is defined as ordinary addition of polynomials and scalar multiplication is defined as multiplication of a polynomial by a real number.

10. Is the following statement true or false? Explain your answer. If V is a vector space and U and W are subspaces of V, then $U \cap W$ is a subspace of V.

11. Is the following statement true or false? Explain your answer. If V is a vector space and U and W are subspaces of V, then $U \cup W$ is a subspace of V.

12. Let $v = (x, y)$ where $x + y = 2$. Let $S = \{kv: k \in \mathcal{R}\}$. Explain why S cannot be a subspace of the space of all vectors in the plane.

13. Let $v = (x, y)$ where $ax + by = c$ and a, b, and c are real numbers. Let $S = \{kv: k \in \mathcal{R}\}$. Under what conditions placed on a, b, and c will S be a subspace of the space of all vectors in the plane?

APPLICATION

Uniform Circular Motion

Vectors can be used to study the motion of a particle traveling with constant angular speed in a circular path.

To do so, suppose that P travels counterclockwise around a circle of radius d centered at the origin. Suppose further that it has constant angular speed ω radians per second and at time $t = 0$, P is at $(d, 0)$. Then at time t, the coordinates of P are $(d \cos \omega t, d \sin \omega t)$, and the position vector of P, $\mathbf{r}(t)$, is

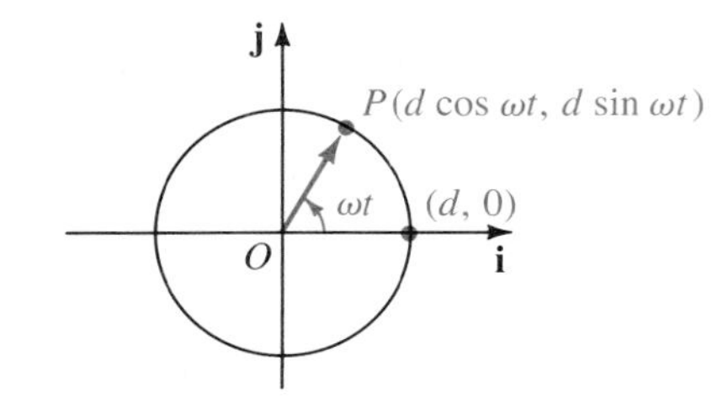

$$\mathbf{r}(t) = (d \cos \omega t)\mathbf{i} + (d \sin \omega t)\mathbf{j}.$$

If, for example, the circle has radius 1 and P has angular speed $\omega = \frac{\pi}{9}$ radians per second, then after 3 seconds, P has position vector

$$\mathbf{r}(3) = \cos\left(\frac{\pi}{9} \cdot 3\right)\mathbf{i} + \sin\left(\frac{\pi}{9} \cdot 3\right)\mathbf{j} = \frac{1}{2}\mathbf{i} + \frac{\sqrt{3}}{2}\mathbf{j}.$$

We define the velocity vector, $\mathbf{v}(t)$, as the derivative of the position vector $\mathbf{r}(t)$, and the acceleration vector, $\mathbf{a}(t)$, as the derivative of $\mathbf{v}(t)$. In the case of uniform circular motion, we have

$$\mathbf{v}(t) = \frac{d}{dt}(\mathbf{r}(t)) = \frac{d}{dt}(d \cos \omega t)\mathbf{i} + \frac{d}{dt}(d \sin \omega t)\mathbf{j}.$$

Using the formulas for the derivatives of trigonometric functions from Section 9-7,

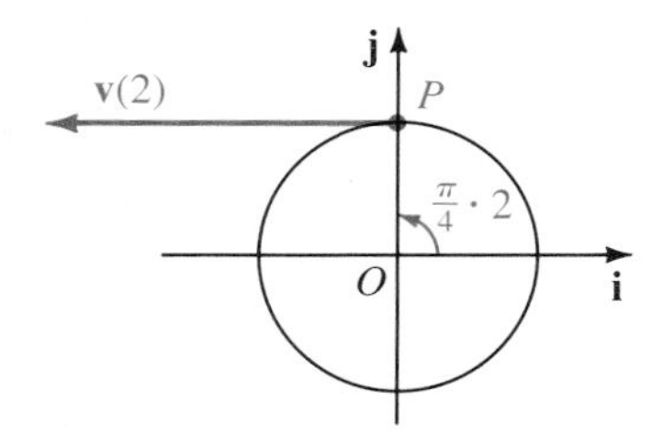

$$\mathbf{v}(t) = (-\omega d \sin \omega t)\mathbf{i} + (\omega d \cos \omega t)\mathbf{j}.$$

If, for example, $d = 2$, $\omega = \frac{\pi}{4}$ radians per second, and $t = 2$ seconds, then

$$\mathbf{v}(2) = (-2)\left(\frac{\pi}{4}\right)\sin\left(\frac{\pi}{4} \cdot 2\right)\mathbf{i} + 2 \cdot \frac{\pi}{4}\cos\left(\frac{\pi}{4} \cdot 2\right)\mathbf{j} = -\frac{\pi}{2}\mathbf{i}.$$

Thus P has a speed of $\frac{\pi}{2}$ radians per second and direction opposite that of $\mathbf{i}$.

EXERCISES

In Exercises 1–5 suppose that P travels counterclockwise around a circle of radius d centered at the origin and suppose that P has constant angular velocity ω radians per second.

1. Show that the speed, $\|\mathbf{v}(t)\|$, of P is constant and equal to $d\omega$.
2. a. Show that $\mathbf{a}(t)$ is given by $\mathbf{a}(t) = (-d\omega^2 \cos \omega t)\mathbf{i} + (-d\omega^2 \sin \omega t)\mathbf{j}$.
   b. Show that $\|\mathbf{a}(t)\|$ is constant and equal to $d\omega^2$.
3. Show that velocity and acceleration are perpendicular by computation of $\mathbf{a}(t) \cdot \mathbf{v}(t)$.
4. Show that $\mathbf{a}(t) = -\omega^2\, \mathbf{r}(t)$ and therefore that acceleration has direction opposite that of $\mathbf{r}(t)$.
5. Newton's second law states that $\mathbf{F} = m\mathbf{a}$ where m is the mass of an object and $\mathbf{a}$ is its acceleration. If an object with mass m is in uniform circular motion, then $\mathbf{F}$ is called **centripetal force**, the force tending to pull the object toward the center of the circle. Show that an object with mass m and constant *linear* speed c traveling in uniform circular motion has a centripetal force $\mathbf{F}$ with magnitude $\|\mathbf{F}\| = \dfrac{mc^2}{d}$.
6. Show that the velocity and acceleration of a point P traveling in uniform circular motion do not depend on the center of the circle being located at the origin. (*Hint:* Suppose that the circle has center (r_1, r_2). Find $\mathbf{r}(t)$. Then find $\mathbf{v}(t)$ and $\mathbf{a}(t)$.)

A centrifuge is used to test the effects of high-speed circular motion on astronauts.

Chapter 12

Matrices, Determinants, and Systems of Linear Equations

In this chapter linear transformations and matrices are discussed. The discussion of conic sections is extended. Methods of solving systems of linear equations are investigated.

Transformations and Matrices

■ 12-1 LINEAR TRANSFORMATIONS

An important concept in mathematics and its applications is the concept of *linear transformation*.

DEFINITION

A **linear transformation** of a vector space V into a vector space V' is a function T having domain V and range in V' such that

$$T(\mathbf{u} + \mathbf{v}) = T(\mathbf{u}) + T(\mathbf{v}) \quad \text{and} \quad T(a\mathbf{u}) = aT(\mathbf{u})$$

for all $\mathbf{u}, \mathbf{v} \in V$ and $a \in \mathcal{R}$.

The two equations in the definition can be replaced by the single equation:

$$T(a\mathbf{u} + b\mathbf{v}) = aT(\mathbf{u}) + bT(\mathbf{v}).$$

The photograph shows a grain ship at its port of call. Systems of linear equations are often used to solve problems involving transportation and cargo shipments.

It is customary to refer to $T(\mathbf{u})$ as the **image** of $\mathbf{u}$ under T. Thus T is linear if and only if the image of each linear combination of vectors is the same linear combination of their images.

EXAMPLE 1 Let V be any vector space and k be a scalar. Show that the transformation defined by $T(\mathbf{v}) = k\mathbf{v}$ is linear.

Solution

$$\begin{aligned} T(a\mathbf{u} + b\mathbf{v}) &= k(a\mathbf{u} + b\mathbf{v}) && \text{definition of } T \\ &= ka\mathbf{u} + kb\mathbf{v} \\ &= ak\mathbf{u} + bk\mathbf{v} \\ &= a(k\mathbf{u}) + b(k\mathbf{v}) \\ &= aT(\mathbf{u}) + bT(\mathbf{v}) && \text{definition of } T \end{aligned}$$

In Example 1 we showed that scalar multiplication is linear. The following figures illustrate the linearity. For simplicity $a = b = 1$ and $k = 2$.

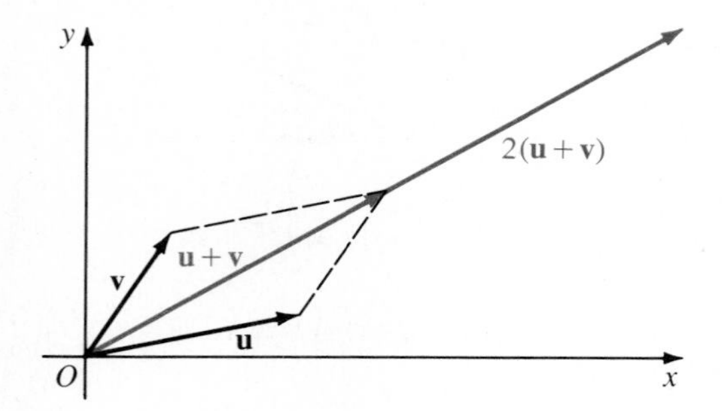

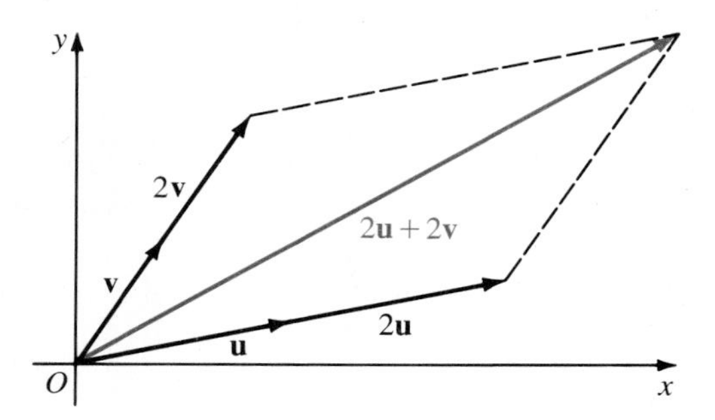

The next example illustrates how to find the image of a vector under a linear transformation T when the images under T of the basis vectors are known.

EXAMPLE 2 If T is a linear transformation of the plane, $T(\mathbf{i}) = (2, 5)$, and $T(\mathbf{j}) = (-4, 7)$, find $T(\mathbf{v})$ where $\mathbf{v} = (1, 3)$.

Solution $\mathbf{v} = (1, 3) = \mathbf{i} + 3\mathbf{j}$. Hence

$$\begin{aligned} T(\mathbf{v}) &= T(\mathbf{i} + 3\mathbf{j}) \\ &= T(\mathbf{i}) + 3T(\mathbf{j}) \\ &= (2, 5) + 3(-4, 7) \\ &= (2, 5) + (-12, 21) \\ &= (-10, 26) \quad \textbf{Answer} \end{aligned}$$

A linear transformation is completely determined by its action on the standard basis vectors. The following theorem states that a linear transformation T of the plane is completely determined by its action on $\mathbf{i} = (1, 0)$ and $\mathbf{j} = (0, 1)$. You are asked to prove the theorem in Exercise 27.

THEOREM 1 Let T be a linear transformation of the plane and let

$$T(\mathbf{i}) = (p, q) \quad \text{and} \quad T(\mathbf{j}) = (r, s).$$

If $\mathbf{v} = (x, y)$, then

$$T(\mathbf{v}) = T((x, y)) = (px + ry, qx + sy).$$

That is, if $T((x, y)) = (x', y')$, then

$$x' = px + ry \quad \text{and} \quad y' = qx + sy.$$

An important linear transformation of the plane is the rotation, T_α, of the plane about the origin through an angle α. See the figure at the left below.

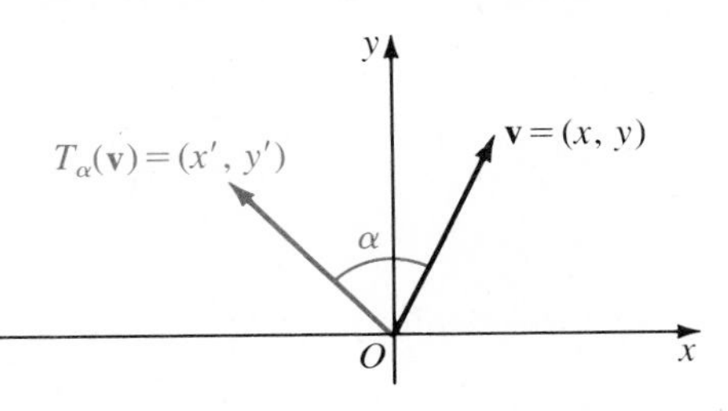

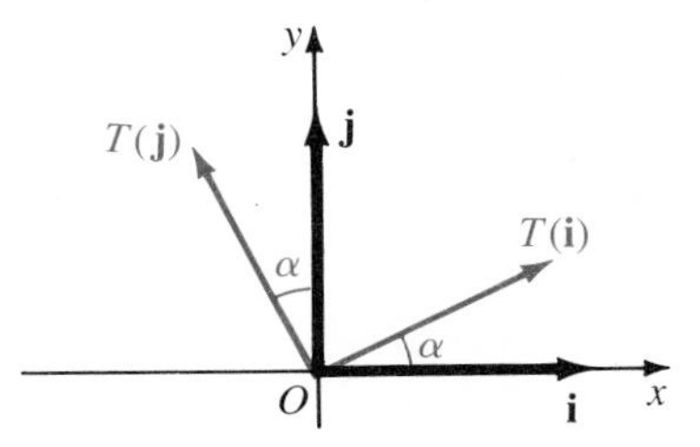

Since the norm of both $T_\alpha(\mathbf{i})$ and $T_\alpha(\mathbf{j})$ is 1, we see from the figure at the right above that

$$T_\alpha(\mathbf{i}) = (\cos\alpha, \sin\alpha) \quad \text{and} \quad T_\alpha(\mathbf{j}) = (\cos(\alpha + 90°), \sin(\alpha + 90°)) = (-\sin\alpha, \cos\alpha).$$

In the notation of Theorem 1, we have

$$(p, q) = (\cos\alpha, \sin\alpha) \quad \text{and} \quad (r, s) = (-\sin\alpha, \cos\alpha).$$

Therefore $\quad x' = x\cos\alpha - y\sin\alpha, \quad y' = x\sin\alpha + y\cos\alpha.$

EXAMPLE 3 The plane is rotated about the origin through a 120° angle.

a. Find the image of $(-4, 0)$. **b.** Find the image of $(6, 2)$.

Solution Equations of the rotation $T_{120°}$ are:

$$x' = x\cos 120° - y\sin 120° = -\frac{1}{2}x - \frac{\sqrt{3}}{2}y \quad \Big| \quad y' = x\sin 120° + y\cos 120° = \frac{\sqrt{3}}{2}x - \frac{1}{2}y$$

a. If $(x, y) = (-4, 0)$, then

$$x' = -\frac{1}{2}(-4) - \frac{\sqrt{3}}{2}(0) = 2 \quad \Big| \quad y' = \frac{\sqrt{3}}{2}(-4) - \frac{1}{2}(0) = -2\sqrt{3}$$

Therefore the image of $(-4, 0)$ is $(2, -2\sqrt{3})$.

b. If $(x, y) = (6, 2)$,

$$x' = -\frac{1}{2}\cdot 6 - \frac{\sqrt{3}}{2}\cdot 2 = -3 - \sqrt{3} \approx -4.73 \quad \Big| \quad y' = \frac{\sqrt{3}}{2}\cdot 6 - \frac{1}{2}\cdot 2 = 3\sqrt{3} - 1 \approx 4.20$$

We can also consider linear transformations from a vector space, such as three-space, into a vector space, such as two-space.

EXAMPLE 4 Show that the transformation defined by $T((x, y, z)) = (x + y, y - z)$ is linear.

Solution Let $\mathbf{u} = (u_1, u_2, u_3)$ and $\mathbf{v} = (v_1, v_2, v_3)$. Then

$$\begin{aligned}
T(\mathbf{u} + \mathbf{v}) &= T((u_1, u_2, u_3) + (v_1, v_2, v_3)) \\
&= T((u_1 + v_1, u_2 + v_2, u_3 + v_3)) \\
&= ((u_1 + v_1) + (u_2 + v_2), (u_2 + v_2) - (u_3 + v_3)) && \text{definition of } T \\
&= ((u_1 + u_2) + (v_1 + v_2), (u_2 - u_3) + (v_2 - v_3)) \\
&= (u_1 + u_2, u_2 - u_3) + (v_1 + v_2, v_2 - v_3) \\
&= T(\mathbf{u}) + T(\mathbf{v}) && \text{definition of } T
\end{aligned}$$

$$\begin{aligned}
T(a\mathbf{u}) &= T(a(u_1, u_2, u_3)) \\
&= T((au_1, au_2, au_3)) \\
&= (au_1 + au_2, au_2 - au_3) && \text{definition of } T \\
&= a(u_1 + u_2, u_2 - u_3) \\
&= aT(\mathbf{u}) && \text{definition of } T
\end{aligned}$$

EXAMPLE 5 If T is a linear transformation from three-space into two-space and $T(\mathbf{i}) = (1, 0)$, $T(\mathbf{j}) = (1, 1)$, $T(\mathbf{k}) = (0, -1)$, find **(a)** $T((3, -1, 2))$ and **(b)** $T((x, y, z))$.

Solution **a.**
$$\begin{aligned}
T((3, -1, 2)) &= T(3\mathbf{i} - \mathbf{j} + 2\mathbf{k}) \\
&= 3T(\mathbf{i}) - T(\mathbf{j}) + 2T(\mathbf{k}) \\
&= 3(1, 0) - (1, 1) + 2(0, -1) \\
&= (3, 0) + (-1, -1) + (0, -2) \\
&= (3 - 1 + 0, 0 - 1 - 2) = (2, -3)
\end{aligned}$$

Therefore $T((3, -1, 2)) = (2, -3)$. **Answer**

b.
$$\begin{aligned}
T((x, y, z)) &= T(x\mathbf{i} + y\mathbf{j} + z\mathbf{k}) \\
&= xT(\mathbf{i}) + yT(\mathbf{j}) + zT(\mathbf{k}) \\
&= x(1, 0) + y(1, 1) + z(0, -1) \\
&= (x, 0) + (y, y) + (0, -z) \\
&= (x + y + 0, 0 + y - z) = (x + y, y - z)
\end{aligned}$$

Therefore $T((x, y, z)) = (x + y, y - z)$. **Answer**

EXERCISES

Find the image of $\mathbf{v}$ under T.

A

1. $\mathbf{v} = (1, 0)$; $T((x, y)) = (x + y, x - y)$
2. $\mathbf{v} = (2, 3)$; $T((x, y)) = (2x - y, x + 3y)$
3. $\mathbf{v} = (-3, -3)$; $T((x, y)) = (3x + 4y, 4x + 3y)$
4. $\mathbf{v} = (5, 2)$; $T((x, y)) = (-6x - 5y, 3x + 4y)$

Find the image of $\mathbf{v}$ under T_α.

5. $\mathbf{v} = (1, 0)$; $\alpha = -45°$

6. $\mathbf{v} = (1, 1)$; $\alpha = 60°$

7. $\mathbf{v} = (-2, 2)$; $\alpha = 90°$

8. $\mathbf{v} = (4, 5)$; $\alpha = 135°$

Find $\mathbf{v}$ given its image $T(\mathbf{v})$ and T.

9. $T(\mathbf{v}) = (0, 0)$; $T((x, y)) = (3x - 7y, 8x + 2y)$

10. $T(\mathbf{v}) = (4, 20)$; $T((x, y)) = (2x - 4y, 5x + y)$

11. $T(\mathbf{v}) = (-3, 8)$; $T((x, y)) = (3x - 7y, 4x + 3y)$

12. $T(\mathbf{v}) = (0, 0)$; $T((x, y)) = (5x + 6y, -3x - 6y)$

Let $\mathbf{v} = (x, y)$. Use Theorem 1 as a guide to write $T(\mathbf{v})$.

13. $T(\mathbf{i}) = (3, 4)$; $T(\mathbf{j}) = (-2, 2)$

14. $T(\mathbf{i}) = (-3, 6)$; $T(\mathbf{j}) = (10, 10)$

15. $T(\mathbf{i}) = (-1, 0)$; $T(\mathbf{j}) = (0, 1)$

16. $T(\mathbf{i}) = (0, 1)$; $T(\mathbf{j}) = (1, 0)$

17. The **identity transformation** I of the plane is defined by $I((x, y)) = (x, y)$. Show that I is linear. (*Hint*: Use an argument similar to the one that shows scalar multiplication is linear.)

B

18. A reflection R about the $\mathbf{j}$-axis is defined by $R((x, y)) = (-x, y)$. Show that R is a linear transformation of the plane.

19. A reflection R about the $\mathbf{i}$-axis is defined by $R((x, y)) = (x, -y)$. Show that R is linear.

20. A reflection R about the line containing $\mathbf{v} = (1, 1)$ is defined by $R((x, y)) = (y, x)$. Show that R is linear.

21. Show that the transformation R defined by $R((x, y)) = (k, k)$, where k is a nonzero scalar, is not linear.

22. Show that the transformation $R((x, y)) = (x^2, y^2)$ is not linear.

Show that each transformation T is linear.

23. $T((x, y)) = (3x - 7y, 8x + 2y)$

24. $T((x, y)) = (-5y, 3x + 9y)$

25. $T((x, y, z)) = (-2x + 4y - 11z, 3x + 7z, z)$

26. $T((x, y, z)) = (13y - 9z, 8x, 5x - z)$

27. Prove Theorem 1.

Find the image of $\mathbf{v}$ under S followed by T.

28. $\mathbf{v} = (-4, 6)$; S a rotation through 45°; T scalar multiplication by 4

29. $\mathbf{v} = (3, -5)$; S a rotation through $-90°$; T reflection in the line $y = x$ (*Hint*: Use Exercise 20.)

30. $\mathbf{v} = (-2, 6)$; S a rotation through 60°; T reflection in the $\mathbf{i}$-axis (*Hint*: Use Exercise 19.)

31. $\mathbf{v} = (4, 9)$; S scalar multiplication by 2; T rotation through 90°

32. $\mathbf{v} = (3, 7)$; S reflection in the $\mathbf{j}$-axis; T scalar multiplication by 5

In Exercises 33–38 write a single linear transformation that is equivalent to the first transformation followed by the second transformation.

33. Rotation through 60° followed by rotation through $-15°$

34. Reflection in the **i**-axis followed by reflection in the **j**-axis

35. Scalar multiplication by 3 followed by rotation through 180°

36. Reflection in the line $y = x$ followed by reflection in the line $y = -x$

37. Rotation through 90° followed by scalar multiplication by -1

38. Reflection in the **i**-axis followed by rotation through 90°

39. Show that if T is a linear transformation of three-space, $\mathbf{v} = (x, y, z)$, $T(\mathbf{i}) = (r, s, t)$, $T(\mathbf{j}) = (m, n, p)$, and $T(\mathbf{k}) = (w, k, q)$, then $T(\mathbf{v})$ is given by $T(\mathbf{v}) = (xr + ym + zw, xs + yn + zk, xt + yp + zq)$.

40. Show that if a vector **v** makes an angle φ with the **i**-axis, then the reflection of **v** in a line making an angle θ with the **i**-axis is a rotation of **v** through an angle $2(\theta - \varphi)$.

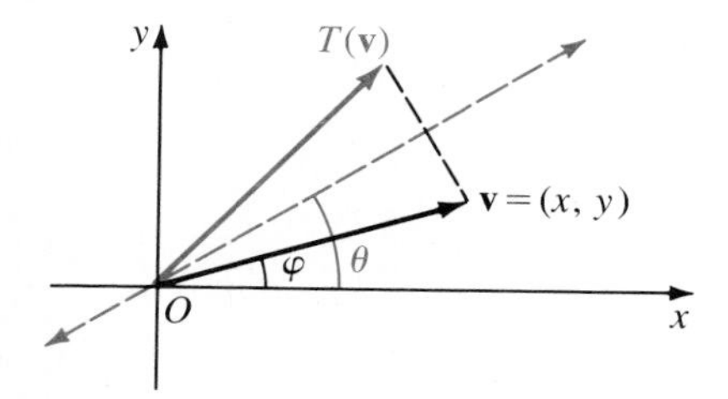

COMPUTER EXERCISES

1. Write a program that will find the image of a vector, **v**, in the plane under a rotation through angle α, scalar multiplication by k, or reflection about an axis. The user should input **v** and α, k, or an axis as appropriate.

2. Run the program to find the image under the transformation given.
   a. $\mathbf{v} = (-1, 0)$; a rotation through 30°
   b. $\mathbf{v} = (3, 2)$; a reflection about the **j**-axis
   c. $\mathbf{v} = (4, -2)$; a scalar multiplication by $\frac{1}{2}$

3. Modify the program of Exercise 1 so that transformations can be performed one after another.

4. Run the program of Exercise 3 to find the image of $(3, -5)$ under a rotation through 135° and then a scalar multiplication by -2.

5. Run the program of Exercise 3 to find the image of $(1, -7)$ under a rotation through 90°, a reflection about the **i**-axis, a scalar multiplication by -1, a reflection about the origin, and finally a scalar multiplication by 5.

12-2 MATRICES

In Section 12–1 we saw that a linear transformation is completely determined by its action on a basis. We can describe such actions by using a *matrix*. The transformation T in Example 5 has the following action on $\{\mathbf{i}, \mathbf{j}, \mathbf{k}\}$.

$$T(\mathbf{i}) = (1, 0) \qquad T(\mathbf{j}) = (1, 1) \qquad T(\mathbf{k}) = (0, -1)$$

We can write this information in a rectangular array, called a matrix:

$$\begin{bmatrix} 1 & 1 & 0 \\ 0 & 1 & -1 \end{bmatrix}$$

(The reason for writing the vectors (1, 0), (1, 1) and (0, −1) as columns rather than rows will appear in the next section.)

DEFINITION

A **matrix** is a rectangular array of numbers, which are called its **entries**, or **elements**. If a matrix has m rows and n columns, we say that it has **dimensions** m by n, written $m \times n$. If $m = n$, the matrix is **square**. A matrix with only one row (column) is called a **row (column) matrix**. Two matrices are **equal** if their corresponding entries are equal.

$$\begin{bmatrix} 2 & -1 \\ -2 & 0 \\ 3 & 2 \end{bmatrix} \qquad \begin{bmatrix} 1 & 0 \\ 0 & 0 \end{bmatrix} \qquad \begin{bmatrix} -2 \\ 1 \\ 3 \end{bmatrix}$$

3×2 matrix $\qquad$ 2×2 (square) matrix $\qquad$ 3×1 (column) matrix

DEFINITION

The **sum** of two matrices A and B of the same dimensions is the matrix $A + B$, each entry of which is the sum of the corresponding entries in A and B. The **product of a scalar *r* and a matrix** A is the matrix rA, each entry of which is r times the corresponding entry of A.

For example, if $\quad A = \begin{bmatrix} 3 & 1 \\ -2 & 0 \end{bmatrix}$ and $B = \begin{bmatrix} -1 & 7 \\ 6 & 1 \end{bmatrix}$,

then $A + B = \begin{bmatrix} 3 & 1 \\ -2 & 0 \end{bmatrix} + \begin{bmatrix} -1 & 7 \\ 6 & 1 \end{bmatrix} = \begin{bmatrix} 3 + (-1) & 1 + 7 \\ -2 + 6 & 0 + 1 \end{bmatrix} = \begin{bmatrix} 2 & 8 \\ 4 & 1 \end{bmatrix}$.

The **additive inverse** of a matrix A, denoted $-A$, is the matrix whose entries are the additive inverses of the corresponding entries of A. The **difference** of two matrices A and B, written $A - B$, is defined by $A + (-B)$. The difference is defined only for matrices of the same dimensions.

EXAMPLE 1 Let $A = \begin{bmatrix} 2 & -1 \\ -2 & 0 \\ 3 & 2 \end{bmatrix}$ and $B = \begin{bmatrix} 3 & 1 \\ 2 & 0 \\ 1 & 2 \end{bmatrix}$.

Find $2A - 3B$.

Solution First find $2A$ and $-3B$.

$$2A = \begin{bmatrix} 2 \cdot 2 & (2)(-1) \\ (2)(-2) & 2 \cdot 0 \\ 2 \cdot 3 & 2 \cdot 2 \end{bmatrix} = \begin{bmatrix} 4 & -2 \\ -4 & 0 \\ 6 & 4 \end{bmatrix}$$

$$-3B = \begin{bmatrix} (-3)(3) & (-3)(1) \\ (-3)(2) & (-3)(0) \\ (-3)(1) & (-3)(2) \end{bmatrix} = \begin{bmatrix} -9 & -3 \\ -6 & 0 \\ -3 & -6 \end{bmatrix}$$

Then add the resulting matrices.

$$2A - 3B = \begin{bmatrix} 4 & -2 \\ -4 & 0 \\ 6 & 4 \end{bmatrix} + \begin{bmatrix} -9 & -3 \\ -6 & 0 \\ -3 & -6 \end{bmatrix} = \begin{bmatrix} -5 & -5 \\ -10 & 0 \\ 3 & -2 \end{bmatrix}$$

A **zero** matrix, denoted O, is one in which all entries are 0. The $n \times n$ **identity** matrix, I, has 1 in row i and column i for $1 \leq i \leq n$ and 0 elsewhere.

$$\begin{bmatrix} 0 & 0 & 0 \\ 0 & 0 & 0 \end{bmatrix} \qquad \begin{bmatrix} 1 & 0 & 0 \\ 0 & 1 & 0 \\ 0 & 0 & 1 \end{bmatrix}$$

2×3 zero matrix $\qquad$ 3×3 identity matrix

It can be shown in a straightforward way that the set of all $m \times n$ matrices with addition and scalar multiplication is a vector space. We shall use this fact. In particular, addition is commutative, and multiplication by a scalar distributes over sums.

We often regard a row matrix as a vector and a column matrix as a vector in column form.

DEFINITION

Let A be an $m \times n$ matrix and B be an $n \times k$ matrix. Their **product**, AB, is the $m \times k$ matrix whose entry in the ith row and jth column is the dot product of the ith row of A and the jth column of B.

Powers of square matrices are defined as for real numbers. That is, $A^2 = AA$, and $A^3 = AAA$. In general,

$$A^n = \underbrace{AAA \cdots A}_{n \text{ factors}}$$

EXAMPLE 2 Let $A = \begin{bmatrix} 1 & 0 & 3 \\ 2 & -1 & 4 \end{bmatrix}$ and $B = \begin{bmatrix} 3 \\ 1 \\ -2 \end{bmatrix}$. Find AB.

Solution Since A is a 2×3 matrix and B is a 3×1 matrix, the product exists and is a 2×1 matrix.

The dot product of the first row of A and the first column of B gives the entry in the first row and first column of AB. This is shown in red.

$$AB = \begin{bmatrix} 1 & 0 & 3 \\ 2 & -1 & 4 \end{bmatrix} \begin{bmatrix} 3 \\ 1 \\ -2 \end{bmatrix} = \begin{bmatrix} 1 \cdot 3 + 0 \cdot 1 + 3 \cdot (-2) \\ 2 \cdot 3 + (-1)(1) + 4(-2) \end{bmatrix} = \begin{bmatrix} -3 \\ -3 \end{bmatrix}$$

Answer

Notice that the product BA is not defined since the number of columns of B does not equal the number of rows of A.

EXAMPLE 3 Let $A = \begin{bmatrix} 2 & -1 \\ 4 & -2 \end{bmatrix}$, $B = \begin{bmatrix} 3 & 0 \\ 1 & 2 \end{bmatrix}$, and $C = \begin{bmatrix} 1 & -2 \\ 2 & -4 \end{bmatrix}$.

Find AB, BA, and AC.

Solution Each product will be a 2×2 matrix.

$$AB = \begin{bmatrix} 2 & -1 \\ 4 & -2 \end{bmatrix} \begin{bmatrix} 3 & 0 \\ 1 & 2 \end{bmatrix}$$

$$= \begin{bmatrix} 2 \cdot 3 + (-1) \cdot 1 & 2 \cdot 0 + (-1) \cdot 2 \\ 4 \cdot 3 + (-2) \cdot 1 & 4 \cdot 0 + (-2) \cdot 2 \end{bmatrix} = \begin{bmatrix} 5 & -2 \\ 10 & -4 \end{bmatrix}$$ **Answer**

$$BA = \begin{bmatrix} 3 & 0 \\ 1 & 2 \end{bmatrix} \begin{bmatrix} 2 & -1 \\ 4 & -2 \end{bmatrix}$$

$$= \begin{bmatrix} 3 \cdot 2 + 0 \cdot 4 & 3 \cdot (-1) + 0 \cdot (-2) \\ 1 \cdot 2 + 2 \cdot 4 & 1 \cdot (-1) + 2 \cdot (-2) \end{bmatrix} = \begin{bmatrix} 6 & -3 \\ 10 & -5 \end{bmatrix}$$ **Answer**

$$AC = \begin{bmatrix} 2 & -1 \\ 4 & -2 \end{bmatrix} \begin{bmatrix} 1 & -2 \\ 2 & -4 \end{bmatrix}$$

$$= \begin{bmatrix} 2 \cdot 1 + (-1) \cdot 2 & 2 \cdot (-2) + (-1)(-4) \\ 4 \cdot 1 + (-2) \cdot 2 & 4 \cdot (-2) + (-2)(-4) \end{bmatrix} = \begin{bmatrix} 0 & 0 \\ 0 & 0 \end{bmatrix}$$ **Answer**

Example 3 shows that matrix multiplication is not, in general, commutative even when both products are defined. It also shows that the product of two nonzero matrices can be the zero matrix. For square matrices we have the results described by the following theorem. In the exercises you are asked to prove the theorem in the case where $n = 2$.

THEOREM 2 Let A, B, and C be $n \times n$ matrices and let r be any real number. Then:

1. $A(BC) = (AB)C$ Multiplication is associative.
2. $A(B + C) = AB + AC$ Multiplication distributes over addition.
3. $(rA)B = A(rB)$

EXERCISES

Let $A = \begin{bmatrix} 5 & -7 \\ 6 & -1 \end{bmatrix}$, $B = \begin{bmatrix} 5 & -1 \\ -9 & 6 \end{bmatrix}$, $C = \begin{bmatrix} 8 & 6 & -1 \\ 0 & -2 & 5 \\ 4 & 3 & 1 \end{bmatrix}$, $D = \begin{bmatrix} 14 & -3 & 2 \\ -4 & 0 & 9 \\ -2 & 0 & 6 \end{bmatrix}$,

$E = \begin{bmatrix} 5 & 8 & -3 \\ -6 & 0 & -4 \end{bmatrix}$, $F = \begin{bmatrix} 2 & 4 & 1 \\ -1 & -3 & 0 \end{bmatrix}$, $G = \begin{bmatrix} 2 & -1 \\ 3 & 5 \\ -1 & 6 \end{bmatrix}$.

Find each matrix.

A

1. $A + B$
2. $A - B$
3. $A - 3B$
4. $2A + B$
5. AB
6. DC
7. FG
8. $3GF$
9. D^2
10. $-2(D + C)$
11. $2(C - 2D)$
12. $CD + C^2$
13. $EG - GE$
14. $(E + F)C$
15. $(B + A)(-A)$
16. $(E + F)(E - F)$

In Exercises 17–19, let $A = \begin{bmatrix} -7 & 3 \\ 5 & 2 \end{bmatrix}$ and $B = \begin{bmatrix} 6 & -1 \\ 9 & -8 \end{bmatrix}$.

17. Does $AB = BA$?
18. **a.** Find $(A + B)(A - B)$.
    b. Find $A^2 - B^2$.
    c. Does $(A + B)(A - B) = A^2 - B^2$? Explain your answer.
19. **a.** Find $(A + B)^2$.
    b. Find $A^2 + 2AB + B^2$.
    c. Does $(A + B)^2 = A^2 + 2AB + B^2$? Explain your answer.

B

20. Show that if A is a 2×2 matrix and I is the 2×2 identity matrix, then $AI = IA = A$.
21. Show that if A is a 3×3 matrix and I is the 3×3 identity matrix, then $IA = AI = A$.

Solve each matrix equation for T.

22. $\begin{bmatrix} -3 & 5 \\ 2 & 6 \end{bmatrix} + T = \begin{bmatrix} -5 & 6 \\ 1 & 0 \end{bmatrix} + 2T$
23. $\begin{bmatrix} -3 & 4 & 5 \\ 0 & 6 & 2 \end{bmatrix} + 2T = \begin{bmatrix} 0 & 4 & 7 \\ 1 & -5 & 8 \end{bmatrix}$

A 3×3 **triangular matrix** has the form $\begin{bmatrix} a & b & c \\ 0 & d & e \\ 0 & 0 & f \end{bmatrix}$.

24. Show that the sum of two such matrices is another such matrix.

25. Show that the product of two such matrices is another such matrix.

Let A, B, and C be 2×2 matrices. Prove each part of Theorem 2.

26. Part (1) **27.** Part (2) **28.** Part (3)

29. Show that if A and B are 3×3 matrices and k is a scalar, $k(A + B) = kA + kB$.

A **symmetric matrix** is one in which the entry in the ith row and jth column is equal to the entry in the jth row and ith column. A 2×2 symmetric matrix has the form $\begin{bmatrix} a & b \\ b & c \end{bmatrix}$.

C **30.** Show that the sum of two symmetric 3×3 matrices is symmetric.

31. Find a 2×2 nonzero matrix A such that $A\begin{bmatrix} 3 & 6 \\ 4 & 8 \end{bmatrix} = \begin{bmatrix} 0 & 0 \\ 0 & 0 \end{bmatrix}$.

32. Find all 2×2 matrices A such that $A^2 - 4I = 0$.

33. Let A and B be 2×2 matrices of the form $\begin{bmatrix} p & -q \\ q & p \end{bmatrix}$. Show that AB and BA have this form and that $AB = BA$.

COMPUTER EXERCISES

1. Write a program that will print the product, if defined, of two matrices T and S. T is an $m \times n$ matrix. S is a $p \times q$ matrix. Have the user input m, n, p, q, and the entries of T and S. If the product is not defined, have the program print an appropriate message.
2. Run the program for each pair of matrices.

a. $T = \begin{bmatrix} 4 & -7 \\ 9 & -3 \end{bmatrix}$; $S = \begin{bmatrix} 5 & 8 \\ -9 & 6 \end{bmatrix}$

b. $T = \begin{bmatrix} 2 & 4 & 9 \\ -3 & 1 & 5 \end{bmatrix}$; $S = \begin{bmatrix} -6 & 2 \\ 3 & 5 \\ -2 & 7 \end{bmatrix}$

c. $T = \begin{bmatrix} -2 & 3 & 8 \\ 1 & 0 & -4 \\ 9 & 13 & 2 \end{bmatrix}$; $S = \begin{bmatrix} -3 & 5 \\ 11 & 6 \end{bmatrix}$

12-3 MATRICES OF LINEAR TRANSFORMATIONS

We can associate with each linear transformation T a matrix $\overline{T}$, called the **matrix of the linear transformation**. The images of **i**, **j**, and **k** under

$$T((x, y, z)) = (x + y, y - z)$$

are $\quad T(\mathbf{i}) = (1, 0) \quad T(\mathbf{j}) = (1, 1) \quad T(\mathbf{k}) = (0, -1)$

and
$$\overline{T} = \begin{bmatrix} 1 & 1 & 0 \\ 0 & 1 & -1 \end{bmatrix}.$$

EXAMPLE 1 Find $\overline{T}$ for each linear transformation T of the plane.

a. scalar multiplication of each vector by k

b. reflection of each vector in the **j**-axis

Solution In each solution find the image of **i** and the image of **j**.

a. $T(\mathbf{i}) = k(1, 0) = (k, 0) \qquad T(\mathbf{j}) = k(0, 1) = (0, k)$

Therefore $\overline{T} = \begin{bmatrix} k & 0 \\ 0 & k \end{bmatrix}$. **Answer**

b. Since $\mathbf{i} = (1, 0)$, $T(\mathbf{i}) = (-1, 0)$. Since $\mathbf{j} = (0, 1)$, $T(\mathbf{j}) = (0, 1)$.

Therefore $\overline{T} = \begin{bmatrix} -1 & 0 \\ 0 & 1 \end{bmatrix}$. **Answer**

Notice that the image under T of each basis vector is written as a column in $\overline{T}$.

Matrices define linear transformations too.

EXAMPLE 2 Find a formula for the linear transformation S having the matrix

$$\overline{S} = \begin{bmatrix} 3 & 1 \\ 0 & 2 \\ -1 & 1 \end{bmatrix}.$$

Solution The domain of S is two-space.

$$S((x, y)) = \begin{bmatrix} 3 & 1 \\ 0 & 2 \\ -1 & 1 \end{bmatrix} \begin{bmatrix} x \\ y \end{bmatrix} = \begin{bmatrix} 3x + y \\ 2y \\ -x + y \end{bmatrix}$$

Thus S is given by $S((x, y)) = (3x + y, 2y, -x + y)$. **Answer**

In Section 12–1 we saw that the rotation T_α of the plane about the origin through an angle α has the following action on **i** and **j**.

$$T_\alpha(\mathbf{i}) = (\cos \alpha, \sin \alpha) \text{ and } T_\alpha(\mathbf{j}) = (-\sin \alpha, \cos \alpha)$$

Thus $\overline{T}_\alpha = \begin{bmatrix} \cos \alpha & -\sin \alpha \\ \sin \alpha & \cos \alpha \end{bmatrix}$.

The close connection between composition of linear transformations and multiplication of matrices is given in the following theorem, which we state without proof.

THEOREM 3 Let S be a linear transformation of a vector space U into a vector space V, let T be a linear transformation of V into W, and let $\overline{S}$ and $\overline{T}$ be their matrices. Then the matrix of the composition $T \circ S$ is $\overline{T}\,\overline{S}$.

EXAMPLE 3 Let T and S be linear transformations having the matrices

$$\overline{T} = \begin{bmatrix} 1 & -1 \\ 1 & 0 \\ 0 & 2 \end{bmatrix} \text{ and } \overline{S} = \begin{bmatrix} 3 & 1 & -1 \\ 2 & 0 & 1 \end{bmatrix} \text{ respectively.}$$

a. Find the domain and range space of T, S, and $T \circ S$.

b. Find the matrix of $T \circ S$.

c. Find $(T \circ S)((1, 3, 4))$.

Solution **a.** The matrix $\overline{T}$ has two columns and three rows. Thus T has domain two-space and range three-space. The matrix $\overline{S}$ has three columns and two rows. Thus S has domain three-space and range two-space. Both the domain and range space of $T \circ S$ are three-space. **Answer**

b. $$\overline{T}\,\overline{S} = \begin{bmatrix} 1 & -1 \\ 1 & 0 \\ 0 & 2 \end{bmatrix} \begin{bmatrix} 3 & 1 & -1 \\ 2 & 0 & 1 \end{bmatrix} = \begin{bmatrix} 1 & 1 & -2 \\ 3 & 1 & -1 \\ 4 & 0 & 2 \end{bmatrix}$$ **Answer**

c. $$(T \circ S)((1, 3, 4)) = \begin{bmatrix} 1 & 1 & -2 \\ 3 & 1 & -1 \\ 4 & 0 & 2 \end{bmatrix} \begin{bmatrix} 1 \\ 3 \\ 4 \end{bmatrix} = \begin{bmatrix} -4 \\ 2 \\ 12 \end{bmatrix} = (-4, 2, 12)$$ **Answer**

EXAMPLE 4 Let S be the rotation of the plane through 90°, and let T be the reflection of the plane in the x-axis.

a. Find the matrices $\overline{S}$ and $\overline{T}$ of S and T.

b. Find the matrices of $S \circ T$ and $T \circ S$.

c. Find $(S \circ T)((1, 2))$ and $(T \circ S)((1, 2))$.

Solution **a.** Substitute 90° for α in $\begin{bmatrix} \cos\alpha & -\sin\alpha \\ \sin\alpha & \cos\alpha \end{bmatrix}$ to obtain $\overline{S} = \begin{bmatrix} 0 & -1 \\ 1 & 0 \end{bmatrix}$.

Reflection T in the x-axis is defined by $T((x, y)) = (x, -y)$. Hence $T(\mathbf{i}) = (1, 0)$, $T(\mathbf{j}) = (0, -1)$, and $\overline{T} = \begin{bmatrix} 1 & 0 \\ 0 & -1 \end{bmatrix}$.

Solution continued on following page

$$\textbf{b. } \overline{ST} = \begin{bmatrix} 0 & -1 \\ 1 & 0 \end{bmatrix}\begin{bmatrix} 1 & 0 \\ 0 & -1 \end{bmatrix} = \begin{bmatrix} 0 & 1 \\ 1 & 0 \end{bmatrix}$$

$$\overline{TS} = \begin{bmatrix} 1 & 0 \\ 0 & -1 \end{bmatrix}\begin{bmatrix} 0 & -1 \\ 1 & 0 \end{bmatrix} = \begin{bmatrix} 0 & -1 \\ -1 & 0 \end{bmatrix}$$

$$\textbf{c. } (S \circ T)((1, 2)) = \begin{bmatrix} 0 & 1 \\ 1 & 0 \end{bmatrix}\begin{bmatrix} 1 \\ 2 \end{bmatrix} = \begin{bmatrix} 2 \\ 1 \end{bmatrix} = (2, 1)$$

$$(T \circ S)((1, 2)) = \begin{bmatrix} 0 & -1 \\ -1 & 0 \end{bmatrix}\begin{bmatrix} 1 \\ 2 \end{bmatrix} = \begin{bmatrix} -2 \\ -1 \end{bmatrix} = (-2, -1)$$

Part (c) of Example 4 is illustrated below.

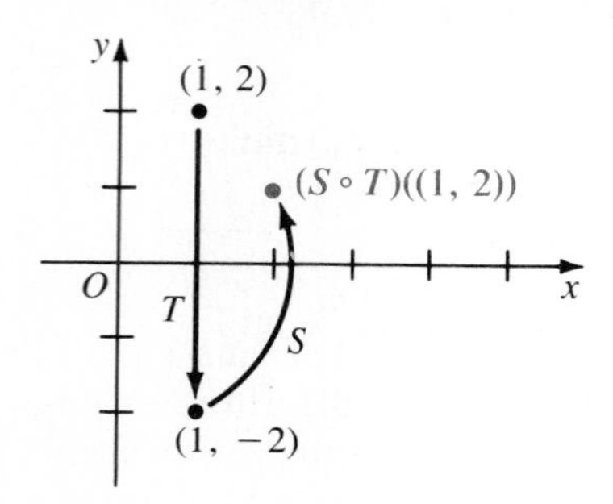

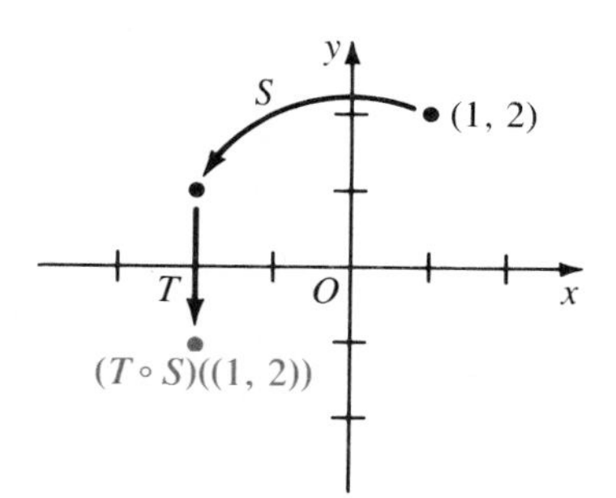

EXERCISES

In Exercises 1–4 **(a)** find $T(\mathbf{i})$ and $T(\mathbf{j})$ and **(b)** write the matrix $\overline{T}$ of T.

A

1. $T((x, y)) = (-x, x + y)$

2. $T((x, y)) = (3x + 2y, -y)$

3. $T((x, y)) = (2y - 3x, 3x - 2y)$

4. $T((x, y)) = (2x + 4y, -2x + y)$

In Exercises 5–9 write the matrix $\overline{T}$ of T.

5. $T((x, y, z)) = (-2x + 4y - 11z, 3x + 7z, x + y - z)$

6. $T((x, y, z)) = (13y - 9z, 8x, 5x - z)$

7. $T((x, y, z)) = (2x - y + z, 3y - 4z, 5x + y)$

8. $T((x, y, z)) = (3x - 7y, 8x + 2y)$

9. $T((x, y, z)) = (-5y, 3x + 9y)$

Find the domain and range space of the linear transformation T having associated matrix $\overline{T}$. Then write the transformation.

10. $\begin{bmatrix} 0 & 4 \\ 3 & 0 \end{bmatrix}$

11. $\begin{bmatrix} 2 & -1 & 3 \\ -1 & 2 & 0 \end{bmatrix}$

12. $\begin{bmatrix} 1 & 2 \\ 3 & 5 \\ -6 & 2 \end{bmatrix}$

13. $\begin{bmatrix} 2 \\ 3 \\ 4 \end{bmatrix}$

Let $S = \begin{bmatrix} k & 0 \\ 0 & k \end{bmatrix}$, $A = \begin{bmatrix} 1 & 0 \\ 0 & -1 \end{bmatrix}$, $B = \begin{bmatrix} 0 & 1 \\ -1 & 0 \end{bmatrix}$, and $C = \begin{bmatrix} 0 & -1 \\ 1 & 0 \end{bmatrix}$.

14. Let $\mathbf{v} = (x, y)$. Find the image of $\mathbf{v}$ under S. Describe geometrically the image of $\mathbf{v}$ if **(a)** $k > 0$ and **(b)** $k < 0$.

15. Let $\mathbf{v} = (x, y)$. Find the image of $\mathbf{v}$ under A. Describe geometrically the action of A.

16. Let $\mathbf{v} = (x, y)$. Find the image of $\mathbf{v}$ under B. Describe geometrically the action of B.

17. Let $\mathbf{v} = (x, y)$. Find the image of $\mathbf{v}$ under C. Describe geometrically the action of C.

18. Describe geometrically the action of $S \circ A$.

19. Describe geometrically the action of $B \circ C$.

B

20. a. Let $\mathbf{v} = (x, y)$. Find the matrix corresponding to scalar multiplication, S, followed by rotation T_α of $S(\mathbf{v})$ through an angle α.

b. Show that $\overline{S}\overline{T_\alpha} = \overline{T_\alpha}\overline{S}$.

21. Show that under any linear transformation T with domain V and range space V' the image of $\mathbf{0} \in V$ is $\mathbf{0} \in V'$.

22. Find a matrix for the linear transformation T such that $T((x, y, z)) = (x, y, 0)$.

C

23. a. Show that $\mathbf{u} = \mathbf{i} + \mathbf{j}$ and $\mathbf{v} = \mathbf{i} - \mathbf{j}$ form a basis for the set of all vectors in the plane.

b. Find a matrix for the linear transformation T such that $T(\mathbf{u}) = 3\mathbf{i} - 7\mathbf{j}$ and $T(\mathbf{v}) = -2\mathbf{i} + 4\mathbf{j}$.

c. Find $T(\mathbf{i})$ and $T(\mathbf{j})$.

The **kernel**, K, of a linear transformation T from V into V' is the set of all vectors $\mathbf{v} \in V$ such that $T(\mathbf{v}) = \mathbf{0}$.

24. Find and describe geometrically the kernel of the linear transformation with matrix $\begin{bmatrix} 3 & -6 \\ -4 & 8 \end{bmatrix}$.

25. Show that if $\mathbf{u}$ and $\mathbf{v}$ are in the kernel K of a linear transformation T, then $a\mathbf{u} + b\mathbf{v}$ is in K for all scalars a and b.

■ 12–4 ROTATION OF AXES

In Section 12–1, when we discussed rotations of the plane about the origin, we kept the coordinate axes fixed and moved the points of the plane. In this section we take a different point of view. We shall keep the points of the plane fixed and rotate the xy-coordinate system to form an $x'y'$-coordinate system. The theorem at the top of the following page describes how the xy-coordinate system and the $x'y'$-coordinate system are related.

THEOREM 4 If the x- and y-axes are rotated through an angle φ into x'- and y'-axes, each point P of the plane has a set of xy-coordinates, (x, y), and a set of $x'y'$-coordinates, (x', y'). These sets of coordinates are related by the equations:

1. $x = x' \cos \varphi - y' \sin \varphi \qquad y = x' \sin \varphi + y' \cos \varphi$

2. $x' = x \cos \varphi + y \sin \varphi \qquad y' = -x \sin \varphi + y \cos \varphi.$

Proof: Let $r = OP$. From triangle OPC, $x' = r \cos \theta$ and $y' = r \sin \theta$.

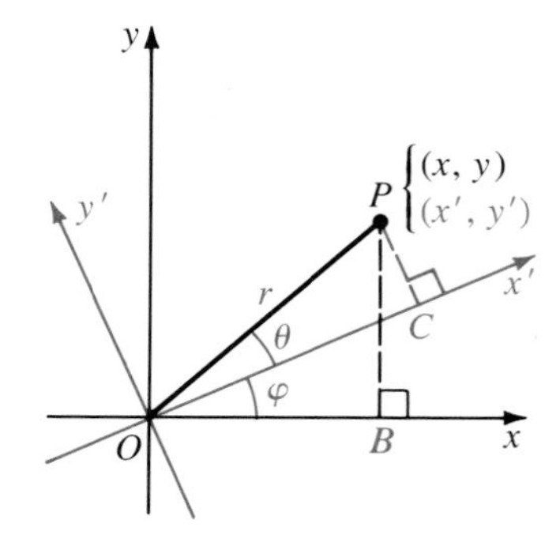

From triangle OPB,

$$\begin{aligned} x &= r \cos (\theta + \varphi) \\ &= r(\cos \theta \cos \varphi - \sin \theta \sin \varphi) \\ &= (r \cos \theta)\cos \varphi - (r \sin \theta)\sin \varphi \end{aligned}$$

and

$$\begin{aligned} y &= r \sin(\theta + \varphi) \\ &= r(\sin \theta \cos \varphi + \cos \theta \sin \varphi) \\ &= (r \cos \theta)\sin \varphi + (r \sin \theta)\cos \varphi \end{aligned}$$

Therefore $x = x' \cos \varphi - y' \sin \varphi$, $y = x' \sin \varphi + y' \cos \varphi$. The other set of equations is obtained by solving these for x' and y' (Exercise 15). ■

EXAMPLE 1 The xy-axes are rotated $-45°$. Find an equation of $x^2 + xy + y^2 = 6$ in the new coordinate system.

Solution Use the equations in part (1) of Theorem 4.

$$x = x' \cos (-45°) - y' \sin (-45°) \qquad y = x' \sin (-45°) + y' \cos (-45°)$$

$$x = \frac{x' + y'}{\sqrt{2}} \qquad y = \frac{y' - x'}{\sqrt{2}}$$

Therefore $x^2 + xy + y^2 = 6$ becomes

$$\left(\frac{x' + y'}{\sqrt{2}}\right)^2 + \left(\frac{x' + y'}{\sqrt{2}}\right)\left(\frac{y' - x'}{\sqrt{2}}\right) + \left(\frac{y' - x'}{\sqrt{2}}\right)^2 = 6.$$

After multiplication and combining of terms:

$$(x')^2 + 3(y')^2 = 12, \quad \text{or} \quad \frac{(x')^2}{12} + \frac{(y')^2}{4} = 1. \qquad \textbf{Answer}$$

The graph of $\dfrac{(x')^2}{12} + \dfrac{(y')^2}{4} = 1$ in the $x'y'$-coordinate system is an ellipse with its major axis along the x'-axis, its minor axis along the y'-axis, and its center at the origin.

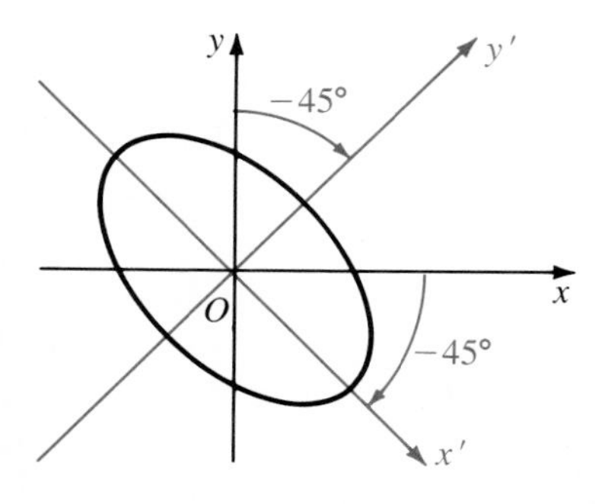

The method of solution in Example 1 illustrates how a second-degree equation involving an xy-term can be simplified by writing it as a new equation in a rotated coordinate system. From the new equation, the graph of the original equation can be identified.

The **general second-degree equation** in two variables is of the form

$$Ax^2 + Bxy + Cy^2 + Dx + Ey + F = 0. \qquad (*)$$

If $B = 0$, we can identify the graph as a conic section by using the methods given in Chapter 2. If $B \neq 0$, we can use a suitable rotation of axes to eliminate the xy-term of (*) and then identify the graph.

THEOREM 5 If the axes are rotated through an angle φ, where $\cot 2\varphi = \dfrac{A - C}{B}$, the transformed equation of the graph of (*) will have no xy-term.

Proof: Use the rotation equations in part (1) of Theorem 4.

$$\begin{aligned} &A(x' \cos \varphi - y' \sin \varphi)^2 \\ &\quad + B(x' \cos \varphi - y' \sin \varphi)(x' \sin \varphi + y' \cos \varphi) \\ &\quad + C(x' \sin \varphi + y' \cos \varphi)^2 + D(x' \cos \varphi - y' \sin \varphi) \\ &\quad + E(x' \sin \varphi + y' \cos \varphi) + F = 0 \end{aligned}$$

After multiplying and combining terms we obtain

$$A'(x')^2 + B'x'y' + C'(y')^2 + D'x' + E'y' + F' = 0,$$

where
$$\begin{aligned} A' &= A \cos^2 \varphi + B \sin \varphi \cos \varphi + C \sin^2 \varphi \\ B' &= (2C - 2A) \sin \varphi \cos \varphi + B(\cos^2 \varphi - \sin^2 \varphi) \\ C' &= A \sin^2 \varphi - B \sin \varphi \cos \varphi + C \cos^2 \varphi \\ D' &= D \cos \varphi + E \sin \varphi \\ E' &= -D \sin \varphi + E \cos \varphi \\ F' &= F \end{aligned}$$

To eliminate the $x'y'$-term, make $B' = 0$ by choosing φ so that

$$(C - A) \cdot 2 \sin \varphi \cos \varphi + B(\cos^2 \varphi - \sin^2 \varphi) = 0.$$

Then
$$(C - A)\sin 2\varphi + B \cos 2\varphi = 0$$
and
$$B \cos 2\varphi = (A - C)\sin 2\varphi.$$

If $\cot 2\varphi = \dfrac{A - C}{B}$, then $B' = 0$. ■

EXAMPLE 2 Identify the graph of $2xy + 2\sqrt{2}x = 1$. Sketch the graph.

Solution From the equation, $A = C = 0$ and $B = 2$. Therefore $\cot 2\varphi = 0$. Choose $\varphi = 45°$. Then the rotation equations to use are

$$x = \frac{x'\sqrt{2} - y'\sqrt{2}}{2} \text{ and } y = \frac{x'\sqrt{2} + y'\sqrt{2}}{2}.$$

The given equation becomes

$$2\left(\frac{x'\sqrt{2} - y'\sqrt{2}}{2}\right)\left(\frac{x'\sqrt{2} + y'\sqrt{2}}{2}\right) + 2\sqrt{2}\left(\frac{x'\sqrt{2} - y'\sqrt{2}}{2}\right) = 1.$$

Solution continued on following page

After simplification, the preceding equation becomes

$$(x')^2 - (y')^2 + 2x' - 2y' = 1.$$

Complete the square in x' and y'.

$$(x' + 1)^2 - (y' + 1)^2 = 1.$$

Therefore $\frac{(x' + 1)^2}{1} - \frac{(y' + 1)^2}{1} = 1.$

The graph is a hyperbola with center $(-1, -1)$ in the $x'y'$-coordinate system and asymptotes $y' + 1 = \pm\frac{1}{1}(x' + 1)$. That is, $y' = x'$ and $y' = -x' - 2$. **Answer**

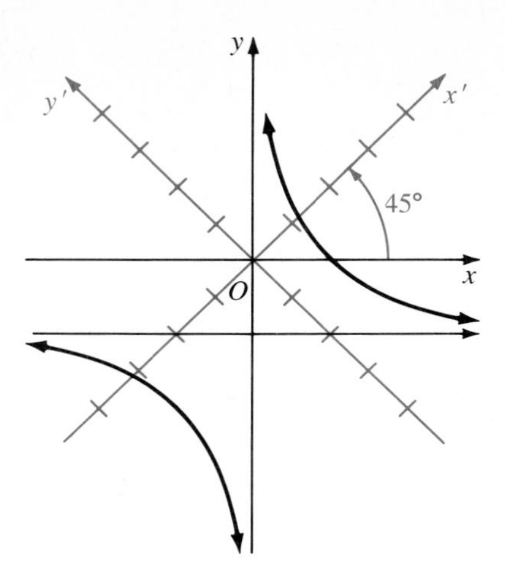

In Exercises 16 and 17 you are asked to show that for all values of φ $A' + C' = A + C$ and that $(B')^2 - 4A'C' = B^2 - 4AC$. We say that $A + C$ and $B^2 - 4AC$ are *invariant* under any rotation of axes. We call $B^2 - 4AC$ the **characteristic**. If φ is the rotation by which the xy-term is eliminated, then

$$B^2 - 4AC = -4A'C'.$$

Using this information and the list by which we can identify conics given in Section 2–7, we have the following theorem.

THEOREM 6 Let $Ax^2 + Bxy + Cy^2 + Dx + Ey + F = 0$ and let at least one of A, B, and C be nonzero.

1. If $B^2 - 4AC = 0$, then the graph is a parabola, two parallel lines, a line, or the empty set.
2. If $B^2 - 4AC < 0$, then the graph is an ellipse, a point, or the empty set.
3. If $B^2 - 4AC > 0$, then the graph is a hyperbola or a pair of intersecting lines.

According to Theorem 6, the graph of the equation in Example 2 is a hyperbola since $B^2 - 4AC = 2^2 - 4 \cdot 0 \cdot 0 = 4 > 0$.

To apply the rotation equations and thus eliminate the xy-term, we may choose 2φ such that $0° < 2\varphi < 180°$, that is, $0° < \varphi < 90°$. From $\cot 2\varphi = \frac{A - C}{B}$, we find $\cos 2\varphi$. Then we may use

$$\sin\varphi = \sqrt{\frac{1 - \cos 2\varphi}{2}} \quad \text{and} \quad \cos\varphi = \sqrt{\frac{1 + \cos 2\varphi}{2}}$$

to find $\sin \varphi$ and $\cos \varphi$. To illustrate finding $\sin \varphi$ and $\cos \varphi$, consider the equation $5x^2 + 3xy + y^2 = 4$. Then $A = 5$, $B = 3$, and $C = 1$. Hence $\cot 2\varphi = \frac{4}{3}$. Thus $\cos 2\varphi = \frac{4}{\sqrt{4^2 + 3^2}} = \frac{4}{5}$.

$$\sin \varphi = \sqrt{\frac{1 - \frac{4}{5}}{2}} = \frac{\sqrt{10}}{10} \quad \text{and} \quad \cos \varphi = \sqrt{\frac{1 + \frac{4}{5}}{2}} = \frac{3\sqrt{10}}{10}$$

EXERCISES

Identify and sketch the graph of each equation.

A

1. $xy - y\sqrt{2} = 2$
2. $7x^2 + 6xy + 15y^2 = 144$
3. $x^2 + xy + y^2 = 8$
4. $2x^2 + 4xy - y^2 - 2x + 3y = 6$
5. $3x^2 + 2xy + 3y^2 = 16$
6. $7x^2 - 4xy + 4y^2 = 240$
7. $9x^2 + 24xy + 16y^2 - 40x - 95y = -75$
8. $3x^2 - 2xy + 3y^2 = 4$
9. $3x^2 + 4\sqrt{3}xy - y^2 = 15$
10. $\sqrt{3}xy + y^2 - 18 = 0$

B

11. $3x^2 + 4xy + 4\sqrt{5}x + 4 = 0$
12. $x^2 + xy + y^2 - 3\sqrt{2}y - 12 = 0$
13. $x^2 + 4xy + 4y^2 - 5\sqrt{5}y = 0$
14. $y^2 - 2xy + x^2 - 5x = 0$
15. Prove the second set of rotation equations in Theorem 4.
16. Prove that in the notation of (*), $A' + C' = A + C$.

C

17. Prove that in the notation of (*), $(B')^2 - 4A'C' = B^2 - 4AC$.

Systems of Linear Equations

12-5 LINEAR SYSTEMS I

Throughout this book we have solved systems of two linear equations in two variables.

In general, a linear equation in n variables is an equation of the form

$$a_1x_1 + a_2x_2 + \cdots + a_nx_n = c.$$

A system of m linear equations in n variables has the form:

$$\begin{aligned} a_{11}x_1 + a_{12}x_2 + \cdots + a_{1n}x_n &= c_1 \\ a_{21}x_1 + a_{22}x_2 + \cdots + a_{2n}x_n &= c_2 \\ &\vdots \\ a_{m1}x_1 + a_{m2}x_2 + \cdots + a_{mn}x_n &= c_m \end{aligned}$$

EXAMPLE 1 Solve the system:

$$\begin{aligned} x + 5y - 6z &= 29 \\ y + 4z &= -10 \\ z &= -3 \end{aligned}$$

Solution Use $z = -3$ to find y from the second equation.

$$y + (4)(-3) = -10$$

Therefore, $y = 2$. Once z and y are known, find x from the first equation by substitution of -3 for z and 2 for y .

$$x + 5 \cdot 2 - 6(-3) = 29$$

Therefore $x = 1$. The solution of the system is $(1, 2, -3)$. **Answer**

In Example 1 we solved the system of equations by *back substitution*. That is, we used the known value of z to find y. Then we used z and y to find x. If we can write a system of equations in a form, called **triangular form**, like that in Example 1, we can easily solve the system.

EXAMPLE 2 Solve the system:

$$(1) \qquad \begin{aligned} x - 3y + z &= 11 \\ 3x + y - 2z &= -2 \\ x - y - z &= 1 \end{aligned}$$

Solution Multiply the first equation of (1) by -3 and add the resulting equation to the second equation to obtain (2).

$$(2) \qquad \begin{aligned} x - 3y + z &= 11 \\ 10y - 5z &= -35 \\ x - y - z &= 1 \end{aligned}$$

Multiply the first equation of (2) by -1 and add the resulting equation to the third equation.

$$(3) \qquad \begin{aligned} x - 3y + z &= 11 \\ 10y - 5z &= -35 \\ 2y - 2z &= -10 \end{aligned}$$

Divide the second equation of (3) by 5 and the third equation by 2.

$$(4) \qquad \begin{aligned} x - 3y + z &= 11 \\ 2y - z &= -7 \\ y - z &= -5 \end{aligned}$$

Interchange the second and third equations of (4).

$$(5) \qquad \begin{aligned} x - 3y + z &= 11 \\ y - z &= -5 \\ 2y - z &= -7 \end{aligned}$$

Multiply the second equation of (5) by -2 and add the resulting equation to the third equation.

$$\begin{aligned} x - 3y + z &= 11 \\ y - z &= -5 \\ z &= 3 \end{aligned}$$

Therefore $z = 3$. Then $y - 3 = -5$ and $y = -2$. Use these values of y and z to obtain $x - (3)(-2) + 3 = 11$. Thus $x = 2$. Therefore the solution of the system is $(2, -2, 3)$. **Answer**

In Example 2, we multiplied one equation by a real number, we added two equations to replace an equation, and we interchanged two equations. These operations are called **elementary operations**.

We can associate with each system of linear equations its **augmented matrix**. The columns of this matrix are the coefficients of the variables, and the last column contains the constants of the system. The system in Example 2 and its augmented matrix are shown below.

$$\begin{aligned} x - 3y + z &= 11 \\ 3x + y - 2z &= -2 \\ x - y - z &= 1 \end{aligned} \qquad \begin{bmatrix} 1 & -3 & 1 & 11 \\ 3 & 1 & -2 & -2 \\ 1 & -1 & -1 & 1 \end{bmatrix}$$

We may operate on the rows of the augmented matrix as if the rows were equations. When we do this, we call the elementary operations **elementary row operations**. Applying these operations to the augmented matrix for Example 2, we obtain

$$W = \begin{bmatrix} 1 & -3 & 1 & 11 \\ 0 & 1 & -1 & -5 \\ 0 & 0 & 1 & 3 \end{bmatrix},$$

from which we can read $z = 3$ and from which we can find y and then x. If we continue to perform elementary row operations on W, we can obtain another matrix W':

$$W' = \begin{bmatrix} 1 & 0 & 0 & 2 \\ 0 & 1 & 0 & -2 \\ 0 & 0 & 1 & 3 \end{bmatrix},$$

from which $x = 2$, $y = -2$, and $z = 3$ can be read immediately.

Systems of linear equations have many applications. For example, if we know that a set B of three vectors in three-space is a basis for that space, then we can write a given vector in the standard basis in terms of the vectors in B.

EXAMPLE 3 Write $\mathbf{v} = (-1, -4, -7)$ as a linear combination of $(1, 1, 0)$, $(1, 0, 1)$, and $(0, 1, 1)$.

Solution Find a, b, and c such that $(-1, -4, -7) = a(1, 1, 0) + b(1, 0, 1) + c(0, 1, 1)$.

$$(-1, -4, -7) = (a + b, a + c, b + c)$$

$$\begin{aligned} a + b &= -1 \\ a + c &= -4 \\ b + c &= -7 \end{aligned}$$

Solution continued on following page

Perform elementary row operations on the augmented matrix.

$$\begin{bmatrix} 1 & 1 & 0 & -1 \\ 1 & 0 & 1 & -4 \\ 0 & 1 & 1 & -7 \end{bmatrix}$$

Multiply row 1 by -1 and add the result to row 2.

$$\begin{bmatrix} 1 & 1 & 0 & -1 \\ 0 & -1 & 1 & -3 \\ 0 & 1 & 1 & -7 \end{bmatrix}$$

Add row 2 to row 1 and add row 2 to row 3.

$$\begin{bmatrix} 1 & 0 & 1 & -4 \\ 0 & -1 & 1 & -3 \\ 0 & 0 & 2 & -10 \end{bmatrix}$$

Multiply row 2 by -1 and row 3 by $\frac{1}{2}$.

$$\begin{bmatrix} 1 & 0 & 1 & -4 \\ 0 & 1 & -1 & 3 \\ 0 & 0 & 1 & -5 \end{bmatrix}$$

Add row 3 to row 2. Multiply row 3 by -1 and add the result to row 1.

$$\begin{bmatrix} 1 & 0 & 0 & 1 \\ 0 & 1 & 0 & -2 \\ 0 & 0 & 1 & -5 \end{bmatrix}$$

Thus $a = 1$, $b = -2$, and $c = -5$. Hence $(-1, -4, -7) = (1, 1, 0) - 2(1, 0, 1) - 5(0, 1, 1)$. **Answer**

Check: $(1, 1, 0) - 2(1, 0, 1) - 5(0, 1, 1) = (1 - 2, 1 - 5, -2 - 5) = (-1, -4, -7)$.

EXERCISES

Solve each system of equations by using elementary operations.

A

1. $2x - 3y = -3$; $x + 3y = 21$

2. $4x + 5y = 9$; $x + 3y = -2$

3. $-2x - 3y = -9$; $3x + 4y = -30$

Solve each system of equations by using back substitution.

4. $2x - 5y + z = 1$; $3y + 2z = 5$; $z = -2$

5. $2x - y + 3z = -13$; $2y - z = -1$; $z = 5$

6. $-x - y + 2z = 21$; $3y - 3z = 7$; $z = -7$

Solve each system of equations by using elementary operations.

7. $2x - y + 2z = 11$; $x - 2y - 3z = -5$; $-3x + z = -3$

8. $4x + y + z = 5$; $x - 2y - z = 3$; $3x + 3y - 2z = 22$

9. $4x - 2y + 5z = 7$; $-x + y - 2z = 0$; $x - 3y - 2z = 4$

10. $x + 2y - z = 12$
$-3x + 2y + z = 0$
$4x - y - z = -3$

11. $5x - 2y + 3z = 6$
$3x + 2y - z = 5$
$3x + 2y + 5z = 5$

12. $-7x - 5y + z = 14$
$4x - 6y + 2z = 2$
$x + 5y - 3z = 0$

13. Find an equation in the form $y = ax^2 + bx + c$ of the parabola containing the points (2, 9), (0, 3), and (−3, 24).

14. Find an equation in the form $y = ax^2 + bx + c$ of the parabola containing the points (1, 2), (2, 7), and (4, 29).

Use substitutions such as $u = \frac{1}{x}$, and so forth, to solve each system of nonlinear equations.

B

15. $-\frac{2}{x} + \frac{3}{y} - \frac{4}{z} = -1$
$\frac{2}{x} + \frac{2}{y} + \frac{2}{z} = 5$
$\frac{1}{x} + \frac{1}{y} + \frac{2}{z} = 3$

16. $\frac{5}{x} + \frac{2}{y} + \frac{2}{z} = 2$
$\frac{1}{x} + \frac{2}{y} - \frac{2}{z} = -1$
$\frac{3}{x} + \frac{1}{y} - \frac{4}{z} = 0$

17. $3x - 2y + \frac{3}{z} = 6$
$x + y + \frac{1}{z} = 2$
$2x - y + \frac{1}{z} = 5$

18. $x - \frac{1}{y} + 2z = 4$
$3x + \frac{1}{y} + 4z = 6$
$x + \frac{2}{y} + z = 3$

12-6 LINEAR SYSTEMS II

A system of linear equations may have a unique solution as was the case in Section 12-5. A system may also have infinitely many solutions or no solution at all.

EXAMPLE 1 Solve the system:

$$x - 2y - z = 2$$
$$2x + y + z = 3$$

Solution Add the equations to obtain $3x - y = 5$, or $y = 3x - 5$. Substitute $3x - 5$ for y in the second equation to obtain

$$2x + (3x - 5) + z = 3.$$

Hence $z = -5x + 8$. The solution set of the system is

$$\{(x, 3x - 5, -5x + 8): x \in \mathcal{R}\}.$$ **Answer**

Let $x = 1$ to obtain $(1, -2, 3)$.

Let $x = 2$ to obtain another solution, $(2, 1, -2)$. There are infinitely many solutions.

EXAMPLE 2 Solve the system:

$$\begin{aligned} x - y + z &= -1 \\ x + y + 2z &= 9 \\ 5x - y + 7z &= 15 \end{aligned}$$

Solution Write the augmented matrix and perform elementary row operations.

$$\begin{bmatrix} 1 & -1 & 1 & -1 \\ 1 & 1 & 2 & 9 \\ 5 & -1 & 7 & 15 \end{bmatrix}$$

Multiply the first row by -1 and add the resulting row to the second row. Multiply the first row by -5 and add the resulting row to the third row.

$$\begin{bmatrix} 1 & -1 & 1 & -1 \\ 0 & 2 & 1 & 10 \\ 0 & 4 & 2 & 20 \end{bmatrix}$$

Multiply the second row by -2 and add the resulting row to the third row.

$$\begin{bmatrix} 1 & -1 & 1 & -1 \\ 0 & 2 & 1 & 10 \\ 0 & 0 & 0 & 0 \end{bmatrix}$$

Multiply the second row by $\frac{1}{2}$.

$$\begin{bmatrix} 1 & -1 & 1 & -1 \\ 0 & 1 & \frac{1}{2} & 5 \\ 0 & 0 & 0 & 0 \end{bmatrix}$$

Add the second row to the first row.

$$\begin{bmatrix} 1 & 0 & \frac{3}{2} & 4 \\ 0 & 1 & \frac{1}{2} & 5 \\ 0 & 0 & 0 & 0 \end{bmatrix}$$

From the matrix finally obtained, $x + \frac{3}{2}z = 4$ and $y + \frac{1}{2}z = 5$. The value of z may be chosen arbitrarily. Solve for x and y in terms of z. The solution set of the system is $\{(4 - \frac{3}{2}z, 5 - \frac{1}{2}z, z): z \in \mathcal{R}\}$. **Answer**

EXAMPLE 3 Solve the system:

$$\begin{aligned} x + 2y - 2z &= 8 \\ 5y - z &= 6 \\ -2x + y + 3z &= -2 \end{aligned}$$

Solution Multiply the first equation by 2 and add the resulting equation to the third equation.

$$\begin{aligned} x + 2y - 2z &= 8 \\ 5y - z &= 6 \\ 5y - z &= 14 \end{aligned}$$

Subtract the second and third equations.

$$\begin{aligned} x + 2y - 2z &= 8 \\ 5y - z &= 6 \\ 0 &= -8 \end{aligned}$$

The last equation is false. The system has no solution. **Answer**

A matrix is said to be in **row-reduced echelon form** when

1. in each row the first nonzero entry is 1,
2. all other entries in the column containing that 1 are 0,
3. the leading 1 in a row is to the right of the leading 1's in all rows above that row, and
4. any row with all zero entries is below all the rows with a leading 1.

The matrix finally obtained in Example 2 is in row-reduced echelon form.

To transform a matrix to row-reduced echelon form, use elementary row operations. Once the matrix of a linear system is in this form, the solutions, if any, can be found.

A system of equations with at least one solution is said to be **consistent**. The systems of equations in Section 12–5 are consistent and each has a unique solution. The systems in Examples 1 and 2 of this section are consistent and each system has infinitely many solutions. In Example 2, the *general solution* is $\{(4 - \frac{3}{2}z, 5 - \frac{1}{2}z, z): z \in \mathcal{R}\}$ and a *particular solution* is $(1, 4, 2)$. When the solution set of a system is infinite, the equations are said to be **dependent**. A system of equations with no solution is said to be **inconsistent**. The system in Example 3 of this section is inconsistent.

EXERCISES

Solve each system of equations. If the solution is unique, find it. If there are infinitely many solutions, write the solution set in the form shown in Example 2. Give two particular solutions. If there is no solution, so state.

A

1. $\begin{aligned} x - 3y &= -5 \\ 2x - 2y &= 6 \end{aligned}$

2. $\begin{aligned} 3x - 3y &= 12 \\ x + y &= 4 \end{aligned}$

3. $\begin{aligned} 3x + 5y &= 10 \\ -6x - 10y &= -20 \end{aligned}$

4. $\begin{aligned} 2x + 5y &= 12 \\ \frac{2}{5}x + y &= 0 \end{aligned}$

5. $\begin{aligned} x + 2y - z &= 3 \\ 2x - y + z &= 7 \end{aligned}$

6. $\begin{aligned} 3x - y + 2z &= 1 \\ 2x + 3y + 2z &= 2 \end{aligned}$

In Exercises 7–9, the augmented matrix of a linear system in x, y, and z is given. Transform each matrix to row-reduced echelon form. Give the unique solution if there is one, write the infinite solution set in the form shown in Example 2, or state that there is no solution.

7. $\begin{bmatrix} 1 & 2 & 5 & 4 \\ 2 & 1 & -2 & -4 \\ -6 & -3 & 4 & 6 \end{bmatrix}$

8. $\begin{bmatrix} 1 & 2 & 7 & -1 \\ 0 & -2 & -6 & 4 \\ 0 & 1 & 4 & -5 \end{bmatrix}$

9. $\begin{bmatrix} 1 & 0 & -2 & -6 \\ 0 & 1 & 3 & 9 \\ 2 & 0 & -1 & 3 \end{bmatrix}$

Solve each system of equations. If the solution is unique, find it. If there are infinitely many solutions, write the solution set in the form shown in Example 2. Give two particular solutions. If there is no solution, so state.

10. $\begin{aligned} a + b + c &= 1 \\ -3a + 7b + 2c &= 0 \\ -2a + 8b + 3c &= 4 \end{aligned}$

11. $\begin{aligned} 2x + 3y - z &= -2 \\ x + 2y + 2z &= 8 \\ 5x + 9y + 5z &= 22 \end{aligned}$

Solve each system of equations. If the solution is unique, find it. If there are infinitely many solutions, write the solution set in the form shown in Example 2. Give two particular solutions. If there is no solution, so state.

12. $\begin{aligned} x + y - z &= 3 \\ x + y + z &= 3 \\ 2x + 2y &= 6 \end{aligned}$

13. $\begin{aligned} 4a - 4b - 3c &= 2 \\ 4a + 3c &= 3 \\ 4b + 6c &= 3 \end{aligned}$

14. $\begin{aligned} 2x + 5y - 2z &= -6 \\ 7x + 2y - 5z &= -4 \\ -2x + 3y + 2z &= -2 \end{aligned}$

15. $\begin{aligned} 2x + 3y + 3z &= 8 \\ 3x + 2y - 2z &= 14 \\ 4x + 5y + 3z &= 38 \end{aligned}$

16. $\begin{aligned} x + y + 9z &= 8 \\ x + 3z &= 1 \\ y + 6z &= 7 \end{aligned}$

17. $\begin{aligned} x + 2y + 3z &= 4 \\ x + 2z &= 3 \\ 2y + z &= 2 \end{aligned}$

B

18. $\begin{aligned} 3x - 2y &= 7 \\ 2x - 5y &= 12 \\ x + 3y &= -5 \end{aligned}$

19. $\begin{aligned} 3x + 4y &= 0 \\ 2x + y &= 3 \\ x - y &= 2 \end{aligned}$

20. $\begin{aligned} x - 2z &= 0 \\ 2x - y + z &= 0 \\ x - 3y + 3z &= 0 \end{aligned}$

21. $\begin{aligned} x - 2y + 3z &= 0 \\ x + 2z &= 0 \\ 3x - 4y + 5z &= 0 \end{aligned}$

22. $\begin{aligned} x + 2y + z &= 0 \\ -2x + y - 2z &= 0 \\ 3x + y - 5z &= 0 \end{aligned}$

23. $\begin{aligned} x - 2y - z &= 0 \\ 3x - 6y &= 0 \\ 4x - 6y - 6z &= 0 \end{aligned}$

24. Use $x^2 + y^2 + ax + by + c = 0$ where a, b, and c are real numbers to find an equation of the circle containing $(-2, 2)$, $(4, 2)$, and $(2, -2)$.

25. Use $x^2 + y^2 + ax + by + c = 0$ where a, b, and c are real numbers to find an equation for the set of all circles containing $(-4, 4)$ and $(4, 4)$. What is an equation for the circle containing these points and with center at the origin?

Solve each system of linear equations.

26. $\begin{aligned} -w + x - y + z &= 0 \\ w + y &= 4 \\ x - z &= 0 \\ -w + y &= 0 \end{aligned}$

27. $\begin{aligned} 3w + 3x + y + 2z &= 6 \\ 2w - 5x + y + 5z &= 0 \\ w + 7x + 2y + 6z &= 10 \\ w - x + 4y + 8z &= 0 \end{aligned}$

The linear system:

$$\begin{aligned} a_{11}x_1 + a_{12}x_2 + \cdots + a_{1n}x_n &= c_1 \\ a_{21}x_1 + a_{22}x_2 + \cdots + a_{2n}x_n &= c_2 \\ &\vdots \\ a_{m1}x_1 + a_{m2}x_2 + \cdots + a_{mn}x_n &= c_m \end{aligned}$$

is **homogeneous** if $c_1 = c_2 = \cdots = c_m = 0$. In Exercises 28–30, find all solutions of each homogeneous system of linear equations. Clearly each system has $(0, 0, 0)$ as a solution.

28. $\begin{aligned} 2x - 3y + z &= 0 \\ -x \qquad - 2z &= 0 \\ 2x - 6y + 2z &= 0 \end{aligned}$

29. $\begin{aligned} 2x - 3y + z &= 0 \\ 5x + 3y - 2z &= 0 \\ 6x - y + 3z &= 0 \end{aligned}$

30. $\begin{aligned} 5x - 2y - 2z &= 0 \\ -x + 3y + 3z &= 0 \\ 4x - 5y - 4z &= 0 \end{aligned}$

The vectors $\mathbf{v}_1, \mathbf{v}_2, \mathbf{v}_3, \ldots, \mathbf{v}_n$ are said to be **linearly independent** if $a_1\mathbf{v}_1 + a_2\mathbf{v}_2 + a_3\mathbf{v}_3 + \cdots + a_n\mathbf{v}_n = \mathbf{0}$ has as its only solution $a_1 = a_2 = a_3 = \cdots = a_n = 0$. Otherwise they are said to be **linearly dependent**.
Determine whether the vectors are linearly independent or linearly dependent.

31. $\mathbf{v}_1 = (1, 0, 0)$, $\mathbf{v}_2 = (0, 1, 0)$, $\mathbf{v}_3 = (0, 0, 1)$

32. $\mathbf{v}_1 = (0, 1, 1)$, $\mathbf{v}_2 = (1, 1, 0)$, $\mathbf{v}_3 = (0, 1, 0)$

33. $\mathbf{v}_1 = (-2, 1, 3)$, $\mathbf{v}_2 = (1, 5, 2)$, $\mathbf{v}_3 = (4, -13, -13)$

34. **a.** Show that the set $B = \{(1, 1, 0), (1, 0, 1), (0, 1, 1)\}$ is a basis for the set of all vectors in three-space.
b. Show that the set $B' = \{(1, -1, 1), (2, 0, 3), (3, 1, 0)\}$ is a basis for the set of all vectors in three-space.
c. Find a matrix $\overline{T}$ for the linear transformation T such that

$$T((1, 1, 0)) = (1, -1, 1),$$
$$T((1, 0, 1)) = (2, 0, 3),$$

and

$$T((0, 1, 1)) = (3, 1, 0).$$

Find all vectors $\mathbf{v} = (x, y, z)$ such that $T(\mathbf{v}) = \mathbf{0}$.

C **35.** $T(\mathbf{v}) = (3x - 2y + z, -x + 4y + 2z, 2x - y + 5z)$

36. $T(\mathbf{v}) = (x + y + z, 2x - y + 2z, 4x + y + 4z)$

37. Let $\begin{aligned} -2x + 3y + 2z &= 2 \\ -x + 4y + z &= 1 \end{aligned} \qquad (*)$

a. Show that $(1, 0, 2)$ is a particular solution of $(*)$.
b. Show that $\{(z, 0, z)\colon z \in \mathcal{R}\}$ is the general solution of

$$\begin{aligned} -2x + 3y + 2z &= 0 \\ -x + 4y + z &= 0 \end{aligned}$$

c. Show that the general solution of $(*)$ can be written $(1, 0, 2) + (z, 0, z)$, or $\{(1 + z, 0, 2 + z)\colon z \in \mathcal{R}\}$.

12-7 DETERMINANTS

Associated with each square matrix is a number, called the *determinant* of the matrix. In Sections 12-8 and 12-9 we shall discuss some of the uses of determinants.

DEFINITION

Let $T = \begin{bmatrix} a & b \\ c & d \end{bmatrix}$. The **determinant** of T, denoted det T or $\begin{vmatrix} a & b \\ c & d \end{vmatrix}$, is defined by:

$$\det T = \begin{vmatrix} a & b \\ c & d \end{vmatrix} = ad - bc.$$

For example, if $A = \begin{bmatrix} 3 & 5 \\ 2 & 8 \end{bmatrix}$, then $\det A = \begin{vmatrix} 3 & 5 \\ 2 & 8 \end{vmatrix} = 3 \cdot 8 - 2 \cdot 5 = 14$.

The number of rows (or columns) of a determinant array is called its **order**. We shall use a_{ij} to denote the entry in the ith row and jth column of a determinant. The **minor** of a_{ij}, denoted M_{ij}, is the determinant obtained by striking out the ith row and the jth column of the determinant. The **cofactor** of a_{ij}, denoted A_{ij}, is $(-1)^{i+j}M_{ij}$.

EXAMPLE 1 Find the cofactor of $a_{22} = -3$ of det T if

$$T = \begin{bmatrix} 1 & 2 & 3 \\ 2 & -3 & 1 \\ 1 & 2 & 4 \end{bmatrix}.$$

Solution To find the minor of -3, strike out the second row and the second column:

$M_{22} = \begin{vmatrix} 1 & 3 \\ 1 & 4 \end{vmatrix}$. The cofactor of -3 is $(-1)^{2+2}\begin{vmatrix} 1 & 3 \\ 1 & 4 \end{vmatrix}$, or $\begin{vmatrix} 1 & 3 \\ 1 & 4 \end{vmatrix}$.

Determinants of order 3 are defined in terms of cofactors and determinants of order 2.

DEFINITION

Let T be an $n \times n$ matrix, let a_{ij} be the entry in the ith row and jth column, and let A_{ij} be its cofactor. Det T is defined by

$$\det T = a_{11}A_{11} + a_{12}A_{12} + \cdots + a_{1j}A_{1j} + \cdots + a_{1n}A_{1n}.$$

If $T = \begin{bmatrix} a_{11} & a_{12} & a_{13} \\ a_{21} & a_{22} & a_{23} \\ a_{31} & a_{32} & a_{33} \end{bmatrix}$, $\det T = a_{11}A_{11} + a_{12}A_{12} + a_{13}A_{13}$

$$= a_{11}\begin{vmatrix} a_{22} & a_{23} \\ a_{32} & a_{33} \end{vmatrix} - a_{12}\begin{vmatrix} a_{21} & a_{23} \\ a_{31} & a_{33} \end{vmatrix} + a_{13}\begin{vmatrix} a_{21} & a_{22} \\ a_{31} & a_{32} \end{vmatrix}$$

EXAMPLE 2 Find the value of $\begin{vmatrix} 3 & 0 & 2 \\ -1 & 1 & -3 \\ 5 & 2 & 1 \end{vmatrix}$.

Solution Expand the determinant along the first row.

$$\begin{vmatrix} 3 & 0 & 2 \\ -1 & 1 & -3 \\ 5 & 2 & 1 \end{vmatrix} = 3\begin{vmatrix} 1 & -3 \\ 2 & 1 \end{vmatrix} - 0\begin{vmatrix} -1 & -3 \\ 5 & 1 \end{vmatrix} + 2\begin{vmatrix} -1 & 1 \\ 5 & 2 \end{vmatrix}$$

$$= 3(7) - 0(14) + 2(-7) = 7 \quad \textbf{Answer}$$

It can be shown that any other row or column can be used instead of the first row and the same value obtained. For example, if we choose the second column to evaluate the determinant in Example 2, we obtain the following:

$$-0\begin{vmatrix} -1 & -3 \\ 5 & 1 \end{vmatrix} + 1\begin{vmatrix} 3 & 2 \\ 5 & 1 \end{vmatrix} - 2\begin{vmatrix} 3 & 2 \\ -1 & -3 \end{vmatrix} = (1)(-7) - (2)(-7) = 7$$

The following theorem (stated without proof) is a great help in evaluating determinants.

THEOREM 7 If the entries of any row (column) of a determinant are multiplied by the same constant and the result is added to any different row (column), the value of the determinant is unchanged.

EXAMPLE 3 Evaluate $\begin{vmatrix} 3 & -1 & 2 \\ 2 & -1 & 1 \\ -2 & 3 & 3 \end{vmatrix}$.

Solution Use Theorem 7 to obtain two zeros in the second row.

$$\begin{vmatrix} 3 & -1 & 2 \\ 2 & -1 & 1 \\ -2 & 3 & 3 \end{vmatrix} \underbrace{=}_{\text{(col 1)} - 2\text{(col 3)}} \begin{vmatrix} -1 & -1 & 2 \\ 0 & -1 & 1 \\ -8 & 3 & 3 \end{vmatrix} \underbrace{=}_{\text{(col 2)} + \text{(col 3)}} \begin{vmatrix} -1 & 1 & 2 \\ 0 & 0 & 1 \\ -8 & 6 & 3 \end{vmatrix}$$

$$= -0\begin{vmatrix} 1 & 2 \\ 6 & 3 \end{vmatrix} + 0\begin{vmatrix} -1 & 2 \\ -8 & 3 \end{vmatrix} - 1\begin{vmatrix} -1 & 1 \\ -8 & 6 \end{vmatrix}$$

$$= -(-6 + 8)$$

$$= -2 \quad \textbf{Answer}$$

EXAMPLE 4 Evaluate $\begin{vmatrix} 3 & 1 & 2 & 4 \\ 2 & 0 & 1 & 0 \\ -1 & 2 & -4 & 0 \\ 3 & 2 & 1 & 1 \end{vmatrix}$.

Solution A third 0 can be obtained in the second row by multiplying each entry in the third column by -2 and adding the results to the entries of the first column.

$$\begin{vmatrix} 3 & 1 & 2 & 4 \\ 2 & 0 & 1 & 0 \\ -1 & 2 & -4 & 0 \\ 3 & 2 & 1 & 1 \end{vmatrix} = \begin{vmatrix} -1 & 1 & 2 & 4 \\ 0 & 0 & 1 & 0 \\ 7 & 2 & -4 & 0 \\ 1 & 2 & 1 & 1 \end{vmatrix}$$

Expand along the second row. Obtain another 0 in the third column.

$$-0 + 0 - 1\begin{vmatrix} -1 & 1 & 4 \\ 7 & 2 & 0 \\ 1 & 2 & 1 \end{vmatrix} + 0 = -\begin{vmatrix} -5 & -7 & 0 \\ 7 & 2 & 0 \\ 1 & 2 & 1 \end{vmatrix}$$

Expand the third-order determinant that results along the third column.

$$-\left(0 - 0 + 1\begin{vmatrix} -5 & -7 \\ 7 & 2 \end{vmatrix}\right) = -(-10 - (-49)) = -39 \quad \textbf{Answer}$$

The diagonal method is a convenient way to evaluate a 3×3 determinant. The following diagrams show how the method is used.

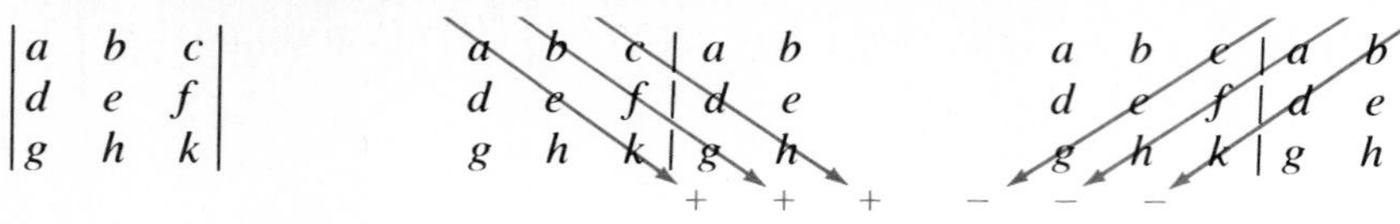

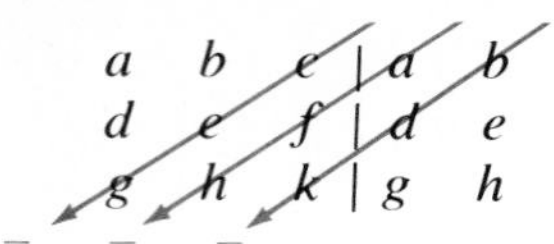

The value of the determinant is

$$aek + bfg + cdh - ceg - afh - bdk.$$

In Exercise 20 you are asked to show that this value agrees with the value obtained by evaluating by cofactors.

EXERCISES

Find the determinant of each matrix.

A

1. $\begin{bmatrix} -3 & 7 \\ -2 & 4 \end{bmatrix}$ **2.** $\begin{bmatrix} 6 & -5 \\ -9 & 8 \end{bmatrix}$ **3.** $\begin{bmatrix} 2 & -1 \\ 6 & -3 \end{bmatrix}$ **4.** $\begin{bmatrix} 1 & 0 \\ 0 & 1 \end{bmatrix}$

5. $\begin{bmatrix} a & 0 & 0 \\ 0 & b & 0 \\ 0 & 0 & c \end{bmatrix}$ **6.** $\begin{bmatrix} 0 & 0 & a \\ 0 & b & 0 \\ c & 0 & 0 \end{bmatrix}$ **7.** $\begin{bmatrix} 0 & 1 & 2 \\ 0 & 2 & 5 \\ 4 & -1 & -3 \end{bmatrix}$

8. $\begin{bmatrix} 0 & 1 & -5 \\ -2 & 17 & 18 \\ 0 & 1 & -4 \end{bmatrix}$ **9.** $\begin{bmatrix} -3 & 27 & -6 \\ 2 & 19 & 7 \\ 0 & -1 & 0 \end{bmatrix}$ **10.** $\begin{bmatrix} 0 & 0 & 0 \\ 1 & 2 & 3 \\ 4 & 5 & 6 \end{bmatrix}$

Evaluate each determinant.

11. $\begin{vmatrix} 8 & -6 & 6 \\ 1 & 0 & -1 \\ -7 & 2 & -3 \end{vmatrix}$ **12.** $\begin{vmatrix} 7 & 6 & 8 \\ -3 & 4 & 6 \\ 1 & -2 & 4 \end{vmatrix}$ **13.** $\begin{vmatrix} 10 & -2 & 4 \\ -7 & 5 & 1 \\ 2 & 3 & 3 \end{vmatrix}$

14. $\begin{vmatrix} 1 & -3 & -5 & 6 \\ 0 & 1 & -8 & 1 \\ 0 & 9 & 2 & 1 \\ 5 & 0 & -1 & 7 \end{vmatrix}$ **15.** $\begin{vmatrix} 2 & 0 & 0 & 10 \\ -8 & 3 & 4 & -2 \\ 5 & -6 & 1 & 4 \\ 9 & -1 & -3 & 2 \end{vmatrix}$ **16.** $\begin{vmatrix} -4 & 6 & 2 & 1 \\ 1 & 0 & -1 & 0 \\ 3 & -2 & 2 & 0 \\ 4 & -5 & 6 & 3 \end{vmatrix}$

17. $\begin{vmatrix} 2 & 0 & 1 & 5 \\ 1 & 2 & 3 & 4 \\ 3 & 4 & -2 & 1 \\ 1 & 7 & -3 & 2 \end{vmatrix}$ **18.** $\begin{vmatrix} 8 & 2 & 5 & 6 \\ -2 & 4 & 1 & 3 \\ 8 & 1 & -2 & 0 \\ 4 & 3 & -1 & 2 \end{vmatrix}$ **19.** $\begin{vmatrix} 1 & 3 & 2 & -4 \\ 3 & 0 & 2 & 1 \\ -7 & 0 & 1 & 5 \\ 6 & -6 & 10 & 2 \end{vmatrix}$

20. Prove that the value of any third-order determinant obtained by the diagonal method agrees with the value obtained by expanding along row 1.

B

21. Show that if one row of a third-order determinant consists entirely of 0 entries, then its value is 0.

22. Let A be a 2×2 matrix and r be a real number. Show that $\det rA = r^2 \det A$.

23. Let A be a 3×3 matrix and r be any real number. Write a formula that relates $\det rA$ and $\det A$.

24. Let $T = \begin{bmatrix} a & b & c & d \\ 0 & e & f & g \\ 0 & 0 & h & j \\ 0 & 0 & 0 & k \end{bmatrix}$. Show that $\det T = aehk$.

25. Show that if two rows of a third-order determinant are the same, then the value of the determinant is 0.

26. Show that if two columns of a third-order determinant are the same, then the value of the determinant is 0.

27. Show that if one row of a determinant is multiplied by a real number r, then the value of the determinant is multiplied by r.

28. Let $\begin{vmatrix} a & b & c \\ d & e & f \\ g & h & k \end{vmatrix}$ be a third-order determinant and r be a real number.

Show that $\begin{vmatrix} a & b & c \\ d & e & f \\ g & h & k \end{vmatrix} = \begin{vmatrix} a & b & c \\ ra + d & rb + e & rc + f \\ g & h & k \end{vmatrix}$.

29. Show that $\begin{vmatrix} a & b & c \\ d & e & f \\ g & h & k \end{vmatrix} = -\begin{vmatrix} d & e & f \\ a & b & c \\ g & h & k \end{vmatrix} = -\begin{vmatrix} b & a & c \\ e & d & f \\ h & g & k \end{vmatrix}$.

30. Show that if A and B are 2×2 matrices, then $\det AB = \det A \det B$.

C

31. Let T be a linear transformation of the plane, $T(\mathbf{v}) = (ax + by, cx + dy)$ where a, b, c, and d are positive real numbers.

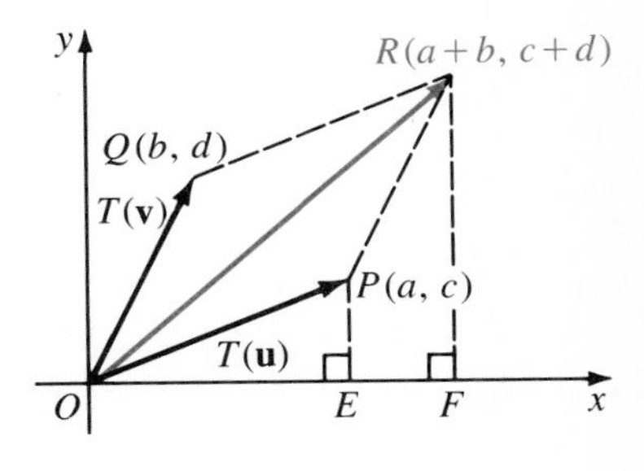

a. Show that the matrix of $T = \begin{bmatrix} a & b \\ c & d \end{bmatrix}$.

b. Show that $\det T$ = area parallelogram $OPRQ$. (*Hint:* Area triangle OPR = area triangle ORF − area triangle OPE − area trapezoid $EPRF$.)

32. Show that $\begin{vmatrix} a & b & 0 & 0 \\ c & d & 0 & 0 \\ 0 & 0 & e & f \\ 0 & 0 & g & h \end{vmatrix} = \begin{vmatrix} a & b \\ c & d \end{vmatrix} \cdot \begin{vmatrix} e & f \\ g & h \end{vmatrix}$.

33. Show that for all real numbers r and s $\begin{vmatrix} a & b & c \\ d & e & f \\ ra + sd & rb + se & rc + sf \end{vmatrix} = 0$.

12-8 THE INVERSE OF A MATRIX

Let R be the rotation of the plane through the angle α and S be the rotation through the angle $-\alpha$. Then S is the *inverse* of R, and

$$S \circ R = \text{the identity transformation.}$$

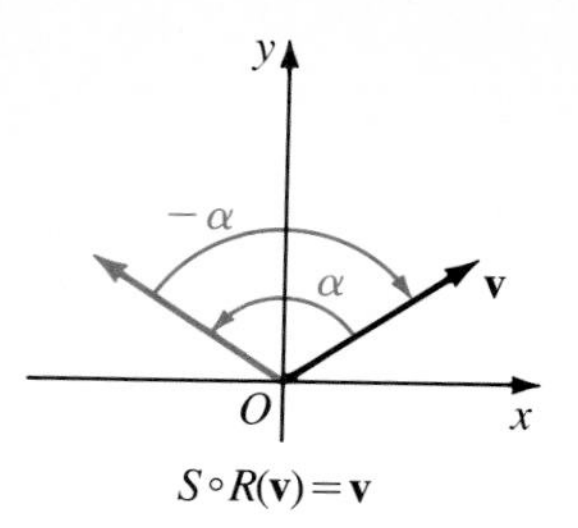

$S \circ R(\mathbf{v}) = \mathbf{v}$

This is equivalent to saying that the product of the matrices of S and R is the identity matrix.

DEFINITION

If A and B are matrices and $AB = BA = I$ where I is the identity matrix, then A and B are **inverses** of each other. The inverse of A is denoted A^{-1}.

EXAMPLE 1 Show that A and B are inverses of each other.

$$A = \begin{bmatrix} -1 & 3 \\ 1 & -2 \end{bmatrix} \qquad B = \begin{bmatrix} 2 & 3 \\ 1 & 1 \end{bmatrix}$$

Solution Find AB and BA.

$$AB = \begin{bmatrix} -1 & 3 \\ 1 & -2 \end{bmatrix}\begin{bmatrix} 2 & 3 \\ 1 & 1 \end{bmatrix}$$

$$= \begin{bmatrix} (-1)(2) + 3 \cdot 1 & (-1)(3) + 3 \cdot 1 \\ 1 \cdot 2 + (-2)(1) & 1 \cdot 3 + (-2)(1) \end{bmatrix} = \begin{bmatrix} 1 & 0 \\ 0 & 1 \end{bmatrix}$$

$$BA = \begin{bmatrix} 2 & 3 \\ 1 & 1 \end{bmatrix}\begin{bmatrix} -1 & 3 \\ 1 & -2 \end{bmatrix}$$

$$= \begin{bmatrix} (2)(-1) + 3 \cdot 1 & 2 \cdot 3 + (3)(-2) \\ (1)(-1) + 1 \cdot 1 & 1 \cdot 3 + (1)(-2) \end{bmatrix} = \begin{bmatrix} 1 & 0 \\ 0 & 1 \end{bmatrix}$$

Therefore $AB = BA = I$. Thus A and B are inverses of each other.

EXAMPLE 2 Determine whether $A = \begin{bmatrix} 1 & 2 \\ 3 & 4 \end{bmatrix}$ has an inverse. If it does, find A^{-1}.

Solution The matrix A has an inverse if there is a matrix $B = \begin{bmatrix} r & s \\ m & n \end{bmatrix}$ such that $AB = I$. Solve $\begin{bmatrix} 1 & 2 \\ 3 & 4 \end{bmatrix}\begin{bmatrix} r & s \\ m & n \end{bmatrix} = \begin{bmatrix} 1 & 0 \\ 0 & 1 \end{bmatrix}$.

This matrix equation is equivalent to the following system:

$$\begin{aligned} r + 2m &= 1 \\ 3r + 4m &= 0 \end{aligned} \qquad \begin{aligned} s + 2n &= 0 \\ 3s + 4n &= 1 \end{aligned}$$

Solve the system to find $r = -2$, $m = \frac{3}{2}$, $s = 1$, $n = -\frac{1}{2}$.

Hence $A^{-1} = \begin{bmatrix} -2 & 1 \\ \frac{3}{2} & -\frac{1}{2} \end{bmatrix}$.

Notice that the determinant of the matrix A in Example 2 has the value $1 \cdot 4 - 3 \cdot 2 = -2$. That is, $\det A = -2 \neq 0$.

THEOREM 8 If A is a matrix and $\det A \neq 0$, then A^{-1} exists.

$$\text{If } A = \begin{bmatrix} a & b \\ c & d \end{bmatrix}, \text{ then } A^{-1} = \frac{1}{\det A}\begin{bmatrix} d & -b \\ -c & a \end{bmatrix}.$$

Applying Theorem 8 to matrix $A = \begin{bmatrix} 1 & 2 \\ 3 & 4 \end{bmatrix}$ in Example 2, we have that

$$A^{-1} = -\frac{1}{2}\begin{bmatrix} 4 & -2 \\ -3 & 1 \end{bmatrix} = \begin{bmatrix} -2 & 1 \\ \frac{3}{2} & -\frac{1}{2} \end{bmatrix}.$$

We can generalize Theorem 8 to find the inverse of any square matrix A with $\det A \neq 0$. The proof of the general result is beyond the scope of this book. The following definitions are needed to state the general method.

DEFINITION

The **transpose** of a matrix A, denoted A^T, is that matrix whose entry $t_{ij} = a_{ji}$.

The **adjoint** of A, denoted adj A, is the transpose of the matrix whose entry in the ith row and jth column is the cofactor of a_{ij}.

THEOREM 9 If A is a matrix and $\det A \neq 0$, then $A^{-1} = \frac{1}{\det A} \operatorname{adj} A$.

EXAMPLE 3 Find the inverse of A if it exists.

$$A = \begin{bmatrix} 2 & -1 & 1 \\ -1 & 3 & 4 \\ -2 & 1 & 0 \end{bmatrix}$$

Solution To determine whether A^{-1} exists find $\det A$. Use column 3 for the expansion.

$$\det A = (1)\begin{vmatrix} -1 & 3 \\ -2 & 1 \end{vmatrix} - 4\begin{vmatrix} 2 & -1 \\ -2 & 1 \end{vmatrix}$$

$$= 5 \neq 0.$$

Thus A^{-1} exists. To find A^{-1}, apply Theorem 9.

Solution continued on following page

Find the matrix of cofactors. The cofactor of a_{11} is $3 \cdot 0 - 1 \cdot 4 = -4$. The cofactor of a_{12} is $(-1)[(-1)(0) - (-2)(4)] = -8$, and so on.

$$\begin{bmatrix} -4 & -8 & 5 \\ 1 & 2 & 0 \\ -7 & -9 & 5 \end{bmatrix}$$

Find the transpose of the matrix of cofactors.

$$\begin{bmatrix} -4 & 1 & -7 \\ -8 & 2 & -9 \\ 5 & 0 & 5 \end{bmatrix}$$

Therefore $A^{-1} = \frac{1}{5}\begin{bmatrix} -4 & 1 & -7 \\ -8 & 2 & -9 \\ 5 & 0 & 5 \end{bmatrix}$. **Answer**

The system of equations

$$\begin{aligned} a_{11}x_1 + a_{12}x_2 + \cdots + a_{1n}x_n &= c_1 \\ a_{21}x_1 + a_{22}x_2 + \cdots + a_{2n}x_n &= c_2 \\ &\vdots \\ a_{m1}x_1 + a_{m2}x_2 + \cdots + a_{mn}x_n &= c_m \end{aligned}$$

can be written in matrix form: $AX = C$,

where $A = \begin{bmatrix} a_{11} & a_{12} & \cdots & a_{1n} \\ a_{21} & a_{22} & \cdots & a_{2n} \\ \vdots & \vdots & & \vdots \\ a_{m1} & a_{m2} & \cdots & a_{mn} \end{bmatrix}$, $X = \begin{bmatrix} x_1 \\ x_2 \\ \vdots \\ x_n \end{bmatrix}$, and $C = \begin{bmatrix} c_1 \\ c_2 \\ \vdots \\ c_m \end{bmatrix}$.

We call the $m \times n$ matrix A the **coefficient matrix** of the system and the $m \times 1$ column matrix C the **matrix of constants**. If $m = n$, then A is square, and if $\det A \neq 0$, then A^{-1} exists. We then have:

$$\begin{aligned} A^{-1}(AX) &= A^{-1}C \\ (A^{-1}A)X &= A^{-1}C \\ IX &= A^{-1}C \\ X &= A^{-1}C \end{aligned}$$

EXAMPLE 4 Solve the system:

$$\begin{aligned} 2x - y + z &= 2 \\ -x + 3y + 4z &= 0 \\ -2x + y \quad\quad &= 1 \end{aligned}$$

Solution The coefficient matrix of the system is matrix A in Example 3. From that example,

$$A^{-1} = \frac{1}{5}\begin{bmatrix} -4 & 1 & -7 \\ -8 & 2 & -9 \\ 5 & 0 & 5 \end{bmatrix}.$$

Multiply the constant matrix by A^{-1} on the left.

$$\begin{bmatrix} x \\ y \\ z \end{bmatrix} = \frac{1}{5}\begin{bmatrix} -4 & 1 & -7 \\ -8 & 2 & -9 \\ 5 & 0 & 5 \end{bmatrix}\begin{bmatrix} 2 \\ 0 \\ 1 \end{bmatrix} = \frac{1}{5}\begin{bmatrix} -15 \\ -25 \\ 15 \end{bmatrix} = \begin{bmatrix} -3 \\ -5 \\ 3 \end{bmatrix}$$

Therefore the solution is $(-3, -5, 3)$. **Answer**

Finding the inverse of a matrix is particularly useful when two or more systems of equations have the same coefficient matrix. In particular, we can solve the systems

$$\begin{aligned} x_1 + 2y_1 &= 7 \\ 3x_1 + 4y_1 &= 17 \end{aligned} \qquad \begin{aligned} x_2 + 2y_2 &= 25 \\ 3x_2 + 4y_2 &= 55 \end{aligned}$$

by using the same coefficient matrix A from Example 2, the inverse A^{-1} found in Example 2, and the discussion preceding Example 4. In Exercise 10 you are asked to solve this pair of systems to find (x_1, y_1) and (x_2, y_2).

EXERCISES

Determine whether the matrices are inverses of each other.

A

1. $\begin{bmatrix} 0 & -1 \\ 1 & 0 \end{bmatrix}; \begin{bmatrix} 0 & 1 \\ -1 & 0 \end{bmatrix}$

2. $\begin{bmatrix} -6 & 5 \\ 4 & -3 \end{bmatrix}; \begin{bmatrix} 1.5 & 2.5 \\ 2 & 3 \end{bmatrix}$

3. $\begin{bmatrix} 5 & 2 & 0 \\ 0 & 0 & 1 \\ 3 & 1 & 0 \end{bmatrix}; \begin{bmatrix} -1 & 3 & 0 \\ 0 & 0 & 1 \\ 0 & -5 & 0 \end{bmatrix}$

4. $\begin{bmatrix} 4 & 1 & 3 \\ 0 & 2 & 0 \\ 1 & 0 & 1 \end{bmatrix}; \begin{bmatrix} 1 & -3 & 0 \\ 0 & 0 & 3 \\ -1 & 4 & -1 \end{bmatrix}$

Find the inverse of each matrix, if it exists. If it does not exist, so state.

5. $\begin{bmatrix} -3 & -7 \\ 2 & 5 \end{bmatrix}$

6. $\begin{bmatrix} 6 & -4 \\ -12 & 8 \end{bmatrix}$

7. $\begin{bmatrix} 8 & -9 \\ 6 & -7 \end{bmatrix}$

8. $\begin{bmatrix} 6 & 4 & 1 \\ -3 & -2 & 0 \\ 2 & 1 & 0 \end{bmatrix}$

9. $\begin{bmatrix} 3 & 4 & -5 \\ -1 & -2 & 2 \\ 0 & 1 & 0 \end{bmatrix}$

10. Use inverses to solve the pair of systems of equations on page 513.

Use inverses to solve each matrix equation for x and y.

11. $\begin{bmatrix} 2 & 1 \\ 1 & 1 \end{bmatrix}\begin{bmatrix} x \\ y \end{bmatrix} = \begin{bmatrix} 3 \\ 2 \end{bmatrix}$

12. $\begin{bmatrix} 1 & 3 \\ 2 & 3 \end{bmatrix}\begin{bmatrix} x \\ y \end{bmatrix} = \begin{bmatrix} 2 \\ 7 \end{bmatrix}$

13. $\begin{bmatrix} 1 & 2 \\ 3 & 2 \end{bmatrix}\begin{bmatrix} x \\ y \end{bmatrix} = \begin{bmatrix} -1 \\ 1 \end{bmatrix}$

14. $\begin{bmatrix} 3 & 2 \\ 1 & 2 \end{bmatrix}\begin{bmatrix} x \\ y \end{bmatrix} = \begin{bmatrix} -12 \\ -4 \end{bmatrix}$

Solve each matrix equation for the matrix A.

B

15. $\begin{bmatrix} 5 & 3 \\ 3 & 2 \end{bmatrix}A - \begin{bmatrix} 2 & 4 \\ 3 & 2 \end{bmatrix} = \begin{bmatrix} 0 & 1 \\ -2 & 2 \end{bmatrix}$

16. $\begin{bmatrix} 3 & 4 \\ 2 & 3 \end{bmatrix}A + \begin{bmatrix} 4 & 0 \\ 2 & 3 \end{bmatrix} = \begin{bmatrix} -1 & 1 \\ -2 & 0 \end{bmatrix}$

17. Show that the matrix of the rotation of the plane through the angle α has an inverse and find it.

Solve each system of equations by using inverses of matrices.

18. $\begin{aligned} x + y + z &= 6 \\ 2x - y - z &= -3 \\ x - 3y + 2z &= 1 \end{aligned}$

19. $\begin{aligned} x - 2y - 3z &= 3 \\ x + y - z &= 2 \\ 2x - 3y - 5z &= 5 \end{aligned}$

20. Show that if $A = \begin{bmatrix} a_1 & b_1 \\ a_2 & b_2 \end{bmatrix}$ and $\begin{bmatrix} a_1 & b_1 \\ a_2 & b_2 \end{bmatrix}\begin{bmatrix} u_1 & v_1 \\ u_2 & v_2 \end{bmatrix} = \begin{bmatrix} 1 & 0 \\ 0 & 1 \end{bmatrix}$, then

$$u_1 = \frac{b_2}{a_1b_2 - a_2b_1} \qquad v_1 = \frac{-b_1}{a_1b_2 - a_2b_1}$$

$$u_2 = \frac{-a_2}{a_1b_2 - a_2b_1} \qquad v_2 = \frac{a_1}{a_1b_2 - a_2b_1}$$

provided $a_1b_2 - a_2b_1 \neq 0$. (*Hint:* Use matrix multiplication and matrix equality to obtain the following two systems.)

$$(1)\quad \begin{aligned} a_1u_1 + b_1u_2 &= 1 \\ a_2u_1 + b_2u_2 &= 0 \end{aligned} \qquad (2)\quad \begin{aligned} a_1v_1 + b_1v_2 &= 0 \\ a_2v_1 + b_2v_2 &= 1 \end{aligned}$$

21. **a.** If $T = \begin{bmatrix} a & b & c \\ 0 & d & e \\ 0 & 0 & f \end{bmatrix}$, under what conditions will T^{-1} exist?

b. Find T^{-1}.

22. A transformation T of three-space is represented by the matrix

$$\overline{T} = \begin{bmatrix} 3 & 2 & -2 \\ 4 & 2 & -3 \\ 0 & 1 & -1 \end{bmatrix}.$$

If $T(\mathbf{v}) = (7, 9, -4)$, find $\mathbf{v} = (x, y, z)$.

23. Show that if a matrix A has an inverse, then A^{-1} is unique. (*Hint:* Suppose that A has two distinct inverses B and C. Show that $AB = I$ and $AC = I$ leads to a contradiction.)

C 24. Show that if A and B are $n \times n$ matrices whose inverses exist, then $(AB)^{-1}$ exists and $(AB)^{-1} = B^{-1}A^{-1}$.

COMPUTER EXERCISES

1. Write a program that will compute the inverse of a 2×2 matrix, if it exists. The program should print A^{-1} for an input matrix A or a message if appropriate.
2. Run the program for each matrix.

a. $A = \begin{bmatrix} 6 & -2 \\ 1 & -10 \end{bmatrix}$ **b.** $A = \begin{bmatrix} 5 & -9 \\ -16 & 72 \end{bmatrix}$ **c.** $A = \begin{bmatrix} 28 & -21 \\ -16 & 12 \end{bmatrix}$

12-9 CRAMER'S RULE

While reduction to echelon form can be used to solve a system of linear equations, *Cramer's rule*, a method of using determinants, can also be used if the coefficient matrix is square and its determinant is not zero.

If the system

$$\begin{aligned} ax + by &= e \\ cx + dy &= f \end{aligned}$$

is solved by using elementary operations, the result will be

$$x = \frac{ed - bf}{ad - bc} \qquad y = \frac{af - ec}{ad - bc}$$

provided that $ad - bc \neq 0$. The numerators and denominators of these fractions can be written as determinants to give

$$x = \frac{\begin{vmatrix} e & b \\ f & d \end{vmatrix}}{\begin{vmatrix} a & b \\ c & d \end{vmatrix}} \qquad y = \frac{\begin{vmatrix} a & e \\ c & f \end{vmatrix}}{\begin{vmatrix} a & b \\ c & d \end{vmatrix}}$$

These are the Cramer's-rule formulas for the two-variable case.

EXAMPLE 1 Use Cramer's rule to solve the system:

$$\begin{aligned} 7x - 11y &= 6 \\ -3x + 5y &= -2 \end{aligned}$$

Solution Find the determinant of the coefficient matrix.

$$\det\begin{bmatrix} 7 & -11 \\ -3 & 5 \end{bmatrix} = 7 \cdot 5 - (-3)(-11)$$
$$= 2$$

To find x, find the value of the determinant of the coefficient matrix with the first column replaced by the constant matrix.

$$\begin{vmatrix} 6 & -11 \\ -2 & 5 \end{vmatrix} = 6 \cdot 5 - (-2)(-11)$$
$$= 8$$

Therefore $x = \frac{8}{2} = 4$.

To find y, find the value of the determinant of the coefficient matrix with the second column replaced by the constant matrix.

$$\begin{vmatrix} 7 & 6 \\ -3 & -2 \end{vmatrix} = (7)(-2) - (-3)(6)$$
$$= 4$$

Therefore $y = \frac{4}{2} = 2$.

The solution of the system is $(4, 2)$. **Answer**

In general, we have **Cramer's rule:**

Given a system of n linear equations in n variables $x_1, x_2, \ldots, x_n$, let A be the coefficient matrix, and let A_j be the matrix obtained from A by replacing column j by the column of constants. Then, if $\det A \neq 0$,

$$x_j = \frac{\det A_j}{\det A} \qquad j = 1, 2, \ldots, n$$

EXAMPLE 2 Use Cramer's rule to solve:

$$\begin{aligned} 5x - 2y + z &= 20 \\ 4x - 4y - 5z &= 0 \\ 3x - 2y &= 12 \end{aligned}$$

Solution Find $\det A$, $\det A_1$, $\det A_2$, and $\det A_3$.

$$\det A = \begin{vmatrix} 5 & -2 & 1 \\ 4 & -4 & -5 \\ 3 & -2 & 0 \end{vmatrix} = -16$$

$$\det A_1 = \begin{vmatrix} 20 & -2 & 1 \\ 0 & -4 & -5 \\ 12 & -2 & 0 \end{vmatrix} = -32$$

$$\det A_2 = \begin{vmatrix} 5 & 20 & 1 \\ 4 & 0 & -5 \\ 3 & 12 & 0 \end{vmatrix} = 48$$

$$\det A_3 = \begin{vmatrix} 5 & -2 & 20 \\ 4 & -4 & 0 \\ 3 & -2 & 12 \end{vmatrix} = -64$$

Therefore $x = \frac{-32}{-16} = 2$, $y = \frac{48}{-16} = -3$, and $z = \frac{-64}{-16} = 4$.

The solution of the system is $(2, -3, 4)$. **Answer**

EXERCISES

Solve each system by Cramer's rule. If the determinant of the coefficient matrix is 0, so state.

A

1. $5x - 11y = 3$
   $4x - 9y = 2$
2. $-9x + 7y = -10$
   $5x - 4y = 4$
3. $3x - 9y = 4$
   $-5x + 16y = -7$
4. $-13x - 7y = 8$
   $9x + 5y = -6$
5. $-3x + 5y = 10$
   $7x - 14y = 9$
6. $7x + 9y = 12$
   $6x + 8y = 9$

7. $\begin{aligned} 3x - y + 2z &= -1 \\ 5x + 4y + z &= 11 \\ -4x + y + 3z &= 2 \end{aligned}$

8. $\begin{aligned} x - 2y + 4z &= 8 \\ 6x - y - 3z &= 2 \\ -x + y - 5z &= -9 \end{aligned}$

9. $\begin{aligned} x - 3y + 4z &= 7 \\ 3x + 4y + z &= 7 \\ 2x + 7y - 4z &= 2 \end{aligned}$

10. $\begin{aligned} 2x - y + 5z &= -1 \\ 2x - 3z &= 4 \\ 6x - 2y + z &= 8 \end{aligned}$

11. $\begin{aligned} 3x + 7y &= 11 \\ 2x + 5y + z &= 6 \\ 2y + 4z &= 7 \end{aligned}$

12. $\begin{aligned} x + 2y - z &= -1 \\ 7x + 2y - 3z &= 5 \\ 2x - 5y + 3z &= 5 \end{aligned}$

In Exercises 13–15 sketch the graph of each pair of equations in the system. Determine whether the lines intersect in a point, are coincident, or are parallel.

13. $\begin{aligned} 2x + 5y &= 0 \\ 4x - 10y &= 0 \end{aligned}$

14. $\begin{aligned} 2x - 5y &= 0 \\ 4x - 10y &= 0 \end{aligned}$

15. $\begin{aligned} 2x - 5y &= 0 \\ 4x - 10y &= 10 \end{aligned}$

B

16. a. Given the system:

$$\begin{aligned} ax + by &= e \\ cx + dy &= f \end{aligned}$$

Write an equation involving a, b, c, and d that gives a necessary and sufficient condition for the graphs of the equations to be either parallel or coincident.

b. Show that the result of part (a) is equivalent to $\begin{vmatrix} a & b \\ c & d \end{vmatrix} = 0$.

Use Cramer's rule to find x.

17. $\begin{aligned} 3x + 2y - 4w &= -2 \\ x - y - z &= 6 \\ 2x + y - 3w &= -1 \\ -x + y + w &= -5 \end{aligned}$

18. $\begin{aligned} 3x - y + 2z + w &= 5 \\ x - z + 4w &= 4 \\ -2x + y + z - 3w &= -1 \\ x + 3z - w &= 6 \end{aligned}$

C

19. Given the system

$$\begin{aligned} a_1x_1 + b_1x_2 + c_1x_3 &= d_1 \\ a_2x_1 + b_2x_2 + c_2x_3 &= d_2 \\ a_3x_1 + b_3x_2 + c_3x_3 &= d_3 \end{aligned}$$

with coefficient matrix A and $\det A \neq 0$. Let A_1, A_2, and A_3 be the cofactors of a_1, a_2, and a_3 respectively. Complete parts (a)–(f) to justify Cramer's rule for finding x_1.

a. Show that $(A_1a_1 + A_2a_2 + A_3a_3)x_1 + (A_1b_1 + A_2b_2 + A_3b_3)x_2 + (A_1c_1 + A_2c_2 + A_3c_3)x_3 = (A_1d_1 + A_2d_2 + A_3d_3)$.

b. Show that $A_1a_1 + A_2a_2 + A_3a_3 = \det A$.

c. Show that $A_1b_1 + A_2b_2 + A_3b_3 = 0$ by expanding A_1, A_2, and A_3.

d. Show that $A_1c_1 + A_2c_2 + A_3c_3 = 0$ by expanding A_1, A_2, and A_3.

e. Show that $A_1d_1 + A_2d_2 + A_3d_3$ is the value of the determinant of the coefficient matrix with the first column replaced by the constant matrix.

f. Use parts (a)–(e) to complete the justification of Cramer's rule for finding x_1.

20. To justify Cramer's rule for finding x_2, what cofactors would be chosen as multipliers?

Chapter Summary

1. A *linear transformation* of a vector space V into a vector space V' is a function T with domain V and range in V' such that
$$T(a\mathbf{u} + b\mathbf{v}) = aT(\mathbf{u}) + bT(\mathbf{v})$$
for all vectors $\mathbf{u}$ and $\mathbf{v}$ in V and all scalars a and $b \in \mathcal{R}$. The *image* (x', y') of a vector (x, y) under a *rotation* of the plane through α, T_α, is given by
$$x' = x \cos \alpha - y \sin \alpha \qquad y' = x \sin \alpha + y \cos \alpha.$$

2. A *matrix* is a rectangular array of numbers, called *entries*. A matrix with m rows and n columns has *dimensions* $m \times n$. The *sum* of two such matrices is obtained by adding corresponding entries. *Scalar multiplication* of one such matrix by a scalar k is obtained by multiplying each entry by k. The set of all $m \times n$ matrices with these operations is a vector space.

   The *product* of an $m \times n$ matrix, A, and an $n \times k$ matrix, B, is the matrix whose entry in the ith row and jth column is the dot product of the ith row of A and jth column of B.

3. Associated with each linear transformation is a matrix. If S and T are linear transformations: T: $U \to V$ and S: $V \to W$ with associated matrices $\overline{S}$ and $\overline{T}$, then the matrix of $S \circ T$ is $\overline{S}\,\overline{T}$.

4. If the x- and y-axes are rotated through an angle φ, each point P has a set of xy-coordinates, (x, y), and a set of $x'y'$-coordinates, (x', y'). These are related by
$$1.\ x = x' \cos \varphi - y' \sin \varphi \qquad y = x' \sin \varphi + y' \cos \varphi$$
$$2.\ x' = x \cos \varphi + y \sin \varphi \qquad y' = -x \sin \varphi + y \cos \varphi$$
   The *general second-degree equation* in two variables is of the form
$$Ax^2 + Bxy + Cy^2 + Dx + Ey + F = 0.$$
   If $B \neq 0$, the xy-term can be eliminated by rotating the xy-axes through an angle φ such that $\cot 2\varphi = \dfrac{A - C}{B}$. The graph of a second-degree equation is a conic section.

5. *Elementary operations* can be used to put a system of linear equations into *triangular form*, from which the solutions can be found by *back substitution*. A system of linear equations can be represented by an *augmented matrix*, from which solutions can be obtained by carrying out *elementary row operations*.

6. A matrix can be transformed to *row-reduced echelon form*. A system having at least one solution is *consistent*. If no solution exists, the system is *inconsistent*.

7. The *determinant* of a 2×2 matrix $T = \begin{bmatrix} a & b \\ c & d \end{bmatrix}$ is defined by
$$\det T = \begin{vmatrix} a & b \\ c & d \end{vmatrix} = ad - bc.$$

Determinants of higher order are defined and evaluated by use of minors and cofactors.

8. If A and B are matrices and $AB = BA = I$ where I is the identity matrix, then A and B are *inverses* of each other. If A has an inverse, A^{-1}, then

$$A^{-1} = \frac{1}{\det A} \operatorname{adj} A.$$

9. *Cramer's rule* is a method that uses determinants to solve a system of n linear equations in n variables.

Chapter Test

12-1 **1.** Show that $T((x, y)) = (3x - y, 5x)$ is a linear transformation.

2. Find the image of $(-5, 2)$ under a transformation of scalar multiplication by 2 followed by a rotation through 90°.

12-2 Let $A = \begin{bmatrix} -3 & 1 \\ 4 & 0 \end{bmatrix}$ and $B = \begin{bmatrix} -1 & -2 \\ 5 & 8 \end{bmatrix}$. Find each matrix.

3. $2A - B$ **4.** A^2 **5.** AB **6.** $BA - A$

12-3 **7.** Let transformations S and T have matrices

$$\overline{S} = \begin{bmatrix} 0 & 5 & 2 \\ 3 & 2 & 0 \\ 1 & 4 & -1 \end{bmatrix} \text{ and } \overline{T} = \begin{bmatrix} -2 & 3 & 1 \\ 1 & 6 & 0 \\ 0 & 4 & 5 \end{bmatrix}.$$

Find the matrix associated with $S \circ T$.

8. Write an expression for $T((x, y))$ if the matrix of T is $\begin{bmatrix} 5 & -9 \\ 0 & 4 \\ -2 & 3 \end{bmatrix}$.

12-4 **9.** Identify the graph of $4x^2 + 2\sqrt{3}xy + 2y^2 - 20 = 0$.

10. With a suitable rotation of axes transform the equation in Exercise 9 into an equation in x' and y' having no $x'y'$-term.

12-5 **11.** Solve the system:

$$\begin{aligned} x - 2y - 4z &= 3 \\ 3x + y + 2z &= 2 \\ -x + 3y + 5z &= -1 \end{aligned}$$

12-6 **12.** Solve the system:

$$\begin{aligned} x + y + z &= 4 \\ 2x - y + 2z &= 5 \\ 7x + y + 7z &= 22 \end{aligned}$$

12-7 Find the value of each determinant.

13. $\begin{vmatrix} -3 & 1 & -2 \\ 5 & -4 & 3 \\ 0 & 0 & 1 \end{vmatrix}$ **14.** $\begin{vmatrix} 1 & -3 & 4 \\ -2 & 5 & -1 \\ 6 & -7 & 2 \end{vmatrix}$

12-8 **15.** Find the inverse of $\begin{bmatrix} 0 & 0 & 2 \\ 1 & 3 & -2 \\ 3 & -2 & 1 \end{bmatrix}$.

12-9 **16.** Find y in Exercise 11 by using Cramer's rule.

COMPUTER APPLICATION

LINEAR SYSTEMS

Example 2 of Section 12–5 and the discussion following that example can serve as the basis for a computer program designed to solve a system of n linear equations in n variables. The strategy is as follows:

1. Reduce, if possible, each diagonal entry, a_{ii}, of the augmented matrix to 1. If $a_{ii} = 0$ for some i, then, in lines 100–200, rows are interchanged to obtain the next nonzero diagonal entry.
2. Transform each other entry in column i to 0. In performing this step, each entry in row i is also changed. See in particular line 310.
3. If the solution is unique, the solution is printed in lines 350–400. If the solution is not unique, either because there is no solution or because there are infinitely many solutions, the fact that the solution is not unique is reported in line 140.

```
10  INPUT N
20  DIM A(N + 1,N + 1)
30  FOR L = 1 TO N
40  FOR C = 1 TO N + 1
50  INPUT A(L,C)
60  NEXT C
70  NEXT L
80  FOR L = 1 TO N
90  IF A(L,L) < > 0 THEN 210
100  FOR K = L + 1 TO N
110  IF A(K,K) < > 0 THEN 160
120  NEXT K
130  PRINT
140  PRINT "NO UNIQUE SOLUTION"
150  GOTO 410
160  FOR I = 1 TO N + 1
170  LET M = A(L,I)
180  LET A(L,I) = A(K,I)
190  LET A(K,I) = M
200  NEXT I
210  IF A(L,L) = 1 THEN 260
220  LET D = A(L,L)
230  FOR C = 1 TO N + 1
240  LET A(L,C) = A(L,C) / D
```

```
250 NEXT C
260 FOR R = 1 TO N
270 IF R = L THEN 330
280 IF A(R,L) = 0 THEN 330
290 LET R1 = A(R,L)
300 FOR C = L TO N + 1
310 LET A(R,C) = A(R,C) - R1 * A(L,C)
320 NEXT C
330 NEXT R
340 NEXT L
350 FOR R = 1 TO N
360 FOR C = 1 TO N + 1
370 PRINT A(R,C);" ";
380 NEXT C
390 PRINT
400 NEXT R
410 END
```

EXERCISES *(You may use either BASIC or Pascal.)*

Run the program to solve each system of linear equations.

1.
$$\begin{aligned} 2x + y - 3z + w &= 10 \\ x + 2y + z - w &= -2 \\ 3x + 5y - 2z + 2w &= 14 \\ 2x + 3y + 2z - 2w &= -4 \end{aligned}$$

2.
$$\begin{aligned} x + y + z + w + u + v &= 21 \\ 2x - 2y - z + w - u + 2v &= 6 \\ x - y + 3z - w + 2u - v &= 8 \\ x + 3y - z + 2w - u - 3v &= -11 \\ 3x - y + 2z - 3w + 3u - v &= 4 \\ x + 4y - 2z + 4w + 2u - 3v &= 11 \end{aligned}$$

3.
$$\begin{aligned} x + y - z + w &= 4 \\ x - y + z - w &= -2 \\ x + y + z - w &= 2 \\ x - y - z + w &= 0 \end{aligned}$$

4.
$$\begin{aligned} x + 2y - z - 2w &= 3 \\ 4z + w &= 5 \\ 2x - y + w &= 0 \\ x + y + z &= 2 \end{aligned}$$

5. The table at the right lists the quantities of five kinds of articles purchased by a craft shop over five weeks. It also lists the total bill for each week. Find the cost of each kind of article.

Week	Quantity					Bill
1	6	7	10	2	1	320
2	10	12	8	3	1	380
3	7	8	6	1	1	255
4	8	10	7	2	1	315
5	10	8	5	3	1	295

APPLICATION

Coding

A major part of the art of encoding and decoding messages is the design of a coding scheme. Matrices and vectors are useful in such designs. Suppose that we wish to send AH YES! in code. We might proceed as follows.

1. Assign to each character a number. Use a 0 for each space.

   A $\leftrightarrow$ 3 H $\leftrightarrow$ 2 space $\leftrightarrow$ 0 Y $\leftrightarrow$ 25 E $\leftrightarrow$ 9 S $\leftrightarrow$ 11 ! $\leftrightarrow$ 99

2. Make each block of successive characters into 2×1 vectors.

$$\begin{bmatrix} 3 \\ 2 \end{bmatrix} \qquad \begin{bmatrix} 0 \\ 25 \end{bmatrix} \qquad \begin{bmatrix} 9 \\ 11 \end{bmatrix} \qquad \begin{bmatrix} 99 \\ 0 \end{bmatrix}$$

3. Choose a 2×2 matrix, *CODE*, whose inverse exists.

$$CODE = \begin{bmatrix} -2 & 4 \\ -1 & 5 \end{bmatrix} \qquad (CODE)^{-1} = \begin{bmatrix} -\frac{5}{6} & \frac{2}{3} \\ -\frac{1}{6} & \frac{1}{3} \end{bmatrix}$$

4. To encode AH, multiply $\begin{bmatrix} 3 \\ 2 \end{bmatrix}$ on the left by *CODE*.

$$\begin{bmatrix} -2 & 4 \\ -1 & 5 \end{bmatrix} \begin{bmatrix} 3 \\ 2 \end{bmatrix} = \begin{bmatrix} (-2)(3) + (4)(2) \\ (-1)(3) + (5)(2) \end{bmatrix} = \begin{bmatrix} 2 \\ 7 \end{bmatrix}$$

5. To decode $\begin{bmatrix} 2 \\ 7 \end{bmatrix}$, multiply $\begin{bmatrix} 2 \\ 7 \end{bmatrix}$ on the left by $(CODE)^{-1}$.

$$\begin{bmatrix} -\frac{5}{6} & \frac{2}{3} \\ -\frac{1}{6} & \frac{1}{3} \end{bmatrix} \begin{bmatrix} 2 \\ 7 \end{bmatrix} = \begin{bmatrix} (-\frac{5}{6})(2) + (\frac{2}{3})(7) \\ (-\frac{1}{6})(2) + (\frac{1}{3})(7) \end{bmatrix} = \begin{bmatrix} 3 \\ 2 \end{bmatrix}$$

   From $\begin{bmatrix} 3 \\ 2 \end{bmatrix}$, 3 $\leftrightarrow$ A and 2 $\leftrightarrow$ H. Thus AH is recovered.

Notes: The code designer is free to choose any character-to-number assignment scheme as long as the scheme is one-to-one. The designer is free to divide successive characters into blocks other than two characters long and is free to choose any coding matrix as long as its dimensions conform to the choice made in step 2 and the matrix has an inverse.

EXERCISES

1. Explain: The assignment of numbers to characters must be one-to-one.
2. Explain: The matrix chosen for coding must have an inverse.
3. **a.** Use *CODE* to encode ES.
   b. Use $(CODE)^{-1}$ to decode the result of part (a).
4. A message consists of thirteen characters. It is divided into five blocks of characters and spaces are placed at the end of the message so as to make 3×1 vectors. How many blocks are made and how many spaces are needed at the end?
5. What are the dimensions of a coding matrix for Exercise 4?
6. The letters A, B, C, . . . , Z are assigned the numbers 1, 2, 3, . . . , 26 respectively. Then the numbers, n, are assigned numbers, n', according to $n' = 6n - 5$. Is the assignment one-to-one?
7. Must the matrix chosen as a coding matrix be a square matrix?
8. May a code designer choose the matrix

$$\begin{bmatrix} 3 & 9 \\ 2 & 6 \end{bmatrix}$$

   as a coding matrix?
9. Must multiplication by *CODE* and $(CODE)^{-1}$ be performed on the left?

Digital switches in modern telephone equipment convert electronic signals into the on-and-off pulses of computer language.

Chapter 13

Further Vector Topics

In this chapter, the cross product, planes, and lines are discussed. Surfaces and other coordinate systems are investigated as well.

Planes and Lines

13-1 THE CROSS PRODUCT

In Chapter 11 we introduced the dot product of two vectors. Recall that

$$\mathbf{u} \cdot \mathbf{v} = \|\mathbf{u}\|\,\|\mathbf{v}\| \cos \theta \quad \text{and} \quad \mathbf{u} \cdot \mathbf{v} = u_1v_1 + u_2v_2 + u_3v_3$$

where $\mathbf{u} = (u_1, u_2, u_3)$, $\mathbf{v} = (v_1, v_2, v_3)$, and θ is the angle between $\mathbf{u}$ and $\mathbf{v}$. We now define another kind of product and, in subsequent sections, use both kinds of products to solve problems in three-space.

DEFINITION

The **cross product**, or **vector product**, of $\mathbf{u} = (u_1, u_2, u_3)$ and $\mathbf{v} = (v_1, v_2, v_3)$, denoted $\mathbf{u} \times \mathbf{v}$, is given by

$$\mathbf{u} \times \mathbf{v} = \left(\begin{vmatrix} u_2 & u_3 \\ v_2 & v_3 \end{vmatrix}, \begin{vmatrix} u_3 & u_1 \\ v_3 & v_1 \end{vmatrix}, \begin{vmatrix} u_1 & u_2 \\ v_1 & v_2 \end{vmatrix} \right).$$

Note that while the dot product of $\mathbf{u}$ and $\mathbf{v}$ is a scalar, the cross product of $\mathbf{u}$ and $\mathbf{v}$ is another *vector*.

If $\mathbf{u} = u_1\mathbf{i} + u_2\mathbf{j} + u_3\mathbf{k}$ and $\mathbf{v} = v_1\mathbf{i} + v_2\mathbf{j} + v_3\mathbf{k}$, then

$$\mathbf{u} \times \mathbf{v} = \begin{vmatrix} u_2 & u_3 \\ v_2 & v_3 \end{vmatrix}\mathbf{i} + \begin{vmatrix} u_3 & u_1 \\ v_3 & v_1 \end{vmatrix}\mathbf{j} + \begin{vmatrix} u_1 & u_2 \\ v_1 & v_2 \end{vmatrix}\mathbf{k}.$$

The ridges in this Death Valley scene show many variations in the slopes of a surface. One concept introduced in this chapter is that of surface.

The preceding formula can be remembered as the expansion of:

$$\begin{vmatrix} \mathbf{i} & \mathbf{j} & \mathbf{k} \\ u_1 & u_2 & u_3 \\ v_1 & v_2 & v_3 \end{vmatrix}$$

EXAMPLE 1 Find the cross product of $\mathbf{u} = (3, -2, 0)$ and $\mathbf{v} = (4, 1, 0)$.

Solution To find $\mathbf{u} \times \mathbf{v}$, expand

$$\begin{vmatrix} \mathbf{i} & \mathbf{j} & \mathbf{k} \\ 3 & -2 & 0 \\ 4 & 1 & 0 \end{vmatrix}.$$

$$\mathbf{u} \times \mathbf{v} = ((-2)0 - 1 \cdot 0)\mathbf{i} - (3 \cdot 0 - 4 \cdot 0)\mathbf{j} + (3 \cdot 1 - 4(-2))\mathbf{k}$$

$$\mathbf{u} \times \mathbf{v} = 0\mathbf{i} - 0\mathbf{j} + 11\mathbf{k}, \text{ or } 11\mathbf{k}, \text{ or } (0, 0, 11)$$ **Answer**

The following theorems state basic properties of the cross product. You are asked to prove Theorem 1 in Exercises 20–23.

THEOREM 1 Let $\mathbf{u}$, $\mathbf{v}$, and $\mathbf{w}$ be vectors and let r and s be scalars.

1. $\mathbf{u} \times \mathbf{v} = -\mathbf{v} \times \mathbf{u}$
2. $(r\mathbf{u}) \times (s\mathbf{v}) = rs(\mathbf{u} \times \mathbf{v})$
   Associative Property
3. $\mathbf{u} \times \mathbf{u} = \mathbf{0}$
4. $\mathbf{u} \times (\mathbf{v} + \mathbf{w}) = \mathbf{u} \times \mathbf{v} + \mathbf{u} \times \mathbf{w}$
   Distributive Property

THEOREM 2

1. $\mathbf{u} \times \mathbf{v}$ is perpendicular to both $\mathbf{u}$ and $\mathbf{v}$.
2. $\|\mathbf{u} \times \mathbf{v}\| = \|\mathbf{u}\|\|\mathbf{v}\| \sin\theta$ where θ, $0° \le \theta < 360°$ is the angle between $\mathbf{u}$ and $\mathbf{v}$.

Proof: Let $\mathbf{u} = (u_1, u_2, u_3)$ and $\mathbf{v} = (v_1, v_2, v_3)$.

1. $(\mathbf{u} \times \mathbf{v}) \cdot \mathbf{u}$

$$= \left(\begin{vmatrix} u_2 & u_3 \\ v_2 & v_3 \end{vmatrix}, \begin{vmatrix} u_3 & u_1 \\ v_3 & v_1 \end{vmatrix}, \begin{vmatrix} u_1 & u_2 \\ v_1 & v_2 \end{vmatrix} \right) \cdot (u_1, u_2, u_3)$$
$$= (u_2v_3 - v_2u_3)u_1 + (u_3v_1 - v_3u_1)u_2 + (u_1v_2 - v_1u_2)u_3$$
$$= u_2v_3u_1 - v_2u_3u_1 + u_3v_1u_2 - v_3u_1u_2 + u_1v_2u_3 - v_1u_2u_3 = 0$$

By similar reasoning, $(\mathbf{u} \times \mathbf{v}) \cdot \mathbf{v} = 0$. Thus $\mathbf{u} \times \mathbf{v}$ is perpendicular to both $\mathbf{u}$ and $\mathbf{v}$. ■

2. First we show that $\|\mathbf{u} \times \mathbf{v}\|^2 = \|\mathbf{u}\|^2\|\mathbf{v}\|^2 - (\mathbf{u} \cdot \mathbf{v})^2$.

$$(\mathbf{u} \cdot \mathbf{v})^2 + \|\mathbf{u} \times \mathbf{v}\|^2 = (u_1v_1 + u_2v_2 + u_3v_3)^2$$
$$+ \left[\sqrt{(u_2v_3 - u_3v_2)^2 + (u_3v_1 - u_1v_3)^2 + (u_1v_2 - u_2v_1)^2}\,\right]^2$$

Expanding the squares on the right side of this equation and combining like terms, we obtain:

$$
\begin{aligned}
(\mathbf{u} \cdot \mathbf{v})^2 + \|\mathbf{u} \times \mathbf{v}\|^2 &= (u_1v_1)^2 + (u_2v_2)^2 + (u_3v_3)^2 + (u_2v_3)^2 + (u_3v_2)^2 \\
&\quad + (u_3v_1)^2 + (u_1v_3)^2 + (u_1v_2)^2 + (u_2v_1)^2 \\
&= (u_1^2 + u_2^2 + u_3^2)(v_1^2 + v_2^2 + v_3^2) \\
&= \|\mathbf{u}\|^2\|\mathbf{v}\|^2
\end{aligned}
$$

Hence $\|\mathbf{u} \times \mathbf{v}\|^2 = \|\mathbf{u}\|^2\|\mathbf{v}\|^2 - (\mathbf{u} \cdot \mathbf{v})^2$.

Next we use the fact that $\mathbf{u} \cdot \mathbf{v} = \|\mathbf{u}\|\,\|\mathbf{v}\| \cos \theta$.

$$
\begin{aligned}
\|\mathbf{u} \times \mathbf{v}\|^2 &= \|\mathbf{u}\|^2\|\mathbf{v}\|^2 - (\|\mathbf{u}\|\,\|\mathbf{v}\| \cos \theta)^2 \\
&= \|\mathbf{u}\|^2\|\mathbf{v}\|^2 (1 - \cos^2 \theta) \\
\|\mathbf{u} \times \mathbf{v}\|^2 &= \|\mathbf{u}\|^2\|\mathbf{v}\|^2 \sin^2 \theta
\end{aligned}
$$

Since the norms and $\sin \theta$ are nonnegative,

$$\|\mathbf{u} \times \mathbf{v}\| = \|\mathbf{u}\|\,\|\mathbf{v}\| \sin \theta. \quad \blacksquare$$

Notice that from part (2), $\|\mathbf{u} \times \mathbf{v}\| = 0$ if and only if $\mathbf{u} = \mathbf{0}$, $\mathbf{v} = \mathbf{0}$, or $\theta = 0$. That is, $\mathbf{u}$ and $\mathbf{v}$ are parallel if and only if $\mathbf{u} \times \mathbf{v} = \mathbf{0}$.

THEOREM 3 The area of the parallelogram having $\mathbf{u}$ and $\mathbf{v}$ as adjacent sides is $\|\mathbf{u} \times \mathbf{v}\|$.

Proof:

$$
\begin{aligned}
\text{Area} &= \text{base} \cdot \text{height} \\
&= \|\mathbf{v}\| \cdot \|\mathbf{u}\| \sin \theta \\
\text{Area} &= \|\mathbf{u} \times \mathbf{v}\| \quad \blacksquare
\end{aligned}
$$

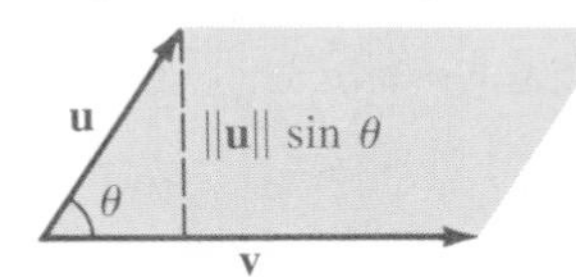

EXAMPLE 2 Find the area of the triangle whose vertices are $A(1, 0, 3)$, $B(1, 3, 4)$, and $C(-2, 5, 1)$.

Solution The area of triangle ABC is one-half the area of the parallelogram formed by $\overrightarrow{AB}$ and $\overrightarrow{AC}$. Find $\frac{1}{2}\|\overrightarrow{AB} \times \overrightarrow{AC}\|$.

$$\overrightarrow{AB} = (0, 3, 1) \qquad \overrightarrow{AC} = (-3, 5, -2)$$

$$\overrightarrow{AB} \times \overrightarrow{AC} = \begin{vmatrix} \mathbf{i} & \mathbf{j} & \mathbf{k} \\ 0 & 3 & 1 \\ -3 & 5 & -2 \end{vmatrix} = -11\mathbf{i} - 3\mathbf{j} + 9\mathbf{k}$$

Hence $\|\overrightarrow{AB} \times \overrightarrow{AC}\| = \sqrt{211}$. Therefore the area of triangle ABC is $\frac{1}{2}\sqrt{211}$, or about 7.3 square units. **Answer**

The **scalar triple product** of $\mathbf{u}$, $\mathbf{v}$, and $\mathbf{w}$ is $\mathbf{u} \cdot (\mathbf{v} \times \mathbf{w})$. The next theorem states how to compute it and gives a geometric application of it.

THEOREM 4 Let $\mathbf{u} = (u_1, u_2, u_3)$, $\mathbf{v} = (v_1, v_2, v_3)$, and $\mathbf{w} = (w_1, w_2, w_3)$.

1. $\mathbf{u} \cdot (\mathbf{v} \times \mathbf{w}) = \begin{vmatrix} u_1 & u_2 & u_3 \\ v_1 & v_2 & v_3 \\ w_1 & w_2 & w_3 \end{vmatrix}$

2. The volume of the parallelepiped with edges $\mathbf{u}$, $\mathbf{v}$, and $\mathbf{w}$ is $|\mathbf{u} \cdot (\mathbf{v} \times \mathbf{w})|$.

We shall prove part (2) and leave the proof of part (1) to Exercise 25.

Proof: Volume = (Height)(Base area)
Base area = $\|\mathbf{v} \times \mathbf{w}\|$
Height = $|\|\mathbf{u}\| \cos \theta|$
Therefore
Volume = $\|\mathbf{u}\|\|\mathbf{v} \times \mathbf{w}\| |\cos \theta|$
$= |\mathbf{u} \cdot (\mathbf{v} \times \mathbf{w})|$
by definition of the dot product. ■

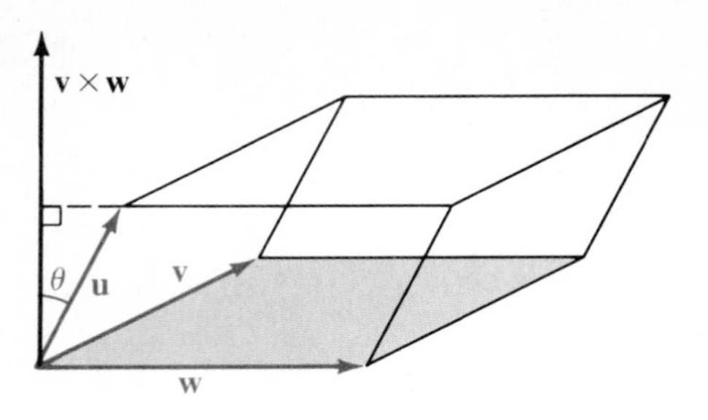

EXAMPLE 3 Find the volume of the parallelepiped with edges $\overrightarrow{OA} = (1, 0, 6)$, $\overrightarrow{OB} = (-1, 4, 1)$, and $\overrightarrow{OC} = (-3, 2, 4)$.

Solution Use part (1) of Theorem 4 to find $\overrightarrow{OC} \cdot (\overrightarrow{OA} \times \overrightarrow{OB})$.

$$\overrightarrow{OC} \cdot (\overrightarrow{OA} \times \overrightarrow{OB}) = \begin{vmatrix} -3 & 2 & 4 \\ 1 & 0 & 6 \\ -1 & 4 & 1 \end{vmatrix} = 74$$

By part (2), the volume is 74 cubic units. **Answer**

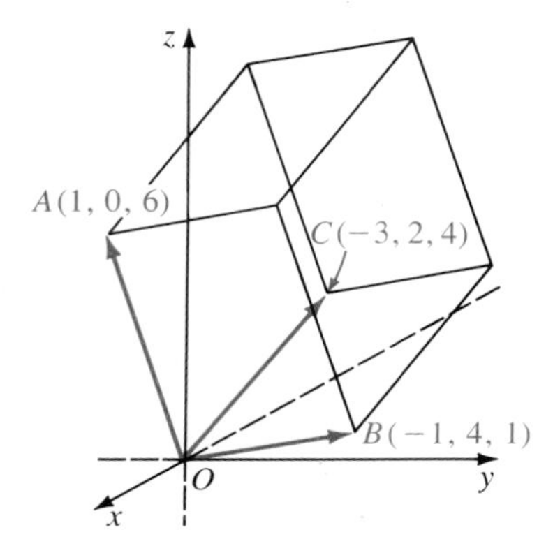

EXERCISES

Find $\mathbf{u} \times \mathbf{v}$.

A

1. $\mathbf{u} = (-3, 5, 2)$; $\mathbf{v} = (1, 6, -4)$
2. $\mathbf{u} = (-1, 2, 7)$; $\mathbf{v} = (0, 5, 8)$
3. $\mathbf{u} = (2, -1, -5)$; $\mathbf{v} = (3, 4, 9)$
4. $\mathbf{u} = (6, -5, 1)$; $\mathbf{v} = (8, -3, -2)$

In Exercises 5–10, prove each statement.

5. $\mathbf{i} \times \mathbf{j} = \mathbf{k}$
6. $\mathbf{j} \times \mathbf{k} = \mathbf{i}$
7. $\mathbf{k} \times \mathbf{i} = \mathbf{j}$
8. $\mathbf{j} \times \mathbf{i} = -\mathbf{k}$
9. $\mathbf{k} \times \mathbf{j} = -\mathbf{i}$
10. $\mathbf{i} \times \mathbf{k} = -\mathbf{j}$

Find a unit vector perpendicular to $\mathbf{u}$ and $\mathbf{v}$.

11. $\mathbf{u} = (1, 1, 1)$; $\mathbf{v} = (2, 3, -1)$
12. $\mathbf{u} = (3, 2, 1)$; $\mathbf{v} = (0, -2, 5)$

Find the area of the parallelogram having $\overrightarrow{AB}$ and $\overrightarrow{AC}$ as adjacent sides.

13. $A(3, 2, 6)$; $B(0, 3, 4)$; $C(5, 4, 0)$
14. $A(4, -2, 5)$; $B(-3, 4, 6)$; $C(5, 4, 5)$

15. Show that $A(3, -2, 4)$, $B(8, 0, 7)$, $C(6, 1, -3)$, and $D(11, 3, 0)$ are vertices of a parallelogram and find its area.

Find the volume of the parallelepiped with edges $\overrightarrow{OA}$, $\overrightarrow{OB}$, and $\overrightarrow{OC}$.

16. $A(3, 2, 3)$; $B(-1, 4, 1)$; $C(-2, -2, -2)$
17. $A(4, 1, 3)$; $B(-1, -1, 5)$; $C(-2, 3, 5)$

18. Find a and b such that $(2, a, 1) \times (1, b, 2) = (3, -3, 1)$.

19. Find a and b such that $(1, 2, a) \times (3, -b, 1) = (10, 5, -10)$.

In Exercises 20–23, prove each part of Theorem 1.

B

20. Part (1) 21. Part (2) 22. Part (3) 23. Part (4)

24. a. Show that if $\mathbf{u}$ and $\mathbf{v}$ are orthogonal, then $\|\mathbf{u} \times \mathbf{v}\| = \|\mathbf{u}\|\|\mathbf{v}\|$.
    b. Show that if $\mathbf{u}$ and $\mathbf{v}$ are also unit vectors, then $\mathbf{u} \times \mathbf{v}$ is also a unit vector.

25. Prove part (1) of Theorem 4.

Find the area of quadrilateral $ABCD$. (*Hint:* Divide the quadrilateral into triangles.)

26. $A(3, 2, 11)$; $B(-5, -6, 3)$; $C(0, 5, 8)$; $D(2, 5, 10)$

27. $A(1, 9, 2)$; $B(3, 8, 3)$; $C(-3, 14, 1)$; $D(4, 6, 3)$

Prove each statement.

28. $(\mathbf{u} + \mathbf{v}) \times (\mathbf{u} - \mathbf{v}) = -2(\mathbf{u} \times \mathbf{v})$

29. $\mathbf{u} \cdot (\mathbf{v} \times \mathbf{w}) = (\mathbf{u} \times \mathbf{v}) \cdot \mathbf{w}$

C

30. $\mathbf{u} \times (\mathbf{v} \times \mathbf{w}) = (\mathbf{u} \cdot \mathbf{w})\,\mathbf{v} - (\mathbf{u} \cdot \mathbf{v})\,\mathbf{w}$

31. $(\mathbf{u} \times \mathbf{v}) \times \mathbf{w} = (\mathbf{w} \cdot \mathbf{u})\,\mathbf{v} - (\mathbf{w} \cdot \mathbf{v})\,\mathbf{u}$

32. $\mathbf{w} \cdot (\mathbf{u} \times \mathbf{v}) = \mathbf{v} \cdot (\mathbf{w} \times \mathbf{u}) = \mathbf{u} \cdot (\mathbf{v} \times \mathbf{w})$

33. Let $\mathbf{u}$ and $\mathbf{v}$ be nonparallel vectors in three-space. Show that every vector $\mathbf{w}$ in three-space can be written as a unique linear combination of $\mathbf{u}$, $\mathbf{v}$, and $\mathbf{u} \times \mathbf{v}$. (*Hint:* Let $\mathbf{u} = (u_1, u_2, u_3)$, $\mathbf{v} = (v_1, v_2, v_3)$, and $\mathbf{w} = (x, y, z)$. Show that the determinant of the matrix of the system resulting from $\mathbf{w} = a\mathbf{u} + b\mathbf{v} + c(\mathbf{u} \times \mathbf{v})$ is nonzero.)

COMPUTER EXERCISES

1. Write a program to find the cross product of two vectors, $\mathbf{u}$ and $\mathbf{v}$, if $\mathbf{u} = (u_1, u_2, u_3)$ and $\mathbf{v} = (v_1, v_2, v_3)$. The vectors $\mathbf{u}$ and $\mathbf{v}$ are input by the user.

2. Run the program in Exercise 1 for the following pairs of vectors.
   a. $\mathbf{u} = (1, 1, 4)$, $\mathbf{v} = (-2, 3, 3)$
   b. $\mathbf{u} = (5, -2, 1)$, $\mathbf{v} = (10, 7, -5)$
   c. $\mathbf{u} = (-9, 4, -1)$, $\mathbf{v} = (3, 2, 1)$
   d. $\mathbf{u} = (5, -2, 3)$, $\mathbf{v} = (1, 4, -1)$

3. Use a program based on a modification of Exercise 1 to find the area of the following figures.
   a. Parallelogram having $\overline{AB}$ and $\overline{AC}$ as adjacent sides; $A(0, 1, -7)$, $B(3, 5, 1)$, $C(-4, 0, -11)$
   b. Triangle whose vertices are $A(9, -3, -3)$, $B(0, 1, 2)$, $C(-5, 1, 1)$

13-2 PLANES

Let P_0 be a point and let **n** be a nonzero vector. The set of all points P such that $\overrightarrow{P_0P}$ is perpendicular to **n**, that is, such that

$$\mathbf{n} \cdot \overrightarrow{P_0P} = 0,$$

is called the plane containing P_0 with normal vector **n**. Letting $\mathbf{r}_0 = \overrightarrow{OP_0}$ and $\mathbf{r} = \overrightarrow{OP}$, we have

$$\mathbf{n} \cdot (\mathbf{r} - \mathbf{r}_0) = 0$$

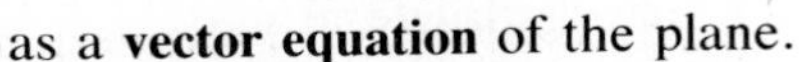

as a **vector equation** of the plane.

If we let $\mathbf{n} = (a, b, c)$, $\mathbf{r}_0 = (x_0, y_0, z_0)$, and $\mathbf{r} = (x, y, z)$, then the equation $\mathbf{n} \cdot (\mathbf{r} - \mathbf{r}_0) = 0$ becomes

$$a(x - x_0) + b(y - y_0) + c(z - z_0) = 0.$$

THEOREM 5 The plane containing $P_0(x_0, y_0, z_0)$ and having normal vector $\mathbf{n} = (a, b, c)$ has an equation of the form

$$a(x - x_0) + b(y - y_0) + c(z - z_0) = 0. \quad (1)$$

The graph of every such equation is a plane containing $P_0(x_0, y_0, z_0)$ and having nonzero normal vector $\mathbf{n} = (a, b, c)$.

You are asked to prove the second part of the theorem in Exercise 26. Equation (1) can also be written in the form

$$ax + by + cz + d = 0 \quad (2)$$

where $d = -(ax_0 + by_0 + cz_0)$. Equation (2) is a **scalar equation** of the plane.

When drawing a plane, first plot its **intercepts**, the points where the plane intersects the coordinate axes. Then join these points in pairs to obtain the **traces** of the plane, the lines in which the plane intersects the coordinate planes. A situation where there are fewer than three intercepts is illustrated in Example 1(b).

EXAMPLE 1 Graph each equation.

a. $2x + 3y + 2z - 12 = 0$ **b.** $x + y - 4 = 0$

Solution In drawing each plane find the traces and sketch the portion of the plane in the first octant.

a. The intercepts of the plane with the coordinate axes are:

$$(6, 0, 0) \qquad (0, 4, 0) \qquad (0, 0, 6)$$

The graph is shown at the left on the following page.

a.

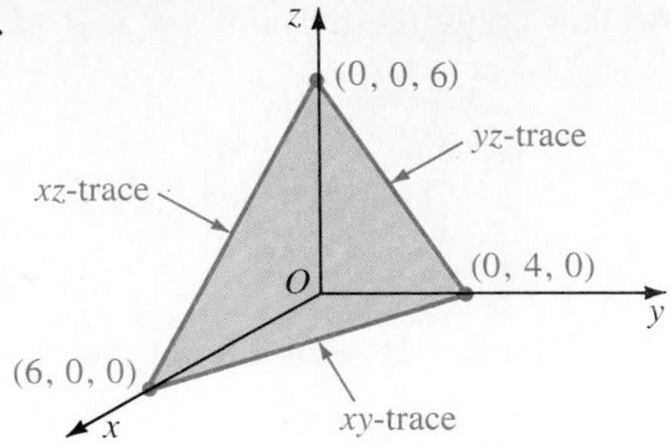

b.

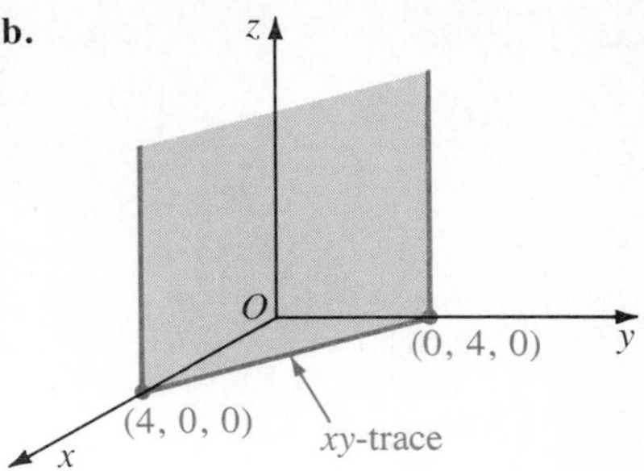

b. The two intercepts of the plane with the coordinate axes are:

$$(4, 0, 0) \qquad (0, 4, 0)$$

The graph is at the right above.

EXAMPLE 2 Find an equation of the plane containing $H(1, -1, 2)$, $J(3, -1, 1)$, and $K(2, 1, 3)$.

Solution The vectors

$$\overrightarrow{HJ} = (2, 0, -1) \quad \text{and} \quad \overrightarrow{HK} = (1, 2, 1)$$

are parallel to the plane. Hence

$$\overrightarrow{HJ} \times \overrightarrow{HK} = \begin{vmatrix} \mathbf{i} & \mathbf{j} & \mathbf{k} \\ 2 & 0 & -1 \\ 1 & 2 & 1 \end{vmatrix} = (2, -3, 4)$$

is perpendicular to both $\overrightarrow{HJ}$ and $\overrightarrow{HK}$ and is normal to the plane. Use (1) with $P_0 = H = (1, -1, 2)$ to obtain

$$2(x - 1) - 3(y + 1) + 4(z - 2) = 0,$$

or

$$2x - 3y + 4z = 13. \qquad \textbf{Answer}$$

We could solve Example 2 by substituting in turn $(1, -1, 2)$, $(3, -1, 1)$, and $(2, 1, 3)$ for (x, y, z) in (2) to obtain the system:

$$\begin{aligned} a - b + 2c + d &= 0 \\ 3a - b + c + d &= 0 \\ 2a + b + 3c + d &= 0 \end{aligned}$$

Then $a = -\frac{2}{13}d$, $b = \frac{3}{13}d$, $c = -\frac{4}{13}d$. Set $d = -13$ and substitute into (2) to obtain

$$2x - 3y + 4z - 13 = 0.$$

The following theorem provides another way to find an equation of the plane containing three noncollinear points. You are asked for a proof in Exercise 27.

THEOREM 6 An equation of the plane that contains the points $P_0(x_0, y_0, z_0)$, $P_1(x_1, y_1, z_1)$, and $P_2(x_2, y_2, z_2)$ is

$$\begin{vmatrix} x - x_0 & y - y_0 & z - z_0 \\ x_1 - x_0 & y_1 - y_0 & z_1 - z_0 \\ x_2 - x_0 & y_2 - y_0 & z_2 - z_0 \end{vmatrix} = 0.$$

The next theorem enables us to find the distance between a point and a plane.

THEOREM 7 The distance D between the point $P_1(x_1, y_1, z_1)$ and the plane with equation $ax + by + cz + d = 0$ is given by

$$D = \frac{|ax_1 + by_1 + cz_1 + d|}{\sqrt{a^2 + b^2 + c^2}}.$$

Proof: Let $P_0(x_0, y_0, z_0)$ be any point in the plane and let $\mathbf{n} = (a, b, c)$ be its normal. Then

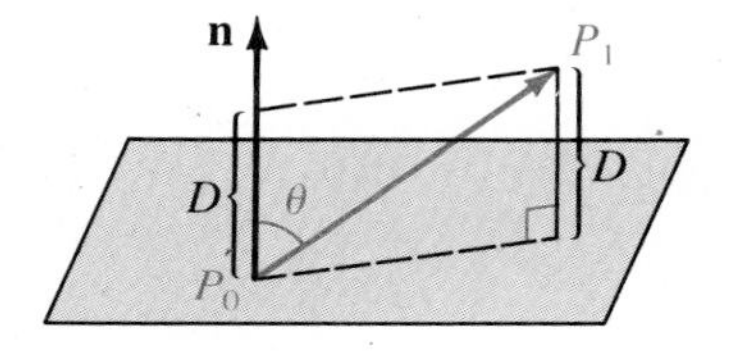

$$D = \left|\|\overrightarrow{P_0P_1}\| \cos\theta\right| = \frac{\left|\|\mathbf{n}\|\,\|\overrightarrow{P_0P_1}\| \cos\theta\right|}{\|\mathbf{n}\|}$$

$$= \frac{|\mathbf{n} \cdot \overrightarrow{P_0P_1}|}{\|\mathbf{n}\|}$$

Since $\overrightarrow{P_0P_1} = (x_1 - x_0, y_1 - y_0, z_1 - z_0)$,

$$D = \frac{|a(x_1 - x_0) + b(y_1 - y_0) + c(z_1 - z_0)|}{\sqrt{a^2 + b^2 + c^2}}$$

$$= \frac{|ax_1 + by_1 + cz_1 - (ax_0 + by_0 + cz_0)|}{\sqrt{a^2 + b^2 + c^2}}$$

Since $P_0(x_0, y_0, z_0)$ is in the plane, $ax_0 + by_0 + cz_0 = -d$. Therefore

$$D = \frac{|ax_1 + by_1 + cz_1 + d|}{\sqrt{a^2 + b^2 + c^2}}. \quad \blacksquare$$

Two planes are said to be **parallel (perpendicular)** if their normal vectors are parallel (perpendicular).

EXAMPLE 3 **a.** Show that the planes X_1: $6x - 4y + 2z + 10 = 0$ and $X_2 : 3x - 2y + z - 4 = 0$ are parallel.

b. Find the distance between them.

Solution **a.** The vectors $\mathbf{n}_1 = (6, -4, 2)$ and $\mathbf{n}_2 = (3, -2, 1)$ are normal to X_1 and X_2 respectively. Since $\mathbf{n}_1 = 2\mathbf{n}_2$, the vectors and hence the planes are parallel.

b. The point $P_1(-1, 1, 0)$ is in X_1. Use Theorem 7 with P_1 and the equation of X_2 to find D.

$$D = \frac{|(3)(-1) + (-2)(1) + 1 \cdot 0 + (-4)|}{\sqrt{3^2 + (-2)^2 + 1^2}} = \frac{9}{\sqrt{14}}, \text{ about } 2.4$$

The distance is about 2.4 units. **Answer**

EXERCISES

Sketch each plane.

A

1. $2x + 3y = 6$
2. $2x + y = 4$
3. $3y - z = 9$
4. $2x - y + z = 6$
5. $2x - 3y + 6z = 12$
6. $3x - 2y + 6z = 24$

Find a scalar equation of the plane containing P and having $\mathbf{n}$ as a normal vector.

7. $P(1, 6, 2)$; $\mathbf{n} = (4, 7, -3)$
8. $P(8, -1, -5)$; $\mathbf{n} = (-2, 5, 9)$
9. $P(3, 6, -7)$; $\mathbf{n} = (-1, -6, 4)$
10. $P(4, 9, 2)$; $\mathbf{n} = (3, 8, -5)$

Find the distance between P and the plane whose equation is given.

11. $P(5, 2, 4)$; $8x + 4y + z - 16 = 0$
12. $P(0, 0, 0)$; $x + y + z - 10 = 0$
13. $P(3, 2, 5)$; $x - y - z - 7 = 0$
14. $P(-2, 5, 1)$; $3x - 2y + 3z - 7 = 0$
15. Show that the planes with equations $3x - y - z - 6 = 0$ and $6x - 2y - 2z - 10 = 0$ are parallel and find the distance between the planes.
16. Show that the planes with equations $9x - 6y + 2z - 11 = 0$ and $18x - 12y + 4z - 1 = 0$ are parallel and find the distance between the planes.

Find an equation of the plane containing A, B, and C.

17. $A(0, 0, 4)$; $B(1, 2, 4)$; $C(-2, 4, 3)$
18. $A(3, 2, 1)$; $B(-2, 4, 3)$; $C(0, 4, -3)$
19. Find a scalar equation of the plane containing $(3, -1, 4)$ and parallel to the plane with equation $3x + 2y - 5z - 6 = 0$.
20. Find an equation of the plane containing $P(2, 3, 5)$ and parallel to the plane containing $A(-2, 3, 1)$, $B(3, 1, 0)$, and $C(5, 1, 2)$.

B

21. Find an equation of the plane containing $(1, -3, 2)$ and perpendicular to the pair of planes that have equations $x + 2y + 3z - 4 = 0$ and $3x - 4y - 4z - 2 = 0$.
22. Find an equation of the plane containing $(3, -1, 2)$ and perpendicular to the pair of planes that have equations $x - 2y + 3z - 6 = 0$ and $4x - 3y - z - 4 = 0$.
23. Find a scalar equation of the plane containing $(3, 2, 1)$ and $(2, 3, 1)$ and perpendicular to the plane with equation $3x + 2y + 6z - 5 = 0$.

24. Find an equation of the plane midway between the planes with equations $x + 3y - z + 3 = 0$ and $x + 3y - z - 5 = 0$.

25. The **angle θ between two planes**, $0° \leq \theta < 180°$, is the angle between their normal vectors. Find the measure of the angle between the planes with equations $x + y - z - 6 = 0$ and $5x - y + z - 7 = 0$.

C

26. Prove the second part of Theorem 5.

27. Prove Theorem 6.

13-3 LINES

In this section we shall derive several forms of equations of lines in space. We shall use the following notation throughout: $P_0(x_0, y_0, z_0)$ and $P_1(x_1, y_1, z_1)$, with position vectors $\mathbf{r}_0 = \overrightarrow{OP_0}$ and $\mathbf{r}_1 = \overrightarrow{OP_1}$, respectively, denote fixed points in space; $P(x, y, z)$, with position vector $\mathbf{r} = \overrightarrow{OP}$, denotes an arbitrary point.

THEOREM 8 An equation of the line L containing P_0 and parallel to a nonzero vector $\mathbf{m} = (a, b, c)$ is

$$\mathbf{r} = \mathbf{r}_0 + t\mathbf{m}. \tag{1}$$

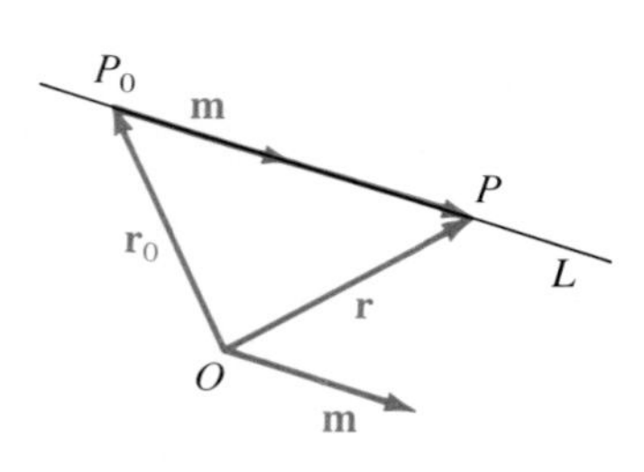

Proof: Point P is on L if and only if $\overrightarrow{P_0P}$ is parallel to $\mathbf{m}$, that is, $\overrightarrow{P_0P} = t\mathbf{m}$ for some scalar t. Since $\overrightarrow{OP} = \overrightarrow{OP_0} + \overrightarrow{P_0P}$, we have $\mathbf{r} = \mathbf{r}_0 + t\mathbf{m}$. ■

Equation (1) is a **vector equation** of L, and $\mathbf{m}$ is called a **direction vector** of L.

Another form of equation (1) is

$$(x, y, z) = (x_0, y_0, z_0) + t(a, b, c), \tag{2}$$

and this in turn is equivalent to the following set of scalar **parametric equations**.

$$x = x_0 + at, \quad y = y_0 + bt, \quad z = z_0 + ct \tag{3}$$

If equations (3) are solved for t and the results equated, we obtain (provided $abc \neq 0$)

$$\frac{x - x_0}{a} = \frac{y - y_0}{b} = \frac{z - z_0}{c}. \tag{4}$$

These are **symmetric-form equations** of L.

EXAMPLE 1 Find **(a)** a vector equation and **(b)** a set of scalar parametric equations for the line L that contains $P_0(2, -1, 3)$ and is perpendicular to the plane with equation $x + 5y + 4z = 7$.

Solution The line L must be parallel to a normal to the plane. Set $\mathbf{m} = (1, 5, 4)$.

a. $(x, y, z) = (2, -1, 3) + t(1, 5, 4)$

b. $x = 2 + t,\ y = -1 + 5t,\ z = 3 + 4t$

Frequently a line is determined by two points, P_0 and P_1, on it. In this case we may take $\mathbf{m} = \mathbf{r}_1 - \mathbf{r}_0$ in Theorem 8 and obtain

$$\mathbf{r} = \mathbf{r}_0 + t(\mathbf{r}_1 - \mathbf{r}_0), \text{ or } \mathbf{r} = (1 - t)\mathbf{r}_0 + t\mathbf{r}_1$$

and

$$x = x_0 + t(x_1 - x_0), \quad y = y_0 + t(y_1 - y_0), \quad z = z_0 + t(z_1 - z_0).$$

If we require that $0 \le t \le 1$, we obtain equations of the line segment $\overline{P_0P_1}$. In particular, the midpoint of $\overline{P_0P_1}$ is obtained when $t = \frac{1}{2}$.

EXAMPLE 2 **a.** Find a set of parametric equations of the line containing $P_0(1, -1, 2)$ and $P_1(3, 4, 2)$.

b. Find the midpoint of $\overline{P_0P_1}$.

Solution **a.** Use $\mathbf{r}_1 - \mathbf{r}_0 = (3 - 1, 4 - (-1), 2 - 2) = (2, 5, 0)$ to obtain

$$(x, y, z) = (1, -1, 2) + t(2, 5, 0).$$

Hence:

$$x = 1 + 2t \qquad y = -1 + 5t \qquad z = 2$$

b. In (a) let $t = \frac{1}{2}$ to obtain $(2, \frac{3}{2}, 2)$.

In three-space, nonparallel lines that do not intersect are called **skew lines**. The angle between any two lines L_1 and L_2 is defined to be the angle θ between vectors $\mathbf{m}_1$ and $\mathbf{m}_2$ parallel to L_1 and L_2, respectively. Recall that

$$\cos\theta = \frac{\mathbf{m}_1 \cdot \mathbf{m}_2}{\|\mathbf{m}_1\|\|\mathbf{m}_2\|}.$$

The distance D between a point P and a line L is the length of the perpendicular segment from P to L. We proceed as follows to find D. Let $\mathbf{m}$ be a direction vector of L and let S be any point on L. From triangle SPT, $\|\overrightarrow{SP}\| \sin\theta = D$.

$$D = \frac{\|\mathbf{m}\|\,\|\overrightarrow{SP}\| \sin\theta}{\|\mathbf{m}\|}$$

$$D = \frac{\|\overrightarrow{SP} \times \mathbf{m}\|}{\|\mathbf{m}\|}$$

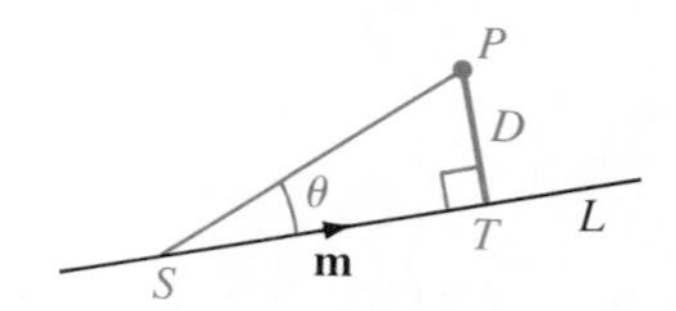

EXAMPLE 3 Given the lines L_1: $\mathbf{r} = (0, -3, 4) + t(2, -1, 2)$ and L_2: $\mathbf{r} = (1, 3, 0) + s(1, -2, 3)$:

a. Find the angle θ between L_1 and L_2.

b. Find scalar parametric equations of the line through the origin perpendicular to both L_1 and L_2.

c. Find the distance between the origin and L_2.

Solution **a.** The vectors $\mathbf{m}_1 = (2, -1, 2)$ and $\mathbf{m}_2 = (1, -2, 3)$ are parallel to L_1 and L_2, respectively.

$$\cos\theta = \frac{2 \cdot 1 + (-1)(-2) + 2 \cdot 3}{\sqrt{9}\sqrt{14}} = \frac{10}{3\sqrt{14}} \approx 0.89087$$

To the nearest degree, $\theta = 27°$. **Answer**

b. The vector

$$\mathbf{m}_1 \times \mathbf{m}_2 = \begin{vmatrix} \mathbf{i} & \mathbf{j} & \mathbf{k} \\ 2 & -1 & 2 \\ 1 & -2 & 3 \end{vmatrix} = \mathbf{i} - 4\mathbf{j} - 3\mathbf{k} = (1, -4, -3)$$

is perpendicular to both $\mathbf{m}_1$ and $\mathbf{m}_2$ and hence to both L_1 and L_2. Use (3) to obtain $x = 0 + 1t$, $y = 0 - 4t$, $z = 0 - 3t$, or

$$x = t, \quad y = -4t, \quad z = -3t \qquad \textbf{Answer}$$

c. Use the formula for D with $S(1, 3, 0)$, $P(0, 0, 0)$ and $\mathbf{m} = (1, -2, 3)$.

$$\overrightarrow{SP} \times \mathbf{m} = \begin{vmatrix} \mathbf{i} & \mathbf{j} & \mathbf{k} \\ -1 & -3 & 0 \\ 1 & -2 & 3 \end{vmatrix} = -9\mathbf{i} + 3\mathbf{j} + 5\mathbf{k} = (-9, 3, 5)$$

Therefore $D = \dfrac{\|\overrightarrow{SP} \times \mathbf{m}\|}{\|\mathbf{m}\|} = \dfrac{\sqrt{115}}{\sqrt{14}} = \sqrt{\dfrac{115}{14}}$ about 2.9 units. **Answer**

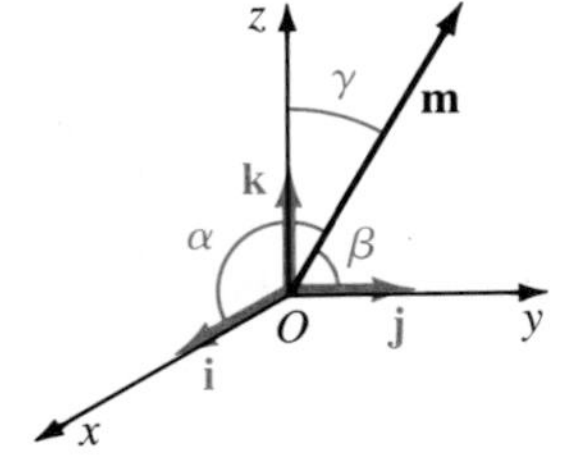

A nonzero vector $\mathbf{m} = a\mathbf{i} + b\mathbf{j} + c\mathbf{k}$ makes angles α, β, and γ with $\mathbf{i}$, $\mathbf{j}$, and $\mathbf{k}$ respectively. These angles are the **direction angles** of $\mathbf{m}$, and their cosines are the **direction cosines** of $\mathbf{m}$. Note that

$$\mathbf{m} \cdot \mathbf{i} = \|\mathbf{m}\|\,\|\mathbf{i}\| \cos\alpha, \qquad \mathbf{m} \cdot \mathbf{j} = \|\mathbf{m}\|\,\|\mathbf{j}\| \cos\beta,$$
$$\mathbf{m} \cdot \mathbf{k} = \|\mathbf{m}\|\,\|\mathbf{k}\| \cos\gamma$$

Since $\|\mathbf{i}\| = \|\mathbf{j}\| = \|\mathbf{k}\| = 1$, and $\mathbf{m} \cdot \mathbf{i} = a$, $\mathbf{m} \cdot \mathbf{j} = b$, $\mathbf{m} \cdot \mathbf{k} = c$, we have

$$\cos\alpha = \frac{a}{\|\mathbf{m}\|}, \quad \cos\beta = \frac{b}{\|\mathbf{m}\|}, \quad \cos\gamma = \frac{c}{\|\mathbf{m}\|} \text{ where } \|\mathbf{m}\| = \sqrt{a^2 + b^2 + c^2}$$

Because $\mathbf{i}\cos\alpha + \mathbf{j}\cos\beta + \mathbf{k}\cos\gamma = \dfrac{a\mathbf{i} + b\mathbf{j} + c\mathbf{k}}{\|\mathbf{m}\|} = \dfrac{\mathbf{m}}{\|\mathbf{m}\|}$, we see that

$$\mathbf{i}\cos\alpha + \mathbf{j}\cos\beta + \mathbf{k}\cos\gamma$$

is the *unit* vector in the direction of $\mathbf{m}$. In particular,

$$\cos^2\alpha + \cos^2\beta + \cos^2\gamma = 1.$$

Direction cosines can be used to find the angle θ between two vectors $\mathbf{m}_1$ and $\mathbf{m}_2$. Let the direction angles of $\mathbf{m}_1$ be α_1, β_1, γ_1, and the direction angles of $\mathbf{m}_2$ be α_2, β_2, γ_2. Then

$$\begin{aligned}\cos\theta &= \frac{\mathbf{m}_1 \cdot \mathbf{m}_2}{\|\mathbf{m}_1\|\,\|\mathbf{m}_2\|} = \frac{\mathbf{m}_1}{\|\mathbf{m}_1\|} \cdot \frac{\mathbf{m}_2}{\|\mathbf{m}_2\|} \\ &= (\mathbf{i}\cos\alpha_1 + \mathbf{j}\cos\beta_1 + \mathbf{k}\cos\gamma_1) \cdot (\mathbf{i}\cos\alpha_2 + \mathbf{j}\cos\beta_2 + \mathbf{k}\cos\gamma_2) \\ &= \cos\alpha_1\cos\alpha_2 + \cos\beta_1\cos\beta_2 + \cos\gamma_1\cos\gamma_2.\end{aligned}$$

EXERCISES

Find a vector equation and a set of parametric equations for the line containing P_0 and having direction vector $\mathbf{m}$.

A

1. $P_0(3, 4, -2)$; $\mathbf{m} = (3, 2, 6)$
2. $P_0(-1, 3, 2)$; $\mathbf{m} = (-1, 2, 5)$

Find a vector equation and a set of parametric equations of $\overleftrightarrow{PQ}$.

3. $P(1, 0, 3)$; $Q(-2, 3, 4)$
4. $P(-1, 5, 7)$; $Q(-4, 1, 3)$

Find the direction cosines of $\mathbf{v}$. Verify that $\cos^2\alpha + \cos^2\beta + \cos^2\gamma = 1$.

5. $\mathbf{v} = (-1, 3, 3)$
6. $\mathbf{v} = (3, 4, 5)$
7. $\mathbf{v} = (-5, 12, 13)$

Find a vector equation of the line containing T and parallel to $\overleftrightarrow{PQ}$.

8. $T(1, 2, 3)$; $P(2, 1, -4)$; $Q(1, 2, 2)$
9. $T(2, -1, -4)$; $P(1, 1, 3)$; $Q(0, 3, -2)$
10. $T(1, 1, 2)$; $P(3, -1, 5)$; $Q(1, 3, 3)$
11. $T(0, 2, -4)$; $P(3, 3, 3)$; $Q(-5, 6, 7)$
12. Find an equation of the line containing $P(-3, 2, 5)$ and perpendicular to the plane with equation $x - y + 4z - 6 = 0$.
13. Find an equation of the line containing $P(-2, 4, 1)$ and perpendicular to the plane containing $A(-3, 1, 0)$, $B(1, 1, 2)$, and $C(4, 0, 5)$.
14. Find the distance from the origin to the line in which planes X_1 and X_2 intersect. X_1: $x + y - z + 4 = 0$ and X_2: $2x - y + z + 1 = 0$.
15. Find the distance from $P(2, 3, 5)$ to the line with equation $\mathbf{r} = (1 + t)\mathbf{i} + 2t\mathbf{j} + (3 - t)\mathbf{k}$.
16. Find the distance from the origin to the line containing $P(-2, 3, 5)$ and $Q(3, 5, 2)$.
17. Find the length of the altitude from A to $\overline{BC}$ if triangle ABC has vertices $A(-2, 3, 4)$, $B(0, 5, 5)$, and $C(3, 7, 2)$.
18. Find an equation of the plane containing $P(-1, 0, 4)$ and perpendicular to the line with equation $\mathbf{r} = 3t\mathbf{i} + (2 - t)\mathbf{j} + (3 + 4t)\mathbf{k}$.
19. Find an equation of the plane that is the perpendicular bisector of the line segment with endpoints $P(-2, 5, 5)$ and $Q(4, 7, 3)$.

In Exercises 20–23, find the coordinates of any point of intersection of the lines. If the lines intersect, find the measure of the angle between them.

B

20. $\mathbf{r}_1 = 3s\mathbf{i} + 4s\mathbf{j}$; $\mathbf{r}_2 = 5t\mathbf{j} - 12t\mathbf{k}$

21. $\mathbf{r}_1 = 12s\mathbf{j} - 5s\mathbf{k}$; $\mathbf{r}_2 = 6t\mathbf{i} + 8t\mathbf{j}$

22. $\mathbf{r}_1 = (-1 + 2s)\mathbf{i} + (-3 + 4s)\mathbf{j} + (9 - 5s)\mathbf{k}$;
$\mathbf{r}_2 = (4 + t)\mathbf{i} + (3t)\mathbf{j} + (-15 + 3t)\mathbf{k}$

23. $\mathbf{r}_1 = (3s)\mathbf{i} + (2 + s)\mathbf{j} + (1 - 3s)\mathbf{k}$;
$\mathbf{r}_2 = (3 + t)\mathbf{i} + (2 + 2t)\mathbf{j} + (1 - 4t)\mathbf{k}$

In Exercises 24–27, P is the point corresponding to $t = 0$ and Q is the point corresponding to $t = 1$. Find the coordinates of the points that divide $\overline{PQ}$ into n segments of equal length.

24. $\mathbf{r} = (2 + t)\mathbf{i} + (3 + 4t)\mathbf{j} + t\mathbf{k}$; $n = 2$

25. $\mathbf{r} = (3 - 2t)\mathbf{i} + (5 + 10t)\mathbf{j} + (t + 1)\mathbf{k}$; $n = 3$

26. $\mathbf{r} = (3 + 4t)\mathbf{i} + (3 - 4t)\mathbf{j}$; $n = 3$

27. $\mathbf{r} = (-1 - t)\mathbf{j} + (3 + 5t)\mathbf{k}$; $n = 4$

28. Find the intersection of the line with equation $\mathbf{r} = (3 + 2t)\mathbf{i} + (2 + t)\mathbf{j} + (1 + 3t)\mathbf{k}$ and the plane with equation $2x - 3y + 4z - 17 = 0$.

29. Find an equation of the line containing $(3, -1, 2)$ and parallel to the planes with equations $x - 2y + 3z - 6 = 0$ and $4x - 3y - z - 4 = 0$.

A set of planes may intersect in a plane, a line, or a point, or may have no common intersection at all. Find a scalar equation of the plane, a set of parametric equations of the line of intersection, or the coordinates of the point of intersection. If the set of planes has no common intersection, so state.

30. $-9x + 12y + 6z - 8 = 0$
$6x - 8y - 4z - 9 = 0$

31. $-12x - 8y + 16z - 4 = 0$
$-6x + 4y - 8z - 2 = 0$

32. $2x - y + 5z - 4 = 0$
$x - \frac{1}{2}y + \frac{5}{2}z - 2 = 0$
$6x - 3y + 15z - 12 = 0$

33. $3x - 2y + 5z - 24 = 0$
$2x + 3y - 4z - 12 = 0$
$x - y + 5z + 15 = 0$

34. $-3x + 12y - 9z = 5$
$-x + 4y - 3z = 10$
$2x - 8y + 6z = 7$

35. $15x - 5y + 10z = 5$
$3x - y - 4z = 1$
$3x - y + 2z = 1$

36. $x - y + z = 5$
$3x - 2y + 5z = 12$
$6x - 4y + 10z = 22$

37. $4x - 2y + 3z = 4$
$2x + 3y - z = 18$
$2x - 5y + 4z = -14$

38. Find symmetric-form equations of the line containing $P(4, 2, -5)$ and having direction angles $\alpha = 60°$, $\beta = 60°$, and γ $(0° < \gamma < 90°)$.

39. Find symmetric-form equations of the line containing $P(4, 2, -5)$ and having direction angles $\alpha = 45°$, $\beta = \text{Cos}^{-1}(\frac{2}{3})$, and $\cos\gamma > 0$.

C **40.** Show that if P, S, and T are three distinct points in space, then S is on $\overline{PT}$ if and only if $\|\overrightarrow{PS}\| + \|\overrightarrow{ST}\| = \|\overrightarrow{PT}\|$.

41. Find a set of parametric equations of the line of intersection of the planes with equations $a_1x + b_1y + c_1z + d_1 = 0$ and $a_2x + b_2y + c_2z + d_2 = 0$.

COMPUTER EXERCISES

1. Write a program to find the distance between a given point and a line whose equation(s) are given. The coordinates of the point and the coordinates of the line's direction vector are input by the user.

2. Run the program in Exercise 1 for each point and line.
a. $(0, 0, 0)$; $x = 3 + 2t$, $y = 2 - t$, $z = 4t$
b. $(-2, 3, 4)$; $\mathbf{r} = (2 - t)\mathbf{i} + (4 + 3t)\mathbf{j} + (5 - t)\mathbf{k}$

3. Run the program to find the length of the altitude from A to $\overline{BC}$ if triangle ABC has vertices $A(0, 0, 0)$, $B(-3, 2, 5)$, and $C(1, 3, 5)$.

Surfaces

13-4 QUADRIC SURFACES

A plane is an example of a *surface*. In this section we investigate *spheres*, *cylinders*, and *quadric surfaces*.

DEFINITION

A **sphere** is the set of all points P at a given distance r, the **radius**, from a fixed point P_0, the **center**.

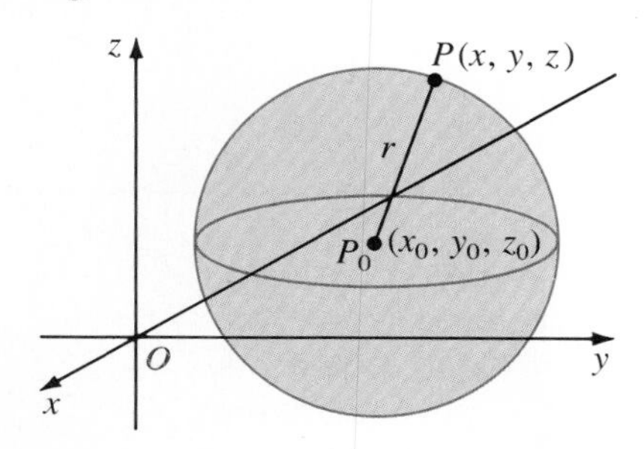

By using the distance formula we see that an equation of the sphere with radius r and center $P_0(x_0, y_0, z_0)$ is

$$(x - x_0)^2 + (y - y_0)^2 + (z - z_0)^2 = r^2.$$

DEFINITION

A **cylinder** is the surface consisting of all lines L, called **rulings**, through a plane curve C, the **directrix**, and parallel to a fixed line L_0.

A cylinder is named for its directrix. This is illustrated in the following figures.

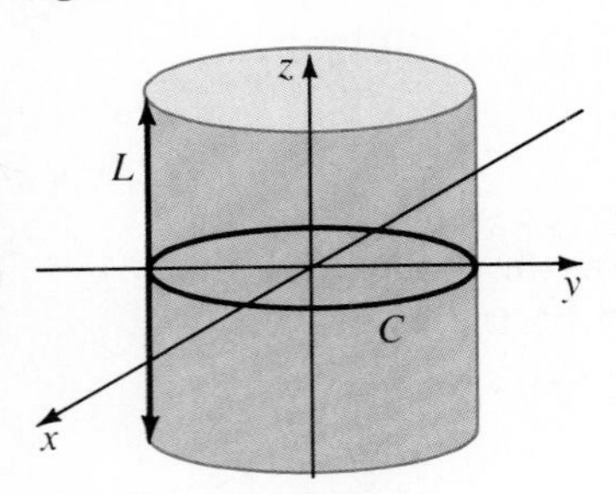

Circular cylinder
Directrix: circle
Rulings: lines parallel to the z-axis

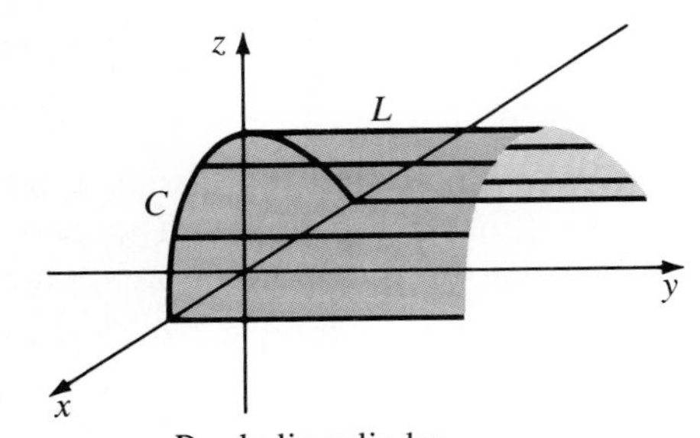

Parabolic cylinder
Directrix: parabola
Rulings: lines parallel to the y-axis

In sketching a surface S, it is often useful to draw *cross sections* and *traces*. A **cross section** of S is the intersection of S with a given plane. A **trace** of a surface S is the intersection of S with a coordinate plane.

EXAMPLE 1 Sketch the graph of $x^2 + (z - 3)^2 = 4$.

Solution Find the traces of the graph.

xy-trace	$z = 0$	$x^2 + (0 - 3)^2 = 4$ $x^2 = -5$	no trace
xz-trace	$y = 0$	$x^2 + (z - 3)^2 = 4$	circle with center (0, 0, 3) and radius 2
yz-trace	$x = 0$	$(z - 3)^2 = 4$ $z = 5$ or 1	two lines parallel to the y-axis.

A cross section of the graph with the plane having equation $y = c$ is a circle with center $(0, c, 3)$ and radius 2. The graph is a cylinder with rulings parallel to the y-axis.

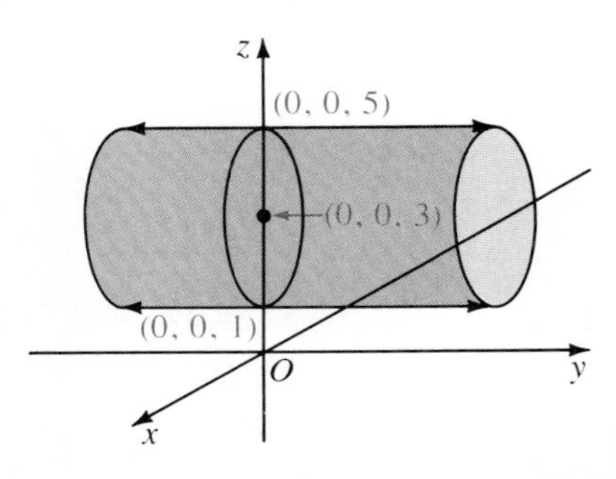

In general, the graph in three-space of an equation in two variables is a cylinder with rulings parallel to the axis of the third variable.

We can use traces and cross sections to graph many equations whether the graphs are cylinders or not.

EXAMPLE 2 Use traces and cross sections to sketch the graph S of

$$\frac{x^2}{4} + \frac{y^2}{9} + \frac{z^2}{4} = 1.$$

Solution Find the traces of S. Use a table to organize the information about the nature of the traces.

xy-trace	$z = 0$	$\frac{x^2}{4} + \frac{y^2}{9} = 1$	ellipse with major axis along the y-axis
xz-trace	$y = 0$	$\frac{x^2}{4} + \frac{z^2}{4} = 1$	circle with radius 2
yz-trace	$x = 0$	$\frac{y^2}{9} + \frac{z^2}{4} = 1$	ellipse with major axis along the y-axis

A cross section of S with a plane parallel to the xy-plane can be found by letting $z = c$. Then

$$\frac{x^2}{4} + \frac{y^2}{9} = 1 - \frac{c^2}{4}.$$

If $|c| > 2$, then $1 - \frac{c^2}{4} < 0$ and the cross section is empty. If $c = 2$ or $c = -2$, the cross section is a single point. If $|c| < 2$, the cross section is an ellipse with major axis directly above the y-axis. Cross sections of S with planes parallel to the other coordinate planes can be found in a similar way. The sketch of S is shown at the right.

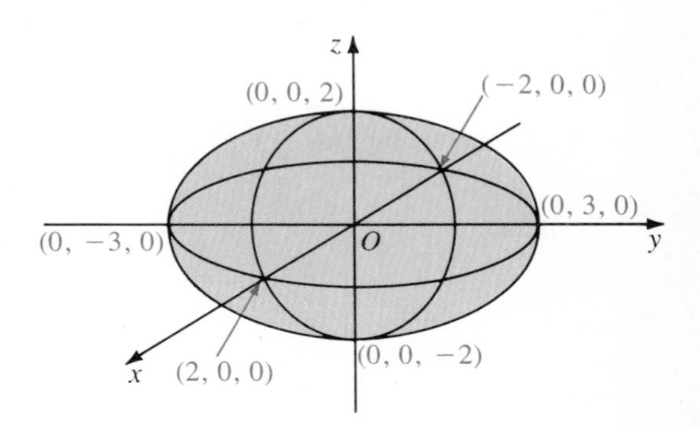

DEFINITION

A **quadric surface** is one obtained as the graph in three-space of a **second-degree equation in three variables**, that is, an equation of the form

$$Ax^2 + By^2 + Cz^2 + Eyz + Fxz + Gxy + Hx + Iy + Jz + K = 0.$$

It can be shown that every nondegenerate quadric surface is either a cylinder or one of the following six types of surfaces. (Such a surface may, of course, be positioned differently with respect to the coordinate axes.)

Ellipsoid $\frac{x^2}{a^2} + \frac{y^2}{b^2} + \frac{z^2}{c^2} = 1$

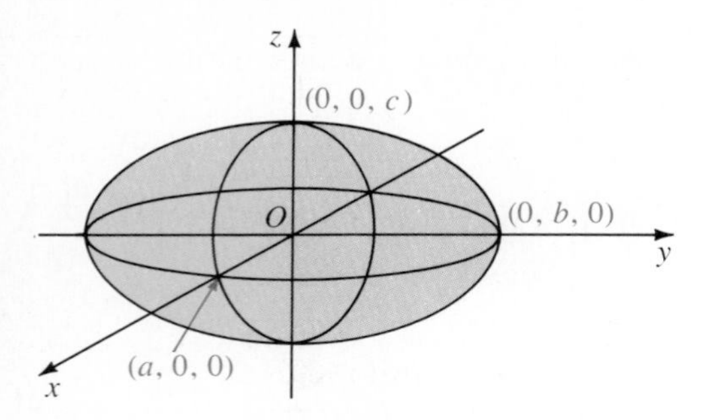

Elliptic Cone $\frac{x^2}{a^2} + \frac{y^2}{b^2} - \frac{z^2}{c^2} = 0$

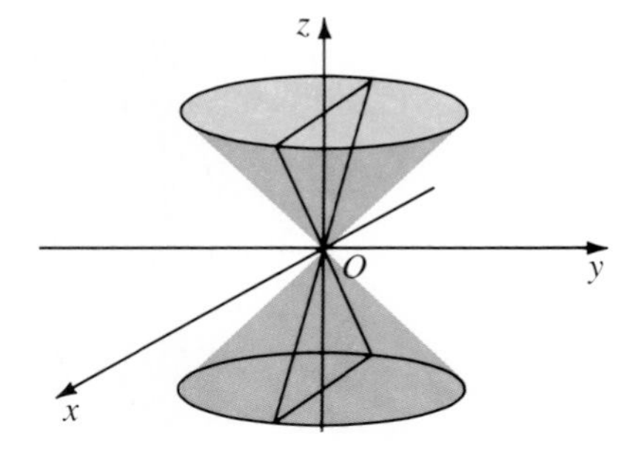

Hyperboloid of one sheet $\frac{x^2}{a^2} + \frac{y^2}{b^2} - \frac{z^2}{c^2} = 1$

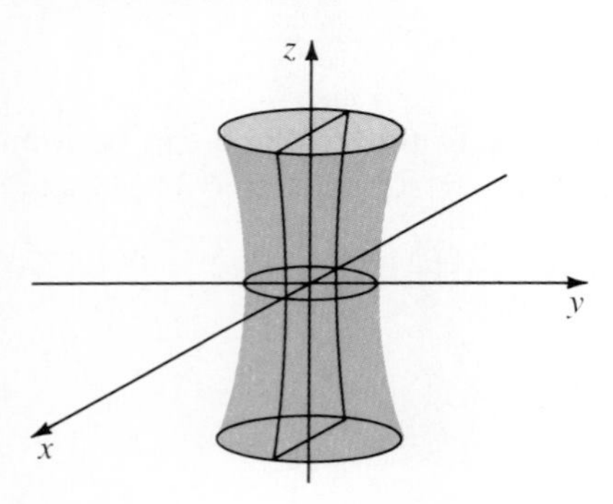

Hyperboloid of two sheets $\frac{x^2}{a^2} - \frac{y^2}{b^2} - \frac{z^2}{c^2} = 1$

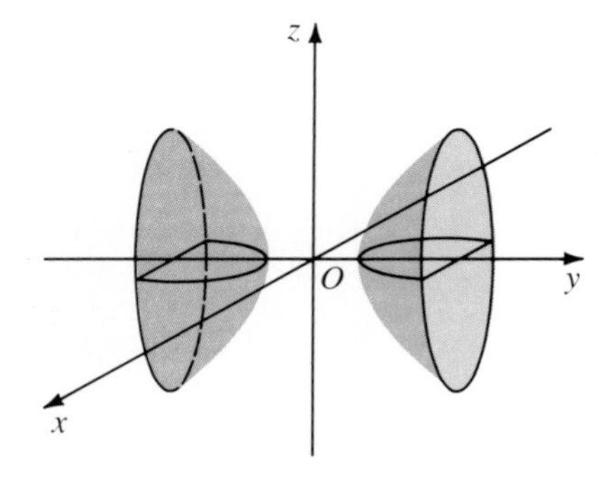

Elliptic Paraboloid $\frac{x^2}{a^2} + \frac{y^2}{b^2} = cz$
$(c > 0)$

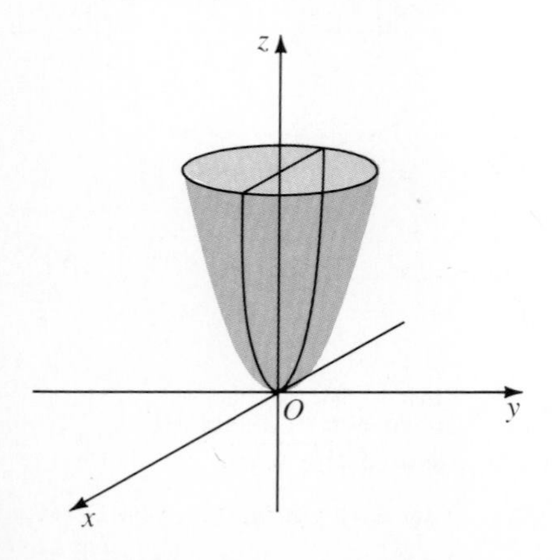

Hyperbolic Paraboloid $\frac{y^2}{b^2} - \frac{x^2}{a^2} = cz$
$(c > 0)$

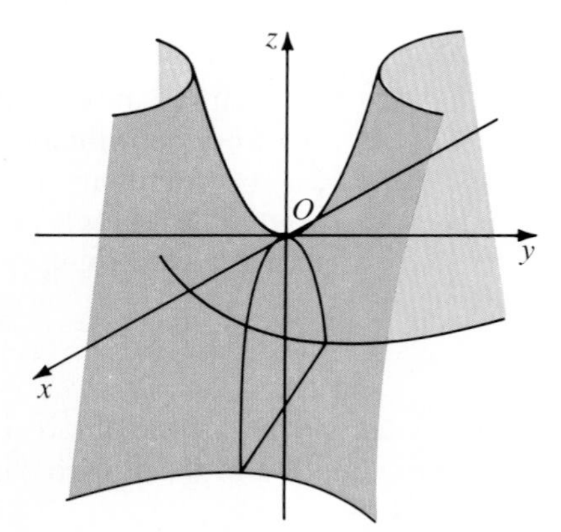

EXAMPLE 3 Identify and sketch the graph of $x^2 + 4y^2 - z^2 = 36$.

Solution Divide both sides of the given equation by 36.

$$\frac{x^2}{36} + \frac{y^2}{9} - \frac{z^2}{36} = 1$$

The graph is a hyperboloid of one sheet. Find the traces of the graph.

xy-trace	$z = 0$	$\frac{x^2}{36} + \frac{y^2}{9} = 1$	ellipse with major axis along the x-axis
xz-trace	$y = 0$	$\frac{x^2}{36} - \frac{z^2}{36} = 1$	hyperbola with transverse axis along the x-axis
yz-trace	$x = 0$	$\frac{y^2}{9} - \frac{z^2}{36} = 1$	hyperbola with transverse axis along the y-axis

Cross sections of the graph with planes parallel to the xy-plane are found by setting $z = c$.

$$\frac{x^2}{36} + \frac{y^2}{9} = 1 + \frac{c^2}{36}$$

If $c > 0$, the graph is an ellipse with major axis directly above the x-axis, and if $c < 0$, the graph is an ellipse with major axis directly below the x-axis.

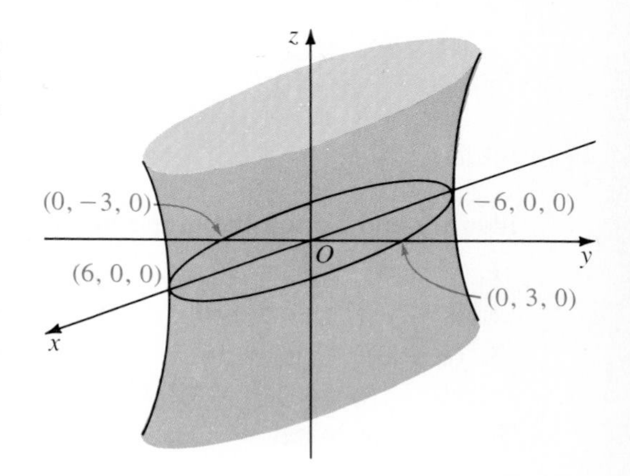

EXAMPLE 4 Identify and sketch the graph of $\frac{x^2}{9} + \frac{z^2}{4} = y$.

Solution Find the traces of the graph.

xy-trace	$z = 0$	$\frac{x^2}{9} = y$	parabola with y-axis as its axis
yz-trace	$x = 0$	$\frac{z^2}{4} = y$	parabola with y-axis as its axis
xz-trace	$y = 0$	$\frac{x^2}{9} + \frac{y^2}{4} = 0$	a single point: (0, 0, 0)

Solution continued on following page

To find the cross sections of the surface parallel to the xz-plane, let $y = c$. Thus $\frac{x^2}{9} + \frac{z^2}{4} = c$. Hence, if $c > 0$, the cross section is elliptical. If $c = 0$, the cross section is a single point, and if $c < 0$ there is no cross section. The surface is an elliptic paraboloid.

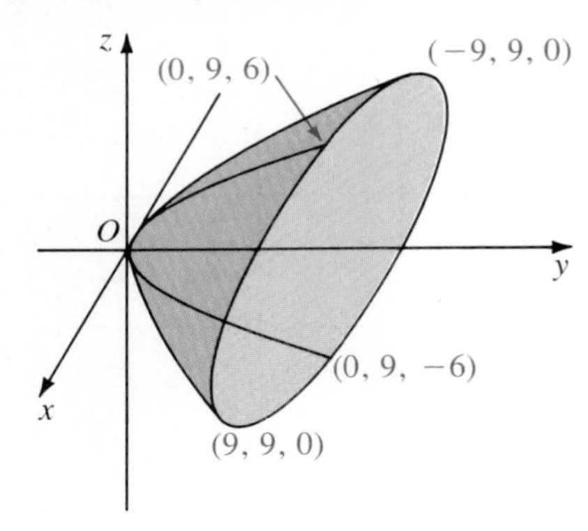

EXERCISES

Identify and sketch the graph of each equation.

A

1. $x^2 + y^2 + z^2 = 25$
2. $x + 2y + 3z - 6 = 0$
3. $x + y + z - 5 = 0$
4. $x^2 + (y - 2)^2 + z^2 = 16$
5. $(x + 3)^2 + y^2 = 16$
6. $y^2 + z^2 = 4$
7. $4(y - 3)^2 + 9(z - 3)^2 = 36$
8. $9(x + 1)^2 + 16(z + 4)^2 = 144$
9. $x^2 + 4y^2 = 4z$
10. $x^2 + y^2 - 4z^2 = 16$
11. $4x^2 + 16y^2 - z^2 = 0$
12. $x^2 - y^2 - z^2 = 1$
13. $25x^2 + 4y^2 + z^2 = 100$
14. $9x^2 + y^2 - z^2 = 0$
15. $4x^2 - y^2 = 16z$
16. $x^2 + 4y^2 + 9z^2 = 36$
17. $y = x^2$
18. $z = \sin y$
19. $xz = 1$
20. $x^2 - y^2 = 0$
21. $x^2 + y^2 - 10y = 0$
22. $x^2 - 6x + z^2 = 0$

In Exercises 23 and 24, find an equation of the plane tangent to the sphere at P. Use the fact that the plane is perpendicular to the radius at that point.

B

23. $x^2 + y^2 + z^2 = 169$; $P(-12, 3, 4)$
24. $x^2 + (y - 1)^2 + (z + 3)^2 = 25$; $P(3, 1, -7)$

25. Show that the intersection of the cone with equation $\frac{x^2}{a^2} + \frac{y^2}{a^2} - \frac{z^2}{a^2} = 0$ and a plane parallel to the xy-plane is a circle.

26. Show that the intersection of the cone with equation $\frac{x^2}{a^2} + \frac{y^2}{a^2} - \frac{z^2}{a^2} = 0$ and a plane parallel to the xz-plane is a hyperbola.

27. Find the intersection of the ellipsoid with equation $\frac{x^2}{a^2} + \frac{y^2}{b^2} + \frac{z^2}{c^2} = 1$ and the elliptic cone with equation $\frac{x^2}{a^2} + \frac{y^2}{b^2} - \frac{z^2}{c^2} = 0$.

28. Find the intersection of the elliptic cone with equation $\frac{x^2}{a^2} + \frac{z^2}{c^2} - \frac{y^2}{b^2} = 0$ with the hyperboloid of one sheet with equation $\frac{x^2}{a^2} + \frac{y^2}{b^2} - \frac{z^2}{c^2} = 1$.

29. Show that the hyperboloid of one sheet with equation $x^2 + y^2 - z^2 = 1$ contains infinitely many lines by completing parts (a)–(c).
   a. Let a and b be real numbers such that $a^2 + b^2 = 1$. Show that $(a, b, 0)$ and $(a + b, b - a, 1)$ are on the hyperboloid.
   b. Find a set of parametric equations of the line containing the two points in part (a).
   c. Show that every point on the line in part (b) lies on the hyperboloid.

30. Find k such that the intersection of the hyperboloid of two sheets with equation $\frac{x^2}{a^2} - \frac{y^2}{b^2} - \frac{z^2}{c^2} = 1$ and the plane with equation $x + ky - 1 = 0$ is **(a)** a hyperbola and **(b)** an ellipse.

13–5 CYLINDRICAL AND SPHERICAL COORDINATES

In addition to rectangular coordinates, two other systems, *cylindrical* and *spherical*, are used in three-space.

The **cylindrical coordinates** of a point $P(x, y, z)$ are $P(r, \theta, z)$ where r and θ are the polar coordinates of the projection of P onto the xy-plane. The following equations and the diagram relate rectangular and cylindrical coordinates.

$$x = r\cos\theta$$
$$y = r\sin\theta$$
$$z = z$$

and

$$\cos\theta = \frac{x}{r} \quad (r \neq 0)$$
$$\sin\theta = \frac{y}{r} \quad (r \neq 0)$$
$$r^2 = x^2 + y^2$$

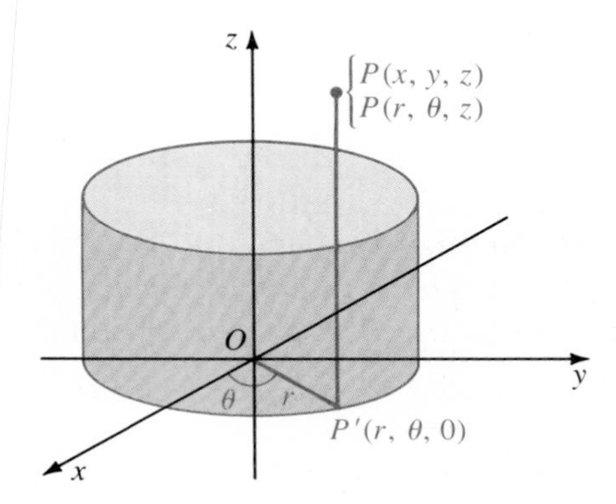

EXAMPLE 1 Describe the graph of each equation.

a. $r = 2$ **b.** $\theta = 60°$

Solution **a.** The variables θ and z are absent from the equation. They may take on all values. The graph is the set of all points that are 2 units from the z-axis, a circular cylinder. The solution can also be obtained by writing $r = 2$ as an equation in rectangular coordinates. If $r = 2$, then $r^2 = x^2 + y^2 = 4$. The graph is a cylinder with circular cross sections, Figure a.

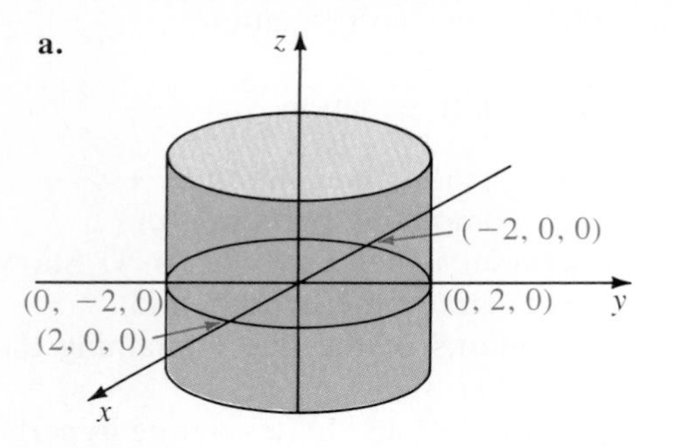

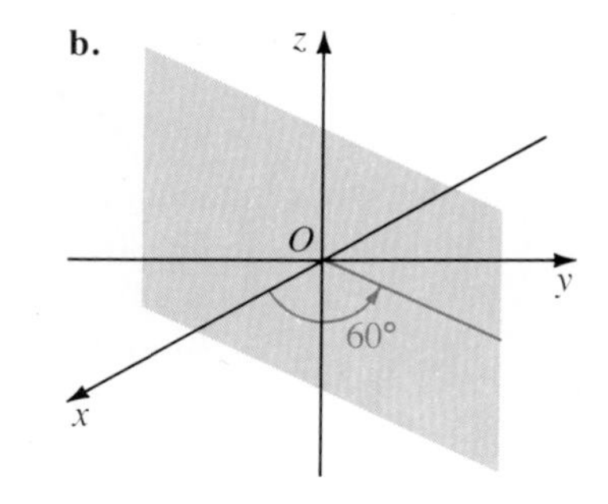

b. Since r and z are absent from the equation they can take on any values. The graph is a plane containing the z-axis and making an angle of 60° with the xz-plane, Figure b.

The **spherical coordinates** of a point P in space are (ρ, θ, φ), where $\rho = OP$, θ is the same as in cylindrical coordinates, and φ is the measure of the angle that $\overrightarrow{OP}$ makes with the positive z-axis, $0 \le \varphi \le \pi$. (See Figure a.)

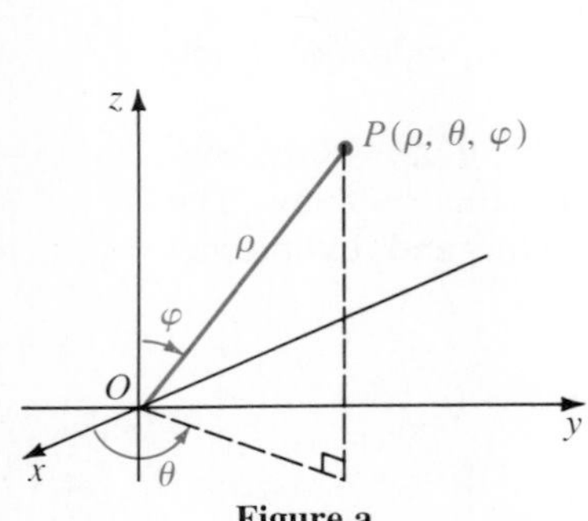

Figure a

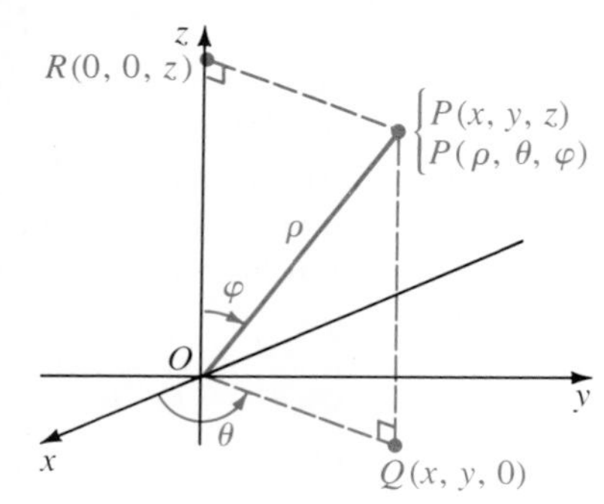

Figure b

By considering the projection R of P onto the z-axis, we obtain $z = \rho \cos \varphi$. (See Figure b.) Letting Q be the projection of P onto the xy-plane, we see that $OQ = RP = \rho \sin \varphi$. Therefore

$$x = OQ \cos \theta = \rho \sin \varphi \cos \theta \quad \text{and} \quad y = OQ \sin \theta = \rho \sin \varphi \sin \theta$$

Using the distance formula, we also have

$$\rho^2 = x^2 + y^2 + z^2.$$

Thus we have the following relationships between spherical and rectangular coordinates.

$$x = \rho \sin \varphi \cos \theta, \quad y = \rho \sin \varphi \sin \theta, \quad z = \rho \cos \varphi$$
$$\rho^2 = x^2 + y^2 + z^2$$

EXAMPLE 2 Describe the graph of each equation.

a. $\rho = 4$ **b.** $\varphi = \frac{3\pi}{4}$

Solution **a.** The graph is the set of all points 4 units from the origin, a sphere of radius 4 and centered at the origin, Figure a.

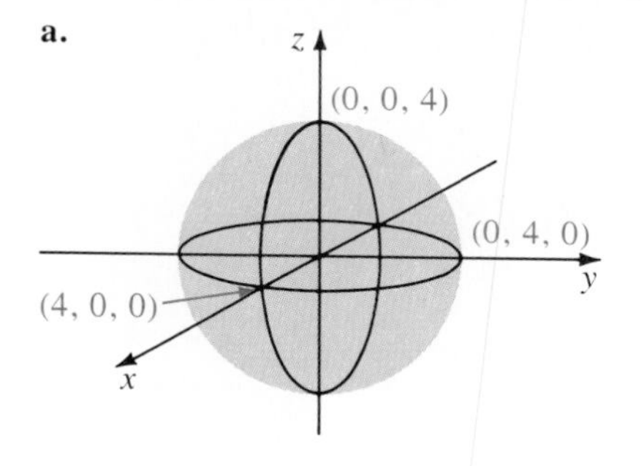

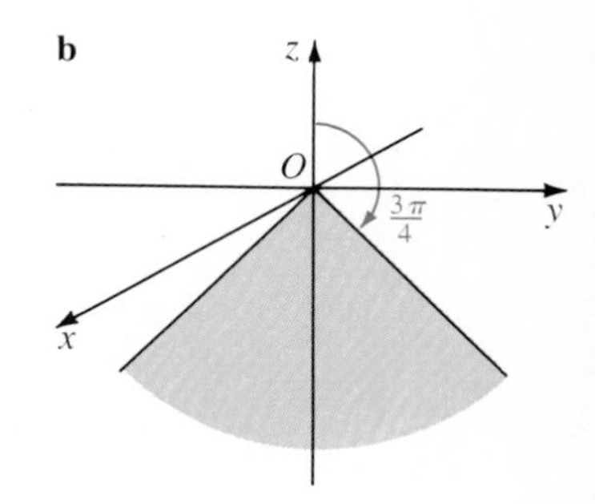

b. The graph is the set of all points P such that $\overline{OP}$ makes an angle of $\frac{3\pi}{4}$ radians with the z-axis. The graph is a cone with vertex at the origin, Figure b.

EXAMPLE 3 Sketch the graph of each equation.

a. $r = 2 \sin \theta$ (in cylindrical coordinates)
b. $\rho = 2 \csc \varphi$ (in spherical coordinates)

Solution In each solution use the transformation equations to obtain an equation in x, y, and z.

a. Since $r = 2 \sin \theta$, $r = 2\left(\frac{y}{r}\right)$.

Therefore $r^2 = 2y$.

$$x^2 + y^2 = 2y$$
$$x^2 + y^2 - 2y + 1 = 1$$
$$x^2 + (y - 1)^2 = 1$$

The trace in the xy-plane is a circle with radius 1 and center (0, 1). Since z is absent from the given equation, z may take on all real values. The graph in three-space is a cylinder.

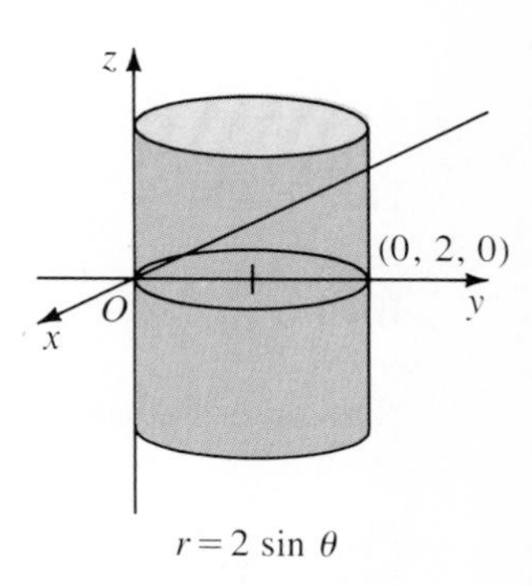

$r = 2 \sin \theta$

Solution continued on following page

b. Since $\rho = 2 \csc \varphi$, $\rho \sin \varphi = 2$. Therefore

$$x = 2 \cos \theta \text{ and } y = 2 \sin \theta.$$

Hence $\quad x^2 + y^2 = 4 \cos^2 \theta + 4 \sin^2 \theta = 4.$

The trace in the xy-plane is a circle of radius 2 centered at the origin. Since z is absent from the given equation, z may take on all real values. The graph in three-space is a cylinder, as shown below.

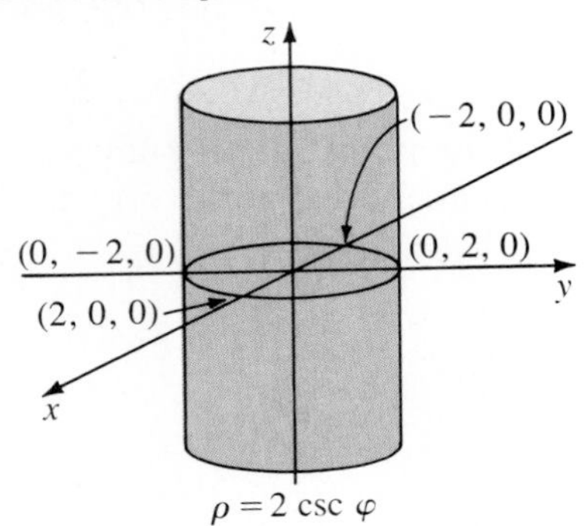

EXERCISES

Use the transformation equations to write the given rectangular coordinates in cylindrical or spherical coordinates.

A

1. $(1, 1, 1)$; cylindrical

2. $(-3\sqrt{3}, 3, 6)$; cylindrical

3. $(-1, 1, \sqrt{2})$; spherical

4. $(\sqrt{3}, -1, 2)$; spherical

5. $(2, 2\sqrt{3}, -4)$; cylindrical

6. $(3, 4, 5)$; spherical

Change the given cylindrical coordinates to rectangular coordinates.

7. $\left(4, \frac{\pi}{2}, 1\right)$

8. $\left(-3, \frac{\pi}{4}, -5\right)$

9. $\left(-3, \frac{\pi}{3}, 4\right)$

10. $\left(2, -\frac{\pi}{2}, 3\right)$

Change the given spherical coordinates to rectangular coordinates.

11. $\left(2, \frac{\pi}{2}, \frac{\pi}{6}\right)$

12. $\left(5, \frac{3\pi}{2}, \frac{\pi}{3}\right)$

13. $\left(2, \frac{\pi}{2}, \frac{\pi}{2}\right)$

14. $\left(3, \frac{\pi}{3}, \frac{3\pi}{4}\right)$

Find an equation in rectangular coordinates of the surface whose equation in cylindrical coordinates is given.

15. $r^2 = 4 - z^2$

16. $r = 9 \cos \theta$

17. $r = 2$

18. $r^2 - z^2 = -4$

Find an equation in rectangular coordinates of the surface whose equation in spherical coordinates is given.

19. $\rho = 2$

20. $\rho = 4 \sin \varphi$

21. $\rho = 2 \sin \varphi \cos \theta$

22. $4 = \rho \cos \varphi$

For each equation in rectangular coordinates, find **(a)** an equation in cylindrical coordinates and **(b)** an equation in spherical coordinates.

23. $x = 4z^2$

24. $x^2 + y^2 + z^2 - 2x = 0$

25. $x^2 + y^2 = 9z$

26. $y = 2x^2 + z$

Sketch each surface.

B

27. $\theta = \pi$; cylindrical coordinates

28. $r = 2 \cos \theta$; cylindrical coordinates

29. $\varphi = \frac{\pi}{4}$; spherical coordinates

30. $r^2 + z^2 = 9$; cylindrical coordinates

31. $\rho = 8 \cos \varphi$; spherical coordinates

32. $\rho = 4 \sin \theta \sin \varphi$; spherical coordinates

Chapter Summary

1. The *cross product* of $\mathbf{u} = (u_1, u_2, u_3)$ and $\mathbf{v} = (v_1, v_2, v_3)$ is

$$\mathbf{u} \times \mathbf{v} = \left(\begin{vmatrix} u_2 & u_3 \\ v_2 & v_3 \end{vmatrix}, \begin{vmatrix} u_3 & u_1 \\ v_3 & v_1 \end{vmatrix}, \begin{vmatrix} u_1 & u_2 \\ v_1 & v_2 \end{vmatrix} \right).$$

Some properties of the cross product are:
a. $\mathbf{u} \times \mathbf{v}$ is perpendicular to both $\mathbf{u}$ and $\mathbf{v}$.
b. $\mathbf{u} \times \mathbf{v} = \|\mathbf{u}\|\|\mathbf{v}\| \sin \theta$ where θ is the angle between $\mathbf{u}$ and $\mathbf{v}$.
The area of the parallelogram with $\mathbf{u}$ and $\mathbf{v}$ as adjacent sides is $\|\mathbf{u} \times \mathbf{v}\|$.
The *scalar triple product* of $\mathbf{u}$, $\mathbf{v}$, and $\mathbf{w}$ is $\mathbf{u} \cdot (\mathbf{v} \times \mathbf{w})$. The parallelepiped with edges $\mathbf{u}$, $\mathbf{v}$, and $\mathbf{w}$ has volume $|\mathbf{u} \cdot (\mathbf{v} \times \mathbf{w})|$.

2. The *plane* containing $P_0(x_0, y_0, z_0)$ and having *normal vector* $\mathbf{n} = (a, b, c)$ has vector equation $\mathbf{n} \cdot (\mathbf{r} - \mathbf{r}_0) = 0$ where $\mathbf{r}_0$ is the position vector of P_0 and $\mathbf{r}$ is the position vector of any point in the plane. A *scalar equation* for the plane is $ax + by + cz + d = 0$.
To sketch a plane, find its *intercepts* and its *traces*.
The distance D between $P_1(x_1, y_1, z_1)$ and the plane with equation $ax + by + cz + d = 0$ is given by

$$D = \frac{|ax_1 + by_1 + cz_1 + d|}{\sqrt{a^2 + b^2 + c^2}}.$$

3. A vector equation of the *line* containing P_0 and having *direction vector* $\mathbf{m} = (a, b, c)$ is $\mathbf{r} = \mathbf{r}_0 + t\mathbf{m}$ where $\mathbf{r}_0$ is the position vector of P_0, $\mathbf{r}$ is the position vector of any point on the line, and t is a scalar variable. A line also has the following forms:

parametric equations $\quad x = x_0 + at \qquad y = y_0 + bt \qquad z = z_0 + ct$

symmetric-form equations $(abc \neq 0)$ $\quad \dfrac{x - x_0}{a} = \dfrac{y - y_0}{b} = \dfrac{z - z_0}{c}$

In three-space, nonparallel lines that do not intersect are called *skew*. The angle between them is defined to be the angle between their direction vectors.

The distance D between a point P and a line containing S and having direction vector $\mathbf{m}$ is given by

$$D = \frac{\|\overrightarrow{SP} \times \mathbf{m}\|}{\|\mathbf{m}\|}.$$

The angles α, β, and γ that a vector $\mathbf{m}$ makes with the coordinate axes are called its *direction angles*, and $\cos \alpha$, $\cos \beta$, and $\cos \gamma$ are its *direction cosines*. The vector $\mathbf{i} \cos \alpha + \mathbf{j} \cos \beta + \mathbf{k} \cos \gamma$ is the unit vector in the direction of $\mathbf{m}$. In particular, $\cos^2 \alpha + \cos^2 \beta + \cos^2 \gamma = 1$. Direction cosines can be used to find the angle θ between two vectors $\mathbf{m}_1$ and $\mathbf{m}_2$.

4. The *sphere* with *radius* r and *center* $P_0(x_0, y_0, z_0)$ has equation

$$(x - x_0)^2 + (y - y_0)^2 + (z - z_0)^2 = r^2.$$

A *cylinder* is the surface consisting of all lines L, called *rulings*, through a plane curve C, the *directrix*, and parallel to a fixed line L_0.

To sketch a surface, draw its *cross sections* and *traces*.

A *quadric surface* is one obtained as the graph of a second-degree equation in three variables:

$$Ax^2 + By^2 + Cz^2 + Eyz + Fxz + Gxy + Hx + Iy + Jz + K = 0.$$

A nondegenerate quadric surface is a cylinder, an *ellipsoid*, an *elliptic cone*, a *hyperboloid of one or two sheets*, an *elliptic paraboloid*, or a *hyperbolic paraboloid*.

5. A point $P(x, y, z)$ has *cylindrical coordinates* $P(r, \theta, z)$ where (r, θ) are the polar coordinates of the projection of P on the xy-plane. Rectangular and cylindrical coordinates are related by

$$x = r \cos \theta \qquad y = r \sin \theta \qquad z = z$$

$$\cos \theta = \frac{x}{r} \qquad r^2 = x^2 + y^2 \qquad \sin \theta = \frac{y}{r} \qquad (r \neq 0)$$

A point $P(x, y, z)$ has *spherical coordinates* $P(\rho, \theta, \varphi)$ where $\rho = OP$, θ is the same as in cylindrical coordinates, and φ is the angle $\overrightarrow{OP}$ makes with the positive z-axis. Rectangular and spherical coordinates are related by

$$x = \rho \sin \varphi \cos \theta \qquad y = \rho \sin \varphi \sin \theta \qquad z = \rho \cos \varphi$$

$$\rho^2 = x^2 + y^2 + z^2$$

Chapter Test

13-1 **1.** Find a unit vector perpendicular to $\mathbf{u} = (3, -2, 0)$ and $\mathbf{v} = (3, -3, -1)$.

2. Find the area of the triangle with vertices $A(2, 1, 3)$, $B(-1, 0, 5)$, and $C(2, 4, 6)$.

3. Find the volume of the parallelepiped with edges $\overrightarrow{OA}$, $\overrightarrow{OB}$, and $\overrightarrow{OC}$ where $O = (0, 0, 0)$, $A = (3, 1, -2)$, $B = (4, 0, -1)$, and $C = (2, 5, 1)$.

13-2 **4.** Graph $2x - 3y + 4z - 12 = 0$.

5. Find a scalar equation of the plane containing $(3, -2, 5)$ and parallel to the plane with equation $x - 4y + 3z + 7 = 0$.

6. Find a scalar equation of the plane containing $(2, 3, -5)$, $(4, -1, 2)$, and $(3, 0, 1)$.

7. Show that the planes with equations $3x - 5y + z = 12$ and $6x - 10y + 2z = 5$ are parallel and find the distance between them.

13-3 **8.** Find **(a)** a vector equation and **(b)** a set of parametric equations for the line containing $(6, -4, 9)$ and $(-3, 5, 7)$.

9. Find the distance between the origin and the line containing $(1, 0, 5)$ and $(7, -3, -1)$.

13-4 Identify and sketch the surface defined by each equation.

10. $4x^2 + 9y^2 - z^2 = 36$

11. $x^2 + 4y^2 = 100z$

13-5 **12.** Find an equation in rectangular coordinates for the surface whose equation in cylindrical coordinates is $3r = 4z$.

13. Find an equation in spherical coordinates for the surface whose equation in rectangular coordinates is $x^2 + y^2 - z^2 + 1 = 0$.

COMPUTER APPLICATION

VECTOR PRODUCTS

Computation of the dot product of two vectors, the cross product of three vectors, and the triple scalar product of three vectors is quite helpful in the solution of many geometric problems. The definitions are summarized below. Let $\mathbf{u} = (u_1, u_2, u_3)$, $\mathbf{v} = (v_1, v_2, v_3)$, and $\mathbf{w} = (w_1, w_2, w_3)$.

$$\mathbf{u} \cdot \mathbf{v} = u_1v_1 + u_2v_2 + u_3v_3$$

$$\mathbf{u} \times \mathbf{v} = \begin{vmatrix} \mathbf{i} & \mathbf{j} & \mathbf{k} \\ u_1 & u_2 & u_3 \\ v_1 & v_2 & v_3 \end{vmatrix} \qquad \mathbf{u} \cdot (\mathbf{v} \times \mathbf{w}) = \begin{vmatrix} u_1 & u_2 & u_3 \\ v_1 & v_2 & v_3 \\ w_1 & w_2 & w_3 \end{vmatrix}$$

Recall that the dot product can be used to find the angle between two vectors, that the cross product can be used to find the area of a parallelogram, and that the triple scalar product can be used to find the volume of a parallelepiped.

EXERCISES *(You may use either BASIC or Pascal.)*

1. Write a computer program to find the dot product of two vectors, the cross product of two vectors, and the triple scalar product of three vectors.

Let $\mathbf{u} = (1, 1, 1)$, $\mathbf{v} = (1, -1, 1)$, $\mathbf{w} = (1, 1, 0)$, and $\mathbf{z} = (-1, 5, -1)$.

2. Find the area of the parallelogram having $\mathbf{u}$ and $\mathbf{v}$ as adjacent sides. Do $\mathbf{u}$ and $\mathbf{v}$ form a basis for the vectors in a plane?
3. Find the volume of the parallelepiped having $\mathbf{u}$, $\mathbf{v}$, and $\mathbf{w}$ as edges. Do $\mathbf{u}$, $\mathbf{v}$, and $\mathbf{w}$ form a basis for the set of all vectors in three-space?
4. Find the volume of the parallelepiped having $\mathbf{u}$, $\mathbf{v}$, and $\mathbf{z}$ as edges. Do $\mathbf{u}$, $\mathbf{v}$, and $\mathbf{z}$ form a basis for the set of all vectors in three-space?
5. Is the parallelepiped in Exercise 3 rectangular?

APPLICATION

SPACE CURVES

The path of the car in the amusement-park ride shown in the photograph is a space curve. The car accelerates downhill and slows as it goes uphill. However, it also twists to the left and the right. The car dips below and rises above the track it covers. This acceleration, deceleration, twisting, and turning makes a thrilling ride.

The path of a spacecraft to its destination is also a space curve, as is the path of an automobile along a hilly road. As a ship changes course continuously along the surface of the sea, it too traces out a space curve.

Engineers and scientists are greatly interested in space curves, and mathematicians have investigated them extensively.

The coordinates of a point $P(x, y, z)$ on a space curve C can be given by equations of the form

$$x = f(t) \qquad y = g(t) \qquad z = h(t)$$

where f, g, and h are differentiable functions of a variable such as time t.

EXERCISES

1. The equations $x = a \cos t$, $y = a \sin t$, and $z = t$ define a space curve C. Show that if P is on C, then P is also on a right circular cylinder having equation $x^2 + y^2 = a^2$.
2. Explain: The points on C corresponding to $t_k = 2\pi k$, k an integer, lie on a ruling of the cylinder in Exercise 1. If k is positive, the point corresponding to t_{k+1} lies directly above the point corresponding to t_k.
3. Sketch the space curve C, known as a *circular helix*, by sketching the cylinder in Exercise 1 and by using your explanation in Exercise 2.

Cumulative Review (*Chapters 11–13*)

1. Let $\mathbf{u} = (-1, 3)$ and $\mathbf{v} = (5, 2)$. Draw each linear combination and give its coordinates.
   a. $-3\mathbf{u}$ **b.** $\mathbf{u} + 2\mathbf{v}$ **c.** $2\mathbf{u} - \mathbf{v}$

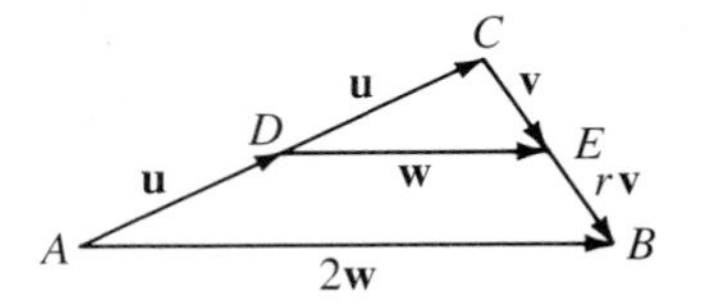

2. Use the vector diagram at the right to show that E is the midpoint of $\overline{CB}$. That is, show that $r = 1$.

3. Find a unit vector having direction opposite that of $\mathbf{v} = (-5, -12, 13)$.

4. Let $A = \begin{bmatrix} -1 & 2 \\ -3 & 1 \end{bmatrix}$ and $B = \begin{bmatrix} 4 & -2 \\ 1 & 3 \end{bmatrix}$. Find each matrix.
   a. $A + 2B$ **b.** AB **c.** B^2 **d.** $3A - 2B$

5. Let $T((x, y, z)) = (x - 2y + z, 4x + 3y)$ and let $\mathbf{v} = (2, -7, 5)$. Find $T(\mathbf{v})$.

6. Let $T((x, y)) = (3x - y, -4x + 2y)$. If $T(\mathbf{v}) = (-2, 1)$, find $\mathbf{v}$.

7. Find the length and the coordinates of the midpoint of the line segment with endpoints $C(-1, 7, 6)$ and $D(3, 2, -8)$.

Solve each system of equations.

8. $$\begin{aligned} x - 2y + z &= -1 \\ 2x + y + 3z &= 0 \\ -x + y - 2z &= 3 \end{aligned}$$

9. $$\begin{aligned} 2x - 3y + z &= 12 \\ x + y + 3z &= 7 \\ 3x - 2y + 3z &= -7 \end{aligned}$$

10. Evaluate
$$\begin{vmatrix} 3 & 2 & -1 & 4 \\ 0 & 1 & 1 & 3 \\ 0 & -2 & 2 & 5 \\ 2 & 1 & 3 & -1 \end{vmatrix}.$$

11. Find the inverse of
$$\begin{bmatrix} 2 & 0 & -1 \\ 1 & 1 & -2 \\ 5 & -3 & 1 \end{bmatrix}.$$

12. Identify and sketch the graph of $x^2 + 2\sqrt{3}xy - y^2 = 8$.

13. Find a scalar equation of the plane containing $(3, -4, 1)$ and parallel to the plane with equation $5x - 2y - 7z + 12 = 0$.

14. Find the coordinates of the vector having the origin as initial point and equivalent to $\overrightarrow{PQ}$, whose endpoints are $P(-2, -3)$ and $Q(-5, 9)$.

15. A plane has a heading of 70° and an air speed of 350 mph. If an east wind is blowing at 25 mph, find the plane's ground speed and true course.

16. Write $\mathbf{t} = (-5, -1)$ as a linear combination of $\mathbf{u} = (2, 3)$ and $\mathbf{v} = (-6, 4)$.

17. Find the volume of the parallelepiped with edges $\overrightarrow{OA}$, $\overrightarrow{OB}$, and $\overrightarrow{OC}$ where O is the origin and $A = (-3, 1, 4)$, $B = (2, 5, -1)$, and $C = (3, -2, 0)$.

18. Find the measure of the angle between $\mathbf{u} = (2, -2, 1)$ and $\mathbf{v} = (2, 5, -14)$.

19. Evaluate $\begin{vmatrix} a & b & c & d \\ 0 & e & f & g \\ 0 & 0 & h & j \\ 0 & 0 & 0 & k \end{vmatrix}$.

20. Let $A = \begin{bmatrix} 3 & -2 \\ 1 & 2 \end{bmatrix}$ and $B = \begin{bmatrix} 1 & 0 \\ -3 & 4 \end{bmatrix}$.

    Find $(A + B)(A - B)$ and $A^2 - B^2$. Are the two resulting matrices equal?

21. Triangle ABC has vertices $A(5, -1, 2)$, $B(-3, 0, 2)$, and $C(-1, 4, 0)$. Determine whether the triangle is equilateral, isosceles, or scalene by finding the lengths of the sides.

22. Identify and sketch the graph of $144x^2 + 16(y - 1)^2 + 9z^2 = 144$.

23. Solve the system for z by using Cramer's rule.

$$\begin{aligned} 3x + 5y - 2z &= -2 \\ y + 4z &= 5 \\ 2x + 3y &= 1 \end{aligned}$$

24. Find the vector components of $\mathbf{t} = (3, -4)$ in the directions of $\mathbf{u} = (7, 4)$ and $\mathbf{v} = (4, -7)$.

25. Let r and s be real numbers and let $\mathbf{u}$ be a complex number. Show that $(rs)\mathbf{u} = r(s\mathbf{u})$.

26. Find a set of parametric equations of the line containing (3, 5, 1) and perpendicular to the plane with equation $x + 3y - 4z + 11 = 0$.

27. Find an equation of the plane containing $(-2, -6, 5)$ and perpendicular to the line with equation $\mathbf{r} = (2 - 3t)\mathbf{i} + 5t\mathbf{j} + (9 + 2t)\mathbf{k}$.

28. Let $\mathbf{u} = (-1, 3, -2)$, $\mathbf{v} = (-5, 0, 7)$, and $\mathbf{w} = (4, -9, -6)$. Find $\mathbf{v} - 2(\mathbf{u} + \mathbf{w})$.

29. Find an equation in rectangular coordinates of the surface whose equation in cylindrical coordinates is $r = 25 \sin \theta$.

30. A force of 36 N is applied at an angle of 47° with the horizontal. It moves an object 20 m up an inclined plane making an angle of 40° with the horizontal. Find the work done.

Chapter 14

Introduction to Integral Calculus

In this chapter basic concepts and methods of integration are discussed. Integration is then applied to geometric and physical problems.

Antiderivatives and Definite Integrals

14-1 ANTIDERIVATIVES

In previous chapters we discussed the operation of differentiation, that is, of finding derivatives of given functions. We shall soon see that many geometric and physical problems can be solved by reversing that operation and using *antiderivatives*.

DEFINITION

The function f has F as an **antiderivative** on an interval I if, for each x in I,

$$F'(x) = f(x), \quad \text{that is,} \quad \frac{d}{dx} F(x) = f(x).$$

For example, $\sin x$ is an antiderivative of $\cos x$ because

$$\frac{d}{dx} \sin x = \cos x.$$

The photograph shows the six generators at the Pickwick Landing plant of the Tennessee Valley Authority. A brief discussion of work and energy, applications of integration, is given in the last section of the chapter.

Notice that for any constant C, $\frac{d}{dx}(\sin x + C) = \cos x$, and thus $\cos x$ has many antiderivatives. However, in view of the following theorem, *every* antiderivative of $\cos x$ is of the form $\sin x + C$.

THEOREM 1 If $F(x)$ is an antiderivative of $f(x)$ on an interval I, then every antiderivative of f on I is of the form $F(x) + C$.

A proof of Theorem 1 is outlined in the Extra on pages 264–265. (Usually we do not need to mention which interval I is involved because the context will make this clear.)

Theorem 1 tells us how to write a general expression for *all* antiderivatives of f once we have found one of them. Such a general expression is called an **indefinite integral** of f and is written $\int f(x)\,dx$. For example,

$$\int \cos x\,dx = \sin x + C.$$

In this expression $\int$ is the **integral sign**, $\cos x$ is the **integrand**, and C is the **constant of integration**. Notice that

$$\frac{d}{dx}F(x) = f(x) \quad \text{and} \quad \int f(x)\,dx = F(x) + C$$

are equivalent statements.

It can be shown that every continuous function has an antiderivative. In this chapter we shall assume without explicit mention that all integrands are continuous.

The process of finding antiderivatives or indefinite integrals is called **integration**. Many functions can be integrated simply by reversing derivative formulas.

EXAMPLE 1 Find each integral: **a.** $\int \sin x\,dx$ **b.** $\int \frac{dx}{x^3}$

Solution In each part find a function whose derivative is the given integrand.

a. Since $\frac{d}{dx}\cos x = -\sin x$, $\frac{d}{dx}(-\cos x) = \sin x$.

Therefore $\int \sin x\,dx = -\cos x + C$. **Answer**

b. $\int \frac{dx}{x^3}$ is another way of writing $\int \frac{1}{x^3}\,dx$, or $\int x^{-3}\,dx$.

Try x^{-2}: $\frac{d}{dx}x^{-2} = -2x^{-3}$. This is exactly -2 times the given integrand.

To obtain x^{-3}, try $-\frac{1}{2}x^{-2}$: $\frac{d}{dx}\left(-\frac{1}{2}x^{-2}\right) = \left(-\frac{1}{2}\right)(-2)x^{-3} = x^{-3}$.

Therefore $\int \frac{dx}{x^3} = -\frac{1}{2}x^{-2} + C$. **Answer**

The following integral formulas are obtained by reversing derivative formulas.

THEOREM 2 1. $\int x^n\,dx = \dfrac{x^{n+1}}{n+1} + C \quad (n \neq -1)$ 2. $\int x^{-1}\,dx = \int \dfrac{dx}{x} = \ln|x| + C$

3. $\int \sin ax\,dx = -\dfrac{1}{a}\cos ax + C$ 4. $\int \cos ax\,dx = \dfrac{1}{a}\sin ax + C$

5. $\int e^{ax}\,dx = \dfrac{1}{a}e^{ax} + C$

We shall prove parts (2) and (5). The proofs of parts (1), (3), and (4) are left as Exercises 29–31.

Proof: 2. We need only show that $\dfrac{d}{dx}\ln|x| = \dfrac{1}{x}$ for all $x \neq 0$.

If $x > 0$, then $|x| = x$, and by Theorem 7, page 423,

$$\frac{d}{dx}\ln|x| = \frac{d}{dx}\ln x = \frac{1}{x}.$$

If $x < 0$, then $|x| = -x$, and, using the chain rule, we have

$$\frac{d}{dx}\ln|x| = \frac{d}{dx}\ln(-x) = \frac{1}{-x}\frac{d}{dx}(-x) = \frac{1}{-x}(-1) = \frac{1}{x}. \quad \blacksquare$$

5. $\dfrac{d}{dx}\left(\dfrac{1}{a}e^{ax}\right) = \dfrac{1}{a}\dfrac{d}{dx}e^{ax} = \dfrac{1}{a}\left(ae^{ax}\right) = e^{ax}. \quad \blacksquare$

Using Theorem 2, we find, for example, that

$$\int e^{-t}\,dt = \frac{1}{-1}e^{-t} + C = -e^{-t} + C.$$

The following theorem enables us to integrate linear combinations of functions when we know the integrals of the individual functions. You are asked to prove the theorem in Exercise 32.

THEOREM 3 Let f and g have antiderivatives. Then for any two constants, a and b,

$$\int [af(x) + bg(x)]\,dx = a\int f(x)\,dx + b\int g(x)\,dx.$$

EXAMPLE 2 Find each integral.

a. $\int (3x^2 - 4x + 5)\,dx$ **b.** $\int \dfrac{x^2+2}{x}\,dx$ **c.** $\int (\sin\theta + \cos\theta)^2\,d\theta$

Solution **a.**

$$\int (3x^2 - 4x + 5)\,dx = 3\int x^2\,dx - 4\int x\,dx + 5\int dx$$
$$= 3\cdot\frac{x^3}{3} - 4\cdot\frac{x^2}{2} + 5x + C$$
$$= x^3 - 2x^2 + 5x + C \quad \textbf{Answer}$$

In parts (b) and (c) use algebra to obtain a linear combination.

Solution continued on following page

b. $\dfrac{x^2+2}{x} = x + \dfrac{2}{x} = x + 2\left(\dfrac{1}{x}\right)$

Therefore $\displaystyle\int \frac{x^2+2}{x}\,dx = \int x\,dx + 2\int \frac{1}{x}\,dx = \frac{x^2}{2} + 2\ln|x| + C$

$\displaystyle = \frac{x^2}{2} + \ln x^2 + C.$ **Answer**

c. $(\sin\theta + \cos\theta)^2 = \sin^2\theta + 2\sin\theta\cos\theta + \cos^2\theta$
$= 1 + \sin 2\theta$

Therefore $\displaystyle\int (\sin\theta + \cos\theta)^2\,d\theta = \int 1\,d\theta + \int \sin 2\theta\,d\theta$

$\displaystyle = \theta - \frac{1}{2}\cos 2\theta + C.$ **Answer**

EXERCISES

Use Theorems 2 and 3 to find the following indefinite integrals.

A

1. $\int 6x^2\,dx$
2. $\int 6t\,dt$
3. $\int (2x-3)\,dx$
4. $\int (1-4x)\,dx$
5. $\int (x^2-2x-3)\,dx$
6. $\int (3z^2-3z+1)\,dz$
7. $\int (1-x)^2\,dx$
8. $\int (2x+1)^2\,dx$
9. $\int \frac{dx}{x^2}$
10. $\int \frac{dt}{t}$
11. $\int \frac{2\,du}{u}$
12. $\int x^{-9}\,dx$
13. $\int \frac{x^2+4}{x}\,dx$
14. $\int \sqrt[3]{x^4}\,dx$
15. $\int (x^{-1}+2)^2\,dx$
16. $\int \frac{9-t^2}{t^2}\,dt$
17. $\int (t^{-1}+2)(t^{-1}-2)\,dt$
18. $\int \frac{z^2-9}{z-3}\,dz$
19. $\int \sin 2\theta\,d\theta$
20. $\int \cos\frac{t}{2}\,dt$
21. $\int e^{2x}\,dx$
22. $\int e^{-\frac{t}{2}}\,dt$
23. $\int (e^t+e^{-t})\,dt$
24. $\int (e^t-e^{-t})\,dt$

B

25. $\int (e^t-e^{-t})^2\,dt$
26. $\int (e^t+e^{-t})^2\,dt$
27. $\int (\cos t - \sin t)^2\,dt$
28. $\int (\cos^2 t - \sin^2 t)\,dt$

Prove each part of Theorem 2.

29. Part (1)
30. Part (3)
31. Part (4)
32. Prove Theorem 3.

Find each indefinite integral.

33. $\int (\cos^4\theta - \sin^4\theta)\,d\theta$ (*Hint*: Write the integrand in terms of cos.)

34. $\int 2 \sin 2x \sin x \, dx$ (*Hint*: Use a product-to-sum identity.)

C 35. Use a product-to-sum identity to find $\int \cos mx \sin nx \, dx$.

36. Use the definition of antiderivative to show that $\int \frac{1}{x^2 + 5x + 6} \, dx = \ln\left(\frac{x+2}{x+3}\right) + C$ for all $x > -2$.

14-2 INTEGRATION BY SUBSTITUTION

Many integrals can be found by using an appropriate substitution. We shall introduce the method by working an example in a purely formal way, leaving the justification of what we do until later.

EXAMPLE 1 Find $\int 2x \cos x^2 \, dx$.

Solution Organize the work by steps.

1. Choose a substitution by which the integral can be rewritten.

Let $u = x^2$. (*)

Then $\frac{du}{dx} = 2x$ and $du = 2x \, dx$. (†)

Make the substitution using (*) and (†).

$$\int \cos x^2 \cdot 2x \, dx = \int \cos u \, du.$$

2. Find the resulting integral.

$$\int \cos u \, du = \sin u + C \quad \text{by Theorem 2, part (4)}$$

3. Return to the original variable x.

$$\int 2x \cos x^2 \, dx = \sin x^2 + C. \quad \textbf{Answer}$$

The substitution method is justified by the following theorem.

THEOREM 4 If $u = g(x)$, then $\int f(g(x))g'(x) \, dx = \int f(u) \, du$.

In Example 1 we used this theorem with $u = g(x) = x^2$, and $f(u) = \cos u$. Then $f(g(x)) = \cos x^2$ and $g'(x) = 2x$.

Proof: (Optional) Let F be an antiderivative of f. Then

$$\int f(u) \, du = F(u) + C,$$

and since $u = g(x)$, $\int f(u) \, du = F(g(x)) + C.$ (1)

Proof continued on following page

On the other hand:

$$\begin{aligned}
\frac{d}{dx}F(g(x)) &= \frac{d}{dx}F(u) && \text{because } u = g(x)\\
&= \frac{d}{du}F(u)\cdot\frac{du}{dx} && \text{by the chain rule}\\
&= F'(u)\cdot g'(x) && \text{because } \frac{du}{dx} = g'(x)\\
&= f(u)\cdot g'(x) && \text{because } F'(u) = f(u)\\
&= f(g(x))\cdot g'(x) && \text{because } u = g(x)
\end{aligned}$$

Thus, $F(g(x))$ is an antiderivative of $f(g(x))g'(x)$. Therefore

$$\int f(g(x))g'(x)\,dx = F(g(x)) + C. \tag{2}$$

Together (1) and (2) imply the conclusion of the theorem. ■

EXAMPLE 2 Find $\int \frac{x\,dx}{1 + x^2}$.

Solution If $u = 1 + x^2$, then $\frac{du}{dx} = 2x$, $du = 2x\,dx$, and $x\,dx = \frac{1}{2}\,du$.

$$\int \frac{x\,dx}{1 + x^2} = \int \frac{\frac{1}{2}\,du}{u} = \frac{1}{2}\int \frac{du}{u} = \frac{1}{2}\ln u + C$$

Therefore $\int \frac{x\,dx}{1 + x^2} = \frac{1}{2}\ln(1 + x^2) + C$, or $\ln\sqrt{1 + x^2} + C$. **Answer**

When the integrand involves $\sqrt{x^2 \pm a^2}$ or $\sqrt{a^2 - x^2}$, look for the combination $x\,dx$.

EXAMPLE 3 Find $\int x\sqrt{4 - x^2}\,dx$.

Solution Let $u = 4 - x^2$. Then $du = -2x\,dx$, and $x\,dx = -\frac{1}{2}\,du$.

$$\int x\sqrt{4 - x^2}\,dx = -\frac{1}{2}\int u^{\frac{1}{2}}\,du = -\frac{1}{2}\frac{u^{\frac{3}{2}}}{\frac{3}{2}} + C = -\frac{1}{3}u^{\frac{3}{2}} + C$$

Therefore $\int x\sqrt{4 - x^2}\,dx = -\frac{1}{3}(4 - x^2)^{\frac{3}{2}} + C$. **Answer**

The following theorem enables us to differentiate functions involving $\mathrm{Sin}^{-1}\,x$ and $\mathrm{Tan}^{-1}\,x$ and integrate functions involving $\frac{1}{\sqrt{1 - x^2}}$ or $\frac{1}{1 + x^2}$.

THEOREM 5 **1.** If $-1 < x < 1$,

$$\frac{d}{dx}\mathrm{Sin}^{-1}\,x = \frac{1}{\sqrt{1 - x^2}} \quad\text{and}\quad \int \frac{dx}{\sqrt{1 - x^2}} = \mathrm{Sin}^{-1}\,x + C.$$

2. $$\frac{d}{dx}\mathrm{Tan}^{-1}\,x = \frac{1}{1 + x^2} \quad\text{and}\quad \int \frac{dx}{1 + x^2} = \mathrm{Tan}^{-1}\,x + C.$$

The following (optional) proof establishes part (1). The proof of part (2) is left as Exercise 29.

Proof: Assuming that $\mathrm{Sin}^{-1} x$ is differentiable on $-1 < x < 1$, let $y = \mathrm{Sin}^{-1} x$. Then $\sin y = x$, $-\frac{\pi}{2} < y < \frac{\pi}{2}$. Differentiating both sides of $\sin y = x$ with respect to x and using the chain rule on the left, we obtain

$$\cos y \frac{dy}{dx} = 1, \quad \text{or} \quad \frac{dy}{dx} = \frac{1}{\cos y}.$$

Since $-\frac{\pi}{2} < y < \frac{\pi}{2}$, $\cos y > 0$. Therefore

$$\begin{aligned}\cos y &= +\sqrt{1 - \sin^2 y}\\ &= \sqrt{1 - x^2}.\end{aligned}$$

Therefore $\frac{dy}{dx} = \frac{d}{dx}\mathrm{Sin}^{-1} x = \frac{1}{\sqrt{1 - x^2}}$. Since $\mathrm{Sin}^{-1} x$ is an antiderivative of $\frac{1}{\sqrt{1 - x^2}}$, we also have $\int \frac{dx}{\sqrt{1 - x^2}} = \mathrm{Sin}^{-1} x + C$. ■

EXAMPLE 4 Find $\int \frac{\cos \theta}{1 + \sin^2 \theta}\, d\theta$.

Solution If $u = \sin \theta$, then $\frac{du}{d\theta} = \cos \theta$, and $\cos \theta\, d\theta = du$.

$$\begin{aligned}\int \frac{\cos \theta}{1 + \sin^2 \theta}\, d\theta &= \int \frac{du}{1 + u^2}\\ &= \mathrm{Tan}^{-1} u + C \quad \text{by Theorem 5, part (2)}\end{aligned}$$

Therefore $\int \frac{\cos \theta}{1 + \sin^2 \theta}\, d\theta = \mathrm{Tan}^{-1} (\sin \theta) + C$. **Answer**

In the next example some algebra is used before a substitution is chosen.

EXAMPLE 5 Find $\int \frac{dt}{\sqrt{9 - t^2}}$.

Solution

$$\int \frac{dt}{\sqrt{9 - t^2}} = \int \frac{dt}{\sqrt{9\left(1 - \left(\frac{t}{3}\right)^2\right)}} = \int \frac{\frac{1}{3}\, dt}{\sqrt{1 - \left(\frac{t}{3}\right)^2}}$$

Use the substitution $u = \frac{t}{3}$. Then $du = \frac{1}{3}\, dt$.

$$\int \frac{\frac{1}{3}\, dt}{\sqrt{1 - \left(\frac{t}{3}\right)^2}} = \int \frac{du}{\sqrt{1 - u^2}} = \mathrm{Sin}^{-1} u + C$$

Therefore $\int \frac{dt}{\sqrt{9 - t^2}} = \mathrm{Sin}^{-1}\left(\frac{t}{3}\right) + C$. **Answer**

The method used to solve Example 5 can be used to prove the following corollary of Theorem 5.

COROLLARY 1. $\displaystyle\int \frac{dx}{\sqrt{a^2 - x^2}} = \text{Sin}^{-1}\frac{x}{a} + C \quad (a > 0)$

2. $\displaystyle\int \frac{dx}{a^2 + x^2} = \frac{1}{a}\,\text{Tan}^{-1}\frac{x}{a} + C$

EXERCISES

Use the suggested substitution to find the following indefinite integrals.

A

1. $\displaystyle\int (x + 2)^4\,dx \quad [u = x + 2]$
2. $\displaystyle\int (1 - x)^9\,dx \quad [u = 1 - x]$
3. $\displaystyle\int \frac{dx}{(4 - x)^2} \quad [u = 4 - x]$
4. $\displaystyle\int \frac{dx}{x + 2} \quad [u = x + 2]$
5. $\displaystyle\int \frac{dx}{2x - 1} \quad [u = 2x - 1]$
6. $\displaystyle\int \frac{dx}{(3x + 2)^2} \quad [u = 3x + 2]$
7. $\displaystyle\int x(x^2 - 1)^4\,dx \quad [u = x^2 - 1]$
8. $\displaystyle\int \frac{x\,dx}{1 + x^2} \quad [u = 1 + x^2]$
9. $\displaystyle\int \frac{x\,dx}{\sqrt{1 - x^2}} \quad [u = 1 - x^2]$
10. $\displaystyle\int x\sqrt{x^2 - 1}\,dx \quad [u = x^2 - 1]$
11. $\displaystyle\int x\,e^{x^2}\,dx \quad [u = x^2]$
12. $\displaystyle\int x \sin x^2\,dx \quad [u = x^2]$
13. $\displaystyle\int e^x \cos e^x\,dx \quad [u = e^x]$
14. $\displaystyle\int \cos x\,e^{\sin x}\,dx \quad [u = \sin x]$

Use substitutions to find the following indefinite integrals.

15. $\displaystyle\int (2x - 1)^4\,dx$
16. $\displaystyle\int \frac{dx}{1 - x}$
17. $\displaystyle\int \frac{x\,dx}{1 - x^2}$
18. $\displaystyle\int x(x^2 + 1)^9\,dx$

B

19. $\displaystyle\int x\sqrt{1 - x^2}\,dx$
20. $\displaystyle\int \frac{x\,dx}{\sqrt{x^2 + 4}}$
21. $\displaystyle\int \frac{x^2\,dx}{x^3 + 1}$
22. $\displaystyle\int x^2\sqrt{x^3 + 1}\,dx$
23. $\displaystyle\int \frac{\cos x\,dx}{1 + \sin x}$
24. $\displaystyle\int \frac{e^x\,dx}{1 + e^x}$

Use the corollary of Theorem 5 to find each indefinite integral.

25. $\displaystyle\int \frac{dx}{1 + 9x^2}$
26. $\displaystyle\int \frac{dx}{\sqrt{1 - 4x^2}}$

Use substitutions to find each indefinite integral. (*Hint:* First write each integrand in terms of sin and cos.)

C **27.** $\int \tan x \, dx$ **28.** $\int \cot x \, dx$

29. Prove Theorem 5, part (2).

30. Prove that $\int \frac{1}{ax + b} \, dx = \frac{1}{a} \ln |ax + b| + C, a \neq 0$. (*Hint:* Consider two cases: $ax + b > 0$ and $ax + b < 0$.)

14-3 DEFINITE INTEGRALS

Let F and G be any two antiderivatives of the function f on the interval $[a, b]$. Then

$$G(b) - G(a) = F(b) - F(a).$$

This fact follows directly from Theorem 1. For some constant C, $G(x) = F(x) + C$, and therefore

$$G(b) - G(a) = [F(b) + C] - [F(a) + C] = F(b) - F(a).$$

Thus, the number $F(b) - F(a)$ depends only on f, a, and b, and not on *which* antiderivative is used. We may therefore make this definition.

DEFINITION

Let F be any antiderivative of f on $[a, b]$. Then the **definite integral** of f from a to b, denoted by $\int_a^b f(x) \, dx$, is defined by

$$\int_a^b f(x) \, dx = F(b) - F(a).$$

The numbers a and b are called the **limits of integration**.

EXAMPLE 1 Evaluate $\int_1^3 (x^2 - 2x + 4) \, dx$.

Solution An antiderivative of $x^2 - 2x + 4$ is $\frac{1}{3}x^3 - x^2 + 4x$.

Therefore:

$$\int_1^3 (x^2 - 2x + 4) \, dx = \left(\frac{1}{3} \cdot 3^3 - 3^2 + 4 \cdot 3\right) - \left(\frac{1}{3} \cdot 1^3 - 1^2 + 4 \cdot 1\right)$$

$$= (12) - \left(\frac{10}{3}\right) = \frac{26}{3} \quad \textbf{Answer}$$

Changing x in $\int_a^b f(x) \, dx$ to any other letter does not change the value of the definite integral. For this reason, x is called a **dummy variable**.

Example 2 illustrates how to use the notation

$$\Big[F(x)\Big]_a^b = F(b) - F(a).$$

EXAMPLE 2 Evaluate **(a)** $\int_0^{\pi/2} \sin\theta\, d\theta$; **(b)** $\int_{-1}^{\sqrt{3}} \frac{dt}{1+t^2}$

Solution **a.** $\int_0^{\pi/2} \sin\theta\, d\theta = \Big[-\cos\theta\Big]_0^{\frac{\pi}{2}} = \left(-\cos\frac{\pi}{2}\right) - (-\cos 0)$

$= 0 - (-1) = 1$ **Answer**

b. $\int_{-1}^{\sqrt{3}} \frac{dt}{1+t^2} = \Big[\text{Tan}^{-1} t\Big]_{-1}^{\sqrt{3}} = \text{Tan}^{-1}\sqrt{3} - \text{Tan}^{-1}(-1)$

$= \frac{\pi}{3} - \left(-\frac{\pi}{4}\right) = \frac{7\pi}{12}$ **Answer**

EXAMPLE 3 Evaluate $\int_0^4 x\sqrt{25 - x^2}\, dx$.

Solution First use the substitution method to find an antiderivative of $x\sqrt{25 - x^2}$. Let $u = 25 - x^2$. Then $du = -2x\, dx$, and $x\, dx = -\frac{1}{2}du$.

$$\int x\sqrt{25 - x^2}\, dx = \int \sqrt{u}\left(-\frac{1}{2}du\right) = -\frac{1}{2}\int u^{\frac{1}{2}}\, du$$

$$= -\frac{1}{2}\,\frac{u^{\frac{3}{2}}}{\frac{3}{2}} = -\frac{1}{3}u^{\frac{3}{2}}$$

Therefore $\int x\sqrt{25 - x^2}\, dx = -\frac{1}{3}(25 - x^2)^{\frac{3}{2}}$.

Next find the value of the definite integral.

$$\int_0^4 x\sqrt{25 - x^2}\, dx = -\frac{1}{3}\Big[(25 - x^2)^{\frac{3}{2}}\Big]_0^4$$

$$= -\frac{1}{3}\Big[(25 - 16)^{\frac{3}{2}} - (25 - 0)^{\frac{3}{2}}\Big]$$

$$= -\frac{1}{3}\,(9^{\frac{3}{2}} - 25^{\frac{3}{2}}) = \frac{98}{3} \quad \textbf{Answer}$$

The solution of Example 3 can be shortened by changing the "x limits," 0 and 4, to the corresponding "u limits" when making the substitution: Since $u = 25 - x^2$, $u = 25$ when $x = 0$, and $u = 9$ when $x = 4$. Thus

$$\int_0^4 x\sqrt{25 - x^2}\, dx = -\frac{1}{2}\int_{25}^{9} u^{\frac{1}{2}}\, du$$

$$= -\frac{1}{2}\cdot\frac{2}{3}\Big[u^{\frac{3}{2}}\Big]_{25}^{9}$$

$$= -\frac{1}{3}\Big[u^{\frac{3}{2}}\Big]_{25}^{9} = -\frac{1}{3}(9^{\frac{3}{2}} - 25^{\frac{3}{2}}) = \frac{98}{3}$$

The procedure just illustrated is justified by the following definite-integral counterpart of Theorem 4.

THEOREM 6 $\displaystyle\int_a^b f(g(x))g'(x)\,dx = \int_{g(a)}^{g(b)} f(u)\,du$

Proof: Let F be an antiderivative of f. Then

$$\int_{g(a)}^{g(b)} f(u)\,du = \Big[F(u)\Big]_{g(a)}^{g(b)} = F(g(b)) - F(g(a)). \qquad (1)$$

In the proof on pages 561–562, we showed that $F(g(x))$ is an antiderivative of $f(g(x))g'(x)$. Therefore

$$\int_a^b f(g(x))g'(x)\,dx = \Big[F(g(x))\Big]_a^b = F(g(b)) - F(g(a)). \qquad (2)$$

Together, (1) and (2) prove the theorem. ■

EXERCISES

Evaluate the following definite integrals.

A

1. $\displaystyle\int_0^3 (2x - 1)\,dx$
2. $\displaystyle\int_0^2 (4 - x)\,dx$
3. $\displaystyle\int_1^2 (4t - 5)\,dt$
4. $\displaystyle\int_1^3 (2x + 3)\,dx$
5. $\displaystyle\int_{-3}^0 (x^2 + 2x - 2)\,dx$
6. $\displaystyle\int_{-2}^2 (s^3 - 3s^2 + 2)\,ds$
7. $\displaystyle\int_1^2 \frac{dx}{x}$
8. $\displaystyle\int_1^e \frac{dx}{x}$
9. $\displaystyle\int_e^{e^2} \frac{du}{u}$
10. $\displaystyle\int_{1/e}^e \frac{dt}{t}$
11. $\displaystyle\int_0^1 e^x\,dx$
12. $\displaystyle\int_0^{\ln 2} e^u\,du$
13. $\displaystyle\int_{-\pi/2}^{\pi/2} \cos\theta\,d\theta$
14. $\displaystyle\int_0^{\pi} \sin\alpha\,d\alpha$
15. $\displaystyle\int_0^{\ln 2} e^{-t}\,dt$

B

16. $\displaystyle\int_0^2 \frac{dx}{4 - x}$
17. $\displaystyle\int_{-2}^0 \frac{dx}{1 - x}$
18. $\displaystyle\int_0^{\pi/4} \cos 2\theta\,d\theta$
19. $\displaystyle\int_0^{\pi/6} \sin 2\theta\,d\theta$
20. $\displaystyle\int_0^{\pi/4} \tan t\,dt$
21. $\displaystyle\int_0^{\pi/2} \frac{\cos\theta}{1 + \sin\theta}\,d\theta$
22. $\displaystyle\int_0^{\pi} \sin^2\theta\,d\theta$
(*Hint:* $\sin^2\theta = \frac{1}{2}(1 - \cos 2\theta)$)
23. $\displaystyle\int_0^{\pi/2} \cos^2 t\,dt$
(*Hint:* $\cos^2 t = \frac{1}{2}(1 + \cos 2t)$)

Evaluate the following definite integrals.

24. $\int_0^{\sqrt{3}} \frac{dx}{9 + x^2}$

25. $\int_0^1 \frac{dx}{\sqrt{4 - x^2}}$

26. $\int_0^{\sqrt{3}} \frac{x\,dx}{9 + x^2}$

27. $\int_0^1 \frac{x\,dx}{\sqrt{4 - x^2}}$

28. $\int_0^4 t\sqrt{9 + t^2}\,dt$

29. $\int_0^3 x\sqrt{16 + x^2}\,dx$

Prove each statement where $a < c < b$.

30. $\int_b^a f(x)\,dx = -\int_a^b f(x)\,dx.$

31. $\int_a^b f(x)\,dx = \int_a^c f(x)\,dx + \int_c^b f(x)\,dx.$

In Exercises 32 and 33, use the fact that every even (odd) continuous function has an odd (even) antiderivative. Prove each statement.

C 32. Let f be an odd continuous function. $\int_{-a}^{a} f(x)\,dx = 0.$

33. Let f be an even continuous function. $\int_{-a}^{a} f(x)\,dx = 2\int_0^a f(x)\,dx.$

Applications of Integration

14-4 AREAS OF PLANE REGIONS

Let f be a continuous function that is nonnegative on the interval $[a, b]$. By the **region under the graph** of f from a to b we mean the set,

$$\{(x, y)\colon a \le x \le b,\ 0 \le y \le f(x)\},$$

shown shaded in Figure a. We shall assume that every such region has a definite area, A, which we shall call the **area under the graph** from a to b. Our immediate objective is to find A, given f, a, and b.

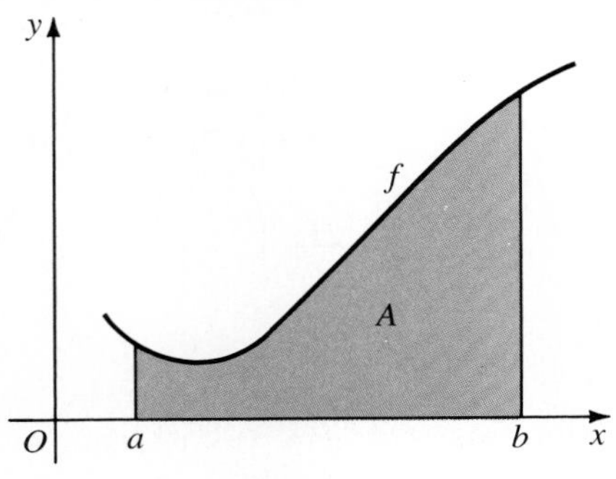

Figure a

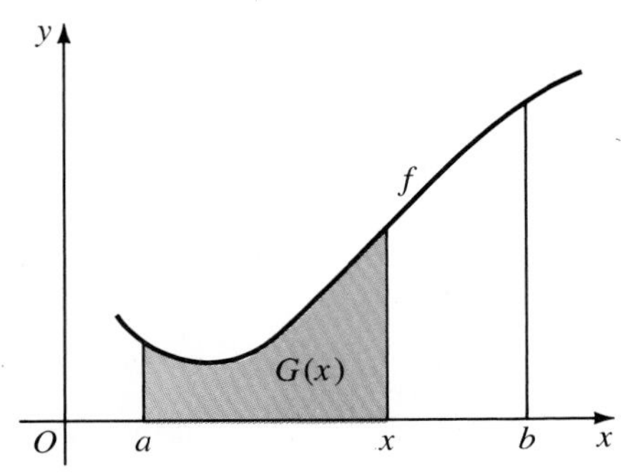

Figure b

We start by introducing the "area function," G, where $G(x)$ is the area under the graph of f from a to x (Figure b). Notice that $G(a) = 0$ and that $G(b) = A$, the area we wish to find.

We next show that G is an antiderivative of f, that is,

$$G'(x) = f(x). \tag{1}$$

The area of the shaded strip in Figure c, shown in more detail in Figure d, is given by $G(x + h) - G(x)$ as well as by $f(\bar{x}) \cdot h$ for some $\bar{x}$ between x and $x + h$.

$$G(x + h) - G(x) = f(\bar{x}) \cdot h$$

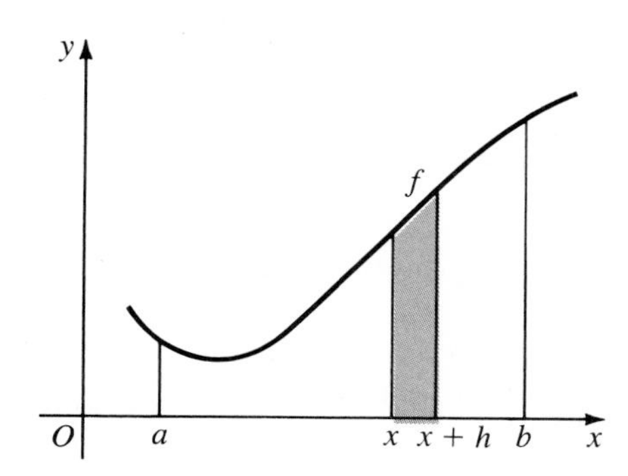

Figure c

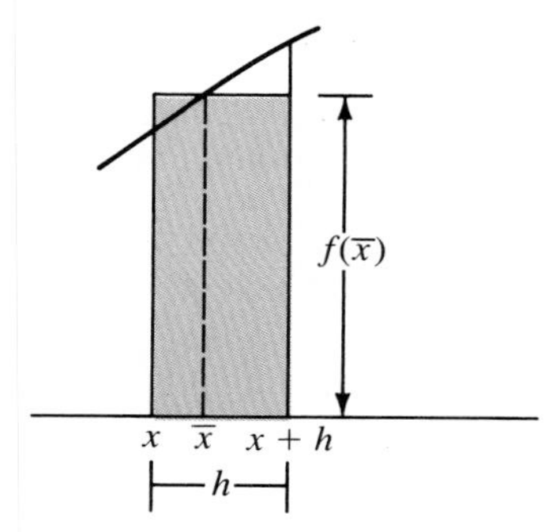

Figure d

Hence
$$\frac{G(x + h) - G(x)}{h} = f(\bar{x}).$$

As $h \to 0$, $\bar{x} \to x$.

Therefore
$$G'(x) = \lim_{h \to 0} \frac{G(x + h) - G(x)}{h}$$
$$= \lim_{\bar{x} \to x} f(\bar{x}).$$

Since f is continuous, $\lim_{\bar{x} \to x} f(\bar{x}) = f(x)$. Therefore $G'(x) = f(x)$, and G is an antiderivative of f. This proves formula (1).

Since G is an antiderivative of f,

$$\begin{aligned} \int_a^b f(x)\, dx &= G(b) - G(a) \\ &= G(b) && \text{since } G(a) = 0 \\ &= A && \text{since } G(b) = A \end{aligned}$$

This proves the following result.

THEOREM 7 Let the function f be continuous and nonnegative on $[a, b]$. The area under the graph of f from a to b is $\int_a^b f(x)\, dx$.

Note that in applying this theorem, we may use *any* antiderivative of f to evaluate the integral, not just the "area function" G.

EXAMPLE 1 Find the area under the graph of $f(x) = \sin x$ from 0 to $\frac{\pi}{3}$.

Solution

$$\text{Area} = \int_0^{\pi/3} \sin x\, dx$$

$$= \Big[-\cos x\Big]_0^{\pi/3}$$

$$= \left(-\cos \frac{\pi}{3}\right) - (-\cos 0)$$

$$= \left(-\frac{1}{2}\right) - (-1) = \frac{1}{2}$$

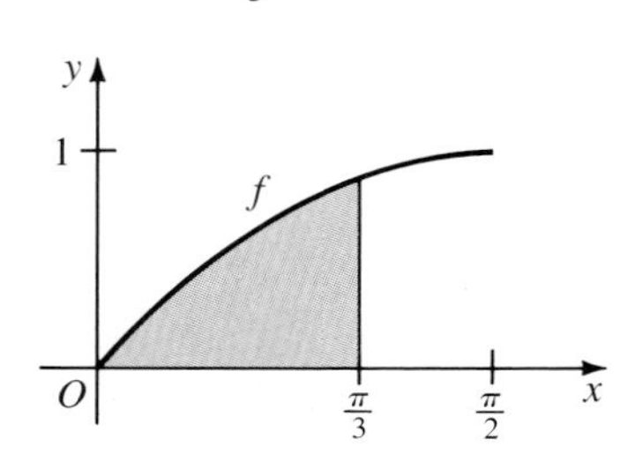

Thus the area is $\frac{1}{2}$ square unit. **Answer**

We can also use definite integrals to find the area of a region bounded by the graphs of two functions. To do so, we use two integrals and their difference.

EXAMPLE 2 Find the area of the region bounded by the graphs of $f(x) = 5 - x^2$ and $g(x) = (x - 1)^2$.

Solution

1. The sketch shows that the region lies below the parabola $y = 5 - x^2$ and above the parabola $y = (x - 1)^2$.
2. Find where the curves intersect. (The x-coordinates of these points will become the needed limits of integration.) Solve the equation
$$(x - 1)^2 = 5 - x^2$$
to obtain $x = -1$ and $x = 2$.

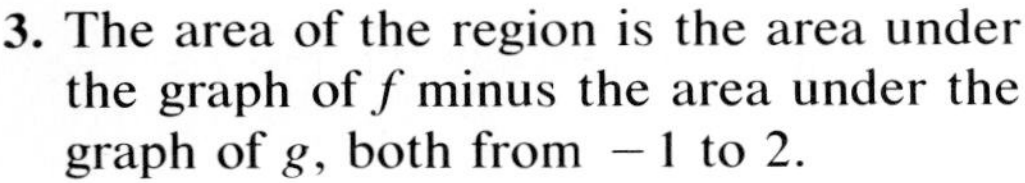

3. The area of the region is the area under the graph of f minus the area under the graph of g, both from -1 to 2.

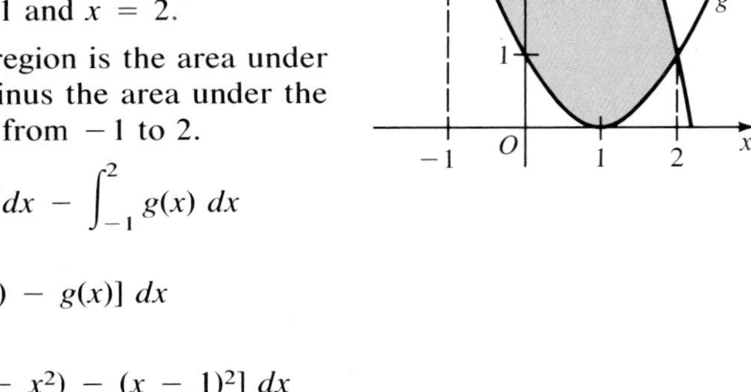

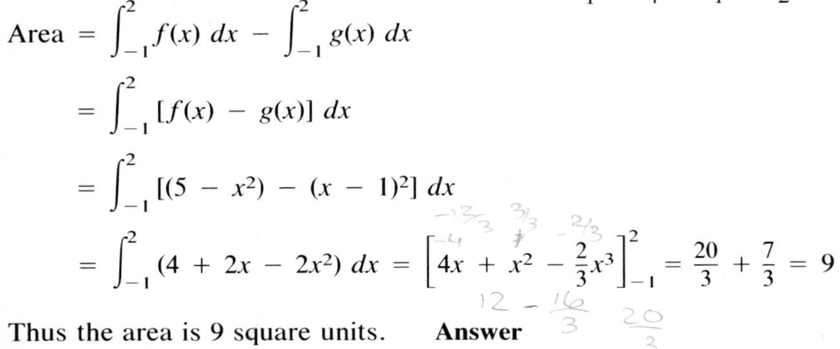

$$\text{Area} = \int_{-1}^{2} f(x)\, dx - \int_{-1}^{2} g(x)\, dx$$

$$= \int_{-1}^{2} [f(x) - g(x)]\, dx$$

$$= \int_{-1}^{2} [(5 - x^2) - (x - 1)^2]\, dx$$

$$= \int_{-1}^{2} (4 + 2x - 2x^2)\, dx = \left[4x + x^2 - \frac{2}{3}x^3\right]_{-1}^{2} = \frac{20}{3} + \frac{7}{3} = 9$$

Thus the area is 9 square units. **Answer**

The method used in solving Example 2 can be adapted to prove Theorem 8.

THEOREM 8 Let the functions f and g be continuous on $[a, b]$ with $f(x) \geq g(x)$ on $[a, b]$. Then the area between the graphs of f and g from a to b is

$$\int_a^b [f(x) - g(x)]\, dx.$$

Note that in Theorem 8 we remove the restriction that f and g be nonnegative on $[a, b]$. You are asked to prove the theorem in Exercise 31. In applying Theorem 8 it may be helpful to indicate an "element of area," a strip extending from one part of the boundary to the other. This has been done in the sketches for Examples 3 and 4.

EXAMPLE 3 Find the area of the region bounded by the graphs of $f(x) = x^3 + x^2$ and $g(x) = 2x$.

Solution **1.** The region along with elements of area is sketched in the figure below.

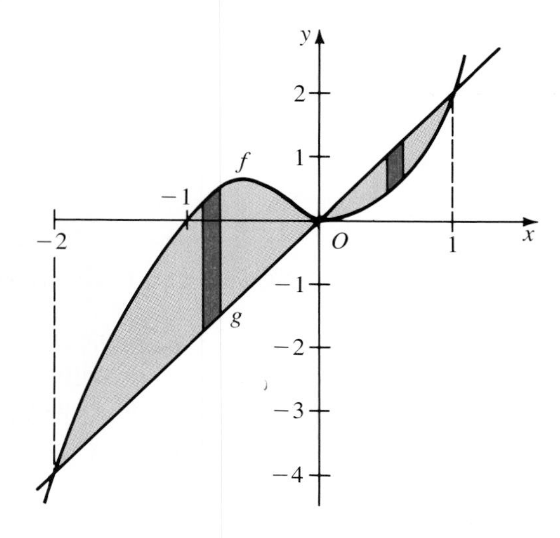

2. Find the x-coordinates of the points where the curve and the line intersect. Solve $x^3 + x^2 = 2x$ and obtain $x = -2$, 0, and 1.

3. Apply Theorem 8. Use the fact that $f(x) \geq g(x)$ on $[-2, 0]$ and that $g(x) \geq f(x)$ on $[0, 1]$.

$$\text{Area} = \int_{-2}^{0} [(x^3 + x^2) - (2x)]\, dx + \int_0^1 [(2x) - (x^3 + x^2)]\, dx$$

$$= \frac{8}{3} + \frac{5}{12} = \frac{37}{12}$$

Thus the area is $\frac{37}{12}$ square units. **Answer**

EXAMPLE 4 Find the area of the region bounded by the graphs of $y^2 = x + 1$ and $x - y = 1$.

Solution The region is sketched in the figure below. It is easier to use a horizontal element of area and find the area of the region by integrating a function of y. To find the limits of integration, simultaneously solve $x = y^2 - 1$ and $x = y + 1$ to obtain $y = -1$ and $y = 2$.

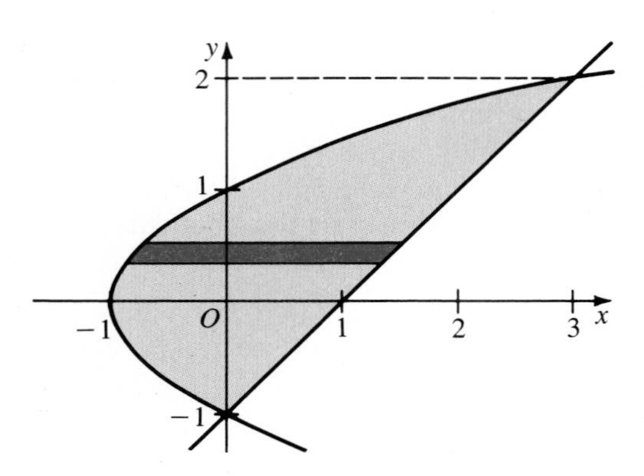

$$\text{Area} = \int_{-1}^{2} [(y + 1) - (y^2 - 1)]\, dy$$

$$= \int_{-1}^{2} [2 + y - y^2]\, dy$$

$$= \left[2y + \frac{y^2}{2} - \frac{y^3}{3}\right]_{-1}^{2} = \left[4 + 2 - \frac{8}{3}\right] - \left[-2 + \frac{1}{2} + \frac{1}{3}\right] = \frac{9}{2}.$$

Thus the area is $\frac{9}{2}$ square units. **Answer**

In solving area problems, sketch the graphs of the equations involved. Use the vertical- or horizontal-line test mentally to see if the equations define functions of x, as in Example 3, or of y, as in Example 4. Find the appropriate x or y limits of integration.

EXERCISES

In Exercises 1–8 find the area under the graph of f from a to b.

A

1. $f(x) = x + 1;\ a = 0,\ b = 3$

2. $f(x) = 4 - x;\ a = -1,\ b = 2$

3. $f(x) = 4 - x^2;\ a = -2,\ b = 2$

4. $f(x) = 4x - x^2;\ a = 0,\ b = 4$

5. $f(x) = \cos x;\ a = -\frac{\pi}{2},\ b = \frac{\pi}{2}$

6. $f(x) = \sin x;\ a = \frac{\pi}{6},\ b = \frac{\pi}{3}$

7. $f(x) = e^{2x};\ a = 0,\ b = 1$

8. $f(x) = e^{x};\ a = -1,\ b = 1$

Each of Exercises 9–28 describes a region in the plane. Sketch the region and find its area.

9. Bounded by the x-axis and the parabola $y = 4 - x^2$

10. Bounded by the x-axis and the parabola $y = 4x - x^2$

11. Bounded by the curve $y = \frac{1}{1 + x^2}$ and the lines $x = 0$, $x = 1$, and $y = 0$

12. Bounded by the curve $y = \sqrt{x}$ and the lines $x = 4$ and $y = 0$

13. Bounded by the curve $y = e^x$ and the lines $x = 0$ and $y = e$

14. Lying in the first quadrant and bounded by the curves $y = \sin x$, $y = 1$, and $x = 0$

B **15.** Bounded by the parabola $y = x^2$ and the line $y = x + 2$

16. Bounded by the parabola $y = 1 - x^2$ and the line $x + y + 1 = 0$

17. Bounded by the parabola $x = y^2 - 2y$ and the line $y = x$

18. Bounded by the parabola $y^2 = x$ and the line $x + y = 2$

19. Bounded by the parabolas $3y = x^2$ and $y = 4x - x^2$

20. Bounded by the parabolas $y = x^2$ and $y + 4x + x^2 = 0$

21. Bounded by the hyperbola $xy = 4$ and the line $x + y = 5$

22. Bounded by the curves $y = \frac{1}{1 + x^2}$ and $y = \frac{x^2}{2}$

23. Bounded by the curves $y = x^3$ and $y = 2x - x^2$ (The region consists of two pieces.)

24. Bounded by the curves $y = x^3 - 2x^2$ and $y = 2x$ (The region consists of two pieces.)

C **25.** Bounded by the curve $x^{\frac{1}{2}} + y^{\frac{1}{2}} = 1$ and the coordinate axes (The curve is an arc of the parabola having focus $(\frac{1}{2}, \frac{1}{2})$ and directrix $x + y = 0$.)

26. Lying in the first quadrant and bounded by the parabolas $y = x^2 - 1$ and $2y^2 = 9x$ (The parabolas intersect at $(2, 3)$.)

27. Lying in the first quadrant and bounded by the curves $y = \sin x$ and $y = \cos x$ and the line $x = 2\pi$.

28. Lying in the first quadrant and bounded by the curves $y = e^x$ and $y = x^e$ (*Hint:* Find a solution of $e^x = x^e$ by inspection.)

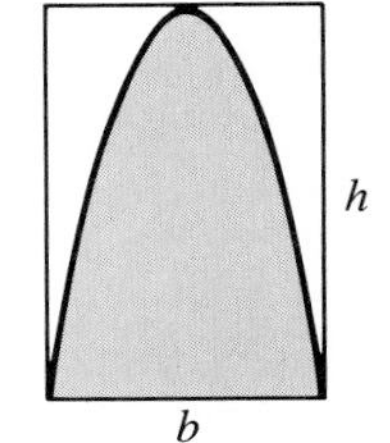

Exercise 29

29. Show that the area of the parabolic segment is two-thirds the area of the rectangle in which it is inscribed.

30. Evaluate $\int_0^a \sqrt{a^2 - x^2}\, dx$ by interpreting the integral as an area.

31. Assuming that Theorem 8 is true for nonnegative functions f and g, prove the theorem for the case where one or both of f and g may be negative. (*Hint:* Use the fact that there is a constant k such that $0 \leq g(x) + k$ for all $x \in [a, b]$.)

32. Show that the formula $G(x + h) - G(x) = f(\bar{x}) \cdot h$ holds when $h < 0$ by justifying the following steps.

a. The area of the shaded strip in Figure d, page 569, is $G(x) - G(x + h)$.

b. The rectangle in Figure d has width $-h$ and therefore has area $f(\bar{x}) \cdot (-h)$ for some $\bar{x}$ between $x + h$ and x.

c. Since the areas in a and b are equal, $G(x + h) - G(x) = f(\bar{x}) \cdot h$.

14-5 VOLUMES OF SOLIDS

Definite integrals can be used to find many geometric and physical quantities besides areas of plane regions. As a preliminary step, we reexamine the area problem from a slightly different point of view.

Given a nonnegative function f, continuous on $[a, b]$, let A be the area of the region under the graph of f from a to b. Let us slice the region into n vertical strips, each of width $\Delta x = \frac{b - a}{n}$ (Δx is read "delta x"). The area of the ith strip is approximated by the area of the shaded rectangle, namely $f(x_i) \cdot \Delta x$, where x_i is in the ith subinterval of $[a, b]$.

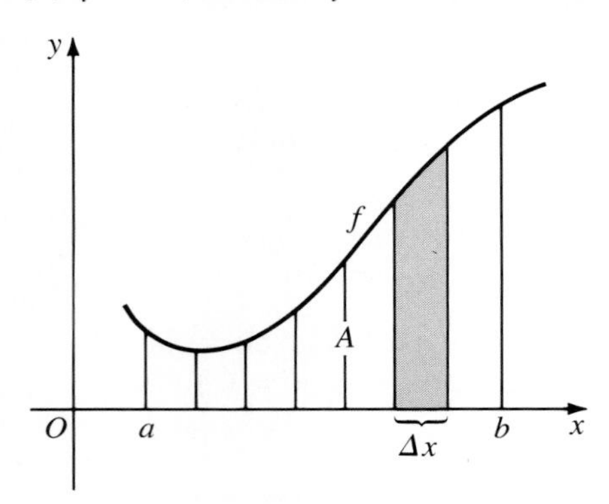

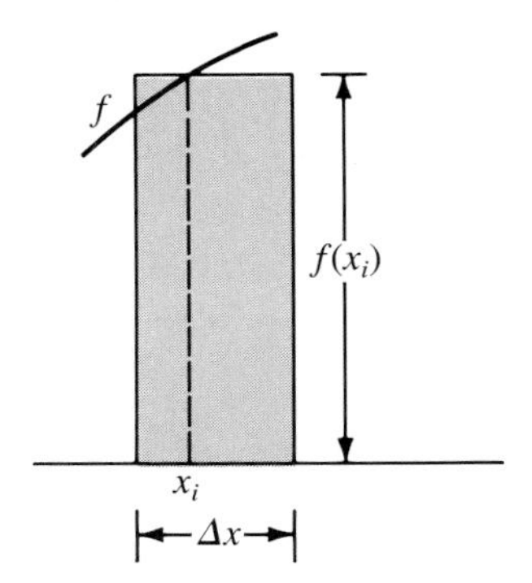

Thus A is approximated by

$$f(x_1)\,\Delta x + f(x_2)\,\Delta x + \cdots + f(x_n)\,\Delta x, \quad \text{or} \quad \sum_{i=1}^{n} f(x_i)\,\Delta x.$$

The narrower the strips are, that is, the larger n is, the better the approximation. It therefore seems reasonable that

$$A = \lim_{n\to\infty} \sum_{i=1}^{n} f(x_i)\,\Delta x.$$

From Section 14-4, $A = \int_a^b f(x)\,dx$, and therefore

$$\lim_{n\to\infty} \sum_{i=1}^{n} f(x_i)\,\Delta x = \int_a^b f(x)\,dx. \tag{1}$$

Notice that (1) makes no reference to areas (although areas were used in its derivation). We can therefore use a definite integral to evaluate *any* quantity that can be expressed as the limit of a sum of the type appearing in (1).

EXAMPLE 1 The region under the curve $y = \frac{1}{x}$ from $x = 1$ to $x = 3$ is revolved about the x-axis. Find the volume V of the solid swept out.

Solution The following sketches show the planar region, the solid obtained by revolving the region about the x-axis, and a circular disk.

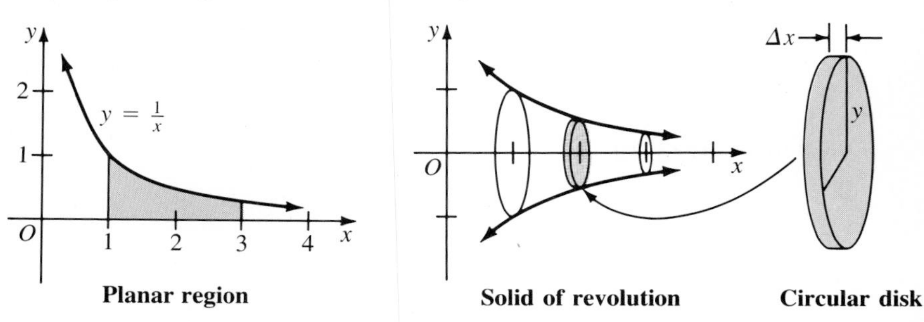

Planar region **Solid of revolution** **Circular disk**

The circular disk has thickness Δx, and radius y. Its volume, therefore, is $\pi y^2 \, \Delta x$, and the total volume V is approximated by a sum of the form $\Sigma \, \pi y^2 \, \Delta x$. Thus

$$V = \lim \Sigma \, \pi y^2 \, \Delta x = \int_1^3 \pi y^2 \, dx.$$

Since $y = \frac{1}{x}$,

$$V = \pi \int_1^3 \frac{1}{x^2} \, dx = \pi \left[-\frac{1}{x} \right]_1^3$$

$$= \pi \left[\left(-\frac{1}{3} \right) - (-1) \right] = \frac{2\pi}{3}.$$

Thus the volume is $\frac{2\pi}{3}$ cubic units. **Answer**

The method used in Example 1 can be extended to solids that are not solids of revolution. If we can find a formula, $A = A(x)$, for the areas of cross sections of the solid perpendicular to some x-axis, then the volume V of the solid is given by

$$V = \int_a^b A(x) \, dx, \qquad (2)$$

where the limits a and b are determined by the extent of the solid. The figures at the top of the next page illustrate such a solid and a cross section from

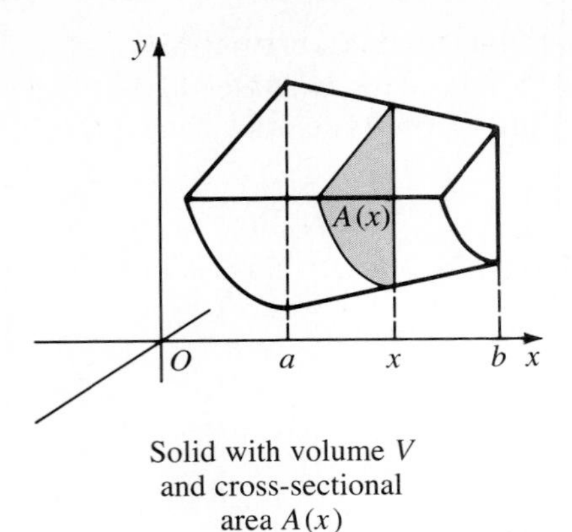

Solid with volume V and cross-sectional area $A(x)$

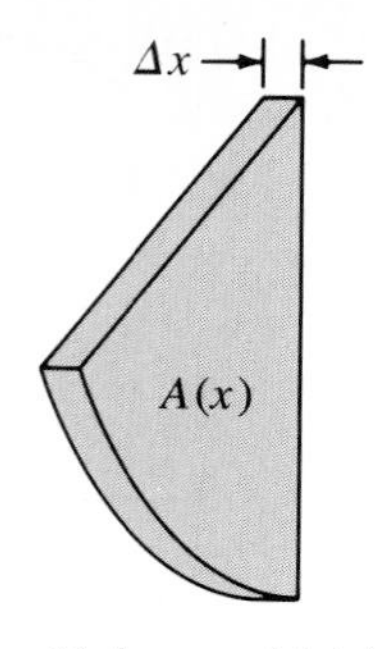

Volume $= A(x)\Delta x$

Solid with base area $A(x)$ and thickness Δx

which an area formula, $A = A(x)$, can be obtained. The use of formula (2),

$$V = \int_a^b A(x)\,dx,$$

is illustrated in the following example. Note in the example that the area formula for an equilateral triangle is used to get $A = A(x)$.

EXAMPLE 2 A solid has a circular base of fixed radius a. Cross sections perpendicular to a fixed diameter are equilateral triangles. Find the volume of the solid.

Solution Introduce an xy-coordinate system in the plane of the base with the origin at the center and the x-axis along the fixed diameter. The following figures show half of the solid and an equilateral cross section.

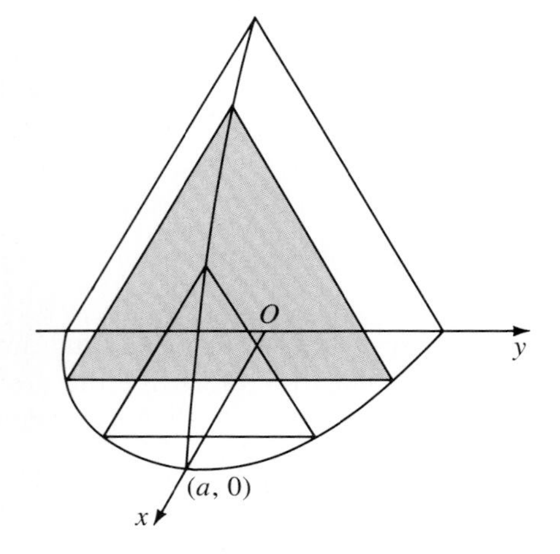

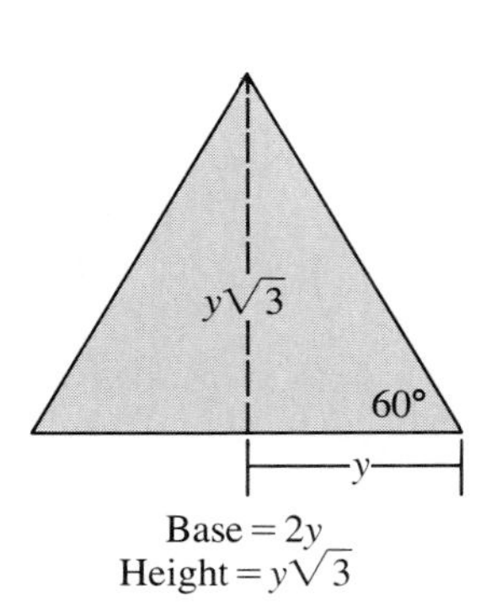

Base $= 2y$
Height $= y\sqrt{3}$

The cross section through the point (x, y) has area

$$A(x) = \frac{1}{2} \cdot 2y \cdot \sqrt{3}y = \sqrt{3}y^2.$$

Since an equation of the circle is $x^2 + y^2 = a^2$, $y^2 = a^2 - x^2$. Thus the area function is given by $A(x) = \sqrt{3}(a^2 - x^2)$. Due to the symmetric shape of the solid, the total volume is twice the volume of the solid in the sketch. Use formula (2). Thus

$$V = 2\int_0^a A(x)\,dx = 2\sqrt{3}\int_0^a (a^2 - x^2)\,dx = 2\sqrt{3}\left[a^2 x - \frac{x^3}{3}\right]_0^a$$

$$= 2\sqrt{3}\left[\left(a^3 - \frac{a^3}{3}\right) - (0)\right] = \frac{4\sqrt{3}}{3}a^3.$$

Thus the volume is $\frac{4\sqrt{3}}{3}a^3$ cubic units. **Answer**

EXAMPLE 3 The region bounded by the parabola P: $y = x^2$, the line L: $2x + y = 8$, and the x-axis is revolved about the y-axis. Find the volume of the solid swept out.

Solution Sketch the planar region and the solid swept out.

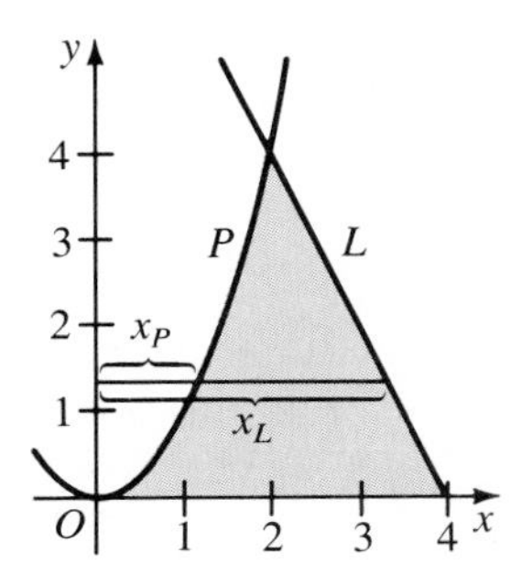

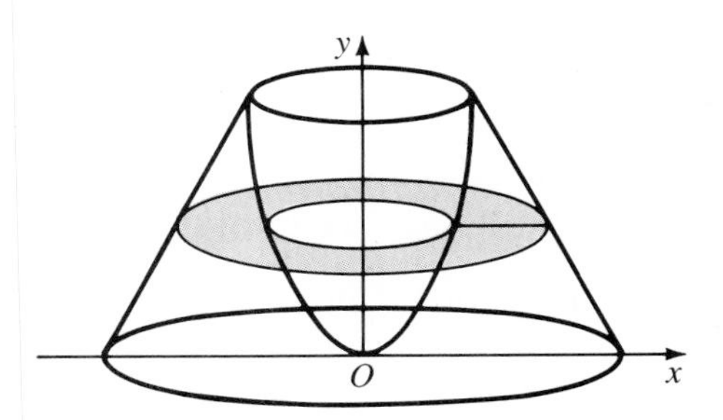

Find the area of a typical cross section. Each cross section perpendicular to the y-axis is an annulus (a washer-shaped figure) having inner radius x_P and outer radius x_L, where x_P and x_L are as indicated below. The cross-sectional area is therefore

$$A(y) = \pi x_L^2 - \pi x_P^2 = \pi(x_L^2 - x_P^2).$$

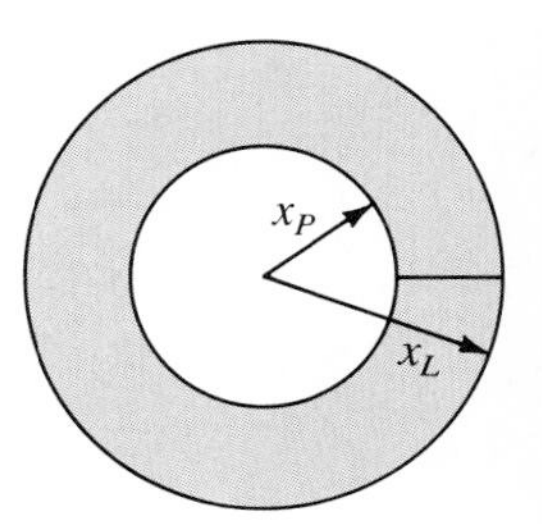

Write x_L and x_P in terms of y. From L: $2x + y = 8$ and P: $y = x^2$,

$$x_L = 4 - \frac{1}{2}y \quad \text{and} \quad x_P = \sqrt{y}.$$

Thus:

$$A(y) = \pi\left[\left(4 - \frac{1}{2}y\right)^2 - \left(\sqrt{y}\right)^2\right]$$

$$= \pi\left(16 - 5y + \frac{y^2}{4}\right)$$

Solution continued on following page

Find the limits of integration and integrate $A(y)$. Since L and P intersect at (2, 4), the upper limit is $y = 4$.

$$\begin{aligned}\text{Volume} &= \pi \int_0^4 \left(16 - 5y + \frac{y^2}{4}\right) dy \\ &= \pi\left[16y - \frac{5y^2}{2} + \frac{y^3}{12}\right]_0^4 \\ &= \frac{88\pi}{3}.\end{aligned}$$

Thus the volume is $\frac{88\pi}{3}$ cubic units. **Answer**

Note: In many books a discussion of area and limits of sums, called *Riemann sums*, motivates the *definition* of the definite integral to be

$$\int_a^b f(x)\,dx = \lim_{n\to\infty} \sum_{i=1}^{n} f(x_i)\,\Delta x.$$

It is then *shown* that $\lim_{n\to\infty} \sum_{i=1}^{n} f(x_i)\,\Delta x = F(b) - F(a)$ where F is any antiderivative of f. The statement

$$\int_a^b f(x)\,dx = F(b) - F(a)$$

is then a theorem, called the *fundamental theorem of calculus.*

EXERCISES

Find the volume of the solid swept out when the region described is revolved about **(a)** the x-axis; **(b)** the y-axis.

A

1. Bounded by the lines $2x + y = 4$, $y = 0$, and $x = 0$
2. Bounded by the lines $2x + 3y = 6$, $y = 0$, and $x = 0$
3. Lying in the first quadrant and bounded by the curves $y = 4 - x^2$, $y = 0$, and $x = 0$
4. Bounded by the curves $y = \sqrt{4 - x}$, $y = 0$, and $x = 0$
5. Bounded by the lines $y = x$, $y = 0$, and $x = 1$
6. Bounded by the lines $x + y = 1$, $x = 1$, and $y = 1$
7. Bounded by the curves $y = x^2$, $y = 0$ and $x = 2$
8. Bounded by the curves $y = x^2$, $x = 0$, and $y = 4$
9. Bounded by the parabola $y^2 = 4 - x$ and the line $x + 2y = 4$
10. Bounded by the parabola $y = x^2$ and the line $y = 2x$
11. Bounded by the hyperbola $xy = 4$ and the line $x + y = 5$

12. Lying in the first quadrant and bounded by the hyperbola $x^2 - y^2 = 9$ and the lines $y = 0$ and $x = 5$

13. The smaller of the two regions into which the line $x + y = a$ divides the disk $x^2 + y^2 \leq a^2$

14. The first-quadrant part of the ellipse $\frac{x^2}{a^2} + \frac{y^2}{b^2} = 1$

In Exercises 15–20, find the volume of the solid swept out when the region described is revolved about the specified line.

B

15. The region under the curve $y = e^x$ from 0 to 1; about the x-axis

16. The region under the curve $y = \sin x$ from 0 to π; about the x-axis (*Hint:* Recall that $\sin^2 x = \frac{1}{2}(1 - \cos 2x)$.)

17. The region of Exercise 7; about the line $x = 2$

18. The region of Exercise 8; about the line $y = 4$

19. The region of Exercise 9; about the line $y = 2$

20. The region of Exercise 10; about the line $x = 2$

A solid has a circular base of radius a. Find its volume if cross sections perpendicular to a fixed diameter of this base are

21. squares. **22.** semicircles.

C

23. The axes of two circular cylinders each of radius a intersect at right angles. Find the volume of the solid region inside both cylinders. (*Hint:* Take the axes of the cylinders to be the x- and y-axes. Find the volume in the first octant by using cross sections perpendicular to the z-axis.)

24. When the disk $x^2 + y^2 \leq a^2$ is revolved about the line $x = b$ $(0 < a < b)$, a doughnut-shaped solid called a *torus* is swept out. Show that its volume is $2\pi^2 a^2 b$. (*Hint:* Use washer-shaped cross sections having inner and outer radii $b - \sqrt{a^2 - x^2}$ and $b + \sqrt{a^2 - x^2}$, respectively. You will need the fact that $\int_0^a \sqrt{a^2 - x^2}\, dx = \frac{\pi a^2}{4}$. (See Exercise 30, page 573.))

COMPUTER EXERCISES

1. Let $f(x) = 2x$, $a = 0$, $b = 5$, and $n = 20$. Write a computer program to find $\sum_{i=1}^{n} f(x_i)\, \Delta x$ where $\Delta x = \frac{b - a}{n}$ and x_i is the midpoint of $[a + (i - 1)\, \Delta x,\ a + i\, \Delta x]$.

2. Modify and run the program for the following sets of information.

a. $f(x) = x$; $a = 0$, $b = 6$, $n = 100$
b. $f(x) = x^2$; $a = 2$, $b = 5$, $n = 100$
c. $f(x) = \sin x$; $a = 0$, $b = \pi$, $n = 100$

14-6 WORK AND ENERGY

Suppose that a constant force of magnitude F directed along the s-axis displaces a particle from a to b. The **work** done by the force, and the **energy** it expends, are given by

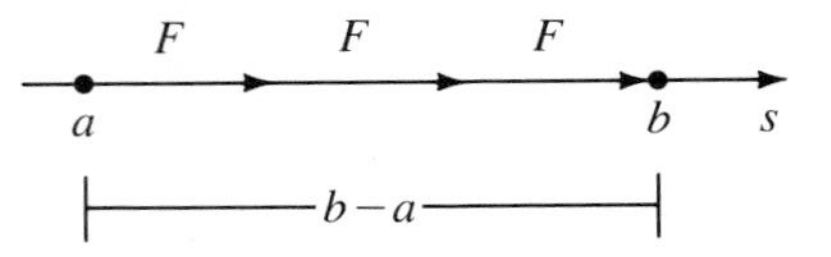

$$F \cdot (\text{displacement}) = F(b - a).$$

Refer to the discussion of work and energy in Section 11–6.

Now suppose that F is no longer constant but is a continuous function of position: $F = F(s)$. We reason that over a small displacement Δs, $F(s)$ is almost constant, and so the work done is about $F(s) \cdot \Delta s$ for some s. Thus, the total work, W, is approximated by a sum

$$W \approx F(s_1)\,\Delta s + F(s_2)\,\Delta s + \cdots + F(s_n)\,\Delta s = \sum_{i=1}^{n} F(s_i)\,\Delta s.$$

Let $\Delta s \to 0$.

$$W = \lim_{n\to\infty} \sum_{i=1}^{n} F(s_i)\,\Delta s = \int_a^b F(s)\,ds$$

In Example 1 we shall use the following facts.

1. When an elastic object, such as a coil spring or a rubber band, is subjected to a distorting force, or **stress**, a deformation, or **strain**, results.
2. In such cases Hooke's law holds.

   Stress (F) is proportional to strain (s).

   $F = ks, \quad k$ a constant
3. One unit for measuring work done is the **foot-pound**, the energy needed to lift a one-pound object one foot.

EXAMPLE 1 A coil spring has natural length 4 feet, and a force of 60 pounds stretches it to length 6 feet. How much work is necessary to stretch it from length 7 feet to length 8 feet?

Solution Let s be the elongation (strain) in the spring when stress force F is applied. When $F = 60$, $s = 6 - 4 = 2$. Hooke's law, $F = ks$, becomes $60 = k \cdot 2$. Thus $k = 30$ and

$$F(s) = 30s.$$

Next find the limits of integration and integrate $F(s) = 30s$ between them. When the spring's length is 7 feet, $s = 3$, and when its length is 8 feet, $s = 4$. The work W done is

$$W = \int_3^4 F(s)\,ds = \int_3^4 30s\,ds$$

$$= \Big[15s^2\Big]_3^4 = 15 \cdot 4^2 - 15 \cdot 3^2 = 105.$$

Therefore the work done is 105 foot-pounds. **Answer**

In Example 2 water is to be pumped out of a reservoir. The following facts are used.

1. Density $= \dfrac{\text{mass}}{\text{volume}}$, and the density δ of water is $\delta = 1000\ \text{kg/m}^3$.
2. Newton's second law states that force = mass × acceleration. The acceleration g due to gravity is $g = 9.8\ \text{m/s}^2$.
3. In the metric system, the basic unit for measuring work and energy is the **joule** (J), one newton-meter.

EXAMPLE 2 A reservoir in the shape of an inverted cone 20 m across the top and 5 m deep is full of water. How much energy is needed to pump all the water to the top of the reservoir?

Solution The sketch shows a horizontal slab of water h meters above the bottom of the reservoir and Δh meters thick.

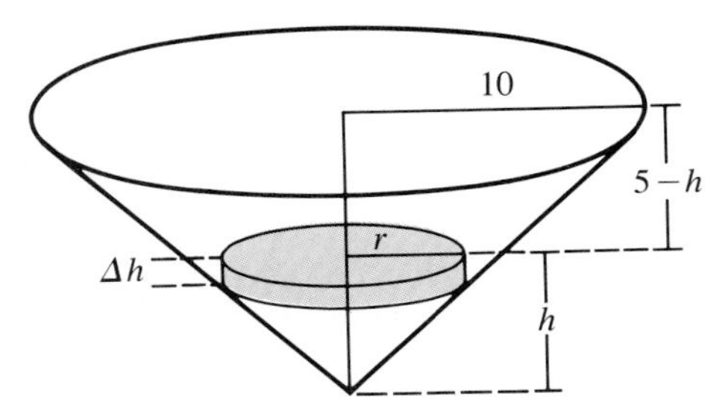

First find an expression for the energy needed to raise the slab to the top of the reservoir. If the radius of the slab is r meters,

Volume $= \pi r^2\, \Delta h;$	m^3
Mass $= \delta\pi r^2\, \Delta h.$	kg
Force $= g \cdot \delta\pi r^2\, \Delta h$	newtons
Energy $= g\delta\pi r^2\, \Delta h \cdot (5 - h)$	joules

The energy needed to raise all such slabs is the integral of the energy function from $h = 0$ to $h = 5$.

$$E = \pi\delta g \int_0^5 r^2(5 - h)\, dh$$

To evaluate the integral, write r in terms of h. From similar triangles,

$$\frac{r}{10} = \frac{h}{5}, \quad \text{or} \quad r = 2h.$$

Thus, $E = \pi\delta g \displaystyle\int_0^5 (2h)^2(5 - h)\, dh = 4\pi\delta g \int_0^5 (5h^2 - h^3)\, dh$

$$= 4\pi\delta g \left[\frac{5}{3}h^3 - \frac{1}{4}h^4\right]_0^5 = \frac{625}{3}\pi\delta g$$

Thus the energy needed is 6,414,085 J. **Answer**

Both the foot-pound and the joule (about 0.738 foot-pound) are small units. Thus the **kilojoule** (kJ), 10^3 J, and **megajoule** (MJ), 10^6 J, are more widely used. Because electricity is such a common source of energy, the **kilowatt-hour** (kW·h), 3.6 MJ, is a popular unit. A frequently used unit for measuring energy derived from foods is the Calorie ($\approx$ 4.2 kJ $\approx$ 3100 foot-pounds). In Example 2, we write

$$E \approx 6.414 \text{ MJ} \approx 1.78 \text{ kW·h}.$$

EXERCISES

Note: Exercises 1–4 require no calculus.

A

1. The mass of a loaded helicopter is 1500 kg. How much energy does its engine expend in ascending vertically for 120 m?
2. The mass of an elevator and its passengers is 3000 kg. How many kilowatt-hours of energy are used in raising it 50 m?
3. How many kilowatt-hours of energy does a crane use in lifting 25 buckets of concrete, each weighing 3000 pounds, to a point 60 feet above street level?
4. How many Calories does a person weighing 150 pounds use up in climbing a hill 1000 feet high? (Ignore the horizontal component of the person's displacement.)
5. How much work is done in stretching the spring of Example 1 **(a)** from its natural length to length 6 feet? **(b)** from length 6 feet to length 7 feet?
6. A force of 800 kN compresses the spring on a railroad yard bumping post by 2 cm. How much energy does the spring absorb by being compressed 10 cm? (*Hint:* Use Hooke's law.)

In Exercises 7 and 8 the J. and M. Power Company wishes to pump water from a cylindrical tank 6 m in diameter and 5 m high.

7. How much energy is needed to pump all the water to the top of the tank?
8. How much energy is needed to pump all the water to a point 3 m above the top of the tank?

B

9. When a steel rod is subjected to a compressive force of 1200 kN, it is shortened by 3 mm. How much work will shorten it by another 3 mm?
10. One hundred foot-pounds of energy is required to stretch a spring from its natural length of 24 inches to 26 inches. How much more would the spring be stretched if 300 foot-pounds more energy were used?

In Exercises 11 and 12 a hemispherical bowl of radius 2 m is full of water.

11. How much energy is needed to pump all of the water to the top?
12. How much energy is needed to pump all of the water to a point 3 m above the top of the bowl?

13. A 100-foot cable weighing 4 pounds per foot hangs from a windlass. How much work is done in winding up all the cable? (*Hint*: The weight of a small portion of the cable Δx feet long is $4\Delta x$, and if this portion is x feet from the top, the work done in raising it to the top is $x \cdot 4\Delta x$ foot-pounds.)

14. In Exercise 13, how much of the cable is still out when half of the work to wind it all in has been done?

15. Work Exercise 13 given that a weightless bucket filled with 500 pounds of sand is attached to the lower end of the cable.

C **16.** Work Exercise 15 given that the cable is reeled in at a uniform rate and that sand is leaking out of the bucket at a constant rate in such a way that half of it is gone when the bucket reaches the top.

17. The repulsive force between two electrically charged particles is inversely proportional to the square of the distance between them and is 3 N when the particles are 10 cm apart. How much work is needed to decrease the distance between them from 10 cm to 4 cm?

18. The force F of gravity on an object above Earth's surface is inversely proportional to the square of the distance x from the center of Earth to the object. How many kilowatt-hours of energy are needed to lift a 5-ton payload to a point 200 miles above the surface of Earth?

$\Big($*Hints*: Evaluate k in the formula $F = \dfrac{k}{x^2}$ by using the fact that when $x = 4000$ miles (the radius of Earth), $F = 5$ tons. Use the fact that 1 mile-ton $= 3.977$ kW·h.$\Big)$

COMPUTER EXERCISES

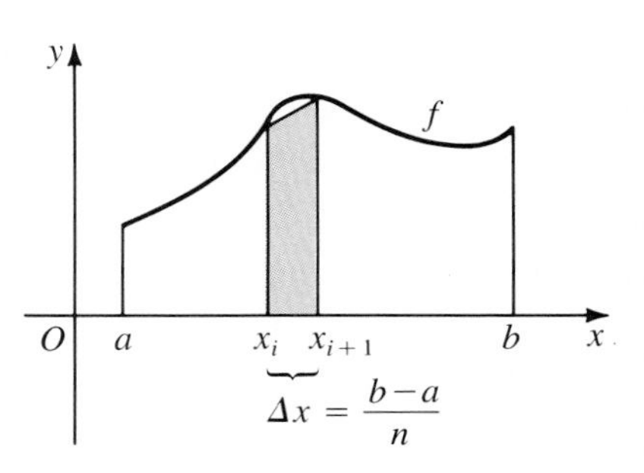

1. Use the fact that the area of the shaded trapezoid is

$$\tfrac{1}{2}(\Delta x)(f(x_i) + f(x_{i+1}))$$

to write a computer program to approximate the area under the graph of $f(x) = x^2$ from $a = 0$ to $b = 4$. Use $n = 100$.

2. Modify and run the program in Exercise 1 for the following sets of information.

a. $f(x) = \sqrt{1 - x^2}$; $a = 0$, $b = 1$, $n = 100$

b. $f(x) = x^3 + x^2$; $a = 1$, $b = 2$, $n = 100$

c. $f(x) = \dfrac{1}{x}$; $a = 1$, $b = 2$, $n = 100$

d. $f(x) = \sin x$; $a = 0$, $b = \pi$, $n = 100$

Chapter Summary

1. If $F'(x) = f(x)$ in an interval I, then F is an *antiderivative* of f on I, and the *indefinite integral* of f is

$$\int f(x)\,dx = F(x) + C,$$

where C is an arbitrary constant. In the notation $\int f(x)\,dx$, $\int$ is the *integral sign* and $f(x)$ is the *integrand.*

2. Some functions can be integrated by simply reversing derivative formulas. In case the integrand is of the form $f(g(x))g'(x)$, the integral can be simplified by making a *substitution*:

If $u = g(x)$, then $\quad \int f(g(x))g'(x)\,dx = \int f(u)\,du.$

3. For the purposes of this book the *definite integral* of f from a to b is defined by

$$\int_a^b f(x)\,dx = F(b) - F(a),$$

where F is any antiderivative of f. For definite integrals the substitution formula becomes

$$\int_a^b f(g(x))g'(x)\,dx = \int_{g(a)}^{g(b)} f(u)\,du.$$

4. If $f(x) \geq 0$ on the interval $[a, b]$, the *region under the graph* of $y = f(x)$ from a to b is $\{(x, y): a \leq x \leq b, 0 \leq y \leq f(x)\}$. The area of this region is given by

$$A = \int_a^b f(x)\,dx.$$

This formula can be modified to find the areas of more complicated regions.

5. If the region under the graph of $y = f(x)$ from a to b is revolved about the x-axis, the *volume* of the resulting solid of revolution is given by

$$V = \int_a^b \pi y^2\,dx = \pi \int_a^b [f(x)]^2\,dx.$$

More generally, if the area of the general cross section of a solid is $A(x)$, then its volume is

$$V = \int_a^b A(x)\,dx.$$

6. If a variable force of magnitude $F(s)$ in the line of an s-axis moves a particle from $s = a$ to $s = b$, then the *work* the force does, and the *energy* it expends, are given by

$$W = \int_a^b F(s)\,ds.$$

Chapter Test

Find the following indefinite integrals.

14-1 **1.** $\int e^{-x}\,dx$ **2.** $\int \frac{4x^2 + 1}{x}\,dx$

3. $\int 3x^2 - 4x^5\,dx$ **4.** $\int \frac{x^2 - 4}{x + 2}\,dx$

14-2 **5.** $\int \frac{x\,dx}{\sqrt{1 + x^2}}$ **6.** $\int \frac{e^x\,dx}{1 + e^x}$

7. $\int x \cos x^2\,dx$ **8.** $\int x(x^2 - 5)^3\,dx$

Evaluate the following definite integrals.

14-3 **9.** $\int_0^4 \frac{x\,dx}{x^2 + 9}$ **10.** $\int_0^{\pi/2} \sin\theta \cos\theta\,d\theta$

11. $\int_0^{\pi/3} \sin 3\theta\,d\theta$ **12.** $\int_{\pi/4}^{\pi/2} \frac{\sin\theta}{1 - \cos\theta}\,d\theta$

14-4 **13.** Find the area of the region under the curve $y = \sqrt{x}$ from $x = 1$ to $x = 4$.

14. Find the area of the region lying above the parabola $y = x^2$ and below the line $y = 2x$.

14-5 **15.** The region of Exercise 13 is revolved about the x-axis. Find the volume of the solid swept out.

16. Find the volume of the solid swept out when the region of Exercise 14 is revolved about **(a)** the x-axis; **(b)** the y-axis.

14-6 **17.** A force of 20 pounds stretches a coil spring from its natural length of 3 feet to length 5 feet. How much work is necessary to stretch the spring an additional foot?

18. A V-shaped trough 2 m across the top, 2 m deep, and 3 m long is full of water. How much energy is necessary to pump all the water to the top of the trough?

COMPUTER APPLICATION

NUMERICAL INTEGRATION

The area A under a curve from x_i to x_{i+1} can be approximated by rectangles, trapezoids, or parabolic segments. The figure at the right shows such a parabolic approximation to the area under the graph of f from x_i to x_{i+1}. It can be shown that the parabolic approximation is given by

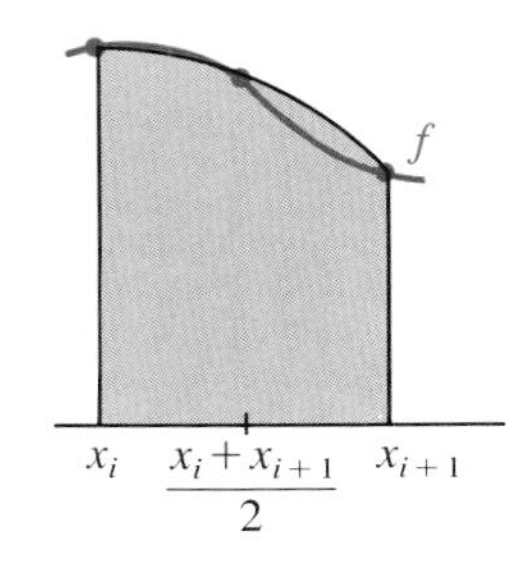

$$A \approx \left(\frac{x_{i+1} - x_i}{6}\right)\left[f(x_i) + f(x_{i+1}) + 4f\left(\frac{x_{i+1} + x_i}{2}\right)\right].$$

This is *Simpson's approximation formula.* Its derivation is somewhat lengthy and involved.

To find the area under a curve from a to b, divide $[a, b]$ into n subintervals of lengths $\frac{b - a}{n}$. Then $x_{i+1} - x_i = \frac{b - a}{n}$ for $i = 0, 1, 2, \ldots, n - 1$. Then apply Simpson's approximation formula above.

EXERCISES *(You may use either BASIC or Pascal.)*

1. Write a program using Simpson's formula to approximate the area under the graph of $y = x$ from $a = 0$ to $b = 12$. Use $n = 5$, 10, and 20.

2. Let $a = 1$, $b = 2$, and $n = 5$, 10, and 20. Modify and run the program in Exercise 1 for each function.

a. $f(x) = 4x^3 - x^4$

b. $f(x) = x^2$

c. $f(x) = 10^x$

3. Write a program using trapezoidal and parabolic formulas to approximate the area under $f(x) = \sin x$ from $a = 0$ to $b = \pi$. Use $n = 50$. See the computer exercises on page 583.

4. Run the program in Exercise 3 to approximate one quarter of the area of the unit circle. Use $f(x) = \sqrt{4 - x^2}$ with $a = 0$, $b = 2$, and $n = 50$. The result should approximate π.

APPLICATION

NEWTON'S LAW OF COOLING

Icebergs stay frozen. However, an ice cube in a glass of tap water melts. Why does one object melt while another object stays frozen? In the Arctic Zone, the difference between the temperature of the iceberg and the temperature of the surrounding air and water is 0. In the kitchen, however, the difference between the ice-cube temperature and the room temperature is not close to 0. It seems that the rate at which an object cools is related to the difference between its temperature and the temperature of its surroundings. Newton's law of cooling makes this relationship more precise:

$$\frac{d}{dt} T(t) = k(T(t) - M)$$

where $T(t)$ is the temperature of the object at time t, M is the temperature of the surroundings, and k is a proportionality constant.

The study of cooling rates is very important to engineers who design spacecraft, as the temperature difference between ice-cold fuel tanks and fiery-hot engines is very great.

EXERCISES

1. Let $T(t) = (T(0) - M)e^{kt} + M$.

a. Show: $\frac{d}{dt} T(t) = k(T(0) - M)e^{kt}$

b. Show: $e^{kt} = \frac{T(t) - M}{T(0) - M}$

c. Show: $\frac{d}{dt} T(t) = k(T(t) - M)$

2. A steel ball having an initial temperature of 300°C is placed in a freezer whose temperature is 0°C. After 10 min the ball has temperature 240°C.

a. Find an equation for the temperature $T(t)$ of the steel ball. Use the given information to find k.

b. Find the temperature of the ball after 60 minutes.

Chapter 15

Probability and Statistics

The first half of this chapter is devoted to some fundamental notions in probability, including conditional probability and Bernoulli trials. The second half is devoted to statistics, including the normal distribution and correlation.

Probability

15-1 PERMUTATIONS

How many ways are there of choosing one member from $S_1 = \{1, 2\}$ and one member from $S_2 = \{3, 4, 5\}$? The *tree diagram* shown at the right illustrates all of the possible selections from S_1 and S_2. For each of the two selections from S_1 there are three selections from S_2. In all there are six selections.

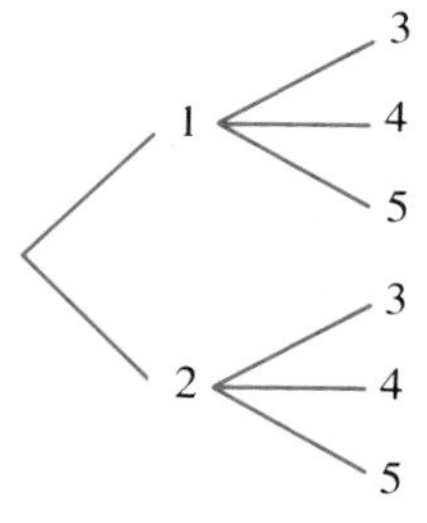

If there is another set $S_3 = \{6, 7, 8, 9\}$, then there are $2 \cdot 3 \cdot 4 = 24$ ways of selecting a member from each of S_1, S_2, and S_3. In general, we have the following principle.

Fundamental Counting Principle
If n sets $S_1, S_2, \ldots, S_n$ contain $m_1, m_2, \ldots, m_n$ members respectively, then the number of ways of selecting a member from each of them is

$$m_1 m_2 \cdots m_n.$$

The statistical analysis of traffic patterns is becoming increasingly important, as suggested by this view of the approach to Brooklyn Bridge.

EXAMPLE 1 How many license plates can be made if each plate consists of three letters followed by two digits?

Solution There are 26 possible letters for each of the first three slots.

$$\underline{26}\quad \underline{26}\quad \underline{26}\quad \underline{\ \ }\quad \underline{\ \ }$$

There are 10 possible digits for each of the last two slots.

$$\underline{26}\quad \underline{26}\quad \underline{26}\quad \underline{10}\quad \underline{10}$$

By the fundamental counting principle, $26 \cdot 26 \cdot 26 \cdot 10 \cdot 10 = 1{,}757{,}600$ license plates can be made. **Answer**

EXAMPLE 2 How many positive odd integers less than 1000 can be formed using 3, 4, 5, and 6?

Solution Such an integer might have 1, 2, or 3 digits. Since the number is odd, the units' digit must be 3 or 5.

	H	T	U		
1 digit			2	2	possible integers
2 digits		4	2	$4 \cdot 2 = 8$	possible integers
3 digits	4	4	2	$4 \cdot 4 \cdot 2 = 32$	possible integers

The total number of positive odd integers that can be so formed is the sum $2 + 8 + 32 = 42$. **Answer**

How many ways are there to arrange the letters a, b, and c *in a definite order*? One such arrangement is abc. In all there are six such arrangements, and they are listed in the following array. Each arrangement is called a *permutation* of a, b, and c.

abc	bac	cab
acb	bca	cba

DEFINITION

An arrangement of the members of a set in a definite order is a **permutation** of the members.

EXAMPLE 3 How many permutations of a, b, c, and d are possible?

Solution Any of the four letters can be used to fill the first slot.

$$\underline{4}\quad \underline{\ \ }\quad \underline{\ \ }\quad \underline{\ \ }$$

The second slot can be filled by any of the three remaining letters, and so forth.

$$\underline{4}\quad \underline{3}\quad \underline{2}\quad \underline{1}$$

Thus there are $4 \cdot 3 \cdot 2 \cdot 1 = 24$ permutations.

If instead we count the permutations of a, b, c, and d taking only two letters at a time, we will obtain $4 \cdot 3 = 12$ permutations. That is, there are four possibilities for the first slot and three possibilities for the second (last) slot.

This discussion can be generalized to any set with n members. In Theorem 1 and all that follows, the number of permutations of n objects taken r at a time will be denoted ${}_nP_r$. (Some texts use the notation $P(n, r)$.)

THEOREM 1 Let S be a set with n members and let r be a positive integer, $r \leq n$.

1. The number of permutations of n objects is

$$ {}_nP_n = n(n - 1)(n - 2) \cdots 1 = n!. $$

2. The number of permutations of n objects taken r at a time is

$$ {}_nP_r = n(n - 1)(n - 2) \cdots [(n - (r - 1)]. $$

Alternatively,

$$ {}_nP_r = \frac{n!}{(n - r)!}. $$

You are asked to prove that the first formula for ${}_nP_r$ implies the second formula for ${}_nP_r$ in Exercise 29.

EXAMPLE 4 How many permutations of the letters in DECIMAL taken 3 at a time are possible?

Solution Use the second formula for ${}_nP_r$ with $n = 7$ and $r = 3$.

$$ {}_7P_3 = \frac{7!}{(7 - 3)!} = \frac{7!}{4!} = 7 \cdot 6 \cdot 5 = 210 \quad \textbf{Answer} $$

EXAMPLE 5 If 5 books have red covers, 7 books have green covers, and 4 books have brown covers, find the number of ways in which the 16 books can be arranged on a shelf so that books of the same color are kept together.

Solution First find the number of ways the three *colors*, red, green, and brown, can be arranged.

$$ {}_3P_3 = 3! = 6 $$

Next find the number of ways the *books of each color* can be arranged.

Red	${}_5P_5 = 5! = 120$
Green	${}_7P_7 = 7! = 5040$
Brown	${}_4P_4 = 4! = 24$

The total number of arrangements is $6(120 \cdot 5040 \cdot 24) = 87{,}091{,}200$. **Answer**

The following sequence of six flags communicates a certain message.

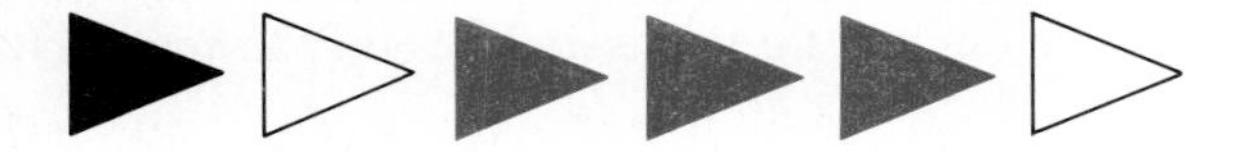

If the two white flags are interchanged, the same message is communicated. This permutation of the flags is *indistinguishable* from the original sequence. The 2! rearrangements of the white flags among themselves and the $3! = 6$ rearrangements of the red flags among themselves yield no new, *distinguishable*, permutation. Of the $6! = 720$ total permutations, only $\frac{720}{2 \cdot 6} = 60$ distinguishable messages can be sent. In general,

> If a set has n members with n_1 alike, n_2 alike, and so on, the number of distinguishable permutations of the n objects is
>
> $$\frac{n!}{n_1!n_2! \cdots}.$$

EXAMPLE 6 A child has 1 red, 3 blue, 5 yellow, 1 green, and 4 orange blocks. In how many different ways can the child arrange the blocks in a row such that the red and green blocks are at the two ends?

Solution There are two ways to place the red and green blocks: R . . . G or G . . . R. Then the 12 blue, yellow, and orange blocks remain. The number of permutations of these blocks is

$$\frac{12!}{3!5!4!},$$

which, after canceling common factors, is 27,720.
Therefore, by the fundamental counting principle, the total number of block arrangements is $2 \cdot 27{,}720 = 55{,}440$. **Answer**

EXERCISES

Evaluate.

A

1. ${}_{12}P_3$ 2. ${}_{8}P_0$ 3. ${}_{6}P_6$ 4. ${}_{9}P_4$ 5. ${}_{7}P_5$ 6. ${}_{20}P_2$
7. In how many ways can 6 reports be stacked in a pile?
8. In how many ways can 5 players be assigned 5 different positions on a basketball team?
9. In how many ways can a tie and a shirt be chosen from 6 ties and 8 shirts?

10. A menu lists 3 kinds of soup, 8 main courses, and 4 desserts. How many different dinners are possible if a diner chooses a soup, a main course, and a dessert?

11. In how many ways can 5 students be seated in a row of 10 seats? (*Hint:* How many choices of seats does the first student have?)

12. How many six-letter permutations can be formed from the letters in BIRTHDAY?

13. How many positive four-digit integers can be formed using 1, 2, 3, 4, and 5 if **(a)** no digit is repeated and **(b)** repetition of digits is allowed?

14. Repeat Exercise 13 using the digits 0, 1, 2, 3, 4, and 5. (*Hint*: The first digit cannot be zero.)

15. In Morse code a digit is represented as a sequence of five dots or dashes. In how many ways can a digit be represented?

16. In how many ways can a president, a vice president, and a treasurer be chosen from 12 council members?

Find the number of permutations of the letters in each word.

17. SUNSHINE 18. ENGINEER 19. TEETER 20. ELLIPTICAL

B

21. How many positive even integers with 5 digits are there?

22. How many positive six-digit multiples of 5 contain only odd digits?

In Exercises 23–25 find the specified number of permutations of the letters in the word QUIET.

23. All letters are used and Q is used first.

24. All letters are used and the vowels are used first.

25. The letter U must follow Q.

26. How many seven-digit phone numbers can be formed if no phone number begins with 0 or 1 and no phone number begins with 800?

27. A license plate consists of 6 symbols of which at least 4 symbols are digits. The remaining symbols, if any, are letters and occur at the beginning of the sequence of symbols. How many such license plates can be made?

28. A debate team consists of 3 historians, 4 economists, and 2 political scientists.
    a. In how many ways can the debate team be seated in a row?
    b. In how many ways can the debate team be seated in a row if a historian is seated at each end?
    c. In how many ways can the debate team be seated in a row if those in the same field must be seated next to one another?

29. Show that ${}_nP_r = n(n-1)\cdots[n-(r-1)] = \dfrac{n!}{(n-r)!}$, $1 \le r \le n$.

30. Find n if ${}_nP_3 = {}_{n-1}P_4$, $n \ge 5$.

31. Show that ${}_nP_r - {}_nP_{r-1} = (n - r)_nP_{r-1}$, $2 \le r \le n$.

32. Show that ${}_nP_r + {}_{n+1}P_{r+1} = (n + 2)_nP_r$, $1 \le r \le n$.

33. How many distinguishable four-letter permutations can be formed using the letters in CHEESE? (*Hint*: The permutation may contain one, two, or three E's.)

34. How many distinguishable five-letter permutations can be formed using the letters in POSSESS?

C 35. The diagram at the right shows one arrangement of three people at a table. Person A has C on the left and B on the right. **a.** Draw two diagrams that maintain the permutation but assign each person a different seat. **b.** Draw a diagram showing a different arrangement of the three people.

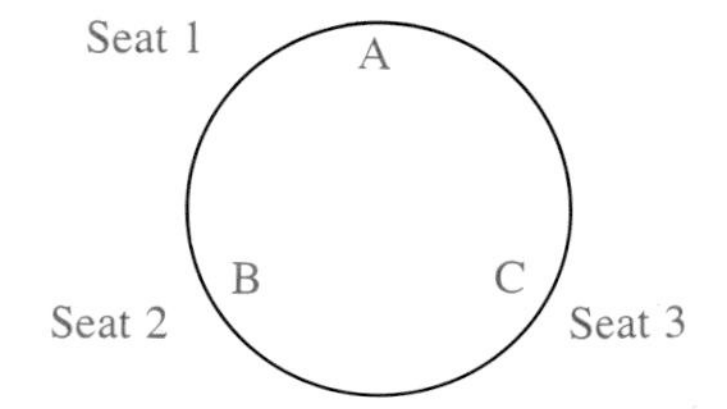

36. In how many different ways can four people, A, B, C, and D, be seated around a circular table?

37. Find a formula for the number of different ways that n people can be seated around a circular table.

15-2 COMBINATIONS

The set $S = \{a, b, c, d\}$ has 6 subsets having 2 members each:

$$\{a, b\} \quad \{a, c\} \quad \{a, d\} \quad \{b, c\} \quad \{b, d\} \quad \{c, d\}$$

Each of these subsets is called a *combination* of a, b, c, and d taken 2 at a time.

DEFINITION

An r-member subset of a set with n members is called a **combination** of n objects taken r at a time. The number of combinations of n objects taken r at a time is denoted ${}_nC_r$ and is read "n choose r." (Some texts use the notation $C(n, r)$.)

Since a set is an unordered collection, the combination $\{a, b\}$ is the same as $\{b, a\}$. The permutation ab, however, is different from the permutation ba since a permutation is ordered. How is the number of combinations related to the number of permutations?

The 24 permutations of a, b, c, and d taken 3 at a time are listed below.

abc	*abd*	*acd*	*bcd*
acb	*adb*	*adc*	*bdc*
bac	*bad*	*cad*	*cbd*
bca	*bda*	*cda*	*cdb*
cab	*dab*	*dac*	*dbc*
cba	*dba*	*dca*	*dcb*

The 4 combinations of a, b, c, and d taken 3 at a time are

$$\{a, b, c\} \qquad \{a, b, d\} \qquad \{a, c, d\} \qquad \{b, c, d\}.$$

Notice that column 1 represents the 6 permutations of a, b, and c, column 2 represents the 6 permutations of a, b, and d, and so forth. From this discussion,

$$4 = \frac{24}{6}$$

$$_4C_3 = \frac{_4P_3}{_3P_3}$$

In general, the number of combinations of n objects taken r at a time, $0 < r \le n$, is

$$_nC_r = \frac{_nP_r}{_rP_r}.$$

We can obtain a useful formula for $_nC_r$ from part (2) of Theorem 1.

$$_nC_r = \frac{_nP_r}{_rP_r}$$

Since $_nP_r = \dfrac{n!}{(n - r)!}$ and $_rP_r = r!$,

$$_nC_r = \frac{\frac{n!}{(n - r)!}}{r!} = \frac{n!}{r!(n - r)!}.$$

THEOREM 2 The number of combinations of n objects taken r at a time, $r \le n$, is given by

$$_nC_r = \frac{n!}{r!(n - r)!}.$$

The following example illustrates the use of Theorem 2. Note that the order in which committee members are selected is not important.

EXAMPLE 1 In how many ways can a four-person committee be formed from nine people?

Solution The committee is a four-member subset of the set of nine people. Use Theorem 2 with $n = 9$ and $r = 4$.

$$\begin{aligned} _9C_4 &= \frac{9!}{4!(9 - 4)!} \\ &= \frac{9!}{4!5!} \\ &= \frac{9 \cdot 8 \cdot 7 \cdot 6}{4 \cdot 3 \cdot 2 \cdot 1} \\ &= 126 \quad \textbf{Answer} \end{aligned}$$

EXAMPLE 2 A 5-member committee is to be made up of 3 teachers and 2 students. If 8 teachers and 7 students are eligible for selection, in how many ways can the committee be formed?

Solution There are ${}_8C_3$ ways to select 3 teachers from 8 teachers.

$${}_8C_3 = \frac{8!}{3!(8-3)!} = 56$$

There are ${}_7C_2$ ways to select 2 students from 7 students.

$${}_7C_2 = \frac{7!}{2!(7-2)!} = 21$$

There are 21 selections of 2 students for each selection of 3 teachers. By the fundamental counting principle, there are $56 \cdot 21 = 1176$ ways to form the committee. **Answer**

The problem of finding how many ways 3 red marbles and 2 green marbles can be selected from a jar containing 8 red marbles and 7 green marbles is exactly the same as the problem posed in Example 2. We can imagine that each red marble is associated with a particular teacher and that each green marble is associated with a particular student. The number of 5-marble selections is then $({}_8C_3)({}_7C_2) = 56 \cdot 21 = 1176$.

EXAMPLE 3 A salad bar contains nine different foods. In how many ways can seven or more foods be selected?

Solution Seven, eight, or nine foods may be selected. The total number of ways the food can be selected is the sum of combinations:

$${}_9C_7 + {}_9C_8 + {}_9C_9$$

$${}_9C_7 = \frac{9!}{7!(9-7)!} = \frac{9 \cdot 8}{2!} = 36$$

$${}_9C_8 = \frac{9!}{8!(9-8)!} = 9$$

$${}_9C_9 = \frac{9!}{9!(9-9)!} = \frac{1}{0!} = 1$$

Seven or more foods can be selected in $36 + 9 + 1 = 46$ ways. **Answer**

EXAMPLE 4 In how many ways can 12 people form 4 debate teams with each team consisting of 3 people?

Solution One team of 3 people can be formed in ${}_{12}C_3$ ways. Members of the other teams are selected from the 9 people that remain. Thus the second team can be selected in ${}_9C_3$ ways. The third team can be selected in ${}_6C_3$ ways and the fourth team in ${}_3C_3$ ways. The total number of selections is:

$$({}_{12}C_3)({}_9C_3)({}_6C_3)({}_3C_3)$$

That is,

$$\frac{12!}{3!9!} \cdot \frac{9!}{3!6!} \cdot \frac{6!}{3!3!} \cdot \frac{3!}{3!0!}.$$

$$\frac{12!}{3!3!3!3!} = 369{,}600$$ **Answer**

There is a strong connection between ${}_nC_r$ and $\binom{n}{r}$, the binomial coefficient introduced in the section on the binomial theorem, page 94. We can see the connection by examining, for example, the expansion of $(a + b)^3$.

$$(a + b)(a + b)(a + b)$$
$$aaa + baa + aab + aba + bba + bab + abb + bbb$$

The term baa comes from multiplying b in the first binomial, a in the second binomial, and a in the third binomial, as shown by the curved arrows. In all there are ${}_3C_2 = 3$ ways of choosing exactly two a's from the three binomials. Thus the coefficient of a^2b,

$$\binom{3}{2}, \text{ is exactly } {}_3C_2.$$

In general, if n and r are natural numbers, $1 \leq r \leq n$, then the binomial coefficient of $a^r b^{n-r}$ in the expansion of $(a + b)^n$ is exactly ${}_nC_r$:

$$\binom{n}{r} = {}_nC_r$$

EXERCISES

Evaluate.

A

1. ${}_{10}C_4$ 2. ${}_7C_3$ 3. ${}_{100}C_2$ 4. ${}_8C_5$ 5. ${}_{20}C_{17}$ 6. ${}_{12}C_1$

7. Write all three-letter combinations of the letters in DAISY.
8. Write all four-letter combinations of the letters in ORCHID.
9. In how many ways can 3 sweaters be chosen from 9 sweaters on sale?
10. There are 85 students in the junior class. Three juniors are needed to serve on the student council. In how many ways can the students be selected?
11. In how many ways can 10 people be assigned to 2 teams with 5 players each?
12. How many telephone lines are needed to connect 12 towns if there is exactly 1 line connecting any 2 towns?
13. Eight points, no three of which are collinear, are plotted in a plane. How many triangles can be made from them?
14. Four batteries are selected at random from a shipment of 200 batteries and tested for any defect. In how many ways can the 4 batteries be chosen?
15. A jar contains six red marbles and four white marbles.
    a. In how many ways can three marbles be chosen?
    b. In how many ways can three marbles be chosen so that two marbles are red and one marble is white?

16. A standard deck of cards consists of 4 suits (clubs, diamonds, hearts, and spades) and one ace, 2, 3, . . . , 10, jack, queen, and king in each suit.
   a. How many five-card hands can be dealt?
   b. How many five-card hands containing 3 hearts and 2 clubs can be dealt?

B **17.** In a standard deck of cards, jacks, queens, and kings are called face cards. How many five-card hands containing exactly 4 face cards can be dealt?

18. How many five-card hands containing cards of one suit can be dealt?

19. In how many ways can a committee of 5 people be formed from 10 officials if one of the officials chosen must be the mayor?

20. In how many ways can a dance committee of 3 boys, 3 girls, and an alternate of either gender be chosen from 6 boys and 7 girls?

21. Use Theorem 2 to show that ${}_nC_r = {}_nC_{n-r}$.

22. How many ways are there to get eight or more correct answers on a ten-question quiz?

23. How many diagonals does a regular octagon have?

C **24.** Find a formula for the number of k-sided polygons that can be formed from n points lying along an ellipse.

25. How many diagonals does a regular polygon with n sides have?

26. Six geologists, three physicists, and four mathematicians are eligible for a certain research team. In how many ways can a five-member team be chosen if there must be at least two geologists, at least one physicist, and at least one mathematician?

27. How many five-letter sequences can be formed from the letters in COUNTERS if each sequence contains exactly 2 vowels and 3 consonants? (*Hint*: First find combinations and then sequences, or permutations.)

COMPUTER EXERCISES

1. Write a program to find the number of permutations or combinations of n objects taken r at a time.

2. Run the program in Exercise 1 for the following:
   a. $n = 10$; $r = 7$; combinations
   b. $n = 14$; $r = 4$; permutations
   c. $n = 9$; $r = 6$; permutations
   d. $n = 12$; $r = 12$; combinations

3. Use the program to verify that the number of combinations of 10 objects taken 7 at a time is the same as the number of combinations of those objects taken 3 at a time.

15-3 PROBABILITY

Most of us have an intuitive feeling about the concept of probability. We may say, for example, "I will probably pass the test" or "I have little chance of winning the grand prize." Such comments suggest some kind of estimate of the likelihood, or probability, of an event occurring. In order to make these ideas precise we give some definitions.

For our purpose an **experiment** is a procedure which can be repeated indefinitely under the same conditions and which has well-defined **outcomes**. The set of all possible outcomes of an experiment is called its **sample space**. For example, in the experiment of rolling a single die and noting the number that comes up, the sample space is $\{1, 2, 3, 4, 5, 6\}$. We do not know in advance what the particular outcome of the experiment will be, but we *do* know that it will be one of these six numbers.

An **event** is any subset of the sample space, and an **elementary event** is one that contains a single outcome. Thus $\{1, 6\}$ is the event that either a 1 or a 6 comes up, and $\{2\}$ is the elementary event that a 2 shows.

EXAMPLE 1 A coin is tossed three times; each time heads (H) or tails (T) shows.

a. Write the sample space S for this experiment.

b. Write each event.
- A: Exactly two heads appear.
- B: At least two heads appear.
- C: There are equal numbers of heads and tails.
- D: There are at least two heads or two tails.

Solution **a.** The sample space S contains 8 possible outcomes.

$$S = \{\text{HHH, HHT, HTH, THH, HTT, THT, TTH, TTT}\}$$

b. The events A, B, C, and D are subsets of S.

$A = \{\text{HHT, HTH, THH}\}$ $\quad$ $B = \{\text{HHH, HHT, HTH, THH}\}$

$C = \emptyset$ $\quad$ $D = S$

EXAMPLE 2 Two dice are thrown and the numbers showing are recorded.

a. Write the sample space.

b. Write the events:
- E: At least one 2 shows.
- F: The sum of the numbers showing is 6.

Solution **a.** The sample space consists of all pairs (i, j) where $1 \le i \le 6$ and $1 \le j \le 6$.

(1, 1)	(1, 2)	(1, 3)	(1, 4)	(1, 5)	(1, 6)
(2, 1)	(2, 2)	(2, 3)	(2, 4)	(2, 5)	(2, 6)
(3, 1)	(3, 2)	(3, 3)	(3, 4)	(3, 5)	(3, 6)
(4, 1)	(4, 2)	(4, 3)	(4, 4)	(4, 5)	(4, 6)
(5, 1)	(5, 2)	(5, 3)	(5, 4)	(5, 5)	(5, 6)
(6, 1)	(6, 2)	(6, 3)	(6, 4)	(6, 5)	(6, 6)

b. To find E, list all pairs in which at least one 2 shows. $E = \{(1, 2), (2, 2), (3, 2), (4, 2), (5, 2), (6, 2), (2, 1), (2, 3), (2, 4), (2, 5), (2, 6)\}$

To find F, list all pairs in which $i + j = 6$.

$F = \{(1, 5), (2, 4), (3, 3), (4, 2), (5, 1)\}$

We use probability functions to measure the likelihood of occurrences.

DEFINITION

A **probability function** P assigns to each event A in the sample space S of an experiment a real number $P(A)$, read "the probability of A," such that

1. $0 \leq P(A) \leq 1$

2. $P(S) = 1$

3. $P(A \cup B) = P(A) + P(B)$ if A and B are **mutually exclusive**, that is, if they have no outcomes in common.

It is easy to see from the definition that

$$P(\emptyset) = 0$$

and that $P(A') = 1 - P(A)$ where A' is the set of all members of S not in A. You are asked to prove these facts in Exercises 17 and 18.

In the case of a finite sample space with n members, we can define a probability function by assigning probabilities to the elementary events. We then define the probability of an event to be the sum of the probabilities of the elementary events that comprise it. If each elementary event is *equally likely* to occur, we assign the probability $\frac{1}{n}$ to each such event.

EXAMPLE 3 A coin is tossed three times. Find the probabilities of the events A, B, C, and D in Example 1.

Solution The sample space consists of 8 equally likely outcomes (if the coin is fair). Assign the probability $\frac{1}{8}$ to each outcome.

$$P(A) = P(\{\text{HHT, HTH, THH}\})$$
$$= \tfrac{1}{8} + \tfrac{1}{8} + \tfrac{1}{8} = \tfrac{3}{8}$$
$$P(B) = P(\{\text{HHH, HHT, HTH, THH}\})$$
$$= \tfrac{1}{8} + \tfrac{1}{8} + \tfrac{1}{8} + \tfrac{1}{8} = \tfrac{4}{8} = \tfrac{1}{2}$$
$$P(C) = P(\emptyset) = 0 \qquad P(D) = P(S) = 1$$

Example 3 suggests the following theorem, whose proof is straightforward (Exercise 19).

THEOREM 3 If the sample space S of an experiment contains n equally likely outcomes, and event $A \subseteq S$ is made up of k outcomes, then $P(A) = \frac{k}{n}$.

In Example 2, event E consists of 11 outcomes and event F consists of 5 outcomes. Since the outcomes of the experiment are equally likely, then, by Theorem 3, $P(E) = \frac{11}{36}$ and $P(F) = \frac{5}{36}$.

EXAMPLE 4 A jar contains 4 white balls and 3 black balls. Two balls are drawn at random. What is the probability that **(a)** both balls are white, that **(b)** one ball is white and the other ball is black, and **(c)** that at least one ball is black?

Solution First find the number of ways two balls can be drawn. There are ${}_7C_2$ ways of drawing two balls.

$${}_7C_2 = \frac{7!}{2!5!} = 21$$

Furthermore, the 21 outcomes are equally likely.

a. Find the number of ways two white balls can be drawn.

$${}_4C_2 = \frac{4!}{2!2!} = 6$$

Therefore the probability that both balls are white is

$$\frac{6}{21} = \frac{2}{7}. \quad \textbf{Answer}$$

b. Find the number of ways of drawing a white ball and a black ball. For each choice of a white ball there are 3 choices of a black ball. There are $4 \times 3 = 12$ ways of drawing one white and one black ball. Therefore the probability of "one white, one black" is

$$\frac{12}{21} = \frac{4}{7}. \quad \textbf{Answer}$$

c. The event B that at least one ball chosen is black is the complement of the event W that both balls chosen are white. Therefore

$$\begin{aligned} P(B) &= 1 - P(W) \\ &= 1 - \tfrac{2}{7} = \tfrac{5}{7}. \quad \textbf{Answer} \end{aligned}$$

EXERCISES

A

1. Refer to the experiment of Example 2. Specify each event and find each probability.
   a. The sum of the numbers showing is 8.
   b. The numbers showing are equal.
   c. The sum of the numbers showing is greater than or equal to 10.
2. Refer to the experiment of Example 1. Specify each event and find each probability.
   a. Heads shows exactly once.
   b. Heads shows at least once.
   c. Heads shows exactly twice or tails shows exactly twice.

In Exercises 3–5 tell how many members are in each sample space.

3. A coin is flipped twice.
4. A card is drawn at random from a standard deck of cards.
5. A red, a green, and a white die are rolled.

6. Two letters are selected at random from the word PLANETS. How many members are in the sample space?

7. Two marbles are drawn at random from a jar containing five blue marbles and four green marbles. List the sample space, specify each event, and find each probability.
   a. One marble of each color is selected.
   b. At least one blue marble is selected.
   c. Both marbles are green.

8. Two coins are chosen at random from a purse containing 7 quarters and 3 dimes. List the sample space. Then find the probability that the combined value of the coins is:
   a. 50¢ **b.** At least 15¢ **c.** 35¢ **d.** Not equal to 20¢

9. Two cards are drawn at random from a standard deck of cards. Find the probability of each event.
   a. Both cards are aces.
   b. Neither card is an ace.
   c. At least one card is a club.

B **10.** A ticket number consists of five digits.
   a. How many tickets are possible if digits may repeat?
   b. How many tickets have numbers in which digits do not repeat?
   c. Find the probability that a randomly selected ticket number has five different digits.

In Exercises 11–16, find the probability of each event.

11. Seven chips labeled A, B, C, D, E, F, and G are placed in a jar. Five chips are drawn at random.
   a. Chips A and B are selected.
   b. No vowel is selected.
   c. At least one vowel is selected.

12. From five women and five men four people are chosen for a committee.
   a. Two women and two men are chosen.
   b. At least three women are selected.
   c. Four men are selected.

13. A jar contains four blue marbles, three white marbles, and three red marbles. Three marbles are drawn.
   a. All three marbles chosen are blue.
   b. No marble chosen is blue.
   c. At most one marble chosen is white.

14. A coin is flipped five times.
   a. Heads shows exactly once.
   b. Tails shows exactly three times.
   c. Heads does not show at all.
   d. Heads shows at least four times.

15. A five-card hand is dealt from a standard deck of cards.
   a. The hand contains exactly two aces.
   b. The hand contains only hearts.
   c. The hand contains an ace, a two, a three, a four, and a five.
16. A jar contains five green marbles, four yellow marbles, and three white marbles. Four marbles are drawn at random.
   a. No white marble is drawn.
   b. At least two marbles drawn are green.
   c. All marbles drawn are the same color.
17. Show that $P(\emptyset) = 0$.
18. Show that if $A \subseteq S$, then $P(A') = 1 - P(A)$.
19. Prove Theorem 3.

C

20. Let $S = \{s: s = 2, 3, 4, \ldots\}$. Let $P(s) = c(\frac{2}{3})^s$. Find c such that P is a probability function.
21. Let $S = \{s: s = 1, 2, 3, \ldots\}$. Show that $P(s) = \frac{1}{s}$ does not define a probability function.

15–4 COMPOUND EVENTS

The diagram below shows events E and F from the dice-rolling experiment in Example 2 of Section 15–3:

E: At least one 2 shows. F: The sum is 6.

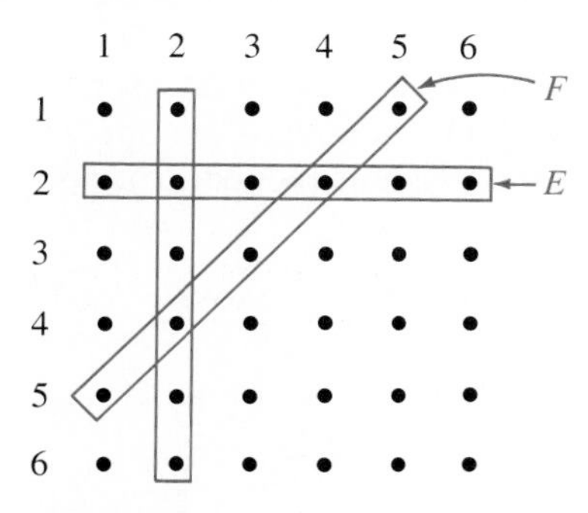

We can form the compound events:

$E \cup F$: At least one 2 shows *or* the sum is 6.
$E \cap F$: At least one 2 shows *and* the sum is 6.

Since the elementary events are equally likely, then by the figure above and Theorem 3,

$$P(E \cup F) = \tfrac{14}{36}, \quad P(E) = \tfrac{11}{36}, \quad P(F) = \tfrac{5}{36}, \quad P(E \cap F) = \tfrac{2}{36}.$$

Notice that $P(E \cup F) = P(E) + P(F) - P(E \cap F)$.

THEOREM 4 **Addition Rule of Probability**

For any two events A and B,

$$P(A \cup B) = P(A) + P(B) - P(A \cap B).$$

Proof: The figure at the right shows $A \cup B$ subdivided into three subsets:

$$A \cup B = (A \cap B') \cup (A \cap B) \cup (B \cap A')$$

Also from the figure:

$$A = (A \cap B') \cup (A \cap B)$$
$$B = (B \cap A') \cup (A \cap B)$$

From part (3) of the definition of a probability function,

$$P(A \cup B) = P(A \cap B') + P(A \cap B) + P(B \cap A'),$$
$$P(A) = P(A \cap B') + P(A \cap B),$$
$$P(B) = P(B \cap A') + P(A \cap B).$$

Hence

$$\begin{aligned} P(A) + P(B) &= P(A \cap B') + P(A \cap B) + P(B \cap A') + P(A \cap B) \\ &= [P(A \cap B') + P(A \cap B) + P(B \cap A')] + P(A \cap B) \\ &= P(A \cup B) + P(A \cap B) \end{aligned}$$

Therefore $P(A \cup B) = P(A) + P(B) - P(A \cap B)$. ■

EXAMPLE 1 A high school has 1200 students, of whom 640 are young women. The junior class has 360 students, of whom 200 are young women. What is the probability that a student picked at random is either a junior or a young woman?

Solution Let W be the event "woman chosen" and J be the event "junior chosen." Find $P(W \cup J)$.

$$P(W) = \frac{640}{1200} = \frac{8}{15} \qquad P(J) = \frac{360}{1200} = \frac{3}{10}$$

$$P(W \cap J) = \frac{200}{1200} = \frac{1}{6}$$

$$\begin{aligned} P(W \cup J) &= P(W) + P(J) - P(W \cap J) \\ &= \frac{8}{15} + \frac{3}{10} - \frac{1}{6} = \frac{2}{3} \end{aligned}$$ **Answer**

EXAMPLE 2 A committee of 4 students is to be formed from 6 seniors and 7 juniors. Find the probability that the committee includes *at least* 2 seniors.

Solution The desired probability is the sum of the probabilities of events A "exactly 2 seniors," B "exactly 3 seniors," and C "exactly 4 seniors." This is the case since the events cannot overlap. Use combinations to find $P(A)$, $P(B)$, and $P(C)$.

$$P(A) = \frac{({}_6C_2)({}_7C_2)}{{}_{13}C_4} = \frac{63}{143} \qquad P(B) = \frac{({}_6C_3)({}_7C_1)}{{}_{13}C_4} = \frac{28}{143} \qquad P(C) = \frac{({}_6C_4)({}_7C_0)}{{}_{13}C_4} = \frac{3}{143}$$

$P(A \cup B \cup C) = P(A) + P(B) + P(C) = \frac{63}{143} + \frac{28}{143} + \frac{3}{143} = \frac{94}{143}$, or about 0.66. **Answer**

Let us return to the experiment of throwing two dice and the events:

E: At least one 2 shows.
F: The sum is 6.

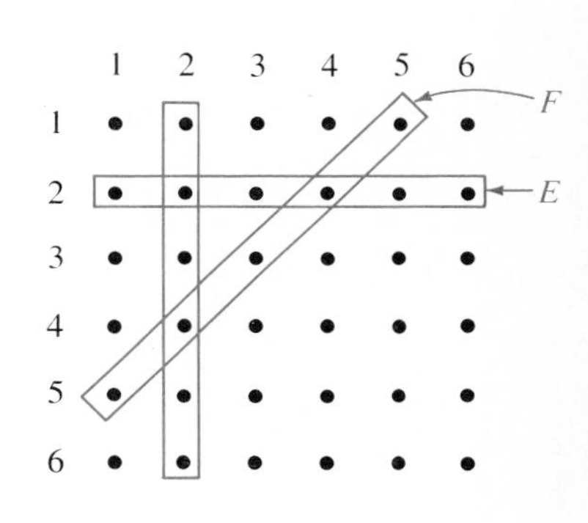

We know that $P(F) = \frac{5}{36}$. Suppose, however, that after the dice are thrown, but before we look at them, we are told that E has occurred, that is, that at least one 2 shows. This information requires us to change our thinking about the probability of F. We now know that the outcome of the experiment lies in E, so that E now plays the role of sample space. Moreover, the event "the sum is 6" is now $E \cap F$. Since the number of outcomes in E is 11 and in $E \cap F$ is 2, the probability of "the sum is 6" is now $\frac{2}{11}$, and we write $P(F \mid E) = \frac{2}{11}$.

The symbol $P(B \mid A)$, read "the probability of B given A," denotes the **conditional probability** that B will occur, given that A has occurred (or is certain to occur). The second formula in Theorem 5 is called the **multiplication rule of probability**.

THEOREM 5 For any two events A and B, with $P(A) \neq 0$,

$$P(B \mid A) = \frac{P(A \cap B)}{P(A)}, \quad \text{or} \quad P(A \cap B) = P(A) \cdot P(B \mid A).$$

Proof: The figure at the right shows S, A, and B. Let S have n members, $A \cap B$ have h members, and A have k members. We assume that all elementary outcomes are equally likely. If event A has occurred, then A becomes the sample space and B becomes $A \cap B$.

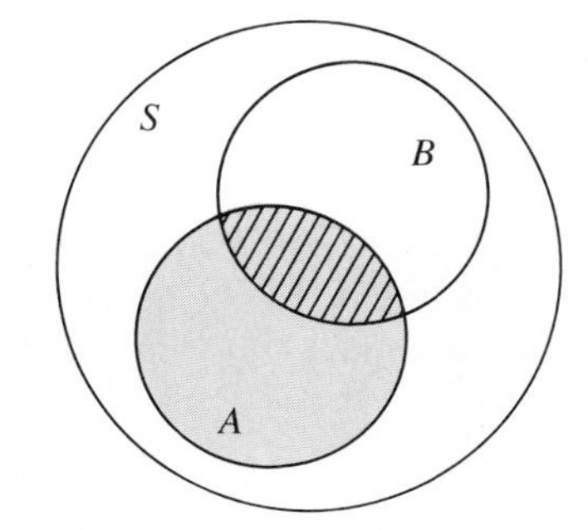

$$P(B \mid A) = \frac{h}{k} = \frac{h}{n} \div \frac{k}{n} = \frac{P(A \cap B)}{P(A)} \quad ■$$

EXAMPLE 3 A box contains 4 red chips and 2 blue chips.

a. A chip is drawn, not replaced, and then a second chip is drawn. Find the probability that the first chip is red and the second chip is blue.

b. Two chips are drawn simultaneously. Find the probability that the chips have different colors.

Solution **a.** Let R be the event "first chip red" and B be the event "second chip blue." Find $P(\text{red, then blue}) = P(R \cap B)$. The box has 6 chips, of which 4 chips are red. By Theorem 3,

$$P(R) = \tfrac{4}{6} = \tfrac{2}{3}.$$

If the first chip drawn is red, there remain 5 chips in the box, of which 2 are blue. Hence

$$P(B \mid R) = \tfrac{2}{5}.$$

Therefore

$$\begin{aligned} P(R \cap B) &= P(R) \cdot P(B \mid R) \\ &= \tfrac{2}{3} \cdot \tfrac{2}{5} = \tfrac{4}{15}. \end{aligned}$$ **Answer**

b. Drawing two chips simultaneously is equivalent to drawing a chip and then drawing another chip without replacing the first one.

$$\begin{aligned} &P((\text{red, then blue}) \text{ or } (\text{blue, then red})) \\ &\quad = P(\text{red, then blue}) + P(\text{blue, then red}) \\ &\quad = P(R \cap B) + P(B \cap R) \end{aligned}$$

since "red, then blue" and "blue, then red" are mutually exclusive.

From part (a), $P(R \cap B) = \frac{4}{15}$. By similar reasoning,

$$\begin{aligned} P(B \cap R) &= P(B) \cdot P(R \mid B) \\ &= \tfrac{2}{6} \cdot \tfrac{4}{5} = \tfrac{4}{15}. \end{aligned}$$

Therefore the probability of drawing two chips of different colors is $\frac{4}{15} + \frac{4}{15} = \frac{8}{15}$. **Answer**

Suppose that a chip is drawn from the box and that the chip is replaced before a second chip is drawn. The outcome of a second draw is *independent* of the outcome of the first draw. For example, suppose a red chip is drawn and returned to the box. The probability $P(\text{blue given red})$ is $\frac{2}{6}$ since there are 2 blue chips among the 6 chips in the box. This probability is exactly the same as $P(B)$.

$$P(B \mid R) = P(B)$$

By Theorem 5, $$P(B \mid R) = \frac{P(R \cap B)}{P(R)}.$$

Hence $$P(R \cap B) = P(R) \cdot P(B).$$

This discussion suggests the following definition of independence.

DEFINITION

Events A and B are **independent** if $P(A \cap B) = P(A) \cdot P(B)$.

In many cases we can immediately tell, without using the definition, that two events A and B are independent. We can then use the definition to find $P(A \cap B)$.

EXAMPLE 4 A coin is tossed and a die thrown. Find the probability that the coin comes up heads and the die shows a 1 or a 6.

Solution Events A (heads) and B (die shows 1 or 6) are clearly independent.

$$\begin{aligned} P(A \cap B) &= P(A) \cdot P(B) \\ &= \tfrac{1}{2} \cdot \tfrac{2}{6} = \tfrac{1}{6} \end{aligned}$$

Sometimes the definition can be used to decide whether or not two events are independent.

EXAMPLE 5 In a certain suburb a survey of family income and home ownership gave the following results.

	Home rented	Home owned
Income ≤ \$50,000	450	150
Income > \$50,000	100	300

Is home ownership independent of whether the family income is greater than \$50,000?

Solution The table shows 4 mutually exclusive groups of people in the survey. Let A be the event "home owned" and B be the event "income > \$50,000." Compare the probability, $P(A \cap B)$, with the probability, $P(A) \cdot P(B)$. The total number of families surveyed is 1000. Hence

$$P(A \cap B) = \frac{300}{1000} = 0.3.$$

$$P(A) = \frac{150}{1000} + \frac{300}{1000} = \frac{450}{1000} = 0.45$$

$$P(B) = \frac{100}{1000} + \frac{300}{1000} = \frac{400}{1000} = 0.4$$

Since 0.3 and $(0.45)(0.4) = 0.18$ are significantly different, the events are *not* independent. **Answer**

EXERCISES

In Exercises 1–6, classify each pair of events as independent, mutually exclusive, or neither. A card is drawn from a standard deck of cards.

A: The card drawn is black. B: The card drawn is a face card.
C: The card drawn is a heart. D: The card drawn is not a king.

A

1. A and D **2.** B and C **3.** A and C
4. A and B **5.** B and D **6.** C and D

7. A card is drawn at random from a standard deck of cards. Find each probability.
 a. The card drawn is a queen given that it is a face card.
 b. The card drawn is a jack given that it is a diamond.
 c. The card drawn is a diamond given that it is a jack.

8. Two dice are rolled. Find each probability.
 a. The sum of the numbers showing is greater than 9 given that at least one number showing is a 5.
 b. One of the numbers showing is a 5 given that the sum of the numbers is 9.
 c. One number showing is twice the other number showing given that at least one number is even.

9. A student's chances of passing chemistry and French are 0.8 and 0.6 respectively. The chance of passing both courses is 0.5. Are passing chemistry and passing French independent events? Explain.

10. If A and B are independent events, $P(B') = 0.7$, and $P(A \cap B) = 0.18$, find $P(A)$.

11. A jar contains three blue chips and two white chips. A second jar contains two blue chips and four white chips. One jar is selected at random and then one chip is selected from it. Find the probability that the chip selected is blue given that it came from the second jar.

12. A die is rolled and a coin is tossed. Find each probability.
 a. The die shows a 2 and the coin shows tails.
 b. The die shows 4 or 5 and the coin shows heads.

13. A jar contains two yellow chips and five red chips. A second jar contains three yellow chips and five green chips. A chip from each jar is selected. Find each probability.
 a. A red chip is drawn from the first jar and a yellow chip is drawn from the second jar.
 b. Both chips drawn are yellow.

14. Three cards are drawn at random from a standard deck of cards. Find the probability that exactly two aces are drawn under each condition.
 a. Selection is made with replacement.
 b. Selection is made without replacement.

B

15. A basket contains eleven tickets numbered from 1 to 11. One ticket is drawn at random and its number noted. Let A be the event that the number is a multiple of 2 and let B be the event that the number is a multiple of 3.
    a. Show that A and B are not independent.
    b. Find $P(A \cup B)$ and $P(A \cup B')$.
    c. Find $P(A \mid B)$ and $P(B \mid A)$.

16. A jar contains four green balls, three yellow balls, and one white ball. Two balls are randomly drawn. Let C be the event that at least one ball drawn is green. Let D be the event that the second ball drawn is yellow.
    a. Show that C and D are not independent.
    b. Find $P(C \cup D)$ and $P(C' \cup D')$.
    c. Find $P(C \mid D)$ and $P(D \mid C)$.

17. A coin is tossed four times. Find each probability.
    a. All four tosses are heads given that the first two tosses are heads.
    b. Exactly two tosses are tails given that at least one toss is tails.
    c. Exactly three tosses are heads given that at least one toss is heads.

18. Let A and B be events.
    a. If $P(A) = 0.35$, $P(B) = 0.55$, and $P(A \cup B) = 0.75$, find $P(A \cap B)$.
    b. Use part (a) to find the probability of A or B but not both.

19. A convention has one morning seminar and one afternoon seminar. Of 60 people attending, 31 people went to the morning seminar and 40 people went to the afternoon seminar. Fourteen people went to neither one. Find the probability that a person selected at random attended both seminars.

20. Of thirty mathematics students, fifteen students study physics and eighteen students study Spanish. A student is selected randomly from the mathematics students. Find each probability.
    a. The student studies either physics or Spanish but not both.
    b. The student studies Spanish but not physics.

21. Show that if A and B are mutually exclusive and $P(A) \neq 0$ and $P(B) \neq 0$, then A and B are not independent.

22. Show that if $P(B) \neq 0$, then $P(A' \mid B) = 1 - P(A \mid B)$.

C

23. Show that if A and B are independent, then A and B' are independent.

24. The probability that Central High School will win this week's football game is $\frac{1}{2}$, that it will win this week's basketball game is $\frac{2}{3}$, and that it will win this week's hockey game is $\frac{3}{5}$. Find the probability that at most one of the three games is lost.

25. A jar contains one blue marble and one red marble. Another jar contains one blue marble and three red marbles. One jar is selected at random and then one marble from it is selected. Find the probability that the marble is drawn from the second jar given that the marble is red.

15-5 RANDOM VARIABLES AND PROBABILITY DISTRIBUTIONS

When we perform experiments, we are frequently interested in some characteristic of an outcome rather than the outcome itself. Suppose, for example, we flip 4 coins. The 16 possible outcomes are:

		HHHH			
	HTTT	THTT	TTHT	TTTH	
HHTT	HTTH	THHT	TTHH	THTH	HTHT
	THHH	HTHH	HHTH	HHHT	
		TTTT			

We might be interested in the characteristic "heads occurs x times in the outcome." To each outcome we can assign one of the numbers, 0, 1, 2, 3, or 4. For example,

$$\begin{aligned} TTTT &\longrightarrow 0 \\ HTTT &\longrightarrow 1 \\ HTHT &\longrightarrow 2 \\ HTHH &\longrightarrow 3 \\ HHHH &\longrightarrow 4 \end{aligned}$$

Such an assignment is called a *random variable*.

DEFINITION

A **random variable** is a function that assigns a numerical value to each outcome of an experiment.

Given a random variable, X, we usually are interested in the probability that X takes on a particular value, x, of its range. We denote this probability by $P(X = x)$. A listing of these probabilities as x runs through the range of X is the **probability distribution** of the random variable X. Associated with each such distribution is a **probability histogram**. The probability distribution and histogram for the 4-coin-toss experiment are shown below.

Outcome	x	$P(X = x)$
TTTT	0	$\frac{1}{16}$
HTTT THTT TTHT TTTH	1	$\frac{4}{16}$
HHTT HTTH THHT TTHH THTH HTHT	2	$\frac{6}{16}$
THHH HTHH HHTH HHHT	3	$\frac{4}{16}$
HHHH	4	$\frac{1}{16}$

Probability distribution

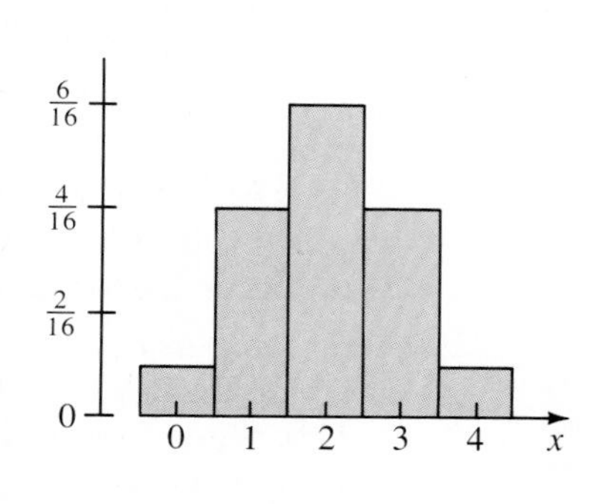

Probability histogram

EXAMPLE 1 Three boxes labeled A, B, and C are placed on a table. Then the labels are covered. Three cards labeled A, B, and C are placed at random in the boxes, one card to a box. Find a probability distribution for matching the card and box labels.

Solution Let Y be the number of boxes to which a card is matched correctly. There are $3! = 6$ ways of distributing the 3 cards:

ABC ACB BAC BCA CAB CBA

One of these, say ABC, is the complete match. In the placement ACB, card A and box A are correctly matched. The probability distribution and histogram are shown below.

Outcome	x	$P(Y = x)$
BCA CAB	0	$\frac{2}{6} = \frac{1}{3}$
CBA ACB BAC	1	$\frac{3}{6} = \frac{1}{2}$
	2	0
ABC	3	$\frac{1}{6}$

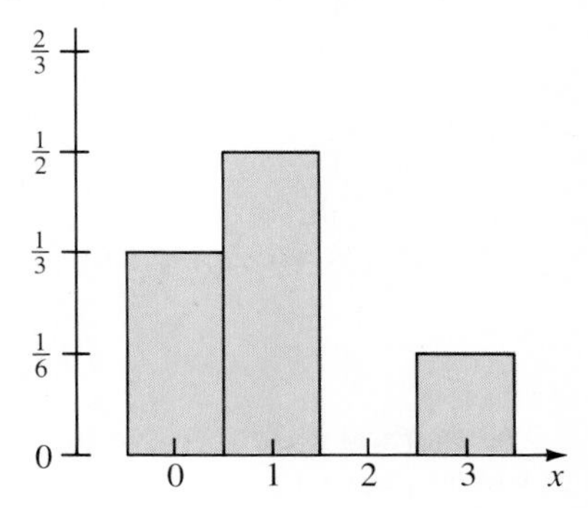

In many experiments, we consider each outcome to be either a **success** or a **failure**. For example, if an urn contains 2 white marbles and 3 red marbles, we might consider drawing a white marble a success and therefore drawing a red marble a failure. We might consider rolling a 1 or 6 on a die to be a success and therefore rolling a 2, 3, 4, or 5 a failure.

The probability of rolling a 1 or 6 on a die is $\frac{2}{6}$, and the probability of rolling 2, 3, 4, or 5 is $\frac{4}{6}$. Thus

$$P(\text{Success}) = \tfrac{2}{6} \text{ and } P(\text{Failure}) = \tfrac{4}{6}.$$

Notice that $P(\text{Failure}) = 1 - \frac{2}{6} = 1 - P(\text{Success})$. In general, if the outcome of an experiment is classified as success or failure and the probability of success is p, then the probability, q, of failure is $q = 1 - p$.

What is the probability of obtaining a specified number of successes in n repetitions, or **trials**, of such an experiment? To answer the question, we first define a *binomial experiment*.

DEFINITION

A **binomial experiment** is one having these properties:

1. There is a fixed number n of repeated trials.
2. The outcome of each trial can be classified as a success or a failure.
3. The probability p of success is the same for all trials.
4. The trials are independent, that is, the outcome of any trial is not affected by the outcomes of preceding trials. The trials themselves are called **Bernoulli trials**.

EXAMPLE 2 Let success, S, be rolling a 1 or 6 on a die and let failure, F, be rolling a 2, 3, 4, or 5. What is the probability of having exactly 2 successes in 5 repetitions of the experiment?

Solution Find the probability of success and the probability of failure on one trial.

$$P(\text{S}) = \tfrac{1}{3} \qquad P(\text{F}) = \tfrac{2}{3}$$

One sequence of 2 successes in 5 repetitions is SFSFF. Since the trials are independent, the multiplication rule of probability applies to find $P(\text{SFSFF})$.

$$P(\text{SFSFF}) = \tfrac{1}{3} \cdot \tfrac{2}{3} \cdot \tfrac{1}{3} \cdot \tfrac{2}{3} \cdot \tfrac{2}{3} = (\tfrac{1}{3})^2(\tfrac{2}{3})^3$$

There are ${}_5C_2 = \binom{5}{2} = 10$ ways of choosing sequences with S appearing exactly twice.

SSFFF, SFSFF, SFFSF, SFFFS, FSSFF
FSFSF, FSFFS, FFSSF, FFSFS, FFFSS

Since the ten events are mutually exclusive, the probability that exactly 2 successes occur in 5 repetitions is the sum of the individual probabilities.

$$\binom{5}{2}\left(\frac{1}{3}\right)^2\left(\frac{2}{3}\right)^3 = 10 \cdot \frac{2^3}{3^5} \approx 0.3292 \qquad \textbf{Answer}$$

The method used to solve the single-die problem discussed above can be adapted to prove the following result.

THEOREM 6 Consider a binomial experiment of n trials with probability of success p and of failure $q = 1 - p$, and let X denote the number of successes. Then the probability distribution of X is given by

$$P(X = k) = \binom{n}{k}p^k q^{n-k}.$$

Since $\binom{n}{k}q^{n-k}p^k$ is the $(k + 1)$st term of the expansion of $(q + p)^n$, the distribution given in Theorem 6 is called a **binomial distribution**, and X is called a **binomial random variable**.

EXAMPLE 3 An urn contains 2 white balls and 3 red balls. A ball is drawn at random and returned to the urn. Let Z denote the number of times a white ball is drawn in 4 trials.

a. Find the probability distribution of Z.
b. Draw the associated probability histogram.
c. Find the probability that a white ball will be drawn at least twice in the 4 trials.

Solution Let success be drawing a white ball from the urn. Since the ball drawn is replaced, the trials are independent. The experiment is binomial with $n = 4$, $p = \tfrac{2}{5}$, and $q = \tfrac{3}{5}$. Apply Theorem 6.

$$P(Z = 0) = \binom{4}{0}\left(\frac{2}{5}\right)^0\left(\frac{3}{5}\right)^4 = 1 \cdot \frac{3^4}{5^4} = \frac{81}{625} = 0.1296$$

$$P(Z = 1) = \binom{4}{1}\left(\frac{2}{5}\right)^1\left(\frac{3}{5}\right)^3 = 4 \cdot \frac{2 \cdot 3^3}{5^4} = \frac{216}{625} = 0.3456$$

$$P(Z = 2) = \binom{4}{2}\left(\frac{2}{5}\right)^2\left(\frac{3}{5}\right)^2 = 6 \cdot \frac{2^2 \cdot 3^2}{5^4} = \frac{216}{625} = 0.3456$$

$$P(Z = 3) = \binom{4}{3}\left(\frac{2}{5}\right)^3\left(\frac{3}{5}\right)^1 = 4 \cdot \frac{2^3 \cdot 3}{5^4} = \frac{96}{625} = 0.1536$$

$$P(Z = 4) = \binom{4}{4}\left(\frac{2}{5}\right)^4\left(\frac{3}{5}\right)^0 = 1 \cdot \frac{2^4}{5^4} = \frac{16}{625} = 0.0256$$

a.

k	$P(Z = k)$
0	0.1296
1	0.3456
2	0.3456
3	0.1536
4	0.0256

b.

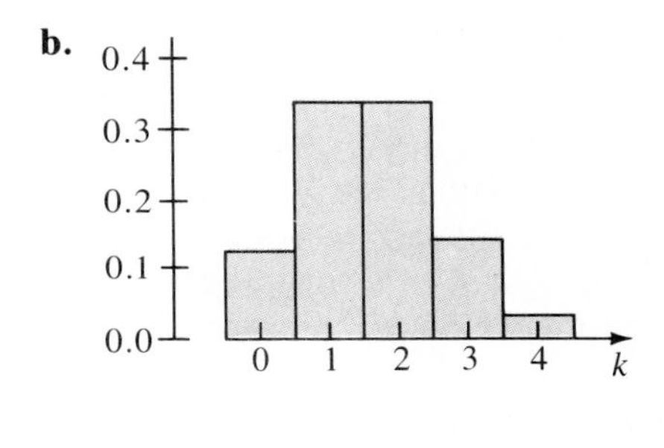

c. $P(Z \geq 2) = P(Z = 2) + P(Z = 3) + P(Z = 4)$
$= 0.3456 + 0.1536 + 0.0256$
$= 0.5248$ **Answer**

The probability distribution of a random variable X provides important information about X. Suppose that an experiment has n equally likely outcomes $x_1, x_2, \ldots, x_n$. Then $P(x_i) = \frac{1}{n}$. The sum

$$x_1\left(\frac{1}{n}\right) + x_2\left(\frac{1}{n}\right) + \cdots + x_n\left(\frac{1}{n}\right) = \frac{x_1 + x_2 + \cdots + x_n}{n}$$

gives the average value, or *expected value*, of $x_1, x_2, \ldots, x_n$. In many experiments, however, the outcomes are not equally likely. This is the case with problems posed in this section. Still the sum of products $x_i \cdot P(X = x_i)$ provides some measure of what value of X can be expected.

DEFINITION

Let the values of a random variable X be $x_1, x_2, \ldots, x_n$. Then the **expected value** of X is

$$E(X) = x_1 \cdot P(X = x_1) + x_2 \cdot P(X = x_2) + \cdots + x_n \cdot P(X = x_n)$$
$$= \sum_{i=1}^{n} x_i \cdot P(X = x_i).$$

EXAMPLE 4 The probabilities are 0.12, 0.20, 0.45, and 0.23 that a customer will buy 1, 2, 3, or 4 units of a certain product on any one day.

a. Find the number of units the store managers expect a customer to buy on any one day.

b. If the managers expect 10 customers a day for this product, how many units might the store keep in stock?

Solution **a.** Let X = the number of units a customer buys.

$$E(X) = 1 \cdot (0.12) + 2 \cdot (0.20) + 3 \cdot (0.45) + 4 \cdot (0.23)$$
$$= 2.79 \quad \textbf{Answer}$$

The managers expect each customer to buy $2.79 \approx 3$ units.

b. The managers might keep $3 \cdot 10 = 30$ units in stock each day. **Answer**

Note: A game of chance is called *fair* if the expected value is 0.

In a binomial experiment, the expected value of a random variable X is given by the following theorem. A proof is outlined in Exercises 17–19.

THEOREM 7 Consider a binomial experiment with n trials and probability of success p and of failure $q = 1 - p$. Let X denote the number of successes. Then $E(X) = np$.

EXERCISES

In Exercises 1–8, **(a)** show the probability distribution in a table; **(b)** draw the associated probability histogram; and **(c)** compute the expected value of the random variable.

A

1. A coin is flipped three times. Let X = number of tails showing.

2. Two dice are rolled. Let Y = the sum of the numbers showing.

3. Five chips numbered 1, 2, 3, 4, and 5 are placed in a jar. Two chips are drawn at random. Let Z = the sum of the numbers on the chips drawn.

4. Two tetrahedral (4-faced) dice, each with faces numbered 1, 2, 3, and 4, are tossed on a table. Let X = the sum of the numbers on the faces in contact with the table.

5. Two marbles are drawn at random from a jar containing three blue marbles, two green marbles, and one white marble. Let Y = the number of blue marbles drawn.

6. Four one-dollar bills and two five-dollar bills are placed in a hat. Three bills are drawn at random. Let Z = the dollar amount drawn.

7. Two cards are drawn from a standard deck of cards. Let X = the number of hearts drawn.

8. Four transistors are selected at random from a box containing twenty transistors, four of which are defective. Let X = the number of defective transistors chosen.

In Exercises 9–14, **(a)** use Theorem 6 to write a formula for the probability distribution of the specified binomial random variable; **(b)** find the probability of event E to the nearest thousandth; and **(c)** find the expected value of the random variable.

9. A die is rolled four times. Let X = the number of times that 5 shows. E: Five shows exactly twice.

10. A weighted coin with probability of heads equal to 0.4 and probability of tails equal to 0.6 is flipped five times. Let H = the number of heads showing. E: Heads shows exactly three times.

11. A card is drawn at random from a standard deck of cards. Let Y = the number of aces drawn in five trials. E: An ace is drawn exactly twice.

12. One chip is drawn at random from a jar containing six chips numbered 2, 3, 4, 5, 6, and 7, its number is noted, and then it is replaced. Let X = the number of times an even-numbered chip is drawn in four trials. E: Exactly three chips drawn are even-numbered.

13. A baseball player's batting average is 0.400. Let B = the number of hits in six times at bat. E: The player gets exactly three hits.

14. A tennis player who serves successfully 80% of the time makes ten practice serves. Let C = the number of successful serves in ten tries. E: Exactly nine of the ten serves are successful.

B

15. A multiple-choice test consists of eight questions. Each question has four choices of answer, only one of which is correct. Find the probability that a student who chooses each answer at random will get five or more answers correct.

16. A garden-supply distributor guarantees that 98% of the seeds in each package of seeds will germinate. Find the probability that 49 or more of the 50 seeds in one package will germinate when planted.

C

17. Show that $k\binom{n}{k} = n\binom{n-1}{k-1}$ for all integers n and k such that $1 \le k \le n$.

18. Show that if $p + q = 1$, then $\sum_{k=0}^{n} \binom{n}{k} p^k q^{n-k} = 1$ for all integers n and k such that $0 \le k \le n$.

19. Use Exercises 17 and 18 to prove Theorem 7. That is, show that if $E(X) = \sum_{k=0}^{n} k\binom{n}{k} p^k q^{n-k}$, then $E(X) = np$. (*Hint*: Factor out p and use the fact that $n - k = n - 1 - k + 1$.)

15-6 STATISTICAL MEASURES

Statistics is the area of mathematics concerned with gathering, organizing, and analyzing numerical *data*. Probability is used for analysis when the amount of data is large. In this section we consider ways of organizing data and we introduce some basic statistical measures.

Suppose that 36 students received the following scores on a test:

100	78	87	73	94	82	56	93	80
83	76	85	98	68	93	90	77	90
74	87	65	75	80	83	83	80	82
88	85	75	85	95	83	65	85	88

We shall find it easier to analyze the data if we group them by frequency. In Table 1 the distinct scores are listed with the frequency of each. In Table 2 the scores are grouped in class intervals of equal width. In this table we choose class intervals [50, 54], [55, 59], [60, 64], and so on. Each table is a form of frequency table, or **frequency distribution**.

x_i (score)	f_i (frequency)
100	1
98	1
95	1
94	1
93	2
90	2
88	2
87	2
85	4
83	4
82	2
80	3
78	1
77	1
76	1
75	2
74	1
73	1
68	1
65	2
56	1

Table 1

Class Interval	f_i (frequency)
50–54	0
55–59	1
60–64	0
65–69	3
70–74	2
75–79	5
80–84	9
85–89	8
90–94	5
95–100	3

Table 2

In constructing Table 2 we divided the scores, which range from about 50 to 100, into 10 non-overlapping intervals each of length 5 (with the exception of the last interval, [95, 100]). Note that a different frequency table will be obtained if a different interval length is chosen.

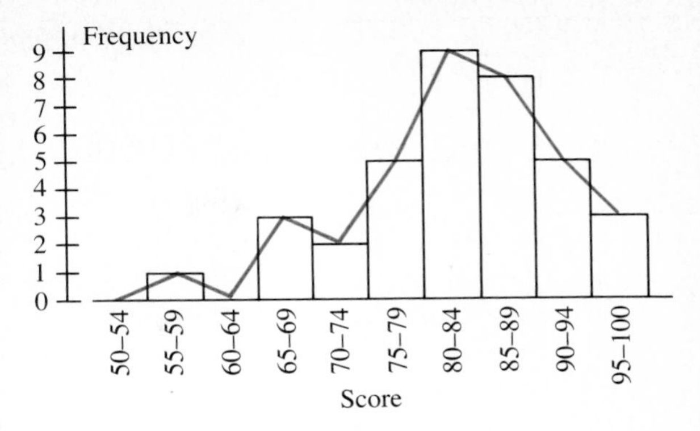

The grouped data in the frequency table can be represented graphically. Both the **histogram** and the **frequency polygon**, shown in red, can be used. The polygon is drawn by joining the midpoints of the tops of the bars.

Three measures of **central tendency** are *mean*, *median*, and *mode*.

DEFINITION

The **mean** $\overline{x}$ of the data $x_1, x_2, \ldots, x_n$ is

$$\overline{x} = \frac{x_1 + x_2 + \cdots + x_n}{n} = \frac{1}{n}\sum_{i=1}^{n} x_i. \qquad (*)$$

It is easy to see that if there are r data values $x_1, x_2, \ldots, x_r$ with frequencies $f_1, f_2, \ldots, f_r$, respectively, then

$$\overline{x} = \frac{x_1 f_1 + x_2 f_2 + \cdots + x_r f_r}{n} = \frac{1}{n}\sum_{i=1}^{r} x_i f_i. \qquad (\dagger)$$

Using Table 1 and (†), we find the mean of the test scores to be

$$\overline{x} = \frac{100 \cdot 1 + 98 \cdot 1 + \cdots + 65 \cdot 2 + 56 \cdot 1}{36} = \frac{1}{36}(2961) = 82.25.$$

Using Table 2 and (†) with x_i being the midpoint of each class interval, we can approximate the mean by

$$\overline{x} = \tfrac{1}{36}(57 \cdot 1 + 67 \cdot 3 + 72 \cdot 2 + 77 \cdot 5 + 82 \cdot 9 + 87 \cdot 8 + 92 \cdot 5 + 97 \cdot 3)$$
$$\overline{x} = \tfrac{1}{36}(2972) = 82.56$$

This is very close to the true mean.

We can find the **median** of a set of n data values by first writing them in order from least to greatest. If n is odd, the median is the middle value. If n is even, the median is the mean of the two middle values. The **mode** is the data value(s) having the greatest frequency.

EXAMPLE On Monday, 25 students were asked how many hours of television each had watched the day before. The answers were:

7, 3, 4, 0, 2, 0, 5, 4, 4, 6, 1, 7, 2, 3, 3, 5, 1, 4, 0, 3, 6, 4, 0, 2, 5

Find the mean, median, and mode of these data.

Solution on following page

Solution Make a frequency distribution and use (†).

Hours x_i	Students f_i	$x_i f_i$
0	4	0
1	2	2
2	3	6
3	4	12
4	5	20
5	3	15
6	2	12
7	2	14

$$n = \sum f_i = 25 \qquad \sum x_i f_i = 81$$

The mean number of hours watched is $\bar{x} = \frac{1}{n}\sum x_i f_i = \frac{81}{25} = 3.24.$

There are 25 data values. Thus the median is the thirteenth value, 3. The mode is 4 since that value has greatest frequency. **Answer**

Note: When there is no chance of confusion, we omit the limits on the summation sign.

The **relative frequency** of a score or class of scores is obtained by dividing its actual frequency by the total number of scores. Thus, in the test-score example, the relative frequency of scores in the 80–84 interval is

$\frac{9}{36}$, or 0.25, or 25%.

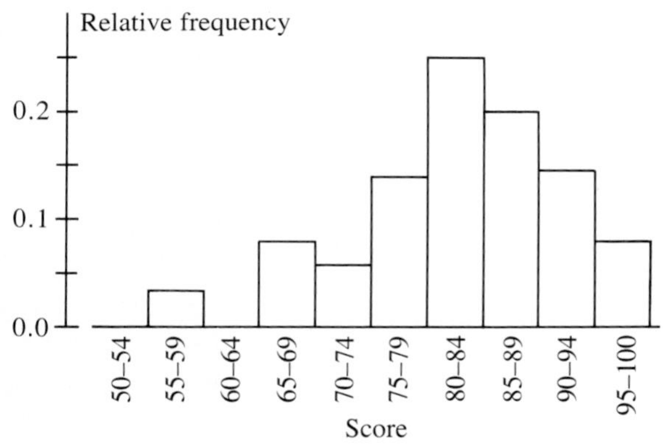

The relative frequency histogram shown at the right is the same as the frequency histogram except that the vertical axis is labeled with relative frequencies.

Relative frequency is related to probability. The probability that a test paper picked at random will have its score in the 80–84 interval is $\frac{9}{36}$, or 0.25, the same as the relative frequency of scores in that interval.

Consider now a set of n items of data having r values $x_1, x_2, \ldots, x_r$ occurring with frequencies $f_1, f_2, \ldots, f_r$, respectively. Let X denote the value of an item of data chosen at random. Then $\frac{f_i}{n} = P(X = x_i)$, and formula (†) can be written as

$$\bar{x} = \frac{1}{n}\sum_{i=1}^{r} x_i f_i = \sum_{i=1}^{r} x_i \frac{f_i}{n} = \sum_{i=1}^{r} x_i P(X = x_i).$$

This suggests the following:

DEFINITION

Let X be a random variable that takes on the values $x_1, x_2, \ldots, x_r$. Then the **mean**, μ, of the probability distribution of X is

$$\mu = \sum_{i=1}^{r} x_i P(X = x_i).$$

(The Greek letter μ is pronounced "mew.")

Note that this mean is the same as the expected value of X defined on page 613.

EXERCISES

In Exercises 1–4, find the mean, the median, and each mode for each set of data.

A **1.** The temperatures recorded at one location at three-hour intervals over a one-day period were (in degrees Celsius):

4, 5, 7, 10, 6, 5, 5, 0

2. Average monthly precipitation (in inches) for Albuquerque is:

0.4, 0.4, 0.5, 0.4, 0.5, 0.5, 1.3, 1.5, 0.9, 0.9, 0.4, 0.5

3. In one hockey season the players for the Calgary Flames scored the following numbers of goals:

37 37 24 23 28 14 20 20 16 22 14 9 19 13 19 3 11 5 2 2

4. Nine tall buildings in Columbus, Ohio, have the following heights (in feet):

624 555 512 485 456 438 357 348 346

In Exercises 5 and 6, draw a frequency histogram for each set of data.

5. A class of history students received the following quiz scores.

Score	0	1	2	3	4	5	6	7	8	9	10
Frequency	0	0	0	1	2	4	8	9	6	2	2

6. A certain machine part is said to have length 25 cm. The actual measurements of 100 parts have the following frequency distribution:

Measurement	24.8	24.9	25.0	25.1	25.2	25.3
Frequency	4	17	45	27	5	2

In Exercises 7 and 8, draw a frequency polygon for each set of data.

7. The atomic weights of 105 elements are grouped in class intervals as follows:

Atomic weight	1–25	26–50	51–75	76–100	101–125	126–150	151–175
Frequency	12	10	11	10	8	11	9

Atomic weight	176–200	201–225	226–250	251–275
Frequency	8	8	10	8

8. The batting averages of the batting champions of the National League from 1926 to 1985 are grouped in class intervals as shown at the right.

Batting average	Frequency
0.315–0.324	2
0.325–0.334	8
0.335–0.344	14
0.345–0.354	14
0.355–0.364	10
0.365–0.374	6
0.375–0.384	2
0.385–0.394	2
0.395–0.404	2

In Exercises 9 and 10, draw a relative frequency histogram for each frequency distribution.

9. Exercise 6 **10.** Exercise 7

In Exercises 11–14, find the mean, the median, and each mode for the data in the specified exercise. In Exercises 13 and 14 use the midpoints of the intervals as the data values.

11. Exercise 5 **12.** Exercise 6

13. Exercise 7 **14.** Exercise 8

15. The following list gives record wind speeds (in km/h) for 27 Canadian cities:

127	177	151	57	117	132	156	132	133
128	161	161	135	177	153	146	193	151
119	122	124	129	145	106	148	129	105

Construct a frequency distribution with class intervals 1–25, 26–50, and so on.

16. Use the frequency distribution in Exercise 15 to find the class interval with the largest number of record speeds and the class interval that contains the median record speed.

B **17.** Let X be the random variable X = record wind speed in Exercise 15. Find the mean of the probability distribution of X.

18. At high school A, the mean score of 40 students on a standardized test is 80. At high school B, the mean score of 50 students on the test is 78.

a. Find the sum of all forty scores at school A.
b. Find the sum of all fifty scores at school B.
c. Find the mean score of all ninety students.

19. At three high schools, the mean scores on a standardized test were as shown in the table. Find the mean score of all 150 students.

School	Students	Mean score
A	40	80
B	50	78
C	60	83

20. Twenty students reported the numbers of children in their families. The mean number was 1.6, the median was 1.5, and the mode was 1.0. A student from a family with three children joins the group. Find the new mean, median, and mode.

21. Find five positive numbers with mean 8, median 9, and mode 12.

C 22. The mean of n scores is M_1. The mean of another m scores is M_2.
  a. Find the mean of the $n + m$ scores.
  b. Under what conditions is the mean of the $n + m$ scores $\frac{M_1 + M_2}{2}$?

23. Show that if the values of a random variable are each increased by k, then the mean of the variable is increased by k.

24. Show that if the values of a random variable are each multiplied by k, then the mean of the variable is multiplied by k.

15-7 VARIANCE AND STANDARD DEVIATION

It is often necessary to have a measure of how much a set of data is spread out, or dispersed. One such measure is the **range**, which is the difference between the largest and smallest data values. In the test-score example on page 616, the range of the data is 100 − 56, or 44.

That the range is often an unsatisfactory measure of dispersion is illustrated by the following three sets of data:

A: 1, 1, 1, 1, 3, 7, 9, 9, 9, 9 mean = 5; range = 8
B: 1, 3, 3, 3, 5, 5, 7, 7, 7, 9 mean = 5; range = 8
C: 1, 3, 5, 5, 5, 5, 5, 6, 6, 9 mean = 5; range = 8

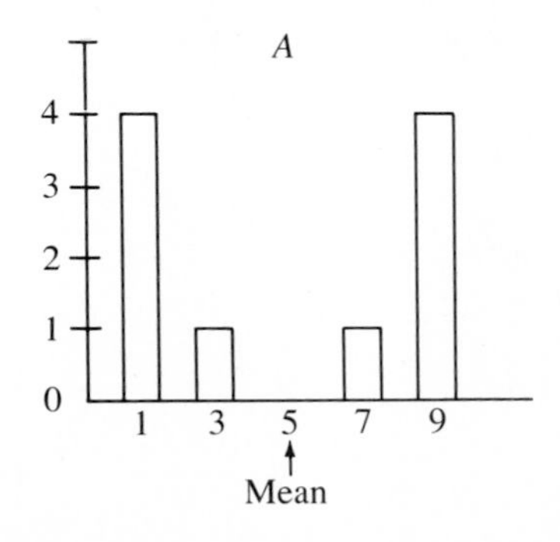

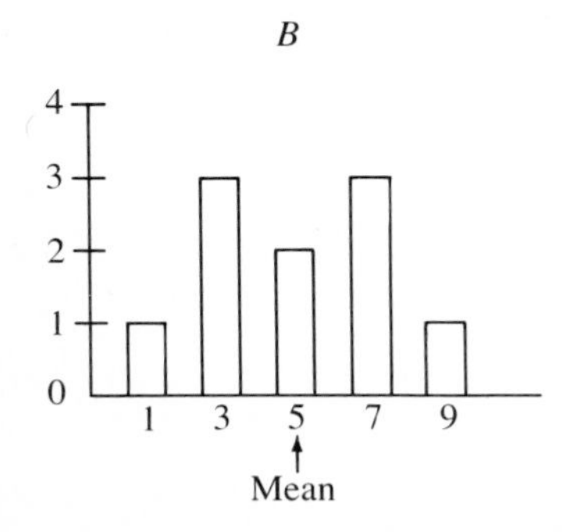

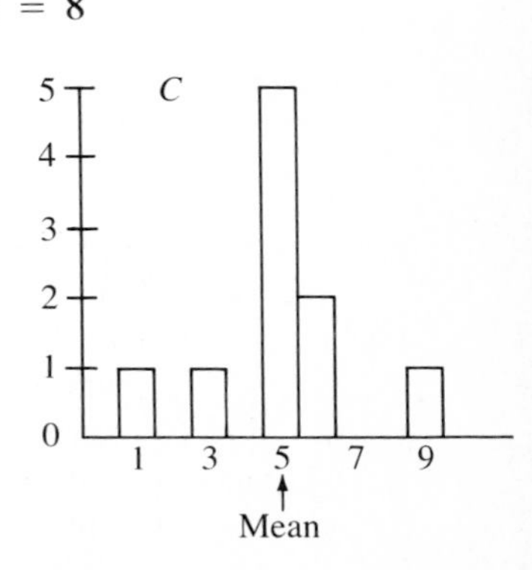

Although the three sets have the same range, A is more dispersed from its mean than B and C are. This dispersion can be detected by measures called the *variance* and the *standard deviation*. Both measures rely on the difference between a data value x_i and the mean of the data values $\bar{x}$.

DEFINITION

Let the mean of the data $x_1, x_2, \ldots, x_n$ be $\bar{x}$. The **variance** of the data is

$$\sigma^2 = \frac{1}{n} \sum_{i=1}^{n} (x_i - \bar{x})^2, \qquad (*)$$

and the **standard deviation** is σ, the principal square root of the variance. (The letter σ is the lower-case Greek sigma.)

If the given data have values $x_1, x_2, \ldots, x_r$ with frequencies $f_1, f_2, \ldots, f_r$ respectively, then the variance and the standard deviation can be calculated from

$$\sigma^2 = \frac{1}{n} \sum_{i=1}^{r} (x_i - \bar{x})^2 f_i. \qquad (\dagger)$$

The variance and standard deviation of the data sets A, B, and C are calculated from the following three tables. Recall that in all three cases the mean is 5.

x_i	f_i	$(x_i - \bar{x})^2 f_i$
1	4	$(1 - 5)^2 4 = 64$
3	1	$(3 - 5)^2 1 = 4$
5	0	$(5 - 5)^2 0 = 0$
7	1	$(7 - 5)^2 1 = 4$
9	4	$(9 - 5)^2 4 = 64$
		Sum = 136

x_i	f_i	$(x_i - \bar{x})^2 f_i$
1	1	$(1 - 5)^2 1 = 16$
3	3	$(3 - 5)^2 3 = 12$
5	2	$(5 - 5)^2 2 = 0$
7	3	$(7 - 5)^2 3 = 12$
9	1	$(9 - 5)^2 1 = 16$
		Sum = 56

x_i	f_i	$(x_i - \bar{x})^2 f_i$
1	1	$(1 - 5)^2 1 = 16$
3	1	$(3 - 5)^2 1 = 4$
5	5	$(5 - 5)^2 5 = 0$
6	2	$(6 - 5)^2 2 = 2$
9	1	$(9 - 5)^2 1 = 16$
		Sum = 38

Thus:

Variance $= \frac{136}{10} = 13.6$

Standard deviation $= \sqrt{13.6} = 3.7$

Variance $= \frac{56}{10} = 5.6$

Standard deviation $= \sqrt{5.6} = 2.4$

Variance $= \frac{38}{10} = 3.8$

Standard deviation $= \sqrt{3.8} = 1.9$

Notice that the more a set of data is spread out from its mean, the larger are its variance and standard deviation.

We sometimes find it convenient to use the following formulas for (*) and (†).

THEOREM 8 $\sigma^2 = \frac{1}{n} \sum_{i=1}^{n} x_i^2 - \bar{x}^2 \quad (**)$ $\qquad \sigma^2 = \frac{1}{n} \sum_{i=1}^{r} x_i^2 f_i - \bar{x}^2 \quad (\dagger\dagger)$

Proof: By definition $\sigma^2 = \frac{1}{n}\sum_{i=1}^{n}(x_i - \bar{x})^2$. We expand the sum to obtain

$$\sigma^2 = \frac{1}{n}\left(x_1^2 - 2x_1\bar{x} + \bar{x}^2 + \cdots + x_n^2 - 2x_n\bar{x} + \bar{x}^2\right)$$

$$= \frac{1}{n}(x_1^2 + \cdots + x_n^2) + \frac{1}{n}(-2x_1\bar{x} + \cdots - 2x_n\bar{x}) + \frac{1}{n}(n\bar{x}^2)$$

$$= \frac{1}{n}\sum_{i=1}^{n}(x_i^2) - 2\bar{x}\left(\frac{x_1 + \cdots + x_n}{n}\right) + \bar{x}^2$$

$$\sigma^2 = \frac{1}{n}\sum_{i=1}^{n}(x_i^2) - 2\bar{x}^2 + \bar{x}^2 = \frac{1}{n}\sum_{i=1}^{n}(x_i^2) - \bar{x}^2 \quad ■$$

EXAMPLE 1 In its first 20 games, a baseball team scored the following numbers of runs.

Number of runs	0	1	2	3	4	5
Frequency	2	5	6	3	3	1

Find the mean $\bar{x}$, variance σ^2, and standard deviation σ.

Solution Arrange the work in a table and use (††).

x_i	f_i	$x_i f_i$	x_i^2	$x_i^2 f_i$
0	2	0	0	0
1	5	5	1	5
2	6	12	4	24
3	3	9	9	27
4	3	12	16	48
5	1	5	25	25
	20	43		129 Totals

$$\bar{x} = \frac{1}{n}\sum x_i f_i = \tfrac{1}{20} \cdot 43 = 2.15$$

$$\sigma^2 = \frac{1}{n}\sum x_i^2 f_i - \bar{x}^2$$

$$= \tfrac{1}{20} \cdot 129 - (2.15)^2$$

$$= 1.83$$

$$\sigma = \sqrt{1.83} = 1.35$$

We can extend the concepts of variance and standard deviation to probability distributions in the same way we treated the mean on page 619. When we replace $\frac{f_i}{n}$ by $P(X = x_i)$ in (†) and (††), we obtain the formulas in the following definition.

DEFINITION

Let X be a random variable that takes on the values $x_1, x_2, \ldots, x_r$ and has mean μ. Then the **variance** of the probability distribution of X is given by

$$\sigma^2 = \sum_{i=1}^{r}(x_i - \mu)^2 P(X = x_i) \quad \text{or} \quad \sigma^2 = \sum_{i=1}^{r} x_i^2 P(X = x_i) - \mu^2.$$

The **standard deviation** of the probability distributions is σ, the principal square root of the variance.

In a *binomial* experiment with n trials and probability of success p, the probability distribution of the number of successes can be shown to have these properties:

$$\begin{aligned} \text{mean} &= \mu = np \\ \text{variance} &= \sigma^2 = np(1-p) \\ \text{standard deviation} &= \sigma = \sqrt{np(1-p)} \end{aligned}$$

EXAMPLE 2 Find the mean, variance, and standard deviation of the probability distribution of Example 3, Section 15-5.

Solution In this binomial experiment $n = 4$ and $p = \frac{2}{5}$.

$$\begin{aligned} \text{mean} &= \mu \\ &= 4 \cdot \tfrac{2}{5} = 1.6 \\ \text{variance} &= \sigma^2 \\ &= 4 \cdot \tfrac{2}{5} \cdot \tfrac{3}{5} = 0.96 \\ \text{standard deviation} &= \sigma \\ &= \sqrt{0.96} \approx 0.98 \end{aligned}$$

EXERCISES

In Exercises 1–6, find the range, variance, and standard deviation of each set of data.

A

1. Air samples containing a certain pollutant (in parts per million) were taken at noon for ten days in a city. Measures of the pollutant were:

15 6 18 7 9 12 14 16 13 10

2. Average monthly temperatures (in degrees Celsius) in Nova Scotia are:

−6 −6 −2 3 9 15 18 18 14 9 3 −3

3. Seven notable I-beam girder bridges have the following lengths (in feet):

438 340 316 316 240 240 224

4. The following table shows points scored by leading touchdown scorers in the American Football League in one season:

Number of points	54	60	66	72	108
Number of players	1	2	2	2	2

5. The following table shows actual speedometer readings of manufactured speedometers compared to a test speedometer at 50 miles per hour:

Reading	48	49	50	51	52	53	54
Number of speedometers	5	4	12	8	2	6	3

6. The following table shows the numbers of games won in one season by Ivy League basketball teams:

Games won	5	8	11	12	13	15
Teams	1	1	2	1	2	1

In Exercises 7–12, find the mean, variance, and standard deviation of the probability distribution of the data in each exercise.

7. Exercise 9, page 615
8. Exercise 10, page 615
9. Exercise 11, page 615
10. Exercise 12, page 615
11. Exercise 13, page 615
12. Exercise 14, page 615

In Exercises 13–18, find the variance and standard deviation of the probability distribution in the specified exercise.

13. Exercise 1, page 614
14. Exercise 2, page 614
15. Exercise 3, page 614
16. Exercise 4, page 614
17. Exercise 5, page 614
18. Exercise 6, page 614

B

19. The following table gives bank rates for short-term loans. Were the rates more variable for the smaller loans or the larger loans? Explain.

	5/81	5/82	5/83	5/84	5/85
\$1000–\$24,000	19.45	18.51	13.86	14.93	13.80
\$500,000–\$999,000	19.58	17.98	11.34	13.37	11.18

A fair coin is flipped n times. For each value of n, find **(a)** the mean number of heads expected, **(b)** the standard deviation in the number of heads expected, and **(c)** the probability that the number of heads expected will exceed the mean plus two standard deviations.

20. $n = 5$

21. $n = 6$

22. $n = 10$

23. $n = 12$

24. Show that if the values of a random variable are each increased by k, then the variance and standard deviation are unchanged.

25. Show that if the values of a random variable are each multiplied by k, then the variance is multiplied by k^2 and the standard deviation is multiplied by $|k|$.

C

26. Let X and Y be random variables such that $Y = aX + b$ where a and b are real. Let $\overline{X}$ and $\overline{Y}$ be the means of X and Y respectively. Show that if $\sigma_X{}^2$ is the variance of X, then the variance of Y, $\sigma_Y{}^2$, is $a^2\sigma_X{}^2$.

COMPUTER EXERCISES

1. Write a program to find the mean, variance, and standard deviation of a set of data values supplied by the user.
2. Run the program for each set of data values.
   a. Exercise 3, Section 15–7 b. Exercise 15, Section 15–6
3. Modify the program in Exercise 1 to find the mean, variance, and standard deviation of a given set of data with given frequencies.
4. Run the program in Exercise 3 for each set of data values.
   a. Exercise 5, Section 15–7 b. Exercise 6, Section 15–6

15-8 NORMAL DISTRIBUTIONS

Statisticians have found that many distributions arising in practice can be approximated by normal distributions. Among these are test scores, dimensions of manufactured parts such as ball bearings, physical characteristics such as height and weight, and annual rainfall over a long span of years.

We shall introduce normal distributions with an example. The heights of some randomly selected two-year-old trees in a reforestation project are recorded. Relative frequency histogram A gives an idea of the height distribution. For example, about 38% of the trees are between 1 m and 2 m tall, and the probability is 0.38 that a tree picked at random will be between 1 m and 2 m tall. Notice that each rectangle has width 1 and therefore the relative frequency is given by both the rectangle's area and its height.

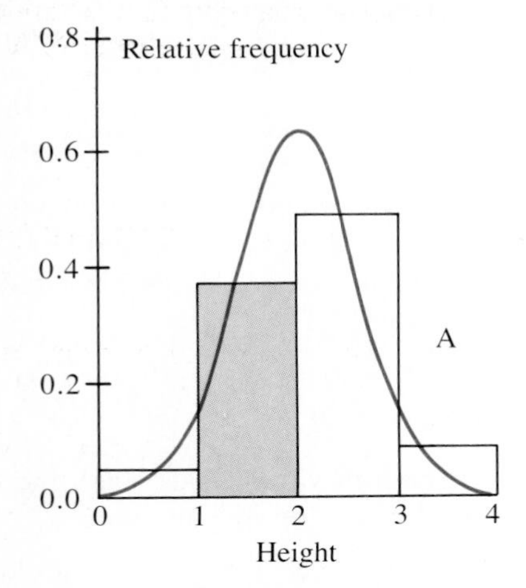

We obtain more accurate pictures of the distribution of height if we use shorter class intervals and therefore narrower rectangles. In histograms B and C, however, the heights of the rectangles no longer give relative frequencies, but the areas still do. In histogram B, for example, the relative frequency of trees between 1.5 m and 2.0 m tall is $(0.52)(0.5) = 0.26$. Thus 26% of the trees are between 1.5 m and 2.0 m tall.

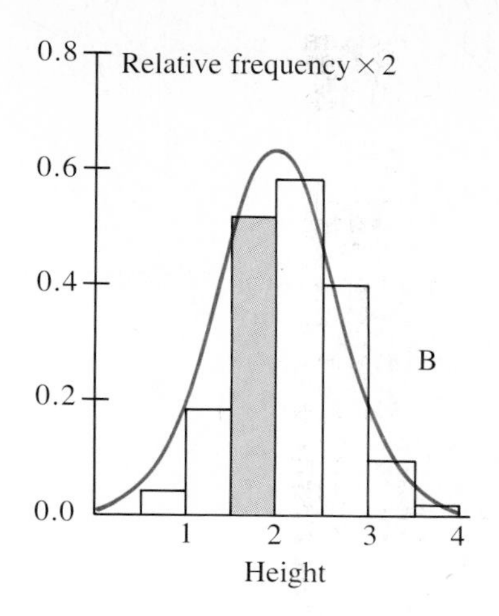

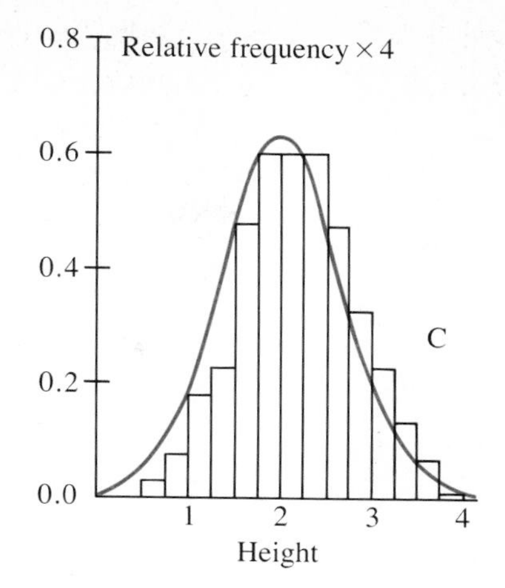

Histograms A, B, and C suggest that as the widths of the rectangles approach zero, the contours of the histograms approach the smooth bell-shaped curve shown in red in the three figures. This is an example of a *normal curve*. The corresponding theoretical frequency or probability distribution is called a *normal distribution*.

A **normal distribution** is determined by its mean μ and standard deviation σ. The corresponding **normal curve** has the equation

$$y = \frac{1}{\sigma\sqrt{2\pi}}\, e^{-\frac{1}{2}z^2} \text{ where } z = \frac{x - \mu}{\sigma}.$$

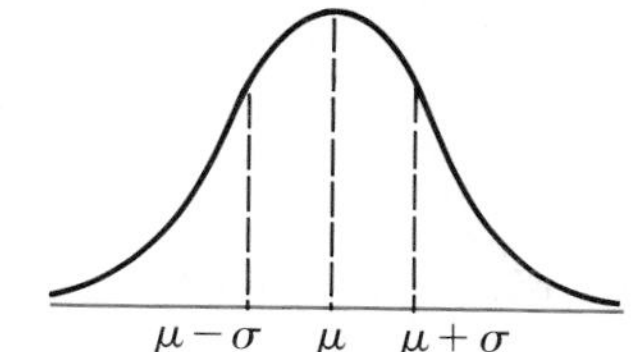

(This equation need not be memorized.)

The **standard normal distribution** is the one with mean $\mu = 0$ and standard deviation $\sigma = 1$. The **standard normal curve** has the equation

$$y = \frac{1}{\sqrt{2\pi}}\, e^{-\frac{1}{2}x^2}.$$

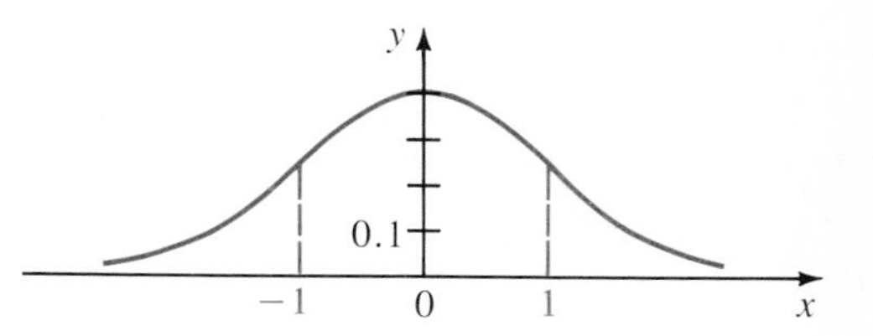

Several normal curves, along with the standard normal curve, are shown in the following figures.

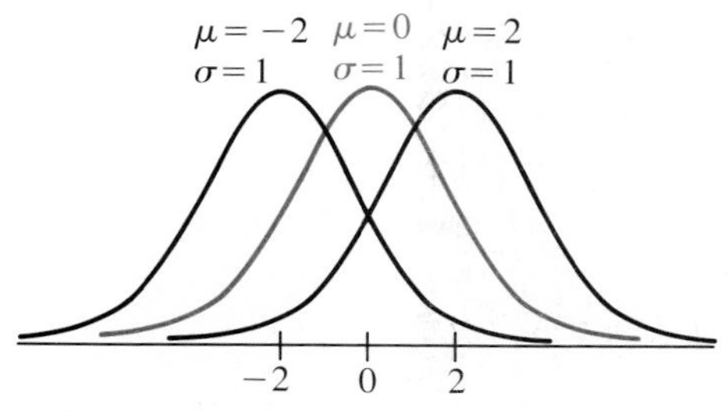

Normal distributions with same standard deviations but different means

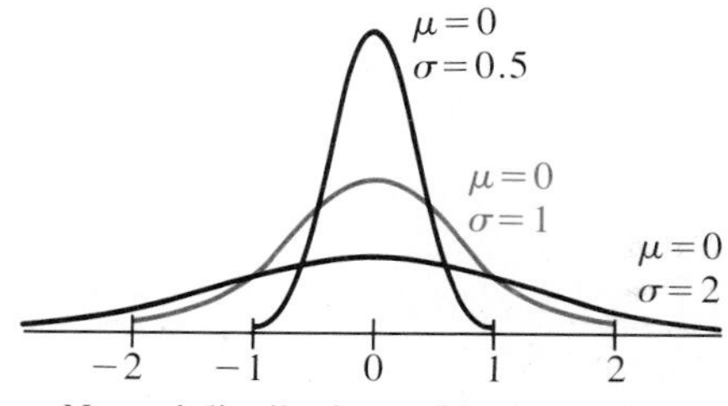

Normal distributions with same means but different standard deviations

A normal curve has the following characteristics.

1. Its maximum occurs at the mean, μ.
2. It has inflection points at $\mu \pm \sigma$.
3. It is symmetric about the line $x = \mu$ and has the x-axis as asymptote.
4. The total area under the curve is 1.

The fourth characteristic requires further explanation. The sum of the areas of the rectangles in any probability histogram is 1 since the sum of all the probabilities is 1. It can be shown, using calculus, that as the rectangles become narrower the sequence of sums approaches the total area and that this area is 1.

An additional important property can be stated in two ways.

1. The proportion (percent) of a normally distributed set of data that lies in the interval $a < x < b$ equals the area under the associated normal curve from a to b. (See the figure below.)

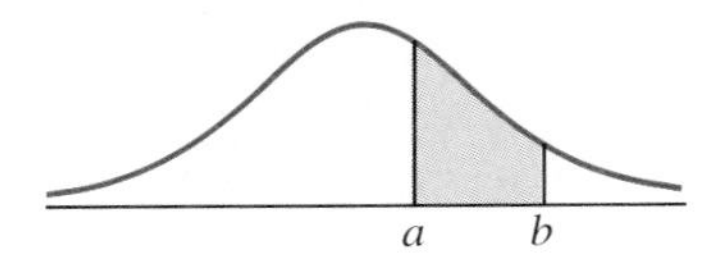

2. The probability that a normally distributed random variable takes on values between a and b equals the area under the associated normal curve from a to b. (See the figure above.)

Tables giving the area A_x under the standard normal curve from 0 to x ($x > 0$) have been compiled. A portion of such a table follows.

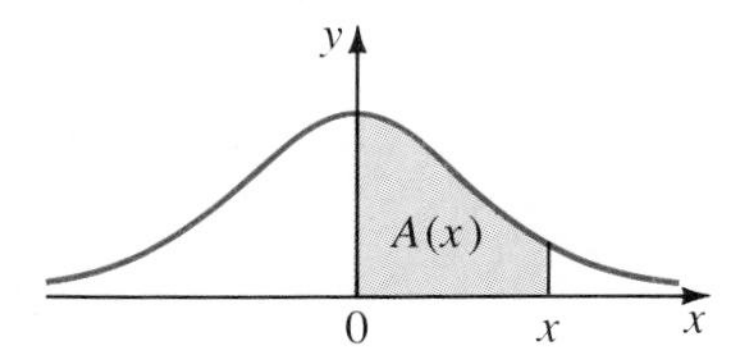

x	Area, $A(x)$	x	Area, $A(x)$	x	Area, $A(x)$	x	Area, $A(x)$
0.0	0.0000	1.0	0.3413	2.0	0.4772	3.0	0.4987
0.1	0.0398	1.1	0.3643	2.1	0.4821	3.1	0.4990
0.2	0.0793	1.2	0.3849	2.2	0.4861	3.2	0.4993
0.3	0.1179	1.3	0.4032	2.3	0.4893	3.3	0.4995
0.4	0.1554	1.4	0.4192	2.4	0.4918	3.4	0.4997
0.5	0.1915	1.5	0.4332	2.5	0.4938	3.5	0.4998
0.6	0.2257	1.6	0.4452	2.6	0.4953	3.6	0.4998
0.7	0.2580	1.7	0.4554	2.7	0.4965	3.7	0.4999
0.8	0.2881	1.8	0.4641	2.8	0.4974	3.8	0.4999
0.9	0.3159	1.9	0.4713	2.9	0.4981	3.9	0.5000
						4.0	0.5000

EXAMPLE 1 Let X be a random variable with *standard* normal distribution. Find each probability.

a. $P(-1 < X < 2)$ **b.** $P(X > 1.5)$ **c.** $P(X < 1.5)$

Solution

a. Use addition and symmetry.

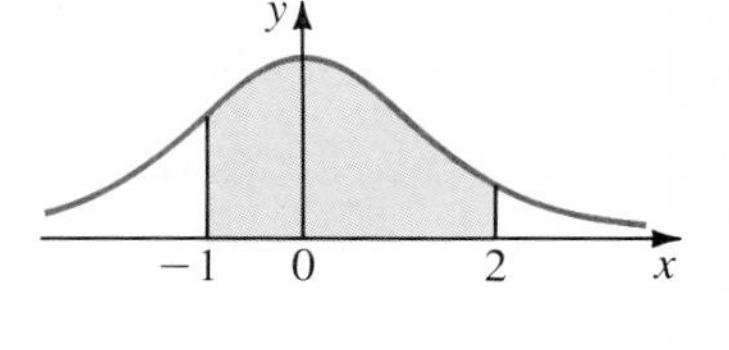

$$\begin{aligned} P(-1 < X < 2) &= P(-1 < X < 0) + P(0 < X < 2) \\ &= P(0 < X < 1) + P(0 < X < 2) \\ &= A(1) + A(2) \\ &= 0.3413 + 0.4772 \\ &= 0.8185 \end{aligned}$$

b. Because of symmetry, the area under the curve and to the right of the y-axis is 0.5.

$$\begin{aligned} P(X > 1.5) &= 0.5 - P(0 < X < 1.5) \\ &= 0.5 - A(1.5) \\ &= 0.5000 - 0.4332 = 0.0668 \end{aligned}$$

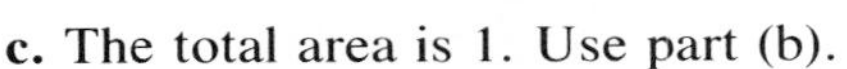

c. The total area is 1. Use part (b).

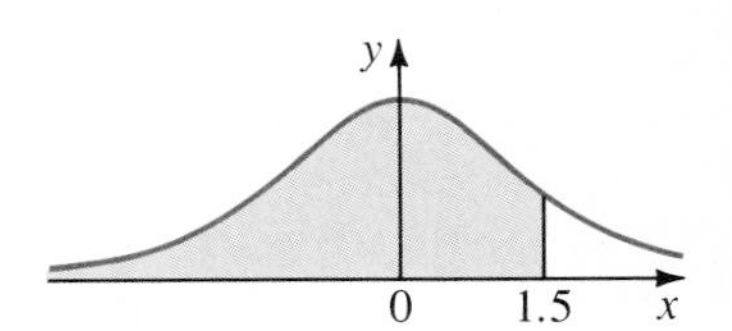

$$\begin{aligned} P(X < 1.5) &= 1 - P(X > 1.5) \\ &= 1.0000 - 0.0668 = 0.9332 \end{aligned}$$

Note that part (c) could be solved by computing $0.5 + P(0 < X < 1.5) = 0.5 + 0.4332 = 0.9332$. Note also that throughout Example 1 we used $<$ and $>$ rather than $\leq$ and $\geq$. We may do this because for any single number c, $P(X = c) = 0$.

The following fact enables us to work with *arbitrary* normal distributions.

> The area under the normal curve with mean μ and standard deviation σ from a to b equals the area under the standard normal curve from $\dfrac{a-\mu}{\sigma}$ to $\dfrac{b-\mu}{\sigma}$.

Arbitrary normal curve

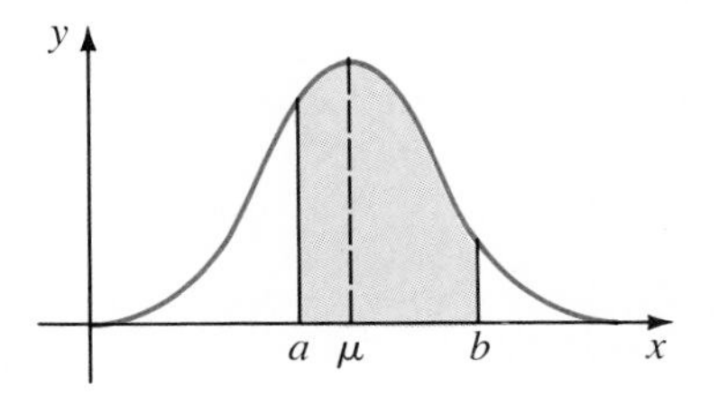

Standard normal curve

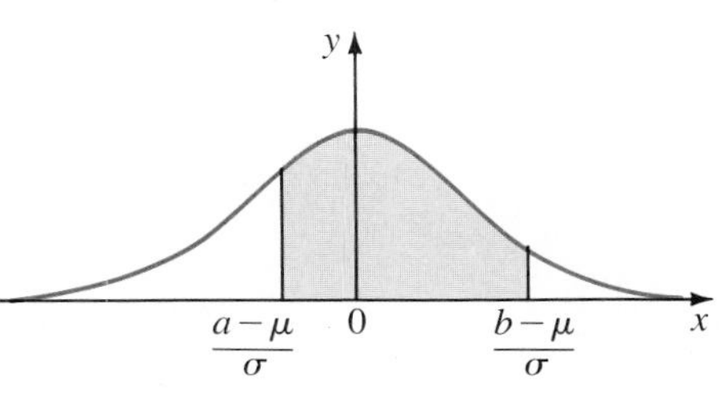

The shaded areas are equal.

In other words, the introduction of so-called **z-scores** by means of the change of scale

$$z = \frac{x - \mu}{\sigma}$$

enables us to use standard normal tables to calculate probabilities and relative frequencies.

EXAMPLE 2 The scores on a nationwide test are normally distributed with mean 75 and standard deviation 10. What percent of the scores lie between 70 and 90?

Solution Find the z-scores z_1 and z_2 corresponding to 70 and 90.

$$z = \frac{x - 75}{10}$$

$$z_1 = \frac{70 - 75}{10} = \frac{-5}{10} = -0.5$$

$$z_2 = \frac{90 - 75}{10} = \frac{15}{10} = 1.5$$

Find the area under the standard normal curve from -0.5 to 1.5.

$$\begin{aligned} P(-0.5 < z < 1.5) &= P(0 < z < 0.5) + P(0 < z < 1.5) \\ &= A(0.5) + A(1.5) \\ &= 0.1915 + 0.4332 = 0.6247 \end{aligned}$$

About 62.5% of the scores lie between 70 and 90. **Answer**

EXAMPLE 3 In Example 2, 90% of the scores are less than what value?

Solution From the standard normal curve shown at the right,

$$0.9 = 0.5 + A(z), \text{ or } A(z) = 0.4$$

for some positive z-score, z. Since 0.4 is between 0.3849 and 0.4032 in the table, z is between 1.2 and 1.3. Thus, as explained in the note below, $z = 1.28$.

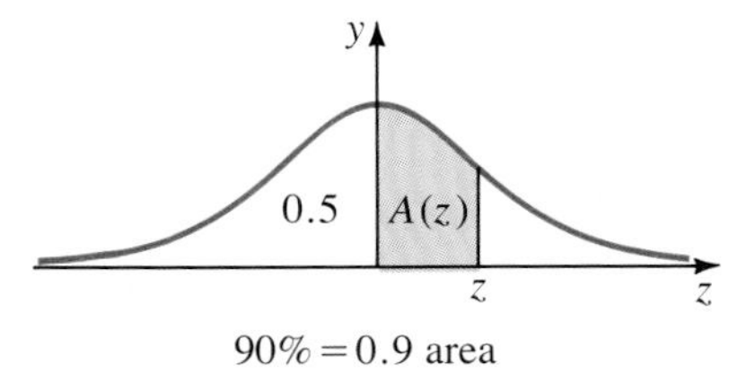

90% = 0.9 area

To find the desired test score, use the z-score formula.

$$1.28 = \frac{x - 75}{10}$$

Therefore $x = 87.8$. Hence 90% of the scores are less than 87.8. **Answer**

Note: A simple proportion was used to find the value of z in Example 3.

$$0.1\left[\begin{array}{l} a\left[\begin{array}{l} 1.2 \\ z \end{array}\right. \\ \;\; 1.3 \end{array}\right. \quad \left.\begin{array}{l} \left.\begin{array}{l} 0.3849 \\ 0.4 \end{array}\right] 0.0151 \\ 0.4032 \end{array}\right] 0.0183$$

$\frac{a}{0.1} = \frac{0.0151}{0.0183}$, from which $a = 0.08$ and $z = 1.2 + 0.08 = 1.28$

EXERCISES

A set of data has mean 20 and standard deviation 2. Find the z-score of each data value.

A **1.** 24 **2.** 18.6 **3.** 17.2 **4.** 27

A set of data has mean 57 and standard deviation 6. Find the data value of each given z-score.

5. $z = 2$ **6.** $z = 1.3$ **7.** $z = -2.2$ **8.** $z = -0.5$

In Exercises 9–14, let Y be a random variable with standard normal distribution. Use the table on page 628 to find each probability.

9. $P(Y < 1.3)$ **10.** $P(Y < -0.9)$ **11.** $P(Y > -1.7)$

12. $P(Y > 2.5)$ **13.** $P(-0.5 < Y < 3)$ **14.** $P(-2 < Y < 2)$

In Exercises 15–26, assume that the specified data are normally distributed. Use the table on page 628 as needed. Use proportions, if necessary, to approximate $A(z)$ to the nearest ten-thousandth.

15. A set of biology-test scores has mean 78 and standard deviation 8. Approximate the percent of scores that can be expected to be **(a)** above 70, **(b)** above 90, and **(c)** between 70 and 90.

16. A certain brand of washing machine has mean life 12 years and standard deviation 2.5 years. Find the probability that a machine chosen at random will last **(a)** more than 14 years, **(b)** more than 9 years, and **(c)** between 11 years and 13 years.

17. A scientist measures weight gain in a certain group of laboratory animals. The mean gain is 12.4 g and the standard deviation is 3.2 g. Find the percent of animals whose weight gain can be expected to be **(a)** less than 14 g, **(b)** more than 14 g, and **(c)** less than 6 g or more than 18 g.

18. On the verbal portion of a standardized test a group of students had a mean score of 525 and a standard deviation 75. Find the probability that a randomly chosen score is **(a)** greater than 600, **(b)** less than 300, and **(c)** between 450 and 650.

B **19.** Find the percent of data in a standard normal distribution within **(a)** one standard deviation of the mean, **(b)** two standard deviations of the mean, and **(c)** three standard deviations of the mean.

20. Annual sales of sales representatives at a certain corporation are normally distributed with mean $120,000 and standard deviation $20,000. If sales representatives in the top 10% are awarded bonuses, what level of sales is needed to get a bonus?

21. A manufacturer of 60-watt bulbs finds that the bulbs have a mean life of 1000 hours and a standard deviation of 50 hours.

a. About 80% of the bulbs can be expected to last more than __?__ hours.

b. About 5% of the bulbs can be expected to last less than __?__ hours.

22. A statistics teacher with a large class found that students achieved a mean score of 73 and a standard deviation of 6. If the teacher gives the top 35% of the students an A or a B, what is the minimum score needed to achieve at least a B?

23. Refer to Exercise 18. **a.** About 17% of the scores are less than a score of __?__. **b.** About 75% of the scores are less than a score of __?__.

24. An automobile manufacturer finds that the mean body height of a seated driver (measured from the seat) is 31 inches and has a standard deviation of 3.2 inches. What minimum height should the manufacturer allow from seat to ceiling if 97% of the drivers are to have at least 1 inch between head and ceiling?

25. A manufacturer finds that 77% of the parts manufactured by a certain machine have diameters between 1.47 cm and 1.53 cm. The mean diameter is 1.5 cm. Find the standard deviation.

26. The mean score on a college entrance examination is 500. If 800 scores of 10,000 scores are 640 or better, find the standard deviation of the scores.

15–9 CONFIDENCE INTERVALS

In statistics, a **population** is any set of individuals or objects under study. A **sample** is a subset of the population. A pollster might sample 500 voters in an area having a 400,000-voter population to estimate how many voters will vote "yes" on a proposition. In many statistical problems information about a sample is used to obtain reliable information about the total population from which the sample is taken.

If, for example, 280 voters indicate a "yes" vote, the pollster might conclude that

$$\tfrac{280}{500} = 0.56, \text{ or } 56\%,$$

of the total voter population will vote "yes". Such a *point estimate* is not necessarily a reliable indicator.

The point estimate,

$$\frac{\text{sample number favorable}}{\text{sample size}},$$

will most likely vary from sample to sample. The larger the sample size, the more accurate the estimator will be. Since most populations are large, usually we must be content with an *interval estimate*, an interval in which the desired population characteristic most likely lies. Let

$$p = \frac{\text{total number favorable}}{\text{population size}} \text{ and let } \bar{p} = \frac{\text{sample number favorable}}{\text{sample size}}.$$

With the aid of the theorem on the following page, we can obtain an interval estimate of p from $\bar{p}$. (The proof of the theorem is beyond the scope of this book.)

THEOREM 9 Let n be the sample size.

1. The values of $\overline{p}$ are normally distributed.
2. The mean of this distribution is p.
3. The standard deviation, $\sigma_{\overline{p}}$, of this distribution with mean p is
$$\sqrt{\frac{p(1-p)}{n}}.$$

Although the value of p is unknown (we wish to estimate it), we may approximate $\sigma_{\overline{p}}$ by

$$\sqrt{\frac{\overline{p}(1-\overline{p})}{n}}.$$

The following figures show the normal distribution of the values of $\overline{p}$ and the area under the normal curve over the intervals $[p - \sigma_{\overline{p}}, p + \sigma_{\overline{p}}]$, $[p - 2\sigma_{\overline{p}}, p + 2\sigma_{\overline{p}}]$, and $[p - 3\sigma_{\overline{p}}, p + 3\sigma_{\overline{p}}]$.

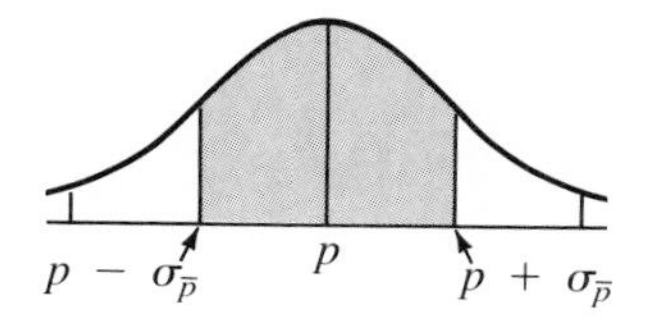

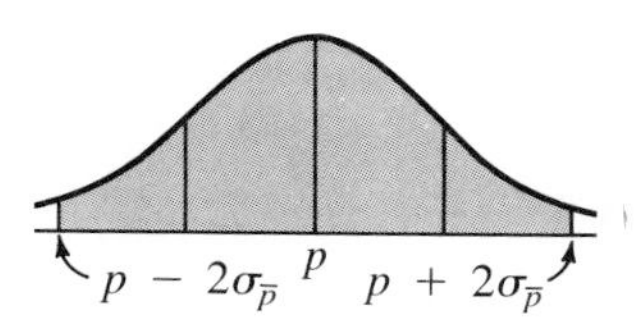

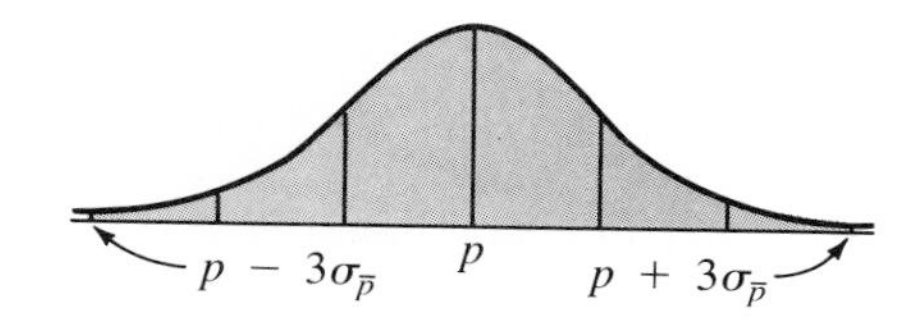

From the table on page 628, the area under the curve between $-\sigma_{\overline{p}}$ and $\sigma_{\overline{p}}$ is

$$2 \cdot A(1) = 2 \cdot 0.3413 = 0.6826.$$

Thus the probability that p lies in $[\overline{p} - \sigma_{\overline{p}}, \overline{p} + \sigma_{\overline{p}}]$ is about 68%. By similar reasoning, the probability that p lies in $[\overline{p} - 2\sigma_{\overline{p}}, \overline{p} + 2\sigma_{\overline{p}}]$ is about 95% and that p lies in $[\overline{p} - 3\sigma_{\overline{p}}, \overline{p} + 3\sigma_{\overline{p}}]$ is about 99%.

DEFINITION

A **68% confidence interval** for p is $[\overline{p} - \sigma_{\overline{p}}, \overline{p} + \sigma_{\overline{p}}]$.
A **95% confidence interval** for p is $[\overline{p} - 2\sigma_{\overline{p}}, \overline{p} + 2\sigma_{\overline{p}}]$.
A **99% confidence interval** for p is $[\overline{p} - 3\sigma_{\overline{p}}, \overline{p} + 3\sigma_{\overline{p}}]$.

EXAMPLE Out of a sample of 120 voters, a pollster finds that 70 voters favor candidate A. Find a 95% confidence interval for the fraction of all voters favoring candidate A.

Solution First find $\overline{p}$: $\overline{p} = \frac{70}{120} = 0.58$

Solution continued on following page

Next find $\sigma_{\bar{p}}$: $\sqrt{\frac{\bar{p}(1 - \bar{p})}{n}}$

$$\sqrt{\frac{(0.58)(0.42)}{120}} = 0.05$$

A 95% confidence interval is [0.58 − 2(0.05), 0.58 + 2(0.05)] = [0.48, 0.68]. The pollster can be 95% confident that between 48% and 68% of the voters favor candidate A. **Answer**

If instead 1000 voters are polled and it is found that 570 voters favor candidate A, then a 95% confidence interval is [0.57 − 2(0.02), 0.57 + 2(0.02)] = [0.53, 0.61]. This interval is narrower than the interval obtained from polling 120 voters. In general, the larger the sample size, the narrower the confidence interval. However, larger samples are more costly and time consuming.

EXERCISES

In Exercises 1–6, y is the number of successes in a sample with size n. **a.** Find the standard deviation. **b.** Find a 68% confidence interval. **c.** Find a 95% confidence interval.

A

1. $y = 80$; $n = 100$ **2.** $y = 240$; $n = 600$ **3.** $y = 9$; $n = 25$

4. $y = 300$; $n = 1200$ **5.** $y = 45$; $n = 100$ **6.** $y = 24$; $n = 150$

7. A pollster reported that out of a sample of 90 people, 60 people preferred Brand A peanut butter to Brand B. Find a 95% confidence interval for the true proportion of people who prefer Brand A to Brand B.

8. Two weeks before an election in Midtown the results of a poll of 300 people showed that 180 people would vote for candidate X. Find a 95% confidence interval for the actual proportion of people who will vote for candidate X.

9. A quality-control inspector found that 24 flashbulbs out of a sample of 500 flashbulbs were defective. Find a 95% confidence interval for the actual proportion of defective bulbs in the factory.

10. A pollster reported that 78 students out of 150 students in a school prefer playing basketball to playing football. Find a 68% confidence interval for the proportion of students in the school who prefer playing basketball.

B

11. A coin is flipped 80 times and shows heads 28 times. Use a 95% confidence interval to explain that it is unlikely the coin is fair. (*Hint*: Is 0.5 in the confidence interval?)

12. A survey found that 75 out of 250 households watched the Channel 3 news at least four nights each week. Can Channel 3 justifiably claim that 2 out of every 5 households watch its evening news?

Let $\overline{p}$ be a point estimate of p and let $\sigma_{\overline{p}}$ be the standard deviation of the set of all point estimates of p of size n.

13. Find a 90% confidence interval for p in terms of $\overline{p}$ and $\sigma_{\overline{p}}$. (*Hint*: Find x such that $A(x) = 0.45$.)

14. Find a 98% confidence interval for p in terms of $\overline{p}$ and $\sigma_{\overline{p}}$.

In Exercises 15 and 16, use the results of Exercises 13 and 14.

15. The experimental medicine ENDO-35 was found to be successful in treating VIRUS-WI in 392 out of 400 cases. Find **(a)** a 90% confidence interval and **(b)** a 98% confidence interval for the true proportion of successful treatments.

16. A thumbtack is tossed 150 times and lands point down 60 times. Find **(a)** a 90% confidence interval and **(b)** a 98% confidence interval.

15-10 CORRELATION

Statisticians often find it necessary to ask whether two variables are related, that is, whether a change in one variable produces a change in the other variable. When two variables are related, we say that there is a **correlation** between them.

The following table shows data taken at noon at a certain beach on six randomly chosen days.

Number of people	5	2	12	17	42	48
Temperature (°C)	15	18	20	24	28	33

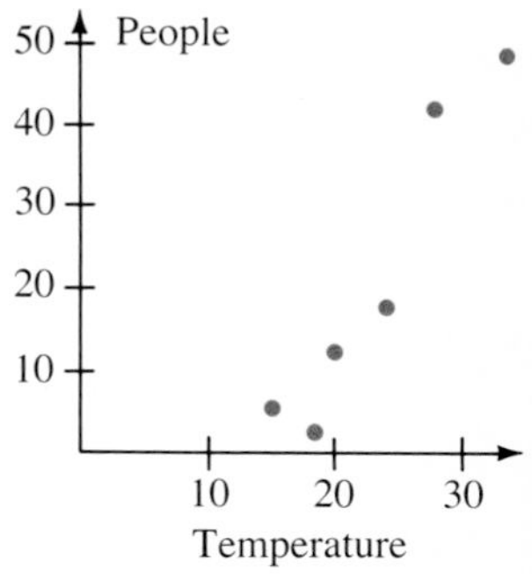

The *scatter diagram* at the right shows the data with temperature plotted along the horizontal axis and the number of people plotted along the vertical axis. From both the table and the scatter diagram, it appears that as temperature rises, the number of people at the beach increases. We say that these variables are *positively correlated*. If, on the other hand, two variables are related in such a way that as one increases the other decreases, we say the variables are *negatively correlated*. The following figures illustrate possible correlation situations.

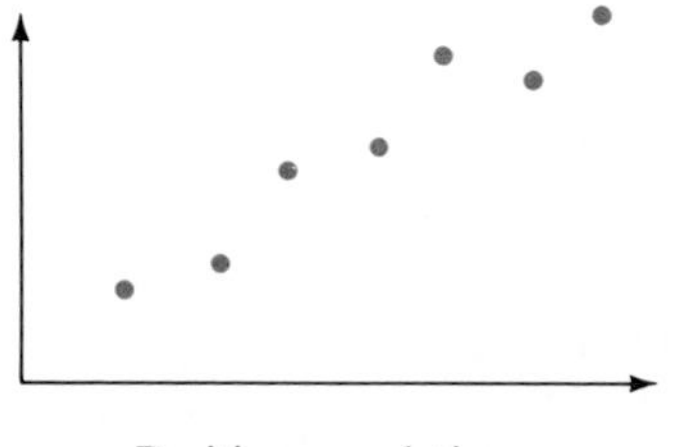

Positive correlation

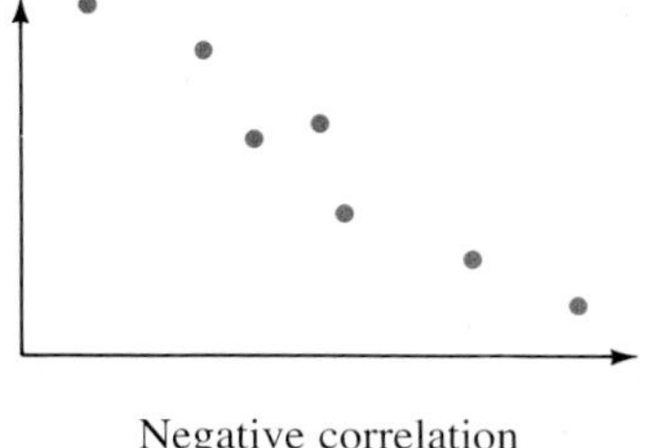

Negative correlation

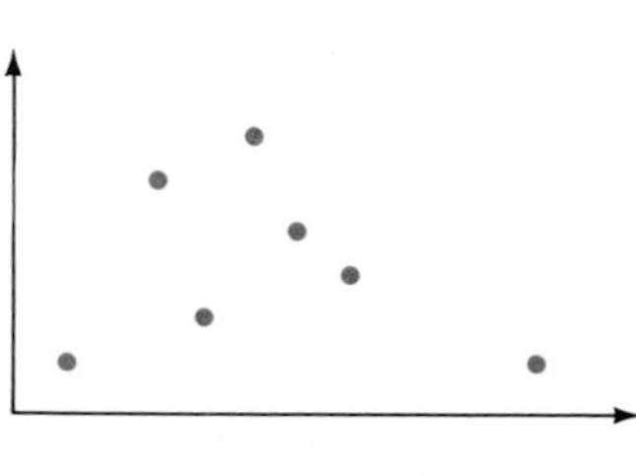

No correlation

We can obtain a numerical measurement of correlation of the beach data by transforming the data to z-scores (page 630).

No. of people (z-score)	-0.90	-1.07	-0.51	-0.23	1.19	1.52
Temperature (°C) (z-score)	-1.31	-0.82	-0.49	0.16	0.82	1.64

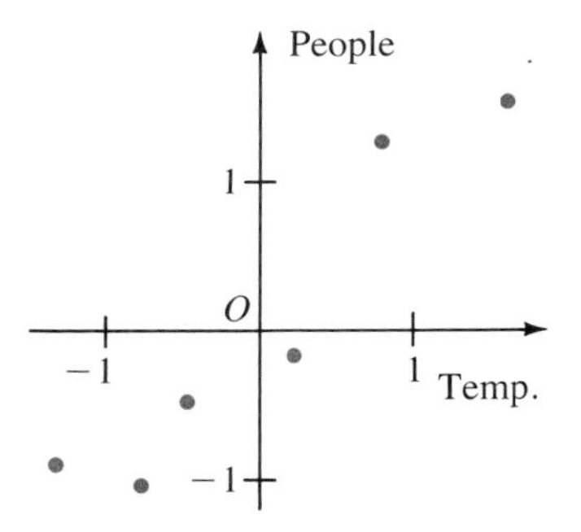

The scatter diagram of the z-scores table, shown at the right, reflects the same pattern as the original scatter diagram. Notice that for the most part the points are in the first and third quadrants, where both coordinates have the same sign. Thus the sum of the products of coordinates will be positive.

In the case where two variables are negatively correlated, the scatter diagram of the z-scores table will have points predominantly in the second and fourth quadrants, where coordinates have opposite signs. In this case, the sum of products is negative.

DEFINITION

Let x be a variable with values $x_1, x_2, \ldots, x_n$ and let y be a variable with values $y_1, y_2, \ldots, y_n$. The **correlation coefficient** r of x and y is defined by

$$r = \frac{1}{n}\left[\sum_{i=1}^{n} z_{x_i} z_{y_i}\right]$$

where z_{x_i} is the z-score of x_i and z_{y_i} is the z-score of y_i.

Application of the formula in the definition involves tedious calculations since z-scores must be computed. The following formula enables us to find the correlation coefficient directly from the values of x and y. Its proof is left as Exercise 15.

THEOREM 10 Let x be a variable with values $x_1, x_2, \ldots, x_n$ and let y be a variable with values $y_1, y_2, \ldots, y_n$.

$$r = \frac{\overline{xy} - \bar{x}\,\bar{y}}{\sigma_x \sigma_y},$$

where $\overline{xy}$ is the mean of the products $x_i y_i$, $1 \leq i \leq n$, and σ_x and σ_y are the standard deviations of x and y from $\bar{x}$ and $\bar{y}$ respectively.

EXAMPLE Refer to the beach data presented in the table on the preceding page. Find the correlation coefficient for the variables x, the number of people on the beach, and y, the noon temperature.

Solution Construct a table for x_i, y_i, x_iy_i, x_i^2, and y_i^2.

x_i	y_i	x_iy_i	x_i^2	y_i^2	
5	15	75	25	225	
2	18	36	4	324	
12	20	240	144	400	
17	24	408	289	576	
42	28	1176	1764	784	
48	33	1584	2304	1089	
126	138	3519	4530	3398	Totals

Use the totals from columns 1, 2, and 3 with $n = 6$ to find $\bar{x}$, $\bar{y}$, and $\overline{xy}$.

$$\bar{x} = \frac{126}{6} = 21 \qquad \bar{y} = \frac{138}{6} = 23 \qquad \overline{xy} = \frac{3519}{6} = 586.5$$

Use the totals from columns 4 and 5 along with the formula for σ^2, page 622, to find σ_x and σ_y:

$$\sigma_x = \sqrt{\tfrac{1}{6}(4530) - (21)^2} = 17.7 \qquad \sigma_y = \sqrt{\tfrac{1}{6}(3398) - (23)^2} = 6.1$$

Since $r = \dfrac{\overline{xy} - \bar{x}\,\bar{y}}{\sigma_x\,\sigma_y}$, $r = \dfrac{586.5 - (21)(23)}{(17.7)(6.1)} \approx 0.96$ **Answer**

It can be shown that the correlation coefficient r satisfies

$$-1 \le r \le 1.$$

In view of this fact, the coefficient $r = 0.96$ is quite high. The variables x and y are highly positively correlated. If r is close to -1, the variables in question are highly negatively correlated. A coefficient close to 0 indicates little correlation.

We should be careful not to infer a cause-effect relationship between two variables with a high correlation coefficient. We may not conclude, for example, that hot weather causes more people to go to the beach.

EXERCISES

In Exercises 1–4, state whether any correlation is likely to be positive or negative. If there seems to be no correlation, so state.

A

1. The price of an item and the demand for it
2. A nurse's salary and the number of years of work experience the nurse has
3. A student's grade-point average and the student's running speed
4. Time and speed needed to travel a given distance

In Exercises 5–10, **(a)** make a scatter diagram for the given data and **(b)** find the correlation coefficient.

5. Engine capacity and fuel economy rating

Engine capacity	61	61	90	91	113
Highway miles per gallon	50	60	42	57	41

6. Canadian economic indicators

Unemployment (%)	4.4	7.0	3.9	5.7	6.9	7.5
Inflation (%)	0.1	1.4	2.4	3.3	10.8	10.2

7. College entrance examination scores and first-year grade-point averages

Score	420	530	530	540	580	600	620	640	660	700
G.P.A.	2.3	2.8	2.2	3.1	3.2	3.4	3.6	3.5	3.8	3.9

8. Height of a basketball player and average number of points per game last season

Height (inches)	68	70	71	72	75	76	78
Points	8	8	6	10	14	7	12

9. The amount of a certain insect repellent applied to a glass surface and the number of insects landing on that surface

Amount	0	2	5	8	10	12	15	20
Insects	52	50	43	30	12	5	4	6

10. Elevation and average annual precipitation

	Amsterdam	Berlin	London	Paris	Rome
Elevation (feet)	5	187	149	164	377
Precipitation (inches)	25.6	23.1	22.9	22.3	29.5

Use the following fact in Exercises 11 and 12. When two variables x and y are strongly correlated, the data cluster around a *regression line* containing $(\bar{x}, \bar{y})$ and having slope $r\left(\frac{\sigma_y}{\sigma_x}\right)$. Find an equation of the regression line for the data in the specified exercise. Then use the equation to estimate the unknown value.

B **11.** Exercise 7. If a student has a score of 680, the student's G.P.A. is approximately __?__.

12. Exercise 9. If 35 insects land on the glass, the amount of repellent applied is approximately __?__.

Find an equation of the regression line. Use it to estimate the missing entry.

13. Boston Celtics individual statistics: minutes played and points scored

Minutes	3161	2653	2850	2976	2564	1495	1127	562	858	?
Points	2295	1565	1394	1254	971	633	499	169	180	1000

14. A scientist separated 140 plants into 7 groups of 20 plants each, exposed the plants to a particular type of bacteria for a certain time, and then gave each plant a specified dosage of an antibacterial treatment. The number of plants damaged in each group was noted.

Dosage (mL)	1.5	2.0	2.5	3.0	3.5	4.0	4.5	0.5
Plants damaged	16	14	9	8	6	2	4	?

15. Prove Theorem 10.

In Exercises 16 and 17 use the results of Exercises 23 and 24 in Section 15–6 and Exercises 24 and 25 in Section 15–7.

16. Show that the correlation coefficient is unchanged if **(a)** all values of one variable are increased by a constant k and **(b)** all values of both variables are increased by k.

17. Show that the correlation coefficient is unchanged if **(a)** all values of one variable are multiplied by a positive constant k and **(b)** all values of both variables are multiplied by k.

Chapter Summary

1. The *fundamental counting principle* states: If sets $S_1, S_2, \ldots, S_n$ contain $m_1, m_2, \ldots, m_n$ members respectively, then there are exactly $m_1 m_2 \cdots m_n$ ways of selecting a member from each set.

   There are $\frac{n!}{(n-r)!}$ *permutations* of n objects taken r at a time. There are $\frac{n!}{r!(n-r)!}$ *combinations* of n objects taken r at a time.

2. A *probability function* P assigns to each event A in a sample space S a real number $P(A)$, the probability of A, such that

   $0 \le P(A) \le 1 \qquad P(S) = 1$

   $P(A \cup B) = P(A) + P(B)$ if A and B are mutually exclusive.

The *addition rule* states:
For any events A and B, $P(A \cup B) = P(A) + P(B) - P(A \cap B)$.

The *multiplication rule* states:
For any events A and B with $P(A) \neq 0$, $P(A \cap B) = P(A) \cdot P(B \mid A)$.

The symbol $P(B \mid A)$ denotes the *conditional probability* of B given A. Events A and B are *independent* if $P(A \cap B) = P(A)P(B)$.

3. To each outcome x of an experiment, a *random variable* X assigns a probability, denoted $P(X = x)$. A listing of all such values is a *probability distribution*. The probability distribution of a random variable for a *binomial experiment* is given by

$$P(X = k) = \binom{n}{k} p^k q^{n-k}$$

where n is the number of trials and p is the probability of success. The *expected value* of a random variable with values $x_1, x_2, \ldots, x_n$ is

$$E(X) = \sum_{i=1}^{n} x_i P(X = x_i).$$

In a binomial experiment, $E(X) = np$.

4. Numerical data can be organized in a *frequency distribution*, a *histogram*, or a *frequency polygon*. The *mean* of $x_1, x_2, \ldots, x_r$ with frequencies $f_1, f_2, \ldots, f_r$ is

$$\bar{x} = \frac{1}{n} \sum_{i=1}^{r} x_i f_i.$$

The *median* and *mode* are other indicators of central tendency. The *relative frequency* of x_i is $\frac{f_i}{n}$. The *mean of a probability distribution* is the expected value of the associated random variable.

5. The *variance* σ^2 of $x_1, x_2, \ldots, x_n$ is given by

$$\sigma^2 = \frac{1}{n} \sum_{i=1}^{n} (x_i - \bar{x})^2$$

$$= \frac{1}{n} \sum_{i=1}^{n} x_i^2 - \bar{x}^2$$

or by

$$\sigma^2 = \frac{1}{n} \sum_{i=1}^{r} (x_i - \bar{x})^2 f_i$$

$$= \frac{1}{n} \sum_{i=1}^{r} x_i^2 f_i - \bar{x}^2$$

where x_i has frequency f_i.
The *standard deviation* of $x_1, x_2, \ldots, x_n$ is found by taking the principal square root of the variance.

6. Tables of areas under the *standard normal curve* are used to find probabilities of a random variable having a *standard normal distribution*. *Z-scores* are used to find probabilities of random variables whose normal distribution has a nonzero mean or a standard deviation other than 1.
7. *Confidence intervals* are obtained by computation of

$$\frac{\overline{p}(1 - \overline{p})}{n},$$

where $\overline{p}$ is the proportion of a sample with size n having a certain characteristic.
8. The *correlation coefficient* r of x and y with values $x_1, x_2, \ldots, x_n$ and $y_1, y_2, \ldots, y_n$ respectively is given by

$$r = \frac{\overline{xy} - \overline{x}\,\overline{y}}{\sigma_x\,\sigma_y},$$

where $\overline{xy}$ is the mean of the products $x_i y_i$, $\overline{x}$ is the mean of the values of x, $\overline{y}$ is the mean of the values of y, and σ_x and σ_y are the standard deviations of x and y.

Chapter Test

15-1 1. Four balls are drawn at random from a drum containing sixteen balls numbered 1, 2, . . . , 16. Balls drawn are not replaced. How many different four-ball drawings are possible?

2. Of eight dinner guests, four guests are seated on one side of a table and the other four guests are seated on the other side. How many seating arrangements are possible?

15-2 3. A wallet contains one bill in each denomination: \$1, \$5, \$10, \$20, \$50. How many different amounts are possible if three or more bills are chosen?

4. A subcommittee of four people is to be formed from a committee of nine people. A particular committee member must be on the subcommittee. A second committee member is not eligible to be on the subcommittee. How many ways are there to form the subcommittee under the given conditions?

15-3 5. Four socks are chosen at random from a drawer that contains six red socks and four green socks. Find the probability of choosing exactly two red socks from the drawer.

6. A coin is flipped three times and a die is rolled. Find the probability of tails showing twice and the die showing 6.

15-4 7. Two dice are rolled. Find the probability that at least one die shows 3 or both dice show the same number.

8. A box contains six red chips and two blue chips. A second box contains four red chips and four blue chips. One box is selected at random and then one chip is selected at random from it. Find the probability that the chip selected is red given that it came from the second box.

15-5 9. The probability of a certain thumbtack landing point down when tossed is $\frac{1}{4}$. Let X = the number of times it lands point down in three tosses. Make a probability distribution for the experiment.

10. A coin is flipped four times. Let X = the number of times that heads shows. Find the expected value of X.

Exercises 11–13 refer to the following frequency distribution of car-braking distances in meters at a fixed speed.

Braking distance	24	30	32	36	38	42	50
Frequency	2	1	4	8	5	4	1

15-6 11. Find the mean.

12. Make a relative frequency distribution.

15-7 13. Find the variance and standard deviation.

15-8 14. In a survey taken to determine the optimum size for a pair of one-size-fits-all men's socks, sock sizes were found to have a mean of 12 and a standard deviation of 1.25. Based on this survey, what percent of men can be expected to have sock sizes between 11 and 14?

15-9 15. Out of 125 workers at a certain factory, 75 workers favored a change in work schedule. Find a 95% confidence interval for the fraction of all workers at that factory favoring such a change.

15-10 16. Find the correlation coefficient of the variables "hours of television viewing" and "grade point average."

Hours of viewing	0.5	1	1.5	2	3	4
Grade point av.	3.6	3.2	2.6	2.8	1.6	1.2

COMPUTER APPLICATION

A NORMAL DISTRIBUTION

The following program can be used to simulate a normal distribution experiment such as the collection into $P + 1$ bins of N balls bouncing to the left ($-$) or the right ($+$) off equally spaced pegs in P rows of pegs. For each ball dropped the computer randomly generates a $+$ or $-$ for each row. The sum S of the number of $+$'s and $-$'s determines the bin into which the ball drops.

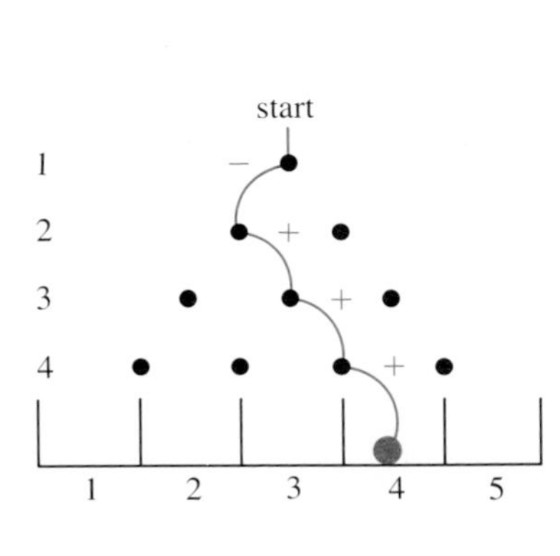


```
10  INPUT N
20  INPUT P
30  FOR I = 1 TO P + 1
40  LET B(I) = 0
50  NEXT I
60  FOR I = 1 TO N
70  LET S = 0
80  FOR J = 1 TO P
90  LET A = 2 * ( INT (RND (1) + .5) - .5)
100  IF A > 0 THEN PRINT "+";
110  IF A < 0 THEN PRINT "-";
120  LET S = S + A
130  NEXT J
140  PRINT "S = " ;S
150  LET V = (S + P + 2) / 2
160  LET B(V) = B(V) + 1
170  NEXT I
180  FOR I = 1 TO P + 1
190  PRINT B(I);" ";
200  NEXT I
210  END
```

EXERCISES

1. Run the program four times for $N = 24$ and $P = 6$. Construct a histogram for the final line of each printout.
2. Run the program four times for $N = 120$ and $P = 8$. Construct a histogram for the final line of each printout.
3. To see that the probability of flipping heads on a fair coin is about 0.5, run the program with $N = 120$ and $P = 1$.

APPLICATION

WEATHER STUDIES

A great deal of the study of weather and weather patterns is empirical. That is, the study of weather depends to a great extent on the gathering, organizing, and analyzing of numerical data. Such data involve temperature, precipitation, humidity, wind velocity, and other variables. In addition to numerical data gathered by measurement, satellite photographs often enhanced by computers are used to study storm movement.

Some meteorologists dedicate themselves to the study of hurricane formation and hurricane paths, others to the study of tornadoes. These meteorologists sometimes use aircraft to examine storms close up.

Of particular concern to meteorologists is the extent to which weather variables are correlated. The following table gives data on temperature and precipitation for ten cities. Annual average temperature T is given in degrees Fahrenheit and precipitation P is given in inches.

	T	P	TP	T^2	P^2
Portland, Me.	45	44	1980	2025	1936
Boston, Mass.	52	44	2288	2704	1936
New York, N.Y.	55	44	2420	3025	1936
Philadelphia, Pa.	54	41	2214	2916	1681
Washington, D.C.	58	39	2262	3364	1521
Richmond, Va.	58	44	2552	3364	1936
Charlotte, N.C.	60	43	2580	3600	1849
Atlanta, Ga.	61	49	2989	3721	2401
Jacksonville, Fla.	68	53	3604	4624	2809
Miami, Fla.	76	58	4408	5776	3364

To see how well T and P are correlated, σ_T and σ_P need to be computed. Columns 3, 4, and 5 of the first table provide the needed products and squares. The table below provides the needed sums and means.

	T	P	TP	T^2	P^2
Sum	587	459	27297	35119	21369
Mean	58.7	45.9	2729.7	3511.9	2136.9

Use of a formula for the standard deviation given in Section 15–7 and a formula for the correlation coefficient r given in Section 15–10 gives:

$$\sigma_T = \sqrt{3511.9 - (58.7)^2} \approx 8.1 \qquad \sigma_P = \sqrt{2136.9 - (45.9)^2} \approx 5.5$$

$$r = \frac{2729.7 - (58.7)(45.9)}{(8.1)(5.5)} \approx 0.8$$

EXERCISES

1. Construct a scatter diagram for the temperature and precipitation data given on the preceding page.

In Exercises 2 and 3, find the correlation coefficient for temperature and precipitation. (You might use a computer program to make the necessary computation.) Then draw a scatter diagram to display each set of data.

2.

Minneapolis, Minn.	45	26
Milwaukee, Wis.	46	31
Chicago, Ill.	49	33
Des Moines, Ia.	50	31
Indianapolis, Ind.	52	39
Kansas City, Mo.	54	35
St. Louis, Mo.	55	34
Memphis, Tenn.	62	52
Jackson, Miss.	65	53
New Orleans, La.	68	60

3.

Juneau, Alas.	40	53
Seattle, Wash.	51	39
Spokane, Wash.	47	17
Portland, Ore.	53	37
Boise, Idaho	51	12
Sacramento, Cal.	61	17
San Francisco, Cal.	57	20
Los Angeles, Cal.	63	12
Phoenix, Ariz.	71	7
Honolulu, H.I.	77	23

Protection from storm damage depends in large measure upon early warning, storm tracking, and both probability and statistics.

Using the Tables

Historically, mathematicians compiled tables of values of trigonometric functions, exponential and logarithmic functions, and even square root and cube root functions. This compilation was time consuming and far from complete. Notice, for example, that Table 2 includes sin $36°10'$ and sin $36°20'$ but not sin $36°16'$.

To find sin $36°16'$ by use of the table only, we need a way to approximate sin $36°16'$ from sin $36°10'$ and sin $36°20'$. To do so, we use *linear interpolation* as illustrated below.

$$10\left[\begin{array}{ll} 6\left[\begin{array}{l} 36°10' \\ 36°16' \end{array}\right. & \left.\begin{array}{r} 0.5901 \\ ? \end{array}\right] a \\ \quad 36°20' & \quad 0.5925 \end{array}\right] 0.0024$$

$$\frac{6}{10} = \frac{a}{0.0024}$$

Solve the proportion to find $a = 0.0014$. Thus sin $36°16' = 0.5901 + 0.0014 = 0.5915$.

The process of linear interpolation can also be used in reverse. To illustrate this, we shall use Table 5 to find the number whose common logarithm is 0.8594. First find the table entries closest to 0.8594.

$$0.01\left[\begin{array}{ll} a\left[\begin{array}{l} 7.23 \\ ? \end{array}\right. & \left.\begin{array}{r} 0.8591 \\ 0.8594 \end{array}\right] 3 \\ \quad 7.24 & \quad 0.8597 \end{array}\right] 6$$

$$\frac{a}{0.01} = \frac{3}{6}$$

Next solve the proportion to find $a = 0.005$. Thus $? = 7.23 + 0.005 = 7.235$.

Note:
The left-hand column of the logarithm table gives numbers between 1 and 10. That is, 72 means 7.2. The top row of the table gives the hundredths place. Thus we took 7.23 and 7.24. Note also that each table entry is a decimal between 0 and 1. Thus we took 0.8591 and 0.8597.

Trigonometric Functions of θ: decimal degrees

TABLE 1

θ Degrees	θ Radians	sin θ	cos θ	tan θ	cot θ	sec θ	csc θ		
0.0	.0000	.0000	1.0000	.0000	undefined	1.000	undefined	1.5708	**90.0**
0.1	.0017	.0017	1.0000	.0017	573.0	1.000	573.0	1.5691	89.9
0.2	.0035	.0035	1.0000	.0035	286.5	1.000	286.5	1.5673	89.8
0.3	.0052	.0052	1.0000	.0052	191.0	1.000	191.0	1.5656	89.7
0.4	.0070	.0070	1.0000	.0070	143.2	1.000	143.2	1.5638	89.6
0.5	.0087	.0087	1.0000	.0087	114.6	1.000	114.6	1.5621	89.5
0.6	.0105	.0105	.9999	.0105	95.49	1.000	95.49	1.5603	89.4
0.7	.0122	.0122	.9999	.0122	81.85	1.000	81.85	1.5586	89.3
0.8	.0140	.0140	.9999	.0140	71.62	1.000	71.62	1.5568	89.2
0.9	.0157	.0157	.9999	.0157	63.66	1.000	63.66	1.5551	89.1
1.0	.0175	.0175	.9998	.0175	57.29	1.000	57.30	1.5533	**89.0**
1.1	.0192	.0192	.9998	.0192	52.08	1.000	52.09	1.5516	88.9
1.2	.0209	.0209	.9998	.0209	47.74	1.000	47.75	1.5499	88.8
1.3	.0227	.0227	.9997	.0227	44.07	1.000	44.08	1.5481	88.7
1.4	.0244	.0244	.9997	.0244	40.92	1.000	40.93	1.5464	88.6
1.5	.0262	.0262	.9997	.0262	38.19	1.000	38.20	1.5446	88.5
1.6	.0279	.0279	.9996	.0279	35.80	1.000	35.81	1.5429	88.4
1.7	.0297	.0297	.9996	.0297	33.69	1.000	33.71	1.5411	88.3
1.8	.0314	.0314	.9995	.0314	31.82	1.000	31.84	1.5394	88.2
1.9	.0332	.0332	.9995	.0332	30.14	1.001	30.16	1.5376	88.1
2.0	.0349	.0349	.9994	.0349	28.64	1.001	28.65	1.5359	**88.0**
2.1	.0367	.0366	.9993	.0367	27.27	1.001	27.29	1.5341	87.9
2.2	.0384	.0384	.9993	.0384	26.03	1.001	26.05	1.5324	87.8
2.3	.0401	.0401	.9992	.0402	24.90	1.001	24.92	1.5307	87.7
2.4	.0419	.0419	.9991	.0419	23.86	1.001	23.88	1.5289	87.6
2.5	.0436	.0436	.9990	.0437	22.90	1.001	22.93	1.5272	87.5
2.6	.0454	.0454	.9990	.0454	22.02	1.001	22.04	1.5254	87.4
2.7	.0471	.0471	.9989	.0472	21.20	1.001	21.23	1.5237	87.3
2.8	.0489	.0488	.9988	.0489	20.45	1.001	20.47	1.5219	87.2
2.9	.0506	.0506	.9987	.0507	19.74	1.001	19.77	1.5202	87.1
3.0	.0524	.0523	.9986	.0524	19.08	1.001	19.11	1.5184	**87.0**
3.1	.0541	.0541	.9985	.0542	18.46	1.001	18.49	1.5167	86.9
3.2	.0559	.0558	.9984	.0559	17.89	1.002	17.91	1.5149	86.8
3.3	.0576	.0576	.9983	.0577	17.34	1.002	17.37	1.5132	86.7
3.4	.0593	.0593	.9982	.0594	16.83	1.002	16.86	1.5115	86.6
3.5	.0611	.0610	.9981	.0612	16.35	1.002	16.38	1.5097	86.5
3.6	.0628	.0628	.9980	.0629	15.89	1.002	15.93	1.5080	86.4
3.7	.0646	.0645	.9979	.0647	15.46	1.002	15.50	1.5062	86.3
3.8	.0663	.0663	.9978	.0664	15.06	1.002	15.09	1.5045	86.2
3.9	.0681	.0680	.9977	.0682	14.67	1.002	14.70	1.5027	86.1
4.0	.0698	.0698	.9976	.0699	14.30	1.002	14.34	1.5010	**86.0**
4.1	.0716	.0715	.9974	.0717	13.95	1.003	13.99	1.4992	85.9
4.2	.0733	.0732	.9973	.0734	13.62	1.003	13.65	1.4975	85.8
4.3	.0750	.0750	.9972	.0752	13.30	1.003	13.34	1.4957	85.7
4.4	.0768	.0767	.9971	.0769	13.00	1.003	13.03	1.4940	85.6
4.5	.0785	.0785	.9969	.0787	12.71	1.003	12.75	1.4923	85.5
4.6	.0803	.0802	.9968	.0805	12.43	1.003	12.47	1.4905	85.4
4.7	.0820	.0819	.9966	.0822	12.16	1.003	12.20	1.4888	85.3
4.8	.0838	.0837	.9965	.0840	11.91	1.004	11.95	1.4870	85.2
4.9	.0855	.0854	.9963	.0857	11.66	1.004	11.71	1.4853	85.1
5.0	.0873	.0872	.9962	.0875	11.43	1.004	11.47	1.4835	**85.0**
5.1	.0890	.0889	.9960	.0892	11.20	1.004	11.25	1.4818	84.9
5.2	.0908	.0906	.9959	.0910	10.99	1.004	11.03	1.4800	84.8
5.3	.0925	.0924	.9957	.0928	10.78	1.004	10.83	1.4783	84.7
5.4	.0942	.0941	.9956	.0945	10.58	1.004	10.63	1.4765	84.6
5.5	.0960	.0958	.9954	.0963	10.39	1.005	10.43	1.4748	84.5
5.6	.0977	.0976	.9952	.0981	10.20	1.005	10.25	1.4731	84.4
5.7	.0995	.0993	.9951	.0998	10.02	1.005	10.07	1.4713	84.3
5.8	.1012	.1011	.9949	.1016	9.845	1.005	9.895	1.4696	84.2
5.9	.1030	.1028	.9947	.1033	9.677	1.005	9.728	1.4678	84.1
6.0	.1047	.1045	.9945	.1051	9.514	1.006	9.567	1.4661	**84.0**
		cos θ	sin θ	cot θ	tan θ	csc θ	sec θ	θ Radians	θ Degrees

TABLE 1

Trigonometric Functions of θ: decimal degrees

θ Degrees	*θ Radians*	*sin θ*	*cos θ*	*tan θ*	*cot θ*	*sec θ*	*csc θ*		
6.0	.1047	.1045	.9945	.1051	9.514	1.006	9.567	1.4661	**84.0**
6.1	.1065	.1063	.9943	.1069	9.357	1.006	9.411	1.4643	83.9
6.2	.1082	.1080	.9942	.1086	9.205	1.006	9.259	1.4626	83.8
6.3	.1100	.1097	.9940	.1104	9.058	1.006	9.113	1.4608	83.7
6.4	.1117	.1115	.9938	.1122	8.915	1.006	8.971	1.4591	83.6
6.5	.1134	.1132	.9936	.1139	8.777	1.006	8.834	1.4574	83.5
6.6	.1152	.1149	.9934	.1157	8.643	1.007	8.700	1.4556	83.4
6.7	.1169	.1167	.9932	.1175	8.513	1.007	8.571	1.4539	83.3
6.8	.1187	.1184	.9930	.1192	8.386	1.007	8.446	1.4521	83.2
6.9	.1204	.1201	.9928	.1210	8.264	1.007	8.324	1.4504	83.1
7.0	.1222	.1219	.9925	.1228	8.144	1.008	8.206	1.4486	**83.0**
7.1	.1239	.1236	.9923	.1246	8.028	1.008	8.091	1.4469	82.9
7.2	.1257	.1253	.9921	.1263	7.916	1.008	7.979	1.4451	82.8
7.3	.1274	.1271	.9919	.1281	7.806	1.008	7.870	1.4434	82.7
7.4	.1292	.1288	.9917	.1299	7.700	1.008	7.764	1.4416	82.6
7.5	.1309	.1305	.9914	.1317	7.596	1.009	7.661	1.4399	82.5
7.6	.1326	.1323	.9912	.1334	7.495	1.009	7.561	1.4382	82.4
7.7	.1344	.1340	.9910	.1352	7.396	1.009	7.463	1.4364	82.3
7.8	.1361	.1357	.9907	.1370	7.300	1.009	7.368	1.4347	82.2
7.9	.1379	.1374	.9905	.1388	7.207	1.010	7.276	1.4329	82.1
8.0	.1396	.1392	.9903	.1405	7.115	1.010	7.185	1.4312	**82.0**
8.1	.1414	.1409	.9900	.1423	7.026	1.010	7.097	1.4294	81.9
8.2	.1431	.1426	.9898	.1441	6.940	1.010	7.011	1.4277	81.8
8.3	.1449	.1444	.9895	.1459	6.855	1.011	6.927	1.4259	81.7
8.4	.1466	.1461	.9893	.1477	6.772	1.011	6.845	1.4242	81.6
8.5	.1484	.1478	.9890	.1495	6.691	1.011	6.765	1.4224	81.5
8.6	.1501	.1495	.9888	.1512	6.612	1.011	6.687	1.4207	81.4
8.7	.1518	.1513	.9885	.1530	6.535	1.012	6.611	1.4190	81.3
8.8	.1536	.1530	.9882	.1548	6.460	1.012	6.537	1.4172	81.2
8.9	.1553	.1547	.9880	.1566	6.386	1.012	6.464	1.4155	81.1
9.0	.1571	.1564	.9877	.1584	6.314	1.012	6.392	1.4137	**81.0**
9.1	.1588	.1582	.9874	.1602	6.243	1.013	6.323	1.4120	80.9
9.2	.1606	.1599	.9871	.1620	6.174	1.013	6.255	1.4102	80.8
9.3	.1623	.1616	.9869	.1638	6.107	1.013	6.188	1.4085	80.7
9.4	.1641	.1633	.9866	.1655	6.041	1.014	6.123	1.4067	80.6
9.5	.1658	.1650	.9863	.1673	5.976	1.014	6.059	1.4050	80.5
9.6	.1676	.1668	.9860	.1691	5.912	1.014	5.996	1.4032	80.4
9.7	.1693	.1685	.9857	.1709	5.850	1.015	5.935	1.4015	80.3
9.8	.1710	.1702	.9854	.1727	5.789	1.015	5.875	1.3998	80.2
9.9	.1728	.1719	.9851	.1745	5.730	1.015	5.816	1.3980	80.1
10.0	.1745	.1736	.9848	.1763	5.671	1.015	5.759	1.3963	**80.0**
10.1	.1763	.1754	.9845	.1781	5.614	1.016	5.702	1.3945	79.9
10.2	.1780	.1771	.9842	.1799	5.558	1.016	5.647	1.3928	79.8
10.3	.1798	.1788	.9839	.1817	5.503	1.016	5.593	1.3910	79.7
10.4	.1815	.1805	.9836	.1835	5.449	1.017	5.540	1.3893	79.6
10.5	.1833	.1822	.9833	.1853	5.396	1.017	5.487	1.3875	79.5
10.6	.1850	.1840	.9829	.1871	5.343	1.017	5.436	1.3858	79.4
10.7	.1868	.1857	.9826	.1890	5.292	1.018	5.386	1.3840	79.3
10.8	.1885	.1874	.9823	.1908	5.242	1.018	5.337	1.3823	79.2
10.9	.1902	.1891	.9820	.1926	5.193	1.018	5.288	1.3806	79.1
11.0	.1920	.1908	.9816	.1944	5.145	1.019	5.241	1.3788	**79.0**
11.1	.1937	.1925	.9813	.1962	5.097	1.019	5.194	1.3771	78.9
11.2	.1955	.1942	.9810	.1980	5.050	1.019	5.148	1.3753	78.8
11.3	.1972	.1959	.9806	.1998	5.005	1.020	5.103	1.3736	78.7
11.4	.1990	.1977	.9803	.2016	4.959	1.020	5.059	1.3718	78.6
11.5	.2007	.1994	.9799	.2035	4.915	1.020	5.016	1.3701	78.5
11.6	.2025	.2011	.9796	.2053	4.872	1.021	4.973	1.3683	78.4
11.7	.2042	.2028	.9792	.2071	4.829	1.021	4.931	1.3666	78.3
11.8	.2059	.2045	.9789	.2089	4.787	1.022	4.890	1.3648	78.2
11.9	.2077	.2062	.9785	.2107	4.745	1.022	4.850	1.3631	78.1
12.0	.2094	.2079	.9781	.2126	4.705	1.022	4.810	1.3614	**78.0**
		cos θ	*sin θ*	*cot θ*	*tan θ*	*csc θ*	*sec θ*	*θ Radians*	*θ Degrees*

Trigonometric Functions of θ: decimal degrees

θ Degrees	θ Radians	$\sin\theta$	$\cos\theta$	$\tan\theta$	$\cot\theta$	$\sec\theta$	$\csc\theta$		
12.0	.2094	.2079	.9781	.2126	4.705	1.022	4.810	1.3614	**78.0**
12.1	.2112	.2096	.9778	.2144	4.665	1.023	4.771	1.3596	77.9
12.2	.2129	.2113	.9774	.2162	4.625	1.023	4.732	1.3579	77.8
12.3	.2147	.2130	.9770	.2180	4.586	1.023	4.694	1.3561	77.7
12.4	.2164	.2147	.9767	.2199	4.548	1.024	4.657	1.3544	77.6
12.5	.2182	.2164	.9763	.2217	4.511	1.024	4.620	1.3526	77.5
12.6	.2199	.2181	.9759	.2235	4.474	1.025	4.584	1.3509	77.4
12.7	.2217	.2198	.9755	.2254	4.437	1.025	4.549	1.3491	77.3
12.8	.2234	.2215	.9751	.2272	4.402	1.025	4.514	1.3474	77.2
12.9	.2251	.2233	.9748	.2290	4.366	1.026	4.479	1.3456	77.1
13.0	.2269	.2250	.9744	.2309	4.331	1.026	4.445	1.3439	**77.0**
13.1	.2286	.2267	.9740	.2327	4.297	1.027	4.412	1.3422	76.9
13.2	.2304	.2284	.9736	.2345	4.264	1.027	4.379	1.3404	76.8
13.3	.2321	.2300	.9732	.2364	4.230	1.028	4.347	1.3387	76.7
13.4	.2339	.2317	.9728	.2382	4.198	1.028	4.315	1.3369	76.6
13.5	.2356	.2334	.9724	.2401	4.165	1.028	4.284	1.3352	76.5
13.6	.2374	.2351	.9720	.2419	4.134	1.029	4.253	1.3334	76.4
13.7	.2391	.2368	.9715	.2438	4.102	1.029	4.222	1.3317	76.3
13.8	.2409	.2385	.9711	.2456	4.071	1.030	4.192	1.3299	76.2
13.9	.2426	.2402	.9707	.2475	4.041	1.030	4.163	1.3282	76.1
14.0	.2443	.2419	.9703	.2493	4.011	1.031	4.134	1.3265	**76.0**
14.1	.2461	.2436	.9699	.2512	3.981	1.031	4.105	1.3247	75.9
14.2	.2478	.2453	.9694	.2530	3.952	1.032	4.077	1.3230	75.8
14.3	.2496	.2470	.9690	.2549	3.923	1.032	4.049	1.3212	75.7
14.4	.2513	.2487	.9686	.2568	3.895	1.032	4.021	1.3195	75.6
14.5	.2531	.2504	.9681	.2586	3.867	1.033	3.994	1.3177	75.5
14.6	.2548	.2521	.9677	.2605	3.839	1.033	3.967	1.3160	75.4
14.7	.2566	.2538	.9673	.2623	3.812	1.034	3.941	1.3142	75.3
14.8	.2583	.2554	.9668	.2642	3.785	1.034	3.915	1.3125	75.2
14.9	.2601	.2571	.9664	.2661	3.758	1.035	3.889	1.3107	75.1
15.0	.2618	.2588	.9659	.2679	3.732	1.035	3.864	1.3090	**75.0**
15.1	.2635	.2605	.9655	.2698	3.706	1.036	3.839	1.3073	74.9
15.2	.2653	.2622	.9650	.2717	3.681	1.036	3.814	1.3055	74.8
15.3	.2670	.2639	.9646	.2736	3.655	1.037	3.790	1.3038	74.7
15.4	.2688	.2656	.9641	.2754	3.630	1.037	3.766	1.3020	74.6
15.5	.2705	.2672	.9636	.2773	3.606	1.038	3.742	1.3003	74.5
15.6	.2723	.2689	.9632	.2792	3.582	1.038	3.719	1.2985	74.4
15.7	.2740	.2706	.9627	.2811	3.558	1.039	3.695	1.2968	74.3
15.8	.2758	.2723	.9622	.2830	3.534	1.039	3.673	1.2950	74.2
15.9	.2775	.2740	.9617	.2849	3.511	1.040	3.650	1.2933	74.1
16.0	.2793	.2756	.9613	.2867	3.487	1.040	3.628	1.2915	**74.0**
16.1	.2810	.2773	.9608	.2886	3.465	1.041	3.606	1.2898	73.9
16.2	.2827	.2790	.9603	.2905	3.442	1.041	3.584	1.2881	73.8
16.3	.2845	.2807	.9598	.2924	3.420	1.042	3.563	1.2863	73.7
16.4	.2862	.2823	.9593	.2943	3.398	1.042	3.542	1.2846	73.6
16.5	.2880	.2840	.9588	.2962	3.376	1.043	3.521	1.2828	73.5
16.6	.2897	.2857	.9583	.2981	3.354	1.043	3.500	1.2811	73.4
16.7	.2915	.2874	.9578	.3000	3.333	1.044	3.480	1.2793	73.3
16.8	.2932	.2890	.9573	.3019	3.312	1.045	3.460	1.2776	73.2
16.9	.2950	.2907	.9568	.3038	3.291	1.045	3.440	1.2758	73.1
17.0	.2967	.2924	.9563	.3057	3.271	1.046	3.420	1.2741	**73.0**
17.1	.2985	.2940	.9558	.3076	3.251	1.046	3.401	1.2723	72.9
17.2	.3002	.2957	.9553	.3096	3.230	1.047	3.382	1.2706	72.8
17.3	.3019	.2974	.9548	.3115	3.211	1.047	3.363	1.2689	72.7
17.4	.3037	.2990	.9542	.3134	3.191	1.048	3.344	1.2671	72.6
17.5	.3054	.3007	.9537	.3153	3.172	1.049	3.326	1.2654	72.5
17.6	.3072	.3024	.9532	.3172	3.152	1.049	3.307	1.2636	72.4
17.7	.3089	.3040	.9527	.3191	3.133	1.050	3.289	1.2619	72.3
17.8	.3107	.3057	.9521	.3211	3.115	1.050	3.271	1.2601	72.2
17.9	.3124	.3074	.9516	.3230	3.096	1.051	3.254	1.2584	72.1
18.0	.3142	.3090	.9511	.3249	3.078	1.051	3.236	1.2566	**72.0**
		$\cos\theta$	$\sin\theta$	$\cot\theta$	$\tan\theta$	$\csc\theta$	$\sec\theta$	θ Radians	θ Degrees

TABLE 1

Trigonometric Functions of θ: decimal degrees

θ Degrees	θ Radians	sin θ	cos θ	tan θ	cot θ	sec θ	csc θ		
18.0	.3142	.3090	.9511	.3249	3.078	1.051	3.236	1.2566	**72.0**
18.1	.3159	.3107	.9505	.3269	3.060	1.052	3.219	1.2549	71.9
18.2	.3177	.3123	.9500	.3288	3.042	1.053	3.202	1.2531	71.8
18.3	.3194	.3140	.9494	.3307	3.024	1.053	3.185	1.2514	71.7
18.4	.3211	.3156	.9489	.3327	3.006	1.054	3.168	1.2497	71.6
18.5	.3229	.3173	.9483	.3346	2.989	1.054	3.152	1.2479	71.5
18.6	.3246	.3190	.9478	.3365	2.971	1.055	3.135	1.2462	71.4
18.7	.3264	.3206	.9472	.3385	2.954	1.056	3.119	1.2444	71.3
18.8	.3281	.3223	.9466	.3404	2.937	1.056	3.103	1.2427	71.2
18.9	.3299	.3239	.9461	.3424	2.921	1.057	3.087	1.2409	71.1
19.0	.3316	.3256	.9455	.3443	2.904	1.058	3.072	1.2392	**71.0**
19.1	.3334	.3272	.9449	.3463	2.888	1.058	3.056	1.2374	70.9
19.2	.3351	.3289	.9444	.3482	2.872	1.059	3.041	1.2357	70.8
19.3	.3368	.3305	.9438	.3502	2.856	1.060	3.026	1.2339	70.7
19.4	.3386	.3322	.9432	.3522	2.840	1.060	3.011	1.2322	70.6
19.5	.3403	.3338	.9426	.3541	2.824	1.061	2.996	1.2305	70.5
19.6	.3421	.3355	.9421	.3561	2.808	1.062	2.981	1.2287	70.4
19.7	.3438	.3371	.9415	.3581	2.793	1.062	2.967	1.2270	70.3
19.8	.3456	.3387	.9409	.3600	2.778	1.063	2.952	1.2252	70.2
19.9	.3473	.3404	.9403	.3620	2.762	1.064	2.938	1.2235	70.1
20.0	.3491	.3420	.9397	.3640	2.747	1.064	2.924	1.2217	**70.0**
20.1	.3508	.3437	.9391	.3659	2.733	1.065	2.910	1.2200	69.9
20.2	.3526	.3453	.9385	.3679	2.718	1.066	2.896	1.2182	69.8
20.3	.3543	.3469	.9379	.3699	2.703	1.066	2.882	1.2165	69.7
20.4	.3560	.3486	.9373	.3719	2.689	1.067	2.869	1.2147	69.6
20.5	.3578	.3502	.9367	.3739	2.675	1.068	2.855	1.2130	69.5
20.6	.3595	.3518	.9361	.3759	2.660	1.068	2.842	1.2113	69.4
20.7	.3613	.3535	.9354	.3779	2.646	1.069	2.829	1.2095	69.3
20.8	.3630	.3551	.9348	.3799	2.633	1.070	2.816	1.2078	69.2
20.9	.3648	.3567	.9342	.3819	2.619	1.070	2.803	1.2060	69.1
21.0	.3665	.3584	.9336	.3839	2.605	1.071	2.790	1.2043	**69.0**
21.1	.3683	.3600	.9330	.3859	2.592	1.072	2.778	1.2025	68.9
21.2	.3700	.3616	.9323	.3879	2.578	1.073	2.765	1.2008	68.8
21.3	.3718	.3633	.9317	.3899	2.565	1.073	2.753	1.1991	68.7
21.4	.3735	.3649	.9311	.3919	2.552	1.074	2.741	1.1973	68.6
21.5	.3752	.3665	.9304	.3939	2.539	1.075	2.729	1.1956	68.5
21.6	.3770	.3681	.9298	.3959	2.526	1.076	2.716	1.1938	68.4
21.7	.3787	.3697	.9291	.3979	2.513	1.076	2.705	1.1921	68.3
21.8	.3805	.3714	.9285	.4000	2.500	1.077	2.693	1.1903	68.2
21.9	.3822	.3730	.9278	.4020	2.488	1.078	2.681	1.1886	68.1
22.0	.3840	.3746	.9272	.4040	2.475	1.079	2.669	1.1868	**68.0**
22.1	.3857	.3762	.9265	.4061	2.463	1.079	2.658	1.1851	67.9
22.2	.3875	.3778	.9259	.4081	2.450	1.080	2.647	1.1833	67.8
22.3	.3892	.3795	.9252	.4101	2.438	1.081	2.635	1.1816	67.7
22.4	.3910	.3811	.9245	.4122	2.426	1.082	2.624	1.1798	67.6
22.5	.3927	.3827	.9239	.4142	2.414	1.082	2.613	1.1781	67.5
22.6	.3944	.3843	.9232	.4163	2.402	1.083	2.602	1.1764	67.4
22.7	.3962	.3859	.9225	.4183	2.391	1.084	2.591	1.1746	67.3
22.8	.3979	.3875	.9219	.4204	2.379	1.085	2.581	1.1729	67.2
22.9	.3997	.3891	.9212	.4224	2.367	1.086	2.570	1.1711	67.1
23.0	.4014	.3907	.9205	.4245	2.356	1.086	2.559	1.1694	**67.0**
23.1	.4032	.3923	.9198	.4265	2.344	1.087	2.549	1.1676	66.9
23.2	.4049	.3939	.9191	.4286	2.333	1.088	2.538	1.1659	66.8
23.3	.4067	.3955	.9184	.4307	2.322	1.089	2.528	1.1641	66.7
23.4	.4084	.3971	.9178	.4327	2.311	1.090	2.518	1.1624	66.6
23.5	.4102	.3987	.9171	.4348	2.300	1.090	2.508	1.1606	66.5
23.6	.4119	.4003	.9164	.4369	2.289	1.091	2.498	1.1589	66.4
23.7	.4136	.4019	.9157	.4390	2.278	1.092	2.488	1.1572	66.3
23.8	.4154	.4035	.9150	.4411	2.267	1.093	2.478	1.1554	66.2
23.9	.4171	.4051	.9143	.4431	2.257	1.094	2.468	1.1537	66.1
24.0	.4189	.4067	.9135	.4452	2.246	1.095	2.459	1.1519	**66.0**
		cos θ	sin θ	cot θ	tan θ	csc θ	sec θ	θ Radians	θ Degrees

Trigonometric Functions of θ: decimal degrees

TABLE 1

θ Degrees	θ Radians	$\sin\theta$	$\cos\theta$	$\tan\theta$	$\cot\theta$	$\sec\theta$	$\csc\theta$		
24.0	.4189	.4067	.9135	.4452	2.246	1.095	2.459	1.1519	**66.0**
24.1	.4206	.4083	.9128	.4473	2.236	1.095	2.449	1.1502	65.9
24.2	.4224	.4099	.9121	.4494	2.225	1.096	2.439	1.1484	65.8
24.3	.4241	.4115	.9114	.4515	2.215	1.097	2.430	1.1467	65.7
24.4	.4259	.4131	.9107	.4536	2.204	1.098	2.421	1.1449	65.6
24.5	.4276	.4147	.9100	.4557	2.194	1.099	2.411	1.1432	65.5
24.6	.4294	.4163	.9092	.4578	2.184	1.100	2.402	1.1414	65.4
24.7	.4311	.4179	.9085	.4599	2.174	1.101	2.393	1.1397	65.3
24.8	.4328	.4195	.9078	.4621	2.164	1.102	2.384	1.1380	65.2
24.9	.4346	.4210	.9070	.4642	2.154	1.102	2.375	1.1362	65.1
25.0	.4363	.4226	.9063	.4663	2.145	1.103	2.366	1.1345	**65.0**
25.1	.4381	.4242	.9056	.4684	2.135	1.104	2.357	1.1327	64.9
25.2	.4398	.4258	.9048	.4706	2.125	1.105	2.349	1.1310	64.8
25.3	.4416	.4274	.9041	.4727	2.116	1.106	2.340	1.1292	64.7
25.4	.4433	.4289	.9033	.4748	2.106	1.107	2.331	1.1275	64.6
25.5	.4451	.4305	.9026	.4770	2.097	1.108	2.323	1.1257	64.5
25.6	.4468	.4321	.9018	.4791	2.087	1.109	2.314	1.1240	64.4
25.7	.4485	.4337	.9011	.4813	2.078	1.110	2.306	1.1222	64.3
25.8	.4503	.4352	.9003	.4834	2.069	1.111	2.298	1.1205	64.2
25.9	.4520	.4368	.8996	.4856	2.059	1.112	2.289	1.1188	64.1
26.0	.4538	.4384	.8988	.4877	2.050	1.113	2.281	1.1170	**64.0**
26.1	.4555	.4399	.8980	.4899	2.041	1.114	2.273	1.1153	63.9
26.2	.4573	.4415	.8973	.4921	2.032	1.115	2.265	1.1135	63.8
26.3	.4590	.4431	.8965	.4942	2.023	1.115	2.257	1.1118	63.7
26.4	.4608	.4446	.8957	.4964	2.014	1.116	2.249	1.1100	63.6
26.5	.4625	.4462	.8949	.4986	2.006	1.117	2.241	1.1083	63.5
26.6	.4643	.4478	.8942	.5008	1.997	1.118	2.233	1.1065	63.4
26.7	.4660	.4493	.8934	.5029	1.988	1.119	2.226	1.1048	63.3
26.8	.4677	.4509	.8926	.5051	1.980	1.120	2.218	1.1030	63.2
26.9	.4695	.4524	.8918	.5073	1.971	1.121	2.210	1.1013	63.1
27.0	.4712	.4540	.8910	.5095	1.963	1.122	2.203	1.0996	**63.0**
27.1	.4730	.4555	.8902	.5117	1.954	1.123	2.195	1.0978	62.9
27.2	.4747	.4571	.8894	.5139	1.946	1.124	2.188	1.0961	62.8
27.3	.4765	.4586	.8886	.5161	1.937	1.125	2.180	1.0943	62.7
27.4	.4782	.4602	.8878	.5184	1.929	1.126	2.173	1.0926	62.6
27.5	.4800	.4617	.8870	.5206	1.921	1.127	2.166	1.0908	62.5
27.6	.4817	.4633	.8862	.5228	1.913	1.128	2.158	1.0891	62.4
27.7	.4835	.4648	.8854	.5250	1.905	1.129	2.151	1.0873	62.3
27.8	.4852	.4664	.8846	.5272	1.897	1.130	2.144	1.0856	62.2
27.9	.4869	.4679	.8838	.5295	1.889	1.132	2.137	1.0838	62.1
28.0	.4887	.4695	.8829	.5317	1.881	1.133	2.130	1.0821	**62.0**
28.1	.4904	.4710	.8821	.5340	1.873	1.134	2.123	1.0804	61.9
28.2	.4922	.4726	.8813	.5362	1.865	1.135	2.116	1.0786	61.8
28.3	.4939	.4741	.8805	.5384	1.857	1.136	2.109	1.0769	61.7
28.4	.4957	.4756	.8796	.5407	1.849	1.137	2.103	1.0751	61.6
28.5	.4974	.4772	.8788	.5430	1.842	1.138	2.096	1.0734	61.5
28.6	.4992	.4787	.8780	.5452	1.834	1.139	2.089	1.0716	61.4
28.7	.5009	.4802	.8771	.5475	1.827	1.140	2.082	1.0699	61.3
28.8	.5027	.4818	.8763	.5498	1.819	1.141	2.076	1.0681	61.2
28.9	.5044	.4833	.8755	.5520	1.811	1.142	2.069	1.0664	61.1
29.0	.5061	.4848	.8746	.5543	1.804	1.143	2.063	1.0647	**61.0**
29.1	.5079	.4863	.8738	.5566	1.797	1.144	2.056	1.0629	60.9
29.2	.5096	.4879	.8729	.5589	1.789	1.146	2.050	1.0612	60.8
29.3	.5114	.4894	.8721	.5612	1.782	1.147	2.043	1.0594	60.7
29.4	.5131	.4909	.8712	.5635	1.775	1.148	2.037	1.0577	60.6
29.5	.5149	.4924	.8704	.5658	1.767	1.149	2.031	1.0559	60.5
29.6	.5166	.4939	.8695	.5681	1.760	1.150	2.025	1.0542	60.4
29.7	.5184	.4955	.8686	.5704	1.753	1.151	2.018	1.0524	60.3
29.8	.5201	.4970	.8678	.5727	1.746	1.152	2.012	1.0507	60.2
29.9	.5219	.4985	.8669	.5750	1.739	1.154	2.006	1.0489	60.1
30.0	.5236	.5000	.8660	.5774	1.732	1.155	2.000	1.0472	**60.0**
		$\cos\theta$	$\sin\theta$	$\cot\theta$	$\tan\theta$	$\csc\theta$	$\sec\theta$	θ Radians	θ Degrees

TABLE 1 Trigonometric Functions of θ: decimal degrees

θ Degrees	θ Radians	sin θ	cos θ	tan θ	cot θ	sec θ	csc θ		
30.0	.5236	.5000	.8660	.5774	1.732	1.155	2.000	1.0472	**60.0**
30.1	.5253	.5015	.8652	.5797	1.725	1.156	1.994	1.0455	59.9
30.2	.5271	.5030	.8643	.5820	1.718	1.157	1.988	1.0437	59.8
30.3	.5288	.5045	.8634	.5844	1.711	1.158	1.982	1.0420	59.7
30.4	.5306	.5060	.8625	.5867	1.704	1.159	1.976	1.0402	59.6
30.5	.5323	.5075	.8616	.5890	1.698	1.161	1.970	1.0385	59.5
30.6	.5341	.5090	.8607	.5914	1.691	1.162	1.964	1.0367	59.4
30.7	.5358	.5105	.8599	.5938	1.684	1.163	1.959	1.0350	59.3
30.8	.5376	.5120	.8590	.5961	1.678	1.164	1.953	1.0332	59.2
30.9	.5393	.5135	.8581	.5985	1.671	1.165	1.947	1.0315	59.1
31.0	.5411	.5150	.8572	.6009	1.664	1.167	1.942	1.0297	**59.0**
31.1	.5428	.5165	.8563	.6032	1.658	1.168	1.936	1.0280	58.9
31.2	.5445	.5180	.8554	.6056	1.651	1.169	1.930	1.0263	58.8
31.3	.5463	.5195	.8545	.6080	1.645	1.170	1.925	1.0245	58.7
31.4	.5480	.5210	.8535	.6104	1.638	1.172	1.919	1.0228	58.6
31.5	.5498	.5225	.8526	.6128	1.632	1.173	1.914	1.0210	58.5
31.6	.5515	.5240	.8517	.6152	1.625	1.174	1.908	1.0193	58.4
31.7	.5533	.5255	.8508	.6176	1.619	1.175	1.903	1.0175	58.3
31.8	.5550	.5270	.8499	.6200	1.613	1.177	1.898	1.0158	58.2
31.9	.5568	.5284	.8490	.6224	1.607	1.178	1.892	1.0140	58.1
32.0	.5585	.5299	.8480	.6249	1.600	1.179	1.887	1.0123	**58.0**
32.1	.5603	.5314	.8471	.6273	1.594	1.180	1.882	1.0105	57.9
32.2	.5620	.5329	.8462	.6297	1.588	1.182	1.877	1.0088	57.8
32.3	.5637	.5344	.8453	.6322	1.582	1.183	1.871	1.0071	57.7
32.4	.5655	.5358	.8443	.6346	1.576	1.184	1.866	1.0053	57.6
32.5	.5672	.5373	.8434	.6371	1.570	1.186	1.861	1.0036	57.5
32.6	.5690	.5388	.8425	.6395	1.564	1.187	1.856	1.0018	57.4
32.7	.5707	.5402	.8415	.6420	1.558	1.188	1.851	1.0001	57.3
32.8	.5725	.5417	.8406	.6445	1.552	1.190	1.846	.9983	57.2
32.9	.5742	.5432	.8396	.6469	1.546	1.191	1.841	.9966	57.1
33.0	.5760	.5446	.8387	.6494	1.540	1.192	1.836	.9948	**57.0**
33.1	.5777	.5461	.8377	.6519	1.534	1.194	1.831	.9931	56.9
33.2	.5794	.5476	.8368	.6544	1.528	1.195	1.826	.9913	56.8
33.3	.5812	.5490	.8358	.6569	1.522	1.196	1.821	.9896	56.7
33.4	.5829	.5505	.8348	.6594	1.517	1.198	1.817	.9879	56.6
33.5	.5847	.5519	.8339	.6619	1.511	1.199	1.812	.9861	56.5
33.6	.5864	.5534	.8329	.6644	1.505	1.201	1.807	.9844	56.4
33.7	.5882	.5548	.8320	.6669	1.499	1.202	1.802	.9826	56.3
33.8	.5899	.5563	.8310	.6694	1.494	1.203	1.798	.9809	56.2
33.9	.5917	.5577	.8300	.6720	1.488	1.205	1.793	.9791	56.1
34.0	.5934	.5592	.8290	.6745	1.483	1.206	1.788	.9774	**56.0**
34.1	.5952	.5606	.8281	.6771	1.477	1.208	1.784	.9756	55.9
34.2	.5969	.5621	.8271	.6796	1.471	1.209	1.779	.9739	55.8
34.3	.5986	.5635	.8261	.6822	1.466	1.211	1.775	.9721	55.7
34.4	.6004	.5650	.8251	.6847	1.460	1.212	1.770	.9704	55.6
34.5	.6021	.5664	.8241	.6873	1.455	1.213	1.766	.9687	55.5
34.6	.6039	.5678	.8231	.6899	1.450	1.215	1.761	.9669	55.4
34.7	.6056	.5693	.8221	.6924	1.444	1.216	1.757	.9652	55.3
34.8	.6074	.5707	.8211	.6950	1.439	1.218	1.752	.9634	55.2
34.9	.6091	.5721	.8202	.6976	1.433	1.219	1.748	.9617	55.1
35.0	.6109	.5736	.8192	.7002	1.428	1.221	1.743	.9599	**55.0**
35.1	.6126	.5750	.8181	.7028	1.423	1.222	1.739	.9582	54.9
35.2	.6144	.5764	.8171	.7054	1.418	1.224	1.735	.9564	54.8
35.3	.6161	.5779	.8161	.7080	1.412	1.225	1.731	.9547	54.7
35.4	.6178	.5793	.8151	.7107	1.407	1.227	1.726	.9529	54.6
35.5	.6196	.5807	.8141	.7133	1.402	1.228	1.722	.9512	54.5
35.6	.6213	.5821	.8131	.7159	1.397	1.230	1.718	.9495	54.4
35.7	.6231	.5835	.8121	.7186	1.392	1.231	1.714	.9477	54.3
35.8	.6248	.5850	.8111	.7212	1.387	1.233	1.710	.9460	54.2
35.9	.6266	.5864	.8100	.7239	1.381	1.235	1.705	.9442	54.1
36.0	.6283	.5878	.8090	.7265	1.376	1.236	1.701	.9425	**54.0**
		cos θ	sin θ	cot θ	tan θ	csc θ	sec θ	θ Radians	θ Degrees

Trigonometric Functions of θ: decimal degrees

TABLE 1

θ Degrees	θ Radians	sin θ	cos θ	tan θ	cot θ	sec θ	csc θ		
36.0	.6283	.5878	.8090	.7265	1.376	1.236	1.701	.9425	**54.0**
36.1	.6301	.5892	.8080	.7292	1.371	1.238	1.697	.9407	53.9
36.2	.6318	.5906	.8070	.7319	1.366	1.239	1.693	.9390	53.8
36.3	.6336	.5920	.8059	.7346	1.361	1.241	1.689	.9372	53.7
36.4	.6353	.5934	.8049	.7373	1.356	1.242	1.685	.9355	53.6
36.5	.6370	.5948	.8039	.7400	1.351	1.244	1.681	.9338	53.5
36.6	.6388	.5962	.8028	.7427	1.347	1.246	1.677	.9320	53.4
36.7	.6405	.5976	.8018	.7454	1.342	1.247	1.673	.9303	53.3
36.8	.6423	.5990	.8007	.7481	1.337	1.249	1.669	.9285	53.2
36.9	.6440	.6004	.7997	.7508	1.332	1.250	1.666	.9268	53.1
37.0	.6458	.6018	.7986	.7536	1.327	1.252	1.662	.9250	**53.0**
37.1	.6475	.6032	.7976	.7563	1.322	1.254	1.658	.9233	52.9
37.2	.6493	.6046	.7965	.7590	1.317	1.255	1.654	.9215	52.8
37.3	.6510	.6060	.7955	.7618	1.313	1.257	1.650	.9198	52.7
37.4	.6528	.6074	.7944	.7646	1.308	1.259	1.646	.9180	52.6
37.5	.6545	.6088	.7934	.7673	1.303	1.260	1.643	.9163	52.5
37.6	.6562	.6101	.7923	.7701	1.299	1.262	1.639	.9146	52.4
37.7	.6580	.6115	.7912	.7729	1.294	1.264	1.635	.9128	52.3
37.8	.6597	.6129	.7902	.7757	1.289	1.266	1.632	.9111	52.2
37.9	.6615	.6143	.7891	.7785	1.285	1.267	1.628	.9093	52.1
38.0	.6632	.6157	.7880	.7813	1.280	1.269	1.624	.9076	**52.0**
38.1	.6650	.6170	.7869	.7841	1.275	1.271	1.621	.9058	51.9
38.2	.6667	.6184	.7859	.7869	1.271	1.272	1.617	.9041	51.8
38.3	.6685	.6198	.7848	.7898	1.266	1.274	1.613	.9023	51.7
38.4	.6702	.6211	.7837	.7926	1.262	1.276	1.610	.9006	51.6
38.5	.6720	.6225	.7826	.7954	1.257	1.278	1.606	.8988	51.5
38.6	.6737	.6239	.7815	.7983	1.253	1.280	1.603	.8971	51.4
38.7	.6754	.6252	.7804	.8012	1.248	1.281	1.599	.8954	51.3
38.8	.6772	.6266	.7793	.8040	1.244	1.283	1.596	.8936	51.2
38.9	.6789	.6280	.7782	.8069	1.239	1.285	1.592	.8919	51.1
39.0	.6807	.6293	.7771	.8098	1.235	1.287	1.589	.8901	**51.0**
39.1	.6824	.6307	.7760	.8127	1.230	1.289	1.586	.8884	50.9
39.2	.6842	.6320	.7749	.8156	1.226	1.290	1.582	.8866	50.8
39.3	.6859	.6334	.7738	.8185	1.222	1.292	1.579	.8849	50.7
39.4	.6877	.6347	.7727	.8214	1.217	1.294	1.575	.8831	50.6
39.5	.6894	.6361	.7716	.8243	1.213	1.296	1.572	.8814	50.5
39.6	.6912	.6374	.7705	.8273	1.209	1.298	1.569	.8796	50.4
39.7	.6929	.6388	.7694	.8302	1.205	1.300	1.566	.8779	50.3
39.8	.6946	.6401	.7683	.8332	1.200	1.302	1.562	.8762	50.2
39.9	.6964	.6414	.7672	.8361	1.196	1.304	1.559	.8744	50.1
40.0	.6981	.6428	.7660	.8391	1.192	1.305	1.556	.8727	**50.0**
40.1	.6999	.6441	.7649	.8421	1.188	1.307	1.552	.8709	49.9
40.2	.7016	.6455	.7638	.8451	1.183	1.309	1.549	.8692	49.8
40.3	.7034	.6468	.7627	.8481	1.179	1.311	1.546	.8674	49.7
40.4	.7051	.6481	.7615	.8511	1.175	1.313	1.543	.8657	49.6
40.5	.7069	.6494	.7604	.8541	1.171	1.315	1.540	.8639	49.5
40.6	.7086	.6508	.7593	.8571	1.167	1.317	1.537	.8622	49.4
40.7	.7103	.6521	.7581	.8601	1.163	1.319	1.534	.8604	49.3
40.8	.7121	.6534	.7570	.8632	1.159	1.321	1.530	.8587	49.2
40.9	.7138	.6547	.7559	.8662	1.154	1.323	1.527	.8570	49.1
41.0	.7156	.6561	.7547	.8693	1.150	1.325	1.524	.8552	**49.0**
41.1	.7173	.6574	.7536	.8724	1.146	1.327	1.521	.8535	48.9
41.2	.7191	.6587	.7524	.8754	1.142	1.329	1.518	.8517	48.8
41.3	.7208	.6600	.7513	.8785	1.138	1.331	1.515	.8500	48.7
41.4	.7226	.6613	.7501	.8816	1.134	1.333	1.512	.8482	48.6
41.5	.7243	.6626	.7490	.8847	1.130	1.335	1.509	.8465	48.5
41.6	.7261	.6639	.7478	.8878	1.126	1.337	1.506	.8447	48.4
41.7	.7278	.6652	.7466	.8910	1.122	1.339	1.503	.8430	48.3
41.8	.7295	.6665	.7455	.8941	1.118	1.341	1.500	.8412	48.2
41.9	.7313	.6678	.7443	.8972	1.115	1.344	1.497	.8395	48.1
42.0	.7330	.6691	.7431	.9004	1.111	1.346	1.494	.8378	**48.0**
		cos θ	sin θ	cot θ	tan θ	csc θ	sec θ	θ Radians	θ Degrees

TABLE 1

Trigonometric Functions of θ: decimal degrees

θ Degrees	θ Radians	sin θ	cos θ	tan θ	cot θ	sec θ	csc θ		
42.0	.7330	.6691	.7431	.9004	1.111	1.346	1.494	.8378	**48.0**
42.1	.7348	.6704	.7420	.9036	1.107	1.348	1.492	.8360	47.9
42.2	.7365	.6717	.7408	.9067	1.103	1.350	1.489	.8343	47.8
42.3	.7383	.6730	.7396	.9099	1.099	1.352	1.486	.8325	47.7
42.4	.7400	.6743	.7385	.9131	1.095	1.354	1.483	.8308	47.6
42.5	.7418	.6756	.7373	.9163	1.091	1.356	1.480	.8290	47.5
42.6	.7435	.6769	.7361	.9195	1.087	1.359	1.477	.8273	47.4
42.7	.7453	.6782	.7349	.9228	1.084	1.361	1.475	.8255	47.3
42.8	.7470	.6794	.7337	.9260	1.080	1.363	1.472	.8238	47.2
42.9	.7487	.6807	.7325	.9293	1.076	1.365	1.469	.8221	47.1
43.0	.7505	.6820	.7314	.9325	1.072	1.367	1.466	.8203	**47.0**
43.1	.7522	.6833	.7302	.9358	1.069	1.370	1.464	.8186	46.9
43.2	.7540	.6845	.7290	.9391	1.065	1.372	1.461	.8168	46.8
43.3	.7557	.6858	.7278	.9424	1.061	1.374	1.458	.8151	46.7
43.4	.7575	.6871	.7266	.9457	1.057	1.376	1.455	.8133	46.6
43.5	.7592	.6884	.7254	.9490	1.054	1.379	1.453	.8116	46.5
43.6	.7610	.6896	.7242	.9523	1.050	1.381	1.450	.8098	46.4
43.7	.7627	.6909	.7230	.9556	1.046	1.383	1.447	.8081	46.3
43.8	.7645	.6921	.7218	.9590	1.043	1.386	1.445	.8063	46.2
43.9	.7662	.6934	.7206	.9623	1.039	1.388	1.442	.8046	46.1
44.0	.7679	.6947	.7193	.9657	1.036	1.390	1.440	.8029	**46.0**
44.1	.7697	.6959	.7181	.9691	1.032	1.393	1.437	.8011	45.9
44.2	.7714	.6972	.7169	.9725	1.028	1.395	1.434	.7994	45.8
44.3	.7732	.6984	.7157	.9759	1.025	1.397	1.432	.7976	45.7
44.4	.7749	.6997	.7145	.9793	1.021	1.400	1.429	.7959	45.6
44.5	.7767	.7009	.7133	.9827	1.018	1.402	1.427	.7941	45.5
44.6	.7784	.7022	.7120	.9861	1.014	1.404	1.424	.7924	45.4
44.7	.7802	.7034	.7108	.9896	1.011	1.407	1.422	.7906	45.3
44.8	.7819	.7046	.7096	.9930	1.007	1.409	1.419	.7889	45.2
44.9	.7837	.7059	.7083	.9965	1.003	1.412	1.417	.7871	45.1
45.0	.7854	.7071	.7071	1.0000	1.000	1.414	1.414	.7854	**45.0**
		cos θ	sin θ	cot θ	tan θ	csc θ	sec θ	θ Radians	θ Degrees

Trigonometric Functions of θ: degrees and minutes

TABLE 2

θ Degrees	θ Radians	sin θ	csc θ	tan θ	cot θ	sec θ	cos θ		
0° 00′	.0000	.0000	Undefined	.0000	Undefined	1.000	1.0000	1.5708	90° 00′
10′	.0029	.0029	343.8	.0029	343.8	1.000	1.0000	1.5679	50′
20′	.0058	.0058	171.9	.0058	171.9	1.000	1.0000	1.5650	40′
30′	.0087	.0087	114.6	.0087	114.6	1.000	1.0000	1.5621	30′
40′	.0116	.0116	85.95	.0116	85.94	1.000	.9999	1.5592	20′
50′	.0145	.0145	68.76	.0145	68.75	1.000	.9999	1.5563	10′
1° 00′	.0175	.0175	57.30	.0175	57.29	1.000	.9998	1.5533	89° 00′
10′	.0204	.0204	49.11	.0204	49.10	1.000	.9998	1.5504	50′
20′	.0233	.0233	42.98	.0233	42.96	1.000	.9997	1.5475	40′
30′	.0262	.0262	38.20	.0262	38.19	1.000	.9997	1.5446	30′
40′	.0291	.0291	34.38	.0291	34.37	1.000	.9996	1.5417	20′
50′	.0320	.0320	31.26	.0320	31.24	1.001	.9995	1.5388	10′
2° 00′	.0349	.0349	28.65	.0349	28.64	1.001	.9994	1.5359	88° 00′
10′	.0378	.0378	26.45	.0378	26.43	1.001	.9993	1.5330	50′
20′	.0407	.0407	24.56	.0407	24.54	1.001	.9992	1.5301	40′
30′	.0436	.0436	22.93	.0437	22.90	1.001	.9990	1.5272	30′
40′	.0465	.0465	21.49	.0466	21.47	1.001	.9989	1.5243	20′
50′	.0495	.0494	20.23	.0495	20.21	1.001	.9988	1.5213	10′
3° 00′	.0524	.0523	19.11	.0524	19.08	1.001	.9986	1.5184	87° 00′
10′	.0553	.0552	18.10	.0553	18.07	1.002	.9985	1.5155	50′
20′	.0582	.0581	17.20	.0582	17.17	1.002	.9983	1.5126	40′
30′	.0611	.0610	16.38	.0612	16.35	1.002	.9981	1.5097	30′
40′	.0640	.0640	15.64	.0641	15.60	1.002	.9980	1.5068	20′
50′	.0669	.0669	14.96	.0670	14.92	1.002	.9978	1.5039	10′
4° 00′	.0698	.0698	14.34	.0699	14.30	1.002	.9976	1.5010	86° 00′
10′	.0727	.0727	13.76	.0729	13.73	1.003	.9974	1.4981	50′
20′	.0756	.0756	13.23	.0758	13.20	1.003	.9971	1.4952	40′
30′	.0785	.0785	12.75	.0787	12.71	1.003	.9969	1.4923	30′
40′	.0814	.0814	12.29	.0816	12.25	1.003	.9967	1.4893	20′
50′	.0844	.0843	11.87	.0846	11.83	1.004	.9964	1.4864	10′
5° 00′	.0873	.0872	11.47	.0875	11.43	1.004	.9962	1.4835	85° 00′
10′	.0902	.0901	11.10	.0904	11.06	1.004	.9959	1.4806	50′
20′	.0931	.0929	10.76	.0934	10.71	1.004	.9957	1.4777	40′
30′	.0960	.0958	10.43	.0963	10.39	1.005	.9954	1.4748	30′
40′	.0989	.0987	10.13	.0992	10.08	1.005	.9951	1.4719	20′
50′	.1018	.1016	9.839	.1022	9.788	1.005	.9948	1.4690	10′
6° 00′	.1047	.1045	9.567	.1051	9.514	1.006	.9945	1.4661	84° 00′
10′	.1076	.1074	9.309	.1080	9.255	1.006	.9942	1.4632	50′
20′	.1105	.1103	9.065	.1110	9.010	1.006	.9939	1.4603	40′
30′	.1134	.1132	8.834	.1139	8.777	1.006	.9936	1.4573	30′
40′	.1164	.1161	8.614	.1169	8.556	1.007	.9932	1.4544	20′
50′	.1193	.1190	8.405	.1198	8.345	1.007	.9929	1.4515	10′
7° 00′	.1222	.1219	8.206	.1228	8.144	1.008	.9925	1.4486	83° 00′
10′	.1251	.1248	8.016	.1257	7.953	1.008	.9922	1.4457	50′
20′	.1280	.1276	7.834	.1287	7.770	1.008	.9918	1.4428	40′
30′	.1309	.1305	7.661	.1317	7.596	1.009	.9914	1.4399	30′
40′	.1338	.1334	7.496	.1346	7.429	1.009	.9911	1.4370	20′
50′	.1367	.1363	7.337	.1376	7.269	1.009	.9907	1.4341	10′
8° 00′	.1396	.1392	7.185	.1405	7.115	1.010	.9903	1.4312	82° 00′
10′	.1425	.1421	7.040	.1435	6.968	1.010	.9899	1.4283	50′
20′	.1454	.1449	6.900	.1465	6.827	1.011	.9894	1.4254	40′
30′	.1484	.1478	6.765	.1495	6.691	1.011	.9890	1.4224	30′
40′	.1513	.1507	6.636	.1524	6.561	1.012	.9886	1.4195	20′
50′	.1542	.1536	6.512	.1554	6.435	1.012	.9881	1.4166	10′
9° 00′	.1571	.1564	6.392	.1584	6.314	1.012	.9877	1.4137	81° 00′
		cos θ	sec θ	cot θ	tan θ	csc θ	sin θ	θ Radians	θ Degrees

TABLE 2

Trigonometric Functions of θ:
degrees and minutes

θ Degrees	θ Radians	sin θ	csc θ	tan θ	cot θ	sec θ	cos θ		
9° 00′	.1571	.1564	6.392	.1584	6.314	1.012	.9877	1.4137	81° 00′
10′	.1600	.1593	6.277	.1614	6.197	1.013	.9872	1.4108	50′
20′	.1629	.1622	6.166	.1644	6.084	1.013	.9868	1.4079	40′
30′	.1658	.1650	6.059	.1673	5.976	1.014	.9863	1.4050	30′
40′	.1687	.1679	5.955	.1703	5.871	1.014	.9858	1.4021	20′
50′	.1716	.1708	5.855	.1733	5.769	1.015	.9853	1.3992	10′
10° 00′	.1745	.1736	5.759	.1763	5.671	1.015	.9848	1.3963	80° 00′
10′	.1774	.1765	5.665	.1793	5.576	1.016	.9843	1.3934	50′
20′	.1804	.1794	5.575	.1823	5.485	1.016	.9838	1.3904	40′
30′	.1833	.1822	5.487	.1853	5.396	1.017	.9833	1.3875	30′
40′	.1862	.1851	5.403	.1883	5.309	1.018	.9827	1.3846	20′
50′	.1891	.1880	5.320	.1914	5.226	1.018	.9822	1.3817	10′
11° 00′	.1920	.1908	5.241	.1944	5.145	1.019	.9816	1.3788	79° 00′
10′	.1949	.1937	5.164	.1974	5.066	1.019	.9811	1.3759	50′
20′	.1978	.1965	5.089	.2004	4.989	1.020	.9805	1.3730	40′
30′	.2007	.1994	5.016	.2035	4.915	1.020	.9799	1.3701	30′
40′	.2036	.2022	4.945	.2065	4.843	1.021	.9793	1.3672	20′
50′	.2065	.2051	4.876	.2095	4.773	1.022	.9787	1.3643	10′
12° 00′	.2094	.2079	4.810	.2126	4.705	1.022	.9781	1.3614	78° 00′
10′	.2123	.2108	4.745	.2156	4.638	1.023	.9775	1.3584	50′
20′	.2153	.2136	4.682	.2186	4.574	1.024	.9769	1.3555	40′
30′	.2182	.2164	4.620	.2217	4.511	1.024	.9763	1.3526	30′
40′	.2211	.2193	4.560	.2247	4.449	1.025	.9757	1.3497	20′
50′	.2240	.2221	4.502	.2278	4.390	1.026	.9750	1.3468	10′
13° 00′	.2269	.2250	4.445	.2309	4.331	1.026	.9744	1.3439	77° 00′
10′	.2298	.2278	4.390	.2339	4.275	1.027	.9737	1.3410	50′
20′	.2327	.2306	4.336	.2370	4.219	1.028	.9730	1.3381	40′
30′	.2356	.2334	4.284	.2401	4.165	1.028	.9724	1.3352	30′
40′	.2385	.2363	4.232	.2432	4.113	1.029	.9717	1.3323	20′
50′	.2414	.2391	4.182	.2462	4.061	1.030	.9710	1.3294	10′
14° 00′	.2443	.2419	4.134	.2493	4.011	1.031	.9703	1.3265	76° 00′
10′	.2473	.2447	4.086	.2524	3.962	1.031	.9696	1.3235	50′
20′	.2502	.2476	4.039	.2555	3.914	1.032	.9689	1.3206	40′
30′	.2531	.2504	3.994	.2586	3.867	1.033	.9681	1.3177	30′
40′	.2560	.2532	3.950	.2617	3.821	1.034	.9674	1.3148	20′
50′	.2589	.2560	3.906	.2648	3.776	1.034	.9667	1.3119	10′
15° 00′	.2618	.2588	3.864	.2679	3.732	1.035	.9659	1.3090	75° 00′
10′	.2647	.2616	3.822	.2711	3.689	1.036	.9652	1.3061	50′
20′	.2676	.2644	3.782	.2742	3.647	1.037	.9644	1.3032	40′
30′	.2705	.2672	3.742	.2773	3.606	1.038	.9636	1.3003	30′
40′	.2734	.2700	3.703	.2805	3.566	1.039	.9628	1.2974	20′
50′	.2763	.2728	3.665	.2836	3.526	1.039	.9621	1.2945	10′
16° 00′	.2793	.2756	3.628	.2867	3.487	1.040	.9613	1.2915	74° 00′
10′	.2822	.2784	3.592	.2899	3.450	1.041	.9605	1.2886	50′
20′	.2851	.2812	3.556	.2931	3.412	1.042	.9596	1.2857	40′
30′	.2880	.2840	3.521	.2962	3.376	1.043	.9588	1.2828	30′
40′	.2909	.2868	3.487	.2994	3.340	1.044	.9580	1.2799	20′
50′	.2938	.2896	3.453	.3026	3.305	1.045	.9572	1.2770	10′
17° 00′	.2967	.2924	3.420	.3057	3.271	1.046	.9563	1.2741	73° 00′
10′	.2996	.2952	3.388	.3089	3.237	1.047	.9555	1.2712	50′
20′	.3025	.2979	3.357	.3121	3.204	1.048	.9546	1.2683	40′
30′	.3054	.3007	3.326	.3153	3.172	1.049	.9537	1.2654	30′
40′	.3083	.3035	3.295	.3185	3.140	1.049	.9528	1.2625	20′
50′	.3113	.3062	3.265	.3217	3.108	1.050	.9520	1.2595	10′
18° 00′	.3142	.3090	3.236	.3249	3.078	1.051	.9511	1.2566	72° 00′
		cos θ	sec θ	cot θ	tan θ	csc θ	sin θ	θ Radians	θ Degrees

Trigonometric Functions of θ: degrees and minutes

TABLE 2

θ Degrees	θ Radians	sin θ	csc θ	tan θ	cot θ	sec θ	cos θ		
18° 00′	.3142	.3090	3.236	.3249	3.078	1.051	.9511	1.2566	72° 00′
10′	.3171	.3118	3.207	.3281	3.047	1.052	.9502	1.2537	50′
20′	.3200	.3145	3.179	.3314	3.018	1.053	.9492	1.2508	40′
30′	.3229	.3173	3.152	.3346	2.989	1.054	.9483	1.2479	30′
40′	.3258	.3201	3.124	.3378	2.960	1.056	.9474	1.2450	20′
50′	.3287	.3228	3.098	.3411	2.932	1.057	.9465	1.2421	10′
19° 00′	.3316	.3256	3.072	.3443	2.904	1.058	.9455	1.2392	71° 00′
10′	.3345	.3283	3.046	.3476	2.877	1.059	.9446	1.2363	50′
20′	.3374	.3311	3.021	.3508	2.850	1.060	.9436	1.2334	40′
30′	.3403	.3338	2.996	.3541	2.824	1.061	.9426	1.2305	30′
40′	.3432	.3365	2.971	.3574	2.798	1.062	.9417	1.2275	20′
50′	.3462	.3393	2.947	.3607	2.773	1.063	.9407	1.2246	10′
20° 00′	.3491	.3420	2.924	.3640	2.747	1.064	.9397	1.2217	70° 00′
10′	.3520	.3448	2.901	.3673	2.723	1.065	.9387	1.2188	50′
20′	.3549	.3475	2.878	.3706	2.699	1.066	.9377	1.2159	40′
30′	.3578	.3502	2.855	.3739	2.675	1.068	.9367	1.2130	30′
40′	.3607	.3529	2.833	.3772	2.651	1.069	.9356	1.2101	20′
50′	.3636	.3557	2.812	.3805	2.628	1.070	.9346	1.2072	10′
21° 00′	.3665	.3584	2.790	.3839	2.605	1.071	.9336	1.2043	69° 00′
10′	.3694	.3611	2.769	.3872	2.583	1.072	.9325	1.2014	50′
20′	.3723	.3638	2.749	.3906	2.560	1.074	.9315	1.1985	40′
30′	.3752	.3665	2.729	.3939	2.539	1.075	.9304	1.1956	30′
40′	.3782	.3692	2.709	.3973	2.517	1.076	.9293	1.1926	20′
50′	.3811	.3719	2.689	.4006	2.496	1.077	.9283	1.1897	10′
22° 00′	.3840	.3746	2.669	.4040	2.475	1.079	.9272	1.1868	68° 00′
10′	.3869	.3773	2.650	.4074	2.455	1.080	.9261	1.1839	50′
20′	.3898	.3800	2.632	.4108	2.434	1.081	.9250	1.1810	40′
30′	.3927	.3827	2.613	.4142	2.414	1.082	.9239	1.1781	30′
40′	.3956	.3854	2.595	.4176	2.394	1.084	.9228	1.1752	20′
50′	.3985	.3881	2.577	.4210	2.375	1.085	.9216	1.1723	10′
23° 00′	.4014	.3907	2.559	.4245	2.356	1.086	.9205	1.1694	67° 00′
10′	.4043	.3934	2.542	.4279	2.337	1.088	.9194	1.1665	50′
20′	.4072	.3961	2.525	.4314	2.318	1.089	.9182	1.1636	40′
30′	.4102	.3987	2.508	.4348	2.300	1.090	.9171	1.1606	30′
40′	.4131	.4014	2.491	.4383	2.282	1.092	.9159	1.1577	20′
50′	.4160	.4041	2.475	.4417	2.264	1.093	.9147	1.1548	10′
24° 00′	.4189	.4067	2.459	.4452	2.246	1.095	.9135	1.1519	66° 00′
10′	.4218	.4094	2.443	.4487	2.229	1.096	.9124	1.1490	50′
20′	.4247	.4120	2.427	.4522	2.211	1.097	.9112	1.1461	40′
30′	.4276	.4147	2.411	.4557	2.194	1.099	.9100	1.1432	30′
40′	.4305	.4173	2.396	.4592	2.177	1.100	.9088	1.1403	20′
50′	.4334	.4200	2.381	.4628	2.161	1.102	.9075	1.1374	10′
25° 00′	.4363	.4226	2.366	.4663	2.145	1.103	.9063	1.1345	65° 00′
10′	.4392	.4253	2.352	.4699	2.128	1.105	.9051	1.1316	50′
20′	.4422	.4279	2.337	.4734	2.112	1.106	.9038	1.1286	40′
30′	.4451	.4305	2.323	.4770	2.097	1.108	.9026	1.1257	30′
40′	.4480	.4331	2.309	.4806	2.081	1.109	.9013	1.1228	20′
50′	.4509	.4358	2.295	.4841	2.066	1.111	.9001	1.1199	10′
26° 00′	.4538	.4384	2.281	.4877	2.050	1.113	.8988	1.1170	64° 00′
10′	.4567	.4410	2.268	.4913	2.035	1.114	.8975	1.1141	50′
20′	.4596	.4436	2.254	.4950	2.020	1.116	.8962	1.1112	40′
30′	.4625	.4462	2.241	.4986	2.006	1.117	.8949	1.1083	30′
40′	.4654	.4488	2.228	.5022	1.991	1.119	.8936	1.1054	20′
50′	.4683	.4514	2.215	.5059	1.977	1.121	.8923	1.1025	10′
27° 00′	.4712	.4540	2.203	.5095	1.963	1.122	.8910	1.0996	63° 00′
		cos θ	sec θ	cot θ	tan θ	csc θ	sin θ	θ Radians	θ Degrees

TABLE 2 Trigonometric Functions of θ: degrees and minutes

θ Degrees	θ Radians	sin θ	csc θ	tan θ	cot θ	sec θ	cos θ		
27° 00′	.4712	.4540	2.203	.5095	1.963	1.122	.8910	1.0996	63° 00′
10′	.4741	.4566	2.190	.5132	1.949	1.124	.8897	1.0966	50′
20′	.4771	.4592	2.178	.5169	1.935	1.126	.8884	1.0937	40′
30′	.4800	.4617	2.166	.5206	1.921	1.127	.8870	1.0908	30′
40′	.4829	.4643	2.154	.5243	1.907	1.129	.8857	1.0879	20′
50′	.4858	.4669	2.142	.5280	1.894	1.131	.8843	1.0850	10′
28° 00′	.4887	.4695	2.130	.5317	1.881	1.133	.8829	1.0821	62° 00′
10′	.4916	.4720	2.118	.5354	1.868	1.134	.8816	1.0792	50′
20′	.4945	.4746	2.107	.5392	1.855	1.136	.8802	1.0763	40′
30′	.4974	.4772	2.096	.5430	1.842	1.138	.8788	1.0734	30′
40′	.5003	.4797	2.085	.5467	1.829	1.140	.8774	1.0705	20′
50′	.5032	.4823	2.074	.5505	1.816	1.142	.8760	1.0676	10′
29° 00′	.5061	.4848	2.063	.5543	1.804	1.143	.8746	1.0647	61° 00′
10′	.5091	.4874	2.052	.5581	1.792	1.145	.8732	1.0617	50′
20′	.5120	.4899	2.041	.5619	1.780	1.147	.8718	1.0588	40′
30′	.5149	.4924	2.031	.5658	1.767	1.149	.8704	1.0559	30′
40′	.5178	.4950	2.020	.5696	1.756	1.151	.8689	1.0530	20′
50′	.5207	.4975	2.010	.5735	1.744	1.153	.8675	1.0501	10′
30° 00′	.5236	.5000	2.000	.5774	1.732	1.155	.8660	1.0472	60° 00′
10′	.5265	.5025	1.990	.5812	1.720	1.157	.8646	1.0443	50′
20′	.5294	.5050	1.980	.5851	1.709	1.159	.8631	1.0414	40′
30′	.5323	.5075	1.970	.5890	1.698	1.161	.8616	1.0385	30′
40′	.5352	.5100	1.961	.5930	1.686	1.163	.8601	1.0356	20′
50′	.5381	.5125	1.951	.5969	1.675	1.165	.8587	1.0327	10′
31° 00′	.5411	.5150	1.942	.6009	1.664	1.167	.8572	1.0297	59° 00′
10′	.5440	.5175	1.932	.6048	1.653	1.169	.8557	1.0268	50′
20′	.5469	.5200	1.923	.6088	1.643	1.171	.8542	1.0239	40′
30′	.5498	.5225	1.914	.6128	1.632	1.173	.8526	1.0210	30′
40′	.5527	.5250	1.905	.6168	1.621	1.175	.8511	1.0181	20′
50′	.5556	.5275	1.896	.6208	1.611	1.177	.8496	1.0152	10′
32° 00′	.5585	.5299	1.887	.6249	1.600	1.179	.8480	1.0123	58° 00′
10′	.5614	.5324	1.878	.6289	1.590	1.181	.8465	1.0094	50′
20′	.5643	.5348	1.870	.6330	1.580	1.184	.8450	1.0065	40′
30′	.5672	.5373	1.861	.6371	1.570	1.186	.8434	1.0036	30′
40′	.5701	.5398	1.853	.6412	1.560	1.188	.8418	1.0007	20′
50′	.5730	.5422	1.844	.6453	1.550	1.190	.8403	.9977	10′
33° 00′	.5760	.5446	1.836	.6494	1.540	1.192	.8387	.9948	57° 00′
10′	.5789	.5471	1.828	.6536	1.530	1.195	.8371	.9919	50′
20′	.5818	.5495	1.820	.6577	1.520	1.197	.8355	.9890	40′
30′	.5847	.5519	1.812	.6619	1.511	1.199	.8339	.9861	30′
40′	.5876	.5544	1.804	.6661	1.501	1.202	.8323	.9832	20′
50′	.5905	.5568	1.796	.6703	1.492	1.204	.8307	.9803	10′
34° 00′	.5934	.5592	1.788	.6745	1.483	1.206	.8290	.9774	56° 00′
10′	.5963	.5616	1.781	.6787	1.473	1.209	.8274	.9745	50′
20′	.5992	.5640	1.773	.6830	1.464	1.211	.8258	.9716	40′
30′	.6021	.5664	1.766	.6873	1.455	1.213	.8241	.9687	30′
40′	.6050	.5688	1.758	.6916	1.446	1.216	.8225	.9657	20′
50′	.6080	.5712	1.751	.6959	1.437	1.218	.8208	.9628	10′
35° 00′	.6109	.5736	1.743	.7002	1.428	1.221	.8192	.9599	55° 00′
10′	.6138	.5760	1.736	.7046	1.419	1.223	.8175	.9570	50′
20′	.6167	.5783	1.729	.7089	1.411	1.226	.8158	.9541	40′
30′	.6196	.5807	1.722	.7133	1.402	1.228	.8141	.9512	30′
40′	.6225	.5831	1.715	.7177	1.393	1.231	.8124	.9483	20′
50′	.6254	.5854	1.708	.7221	1.385	1.233	.8107	.9454	10′
36° 00′	.6283	.5878	1.701	.7265	1.376	1.236	.8090	.9425	54° 00′
		cos θ	sec θ	cot θ	tan θ	csc θ	sin θ	θ Radians	θ Degrees

Trigonometric Functions of θ: degrees and minutes

TABLE 2

θ									
Degrees	Radians	sin θ	csc θ	tan θ	cot θ	sec θ	cos θ		
36° 00′	.6283	.5878	1.701	.7265	1.376	1.236	.8090	.9425	54° 00′
10′	.6312	.5901	1.695	.7310	1.368	1.239	.8073	.9396	50′
20′	.6341	.5925	1.688	.7355	1.360	1.241	.8056	.9367	40′
30′	.6370	.5948	1.681	.7400	1.351	1.244	.8039	.9338	30′
40′	.6400	.5972	1.675	.7445	1.343	1.247	.8021	.9308	20′
50′	.6429	.5995	1.668	.7490	1.335	1.249	.8004	.9279	10′
37° 00′	.6458	.6018	1.662	.7536	1.327	1.252	.7986	.9250	53° 00′
10′	.6487	.6041	1.655	.7581	1.319	1.255	.7969	.9221	50′
20′	.6516	.6065	1.649	.7627	1.311	1.258	.7951	.9192	40′
30′	.6545	.6088	1.643	.7673	1.303	1.260	.7934	.9163	30′
40′	.6574	.6111	1.636	.7720	1.295	1.263	.7916	.9134	20′
50′	.6603	.6134	1.630	.7766	1.288	1.266	.7898	.9105	10′
38° 00′	.6632	.6157	1.624	.7813	1.280	1.269	.7880	.9076	52° 00′
10′	.6661	.6180	1.618	.7860	1.272	1.272	.7862	.9047	50′
20′	.6690	.6202	1.612	.7907	1.265	1.275	.7844	.9018	40′
30′	.6720	.6225	1.606	.7954	1.257	1.278	.7826	.8988	30′
40′	.6749	.6248	1.601	.8002	1.250	1.281	.7808	.8959	20′
50′	.6778	.6271	1.595	.8050	1.242	1.284	.7790	.8930	10′
39° 00′	.6807	.6293	1.589	.8098	1.235	1.287	.7771	.8901	51° 00′
10′	.6836	.6316	1.583	.8146	1.228	1.290	.7753	.8872	50′
20′	.6865	.6338	1.578	.8195	1.220	1.293	.7735	.8843	40′
30′	.6894	.6361	1.572	.8243	1.213	1.296	.7716	.8814	30′
40′	.6923	.6383	1.567	.8292	1.206	1.299	.7698	.8785	20′
50′	.6952	.6406	1.561	.8342	1.199	1.302	.7679	.8756	10′
40° 00′	.6981	.6428	1.556	.8391	1.192	1.305	.7660	.8727	50° 00′
10′	.7010	.6450	1.550	.8441	1.185	1.309	.7642	.8698	50′
20′	.7039	.6472	1.545	.8491	1.178	1.312	.7623	.8668	40′
30′	.7069	.6494	1.540	.8541	1.171	1.315	.7604	.8639	30′
40′	.7098	.6517	1.535	.8591	1.164	1.318	.7585	.8610	20′
50′	.7127	.6539	1.529	.8642	1.157	1.322	.7566	.8581	10′
41° 00′	.7156	.6561	1.524	.8693	1.150	1.325	.7547	.8552	49° 00′
10′	.7185	.6583	1.519	.8744	1.144	1.328	.7528	.8523	50′
20′	.7214	.6604	1.514	.8796	1.137	1.332	.7509	.8494	40′
30′	.7243	.6626	1.509	.8847	1.130	1.335	.7490	.8465	30′
40′	.7272	.6648	1.504	.8899	1.124	1.339	.7470	.8436	20′
50′	.7301	.6670	1.499	.8952	1.117	1.342	.7451	.8407	10′
42° 00′	.7330	.6691	1.494	.9004	1.111	1.346	.7431	.8378	48° 00′
10′	.7359	.6713	1.490	.9057	1.104	1.349	.7412	.8348	50′
20′	.7389	.6734	1.485	.9110	1.098	1.353	.7392	.8319	40′
30′	.7418	.6756	1.480	.9163	1.091	1.356	.7373	.8290	30′
40′	.7447	.6777	1.476	.9217	1.085	1.360	.7353	.8261	20′
50′	.7476	.6799	1.471	.9271	1.079	1.364	.7333	.8232	10′
43° 00′	.7505	.6820	1.466	.9325	1.072	1.367	.7314	.8203	47° 00′
10′	.7534	.6841	1.462	.9380	1.066	1.371	.7294	.8174	50′
20′	.7563	.6862	1.457	.9435	1.060	1.375	.7274	.8145	40′
30′	.7592	.6884	1.453	.9490	1.054	1.379	.7254	.8116	30′
40′	.7621	.6905	1.448	.9545	1.048	1.382	.7234	.8087	20′
50′	.7650	.6926	1.444	.9601	1.042	1.386	.7214	.8058	10′
44° 00′	.7679	.6947	1.440	.9657	1.036	1.390	.7193	.8029	46° 00′
10′	.7709	.6967	1.435	.9713	1.030	1.394	.7173	.7999	50′
20′	.7738	.6988	1.431	.9770	1.024	1.398	.7153	.7970	40′
30′	.7767	.7009	1.427	.9827	1.018	1.402	.7133	.7941	30′
40′	.7796	.7030	1.423	.9884	1.012	1.406	.7112	.7912	20′
50′	.7825	.7050	1.418	.9942	1.006	1.410	.7092	.7883	10′
45° 00′	.7854	.7071	1.414	1.000	1.000	1.414	.7071	.7854	45° 00′
		cos θ	sec θ	cot θ	tan θ	csc θ	sin θ	Radians	Degrees
								θ	

TABLE 3

Trigonometric Functions of θ: radians

rad.	deg.	sin θ	csc θ	tan θ	cot θ	sec θ	cos θ
0.00	0° 00′	0.0000	Undefined	0.0000	Undefined	1.000	1.000
.01	0° 34′	.0100	100.0	.0100	100.0	1.000	1.000
.02	1° 09′	.0200	50.00	.0200	49.99	1.000	0.9998
.03	1° 43′	.0300	33.34	.0300	33.32	1.000	0.9996
.04	2° 18′	.0400	25.01	.0400	24.99	1.001	0.9992
0.05	2° 52′	0.0500	20.01	0.0500	19.98	1.001	0.9988
.06	3° 26′	.0600	16.68	.0601	16.65	1.002	.9982
.07	4° 01′	.0699	14.30	.0701	14.26	1.002	.9976
.08	4° 35′	.0799	12.51	.0802	12.47	1.003	.9968
.09	5° 09′	.0899	11.13	.0902	11.08	1.004	.9960
0.10	5° 44′	0.0998	10.02	0.1003	9.967	1.005	0.9950
.11	6° 18′	.1098	9.109	.1104	9.054	1.006	.9940
.12	6° 53′	.1197	8.353	.1206	8.293	1.007	.9928
.13	7° 27′	.1296	7.714	.1307	7.649	1.009	.9916
.14	8° 01′	.1395	7.166	.1409	7.096	1.010	.9902
0.15	8° 36′	0.1494	6.692	0.1511	6.617	1.011	0.9888
.16	9° 10′	.1593	6.277	.1614	6.197	1.013	.9872
.17	9° 44′	.1692	5.911	.1717	5.826	1.015	.9856
.18	10° 19′	.1790	5.586	.1820	5.495	1.016	.9838
.19	10° 53′	.1889	5.295	.1923	5.200	1.018	.9820
0.20	11° 28′	0.1987	5.033	0.2027	4.933	1.020	0.9801
.21	12° 02′	.2085	4.797	.2131	4.692	1.022	.9780
.22	12° 36′	.2182	4.582	.2236	4.472	1.025	.9759
.23	13° 11′	.2280	4.386	.2341	4.271	1.027	.9737
.24	13° 45′	.2377	4.207	.2447	4.086	1.030	.9713
0.25	14° 19′	0.2474	4.042	0.2553	3.916	1.032	0.9689
.26	14° 54′	.2571	3.890	.2660	3.759	1.035	.9664
.27	15° 28′	.2667	3.749	.2768	3.613	1.038	.9638
.28	16° 03′	.2764	3.619	.2876	3.478	1.041	.9611
.29	16° 37′	.2860	3.497	.2984	3.351	1.044	.9582
0.30	17° 11′	0.2955	3.384	0.3093	3.233	1.047	0.9553
.31	17° 46′	.3051	3.278	.3203	3.122	1.050	.9523
.32	18° 20′	.3146	3.179	.3314	3.018	1.053	.9492
.33	18° 55′	.3240	3.086	.3425	2.920	1.057	.9460
.34	19° 29′	.3335	2.999	.3537	2.827	1.061	.9428
0.35	20° 03′	0.3429	2.916	0.3650	2.740	1.065	0.9394
.36	20° 38′	.3523	2.839	.3764	2.657	1.068	.9359
.37	21° 12′	.3616	2.765	.3879	2.578	1.073	.9323
.38	21° 46′	.3709	2.696	.3994	2.504	1.077	.9287
.39	22° 21′	.3802	2.630	.4111	2.433	1.081	.9249
0.40	22° 55′	0.3894	2.568	0.4228	2.365	1.086	0.9211
.41	23° 30′	.3986	2.509	.4346	2.301	1.090	.9171
.42	24° 04′	.4078	2.452	.4466	2.239	1.095	.9131
.43	24° 38′	.4169	2.399	.4586	2.180	1.100	.9090
.44	25° 13′	.4259	2.348	.4708	2.124	1.105	.9048
0.45	25° 47′	0.4350	2.299	0.4831	2.070	1.111	0.9004
.46	26° 21′	.4439	2.253	.4954	2.018	1.116	.8961
.47	26° 56′	.4529	2.208	.5080	1.969	1.122	.8916
.48	27° 30′	.4618	2.166	.5206	1.921	1.127	.8870
.49	28° 05′	.4706	2.125	.5334	1.875	1.133	.8823

Trigonometric Functions of θ: radians

TABLE 3

rad.	deg.	sin θ	csc θ	tan θ	cot θ	sec θ	cos θ
0.50	28° 39′	0.4794	2.086	0.5463	1.830	1.139	0.8776
.51	29° 13′	.4882	2.048	.5594	1.788	1.146	.8727
.52	29° 48′	.4969	2.013	.5726	1.747	1.152	.8678
.53	30° 22′	.5055	1.978	.5859	1.707	1.159	.8628
.54	30° 56′	.5141	1.945	.5994	1.668	1.166	.8577
0.55	31° 31′	0.5227	1.913	0.6131	1.631	1.173	0.8525
.56	32° 05′	.5312	1.883	.6269	1.595	1.180	.8473
.57	32° 40′	.5396	1.853	.6410	1.560	1.188	.8419
.58	33° 14′	.5480	1.825	.6552	1.526	1.196	.8365
.59	33° 48′	.5564	1.797	.6696	1.494	1.203	.8309
0.60	34° 23′	0.5646	1.771	0.6841	1.462	1.212	0.8253
.61	34° 57′	.5729	1.746	.6989	1.431	1.220	.8196
.62	35° 31′	.5810	1.721	.7139	1.401	1.229	.8139
.63	36° 06′	.5891	1.697	.7291	1.372	1.238	.8080
.64	36° 40′	.5972	1.674	.7445	1.343	1.247	.8021
0.65	37° 15′	0.6052	1.652	0.7602	1.315	1.256	0.7961
.66	37° 49′	.6131	1.631	.7761	1.288	1.266	.7900
.67	38° 23′	.6210	1.610	.7923	1.262	1.276	.7838
.68	38° 58′	.6288	1.590	.8087	1.237	1.286	.7776
.69	39° 32′	.6365	1.571	.8253	1.212	1.297	.7712
0.70	40° 06′	0.6442	1.552	0.8423	1.187	1.307	0.7648
.71	40° 41′	.6518	1.534	.8595	1.163	1.319	.7584
.72	41° 15′	.6594	1.517	.8771	1.140	1.330	.7518
.73	41° 50′	.6669	1.500	.8949	1.117	1.342	.7452
.74	42° 24′	.6743	1.483	.9131	1.095	1.354	.7385
0.75	42° 58′	0.6816	1.467	0.9316	1.073	1.367	0.7317
.76	43° 33′	.6889	1.452	.9505	1.052	1.380	.7248
.77	44° 07′	.6961	1.437	.9697	1.031	1.393	.7179
.78	44° 41′	.7033	1.422	.9893	1.011	1.407	.7109
.79	45° 16′	.7104	1.408	1.009	.9908	1.421	.7038
0.80	45° 50′	0.7174	1.394	1.030	0.9712	1.435	0.6967
.81	46° 25′	.7243	1.381	1.050	.9520	1.450	.6895
.82	46° 59′	.7311	1.368	1.072	.9331	1.466	.6822
.83	47° 33′	.7379	1.355	1.093	.9146	1.482	.6749
.84	48° 08′	.7446	1.343	1.116	.8964	1.498	.6675
0.85	48° 42′	0.7513	1.331	1.138	0.8785	1.515	0.6600
.86	49° 17′	.7578	1.320	1.162	.8609	1.533	.6524
.87	49° 51′	.7643	1.308	1.185	.8437	1.551	.6448
.88	50° 25′	.7707	1.297	1.210	.8267	1.569	.6372
.89	51° 00′	.7771	1.287	1.235	.8100	1.589	.6294
0.90	51° 34′	0.7833	1.277	1.260	0.7936	1.609	0.6216
.91	52° 08′	.7895	1.267	1.286	.7774	1.629	.6137
.92	52° 43′	.7956	1.257	1.313	.7615	1.651	.6058
.93	53° 17′	.8016	1.247	1.341	.7458	1.673	.5978
.94	53° 52′	.8076	1.238	1.369	.7303	1.696	.5898
0.95	54° 26′	0.8134	1.229	1.398	0.7151	1.719	0.5817
.96	55° 00′	.8192	1.221	1.428	.7001	1.744	.5735
.97	55° 35′	.8249	1.212	1.459	.6853	1.769	.5653
.98	56° 09′	.8305	1.204	1.491	.6707	1.795	.5570
.99	56° 43′	.8360	1.196	1.524	.6563	1.823	.5487

TABLE 3

Trigonometric Functions of θ: radians

rad.	deg.	sin θ	csc θ	tan θ	cot θ	sec θ	cos θ
1.00	57° 18′	0.8415	1.188	1.557	0.6421	1.851	0.5403
1.01	57° 52′	.8468	1.181	1.592	.6281	1.880	.5319
1.02	58° 27′	.8521	1.174	1.628	.6142	1.911	.5234
1.03	59° 01′	.8573	1.166	1.665	.6005	1.942	.5148
1.04	59° 35′	.8624	1.160	1.704	.5870	1.975	.5062
1.05	60° 10′	0.8674	1.153	1.743	0.5736	2.010	0.4976
1.06	60° 44′	.8724	1.146	1.784	.5604	2.046	.4889
1.07	61° 18′	.8772	1.140	1.827	.5473	2.083	.4801
1.08	61° 53′	.8820	1.134	1.871	.5344	2.122	.4713
1.09	62° 27′	.8866	1.128	1.917	.5216	2.162	.4625
1.10	63° 02′	0.8912	1.122	1.965	0.5090	2.205	0.4536
1.11	63° 36′	.8957	1.116	2.014	.4964	2.249	.4447
1.12	64° 10′	.9001	1.111	2.066	.4840	2.295	.4357
1.13	64° 45′	.9044	1.106	2.120	.4718	2.344	.4267
1.14	65° 19′	.9086	1.101	2.176	.4596	2.395	.4176
1.15	65° 53′	0.9128	1.096	2.234	0.4475	2.448	0.4085
1.16	66° 28′	.9168	1.091	2.296	.4356	2.504	.3993
1.17	67° 02′	.9208	1.086	2.360	.4237	2.563	.3902
1.18	67° 37′	.9246	1.082	2.428	.4120	2.625	.3809
1.19	68° 11′	.9284	1.077	2.498	.4003	2.691	3717
1.20	68° 45′	0.9320	1.073	2.572	0.3888	2.760	0.3624
1.21	69° 20′	.9356	1.069	2.650	.3773	2.833	.3530
1.22	69° 54′	.9391	1.065	2.733	.3659	2.910	.3436
1.23	70° 28′	.9425	1.061	2.820	.3546	2.992	.3342
1.24	71° 03′	.9458	1.057	2.912	.3434	3.079	.3248
1.25	71° 37′	0.9490	1.054	3.010	0.3323	3.171	0.3153
1.26	72° 12′	.9521	1.050	3.113	.3212	3.270	.3058
1.27	72° 46′	.9551	1.047	3.224	.3102	3.375	.2963
1.28	73° 20′	.9580	1.044	3.341	.2993	3.488	.2867
1.29	73° 55′	.9608	1.041	3.467	.2884	3.609	.2771
1.30	74° 29′	0.9636	1.038	3.602	0.2776	3.738	0.2675
1.31	75° 03′	.9662	1.035	3.747	.2669	3.878	.2579
1.32	75° 38′	.9687	1.032	3.903	.2562	4.029	.2482
1.33	76° 12′	.9711	1.030	4.072	.2456	4.193	.2385
1.34	76° 47′	.9735	1.027	4.256	.2350	4.372	.2288
1.35	77° 21′	0.9757	1.025	4.455	0.2245	4.566	0.2190
1.36	77° 55′	.9779	1.023	4.673	.2140	4.779	.2092
1.37	78° 30′	.9799	1.021	4.913	.2035	5.014	.1994
1.38	79° 04′	.9819	1.018	5.177	.1931	5.273	.1896
1.39	79° 39′	.9837	1.017	5.471	.1828	5.561	.1798
1.40	80° 13′	0.9854	1.015	5.798	0.1725	5.883	0.1700
1.41	80° 47′	.9871	1.013	6.165	.1622	6.246	.1601
1.42	81° 22′	.9887	1.011	6.581	.1519	6.657	.1502
1.43	81° 56′	.9901	1.010	7.055	.1417	7.126	.1403
1.44	82° 30′	.9915	1.009	7.602	.1315	7.667	.1304
1.45	83° 05′	0.9927	1.007	8.238	0.1214	8.299	0.1205
1.46	83° 39′	.9939	1.006	8.989	.1113	9.044	.1106
1.47	84° 14′	.9949	1.005	9.887	.1011	9.938	.1006
1.48	84° 48′	.9959	1.004	10.98	.0911	11.03	.0907
1.49	85° 22′	.9967	1.003	12.35	.0810	12.39	.0807

Trigonometric Functions of θ: radians

TABLE 3

rad.	deg.	sin θ	csc θ	tan θ	cot θ	sec θ	cos θ
1.50	85° 57′	0.9975	1.003	14.10	0.0709	14.14	0.0707
1.51	86° 31′	.9982	1.002	16.43	.0609	16.46	.0608
1.52	87° 05′	.9987	1.001	19.67	.0508	19.70	.0508
1.53	87° 40′	.9992	1.001	24.50	.0408	24.52	.0408
1.54	88° 14′	.9995	1.000	32.46	.0308	32.48	.0308
1.55	88° 49′	0.9998	1.000	48.08	0.0208	48.09	0.0208
1.56	89° 23′	.9999	1.000	92.62	.0108	92.63	.0108
1.57	89° 57′	1.000	1.000	1256	.0008	1256	.0008

TABLE 4

Squares and Square Roots

N	N^2	$\sqrt{N}$	$\sqrt{10N}$	N	N^2	$\sqrt{N}$	$\sqrt{10N}$
1.0	1.00	1.000	3.162	**5.5**	30.25	2.345	7.416
1.1	1.21	1.049	3.317	**5.6**	31.36	2.366	7.483
1.2	1.44	1.095	3.464	**5.7**	32.49	2.387	7.550
1.3	1.69	1.140	3.606	**5.8**	33.64	2.408	7.616
1.4	1.96	1.183	3.742	**5.9**	34.81	2.429	7.681
1.5	2.25	1.225	3.873	**6.0**	36.00	2.449	7.746
1.6	2.56	1.265	4.000	**6.1**	37.21	2.470	7.810
1.7	2.89	1.304	4.123	**6.2**	38.44	2.490	7.874
1.8	3.24	1.342	4.243	**6.3**	39.69	2.510	7.937
1.9	3.61	1.378	4.359	**6.4**	40.96	2.530	8.000
2.0	4.00	1.414	4.472	**6.5**	42.25	2.550	8.062
2.1	4.41	1.449	4.583	**6.6**	43.56	2.569	8.124
2.2	4.84	1.483	4.690	**6.7**	44.89	2.588	8.185
2.3	5.29	1.517	4.796	**6.8**	46.24	2.608	8.246
2.4	5.76	1.549	4.899	**6.9**	47.61	2.627	8.307
2.5	6.25	1.581	5.000	**7.0**	49.00	2.646	8.367
2.6	6.76	1.612	5.099	**7.1**	50.41	2.665	8.426
2.7	7.29	1.643	5.196	**7.2**	51.84	2.683	8.485
2.8	7.84	1.673	5.292	**7.3**	53.29	2.702	8.544
2.9	8.41	1.703	5.385	**7.4**	54.76	2.720	8.602
3.0	9.00	1.732	5.477	**7.5**	56.25	2.739	8.660
3.1	9.61	1.761	5.568	**7.6**	57.76	2.757	8.718
3.2	10.24	1.789	5.657	**7.7**	59.29	2.775	8.775
3.3	10.89	1.817	5.745	**7.8**	60.84	2.793	8.832
3.4	11.56	1.844	5.831	**7.9**	62.41	2.811	8.888
3.5	12.25	1.871	5.916	**8.0**	64.00	2.828	8.944
3.6	12.96	1.897	6.000	**8.1**	65.61	2.846	9.000
3.7	13.69	1.924	6.083	**8.2**	67.24	2.864	9.055
3.8	14.44	1.949	6.164	**8.3**	68.89	2.881	9.110
3.9	15.21	1.975	6.245	**8.4**	70.56	2.898	9.165
4.0	16.00	2.000	6.325	**8.5**	72.25	2.915	9.220
4.1	16.81	2.025	6.403	**8.6**	73.96	2.933	9.274
4.2	17.64	2.049	6.481	**8.7**	75.69	2.950	9.327
4.3	18.49	2.074	6.557	**8.8**	77.44	2.966	9.381
4.4	19.36	2.098	6.633	**8.9**	79.21	2.983	9.434
4.5	20.25	2.121	6.708	**9.0**	81.00	3.000	9.487
4.6	21.16	2.145	6.782	**9.1**	82.81	3.017	9.539
4.7	22.09	2.168	6.856	**9.2**	84.64	3.033	9.592
4.8	23.04	2.191	6.928	**9.3**	86.49	3.050	9.644
4.9	24.01	2.214	7.000	**9.4**	88.36	3.066	9.695
5.0	25.00	2.236	7.071	**9.5**	90.25	3.082	9.747
5.1	26.01	2.258	7.141	**9.6**	92.16	3.098	9.798
5.2	27.04	2.280	7.211	**9.7**	94.09	3.114	9.849
5.3	28.09	2.302	7.280	**9.8**	96.04	3.130	9.899
5.4	29.16	2.324	7.348	**9.9**	98.01	3.146	9.950
5.5	30.25	2.345	7.416	**10**	100.00	3.162	10.000

Common Logarithms

TABLE 5

N	0	1	2	3	4	5	6	7	8	9
10	0000	0043	0086	0128	0170	0212	0253	0294	0334	0374
11	0414	0453	0492	0531	0569	0607	0645	0682	0719	0755
12	0792	0828	0864	0899	0934	0969	1004	1038	1072	1106
13	1139	1173	1206	1239	1271	1303	1335	1367	1399	1430
14	1461	1492	1523	1553	1584	1614	1644	1673	1703	1732
15	1761	1790	1818	1847	1875	1903	1931	1959	1987	2014
16	2041	2068	2095	2122	2148	2175	2201	2227	2253	2279
17	2304	2330	2355	2380	2405	2430	2455	2480	2504	2529
18	2553	2577	2601	2625	2648	2672	2695	2718	2742	2765
19	2788	2810	2833	2856	2878	2900	2923	2945	2967	2989
20	3010	3032	3054	3075	3096	3118	3139	3160	3181	3201
21	3222	3243	3263	3284	3304	3324	3345	3365	3385	3404
22	3424	3444	3464	3483	3502	3522	3541	3560	3579	3598
23	3617	3636	3655	3674	3692	3711	3729	3747	3766	3784
24	3802	3820	3838	3856	3874	3892	3909	3927	3945	3962
25	3979	3997	4014	4031	4048	4065	4082	4099	4116	4133
26	4150	4166	4183	4200	4216	4232	4249	4265	4281	4298
27	4314	4330	4346	4362	4378	4393	4409	4425	4440	4456
28	4472	4487	4502	4518	4533	4548	4564	4579	4594	4609
29	4624	4639	4654	4669	4683	4698	4713	4728	4742	4757
30	4771	4786	4800	4814	4829	4843	4857	4871	4886	4900
31	4914	4928	4942	4955	4969	4983	4997	5011	5024	5038
32	5051	5065	5079	5092	5105	5119	5132	5145	5159	5172
33	5185	5198	5211	5224	5237	5250	5263	5276	5289	5302
34	5315	5328	5340	5353	5366	5378	5391	5403	5416	5428
35	5441	5453	5465	5478	5490	5502	5514	5527	5539	5551
36	5563	5575	5587	5599	5611	5623	5635	5647	5658	5670
37	5682	5694	5705	5717	5729	5740	5752	5763	5775	5786
38	5798	5809	5821	5832	5843	5855	5866	5877	5888	5899
39	5911	5922	5933	5944	5955	5966	5977	5988	5999	6010
40	6021	6031	6042	6053	6064	6075	6085	6096	6107	6117
41	6128	6138	6149	6160	6170	6180	6191	6201	6212	6222
42	6232	6243	6253	6263	6274	6284	6294	6304	6314	6325
43	6335	6345	6355	6365	6375	6385	6395	6405	6415	6425
44	6435	6444	6454	6464	6474	6484	6493	6503	6513	6522
45	6532	6542	6551	6561	6571	6580	6590	6599	6609	6618
46	6628	6637	6646	6656	6665	6675	6684	6693	6702	6712
47	6721	6730	6739	6749	6758	6767	6776	6785	6794	6803
48	6812	6821	6830	6839	6848	6857	6866	6875	6884	6893
49	6902	6911	6920	6928	6937	6946	6955	6964	6972	6981
50	6990	6998	7007	7016	7024	7033	7042	7050	7059	7067
51	7076	7084	7093	7101	7110	7118	7126	7135	7143	7152
52	7160	7168	7177	7185	7193	7202	7210	7218	7226	7235
53	7243	7251	7259	7267	7275	7284	7292	7300	7308	7316
54	7324	7332	7340	7348	7356	7364	7372	7380	7388	7396

*Mantissas; decimal points omitted. Characteristics are found by inspection.

TABLE 5

Common Logarithms

N	0	1	2	3	4	5	6	7	8	9
55	7404	7412	7419	7427	7435	7443	7451	7459	7466	7474
56	7482	7490	7497	7505	7513	7520	7528	7536	7543	7551
57	7559	7566	7574	7582	7589	7597	7604	7612	7619	7627
58	7634	7642	7649	7657	7664	7672	7679	7686	7694	7701
59	7709	7716	7723	7731	7738	7745	7752	7760	7767	7774
60	7782	7789	7796	7803	7810	7818	7825	7832	7839	7846
61	7853	7860	7868	7875	7882	7889	7896	7903	7910	7917
62	7924	7931	7938	7945	7952	7959	7966	7973	7980	7987
63	7993	8000	8007	8014	8021	8028	8035	8041	8048	8055
64	8062	8069	8075	8082	8089	8096	8102	8109	8116	8122
65	8129	8136	8142	8149	8156	8162	8169	8176	8182	8189
66	8195	8202	8209	8215	8222	8228	8235	8241	8248	8254
67	8261	8267	8274	8280	8287	8293	8299	8306	8312	8319
68	8325	8331	8338	8344	8351	8357	8363	8370	8376	8382
69	8388	8395	8401	8407	8414	8420	8426	8432	8439	8445
70	8451	8457	8463	8470	8476	8482	8488	8494	8500	8506
71	8513	8519	8525	8531	8537	8543	8549	8555	8561	8567
72	8573	8579	8585	8591	8597	8603	8609	8615	8621	8627
73	8633	8639	8645	8651	8657	8663	8669	8675	8681	8686
74	8692	8698	8704	8710	8716	8722	8727	8733	8739	8745
75	8751	8756	8762	8768	8774	8779	8785	8791	8797	8802
76	8808	8814	8820	8825	8831	8837	8842	8848	8854	8859
77	8865	8871	8876	8882	8887	8893	8899	8904	8910	8915
78	8921	8927	8932	8938	8943	8949	8954	8960	8965	8971
79	8976	8982	8987	8993	8998	9004	9009	9015	9020	9025
80	9031	9036	9042	9047	9053	9058	9063	9069	9074	9079
81	9085	9090	9096	9101	9106	9112	9117	9122	9128	9133
82	9138	9143	9149	9154	9159	9165	9170	9175	9180	9186
83	9191	9196	9201	9206	9212	9217	9222	9227	9232	9238
84	9243	9248	9253	9258	9263	9269	9274	9279	9284	9289
85	9294	9299	9304	9309	9315	9320	9325	9330	9335	9340
86	9345	9350	9355	9360	9365	9370	9375	9380	9385	9390
87	9395	9400	9405	9410	9415	9420	9425	9430	9435	9440
88	9445	9450	9455	9460	9465	9469	9474	9479	9484	9489
89	9494	9499	9504	9509	9513	9518	9523	9528	9533	9538
90	9542	9547	9552	9557	9562	9566	9571	9576	9581	9586
91	9590	9595	9600	9605	9609	9614	9619	9624	9628	9633
92	9638	9643	9647	9652	9657	9661	9666	9671	9675	9680
93	9685	9689	9694	9699	9703	9708	9713	9717	9722	9727
94	9731	9736	9741	9745	9750	9754	9759	9763	9768	9773
95	9777	9782	9786	9791	9795	9800	9805	9809	9814	9818
96	9823	9827	9832	9836	9841	9845	9850	9854	9859	9863
97	9868	9872	9877	9881	9886	9890	9894	9899	9903	9908
98	9912	9917	9921	9926	9930	9934	9939	9943	9948	9952
99	9956	9961	9965	9969	9974	9978	9983	9987	9991	9996

Values of e^x and e^{-x}

TABLE 6

x	e^x	e^{-x}	x	e^x	e^{-x}
0.00	1.0000	1.0000	**2.5**	12.182	0.0821
0.05	1.0513	0.9512	**2.6**	13.464	0.0743
0.10	1.1052	0.9048	**2.7**	14.880	0.0672
0.15	1.1618	0.8607	**2.8**	16.445	0.0608
0.20	1.2214	0.8187	**2.9**	18.174	0.0550
0.25	1.2840	0.7788	**3.0**	20.086	0.0498
0.30	1.3499	0.7408	**3.1**	22.198	0.0450
0.35	1.4191	0.7047	**3.2**	24.533	0.0408
0.40	1.4918	0.6703	**3.3**	27.113	0.0369
0.45	1.5683	0.6376	**3.4**	29.964	0.0334
0.50	1.6487	0.6065	**3.5**	33.115	0.0302
0.55	1.7333	0.5769	**3.6**	36.598	0.0273
0.60	1.8221	0.5488	**3.7**	40.447	0.0247
0.65	1.9155	0.5220	**3.8**	44.701	0.0224
0.70	2.0138	0.4966	**3.9**	49.402	0.0202
0.75	2.1170	0.4724	**4.0**	54.598	0.0183
0.80	2.2255	0.4493	**4.1**	60.340	0.0166
0.85	2.3396	0.4274	**4.2**	66.686	0.0150
0.90	2.4596	0.4066	**4.3**	73.700	0.0136
0.95	2.5857	0.3867	**4.4**	81.451	0.0123
1.0	2.7183	0.3679	**4.5**	90.017	0.0111
1.1	3.0042	0.3329	**4.6**	99.484	0.0101
1.2	3.3201	0.3012	**4.7**	109.95	0.0091
1.3	3.6693	0.2725	**4.8**	121.51	0.0082
1.4	4.0552	0.2466	**4.9**	134.29	0.0074
1.5	4.4817	0.2231	**5.0**	148.41	0.0067
1.6	4.9530	0.2019	**5.5**	244.69	0.0041
1.7	5.4739	0.1827	**6.0**	403.43	0.0025
1.8	6.0496	0.1653	**6.5**	665.14	0.0015
1.9	6.6859	0.1496	**7.0**	1096.6	0.0009
2.0	7.3891	0.1353	**7.5**	1808.0	0.0006
2.1	8.1662	0.1225	**8.0**	2981.0	0.0003
2.2	9.0250	0.1108	**8.5**	4914.8	0.0002
2.3	9.9742	0.1003	**9.0**	8103.1	0.0001
2.4	11.023	0.0907	**10.0**	22026	0.00005

TABLE 7

Natural Logarithms

x	ln x	x	ln x	x	ln x
		4.5	1.5041	9.0	2.1972
0.1	−2.3026	4.6	1.5261	9.1	2.2083
0.2	−1.6094	4.7	1.5476	9.2	2.2192
0.3	−1.2040	4.8	1.5686	9.3	2.2300
0.4	−0.9163	4.9	1.5892	9.4	2.2407
0.5	−0.6931	5.0	1.6094	9.5	2.2513
0.6	−0.5108	5.1	1.6292	9.6	2.2618
0.7	−0.3567	5.2	1.6487	9.7	2.2721
0.8	−0.2231	5.3	1.6677	9.8	2.2824
0.9	−0.1054	5.4	1.6864	9.9	2.2925
1.0	0.0000	5.5	1.7047	10	2.3026
1.1	0.0953	5.6	1.7228	11	2.3979
1.2	0.1823	5.7	1.7405	12	2.4849
1.3	0.2624	5.8	1.7579	13	2.5649
1.4	0.3365	5.9	1.7750	14	2.6391
1.5	0.4055	6.0	1.7918	15	2.7081
1.6	0.4700	6.1	1.8083	16	2.7726
1.7	0.5306	6.2	1.8245	17	2.8332
1.8	0.5878	6.3	1.8405	18	2.8904
1.9	0.6419	6.4	1.8563	19	2.9444
2.0	0.6931	6.5	1.8718	20	2.9957
2.1	0.7419	6.6	1.8871	25	3.2189
2.2	0.7885	6.7	1.9021	30	3.4012
2.3	0.8329	6.8	1.9169	35	3.5553
2.4	0.8755	6.9	1.9315	40	3.6889
2.5	0.9163	7.0	1.9459	45	3.8067
2.6	0.9555	7.1	1.9601	50	3.9120
2.7	0.9933	7.2	1.9741	55	4.0073
2.8	1.0296	7.3	1.9879	60	4.0943
2.9	1.0647	7.4	2.0015	65	4.1744
3.0	1.0986	7.5	2.0149	70	4.2485
3.1	1.1314	7.6	2.0281	75	4.3175
3.2	1.1632	7.7	2.0412	80	4.3820
3.3	1.1939	7.8	2.0541	85	4.4427
3.4	1.2238	7.9	2.0669	90	4.4998
3.5	1.2528	8.0	2.0794	100	4.6052
3.6	1.2809	8.1	2.0919	110	4.7005
3.7	1.3083	8.2	2.1041	120	4.7875
3.8	1.3350	8.3	2.1163	130	4.8676
3.9	1.3610	8.4	2.1282	140	4.9416
4.0	1.3863	8.5	2.1401	150	5.0106
4.1	1.4110	8.6	2.1518	160	5.0752
4.2	1.4351	8.7	2.1633	170	5.1358
4.3	1.4586	8.8	2.1748	180	5.1930
4.4	1.4816	8.9	2.1861	190	5.2470

REFERENCE MATERIALS

Following is a list of suggested references for further investigation of some of the topics presented in this course. These sources are suitable for individual or group activities such as projects, term papers, and class presentations.

Of General Interest

Bell, Eric Temple. *Men of Mathematics.* New York: Simon and Schuster, 1937.

Courant, Richard, and Herbert Robbins. *What Is Mathematics? An Elementary Approach to Ideas and Methods.* London: Oxford University Press, 1941.

Douglis, Avron. *Ideas in Mathematics.* Philadelphia: W. B. Saunders, 1970.

Jacobs, Harold R. *Mathematics: A Human Endeavor.* San Francisco: W. H. Freeman, 1970.

Kasner, Edward, and James R. Newman. *Mathematics and the Imagination.* New York: Simon and Schuster, 1967.

Kramer, Edna. *The Nature and Growth of Modern Mathematics* (2 vols.). Greenwich, Conn.: Fawcett, 1970.

Mathematics: An Introduction to Its Spirit and Use: Readings from "Scientific American." San Francisco: W. H. Freeman, 1979.

Newman, James R., ed. *The World of Mathematics* (4 vols.). New York: Simon and Schuster, 1956.

Perl, Teri. *Math Equals: Biographies of Women Mathematicians + Related Activities.* Menlo Park, Calif.: Addison-Wesley, 1978.

Singh, Jagjit. *Great Ideas of Modern Mathematics: Their Nature and Use.* New York: Dover, 1959.

Logic (Chapter 1)

Kemeny, John G.; J. Laurie Snell; and Gerald L. Thompson. *Introduction to Finite Mathematics,* 3rd ed. Englewood Cliffs, N.J.: Prentice-Hall, 1974.

Nagel, Ernest. "Symbolic Notation, Haddock's Eyes and the Dog-Walking Ordinance" in *The World of Mathematics,* Volume III, pp. 1878–1900. *Entertaining applications of symbolic logic.*

Roethel, Louis F., and Abraham Weinstein. *Logic, Sets, and Numbers,* 2nd ed. Belmont, Calif.: Wadsworth, 1976.

Trigonometry, Complex Numbers (Chapters 7, 8, 9)

Dewdney, A. K. "Computer Recreations." *Scientific American,* August 1985.
Contains computer displays of fractal curves in the complex plane.

Niven, Ivan. *Maxima and Minima without Calculus.* Washington, D.C.: Mathematical Association of America, 1981, pp. 92–143.
Uses trigonometry to determine polygons of maximal area.

Groups, Rings, Fields

Gudder, Stanley. *A Mathematical Journey.* New York: McGraw-Hill, 1976, pp. 281–321.
A historical development of group theory; many examples and an application to communication systems.

Sawyer, W. W. *Prelude to Mathematics.* Baltimore: Penguin, 1955.

Vectors (Chapters 11, 12, 13)

Lord, N. J. "A Method for Vector Proofs in Geometry." *Mathematics Magazine,* March 1985, pp. 85–89.

Schiffer, M. M., and Leon Bowden. *The Role of Mathematics in Science.* Washington, D.C.: Mathematical Association of America, pp. 104–131.
Relates linear algebra to analytic geometry and trigonometry.

Calculus and Analytic Geometry (Chapters 2–4, 6, 10, 14)

Goodman, A. W. *Analytic Geometry and the Calculus,* 4th ed. New York: Macmillan, 1980.

Grabiner, Judith V. "The Changing Concept of Change: The Derivative from Fermat to Weierstrass." *Mathematics Magazine,* September 1983.

Thomas, George, and Ross L. Finney. *Calculus and Analytic Geometry.* Reading, Mass.: Addison-Wesley, 1979.

Computers and Mathematics

Dromey, R. G. *How to Solve It by Computer.* Englewood Cliffs, N.J.: Prentice-Hall, 1982.

Kepner, Henry S., Jr., and Joseph W. Kmoch. "How Close Is Close?" in *Topics for Mathematics Clubs,* Leroy C. Dalton and Henry D. Snyder, eds. Reston, Va.: National Council of Teachers of Mathematics, 1983, pp. 96–106.

Pavelle, Richard; Michael Rothstein; and John Fitch. "Computer Algebra." *Scientific American,* December 1981.

Reagan, James. "Get the Message? Cryptographs, Mathematics, and Computers." *The Mathematics Teacher,* October 1986, pp. 547–553.

GLOSSARY

Abscissa (p. 42). The first coordinate of a point in the plane.

Absolute, or global, minimum value (p. 240).

Absolute value (p. 23).

$$|x| = \begin{cases} x \text{ if } x \geq 0 \\ -x \text{ if } x < 0 \end{cases}$$

Absolute value function (p. 147). $x \to |x|$.

Absolute value $|z|$ of a complex number z (p. 190). Also called the modulus. If $z = x + yi$, then $|z| = \sqrt{x^2 + y^2}$.

Acceleration (p. 253). Rate of change of velocity with respect to time.

Additive inverse of a matrix A (p. 485). Denoted $-A$. The matrix whose entries are the additive inverses of the corresponding entries of A.

Additive inverse $-\mathbf{u}$ of $\mathbf{u}$ (p. 442). The vector such that $\mathbf{u} + (-\mathbf{u}) = \mathbf{0}$.

Adjacent (p. 292). The sides of a triangle forming an angle are adjacent to the angle and the angle is included by the sides.

Adjoint of a matrix A (p. 511). Denoted adj A. The transpose of the matrix whose entry in the ith row and jth column is the cofactor of a_{ij}.

Air speed (p. 462). Speed in still air.

Algebraically complete, or algebraically closed (p. 202). A field F is algebraically complete if every polynomial equation of positive degree over F has a root in F.

Amplitude (p. 340). If a periodic function has maximum value M and minimum value m, its amplitude is $\frac{M - m}{2}$.

Angle (p. 281). In geometry, the union of two rays (its sides) having a common endpoint (its vertex). In trigonometry, the figure generated by a rotation of one side (the initial side) into the other (the terminal side).

Angle between two planes (p. 534). The angle between their normal vectors.

Angle of depression (p. 293). The angle made by a line of sight with the horizontal, where the object being observed is *below* the observer.

Angle of elevation (p. 293). The angle made by a line of sight with the horizontal, where the object being observed is *above* the observer.

Angular speed (p. 284). Speed of a point P moving in uniform circular motion.

Antiderivative of f (p. 557). F if $F'(x) = f(x)$.

Area under a graph (p. 568). *See* Region under a graph.

Arithmetic means (pp. 104–105). The terms between two given terms of an arithmetic sequence. A single arithmetic mean between two numbers is called *the* arithmetic mean, or the average, of the numbers.

Arithmetic sequence (p. 104). Also called an arithmetic progression. A sequence such that $t_1 = a$, and $t_{n+1} = t_n + d$; d is called the common difference.

Arithmetic series (p. 105). One whose terms form an arithmetic sequence.

Asymptote (p. 173). A line L is an asymptote of a curve C if a point P moving on C becomes arbitrarily close to L as the distance from P to the origin increases without bound.

Augmented matrix (p. 499). Associated with each system of linear equations so that the columns are the coefficients of the variables, and the last column contains the constants of the system. Operations on the rows as if they were equations are called elementary row operations.

Axes (p. 42). Two (or three) perpendicular number lines intersecting in a point, the origin.

Average. *See* Arithmetic means.

Bearing (p. 462). The bearing of a vector $\mathbf{v}$ is the angle θ that $\mathbf{v}$ makes with due north, $0° \leq \theta < 360°$.

Bernoulli trials (p. 611). Trials in a binomial experiment.

Biconditional sentence (p. 10). "p if and only if q" (also written $p \leftrightarrow q$); the conjunction of "if p, then q" and its converse where p and q represent sentences.

Binomial coefficients (p. 94). The coefficients in the expansion of $(a + b)^n$.

Binomial distribution (p. 612). The probability distribution associated with a binomial experiment.

Binomial experiment (p. 611). One having the following properties: (1) There is a fixed number n of repeated trials. (2) The outcome of each trial can be classified as a success or a failure. (3) The probability p of success is the same for all trials. (4) The trials, Bernoulli trials, are independent.

Binomial random variable (p. 612). The number of successes in a binomial experiment.

Bounded sequence (p. 126). $a_1, a_2, a_3, \ldots, a_n, \ldots$ such that $|a_n| \leq B$ for all n.

Cartesian product (p. 41). The set of all ordered pairs of real numbers.

Center of sphere. *See* Sphere.

Central tendency (p. 617). Measured by mean, median, or mode.

Characteristic (p. 496). $B^2 - 4AC$ in a second-degree equation in two variables.

Circle (p. 53). The set of all points in the plane a fixed distance, the radius, from a fixed point, the center.

Circular-arc length (p. 283). The length of an arc of a circle.

Circular functions (p. 272). Trigonometric functions defined by use of the unit circle.

Circular-sector area (p. 283). The area of the region bounded by two radii of a circle and the arc intercepted.

Closed interval (p. 167). $[a, b] = \{x : a \leq x \leq b\}$.

Coefficient matrix (p. 512). Its columns are the coefficients of the variables in a system of linear equations.

Cofactor (p. 506). *See* Minors of a determinant.

Column matrix (p. 485). A matrix with only one column.

Combination (p. 594). An r-member subset of a set with n members where $1 \leq r \leq n$.

Common factors (p. 28). If a, b, and c are integers, $c \mid a$ and $c \mid b$, c is a common factor of a and b. The greatest common factor (GCF), or greatest common divisor, of a and b is the largest such c.

Common logarithms (p. 418). Logarithms with base 10.

Commutative group (p. 15). A group that satisfies the commutative axiom.

Complement of a set A (p. 3). Denoted A'. The set of all members of U (the universal set) not in A.

Conditional sentence (p. 10). "If p, then q" (also written $p \rightarrow q$) where p and q represent sentences.

Complex field, C (pp. 185–186). The set of all (x, y) where x and y are real numbers. If (u, v) and (x, y) are members of C, $(u, v) + (x, y) = (u + x, v + y)$ and $(u, v) \cdot (x, y) = (ux - vy, uy + vx)$.

Complex number (p. 186). Each member of the complex field. In $x + yi$, $i^2 = -1$, x is the real part and y is the imaginary part. If $x = 0$ and $y \neq 0$, $x + yi$ is pure imaginary.

Complex plane (p. 189). The plane used to represent complex numbers. The horizontal (vertical) axis is called the real (imaginary) axis.

Composition of functions (p. 151). $(f \circ g)(x) = f(g(x))$.

Compound interest formula (p. 406). $A = P\left(1 + \frac{r}{n}\right)^{nt}$

Concave upward (downward) (p. 242). The graph of a differentiable function f is concave upward (downward) on an interval I if f' is increasing (decreasing) on I.

Conditional probability (p. 605). The probability that an event will occur, given that another event has occurred or is certain to occur.

Confidence intervals (p. 633). Obtained from $\frac{\bar{p}(1 - \bar{p})}{n}$, where $\bar{p}$ is the proportion of a sample size n.

Conic sections (p. 73). Circles, ellipses, hyperbolas, and parabolas.

Conjugate of $z = x + yi$ (p. 189). $\bar{z} = x - yi$.

Conjunction (p. 6). "p and q" (also written $p \wedge q$) where p and q represent sentences.

Consistent system of equations (p. 503). One with at least one solution.

Constant function (p. 149). $f(x) = c$, c constant.

Constant sequence (p. 115). $a_n = c$, $n \geq 1$.

Continuous function (p. 163). f is continuous at c if and only if $\lim_{x \to c} f(x) = f(c)$. If the condition is not met, the function is discontinuous at c.

Contradiction (p. 5). A sentence whose solution set is empty.

Contrapositive (p. 11). The contrapositive of $p \rightarrow q$ is $q' \rightarrow p'$ where p and q represent sentences.

Convergent sequence (p. 115). A sequence that has a limit.

Convergent series (p. 119). A series whose sequence of partial sums converges.

Converse (p. 10). The converse of "if p, then q" is "if q, then p."

Correlation (p. 635). When two variables are related, there is a correlation between them.

Correlation coefficient (p. 636). For variables x and y, the correlation coefficient r of x and y is

$$r = \frac{1}{n}\left|\sum_{i=1}^{n} z_{x_i} z_{y_i}\right|$$

where z_{x_i} is the z-score of x_i and z_{y_i} is the z-score of y_i.

Coordinate planes (p. 447). The xy-, yz-, and xz-planes.

Cosecant (p. 277). $\csc x = \frac{1}{\sin x}$ $(\sin x \neq 0)$.

Cosine (p. 272). The cosine of x ($\cos x$) is the first coordinate of P_x, a point on the unit circle.

Cotangent (p. 277). $\cot x = \frac{\cos x}{\sin x}$ $(\sin x \neq 0)$.

Cramer's rule (pp. 515–516). A method using determinants to solve a system of linear equations.

Cross product of u and v (p. 525). Also vector product. If $\mathbf{u} = (u_1, u_2, u_3)$ and $\mathbf{v} = (v_1, v_2, v_3)$,

$$\mathbf{u} \times \mathbf{v} = \left(\begin{vmatrix} u_2 & u_3 \\ v_2 & v_3 \end{vmatrix}, \begin{vmatrix} u_3 & u_1 \\ v_3 & v_1 \end{vmatrix}, \begin{vmatrix} u_1 & u_2 \\ v_1 & v_2 \end{vmatrix}\right).$$

Cross section (p. 540). The intersection of a surface with a given plane.

Cycle of the graph of a periodic function (p. 334). The graph from $(c, f(c))$ to $(c + p, f(c + p))$ where p is the period.

Cylinder (p. 540). The surface consisting of all lines, rulings, through a plane curve, the directrix, and parallel to a fixed line.

Cylindrical coordinates (p. 545). Point $P(x, y, z)$ has cylindrical coordinates $P(r, \theta, z)$ where r and θ are the polar coordinates of the projection of P onto the xy-plane.

Decreasing function (p. 238). f is decreasing on an interval I if $f(p) > f(q)$ for all p and q in I with $p < q$.

Definite integral (p. 565). Let F be any antiderivative of f on $[a, b]$. Then the definite integral of f from a to b is defined by $\int_a^b f(x)dx = F(b) - F(a)$. The limits of integration are a and b.

Degrees (p. 282). Used in measuring angles. $360° = 1$ revolution.

Dependent system of equations (p. 503). A system having more than one solution.

Depressed equation (p. 202). Also called reduced equation. $Q(x) = 0$ where r is a root of $P(x)$ and $P(x) = (x - r)Q(x)$.

Derivative of f at c (p. 223).

$$f'(c) = \lim_{x \to c} \frac{f(x) - f(c)}{x - c} \text{ if the limit exists.}$$

Derivative function (p. 224). Also derived function. The function $x \to f'(x)$ obtained by finding the derivative.

Derivative operator notation (p. 230). $\frac{d}{dx}$.

Determinant (p. 505). A number associated with each square matrix. Let $T = \begin{bmatrix} a & b \\ c & d \end{bmatrix}$. The determinant $\left(\text{denoted det } T \text{ or } \begin{vmatrix} a & b \\ c & d \end{vmatrix}\right)$ of T is defined by

$$\det T = \begin{vmatrix} a & b \\ c & d \end{vmatrix} = ad - bc.$$

Difference of functions (p. 147). $(f - g)(x) = f(x) - g(x)$.

Difference of two matrices (p. 485). $A - B = A + (-B)$.

Difference of vectors (p. 442). $\mathbf{u} - \mathbf{v} = \mathbf{u} + (-\mathbf{v})$.

Differentiable at c (p. 223). *See* Derivative.

Differentiation (p. 224). The process of finding a derivative.

Dimensions of a matrix (p. 485). If a matrix has m rows and n columns, it has dimensions $m \times n$, or m by n. If $m = n$, the matrix is square.

Directed line segment (p. 441). One to which a direction has been assigned. For $\overrightarrow{PQ}$, P is its initial point, or tail, and Q is its terminal point, or head.

Direction angles (p. 536). The angles formed by a vector and **i**, **j**, and **k**.

Direction cosines (p. 536). Cosines of direction angles of a nonzero vector.

Direction vector. *See* Vector equation.

Directrix. *See* Cylinder and Parabola.

Disjunction (p. 7). "p or q," also written $p \vee q$, where p and q represent sentences.

Distance between two points. On the number line (p. 25) $|x - y|$. In the plane (p. 43) $\sqrt{(x_2 - x_1)^2 + (y_2 - y_1)^2}$. In three-space (p. 448) $\sqrt{(x_2 - x_1)^2 + (y_2 - y_1)^2 + (z_2 - z_1)^2}$.

Divergent sequence (p. 117). A sequence that has no limit.

Divergent series (p. 119). A series whose sequence of partial sums diverges.

Divide-and-average method (p. 260). A method for finding square roots.

Divisor. *See* Factor.

Domain (1) If x represents any member of a set D, then D is called the domain of x. Also called replacement set (p. 5). (2) The set of all first members of pairs of a function (p. 141).

Dot product of u and v (p. 454). $\mathbf{u} \cdot \mathbf{v} = \|\mathbf{u}\| \|\mathbf{v}\| \cos \theta$ where θ is the angle between **u** and **v**.

Double-angle formulas (p. 322). Identities that are used to find values of trigonometric functions of 2θ from the values of θ.

Dummy variable (p. 565). x in $\int_a^b f(x)dx$.

Elementary event (p. 599). One that contains a single outcome.

Elementary operations (p. 499). Operations used in solving a system of linear equations.

Ellipse (p. 57). The set of all points in the plane such that the sum of the distances from each point to two fixed points (called foci) is a constant.

An ellipse has a major axis, a minor axis, a center, and two vertices (p. 59).

Ellipsoid (p. 542). A graph of $\frac{x^2}{a^2} + \frac{y^2}{b^2} + \frac{z^2}{c^2} = 1$.

Elliptic cone (p. 542). A graph of

$$\frac{x^2}{a^2} + \frac{y^2}{b^2} - \frac{z^2}{c^2} = 0.$$

Elliptic paraboloid (p. 542). A graph of

$$\frac{x^2}{a^2} + \frac{y^2}{b^2} = cz \ (c > 0).$$

Empty set. *See* Null set.

Energy. *See* Work.

Equal matrices (p. 485). Those having equal corresponding entries.

Equal ordered pairs (p. 41). $(a, b) = (c, d)$ if $a = c$ and $b = d$.

Equal sets (p. 2). Sets that have exactly the same members.

Equivalent vectors (p. 442). Vectors having the same magnitude and direction.

Even function (p. 278). f if $f(-x) = f(x)$ for every x in the domain of f.

Event. (p. 599). Any subset of a sample space.

Experiment (p. 599). A procedure that can be repeated indefinitely under the same conditions and that has well-defined outcomes.

Exponential function (p. 403). The function $x \to b^x$, $b > 0$ and $x \in \mathcal{R}$.

Exponential growth model (p. 427). $P(t) = P_0(1 + r)^t$, where P_0 is an initial population and $P(t)$ the population at time t when growing at $100r\%$ ($r > 0$). When $r < 0$, the formula becomes an exponential decay model. The population is said to grow or decay exponentially.

Extremum (p. 239). Either a maximum or a minimum.

Factor (p. 28) For any integers a and b, a is a factor, or divisor, of b (written $a \mid b$ and read "a divides b") if there is an integer c such that $ac = b$.

Factors of $P(x)$ (p. 201). Polynomials $F(x)$ and $G(x)$ if $F(x)G(x) = P(x)$.

Field (p. 15). A set F of objects with two operations such that (1) F is a commutative group under the first operation; (2) its nonzero members form a commutative group under the second operation; and (3) the second operation is distributive over the first.

Finite decimal (p. 121). An expression of the form $a_r a_{r-1} \ldots a_1 a_0 \,.\, b_1 b_2 \ldots b_n$ where a_i and b_i are integers, and $0 \le a_i, b_i \le 9$.

Finite sequence (p. 99). One in which the members of 1, 2, 3, . . . , m are paired with the terms of the sequence.

First-quadrant number (p. 274). A number x is so-called if P_x is in the first quadrant, and so on for numbers in the other quadrants.

Foot-pound (p. 580). A unit for measuring work done.

Frequency of a periodic function (p. 341). The reciprocal of its period.

Frequency distribution (p. 616). A frequency table.

Frequency polygon (p. 617). Graph of data in a frequency table.

Function. (1) A set of ordered pairs in which different pairs have different first members (p. 141).

(2) A rule that assigns to each member of a set D, the domain, a unique member of a set R, the range (p. 142).

(3) Piecewise definition. A function may be defined by different formulas on different parts of its domain (p. 147).

Functional notation (p. 142). Used to denote the value of a function at a value of its domain. $y = f(x)$ is read "y equals f of x."

General second-degree equation in two variables (p. 495). $Ax^2 + Bxy + Cy^2 + Dx + Ey + F = 0$.

General solution (p. 331). One or more formulas giving all solutions of a trigonometric equation.

Geometric means (p. 109). Terms between any two given terms in a geometric sequence. A single positive geometric mean between two positive real numbers is called *the* geometric mean, or mean proportional, of the two numbers.

Geometric sequence (p. 108). A sequence such that $t_1 = a$ and $t_{n+1} = rt_n$ $(n \ge 1)$. The common ratio is r.

Geometric series (p. 110). A series whose terms form a geometric sequence.

Graph. (1) of a function f. The set of all points having coordinates of the form $(x, f(x))$ where x is in the domain of f (p. 143). (2) of an ordered pair (a, b). The point having (a, b) as its coordinates (p. 46). (3) of an open sentence in two variables. The set of all points in the plane whose coordinates satisfy the open sentence (p. 46). (4) of a polar equation. The set of all points whose polar coordinates (r, θ) satisfy an equation in r and θ (p. 363).

Greatest integer function (p. 148). Denoted $[\![x]\!]$. The greatest integer not exceeding x.

Ground speed (p. 462). Actual speed of an object relative to the ground.

Group (p. 15). A set of objects having one operation satisfying the axioms of closure, associativity, identity, and inverses.

Half-angle formulas (p. 323). Identities used to find the values of trigonometric functions of $\frac{\theta}{2}$ from those of θ.

Harmonic sequence (p. 115). $1, \frac{1}{2}, \frac{1}{3}, \frac{1}{4}, \ldots, \frac{1}{n}, \ldots$

Harmonic series (p. 119). $\sum_{n=1}^{\infty} \frac{1}{n}$.

Heading (p. 462). The direction in which a plane is pointed.

Hero's formula (p. 305). A formula for the area of a triangle.

Hertz (p. 341). 1 cycle per second.

Histogram (p. 617). Graph of the data in a frequency table.

Homogeneous linear system (p. 504). *See* System of linear equations. A homogeneous linear system has $c_1 = c_2 = \cdots = c_m = 0$.

Hyperbola (p. 64). The set of all points in the plane such that the absolute value of the differences of the distances from each point in the set to two fixed points (the foci) is constant.

If $\frac{x^2}{a^2} - \frac{y^2}{b^2} = 1$, the hyperbola has $y = \pm\frac{b}{a}x$ as equations of its asymptotes.

A hyperbola has two vertices, a transverse axis, and a center (p. 65).

Hyperbolic paraboloid (p. 542). A graph of

$$\frac{y^2}{b^2} - \frac{x^2}{a^2} = cz \ (c > 0).$$

Hyperboloid of one sheet (p. 542). A graph of

$$\frac{x^2}{a^2} + \frac{y^2}{b^2} - \frac{z^2}{c^2} = 1.$$

Hyperboloid of two sheets (p. 542). A graph of

$$\frac{x^2}{a^2} - \frac{y^2}{b^2} - \frac{z^2}{c^2} = 1.$$

Identity (p. 5). An open sentence whose solution set is the domain of its variables.

Identity function (p. 149). $I(x) = x$ for all $x \in X$.

Identity matrix (p. 486). A square matrix in which $a_{ij} = 1$ if $i = j$ and $a_{ij} = 0$ otherwise.

Identity transformation I of the plane (p. 483). $I((x, y)) = (x, y)$.

Included. *See* Adjacent.

Inconsistent system of equations (p. 503). One with no solution.

Increasing function (p. 238). f is increasing on an interval I if $f(p) < f(q)$ for all p and q in I with $p < q$.

Independent events (p. 607). A and B if $P(A \cap B) = P(A) \cdot P(B)$.

Index of summation (p. 101). k in $\sum_{k=1}^{n} a_k$.

Induction hypothesis (p. 90). The assumption $x \in S$ used in mathematical induction.

Inequality (p. 20). Any sentence involving an inequality symbol.

Infinite sequence (p. 99). One in which the terms of the sequence are paired with the members of N.

Inflection point (p. 242). The point at which the concavity of a graph changes sense.

Intersection (p. 3). The intersection of two sets is the set consisting of all members of both sets.

Intercepts of a plane (p. 530). The points where the plane intersects the coordinate axes.

Inverse cosecant (p. 353). $\text{Csc}^{-1} x = \text{Sin}^{-1} \frac{1}{x}$.

Inverse cosine (p. 351). Cos^{-1} is the inverse of Cos. $y = \text{Cos}^{-1} x$ if and only if $\cos y = x$ and $0 \le y \le \pi$.

Inverse cotangent (p. 352). Cot^{-1} is the inverse of Cot. $y = \text{Cot}^{-1} x$ if and only if $\cot y = x$ and $0 < y < \pi$.

Inverse of a matrix (p. 510). If A and B are matrices and $AB = BA = I$ where I is an identity matrix, then A and B are inverses of each other.

Inverse secant (p. 353). $\text{Sec}^{-1} x = \text{Cos}^{-1} \frac{1}{x}$.

Inverse sine (p. 351). Sin^{-1} is the inverse of Sin. $y = \text{Sin}^{-1} x$ if and only if $\sin y = x$ and $-\frac{\pi}{2} \le y \le \frac{\pi}{2}$.

Inverse tangent (p. 352). Tan^{-1} is the inverse of Tan. $y = \text{Tan}^{-1} x$ if and only if $\tan y = x$ and $-\frac{\pi}{2} < y < \frac{\pi}{2}$.

Iterate (p. 259). Repeat.

Joule (p. 464). The work done by a force of one newton in moving an object one meter; the basic metric unit for measuring work (p. 581).

Kilojoule (p. 582). A unit for measuring work, 1000 J.

Kilowatt-hour (p. 582). A unit for measuring electricity.

Lag (p. 344). The amount by which one sine curve is to the left of a second sine curve.

Latitude (p. 285). The angle formed by a point on Earth, the center of Earth, and a point on the equator.

Lead (p. 344). The amount by which one sine curve is to the right of a second sine curve.

Least common multiple (LCM) (p. 30). The smallest positive integer that is a multiple of given integers.

Limit of a function at c (p. 158). $\lim_{x \to c} f(x) = l$, if and only if $\lim_{n \to \infty} f(x_n) = l$ for every sequence $x_1, x_2, \ldots, x_n, \ldots$ in the domain of f with $\lim_{n \to \infty} x = c$ $(x_n \ne c)$.

Limits of integration. *See* Definite integral.

Limit of a sequence (p. 115). The infinite sequence $a_1, a_2, \ldots, a_n, \ldots$ has A as limit, written $\lim_{n \to \infty} a_n = A$, if for each positive number h, there is a number M such that $|a_n - A| < h$ whenever $n > M$.

Linear combination of functions (p. 227). $(af + bg)(x), = af(x) + bg(x), a, b \in \mathcal{R}$.

Linear combination of vectors (p. 443). $a\mathbf{u} + b\mathbf{v}$, $a, b \in \mathcal{R}$.

Linear transformation (p. 479). T if $T(\mathbf{u} + \mathbf{v}) = T(\mathbf{u}) + T(\mathbf{v})$ and $T(a\mathbf{u}) = aT(\mathbf{u})$; (p. 480) $T(\mathbf{u})$ is the image of $\mathbf{u}$.

Linearly independent vectors (p. 505). $\mathbf{v}_1, \mathbf{v}_2, \mathbf{v}_3, \ldots \mathbf{v}_n$ if $a_1\mathbf{v}_1 + a_2\mathbf{v}_2 + \cdots + a_n\mathbf{v}_n = 0$ has as its only solution $a_1 = a_2 = \cdots = a_n = 0$. Otherwise they are said to be linearly dependent.

Logarithmic function with base b (p. 412). The inverse of the exponential function with base b; $\log_b x = y$ if and only if $b^y = x$.

Lower and upper bounds of roots (p. 208). $[l, m]$ in which all real roots of a polynomial equation over $\mathcal{R}$ lie; l is a lower bound and m is an upper bound.

Mapping diagram (p. 152). A diagram relating domain members and corresponding range members of a function.

Mathematical system (p. 13). A formal mathematical system consists of undefined objects, postulates or axioms, definitions, and theorems.

Matrix (p. 485). A rectangular array of numbers. Each number is called an entry.

Matrix of constants (p. 512). The matrix consisting of the constant terms in a system of linear equations.

Matrix of a linear transformation (p. 490). A matrix associated with a linear transformation.
Maximized (p. 245). Made as large as possible.
Mean. (1) The mean $\bar{x}$ of the data $x_1, x_2, \ldots, x_n$ is $\frac{1}{n}\sum_{i=1}^{n} x_i$. (p. 617)
(2) If X is a random variable that takes on the values $x_1, x_2, \ldots, x_r$, then the mean μ of the probability distribution of X is $\sum_{i=1}^{n} x_i P(X = x_i)$. (p. 619)
Median (p. 617). When a set of n data is written in order from least to greatest, the median is the middle value if n is odd and the mean of the two middle values if n is even.
Megajoule (p. 582). 10^6 J.
Members of a set (p. 1). The objects in a given set.
Minimized (p. 245). Made as small as possible.
Minors of a determinant (p. 506). Using a_{ij} to denote the entry in the ith row and jth column of a determinant, the minor of a_{ij}, denoted M_{ij}, is the determinant obtained by striking out the ith row and the jth column of the determinant. The cofactor of a_{ij}, denoted A_{ij}, is $(-1)^{i+j}M_{ij}$.
Mode (p. 617). Data value with greatest frequency.
Modulus. *See* Absolute value of a complex number.
Multiplication rule of probability (p. 605). $P(A \cap B) = P(A) \cdot P(B \mid A)$.
Multiplicity (p. 202). A root r of $P(x) = 0$ has multiplicity k if $(x - r)$ occurs exactly k times in the complete factorization of $P(x)$.
Mutually exclusive. *See* Probability function.

Natural exponential function (p. 408). $x \to e^x$.
Negation of p (p. 7). "not p" where p represents a sentence.
Negative number (p. 19). c is negative if $c < 0$.
Newton's method (p. 259). A method for approximating zeros of differentiable functions.
Nondecreasing sequence (p. 126). $a_1, a_2, a_3, \ldots, a_n, \ldots$ such that $a_{n+1} \geq a_n$ for all n.
Norm of a vector v (p. 451). Denoted $\|\mathbf{v}\|$. The magnitude of **v**.
Normal curve (p. 627). The graph of

$$y = \frac{1}{\sigma\sqrt{2\pi}} e^{-\frac{1}{2}z^2} \text{ where } z = \frac{x - \mu}{\sigma}.$$

The standard normal curve is the graph of

$$y = \frac{1}{\sqrt{2\pi}} e^{-\frac{1}{2}x^2}$$

Normal distribution (p. 627). Determined by its mean μ and standard deviation σ. The standard normal distribution is the one with $\mu = 0$ and $\sigma = 1$.
Null set (p. 2). Also called the empty set. The set having no members.

***n*th-quadrant angle** (p. 287). θ if it is in standard position and its terminal side lies in the nth quadrant ($n = 1, 2, 3, 4$).

Octants (p. 447). The eight spatial regions into which the coordinate planes divide three-space.
Odd function (p. 278). f if $f(-x) = -f(x)$ for every x in the domain of f.
One-to-one function (p. 153). f if $f(x_1) \neq f(x_2)$ whenever $x_1 \neq x_2$.
Open interval (p. 167). $(a, b) = \{x: a < x < b\}$; (p. 238). $(a, \infty) = \{x: x > a\}$ and $(-\infty, b) = \{x: x < b\}$.
Open sentence (p. 5). Any sentence involving a variable.
Optimization problems (p. 245). Problems dealing with maxima and minima.
Order of a determinant (p. 506). The number of rows (or columns).
Ordinate (p. 42). The second coordinate of a point in the plane.
Origin (p. 42). The point where the coordinate axes intersect. Its coordinates are (0, 0) or (0, 0, 0).
Orthogonal vectors (p. 455). So called if the angle between them is 90° or if either of them is 0.
Outcomes. *See* Experiment.

Parabola (p. 69). A set of all points in the plane that are equidistant from a fixed line (the directrix) and a fixed point (the focus) not on the line. Each parabola has an axis (of symmetry).
Parallel planes (p. 532). Planes with parallel normal vectors.
Parallel vectors (p. 443). **u** and **v** if $\mathbf{u} = k\mathbf{v}$ for a scalar k.
Parameter (p. 375). When a curve in the plane is described by giving the x- and y-coordinates of a point on the curve in terms of a third variable, say t, t is called a parameter.
Parametric equations. Equations in one variable that give the coordinates of points on a curve in the plane (pp. 375, 464), or in three-space (p. 534).
Partial sum (p. 119). The sum of the first n terms of an infinite series.
Periodic function (p. 334). f if for some positive constant p, $f(x + p) = f(x)$ for every x in the domain of f.
Permutation (p. 590). An arrangement of the members of a set in a definite order.
Perpendicular planes (p. 532). Planes having perpendicular normal vectors.
Phase shift (p. 335). A horizontal translation of a periodic function.
Point-slope form (p. 48). The equation $y - y_1 = m(x - x_1)$ for the line containing (x_1, y_1) and having slope m.
Polar coordinates (p. 361). The reference system for polar coordinates consists of a point O, the pole, and

a ray, the polar axis, having O as its endpoint; in (r, θ), $r = OP$ and θ = the measure of an angle from the polar axis to $\overline{OP}$.

Polar (trigonometric) form of a complex number z (p. 369). $z = r(\cos \theta + i \sin \theta)$. r is the modulus, or absolute value, of z. θ is the argument of z (arg z).

Polynomial completely factored (p. 201). When expressed as a product of polynomials irreducible over a field F.

Polynomial over a field F (p. 196). An expression of the form $P(x) = a_0x^n + a_1x^{n-1} + \cdots + a_{n-1}x + a_n$, where n is a nonnegative integer, and the coefficients $a_0, a_1, \ldots, a_n$ are members of F, with the leading coefficient, a_0, not 0; linear: degree 1; quadratic: degree 2; cubic: degree 3; quartic: degree 4.

Polynomial function (p. 164). Any function defined by a polynomial.

Population (p. 632). In statistics, any set of individuals or objects under study.

Positive nth root (p. 128). We write $x = \sqrt[n]{a}$ where x is the positive nth root of a.

Positive number (p. 19). c if $0 < c$.

Position vector of P (p. 464). $\overrightarrow{OP}$.

Prime number, or prime (p. 28). Any integer greater than 1 that has only 1 and itself as factors.

Principal solutions (p. 331). Solutions of a trigonometric equation in a specified interval.

Probability distribution (p. 610). A listing of the probabilities that a random variable takes in a particular value of its range.

Probability function (p. 600). A probability function P assigns to each event A in the sample space S of an experiment a real number $P(A)$, read "the probability of A," such that (1) $0 \le P(A) \le 1$; (2) $P(S) = 1$; (3) $P(A \cup B) = P(A) + P(B)$ if A and B are mutually exclusive, that is, if they have no outcomes in common.

Probability histogram (p. 610). A graph showing the probability distribution of a random variable.

Product of functions (p. 146). $(fg)(x) = f(x) \cdot g(x)$.

Product of matrices (p. 486). Let A be an $m \times n$ matrix and B be an $n \times k$ matrix. AB is the $m \times k$ matrix whose entry in the ith row and jth column is the dot product of the ith row of A and the jth column of B.

Product of a scalar r and a matrix A (p. 485). The matrix rA, each entry of which is r times the corresponding entry of A.

Quadrantal angle (p. 287). An angle in standard position with its terminal side along a coordinate axis.

Quadrantal number (p. 274). A number x if P_x lies on a coordinate axis.

Quadrants (p. 42). The four regions into which the axes divide the plane.

Quadric surface (p. 541). One obtained as the graph of a second-degree equation in three variables.

Quotient of functions (p. 147). $\frac{f}{g}(x) = \frac{f(x)}{g(x)}$ where $g(x) \neq 0$.

Radian measure (p. 282). A system of angle measurement. The radian measure of an angle θ is $\frac{s}{r}$ where r is the radius of circle C centered at the vertex of θ and s is the length of the arc of C intercepted by θ.

Radical (p. 128). The symbol $\sqrt[n]{a}$. a is the radicand, n is the index.

Radius of sphere. *See* Sphere.

Random variable (p. 610). A function that assigns a numerical value to each outcome of an experiment.

Range (p. 141) set of all second members of pairs of a function; (p. 142) each member is called a value of the function; (p. 621) measure of the spread of a set of data, that is, the difference between the largest and smallest data values.

Rational function (p. 172). A function defined by the quotient of polynomials not having common zeros.

Rectangular (Cartesian) coordinate system (p. 42). The coordinate system in which lengths of line segments are used to assign coordinates to points.

Rectilinear motion (p. 253). Motion along a straight line.

Recursion formula (p. 99). A sequence is defined by a recursion formula, or recursively, when rules are given that identify the first term or the first several terms of the sequence and indicate how to find subsequent terms from preceding terms.

Reducible (p. 201). A polynomial over a field F is reducible over F if it is the product of two or more nonconstant polynomials over F; otherwise it is irreducible over F.

Reference angle of θ (p. 288). The acute angle that the terminal side of θ makes with the horizontal axis.

Region under the graph of f (p. 568). The set
$$\{(x, y): a \le x \le b, 0 \le y \le f(x)\}.$$
Area under the graph refers to the area of the region.

Related-rate problems (p. 254). Problems involving rates that are related.

Relation (p. 141). Any set of ordered pairs.

Relative frequency (p. 618). Actual frequency of a score or class of scores divided by the total number of scores.

Relative (local) maximum value (p. 239). f has a relative maximum at c if $f(c) \ge f(x)$ for all x in some open interval containing c.

Relative (local) minimum value (p. 239). f has a relative minimum at c if $f(c) \le f(x)$ for all x in some open interval containing c.

Relatively prime (p. 28). Two integers whose GCF is 1.

Replacement set. *See* Domain.

Resultant (p. 463). The sum of two vectors.

Ring (p. 179). Any set of objects G together with two operations, say $\oplus$ and $\odot$, such that (1) G is a commutative group under $\oplus$, (2) G is closed under $\odot$, and $\odot$ is associative, and (3) $\odot$ is distributive over $\oplus$.

Ring with identity (p. 179). One that has an identity under $\odot$.

Roster, or listing (p. 2). A way of writing a set.

Row matrix (p. 485). A matrix with one row.

Row-reduced echelon form of a matrix (p. 503). So-called when in each row the first nonzero entry is 1, all other entries in the column containing that 1 are 0, the leading 1 in a row is to the right of the leading 1's in all rows above that row, and any row with all zero entries is below all the rows with a leading 1.

Rulings *See* Cylinder.

Sample (p. 632). In statistics, a subset of a population.

Sample space (p. 599). The set of all possible outcomes of an experiment

Scalar equation of a plane (p. 530).

$$ax + by + cz + d = 0.$$

Scalar multiplication (p. 443). $k\mathbf{v}$ is the vector whose magnitude is $|k|$ times the magnitude of $\mathbf{v}$ and whose direction is the same as that of $\mathbf{v}$ if $k > 0$ and opposite that of $\mathbf{v}$ if $k < 0$. If k is 0, $k\mathbf{v}$ has no direction.

Scalar triple product of u, v, w (p. 527). $\mathbf{u} \cdot (\mathbf{v} \times \mathbf{w})$.

Secant (p. 277). $\sec x = \dfrac{1}{\cos x} \quad (\cos x \neq 0)$.

Second-degree equation in three variables (p. 541). $Ax^2 + By^2 + Cz^2 + Eyz + Fxz + Gxy + Hx + Iy + Jz + K = 0$.

Second-degree equation in two variables (pp. 75, 495). Any equation in this form: $Ax^2 + Bxy + Cy^2 + Dx + Ey + F = 0$.

Second derivative (p. 231). If f' is differentiable, the second derivative of f is the derivative of f'.

Sequence (p. 99) A set of real numbers in a specified order. Each number is called a term of the sequence; (p. 142) a function whose domain is N or an initial segment $1, 2, 3, \ldots, n$ of N.

Series (p. 100). An expression for the sum of the terms of a sequence.

Set (p. 1). A collection of objects with a definite criterion determining what is in the set and what is not.

Sigma (Σ) (p. 101). The sum or summation symbol.

Signum or sign function (p. 148).

$$x \to \operatorname{sgn} x = \begin{cases} 1 \text{ if } x > 0 \\ 0 \text{ if } x = 0 \\ -1 \text{ if } x < 0 \end{cases}$$

Simple harmonic motion (p. 344). Occurs when the net force acting on an object is negatively proportional to its displacement from some central position.

Sine (p. 272). The sine of x, $\sin x$, is the second coordinate of P_x, a point on the unit circle.

Sine curve (p. 343). The graph of an equation of the form $y = A \sin \omega x$ or a translation of it.

Sinusoid (p. 343). A sine curve. Also called a sinusoidal curve.

Skew lines (p. 535). In three-space, nonparallel lines that do not intersect.

Slope-intercept form (p. 48). $y = mx + b$, for the line with slope m and y-intercept b.

Slope of a curve at c (p. 223).

$$\lim_{x \to c} \frac{f(x) - f(c)}{x - c} \text{ if the limit exists.}$$

Solving a triangle (p. 292). Finding the measures of its sides and angles.

Solution set (p. 5). The solution set (or truth set) of an open sentence in x is the set of all x in a set D such that the open sentence is true.

Sphere (p. 539). The set of all points at a given distance, the radius, from a fixed point, the center.

Spherical coordinates of *P* (p. 546). (ρ, θ, φ) where $\rho = OP$, θ is the same as in cylindrical coordinates, and φ is the measure of the angle that $\overrightarrow{OP}$ makes with the positive z-axis, $0 \leq \varphi \leq \pi$.

Square root function (p. 164). $x \to \sqrt{x}$ where $x \geq 0$.

Standard basis (p. 457). $\{\mathbf{i}, \mathbf{j}\}$ or $\{\mathbf{i}, \mathbf{j}, \mathbf{k}\}$.

Standard deviation of the probability distribution (p. 623). The principal square root of the variance.

Standard form of an equation of a line (p. 48). $Ax + By = C$.

Standard position of θ (p. 286). θ is in standard position if the initial side of θ coincides with the positive horizontal axis.

Stationary point (p. 239). c if $f'(c) = 0$.

Stress, strain (p. 580). When an elastic object is subjected to a distorting force (stress), a deformation (strain) results.

Subset (p. 2). If every member of a set G is also a member of a set H, then G is a subset of H. If H has a member that G does not have, then G is a proper subset of H.

Success, failure (p. 611). Classifications of outcomes in a binomial experiment.

Sum of two matrices (p. 485). The matrix $A + B$, each entry of which is the sum of the corresponding entries in A and B.

Sum of a finite series (p. 100). The sum of the terms in the sequence.

Sum of functions (p. 146). $(f + g)(x) = f(x) + g(x)$.

Sum of two vectors (p. 442). For vectors $\mathbf{u}$ and $\mathbf{v}$, the vector obtained by placing the tail of $\mathbf{v}$ at the head of $\mathbf{u}$ and then drawing the arrow whose tail is the tail of $\mathbf{u}$ and whose head is the head of $\mathbf{v}$.

Symmetric (p. 58). Used in describing a graph with respect to the axes.

Symmetric-form equations. *See* Parametric equations.

Symmetric matrix (p. 489). One in which $a_{ij} = a_{ji}$.

Synthetic substitution (p. 169). A method for evaluating polynomials.

System of linear equations (p. 497).

$$a_{11}x_1 + a_{12}x_2 + \cdots + a_{1n}x_n = c_1$$
$$\vdots$$
$$a_{m1}x_1 + a_{m2}x_2 + \cdots + a_{mn}x_n = c_n$$

Tangent (p. 277). $\tan x = \dfrac{\sin x}{\cos x}$ $(\cos x \neq 0)$.

Three-space, or three-dimensional space (p. 446). Points located by drawing three mutually perpendicular coordinate axes whose origins intersect in a point, the origin.

Traces of a plane (p. 530). The lines in which the plane intersects the coordinate planes.

Trace of a surface (p. 540). The intersection of the surface with a coordinate plane.

Transpose of a matrix *A* (p. 511). Denoted A^T. The transpose of a matrix A is that matrix whose entry $t_{ij} = a_{ji}$.

Trials (p. 611). Repetitions of an experiment.

Triangle inequality. (p. 26) $|a + b| \le |a| + |b|$; (p. 452) $\|\mathbf{u} + \mathbf{v}\| \le \|\mathbf{u}\| + \|\mathbf{v}\|$.

Triangular form (p. 498). Term applied to a system of equations in a form like

$$\begin{aligned} x + 5y - 6z &= 29 \\ y + 4z &= -10 \\ z &= -3 \end{aligned}$$

Trigonometric cosine (p. 286). For θ in standard position, $\cos\theta = \dfrac{u}{r}$.

Trigonometric sine (p. 286). For θ in standard position, $\sin\theta = \dfrac{v}{r}$.

True course (p. 462). Direction in which an object actually travels.

Truth set. *See* Solution set.

Truth value (p. 6). The truth or falsity of a sentence.

Uniform circular motion (p. 283). Motion described by a point that moves with constant speed in a circular path.

Union (p. 3). The union of two sets is the set consisting of all members of either set.

Unit circle (p. 271). Radius 1 centered at (0, 0).

Unit vector (p. 451). A vector with norm 1.

Variable (p. 5). Any member of the replacement set or domain.

Variance (p. 622). If $\bar{x}$ is the mean of data $x_1, x_2, \ldots, x_n$, $\sigma^2 = \dfrac{1}{n}\sum_{i=1}^{n}(x_i - \bar{x})^2$.

Variance of a probability distribution (p. 623). For a random variable X that takes on values $x_1, x_2, \ldots, x_r$ and has mean μ, the variance of the probability distribution of X is

$$\sigma^2 = \sum_{i=1}^{r}(x_i - \mu)^2 P(X = x_i) \text{ or}$$
$$\sigma^2 = \sum_{i=1}^{r} x_i^2 P(X = x_i) - \mu^2.$$

Variation in sign (p. 209). Term applied to $P(x)$ whenever the coefficients of two adjacent terms have opposite signs.

Vector. (p. 441) A quantity that has both magnitude and direction; (p. 444) when denoted by an ordered pair (a, b), a is the first component, and b the second component.

Vector components (p. 459). $a\mathbf{u}$ and $b\mathbf{v}$ if $\mathbf{t} = a\mathbf{u} + b\mathbf{v}$; scalar components of $\mathbf{t}$ are a and b.

Vector equation (p. 534). $\mathbf{r} = \mathbf{r}_0 + t\mathbf{m}$, an equation of the line L containing P_0 and parallel to a nonzero vector $\mathbf{m} = (a, b, c)$. $\mathbf{m}$ is called a direction vector of L.

Vector product. *See* Cross product.

Velocity (p. 253). Rate of change of position with respect to time.

Work. (p. 463) When a force $\mathbf{F}$ moves an object from A to B along a straight path ($\mathbf{d}$), then the work done by $\mathbf{F}$ is $\|\mathbf{F}\|\,\|\mathbf{d}\|\cos\theta$, where θ is the angle between $\mathbf{F}$ and $\mathbf{d}$; (p. 580) given a constant force of magnitude F directed along the s-axis displacing a particle from a to b. The work done by the force and the energy it expends are given by $F \cdot (\text{displacement}) = F(b - a)$.

***x*-intercept** (p. 48). The x-coordinate of the point where a line crosses the x-axis.

***y*-intercept** (p. 48). The y-coordinate of the point at which a nonvertical line crosses the y-axis.

Zero matrix (p. 486). One in which all entries are 0.

Zero-product property (p. 16). For all real numbers a and b: $ab = 0$ if and only if $a = 0$ or $b = 0$.

Zero vector 0 (p. 442). The vector such that $\mathbf{u} + \mathbf{0} = \mathbf{u}$.

INDEX

Abscissa, 42
Absolute (global) maximum, 240
Absolute (global) minimum, 240
Absolute value
  of a complex number, 190–191, 215, 369
  of real numbers, 23–27, 33
Absolute value function, 147, 165
Acceleration, 253, 263
Addition
  axioms of, 14, 19
  cancellation law for, 15
  of complex numbers, 185–186, 215
  of probabilities, 604, 640
  of real numbers, 14
  of vectors, 442–443, 444, 449, 462, 469, 471
Addition formulas
  for sine and cosine, 318–322, 356
  for tangent, 326, 356
Additive identity, 14
Additive inverse, 14, 15
  of a matrix, 485
  of a vector, 442, 469, 471
Air speed, 462
Alternating current, 358–359
Ambiguous case in solving triangles, 301
Amplitude of a function, 340
Analysis, 157
Analytic geometry, 41–87
Angle(s), 281
  coterminal, 288
  of depression, 293
  of elevation, 293
  generated by rotation, 281–282
  measure of, 282–284, 285, 290, 308
  negative, 282
  positive, 282
  quadrant of an, 287
  quadrantal, 287
  reference, 288, 308
  sides of, 281
  between vectors, 454, 456
  vertex of, 281
Angular speed, 284
Annuities, 136–137
Antiderivative, 557–561, 584
Apogee, 86
Application (feature)
  alternating current, 358–359
  annuities, 136–137
  coding, 522–523
  complex numbers and functions, 218–219
  elasticity, 267
  logic networks, 38–39
  functions in the sciences, 182–183
  magnitudes of stars, 437
  Newton's law of cooling, 587
  orbits, 86–87
  space curves, 553
  triangles in structures, 311
  two laws of optics, 396–397
  uniform circular motion, 476–477
  weather studies, 644–645
Approximation
  of areas, 84–85, 586
  by linear interpolation, 181
  of zeros of functions, 258–261, 263
Arc, length of, 283
Area(s)
  approximating, 84–85, 586
  circular-sector, 283
  under a curve, 568, 584, 586, 628, 641
  of a parallelogram, 527, 549
  of plane regions, 568–574, 584, 586
  of triangles, 297, 305–307, 309, 310
Arithmetic mean(s), 104–105
Arithmetic progression, 104
Arithmetic sequences, 103–108, 134
Arithmetic series, 105–106, 134
Associativity of real numbers, 14
Asymptote(s)
  of a curve C, 173
  of a hyperbola, 65
  in trigonometric graphs, 348, 349
Average, 105
Axes, 42
  of an ellipse, 59
  rotation of, 493–497, 518
  in three-space, 446
Axiom(s)
  of addition, 14
  of comparison, 19
  of completeness, 125–129, 134
  distributive, 15
  of equality, 13
  of multiplication, 14
  of order, 19, 20
  substitution, 14
  of vector space, 469
Axis
  imaginary, 189
  of a parabola, 69
  polar, 361, 362
  real, 189
  transverse, of hyperbola, 65

Base of an exponential function, 403, 408, 432
Basis vectors, 457–461, 480
Bernoulli trials, 611
Bernoulli's inequality, 92
Biconditional sentence, 10
Binomial coefficients, 94, 597
Binomial distribution, 612–613
Binomial experiment, 611–614, 624, 640
Binomial theorem, 94–99, 133
Bound(s)
  for roots of polynomial equations, 208–210
  for a sequence, 126
Boyle's law, 146

Calculus. *See* Differential calculus *and* Integral calculus.
Cancellation law
  for addition, 15
  for multiplication, 15
Cardioid, 368
Cartesian coordinate system, 42–44, 80
Cartesian product, 41
Cauchy-Schwarz inequality, 455–456
Center
  of a circle, 53, 80
  of an ellipse, 59, 80
  of a hyperbola, 65, 80
  of a sphere, 539
Centesimal system of angle measure, 285
Chain rule, 234–237, 262

Chapter summaries, 32–33, 80–81, 133–134, 177–178, 215–216, 262–263, 308–309, 355–356, 393–394, 432–433, 471–472, 518–519, 549–550, 584, 639–641
Chapter tests, 33, 81, 135, 178, 216, 263, 309, 357, 394, 433, 472–473, 519, 551, 585, 641–642
Characteristic, 496
Circle(s), 53–56, 73, 80
center of, 53, 80
equation of, 53, 80
radius of, 53, 80
tangent to a, 54
unit, 271–272, 308
Circular functions, 271–281, 308
Change-of-base formula, 419, 433
Closed interval, 167
Closure for real numbers, 14
Coding, 522–523
Coefficient(s)
binomial, 94, 597
leading, 196
of polynomials, 196
Cofactor, 506
Cofunctions, 321
Column matrix, 485, 486
Combinations, 594–598, 639
Common difference, 104, 134
Common factor, 28
Common logarithms, 418
table of, 665–666
Common ratio, 108, 134
Commutative group, 15
Commutativity of real numbers, 14
Comparison, axiom of, 19
Complement of a set, 3, 32
Completeness, axiom of, 125–129, 134
Complex field, 185–188, 215
Complex number(s), 185–195
absolute value (modulus) of, 190–191, 215, 369
addition of, 185–186, 215
argument of, 369, 370
conjugate of, 189–190, 215
equal, 186
field of, 185–188, 215
and functions, 218–219
graphs of, 189–192
imaginary part of, 186
multiplication of, 185–186, 215, 369–370, 393
polar form of, 369–372, 393
powers of, 217, 372–373
quotient of, 187, 370, 393
real part of, 186
roots of, 193–195, 215, 373–374
Complex plane, 189–192
Components of a vector, 444, 471
Composition of functions, 151–152, 164, 177
Compound interest formula, 406, 409, 432
continuous, 409–410, 432
Computer applications
approximating areas, 84–85
area (of a triangle), 310
integral powers of complex numbers, 217
interpolation, 181
linear systems, 520–521
minimum distance, 266
a normal distribution, 643
numerical integration, 586
powers of *e*, 436
truth tables, 36–37
vector products, 552
Computer exercises, 5, 9, 19, 27, 31, 52, 63, 79, 99, 113, 124, 146, 156, 171, 188, 193, 200, 214–215, 226, 241, 261, 281, 295, 305, 317, 322, 333, 343, 350, 368, 375, 383, 412, 421, 432, 446, 453, 461, 466, 484, 489, 514, 529, 539, 579, 583, 598, 626
Concavity of a graph, 242, 243, 262
Conditional probability, 605, 640
Conditional sentences, 10–11, 32
Confidence intervals, 632–635, 641
Conic sections, 53–79, 80, 81, 494–497, 518
degenerate, 73, 81
eccentricity of, 395
intersections of, 76–79, 81
polar equations of, 395
Conjugate of a complex number, 189–190, 215
Conjunction, 6, 32
negation of, 8
truth table for, 6
Constant function, 149
Constant of integration, 558
Continuity, 163
of exponential function, 406, 432
of polynomial functions, 164
of trigonometric functions, 381
Continuous functions, 163–170, 177, 225, 381, 406, 408, 432
Contradiction, 5
proof by, 11
Contrapositive, 11, 32
Convergent series, 119, 134
Converse of conditional sentence, 10, 32
Conversion formula for angle measure, 282
Coordinate box, 447
Coordinate planes, 446–447
Coordinates
Cartesian, 42, 80, 362, 393
cylindrical, 545–546, 547, 550
of a point in three-space, 446–447
polar, 361–362, 393
spherical, 546–548, 550
Correlation, 635–639
Correlation coefficient, 636–637, 641
Cosecant
as circular function, 277–279, 308
derivative of, 385, 394
graph of, 349
inverse, 353
as trigonometric function, 286, 287
Cosine
as circular function, 271–276, 278, 279, 308
derivative of, 384, 394
graphs of, 338–343
inverse, 350–351, 356
in solving triangles, 292, 296–297
as trigonometric function, 286–288
Cosines, law of, 296, 308
Cotangent
as circular function, 277–279, 308
derivative of, 385, 394
graph of, 349
inverse, 352
as trigonometric function, 286, 287
Cramer's rule, 515–517, 519
Cross product of vectors, 525–529, 549
Cross section, 540, 541, 575–577
Cubic polynomial, 196
Cumulative review, 138–139, 268–269, 438–439, 554–555
Cycle of a graph, 334
Cycloid, 377
Cylinders, 540–541, 550

Decibels, 415

Decimals
finite, 121
infinite, 121, 126–127
repeating, 121–122
Degree measure of an angle, 282, 290, 308
Degree of a polynomial, 196
De Moivre's theorem, 372, 393
Depression, angle of, 293
Derivative(s), 223
applications of, 245–258, 427–431
of e^x, 423, 433
in graphing functions, 238–245
of ln *x*, 422–423, 433
notations for, 230–233, 262
order of, 231
of polynomial functions, 221–226, 262
properties of, 226–230
as rates of change, 253–258
second, 231, 242–244
of trigonometric functions, 384–387, 394
Derivative function, 223, 262
Derivative operator notation, 230
Descartes' rule of signs, 209–210
Determinants, 505–509, 518–519
evaluating, 506–508
order of, 506
Difference
common, 104, 134
of functions, 147, 177
of matrices, 485
of vectors, 442
Differentiable function, 223, 262
Differential calculus, 221–267
applied maxima and minima, 245–252
the chain rule, 234–237, 262
concavity and the second derivative, 242–245, 262
derivatives of functions, 221–226, 262
derivatives as rates of change, 253–258, 262
mean-value theorem, 264–265
Newton's method, 258–261, 263
notations for derivatives, 230–233, 262
properties of derivatives, 226–230, 262
using derivatives in graphing, 238–241, 262
Differentiation, 221–237
rules for, 233
Dimensions of a matrix, 485, 518
Directed line segment, 441
Direction angles, 536, 550
Direction cosines, 536, 537, 550
Direction vector, 534
Directrix
of a cylinder, 540, 550
of a parabola, 69
Discontinuous function, 164
Disjunction, 7, 32
negation of, 8
truth table for, 7
Displacement, 580
Distance
in coordinate plane, 42–43, 80
minimum, 266
on number line, 23, 24, 25, 33, 42
between a point and a line, 535–536, 550
between a point and a plane, 532, 549
in polar coordinates, 364
in three-space, 448, 471
Distance formula, 43, 80
for three-space, 448, 471
Distribution(s)
binomial, 612–613
frequency, 616–617, 640
normal, 626–632, 641, 643
probability, 610–615, 618–619, 623–624, 640
Distributive axiom, 15
Divergent series, 119, 134
Division
of real numbers, 15
synthetic, 197–198, 215
Division algorithm, 29, 33, 196–197, 215
Divisor, greatest common, 28
Domain
of a function, 141, 142, 177
of a variable, 5, 32
Dot product of vectors, 454–457, 472
Double-angle formulas
for sine and cosine, 322–323, 326, 356
for tangent, 326, 356

e (base of natural exponential function), 408, 432
powers of, 436, 667 (table)
Eccentricity of a conic, 395
Elasticity (of demand), 267
Elevation, angle of, 293
Ellipse(s), 56–63, 73, 80, 494, 496
center, 59, 80
equations of, 56–61, 80
focal radius, 57
foci, 57, 80
latus rectum, 62
major axis, 59
minor axis, 59
vertices, 59, 80
Ellipsoid, 542
Elliptic cone, 542
Empty set, 2, 32
Energy, 580–583, 584
Entries (elements) of a matrix, 485, 518
Equality, 13
axioms of, 13
of complex numbers, 186
of matrices, 485
of sets, 2
Equation(s)
of conic sections, 53–81, 494–496
depressed, 202
exponential, 131–132
of lines, 46–50, 80, 534–539, 550
of planes, 530–534, 549
polar, 362–363, 365–368
polynomial, 201–215
of quadric surfaces, 539–545, 550
reduced, 202
of rotation, 494–497
systems of. *See under* Systems.
trigonometric, 330–333
Estimate
interval, 632–633
point, 632
Euclidean algorithm, 29
Euler, Leonard, 408
Euler's formula, 436
Even function, 278, 335–336, 356
Event(s), 599
compound, 603–609
elementary, 599
independent, 606–607, 640
mutually exclusive, 600
Expansion of $(a + b)^n$, 94–99, 597
Expected value, 613, 640
Experiment, 599, 640
binomial, 611–614, 624, 640
Exponential decay, 427–432, 433
Exponential functions, 403–412, 423, 432–433
Exponential growth, 427–432, 433
Exponent(s)
laws of, 90–91, 130, 401

negative, 129–130
positive integral, 90, 133
rational, 130–133
real, 399–403
zero, 91
Extra (feature)
groups and subgroups, 34–35
hyperbolic functions, 434–435
a mean-value theorem, 264–265
polar equations of conics, 395
rings, 179–180
symmetry and graphs, 82–83
vector spaces and subspaces, 474–475
Extremum (extrema), 239, 262, 388–393

Factor(s), 28
common, 28
greatest common, 28, 33
of a polynomial, 201, 216
Factor theorem, 201, 215
Factorial, 94, 133
Factorization, prime, 28–29
Fibonacci sequence, 99
Field, 15, 32
algebraically complete (closed), 202
complex, 185–188, 215
of a polynomial, 196–198, 201–203
Finite sequence, 99
First-derivative test, 240
Focus (foci)
of an ellipse, 57, 80
of a hyperbola, 64
of a parabola, 69, 80
Force acting on an object, 463–466, 580, 584
Formula(s)
change-of-base, 419, 433
compound interest, 406, 409–410, 432
distance, 43, 80, 448, 471
double-angle, 322–323, 326, 356
Euler's, 436
half-angle, 323–324, 326–327, 356
Hero's, 305, 307, 309
midpoint, 43, 80, 448, 471
product-to-sum, 327
quadratic, 194–195
reduction, 320
sum-to-product, 328
trigonometric addition, 318, 356
Frequency, 616
of a periodic function, 341
relative, 618, 640
Frequency distribution, 616–617, 640
Frequency polygon, 617, 640
Function(s), 141–144, 177
absolute value, 147, 165
algebra of, 146–151
circular, 271–281, 308
composite, 234–236
composition of, 151–152, 164, 177
constant, 149
continuous, 163–170, 177, 225, 381, 406, 408, 432
decreasing, 238, 262
derivative (derived), 224, 262
derivatives of, 221–226. *See also* Derivative(s).
difference of, 147, 177
differentiable, 223, 262
discontinuous, 164
domain of, 141, 142, 177
even, 278, 335–336, 356
exponential, 403–412, 423, 433
graphs of. *See under* Graphs.
greatest integer, 148
hyperbolic, 434–435
identity, 149
increasing, 238, 262
inverse of, 153–154, 177
inverse trigonometric, 350–355
limits of, 157–167, 172–175, 177, 406, 410–411
linear fractional, 156
logarithmic, 412–421, 422–423, 433
natural exponential, 408–412, 432
odd, 278, 335–336, 356
one-to-one, 153, 177, 350–351, 408, 412
periodic, 334–337, 356
piecewise definition of, 147
polynomial, 164
powers of, 147
probability, 600
product of, 146, 177
quotient of, 147, 177
range of, 141, 142, 177
rational, 172–177, 178
reciprocal, 277
signum (sign), 148
square root, 164
sum of, 146, 177
trigonometric, 281–307, 308
value of, 142, 177, 239
zero of, 168
Functional notation, 142
Fundamental counting principle, 589, 639
Fundamental theorem
of algebra, 201–204, 216
of arithmetic, 28, 33
of calculus, 578

Gauss, Karl Friedrich, 202
General solution of trigonometric equation, 331, 356
Generalized power rule, 232, 234, 262
Geometric means, 109
Geometric sequences, 108–113, 134
Geometric series, 110–111, 134
Grad, 285
Graph(s)
area under, 568–574, 584, 586
of complex numbers, 189–192, 215, 217–218, 369
of conic sections, 53–79, 80, 81, 395, 494–496
of exponential functions, 403–405, 408–409, 423–425, 432
of functions, 143–144, 147–149, 152–153, 157–158, 165, 166, 167–169, 172–175, 177, 181
of hyperbolic functions, 434–435
illustrating limits, 114, 125, 172–175
of inverse trigonometric functions, 350–355
of logarithmic functions, 412–414
on the number line, 6–7, 21–22, 23–24, 114, 125
of an open sentence in two variables, 46–47, 80
of an ordered pair, 46
of polar equations, 363, 365–368, 376–380, 395
symmetry in, 82–83, 340, 356
in three-space, 446–447, 530–531, 540–549
translation of, 148–149
trigonometric, 334–355, 356
using derivatives, 221–223, 225, 238–245, 258–260, 262, 423–425
Greatest common factor, 28–29, 33
Greatest integer function, 148
Ground speed, 462
Group(s), 15, 34–35, 179
commutative, 15
subgroup of, 34, 35

Half-angle formulas
for sine and cosine, 323–324, 356
for tangent, 326–327, 356
Half-life, 428, 431
Harmonic motion, simple, 344, 356
Heading, 462
Hero's formula, 305, 307, 309
Histogram
frequency, 617, 626–627, 640
probability, 610–611
relative frequency, 618
Hooke's law, 580
Horizontal line, 48
Hyperbola(s), 63–68, 73, 80, 495–496
asymptotes, 65, 80
center, 65, 80
equations of, 64–67, 80
focal radius, 64
foci, 64
latus rectum, 68
transverse axis, 65
vertices, 65, 80
Hyperboloids, 542, 543

i, meaning of, 186, 215
Identities, trigonometric. *See* Trigonometric identities.
Identity, 5
Identity function, 149
Identity matrix, 486, 510
Identity transformation, 483
Image under a rotation, 480, 518
Imaginary axis, 189
Increasing functions, 238, 262
Independent events, 606–607, 640
Index
of a radical, 128
of summation, 101
Indirect proof, 11, 32
Induction, mathematical, 89–93, 133
Induction hypothesis, 90
Inequality, 20
solving an, 20–27
symbols of, 20, 32
Infinite sequence(s), 99, 114–118, 134
Infinite series, 119–124, 134
Infinity, limits involving, 172–177
Inflection point, 242, 262
Initial point of directed line segment, 441
Initial side of an angle, 281
Inner product of vectors, 454–457, 472
Integers
composite, 28
relatively prime, 28
set of, 2, 28–29, 32
Integral(s)
definite, 565–568, 578, 584
indefinite, 558, 584
Integral calculus, 557–585. *See also* Integration.
Integral sign, 558, 584
Integrand, 558, 584
Integration, 558
applications of, 568–583
formulas for, 559
limits of, 565–566
numerical, 586
by substitution, 561–565, 584
Intensity of a sound, 414–415
Intermediate-value theorem, 167–171, 177, 178
Interpolation, linear, 181
Intersection(s)
of conics, 76–79
of planes, 538
of sets, 3, 32
Interval(s)
closed, 167
open, 167, 238
Inverse
of exponential function, 412
of a function, 153–154, 177
of a matrix, 510–514, 519
Inverse trigonometric functions, 350–355, 356
Irrational number(s), 29–30, 33, 127
Irreducible polynomial, 201, 215
Iteration, 259

Kernel of a linear transformation, 493

Latitude, 285
Law(s)
of cooling, Newton's, 587
of cosines, 296, 308
of exponents, 90–91
of logarithms, 416, 433
of optics, 396–397
of sines, 297–298, 308
of tangents, 304
Leading coefficient, 196
Leibniz, 221
Lemniscate, 366, 368
Length, circular-arc, 283
Leonardo of Pisa, 99
Limaçon, 366–367, 368
Limit(s)
in differential calculus, 222–225, 262
of exponential function, 406, 410–411
of functions, 157–167, 172–175, 177
of integration, 565–566
of a sequence, 114–118, 134
of trigonometric functions, 380–383, 394
Line(s)
angle between, 535, 550
direction vector of, 534, 550
equations of, 46–50, 80, 534–539, 550
parallel, 49, 80
perpendicular, 49, 80
scalar parametric equations of, 534, 550
skew, 535, 550
slope of, 47, 80
in space, 534–539
symmetric-form equations of, 534, 550
vector equation of, 534, 550
Line segment
directed, 441, 471
equations of, 535
midpoint of, 43–44, 80, 448, 535
Linear combination
of functions, 227
of vectors, 443, 457–461, 472, 480
Linear equations, systems of, 497–505, 515–517, 518, 520–521
Linear interpolation, 181
Linear polynomial, 196
Linear transformations, 479–484, 518
Location principle, 168, 178
Logarithmic functions, 412–421, 422–423, 433
Logarithms
change of base of, 419, 433
common, 418
table of, 665–666
laws of, 416, 433
natural, 418
table of, 668
Logic networks, 38–39
Lower bounds, 208–209

Magnitude
of a star, 437

of a vector quantity, 441, 471
Mapping diagrams, 152–153
Mathematical induction, 89–99, 133
principle of, 89
Mathematical system, formal, 13
Matrix (matrices), 485–493, 518
additive inverse of, 485
adjoint of, 511
augmented, 499, 518
coefficient, 512
column, 485, 486
of constants, 512
difference of, 485
dimensions of, 485, 518
elementary row operations on, 499–500
entries of, 485, 518
equal, 485
identity, 486, 510
inverse of, 510–514, 519
of linear transformations, 490–493, 518
powers of, 486
product of, 486–487, 518
row, 485, 486
row-reduced echelon form of, 503
scalar multiplication of, 485–486, 518
square, 485, 486
sum of, 485, 518
symmetric, 489
transpose of, 511
zero, 486, 487
See also Determinants.
Maximum value of a function, 239, 240
applications of, 245–252, 389–393
Mean, 617, 640
arithmetic, 104–105
geometric, 109
of a probability distribution, 619, 640
Mean proportional, 109
Mean-value theorem of differential calculus, 264–265
Measurement of angles, 282–284, 308
Median, 617, 640
Members of a set, 1, 2
Midpoint of a line segment, 43–44, 80
in three-space, 448, 471, 535
Minimum value of a function, 239, 240
applications of, 245–252, 389–393
Minor of entry of a determinant, 506
Minute in angle measure, 290
Mode, 617, 640
Modulus of a complex number, 190, 215, 369
Mollweide's equations, 304
Motion
simple harmonic, 344, 356
uniform circular, 283–284, 476–477
Multiplication
axioms of, 14, 15, 20
cancellation law for, 15
of complex numbers, 185–186, 215, 369–370, 393
of probabilities, 605, 640
of real numbers, 14–15
Multiplicative identity, 14
Multiplicative inverse, 14, 15
Multiplicity of a root, 202
Mutually exclusive events, 600

Natural logarithms, 418
table of, 668
Natural numbers, set of, 2, 28, 32
Negation of a sentence, 7, 8, 32
Negative number, 19
Newton, Sir Isaac, 221
Newton's law of cooling, 587
Newton's method, 258–261, 263
Norm of a vector, 451–453, 472
Normal curve, 627–630
area under, 628–630, 641
standard, 627–630, 641
Normal distributions, 626–632, 643
standard, 627, 641
Normal form of equation of a line, 364–365
Normal vector to a plane, 530, 549
Null set, 2, 32
Number(s)
integers, 2, 28–29, 32
irrational, 29–30, 33, 127
natural, 2, 28, 32
prime, 28
rational, 2, 29, 32, 33
real. *See* Real numbers.
related to quadrants, 274, 308
Number line, 19, 114, 125
distance on, 23, 24, 25, 33, 42

Octants, 447
Odd function, 278, 335–336, 356
Ohm's law, 145
One-to-one function, 153, 177, 350–351, 408, 412
Open interval, 167, 238
Open sentence, 5
Optics, laws of, 396–397
Optimization problems, 245–252
Orbits, 86–87
Order
axioms of, 19, 20
of a derivative, 231
of a determinant, 506
Ordered field, 125–126
Ordered pair(s), 41, 146, 185
denoting a vector, 444
equal, 41
Ordered triples, 446, 449
Ordinate, 42
Origin, 42, 446
Orthogonal vectors, 455, 472

Parabola(s), 69–73, 80, 496
axis of symmetry, 69, 80
directrix, 69
equations of, 69–71
focal radius, 72
focus, 69, 80
latus rectum, 72
vertex, 69, 80
Paraboloids, 542, 544
Parallel planes, 532
Parallel vectors, 443, 471
Parallelepiped, 528, 549
Parallelogram rule, 189
Parameter, 375
Parametric equations
of a curve, 375–380, 394
of a line, 464, 534–536, 550
Partial sum of a series, 119, 134
Pascal, Blaise, 94
Pascal's triangle, 94
Perigee, 86
Period of a function, 334, 340
Periodic functions, 334–337, 356
Permutations, 589–594, 639
Perpendicular planes, 532
Phase shift, 335
Plane(s), 446–447, 530–534, 549
angle between, 534
intercepts of, 530–531
parallel, 532
perpendicular, 532
scalar equation of, 530, 549
traces of, 530
vector equation of, 530, 549
Plane regions, areas of, 568–574, 584

Points
coordinates of. *See* Coordinates.
distance between, 23, 24, 25, 33, 42, 80, 364, 448, 471
Polar axis, 361, 362
Polar coordinate system, 361–365
Polar coordinates, 361–362, 393
changing to rectangular, 362, 393
Polar form of complex numbers, 369–371
Pole, 361
Polynomial(s)
coefficients of, 196, 205
completely factored, 201
cubic, 196
dividing, 196–198
factoring, 201–204, 206
over fields, 196–200, 215–216
irreducible, 201, 215
linear, 196
quadratic, 196
quartic, 196
reducible, 201, 215
set of, as vector space, 470
zero, 196, 197
Polynomial equations, 185, 201–215, 216
bounds for real roots, 208–211
conjugate imaginary roots, 205–208
double root, 202
multiplicity of a root, 202
rational roots, 211–215, 216
Polynomial function, 164
Population, 632
Positive number, 19
Power rule, 226–227, 262
generalized, 232, 234, 262
Powers
of complex numbers, 217–218, 372–373
of functions, 147
See also Exponents.
Prime number, 28
Principal solutions of trigonometric equation, 331, 356
Probability(ies), 599–615, 639–640
addition rule of, 604, 640
conditional, 605, 640
multiplication rule of, 605, 640
mutually exclusive, 600
Probability distributions, 610–615, 618–619, 623–624, 640
Probability function, 600, 639–640
Probability histogram, 610–611
Product
Cartesian, 41
of functions, 146, 177
of matrices, 486–487, 518
of a scalar and a matrix, 485, 518
of vectors, 454–457, 472, 525–529, 549, 552
Progression, 104
See also Sequence(s) *and* Series.
Projection of a point onto a plane, 546
Proof
by contradiction, 11
by contraposition, 11
indirect, 11, 32
by induction, 90–92
using analytic geometry, 44
using vectors, 466–469
Proper subset, 2, 32
Properties. *See* Axiom(s).
Pure imaginary number, 186
Pythagorean identities, 314, 355
Pythagorean theorem, 42–43, 454

Quadrantal angle, 287
Quadrantal number, 274
Quadrants, 42
Quadratic equations with complex coefficients, 194–195
Quadratic formula, 194–195
Quadratic polynomial, 196
Quadric surfaces, 539–545, 550
Quartic polynomial, 196
Quotient
of complex numbers, 187, 370, 393
of functions, 147, 177

Radian measure of an angle, 282–283, 308
Radical, 128
Radicand, 128
Radius
of a circle, 53, 80
of a sphere, 539
Random variable, 610, 640
binomial, 612
Range
of a function, 141, 142, 177
of a set of data, 621–622
Rates of change, 253–258, 388, 391–393, 429–430
Ratio, common, 108, 134
Rational functions, 172–177, 178
Rational numbers
represented by decimals, 121–122
set of, 2, 29, 32, 33
Rational root theorem, 211, 216
Real axis, 189
Real numbers
absolute value of, 23
axioms of, 13–15, 19–20
order of, 19–20
ordered pairs of, 41
ordered triples of, 446, 449
represented by decimals, 126–127, 134
roots of, 130
set of, 2, 28–30, 32, 272
and the unit circle, 272, 308
Reciprocal functions, 277
Reciprocal identities, 314, 355
Rectangular coordinate system, 42
Rectangular coordinates, 42
changing to polar, 362, 393
Rectilinear motion, 253–258
Recursion formula, 99–100
Reducible polynomial, 201, 215
Reduction formulas, 320
Reflexive property of equality, 13
Region under a graph, 568–574, 584, 586
Related-rate problems, 254–256
Relation, 141
Relative (local) maximum, 239, 262
Relative (local) minimum, 239, 262
Remainder theorem, 197, 215
Repeating decimals, 121–122
Replacement set of a variable, 5, 32
Resultant, 463, 472
Review, cumulative, 138–139, 268–269, 438–439, 554–555
Richter scale, 415
Right triangles, solving, 292–295
Ring(s), 179–180
commutative, 179
with identity, 179
Rolle's theorem, 265
Root(s)
bounds for, 208–210
of complex numbers, 193–195, 215, 373–374
conjugate imaginary, 205–208, 216
double, 202
of multiplicity *k*, 202
of polynomial equations, 201–215, 216
rational, 211–213
of real numbers, 130, 134
Rose graphs, 365, 368
Roster of a set, 2

Rotation
  angles generated by, 281–282
  of axes, 493–497, 518
  equations of, 494–497, 518
  of the plane, 481, 490
Row matrix, 485, 486
Rulings, 540, 550

Sample of a population, 632
Sample space, 599
Scalar, 443
Scalar multiplication
  of a matrix, 485–486, 518
  of vectors, 443, 449, 469, 471, 480
Scalar product, 454–457, 472
  triple, 527–528, 549
Scatter diagram, 635, 636
Scientific notation, 418
Secant
  circular function, 277, 279, 308
  derivative of, 385, 394
  graph of, 349
  inverse, 353
  trigonometric function, 286, 287
Second in angle measure, 290
Second derivative, 231
Second-derivative test, 243, 262
Sector, area of, 283
Segment. *See* Line segment.
Sentence(s), 5
  biconditional, 10, 32
  conditional, 10–13, 32
  logically equivalent, 36
  open, 5–9, 32
  truth value of, 6, 36
Sequence(s), 99, 134
  arithmetic, 103–108, 134
  bounded, 126, 134
  constant, 115
  convergent, 115
  defined as a function, 142
  divergent, 117
  finite, 99
  geometric, 108–113, 134
  harmonic, 115
  infinite, 99, 114–118, 134
  limits of, 114–118, 134
  nondecreasing, 126, 134
  recursive definition of, 99–100, 108, 134
  term of a, 99
Series, 100, 134
  arithmetic, 105–106, 134
  convergent, 119, 134
  divergent, 119, 134
  geometric, 110–111, 134
  harmonic, 119
  infinite, 119–124, 134
  partial sum of a, 119, 134
  sum of a, 100–101, 105–106, 134
Set(s), 1–5, 32
  complement of, 3, 32
  equal, 2
  intersection of, 3, 32
  members of, 1, 2, 32
  null (empty), 2, 32
  subset of, 2, 32
  union of, 3, 32
  universal, 3
Set-builder notation, 5–6
Sides of an angle, 281
Sigma (summation symbol), 101
Signum (sign) function, 148
Simple harmonic motion, 344, 356
Simpson's approximation formula, 586
Sine
  circular function, 271–276, 278, 279, 308
  derivative of, 384, 394
  graphs of, 338–347, 356
  inverse, 351, 356
  in solving triangles, 292–293, 297–298, 308
  trigonometric function, 286–288
Sine curve, 343, 356
Sines, law of, 297–298, 308
Singularity of a function, 172, 178
Sinusoid, 343, 356
Sinusoidal curve, 343
Skew lines, 535, 550
Slope
  of a curve, 221–223, 238, 242, 262
  of a line, 47, 80
Solids of revolution, 575, 577–578, 584
Solution set of an open sentence, 5, 32
Solving triangles, 292–305, 308
Space curves, 553
Speed, 253
Sphere, 539, 550
Spiral, 367, 368
Square matrix, 485
Square root function, 164
Square roots
  of complex numbers, 193–195, 215
  finding by divide-and-average method, 260
  of integers, 29
  table of, 664
Squares, table of, 664
"Squeeze" theorem, 160–161
Standard basis vectors, 457, 480
Standard deviation, 622–624, 640
Standard position of an angle, 286
Stationary point, 239, 262
Statistics, 616–639
Strain and stress, 580
Subgroup, 34, 35
Subset, 2, 32
  proper, 2, 32
Subspace of vector space, 474–475
Substitution, synthetic, 169–170, 197–198, 200
Substitution axiom, 14
Subtraction
  of real numbers, 15
  of vectors, 442
Sum
  of an arithmetic series, 105–106, 134
  of a finite series, 100–101, 105–106, 110–111, 134
  of functions, 146, 177
  of a geometric series, 105–106, 134
  of an infinite series, 119–122, 125–126, 134
  of matrices, 485, 518
  partial, 119, 125–126, 134
  of vectors, 442, 443, 444, 449, 462, 469, 471
Summaries. *See* Chapter summaries.
Summation symbol, 101
Symmetric-form equations of a line, 534, 550
Symmetric property of equality, 13
Symmetry
  axis of, 69, 80
  used in graphing, 82–83
Synthetic division, 197–198, 215
Synthetic substitution, 169–170, 197–198, 200
Systems of equations
  consistent and inconsistent, 503, 518
  dependent, 503
  elementary operations on, 499
  linear, in *n* variables, 497–505, 518, 520–521
  in matrix form, 512–513
  solving by Cramer's rule, 515–517
  in triangular form, 498, 518

Tables
area under standard normal curve, 628
common logarithms, 665–666
natural logarithms, 668
squares and square roots, 664
trigonometric functions of θ: decimal degrees, 647–654
trigonometric functions of θ: degrees and minutes, 655–659
trigonometric functions of θ: radians, 660–663
using, 290–291, 646
values of e^x and e^{-x}, 667
Tangent
circular function, 277–279, 308
derivative of, 385, 394
graph of, 348
inverse, 352
in solving triangles, 292–293
trigonometric function, 286–288
Tangents
to a circle, 54
to curves, 221–223, 238–239, 258, 264
to graphs of exp and log, 423, 424
Tangents, law of, 304
Taylor's theorem, 436
Term of a sequence, 99
Terminal point of directed line segment, 441
Terminal side of an angle, 281
Tests. *See* Chapter tests.
Theorems. *See under names of specific theorems.*
Three-space, 446
Traces of a surface, 540
Transformations, linear, 479–484, 518
Transitive property
of equality, 13
of order, 19
Translation of a graph, 148–149
Transpose of a matrix, 511
Triangle(s)
angles of, 292
areas of, 297, 305–307, 309, 310
right, 292–295
sides of, 292
solving, 292–305, 308
in structures, 311
Triangle inequality, 190, 191
for vectors, 452, 453, 472
Trigonometric equations, 330–333
Trigonometric form of a complex number, 369
Trigonometric functions, 281–307, 308
Trigonometric identities, 313–330, 355–356
addition formulas, 318, 356
double-angle formulas, 322, 326, 356
half-angle formulas, 323, 326, 356
product-to-sum formulas, 327
Pythagorean, 314, 355
reciprocal, 314, 355
sum-to-product formulas, 328
Trigonometric tables. *See* Tables.
True course, 462
Truth set of open sentence, 5
Truth table, 6
for biconditional sentence, 10
for conditional sentence, 10
for conjunction, 6
constructing, 36–37
for disjunction, 7
Truth value of a sentence, 6
See also Truth table.

Uniform circular motion, 283–284, 476–477
Union of sets, 3, 32
Unit circle, 271–272, 308
Unit vector, 451, 536, 550
Unity, *n*th roots of, 374, 375
Universal set, 3
Upper bound, 208–209

Value(s)
of a function, 142, 177
of trigonometric functions, 290–291
Variable, 5, 32
dummy, 565
random, 610, 612, 640
Variance, 622–624, 640
Variation in sign, 209
Vector(s), 441–446, 469–473, 525–539, 549–550
additive inverse of, 442, 469, 471
angle between, 454, 456
applications of, 462–465
basis, 457–461, 480
components of, 444, 471
cross product of, 525–529, 549
difference of, 442
direction, 534
direction angles of, 536, 550
direction cosines of, 536, 537, 550
dot product of, 454–457, 472
equivalent, 442, 444
in geometric proofs, 466–469
image of, 480, 518
linear combination of, 443, 457–461, 472, 480
norm of, 451–453, 472
notation for, 442, 444
orthogonal, 455, 472
parallel, 443, 471
position, 464, 472
resultant of, 463, 472
scalar multiplication of, 443, 449, 469, 471, 480
scalar triple product of, 527–528, 549
standard basis, 457, 480
subtraction of, 442
sum of, 442, 443, 444, 449, 462, 469, 471
unit, 451, 536, 550
vector product of, 525
zero, 442, 469, 471
Vector quantities, 441, 471
Vector spaces, 469–471, 472, 474–475
linear transformation of, 479–484, 518
Velocity, 253, 262, 441
Vertex (vertices)
of an angle, 281
of an ellipse, 59, 80
of a hyperbola, 65, 80
of a parabola, 69, 80
Vertical line, 47, 48
Vertical-line test, 144
Volumes of solids, 574–579, 584

Work, 463–466, 472, 580–583, 584

x-axis, 42
x-intercept, 48

y-axis, 42
y-intercept, 48

Zero(s)
as an exponent, 91
of a function, 168–171, 258–261
Zero matrix, 486, 487
Zero polynomial, 196, 197
Zero-product property, 16–17
Zero vector, 442, 469, 471
z-scores, 630, 641

Credits

Book designed by The Quarasan Group, Inc.

Cover concept and design by Richard C. Bartlett

Cover photograph by David F. Hughes/The Picture Cube

Mechanical art by Precision Graphics

Photographs

page xiv: Adapted from *The Fractal Geometry of Nature* by Benoit B. Mandelbrot. W. H. Freeman, 1982
page 39: Jean-Pierre Ragot/Stock Boston
page 40: NASA/Grant Heilman Photography
page 87: NASA/FPG
page 88: Algimantas Kezys/Click Chicago
page 137: Devaney Stock Photos
page 140: ©Tom McHugh/Photo Researchers, Inc.
page 183: Bohdan Hrynewych/Stock Boston
page 184: Alan Pitcairn/Grant Heilman Photography
page 219: Dr. Robert E. Hager/FPG
page 220: Dr. Harold E. Edgerton, MIT, Cambridge, Mass.
page 267: Peter Menzel/Stock Boston
page 270: ©Joel Gordon, 1982
page 311: ©Ulrike Welsch/Nawrocki Stock Photos
page 312: Jack Zehrt/FPG
page 359: H. Armstrong Roberts
page 360: Jim and Mary Whitmer
page 396: H. Armstrong Roberts
page 398: Richard Harrington/FPG
page 437: ©Herb Levart/Photo Researchers, Inc.
page 440: Paul Fortin/Stock Boston
page 477: Devaney Stock Photos
page 478: Grant Heilman/Grant Heilman Photography
page 523: Bob Frett/Courtesy of Illinois Bell
page 524: Jean-Claude LeJeune/Stock Boston
page 553: H. Armstrong Roberts
page 556: Courtesy of TVA
page 587: Ira Kirschenbaum/Stock Boston
page 588: Mark Antman/Stock Boston
page 645: Michael Weisbrot/Stock Boston

ANSWERS TO SELECTED EXERCISES

EXERCISES 1-1, Pages 4–5

1. $D \subset A$ **3.** $(C \cup D) \subseteq A; A \subseteq (C \cup D);$ $A = (C \cup D)$ **5.** O **7.** Z **9.** false **11.** true **13.** $\{1, 2, 3, 4, 6, 8\}$ **15.** $\{1, 2, 3, 4, 5, 6, 8, 10\}$ **17.** $\{5, 6, 7, 8, 9, 10, 11, 12, 13\}$ **19.** $\{2, 3, 5, 6, 7, 8, 9, 10, 11, 12, 13\}$ **21.** $\{1, 3, 6\}$ **23.** $\{7, 9, 11, 12, 13\}$ **25.** A **27.** $\{7, 9, 11, 12, 13\}$ **29.** $\emptyset, \{4\}$ **31.** $\emptyset, \{1\}, \{2\}, \{3\}, \{1, 2\}, \{1, 3\}, \{2, 3\}, \{1, 2, 3\}$ **33.** 2 **35.** 8 **37.** 2^n **39.** A, 3; B, 4

EXERCISES 1-2, Pages 8–9

1. true **3.** false **5. a.** $\{x: x > 0\}$ **b.** $\{x: x < 3\}$ **c.** $\{x: 0 < x < 3\}$ **7. a.** $\{x: x \leq 3\}$ **b.** $\{x: x \geq 3\}$ **c.** $\{3\}$ **9. a.** $\{t: t \geq 2\}$ **b.** $\{t: t \leq 7\}$ **c.** $\{t: 2 \leq t \leq 7\}$ **11. a.** $\{x: x > 0\}$ **b.** $\{x: x < -2\}$ **c.** $\{x: x < -2$ or $x > 0\}$ **13. a.** $\{w: w > 2\}$ **b.** $\{w: w \leq -1\}$ **c.** $\{w: w \leq -1$ or $w > 2\}$ **15. a.** $\{x: x \geq 2\}$ **b.** $\{x: x \leq 2\}$ **c.** $\mathcal{R}$ **17.** There is no positive square root of 2. **19.** There is at least one real number that is not less than, equal to, or greater than 5. **21.** $2 \neq \frac{6}{3}$ or $\frac{6}{3} \neq \frac{12}{16}$. **23.** $2x \geq -4$ and $3x \leq 6$ **25.** $\{x: -2 \leq x \leq 2\}$ **27.** $\{x: 2 < x \leq 5\}$ **29.** identity **31.** sometimes true, sometimes false **33.** sometimes true, sometimes false **35.** true **37.** true **39.** false **41.** true **43.** $\{x: 3 < x < 5\} \cup \{x: x > 6\}$ **45.** $\mathcal{R}$

EXERCISES 1-3, Pages 12–13

1. true **3.** false **5.** If the corresponding angles of two triangles are congruent, then the triangles are congruent; sometimes true **7.** If x^2 is an odd integer, then x is an odd integer; always true. **9.** If $b < c$, then $ab < ac$; sometimes true **11.** If violets are not green, then roses are not black. **13.** If a parallelogram is not a rectangle, then no angle of the parallelogram is a right angle. **15.** If $a \neq 1$, then $ab \neq b$ or $b = 0$. **17.** $x^2 \neq y^2$ and $x = y$. **19.** $1 = 2$ **21.** If R and S are sets and $R \subset S$, then $R \cap S = R$. **23.** If $a, b \in \mathcal{R}$, $a \leq b$, and $b \leq a$, then $a = b$. **25.** If $a > 0$, $b > 0$, $m \neq 0$, and $a^m = b^m$, then $a = b$. **27.** If x and y are real numbers, then $a^x a^y = a^{x+y}$. **29.** $p' \vee q$ **31.** $p' \wedge q'$

EXERCISES 1-4, Pages 17–19

1. definition of subtraction; associative axiom of addition; additive inverses; additive identity; transitive property **3.** Theorem 1, part (8); Theorem 1, part (8) again; Theorem 1, part (4); transitive property **5.** $x = 25$ **7.** $x = -\frac{4}{5}$ **9.** $x = -\frac{1}{2}$ or $x = -\frac{1}{3}$ **11. a.** closed **b.** closed **13. a.** closed **b.** closed **15. a.** not closed; $3 + 3 = 6$ **b.** not closed; $3 \cdot 3 = 9$ **17. a.** 6 **b.** closed **c.** commutative **d.** associative **19. a.** 12 **b.** closed **c.** commutative **d.** associative **45.** n is a prime number.

EXERCISES 1-5, Pages 22–23

1. $\left\{t: t \geq \frac{3}{2}\right\}$ **3.** $\{x: x \geq 0\}$ **5.** $\{r: -1 \leq r < 5\}$ **7.** $\{x: 4 < x < 5\}$ **9.** $\emptyset$ **11.** $\left\{t: t \leq -3 \text{ or } t \geq \frac{1}{2}\right\}$ **13.** $\{r: r \neq -5\}$ **15.** $\{x: -2 < x < 3\}$ **17.** $\emptyset$ **19. a.** $a < 0$ and $b > 0$ **33.** $\{x: x \leq -1$ or $-\frac{1}{2} < x \leq 1\}$ **35.** $\{x: -1 < x \leq 2$ or $x \geq 3\}$

EXERCISES 1-6, Pages 26–27

1. $\{a: -2.5 \leq a \leq 2.5\}$ **3.** $\{1\}$ **5.** $\emptyset$ **7.** $\{x: x \leq -5$ or $x \geq -2\}$ **9.** $\{k: -3.5 < k < 4\}$ **11.** $\{y: y \neq -6\}$ **13.** $\{-8, 4\}$ **15.** $\{y: y < -2$ or $y > 5\}$ **17.** $\{3\}$ **19.** $\left\{n: n \geq -\frac{1}{2}\right\}$ **21.** $\{x: x < 0$ or $0 < x < \frac{1}{5}$ or $x > 1\}$ **23.** $\left\{\frac{5}{3}, -\frac{5}{4}\right\}$ **25.** $\{y: y \neq -3\}$ **27.** $\left\{a: a < -\frac{2}{3} \text{ or } a > 4\right\}$ **29.** $\{d: d \leq -8$ or $d \geq 1.2\}$ **31.** $\left\{-\frac{3}{7}, \frac{3}{5}\right\}$ **33.** $\emptyset$ **35.** $\{x: x \neq 3$ and $x \neq 4\}$ **37.** $\{x: x \leq 0$ or $x = 1$ or $x \geq 2\}$

EXERCISES 1-7, Pages 30–31

1. 3^6 **3.** $3^2 \cdot 17$ **5.** 151 **7.** $2^3 \cdot 11^2$ **9.** 15 **11.** 97 **13.** 96 **15.** 65 **17.** 8 **19.** 1 **21.** yes **23.** no

EXERCISES 2-1, Pages 45–46

1. $A(0, 5), B(3, 4), E(5, -3), G(-2, 0), H(0, 0)$ **3.** 3 **5.** $2\sqrt{37}$ **7.** $2\sqrt{2}$ **9.** $(1, 5)$ **11.** $\left(\frac{13}{4}, \frac{13}{15}\right)$ **13.** $\left(\frac{11 - \sqrt{2}}{12}, \frac{7 - 2\sqrt{2}}{6}\right)$ **15.** $x = -2$ **17.** $x = 1$ or $x = -\frac{1}{2}$; $y = 4$ or $y = -1$ **19.** $x = \frac{1}{2}, y = \frac{11}{2}$ **21.** $\sqrt{a^2 + 4b^2}$ **23.** $2\sqrt{b^2 + d^2}$ **25.** $(-2, 2\sqrt{2})$ **27.** examples: $(-3, 4), (-4, 3), (-2, \sqrt{21}), (-\sqrt{5}, 2\sqrt{5})$ **29.** $(8, 0)$ **31.** $\frac{1}{2}\sqrt{53}$ **33.** $(AB)^2 + (BC)^2 = (AC)^2$; $\frac{25}{2}$ square units **43.** no

EXERCISES 2-2, Pages 50–52

1. positive **3.** negative **5.** negative **7.** the line through $(0, -3)$ and $(3, 0)$ **9.** the line through

(0, −2) and (2, 0) **11.** the line through (0, −2) and (1, 1) **13.** the vertical line with x-intercept 3 **15.** the line through (0, −7) and (−7, 0) **17.** 0 **19.** $\frac{3}{2}$ **21.** $\frac{1}{6}\sqrt{6}$ **23.** $-\frac{1}{4}$; $\left(\frac{3}{2}, 0\right)$; $\left(0, \frac{3}{8}\right)$ **25.** 0; each point lies on the x-axis. **27.** 5; (0, 13); $\left(-\frac{13}{5}, 0\right)$ **29.** $2x - 3y = 23$ **31.** $5x + 4y = 24$ **33.** $3x - 2y = 1$ **35.** $x + 4y = 9$ **37.** $y = 17$ **39.** $x = -5$ **41.** parallel **43.** neither **45.** perpendicular **47.** Answers will vary. Examples are given. **a.** (5, 1), (6, 4) **b.** (6, 3), (5, 5) **c.** (1, −1), (5, 1) **d.** (1, 7), (4, 11) **49.** $Bx - Ay = Bu - Av$

51.

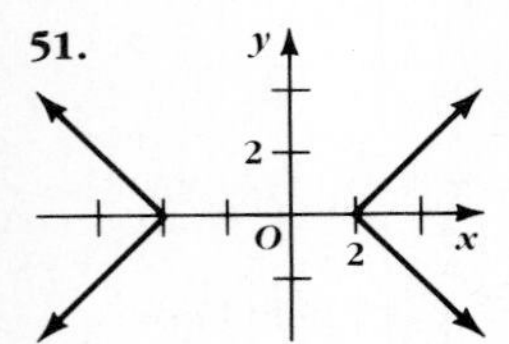

53. $2x - y = -9$ **55.** $12x + 14y = 167$ **57. b.** $AB + BC = k\sqrt{a^2 + b^2} = AC$

EXERCISES 2-3, Pages 55–56

1. $(x - 2)^2 + (y - 3)^2 = 9$ **3.** $x^2 + (y - 3)^2 = 144$ **5.** the circle with center (−2, 3) and radius 1 **7.** the circle with center (4, 0) and radius 2 **9.** the circle with center (0, 1) and radius 1 **11.** the circle with center $\left(-\frac{5}{2}, -\frac{7}{2}\right)$ and radius $\frac{1}{2}\sqrt{74}$ **13.** The graph is empty. **15.** (−4, 0) **17.** $x + 3y = -8$ **19.** outside **21.** inside **23.** $(x-3)^2 + (y-5)^2 = 25$ **25.** $\left(x-\frac{5}{2}\right)^2 + \left(y-\frac{5}{2}\right)^2 = \frac{25}{4}$ or $\left(x+\frac{5}{2}\right)^2 + \left(y-\frac{5}{2}\right)^2 = \frac{25}{4}$ **27.** $\left(x+\frac{3}{2}\right)^2 + (y-3)^2 = \frac{9}{4}$ **29.** $(x + 5)^2 + (y-4)^2 = 20$ **31.** $(x-6)^2 + y^2 = 18$ **33.** (400, 237.5), or 400 m east and 237.5 m north of the entrance

EXERCISES 2-4, Pages 61–63

1. $C(0, 0)$; (0, ±3), (±2, 0) **3.** $C(-4, -2)$; (−4, 3), (−4, −7), (−7, −2), (−1, −2) **5.** $C(0, 0)$; (0, ±3), (±5, 0) **7.** $C(3, -5)$; radius of circle is 9. **9.** $C(0, 1)$; (0, 4), (0, −2), (±4, 1) **11.** $C(2, -3)$; (2, −4), (2, −2), (0, −3), (4, −3) **13.** $C(-3, 2)$; (−4, 2), (−2, 2), (−3, 0), (−3, 4) **15.** $\frac{x^2}{16} + \frac{y^2}{1} = 1$ **17.** $\frac{(x + 6)^2}{100} + \frac{(y - 4)^2}{64} = 1$ **19.** $\frac{(x + 7)^2}{56} + \frac{(y - 6)^2}{225} = 1$ **21.** $\frac{x^2}{36} + \frac{y^2}{20} = 1$ **23.** $\frac{x^2}{18} + \frac{y^2}{9} = 1$ **25.** (±1, 0), (0, ±2) **27.** $\left(0, 2\pm\frac{4}{3}\sqrt{2}\right)$, (1, 0) **29.** $\left(0, \frac{-3 \pm \sqrt{13}}{4}\right)$, $(-1 \pm \sqrt{2}, 0)$ **31.** outside **33.** on **35.** $\frac{(x - 4)^2}{13} + \frac{(y - 4)^2}{4} = 1$ or $\frac{(x - 1)^2}{9} + \frac{(y - 6)^2}{13} = 1$ **39.** $3x^2 - 2xy + 3y^2 = 8$ **43.** $4x^2 + 3y^2 - 12y = 36$

EXERCISES 2-5, Pages 67–68

1. $C(0, 0)$; (±5, 0) **3.** $C(-2, 2)$; (−5, 2), (1, 2)

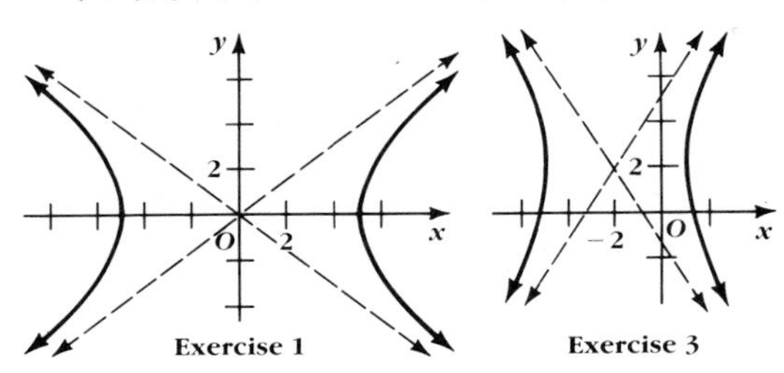

5. $C(3, 4)$; (1, 4), (5, 4) **7.** $C(-1, 1)$; (−3, 1), (1, 1)

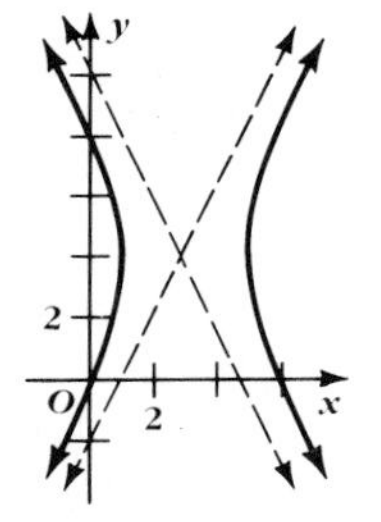

Exercise 5

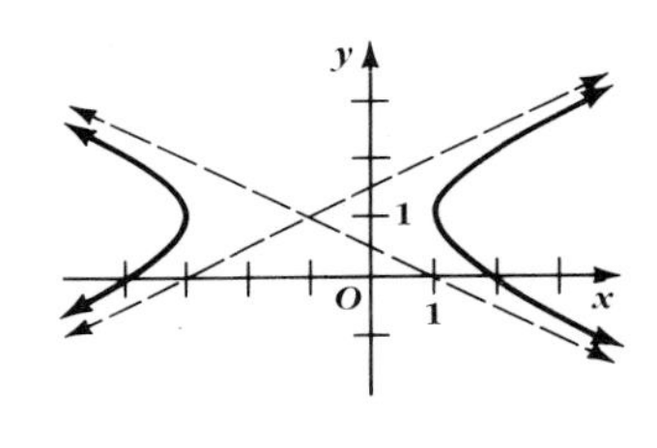

Exercise 7

9. $C(3, 0)$; (3, ±2) **11.** $C(1, -2)$; (1, 0), (1, −4)

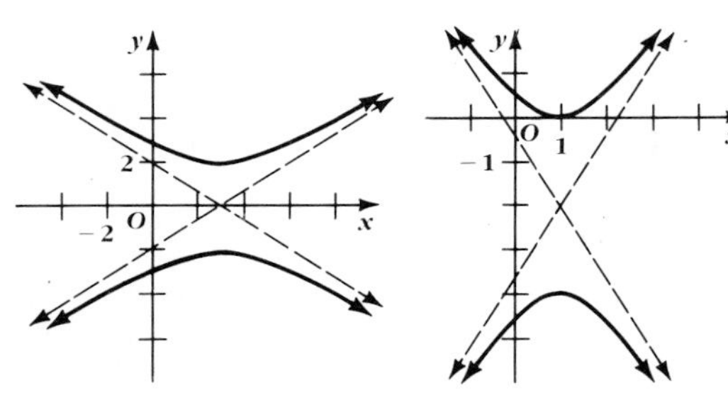

13. $C\left(-\frac{1}{2}, \frac{3}{2}\right)$; $\left(-\frac{7}{2}, \frac{3}{2}\right)$, $\left(\frac{5}{2}, \frac{3}{2}\right)$

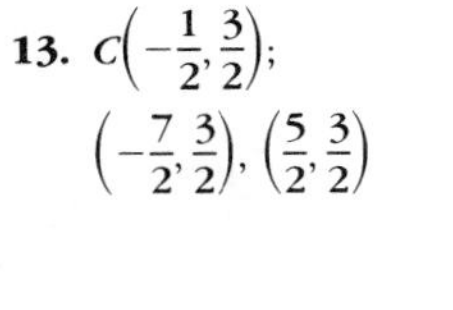

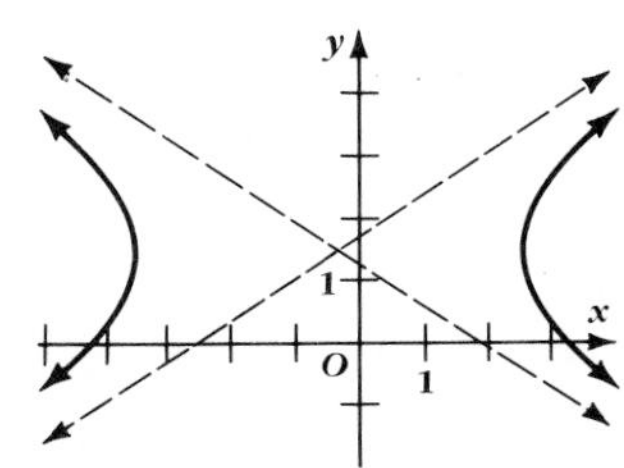

15. $\frac{x^2}{1} - \frac{y^2}{8} = 1$ 17. $\frac{(y-9.5)^2}{2.25} - \frac{(x+9)^2}{18} = 1$

19. $\frac{(y-5)^2}{-1} - \frac{(x+1)^2}{15} = 1$ 21. $\frac{(x+4)^2}{64} - \frac{(y+1)^2}{4} = 1$

23. $\frac{(y-7)^2}{2} - \frac{(x+3)^2}{2} = 1$

25. $\frac{(x-6)^2}{\frac{2}{5}} - \frac{(y-6)^2}{\frac{18}{5}} = 1$

27. the graph is the lines $y = \frac{1}{3}x$ and $y = -\frac{1}{3}x$.

29. the graph is the lines $4x-y = -9$ and $4x+y = -7$.

EXERCISES 2-6, Pages 71–73

1. $V(0, 0)$; $x = 0$; $F\left(0, \frac{1}{16}\right)$; $y = -\frac{1}{16}$; $(0, 0)$

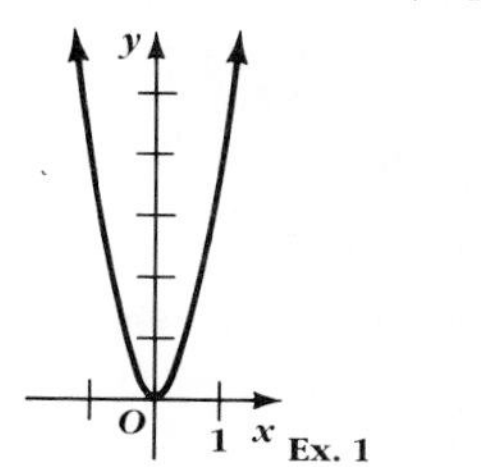

Ex. 1

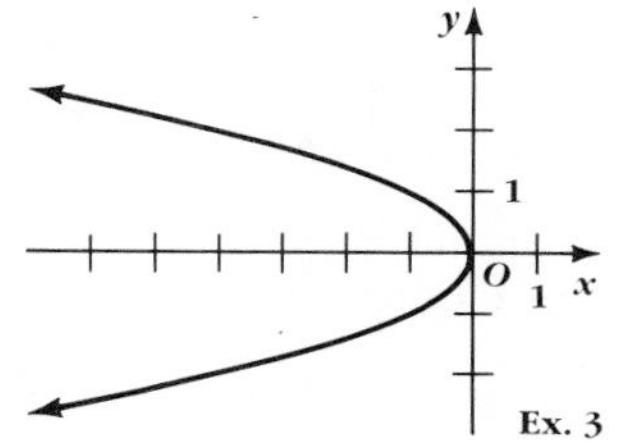

Ex. 3

3. $V(0, 0)$; $y = 0$; $F\left(-\frac{1}{4}, 0\right)$; $x = \frac{1}{4}$; $(0, 0)$

5. $V(2, 3)$; $x = 2$; $F\left(2, 3\frac{1}{4}\right)$; $y = 2\frac{3}{4}$; $(0, 7)$

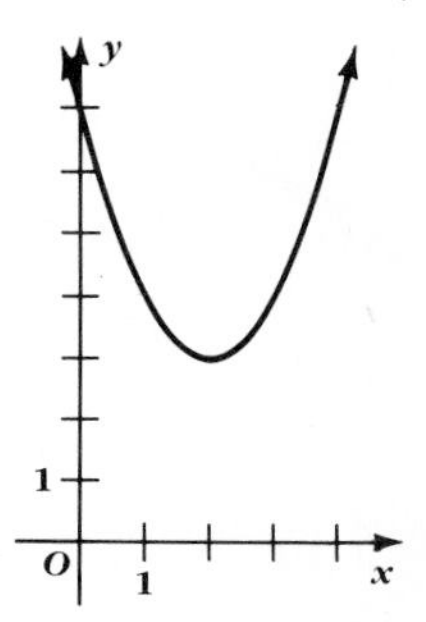

7. $V(3, -1)$; $y = -1$; $F\left(2\frac{3}{4}, -1\right)$; $x = 3\frac{1}{4}$; $(0, -1\pm\sqrt{3})$; $(2, 0)$

9. $V(2, 2)$; $x = 2$; $F\left(2, \frac{7}{4}\right)$; $y = \frac{9}{4}$; $(0, -2)$; $(2\pm\sqrt{2}, 0)$

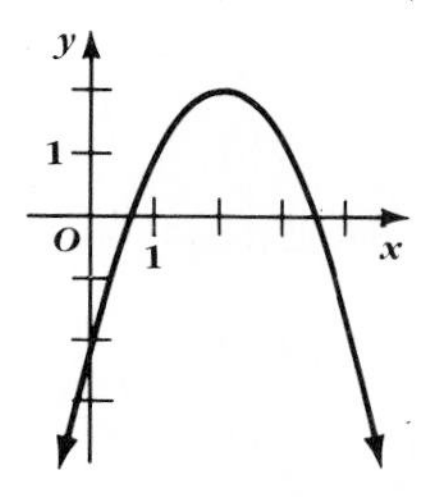

11. $V\left(-\frac{13}{4}, \frac{1}{4}\right)$; $y = \frac{1}{4}$; $F\left(-\frac{51}{16}, \frac{1}{4}\right)$; $x = -\frac{53}{16}$; $(-3, 0)$; $\left(0, \frac{1 \pm \sqrt{13}}{4}\right)$

13. $V(1, 1)$; $x = 1$; $F\left(1, \frac{13}{12}\right)$; $y = \frac{11}{12}$; $(0, 4)$

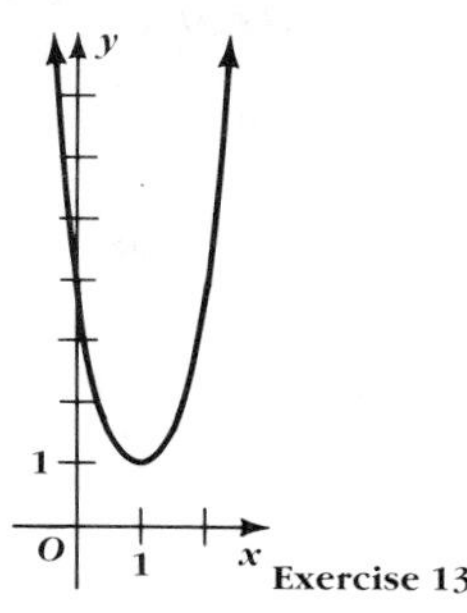

Exercise 13

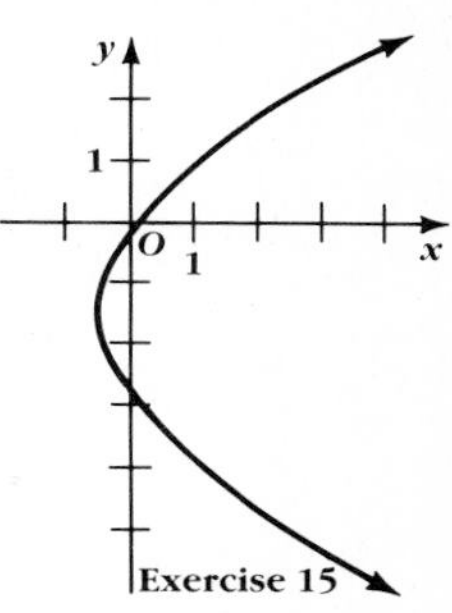

Exercise 15

15. $V\left(-\frac{1}{2}, -\frac{3}{2}\right)$; $y = -\frac{3}{2}$; $F\left(\frac{1}{2}, -\frac{3}{2}\right)$; $x = -\frac{3}{2}$; $\left(\frac{1}{16}, 0\right)$; $\left(0, \frac{-3 \pm 2\sqrt{2}}{2}\right)$

17. $y = \frac{1}{48}x^2$ 19. $x = \frac{1}{8}y^2$ 21. $y + 5 = \frac{5}{4}x^2$

23. $x - 6 = -\frac{4}{9}(y - 6)^2$ 31. 40 miles per hour; 20 miles per gallon 33. -18 and 18 35. $\frac{6}{5}\sqrt{5}$

EXERCISES 2-7, Page 76

1. $\frac{(x-3)^2}{4} + \frac{(y-1)^2}{1} = 1$; an ellipse containing $(1, 1)$, $(5, 1)$, $(3, 0)$, and $(3, 2)$ 3. $\frac{(x+1)^2}{16} - \frac{y^2}{1} = 1$; a hyperbola with vertices $(3, 0)$ and $(-5, 0)$

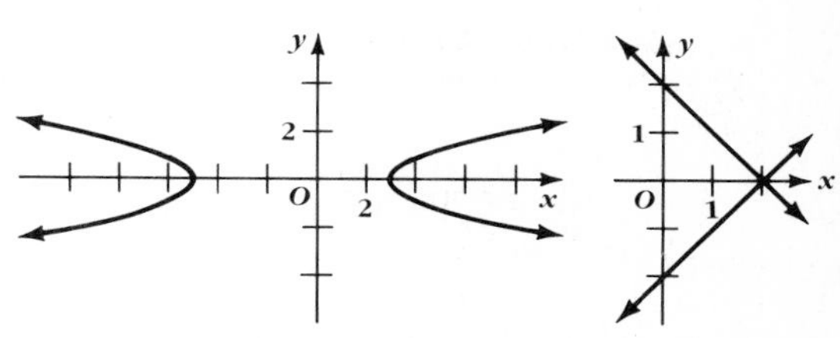

Exercise 3 Exercise 5

5. the intersecting lines $y = x-2$ and $y = -x+2$

7. the point $(0, -1)$ 9. the parallel lines $x = -1$ and $x = -2$ 11. the line $x = -2$

13. $x + 18 = (y+4)^2$; a parabola 15. a. the point $(0, 0)$ b. a pair of intersecting lines

17. the line $x = -1$

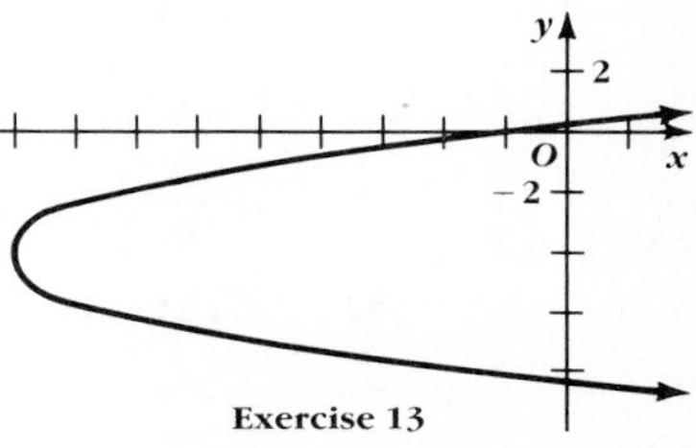

Exercise 13

EXERCISES 2-8, Pages 78–79

1. $\left(\frac{19}{16}, -\frac{9}{8}\right)$ **3.** $(1, -2)$ **5.** $(1 + \sqrt{3}, 2 + \sqrt{3})$, $(1 - \sqrt{3}, 2 - \sqrt{3})$ **7.** no intersection points **9.** $(1, \sqrt{19}), (1, -\sqrt{19})$ **11.** $(-3, -2), (-2, -3), (2, 3), (3, 2)$ **13.** $(1, \sqrt{2}), (1, -\sqrt{2}), (-1, -\sqrt{2}), (-1, \sqrt{2})$ **15.** $\left(\frac{5}{4}\sqrt{2}, \frac{5}{4}\sqrt{14}\right), \left(\frac{5}{4}\sqrt{2}, -\frac{5}{4}\sqrt{14}\right), \left(-\frac{5}{4}\sqrt{2}, -\frac{5}{4}\sqrt{14}\right), \left(-\frac{5}{4}\sqrt{2}, \frac{5}{4}\sqrt{14}\right)$
17. $(-1 + \sqrt{2}, 2), (-1 - \sqrt{2}, 2)$ **19.** $(2\sqrt{3}, 0), (-2\sqrt{3}, 0)$ **21.** $\left(\frac{\sqrt{5}}{2}, \frac{3}{2}\right), \left(\frac{\sqrt{5}}{2}, -\frac{3}{2}\right), \left(-\frac{\sqrt{5}}{2}, -\frac{3}{2}\right), \left(-\frac{\sqrt{5}}{2}, \frac{3}{2}\right)$ **23.** $x = \frac{3}{2}$ **25.** $6x - 4y = 5$ **27.** $x = 0$

EXERCISES 3-1, Pages 92–93

1. $-2304x^8y^8$ **3.** $1728m^9n^{12}$ **5.** $x^{m+n} + x^my^n + x^ny^m + y^{m+n}$ **7.** $\frac{x^{2q}}{13}$ **9.** $2 + 4 + 6 + \cdots + 18 = 90 = 9(10)$ **29.** If the first flower is removed, then the remaining flowers $\{f_2, f_3, \ldots, f_{x+1}\}$ are not necessarily the same color. Thus, the second part of the proof cannot be proved.

EXERCISES 3-2, Pages 97–98

1. $(a + b)(a^2 + 2ab + b^2) = a^3 + 3a^2b + 3ab^2 + b^3$ **3.** 7 **5.** 126 **7.** 20 **9.** $84 + 126 = 210$
11. $(n + 1)!$ **13.** $\frac{100!}{97!}$ **15.** $(n - r)!$
17. $\binom{10}{0}a^{10}b^0 + \binom{10}{1}a^9b^1 + \binom{10}{2}a^8b^2 = a^{10} + 10a^9b + 45a^8b^2$ **19.** $\binom{5}{0}(3c)^5(-d^2)^0 + \binom{5}{1}(3c)^4(-d^2)^1 + \binom{5}{2}(3c)^3(-d^2)^2 = 243c^5 - 405c^4d^2 + 270c^3d^4$
21. $\binom{4}{0}(6y)^4\left(-\frac{z}{2}\right)^0 + \binom{4}{1}(6y)^3\left(-\frac{z}{2}\right)^1 + \binom{4}{2}(6y)^2\left(-\frac{z}{2}\right)^2 = 1296y^4 - 432y^3z + 54y^2z^2$ **23.** $462a^5b^6$
25. $20y^3$ **27.** $\frac{n(n-1)(n-2)(n-3)}{24} \cdot a^4b^{n-4}, n > 3$
29. $-3584x^{10}y^3$ **31.** $3432r^7s^7$ **33.** $455a^{12}b^3$ and $455a^3b^{12}$ **35.** $n = 5$ **37.** $n = 9$ **39.** $n = 6$ or $n = 7$ **41.** $x^5 - 10x^3 + 40x - \frac{80}{x} + \frac{80}{x^3} - \frac{32}{x^5}$
43. $8s^6 + 12s^4t + 6s^2t^2 + t^3$ **45.** 0.922 **47.** 1.158
49. $1 + 4 + 6 + 4 + 1 = 16 = 2^4$

EXERCISES 3-3, Pages 102–103

1. 1, 1, 2, 3, 5 **3.** $-20, -14, -8, -2, 4, 10$
5. $32, -16, 8, -4, 2, -1$ **7.** $\frac{1}{3}, \frac{1}{3}, -\frac{1}{3}, -\frac{1}{3}, \frac{1}{3}, \frac{1}{3}$ **9.** $1, \frac{1}{3}, \frac{2}{15}, \frac{2}{35}, \frac{8}{315}, \frac{8}{693}$ **11.** $-1, \frac{1}{2}, -\frac{1}{3}, \frac{1}{4}, -\frac{1}{5}$ **13.** $\frac{1}{2}, \frac{\sqrt{2}}{3}, \frac{\sqrt{3}}{4}, \frac{2}{5}, \frac{\sqrt{5}}{6}$ **15.** $1, -\sqrt{2}, \sqrt{3}, -2, \sqrt{5}$ **17.** $a_n = 4n - 6$ **19.** $a_n = (n - 1)!$ **21.** $a_n = 6n$
23. $a_n = \frac{1}{n!}$ **25.** $a_n = \frac{n^2 + 1}{2}$ **27.** $0 + 1 + 4 + 9 + 16 + 25$ **29.** $2 + 5 + 8 + 11 + 14$ **31.** $2 - 6 + 12 - 20$ **33.** $\sum_{m=1}^{3} (x_m)^3$ **35.** example: $\sum_{n=1}^{4} 36 \cdot (-\frac{1}{3})^{n-1}$ **37.** example: $\sum_{k=1}^{10}(-1)^{k+1} \cdot \frac{1}{k(k+1)}$
39. 100 **47.** $S_n = \frac{n(n+1)(2n-17)}{6}$

EXERCISES 3-4, Pages 106–108

1. 47 **3.** $a + 12b$ **5.** $a = -2; d = 5$ **7.** $a = 4.5; d = 4.5$ **9.** 11; 17; 20 **11.** $12\frac{1}{3}; 5\frac{2}{3}; 2\frac{1}{3}$ **13.** $\frac{5}{2}x; -\frac{1}{2}x; -\frac{7}{2}x$ **15.** the 32nd term **17.** 2115 **19.** 2
21. 5 **23.** -475 **25.** 16 **27.** 40, 38, 36
29. $k = -\frac{4}{5}$ **31.** yes; The first and last terms must be opposites. **33.** no **37.** $\frac{n(a + t_n) - k(a + t_k) + t_k}{2}$
39. $\frac{n(n+1)}{2}$ **41.** $\frac{[a + (m-1)d] - [a + (n-1)d]}{m - n} = d$
43. $\frac{n^2 + n + 2}{2}$ regions

EXERCISES 3-5, Pages 111–113

1. -3 **3.** $\frac{8}{2401}$ **5.** 192 **7.** 6 **9.** $\frac{1}{3}$
11. $-\frac{635}{64}$ **13.** 25.5 **15.** 1093 **17.** $\frac{15}{4}$ **19.** 25, 125 **21.** $3\sqrt{3}, 9, 9\sqrt{3}$ or $-3\sqrt{3}, 9, -9\sqrt{3}$
23. 1920 or -1920 **27.** $t_3 = -8; r = -1$ **29.** $n = 3; t_3 = -6$ **31.** $n = 6; S_6 = \frac{665}{12}$ **33.** 3 **35.** 4096 times **37.** about 14.4 million people **39. a.** 6.5536 ft **b.** $h \cdot (0.8)^n$ ft **41.** $\frac{32}{243} \approx 13\%$ **43.** $(10121)_3$
47. 0.5; 0.75; 0.875; 0.9375; 0.96875; 0.984375; 0.9921875; 0.99609375; 0.99804687; 0.99902344

EXERCISES 3-6, Pages 117–118

1. $\frac{1}{2}$ **3.** does not exist **5.** $\frac{3}{4}$ **7.** does not exist
9. $\frac{2}{3}$ **11.** 0 **13.** true **15.** false; example: $1, -\frac{1}{2}, \frac{1}{4}, -\frac{1}{8}, \ldots, \left(-\frac{1}{2}\right)^{n-1}$ has a limit of 0. **17.** false; example: If $a_n = \left(-\frac{1}{2}\right)^{n-1}$, then $\lim_{n\to\infty} a_n = 0$ but $a_n \neq 0$ for any n.
19. false; example: If $a_n = (-1)^n$, then $\lim_{n\to\infty} |a_n| = 1$,

but $\lim_{n\to\infty} a_n$ does not exist. **21.** $n = 2$
23. example: $a_n = n$ and $b_n = 3 - n$ **25.** example: $a_n = \frac{1}{n^2}$, $b_n = 2n$ **29.** A sequence $a_1, a_2, \ldots,$ $a_n, \ldots$ decreases without bound if for every T (including negative numbers with large absolute values) there is an $M \in N$ such that $a_n < T$ for all $n \geq M$.

EXERCISES 3-7, Pages 122–124

1. $\frac{21}{2}$ **3.** $-12 - 2\sqrt{3}$ **5.** $\frac{16}{3}$ **7.** $\frac{3}{11}$ **9.** $42 - 7\sqrt{6}$ **11.** $\frac{9}{2}, \frac{9}{5}, \frac{18}{25}$ **13.** $28, -21, \frac{63}{4}$ **15.** $0.3958\overline{3}$
17. $1.3\overline{20}$ **19.** $\frac{14}{99}$ **21.** $\frac{170}{27}$ **23.** no; example: $\sum_{n=1}^{\infty}\left(\frac{1}{2}\right)^n = 1 = \sum_{n=1}^{\infty}\frac{4}{5}\left(\frac{1}{5}\right)^{n-1}$ **25.** no; the set of rational numbers is closed under multiplication.
27. convergent; 1 **29.** divergent **31.** convergent; 0
33. $-\frac{4}{3} < x < -\frac{2}{3}$ **35.** $0 < x < 4$ **37.** $x = \frac{1}{2}$
39. $x = -\frac{1}{2}$ or $x = \frac{2}{5}$ **43.** $80(2+\sqrt{2})$ in.; 800 sq. in.

EXERCISES 3-8, Pages 128–129

1. false; example: $a_n = 2^n$ is a nondecreasing sequence that has no limit. **3.** false; example: $\sum_{n=1}^{\infty}\left(-\frac{1}{3}\right)^{n-1} = \frac{1}{1-\left(-\frac{1}{3}\right)} = \frac{3}{4}$ **5.** false; example: $\sum_{k=1}^{\infty} k$ has a nondecreasing sequence of partial sums that is not bounded.
7. 6; rational **9.** $12\sqrt{5}$; irrational **11.** 2; rational
13. examples: $r = 2.62$; $s = 2.61234567\ldots$
15. examples: $r = \frac{16}{5}$; $s = \sqrt{10}$ **17.** 1.5 **19.** 0.5
25. example: the sequence $a_n = (-1)^n$ is bounded since $|a_n| \leq 1$ for all n, but the sequence diverges.
29. If for some n the finite decimal S_n is such that $S_n{}^2 = 2$, then $\sqrt{2} = S_n$ and $\sqrt{2}$ is rational. This would contradict Theorem 11, page 30. $S_n{}^2 \neq 2$ for all n.
33. example: $(x-2)^3 = 3$, or $x^3 - 6x^2 + 12x - 11 = 0$
35. $\frac{2 + 5\sqrt{2} + 4\sqrt{3} + \sqrt{6}}{4}$ **37.** The converse is false; example: if $a_n = \frac{1}{n}$, then $\lim_{n\to\infty} a_n = 0$, but $\sum_{k=1}^{\infty} a_k$ diverges.

EXERCISES 3-9, Pages 132–133

1. 27 **3.** $\frac{1}{5}$ **5.** $32t^5$ **7.** x^8y^{12} **9.** $-\frac{1}{4}r^4s^5$
11. $-\frac{1}{4}$ **13.** 6 **15.** $-\frac{x^2 + xy + y^2}{xy(x+y)}$ **17.** $\frac{1}{4}$
19. $\frac{8}{3}$ **21.** $\sqrt[6]{x^5}$ **23.** $\sqrt[6]{2^{11}}$ **25.** $\sqrt[4]{125x}$

27. $\frac{2}{3}$ **29.** $-\frac{8}{9}$ **31.** $\frac{1}{16}$ **33.** $\frac{9}{2}$ **41.** 2 **43.** 2
45. 3 **53.** approximate values: 1.0049876; 1.024695; 1.0488 **55.** $1 - \frac{2}{3}x^2 - \frac{1}{9}x^4 - \frac{4}{81}x^6 - \frac{7}{243}x^8$

CUMULATIVE REVIEW (Chapters 1–3), Pages 138–139

1. $(3, 0)$ satisfies $(x+2)^2 + (y-5)^2 = 50$ **3.** $\frac{1}{3}$
5. a. 35 **b.** 1960 **7.** $\left\{n: -3 < n < \frac{1}{2}\right\}$ **9.** an ellipse with center $(-1, 0)$ and containing $(1, 0), (-3, 0)$, and $(-1, \pm 1)$ **13.** $y - 3 = 2(x+1)^2$ **17.** $(0, 3)$, $(0, -5)$ **19.** 0 **21.** $\frac{x^2}{25} + \frac{y^2}{169} = 1$
23. $\frac{(x+2)^2}{4} - \frac{(y-1)^2}{4} = 1$ **25.** $y + 3 = \frac{3}{4}(x + 5)$

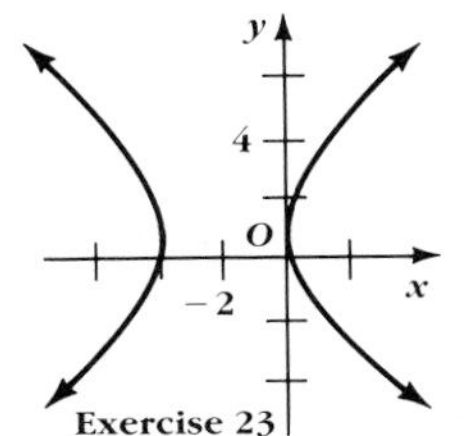

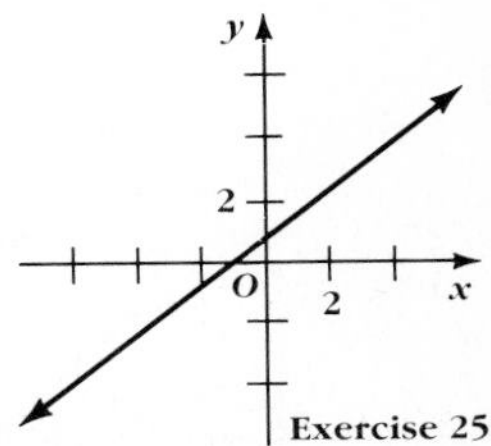

27. never **31.** the 72nd **33. a.** $\mathcal{R}$ **b.** $\{x: -1 < x < 4\}$ **35.** $\{3, -2\}$ **37.** the intersecting lines $y = \frac{1}{2}x$ and $y = -\frac{1}{2}x$

39.

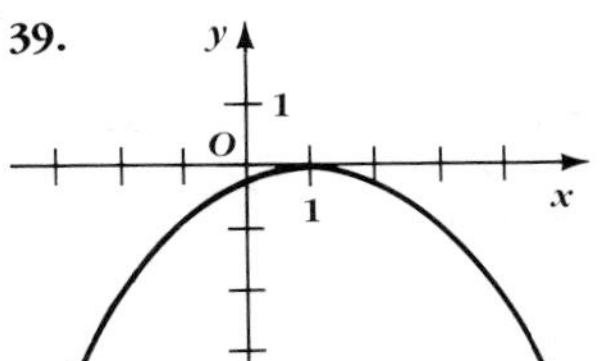

$F(1, -1)$; directrix, $y = 1$; $V(1, 0)$; $(1, 0)$ and $\left(0, -\frac{1}{4}\right)$

41. 2, 3, 5, 9, 17, 33 **43.** $BC = 10 = AC$ **45.** If $|x| = 2$, then $x = -2$; not true

EXERCISES 4-1, Pages 144–146

1. function **3.** not a function **5.** not a function
7. function **9.** not a function **11.** not a function
13. $\{x: x \neq \pm 1\}$ **15.** $\{x: x \leq 4\}$ **17.** $\{x: x \leq -3$ or $x \geq 3\}$ **19.** $\{x: 0 \leq x < 1$ or $x > 1\}$ **21.** $\{x: x \leq -1$ or $x > 1\}$ **23.** $\{y: y \geq 0\}$ **25.** $\{y: y \geq -1\}$
27. $\{y: 0 < y \leq 1\}$ **29.** $\{y: 0 < y \leq 1\}$ **31. a.** 0 **b.** 1 **c.** 3 **d.** $2x + 5$ **e.** 2 **33. a.** 2 **b.** 1 **c.** 3 **d.** $\frac{2}{x+2}$ **e.** $\frac{-2}{x(x+h)}$ **35. a.** 0 **b.** 12 **c.** 2 **d.** $x^2 + 4x$ **e.** $2x + h$

37. 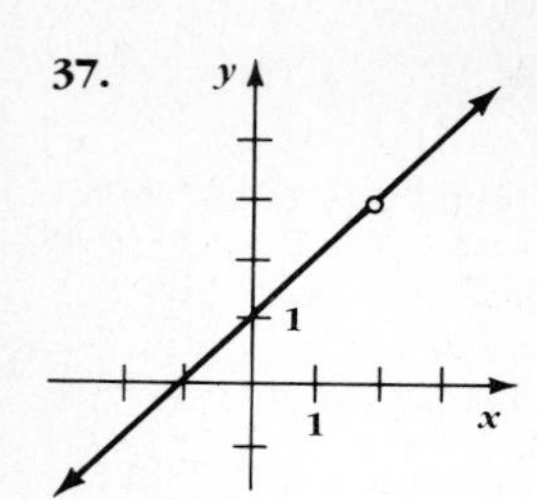

39.  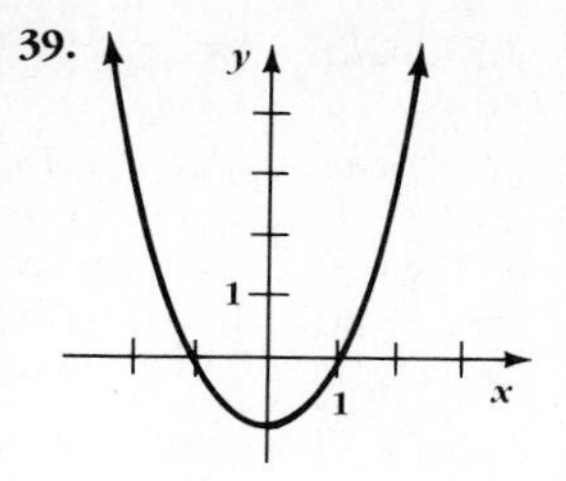

41. 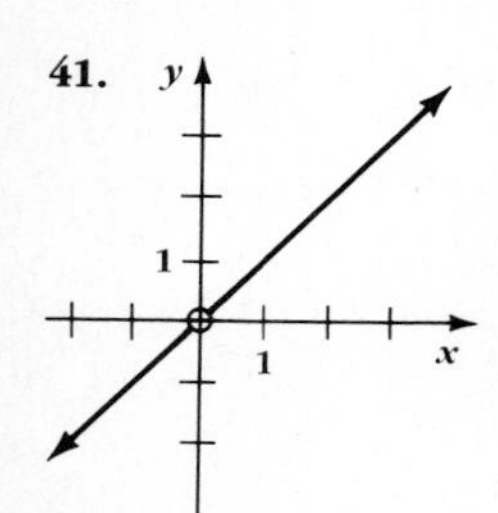

43. 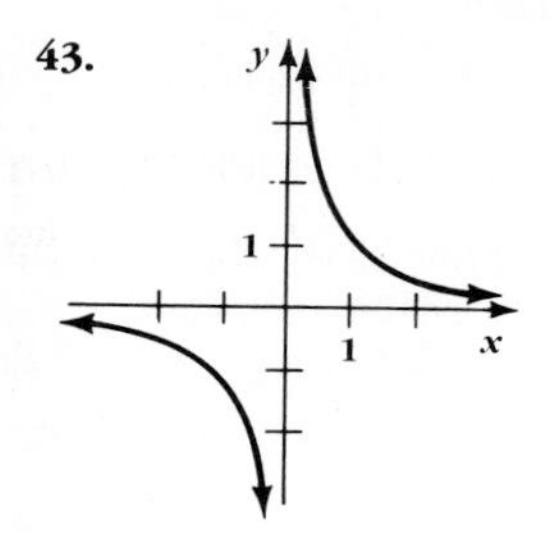

45. 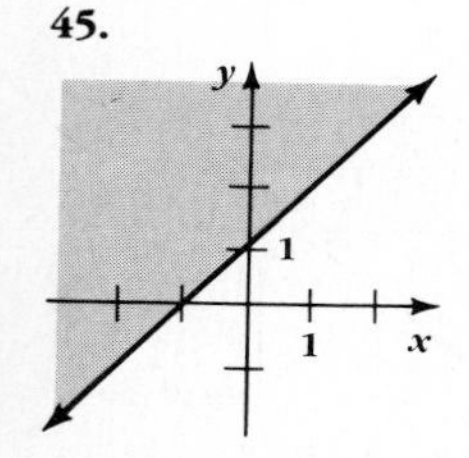

47. 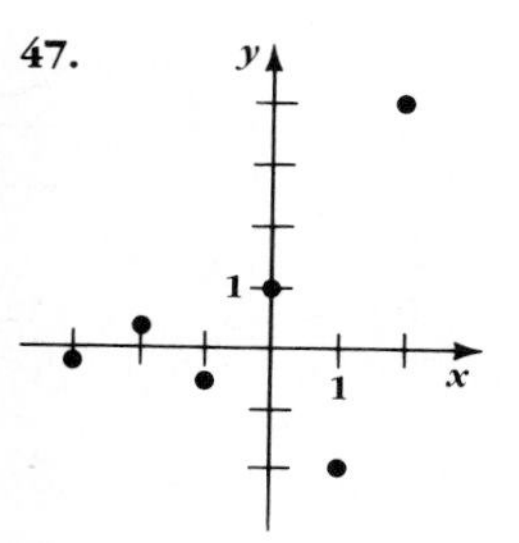

61. $\{I: 4.8 \leq I \leq 12\}$
63. $\{5, \{3, 5\}\}, \{5, \{5, 3\}\}, \{\{5, 3\}, 5\}$
65. $\{a, \{b, a\}\}, \{\{a, b\}, a\}$

EXERCISES 4-2, Pages 150–151

1. a. $2x + 1, \mathcal{R}$ **b.** $x^2 + x, \mathcal{R}$ **c.** $\frac{x+1}{x}, \{x: x \neq 0\}$ **d.** $2x + 1, \mathcal{R}$ **3. a.** $\sqrt{1+x} + \sqrt{1-x}, \{x: -1 \leq x \leq 1\}$ **b.** $\sqrt{1-x^2}, \{x: -1 \leq x \leq 1\}$ **c.** $\sqrt{\frac{1+x}{1-x}}, \{x: -1 \leq x < 1\}$ **d.** $2x, \{x: -1 \leq x \leq 1\}$
5. a. $\frac{2x^2}{x^2-1}, \{x: x \neq \pm 1\}$ **b.** $\frac{x^2}{x^2-1}, \{x: x \neq \pm 1\}$ **c.** $\frac{x-1}{x+1}, \{x: x \neq -1, 0, 1\}$ **d.** $\frac{-4x^3}{(x+1)^2(x-1)^2}, \{x: x \neq \pm 1\}$ **7. a.** $\frac{\sqrt{1+x} + 1 - x}{\sqrt{1-x}}, \{x: -1 \leq x < 1\}$ **b.** $\sqrt{1+x}, \{x: -1 \leq x < 1\}$ **c.** $\frac{\sqrt{1+x}}{1-x}, \{x: -1 \leq x < 1\}$ **d.** $\frac{x(3-x)}{1-x}, \{x: -1 \leq x < 1\}$ **9.** 0; 0 **11.** $0; \frac{3}{2}$
13. 1; 2 **15.** 0; −1 **17.** $f \neq g$; f is not defined when $x = 0$.

19.

a. 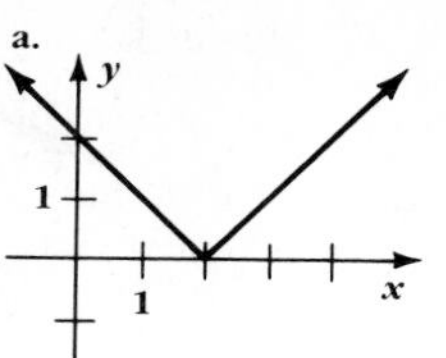

b. 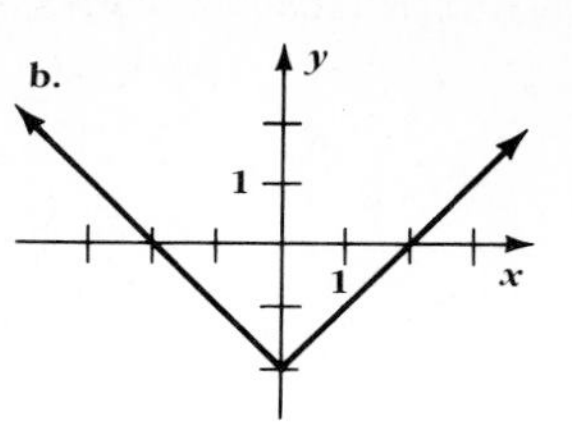

21.

a. 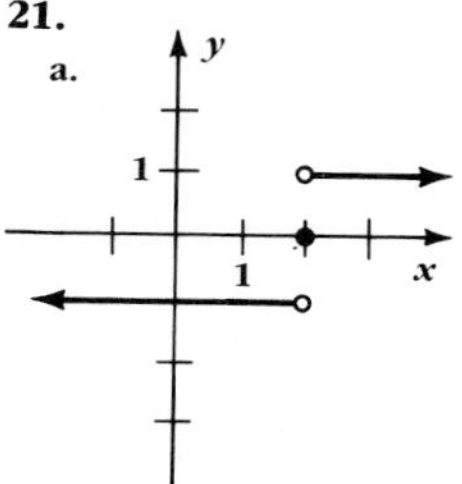

b. 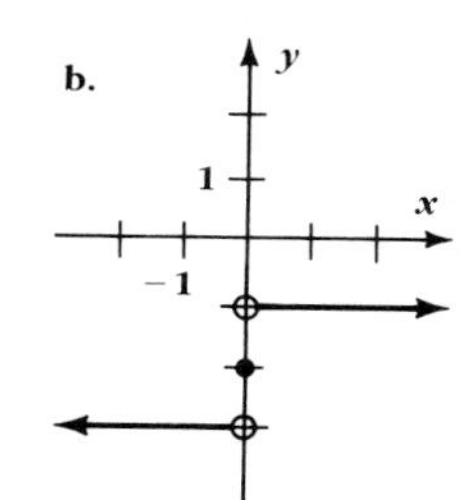 

23. 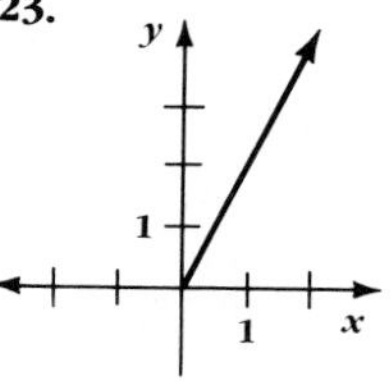

25. 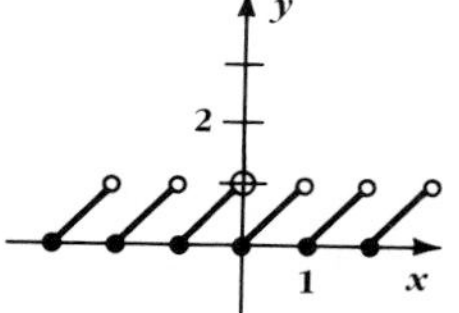

27. 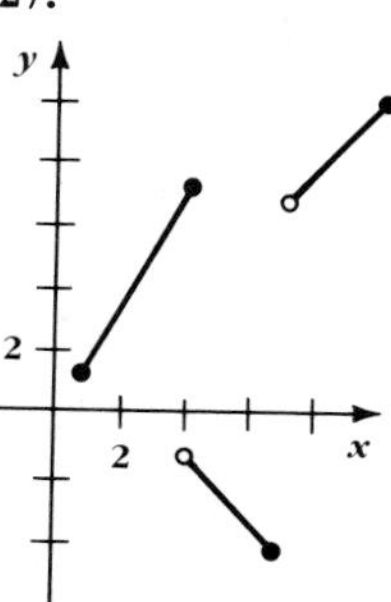

29. 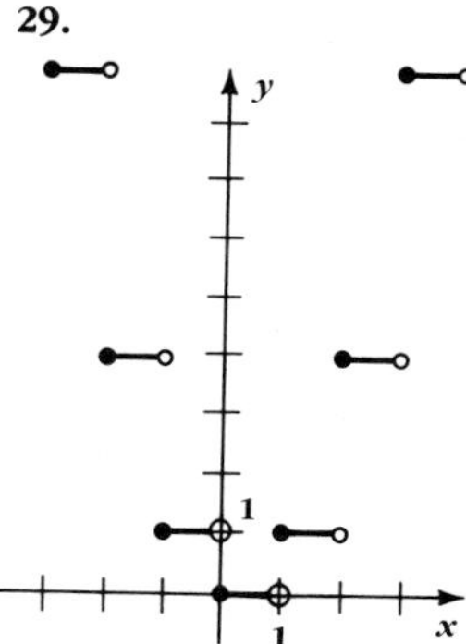

31. the ellipse with center (2, 3) and containing points (0, 3), (4, 3), (2, 0), and (2, 6)

33. 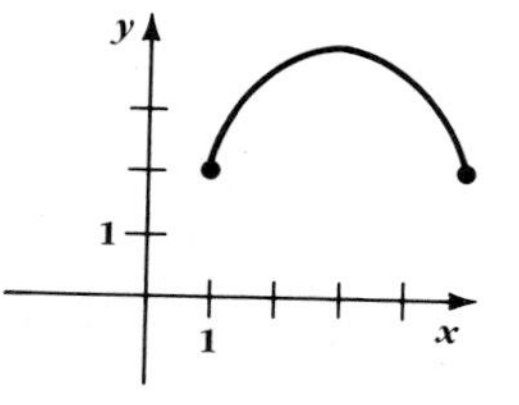

35. 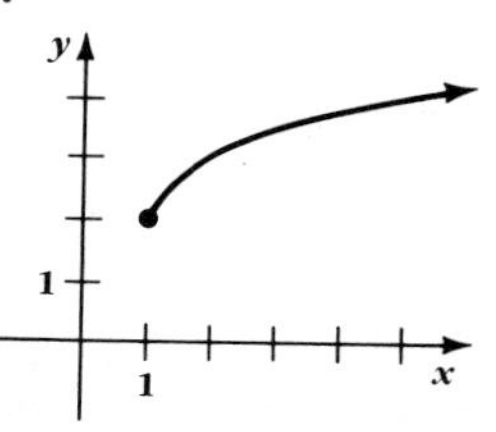

EXERCISES 4-3, Pages 155–156

1. $(f \circ g)(x) = 2x$, $\mathcal{R}$; $(g \circ f)(x) = 2x - 1$, $\mathcal{R}$
3. $(f \circ g)(x) = x^2 + 2x$, $\mathcal{R}$; $(g \circ f)(x) = x^2$, $\mathcal{R}$
5. $(f \circ g)(x) = \dfrac{x}{1 - x}$, $\{x: x \neq 0, 1\}$; $(g \circ f)(x) = x - 1$, $\{x: x \neq 1\}$ **7.** $(f \circ g)(x) = \sqrt{-x^2}$, $\{0\}$; $(g \circ f)(x) = 2 - x$, $\{x: x \leq 1\}$ **9.** Let $u = g(x) = x + 1$ and $f(u) = \sqrt{u}$. **11.** Let $u = g(x) = x^2 - 4$ and $f(u) = u^8$. **13.** Let $u = g(x) = \operatorname{sgn} x$ and $f(u) = u^2$. **15.** Let $u = g(x) = x + 1$ and $f(u) = u^2$. **17.** $f^{-1}(x) = 2x + 1$
19. $f^{-1}(x) = x^2 + 1, x \geq 0$ **21.** $f^{-1}(x) = \dfrac{1 - x}{x}$
23. $f^{-1}(x) = 1 + \sqrt[3]{x}$

25.

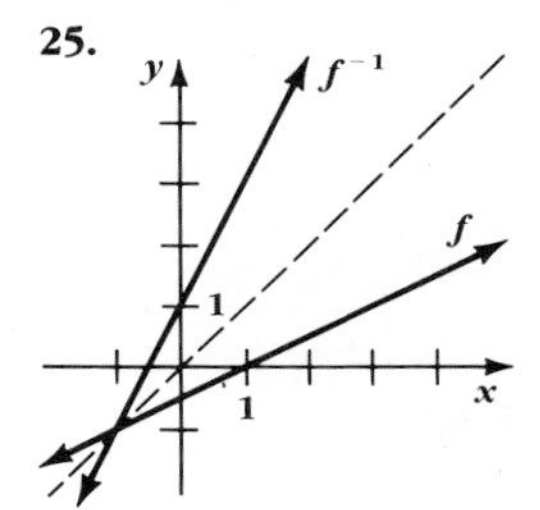

27.

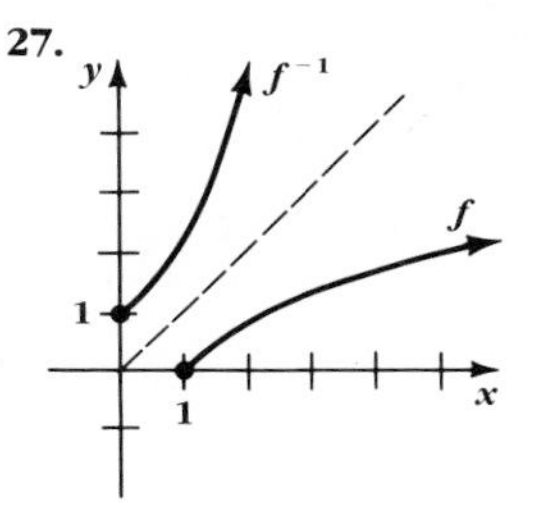

31. $f^{-1}(x) = \dfrac{x - b}{m}$ **33.** $V = \dfrac{4}{3}\pi(4 + 0.5t)^3$; $\dfrac{157{,}216\pi}{3}$ cm^3 **35.** $f^{-1}(x) = \dfrac{-dx + b}{cx - a}, x \neq \dfrac{a}{c}$
37. $(f \circ f)(x) = \dfrac{a^2x + ab + bcx + bd}{acx + bc + cdx + d^2}$; if $f \circ f = I$, then $a = -d$.

EXERCISES 4-4, Pages 161–163

1. yes, 4 **3.** no **5.** no **7.** no **9.** yes, 0 **11.** no **13.** yes, 3 **15.** yes, 3 **27.** 51 **29.** $\sqrt{2}$
31. $4 - \sqrt{2}$ **33.** 0 **35.** 2 **37.** -1 **39.** $-\dfrac{1}{2}$

EXERCISES 4-5, Pages 165–167

11.

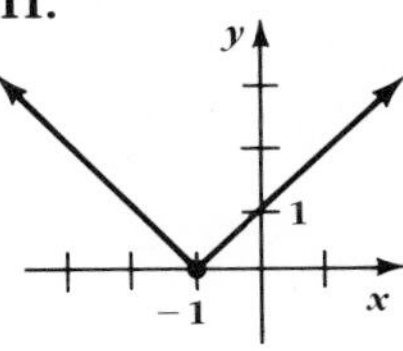

13.

15.

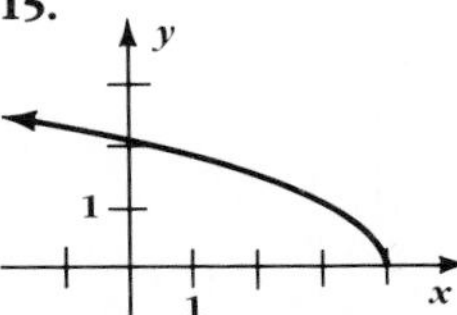

17.

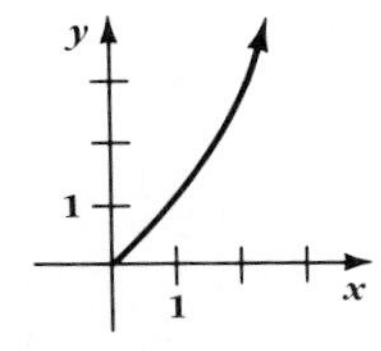

19. discontinuous at 5; continuous at 4.5 **21.** continuous at 3 and at 0 **35.** yes, at 2

EXERCISES 4-6, Pages 170–171

1. 3.5 **3.** 1.5 **5.** 1 **7.** $1 + \sqrt{3}$ **9.** 3.5 **11.** between 2 and 3 **13.** between 0 and -1 and between -1 and -2 **15.** between 0 and 1, 2 and 3, and -2 and -3 **17.** between 0 and 1, 2 and 3, 0 and -1, and -2 and -3 **19.** $f(1) > 0$ and $f(2) > 0$ but, for example, $f(1.5) < 0$ **21.** $f(1) > 0$ and $f(2) > 0$ but, for example, $f(1.5) < 0$ **23.** $[1.4, 1.5]$ **25.** $f(x) =$

$$\begin{cases} 2x & \text{if } 0 \leq x \leq \frac{1}{2} \\ 2x - 1 & \text{if } \frac{1}{2} < x \leq 1 \end{cases}$$

EXERCISES 4-7, Pages 176–177

1. $\lim\limits_{x \to -\infty} f(x) = 2$ **3.** $\lim\limits_{x \to 0} \varphi(x) = -\infty$
5. $\lim\limits_{x \to \infty} H(x) = -x$

7.

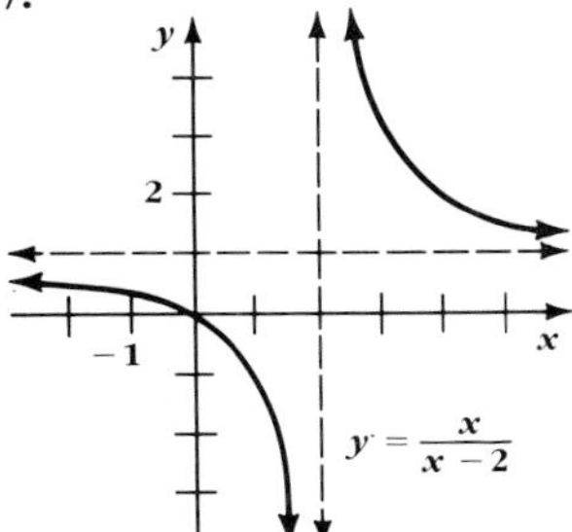

9.

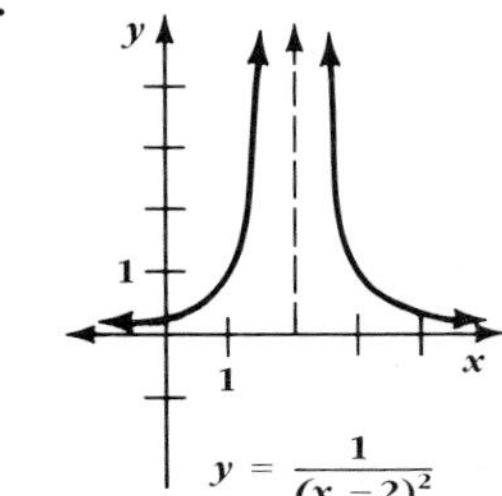

11.

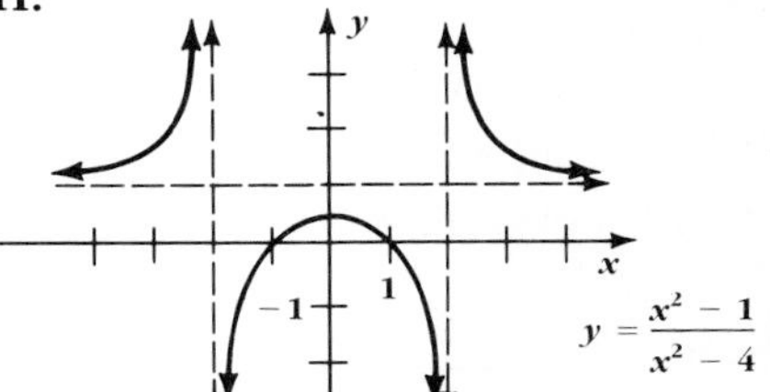

13.

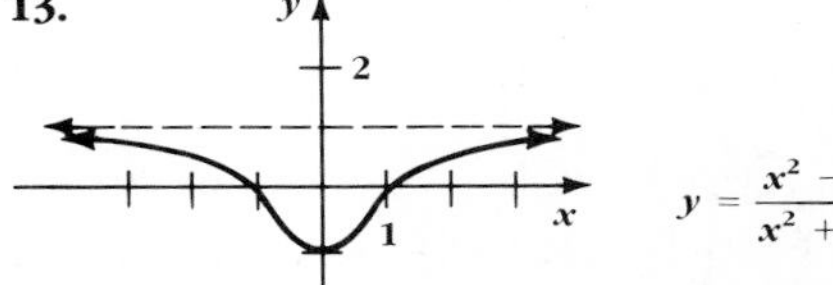

15.

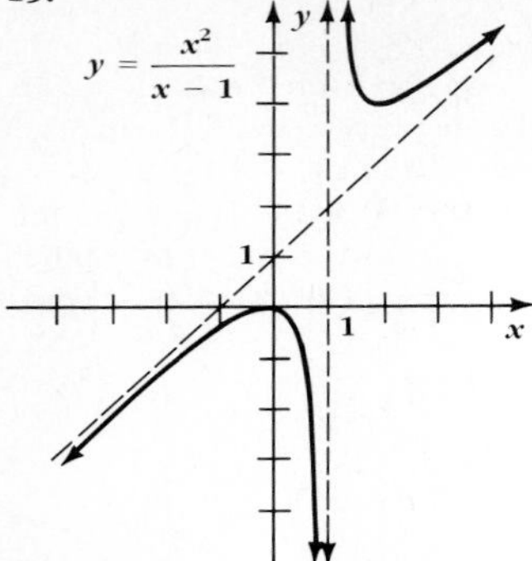

17.

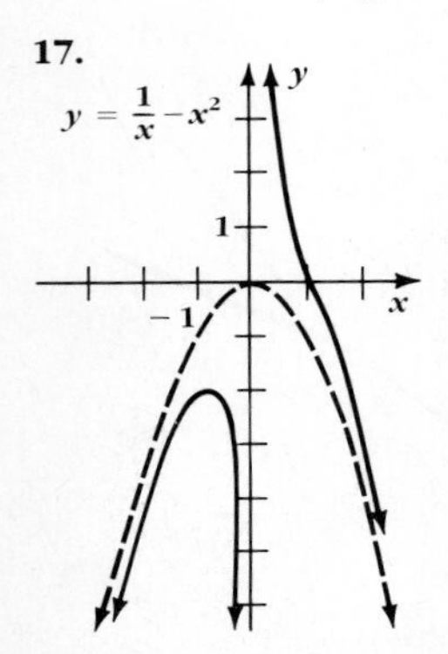

19. For each number M (no matter how large $-M$ is), there is a number N such that $x_n < M$ provided $n > N$.

EXERCISES 5-1, Pages 187–188

1. $2 - i; 3 - i$ **3.** $3 + i; 2 + 6i$ **5.** $5 - 4i; 1 - 7i$ **7.** $-3i; -\frac{1}{2} - \frac{3}{2}i$ **9.** $-3 + 3i; -\frac{1}{5} + \frac{3}{5}i$ **11.** $-3 + 2i; \frac{7}{25} - \frac{1}{25}i$ **13.** $-\frac{1}{2}i$ **15.** $-\frac{1}{2} + \frac{1}{2}i$ **17.** $-i$ **19.** $-i$ **21.** -1 **23.** 200 **25.** $\frac{248}{305} - \frac{44}{305}i$ **27.** $12 + 18i$ **29.** $r = 4, s = 3$ **31.** $r = 2, s = 3$ or $r = 3, s = 2$ **33.** $z = i$ **35.** $z = -i$ **37.** Show that $(1 + i)^2 - 2(1 + i) + 2 = 0$ and $(1 - i)^2 - 2(1 - i) + 2 = 0$. **39.** yes; no

EXERCISES 5-2, Pages 192–193

1. $w = 2 + i, z = -1 + i, w + z = 1 + 2i, w - z = 3$ **3.** $w = -2 + i, z = -2i, w + z = -2 - i, w - z = -2 + 3i$ **5.** $w = 12 + 16i, z = 9 + 12i, w + z = 21 + 28i, w - z = 3 + 4i$ **7.** $\overline{w} = 2 - i, \overline{z} = -1 - i, \overline{w + z} = 1 - 2i$ **9.** $\overline{w} = -2 - i, \overline{z} = 2i, \overline{w + z} = -2 + i$ **11.** $\overline{w} = 12 - 16i, \overline{z} = 9 - 12i, \overline{w + z} = 21 - 28i$ **13.** $1 - 2i = (2 - i) + (-1 - i)$ **15.** $-2 + i = (-2 - i) + 2i$ **17.** $21 - 28i = (12 - 16i) + (9 - 12i)$ **19.** $\sqrt{5} \le \sqrt{5} + \sqrt{2}$ **21.** $\sqrt{5} \le \sqrt{5} + 2$ **23.** $35 \le 20 + 15$ **33.** *Note:* Consider cases one of which is $0 < t < 1$. **35.** Show that all three complex numbers satisfy $|z| = 1$. **37.** w, z, and $w + z$ are collinear points and $w + z$ does not lie between w and z **39.** a circle with center $(1, 0)$ and radius 1 **41.** the horizontal line $y = 2$.

43.

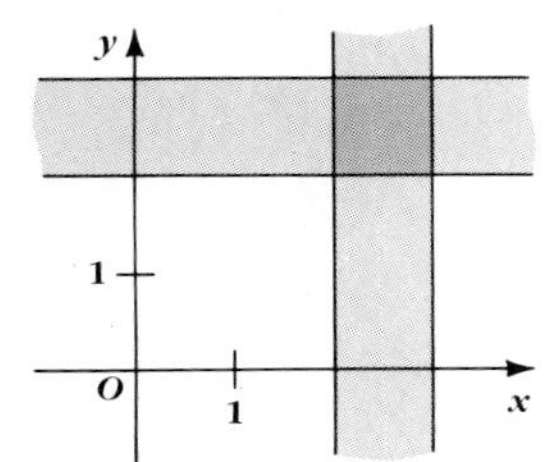

45. the ellipse with center at the origin, foci $(\pm 1, 0)$, and major axis of length $2\sqrt{10}$, that is, the ellipse with equation $\frac{x^2}{10} + \frac{y^2}{9} = 1$ **53.** Ex. 50; Thm. 2, part (1); Ex. 51; Thm. 3, part (2); Ex. 52

EXERCISES 5-3, Page 195

1. $1 \pm i$ **3.** $-2i \pm 2$ **5.** $-3 \pm 2i\sqrt{3}$ **7.** $0, 2i$ **9.** $1 \pm 3i$ **11. a.** -6 **b.** 6 **13. a.** $2i$ **b.** $2i$ **15.** $1 + i, -1 - i$ **17.** $1 + 2i, -1 - 2i$ **19.** $\sqrt{3} - i, -\sqrt{3} + i$ **21.** $2 + i, -i$ **23.** $i, -1 - i$ **25.** $1 + i, 1 - i, -1 + i, -1 - i$ **27.** If $c = \sqrt{\frac{p + \sqrt{p^2 + q^2}}{2}}$, then the roots are $\pm c \pm \frac{q}{2c}i$ if $q > 0$; $\pm c \mp \frac{q}{2c}i$ if $q < 0$.

EXERCISES 5-4, Pages 199–200

1. $h = 3, k = -2$ **3.** $h = 2, k = -1$ **5.** $9x^3 + 9x^2 - 3x + 4; 3x^3 + 5x^2 - 3x; 18x^6 + 33x^5 + 5x^4 + 12x^3 + 18x^2 - 6x + 4$ **7.** $x^4 - x^2 + 11x + 3; x^4 - 3x^2 + 7x + 5; x^6 + 2x^5 - 3x^4 + 5x^3 + 24x^2 - x - 4$ **9.** $Q(x) = 2; R(x) = 3x^2 - 3x - 2$ **11.** $Q(x) = x^2 - 2x + 3; R(x) = x + 7$ **13.** $2x^3 + 4x^2 - x - 3 - \frac{4}{x - 2}$ **15.** $2x^3 - x + 3 - \frac{4}{x + 3}$ **17.** $x^3 + 3x^2 + 2x - 1$ **19.** $x^2 + (2 + 3i)x + 6i - \frac{8}{x - 3i}$ **21.** $k = -15$ **23.** $k = 5$ **25.** example: $P(x) = x^3 - 16x$ **29.** $2abc$

EXERCISES 5-5, Pages 203–204

1. yes **3.** no **5.** yes **7.** $k = -3$ **9.** $3, \pm i\sqrt{2}$ **11.** $2, \pm 2i$ **13.** $-\frac{2}{3}, -\frac{2}{3}, 2$ **15.** example: $2x^3 - x^2 - 8x + 4 = 0$ **17.** example: $4x^3 - 13x - 6 = 0$

19. example: $x^3 - 6x^2 + 13x - 10 = 0$ **21.** $1, -2, \pm i\sqrt{2}$ **23.** $2, -\frac{1}{2}, \pm i$ **25.** $-2, -2, -2, 2$

27. a. $(x^2 - 3)(x^2 + 1)$
b. $(x + \sqrt{3})(x - \sqrt{3})(x^2 + 1)$
c. $(x + \sqrt{3})(x - \sqrt{3})(x + i)(x - i)$

29. a. $(x^2 + 2)(x^2 - 2)$
b. $(x^2 + 2)(x + \sqrt{2})(x - \sqrt{2})$
c. $(x + i\sqrt{2})(x - i\sqrt{2})(x + \sqrt{2})(x - \sqrt{2})$

31. a. $(x^2 - 2)^2$
b. $(x + \sqrt{2})^2 (x - \sqrt{2})^2$
c. $(x + \sqrt{2})^2 (x - \sqrt{2})^2$

33. $\pm 3, 1; (x + 3)(x - 3)(x - 1)$ **35.** $-2, 2, 2; (x + 2)(x - 2)^2$ **37.** $\pm 1, -\frac{2}{3}; (x + 1)(x - 1)(3x + 2)$

39. $3, -2, \pm 3i; (x - 3)(x + 2)(x + 3i)(x - 3i)$

EXERCISES 5-6, Pages 207–208

1. example: $x^3 - 4x^2 + 6x - 4 = 0$ **3.** example: $4x^4 + 3x^2 - 1 = 0$ **5.** example: $x^4 + 4 = 0$ **7.** example: $x^4 - 4x^3 + 8x^2 - 8x + 4 = 0$ **9.** $1 \pm i, -2$

11. $-2 \pm i, \pm i\sqrt{2}$ **13.** $\pm i, -1, \frac{1}{2} \pm \frac{\sqrt{3}}{2}i$

15. $(x + 2)(x^2 - 2x + 2)$

17. $(x^2 + 2)(x^2 + 4x + 5)$ **19.** $(x + 1)(x^2 + 1) \cdot (x^2 - x + 1)$ **21. a.** $\pm\sqrt{3}, -\frac{1}{2}$

b. $(2x + 1)(x^2 - 3)$ **23. a.** $1 \pm \sqrt{3}, -1 \pm i$
b. $(x^2 - 2x - 2)(x^2 + 2x + 2)$

EXERCISES 5-7, Pages 210–211

In Exercises 1–11, let (p, n, i) be the possible numbers of positive, negative, and imaginary roots. **1.** $(1, 0, 2)$ **3.** $(0, 3, 0)$ or $(0, 1, 2)$ **5.** $(1, 1, 2)$ **7.** $(1, 3, 0)$ or $(1, 1, 2)$ **9.** $(1, 2, 2)$ or $(1, 0, 4)$ **11.** $(3, 2, 0)$, $(3, 0, 2)$, $(1, 2, 2)$, or $(1, 0, 4)$ **13. a.** 3 **b.** -2 **15. a.** 4 **b.** -3 **17. a.** 3 **b.** -5 **19. a.** 3 **b.** -2

EXERCISES 5-8, Pages 213–214

1. $3, 1 \pm i$ **3.** no rational roots **5.** $-\frac{1}{2}, -1 \pm i$

7. $\frac{2}{3}, \pm\frac{\sqrt{2}}{2}i$ **9.** $-2, 3, 1 \pm \sqrt{2}$ **11.** $\frac{2}{3}, -4, -1 \pm \sqrt{2}$ **13.** $\frac{1}{2}, \frac{1}{2}, \pm\sqrt{3}$ **15.** $\frac{2}{3}, \frac{3}{2}$ **23.** $-\sqrt{p} + \sqrt{q}$ and $-\sqrt{p} - \sqrt{q}$

EXERCISES 6-1, Pages 225–226

1. a. 4 **b.** $y = 4x - 4$ **3. a.** 0 **b.** $y = 1$
5. a. 1 **b.** $y = x - 3$ **7. a.** -3 **b.** $y = -3x - 1$

9. $f'(x) = 4x^3$ **11.** $f'(x) = -\frac{1}{2x\sqrt{x}}$ **13.** $f'(x) = \frac{1}{(x + 1)^2}$ **15.** $x = 0$ **17.** $x = -3, x = 3$

21. $f'(0) = 0$ **23.** $f'(0)$ does not exist.

EXERCISES 6-2, Pages 229–230

1. a. $P'(x) = 4x + 5$ **b.** $y = -7x - 19$
3. a. $P'(x) = 3x^2 - 6x$ **b.** $y = 4$ **5. a.** $P'(x) = 4x^3 - 2x + 2$ **b.** $y = 4x - 1$ **7.** $F'(x) = \frac{3}{2}x^{\frac{1}{2}} - \frac{1}{2}x^{-\frac{1}{2}}$ **9.** $\varphi'(x) = \frac{-2x}{(x^2 + 1)^2}$ **11.** $g'(x) = \frac{5}{2}x^{\frac{3}{2}} - 3x^{\frac{1}{2}} + \frac{1}{2}x^{-\frac{1}{2}}$ **13.** $G'(x) = \frac{2x}{(1 - x^2)^2}$

15. $f'(x) = x^{-\frac{1}{2}}(1 - x^{\frac{1}{2}})^{-2}$ **17.** $\varphi'(x) = 3x^2 - 2 - x^{-2}$

EXERCISES 6-3, Page 233

1. $20(2x + 3)^9$ **3.** $-8t(1 - t^2)^3$
5. $-6x(x^2 + 1)^{-4}$ **7.** $(1 - 7u^2)(1 - u^2)^2$
9. $(1 - 3x^2)(1 + x^2)^{-3}$ **11.** $(x - 3)(x + 1)^3 x^{-4}$
13. $y' = 3x^2 - 6x; y'' = 6x - 6$
15. $y' = 10(2x - 1)^4; y'' = 80(2x - 1)^3$
17. $y' = 8x(x^2 - 1)^3; y'' = 8(7x^2 - 1)(x^2 - 1)^2$
19. $y' = (1 - x)^{-2} = y^2$ **21.** $y' = \pm 2\sqrt{y}$

EXERCISES 6-4, Pages 236–237

1. $10(1 - x)(2x - x^2)^4$ **3.** $(2x + 3)^{-\frac{1}{2}}$
5. $-x(4 - x^2)^{-\frac{1}{2}}$ **7.** $-x(x^2 + 1)^{-\frac{3}{2}}$
9. $\frac{1}{2}x^{-\frac{1}{2}}(1 - x)^{-\frac{3}{2}}$ **11.** $\frac{1}{2}x(16 - 5x)(4 - x)^{-\frac{1}{2}}$
13. $3x(3x^2 - 9)^{-\frac{1}{2}}$ **15.** -1
17. $16x(x^2 - 1)^7(1 - (x^2 - 1)^4)^{-3}$
21. $2(2 - x^2)(4 - x^2)^{-\frac{1}{2}}$
23. $(x^3 - 2x)(x^2 - 1)^{-\frac{3}{2}}$
27. $6x(4x^2 + 23)(3x^2 - 9)^4(2x^2 + 1)^{-4}$
29. $4(5x^2 + 2x + 1)^9(3x^2 + 6)^3 \cdot (105x^3 + 27x^2 + 156x + 30)$

EXERCISES 6-5, Pages 240–241

1. a. increasing on $(2, \infty)$; decreasing on $(-\infty, 2)$
b. relative minimum at 2 **c.** See below.

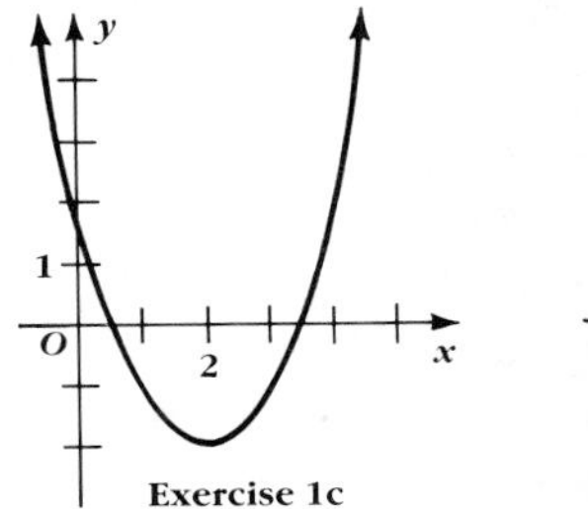

Exercise 1c

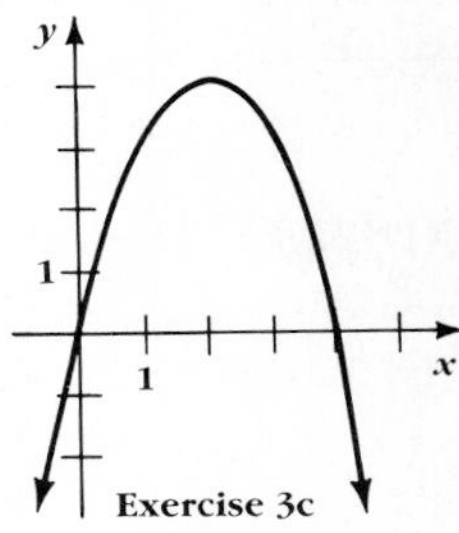

Exercise 3c

3. a. increasing on $(-\infty, 2)$; decreasing on $(2, \infty)$
b. relative maximum at 2 **c.** See above.
5. a. increasing on $(-\infty, 0)$ and on $(2, \infty)$; decreasing on $(0, 2)$ **b.** relative maximum at 0; relative minimum at 2

5. c. See below.

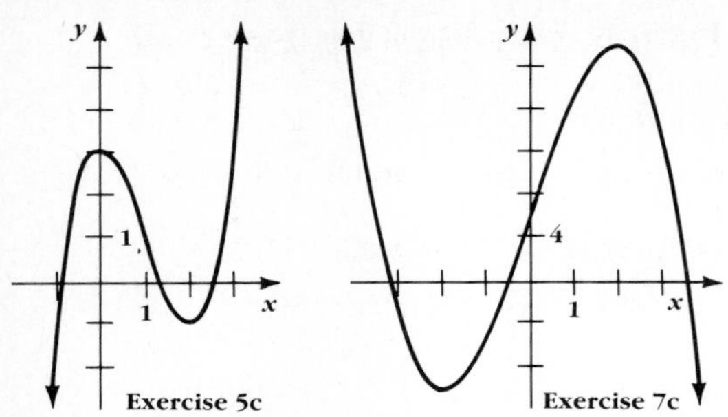

Exercise 5c Exercise 7c

7. a. increasing on $(-2, 2)$; decreasing on $(-\infty, -2)$ and on $(2, \infty)$ **b.** relative minimum at -2; relative maximum at 2 **c.** See above. **9. a.** increasing on $(-\infty, -2)$ and on $(0, 2)$; decreasing on $(-2, 0)$ and on $(2, \infty)$ **b.** relative maximum at -2 and at 2; relative minimum at 0 **c.** See below.

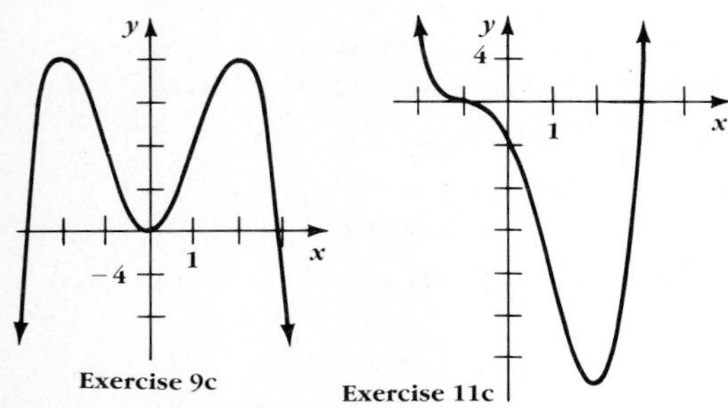

Exercise 9c Exercise 11c

11. a. increasing on $(2, \infty)$; decreasing on $(-\infty, -1)$ and $(-1, 2)$ **b.** relative minimum at 2 **c.** See above. **13.** no relative extrema **15.** relative minimum at -1; relative maximum at 1 **17.** relative minimum at 2 **21.** See below. The "corners" at -2, 0, and 2 show that g has no slope at these values.

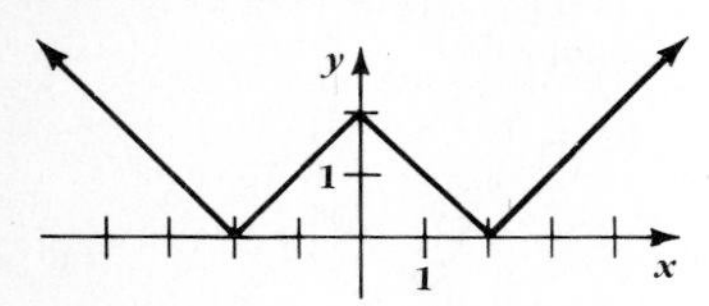

EXERCISES 6-6, Pages 244–245

1. a. $(2, \infty)$ **b.** $(-\infty, 2)$ **c.** $(2, 2)$

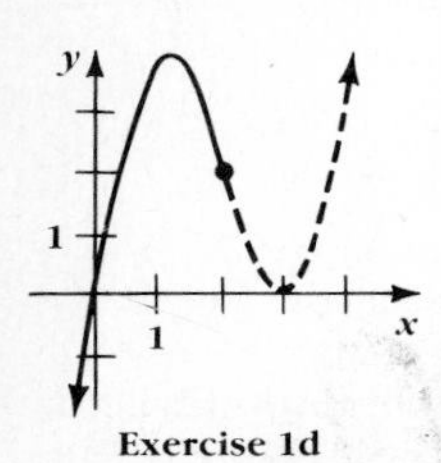

Exercise 1d

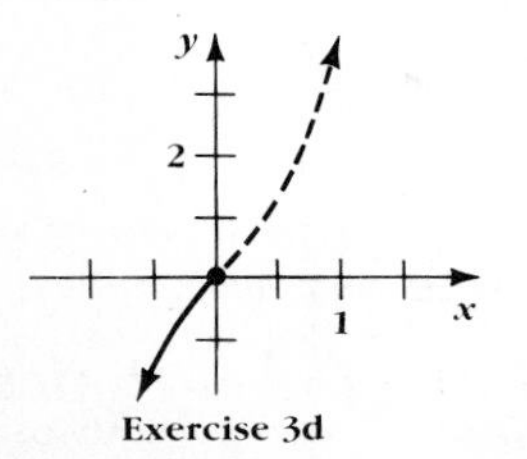

Exercise 3d

3. a. $(0, \infty)$ **b.** $(-\infty, 0)$ **c.** $(0, 0)$
5. a. $(-1, 1)$ **b.** $(-\infty, -1)$ and $(1, \infty)$
c. $(-1, 5)$ and $(1, 5)$
d. See below.

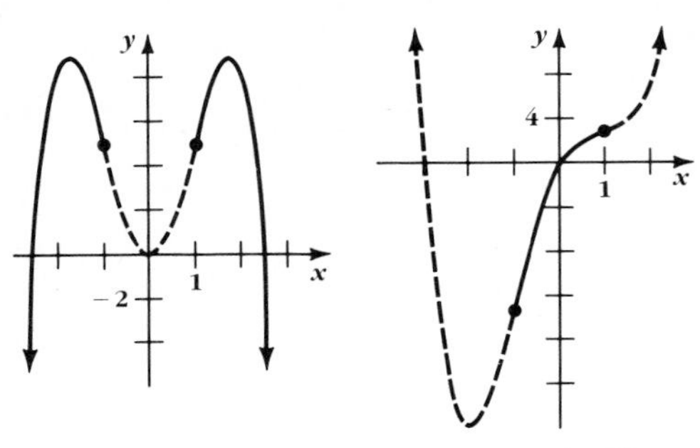

Exercise 5d Exercise 7d

7. a. $(-\infty, -1)$ and $(1, \infty)$ **b.** $(-1, 1)$
c. $(-1, -13)$ and $(1, 3)$ **d.** See above.
9. a. $(-1, 0)$ and $(1, \infty)$
b. $(-\infty, -1)$ and $(0, 1)$
c. $(-1, 7)$, $(0, 0)$, and $(1, -7)$
d. See right.
11. relative minimum at 1
13. relative minimum at 2
15. $f'(1) = 0$; $f''(1) = 2 + \frac{4}{1^3} = 6$

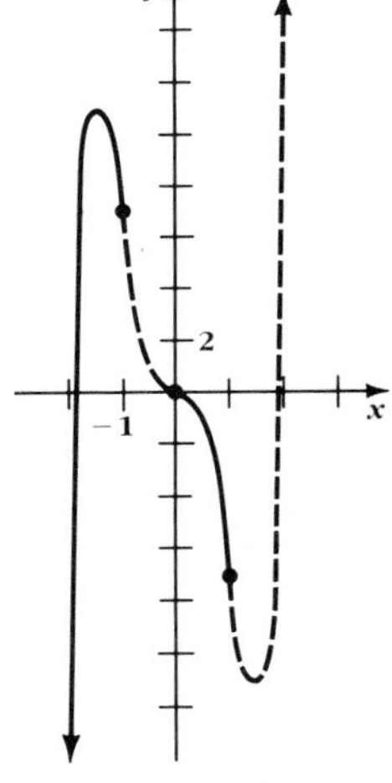

Exercise 9d

17.

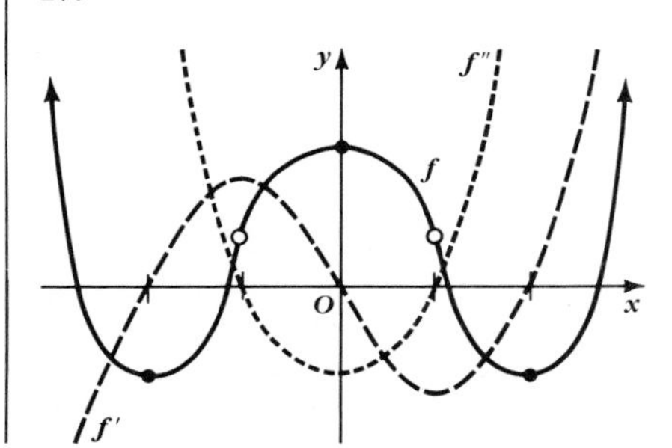

19.

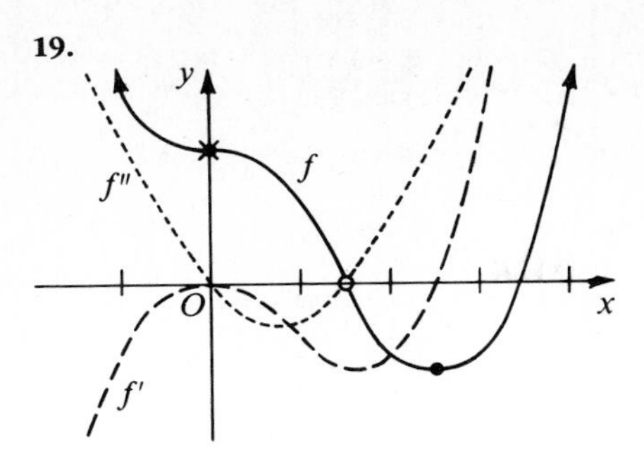

21. **a.** minimum at 0 **b.** neither a maximum nor a minimum at 0 **c.** maximum at 0 **25.** examples: two inflection points: $y = 2x^3 - x^4$; no inflection point: $y = x^4$

EXERCISES 6-7, Pages 250–252

1. 161 ft **3.** 6 cm by 6 cm **5.** 4 and 8, where 8 is to be squared **7.** about 490 m **9.** 3750 yd^2 **11.** 12:02 P.M.; 15 in. **13.** $w \approx 7$ in.; $h \approx 10$ in. **15.** 1st number is $\frac{3}{5}$; 2nd number is $\frac{2}{5}$. **17.** 2 units by 8 units **19.** (1, 1) **21.** 125 passengers **23.** 48π cubic units **25.** $\frac{8}{3}\pi a^3$ cubic units **27.** $w = 6$ in.; $h \approx 10$ in. **29.** $\frac{3a^2\sqrt{3}}{4}$ square units

EXERCISES 6-8, Pages 256–258

1. a. $(4 - 2t)$ m/s **b.** -2 m/s^2 **c.** right: $0 < t < 2$; left: $t > 2$ **d.** 6 m; -2 m/s^2
e.

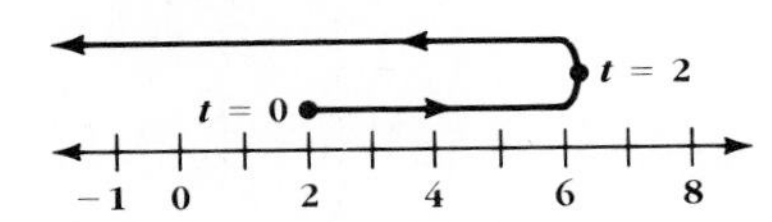

3. a. $(6t^2 - 18t + 12)$ m/s **b.** $(12t - 18)$ m/s^2 **c.** right: $0 < t < 1$ and $t > 2$; left: $1 < t < 2$ **d.** At $t = 1$, $s = 5$ m and $a = -6$ m/s^2; at $t = 2$, $s = 4$ m and $a = 6$ m/s^2
e.

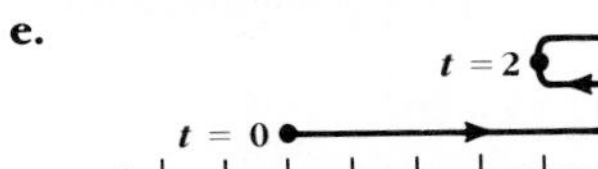

5. a. $4(t - 1)(t - 2)(t - 3)$ m/s **b.** $4(3t^2 - 12t + 11)$ m/s^2 **c.** right: $1 < t < 2$ and $t > 3$; left: $0 < t < 1$ and $2 < t < 3$ **d.** At $t = 1$, $s = -1$ m and $a = 8$ m/s^2; at $t = 2$, $s = 0$ m and $a = -4$ m/s^2; at $t = 3$, $s = -1$ m and $a = 8$ m/s^2
e.

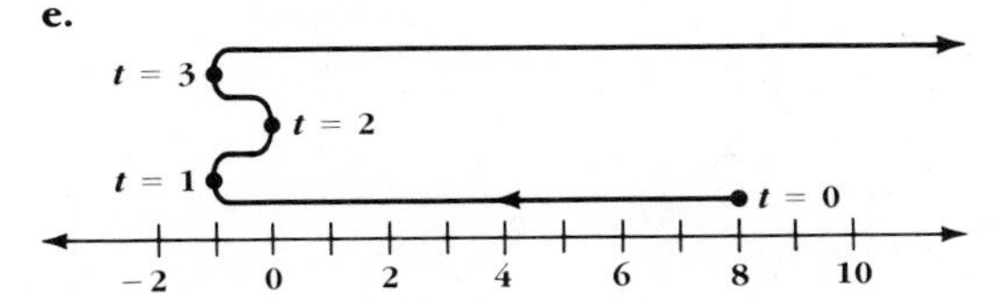

7. 12π ft^2/s **9.** 7.5 ft/s **11.** increasing at 32 mph **13.** $-\frac{8}{3}$ units/s **15. a.** 12 ft/s **b.** 4 ft/s **17.** $V = \frac{4}{3}\pi r^3$; $\frac{dV}{dt} = \frac{4}{3}\pi \cdot 3r^2 \cdot \frac{dr}{dt} = 4\pi r^2 \cdot \frac{dr}{dt} = S \cdot \frac{dr}{dt}$ **19.** $\frac{1}{3\pi}$ ft/min **21. a.** 1 inch/min **b.** $\frac{5}{8}$ inch/min

EXERCISES 6-9, Page 261

1. 4.3589 **3.** 2.1544 **5.** 1.4953 **7.** -1.5849 **9.** 1.4562 **11.** -0.8393 **13.** 1.5747 **15.** $-\frac{b}{a}$ **17.** 3

CUMULATIVE REVIEW (Chapters 4–6), Pages 268–269

1. $-i$ **3. a.** $-5 + 10i$ **b.** $-6 - 4i$ **c.** $2 + i$ **d.** $-\frac{11}{5} + \frac{2}{5}i$ **5.** $3 - 2i$ and $-3 + 2i$ **7.** $y = 2x + 5$ **9.** For example, $2x^3 - 5x^2 - x + 6 = 0$ **11.** 2 positive, 3 negative, and 0 imaginary; or 2 positive, 1 negative, and 2 imaginary; or 0 positive, 3 negative, and 2 imaginary; or 0 positive, 1 negative, and 4 imaginary **13.** $\frac{1}{2}$, $-1 + 2i$, $-1 - 2i$ **15. a.** $x^2 + 4x$ **b.** $x^2 - 2$ **c.** $x^2 + 2x - 3$ **d.** $2x + h$ **17.** $f^{-1}(x) = \frac{1}{1-x}$ **19.** See below. **21.** yes; at $x = -1$ **23.** at $x = 0$ and at $x = 1$

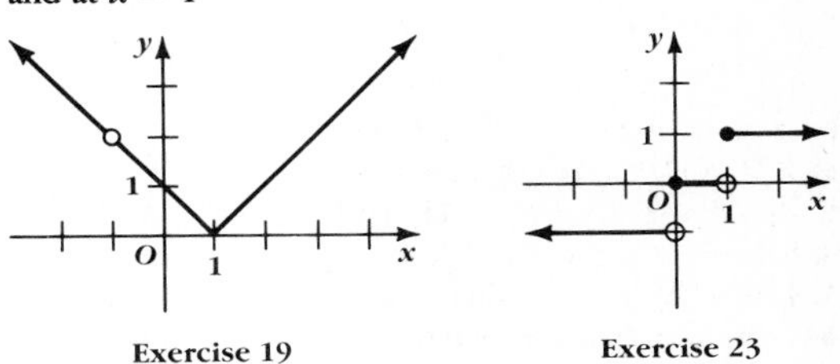

Exercise 19 Exercise 23

25. a. 1 **b.** 0 **c.** does not exist

27.

29. $(1 - x)^{-2}$ **31.** $\frac{1 + x^2}{(1 - x^2)^2}$ **33.** $\frac{1 - 2x^2}{\sqrt{1 - x^2}}$ **35. a.** $(-\infty, -1)$ and $(3, \infty)$ **b.** $(-1, 3)$ **c.** $(1, \infty)$ **d.** $(-\infty, 1)$ **e.** rel. maximum at -1; rel. minimum at 3

35. f.

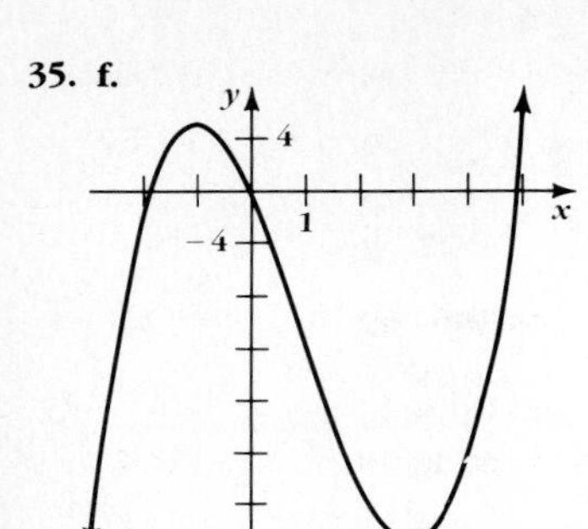

37. relative minimum at -1 **39.** radius = 3 inches, height = 6 inches

EXERCISES 7-1, Pages 275–276

1. $\cos x = -\frac{3}{5}$; $\sin x = \frac{4}{5}$ **3.** $\cos x = -1$; $\sin x = 0$ **5.** $\cos x = \frac{5}{13}$; $\sin x = \frac{12}{13}$ **7.** $\cos x = -\frac{1}{3}$; $\sin x = -\frac{2\sqrt{2}}{3}$ **9.** $\cos x = -\frac{\sqrt{2}}{2}$; $\sin x = \frac{\sqrt{2}}{2}$ **11.** $\cos x = \frac{\sqrt{3}}{2}$; $\sin x = -\frac{1}{2}$ **13.** $\cos x = \frac{1}{2}$; $\sin x = -\frac{\sqrt{3}}{2}$ **15.** $\cos x = \frac{\sqrt{3}}{2}$; $\sin x = \frac{1}{2}$ **17.** $\cos x = -\frac{3}{5}$ **19.** $\sin x = -\frac{1}{\sqrt{5}}$ **21.** $\cos x = \frac{\sqrt{5}}{3}$ **23.** $\cos x = \frac{2\sqrt{2}}{3}$ **25.** $\sin x = -\frac{2}{7}$ **27. a.** $\frac{\sqrt{3}}{2} = 2 \cdot \frac{1}{2} \cdot \frac{\sqrt{3}}{2}$ **b.** $\frac{\sqrt{3}}{2} = 2 \cdot \frac{\sqrt{3}}{2} \cdot \frac{1}{2}$ **29. a.** $\frac{1}{2} = 2\left(-\frac{\sqrt{3}}{2}\right)^2 - 1$ **b.** $0 = 2\left(-\frac{\sqrt{2}}{2}\right)^2 - 1$ **31.** $QP = VP$ (distance formula); squaring a monomial and a binomial, and $u^2 + v^2 = 1$ because P lies on the unit circle; addition and multiplication axioms; factoring and the zero product rule; substitution axiom and definition of sine function

EXERCISES 7-2, Pages 279–281

1. III **3.** II **5.** IV **7.** II **9.** III **11.** $\tan x = -\sqrt{3}$; $\cot x = -\frac{\sqrt{3}}{3}$; $\sec x = 2$; $\csc x = -\frac{2\sqrt{3}}{3}$ **13.** $\tan x = \frac{\sqrt{3}}{3}$; $\cot x = \sqrt{3}$; $\sec x = -\frac{2\sqrt{3}}{3}$; $\csc x = -2$ **15.** $\tan x = -1 = \cot x$; $\sec x = -\sqrt{2}$; $\csc x = \sqrt{2}$ **17.** $\tan x = \sqrt{3}$; $\cot x = \frac{\sqrt{3}}{3}$; $\sec x = -2$; $\csc x = -\frac{2\sqrt{3}}{3}$ **19.** $\sqrt{3} = \frac{2(-\sqrt{3})}{1 - (-\sqrt{3})^2}$ **21.** The given statement is false. **23.** $\cos x = \frac{3}{5}$; $\sin x = -\frac{4}{5}$; $\tan x = -\frac{4}{3}$; $\csc x = -\frac{5}{4}$; $\cot x = -\frac{3}{4}$ **25.** $\cos x = -\frac{24}{25}$; $\tan x = -\frac{7}{24}$; $\csc x = \frac{25}{7}$; $\sec x = -\frac{25}{24}$; $\cot x = -\frac{24}{7}$ **27.** $\sin x = -\frac{\sqrt{10}}{10}$; $\cos x = -\frac{3\sqrt{10}}{10}$; $\tan x = \frac{1}{3}$; $\cot x = 3$; $\sec x = -\frac{\sqrt{10}}{3}$ **29.** $\cot x = -\frac{1}{2}$; $\sin x = \frac{2\sqrt{5}}{5}$; $\cos x = -\frac{\sqrt{5}}{5}$; $\csc x = \frac{\sqrt{5}}{2}$; $\sec x = -\sqrt{5}$

31. a. neither **b.** odd **33. a.** odd **b.** even

EXERCISES 7-3, Pages 284–286

1. a. $\frac{\pi}{6}$ **b.** $\frac{3\pi}{4}$ **c.** $\frac{7\pi}{6}$ **d.** $-\frac{\pi}{3}$ **e.** $\frac{5\pi}{3}$ **3. a.** 114.6° **b.** −57.3° **c.** 22.9° **d.** 412.5° **e.** 2.9° **5.** $s = 3$; $A = 3$ **7.** $\theta = 2$; $A = 9$ **9.** $\theta = 0.4$; $s = 4$ **11.** $r = 1.5$; $A = 1.35$ **13.** $r = 5$; $\theta = 1.2$ **15.** $r = 6$; $s = \pi$ **19.** example: 2π radians = 400 grads **21.** example: 1 revolution = 360 degrees **23.** 250 rpm; 3000π cm/min **25.** 2918.2 miles **27.** 19.6% **29.** 70.3 ft/s

EXERCISES 7-4, Page 289

1. $\cos\theta = \frac{4}{5}$; $\sin\theta = \frac{3}{5}$; $\tan\theta = \frac{3}{4}$; $\cot\theta = \frac{4}{3}$; $\sec\theta = \frac{5}{4}$; $\csc\theta = \frac{5}{3}$ **3.** $\cos\theta = -\frac{5}{13}$; $\sin\theta = \frac{12}{13}$; $\tan\theta = -\frac{12}{5}$; $\cot\theta = -\frac{5}{12}$; $\sec\theta = -\frac{13}{5}$; $\csc\theta = \frac{13}{12}$ **5.** $\cos\theta = -1$; $\sin\theta = 0$; $\tan\theta = 0$; $\sec\theta = -1$; $\cot\theta$ and $\csc\theta$ are not defined. **7.** $\cos\theta = \frac{15}{17}$; $\sin\theta = -\frac{8}{17}$; $\tan\theta = -\frac{8}{15}$; $\cot\theta = -\frac{15}{8}$; $\sec\theta = \frac{17}{15}$; $\csc\theta = -\frac{17}{8}$ **9.** $\cos\theta = -\frac{2\sqrt{5}}{5}$; $\sin\theta = \frac{\sqrt{5}}{5}$; $\tan\theta = -\frac{1}{2}$; $\cot\theta = -2$; $\sec\theta = -\frac{\sqrt{5}}{2}$; $\csc\theta = \sqrt{5}$ **11.** $\cos\theta = -\frac{2}{3}$; $\sin\theta = \frac{\sqrt{5}}{3}$; $\tan\theta = -\frac{\sqrt{5}}{2}$; $\cot\theta = -\frac{2\sqrt{5}}{5}$; $\sec\theta = -\frac{3}{2}$; $\csc\theta = \frac{3\sqrt{5}}{5}$ **13.** I **15.** III **17.** II **19.** III **21.** $\sin 0° = \tan 0° = 0$; $\cos 0° = \sec 0° = 1$; $\cot 0°$ and $\csc 0°$, not defined **23.** $\sin 180° = \tan 180° = 0$; $\cos 180° = \sec 180° = -1$; $\cot 180°$ and $\csc 180°$, not defined **25.** $\sin 120° = \frac{\sqrt{3}}{2}$; $\cos 120° = -\frac{1}{2}$; $\tan 120° = -\sqrt{3}$; $\cot 120° = -\frac{\sqrt{3}}{3}$; $\sec 120° = -2$; $\csc 120° = \frac{2\sqrt{3}}{3}$ **27.** $\sin 150° = \frac{1}{2}$; $\cos 150° = -\frac{\sqrt{3}}{2}$; $\tan 150° = -\frac{\sqrt{3}}{3}$; $\cot 150° = -\sqrt{3}$; $\sec 150° = -\frac{2\sqrt{3}}{3}$; $\csc 150° = 2$ **29.** $\sin(-135°) = -\frac{\sqrt{2}}{2} = \cos(-135°)$;

$\tan(-135°) = 1 = \cot(-135°)$; $\csc(-135°) = -\sqrt{2} = \sec(-135°)$ **31.** $\cos\varphi = -\frac{4}{5}$; $\tan\varphi = -\frac{3}{4}$; $\cot\varphi = -\frac{4}{3}$; $\sec\varphi = -\frac{5}{4}$; $\csc\varphi = \frac{5}{3}$ **33.** $\cos\varphi = \frac{8}{17}$; $\sin\varphi = -\frac{15}{17}$; $\tan\varphi = -\frac{15}{8}$; $\sec\varphi = \frac{17}{8}$; $\csc\varphi = -\frac{17}{15}$

35. $\sin\varphi = \cos\varphi = -\frac{\sqrt{2}}{2}$; $\cot\varphi = 1$; $\sec\varphi = \csc\varphi = -\sqrt{2}$ **37.** $\sin\varphi = \frac{-2\sqrt{2}}{3}$; $\tan\varphi = -2\sqrt{2}$; $\cot\varphi = -\frac{\sqrt{2}}{4}$; $\sec\varphi = 3$; $\csc\varphi = -\frac{3\sqrt{2}}{4}$ **39.** $\sin\varphi = \frac{2\sqrt{2}}{3}$; $\tan\varphi = -2\sqrt{2}$; $\cot\varphi = -\frac{\sqrt{2}}{4}$; $\sec\varphi = -3$; $\csc\varphi = \frac{3\sqrt{2}}{4}$

EXERCISES 7-5, Page 291

1. 0.8059 **3.** 2.006 **5.** 0.2571 **7.** 0.2473 **9.** 0.2063 **11.** −0.9583 **13.** −0.7431 **15.** −0.8280 **17.** 32.52° **19.** 55.30° **21.** 0.451 **23.** 34.35° **25.** 50.44° **27.** 0.983 **29.** 38.54°, 141.46° **31.** 148.04°, 211.96° **33.** 3.0380, 6.1796

EXERCISES 7-6, Pages 294–295

1. $\beta = 48.0°$; $b = 26.7$; $c = 35.9$ **3.** $\beta = 23.5°$; $a = 18.7$; $c = 20.4$ **5.** $\alpha = 32.7°$; $\beta = 57.3°$; $c = 630$ **7.** $\alpha = 36.6°$; $\beta = 53.4°$; $b = 29.0$ **9.** $\alpha = 73.0°$; $\gamma = 34.0°$; $a = 16.0$; $c = 9.36$ **11.** $\alpha = \beta = 23.0°$; $b = 2.65$; $c = 4.88$ **13.** The triangle cannot be solved. **15.** $\alpha = \beta = 71.7°$; $a = b = 420$ **17.** 30.8° **19.** 692.5 m **21.** 81.6° **23.** 13.1 m **25.** 8065.5 ft; 27,586.5 ft **27.** $\sqrt{ab}$ ft **29. a.** 6.18 **b.** 6.50 **33.** π, π

EXERCISES 7-7, Pages 298–300

1. 284 **3.** 48.3° **5.** 0.0759 **7.** 125.4° **9.** 10.2 **11.** 83.1 **13.** 2.09 miles **15.** 39,700 ft **17.** 93.8° **19.** 6.45 and 10.1 **25.** 11.8 m

EXERCISES 7-8, Pages 303–304

1. Use $\alpha = 180° - \beta - \gamma$ to find α; use the law of sines to find b; use the law of sines to find c. **3.** Use the law of cosines to find γ; use the law of sines to find β; use $\alpha + \beta + \gamma = 180°$ to find the third angle. **5.** Use the law of cosines to find a; use the law of sines to find β or γ; use $\alpha + \beta + \gamma = 180°$ to find the third angle. **7.** Since α is obtuse, there is one solution. Use the law of sines to find β; use $\gamma = 180° - \alpha - \beta$ to find γ; use the law of sines to find c. **9.** Use the law of sines to find each possible value for α; for each value of α, use $\gamma = 180° - \alpha - \beta$ and the law of sines to find γ and c. **11.** Use the law of sines to find each possible value for γ; for each value of γ, use $\alpha = 180° - \beta - \gamma$ and the law of sines to find α and a. **13.** $\alpha = 98.0°$; $b = 8.56$; $c = 12.4$ **15.** $\gamma = 97.9°$; $\beta = 52.4°$; $\alpha = 29.7°$ **17.** $a = 204$; $\beta = 35.5°$; $\gamma = 44.5°$ **19.** $\beta = 30.4°$; $\gamma = 19.6°$; $c = 14.2$ **21.** First sol.: $\alpha = 61.9°$; $\gamma = 83.1°$; $c = 13.5$; second sol.: $\alpha = 118.1°$; $\gamma = 26.9°$; $c = 6.16$ **23.** no solution **25.** $\sqrt{7}$ **27.** 3.86; 4.33 **29.** 55.4°

EXERCISES 7-9, Pages 306–307

1. 52.5 sq. units **3.** 19.8 sq. units **5.** 8570 sq. units **7.** 117 sq. units **9.** $K_1 = 46.5$ sq. units; $K_2 = 21.2$ sq. units **11.** No triangle exists. **13.** 75.2 cm² **15.** 136.7°

EXERCISES 8-1, Pages 316–317

1. $\cot^2 x$ **3.** $\tan\alpha$ **5.** 1 **7.** 1 **9.** $-\sin^2 u$ **11.** $\csc^2\varphi$ **13.** $\cos x$ **15.** $\cos x$ **17.** $\cos x = \pm\sqrt{1 - \sin^2 x}$; $\tan x = \pm\frac{\sin x}{\sqrt{1 - \sin^2 x}}$; $\cot x = \pm\frac{\sqrt{1 - \sin^2 x}}{\sin x}$; $\sec x = \pm\frac{1}{\sqrt{1 - \sin^2 x}}$; $\csc x = \frac{1}{\sin x}$

19. $\sin x = \pm\frac{\sqrt{\sec^2 x - 1}}{\sec x}$; $\cos x = \frac{1}{\sec x}$; $\tan x = \pm\sqrt{\sec^2 x - 1}$; $\cot x = \pm\frac{1}{\sqrt{\sec^2 x - 1}}$; $\csc x = \pm\frac{\sec x}{\sqrt{\sec^2 x - 1}}$ **21.** $\sin x = \pm\frac{1}{\sqrt{1 + \cot^2 x}}$; $\cos x = \pm\frac{\cot x}{\sqrt{1 + \cot^2 x}}$; $\tan x = \frac{1}{\cot x}$; $\sec x = \pm\frac{\sqrt{1 + \cot^2 x}}{\cot x}$; $\csc x = \pm\sqrt{1 + \cot^2 x}$ **33.** $\cot t$ **35.** $\cos x$ **37.** $\cos\theta$ **39.** $2\cos x$ **41.** $\sin x$ **43.** 2 **45.** $10\csc^2 x$

EXERCISES 8-2, Pages 320–321

1. 1 **3.** $\frac{1}{2}$ **5.** $-\frac{1}{2}$ **7.** $\cos 3\alpha$ **9.** $\sin 5t$ **11.** $\frac{\sqrt{6} - \sqrt{2}}{4}$ **13.** $\frac{\sqrt{6} - \sqrt{2}}{4}$ **15.** $\frac{\sqrt{2} - \sqrt{6}}{4}$ **17.** $\frac{\sqrt{6} - \sqrt{2}}{4}$ **27. a.** $-\frac{4}{5}$ **b.** $-\frac{3}{5}$ **c.** $p-q$ is a third-quadrant number. **29. a.** $\frac{56}{65}$ **b.** $-\frac{33}{65}$ **c.** $p + q$ is a second-quadrant number. **31.** $-\sin 20°$ **33.** $\sin 25°$ **35.** $-\sec 35°$ **37.** Answers will vary. For example, let $\alpha = \beta = 45°$.

EXERCISES 8-3, Pages 324–325

1. $-\frac{24}{25}$ **3.** $\frac{\sqrt{10}}{10}$ **5.** $-\frac{119}{169}$ **7.** $\frac{2\sqrt{13}}{13}$

9. a. $\sin 2\theta = \frac{24}{25}$; $\cos 2\theta = -\frac{7}{25}$ **b.** II

11. a. $\sin 2\theta = -\frac{120}{169}$; $\cos 2\theta = -\frac{119}{169}$ **b.** III

13. a. $\sin 2\theta = \frac{4\sqrt{5}}{9}$; $\cos 2\theta = -\frac{1}{9}$ **b.** II

15. **a.** $\frac{\sqrt{2+\sqrt{3}}}{2}$ **b.** $\frac{\sqrt{6}+\sqrt{2}}{4}$

17. **a.** $-\frac{\sqrt{2-\sqrt{3}}}{2}$ **b.** $\frac{\sqrt{2}-\sqrt{6}}{4}$

19. **a.** $-\frac{\sqrt{2-\sqrt{3}}}{2}$ **b.** $\frac{\sqrt{2}-\sqrt{6}}{4}$

21. **a.** $\frac{\sqrt{2+\sqrt{3}}}{2}$ **b.** $\frac{\sqrt{2}+\sqrt{6}}{4}$ **31.** $3 \sin x - 4 \sin^3 x$

EXERCISES 8-4, Pages 329–330

1. $\sqrt{3}$ **3.** $\frac{\sqrt{3}}{3}$ **5.** $2-\sqrt{3}$ **7.** $\sqrt{3}-2$

9. $\sqrt{2}-1$ **11.** $\sqrt{2}+1$ **13.** $-\frac{33}{56}$ **15.** 2

17. $\frac{\cot s \cot t - 1}{\cot t + \cot s}$ **25.** $\cos t + \cos 3t$

27. $\frac{1}{2} \sin 6\alpha - \frac{1}{2} \sin 2\alpha$ **29.** $\frac{1}{2} \cos 2x$

31. $2 \sin 2\alpha \cos \alpha$ **33.** $2 \cos 5x \cos x$ **35.** $\cos \alpha$

EXERCISES 8-5, Pages 332–333

1. 45°, 225° **3.** $\frac{\pi}{3}, \frac{2\pi}{3}, \frac{4\pi}{3}, \frac{5\pi}{3}$ **5.** 60°, 120°, 240°, 300° **7.** 40°, 160° **9.** $\frac{\pi}{18}, \frac{31\pi}{18}$

In Exercises 11–19, k represents an integer.

11. $90° + k \cdot 360°$ **13.** $\frac{\pi}{4} + k \cdot \frac{\pi}{2}$

15. $60° + k \cdot 180°, 120° + k \cdot 180°$

17. $\frac{\pi}{6} + k \cdot \pi, \frac{5\pi}{6} + k \cdot \pi$ **19.** $35° + k \cdot 180°$

21. 0°, 60°, 180°, 300° **23.** $\frac{3\pi}{8}, \frac{7\pi}{8}, \frac{11\pi}{8}, \frac{15\pi}{8}$

25. $\frac{\pi}{2}, \frac{7\pi}{6}, \frac{3\pi}{2}, \frac{11\pi}{6}$ **27.** 0°, 120°, 240° **29.** $\frac{2\pi}{3}, \frac{4\pi}{3}$ **31.** 54.7°, 125.3°, 234.7°, 305.3° **33.** 0.253, 1.571, 2.889, 4.712 **35.** 111.5°, 248.5°

37. 32.5°, 122.5°, 212.5°, 302.5°

39. 0.087, 0.785, 2.182, 2.880, 4.276, 4.974

41. 30°, 60°, 210°, 240° **43.** 10°, 50°, 130°, 170°, 250°, 290° **45.** 15°, 75°

EXERCISES 8-6, Pages 336–338

1. **a.** 2 **b.** odd **3.** **a.** 2 **b.** neither

5. **a.**

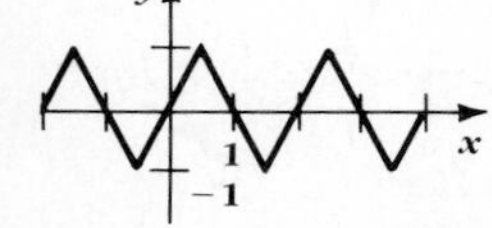

b. odd

7. **a.**

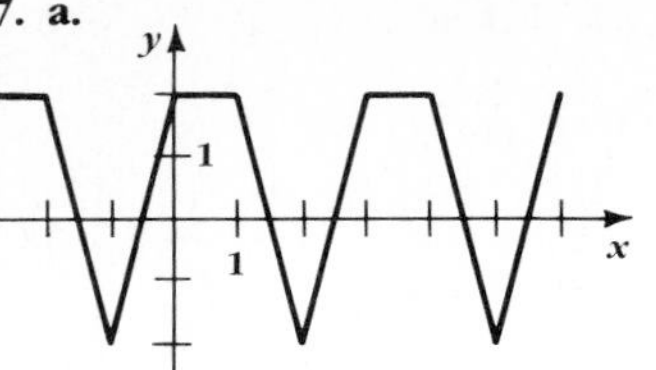

b. neither

9. **a.**

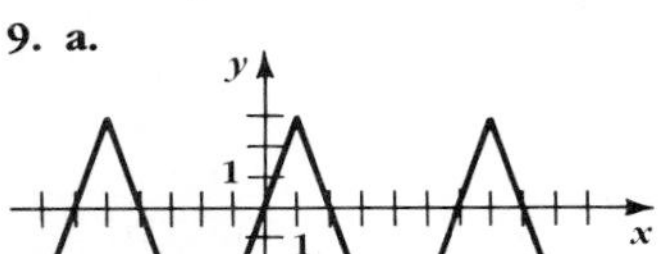
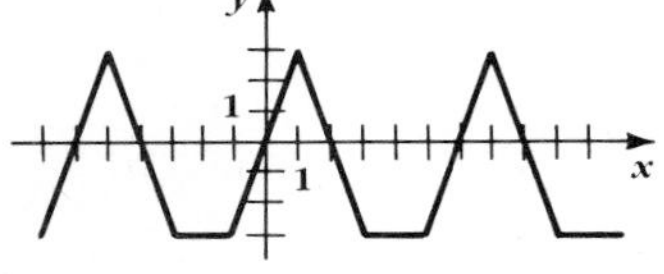

b. neither

11. **a.**

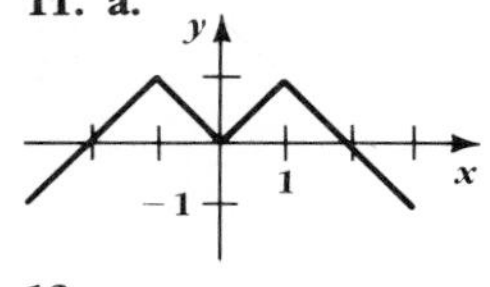

b.

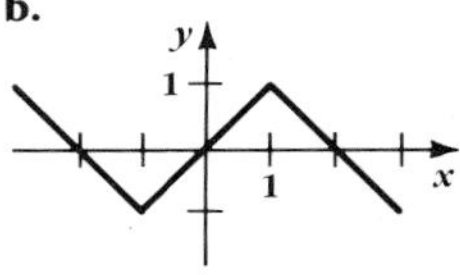

13. **a.**

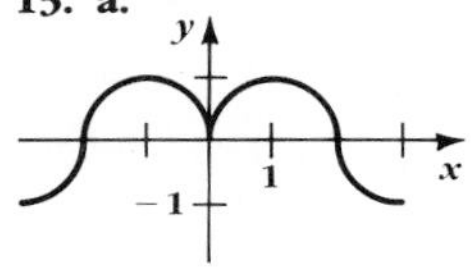

b.

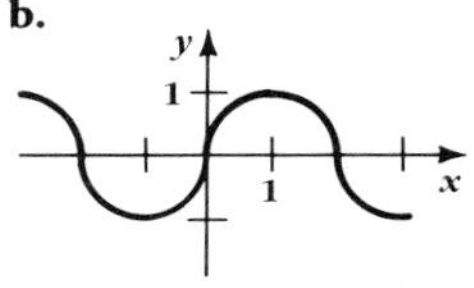

15. **a.**

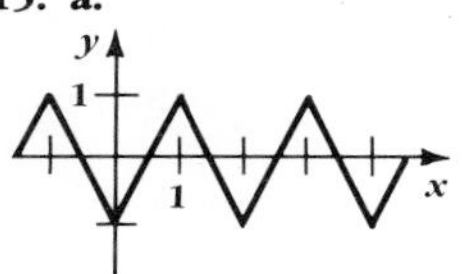

b.

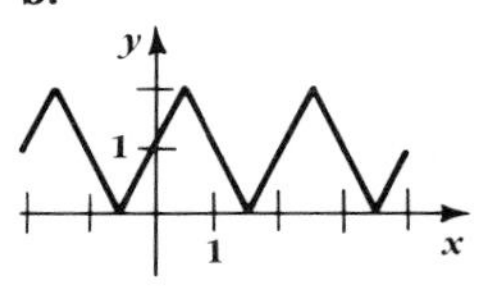

17. **a.**

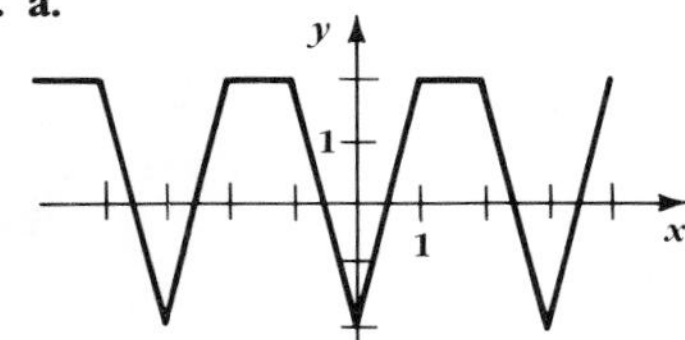

b.

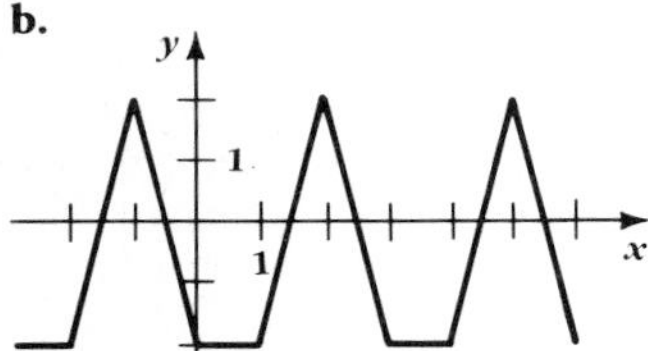

19. **a.**

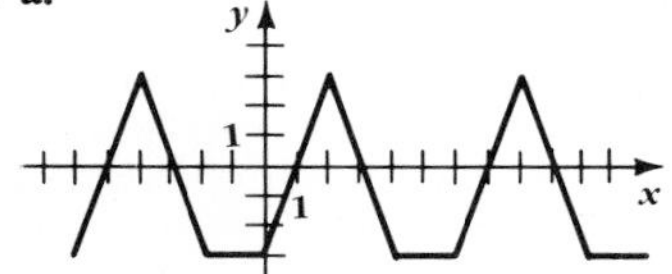

19. b.

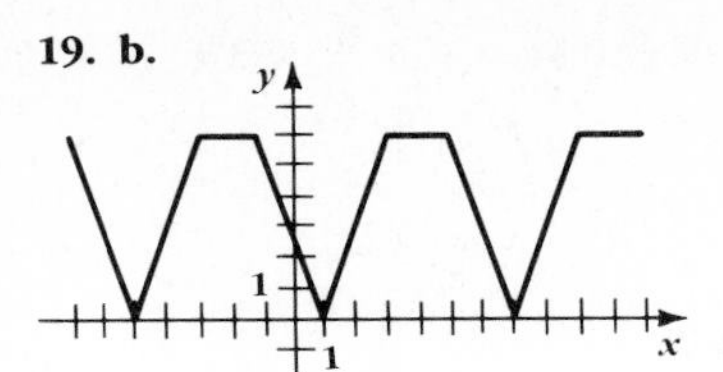

21. even **23.** odd **25.** odd

EXERCISES 8-7, Pages 341–342

1–13. Answers vary. Examples are given. **1.** $g(x) = 3 \sin x$ **3.** $g(x) = \cos 2x$ **5.** $g(x) = 5 \sin 4x$

7. $g(x) = -4 \cos \frac{2\pi}{3}x$ **9.** $g(x) = 20 \sin 20\pi x$

11. $g(x) = 6 \sin 2x$ **13.** $g(x) = 4 \sin 2\pi x$

15.

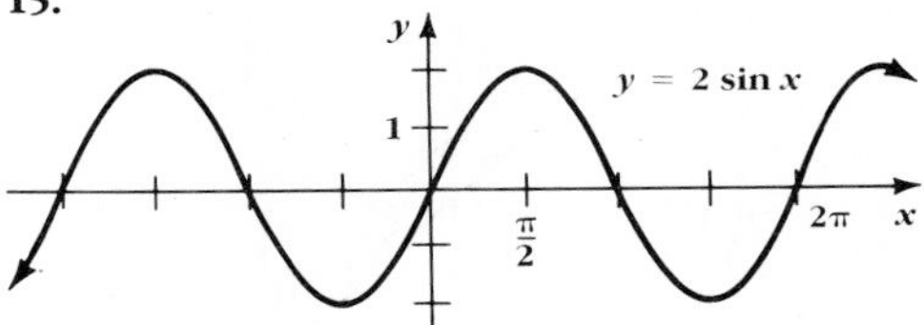

17.

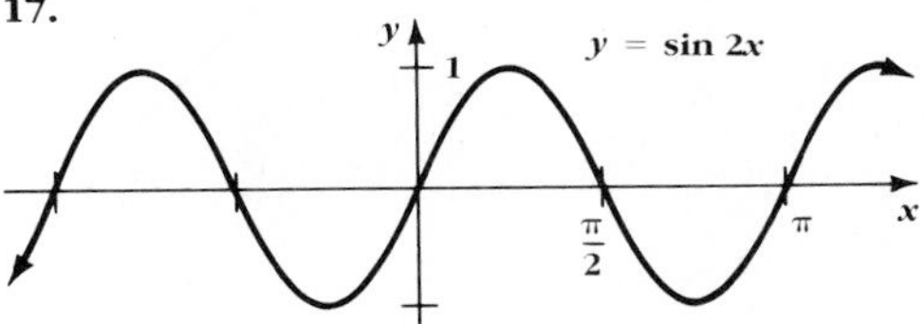

19.

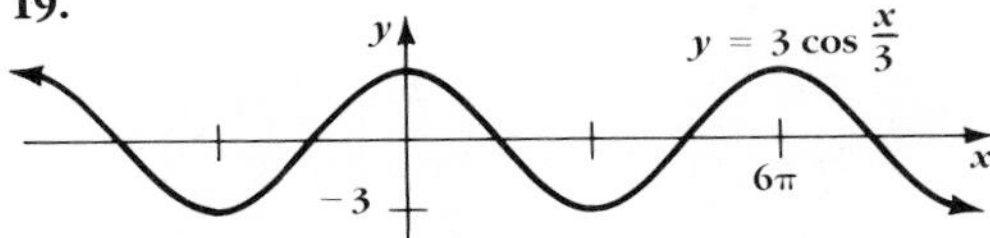

21. **23.**

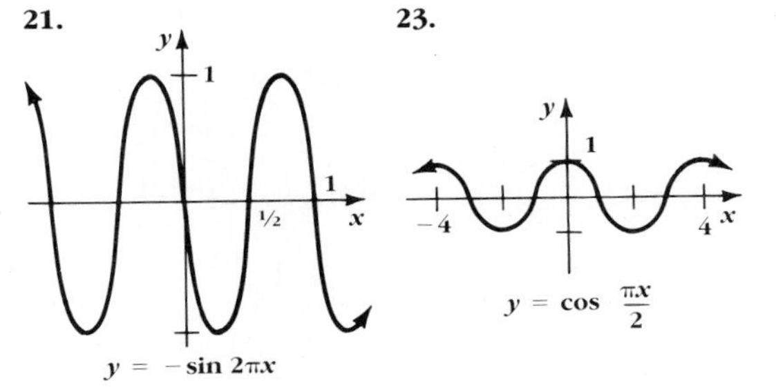

25–29. Answers may vary.

25. $y = 2 \sin x$

27. $y = \frac{3}{2} \cos \frac{x}{2}$ **29.** $y = \sin \pi x$

31.

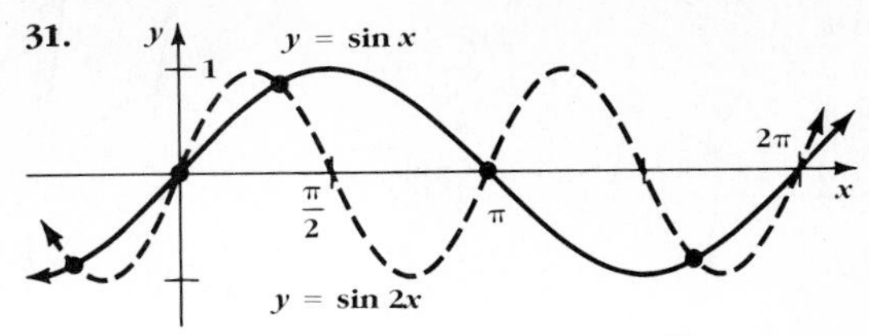

$(k\pi, 0)$, $\left(\frac{\pi}{3} + 2k\pi, \frac{\sqrt{3}}{2}\right)$, $\left(\frac{5\pi}{3} + 2k\pi, -\frac{\sqrt{3}}{2}\right)$ where k is an integer

33.

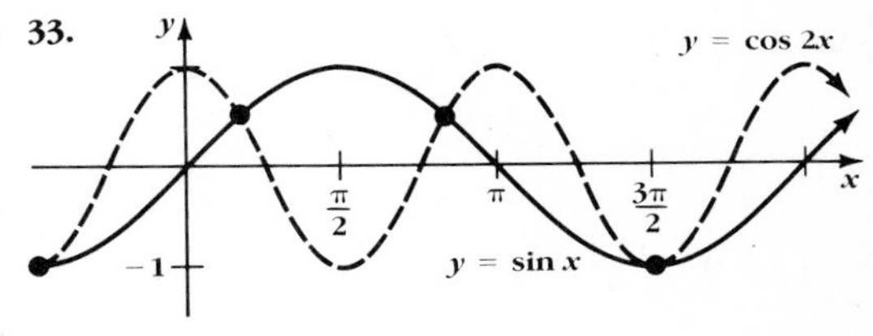

$\left(\frac{\pi}{6} + 2k\pi, \frac{1}{2}\right)$, $\left(\frac{3\pi}{2} + 2k\pi, -1\right)$, $\left(\frac{5\pi}{6} + 2k\pi, \frac{1}{2}\right)$ where k is an integer

EXERCISES 8-8, Pages 346–347

1. leads by $\frac{\pi}{6}$ units **3.** lags by $\frac{\pi}{3}$ units **5.** leads by $\frac{\pi}{4}$ units **7.** leads by 1 unit **9.** lags by 1 unit

11.

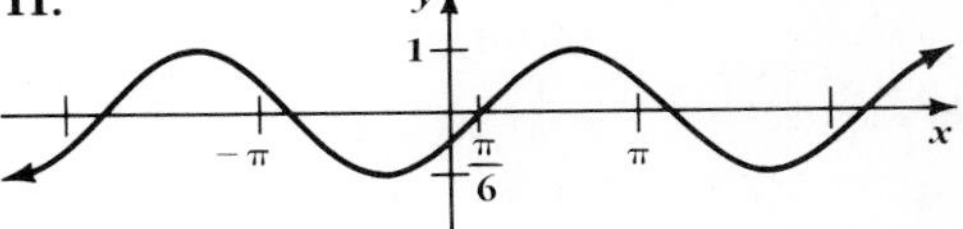

13.

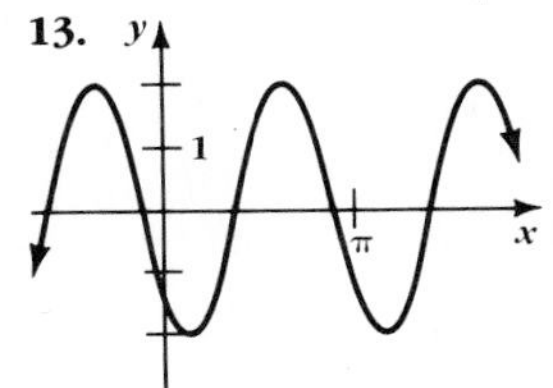

15.

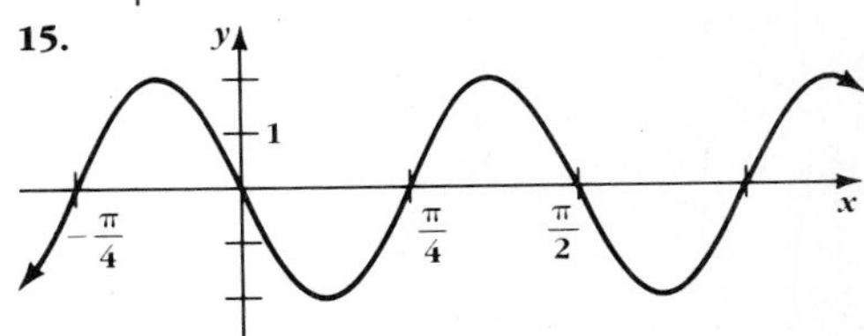

17.

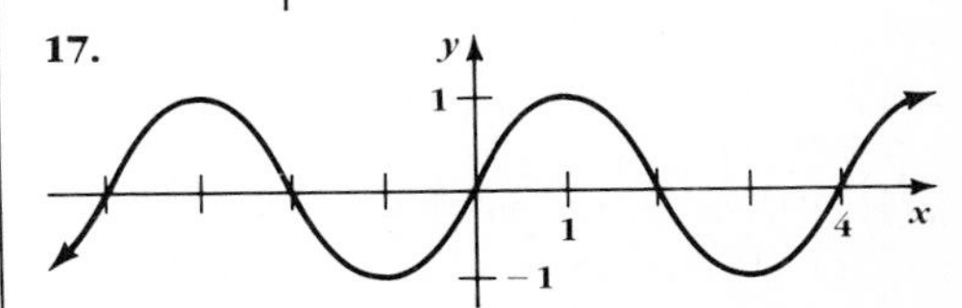

19.

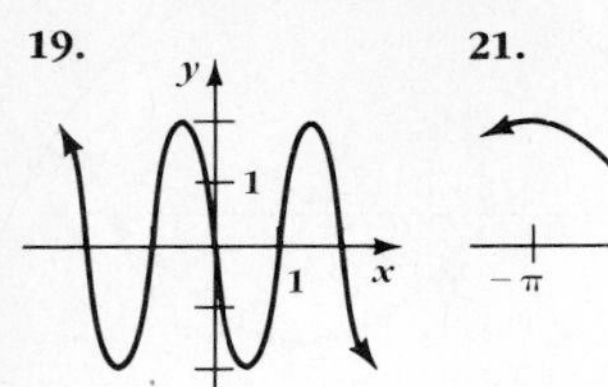

21.

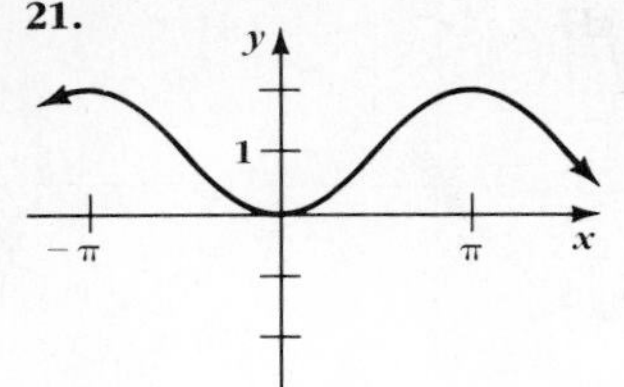

23.

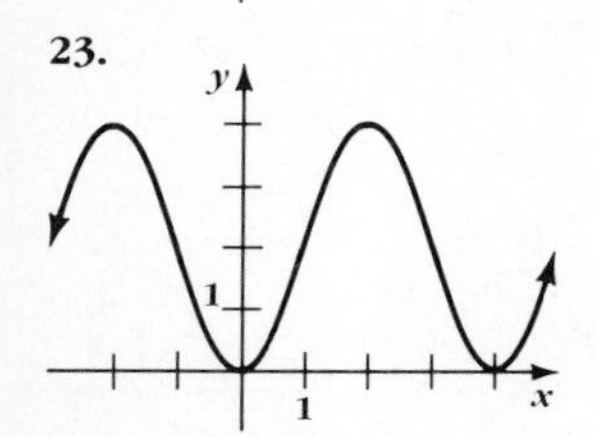

25. 0°, 120°
27. 2.88, 1.83
29. 69.2°, 327.7°
31. 1.85, 3.52

33.

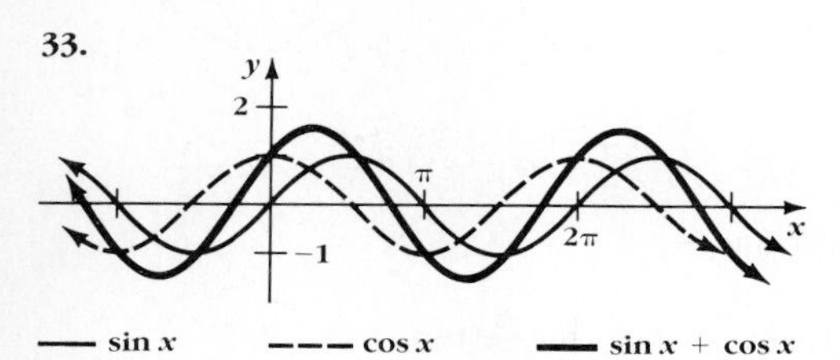

35.

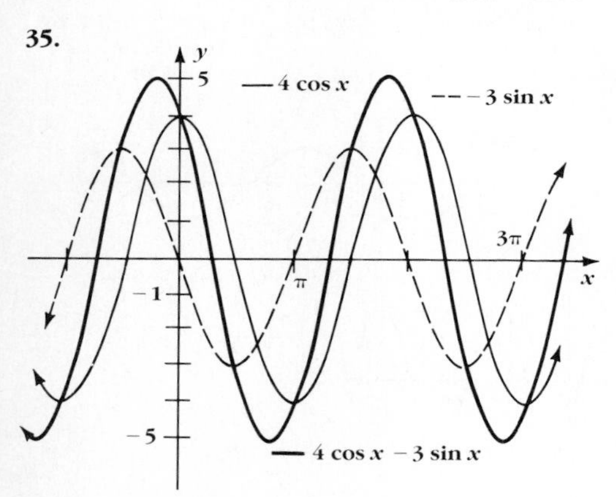

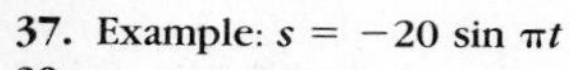

37. Example: $s = -20 \sin \pi t$

39.

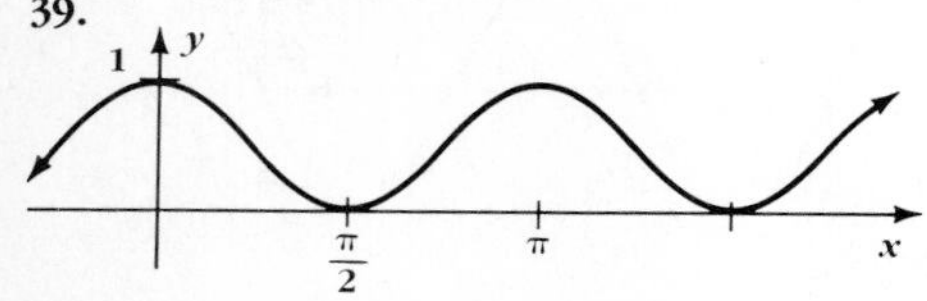

41. Answers may vary. For example, If s = the distance in miles from Earth's center and t = the number of minutes elapsed since the golf ball was dropped, then $s = 4000 \sin\left(\frac{2\pi t}{85} + \frac{\pi}{2}\right)$.

EXERCISES 8-9, Pages 349–350

1. 2π 3. 2π 5. 1 7. 4π 9. π

11.

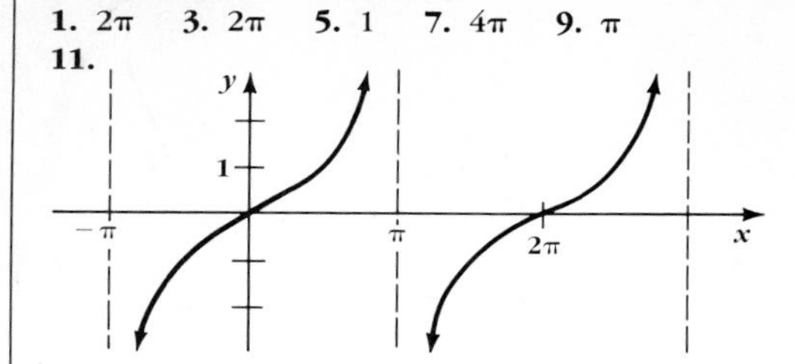

13. 15.

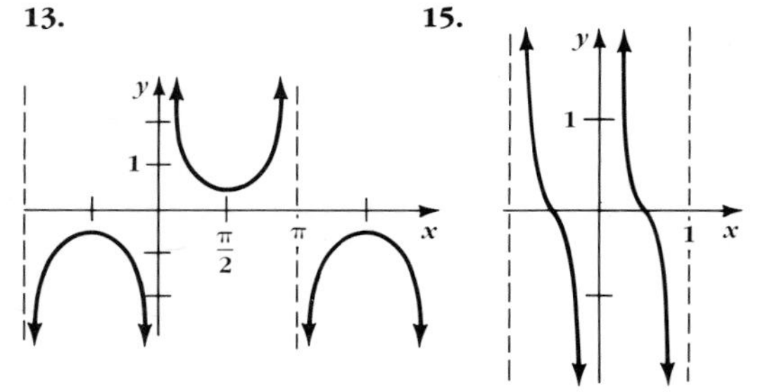

17. 19.

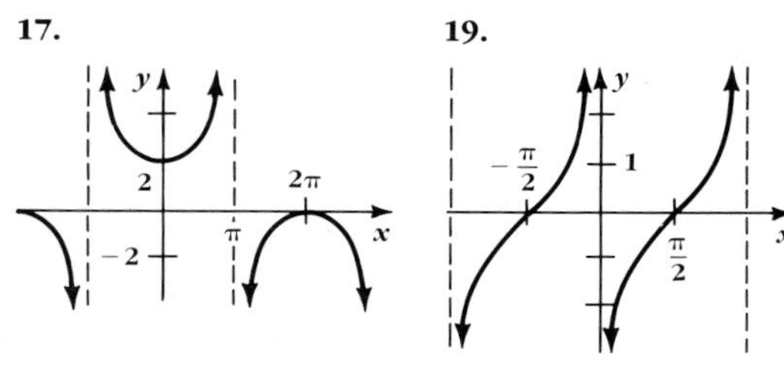

21. 23.

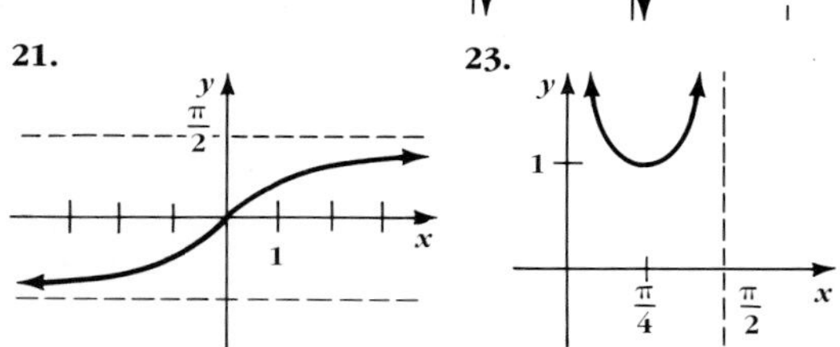

The two graphs are identical.

EXERCISES 8-10, Pages 354–355

1. $\frac{\pi}{3}$ 3. $\frac{\pi}{4}$ 5. π 7. $\frac{3\pi}{4}$ 9. $\frac{\pi}{3}$ 11. $\frac{3\pi}{4}$

13. $\frac{\pi}{3}$ 15. $\frac{4}{5}$ 17. $\frac{\sqrt{5}}{2}$ 19. $\frac{24}{25}$ 21. $-\frac{4}{3}$

23. -1 25. $-\frac{33}{65}$ 27. $\frac{2 + 2\sqrt{10}}{\sqrt{5} - 4\sqrt{2}} = -\frac{2}{3}(\sqrt{5} + \sqrt{2})$ 29. $\frac{\sqrt{2}}{2}$ 31. $\sqrt{1-x^2}$

33. x 35. $2x^2 - 1$ 37. $\frac{2x}{x^2-1}$ 39. $\frac{2-x^2}{x^2}$

41. $\frac{v-u}{uv+1}$ 43. $u\sqrt{1-v^2} + v\sqrt{1-u^2}$

45.

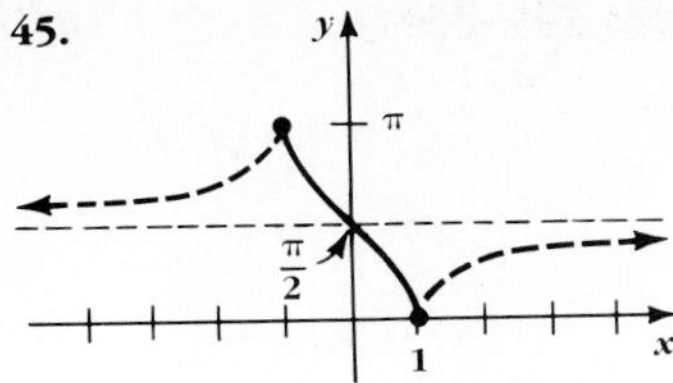

EXERCISES 9-1, Pages 364–365

1. $(1, \sqrt{3})$

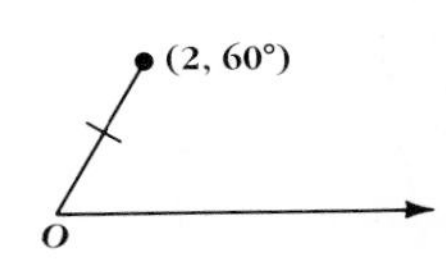

3. $(2\sqrt{2}, -2\sqrt{2})$

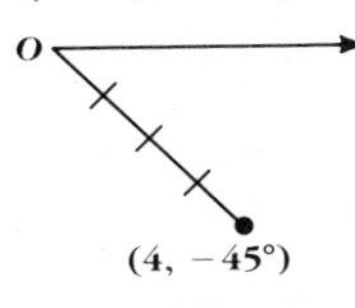

5. $(-2, 2\sqrt{3})$

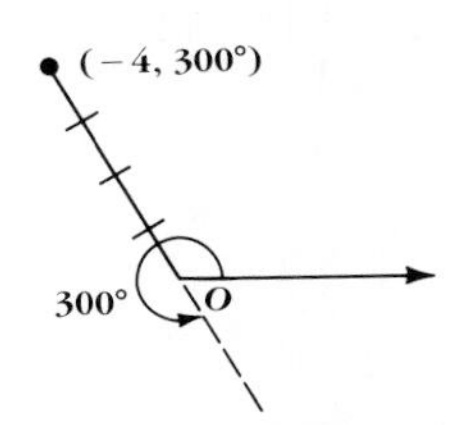

7. $(-3, 3\sqrt{3})$

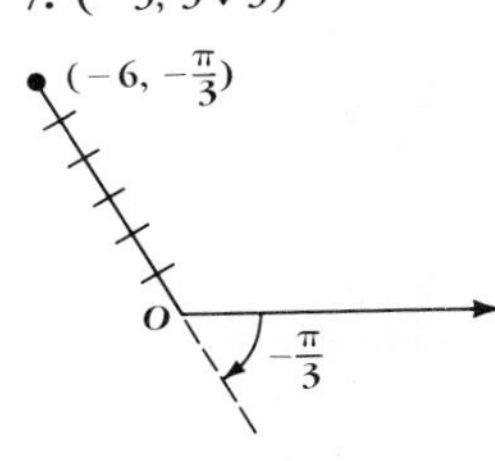

9–15. Answers may vary. Examples are given.
9. (2, 60°) 11. (5, π) 13. (2, 135°)
15. $\left(2\sqrt{2}, \frac{\pi}{3}\right)$ 17. $r \sin \theta = 3$ 19. $\theta = \frac{\pi}{4}$
21. $\theta = \frac{3\pi}{4}$ 23. $r = 1$

25.

27.

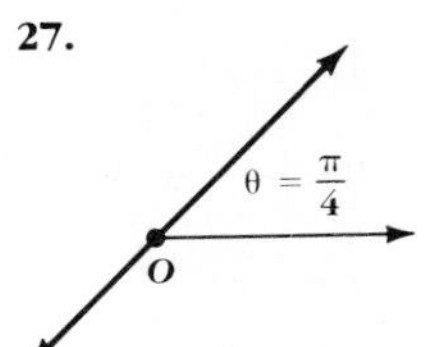

29.

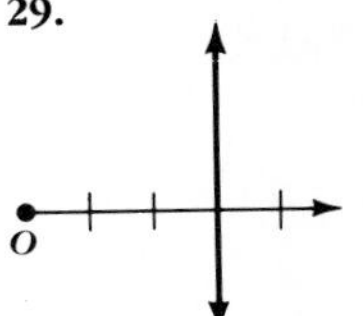

31.

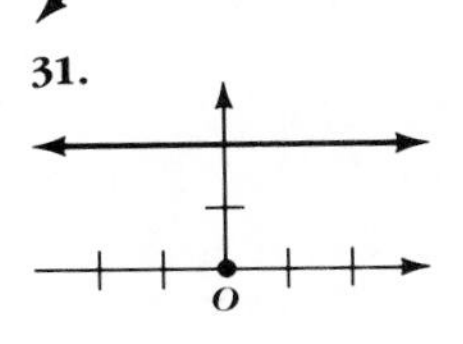

33. $x^2 + (y - 1)^2 = 1$

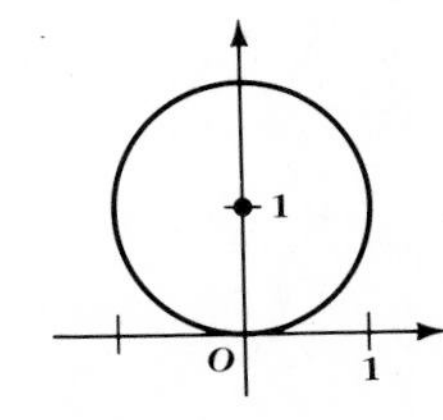

35. $(x - 3)^2 + (y - 4)^2 = 25$

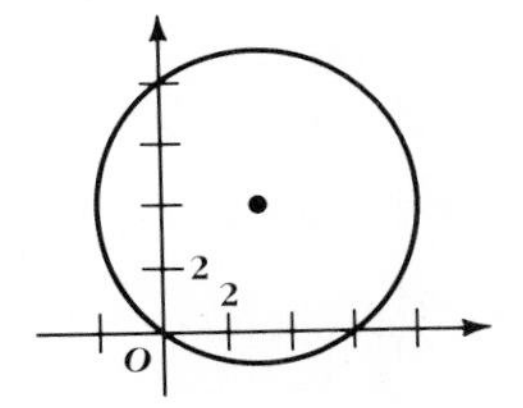

EXERCISES 9-2, Page 368

1.

3.

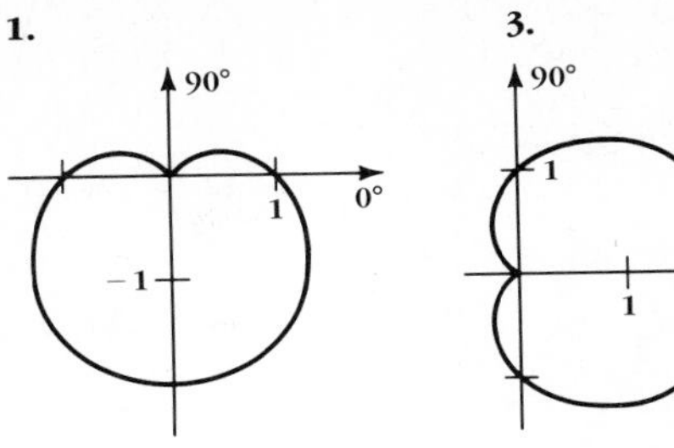

5.

7.

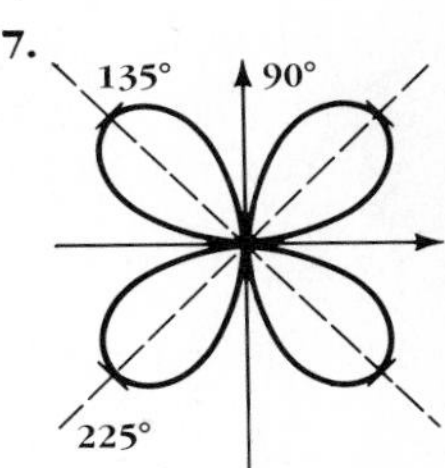

9.

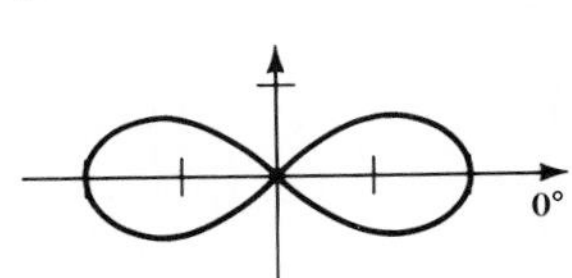

11.

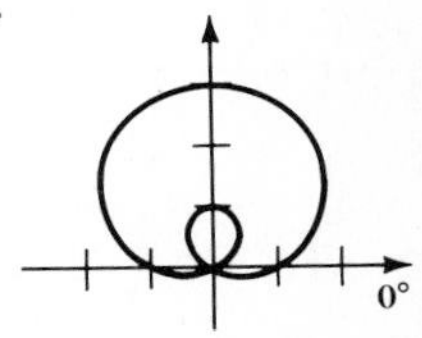

13.

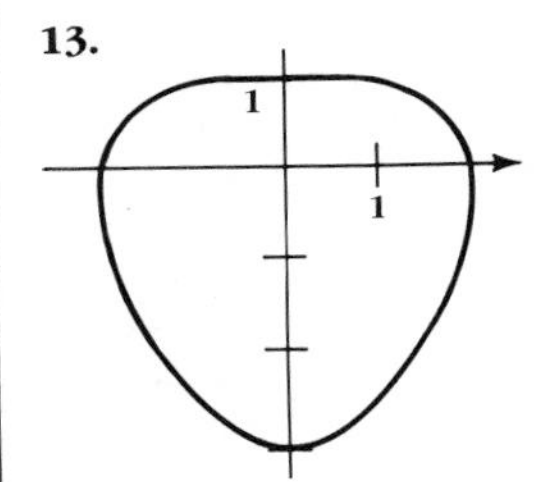

15.

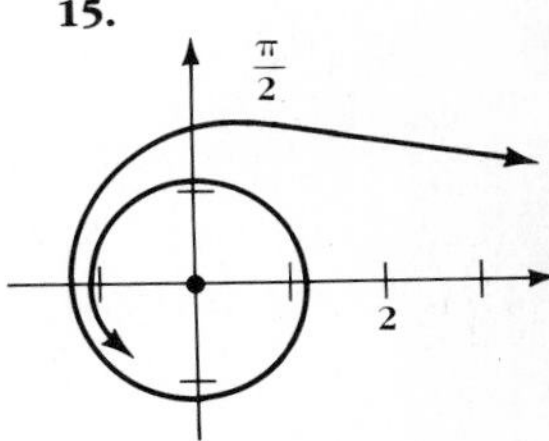

17.

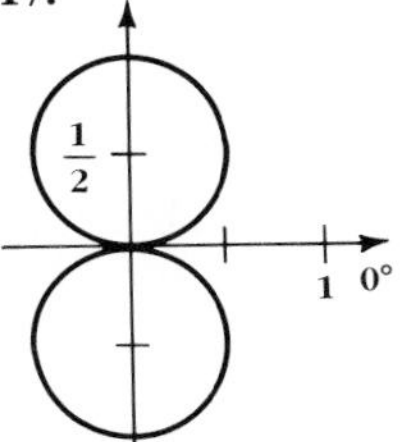

19.

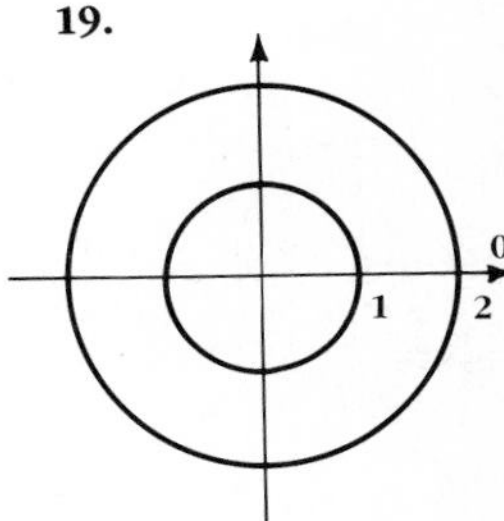

21.

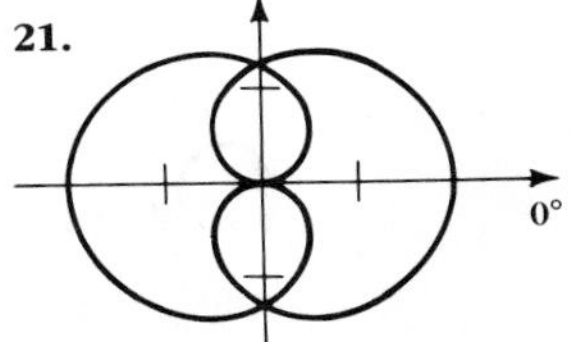

EXERCISES 9-3, Pages 370–371

1. 5 **3.** $-3\sqrt{2} + 3i\sqrt{2}$ **5.** $-5 - 5i\sqrt{3}$
7. $-0.35 + 1.97i$ **9.** $4\sqrt{2}(\cos 315° + i \sin 315°)$
11. $4(\cos 180° + i \sin 180°)$
13. $2(\cos 330° + i \sin 330°)$
15. a. $18(\cos 270° + i \sin 270°)$ **b.** $-18i$
c. $w = -3 + 3i\sqrt{3}; z = -\frac{3}{2}\sqrt{3} + \frac{3}{2}i$ **d.** $-18i$
17. a. $8(\cos 90° + i \sin 90°)$ **b.** $8i$ **c.** $w = -2\sqrt{2} + 2i\sqrt{2}; z = \sqrt{2} - i\sqrt{2}$ **d.** $8i$
19. a. $2(\cos 330° + i \sin 330°)$
b. $\sqrt{3} - i$ **c.** $w = -3 + 3i\sqrt{3}; z = -\frac{3}{2}\sqrt{3} + \frac{3}{2}i$
d. $\sqrt{3} - i$ **21. a.** $2(\cos 180° + i \sin 180°)$
b. -2 **c.** $w = -2\sqrt{2} + 2i\sqrt{2}; z = \sqrt{2} - i\sqrt{2}$
d. -2
27.

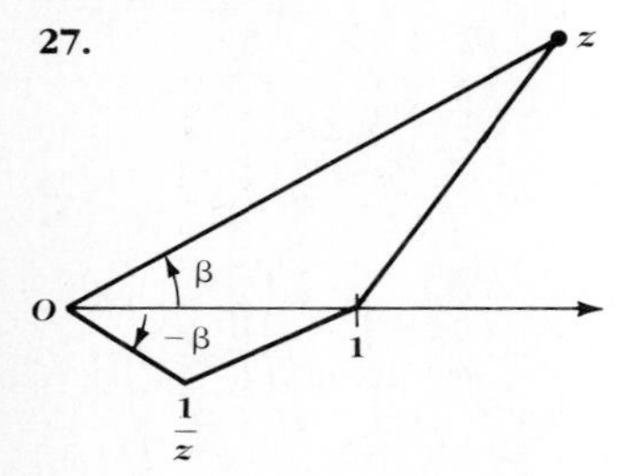

EXERCISES 9-4, Pages 374–375

1. 16 **3.** $8i$ **5.** $-64\sqrt{3} + 64i$ **7.** 512 **9.** $1, i, -1, -i$ **11.** $1, \frac{1}{2} + \frac{\sqrt{3}}{2}i, -\frac{1}{2} + \frac{\sqrt{3}}{2}i, -1, -\frac{1}{2} - \frac{\sqrt{3}}{2}i, \frac{1}{2} - \frac{\sqrt{3}}{2}i$ **13.** $\cos 0° + i \sin 0°$; $\cos 72° + i \sin 72°$, $\cos 144° + i \sin 144°$, $\cos 216° + i \sin 216°$, $\cos 288° + i \sin 288°$ **15.** $-1, \frac{1}{2} - \frac{\sqrt{3}}{2}i, \frac{1}{2} + \frac{\sqrt{3}}{2}i$

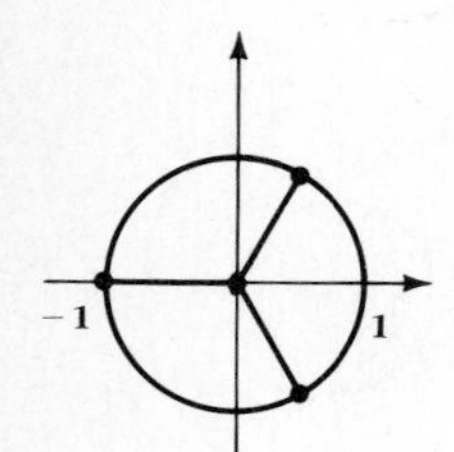

Exercise 15

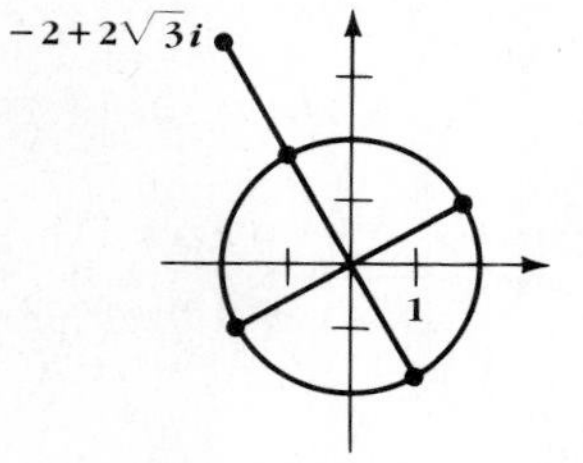

Exercise 17

17. $\frac{\sqrt{6}}{2} + \frac{\sqrt{2}}{2}i, -\frac{\sqrt{2}}{2} + \frac{\sqrt{6}}{2}i, -\frac{\sqrt{6}}{2} - \frac{\sqrt{2}}{2}i, \frac{\sqrt{2}}{2} - \frac{\sqrt{6}}{2}i$
19. $\omega^7 = (\omega^3)^5$ **21.** $\omega^2 = (\omega^5)^4$
23. $\cos 3\theta = \cos\theta(4\cos^2\theta - 3)$; $\sin 3\theta = \sin\theta(3 - 4\sin^2\theta)$ **29.** $k = 1 + 8n, 3 + 8n, 5 + 8n, 7 + 8n, n \in \{0, 1, 2, \ldots\}$

EXERCISES 9-5, Pages 378–380

1. $y = -x + 2$

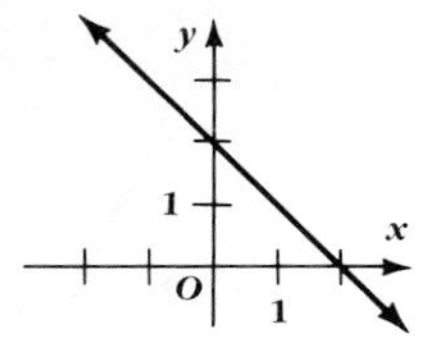

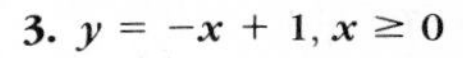

3. $y = -x + 1, x \geq 0$

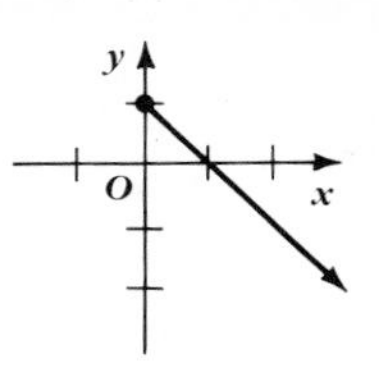

5. $y = -x + 2;$ $x \geq 0, y \geq 0$

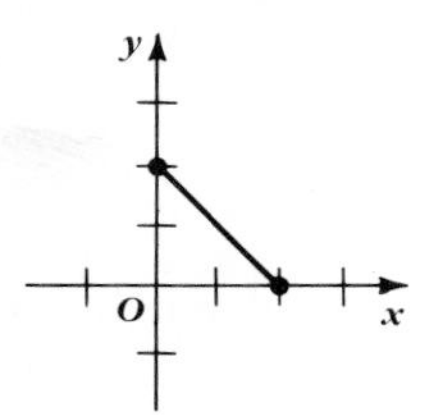

7. $x^2 + y^2 = 4$

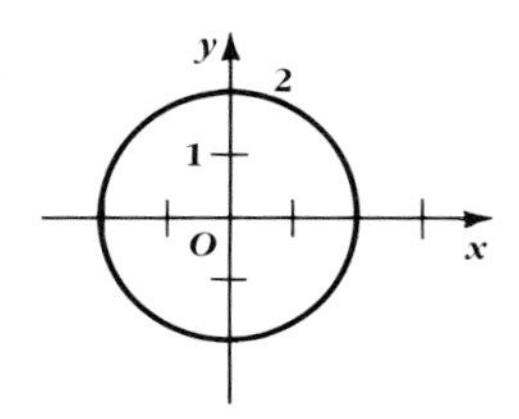

9. $y^2 - x^2 = 1;$ $x \geq 0; y \geq 1$

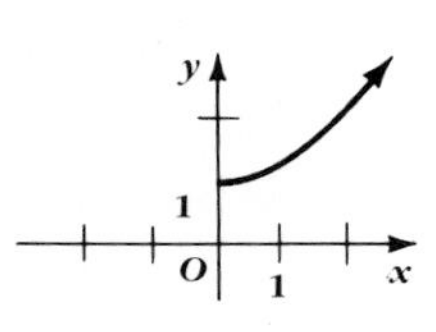

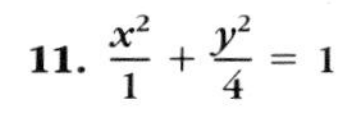

11. $\frac{x^2}{1} + \frac{y^2}{4} = 1$

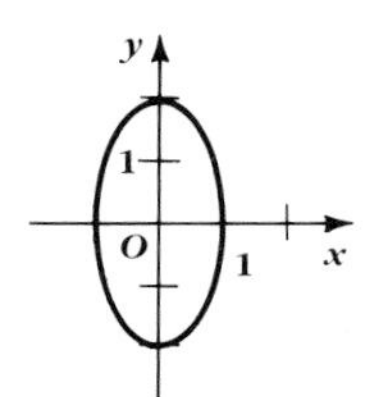

13. $y^2 - x^2 = 1$

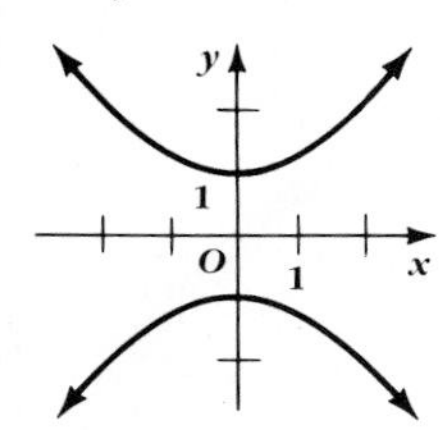

15. $(x - 1)^2 + (y - 1)^2 = 1$

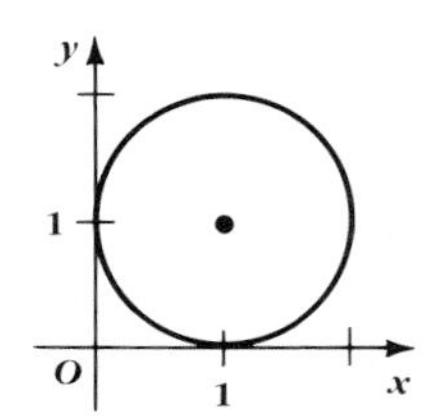

17. $xy = 1; -1 \leq x < 0$ or $0 < x \leq 1$

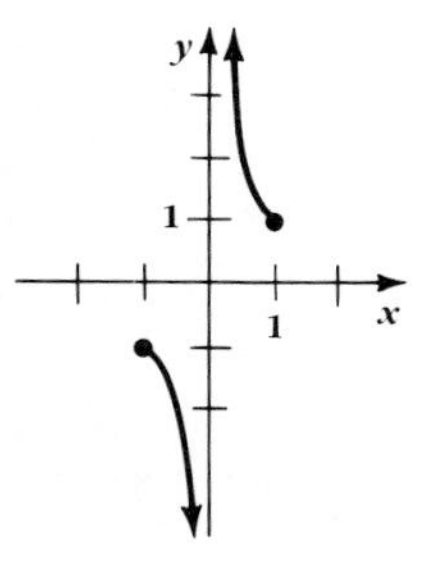

19. $x^2 + (y - 1)^2 = 1$

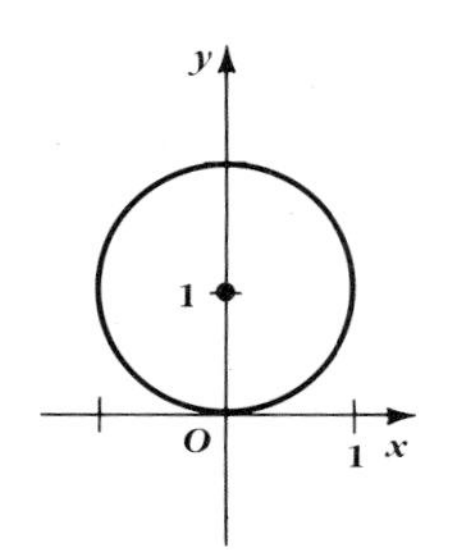

21. $x = 1 + r\cos\theta$; $y = 1 + r\sin\theta$
23. $x = \cos\theta(\cos^2\theta - \sin^2\theta)$;
$y = \sin\theta(\cos^2\theta - \sin^2\theta)$
25. $r = \theta$; spiral **27.** See right.
29.

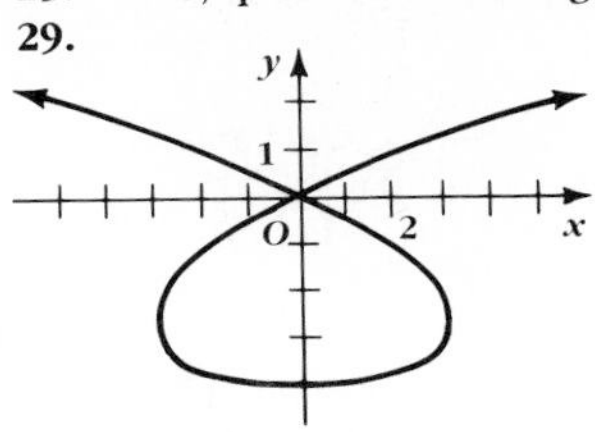

Exercise 27

31. a. $x = a\cos\theta$; $y = b\sin\theta$
b. $\frac{x^2}{a^2} + \frac{y^2}{b^2} = 1$; ellipse

EXERCISES 9-6, Page 383

1. $\frac{1}{2}$ **3.** 2 **5.** 1

EXERCISES 9-7, Page 387

5. $x + 2y = \sqrt{3} + \frac{\pi}{6}$ **7.** $y = \frac{3}{2}$ **9.** $3\sec 3x\tan 3x$
11. $a\cos at$ **13.** $-\frac{\pi}{180}\sin x^\circ$ **15.** $2x\sec^2 x^2$
17. $2\tan x\sec^2 x$ **19.** $\cos 2\theta$ **21.** $t\sin t$
23. $x^2\cos x$ **25.** $4x(1 - \cos 2x)$, or $8x\sin^2 x$
27. $2\tan x\sec^2 x$ **29.** $\frac{3}{2}$; -3 **31.** $\frac{3}{2}\sqrt{3}$; $-\frac{3}{2}\sqrt{3}$
35. a. $f'(x) = -\sin x$; $f''(x) = -\cos x$; $f'''(x) = \sin x$; $f^{(4)}(x) = \cos x$ **b.** $f^{(n)}(0) = 0$ if n is an odd integer, $f^{(n)}(0) = 1$ if $\frac{1}{2}n$ is an even integer, and $f^{(n)}(0) = -1$ if $\frac{1}{2}n$ is an odd integer

EXERCISES 9-8, Pages 391–393

1. a. 2 radians/min **b.** 0.4 radians/min **3.** $\theta = 60^\circ$
5. $3\pi\sqrt{3}$ ft/s **7.** $\theta = 35.26^\circ$ **9.** $\theta = 45^\circ$
11. π miles/s **13.** 6 ft **15.** 23.44 ft

EXERCISES 10-1, Pages 402–403

1, 3. Answers may vary slightly. **1.** 2; 2.63902; 2.65737; 2.66475 **3.** 4; 1.14870; 1.01396; 1.00139
5. $3^{2\sqrt{2}}$ **7.** $2^{2\sqrt{2}}$ **9.** $x^{3\sqrt[3]{3}}$ **11.** $x^{\sqrt{2}-\sqrt{3}}$
13. $x^{2\sqrt{2}} - y^{2\sqrt{2}}$ **15.** $\frac{x^5}{y^{10}}$ **17.** 8 **19.** 17
21. $\pm\sqrt{33}$ **23.** $\frac{41}{8}, \frac{39}{8}$

EXERCISES 10-2, Pages 407–408

1.

3.

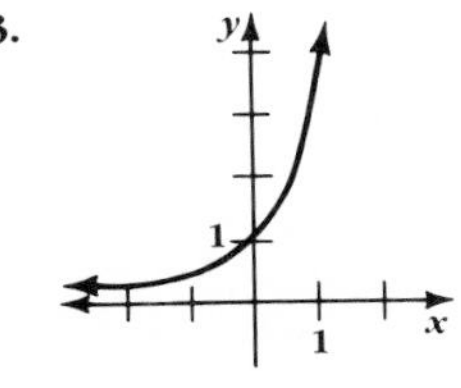

5.

7.

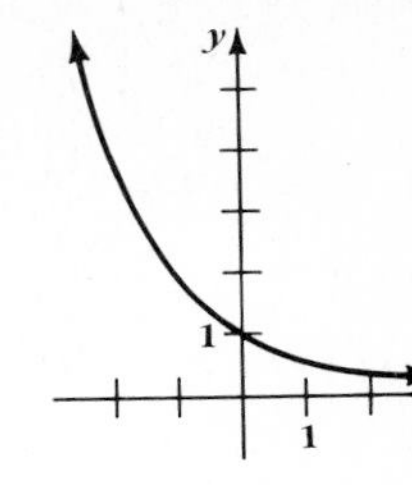

9.

11. $\pm\sqrt{2}$ **13.** $\frac{3}{2}$
15. $\frac{1}{2}$ **17.** $2^{\frac{\sqrt{3}}{3}}$
19. 1 **21.** $\frac{1}{4}$ **23.** 27
25. $2^{\sqrt{2}-1}$ **27.** 2, 3
29. 2 **31.** \$1435.25

33. \$1200.00 **35.** 0.07
37. 14,773 bacteria
39. 17,831 bacteria
41. See right.

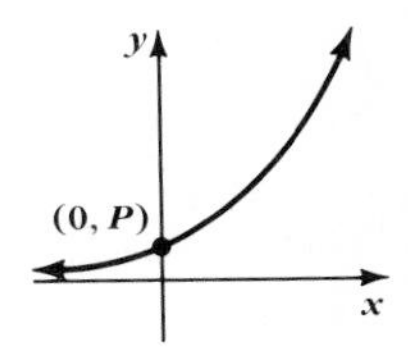

Exercise 41

EXERCISES 10-3, Pages 411–412

1. $2e$ **3.** e^3 **5.** e^3 **7.** e
9.

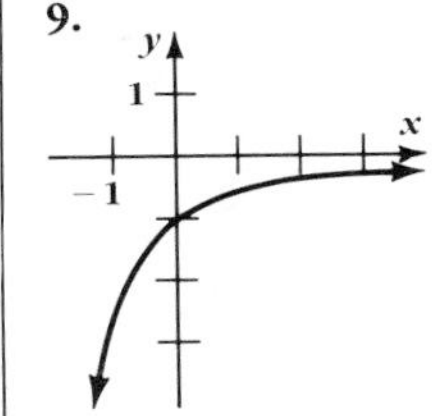

11.

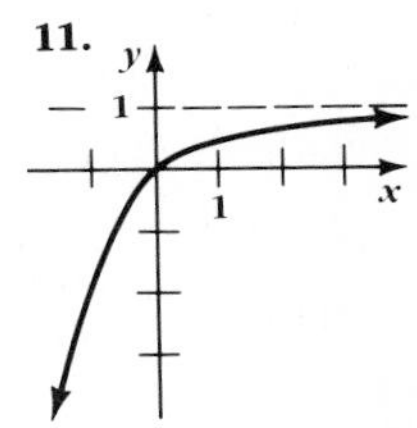

13. \$1377.13 **15.** \$235.18 **17. a.** 1.71828 **b.** 1.05171 **c.** 1.0050 **d.** 1.0005 **e.** 1.00005

EXERCISES 10-4, Pages 415–416

1. 4 **3.** 1 **5.** $\frac{1}{2}$ **7.** $\sqrt{3}$ **9.** -1 **11.** -5
13.

15.

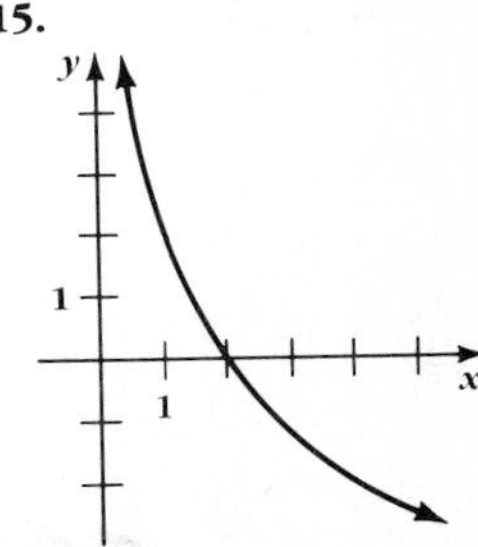

17.

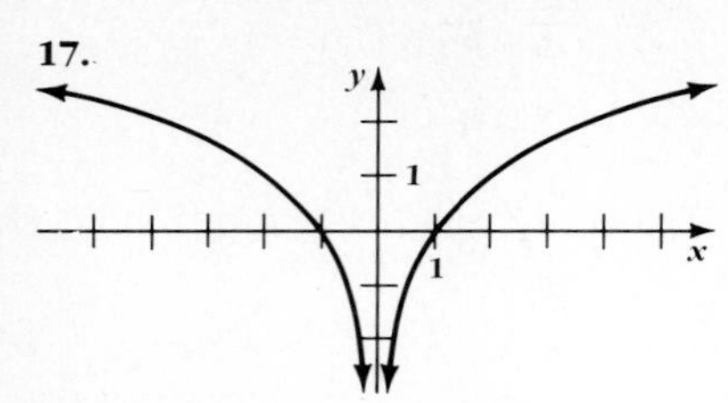

19. $\frac{1}{4}$ or 8   21. 16 or 1024   23. $\sqrt{2}$ 25. $\frac{1}{81}$
27. 2^{-32} 29. $f^{-1}(x) = 2^{x-1}$
31.

a.

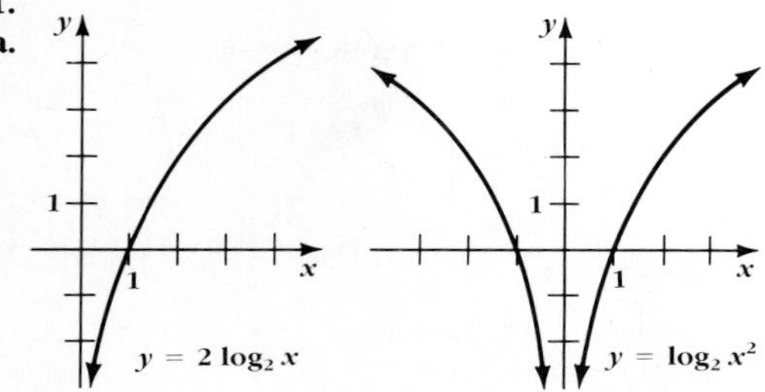

b. no; true only for $x > 0$ 33. 10^{12} times I_0
35. 199,530,000 times I_0 37. 7   39. yes; pH is 3.

EXERCISES 10-5, Pages 420–421

1. log 1080   3. $\log \frac{\sqrt{3}}{2}$ 5. $\log_5 t^{\frac{3}{2}}$ 7. $\log I - \log I_0$
9. $t + \log_2 N$ 11. 64   13. 25   15. 6.9103
17. 69.1035   19. 1967.8   21. 4.1734   23. 1.4833
25. e^2 or e^3 27. 1.7712   29. 1.7925   31. 4.1779
35. Answers may vary. Examples are given.   a. Let a = 10 and c = 1.   b. Let $a = c = 2$.   37. a. $\log_{\frac{1}{b}} x = -\log_b x$ 41. 13.87 years   43. 6.93 years
45. 230 months   47. increasing

EXERCISES 10-6, Pages 425–427

1. $-5e^{-5x}$ 3. $\frac{1}{x}$ 5. $e^x(x^5 + 5x^4)$

7. $\frac{e^x(x^2 - 2x - 1)}{(x^2 - 1)^2}$ 9. $\frac{x(2 \ln x - 1)}{(\ln x)^2}$

11. $\frac{4x + 3}{2x^2 + 3x + 1}$ 13. $\frac{4e^{2x}}{(e^{2x} + 1)^2}$

15. $\frac{(x - 1)\ln(x - 1) - (x + 1)\ln(x + 1)}{(x^2 - 1)(\ln(x - 1))^2}$

17. $\frac{10x^4 + 4}{2x^5 + 4x + 1}$ 19. $\frac{1}{2x\sqrt{\ln x}}$ 21. $\frac{-2}{x(\ln x)^3}$

23. $\frac{e^x(x + 1)}{2\sqrt{xe^x}}$ 25. $\frac{4}{e^2}$ 27. $\frac{e}{2}$ 29. $3e$ 31. $x - 2y = 1$ 33. $y - e = 3e(x - 1)$ 35. $x - y = 0$

37. $\cot x$ 39. $e^{-x}(\cos x - \sin x)$ 41. $\frac{1}{x} \cdot \cos x - \ln x \cdot \sin x$ 45. $\frac{x(1 + 2 \ln x)}{\ln 3}$

47. $\frac{10^x \ln 10}{\ln x}\left[\ln 10 - \frac{1}{x \ln x}\right]$

49. $(2x - 3) \cdot \ln 10 \cdot 10^{x^2 - 3x}$ 51. no relative maxima, minima, or points of inflection

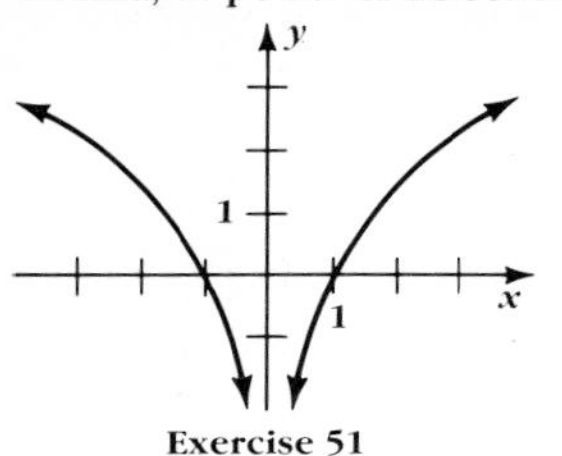

Exercise 51

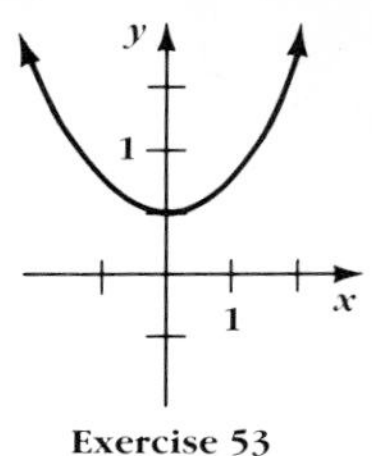

Exercise 53

53. a relative minimum at 0

59. $8x^{4 \ln x - 1} \ln x$

EXERCISES 10-7, Pages 430–431

1. \$3617.57   3. 2.37 years   5. 1.87 years
7. 0.019   9. 3.1623×10^{10} times I_0 11. 0 decibels to 120 decibels   13. no   15. 3261 trees; 4906 trees
17. 53.7 years   19. 63.4% decrease   21. 38 years
23. $k = \ln 2$ 25. $k = -0.2$ 27. $2xe^{x^2} \neq ke^{x^2}$
29. a. no
b.

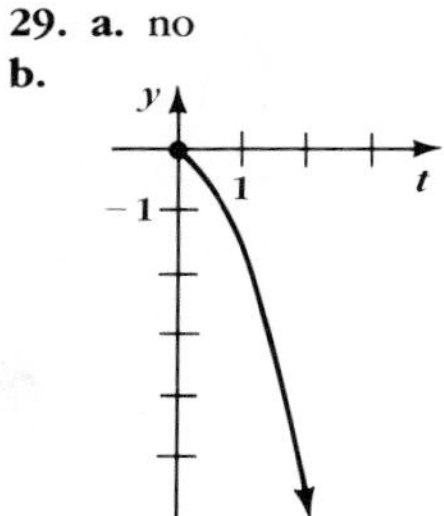

31.

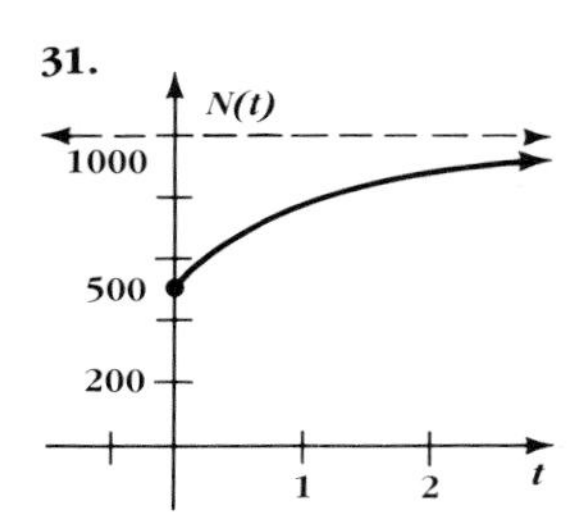

33.

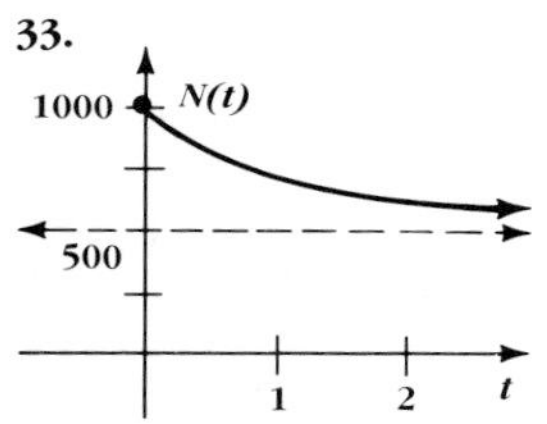

CUMULATIVE REVIEW, Pages 438–439

1. $\sin x = \cos x = -\frac{\sqrt{2}}{2}$; $\tan x = \cot x = 1$; $\csc x = \sec x = -\sqrt{2}$ 3. a. $\frac{7\pi}{3}$ radians   b. $\left(\frac{720}{\pi}\right)^\circ \approx 229.2^\circ$

5. $\sin \theta = -\frac{2\sqrt{5}}{5}$; $\cos \theta = \frac{\sqrt{5}}{5}$; $\tan \theta = -2$; $\cot \theta = -\frac{1}{2}$; $\csc \theta = -\frac{\sqrt{5}}{2}$; $\sec \theta = \sqrt{5}$ 7. 2982 miles

9. 6.04 square units   11. $\frac{1 - 2\cos^2 \alpha}{\cos^2 \alpha(1 - \cos^2 \alpha)}$

13. $-\frac{1}{2}$ 15. 60°, 150°, 240°, 330°

17.

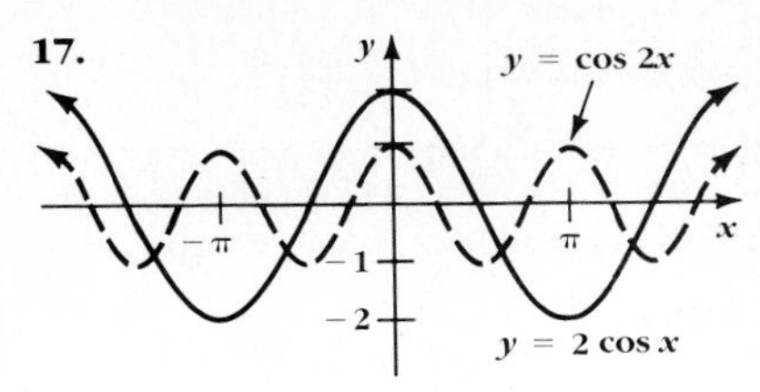

19.

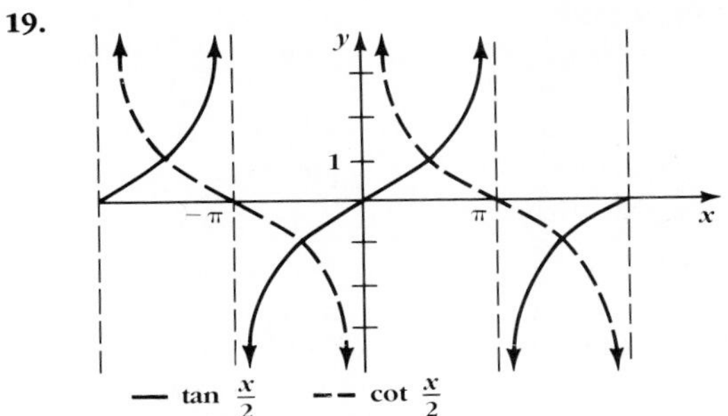

21. $\frac{2x}{x^2 + 1}$ 23. $x + y = 4$

25. $4(\cos 240° + i \sin 240°)$
27. **a.** $2(\cos 180° + i \sin 180°)$
b. -2
29. the ray $y = 1 - 2x, x \geq 0$; see right 31. $\cos 2x$

33. 18.4 degrees/minute

35. \$4529 37. **a.** $\frac{1}{4}$ **b.** 3

39. **a.** 8.75 years **b.** 8.66 years

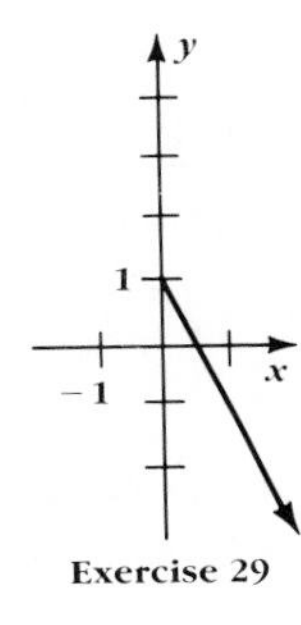

Exercise 29

EXERCISES 11-1, Pages 445–446

1. $(-2, 4)$; $(9, 1)$ 3. $(-1, 6)$; $(-9, 2)$ 5. $(1, -5)$
7. $(3, 0)$ 9. $(3, 3)$

11.

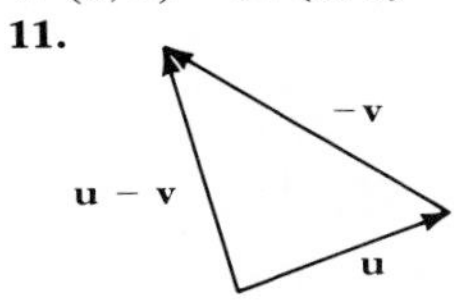

13.

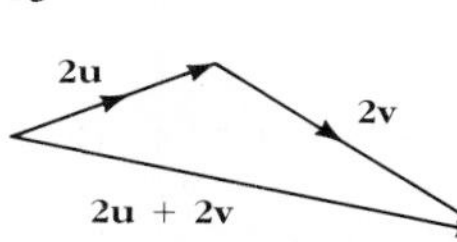

15.

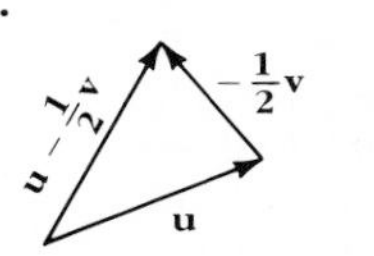

17.

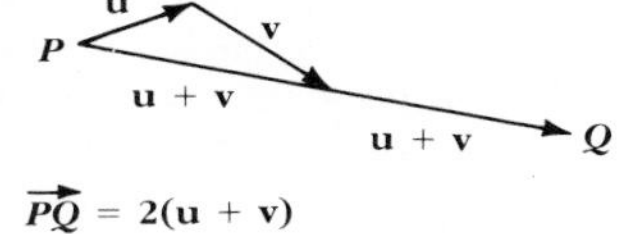

$\overrightarrow{PQ} = 2(\mathbf{u} + \mathbf{v})$

19. $(2, 8)$ 21. $(-12, 9)$
23. $(10, 8)$ 25. $(-5, -4)$
27. $(58, 62)$ 29. $(2, 7)$
31. $(-43, -33)$ 33. $(-58, -28)$

35. $\overrightarrow{AC} = 2(\mathbf{u} + \mathbf{v}) = \overrightarrow{AB} + \overrightarrow{BC} = 2\mathbf{u} + 2\mathbf{v}$

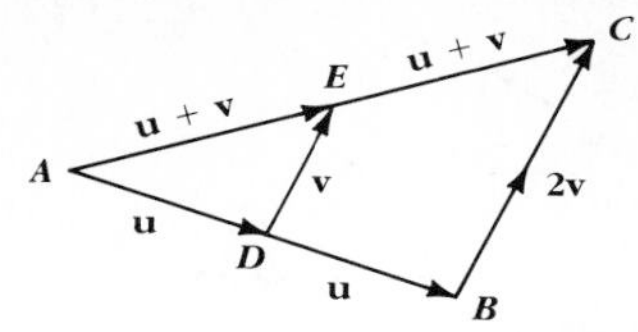

37. $\mathbf{0}$ 39. $\overrightarrow{AB} + \overrightarrow{AD}$, or $\overrightarrow{AD} + \overrightarrow{AB}$ 41. $\frac{1}{2}\overrightarrow{AB} + \overrightarrow{AD}$
43. $a = 2, b = -3$ 45. $a = b = 0$

EXERCISES 11-2, Pages 449–450

1.

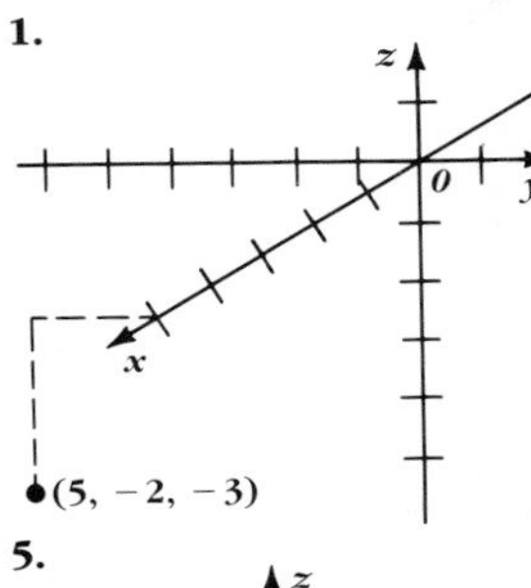

3.

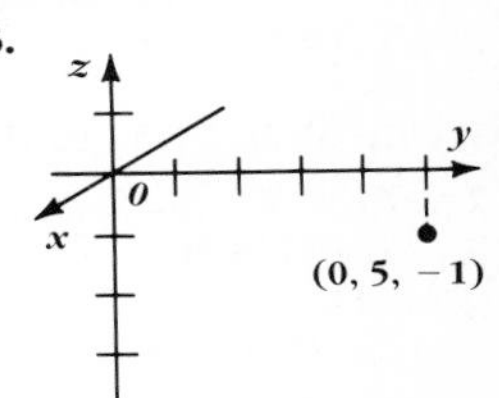

5.

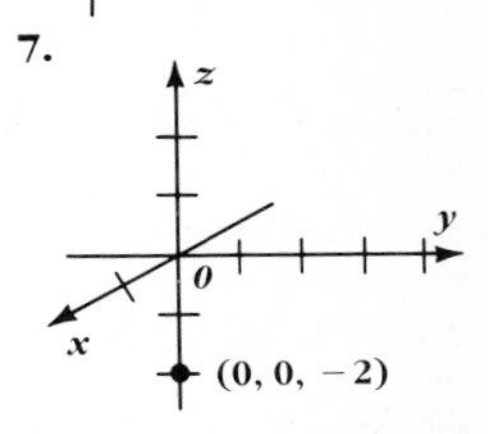

7.

(0, 0, −2)

9.

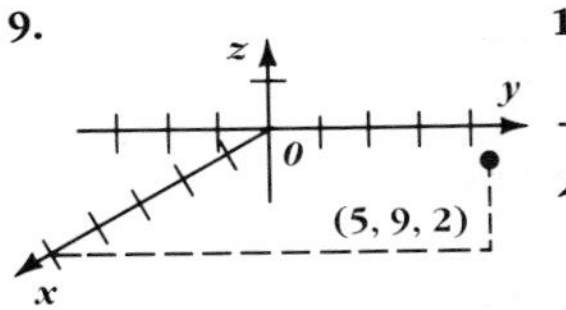

11.

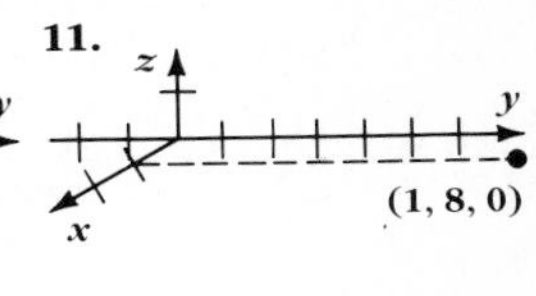

13.

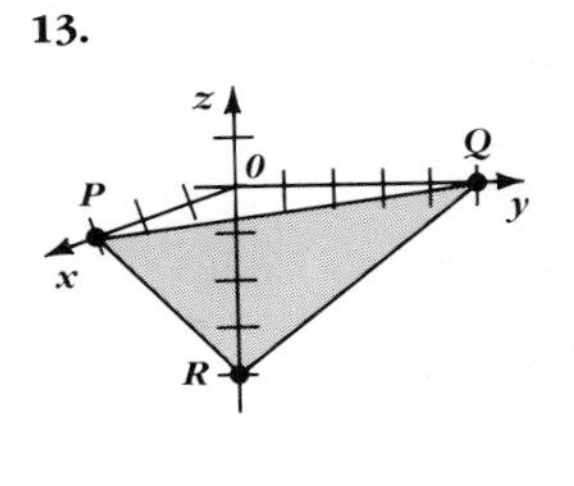

15.

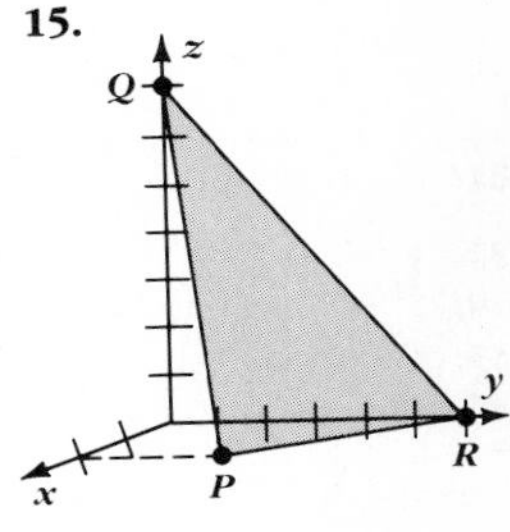

17. 6; $(4, -2, 2)$ 19. $\sqrt{101}$; $\left(2, -2, \frac{15}{2}\right)$

21. $\sqrt{362}$; $\left(7, -\frac{1}{2}, \frac{5}{2}\right)$ 23. $(-3, -6, 8)$

25. $(12, 3, 3)$ 27. $(22, -10, 2)$ 29. $(2, 14, 2)$
31. scalene 33. equilateral 35. a right triangle
39. $(0, 5, 0)$ and $(0, 9, 0)$

EXERCISES 11-3, Pages 452–453

1. 13 **3.** $\sqrt{29}$ **5.** 15 **7.** $5\sqrt{2}$ **9.** $\frac{1}{2}\sqrt{5713}$ **11.** 1 **13.** $15\sqrt{6}$ **15.** $18\sqrt{5}$ **17.** $\|2\mathbf{u}\| = \|(6, 10, -16)\| = 14\sqrt{2}$; $2\|\mathbf{u}\| = 2\cdot 7\sqrt{2} = 14\sqrt{2}$ **19.** $\left(\frac{5}{13}, \frac{12}{13}\right)$

EXERCISES 11-4, Pages 456–457

1. 8; not orthogonal **3.** -144; not orthogonal **5.** 0; orthogonal **7.** -40; not orthogonal **9.** 0; orthogonal **11.** 0; orthogonal **13.** 2; not orthogonal **15.** 44° **17.** 90° **19.** $|-36| \le 25\cdot 13$ **21.** $|11| \le 3\cdot 7$ **23. a.** -14 **b.** $\frac{32}{7}$ **25. a.** $-\frac{5}{2}$ **b.** 10 **27. a.** 0 **b.** ± 6

EXERCISES 11-5, Pages 460–461

1. $-5\mathbf{i} + 3\mathbf{j}$ **3.** $2\mathbf{i} + \mathbf{j}$ **5.** $5\mathbf{i} - 11\mathbf{j}$ **7.** $-\mathbf{i} + 0\mathbf{j} + 5\mathbf{k}$ **9.** $8\mathbf{i} + 9\mathbf{j} + 0\mathbf{k}$ **11.** $a\mathbf{i} + b\mathbf{j} + 0\mathbf{k}$ **13.** $(3, 4, -1)$ **15.** $(13, 0, -6)$ **19.** $\mathbf{t} = -5\mathbf{u} + 2\mathbf{v}$ **21.** $\mathbf{t} = 8\mathbf{u} + 0\mathbf{v}$ **23.** $\mathbf{t} = 2\mathbf{u} + \mathbf{v}$ **25.** $-\frac{11}{13}\mathbf{u}, \frac{3}{26}\mathbf{v}$ **27.** $\frac{-3a + b}{a^2 + b^2}\mathbf{u}, \frac{a + 3b}{a^2 + b^2}\mathbf{v}$ **29.** $5\mathbf{u}, \frac{1}{4}\mathbf{v}$ **31.** $\frac{7}{6}\mathbf{u}, \frac{1}{2}\mathbf{v}$

33–37. Answers may vary. Examples are given.

33. Let $\mathbf{v} = (3, 1)$; $\mathbf{t} = -\frac{13}{20}\mathbf{u} + \frac{9}{10}\mathbf{v}$

35. Let $\mathbf{v} = (0, 1)$; $\mathbf{t} = -\mathbf{u} - 3\mathbf{v}$

37. Let $\mathbf{v} = (1, -1)$; $\mathbf{t} = \frac{1}{10}\mathbf{u} + \frac{7}{2}\mathbf{v}$

EXERCISES 11-6, Pages 464–466

1. (1.0419, 5.9088) **3.** (−2.7500, 4.7631) **5.** (−8, 0) **7.** (7.3861, 1.3024) **9.** $x = 4 + t$, $y = 7 + 3t$ **11.** $x = -3 + 3t, y = -3 + 3t$ **13.** $x = -6t, y = 8t$ **15.** $\frac{35\sqrt{2}}{2}\mathbf{i} + \frac{35\sqrt{2}}{2}\mathbf{j}$ **17.** $32.7\sqrt{3}\mathbf{i} - 32.7\mathbf{j}$

19. $\mathbf{t} = 16.38\left(\frac{\sqrt{2}}{2}, \frac{\sqrt{2}}{2}\right) + 11.47\left(-\frac{\sqrt{2}}{2}, \frac{\sqrt{2}}{2}\right)$

21. $\mathbf{t} = 0.5075\left(-\frac{1}{2}, \frac{\sqrt{3}}{2}\right) + 0.8616\left(\frac{\sqrt{3}}{2}, \frac{1}{2}\right)$ **23.** 869 J **25.** 153 J **27.** 26.7 J **29.** $x = -3 + 7t, y = 6 + 10t$ **31.** example: $x = 4t$ and $y = 6 + t$ **33.** 378 mph; 54° **35.** 3.49 knots; 118° **37.** 823,200 J

EXERCISES 11-7, Pages 468–469

1. $\mathbf{w} + \mathbf{x}$ **3.** $\frac{1}{2}(\mathbf{x} - \mathbf{w})$ **5.** $\mathbf{w} - \mathbf{x}$ **7.** $3(\mathbf{u} + \mathbf{v})$ **9.** $\mathbf{u} + 3\mathbf{v}$

EXERCISES 12-1, Pages 482–484

1. (1, 1) **3.** (−21, −21) **5.** $\left(\frac{\sqrt{2}}{2}, -\frac{\sqrt{2}}{2}\right)$ **7.** (−2, −2) **9.** (0, 0) **11.** $\left(\frac{47}{37}, \frac{36}{37}\right)$ **13.** $(3x - 2y, 4x + 2y)$ **15.** $(-x, y)$ **29.** (−3, −5) **31.** (−18, 8) **33.** rotation through 45° **35.** scalar multiplication by −3 **37.** rotation through −90°

EXERCISES 12-2, Pages 488–489

1. $\begin{bmatrix} 10 & -8 \\ -3 & 5 \end{bmatrix}$ **3.** $\begin{bmatrix} -10 & -4 \\ 33 & -19 \end{bmatrix}$ **5.** $\begin{bmatrix} 88 & -47 \\ 39 & -12 \end{bmatrix}$ **7.** $\begin{bmatrix} 15 & 24 \\ -11 & -14 \end{bmatrix}$

9. $\begin{bmatrix} 204 & -42 & 13 \\ -74 & 12 & 46 \\ -40 & 6 & 32 \end{bmatrix}$

11. $\begin{bmatrix} -40 & 24 & -10 \\ 16 & -4 & -26 \\ 16 & 6 & -22 \end{bmatrix}$

13. The difference isn't defined because EG and GE do not have the same dimensions.

15. $\begin{bmatrix} -2 & 62 \\ -15 & -16 \end{bmatrix}$ **17.** no **19. a.** $\begin{bmatrix} 29 & -14 \\ -98 & 64 \end{bmatrix}$ **b.** $\begin{bmatrix} 61 & -47 \\ 53 & 32 \end{bmatrix}$ **c.** no; since matrix multiplication is not usually commutative, $A^2 + AB + BA + B^2$ is not usually equal to $A^2 + 2AB + B^2$. **23.** $\begin{bmatrix} \frac{3}{2} & 0 & 1 \\ \frac{1}{2} & -\frac{11}{2} & 3 \end{bmatrix}$

31. any matrix of the form $\begin{bmatrix} 4x & -3x \\ 4y & -3y \end{bmatrix}$, for example, $\begin{bmatrix} 4 & -3 \\ 8 & -6 \end{bmatrix}$

EXERCISES 12-3, Pages 492–493

1. a. (−1, 1); (0, 1) **b.** $\begin{bmatrix} -1 & 0 \\ 1 & 1 \end{bmatrix}$ **3. a.** (−3, 3); (2, −2) **b.** $\begin{bmatrix} -3 & 2 \\ 3 & -2 \end{bmatrix}$ **5.** $\begin{bmatrix} -2 & 4 & -11 \\ 3 & 0 & 7 \\ 1 & 1 & -1 \end{bmatrix}$

7. $\begin{bmatrix} 2 & -1 & 1 \\ 0 & 3 & -4 \\ 5 & 1 & 0 \end{bmatrix}$ **9.** $\begin{bmatrix} 0 & -5 & 0 \\ 3 & 9 & 0 \end{bmatrix}$ **11.** domain, three-space; range, two-space; $T((x, y, z)) = (2x - y + 3z, -x + 2y)$ **13.** domain, one-space; range, three-space; $T(x) = (2x, 3x, 4x)$ **15.** $(x, -y)$; reflection of the plane in the x-axis **17.** $(-y, x)$; rotation of the plane through 90° **19.** $B \circ C$ is the identity transformation, mapping each vector of the plane into itself. **23. b.** $\begin{bmatrix} \frac{1}{2} & \frac{5}{2} \\ -\frac{3}{2} & -\frac{11}{2} \end{bmatrix}$ **c.** $\left(\frac{1}{2}, -\frac{3}{2}\right); \left(\frac{5}{2}, -\frac{11}{2}\right)$

EXERCISES 12-4, Page 497

1. $\frac{(x'-1)^2}{4} - \frac{(y'+1)^2}{4} = 1$; a hyperbola with center $(1, -1)$ in the $x'y'$ system

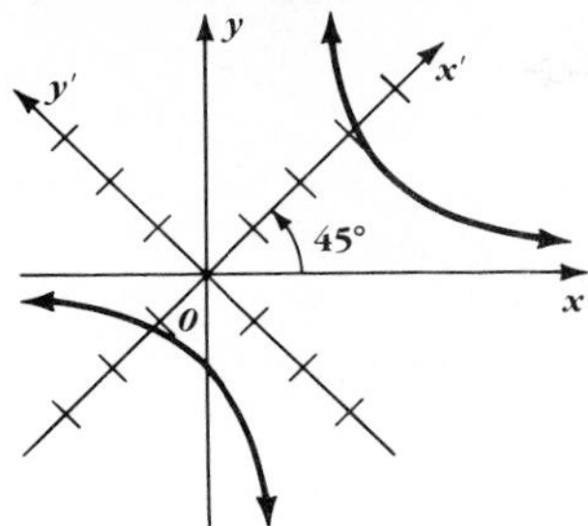

3. $\frac{(x')^2}{\frac{16}{3}} + \frac{(y')^2}{16} = 1$; an ellipse with center $(0, 0)$

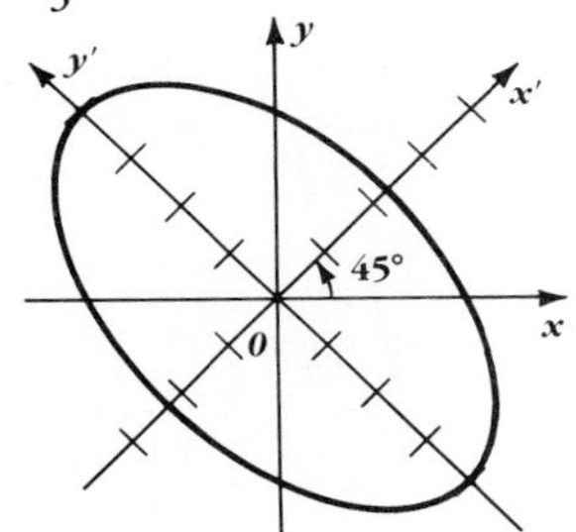

5. $\frac{(x')^2}{4} + \frac{(y')^2}{8} = 1$; an ellipse with center $(0, 0)$

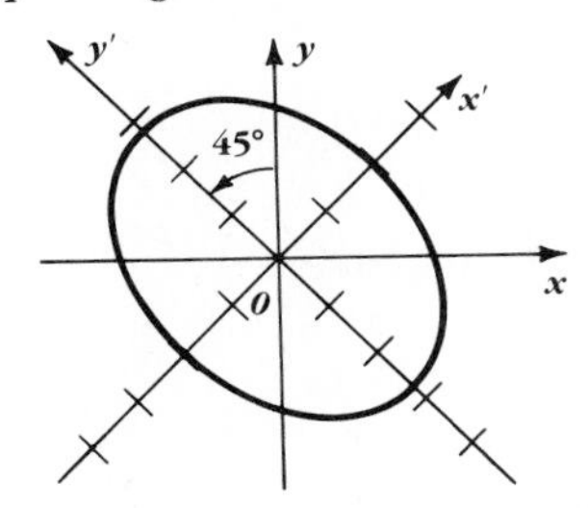

7. $(x'-2)^2 = y' + 1$; a parabola with vertex at $(2, -1)$ in the $x'y'$ system

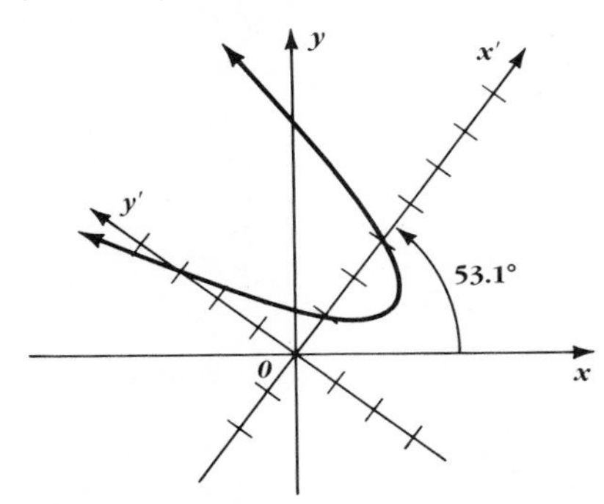

9. $\frac{(x')^2}{3} - \frac{(y')^2}{5} = 1$; a hyperbola with center $(0, 0)$

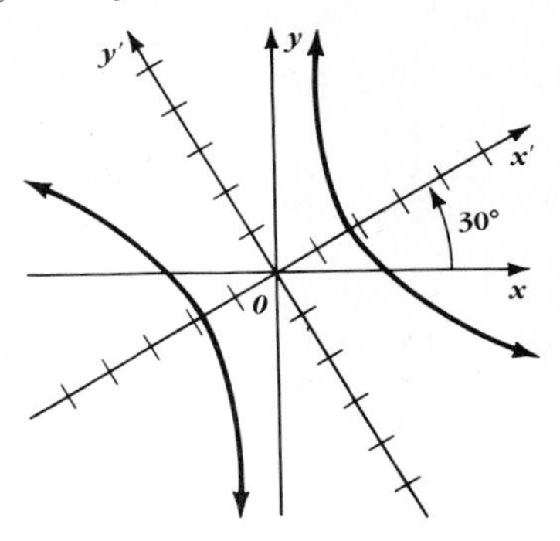

11. $\frac{(y'+2)^2}{4} - \frac{(x'+1)^2}{1} = 1$; a hyperbola with center $(-1, -2)$ in the $x'y'$ system

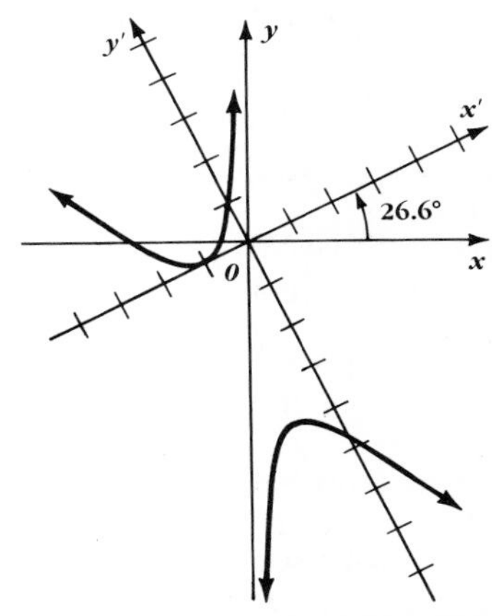

13. $(x'-1)^2 = y' + 1$; a parabola with vertex $(1, -1)$ in the $x'y'$ system

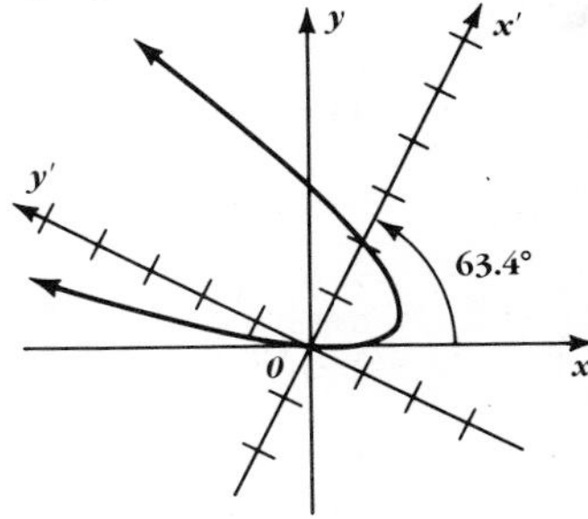

EXERCISES 12-5, Pages 500–501

1. $(6, 5)$ **3.** $(-126, 87)$ **5.** $(-13, 2, 5)$

7. $(2, -1, 3)$ **9.** $\left(\frac{30}{7}, \frac{8}{7}, -\frac{11}{7}\right)$ **11.** $\left(\frac{11}{8}, \frac{7}{16}, 0\right)$

13. $y = 2x^2 - x + 3$ **15.** $(1, 1, 2)$

17. $(3, 0, -1)$

EXERCISES 12-6, Pages 503–505

1. (7, 4) **3.** $\{(x, 2 - \frac{3}{5}x) : x \in \mathcal{R}\}$ or $\{(\frac{10}{3} - \frac{5}{3}y, y) : y \in \mathcal{R}\}$; examples: (0, 2) and $(\frac{10}{3}, 0)$ **5.** $\{(x, 10 - 3x, 17 - 5x): x \in \mathcal{R}\}$, or $\{(\frac{10}{3} - \frac{1}{3}y, y, \frac{5}{3}y + \frac{1}{3}) : y \in \mathcal{R}\}$, or $\{(\frac{17}{5} - \frac{1}{5}z, \frac{3}{5}z - \frac{1}{5}, z): z \in \mathcal{R}\}$; examples: (0, 10, 17) and (1, 7, 12) **7.** $\begin{bmatrix} 1 & 0 & 0 & 5 \\ 0 & 1 & 0 & -8 \\ 0 & 0 & 1 & 3 \end{bmatrix}$; (5, −8, 3)

9. $\begin{bmatrix} 1 & 0 & 0 & 4 \\ 0 & 1 & 0 & -6 \\ 0 & 0 & 1 & 5 \end{bmatrix}$; (4, −6, 5)

11. $\{(x, \frac{1}{2} - \frac{5}{8}x, \frac{1}{8}x + \frac{7}{2}): x \in \mathcal{R}\}$, or $\{(\frac{4}{5} - \frac{8}{5}y, y, \frac{18}{5} - \frac{1}{5}y): y \in \mathcal{R}\}$, or $\{(8z - 28, 18 - 5z, z): z \in \mathcal{R}\}$; examples: (4, −2, 4) and (−20, 13, 1) **13.** no solution **15.** (−122, 137, −53) **17.** no solution **19.** no solution **21.** (0, 0, 0) **23.** (0, 0, 0) **25.** $\{x^2 + y^2 + by - (4b + 32) = 0: b \in \mathcal{R}\}$; $x^2 + y^2 = 32$ **27.** (1, 1, −2, 1) **29.** (0, 0, 0) **31.** independent **33.** dependent **35.** (0, 0, 0) **37. a.** $-2 \cdot 1 + 3 \cdot 0 + 2 \cdot 2 = 2; -1 + 4 \cdot 0 + 2 = 1$

EXERCISES 12-7, Pages 508–509

1. 2 **3.** 0 **5.** abc **7.** 4 **9.** −9 **11.** −32 **13.** −50 **15.** −1088 **17.** −270 **19.** 1014 **23.** $\det(rA) = r^3 \det A$

EXERCISES 12-8, Pages 513–514

1. inverses **3.** not inverses **5.** $\begin{bmatrix} -5 & -7 \\ 2 & 3 \end{bmatrix}$

7. $\begin{bmatrix} \frac{7}{2} & -\frac{9}{2} \\ 3 & -4 \end{bmatrix}$ **9.** $\begin{bmatrix} 2 & 5 & 2 \\ 0 & 0 & 1 \\ 1 & 3 & 2 \end{bmatrix}$ **11.** (1, 1) **13.** (1, −1)

15. $\begin{bmatrix} 1 & -2 \\ -1 & 5 \end{bmatrix}$ **17.** The matrix $\begin{bmatrix} \cos\alpha & -\sin\alpha \\ \sin\alpha & \cos\alpha \end{bmatrix}$ has determinant 1 and inverse $\begin{bmatrix} \cos\alpha & \sin\alpha \\ -\sin\alpha & \cos\alpha \end{bmatrix}$

19. (−1, 1, −2) **21. a.** a, d, and f are nonzero.

b. $T^{-1} = \frac{1}{adf}\begin{bmatrix} df & -bf & be - cd \\ 0 & af & -ae \\ 0 & 0 & ad \end{bmatrix}$

EXERCISES 12-9, Pages 516–517

1. (5, 2) **3.** $\left(\frac{1}{3}, -\frac{1}{3}\right)$ **5.** $\left(-\frac{185}{7}, -\frac{97}{7}\right)$

7. $\left(\frac{13}{47}, \frac{110}{47}, \frac{12}{47}\right)$ **9.** $\left(\frac{87}{13}, -\frac{36}{13}, -2\right)$

11. $\left(-\frac{79}{2}, \frac{37}{2}, -\frac{15}{2}\right)$

13. intersect in a point **15.** are parallel

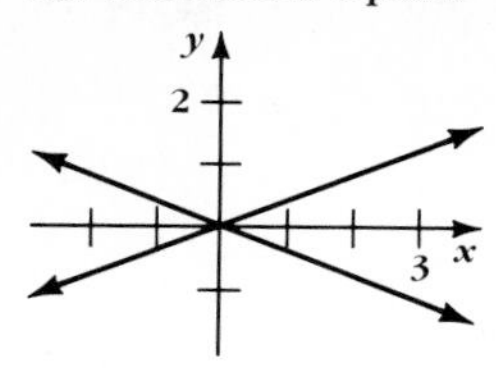

Exercise 13

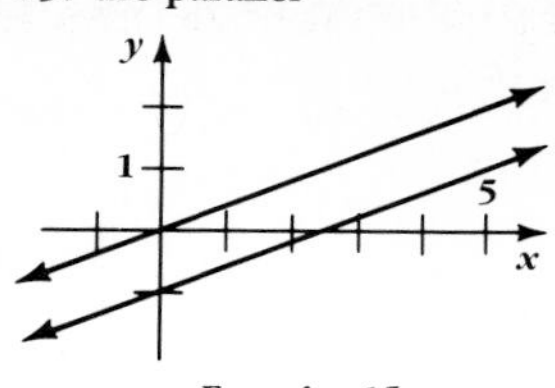

Exercise 15

17. $x = 4$

EXERCISES 13-1, Pages 528–529

1. $-32\mathbf{i} - 10\mathbf{j} - 23\mathbf{k}$ **3.** $11\mathbf{i} - 33\mathbf{j} + 11\mathbf{k}$

5. $\begin{vmatrix} \mathbf{i} & \mathbf{j} & \mathbf{k} \\ 1 & 0 & 0 \\ 0 & 1 & 0 \end{vmatrix} = \mathbf{k}$ **7.** $\begin{vmatrix} \mathbf{i} & \mathbf{j} & \mathbf{k} \\ 0 & 0 & 1 \\ 1 & 0 & 0 \end{vmatrix} = \mathbf{j}$

9. $\begin{vmatrix} \mathbf{i} & \mathbf{j} & \mathbf{k} \\ 0 & 0 & 1 \\ 0 & 1 & 0 \end{vmatrix} = -\mathbf{i}$

11. $-\frac{2\sqrt{26}}{13}\mathbf{i} + \frac{3\sqrt{26}}{26}\mathbf{j} + \frac{\sqrt{26}}{26}\mathbf{k}$ or $\frac{2\sqrt{26}}{13}\mathbf{i} - \frac{3\sqrt{26}}{26}\mathbf{j} - \frac{\sqrt{26}}{26}\mathbf{k}$ **13.** $2\sqrt{138}$ sq. units ≈ 23.5 sq. units **15.** area = $\sqrt{2546}$ ≈ 50.5 sq. units **17.** 100 cubic units **19.** $a = 2$, $b = 4$ **27.** $\frac{3}{2}\sqrt{14}$ ≈ 5.6 sq. units

EXERCISES 13-2, Pages 533–534

1.

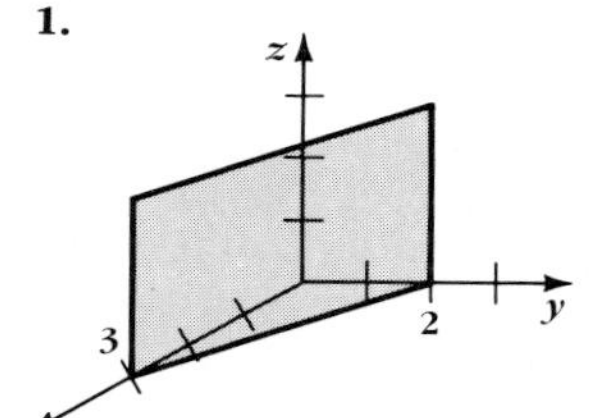

3.

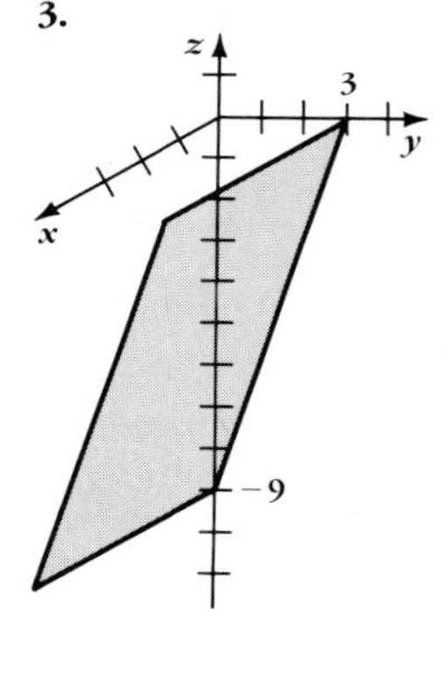

5.

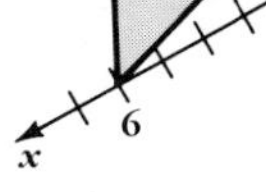

7. $4x + 7y - 3z = 40$ **9.** $x + 6y - 4z = 67$

11. 4 units **13.** $\frac{11\sqrt{3}}{3}$ ≈ 6.4 units

15. $2(3, -1, -1) = (6, -2, -2)$; distance is $\frac{1}{\sqrt{11}}$ ≈ 0.3 units **17.** $2x - y - 8z + 32 = 0$ **19.** $3x + 2y - 5z + 13 = 0$

21. $4x + 13y - 10z + 55 = 0$ **23.** $6x + 6y - 5z - 25 = 0$ **25.** $\theta = \text{Cos}^{-1}\left(\frac{1}{3}\right) \approx 70.5°$

EXERCISES 13-3, Pages 537–539

1. $(x, y, z) = (3, 4, -2) + t(3, 2, 6)$; $x = 3 + 3t$, $y = 4 + 2t$, $z = -2 + 6t$ **3.** Answers may vary. Example: $(x, y, z) = (1, 0, 3) + t(-3, 3, 1)$; $x = 1 - 3t$, $y = 3t$, $z = 3 + t$ **5.** $\cos\alpha = -\frac{\sqrt{19}}{19}$; $\cos\beta = \frac{3\sqrt{19}}{19} = \cos\gamma$
7. $\cos\alpha = -\frac{5\sqrt{2}}{26}$; $\cos\beta = \frac{6\sqrt{2}}{13}$, $\cos\gamma = \frac{\sqrt{2}}{2}$
9. $(x, y, z) = (2, -1, -4) + t(-1, 2, -5)$
11. $(x, y, z) = (0, 2, -4) + t(-8, 3, 4)$
13. $(x, y, z) = (-2, 4, 1) + t(1, -3, -2)$
15. $\frac{1}{6}\sqrt{354}$ **17.** $\frac{1}{22}\sqrt{3278}$ **19.** $3x + y - z - 5 = 0$
21. $(0, 0, 0)$; $\text{Cos}^{-1}\left(\frac{48}{65}\right) \approx 42.4°$ **23.** None
25. $\left(\frac{7}{3}, \frac{25}{3}, \frac{4}{3}\right)$; $\left(\frac{5}{3}, \frac{35}{3}, \frac{5}{3}\right)$
27. $\left(0, -\frac{5}{4}, \frac{17}{4}\right)$, $\left(0, -\frac{3}{2}, \frac{11}{2}\right)$, $\left(0, -\frac{7}{4}, \frac{27}{4}\right)$
29. $\mathbf{r} = (3 + 11t)\mathbf{i} + (-1 + 13t)\mathbf{j} + (2 + 5t)\mathbf{k}$
31. a line: $x = -\frac{1}{3}$, $y = 2t$, $z = t$
33. $\left(\frac{143}{12}, -\frac{91}{6}, -\frac{101}{12}\right)$ **35.** a line: $x = t$, $y = -1 + 3t$, $z = 0$ **37.** a line: $x = 3 - 7t$, $y = 4 + 10t$, $z = 16t$ **39.** $\frac{x - 4}{3\sqrt{2}} = \frac{y - 2}{4} = \frac{z + 5}{\sqrt{2}}$ **41.** Let $\mathbf{m} = (b_1c_2 - b_2c_1, a_2c_1 - a_1c_2, a_1b_2 - a_2b_1)$; if $\mathbf{m} \neq 0$, then equations of the line of intersection are $x = x_0 + t(b_1c_2 - b_2c_1)$, $y = y_0 + t(a_2c_1 - a_1c_2)$, $z = z_0 + t(a_1b_2 - a_2b_1)$ where (x_0, y_0, z_0) is a common point; for example, when $a_1b_2 \neq a_2b_1$, then $(x_0, y_0, z_0) =$

$$\left(\frac{\begin{vmatrix} -d_1 & b_1 \\ -d_2 & b_2 \end{vmatrix}}{\begin{vmatrix} a_1 & b_1 \\ a_2 & b_2 \end{vmatrix}}, \frac{\begin{vmatrix} a_1 & -d_1 \\ a_2 & -d_2 \end{vmatrix}}{\begin{vmatrix} a_1 & b_1 \\ a_2 & b_2 \end{vmatrix}}, 0\right)$$ is a common point.

EXERCISES 13-4, Pages 544–545

1. sphere

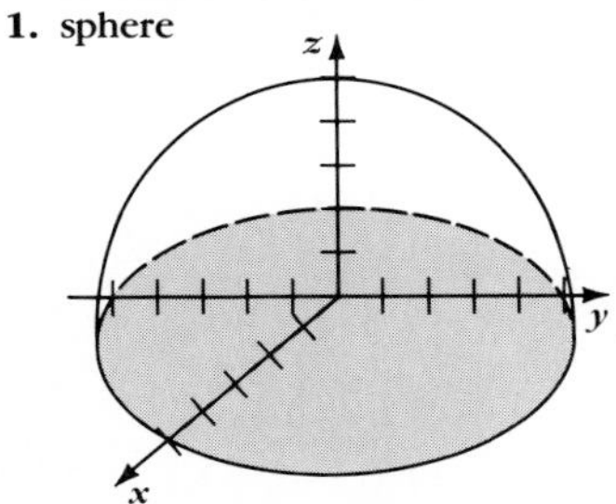

3. plane **5.** circular cylinder

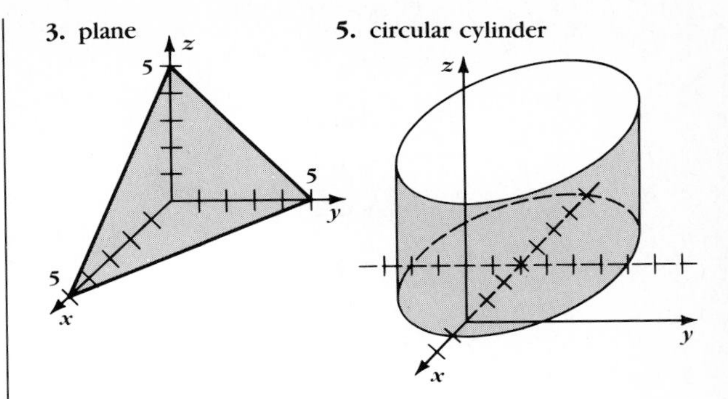

7. elliptic cylinder

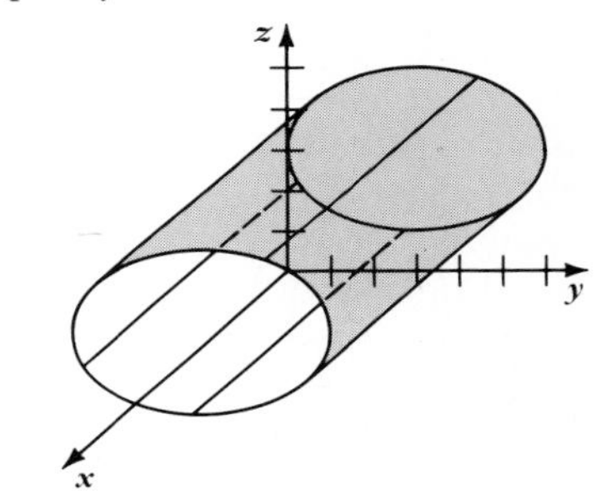

9. elliptic paraboloid **11.** elliptic cone

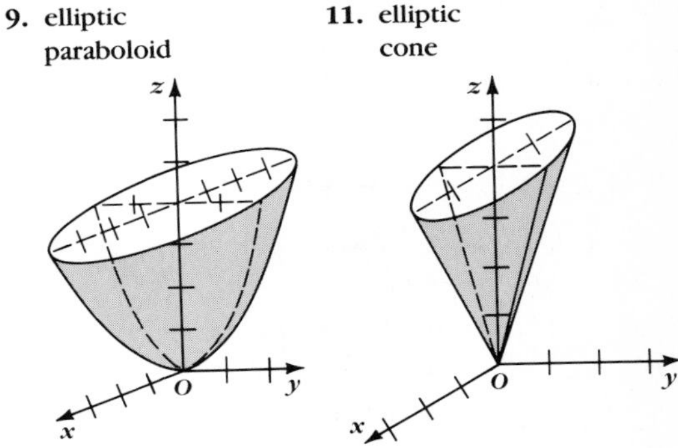

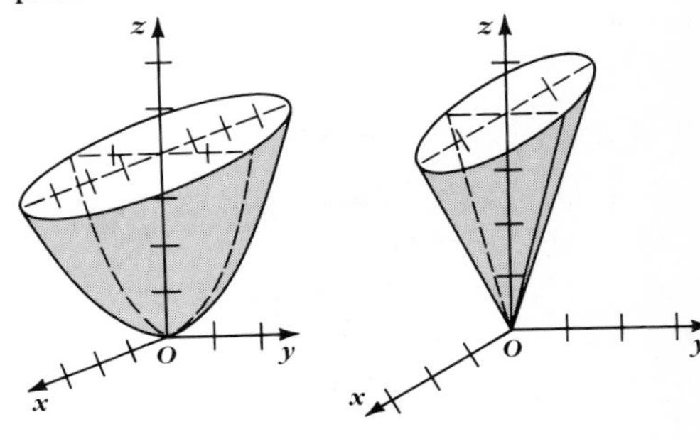

13. ellipsoid

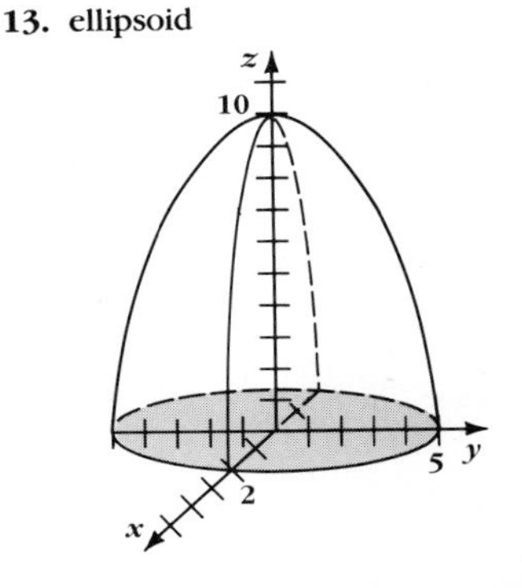

15. hyperbolic paraboloid

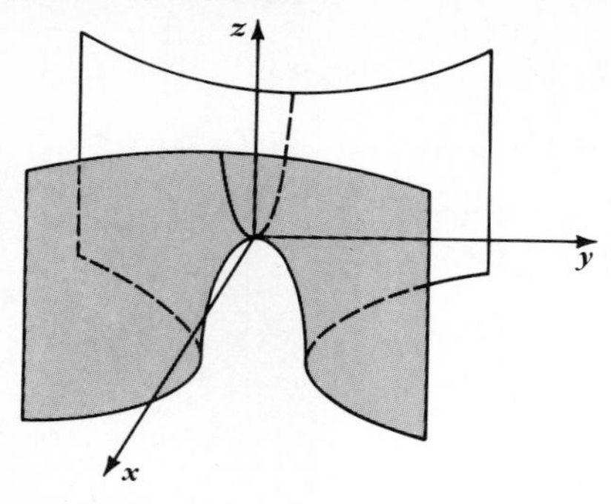

17. parabolic cylinder

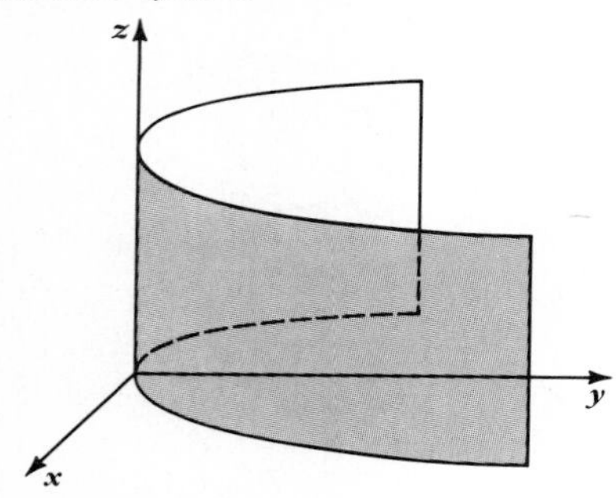

19. hyperbolic cylinder

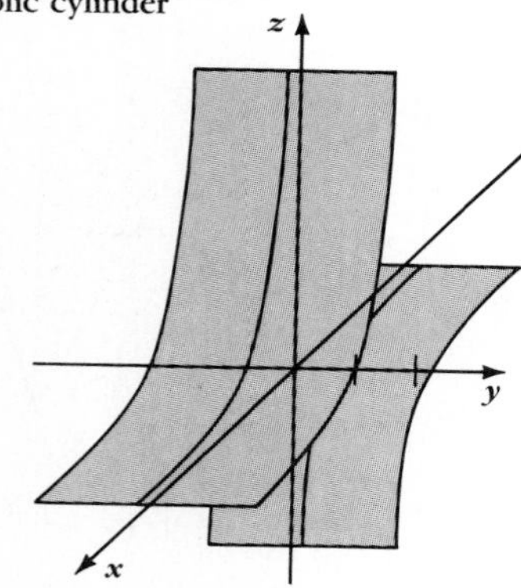

21. circular cylinder

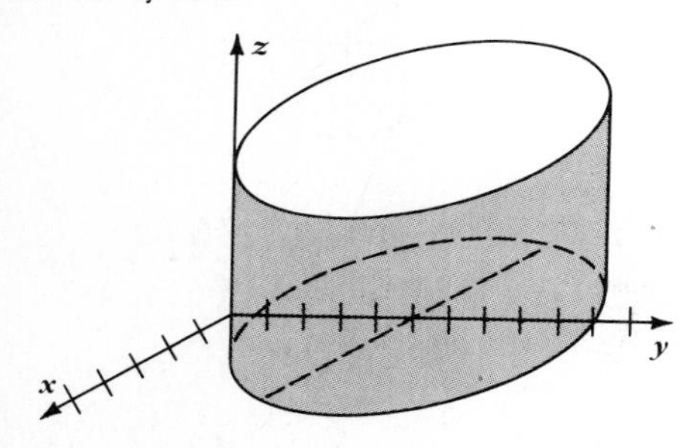

23. $12x - 3y - 4z + 169 = 0$ **27.** a pair of ellipses $\frac{x^2}{a^2} + \frac{y^2}{b^2} = \frac{1}{2}$ with $z = \pm\frac{c\sqrt{2}}{2}$

EXERCISES 13-5, Pages 548–549

1. $\left(\sqrt{2}, \frac{\pi}{4}, 1\right)$ or $\left(-\sqrt{2}, \frac{5\pi}{4}, 1\right)$ **3.** $\left(2, \frac{3\pi}{4}, \frac{\pi}{4}\right)$
5. $\left(4, \frac{\pi}{3}, -4\right)$ or $\left(-4, \frac{4\pi}{3}, -4\right)$ **7.** $(0, 4, 1)$
9. $\left(-\frac{3}{2}, -\frac{3\sqrt{3}}{2}, 4\right)$ **11.** $(0, 1, \sqrt{3})$ **13.** $(0, 2, 0)$
15. $x^2 + y^2 + z^2 = 4$ **17.** $x^2 + y^2 = 4$ **19.** $x^2 + y^2 + z^2 = 4$ **21.** $(x - 1)^2 + y^2 + z^2 = 1$
23. a. $r \cos \theta = 4z^2$ **b.** $\frac{1}{4} \sin \varphi \cos \theta = \rho \cos^2 \varphi$
25. a. $r^2 = 9z$ **b.** $\rho = 9 \cot \varphi \csc \varphi$
27.

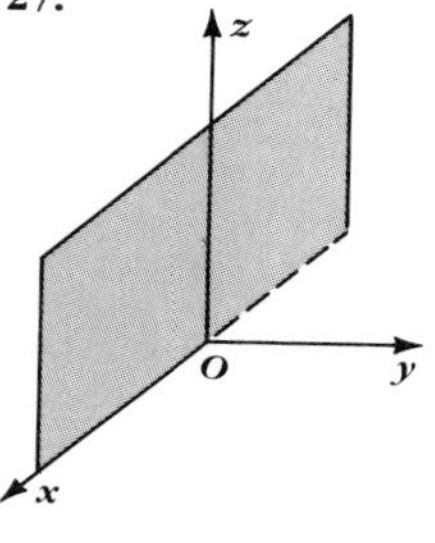

29.

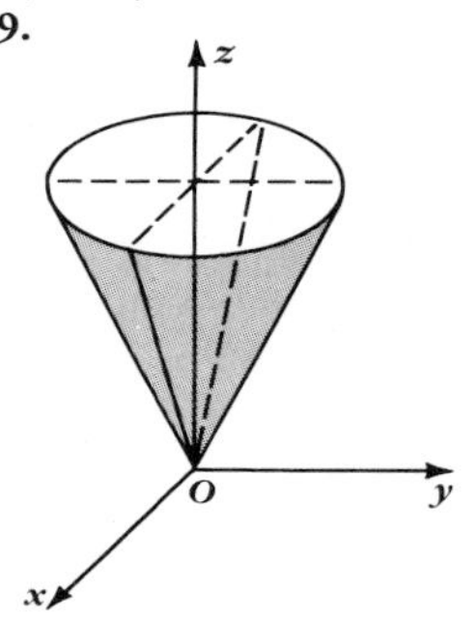

31.

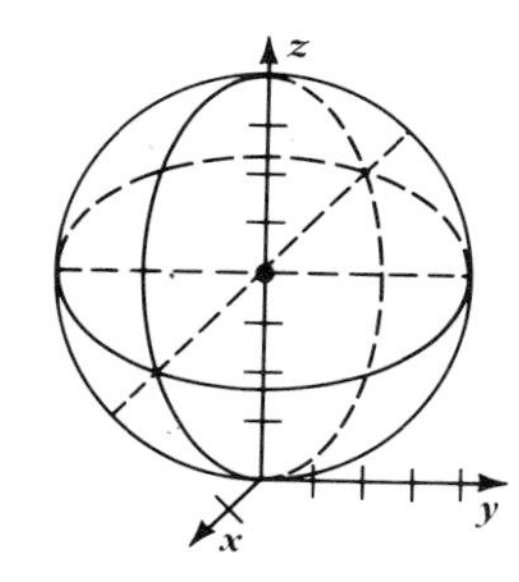

CUMULATIVE REVIEW (Chapters 11–13), Pages 554–555

1. a. $(3, -9)$
b. $(9, 7)$
c. $(-7, 4)$

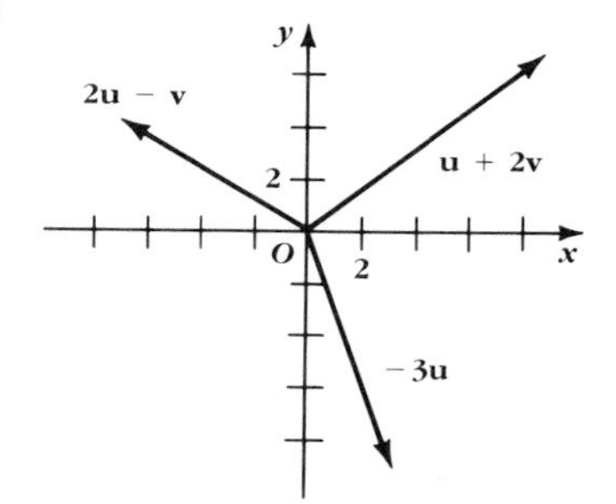

3. $\left(\frac{5\sqrt{2}}{26}, \frac{6\sqrt{2}}{13}, -\frac{\sqrt{2}}{2}\right)$ 5. $(21, -13)$ 7. $\sqrt{237}$; $\left(1, \frac{9}{2}, -1\right)$ 9. $\left(-\frac{227}{5}, -\frac{128}{5}, 26\right)$

11. $\begin{bmatrix} \frac{5}{2} & -\frac{3}{2} & -\frac{1}{2} \\ \frac{11}{2} & -\frac{7}{2} & -\frac{3}{2} \\ 4 & -3 & -1 \end{bmatrix}$ 13. $5x - 2y - 7z - 16 = 0$

15. 327 mph; 68.5° 17. 73 cubic units 19. *aehk*

21. $AB = AC$; isosceles 23. $z = \frac{3}{2}$ 27. $3x - 5y - 2z - 14 = 0$ 29. $x^2 + \left(y - \frac{25}{2}\right)^2 = \frac{625}{4}$

EXERCISES 14-1, Pages 560–561

1. $2x^3 + C$ 3. $x^2 - 3x + C$ 5. $\frac{x^3}{3} - x^2 - 3x + C$

7. $\frac{x^3}{3} - x^2 + x + C$ 9. $-\frac{1}{x} + C$ 11. $\ln u^2 + C$

13. $\frac{x^2}{2} + \ln x^4 + C$ 15. $4x - \frac{1}{x} + \ln x^4 + C$

17. $-4t - t^{-1} + C$ 19. $-\frac{1}{2}\cos 2\theta + C$ 21. $\frac{1}{2}e^{2x} + C$ 23. $e^t - e^{-t} + C$ 25. $\frac{1}{2}e^{2t} - \frac{1}{2}e^{-2t} - 2t + C$

27. $t + \frac{1}{2}\cos 2t + C$ 33. $\frac{1}{2}\sin 2\theta + C$

35. $\frac{(m + n)\cos(m - n)x - (m - n)\cos(m + n)x}{2(m^2 - n^2)} + C$

EXERCISES 14-2, Pages 564–565

1. $\frac{(x + 2)^5}{5} + C$ 3. $\frac{1}{4 - x} + C$

5. $\ln|2x - 1|^{\frac{1}{2}} + C$ 7. $\frac{(x^2 - 1)^5}{10} + C$

9. $-\sqrt{1 - x^2} + C$ 11. $\frac{1}{2}e^{x^2} + C$ 13. $\sin e^x + C$

15. $\frac{(2x - 1)^5}{10} + C$ 17. $\ln|1 - x^2|^{-\frac{1}{2}} + C$

19. $-\frac{1}{3}(1 - x^2)^{\frac{3}{2}} + C$ 21. $\ln|x^3 + 1|^{\frac{1}{3}} + C$

23. $\ln|1 + \sin x| + C$ 25. $\frac{1}{3}\text{Tan}^{-1} 3x + C$

27. $-\ln|\cos x| + C$

EXERCISES 14-3, Pages 567–568

1. 6 3. 1 5. −6 7. ln 2 9. 1 11. $e - 1$

13. 2 15. $\frac{1}{2}$ 17. ln 2 19. $\frac{1}{4}$ 21. ln 2

23. $\frac{\pi}{4}$ 25. $\frac{\pi}{6}$ 27. $2 - \sqrt{3}$ 29. $\frac{61}{3}$

EXERCISES 14-4, Pages 572–574

1. $\frac{15}{2}$ sq. units 3. $\frac{32}{3}$ sq. units 5. 2 square units

7. $\frac{1}{2}(e^2 - 1)$ sq. units 9. $\frac{32}{3}$ sq. units

11. $\frac{\pi}{4}$ sq. units 13. 1 sq. unit

15. $\frac{9}{2}$ sq. units 17. $\frac{9}{2}$ sq. units

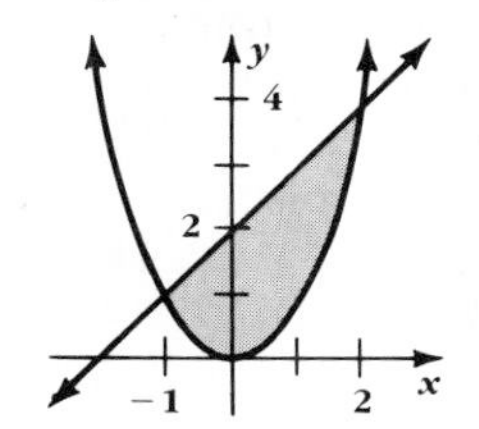

19. 6 sq. units 21. $\frac{15}{2} - \ln 256$ sq. units

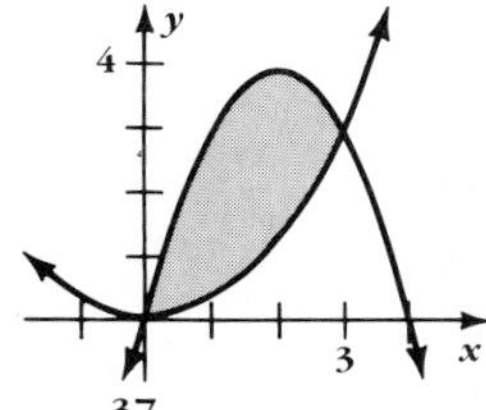

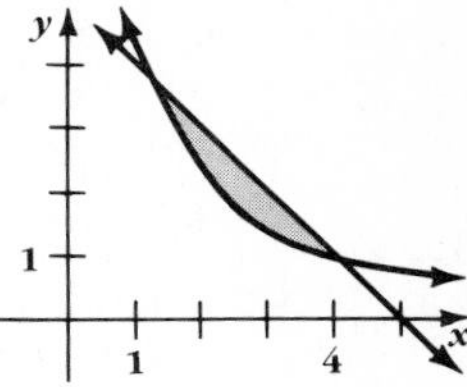

23. $\frac{37}{12}$ sq. units 25. $\frac{1}{6}$ sq. unit

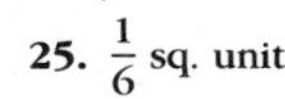

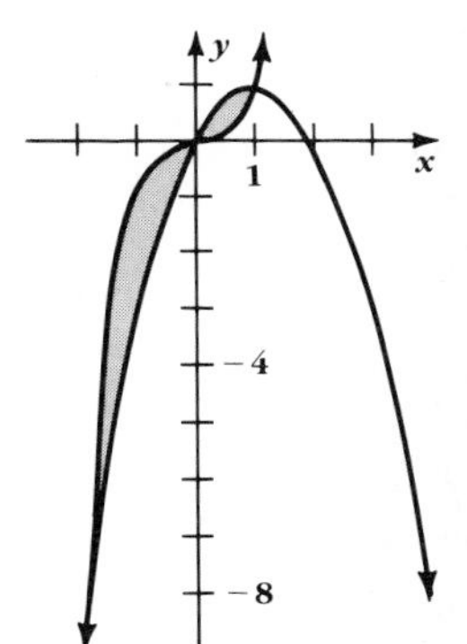

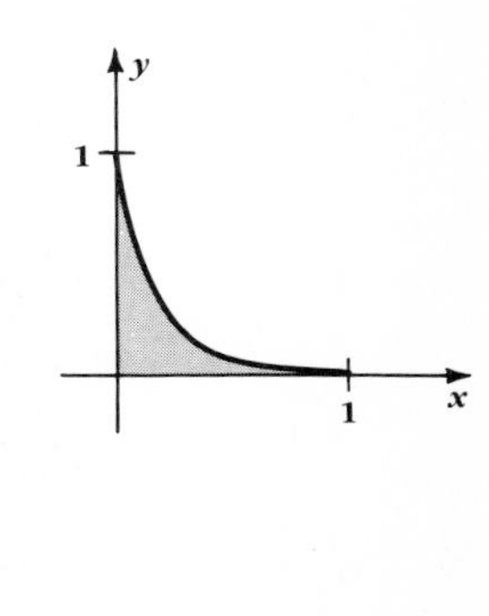

27. Area = $2\sqrt{2}$ sq. units

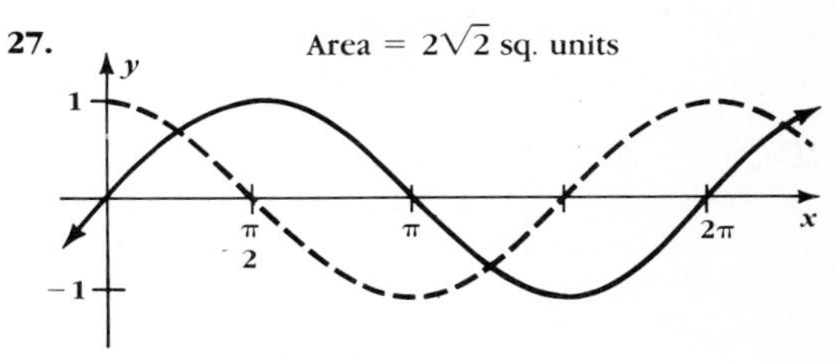

EXERCISES 14-5, Pages 578–579

Each answer is expressed in cubic units.

1. a. $\frac{32\pi}{3}$ b. $\frac{16\pi}{3}$ 3. a. $\frac{256\pi}{15}$ b. 8π

5. a. $\frac{\pi}{3}$ b. $\frac{2\pi}{3}$ 7. a. $\frac{32\pi}{5}$ b. 8π

9. a. $\frac{8\pi}{3}$ **b.** $\frac{32\pi}{5}$ **11. a.** 9π **b.** 9π
13. a. $\frac{1}{3}\pi a^3$ **b.** $\frac{1}{3}\pi a^3$ **15.** $\frac{1}{2}\pi(e^2 - 1)$
17. $\frac{8\pi}{3}$ **19.** $\frac{8\pi}{3}$ **21.** $\frac{16}{3}a^3$ **23.** $\frac{16}{3}a^3$

EXERCISES 14-6, Pages 582–583

1. 1.764 MJ **3.** 1.69 kW · h **5. a.** 60 ft-lb **b.** 75 ft-lb **7.** 3.46 MJ **9.** 5.4 kJ **11.** 123,000 J **13.** 2×10^4 ft-lb **15.** 7×10^4 ft-lb **17.** 0.45 J

EXERCISES 15-1, Pages 592–594

1. 1320 **3.** 720 **5.** 2520 **7.** 720 ways **9.** 48 ways **11.** 30,240 ways **13. a.** 120 integers **b.** 625 integers **15.** 32 ways **17.** 10,080 **19.** 60 **21.** 45,000 **23.** 24 **25.** 24 **27.** 10,360,000 license plates **33.** 72

35. a.

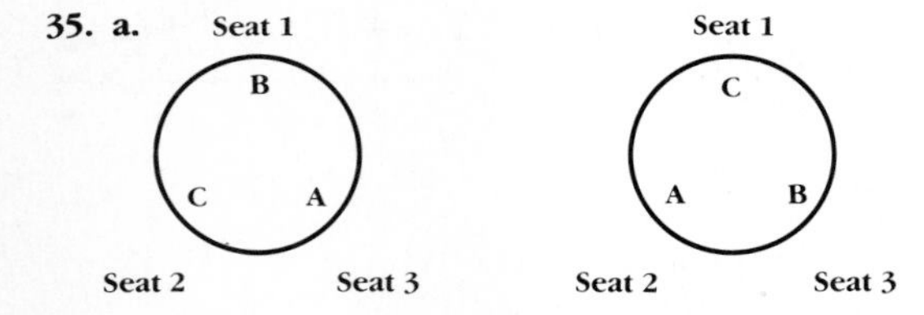

b. Diagrams may vary. An example is given.

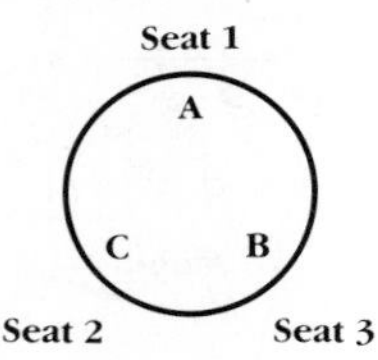

37. $(n - 1)!$

EXERCISES 15-2, Pages 597–598

1. 210 **3.** 4950 **5.** 1140 **7.** {D, A, I}, {D, A, S}, {D, A, Y}, {D, I, S}, {D, I, Y}, {D, S, Y}, {A, I, S}, {A, I, Y}, {A, S, Y}, {I, S, Y} **9.** 84 ways **11.** 252 ways **13.** 56 triangles **15. a.** 120 ways **b.** 60 ways **17.** 19,800 hands **19.** 126 ways **23.** 20 diagonals **25.** $\frac{n(n-3)}{2}$ diagonals **27.** 3600 sequences

EXERCISES 15-3, Pages 601–603

1. a. {(2, 6), (3, 5), (4, 4), (5, 3), (6, 2)}; $\frac{5}{36}$
b. {(1, 1), (2, 2), (3, 3), (4, 4), (5, 5), (6, 6)}; $\frac{1}{6}$
c. {(4, 6), (5, 5), (5, 6), (6, 4), (6, 5), (6, 6)}; $\frac{1}{6}$
3. 4 **5.** 216 **7.** S contains ${}_9C_2 = 36$ elementary events, each of the form {B, B}, {B, G}, or {G, G}. **a.** the events of the form {B, G}; $\frac{5}{9}$ **b.** the events of the forms {B, B} and {B, G}; $\frac{5}{6}$ **c.** the events of the form {G, G}; $\frac{1}{6}$
9. a. $\frac{1}{221}$ **b.** $\frac{188}{221}$ **c.** $\frac{15}{34}$ **11. a.** $\frac{10}{21}$ **b.** $\frac{1}{21}$ **c.** $\frac{20}{21}$
13. a. $\frac{1}{30}$ **b.** $\frac{1}{6}$ **c.** $\frac{49}{60}$ **15. a.** $\frac{2162}{54,145}$ **b.** $\frac{33}{66,640}$
c. $\frac{64}{162,435}$ **21.** $P(S) > 1$

EXERCISES 15-4, Pages 608–609

1. independent **3.** mutually exclusive **5.** neither
7. a. $\frac{1}{3}$ **b.** $\frac{1}{13}$ **c.** $\frac{1}{4}$ **9.** no; $0.5 \neq (0.8)(0.6)$
11. $\frac{1}{3}$ **13. a.** $\frac{15}{56}$ **b.** $\frac{3}{28}$ **15. a.** $P(A \cap B) \neq P(A) \cdot P(B)$ **b.** $\frac{7}{11}, \frac{9}{11}$ **c.** $\frac{1}{3}, \frac{1}{5}$ **17. a.** $\frac{1}{4}$ **b.** $\frac{2}{5}$ **c.** $\frac{4}{15}$
19. $\frac{5}{12}$ **25.** $\frac{3}{5}$

EXERCISES 15-5, Pages 614–615

1. a.

x	$P(X = x)$
0	0.125
1	0.375
2	0.375
3	0.125

b.

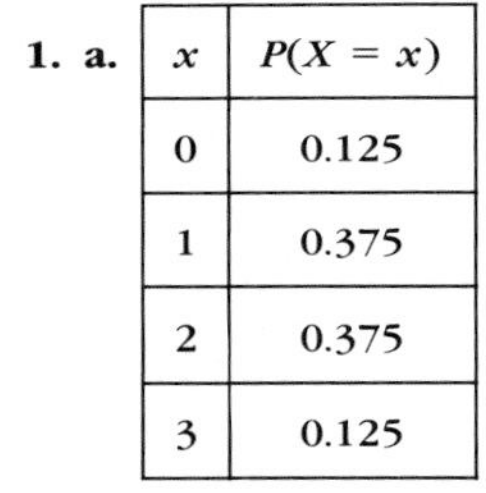

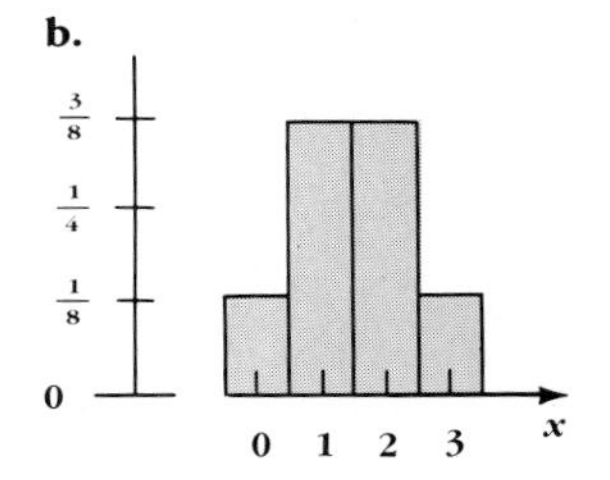

c. 1.5

3. a.

z	$P(Z = z)$
3	0.1
4	0.1
5	0.2
6	0.2
7	0.2
8	0.1
9	0.1

3. b.

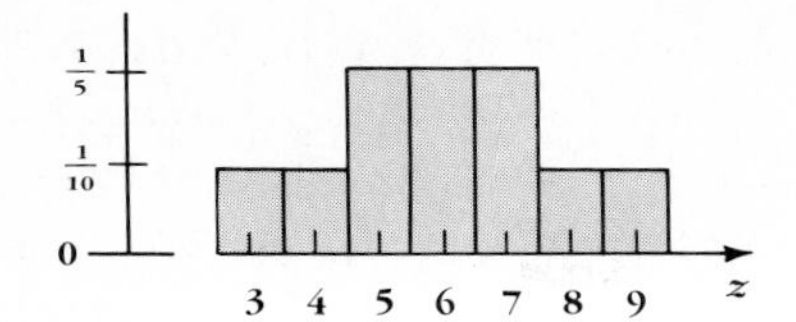

c. 6

5. a.

y	$P(Y = y)$
0	$\frac{1}{5}$
1	$\frac{3}{5}$
2	$\frac{1}{5}$

b.

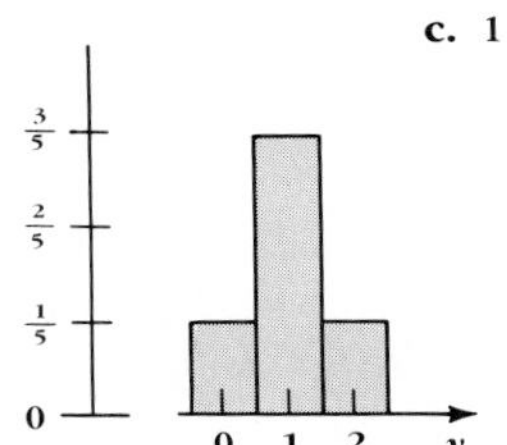

c. 1

7. a.

x	$P(X = x)$
0	$\frac{19}{34}$
1	$\frac{13}{34}$
2	$\frac{2}{34}$

c. $\frac{1}{2}$

b.

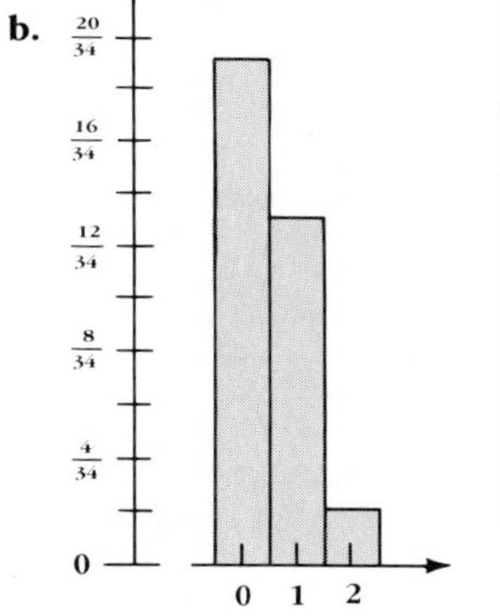

9. a. $P(X = k) = \binom{4}{k}\left(\frac{1}{6}\right)^k\left(\frac{5}{6}\right)^{4-k}$ **b.** 0.116 **c.** $\frac{2}{3}$

11. a. $P(Y = k) = \binom{5}{k}\left(\frac{1}{13}\right)^k\left(\frac{12}{13}\right)^{5-k}$ **b.** 0.047

c. $\frac{5}{13}$ **13. a.** $P(B = k) = \binom{6}{k}(0.4)^k(0.6)^{6-k}$

b. 0.276 **c.** 2.4 **15.** 0.027

EXERCISES 15-6, Pages 619–621

1. 5.25; 5; 5 **3.** 16.9; 17.5; 2, 14, 19, 20, and 37

5.

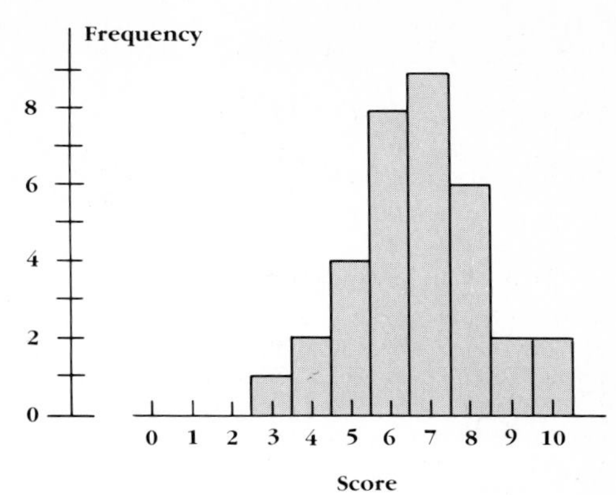

7.

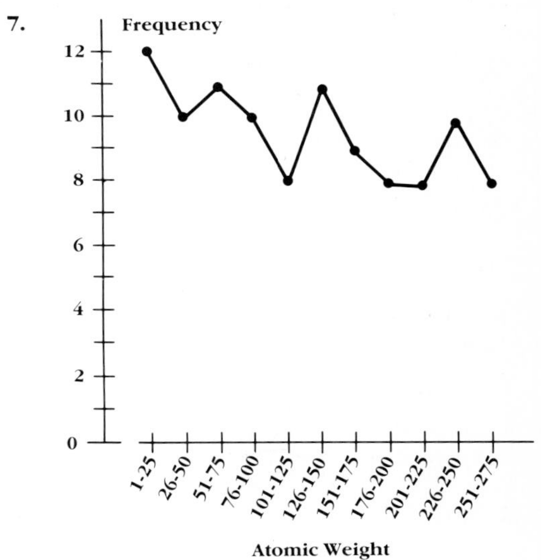

9.

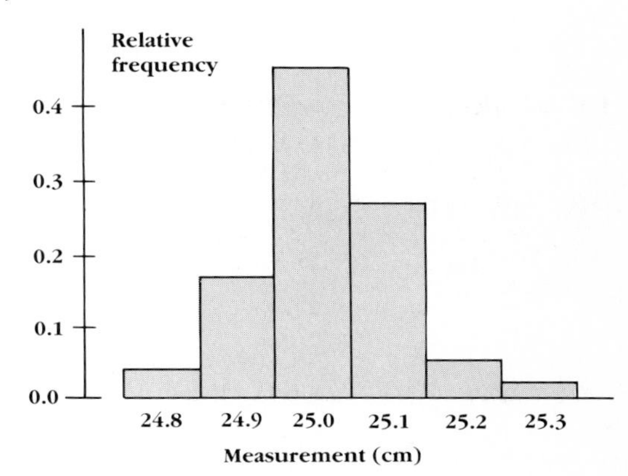

11. 6.71; 7; 7

13. 130.38; 138; 13

15.

Wind Speed (km/h)	Freq.
1–25	0
26–50	0
51–75	1
76–100	0
101–125	6
126–150	11
151–175	6
176–200	3

17. Using class intervals: 140.8 **19.** 80.5
21. Answers may vary. Example: 2, 5, 9, 12, 12

EXERCISES 15-7, Pages 624–625

1. 12; 14; 3.74 **3.** 214; 4872; 69.8 **5.** 6; 3.06; 1.75
7. $\frac{2}{3}; \frac{5}{9}$; 0.745 **9.** $\frac{5}{13}; \frac{60}{169}$; 0.596 **11.** 2.40; 1.44; 1.2
13. 0.75; 0.866 **15.** 3.00; 1.73 **17.** 0.4; 0.632
19. rates for larger loans were more variable (smaller loans: $\sigma^2 \approx 5.74$; larger loans: $\sigma^2 \approx 12.0$) **21. a.** 3
b. 1.22 **c.** $\frac{1}{64} \approx 0.0156$ **23. a.** 6 **b.** 1.73
c. $\frac{79}{4096} \approx 0.0193$

EXERCISES 15-8, Pages 631–632

1. 2 **3.** −1.4 **5.** 69 **7.** 43.8 **9.** 0.9032
11. 0.9554 **13.** 0.6902 **15. a.** 84.13% **b.** 6.68%
c. 77.45% **17. a.** 69.15% **b.** 30.85% **c.** 6.3%
19. a. 68.3% **b.** 95.4% **c.** 99.7% **21. a.** 958
b. 918 **23. a.** 453 **b.** 576 **25.** 0.025

EXERCISES 15-9, Pages 634–635

1. a. 0.04 **b.** [0.76, 0.84] **c.** [0.72, 0.88]
3. a. 0.096 **b.** [0.264, 0.456] **c.** [0.168, 0.552]
5. a. 0.050 **b.** [0.40, 0.50] **c.** [0.35, 0.55]
7. [0.57, 0.77] **9.** [0.029, 0.067] **11.** The 95% confidence interval, [0.24, 0.46], does not contain 0.5, the probability of heads for a fair coin.
13. $[\bar{p} - 1.647\,\sigma_{\bar{p}}, \bar{p} + 1.647\,\sigma_{\bar{p}}]$
15. a. [0.968, 0.992] **b.** [0.964, 0.996]

EXERCISES 15-10, Pages 637–639

1. negative **3.** no correlation

5. a.

mpg
60
40
20
20 40 60 80 100
Engine Capacity

b. −0.64
7. a.

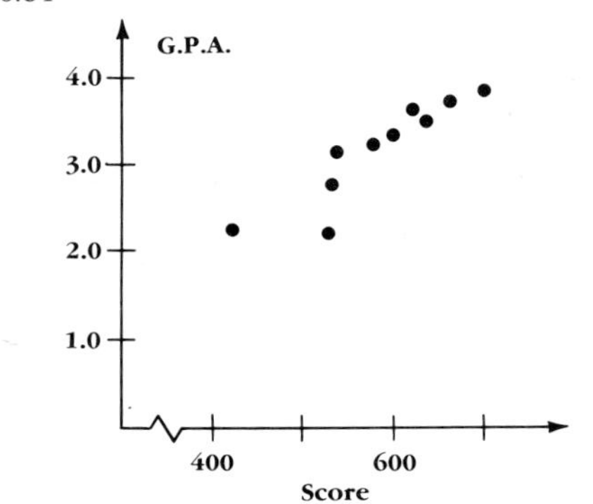

b. 0.91
9. a.

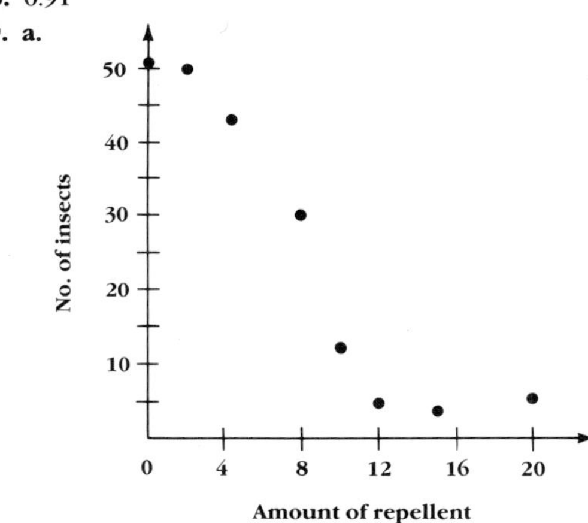

b. −0.92
11. 3.83 **13.** about 2030